Ticks in the genus *Rhipicephalus* include many important vectors of animal and human pathogens, but many species are notoriously difficult to identify, particularly as immature stages. This reference volume provides identification keys for adult ticks from the Afrotropical regions and elsewhere. For the nymphs and larvae, unique plates have been compiled in which line drawings of the capitula of similar species are grouped together to facilitate identification. Brief, well-illustrated, descriptions of the known stages of every species are given, plus information on their hosts, distribution, and disease relationships. Tables providing data on host/parasite relationships and disease transmission are also included, making this the definitive reference source on this group for all those interested in acarology, veterinary or medical parasitology, and entomology for many years to come.

JANE B. WALKER has spent a lifetime studying the taxonomy, hosts, and distribution of African ticks, particularly in the genus *Rhipicephalus.* Although now officially retired, she still works as a scientist at the Onderstepoort Veterinary Institute. She is the author of *The Ixodid Ticks of Kenya* (1974) and co-author of *The Ixodid Ticks of Tanzania* with G.H. Yeoman (1967).

JAMES E. KEIRANS is Professor of Biology and Curator of the U.S. National Tick Collection at the Institute of Arthropodology and Parasitology at Georgia Southern University. His previous books include *The Mallophaga of New England Birds* (1967), *George Henry Falkiner Nuttall and the Nuttall Tick Catalogue* (1985), *Systematics and Ecology of the Subgenus* Ixodiopsis (*Acari: Ixodidae*; Ixodes) (1992, with R.G. Robbins), and *Nymphs of the genus* Ixodes *of the United States* (1996) with L.A. Durden.

IVAN G. HORAK is Emeritus Professor of Veterinary Ectoparasitology at the Faculty of Veterinary Science, University of Pretoria. He is co-editor of *Tick Vector Biology: Medical and Veterinary Aspects* (1992) with B.H. Fivaz and T.N. Petney.

Frontispiece. Cecil Warburton. Photo courtesy of the Master and Fellows of Christ's College, Cambridge.

The Genus *Rhipicephalus* (Acari, Ixodidae):

A Guide to the Brown Ticks of the World

by

JANE B. WALKER

JAMES E. KEIRANS

AND IVAN G. HORAK

CAMBRIDGE UNIVERSITY PRESS
Cambridge, New York, Melbourne, Madrid, Cape Town, Singapore, São Paulo

Cambridge University Press
The Edinburgh Building, Cambridge CB2 2RU, UK

Published in the United States of America by Cambridge University Press, New York

www.cambridge.org
Information on this title: www.cambridge.org/9780521480086

First published 2000
This digitally printed first paperback version 2005

A catalogue record for this publication is available from the British Library

Library of Congress Cataloguing in Publication data

Walker, Jane B. (Jane Brotherton)
The genus *Rhipicephalus* (Acari, Ixodidae): a guide to the brown ticks
of the world / by Jane B. Walker, James E. Keirans, and Ivan G. Horak.
p. cm.
ISBN 0 521 48008 6 (hardback)
1. *Rhipicephalus*. I. Keirans, James E. II. Horak, Ivan G.
III. Title.
QL458.2.I9W337 1999
595.4′29–dc21 98-26534 CIP

ISBN-13 978-0-521-48008-6 hardback
ISBN-10 0-521-48008-6 hardback

ISBN-13 978-0-521-01977-4 paperback
ISBN-10 0-521-01977-X paperback

This contribution to our knowledge of the genus *Rhipicephalus* is respectfully dedicated to the memory of

CECIL WARBURTON
(1854–1958)

In his valuable paper on this genus published in 1912 he commented: 'The systematist has no need to apologise for a want of definiteness the responsibility for which lies with Nature herself.' We therefore feel that he would be sympathetic to the 'want of definiteness' that remains in parts of this book.

Errata

p.31. The correct date of publication for *R. oreotragi* Walker & Horak is 2000, not 1999.

p.38. The correct date of publication for *R. warburtoni* Walker & Horak is 2000, not 1999.

p.45. The Rocky Mountain Laboratory is in Hamilton, Montana (MT), not Massachussetts (MA).

p.408. *Notes on identification.* First sentence should read 'We have not seen the type series of *R. serranoi* (2♂♂, 5♀♀).

Contents

Acknowledgements

This book represents the combined achievements of a large group of people and we are most grateful for the time and effort that so many of our colleagues have expended on our behalf.

We much appreciate the facilities and financial support that have enabled us to carry out the work. These were provided by Dr D.W. Verwoerd, Director, and Dr J.D. Bezuidenhout, Deputy Director, and Dr D.T. de Waal, Onderstepoort Veterinary Institute, Republic of South Africa (J.B.W.); Dr J. H. Oliver Jr, Institute of Arthropodology and Parasitology, Georgia Southern University, Statesboro, Georgia, U.S.A. (J.E.K.), and Professor R.I. Coubrough, Dean, Faculty of Veterinary Science, University of Pretoria, Republic of South Africa (I.G.H.).

The publication of this book was supported in part by grants from the South African Agricultural Research Council and the South African Foundation for Research Development, to whom we are most grateful. In addition portions of this work were supported by National Institute of Allergy and Infectious Diseases grant A1 30026 to J.E.K. He was also the recipient of one Research Fellowship from the South African Department of Agriculture, and two Research Fellowships from the South African Agricultural Research Council, to conduct research for this publication at the Onderstepoort Veterinary Institute.

Several of our co-workers merit our special thanks. Mr A. Olwage, who has worked with J.B.W. for over 10 years, drew the meticulously-detailed illustrations of all the *Rhipicephalus* species adults and the capitula of the immature stages. He also produced the final versions of all the maps as well as providing various other professional services. Mr A.C. Uys assisted I.G.H. with tick surveys; measured all stages of many of the tick species; constructed the host/parasite list of the Afrotropical species, and rendered invaluable help in many other ways. We sincerely thank them both.

We were fortunate to have the services of four skilled electron microscopists, Mr M.D. Corwin, Ms Pat Hill, Mr J.F. Putterill and Dr R.G. Robbins. Between them they produced hundreds of scanning electron photomicrographs of all stages of our ticks. The onerous task of printing almost all these photomicrographs was carried out by Miss Heloise Heyne with commendable care and attention to detail. She also mounted many of them. These photomicrographs form an integral and much valued part of our book.

In addition Miss Heyne contributed numerous unpublished records of rhipicephalids, particularly from Namibia, which we greatly appreciate.

A considerable contribution was made by the four ladies who processed, and time and again revised, the many versions of our manuscript. The major part of this task was carried out by Mrs M. Viljoen, ably assisted by Mesdames

C.M.S. Lubbe, A.S. Meiring and H.M. Serfontein. We extend very warm thanks to them all for their efforts.

I.G.H. wishes to express his thanks to the South African National Parks, the KwaZulu-Natal Parks Board and the Department of Nature Conservation, Namibia for making the wild animals which were examined in numerous surveys available to him, and Dr H.C. Biggs, Professor J. Boomker, Dr L.E.O. Braack, Dr V. de Vos, Professor L.J. Fourie and Mr A.M. Spickett for help during these surveys.

We also appreciate the help we have received from several other colleagues. Complete series of several South African *Rhipicephalus* species were reared for us by the late W.O. Neitz, by Mrs M. Dunsterville and by Dr L. López-Rebollar. The laborious task of entering field collection records into databases was carried out by Mrs D.J. van Wyk and Ms L. Booth, and the calculation of the length of the scale bars on several of the figures by Ms P.J. Reeve, who thereby earned our thanks. Dr M.-L. Penrith is acknowledged with appreciation for her helpful discussions with J.B.W. of various taxonomic problems and the derivation of some specific names. Greatly appreciated help has also been given on many occasions to J.B.W. by Mrs A.E. van der Walt, Mr D.G. de Klerk and Mr T.E. Krecek.

Dr R.G. Robbins, Defense Pest Management Information Analysis Center, Armed Forces Pest Management Board, Forest Glen Section, WRAMC, Washington, DC, U.S.A., contributed important literature references to the genus *Rhipicephalus* which were previously unavailable to us. Other literature references were obtained for us, sometimes from obscure journals overseas at very short notice, by Mr D. Swanepoel and his staff in the library at the Onderstepoort Veterinary Institute and by Mrs E. van der Westhuizen and her staff in the library at the Faculty of Veterinary Science, Onderstepoort.

Much appreciated assistance has been received from a number of our other friends. Both reared specimens and field collections of *Rhipicephalus* species that we would otherwise have lacked, some of them very rare, were donated by Dr J.-L. Camicas, the late Dr P.C. Morel, Dr R.G. Pegram and the late Dr J.A.T. Santos Dias. Up-to-date distribution maps of *R. appendiculatus* and *R. zambeziensis* were provided by Dr B.D. Perry and Mr R. Kruska, to whom we are most grateful. Parts of the manuscript were reviewed by Dr L.A. Durden, Institute of Arthropodology and Parasitology, Georgia Southern University, Statesboro, Georgia, U.S.A., who furnished valuable comments on the keys, and by Dr R.G. Pegram. We thank Dr Lorenza Beati who, with a small collection of *R. zumpti*, brought to light an error in our Afrotropical *Rhipicephalus* key. We are most grateful to Dr V.N. Belozerov for tracing and translating some of the data on human disease transmission published in Russian for us.

We have been fortunate in the cooperation that we have always received from authorities in museums and other institutions overseas when we have sought their help. In particular we have received loans, and in some cases gifts of specimens, from Dr Anne Baker, The Natural History Museum, London; Dr M. Moritz, Museum für Naturkunde der Humboldt Universität zu Berlin; Dr F. Puylaert, Musée Royal de l'Afrique Centrale, Tervuren; Dr M. Judson, Muséum National d'Histoire Naturelle, Paris, and Dr Paula Dias, Instituto Nacional de Investigação Veterinária, Maputo, Mozambique.

For information on *Rhipicephalus* collections under their care we also thank Dr P.J. van Helsdingen, National Museum van Natuurlijke Historié, Leiden, The Netherlands; Dr L.S. Hiregoudar, Hubli, Karnataka, India, and Dr I. Lansbury, Hope Entomological Collections, The University Museum, Oxford.

For permission to utilize copyright material from their publications we offer our sincere thanks to the Editor, *Onderstepoort Journal of Veterinary Research*, and the Agricultural Research Council, Republic of South Africa; the Entomological Society of America for extracts from the *Journal of Medical Entomology*; Kluwer Academic Publishers for extracts from *Systematic Parasitol-*

ogy; the family of the late Dr P.C. Morel; the Editors, *Acarologia*; the Editor-in-Chief, *Parazitologicheskiy Sbornik*, and Dr N.A. Filippova.

Finally we should like to acknowledge in the warmest terms the support we have received from the staff of Cambridge University Press during the production of this book. In particular we sincerely thank Dr Tracey Sanderson (Commissioning Editor: Biological Sciences), Alison Litherland (Copy Editorial Controller), Dr Sharon Erzinçlioğlu and Mrs Sandi Irvine (copy editors), also Mrs Sue Tuck (Production Controller and Ms. Angela Cottingham (Indexer) for all their help.

1

Introduction

'Haba na haba hujaza kibaba'. Little by little fills up the measure! This Swahili proverb encapsulates the contributions made by many people, starting in 1806, towards an understanding of the genus *Rhipicephalus*, an important group of ixodid ticks occurring mainly, but by no means exclusively, in Africa. So far as we know our book represents the first completed attempt to review this knowledge. Almost certainly G.H.F. Nuttall, C. Warburton, W.F. Cooper and L.E. Robinson originally intended to include the genus in their series of monographs on the Ixodoidea but they never managed to do so. The first part, by Nuttall *et al.* on the family Argasidae, was published in 1908. Thereafter three further parts on individual ixodid genera appeared, by Nuttall & Warburton (1911, 1915) on *Ixodes* and *Haemaphysalis*, respectively, and by Robinson (1926) on *Amblyomma*.

In 1939 F. Zumpt, in the first of a series of papers entitled 'Vorstudie zu einer Revision der Gattung *Rhipicephalus*', noted that he planned to revise the genus in collaboration with Dr W. Minning. However, after publishing a key to the known species within the genus in 1949 he apparently abandoned this idea. In 1960 D.R. Arthur, of King's College, London, produced the fifth volume in the monographic series started by Nuttall and his colleagues in which he dealt with the genera *Dermacentor*, *Anocentor*, *Cosmiomma*, *Boophilus* and *Margaropus*. At the same time he gave notice of his intention to complete the series, including a study of *Rhipicephalus* and *Rhipicentor*, but neither this nor his proposed volume on the genus *Hyalomma* ever materialized. Two years later Gertrud Theiler issued a long report in which she dealt with all the ticks known to occur in Africa, including the rhipicephalids, together with their hosts and distribution. P.C. Morel followed this in 1969 with his valuable thesis on all the ixodid ticks occurring in Africa, including maps showing their distribution. The section in that study on the *Rhipicephalus* spp. has been most helpful to us. We have often referred to it, especially in connection with the West African species, of which we ourselves have little first-hand knowledge.

One of us (J.B.W.) has been particularly involved with the rhipicephalids for over 40 years and finally decided to try, with the help of her colleagues, to consolidate available information on the genus. It has been a daunting task and, despite our best efforts, we are well aware of some of the remaining shortcomings of the final result. Wherever possible we have tried to draw attention to outstanding problems and possible mistakes in our interpretation of the existing data. We therefore remain hopeful that our contribution will provide a useful foundation for further studies on these interesting ticks.

We start with a brief account of the relationships of the ticks and a definition of the genus *Rhipicephalus*. This is followed by an explanation of the format used in the accounts of individual species, then a glossary, together with labelled diagrams showing the essential external morphological features of adult rhipicephalids. The last part of this introductory section of the book is a list of the *Rhipicephalus* species names of the world.

Two major sections appear next, the first on species occurring in the Afrotropical region and the second on those found elsewhere. Accounts of the four species that are present in both regions, *R. camicasi*, *R. evertsi evertsi*, *R. sanguineus* and *R. turanicus*, feature in the Afrotropical section. Each of these sections comprises a historical review of research on the *Rhipicephalus* spp. recorded in the region, keys for the identification of the adults, the accounts of individual species listed alphabetically, and a host/parasite list.

The identification of the immature stages of the rhipicephalids has always been particularly difficult because many of them are very much alike in appearance. We have not attempted to produce keys for their identification. Instead we have included a series of plates in which line drawings of the capitula of the nymphs and larvae of morphologically similar species are grouped together. This will facilitate direct comparisons between them and thus, we hope, help readers to identify them.

The last section in the book comprises information on the transmission of various pathogens to animals and humans by *Rhipicephalus* spp.

Pertinent references are listed at appropriate points throughout the text.

2

Relationships of the ticks (Ixodida) and definition of the genus *Rhipicephalus*

Ticks are all obligate blood-feeding parasites of terrestrial vertebrates at some stage of their life cycle. Many species are of considerable interest and importance as vectors of a wide variety of pathogens to both humans and animals.

They are members of the phylum Arthropoda, the jointed-legged animals. Although often referred to as insects, whose adults have six legs, they are in fact members of the class Arachnida. This class, whose adults have eight legs, includes spiders and scorpions as well as the order Acari, a large and diverse group to which the ticks and mites belong. Within the Acari the suborder Ixodida encompasses the three families of ticks, the Argasidae, Nuttalliellidae and Ixodidae. The systematics of the Ixodida were reviewed recently by Keirans (1992).

Members of the family Ixodidae, to which the genus *Rhipicephalus* belongs, are characterized by having a hard sclerotized scutum. This completely covers the dorsal surface of the body in the males but is merely a smaller shield just behind the capitulum in the females and immature stages. The mouthparts of all these ticks are anterior in position; their eyes, when present, are near the lateral margin of the scutum, and their spiracles, which are large, are located behind coxae IV. This combination of characters readily distinguishes ixodid ticks from species in the Nuttalliellidae and the Argasidae. Members of both the latter families lack a hard sclerotized scutum: the main feature characterizing them is the leathery integument that covers their bodies.

Keirans (1992) noted that: 'The family Ixodidae is usually considered to be composed of approximately 13 genera . . .', of which the genus *Rhipicephalus* is one of the largest. Species in this genus have the following morphological features in common: their hypostome and palps are short and their basis capituli is usually hexagonal; they have eyes, festoons and, in the males, adanal plates. With the exception of four species, *R. dux, R. humeralis, R. maculatus* and *R. pulchellus,* they are inornate, i.e. the adults do not have a colour pattern on the scutum, hence their common name 'the brown ticks'.

In this book we recognize 74 *Rhipicephalus* species and 2 subspecies. We believe, however, that further studies may show that several other entities whose precise status is at present uncertain are in fact valid species. It is mainly an African genus. Of the known species one, *R. sanguineus,* occurs practically worldwide between latitudes 50°N and 30°S. Sixty species have been recorded only in the Afrotropical region and one, *R. fulvus,* appears to be confined to parts of north-western Africa. Two species, *R. turanicus* and to a lesser extent *R. camicasi,* are widely distributed both in Africa and further afield. Ten species are known only from outside the Afrotropical region. One subspecies, *R. evertsi evertsi*, which is very widely distributed in the Afrotropical region, has apparently now also gained a foothold on the Arabian peninsula; time will show how far it manages to spread there.

REFERENCES

Keirans, J.E. (1992). Systematics of the Ixodida (Argasidae, Ixodidae, Nuttalliellidae): an overview and some problems. In *Tick Vector Biology, Medical and Veterinary Aspects*, ed. B. Fivaz, T. Petney & I. Horak, pp. 1–21. Heidelberg: Springer-Verlag.

3

Format for the accounts of individual species

SOURCES OF INFORMATION

The data presented in this book have been obtained from many sources. A major source has been the data files of the United States National Tick Collection (USNTC), whose history has been documented by Durden, Keirans & Oliver (1996). It had its origins early in the century at the Rocky Mountain Laboratory, Hamilton, Montana, hence the prefix 'RML' to its collection numbers. In 1983 it was donated to the United States National Museum of Natural History (Smithsonian Institution), where it was curated by J.E.K. Shortly after H. Hoogstraal died in 1986 his tick collection was also sent to the Smithsonian Institution for incorporation into the USNTC. In 1990 the collection was transferred on long-term enhancement loan from the Smithsonian Institution to the Institute of Arthropodology and Parasitology, Georgia Southern University, where it is still curated by J.E.K., assisted by L.A. Durden. It is the world's largest tick collection, including over 300 types and more than 122 500 individual accessioned collections.

Another important source of information has been material from the collections of The Natural History Museum, London, loaned to us by Anne Baker. These include, amongst many others, the Nuttall Tick Collection (Keirans, 1985) and the Tanzanian Tick Survey Collection (Yeoman & Walker, 1967). Among other sources from which we have obtained data are the Onderstepoort Tick Collection, built up largely by Gertrud Theiler; the Namibian Tick Survey Collection, by courtesy of Heloise Heyne; the collection of the Musée Royal de l'Afrique Centrale, Tervuren, Belgium, including many specimens obtained during colonial times in the Democratic Republic of Congo and now curated by F. Puylaert; and numerous collections from East Africa and South Africa accumulated by J.B.W. and I.G.H. respectively.

Details of various individual collections, obtained as either gifts or loans, are detailed elsewhere in the text, as are literature references.

DESCRIPTIONS

Brief descriptions are given of every available stage of all the *Rhipicephalus* species that are presently regarded as valid. Explanations of their specific names are given and their synonyms, if any, are listed. Often the descriptions, especially those of the nymphs and larvae, are based on laboratory-reared specimens. They are illustrated with line drawings of the adults and, in almost every case, with scanning electron micrographs (SEMs) of each of the known stages. The descriptive terms used are defined in the glossary. Some also appear on Figs 1 and 2.

Measurements, quoted in mm, are given simply as guides to the size of the species under discussion. They are not statistically valid.

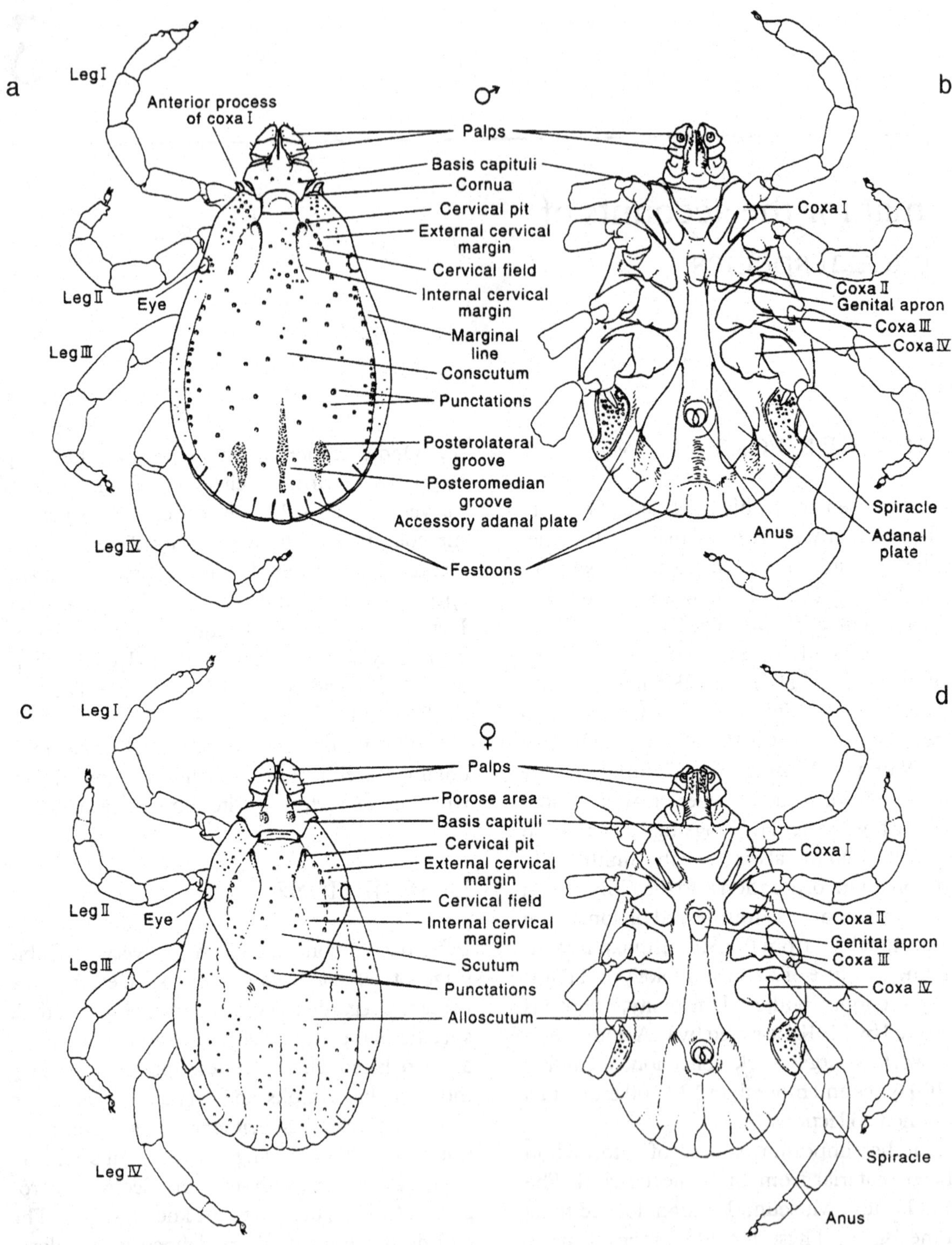

Figure 1. *Rhipicephalus* sp. adults showing morphological features. Male: (a) dorsal view; (b) ventral view. Female: (c) dorsal view; (d) ventral view.

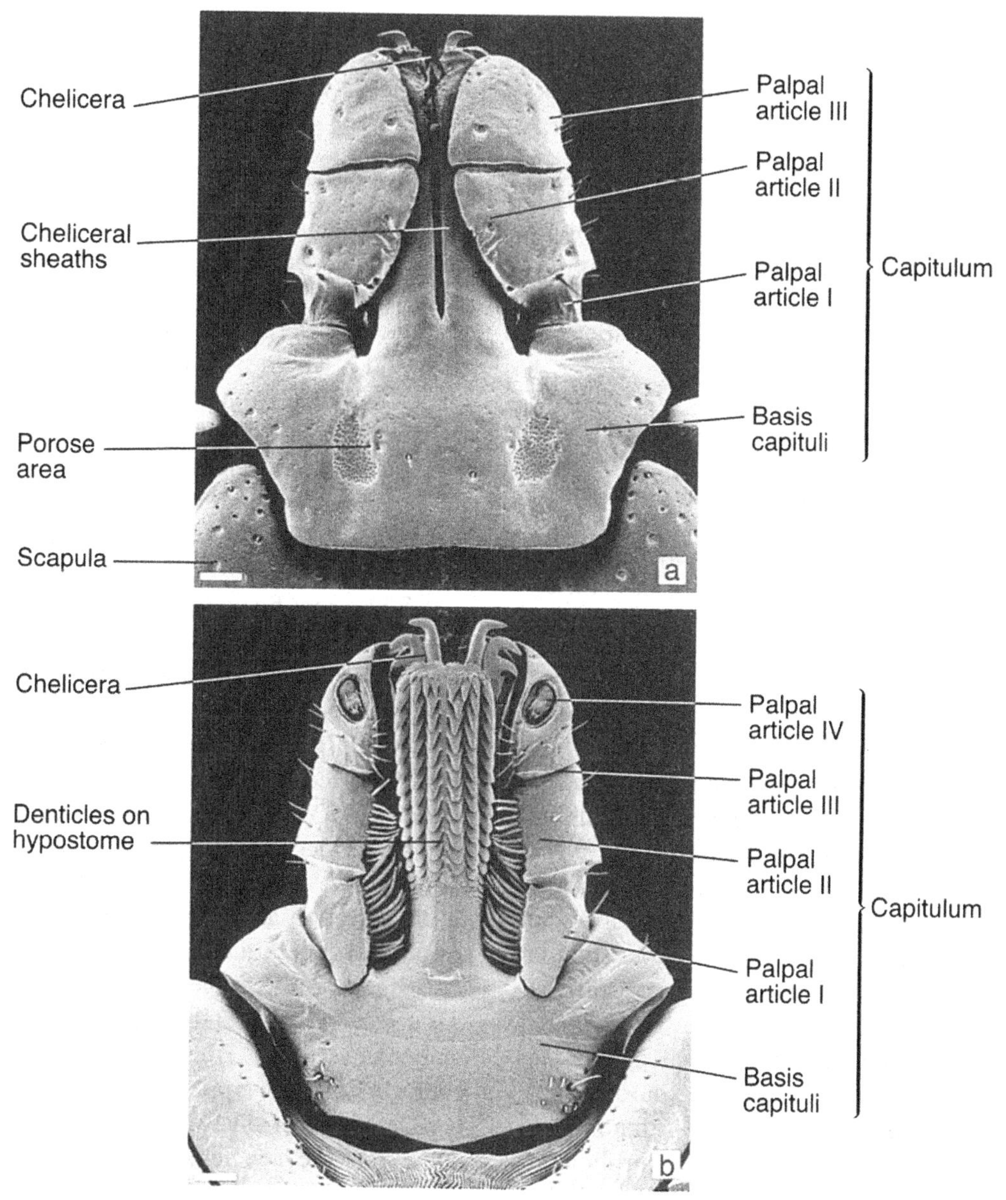

Figure 2. *Rhipicephalus longicoxatus* female capitulum showing morphological features: (a) dorsal view; (b) ventral view. Scale bars represent 0.10 mm. SEMs by J.F. Putterill.

Sometimes it has been possible to give an idea of the potential range of sizes seen in a species, but in some cases very few specimens were available to us for measurement. Rarely, for example for the seldom-collected Saharan species *R. fulvus*, we have merely cited measurements from the literature.

The differences between the length and breadth of various morphological features are described in standardized terms. If, for example, the mean of the smallest and greatest measurements of the length of a morphological feature differs by less than 1% from the mean of the smallest and greatest measurements of its width it is described as being either 'as broad as long' or, conversely, 'as long as broad'. If this difference ranges from 1 % to less than 5% it is described as 'slightly broader than long' or 'slightly longer than broad'. With a difference of 5% to less than 20% it is described simply as 'broader than long' or 'longer than broad'. When the difference is greater than 20% it is described as 'much broader than long' or 'much longer than broad'.

One of the greatest difficulties experienced when attempts are made to identify rhipicephalids is to establish the range of morphological variation shown by individual species. Warburton (1912) began his valuable paper on the genus by saying:

> 'The identification of species of *Rhipicephalus* is likely to give more trouble than is the case with any other genus of Ixodidae, for while, on the one hand, there are few species which depart greatly from the general type, on the other hand the range of variation within the species is extremely great. In no genus is it so dangerous to describe a new species from a single individual, especially if the specimen be a female'.

He went on to discuss the variations seen in some of the species and the consequent problems encountered with their identification. He regarded the following features as the most useful for identifying the males: the shape of the basis capituli, including the position and precise form of its lateral angles; the anterior process of coxae I; the presence or absence, and shapes, of the various grooves and other depressions on the conscutum; the punctation pattern, and the shapes of the spiracles and adanal plates. He commented that, in the females, the porose areas and shape of the scutum are also worth noting.

A major factor influencing this morphological variation is nutrition. In general some species are large while others are small. It has been shown experimentally, though, that within a single species a great range in the size of the adults can occur that is directly related to the degree of engorgement of the immature stages. These differences in size are usually accompanied by other morphological disparities, for example in the shape of the basis capituli or in the punctation pattern.

Our descriptions are followed by notes on identification. These comprise information on aspects such as previous misidentifications of the tick in question, its possible relationships within the genus and comparisons with other morphologically similar species.

HOSTS

The common names used for domestic animals include the various breeds of each of them. Several breeds of the same species occur in different areas and it would be impossible to list the tick species recorded from each of these breeds separately. In the case of cattle, records from the two species that have been described, *Bos taurus* and *Bos indicus*, as well as the records from all the breeds are listed together under the common name 'cattle'.

The scientific nomenclature for wild mammalian hosts that we have used is based mainly, but not exclusively, on that given by the numerous contributors to Wilson & Reeder (1993). For the primate genera *Cercopithecus* and *Papio* we have followed the classification of Napier & Napier (1967). In most cases we have used the host's binomen, i.e. its generic plus its specific name, but we have gone to the subspecific level of classification given by Ansell (1971) for the artiodactyl genera *Alcelaphus* and *Damaliscus* because marked differences exist between the various subspecies of hartebeests. Most of the common names that we have added for the hosts conform to those appearing in Corbet & Hill (1991). Occasionally we have quoted a common name from the new field guide for African mammals by Kingdon (1997), in which recent changes in the nomenclature of African mammals have been incorporated. This useful publication includes illustrations of many species as well as numerous distribution maps for individual species or species groups. Some names for the Palaearctic and Indian mammals have been taken from Ellerman & Morrison-Scott (1951).

For various reasons it has not always been feasible to give the host's binomen. Sometimes it was originally designated only in vague terms, e.g. 'jackal', 'wild pig', 'duiker' or 'hare'. In such cases it has been impossible to determine the exact species involved. Sometimes we have been unable to decide to which host species certain tick records should be assigned when the nomenclature of a previously monotypic entity has been emended to encompass two or more species. This has been the case with both *Phacochoerus* spp. (warthogs) and *Potamochoerus* spp. (bushpigs). These have been regarded as monotypic genera by most authorities in recent years but P. Grubb (in Wilson & Reeder, 1993) has now designated two species in each. He considers that *Phacochoerus aethiopicus* presently occurs in 'N Kenya and Somalia', and *P. africanus* 'outside [the] forest zone of Africa from Senegal to Somalia, south to S Africa'. He regards *Potamochoerus porcus* as an inhabitant of the rainforests of Africa from Senegal to the Democratic Republic of Congo (formerly Zaire) and designates as *P. larvatus* the species occurring in 'Ethiopia, S Sudan and E Zaire south to E and S South Africa, west to N Botswana and Angola'.

We have also had problems with some records from African porcupines. Basically *Hystrix africaeaustralis* occurs in the southern half of the continent, from about the level of the mouth of the Congo River eastwards to Rwanda, Uganda, Kenya, W. and S. Tanzania and thence southwards. *Hystrix cristata* occurs in the northern half of the continent. In parts of central Africa, however, these two species are sympatric. In addition even mammalogists sometimes confess to difficulty in distinguishing between them morphologically. In cases of doubt, therefore, we have assigned records merely to *Hystrix* sp.

References to other publications on African mammals that we have consulted appear in appropriate places in the text.

For birds we have used both the scientific and common names listed by Howard & Moore (1991).

In the host tables that are given for most *Rhipicephalus* species we have tried to indicate the importance or otherwise of each host included by noting the number of collections of the tick in question recorded from it. It should be emphasized that the numbers given in these tables represent numbers of collections of a particular *Rhipicephalus* species from a host, not the number of individual ticks of that species on the host in question. A host may have a burden of several hundred ticks but this would be counted as a single collection. These figures have been compiled from our own data plus information from the literature. Obviously they are often extremely conservative, particularly for the commoner rhipicephalids. In the case of some of the commonest species found on cattle, for example *R. appendiculatus* and *R. duttoni*, we have simply stated that they are 'commonly parasitized' as any attempt to give actual figures for these species would be completely meaningless. For other rhipicephalid species, though, when an author stated that collections were made from 'cattle', without giving any indication of the number of animals examined or noting how many were parasitized by the tick in question, we have added only 'one' to the number of collections recorded. But if, for example, an author stated that collections were made from cattle monthly for a year, and the tick was recorded from March to September, a total of 'seven' would be added to the number of collections recorded. We have also included data on the rhipicephalids collected during various long-term surveys of the parasites of domestic and wild animals. The animals concerned are listed in the introduction to the host/parasite tables. Obviously these data will considerably increase the number of records of the particular ticks found during such surveys and may give a false impression of host preference. If possible we have indicated which hosts are parasitized by the immature stages of the ticks under discussion.

Whenever we consider that a specific tick prefers a certain host that tick's binomen appears in **bold** under the host's name in the host/parasite list.

For a few species there are, for various reasons, doubts about some records that have been included. These are listed as 'unconfirmed'.

Despite the shortcomings that are undoubtedly inherent in these methods we trust that our findings will be more informative than simply listing the names of a tick's known hosts would have been.

ZOOGEOGRAPHY

In the Afrotropical region the distributions of the two most important rhipicephalid vectors of pathogens, *R. appendiculatus* and *R. zambeziensis*, have been mapped as precisely as possible. The locations of almost all the other *Rhipicephalus* species in this region are indicated with symbols mapped on 1° squares. Occasionally, though, it has only been possible to register the presence of a tick in a country, using a single symbol and giving no indication of the areas where it occurs. The distributions of species occurring outside the Afrotropical region have been indicated in very broad terms only.

As in the case of the hosts the information we present is derived partly from our own data and partly from records in the literature.

The current names of African countries are shown on Map 1. Although their boundaries have rarely been altered since colonial times the names of many of these countries have been changed, sometimes more than once. Naturally the older names feature in the earlier literature on African ticks. To avoid confusion we are therefore listing below those current names (in bold) that have been changed together with their earlier names (in parenthesis), and some in other European languages, as follows: **Algeria** (Algérie); **Angola** (Portuguese Congo; Portuguese West Africa); **Benin** (Dahomey); **Botswana** (Bechuanaland Protectorate); **Burkina Faso** (Upper Volta; Haute Volta); **Burundi** (Ruanda-Urundi (in part)); **Cameroon** (British Cameroons; plus Kamerun, later Cameroun); **Central African Republic** (Ubangi Shari; Oubangui Chari); **Chad** (Tchad); **Congo** (Congo Français; Moyen Congo; People's Republic of Congo); **Democratic Republic of Congo** (Congo Free State; Belgian Congo; Congo Belge; Zaire); **Djibouti** (French Somaliland; Afars and Issas); **Egypt** (United Arab Republic); **Equatorial Guinea** (Spanish Guinea; Rio Muni); **Eritrea** (Italian Eritrea; Ethiopia (in part)); **Ethiopia** (Abyssinia); **Ghana** (Gold Coast); **Guinea** (French Guinea; Guinée Français); **Guinea-Bissau** (Portuguese Guinea; Guinée Portugaise; Guiné Portuguesa); **Ivory Coast** (Côte d'Ivoire); **Kenya** (East African Protectorate; British East Africa); **Lesotho** (Basutoland); **Libya** (comprising Tripolitania, Cyrenaica & Fezzan); **Malawi** (Nyasaland Protectorate); **Mali** (French Sudan, Sudan Français); **Mauritania** (Mauritanie); **Morocco** (Maroc; Spanish Morocco (in part)); **Mozambique** (Portuguese East Africa, Moçambique); **Namibia** (Deutsch Südwestafrika, South West Africa); **Republic of South Africa** or **South Africa** (Union of South Africa); **Rwanda** (Ruanda-Urundi (in part)); **Senegal** (Sénégal); **Somalia** (Somaliland Protectorate (British) plus Italian Somaliland (Somalia)); **Sudan** (Anglo-Egyptian Sudan); **Tanzania** (German East Africa; Deutsch Ostafrika; Tanganyika plus Zanzibar); **Tunisia** (Tunisie); **Western Sahara** (Spanish West Africa; Africa Occidental Española; Rio de Oro); **Zambia** (Northern Rhodesia); **Zimbabwe** (Southern Rhodesia; Zimbabwe Rhodesia).

During the colonial era the French territories in the Afrotropical region, with the exception of French Somaliland, were divided into two major groups. French West Africa (Afrique Occidentale Française) comprised Sénégal, Mauritanie, Guinée Française, Côte d'Ivoire, Soudan Français, Haute Volta, Niger, Dahomey and the Trust Territory of Togo. French Equatorial Africa (Afrique Équatoriale Française) comprised Tchad, Cameroun, Oubangui Chari, Moyen Congo and Gabon. Further south in the continent the three British territories of Northern Rhodesia, Southern Rhodesia and Nyasaland were amalgamated into the Federation of Rhodesia and Nyasaland from 1953–1963.

Authors have sometimes listed the geographical coordinates of localities where specific ticks have been recorded. Otherwise these coordinates have either been read from maps of the

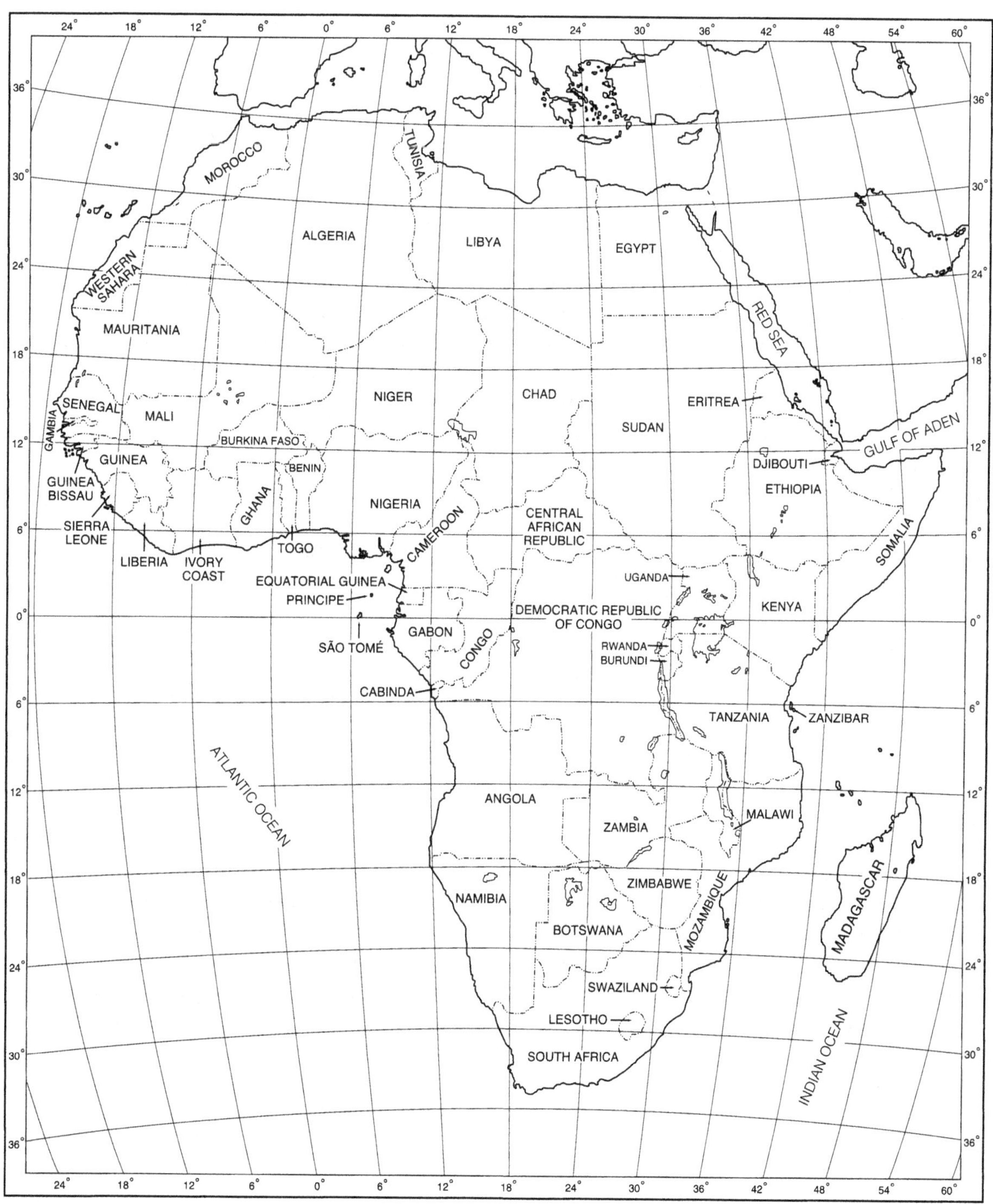

Map 1. Africa: political divisions.

areas concerned or obtained either from *The Times Atlas of the World – Comprehensive Edition* (Bartholomew & Times Books, 1992) or from the excellent gazetteers for individual countries produced by the United States Board on Geographic Names. (At the time of writing these gazetteers were obtainable from the following address: USGS Map Sales, Box 25286, DFC, Denver, CO 80225-0046, U.S.A.). These sources of information have rarely failed us, although it has sometimes been impossible to determine which of the coordinates for several places, or farms, listed under the same name are relevant in the circumstances. We have usually tried to avoid records based on single ticks, especially females, and have not included those of ticks collected from animals at abattoirs unless the origin of these animals was specifically known.

When possible we have included brief notes on the general ecological conditions in the areas favoured by individual tick species, particularly on altitude, rainfall (Jackson, 1961) and vegetation (White, 1983).

Readers must remember that tick distributions, like those of any other living organisms, may vary from time to time in response to changes in the distribution of their hosts and/or expanding human populations. These shifts can sometimes take place rapidly over large areas. In Africa probably the most widespread environmental change during approximately the last 25 years has been the degradation of the fragile ecosystem along the southern borders of the Sahel. This has been caused by persistent drought exacerbated by factors such as deforestation and increasing cultivation. As a result the Sahel has moved southwards, displacing and impoverishing entire communities. There are a few references, mentioned later in the text, to the effects that this had also had on tick populations in northern Senegal. Doubtless such effects are far more widespread but they remain to be documented. We have of necessity relied largely on the findings of Morel (1969) and his colleagues regarding tick distribution in West Africa. This information may well be somewhat out-of-date now, but we have usually had no means of revising it.

DISEASE RELATIONSHIPS

The pathogens known to be transmitted by rhipicephalids have been included in the accounts of the individual tick species. We have also appended disease transmission tables at the end of this book for both animals and humans in which we have listed the diseases and the pathogens that cause them, followed by details regarding their specific tick vectors.

BASIC REFERENCES

Comprehensive lists of the references concerning every *Rhipicephalus* sp. described up to 1969, valid or otherwise, and their hosts and geographical distribution, have been compiled by Doss *et al.* (1974–1978). In these publications literature citations appear under the name(s) of the author(s) and date of publication. Usually, but not invariably, some indication of the contents of the reference in question is also given. When required the complete title of a reference is obtainable either from the author index of the *Index-Catalogue of Medical and Veterinary Zoology* (1932–1982), published by the United States Department of Agriculture, or from the comprehensive bibliography published by Hoogstraal (1970–1988).

We have listed pertinent references at the ends of chapters and after each account of an individual species. These lists commonly appear in two sections. References that apply mainly or exclusively to a particular chapter or species account are given first. They are always quoted in the preceding text and listed in full. Although they may be referred to more than once in the book they are often relatively restricted in scope. Some, indeed, deal with only one species of *Rhipicephalus*. These specific references are usually followed, after the words 'Also see the following Basic References (pp. 12–14)', by a number of basic references, cited under the name(s) of their author(s) and date of publication only. These basic references are sometimes, but not invariably, quoted in the preceding text. They are always wider in scope. They are either books, such as the mammalian checklists and accounts of rainfall and vegetation mentioned above, or publications such as those describing the tick fauna of a particular country or region. With the exception of Warburton (1912), who is quoted in full later as the author of three specific names, the full titles of basic references are listed once only, below.

Aeschlimann, A. (1967). Biologie et écologie des tiques (Ixodoidea) de Côte d'Ivoire. *Acta Tropica*, **24**, 281–405.

Ansell, W.F.H. (1971). Artiodactyla. In *The Mammals of Africa: An Identification Manual*, ed. J. Meester & H.W. Setzer, pp. 1–84. Washington, DC: Smithsonian Institution Press.

Bartholomew & Times Books (1992). *The Times Atlas of the World – Comprehensive Edition.* London: Times Books, a Division of HarperCollins.

Clifford, C.M. & Anastos, G. (1962). Ticks. *Exploration du Parc National de l'Upemba. Mission G.F. de Witte (1946–1949)*, **Fascicule 66**, 45 pp.+ 23 plates.

Clifford, C.M. & Anastos, G. (1964). Ticks. *Exploration du Parc National de la Garamba. Mission H. de Saeger (1949–1952)*, **Fascicule 44**, 40 pp.

Corbet, G.B. & Hill, J.E. (1991). *A World List of Mammalian Species*, 3rd edn. London: Natural History Museum Publications & Oxford University Press.

Doss, M.A., Farr, M.M., Roach, K.F. & Anastos, G. (1974). *Index-Catalogue of Medical and Veterinary Zoology. Special Publication No. 3. Ticks and Tickborne Diseases, I. Genera and Species of Ticks. Part 3. Genera O–X.* University of Maryland & United States Department of Agriculture. Washington, DC: U.S. Government Printing Office.

Doss, M.A., Farr, M.M., Roach, K.F. & Anastos, G. (1974). *Index-Catalogue of Medical and Veterinary Zoology. Special Publication No. 3. Ticks and Tickborne Diseases, II. Hosts. Part 1: A–F. Part 2: G–P. Part 3: Q–Z.* University of Maryland & United States Department of Agriculture. Washington, DC: U.S. Government Printing Office.

Doss, M.A. & Anastos, G. (1977). *Index-Catalogue of Medical and Veterinary Zoology. Special Publication No. 3. Ticks and Tickborne Diseases, III. Checklist of Families, Genera, Species, and Subspecies of Ticks.* University of Maryland & United States Department of Agriculture. Washington, DC: U.S. Government Printing Office.

Doss, M.A., Farr, M.M., Roach, K.F. & Anastos, G. (1978). *Index-Catalogue of Medical and Veterinary Zoology. Special Publication No. 3. Ticks and Tickborne Diseases, IV. Geographical Distribution of Ticks.* University of Maryland & United States Department of Agriculture. Washington, DC: U.S. Government Printing Office.

Durden, L.A., Keirans, J.E. & Oliver, J.H., Jr (1996). The U.S. National Tick Collection: a vital resource for systematics and human and animal welfare. *American Entomologist*, **42**, 239–43.

Elbl, A. & Anastos, G. (1966). Ixodid ticks (Acarina, Ixodidae) of Central Africa. Vol. III. Genus *Rhipicephalus* Koch, 1844. *Annales du Musée Royal de l'Afrique Centrale, Serie in 8°*, **No. 147**, x + 555 pp.

Ellerman, J.R. & Morrison-Scott, T.C.S. (1951). *Checklist of Palaearctic and Indian Mammals 1758–1946.* London: British Museum (Natural History).

Filippova, N.A. (1997). *Ixodid ticks of the subfamily Amblyomminae. Fauna of Russia and Neighbouring Countries. Arachnoidea. Volume IV, issue 5*, 436 pp. + 44 pls. St Petersburg, Russia: Nauka Publishing House. [In Russian, English summary].

Hoogstraal, H. (1956). *African Ixodoidea. I. Ticks of the Sudan (With Special Reference to Equatoria Province and With Preliminary Reviews of the Genera* Boophilus, Margaropus *and* Hyalomma*).* Research Report NM 005 050. 29.07, 1101 pp. Washington, DC: Department of the Navy, Bureau of Medicine and Surgery.

Hoogstraal, H. (1970–1988). *Bibliography of Ticks and Tickborne Diseases from Homer (about 800 B.C.) to 31 December 1984*, Vols. 1–8. Cairo, Egypt: United States Naval Medical Research Unit No. 3.

Howard, R. & Moore, A. (1991). *A Complete Checklist of Birds of the World*, 2nd edn. London: Academic Press.

Jackson, S.P. (1961). *Climatological Atlas of Africa.* [Commission for Technical Co-operation in Africa south of the Sahara (CCTA/CSA), Joint Project No. 1]. Pretoria: Government Printer.

Keirans, J.E. (1985). *George Henry Falkiner Nuttall and the Nuttall Tick Catalogue.* United States Department of Agriculture, Agricultural Research Service, Miscellaneous Publication No. 1438, 1785 pp. Washington, DC: U.S. Government Printing Office.

Kingdon, J. (1997). *The Kingdon Field Guide to African Mammals.* London: Academic Press.

Matthysse, J.G. & Colbo, M.H. (1987). *The Ixodid Ticks of Uganda together with Species pertinent to Uganda because of their Present Known Distribution.* College Park, MD: Entomological Society of America.

Morel, P.C. (1969). *Contribution à la Connaissance de la Distribution des Tiques (Acariens, Ixodidae et Amblyommidae) en Afrique Éthiopienne Continentale,* 388 pp. + annexure cartographique, cartes 1–62. DSc thesis, University of Paris.

Morel, P.C. (1980). *Study on Ethiopian Ticks (Acarida, Ixodida).* Maisons-Alfort, Paris: Institut d'Élevage et de Médecine Vétérinaire des Pays Tropicaux.

Moritz, M. & Fischer, S.-C. (1981). Die Typen der Arachniden-Sammlung des Zoologischen Museums Berlin. IV. Ixodei. *Mitteilungen aus dem Zoologischen Museum in Berlin,* 57, 341–64.

Napier, J.R. & Napier, P.H. (1967). *A Handbook of Living Primates.* London: Academic Press.

Pegram, R.G. (1979). *Ticks (Ixodoidea) of Ethiopia with Special Reference to Cattle and a Critical Review of the Taxonomic Status of Species within the* Rhipicephalus sanguineus *Group,* xi + 169 pp., 10 tables, 36 maps & figures, 123 plates. M. Phil. thesis, Brunel University.

Pegram, R.G. (1984). *Biosystematic Studies on the Genus* Rhipicephalus*: The* R. sanguineus *and* R. simus *Groups (Ixodoidea, Ixodidae),* 160 pp. + 90 plates. PhD thesis, Brunel University.

Santos Dias, J.A.T. (1960). Lista das carraças de Moçambique e respectivos hospedeiros. III. *Anais dos Serviços de Veterinária e Indústria Animal de 1953–1954,* **No. 6**, 213–87.

Santos Dias, J.A.T. (1963). Contribuição para o estudo da sistemática dos Ácaros da subordem Ixodoidea Banks, 1894. I. Família Ixodidae Murray, 1877. *Memórias e Estudos do Museu Zoológico da Universidade de Coimbra,* **No. 285**, 34 pp.

Santos Dias, J.A.T. (1993). Some data concerning the ticks (Acarina-Ixodoidea) presently known in Mozambique. *Garcia de Orta, Sér. Zool., Lisboa,* **18 (1991)**, 27–48.

Scaramella, D. (1988). Studio monografico sugli Ixodidi e gli Argasidi della Somalia. *Acta Medica Veterinaria,* **34**, 91–172.

Skinner, J.D. & Smithers, R.H.N. (1990). *The Mammals of the Southern African Subregion.* Pretoria: University of Pretoria.

Sousa Dias, V. (1950). Subsídios para o estudo dos ixodídeos de Angola. *Pecuaria,* **2**, 127–280 (1947–1948). (Reprint pp. 1–154).

Theiler, G. (1947). Ticks in the South African Zoological Survey Collection. Part VI. Little known African Rhipicephalids. *Onderstepoort Journal of Veterinary Science and Animal Industry,* **21**, 253–300.

Theiler, G. (1962). *The Ixodoidea Parasites of Vertebrates in Africa South of the Sahara (Ethiopian Region).* Project S 9958, 260 pp. Report to the Director of Veterinary Services, Onderstepoort. Mimeographed.

Walker, J.B. (1974). *The Ixodid Ticks of Kenya. A Review of present Knowledge of their Hosts and Distribution.* London: Commonwealth Institute of Entomology.

Walker, J.B., Mehlitz, D. & Jones, G.E. (1978). *Notes on the Ticks of Botswana.* Eschborn: German Agency for Technical Cooperation, Ltd. (GTZ).

Warburton, C. (1912). Notes on the genus *Rhipicephalus,* with the description of new species, and the consideration of some species hitherto described. *Parasitology,* **5**, 1–20.

White, F. (1983). *The Vegetation of Africa. A Descriptive Memoir to Accompany the Unesco/AETFAT/UNSO Vegetation Map of Africa.* Paris: Unesco.

Wilson, D.E. & Reeder, D.M. (1993). *Mammal Species of the World – A Taxonomic and Geographic Reference,* 2nd edn. Washington and London: Smithsonian Institution Press.

Yeoman, G.H. & Walker, J.B. (1967). *The Ixodid Ticks of Tanzania. A Study of the Zoogeography of the Ixodidae of an East African Country.* London: Commonwealth Institute of Entomology.

Zumpt, F. (1949). Preliminary study to a revision of the genus *Rhipicephalus* Koch. Key to the adult ticks of the genus *Rhipicephalus* and description of two new species. *Moçambique,* **No. 60**, 57–123. (The date of issue for this publication is given as 1950 in Hoogstraal's Bibliography of Ticks and Tickborne Diseases (see above)).

4

Glossary

Morphological features used in descriptions of *Rhipicephalus* spp. and taxonomic and other terms.

Accessory adanal plates	Pair of small sclerotized structures lateral to the adanal plates in males. May be present or absent. Also called accessory plates.
Adanal plates	Ventral pair of large sclerotized structures lateral to the anus in males.
Alloscutum	The portion of the body wall in females and immatures posterior to the scutum that expands enormously to accomodate the large volume of blood ingested when these stages feed.
Anterior	Towards the front end.
Anterior process	Referring to anterior process on coxae I. When seen from the dorsal surface it is a sclerotized anterior projection of these coxae protruding in front of the scapulae as in *R. appendiculatus*.
Anus	Posterior opening of the alimentary tract, situated ventrally and medially, posterior to the last pair of legs.
Article	Distinct articulated portion of a jointed appendage.
Basal	Closest to the origin or insertion.
Basis capituli	Basal portion of the capitulum on which the mouthparts are attached; in the genus *Rhipicephalus* the basis capituli is hexagonal.
Bifid	Deeply cleft, forming two long spurs. (*See* coxa).
Capitulum	Anterior movable portion of the body which includes the basis capituli, palps, hypostome and chelicerae. The tick's 'head'.
Caudal process	Distinct projection arising posteromedially from the median posterior end of engorged males of many *Rhipicephalus* species.
Central field	Area on the conscutum or scutum between and delimited by the internal cervical margins.
Cervical fields	Areas anteriorly on the conscutum or scutum delimited by the internal and external cervical margins. These fields are

variously shaped, depending on the species. They are often depressed into the scutal surface and may be flat, roughened or smooth.

Cervical grooves — *See* internal cervical margins.

Cervical pits — Paired anterior depressions in the scuta of both sexes; vary in shape and depth.

Chaetotaxy — Arrangement and nomenclature of setae.

Chelicera (pl. chelicerae) — Paired cutting mouthparts lying dorsal to the hypostome, each terminating in an internal (fixed) and external (moveable) digit.

Clavate — Clubbed; thickening gradually toward the tip as in the alloscutal setae of *R. humeralis* females.

Conscutum — Sclerotized plate posterior to the capitulum that covers almost the entire dorsal surface of male *Rhipicephalus*.

Cornua — Small, paired projections extending from the dorsal, posterolateral angles of the basis capituli.

Coxa (pl. coxae) — Sclerotized plate on the venter representing the first leg segment to which the trochanter is attached. From anterior to posterior the coxae are designated by Roman numerals I, II, III and IV. *Rhipicephalus* species adults have bifid first coxae.

Coxal spurs — Large or small projections extending from the posterior margins of the coxae.

Cusps — Small points or projections extending medially from the internal aspect of the adanal plates.

Denticles — Small, individual, recurved projections or 'teeth' on the ventral surface of the hypostome.

Distal — Farthest from the point of attachment or origin.

Dorsal — Pertaining to the back or top of the body.

Dorsal prolongation — Posterodorsal extension of the spiracle.

Dorsum — The entire dorsal surface of the body.

Emargination — Anterior indentation or excavation in the scutum between the scapulae that receives the basis capituli.

Engorged — Enlargement of a tick following a blood meal. Pronounced distention is seen in females, nymphs and larvae. Males, because they have a sclerotized conscutum, cannot take in such large blood meals as females and immatures, and are therefore incapable of the same degree of engorgement.

et al. [Latin] — (*et alii*); and others.

External cervical margins — Faint or definite ridges or punctations found anterolaterally on the scutum or conscutum. May be present or absent. Also known as lateral carinae.

Eyes — A pair of lens-like structures at the lateral margins of the scutum. May be flat, convex or beady; sometimes delimited by grooves (orbited).

Festoons — Eleven uniform rectangular areas, separated by distinct grooves, located on the posterior margin of the

	tick's body; distinct in males and unengorged females and immatures, but difficult to discern in engorged specimens of the latter.
Genital aperture	External opening of the genital organs; located anteriorly on the ventromedian line, posterior to the basis capituli. In the genus *Rhipicephalus* only the female genital aperture has taxonomic value.
Genital apron	The area within the lateral margins of the female genital aperture.
Goblets	Round structures located in the spiracle; they may be small and numerous or relatively large and few.
Haller's organ	Sensory structure located on the subapical dorsal surface of tarsus I.
Hyaline	Glassy; tending towards the transparent.
Hyaline flaps	External projections from the lateral margins of the female genital aperture.
Hypostome	Median ventral structure of the mouthparts that lies parallel to and between the palps, ventral to the chelicerae, and is immovably attached to the basis capituli. It bears the recurved denticles or 'teeth' and acts as a holdfast organ.
Idiosoma	All of the tick's body exclusive of the capitulum.
Incertae sedis [Latin]	Of uncertain taxonomic position.
in litt. [Latin]	(*in letteris*) in correspondence.
Inornate	Absence of a colour pattern on the conscutum of males and scutum of females.
Internal cervical margins	A pair of grooves or ridges on the conscutum of males or scutum of females extending posteriorly from the inner angles of the cervical pits. May be continuous or interrupted, shallow or deep, faint or absent.
Integument	Outer covering or cuticle of the tick's body.
Lapsus calami [Latin]	Slip of the pen.
Lateral	Towards the side.
Lateral carinae	*See* external cervical margins.
Legs	Segmented appendages of which nymphs and adults have four pairs and larvae have three pairs. From anterior to posterior the legs are identified by Roman numerals I, II, III, IV. The segments from proximal to distal are designated coxa, trochanter, femur, genu, tibia, metatarsus, and tarsus.
Macula	Large sclerotized structure located in the spiracle of adult *Rhipicephalus*. It may be of variable size, shape and location.
Marginal lines	Lines or grooves running along the lateral margins of the male's conscutum. They may be faint, deep, punctate or impunctate, and delimit one or more festoons. (Note: a series of discrete punctations does not constitute a marginal line; a series of contiguous punctations does.)
Medial	Situated in or extending toward the median axis of the body.

Median	Lying in a plane that divides the body into right and left halves.
Morphological	Pertaining to form and structure.
Nomen nudem [Latin]	Naked name. A name that fails the rules for availability then in force.
Orbited	Descriptive term for the eyes of some species of *Rhipicephalus* that are surrounded by grooves in the conscutum or scutum.
Ornamentation	Enamel-like colour pattern that is superimposed on the base colour of the integument in adults of the following four species of *Rhipicephalus*, *R. dux*, *R. humeralis*, *R. maculatus* and *R. pulchellus*.
Ornate	Definite colour pattern on the conscutum of males and scutum of females of four species of *Rhipicephalus*. (*See* ornamentation).
Palps	Paired articulated appendages located anterolaterally upon the basis capituli and lying parallel to the hypostome. Article I (segment I) is proximal, article IV (segment IV) is distal.
Pedicel	A narrow stalk-like first palpal article as in *R. glabroscutatum*.
Periphery	Circumference or outer margin.
Porose areas	A pair of pitted areas, usually depressed and round to oval in shape, on the dorsal surface of the basis capituli of female ixodid ticks; they contain the openings through which the anti-oxidant, produced by Gene's organ for stabilizing the egg lipids, is discharged.
Posterior	Toward the rear end.
Posterolateral grooves	Paired depressions lateral to the posteromedian groove in the conscutum of almost all male *Rhipicephalus*.
Posteromedian groove	A depression situated slightly anterior to the median festoon and between the posterolateral grooves in almost all male *Rhipicephalus*.
Protuberance	Any elevation above the surface.
Proximal	Nearest to the point of attachment or origin.
Pseudoscutum	In males the anterior portion of the conscutum which corresponds to the female scutum.
Punctations	Pits in the surface of the cuticle, present on the conscutum and scutum, and sometimes on the basis capituli of all *Rhipicephalus* adults. These pits vary in number, size and depth, and are often arranged in patterns that are specifically characteristic.
Reticulate	*See* shagreened.
Rugose	Roughened or slightly ridge-like in appearance.
Scapulae	Anterior angles or 'shoulders' of the conscutum or scutum that extend forwards on either side of the emargination.
Sclerotized	Hardened in definite areas by the deposition or formation of organic or inorganic substances in the cuticle.
Scutum	Sclerotized dorsal plate posterior to the capitulum in

	females, nymphs and larvae. It covers one-third to one-half of the dorsum in unengorged specimens.
Segment	Distinct articulated portion of a palp or a leg. Synonym of article.
Sensu lato [Latin]	In the broad sense.
Sensu stricto [Latin]	In the strict sense.
Seta (pl. setae)	Slender hollow extensions of the epidermal layer of the cuticle.
Setiferous punctations	Usually relatively large punctations each bearing a seta on the conscutum of males and the scutum and alloscutum of females and nymphs.
Shagreened	Granular in texture; displaying a pebbled or grainy appearance.
'*Simus*' pattern	Large conscutal punctations roughly arranged in four longitudinal rows, like the punctation pattern found in *R. simus*.
sp. nov. [Latin]	(*Species nova*); new species.
Spiracles	Paired plates located ventrolaterally and posterior to coxae IV in adults and nymphs; may be oval, rounded or comma-shaped. They are the external portion of the respiratory system.
Spurs	Coxal spurs are projections from the posterior margin of the coxae; projections on the inner side are called internal spurs; those on the outer side are called external spurs. Spurs may also be present on the ventral posterolateral border of the basis capituli.
Subterminal	Before the end, or not quite attaining the end.
Stat. nov. [Latin]	(*Status nova*); new status.
Tarsus	The terminal or most distal segment of the leg.
Type	A term used alone, or as part of a compound term, to denote a particular kind of specimen or taxon. Allotype, a designated specimen of opposite sex to the holotype. Holotype, a single specimen designated as the name-bearing type of a species or subspecies when it was established, or the single specimen on which such a taxon was based when no type was specified. Lectotype, a syntype designated as the single name-bearing type specimen subsequent to the establishment of a nominal species or subspecies. Neotype, a single specimen designated as the name-bearing type of a nominal species or subspecies for which no holotype, or lectotype, or syntype(s), or prior neotype is believed to exist. Paralectotype, each specimen of a former syntype series remaining after the designation of a lectotype. Paratype, each specimen of a type series other than the holotype. Syntype, each specimen of a type series from which neither a holotype nor a lectotype has been designated.
Venter	Entire ventral or underside of the body.
Ventral	Pertaining to the underside of the body.
Vide [Latin]	See.

5

Rhipicephalus species names of the world

[Numbered species names in **bold** are considered to be valid at this time].

Rhipicephalus anatolicum (Koch, 1844)

Dönitz, W. 1905. *Sitzungsberichte der Gesellschaft naturforschender Freunde zu Berlin*, (4): 114.

Dönitz, in discussing Koch's naming of new *Hyalomma* species from different areas, mentioned the names *R. hispanum*, *lusitanicum* and *anatolicum*. The generic initial '*R.*' before the three species names is, in all probability, a typographical error and not a change of generic status from *Hyalomma* to *Rhipicephalus*.

[= *Hyalomma anatolicum* Koch]

Rhipicephalus annulatus (Say, 1821)

Neumann, L.G. 1897. *Mémoires de la Société Zoologique de France*, **10**: 407, 419 (key), figs. 37–41.

New combination for *Ixodes annulatus* Say, 1821.

[= *Boophilus annulatus* (Say)]

Rhipicephalus (*Boophilus*) *annulatus argentina* (*sic*) Neumann, 1901

Neumann, L.G. 1904. *Archives de Parasitologie*, 8: 450.

New combination for *Rhipicephalus annulatus argentinus*.

[= *Boophilus microplus* (Canestrini)]

Rhipicephalus annulatus argentinensis Neumann, 1901

Mémoires de la Société Zoologique de France, **14**: 361.

[*Lapsus calami* for *R. annulatus argentinus*]

Rhipicephalus annulatus argentinus Neumann, 1901

Mémoires de la Société Zoologique de France, **14**: 280.

From 160 or so specimens [sex/stage not stated but ♂ unknown], host not stated, Province of Buenos Aires, Argentina. Deposited in the Zoological Museum, Hamburg.

[= *Boophilus microplus* (Canestrini)]

Rhipicephalus annulatus australis (Fuller, 1899)

Neumann, L. G. 1901. *Mémoires de la Société Zoologique de France*, **14**: 280.

Stat. nov. for *Rhipicephalus australis* Fuller, 1899.

Described from a number of specimens ex cattle, Australia; 1 ♂, Borneo ex deer, 1 ♂, Sumatra, ex buffalo in the Oudemans collection; 1 ♀, Sumatra in the Zoological Museum, Hamburg; 4 ♂♂, 2 nymphs, New South Wales, ex cattle, in the Department of Agriculture, New South Wales.

[= *Boophilus microplus* (Canestrini)]

Rhipicephalus (*Boophilus*) *annulatus calcarata* (Birula, 1894)

Neumann, L.G. 1904. *Archives de Parasitologie*, 8: 450.

New combination for *Ixodes calcaratus* Birula.

[= *Boophilus annulatus* (Say)]

Rhipicephalus annulatus caudatus Neumann, 1897

Mémoires de la Société Zoologique de France, **10**: 413, fig. 42.

Described from 23 ♀♀, 2 ♂♂, Miyasaki, Japan ex horse, deposited in the Neumann collection; 250 ♀♀, 6 ♂♂, l'Ile de France ex cattle (probably), in the Natural History Museum, Paris; 22 ♀♀, 1 ♂ from 'Cayenne'; 1 ♂, 1 ♀ from Senegal in A. Railliet collection.

[= *Boophilus microplus* (Canestrini)]

Rhipicephalus annulatus decoloratus Koch, 1844

Neumann, L. G. 1901. *Mémoires de la Société Zoologique de France,* **14**: 279.

New combination for *Rhipicephalus decoloratus* Koch, 1844.

[= *Boophilus decoloratus* (Koch)]

Rhipicephalus annulatus dugesi (Mégnin, 1880)

Neumann, L.G. 1901. *Mémoires de la Société Zoologique de France,* **14**: 279.

Described from 4 ♂♂, 23 ♀♀, Morocco, ex cattle; 3 ♂♂, many ♀♀, Algeria, ex cattle; 2 ♀♀, Blida, one ex deer and one ex cattle; many ♀♀, Egypt, ex cattle.

[= *Boophilus annulatus* (Say)]

Rhipicephalus annulatus microplus (Canestrini, 1888)

Neumann, L.G. 1901. *Mémoires de la Société Zoologique de France,* **14**: 280.

Described from a number of ♂♂ and ♀♀, Argentina; Guadeloupe ex cattle; Antigua ex cattle, sheep and deer; 8 ♀♀, Montevideo, Uruguay; 3 ♀♀, Guatemala; ± 100 ♀♀, 15 ♂♂ Jamaica ex cattle, horses and dogs; many ♀♀ from Brazil, Paraguay and Guatemala, in the Zoological Museum, Hamburg; several ♀♀ Cuba ex cattle, in the Bureau of Animal Industry, Washington.

[= *Boophilus microplus* (Canestrini)]

1. ***Rhipicephalus appendiculatus* Neumann, 1901**

Mémoires de la Société Zoologique de France, **14**: 270.

Described from 3 ♂♂, 7 ♀♀ Cape Colony, South Africa, Lounsbury collection; 3 ♂♂, of which the origin of two is unknown, the other from l'Afrique S.-O; ± 100 ♂♂ and ♀♀ ex *Bos caffer* [= *Syncerus caffer*]. In the Berlin Museum. We herein designate as lectotype ♂ and paralectotype ♀ the syntypes collected from *Bos caffer* [= *Syncerus caffer*], Tangani, German East Africa [Tanzania], received 1900 from Schillings (Nuttall Collection 2893, Neumann 1143), deposited in The Natural History Museum, London (Keirans, 1985). Furthermore, we amend the locality designation for these types to Pangani River, German East Africa, at Masimani Mountain (cf. ZMB 16890, syntypes, in the Zoological Museum, Berlin – see Moritz & Fischer, 1981).

2. ***Rhipicephalus aquatilis* Walker, Keirans & Pegram, 1993**

Onderstepoort Journal of Veterinary Research, **60**: 205, figs. 1–8.

Holotype ♂, allotype ♀, and 4 ♂♂, 2 ♀♀ paratypes collected from a sitatunga (*Tragelaphus spekii*) at Kaisho (01° 19′ S, 30° 37′ E), Karagwe District, Tanzania, 14 August 1959, Mrs. G. Tullock, from Tanzania Tick collection WA99, deposited in The Natural History Museum, London. Other paratypes deposited in the Onderstepoort Tick Collection 3143i, ii, iii, and in the U.S. National Tick Collection [RML 120946].

Rhipicephalus arakeri Hiregoudar, 1975

Mysore Journal of Agricultural Science, **9**: 473, figs. 1–10.

Described from 2 ♂♂, 1 ♀, 8 nymphs, 15 larvae ex *Mus rattus* [= *Rattus rattus*], Anand, Gujarat, India. Types: holotype ♂, allotype ♀, 1 ♂, 8 nymphs, 15 larvae paratypes. The holotype and allotype deposited in the Zoological Survey of India, Calcutta. Depository for nymphal and larval paratypes not stated. Correspondence with both Professor Hiregoudar and the Zoological Survey of India, Calcutta revealed that the types of this species could not be located. Subsequently, Professor Hiregoudar sent the U.S. National Tick Collection slide-mounted specimens of *R. arakeri*, one slide being labelled 'Paratypes'. Although much overcleared, the specimens appear to be *R. ramachandrai*.

[= *Rhipicephalus ramachandrai* Dhanda]

3. ***Rhipicephalus armatus* Pocock, 1900**

Proceedings of the Zoological Society of London, Pt. 1: 50, pl. III, figs. 2–2f.

Described from 4 ♂♂, 2 ♀♀, Bularli, West Somaliland. Type ♂ and two co-types (♂ and ♀) in The Natural History Museum, London. Three co-types (2 ♂♂ and 1 ♀) in the Hope Museum, Oxford.

4. ***Rhipicephalus arnoldi* Theiler & Zumpt, 1949**

In: Zumpt, F. 1949. *Moçambique*, (**60**): 111, figs. 32–36.

Described from F1 generation of a ♀ ex *Lepus* sp. Also from wild hare, *Pronolagus* sp., Transvaal and Cape Province, South Africa. Types: the F1 generation of a ♀ ex *Lepus* sp., Toverwater, Murraysburg, Cape Province, South Africa, deposited in Onderstepoort Tick Collection 2793 and 2845. Three ♀♀ paratypes from *Pronolagus* sp., Murraysburg, South Africa are deposited in the Zoological Museum, Hamburg.

Rhipicephalus attenuatus Neumann, 1908

Archives de Parasitologie, **12**: 12, fig. 8.

Described from 1 ♀ holotype, ex *Equus caballus*, Kansanshi (Congo indépendant) [Democratic Republic of Congo], on the River Lualaba (10° 40′ south latitude). In The Natural History Museum, London.

[= *Rhipicephalus lunulatus* Neumann]

Rhipicephalus aurantiacus Neumann, 1907

Notes from the Leyden Museum, **29**: 90, figs. 3, 4.

Described from 5 ♂♂, 3 ♀♀ ex *Buffelus pumilus* [presumably = *Syncerus caffer nanus*], Liberia. In the National Museum van Natuurlijke Historié, Leiden.

[= ?*Rhipicephalus ziemanni* Neumann]

Rhipicephalus australis Fuller, 1899

Agricultural Journal of the Cape of Good Hope, **14**: 369, 3 figs.

Described from ♂, ♀[number not stated], ex horses, cattle, etc, north-west to north-east Australia.

[Also described as a new species in 1899, *Queensland Agricultural Journal*, **4**: 392].

[= *Boophilus microplus* (Canestrini)]

Rhipicephalus ayrei Lewis, 1933

Parasitology, Cambridge, **25**: 269, figs. 1, 2.

Described from numerous ♂♂, ♀♀ ex buffalo, rhinoceros and lion, Kiagu and Mbeyo, Meru district, Mount Kenya and Sianna Plains, Masai reserve, Kenya. Co-types sent to Imperial Bureau of Entomology, British Museum (Natural History) [now The Natural History Museum, London], and to The Molteno Institute for Research in Parasitology, University of Cambridge. A collection retained at the Veterinary Research Laboratory, Kabete, Kenya and a collection sent to the Coryndon Memorial Natural History Museum, [now the National Museum], Nairobi, Kenya. Those syntypes sent to The Molteno Institute are now in The Natural History Museum, London (Nuttall 3845) – see Keirans (1985).

[= *Rhipicephalus compositus* Neumann]

Rhipicephalus beccarii Pavesi, 1883

Annali del Museo Civico di Storia Naturale di Genova, **20**: 102.

Description based on a ♂ from Bogos, Abyssinia [Ethiopia]. Depository not stated.

[= *Rhipicephalus sanguineus* (Latreille)]

Rhipicephalus belli Sonenshine, 1993

Biology of Ticks, vol. 2, p. 463. Oxford and New York: Oxford University Press. [*Lapsus calami* for *Rickettsia belli*]

5. ***Rhipicephalus bequaerti* Zumpt, 1949**

Moçambique, (**60**): 119, figs. 37–39.

Described from 1 ♂, 1 ♀. 'The pair at my disposal was caught in Central Africa on the buffalo (*Syncerus caffer* Sparrm.). The tag on the ♂ says that it was found in 'Lissenji' (?).' Depository not stated. However, types were deposited in the Institute for Medical Research, Johannesburg.

6. ***Rhipicephalus bergeoni*** **Morel & Balis, 1976**

Revue d'Élevage et de Médecine Vétérinaire des Pays Tropicaux, **29** (nouvelle série): 141, figs. 1 a–h, 2 a–i.

'Type' ♂ [i.e. holotype] and allotype ♀ from Hubeta, Harrar, Ethiopia. Depository not stated.

Rhipicephalus bhamensis Supino, 1897

Atti della Società Veneto-Trentina di Scienze Naturali in Padova, 2nd ser. **3**: 233, pl. V, figs. 1–6.

Description based on a ♂, Bhamò, Burma.

[= *Rhipicephalus sanguineus* (Latreille)]

Rhipicephalus bicornis (Nuttall & Warburton, 1908)

Brumpt, E. 1936. *Précis de Parasitologie,* **2**: 119.

[*Lapsus calami* for *Rhipicentor bicornis* Nuttall & Warburton]

Rhipicephalus bilenus Pavesi, 1883

Annali del Museo Civico di Storia Naturale di Genova, **20**: 102.

Description based on a ♂ from Bogos, Abyssinia [Ethiopia]. [= perhaps *Rhipicephalus bursa* Canestrini and Fanzago. *Fide* Neumann (1911), *Das Tierreich,* p. 46]

7. ***Rhipicephalus boueti*** **Morel, 1957**

Bulletin de la Société de Pathologie Exotique, **50**: 696, fig. 1 (1956).

Described from 2 ♂♂, 3 ♀♀, ex *Procavia ruficeps* subsp., Agouagon, Dahomey [Benin]. Holotype ♂ deposited in the Neumann Collection, École Vétérinaire, Toulouse. Allotypes 2 ♀♀ (*sic*), 1 ♂, 1 ♀ paratypes deposited in Laboratoire Fédéral de l'Élevage, Dakar, Senegal.

Rhipicephalus bovis (Riley, 1869)

Galli-Valerio, B. 1901. *Bulletin de la Société Vaudoise des Sciences Naturelles,* Ser. 4, **37**: 355.

New combination for *Ixodes bovis* Riley, 1869.

[= *Boophilus annulatus* (Say)]

Rhipicephalus breviceps Warburton, 1910

Parasitology, Cambridge, **3**: 398, fig. 3.

Description based on a single ♀ holotype ex *Erinaceus collaris,* Sind, Pakistan. Depository not stated. This specimen is now in the Nuttall Collection (Nuttall 1103) in The Natural History Museum, London. The genital aperture has been removed and mounted on a slide. See Keirans (1985).

[= *Rhipicephalus sanguineus* (Latreille)]

Rhipicephalus brevicollis Neumann, 1897

Mémoires de la Société Zoologique de France, **10**: 402, fig. 35.

Description based on a single ♀ holotype from Mombas, Zanzibar [Mombasa, Kenya]. Deposited in the Zoological Museum, Berlin.

[= *Rhipicephalus sanguineus* (Latreille)]

Rhipicephalus brevicoxatus Morel & Mouchet, 1958

Annales de Parasitologie Humaine et Comparée, **33**: 94, fig. 2 f–h.

Description based on a ♂ from Akok Bekoé, Cameroon, and described as a 'form' of *Rhipicephalus ziemanni* Neumann.

8. ***Rhipicephalus bursa*** **Canestrini & Fanzago, 1878**

Atti della Reale Istituto Veneto di Scienze, Lettere ed Arte, ser. 5, **4**: 190.

Described from ♂ and ♀ collected live on a 'cinghiale' (wild boar). Depository not stated.

Rhipicephalus bursa americanus Neumann, 1897

Mémoires de la Société Zoologique de France, **10**: 393.

Two ♀♀ from Jamaica in the collection of G. Marx, Smithsonian Institution. [These specimens are not in the U.S. National Tick Collection].

[= *Rhipicephalus sanguineus* (Latreille)]

Rhipicephalus bursa pusillus Gil Collado, 1936

Treballs del Museu de Ciències Naturals de Barcelona, Ser. Entomològica, **11**: 4.

Described from 2 ♀♀ ex fox, Barcelona. Depository not stated.

[= *Rhipicephalus pusillus* Gil Collado]

Rhipicephalus calcaratus (Birula, 1894)

Birula, A. 1895. *Izvêstiya Imperatorskoi*

Akademii Nauk, Ser. 5, 2: 137.
? New combination for *Ixodes calcaratus* Birula, 1894.
[= *Boophilus annulatus* (Say)]

Rhipicephalus camelopardalis Walker & Wiley, 1959
Parasitology, Cambridge, **49**: 448, figs. 1–4.
Described from specimens collected from a giraffe, Talek River, Cis-Mara area of West Masai, Kenya. Holotype ♂, allotype ♀ ex *Giraffa camelopardalis tippelskirchi,* locality as above; deposited in The Natural History Museum, London. Paratypes in The Natural History Museum, London; Rocky Mountain Laboratory (now U.S. National Tick Collection, Georgia Southern University); Onderstepoort Tick Collection 3058; Veterinary Research Station, Mpwapwa, Tanzania; in H. Hoogstraal's collection, now in U.S. National Tick Collection [RML 34921, 90316 – see Keirans & Clifford (1984)]; Veterinary Research Laboratory, Kabete; East African Veterinary Research Organization, Kikuyu, Kenya.
[= *R. longicoxatus*]

9. ***Rhipicephalus camicasi* Morel, Mouchet & Rodhain, 1976**
Revue d'Élevage et de Médecine Vétérinaire des Pays Tropicaux, **29** (nouvelle série): 337, fig. 1.
Description based on ♂♂, ♀♀ from Ethiopia and Afars and Issas [Djibouti] on cattle, sheep, camel, goat and *Lepus capensis.* Holotype ♂ and allotype ♀ ex sheep, Randa, Afars and Issas [Djibouti]. Depository not stated.

10. ***Rhipicephalus capensis* Koch, 1844**
Archiv für Naturgeschichte, **10**: 238.
Described from 1 ♂ [holotype], Cape Colony, South Africa. Deposited in the Zoological Museum, Berlin, accession number ZMB 1097 – see Moritz & Fischer (1981).

Rhipicephalus capensis compositus Neumann, 1905
Archives de Parasitologie, **9**: 231.
New subspecific status for *Rhipicephalus compositus* Neumann, 1897.
[= *Rhipicephalus compositus* Neumann]

Rhipicephalus capensis longus Neumann, 1907
Zumpt, F. 1942. *Zeitschrift für Parasitenkunde,* **12**: 484, figs. 6–8.
Stat. nov. for *Rhipicephalus longus* Neumann, 1907.
[= *Rhipicephalus longus* Neumann]

Rhipicephalus capensis pseudolongus Santos Dias, 1953
Memórias e Estudos do Museu Zoológico da Universidade de Coimbra, (**214**): 5, figs. 1–3.
Holotype ♂ and allotype ♀ ex *Bos indicus,* Yaoundé, Cameroon. Depository not stated.
[= *Rhipicephalus longus* Neumann]

Rhipicephalus carinatus Frauenfeld, 1867
Verhandlungen der Kaiserlich-Königlichen Zoologisch-Botanischen Gesellschaft in Wien, **17**: 462. [Publication not seen]
From 1 ♂ collected on the deck of a ship in the China Sea.
[= *Rhipicephalus sanguineus* (Latreille)]

11. ***Rhipicephalus carnivoralis* Walker, 1966**
Parasitology, Cambridge, **56**: 1, figs. 1–15a, 16a.
Described from numerous specimens, both wild caught and reared. Holotype ♂ and allotype ♀ from laboratory reared series, the progeny of a ♀ ex lioness, Muguga, Kiambu District, Kenya [accession number B.S. 913]. Deposited in The Natural History Museum, London. Paratypes in many institutions and collections including the U.S. National Tick Collection [RML 46183 – see Keirans & Clifford (1984)].

Rhipicephalus cliffordi Morel, 1965
Revue d'Élevage et de Médecine Vétérinaire des Pays Tropicaux, **17 for 1964** (nouvelle série): 637, figs. 1–4 (1965).
Holotype ♂, allotype ♀ ex *Syncerus caffer nanus,* Assagni, Ivory Coast. Depository not stated. (Vol. 17 of this journal commenced publication in 1964, but the final

part containing descriptions of this species and *R. moucheti* was published in 1965).
[= *Rhipicephalus pseudolongus*]

12. ***Rhipicephalus complanatus* Neumann, 1911**
Archives de Parasitologie, **14**: 415.
Stat. nov. for *Rhipicephalus planus complanatus* Neumann, 1911. [*nec Rhipicephalus simus planus* Neumann, 1907].

13. ***Rhipicephalus compositus* Neumann, 1897**
Mémoires de la Société Zoologique de France, **10**: 393.
Described from 1 ♂ [holotype], host unknown, Khartoum, Sudan. Deposited in the Natural History Museum, Paris. [*R. compositus* is a highland species; therefore, the type locality of Khartoum is probably incorrect].

Rhipicephalus confusus Santos Dias, 1956
Moçambique, (87): 2, figs. 3–5.
Holotype ♂ ex buffalo, Ile, Zambézia, Mozambique. Deposited in the Laboratory of Veterinary Pathology, Maputo, Mozambique. One ♂, 1♀ paratypes deposited in The Natural History Museum, London.
[= *Rhipicephalus longus*]

Rhipicephalus coriaceus Nuttall & Warburton, 1908
Proceedings of the Cambridge Philosophical Society, **14**: 402, figs. 17–20.
Description based on 2 ♂♂, 6 ♀♀, host unknown, North Nyasa [Malawi]; 1 ♂ host unknown, Benguella, W. Africa [Angola]. Deposited in the Nuttall Collection (Nuttall 320e) in The Natural History Museum, London – see Keirans (1985). Lectotype and paralectotypes designated – see Keirans & Brewster (1981).
[= *Rhipicephalus supertritus* Neumann]

Rhipicephalus cuneatus Neumann, 1908
Notes from the Leyden Museum, **30**: 76, figs. 2, 3.
Described from 3 ♂♂ ex cattle, Ngômo on l'Ogooué, French Congo [Congo]. Deposited in the collection of Professor Galli-Valerio, Lausanne, Switzerland.
[= *Rhipicephalus ziemanni* Neumann]

14. ***Rhipicephalus cuspidatus* Neumann, 1906**
Archives de Parasitologie, **10**: 209, fig. 11.
Described from 1 ♂, 4 ♀♀, ex *Phacochoerus* sp., Senegal. Deposited in The Natural History Museum, London.

Rhipicephalus decoloratus (Birula, 1894)
Velu, H. 1922. *Mémoires de la Société des Sciences Naturelles (et Physiques) du Maroc,* **2**: 187.
New combination for *Boophilus calcaratus* (Birula, 1894).
[= *Boophilus annulatus* (Say)]

15. ***Rhipicephalus deltoideus* Neumann, 1910**
Tijdschrift voor Entomologie, **53**: 13, figs. 3–7.
Described from 1 ♂, 3 ♀♀, host unknown, Basutoland [Lesotho]. Depository not stated. One ♀ syntype (Nuttall 2894) is in The Natural History Museum, London – see Keirans (1985).

16. ***Rhipicephalus distinctus* Bedford, 1932**
Report of the Director of Veterinary Services and Animal Industry, Union of South Africa, (**18**): 523.
New name for *Rhipicephalus punctatus* Bedford, 1929 (preoccupied).
Described from specimens on dassies (rock hyrax) at Onderstepoort and in South-West Africa [Namibia], also on sheep at Victoria West, Cape Province, South Africa. [These Victoria West specimens are *Rhipicephalus neumanni* – see Walker (1990)]. Depository not stated but in the Onderstepoort Tick Collection 2488 ii, iii, iv.

Rhipicephalus dugesi Neumann, 1896
Mémoires de la Société Zoologique de France, **9**: 11.
New combination for *Ixodes bovis* Riley, 1869.
[= *Boophilus annulatus* (Say)]

17. ***Rhipicephalus duttoni* Neumann, 1907**
Annals of Tropical Medicine and Parasitol-

ogy, **1**: 115, figs. 22, 23.
Described from a single ♂, ex bovine [presumably *Bos taurus*], Zambi (*sic*). Depository not stated. Neumann (1911) stated that the locality was 'Congo indépendant [Democratic Republic of Congo], (Zambu)'.

18. ***Rhipicephalus dux*** **Dönitz, 1910**
Sitzungsberichte der Gesellschaft naturforschender Freunde zu Berlin, (**6**): 275, figs. 1–3.
Described from 1 ♂, 1 ♀ ex elephant, oberen Congogebiet [Democratic Republic of Congo]. Depository not stated. One ♂ syntype is in the Zoological Museum, Berlin [ZMB 17714] – see Moritz & Fischer (1981).

Rhipicephalus ecinctus Neumann, 1901
Mémoires de la Société Zoologique de France, **14**: 275.
Described from 6 ♂♂, origin unknown. Deposited in the Berlin Museum. All 6 syntype males are in the Zoological Museum, Berlin [ZMB 17690] – see Moritz & Fischer (1981).
[♂ = *Rhipicephalus maculatus* Neumann]

Rhipicephalus ellipticum Koch, 1847
Uebersicht des Arachnidensystems, (**4**): 135, pl. 30, fig. 111.
New combination for *Rhipistoma ellipticum* Koch, 1844.
[= *Haemaphysalis leachi* (Audouin)]

Rhipicephalus erlangeri Neumann, 1902
Archives de Parasitologie, **6**: 111.
Described from a single ♂ holotype ex horse, near the river Daroli. Depository not stated.
[= *Rhipicephalus simus* group]

19. ***Rhipicephalus evertsi evertsi*** **Neumann, 1897**
Mémoires de la Société Zoologique de France, **10**: 405, fig. 36.
Described from 8 ♂♂, 3 ♀♀, host unknown, Transvaal, South Africa. Deposited in the Oudemans Collection, National Museum van Natuurlijke Historié, Leiden.

Rhipicephalus evertsi albigeniculatus Warburton, 1916
In: Nuttall, G. H. F. & Warburton, C. 1916. *Bulletin of Entomological Research,* **6**: 327.
From Lower Congo; further details not given. For hosts and collecting localities see Keirans (1985). Lectotype and paralectotypes designated by Keirans & Brewster (1981).
[= *Rhipicephalus evertsi mimeticus* Dönitz]

20. ***Rhipicephalus evertsi mimeticus*** **Dönitz, 1910**
Denkschriften der Medizinisch-Naturwissenschaftlichen Gesellschaft zu Jena, **16**: 475.
Described from ♂♂, ex cattle, South-West Africa [Namibia]. Depository not stated. Syntypes in the Zoological Museum, Berlin – see Moritz & Fischer (1981).

21. ***Rhipicephalus exophthalmos*** **Keirans & Walker, 1993**
In: Keirans, J.E., Walker, J.B., Horak, I.G. & Heyne, H. 1993. *Onderstepoort Journal of Veterinary Research,* **60**: 230, figs. 1–16.
Described from laboratory-reared specimens of all stages from ♀ originally collected on a farm 'Kosos 11', Karas region (formerly Bethanien), Namibia, c. 10.III.1970, host not stated. Holotype ♂, allotype ♀ and ♂♂, ♀♀, nymphs, and larvae paratypes deposited in Onderstepoort Tick Collection 3144. Paratypes of all stages deposited in The Natural History Museum, London and in the U.S. National Tick Collection [RML 56729].

Rhipicephalus expositicius Koch, 1877
Abhandlungen der Naturhistorischen Gesellschaft zu Nürnberg, **6**: 196. [Publication not seen].
[= *Haemaphysalis punctata* (in part). *Fide* Neumann, 1911]

Rhipicephalus falcatus Neumann, 1908
Notes from the Leyden Museum, **30**: 77, fig. 4.
Described from 3 ♂♂, 8 ♀♀, host unknown, north of Lake Nyasa, deposited in The Natural History Museum, London; 4 ♂♂, 1 ♀, host unknown, from Liberia, deposited in the Leiden Museum.
[= *Rhipicephalus longus*]

Rhipicephalus flavus Supino, 1897
Atti della Società Veneto-Trentina di Scienze Naturali in Padova, 2nd. ser., **3**: 233, pl. IV, figs. 5–9.
Description based on a ♀, Prome, Burma [Myanmar].
[= *Rhipicephalus sanguineus* (Latreille)]

22. ***Rhipicephalus follis* Dönitz, 1910**
Denkschriften der Medizinisch-Naturwissenschaftlichen Gesellschaft zu Jena, **16**: 481, pl. 16A, fig. 3.
Described from 2 ♂♂, host probably domestic cattle, locality not stated. Depository not stated. Two ♂♂ syntypes (Nuttall 2110) are deposited in The Natural History Museum, London – see Keirans (1985).

23. ***Rhipicephalus fulvus* Neumann, 1913**
Bulletin de la Société Zoologique de France, **38**: 147, figs. 1, 2.
Described from a single ♂ [holotype], host unknown, Matmata, south of Gabès [Tunisia]. Depository not stated.

Rhipicephalus furcosum Neumann, 1901
Mémoires de la Société Zoologique de France, **14**: 361.
Typographical error for *Amblyomma furcosum* Neumann, 1901.
[= *Amblyomma helvolum* Koch]

24. ***Rhipicephalus gertrudae* Feldman-Muhsam, 1960**
Journal of Parasitology, **46**: 104, figs. 3–5.
Described from a batch of ticks collected in Namaqualand. Holotype ♂, allotype ♀ in Feldman-Muhsam collection. Paratypes sent to Dr Theiler [Onderstepoort Tick Collection 3137 ii. Paratypes: Namaqualand. Gen. apert. No. 144–145]. One ♀ paratype [ZMB 1101] in the Zoological Museum, Berlin – see Moritz & Fischer (1981).

25. ***Rhipicephalus glabroscutatum* Du Toit, 1941**
Onderstepoort Journal of Veterinary Science and Animal Industry, **16**: 115, figs. 1, 2.
Described from all stages, ex domestic stock, Fairview District, Aberdeen, Cape Province, South Africa. Depository not stated but in the Onderstepoort Tick Collection 2711 i. Type ♂♂ and ♀♀ ex Angora goats.

Rhipicephalus gladiger Neumann, 1908
Archives de Parasitologie, **12**: 8, figs. 4–7.
Described from 2 ♂♂, 2 ♀♀, ex *Equus caballus,* Kansanshi (Congo indépendant) [Democratic Republic of Congo] on the Lualaba River (10° 40′ south latitude). Deposited in The Natural History Museum, London. Also 1 ♂ ex *Capra hircus,* Pweto, Congo indépendant [Democratic Republic of Congo]. Depository not stated.
[= *Rhipicentor bicornis* Nuttall and Warburton]

Rhipicephalus glyphis Dönitz, 1910
Sitzungsberichte der Gesellschaft naturforschender Freunde zu Berlin, (**6**): 278, fig. 4.
Described from 2 ♂♂, one from a 'Wasserschwein', Lake Tanganyika [Tanzania], the other from Togo, probably on a bovine. Depository not stated. A pinned syntype ♂ (Nuttall 2108) from a 'Wasserschwein' [most likely *Potamochoerus larvatus*], Lake Tanganyika [Tanzania] is in The Natural History Museum, London – see Keirans (1985). The ♂ from Togo is in the Zoological Museum, Berlin [ZMB 17734] – see Moritz & Fischer (1981).
[= *Rhipicephalus lunulatus* Neumann]

26. ***Rhipicephalus guilhoni* Morel & Vassiliades, 1963**
Revue d'Élevage et de Médecine Vétérinaire des Pays Tropicaux, **15 for 1962** (nouvelle série): 378, figs. 7–11.
Described from a collection of ticks; ♀ holotype (gonopore slide-mounted), ♂ allotype ex sheep, Mioro, Mali. Allotype and ♀ paratype deposited in the National Museum of Natural History, Paris.

27. ***Rhipicephalus haemaphysaloides* Supino, 1897**
Neumann, L.G. 1897. *Mémoires de la Société Zoologique de France,* **10**: 417.
New combination for *Rhipicephalus haemaphysaloides niger* Supino, 1897. Two ♂♂, 1 ♀ syntypes (Nuttall 2963, 2965, 2971) are in The Natural History Museum, London –

see Keirans (1985).

Rhipicephalus haemaphysaloides expeditus Neumann, 1904

Archives de Parasitologie, **8**: 454.

Described from ♂, ♀ ex *Buffelus indicus* [*Bubalus bubalis*], China, Sumatra.

[= *Rhipicephalus haemaphysaloides* Supino]

Rhipicephalus haemaphysaloides niger Supino, 1897

Atti della Società Veneto-Trentia di Scienze Naturali in Padova, 2nd Ser., **3**: 234, pl. V, figs. 6–8.

Described from a ♀, host unknown, Yado, Burma [Myanmar], deposited in the Genoa Museum.

[= *Rhipicephalus haemaphysaloides* Supino]

Rhipicephalus haemaphysaloides pilans Schulze, 1935

Zumpt, F. 1940. *Zeitschrift für Parasitenkunde,* **11**: 675.

Stat. nov. for *Rhipicephalus pilans* Schulze, 1935.

[= *Rhipicephalus pilans* Schulze]

Rhipicephalus haemaphysaloides paulopunctata Neumann, 1897

Neumann, L.G. 1904. *Archives de Parasitologie,* **8**: 449.

Stat. nov. for *Rhipicephalus paulopunctatus* Neumann, 1897.

[= *Rhipicephalus pilans* Schulze]

Rhipicephalus haemaphysaloides ruber Supino, 1897

Atti della Società Veneto-Trentia di Scienze Naturali in Padova, 2nd Ser., **3**: 234, pl. VI, figs. 6–10.

Described from ♂, ♀, host unknown, Mt Mooleyit, Tenasserim; Meteléo, Thagatà, Juva, Cagò del Cadù Gianng, north-east of Bomò, Burma [Myanmar]. Deposited in the Genoa Museum.

[= *Rhipicephalus haemaphysaloides* Supino]

Rhipicephalus hilgerti Neumann, 1902

Archives de Parasitologie, **6**: 111.

Described from 1 ♂, 1 ♀ ex *Canis variegatus* [*Canis aureus*], Abyssinia [Ethiopia]. Depository not stated.

[= *Rhipicephalus praetextatus* Gerstäcker]

Rhipicephalus hispanum Dönitz, 1905

Sitzungsberichte der Gesellschaft naturforschender Freunde zu Berlin, (**4**): 114.

[See comment under *Rhipicephalus anatolicum*].

28. ***Rhipicephalus humeralis* Rondelli, 1926**

Zumpt, F. 1949. *Moçambique,* (**60**): 49.

Stat. nov. for *Rhipicephalus pulchellus humeralis* Rondelli, 1926.

29. ***Rhipicephalus hurti* Wilson, 1954**

Parasitology, Cambridge, **44**: 277, figs. 1–6.

Described from ♂♂, ♀♀ ex buffalo, Karati Forest, Naivasha, Kenya (Coll. No. 221). Type ♂ and ♀ deposited in The Natural History Museum, London. Paratypes in the Onderstepoort Tick Collection 2929 ii. Paratypes 2 ♂♂, 2 ♀♀ sent to Dr H. Hoogstraal in Cairo, are no longer present in the Hoogstraal collection, now part of the U.S. National Tick Collection. Paratypes sent to Dr J.T. Santos Dias, Mozambique.

Rhipicephalus inermis (Birula, 1895)

L'Insecte et l'Infection, **1**: *Acarines,* p. 103.

[= *Haemaphysalis inermis* Birula]

Rhipicephalus? (*sic*) *intermedius* (Neumann, 1897)

Neumann, L.G. 1901. *Mémoires de la Société Zoologique de France,* **10**: 416.

Questionable combination for *Phaulixodes intermedius* Neumann, 1897.

[= *Rhipicephalus sanguineus* (Latreille) *fide* Santos Dias (1991)]

30. ***Rhipicephalus interventus* Walker, Pegram & Keirans, 1995**

Onderstepoort Journal of Veterinary Research, **62**: 89, figs. 1a, b; 2a–f.

Described from 28 ♂♂, 69 ♀♀ collected primarily from cattle in Uganda, Tanzania, Rwanda, Malawi, and Zambia. Holotype ♂, allotype ♀ collected at Kawoko-Masaka, Masaka district, Uganda, 2.III.1967, J. G. Matthysse, deposited in the U.S. National Tick Collection (RML 53849). Paratypes 2 ♂♂, 7 ♀♀ data as above; 2 ♂♂, 8 ♀♀ collected at Igula village, Ihimbu Gunguli, Tanzania,

31.XII.1960, Veterinary Assistant Robert, deposited in The Natural History Museum, London, (Tanzania Tick Collection IR/ 47); and 1 ♂, 2 ♀♀ collected at Lutale, Mumbwa, Zambia, XII.1981, R.G. Pegram, deposited in the Onderstepoort Tick Collection 3144 i.

Rhipicephalus janneli Pomerantsev, 1936
Parazitologicheskii Sbornik Institut Akademiya Nauk SSSR, **6**: 15.
[*Lapsus calami* for *Rhipicephalus jeanneli* Neumann, 1913]

Rhipicephalus javanensis Supino, 1897
Atti della Società Veneto-Trentia di Scienze Naturali in Padova, 2nd Ser, **3**: 233.
A nymphal syntype (Nuttall 2955) is deposited in The Natural History Museum, London – see Keirans (1985).
[= *Amblyomma javanense* (Supino)]

31. ***Rhipicephalus jeanneli*** **Neumann, 1913**
Voyage de Ch. Alluaud et R. Jeannel en Afrique Orientale (1911–1912). Résultats Scientifiques. Arachnides II. Ixodidae: 31, figs. 3–5.
Description based on 4 ♂♂, 5 ♀♀, host unknown, Molo, Burgurett River, also 1 ♂, 1 ♀ Station 38, Kenya; 2 ♂♂ upper edge of forest on Kilimandjaro near Bismark Hill, also 1 ♂ Station 71, German East Africa [Tanzania]. Depository not stated.

32. ***Rhipicephalus kochi*** **Dönitz, 1905**
Sitzungsberichten der Gesellschaft naturforschender Freunde zu Berlin, (**4**): 106.
Described from 1 ♂, 5 ♀♀ Sadani, and 3 ♀♀ Lindi, German East Africa [Tanzania], ex cattle. Lectotype ♀ designated by Clifford *et al.* (1983), deposited in the Zoological Museum, Berlin, No. 8490; paralectotypes (Nuttall 2109) deposited in The Natural History Museum, London – see Keirans (1985).

Rhipicephalus leachi Schwetz, 1927
Revue Zoologique Africaine, **15**: 89.
[*Lapsus calami* for *Haemaphysalis leachi*]

33. ***Rhipicephalus leporis*** **Pomerantsev, 1946**
Opredeliteli po Faune SSSR, Izdavaemye Zoologicheskim Institutom Akademii Nauk SSSR, (**26**); 20, fig. 16.
Described by Pomerantsev (*in litt.*) but a valid description and figure given from specimens ex rabbit, Uzbekistan.

Rhipicephalus limbatus Koch, 1844
Archiv für Naturgeschichte, **10**: 239.
Described from 1 ♂, Egypt. Holotype in the Zoological Museum, Berlin – see Moritz & Fischer (1981).
[= *Rhipicephalus sanguineus*]

Rhipicephalus linnei (*sic*) (Audouin, 1826)
Koch, C.L. 1844. *Archiv für Naturgeschichte,* **10**: 238.
New combination for *Ixodes linnaei* Audouin, 1826.
[= Unknown species]

34. ***Rhipicephalus longiceps*** **Warburton, 1912**
Parasitology, Cambridge, **5**: 11, figs. 6, 7.
Described from 18 ♂♂, 3 ♀♀ [Nuttall 351] ex Klipspringer Bok [Klipspringer], Benguella Hinterland, Angola and 19 ♂♂, 2 ♀♀, no host recorded, same locality. Originally deposited in Cambridge. Lectotype and paralectotypes designated – see Keirans & Brewster (1981). Collections now in The Natural History Museum, London.

35. ***Rhipicephalus longicoxatus*** **Neumann, 1905**
Archives de Parasitologie, **9**: 225.
Described from 1 ♂, 2 ♀♀ from German East Africa [Tanzania], deposited in the Zoological Museum, Berlin. There is 1 ♀ paratype, accession no. ZMB 17740 in the collection – see Moritz & Fischer (1981).

Rhipicephalus longoides Zumpt, 1943
Tendeiro, J. 1952. *Anais do Instituto de Medicina Tropical, Lisboa,* **9**: 234.
Stat. nov. for *Rhipicephalus simus longoides* Zumpt, 1943.
[= *Rhipicephalus senegalensis* Koch]

36. ***Rhipicephalus longus*** **Neumann, 1907**
Annals of Tropical Medicine and Parasitology, **1**: 117, figs. 24, 25.

Described from a single ♂ [holotype] ex *Bos taurus*, Kasongo, Nyasaland [Malawi]. Depository not stated. This specimen (Nuttall 1460) is in The Natural History Museum, London – see Keirans (1985).

37. ***Rhipicephalus lounsburyi*** **Walker, 1990**

Onderstepoort Journal of Veterinary Research, **57**: 57, figs. 1–24.

Described from ♂♂, ♀♀, NN, LL, reared from material originally collected at Dordrecht, Eastern Cape Province, South Africa. Holotype ♂, allotype ♀ No. 2820 and paratypes No. 2821, deposited in Onderstepoort Tick Collection. Paratypes [RML 105789] deposited in the U.S. National Tick Collection.

Rhipicephalus lundbladi Schulze, 1939

Arkiv för Zoologi, 31: 2, pl. 1.

Described from 1 ♂, ex vegetation, Paul de Serra, Madeira [Portugal]. Type [i.e. holotype] in the Natural History Museum, Stockholm.

[= *Rhipicephalus* species *incertae sedis*]

38. ***Rhipicephalus lunulatus*** **Neumann, 1907**

Archives de Parasitologie, **11**: 215, fig. 1.

Described from 2 ♂♂ ex horse, near the Lualaba River, Congo Free State [Democratic Republic of Congo]. Deposited in The Natural History Museum, London. Lectotype [BM(NH) 5708; RML 105707] designated by Walker *et al.* (1988).

Rhipicephalus lusitanicum Dönitz, 1905

Sitzungsberichte der Gesellschaft naturforschender Freunde zu Berlin, (**4**): 114.

[See comment under *Rhipicephalus anatolicum*].

Rhipicephalus macropis Schulze, 1936

Zeitschrift für Parasitenkunde, **8**: 521, figs. 1, 2.

Description based on ♂♂, ex dog, Aden, and ♀♀ ex dog, Port Sudan. Type ♂ deposited in the Zoological Museum, Berlin. Specimen lost? Not cited by Moritz & Fischer (1981).

[= *Rhipicephalus sanguineus* (Latreille)]

39. ***Rhipicephalus maculatus*** **Neumann, 1901**

Mémoires de la Société Zoologique de France, **14**: 273.

Described from 1 ♂, 2 ♀♀ ex *Psytalla horrida* (syn. *Platymeris horrida*), Cameroon, deposited in the Berlin Museum. Syntypes [ZMB 17716] in the Zoological Museum, Berlin – see Moritz & Fischer (1981). [The type host listed is an assassin bug, family Reduviidae; obviously this is incorrect].

Rhipicephalus marmoreus Pocock, 1900

Proceedings of the Zoological Society of London, Pt. **1**: 50, pl. III, figs. 1–1d.

From a single ♂ specimen, holotype, Bularli, West Somaliland, deposited in The Natural History Museum, London.

[= *Rhipicephalus pulchellus* Gerstäcker]

40. ***Rhipicephalus masseyi*** **Nuttall & Warburton, 1908**

Proceedings of the Cambridge Philosophical Society, **14**: 404, figs. 21–26.

Described from 31 ♂♂, 21 ♀♀ ex *Bos caffer* [= *Syncerus caffer*], Kansanshi, north-west Rhodesia [Zambia]. Depository not stated. Lectotype and paralectotypes designated – see Keirans & Brewster (1981). Specimens deposited in The Natural History Museum, London.

Rhipicephalus microplus (Canestrini, 1888)

Canestrini, G. 1890. *Prospetto dell' Acarofauna Italiana*, (**4**): 493.

New combination for *Haemaphysalis microplus* Canestrini, 1888.

[= *Boophilus microplus* (Canestrini)]

Rhipicephalus mossambicus Santos Dias, 1950

Moçambique, (**59**): 137, figs. 4–6.

Described from 1 ♂, ex buffalo, Mozambique. Deposited in the Santos Dias collection.

[= *Rhipicephalus pravus* Dönitz]

41. ***Rhipicephalus moucheti*** **Morel, 1965**

Revue d'Élevage et de Médecine Vétérinaire des Pays Tropicaux, **17 for 1964** (nouvelle série): 615, fig. 1 (1965).

Holotype ♂ ex *Erythrocebus patas*, Cameroon; 'Plésiotype' ex cattle, Toui

(Kandi), Dahomey [Benin]. Depository not stated.

42. ***Rhipicephalus muehlensi*** **Zumpt, 1943**
Zeitschrift für Parasitenkunde, **13**: 105, figs. 1–4.
Described from numerous specimens ex bushbuck, giraffe, roan antelope, Kondoa-Irangi, Mikindani and Maliwe-See [Lake Maliwe], German East Africa [Tanzania]. All specimens in the Berlin Museum. Holotype [ZMB 17742] and paratypes are in the Zoological Museum, Berlin – see Moritz & Fischer (1981).

43. ***Rhipicephalus muhsamae*** **Morel & Vassiliades, 1965**
Revue d'Élevage et de Médecine Vétérinaire des Pays Tropicaux, **17 for 1964** (nouvelle série): 619, figs. 1–4 (1965).
Holotype ♀, allotype ♂ ex cattle, Sangalkam, Senegal. Depository not stated.

Rhipicephalus neavei Warburton, 1912
Parasitology, Cambridge, 5: 7, figs. 2, 3.
Described from specimens ex eland near the mouth of the Tasangazi River, Luangwa Valley, north-east Rhodesia [Zambia]. Types were deposited at the British Museum (Natural History) and Cambridge. Lectotype and paralectotypes designated – see Keirans & Brewster (1981). All specimens are now in The Natural History Museum, London.
[= *Rhipicephalus kochi* Dönitz]

Rhipicephalus neavei punctatus Warburton, 1912
Parasitology, Cambridge, 5: 10, figs. 4, 5.
Described from 13 ♂♂, 8 ♀♀ ex kudu, near Fort Mlangeni, Central Angoniland, Nyasaland [Malawi], and 1 ♀ ex reed-buck, valley of the Rukuru River, northern Nyasaland. Types were deposited at the British Museum (Natural History) and Cambridge. Lectotype and paralectotypes designated – see Keirans & Brewster (1981). All specimens are now in The Natural History Museum, London.
[= *Rhipicephalus punctatus* Warburton]

44. ***Rhipicephalus neumanni*** **Walker, 1990**
Onderstepoort Journal of Veterinary Research, **57**: 66, figs. 26–49, 52–54.
Described from specimens collected on sheep and goats, Namibia. Holotype ♂, allotype ♀ ex sheep, farm 'Soutdoringvlei', Bethanien District, Namibia [Onderstepoort Tick Collection 3141 i]. Paratypes [Onderstepoort Tick Collection 3141 ii, iii]. Paratypes also deposited in the U.S. National Tick Collection [RML 119860] and in The Natural History Museum, London [1990.1.18.1–4].

Rhipicephalus niger Rudow, 1870
Zeitschrift für die Gesammte Naturwissenschaften Halle, **35**: 19.
Described from 1 ♂ ex *Boa* sp., locality and depository not stated.
[= species *incertae sedis.* A nymph]

45. ***Rhipicephalus nitens*** **Neumann, 1904**
Archives de Parasitologie, **8**: 462.
Described from 6 ♂♂, 12 ♀♀ ex vegetation, Stellenbosch, Cape Colony, South Africa. [Onderstepoort Tick Collection 2667 i. 1 ♀, Stellenbosch, Dec. 1902, Lounsbury Collection 1456. There is no definite indication that this ♀ is a type specimen]. One ♂ syntype (Nuttall 2896) is in The Natural History Museum, London – see Keirans (1985).

46. ***Rhipicephalus oculatus*** **Neumann, 1901**
Mémoires de la Société Zoologique de France, **14**: 274.
Described from 2 ♂♂, 2 ♀♀ ex *Lepus timidus*, Damaras, [Namibia], and 1 ♀ ex cattle, Kilossa [Tanzania]. [The latter ♀ is probably not *R. oculatus*, which is not now thought to occur in Tanzania]. In Berlin and Hamburg Museums. Syntype ♂ and ♀ in the Zoological Museum, Berlin – see Moritz & Fischer (1981), Keirans *et al.* (1993).

47. ***Rhipicephalus oreotragi*** **Walker & Horak, 1999**
In: Walker, J.B., Keirans, J.E. & Horak, I.G. 1999. *The genus* Rhipicephalus *(Acari: Ixodidae): A Guide to the Brown Ticks of the World,* pp. 330–333. Cambridge: Cambridge University Press.

Described from holotype ♂, allotype ♀, ♂♂ and ♀♀ paratypes collected from klipspringer (*Oreotragus oreotragus*), Sentinel Ranch, 70 km west of Beit Bridge, Zimbabwe, on 4 July 1992. Holotype ♂, allotype ♀, ♂♂, ♀♀ paratypes deposited in Onderstepoort Tick Collection 3145 i, ii. Paratypes in the U.S. National Tick Collection and The Natural History Museum, London.

Rhipicephalus paulopunctatus Neumann, 1897
Mémoires de la Société Zoologique de France, **10**: 397.
Described from 1 ♂, 1 ♀, Indrapura, Sumatra. Deposited in the Hamburg Museum. Anastos (1950) indicated that the types were lost, but one of us (J.E.K.) saw the holotype ♂ in the Zoological Museum, Hamburg in 1995.
[= *Rhipicephalus pilans* Schulze]

Rhipicephalus perpulcher Gerstäcker, 1873
Baron Carl Claus von der Decken's Reisen in Ost-Afrika, **3**: 469.
Described from 1 ♀, Mombassa [Mombasa, Kenya]. Holotype with genital aperture dissected out [ZMB 2373] deposited in the Zoological Museum, Berlin – see Moritz & Fischer (1981) and Pegram *et al.* (1987).
[= *Rhipicephalus praetextatus* Gerstäcker]

Rhipicephalus phthirioides Cooper & Robinson, 1907
Journal of the Linnean Society of London, Zoology, **30**: 35, figs. 1–4, pl. 5 (consisting of photos 1–4). Also inset by Cooper & Robinson (published privately): figs. 1–4 (same as pl. 5, above), figs. 5 and 6.
Described from 1 ♂, 1 ♀ [dried specimens] ex horse, Rhodesia [Zimbabwe]. Depository not stated.
[= *Margaropus winthemi* Karsch]

48. ***Rhipicephalus pilans* Schulze, 1935**
Wissenschaftliche Ergebnisse der Niederländischen Expeditionen in dem Karakorum, (1922–1930): 180.
Described in a key. Specimens from Flores, Dutch East Indies [Indonesia]. Depository not stated. Full description and figures in Schulze P. 1936. *Zeitschrift für Parasitenkunde*, **8**: 521–7, figs. 1–6.

Rhipicephalus piresi Santos Dias, 1950
Moçambique, (**62**): 133, figs. 1, 2.
Described from 1 ♂ ex *Paraxerus cepapi auriventris* from Lumasse (vicinity of Guijá, Gaza district), Mozambique, 15.IX.1947, Dr F. Pires. Depository not stated but in the Veterinary Research Laboratory, Maputo, Mozambique.
[= *Rhipicephalus kochi* Dönitz]

49. ***Rhipicephalus planus* Neumann, 1907**
Morel, P.C. 1976. *Etude sur les tiques d'Ethiopie (Acariens, Ixodides)*, p. 156. Paris: Institut d'Élevage et de Médecine Vétérinaire des Pays Tropicaux.
Stat. nov. for *Rhipicephalus simus planus* Neumann, 1907.
[= *Rhipicephalus planus*]

Rhipicephalus planus Neumann, 1910
Annales des Sciences Naturelles, Zoologie, **12**: 165, figs. 4–7.
Described from 6 ♂♂, 1 ♀ ex wild boar, l'Ivindo basin, Cameroon. Deposited in the Natural History Museum, Paris. [Preoccupied name; *nec Rhipicephalus simus planus* Neumann, 1907].
[= *Rhipicephalus complanatus*]

Rhipicephalus planus complanatus Neumann, 1911
Zumpt, F. 1943. *Zeitschrift für Parasitenkunde*, **13**: 17, figs. 15–17.
Stat. nov. for *Rhipicephalus complanatus* Neumann, 1911.
[= *Rhipicephalus complanatus* Neumann]

Rhipicephalus planus planus Neumann, 1910
Zumpt, F. 1943. *Zeitschrift für Parasitenkunde*, **13**: 15, figs. 13, 14.
Stat. nov. for *Rhipicephalus planus* Neumann, 1910.
[= *Rhipicephalus planus* Neumann]

Rhipicephalus plumbeus (Panzer, 1795)
Neumann, L.G. 1901. *Mémoires de la Société Zoologique de France*, **14**: 353. Also in 1901 by Salmon, D.E. & Stiles, C.W. *17th Annual Report Bureau of Animal Industry*: 419.

[= ? *Ixodes lividus* Koch]

Rhipicephalus pomeranzevi Muratbekov, 1945

Byulleten' Sredne-Aziatskago Gosudarstvennago Universiteta. Tashkent, (23): 147.

Described from specimens ex *Lepus tibetanus*, Kzyl-Kum.

[= *Rhipicephalus* species – publication not seen. Some authors consider this taxon to be a junior synonym of *Rhipicephalus leporis*]

50. ***Rhipicephalus praetextatus* Gerstäcker, 1873**

Baron Carl Claus von der Decken's Reisen in Ost-Afrika, 3: 468.

Described from 1 ♂, Mombassa [Mombasa, Kenya]. Holotype [ZMB 2372] deposited in the Zoological Museum, Berlin – see Moritz & Fischer (1981) and Pegram *et al.* (1987).

51. ***Rhipicephalus pravus* Dönitz, 1910**

Denkschriften der Medicinisch-Naturwissenschaftlichen Gesellschaft zu Jena, 16: 479.

Described from several specimens ex buffalo, giraffe and antelopes, Damaraland, Transvaal and Massai steppe. Depository not stated. Syntypes [ZMB 9844, 17668, 17669] in the Zoological Museum, Berlin – see Moritz & Fischer (1981). Three ♂♂, 2 ♀♀ syntypes (Nuttall 1246) are in The Natural History Museum, London – see Keirans (1985).

Rhipicephalus pravus pravus Dönitz, 1910

Santos Dias, J.A.T. 1951. *Anais do Instituto de Medicina Tropical, Lisboa,* 8: 373.

Stat. nov. for *Rhipicephalus pravus* Dönitz, 1908.

[= *Rhipicephalus pravus* Dönitz]

52. ***Rhipicephalus pseudolongus* Santos Dias, 1953**

Santos Dias, J.A.T. 1955. *Boletim da Sociedade de Estudios de Moçambique,* (92): 110, fig. 2.

Stat. nov. for *Rhipicephalus capensis pseudolongus* Santos Dias, 1953.

53. ***Rhipicephalus pulchellus* (Gerstäcker, 1873)**

Baron Carl Claus von der Decken's Reisen in Ost-Afrika, 3: 467, pl. 18, fig. 2.

Described as a species of *Dermacentor* from a few ♂♂, host not stated, Aruscha [Arusha, Tanzania], Uru [Tanzania], and Lake Jipe [Kenya/Tanzania border]. Depository not stated. There are 3 ♂♂ syntypes [ZMB 2347] deposited in the Zoological Museum, Berlin – see Moritz & Fischer (1981).

Rhipicephalus pulchellus humeralis Rondelli, 1926

Res Biologicae, 1: 34, fig. 1.

Described from 16 ♂♂, host unknown, Mogadiscio, Somalia. Depository not stated.

[= *Rhipicephalus humeralis* Rondelli]

54. ***Rhipicephalus pumilio* Schulze, 1935**

Wissenschaftliche Ergebnisse der Niederländischen Expeditionen in dem Karakorum, 1: 178, fig. 1.

Described from 4 ♂♂, host unknown, Maralbashi. Type 1 ♂, deposited in the Zoological Museum, Amsterdam.

Rhipicephalus punctatissimus Gerstäcker, 1873

Baron Carl Claus von der Decken's Reisen in Ost-Afrika, 3: 470.

Described from 1 ♀, Mombassa [Mombasa, Kenya], deposited in the Zoological Museum, Berlin [ZMB 2370] – see Moritz & Fischer (1981).

[= *Rhipicephalus sanguineus* (Latreille)]

55. ***Rhipicephalus punctatus* Warburton, 1912**

Santos Dias, J.A.T. 1951. *Anais do Instituto de Medicina Tropical, Lisboa,* 8: 383.

Stat. nov. for *Rhipicephalus neavei punctatus* Warburton, 1912.

[= *Rhipicephalus punctatus*]

Rhipicephalus punctatus Bedford, 1929

15th Annual Report of the Director of Veterinary Services, Union of South Africa, 1: 495, pl. III, Figs. 4B, 4D, 5B.

Described from 7 ♂♂, 1 ♀ ex *Procavia capensis coombsi*, near Onderstepoort. Also 2 ♂♂, 2 ♀♀ ex dassie [rock hyrax], Omaruru, South-West Africa [Namibia]. Depository not stated but in the Onderstepoort Tick Collection 2488.

[= *Rhipicephalus distinctus* Bedford]

56. ***Rhipicephalus pusillus* Gil Collado, 1936**

Gil Collado, J. 1938. *Broteria, Lisboa, Serie Trimestral: Ciências Naturais,* 7: 102.
Stat. nov. for *Rhipicephalus bursa pusillus* Gil Collado, 1936.
[= *Rhipicephalus pusillus* Gil Collado]

57. ***Rhipicephalus ramachandrai* Dhanda, 1966**

Journal of Parasitology, **52**: 1025, figs. 1–17.
Described from numerous specimens, holotype ♂, allotype ♀ reared from fed nymph ex *Tatera indica,* Mundhwa, Poona district, Maharashta State, India, deposited in Virus Research Centre, Poona. Paratypes in various institutions, including The Natural History Museum, London with 1 ♂ (1966.12.21.1) and 1 ♀ (1966.12.21.2), and in the U.S. National Tick Collection [RML 47338, 101124, 101125 – see Keirans & Clifford (1984)].

Rhipicephalus reichenowi Zumpt, 1943

Zeitschrift für Parasitenkunde, **13**: 19, figs. 18, 19.
Described from 13 ♂♂ ex *Hystrix africaeaustralis,* near Mikesse, German East Africa [Tanzania]. Depository not stated. Holotype ♂ and 2 paratype ♂♂ are in the Zoological Museum, Berlin, ZMB No. 17687 – see Moritz & Fischer (1981).
[= *Rhipicephalus planus*]

Rhipicephalus rhipicephaloides (Neumann, 1901)

Jacob, E. 1924. *Zeitschrift für Morphologie und Ökologie der Tiere,* **1**: 364.
New combination for *Hyalomma rhipicephaloides* Neumann, 1901.
[= *Hyalomma rhipicephaloides* Neumann]

Rhipicephalus ricinus (Linnaeus, 1758)

Marchoux, E. & Couvy, L. 1912. *Bulletin de la Société de Pathologie Exotique,* **5**: 798.
[Probably a *lapsus calami* for *Ixodes ricinus* (Linnaeus)]

Rhipicephalus rosea (Koch, 1844)

Salmon, D.E. & Stiles, C.W. 1901. *17th Annual Report Bureau of Animal Industry,* pl. LXXXI, figs. 167, 168.
[*Lapsus calami* for *Haemaphysalis rosea* Koch, 1844]

58. ***Rhipicephalus rossicus* Yakimov & Kol-Yakimova, 1911**

Archives de Parasitologie, **14**: 419, figs. 1–4.
Described from ♂, ♀, host not stated, from Government of Saratov, Russia [USSR]. Depository not stated. One ♂, 1 ♀ syntypes (Nuttall 2897) are in The Natural History Museum, London – see Keirans (1985). Filippova (1996) designated a neotype male for *Rhipicephalus rossicus* and deposited it in the Zoological Institute of the Russian Academy of Sciences in St. Petersburg. However, because syntypes exist, this neotype has no validity.

Rhipicephalus ruber Supino, 1897

Neumann, L. G. 1897. *Mémoires de la Société Zoologique de France,* **10**: 418.
Stat. nov. for *Rhipicephalus haemaphysaloides ruber* Supino, 1897.
[= *Rhipicephalus haemaphysaloides* Supino]

Rhipicephalus rubicundus Frauenfeld, 1867

Verhandlungen der Zoologisch-Botanischen Gesellschaft in Wein, **17**: 462.
Described from 1 ♂, found on a ship, Sonde Sea (probably the Sunda Sea). Depository not stated.
[Probably *Rhipicephalus sanguineus* (Latreille)]

Rhipicephalus rufus (Koch, 1844)

Salmon, D.E. & Stiles, C.W. 1901. *17th Annual Report Bureau of Animal Industry*: 419.
New combination for *Ixodes rufus* Koch, 1844.
[= *Ixodes ricinus* (Linnaeus). *Fide* Neumann (1901) who examined the types]

Rhipicephalus rutilus Koch, 1844

Archiv für Naturgeschichte, **10**: 238.
Described from 1 ♀, Damiette, Egypt. Holotype in the Zoological Museum, Berlin – see Moritz & Fischer (1981).
[= *Rhipicephalus sanguineus* (Latreille)]

59. ***Rhipicephalus sanguineus* (Latreille, 1806)**

Koch, C.L. 1844. *Archiv für Naturgeschichte,* **10**: 238.
New combination for *Ixodes sanguineus* Latreille, 1806. [The type specimen of this species is apparently lost].
[= *Rhipicephalus sanguineus* (Latreille)]

Rhipicephalus sanguineus brevicollis Neumann, 1897
Neumann, L.G. 1904. *Archives de Parasitologie,* **9**: 449.
Stat. nov. for *Rhipicephalus sanguineus* (Latreille, 1806).
[= *Rhipicephalus sanguineus* (Latreille)]

Rhipicephalus sanguineus punctatissimus Gerstäcker, 1873
Neumann, L.G. 1904. *Archives de Parasitologie,* **9**: 449.
Stat. nov. for *Rhipicephalus sanguineus* (Latreille, 1806).
[= *Rhipicephalus sanguineus* (Latreille)]

Rhipicephalus sanguineus rossicus Yakimov & Kol-Yakimova, 1911
Zumpt, F. 1939. *Zeitschrift für Parasitenkunde,* **11**: 405, fig. 3.
Stat. nov. for *Rhipicephalus rossicus* Yakimov & Kol-Yakimova, 1911.
[= *Rhipicephalus rossicus* Yakimov & Kol-Yakimova]

Rhipicephalus sanguineus sanguineus (Latreille, 1806)
Neumann, L.G. 1911. *Das Tierreich. Acarina,* p. 35. Berlin: R. Friedländer & Sohn.
Stat. nov. for *Rhipicephalus sanguineus* (Latreille, 1806).
[= *Rhipicephalus sanguineus* (Latreille)]

Rhipicephalus sanguineus schulzei Olenev, 1929
Zumpt, F. 1940. *Zeitschrift für Parasitenkunde,* **11**: 677.
Stat. nov. for *Rhipicephalus schulzei* Olenev, 1929.
[= *Rhipicephalus schulzei* Olenev]

Rhipicephalus sanguineus simus Koch, 1844
Díaz Ungría, C. 1957. *Revista de Sanidad y Asistencia Social, Caracas,* **22**: 463.
Stat. nov. for *Rhipicephalus simus* Koch, 1844.
[= *Rhipicephalus simus* Koch]

Rhipicephalus sanguineus sulcatus Neumann, 1908
King, H.H. 1926. *Sudan Government Entomology Section Bulletin,* (**23**): Appendix I–II, 12.
Stat. nov. for *Rhipicephalus sulcatus* Neumann, 1908.
[= *Rhipicephalus sulcatus* Neumann]

60. ***Rhipicephalus scalpturatus* Santos Dias, 1959**
Memórias e Estudos do Museo Zoológico da Universidade de Coimbra, (**256**): 1, figs. 1–3.
Described from 4 ♂♂, 2 ♀♀, host not stated, Umsaw, Khasi Hills, Assam [India]. Holotype ♂ and 3 ♂♂, 2 ♀♀ paratypes deposited in the Zoological Museum, Hamburg.

61. ***Rhipicephalus schulzei* Olenev, 1929**
Vestnik Sovremennoy Veterinarii, **5**: 192, pl. 1, fig. 6.
Described without details on host or locality. Depository not stated.

Rhipicephalus schwetzi Larousse, 1927
Revue Zoologique Africaine, **15**: 214, figs. 2, 3.
Described from 7 ♂♂, 5 ♀♀ ex *Hylochoerus ituriensis,* Koteli (Bas-Uelé) [Democratic Republic of Congo]. Depository not stated.
[= *Rhipicephalus dux* Dönitz]

62. ***Rhipicephalus sculptus* Warburton, 1912**
Parasitology, Cambridge, **5**: 13, figs. 8, 9.
Described from 11 ♂♂, 5 ♀♀ ex roan antelope, Mpalali River, Nyasaland [Malawi], 1 ♂ from the same locality and host, 3 ♂♂, 1 ♀ ex zebra, S. Rukuru Valley, N. Nyasaland [Malawi]. Types were deposited in the British Museum (Natural History) and Cambridge. Lectotype and paralectotypes designated – see Keirans & Brewster (1981). All specimens are now in The Natural History Museum, London.

Rhipicephalus secundus Feldman-Muhsam, 1952
Bulletin of the Research Council of Israel, **2**: 192, figs. 3A, 4A, 5, 6A, 8.
Described from ♀, nymph and larva, based

on laboratory reared specimens. Depository not stated.

[=*Rhipicephalus turanicus* Pomerantsev]

63. ***Rhipicephalus senegalensis* Koch, 1844**

Archiv für Naturgeschichte, **10**: 238.

Described from ♀♀, host not stated, from Senegal and Egypt. Depository not stated. Three ♀♀ syntypes in the Zoological Museum, Berlin – see Moritz & Fischer (1981).

64. ***Rhipicephalus serranoi* Santos Dias, 1950**

Moçambique, (**63**): 143, figs. 1–3.

Described from 2 ♂♂, 5 ♀♀ ex 2 *Oreotragus oreotragus oreotragus*, Mutuali, Nampula District, Niassa Province, Mozambique, 23.VI.50 and 27.VII.50, Dr António de Melo Serrano. Depository not stated but in the Veterinary Research Laboratory, Maputo, Mozambique.

Rhipicephalus shipleyi Neumann, 1902

Archives de Parasitologie, **6**: 112.

Described from 2 ♂♂, 3 ♀♀ ex *Hyaena* sp.?, from the Sudan. Deposited in Cambridge. One ♀ syntype (Nuttall 1562) is in The Natural History Museum, London – see Keirans (1985).

[= *Rhipicephalus praetextatus* Gerstäcker]

Rhipicephalus siculus Koch, 1844

Archiv für Naturgeschichte, **10**: 239.

Described from 2 ♂♂, 1♀ from Sicily. Syntypes [ZMB 1091] in the Zoological Museum, Berlin – see Moritz & Fischer (1981).

[= *Rhipicephalus sanguineus* (Latreille)]

65. ***Rhipicephalus simpsoni* Nuttall, 1910**

Parasitology, Cambridge, **3**: 413, figs. 6, 7.

Described from 5 ♂♂, 11 ♀♀ ex large rodent, Oshogbo, S. Nigeria. Depository not stated. Lectotype and paralectotypes designated – see Keirans & Brewster (1981). Specimens in The Natural History Museum, London.

66. ***Rhipicephalus simus* Koch, 1844**

Archiv für Naturgeschichte, **10**: 238.

Described from 1 ♂, South Africa. Holotype [ZMB 1098] deposited in the Zoological Museum, Berlin – see Moritz & Fischer (1981) and Pegram *et al.* (1987).

Rhipicephalus simus erlangeri Neumann, 1902

Neumann, L.G. 1904. *Archives de Parasitologie*, **8**: 449.

Stat. nov. for *Rhipicephalus erlangeri* Neumann, 1902.

[= *Rhipicephalus simus* group]

Rhipicephalus simus hilgerti Neumann, 1902

Neumann, L.G. 1904. *Archives de Parasitologie*, **8**: 449.

Stat. nov. for *Rhipicephalus hilgerti* Neumann, 1902.

[= *Rhipicephalus praetextatus* Gerstäcker]

Rhipicephalus simus longoides Zumpt, 1943

Zeitschrift für Parasitenkunde, **13**: 11, figs. 6, 7.

Described from specimens from Cameroon, Sierra Leone, Ivory Coast, Gold Coast [Ghana], Togo, French Congo [Congo] and Belgian Congo [Democratic Republic of Congo]. Type and paratypes (16 ♂♂, 24 ♀♀ in the Museum of Berlin and in my [i.e. Zumpt's] collection) come from Cameroon. Holotype [ZMB 17780] and paratypes [ZMB 17780–17786] in the Zoological Museum, Berlin – see Moritz & Fischer (1981). An additional paratype in the Zumpt collection, South African Institute for Medical Research, Johannesburg. There is also 1 ♂ paratype ex *Bos taurus*, Akra (*sic*), Gold Coast [Ghana] in the Zoological Museum, Hamburg which Neumann identified as *Rhipicephalus bursa* (Neumann, 1897. *Mémoires de la Société Zoologique de France*, **10**: 393), and determined subsequently by Zumpt as *Rhipicephalus simus longoides*.

[= *Rhipicephalus senegalensis* Koch]

Rhipicephalus simus lunulatus Warburton, 1912

Parasitology, Cambridge, **5**: 19, fig. 12.

Stat. nov. for *Rhipicephalus lunulatus* Neumann.

[= *Rhipicephalus lunulatus* Neumann]

Rhipicephalus simus planus Neumann, 1907

Wissenschaftliche Ergebnisse der Schwedischen Zoologische Expedition nach dem Kilimandjaro, dem Meru und den umgeben-

den Massaisteppen, Deutsch-Ostafrikas 1905–06 (Sjöstedt), 3, Abteilung 20: *Arachnoidea*, (2): 20.
Described from 4 collections: 3 ♂♂, 1 ♀ ex *Hystrix africaeaustralis*, Kilimandjaro, Kibonoto; 2 ♀♀, host unknown, Kilimandjaro, Kibonoto; 3 ♂♂, host unknown, Kilimandjaro; 1 ♂, Kilimandjaro, Steppe [Tanzania]. Depository not stated.
[= *Rhipicephalus planus* Neumann]

Rhipicephalus simus senegalensis Koch, 1844
Zumpt, F. 1949. *Moçambique*, (**60**): 91, figs. 23, 24.
Stat. nov. for *Rhipicephalus senegalensis* Koch.
[= *Rhipicephalus senegalensis* Koch]

Rhipicephalus simus shipleyi Neumann, 1902
Neumann, L.G. 1904. *Archives de Parasitologie*, **8**: 449.
Stat. nov. for *Rhipicephalus shipleyi* Neumann, 1902.
[= *Rhipicephalus praetextatus* Gerstäcker]

Rhipicephalus simus simus Koch, 1844
Zumpt, F. 1949. *Moçambique*, (**60**): 90.
Stat. nov. for *Rhipicephalus simus* Koch.
[= *Rhipicephalus simus* Koch]

Rhipicephalus simus tricuspis Dönitz, 1910
Paoli, G. 1916. *Redia*, **11**: 281, pl. 5, figs. 12, 13.
Stat. nov. for *Rhipicephalus tricuspis* Dönitz, 1910.
[= *Rhipicephalus tricuspis* Dönitz]

Rhipicephalus stigmaticus Gerstäcker, 1873
Baron Carl Claus von der Decken's Reisen in Ost-Afrika, **3**: 469.
Described from 1 ♂, Mombassa [Mombasa, Kenya]. Holotype [ZMB 2371] deposited in the Zoological Museum, Berlin – see Moritz & Fischer (1981).
[= *Rhipicephalus sanguineus* (Latreille)]

67. ***Rhipicephalus sulcatus* Neumann, 1908**
Bulletin du Museum National d'Histoire Naturelle, **14**: 352, figs. 1, 2.
Described from 3 ♂♂, 4 ♀♀, host not stated, from Congo [Democratic Republic of Congo]. Deposited in the Museum of Natural History, Paris.

68. ***Rhipicephalus supertritus* Neumann, 1907**
Archives de Parasitologie, **11**: 216, figs. 2, 3.
Described from 2 ♂♂ ex horse, near the Lualaba River, Congo Free State [Democratic Republic of Congo]. Deposited in The Natural History Museum, London.

Rhipicephalus tendeiroi Santos Dias, 1950
Anais do Instituto de Medicina Tropical, Lisboa, 7: 217, figs. 1, 2.
Described from 1 ♂ ex *Felis leo leo*, Mocímboa do Rovuma, Macondes, district of Porto Amélia, Niassa Province, Mozambique, 28.III.50, L. Silveira. Depository not stated but in the Veterinary Research Laboratory, Maputo, Mozambique.
[= *Rhipicephalus masseyi* - see Santos Dias (1989)]

Rhipicephalus tetracornus Kitaoka & Suzuki, 1983
Tropical Medicine, **25**: 210, figs. 6–23.
Described from larvae, nymphs and reared adults. Immatures ex *Rattus surifer, R. nitidus, Bandicota savilei, Mus pahari*, and possibly *Eothenomys melanogaster* and *Anourosorex squamips*, Nakhon Nayok and Doi Inthanon, Thailand. Deposited in the National Science Museum, Natural History Institute, Shinjuku, Tokyo. [Unfortunately, all adults and some of the nymphs and larvae of this species were lost while being sent to Dr Harry Hoogstraal in Cairo, Egypt for illustration. Because we do not have specimens, and until additional material becomes available, we are considering *R. tetracornus* to be a species *incertae sedis*].

Rhipicephalus texanus Banks, 1908
Technical Series Bureau of Entomology, U. S. Department of Agriculture, (**15**): 34, pl. 5, figs. 1–4.
Described from ♂♂, ♀♀ ex dogs and horses, Texas, New Mexico and Mexico. Depository not stated. Syntypes in the U.S. National Tick Collection [RML 21579, 56742 – see Keirans & Clifford (1984)].
[= *Rhipicephalus sanguineus* (Latreille)]

69. ***Rhipicephalus theileri* Bedford & Hewitt, 1925**

South African Journal of Natural History, 5: 263, figs. 7–9.
Described from 1 ♂ holotype, 1 partially engorged ♀ ex *Xerus capensis*, Glen, Orange Free State, Republic of South Africa. Depository not stated but in Onderstepoort Tick Collection [2480 i. ♂, ♀ (type) ex *Geosciurus capensis*, Glen, O.F.S., 15. viii.21, R. Bigalke].

70. ***Rhipicephalus tricuspis* Dönitz, 1906**
Sitzungsberichte der Gesellschaft Naturforschender Freunde zu Berlin, (5): 146, figs. 7–9.
Described from 1 ♂, 1 ♀ found free in the Kalahari [Botswana]. Depository not stated. Syntypes in Zoological Museum, Berlin – see Moritz & Fischer (1981). Lectotype designated – see Walker *et al.* (1988).

Rhipicephalus turamicus Pomerantsev, 1936
Uzakov, U. Ya. 1964. *Trudy Vsesoyuznogo Nauchno-Issledovatel'skogo Instituta Veterinarnoi Sanitarii i Éktoparazitologii,* 24: 341.
[*Lapsus calami* for *Rhipicephalus turanicus* Pomerantsev]

71. ***Rhipicephalus turanicus* Pomerantsev, 1936**
Parazitologicheskii Sbornik, (6): 6
Subgeneric group description which included *Rhipicephalus turanicus* Pomerantsev. Depository not stated.

Rhipicephalus walckenaeri (Gervais, 1842)
Kratz, W. 1940. *Zeitschrift für Parasitenkunde,* 11: 560.
New combination for *Ixodes walckenaerii* Gervais, 1842.
[Species *incertae sedis*]

72. ***Rhipicephalus warburtoni* Walker & Horak, 1999**
In: Walker, J.B., Keirans, J.E. & Horak, I.G. 1999. *The genus* Rhipicephalus *(Acari: Ixodidae): A Guide to the Brown Ticks of the World,* pp. 463–470. Cambridge: Cambridge University Press. Described from holotype ♂, allotype ♀, ♂♂, ♀♀, nymphs and larvae paratypes, laboratory reared from a strain established from a ♀ ex Dorper sheep, Preezfontein, Free State, South Africa, on 14 October 1993 by L.J. Fourie. Holotype ♂, allotype ♀ and some paratypes in the Onderstepoort Tick Collection 3146 i, ii. Paratypes of all stages in the U.S. National Tick Collection and The Natural History Museum, London.

Rhipicephalus zambesiensis (*sic*) Walker, Norval & Corwin, 1981.
Sonenshine, D.E. 1993. *Biology of Ticks,* vol. 2, pp. 151, 463. Oxford and New York: Oxford University Press.
[*Lapsus calami* for *Rhipicephalus zambeziensis*]

Rhipicephalus zambeziensis Lawrence & Norval, 1979
Rhodesian Veterinary Journal, 10: 28.
[= *Nomen nudum*]

73. ***Rhipicephalus zambeziensis* Walker, Norval & Corwin, 1981**
Onderstepoort Journal of Veterinary Research, 48: 87, figs. 1–18, 21–31.
Described from holotype ♂, allotype ♀, ♂♂, ♀♀, nymphs and larvae paratypes, laboratory reared from a strain ex cattle, near West Nicholson, Gwanda District, Zimbabwe. Holotype, allotype and some paratypes in the Onderstepoort Tick Collection 3140 [not 3240 as appeared in the publication]. Other paratypes in Veterinary Research Laboratory, Harare, Zimbabwe; The Natural History Museum, London and U.S. National Tick Collection [RML 105751 – see Keirans & Clifford (1984)].

74. **Rhipicephalus ziemanni Neumann, 1904**
Archives de Parasitologie, 8: 464.
Described from 13 ♂♂, 19 ♀♀ ex cow, Cameroon. Depository not stated. One ♂, 1 ♀ syntypes (Nuttall 2898) are in The Natural History Museum, London – see Keirans (1985).

Rhipicephalus ziemanni aurantiacus Neumann, 1907
Morel, P.C. & Mouchet, J. 1958. *Annales de Parasitologie,* 33: 71.
Stat. nov. for *Rhipicephalus aurantiacus* Neumann, 1907.

[= *Rhipicephalus ziemanni* Neumann]

75. ***Rhipicephalus zumpti* Santos Dias, 1950** *Moçambique,* (**61**): 156, fig. 12.
Described from 1 ♂ ex buffalo, [*Syncerus caffer*], Govuro, Mozambique, 5.I.50, Dr J.M. da Silva. Deposited in the Santos Dias collection.

REFERENCES

Anastos, G. (1950). The scutate ticks, or Ixodidae of Indonesia. *Entomologica Americana (new series),* **30**, 1–144.

Clifford, C.M., Walker, J.B. & Keirans, J.E. (1983). Clarification of the status of *Rhipicephalus kochi* Dönitz, 1905 (Ixodoidea, Ixodidae). *Onderstepoort Journal of Veterinary Research,* **50**, 77–89.

Filippova, N.A. (1996). Designation of the neotypes for two species of ticks family Ixodidae. *Parasitologiya,* **30**, 404–9. [In Russian].

Keirans, J.E. & Brewster, B.E. (1981). The Nuttall and British Museum (Natural History) tick collections: lectotype designations for ticks (Acarina: Ixodoidea) described by Nuttall, Warburton, Cooper & Robinson. *Bulletin of the British Museum of Natural History (Zoology),* **41**, 153–78.

Keirans, J.E. & Clifford, C.M. (1984). A checklist of types of Ixodoidea (Acari) in the collection of the Rocky Mountain Laboratories. *Journal of Medical Entomology,* **21**, 310–20.

Keirans, J.E., Walker, J.B., Horak, I.G. & Heyne, H. (1993). *Rhipicephalus exophthalmos* sp. nov., a new tick species from southern Africa, and redescription of *Rhipicephalus oculatus* Neumann, 1901, with which it has hitherto been confused (Acari: Ixodida: Ixodidae). *Onderstepoort Journal of Veterinary Research,* **60**, 229–46.

Neumann, L.G. (1901). Révision de la famille des Ixodidés (4ᵉ Mémoire). *Mémoires de la Société Zoologique de France,* **14**, 249–372.

Neumann, L.G. (1911). Ixodidae. In *Das Tierreich,* ed. F.E. Schulze, 26 Lief. xvi + 169 pp. Berlin: R. Friedländer & Sohn.

Pegram, R.G., Walker, J.B., Clifford, C.M. & Keirans, J.E. (1987). Comparison of populations of the *Rhipicephalus simus* group: *R. simus*, *R. praetextatus*, and *R. muhsamae* (Acari: Ixodidae). *Journal of Medical Entomology,* **24**, 666–82.

Santos Dias, J.A.T. (1989). Acerca da posição taxonómica de algumas espécies do género *Rhipicephalus* (Acarina – Ixodoidea) da Região Afrotropical. *Garcia de Orta, Serie de Zoologia, Lisboa,* **13**, 107–15.

Santos Dias, J.A.T. (1991). Some data concerning the ticks (Acarina-Ixodoidea) presently known in Mozambique. *Garcia de Orta, Serie de Zoologia, Lisboa,* **18**, 27–48.

Walker, J.B. (1990). Two new species of ticks from southern Africa whose adults parasitize the feet of ungulates: *Rhipicephalus lounsburyi* n. sp. and *Rhipicephalus neumanni* n. sp. (Ixodoidea, Ixodidae). *Onderstepoort Journal of Veterinary Research,* **75**, 57–75.

Walker, J.B., Keirans, J.E., Pegram, R.G. & Clifford, C.M. (1988). Clarification of the status of *Rhipicephalus tricuspis* Dönitz, 1906 and *Rhipicephalus lunulatus* Neumann, 1907 (Ixodoidea, Ixodidae). *Systematic Parasitology,* **12**, 159–86.

Also see the following Basic References (pp. 12–14): Keirans (1985); Moritz & Fischer (1981); Theiler (1962).

6

Rhipicephalus species occurring in the Afrotropical region

HISTORICAL REVIEW

The first rhipicephalid to be recognized was the cosmopolitan species *Rhipicephalus sanguineus*, collected in France and described by Pierre André Latreille (1806). His description was brief in the extreme – '*Sanguineus, punctatus, postice lineolis tribus impressis; dorso antico macula nulla thoracica, distincta.*' Translated from the Latin this means: 'Blood red, punctate posteriorly with three impressed lines; no distinct thoracic spot anterodorsally' (M.-L. Penrith, pers. comm., 1997). Latreille placed this tick in the genus *Ixodes* but Koch (1844) reclassified it as a member of his newly erected genus *Rhipicephalus*, and at the same time described three new species, *R. capensis*, *R. senegalensis* and *R. simus*. Nearly 30 years later A. Gerstäcker (1873), who studied various arthropods collected by Baron C.C. von der Decken during his travels in East Africa, described *R. praetextatus* and *R. pulchellus*. The latter, on account of the ornate scutal pattern of the adults, was originally regarded as a species of *Dermacentor*.

Towards the end of the 19th century and early in this century research on ticks rapidly gathered momentum as these parasites became increasingly recognized as important vectors of various animal pathogens. In 1898 an American, C.P. Lounsbury, who had arrived in South Africa 3 years earlier to take up his appointment as Government Entomologist to the Department of Agriculture, Cape of Good Hope, turned his attention to ticks and tick-borne diseases. He soon established contact with two outstanding authorities in Europe, L.G. Neumann in Toulouse, France, and G.H.F. Nuttall in Cambridge, England. Neumann, who had already described *R. compositus* in 1877 and *R. evertsi* in 1899, based his descriptions of *R. appendiculatus* and *R. nitens* on Lounsbury's collections. By 1911 he had described 19 rhipicephalids that are still regarded as valid from specimens obtained in different parts of Africa. During this period one of the rarest and most distinctive species in the genus, *R. armatus*, was described by R.I. Pocock (1900) from Harar Province in Ethiopia.

Amongst other ticks Lounsbury managed to supply live *R. appendiculatus* to Nuttall for his studies on the variations in size and morphology due to nutrition seen in this species, a remarkable feat considering communications at that time. Many other collectors in Africa also sent ticks to Nuttall. These were all carefully documented, deposited in the Nuttall collection and later donated to The Natural History Museum, London (Keirans, 1985). Nuttall described only one rhipicephalid himself, *R. simpsoni*, and a second, *R. masseyi*, with his colleague C. Warburton. Another three, *R. longiceps*, *R. punctatus* and *R. sculptus*, were erected by Warburton (1912) in a paper that is still extremely relevant for anyone interested in this genus.

During this early period the eminent German taxonomist, W. Dönitz, was also active and was the author of five species, *R. dux*, *R. follis*, *R. kochi*, *R. pravus* and *R. tricuspis*, and one subspecies, *R. evertsi mimeticus*. Although neither he nor Neumann ever co-authored any papers with Nuttall and his colleagues there was evidently a cordial relationship between all these scientists and they exchanged specimens and information.

By 1913, therefore, 36 *Rhipicephalus* species, i.e. just over half of those now recognized from the Afrotropical region, had been described, all of them by authorities working in Europe. But apart from those studied by Lounsbury and by Sir Arnold Theiler at the newly established Veterinary Research Institute at Onderstepoort, both of whom were interested primarily in the life cycles and disease relationships of South African ticks rather than their taxonomy, little was known about many of them. Often only the hosts from which they had been collected and the localities where they were found were recorded, and sometimes not even these basic details. Subsequently the rate at which new species were erected decreased markedly, and most were described by scientists who were living and working in Africa even though their origins were often elsewhere. Of these the first was G.A.H. Bedford, who had arrived in 1912 from England to take up the post of Entomologist at Onderstepoort, where he worked until his death in 1938. Referred to later by Gertrud Theiler (1975) as 'a taxonomist *par excellence*', he described several new species of ticks, among them *R. distinctus* and, with J. Hewitt, *R. theileri*. He also produced a valuable checklist and host list of the external parasites, including ticks, of South African mammals, birds and reptiles. He was succeeded by R. du Toit, who described *R. glabroscutatum* in 1941.

In 1940 Gertrud Theiler was appointed in the Entomology Section at Onderstepoort. She soon took over responsibility for the research on ticks, a task that was to occupy her for over 25 years and for which she became world renowned. Although she was involved, with F. Zumpt, in the description of only one new rhipicephalid, *R. arnoldi*, in 1949, she contributed greatly to our knowledge of this genus, especially by rearing and describing the immature stages of several South African species and studying their host relationships and zoogeography. In 1947 she published a particularly valuable paper in which she collated existing information on many of the lesser known African rhipicephalids. In 1962 she issued a report documenting the hosts and distribution of all the African ticks, including the rhipicephalids. She always maintained close contact with other tick workers, who relied on her for advice and help such as pertinent extracts from the literature and reference specimens of ticks. These were a great boon to those without ready access to such aids.

Fritz Zumpt had become interested in the genus *Rhipicephalus* during the 1940s. While he was still in Germany he wrote eight papers on individual species and groups with the evident aim of revising this genus later. After the Second World War he and his family emigrated to South Africa. In 1949 he published the last of these preliminary studies to his intended revision in the form of a key to the known species, plus a description of *R. bequaerti*. Thereafter he left this field of study to Dr Theiler. He was for many years Head of the Department of Entomology, South African Institute for Medical Research in Johannesburg.

Although most veterinarians and medical authorities in Africa have not been interested in tick taxonomy *per se* they have always been deeply concerned with the various tickborne diseases of animals and man and also realized the necessity of studying their vectors. Consequently research on these parasites often received a high priority during the colonial era. As early as 1907 R.J. Stordy, the Chief Veterinary Officer of the East African Protectorate (as Kenya was then known), had already identified six tick-borne diseases of domestic animals there and had become alarmed about the possible spread of African Coast fever, i.e. East Coast fever (ECF). He listed eight species of ticks occurring in the country, including *R. appendiculatus*, *R. evertsi*, *R. pulchellus* and *R. simus*, and gave details of the localities where they occurred. Work on both the ticks and the diseases they transmitted, especially ECF and Nairobi sheep disease, was continued in Kenya, even during the First World War, by successive researchers including T.J. Anderson, E. Montgomery and, a few years later, by R. Daubney and J.R. Hudson, at the Veterinary Research Laboratory, Kabete. The work was given extra impetus by the appointment in 1930 of E.A. Lewis, who conducted investigations on ticks and tick-borne diseases in Kenya, and also identified some tick collections from Tanzania and Uganda, during the following 22 years.

In general, though, fewer advances in knowledge were made in either Tanzania or Uganda during this early period. Following the outbreak of the First World War in 1914 tick collections from domestic animals in Tanzania virtually ceased for almost 20 years. Thereafter, despite widespread recognition of the importance of ECF in the country, only sporadic investigations were carried out until after the Second World War. Finally in 1955 G.H. Yeoman was able to plan and, with J.P.J. Ross and T. Docker, conduct a well-organized tick survey covering the entire country (Yeoman & Walker, 1967). In Uganda a tick collection that accumulated at the Animal Health Research Centre during the late 1920s and early 1930s was documented by R.W.M. Mettam in his annual reports. Fortunately these specimens were deposited in the Nuttall Collection, and some of Mettam's identifications were later emended by Keirans (1985). The results of a survey by S.G. Wilson during the late 1940s also appeared mainly in the annual reports of the Uganda Veterinary Department. In 1965, with the support of the Uganda Government and the United States Agency for International Development, J.G. Matthysse and M.H. Colbo carried out further field work to fill gaps in the existing knowledge of Uganda ticks. By 1972 their first manuscript documenting this information was completed but it was largely destroyed during the political turmoil in the country at that time. A revised, rewritten and updated version of their work was published only in 1987. It is particularly useful because it is copiously illustrated and includes a number of species that the authors thought might occur in Uganda, plus keys for their identification (Keirans, 1988).

Prior to his service in Uganda, Wilson had worked in Malawi for some years and made valuable contributions to our knowledge of the ticks there, especially *Rhipicephalus* species, in which he was particularly interested. When he left Uganda he served for a short period in Kenya, during which he described *R. hurti* and organized some valuable collections of ticks. He was succeeded by A.J. Wiley, who had trained originally with E.A. Lewis and consequently had much experience with Kenya ticks. Sadly he died just a few years later, in 1959. Research on the hosts and distribution of ticks in Kenya has since been carried out by the senior author (Walker, 1974), and also under the auspices of the United Nations Development Programme (FAO, 1975).

Before the Second World War the ticks occurring in southern Somalia were studied by Maria Tonelli-Rondelli, who described *R. humeralis*, originally as a subspecies of *R. pulchellus*. Subsequently both she and E. Stella, amongst others, extended their investigations to parts of Ethiopia. In 1976 R.G. Pegram reviewed the ixodid ticks occurring in northern Somalia, followed 2 years later by D. Scaramella, who recorded the known tick species present throughout this country. Afterwards Pegram turned his attention to the ticks of Ethiopia, which were also the subject of research during the 1970s by the French workers P. Bergeon, J. Balis and especially P.C. Morel, who had previously worked in West Africa for many years. The review of Ethiopian ticks by Morel (1976) was originally published in French. It was reissued in 1980 in English and we have quoted the English edition throughout this book.

Harry Hoogstraal, who has been described as 'the greatest authority on ticks and tick-borne diseases who ever lived' (Keirans, 1986), was based at the United States Naval Medical Research Unit No. 3 (NAMRU3) in Cairo, Egypt from 1949 until his death in 1986. From 1950 onwards he served as Head of the Medical Zoology Department there. In 1956 he published his monograph on the ticks of the Sudan, based on research he had carried out in Equatoria Province. This book is particularly valuable for its detailed reviews of those ticks he included and its comprehensive bibliography. He was not, however, especially interested in the rhipicephalids, about which both he and his associate Makram Kaiser often consulted either Gertrud Theiler or authorities at the Rocky Mountain Laboratory, Hamilton, Montana, USA, especially Carleton M. Clifford.

During the colonial era West Africa, broadly speaking, was primarily the sphere of interest of a succession of eminent French scientists. Those who made initial contributions to our knowledge of *Rhipicephalus* spp. there include J. Colas-Belcour, who studied *R. fulvus* in North Africa in 1932; J. Rageau, who worked on the ticks of Cameroon in the early 1950s, and R. Rousselot, who had a general interest in the ticks recorded in the French West African territories, also during the 1950s. In 1954 P.C. Morel arrived to take charge of the research on entomology and parasitology at the laboratory operating in Dakar-Hann, Senegal, under the auspices of the Institute d'Élevage et de Médecine Vétérinaire des Pays Tropicaux, Paris (IEMVT). During the following 10 years he, together with his associates P. Finelle, M. Graber, J. Magimel, J. Mouchet and G. Vassiliades, made major contributions to the systematics of West African ticks, to which frequent reference is made later in this book, especially in Chapter 7. Particular mention should be made of the study that he and Vassiliades published on the *R. sanguineus* group in 1963. Unfortunately, though, it has up to now been impossible to integrate some of their conclusions fully with other workers' findings, particularly those of R.G. Pegram. In 1965 Morel returned to Alfort, Paris, as Head of the Parasitology Section at the IEMVT. In 1969 he produced a remarkable thesis, accompanied by distribution maps, on the systematics of all the ticks then known to occur in the Afrotropical region; this has been of material assistance to us during the preparation of this book.

Notable contributions to our knowledge of West African rhipicephalids were also made between 1946 and 1963 by the Portuguese scientist J. Tendeiro, working mainly in Guinea, and by A. Aeschlimann, from Switzerland, who spent three years (1959–1961) collecting ticks in the Ivory Coast (Aeschlimann, 1967). A little earlier, records of *Rhipicephalus* spp. occurring in Nigeria, particularly in the northern territories, were included in the results of a survey on cattle ticks published by K. Unsworth (1952). Further surveys, again with emphasis on the ticks in northern Nigeria, were carried out by K.L. Strickland (1961) and by A.N. Mohammed (1977). More recently investigations in West Africa have centred on the ticks of Mali (Matthysse, 1980); Senegal (Gueye *et al.*, 1986, 1987), and Cameroon (Merlin, Tsangueu & Rousvoal, 1986, 1987).

Since the beginning of this century various publications have featured the *Rhipicephalus* spp. occurring in the Democratic Republic of Congo, Rwanda and Burundi. Among these are the reports published by Nuttall & Warburton (1916); Bequaert (1931); Pierquin & Niemegeers (1957, 1958); Theiler & Robinson (1954); Clifford & Anastos (1962, 1964) and Elbl & Anastos (1966). A useful feature of the latter work is the lists of species from this vast area recorded in the literature and in the collections of the Musée Royal de l'Afrique Centrale, Tervuren, Belgium.

Until about the last 50 years little attention had been paid to the ticks of Zambia. Both the checklist by Theiler & Robinson (1954) and the report by Matthysse (1954) on the tick-borne diseases occurring there included records of various *Rhipicephalus* spp. In 1961 W.J. Gray described his biological studies carried out on *R. evertsi* on the Kafue Flats. From 1966 to 1973 J. MacLeod and his colleagues M.H. Colbo and B. Mwanaumo, working initially under the auspices of the FAO and later the Agricultural Research Council (Zambia), studied tick populations in various parts of the country. Their findings, published in a series of papers cited under the accounts of individual species, provide a foundation for much of our knowledge of Zambian ticks. From 1980 onwards further work was carried out, still under the auspices of the FAO, by R.G. Pegram, including important taxonomic and biological studies on the *R. sanguineus* and *R. simus* groups. Unfortunately, though, we have been unable to trace any recent publications by the Zambian taxonomist, the late F. Zulu, who was involved in research on *Rhipicephalus* species, amongst others, during the 1980s.

From 1910 onwards numerous contributions to our knowledge of ticks in Zimbabwe were made by R.W. Jack, culminating in 1942, to which Kate Jooste (1969) and also J. MacLeod, again working under the auspices of the FAO, added over 25 years later. In 1975 a National Tick Survey was instituted by R.A.I. Norval aided by several co-workers, particularly J. Muchuwe, as described by Mason & Norval (1980). This resulted in the acquisition of much valuable information on the hosts and distribution of the *Rhipicephalus* spp. occurring in that country.

In the former Portuguese colonies of Angola and Mozambique considerable attention has been paid to the tick fauna. In Angola, V.A. Sousa Dias (1950) compiled a detailed account of the tick species present, and a number of collections made there subsequently have been recorded by J.A.T. Santos Dias. Despite the constraints imposed by the devastating civil war in that country further investigations have been organized recently in the Huambo area, in the south of the country, by A. Gomes (see Gomes, Pombal & Venturi, 1994). The main focus of attention by Santos Dias has been the ticks of Mozambique. He was based at the Veterinary School in Maputo (formerly Lourenço Marques) from 1946 to 1981 and published many papers during this period, including descriptions of two new species, *R. serranoi* and *R. pseudolongus*. He then returned to Portugal, where he continued his research at the Instituto de Investigação Cientifica Tropical until his retirement in 1990, though he maintained his interest in ticks until his death 5 years later. Additional contributions to our knowledge of Mozambican ticks were also made by J. Tendeiro.

In 1910 Dönitz described *R. evertsi mimeticus* from Namibia, *R. follis* from 'Südafrika' and *R. tricuspis* from Botswana, and discussed several other rhipicephalids occurring in southern Africa. Ticks occurring in Botswana featured in few publications thereafter until 1955, when Santos Dias published the results of a survey by Zumpt. Available records of the ticks occurring there were consolidated by Walker, Mehlitz & Jones (1978) and G.D. Paine added further information, following a survey that he had undertaken, in 1982. In Namibia early studies on ticks were carried out by Trommsdorff (1914) and Hans Sigwart (1915) but little further information became available until Santos Dias (1955) published the records of a short survey conducted in 1952 by Zumpt. Numerous collections, covering much of the country, have been made since 1970 by J.D. Bezuidenhout and others. These have included a new species, *R. neumanni* Walker, 1990. Many unpublished records of other rhipicephalids identified in these collections have kindly been made available to us by Heloise Heyne for inclusion in our accounts of individual species.

In a short review such as this it is obviously possible to mention only a few of those responsible for existing knowledge of African rhipicephalids. Comprehensive references to those who have contributed further information prior to 1969 are quoted by Doss *et al.* (1974–1978). A brief note should be added about our own involvement in the acquisition of this knowledge. J.B.W. has been concerned with research on African ticks since 1949. After training with Gertrud Theiler she worked until 1966 at the East African Veterinary Research Organization, Muguga, Kenya. She then moved to the Onderstepoort Veterinary Institute, South Africa, where she took over from Dr Theiler. She has been particularly involved in studies on the systematics of the African *Rhipicephalus* spp. and for over 10 years has been working towards a consolidation of information on these ticks. She has been the author or co-author of the following nine species: *R. aquatilis* (with J.E.K. and R.G. Pegram); *R. carnivoralis*; *R. exophthalmos* (with

J.E.K.); *R. interventus* (with R.G. Pegram and J.E.K.); *R. lounsburyi*; *R. neumanni*; *R. oreotragi* sp.nov. (with I.G.H.); *R. warburtoni* sp.nov. (with I.G.H.), and *R. zambeziensis* (with R.A.I. Norval and D. Corwin).

J.E.K. has been curator of the U.S. National Tick Collection for several years, first with the National Institutes of Health at the Rocky Mountain Laboratory, Hamilton, MA (with Dr Carleton M. Clifford), then at the Smithsonian Institution, Washington, DC, and currently at Georgia Southern University, Statesboro, GA. He and J.B.W. have collaborated on the descriptions of *R. aquatilis*, *R. exophthalmos* and *R. interventus* and on systematic studies on other *Rhipicephalus* spp.

I.G.H. has been involved in research on helminths in South Africa since 1961 and on ticks since 1974 when he joined the Faculty of Veterinary Science, University of Pretoria at Onderstepoort as a lecturer in Ectoparasitology. After spending 5 years from 1982 to 1987 at the Tick Research Unit, Rhodes University, Grahamstown he rejoined the staff of the Veterinary Faculty. His research has concentrated on the distribution, host preferences and seasonal abundance of ticks in South Africa and Namibia.

REFERENCES*

*Excluding authors of original descriptions, who are quoted under the species concerned in Chapter 7.

Bequaert, J. (1931). Synopsis des tiques du Congo Belge. *Revue de Zoologie et de Botanique Africaine*, **20**, 209–51.

Colas-Belcour, J. (1932). Contribution à l'étude de *Rhipicephalus (Pterygodes) fulvus* Neumann et de sa biologie. *Archives de l'Institut Pasteur, Tunis*, **20**, 430–43.

FAO (1975). *Research on Tick-borne Cattle Diseases and Tick Control, Kenya. Epizootiological Survey on Tick-borne Cattle Diseases. AG:DP/KEN/70/522. Technical Report 1*, 52 pp. Rome: United Nations Development Programme, Food and Agriculture Organization of the United Nations.

Gomes, A.F., Pombal, A.M. & Venturi, L. (1994). Observations on cattle ticks in Huila Province (Angola). *Veterinary Parasitology*, **51**, 333–6.

Gray, W.J. (1961). *Rhipicephalus evertsi*: notes on free-living phases. *Bulletin of Epizootic Diseases of Africa*, **9**, 25–7.

Gueye, A., Mbengue, Mb., Diouf, A. & Seye, M. (1986). Tiques et hémoparasitoses du betail au Sénégal. I. La region des Niayes. *Revue d'Élevage et de Médecine Vétérinaire des Pays Tropicaux*, **39** (nouvelle série), 381–93.

Gueye, A., Camicas, J.L., Diouf, A. & Mbengue, Mb. (1987). Tiques et hémoparasitoses du betail au Sénégal. II. La zone sahélienne. *Revue d'Élevage et de Médecine Vétérinaire des Pays Tropicaux*, **40** (nouvelle série), 119–25.

Jack, R.W. (1942). Ticks infesting domestic animals in Southern Rhodesia. *Rhodesian Agricultural Journal*, **39**, 95–109.

Jooste, K.F. (1969). The role of Rhodesia in ixodid tick distribution in central and southern Africa. In *Proceedings of a Symposium on the Biology and Control of Ticks in Southern Africa*, convenor G.B. Whitehead, pp. 37–42. Grahamstown, South Africa: Rhodes University.

Keirans, J.E. (1986). Harry Hoogstraal (1917–1986). *Journal of Medical Entomology*, **23**, 342–3.

Keirans, J.E. (1988). Book review. *The Ixodid Ticks of Uganda*. By John G. Matthysse and Murray H. Colbo. College Park, MD: Entomological Society of America. *Proceedings of the Entomological Society of Washington*, **90**, 398–400.

Mason, C.A. & Norval, R.A.I. (1980). The ticks of Zimbabwe. I. The genus *Boophilus*. *Zimbabwe Veterinary Journal*, **11**, 36–43.

Matthysse, J.G. (1954). *Report on Tick-borne Diseases*. Lusaka, Northern Rhodesia: Government Printer. 28 pp.

Matthysse, J.G. (1980). *Research and Training on Vector-borne Hemoparasites of Livestock and their Vectors in Mali. Phase I, Objective 7: Survey to determine Species Diversity and Distribution of Ticks attacking Cattle in Mali*. 109 pp. Technical Assistance Contract between the United States Agency for International Development and the International Division of Texas A & M University. Project AID/afr-c-1262.

Merlin, P., Tsangueu, P. & Rousvoal, D. (1986). Dynamique saisonnière de l'infestation des bovins par les tiques (Ixodoidea) dans les hauts plateaux de l'Ouest du Cameroun. I. Étude de trois sites autour de Bamenda pendant un an. *Revue d'Élevage et de Médecine Vétérinaire des Pays Tropicaux*, **39** (nouvelle série), 367–76.

Merlin, P., Tsangueu, P. & Rousvoal, D. (1987). Dynamique saisonnière de l'infestation des bovins par les tiques (Ixodoidea) dans les hauts plateaux de l'Ouest du Cameroun. II. Élevage extensif traditionnel. *Revue d'Élevage et de Médecine Vétérinaire des Pays Tropicaux*, **40** (nouvelle série), 133–40.

Mohammed, A.N. (1977). The seasonal incidence of ixodid ticks of cattle in Northern Nigeria. *Bulletin of Animal Health and Production in Africa*, **25**, 273–93.

Morel, P.C. & Vassiliades, G. (1963). Les *Rhipicephalus* du groupe *sanguineus*: espèces africaines (Acariens: Ixodoidea). *Revue d'Élevage et de Médecine Vétérinaire des Pays Tropicaux*, **15 for 1962** (nouvelle série), 343–86.

Nuttall, G.H.F. & Warburton, C. (1916). Ticks of the Belgian Congo and the diseases they convey. *Bulletin of Entomological Research*, **6**, 313–52.

Paine, G.D. (1982). Ticks (Acari:Ixodoidea) in Botswana. *Bulletin of Entomological Research*, **72**, 1–16.

Pegram, R.G. (1976). Ticks (Acarina, Ixodoidea) of the northern regions of the Somali Democratic Republic. *Bulletin of Entomological Research*, **66**, 345–63.

Pierquin, L. & Niemegeers, K. (1957). Répertoire et distribution geographique des tiques au Congo Belge et au Ruanda-Urundi. *Bulletin Agricole du Congo Belge*, **48**, 1177–224.

Pierquin, L. & Niemegeers, K. (1958). Tables dichotomiques pour l'identification des tiques adultes du Congo Belge et Ruanda-Urundi. *Bulletin Agricole du Congo Belge*, **49**, 421–60.

Rageau, J. (1951). Ixodidés du Cameroun. *Bulletin de la Société de Pathologie Exotique*, **44**, 441–6.

Rageau, J. (1953). Note complémentaire sur les Ixodidae du Cameroun. *Bulletin de la Société de Pathologie Exotique*, **46**, 1090–8.

Rousselot, R. (1953). Notes de parasitologie tropicale. Tome II. *Ixodes*. 135 pp. Paris: Vigot Frères, Éditeurs.

Santos Dias, J.A.T. (1955). Contribuição para o conhecimento da fauna ixodológica do Sudoeste Africano. *Anais do Instituto de Medicina Tropical*, **12**, 75–100.

Santos Dias, J.A.T. (1955). Subsídios para o estudo da fauna ixodológica da Bechuanalândia. *Memórias e Estudos do Museu Zoológico da Universidade de Coimbra*, **No. 231**, 1–10.

Sigwart, H. (1915). Beitrag zur Zeckenkenntnis von Deutsch-Südwestafrika, unter besonderer Berücksichtigung der Funde in den Bezirken Outjo und Waterberg. *Zeitschrift für Infektkrankheit der Haustiere*, **16**, 434–44.

Stordy, R.J. (1907). Appendix VIII. Report of the Veterinary Department. Cattle diseases. African Coast fever. *Colonial Report No.* **519**. *East African Protectorate* **1905–1906**, 103–11.

Strickland, K.L. (1961). *A Study of the ticks of domesticated animals in Northern Nigeria: a preliminary to disease investigation.* M.Sc. thesis, University of Dublin. 111 pp.

Theiler, Gertrud (1975). Past-workers on tick and tick-borne diseases in southern Africa. *Journal of the South African Veterinary Association*, **46**, 303–10.

Theiler, Gertrud & Robinson, Britha N. (1954). Tick Survey VIII. – Checklists of ticks recorded from the Belgian Congo and Ruanda Urundi, from Angola, and from Northern Rhodesia. *Onderstepoort Journal of Veterinary Research*, **26**, 447–61 + 4 maps.

Trommsdorff, –. (1914). Beitrag zur Kenntnis der in Deutsch-Südwestafrika verkommenden Zeckenarten. *Archiv für Schiffs- und Tropenhygiene*, **18**, 731–47.

Unsworth, K. (1952). The ixodid parasites of cattle in Nigeria, with particular reference to the northern territories. *Annals of Tropical Medicine and Parasitology*, **46**, 331–6.

Also see the following Basic References (pp. 12–14): Aeschlimann (1967); Clifford & Anastos (1962, 1964); Doss *et al.* (1974–78); Hoogstraal (1956); Keirans (1985); Matthysse & Colbo (1987); Morel (1969, 1980); Sousa Dias (1950); Theiler (1947, 1962); Walker, Mehlitz & Jones (1978); Yeoman & Walker (1967).

KEYS

Those readers who have previously attempted to key species of *Rhipicephalus* realize that it can be a daunting and, at times, a frustrating process. We have done our best to make this difficult task as straightforward as possible including, when appropriate, miscellaneous information on hosts and distribution within the couplets themselves. After keying a tick to a particular species we urge the reader not only to compare the specimen with the illustrations provided under that name but also to read the full description, hosts, sites of attachment and zoogeography. Even if you have given a tick a name, this does not necessarily mean you have got it right.

In preparing the keys we owe a considerable debt to those who have travelled this road before: Zumpt, Pomerantsev, and Matthysse & Colbo. Because of the conservative nature of these parasites, we freely admit to borrowing one or more of these authors' couplets.

The host preferences listed are those of the adults. In several cases the immature stages prefer different hosts.

KEY TO THE AFROTROPICAL *RHIPICEPHALUS* SPECIES MALES

These keys were constructed merely to help readers to identify their specimens, not to demonstrate phylogenetic relationships.

1a. Small subadanal plates visible just posterior to adanal plates. (Subgenus *Hyperaspidion* Pomerantsev, 1936). One of the rarer species, recorded most commonly from carnivores in East Africa . . . . ***R. armatus*** (Note: Some male specimens of *R. cuspidatus* have small to minute idiosomal plaques beneath or medial to the adanal plates, but these integumental thickenings do not constitute subadanal plates)

1b. Subadanal plates absent 2

2a. Adanal plates large, curving externally and extended posteriorly into broadly-rounded points; medially projecting cusps absent; accessory adanal plates absent 3

2b. Adanal plates of various shapes, but not produced posteriorly into broadly-rounded points; medially projecting cusps and accessory adanal plates present or absent 4

3a. Spiracles each with a long, narrow dorsal prolongation. Almost exclusively a parasite of yellow mongoose, meercat and Cape ground squirrel in South Africa, Botswana and Namibia . . . . ***R. theileri***

3b. Spiracles each with a long, broad dorsal prolongation. Parasitic mainly on warthogs and porcupines in West Africa, from Senegal eastward to the Sudan . . . . ***R. cuspidatus***

4a. Scapular areas of conscutum bearing a large anterior process on each side of the basis capituli. (Subgenus *Pterygodes* Neumann, 1913). Parasitic on the gundi and some larger animals in North Africa and the Sahara . . . . ***R. fulvus***

4b. Scapular area of conscutum not bearing a large anterior process on each side of the basis capituli 5

5a. Conscutum ornamented with ivory markings 6

5b. Conscutum lacking ornamentation 9

6a. Ivory colouration encircling conscutal margin and often much of the central area. A common tick in dry areas in eastern Africa east of the Rift Valley . . . . . ***R. pulchellus***

6b. Ivory colouration not encircling conscutal margin; markings restricted to scapular or central areas 7

7a. An ivory patch present on each scapula (uncommonly a patch may also be present behind each eye, which rarely may extend forward to the scapular patch). Present in dry areas in southern Somalia, Kenya and Tanzania . . . . ***R.humeralis***

7b. Lacking ornamentation in scapular areas 8

8a. A variable pattern of diffuse ornamentation found on the conscutum often associated with larger punctations. Occurring on a wide range of hosts, mainly in coastal regions from Kenya to northern KwaZulu-Natal, South Africa but also recorded sometimes from Zambia, Zimbabwe and Malawi . . . . *R. maculatus*

8b. A small patch of ornamentation present anteriorly and centrally on the conscutum plus a larger patch posteriorly partially surrounding the posterior grooves; the patches may be joined. Present in western Uganda, and in Rwanda and Democratic Republic of Congo . . . . *R. dux*

9a. Eyes round, beady, deeply orbited 10

9b. Eyes flat, slightly bulging or convex, but not beady and deeply orbited 14

10a. Spiracles surrounded by numerous long setae 11

10b. Long setae absent from circumspiracular area 12

11a. A rather densely punctate tick with larger punctations marking the external cervical margins; legs uniformly bright reddish orange. Very common and widely distributed in the Afrotropical region . . . . . *R. evertsi evertsi*

11b. As in couplet 11a, but with saffron annulations on the legs. Almost exclusively in dry areas from south-western Democratic Republic of Congo south to Namibia and western Botswana . . . . *R. evertsi mimeticus*

12a. Conscutum smooth, shiny, almost devoid of punctations. A parasite of the lower legs and feet of ungulates in the Eastern and Western Cape Provinces, South Africa . . . . *R. glabroscutatum*

12b. Conscutum densely punctate 13

13a. Adanal plates with posterior margins broad and slightly convex; internal margins nearly straight until they lead to small medial points or cusps. Recorded mainly from southern Namibia and South Africa . . . . *R. exophthalmos*

13b. Adanal plates with posterior margins almost straight and without small medial points or cusps. Almost exclusively a parasite of hares in the drier areas of Namibia, Botswana and South Africa . . . . *R. oculatus*

14a. Conscutum with a large depressed area centrally, posteriorly, or in a central posterior trough, giving the conscutum a concave appearance 15

14b. Conscutum convex or flat but without a concave appearance or large depressed area 16

15a. Conscutum length from 3.0 mm to 4.0 mm; depressed from anterior margin to festoons; adanal plates broadly triangular with concave posterior margins . . . . *R. complanatus*

15b. Conscutum length from 2.5 mm to 3.0 mm; depressed in posterior half; adanal plates almost sickle-shaped, with broadly-rounded or convex posterior margins . . . . *R. planus*

16a. Adanal plates either with posterior margins deeply concave between two bluntly-rounded cusps, or with a bluntly-rounded internal and a sharply-pointed external cusp; accessory adanal plates narrowly elongate and pointed; thus giving the adanal plates a 'tricuspid' appearance 17

16b. Adanal plates of various shapes but not deeply concave posteriorly, and not appearing 'tricuspid'; accessory adanal plates present or absent 19

17a. Pedicel of palpal article I elongate and easily visible dorsally, producing a U-shaped indentation in the external margin of the first article. Widespread on numerous hosts in the Afrotropical region . . . . *R. lunulatus*

17b. Pedicel of palpal article I visible dorsally but short and not producing a U-shaped indentation in the external margin of the first article 18

18a. Spiracles each with a short, broad dorsal prolongation. Parasitic on cattle and various mostly small to medium-sized antelopes in East and Central Africa; sometimes sympatric with *R. lunulatus* . . . . *R. interventus*

18b. Spiracles each with a long, narrow dorsal prolongation. Recorded occasionally from domestic animals but more commonly from the smaller wild carnivores, and from antelopes, springhares and hares, mainly in southern Africa, more rarely further north . . . . ***R. tricuspis***

19a. Coxae of the first pair of legs exceptionally long, deeply divided; spiracles unique in shape, oval, each with a short, stubby, dorsal prolongation. A large tick of giraffes in Kenya and Tanzania . . . . ***R. longicoxatus***

19b. Coxae of the first pair of legs not exceptionally long; spiracles of various shapes but not as in couplet 19a 20

20a. Adanal plates tending to be sickle shaped, their external and internal margins roughly parallel and curving around the anus 21

20b. Adanal plates of various shapes, may be somewhat curved, but not sickle shaped 26

21a. External cervical margins marked by a cessation of the punctation pattern in the cervical fields . . . . ***R. pseudolongus***

21b. External cervical margins marked by a row of punctations larger than those of the cervical fields 22

22a. A '*simus*' pattern of punctations obvious on the conscutum 23

22b. A '*simus*' pattern of punctations absent or obscured because the conscutum is too densely punctate 24

23a. Clusters of large punctations present between the posterolateral grooves and the posteromedian groove. Primarily a West African species . . . . ***R. senegalensis***

23b. Clusters of large punctations lacking between the posterolateral grooves and the posteromedian groove, conscutum essentially impunctate except for the '*simus*' pattern. Widely distributed in the Afrotropical region in association with its preferred host, the greater cane rat; rarely recorded from other hosts . . . . ***R. simpsoni***

24a. Conscutum densely punctate; marginal lines deep grooves composed of large punctations . . . . ***R. longus***

24b. Conscutum punctate but not densely so; marginal lines slender, narrow, shallow grooves outlined by small punctations 25

25a. Posteromedian groove quite short, broad, deep; posterolateral grooves small, rounded and deep. A rare West African species . . . . ***R. moucheti***

25b. Posteromedian groove long, narrow and shallow; posterolateral grooves begin at festoons and extend forward as comma-shaped shallow grooves. A parasite of the sitatunga and sometimes other mammals sharing its semi-aquatic habitat . . . . ***R. aquatilis***

26a. Marginal lines either absent, or short and merely indicated by a row of discrete punctations, or by shagreening. Slight grooves may occasionally be seen in some specimens 27

26b. Marginal lines as definite grooves, either narrow or broad, shallow or deep, for at least part of their length; may be formed by depressions in the conscutum or by contiguous punctations, and may also contain discrete punctations 35

27a. Conscutum with large discrete punctations obvious on either a smooth or a finely punctate background 28

27b. Conscutum without large discrete punctations; it may be smooth with a background of fine punctations or have a relatively punctate background of small to medium-sized punctations 30

28a. Conscutum with a background of numerous fine punctations and scattered rows of medium-sized to large punctations; without clusters of large punctations posteriorly; coxae I each with distinct anterior process visible dorsally. A rare tick recorded from East and Central Africa on the African buffalo . . . . ***R. bequaerti***

28b. Conscutum with large punctations on a smooth, mostly impunctate background; clusters of large punctations posteriorly; coxae I each with only a slight anterior process visible dorsally 29

29a. Cervical pits small; posteromedian and posterolateral grooves absent or only faintly indicated. Present in East, Central and southern Africa, almost exclusively on hyraxes . . . . *R. distinctus*

29b. Cervical pits large; posteromedian and posterolateral grooves small but obvious. A parasite of klipspringers in southern Africa . . . . *R. oreotragi*

30a. At high magnification each conscutal punctation is seen to be composed of a cluster of closely packed fine punctations. Recorded mainly from cattle, impala and tragelaphine antelopes in countries along the eastern seaboard of Africa from Somalia southwards . . . . *R. muehlensi*

30b. At high magnification, each conscutal punctation is seen to be a discrete entity 31

31a. Adanal plates each with a medially-projecting cusp or point 32

31b. Adanal plates each without a medially-projecting cusp or point 33

32a. Posteromedian and posterolateral grooves either absent or merely indicated by a slight scoring on the conscutum; accessory adanal plates long and sharply pointed. An East African highland species . . . . *R. jeanneli*

32b. Posteromedian groove long and narrow; posterolateral grooves small and rounded; accessory adanal plates broad and bluntly rounded . . . . *R. masseyi*

33a. Cornua short; adanal plates tend to appear banana shaped. To date recorded twice only from rock hyraxes in West Africa . . . . *R. boueti*

33b. Cornua long; adanal plates either pear shaped or triangular in shape 34

34a. Adanal plates pear shaped; a small tick with a capitulum that superficially appears triangular in shape. Almost exclusively a parasite of red rock rabbits in southern Zimbabwe and in South Africa . . . . *R. arnoldi*

34b. Adanal plates triangular in shape; a moderately-large tick with the typical hexagonal basis capituli . . . . *R. ziemanni*

35a. Cervical fields and/or marginal lines shagreened or reticulated 36

35b. Cervical fields and/or marginal lines not shagreened or reticulated 41

36a. Basis capituli with long sharp lateral angles projecting over the anterior process of coxae I; a single caudal process is seen in replete specimens. A parasite of hyaenas and the larger wild felids, with most records to date from East Africa . . . . *R. carnivoralis*

36b. Basis capituli with short, blunt lateral angles not projecting over the anterior process of coxae I; either no caudal process or three caudal processes present in replete specimens 37

37a. With smooth glossy ridges defining a pseudoscutum; conscutum divided into separate raised areas by shagreened tracts; no caudal process in replete specimens . . . . *R. sculptus*

37b. Without smooth glossy ridges defining a pseudoscutum; conscutum not divided into separate areas; one or three caudal process(es) in replete specimens. 38

38a. Central area of conscutum densely punctate, punctations contiguous giving the conscutum a rugose appearance; marginal lines deep, shagreened grooves lacking punctations; three caudal processes in replete specimens. Most commonly parasitic on cattle, African buffalo and the larger antelopes, especially in Central Africa but with scattered records from other areas . . . . *R. supertritus*

38b. Central area of conscutum not densely punctuate, or not densely punctate enough to give the conscutum a rugose appearance; marginal lines with punctations; a single caudal process in replete specimens . . . 39

39a. Lateral angles of basis capituli quite broad and recurved; conscutal punctations coalesce to form a roughened depressed area just anterior to the posteromedian groove. A parasite of cattle, the African buffalo and probably other ungulates, primarily in Angola . . . . *R. duttoni*

39b. Lateral angles of basis capituli short, only

slightly recurved; conscutal punctations not forming a roughened depressed area just anterior to the posteromedian groove. . 40

40a. Conscutum quite densely punctate . . . . ***R. zambeziensis***

40b. Conscutum quite lightly punctate . . . . ***R. appendiculatus***
(Males of *R. zambeziensis* and *R. appendiculatus* are difficult to differentiate, especially when only one or two specimens, or only small specimens are available. Refer to the descriptions, figures and distributions of these two species).

41a. Posterior grooves on the conscutum composed of four elements; essentially two posteromedian grooves and two posterolateral grooves 42

41b. Posterior grooves three in number; one posteromedian groove and two posterolateral grooves 43

42a. Cornua very short; two posteromedian punctate depressions and two posterolateral aggregates of large punctations. A small species, collected in northern Mozambique and eastern Zambia . . . . ***R. serranoi***

42b. Cornua very long; all posterior grooves composed of four branches which coalesce anteriorly. A large triangular-shaped species, collected originally in Lesotho . . . . ***R. deltoideus***

43a. Marginal lines long, reaching anteriorly almost to eye level 46

43b. Marginal lines short to medium in length, ending well behind eyes. (If a short marginal line ends and is continued anteriorly towards the eye merely by a series of punctations it is still considered to be short) 44

44a. Conscutum heavily punctate with many large punctations posteriorly; broadly rounded posteriorly; caudal process bulbous . . . . ***R. gertrudae***

44b. Conscutum moderately or lightly punctate 45

45a. Marginal lines outlined by large punctations; adanal plates scooped out posterior to anus, posterior margins broadly rounded . . . . ***R. zumpti***

45b. Marginal lines outlined with small punctations and extending anteriorly as a series of punctations; adanal plates approaching the triangular. A highland species, occurring primarily in East Africa . . . . ***R. hurti***

46a. The medial aspect of the conscutum (i.e. internal to the marginal lines and external cervical margins), gives the overall impression that this male is a densely and quite evenly punctate tick regardless of the size of the punctations 47

46b. The medial aspect of the conscutum gives the overall impression that this male is a moderately or lightly-punctate tick regardless of the size of the punctations 52

47a. Anterior projection on coxae I prominent, visible dorsally, often heavily sclerotized, as in *R. appendiculatus* 48

47b. Anterior projection on coxae I when present, not prominent nor visible dorsally although a small process may be seen . . . 49

48a. Conscutum uniformly densely punctate, but less so in scapular areas; most punctations of the same size . . . . . ***R. punctatus***

48b. Conscutum more densely punctate posterior to the pseudoscutal area; punctations of varying sizes. Thus far found almost exclusively in the Free State Province of South Africa . . . . ***R. warburtoni***

49a. Cervical fields long, narrow, depressed, with both their internal and external cervical margins sharply defined. Found almost exclusively in the western regions of the Western Cape Province of South Africa . . . . ***R. capensis***

49b. Limits of cervical fields not clearly defined, at least on the internal aspect; external cervical margins usually indicated by a row of larger punctations or by a cessation of the punctation pattern 50

50a. External cervical margins delimited by a cessation of the punctation pattern; conscutum covered with a dense, even pattern of medium-sized punctations that sometimes coalesce . . . . ***R. compositus***

50b. External cervical margins delimited by a row of larger conscutal punctations. . . 51

51a. Posteromedian and posterolateral grooves indicated by pronounced sunken areas in the conscutum; posterolateral grooves quite broad; dorsal prolongation of spiracles narrow. . . . . ***R. sulcatus***

51b. Posteromedian and posterolateral grooves usually present but shallow and indistinct; dorsal prolongation of spiracles broad. Occurs in eastern South Africa. . ***R. follis***

52a. Anterior projection on coxae I prominent, visible dorsally. 53

52b. Anterior projection on coxae I, when present, not prominent, although a small process may be seen. 55

53a. Posteromedian groove and posterolateral grooves well marked, long and narrow; adanal plates elongated posteromedially into extended rounded points. Found almost exclusively in the Fynbos regions of the Eastern and Western Cape Provinces of South Africa. . . . . ***R. nitens***

53b. Posteromedian groove long and narrow; posterolateral grooves bluntly rounded; adanal plates not elongated posteromedially into extended rounded points. 54

54a. Cervical fields narrow and lanceolate; internal cervical margins marked by rugosity, external cervical margins marked by a row of large punctations; eyes slightly to markedly convex. . . . . ***R. pravus***

54b. Cervical fields narrow, tapering, shallow; internal cervical margins marked by a slight declination, external cervical margins marked by a few small punctations; eyes flat. ***R. kochi***

55a. Posteromedian groove and posterolateral grooves primarily indicated as pronounced sunken areas in the conscutum; posterolateral grooves relatively broad, often subcircular or comma shaped. 56

55b. Posteromedian groove and posterolateral grooves primarily indicated by rugosity or punctations that may be entirely absent or present as shallow grooves, the posterolateral grooves are not broad sunken areas in the conscutum. 60

56a. Marginal lines deep, distinct, heavily punctate. 57

56b. Marginal lines shallow, quite indistinct, lightly punctate. . . . . ***R. camicasi***

57a. Dorsal prolongation of spiracles broad, equal to the breadth of the adjacent festoons. 58

57b. Dorsal prolongation of spiracles narrow, equal to about one-half the breadth of the adjacent festoons. Almost exclusively a parasite of domestic dogs around the world. . . . . ***R. sanguineus***

58a. Posterolateral grooves subcircular, not originating either at, or very near the festoons; punctations present in areas adjacent to the marginal lines. 59

58b. Posterolateral grooves comma shaped, beginning at, or very near the festoons; punctations in the areas adjacent to the marginal lines either sparse or absent. Recorded almost exclusively in Ethiopia . . . . ***R. bergeoni***

59a. A moderately sized tick with small cervical fields and a diffuse pattern of punctations scattered over the conscutal surface . . . . ***R. guilhoni***

59b. A large tick with large cervical fields and a dense pattern of punctations scattered over the conscutal surface. . . . . ***R. turanicus***

60a. Posteromedian groove and posterolateral grooves, when present, shallow and rather inconspicuous; palps short. 61

60b. Posteromedian groove long and narrow, posterolateral grooves shorter and broader; palps somewhat elongated. A rare species in Namibia and Angola. . . . ***R. longiceps***

61a. Adanal plates broadly curved posterior to anus, but never as strongly as the curved sickle-shape of those of *R. senegalensis* 63

61b. Adanal plates broad posteriorly and may be slightly concave on their inner aspect, but not strongly curved posterior to anus. Parasitic on the feet of sheep, goats and antelopes in the southern part of Africa 62

62a. Basis capituli twice as broad as long; posterior margin of basis capituli straight to slightly sinuous between cornua .... ***R. lounsburyi***
62b. Basis capituli approximately one-third broader than long; posterior margin of basis capituli distinctly concave between cornua .... ***R. neumanni***
63a. Marginal lines delimit one festoon; posteromedian and posterolateral grooves either superficial or absent 64
63b. Marginal lines delimit first two festoons; posteromedian groove and posterolateral grooves discernible; posterolateral grooves either linear or slightly curved. Primarily a West African species .... ***R. muhsamae***
64a. '*Simus*' pattern of punctations present on an obvious background of fine interstitial punctations. Occurs in the southern part of Africa .... ***R. simus***
64b. '*Simus*' pattern of punctations present on a smooth shiny background without a field of fine interstitial punctations. Occurs in north-eastern and eastern Africa .... ***R. praetextatus***

KEY TO THE AFROTROPICAL *RHIPICEPHALUS* SPECIES FEMALES

The host preferences listed are those of the adults. In several cases the immature stages prefer different hosts.

1a. Scutum with ivory ornamentation 2
1b. Scutum without ivory ornamentation ... 5
2a. Scutum dark over most of its surface, with ornamentation limited; legs not enamelled dorsally 3
2b. Scutum extensively ornamented, with small dark intrusions on the scutal margins; legs with some dorsal enamelling 4
3a. Scutum with a single posteromedial ivory patch, roughly triangular in shape; numerous small to medium-sized punctations scattered over the scutum. Present in western Uganda, and in Rwanda and Democratic Republic of Congo .... ***R. dux***
3b. Scutum with a few medium-sized punctations; ornamentation semicircular posteriorly, becoming thin and smokey anteriorly. Collected from a wide range of hosts, mainly in coastal regions from Kenya to northern KwaZulu-Natal, South Africa but also recorded sometimes from Zambia, Zimbabwe and Malawi ... ***R. maculatus***
4a. Few clavate alloscutal setae. Present in dry areas in southern Somalia, Kenya and Tanzania .... ***R. humeralis***
4b. Numerous clavate alloscutal setae. A common tick in dry areas in eastern Africa east of the Rift Valley .... ***R. pulchellus***
5a. Eyes round, beady, deeply orbited 6
5b. Eyes flat, slightly bulging or convex, but not beady and deeply orbited 10
6a. Spiracles with numerous circumspiracular setae 7
6b. Spiracles without numerous circumspiracular setae 8
7a. Legs uniformly bright reddish orange. Very common and widely distributed in the Afrotropical region ..... ***R. evertsi evertsi***
7b. Legs with saffron annulations. Almost exclusively in dry areas from south-western Democratic Republic of Congo south to Namibia and western Botswana .... ***R. evertsi mimeticus***
8a. Scutum relatively smooth and impunctate. A parasite of the lower legs and feet of ungulates in the Eastern and Western Cape Provinces, South Africa .... ***R. glabroscutatum***
8b. Scutum not smooth and impunctate, but rather extensively and densely punctate 9
9a. Genital aperture U-shaped, the area within the opening bulging. Recorded mainly from southern Namibia and South Africa .... ***R. exophthalmos***
9b. Genital aperture V-shaped, the area within the opening depressed. Almost exclusively a parasite of hares in the drier areas of Namibia, Botswana and South Africa .... ***R. oculatus***

10a. Palps short, tapering markedly to pointed or very narrowly rounded apices, giving the capitulum a distinctly triangular appearance. Almost exclusively a parasite of red rock rabbits in southern Zimbabwe and in South Africa . . . . ***R. arnoldi***

10b. Palps various but not giving the capitulum a distinctly triangular appearance 11

11a. Palpal article I very long and narrow; genital aperture a long, narrow U-shape. A rare species in Namibia and Angola . . . . ***R. longiceps***

11b. Lacking the combination of very long palpal article I and a long, narrow U-shaped genital aperture 12

12a. Dorsally, alloscutum with four, broad, conspicuous longitudinal bands of white setae; festoons also beset with setae. Almost exclusively a parasite of yellow mongoose, meercat and Cape ground squirrel in South Africa, Botswana and Namibia . . . . ***R. theileri***

12b. Dorsally, alloscutum and festoons not beset with longitudinal bands of white setae 13

13a. External cervical margins of scutum not clearly defined; no declinations or definite punctation pattern outlining their position 14

13b. External cervical margins of the scutum defined by a row of larger punctations which may or may not be contiguous, or by a declination from the raised lateral borders, or by a groove, or by a distinct cessation of the punctation pattern found on the cervical fields, or by a combination of these characters 18

14a. Scutum smooth and glossy with a few medium-sized punctations on the raised lateral margins and in the medial area; no punctations in cervical fields; genital aperture U-shaped with a short anterolateral extension on either side. A large tick of giraffes in Kenya and Tanzania . . . . ***R. longicoxatus***

14b. Scutum punctate including cervical fields; genital aperture U-shaped or V-shaped, but without a short anterolateral extension on either side 15

15a. Basis capituli lacking cornua; scutum longer than broad, sinuous posteriorly. To date recorded twice only from rock hyraxes in West Africa . . . . ***R. boueti***

15b. Basis capituli with cornua present; scutum as broad as long or broader than long . . 16

16a. Scutal punctations uniformly fine to small, evenly dispersed; genital aperture broadly cup shaped . . . . ***R. ziemanni***

16b. Scutum with punctations larger than background interstitial punctations, often on the scapulae or in the area of the external cervical margins; genital aperture not broadly cup shaped 17

17a. Poorly-developed internal spurs on coxae II to IV; genital aperture broadly U-shaped . . . . ***R. masseyi***

17b. Definite internal spurs on coxae II to IV; genital aperture broadly V-shaped. An East African highland species . . . . ***R. jeanneli***

18a. Cervical pits broadly rounded anteriorly. Parasitic on the gundi and some larger mammals in North Africa and the Sahara . . . . ***R. fulvus***

18b. Cervical pits various but not broadly rounded anteriorly 19

19a. Cervical fields long and narrow because external cervical margins tend to be straight rather than curved; central field tends to be straight sided and narrow 20

19b. Cervical fields and central field broad because the external cervical margins are curved outward 37

20a. External cervical margins marked by a distinct sharp declination from the raised lateral borders 21

20b. External cervical margins not marked by a distinct sharp declination from the raised lateral borders 30

21a. Scutum appearing lightly punctate even if small to medium-sized punctations are present, especially in the central field 22

21b. Scutum appearing moderately to densely punctate 24

22a. Scutal central field and cervical fields covered with small to medium-sized dense punctations. A highland species, occurring in East Africa . . . . ***R. hurti***

22b. Scutal central field and cervical fields with few punctations 23

23a. External cervical margins forming a very steep declination; genital aperture with a broadly curved posterior margin. Widely distributed in the Afrotropical region in association with its preferred host, the greater cane rat; rarely recorded from other hosts . . . . ***R. simpsoni***

23b. Declination of external cervical margins distinct but not very steep; genital aperture wide, deeply crescentic. Parasitic on the feet of sheep, goats and antelopes in the southern part of South Africa . . . . ***R. lounsburyi***

24a. Scutum about as broad as long or broader than long 25

24b. Scutum longer than broad 27

25a. Scutum densely punctate obscuring the '*simus*' pattern of punctations. Occurs in eastern South Africa . . . . ***R. follis***

25b. Scutum moderately punctate not obscuring the '*simus*' pattern of punctations 26

26a. Spiracles fairly large, each with a wide and usually angled dorsal prolongation; genital aperture narrowly U-shaped . . . . ***R. turanicus***

26b. Spiracles each with a narrow and curved dorsal prolongation; genital aperture broadly U-shaped. Almost exclusively a parasite of domestic dogs around the world . . . . ***R. sanguineus***

27a. Scutum quite heavily punctate especially in the cervical fields and central field obscuring the '*simus*' pattern of punctations . . 28

27b. All areas of the scutum quite lightly punctate not obscuring the '*simus*' pattern of punctations 29

28a. Raised borders lateral to the external cervical margins relatively impunctate; external cervical margins outlined by small to medium-sized punctations; genital aperture U-shaped without hyaline flaps. A rare West African species . . . . . . ***R. moucheti***

28b. Raised borders lateral to the external cervical margins with small and medium-sized punctations; external cervical margins outlined by many large setiferous punctations; genital aperture U-shaped with hyaline flaps . . . . ***R. sulcatus***

29a. Genital aperture with distinct hyaline flaps . . . . ***R. guilhoni***

29b. Genital aperture without distinct hyaline flaps . . . . ***R. camicasi***

30a. Cornua very long, sharply pointed. A large species, collected originally in Lesotho . . . . ***R. deltoideus***

30b. Cornua short, bluntly rounded 31

31a. Scutal punctations small to medium in size, but uniform in depth; lateral borders anterior to eyes smooth and shiny; genital aperture with area anterior to opening bulging . . . . ***R. kochi***

31b. Scutal punctations uneven in size and depth; lateral borders may be smooth or punctate; genital aperture with area anterior to opening not bulging 32

32a. Genital aperture U-shaped or sharply V-shaped with genital apron depressed 33

32b. Genital aperture broadly V-shaped or U-shaped with genital apron bulging 35

33a. Genital aperture U-shaped. Recorded almost exclusively in Ethiopia . . . . ***R. bergeoni***

33b. Genital aperture sharply V-shaped 34

34a. Areas of cervical fields shagreened, especially along the internal cervical margins; cervical fields and central field lightly punctate. A parasite of cattle, the African buffalo and probably other ungulates, primarily in Angola . . . . ***R. duttoni***

34b. Areas of cervical fields not shagreened; cervical fields and central field quite heavily punctate. Found almost exclusively in the Fynbos regions of the Eastern and Western Cape Provinces of South Africa . . . . ***R. nitens***

35a. Cervical fields and central field lightly punctate; eyes slightly raised to distinctly convex . . . . ***R. pravus***

35b. Cervical fields and central field densely punctate; eyes may not be flush with scutum, but are not convex 36

36a. Genital aperture broadly rounded, U-shaped . . . . ***R. punctatus***

36b. Genital aperture broadly V-shaped. Thus far found almost exclusively in the Free State Province of South Africa . . . . ***R. warburtoni***

37a. Coxae I each with spurs widely separated and not deeply divided; scutum smooth with external cervical margins marked by three or more large, deep punctations; genital aperture very broad, crescent shaped. Parasitic mainly on warthogs and porcupines in West Africa, from Senegal eastward to the Sudan . . . . . . ***R. cuspidatus***

37b. Coxae I each with spurs deeply divided and not widely separated; scutum and genital aperture various 38

38a. Scutal margin posterior to eyes distinctly concave and with a central posterior protrusion, giving the scutum a shield shape 39

38b. Scutum not shield shaped 43

39a. Scutum smooth with external cervical margins marked by several large punctations and a sharp declination; internal cervical margins marked anteriorly by a sharp declination. One of the rarer species, recorded most commonly from carnivores in East Africa . . . . ***R. armatus***

39b. Scutum beset with numerous small to medium-sized punctations giving it a non-smooth appearance; external cervical margins may be marked by large punctations but without a sharp declination; internal cervical margins not marked anteriorly by a sharp declination 40

40a. Porose areas quite small, oval, about one and a half times their own diameter apart; external cervical margins marked by a row of large setiferous punctations 41

40b. Porose areas large, round, less than their own diameter apart; external cervical margins inapparent or marked only by a faint declination in the scutal surface. A rare tick recorded from East and Central Africa on the African buffalo . . . . ***R. bequaerti***

41a. Palpal article I with a stalked pedicle giving the palps a distinctly elongate appearance; genital aperture a very broad and shallow U-shape. Widespread on numerous hosts in the Afrotropical region . . ***R. lunulatus***

41b. Palpal article I with a short pedicle; genital aperture a narrow U-shape with long lateral arms 42

42a. Spiracles each with a short, narrow, dorsal prolongation; genital aperture without hyaline flaps. Recorded occasionally from domestic animals but more commonly from the smaller wild carnivores, and from antelopes, springhares and hares, mainly in southern Africa, more rarely further north . . . . ***R. tricuspis***

42b. Spiracles each with a short, broad, dorsal prolongation; genital aperture with hyaline flaps. Parasitic on cattle and various, mostly small to medium-sized antelopes, in East and Central Africa; sometimes sympatric with *R. lunulatus* . . . . ***R. interventus***

43a. Cervical fields distinctly sunken; genital aperture tripartite in appearance, with a narrow, raised central area flanked on either side by a rounded depression . . ***R. planus***

43b. Cervical fields not distinctly sunken; genital aperture various but not tripartite 44

44a. Cervical fields shagreened anteriorly and external cervical margins a shagreened groove 45

44b. Scutum without shagreening in cervical pits and the external cervical margins not a shagreened groove 46

45a. Punctations with short setae along external cervical margins; genital aperture tongue shaped without hyaline flaps . . . . ***R. sculptus***

45b. Punctations with conspicuous long fine setae along external cervical margins; genital aperture U-shaped with an almost straight posterior margin and hyaline flaps. Most commonly parasitic on cattle, African buffalo and the larger antelopes, especially in Central Africa but with scattered records from other areas . . . . ***R. supertritus***

46a. External cervical margins marked exclusively, or nearly so, by a row of discrete setiferous punctations; declinations absent 47

46b. External cervical margins marked by contiguous punctations, a declination or other features, but not by a row of discrete punctations 53

47a. Basis capituli very broad, and very short in the antero-posterior plane; porose areas small, twice or three times their own diameter apart 48

47b. Basis capituli of the normal *Rhipicephalus* shape. 49

48a. Scutum about as long as broad or slightly longer than broad, smooth and glossy; porose areas very small, about three times their own diameter apart. Present in East, Central and southern Africa, almost exclusively on hyraxes . . . . ***R. distinctus***

48b. Scutum much longer than broad, not smooth and glossy, but with numerous small to medium-sized punctations; porose areas small, about twice their own diameter apart. A small species, collected in northern Mozambique and eastern Zambia . . . . ***R. serranoi***

49a. Porose areas small, at least twice or three times their own diameter apart; scutum longer than broad 50

49b. Porose areas large, only one or one and a half times their own diameter apart; scutum about as broad as long 51

50a. Porose areas three times their own diameter apart. A parasite of klipspringers in southern Africa . . . . ***R. oreotragi***

50b. Porose areas twice their own diameter apart. Parasitic on the feet of sheep, goats and antelopes in the southern part of Namibia and South Africa . . . . ***R. neumanni***

51a. Scutum lightly to moderately and evenly punctate medially with a background of interstitial punctations; genital aperture broadly V-shaped, without hyaline flaps. Primarily a West African species . . . . ***R. senegalensis***

51b. Scutum impunctate except for the '*simus*' pattern and scapular punctations, and with either very light or no interstitial punctations; genital aperture with hyaline flaps 52

52a. No, or at most, only a few interstitial punctations centrally on the scutum; genital aperture with narrow hyaline flaps. Occurs in north-eastern and eastern Africa . . . . ***R. praetextatus***

52b. Light to moderate scattering of interstitial punctations centrally on the scutum; genital aperture with broad hyaline flaps. Primarily a West African species . . . ***R. muhsamae***

53a. External cervical margins marked by a cessation of the punctation pattern, no large punctations or steep declination marking these margins 54

53b. External cervical margins marked by punctations or a declination or both 55

54a. Porose areas three times their own diameter apart; circumspiracular area setose. Collected mainly from cattle, impala and tragelaphine antelopes in countries along the eastern seaboard of Africa from Somalia southwards . . . . ***R. muehlensi***

54b. Porose areas about one and a half times their own diameter apart; circumspiracular area glabrous. A parasite of the sitatunga and sometimes other mammals sharing its semi-aquatic habitat . . . . ***R. aquatilis***

55a. Medial area of scutum internal to the external cervical margins densely and heavily punctate 56

55b. Medial area of scutum not densely and heavily punctate 61

56a. Cervical fields well marked and depressed, cervical pits and especially internal cervical margins shagreened; genital aperture very wide with short lateral arms and a straight posterior margin . . . . . ***R. zambeziensis***

56b. Cervical fields without shagreening; genital aperture various but without a straight posterior margin 57

57a. Palps with internal margins convex, external margins concave; internal cervical margins not well defined 58

57b. Palps with internal margins straight sided or curved, but not convex; internal cervical margins may or may not be well defined 59

58a. Genital aperture narrowly V-shaped . . . . ***R. longus***

58b. Genital aperture broadly V-shaped to U-shaped . . . . ***R. pseudolongus***
(Females of these two species are difficult to separate. *R. longus* is less heavily punctate than *R. pseudolongus*, and its scutum is less evenly rounded posteriorly. See figs 109b and 165b of these two females)

59a. Cervical fields deeply depressed with both internal and external cervical margins sharply delimited; genital aperture broadly V-shaped without sigmoid arms or a tripartite appearance. Found almost exclusively in the western regions of the Western Cape Province of South Africa . . . ***R. capensis***

59b. Cervical fields not deeply depressed, external cervical margins well marked by either punctations or a declination, but internal cervical margins not sharply delimited; genital aperture either with sigmoid arms or tripartite 60

60a. External cervical margins as slight declinations, marked by large punctations; genital aperture tripartite, central area vase shaped, narrow, tapering hyaline flaps laterally . . . . ***R. gertrudae***

60b. External cervical margins with sharp declinations and marked by small punctations; genital aperture a wide V-shape with lateral arms a sigmoid curve . . . . ***R. compositus***

61a. Cervical fields broad and deep, with well-defined internal and external cervical margins 62

61b. Cervical fields broad or narrow but only slightly depressed, external but not internal cervical margins may be well defined . . 63

62a. Cervical fields shagreened anteriorly and rugose laterally because of a contiguous row of slanted punctations along the inner edge of the external cervical margins. A parasite of hyaenas and the larger wild felids, with most records to date from East Africa . . . . ***R. carnivoralis***

62b. Cervical fields without shagreening and rugosity due to slanted punctations; scutum covered with small to fine scattered punctations . . . . ***R. complanatus***

63a. Scutum posteriorly broad and smoothly rounded; cervical fields not shagreened; external cervical margins well marked by either discrete or contiguous punctations 64

63b. Scutum sinuous posteriorly; external cervical margins declinations, not clearly outlined by rows of punctations; cervical fields shagreened . . . . ***R. appendiculatus***

64a. External cervical margins well marked by large contiguous punctations; internal cervical margins poorly defined. Occurs in the southern part of Africa . . . . ***R. simus***

64b. External cervical margins well marked by large discrete punctations; internal cervical margins well defined by a declination . . . . ***R. zumpti***

7

Accounts of individual species occurring in the Afrotropical region

RHIPICEPHALUS APPENDICULATUS NEUMANN, 1901

This specific name *appendiculatus*, from the Latin *appendo* meaning 'to hang', doubtless refers to the appearance of the engorging adults.

Diagnosis

A moderate-sized reddish-brown tick.

Male (Figs 3(a), 4(a) to (c))
Capitulum much longer than broad, length × breadth ranging from 0.69 mm × 0.56 mm to 0.86 mm × 0.71 mm. Basis capituli of smaller males much broader than long but that of larger males only slightly broader than long, with very short obtuse lateral angles at about anterior quarter of its length. Palps short, broad. Conscutum length × breadth ranging from 2.85 mm × 1.85 mm to 3.67 mm × 2.43 mm; large sharp strongly-sclerotized anterior process present on coxae I. In engorged specimens body wall expanded slightly posterolaterally and forming a long slender caudal process posteromedially. Eyes marginal, almost flat, delimited dorsally by a very shallow groove and sometimes one or two large punctations. Cervical fields broad, depressed, with finely-reticulate surfaces. Marginal lines well developed, extending anteriorly nearly to eye level, delimiting one festoon posteriorly. Posteromedian groove long and narrow, posterolateral grooves short and broad, all with finely-reticulate surfaces. Large setiferous punctations present on the scapulae, along the outer margins of the cervical fields and scattered amongst numerous medium-sized punctations medially on the conscutum. Areas surrounding the eyes, adjacent to the marginal lines and anterior to the festoons with fine pinpoint punctations only. In small specimens the pattern of grooves and punctations may be much reduced. Legs increase markedly in size from I to IV. Ventrally spiracles broadly comma shaped, curving gently towards the dorsal surface. Adanal plates large, well sclerotized, tapering posterointernally to broadly rounded points; accessory adanal plates represented by small, short sclerotized points.

Female (Figs 3(b), 4(d) to (f))
Capitulum slightly broader than long, length × breadth ranging from 0.72 mm × 0.75 mm to 0.83 mm × 0.87 mm. Basis capituli with broad lateral angles overlapping the scapulae; porose areas round, well over twice their own diameter apart. Palps short, broad, bluntly rounded apically. Scutum longer than broad, length × breadth ranging from 1.40 mm × 1.32 mm to 1.59 mm × 1.52 mm; posterior margin sinuous. Eyes marginal, at widest point of scutum, almost flat, delimited dorsally by a faint groove and sometimes a few punctations. Cervical fields

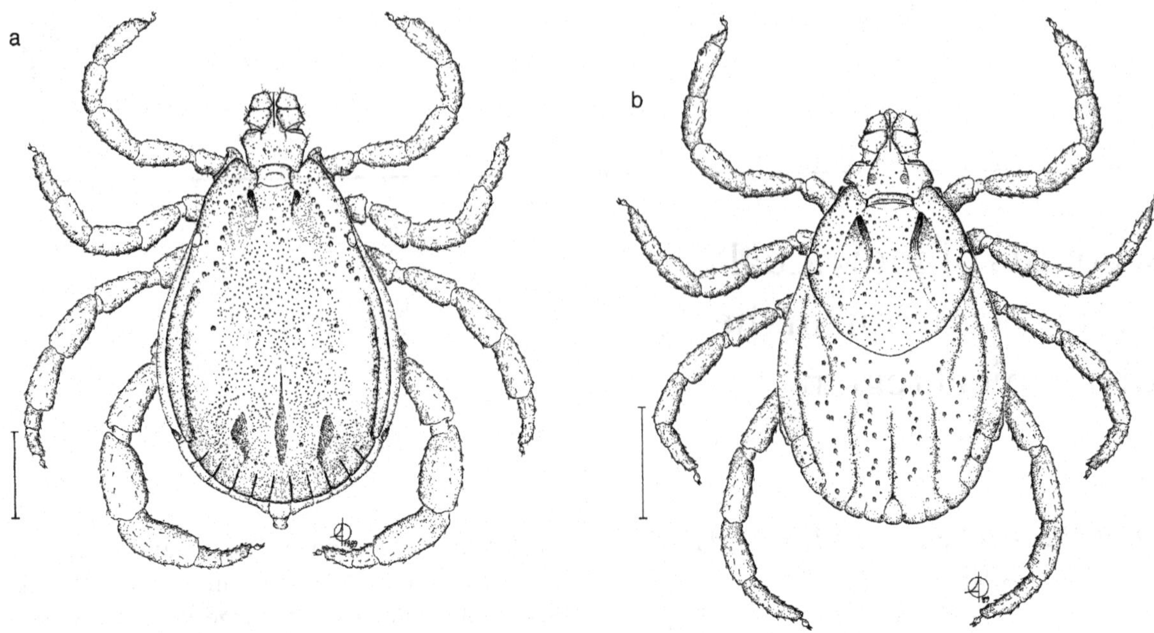

Figure 3. *Rhipicephalus appendiculatus* [collected from calf, 'Bucklands' (farm), near Grahamstown, Eastern Cape Province, South Africa, on 27 January 1987 by I.G. Horak]. (a) Male, dorsal; (b) female, dorsal. Scale bars represent 1 mm. A. Olwage *del.*

broad, depressed, slightly reticulate in places. Large setiferous punctations present on the scapulae, along the outer margins of the cervical fields and scattered medially on the scutum, interspersed with fine punctations. Ventrally genital aperture shaped like the tip of a tongue.

Nymph (Fig. 5)

Capitulum much broader than long, length × breadth ranging from 0.24 mm × 0.31 mm to 0.28 mm × 0.34 mm. Basis capituli well over twice as broad as long, lateral angles in anterior half of its length, short, slightly forwardly curved; ventrally with short, blunt spurs on posterior margin. Palps short, broadly rounded apically. Scutum broader than long, length × breadth ranging from 0.55 mm × 0.64 mm to 0.58 mm × 0.70 mm; posterior margin a broad smooth curve. Eyes at widest point, about halfway back, mildly convex and edged dorsally by a shallow groove. Cervical fields broad, divergent, slightly depressed, almost reaching posterior margin of scutum. Ventrally coxae I each with a long, narrow external spur and a shorter broader internal spur; coxae II to IV each with a short sharp external spur only.

Larva (Fig. 6)

Capitulum broader than long, length × breadth ranging from 0.129 mm × 0.141 mm to 0.138 mm × 0.147 mm. Basis capituli a little over twice as broad as long, with very short blunt lateral angles at about mid-length. Palps constricted proximally, then widening, flattened apically. Scutum much broader than long, length × breadth ranging from 0.233 mm × 0.360 mm to 0.246 mm × 0.373 mm, posterior margin a broad smooth curve. Eyes at widest point, almost flat, delimited dorsally by a faint groove. Cervical grooves short, slightly convergent. Ventrally coxae I each with a broad blunt spur; coxae II and III each with a broad ridge-like spur.

Designation of lectotype

Neumann (1901, p. 270) described *R. appendiculatus* from 3 ♂♂, 7 ♀♀ from the Cape Colony, South Africa, collected by C.P. Lounsbury; from 3 ♂♂, two of whose origins are unknown and one

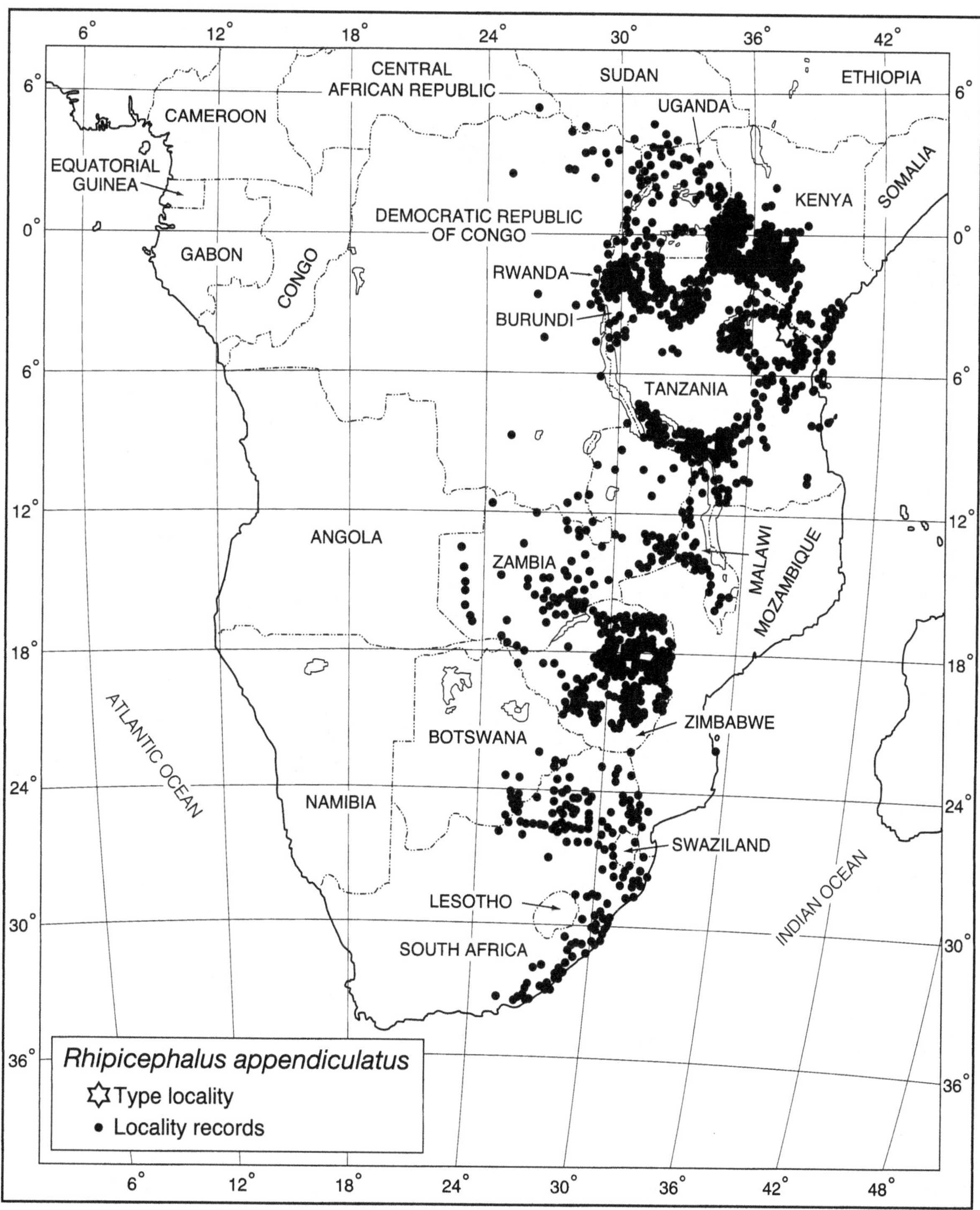

Map 2. *Rhipicephalus appendiculatus*: distribution (with acknowledgements to B.D. Perry and R. Kruska).

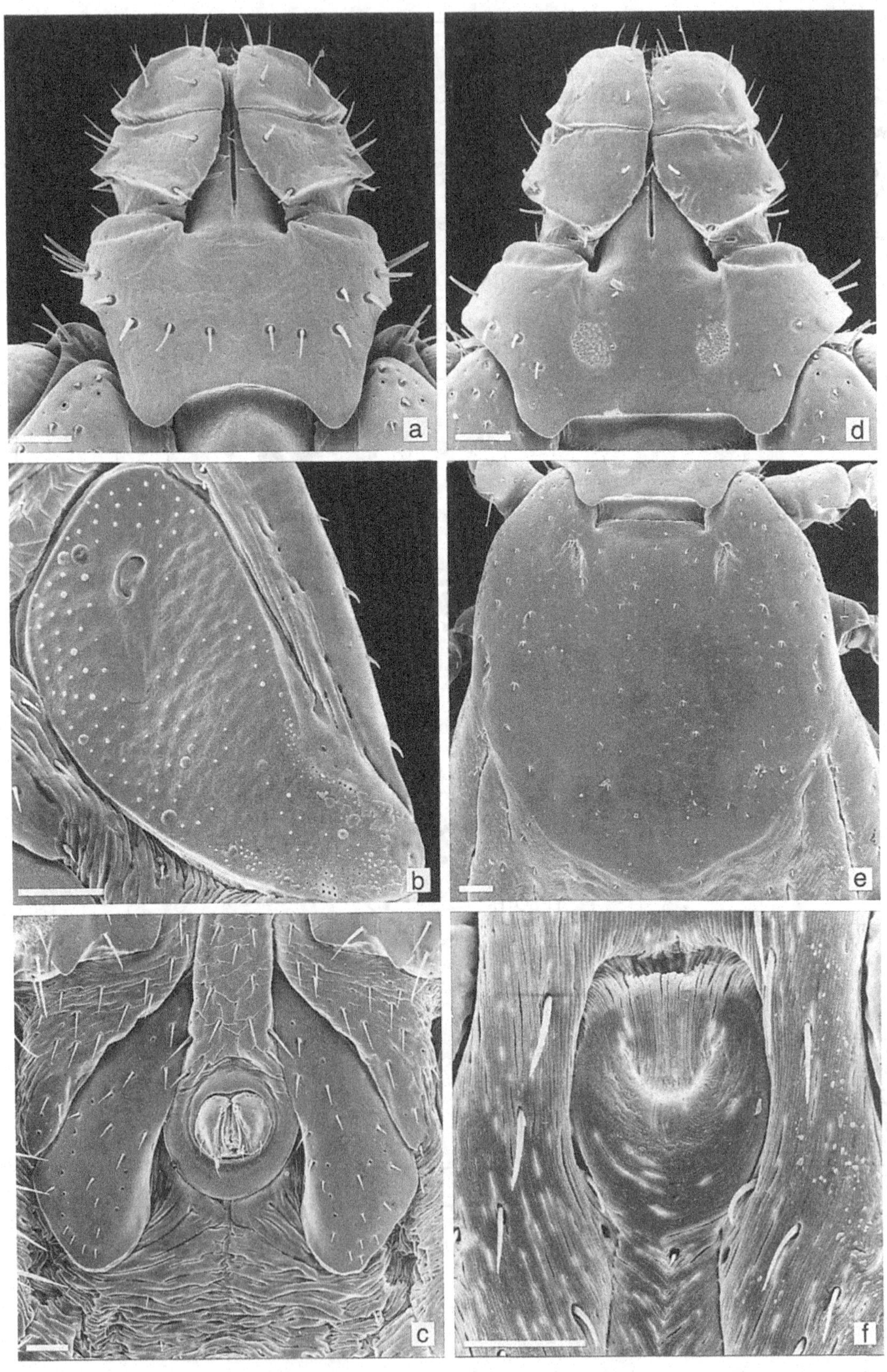
a
d
b
e
c
f

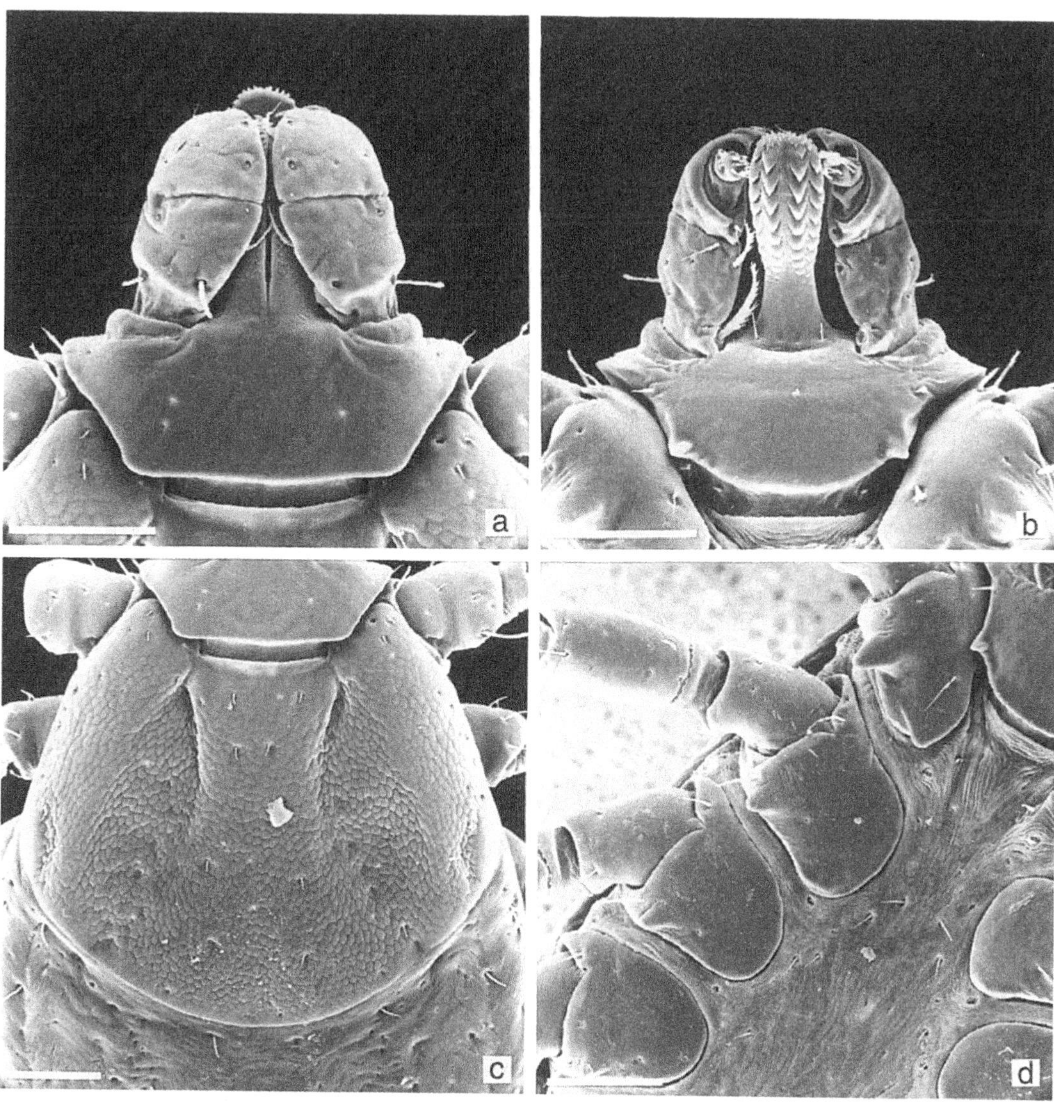

Figure 5 (*above*). *Rhipicephalus appendiculatus* (B.S. 292/-, RML 66302, laboratory reared, original ♀ from a strain maintained at the Veterinary Research Laboratory, Kabete, Kenya, 1951). Nymph: (a) capitulum, dorsal; (b) capitulum, ventral; (c) scutum; (d) coxae. Scale bars represent 0.10 mm. SEMs by M.D. Corwin.

Figure 4 (*opposite*). *Rhipicephalus appendiculatus* (B.S. 292/-, RML 66302, laboratory reared, original ♀ from a strain maintained at the Veterinary Research Laboratory, Kabete, Kenya, 1951). Male: (a) capitulum, dorsal; (b) spiracle; (c) adanal plates. Female: (d) capitulum, dorsal; (e) scutum; (f) genital aperture. Scale bars represent 0.10 mm. SEMs by M.D. Corwin.

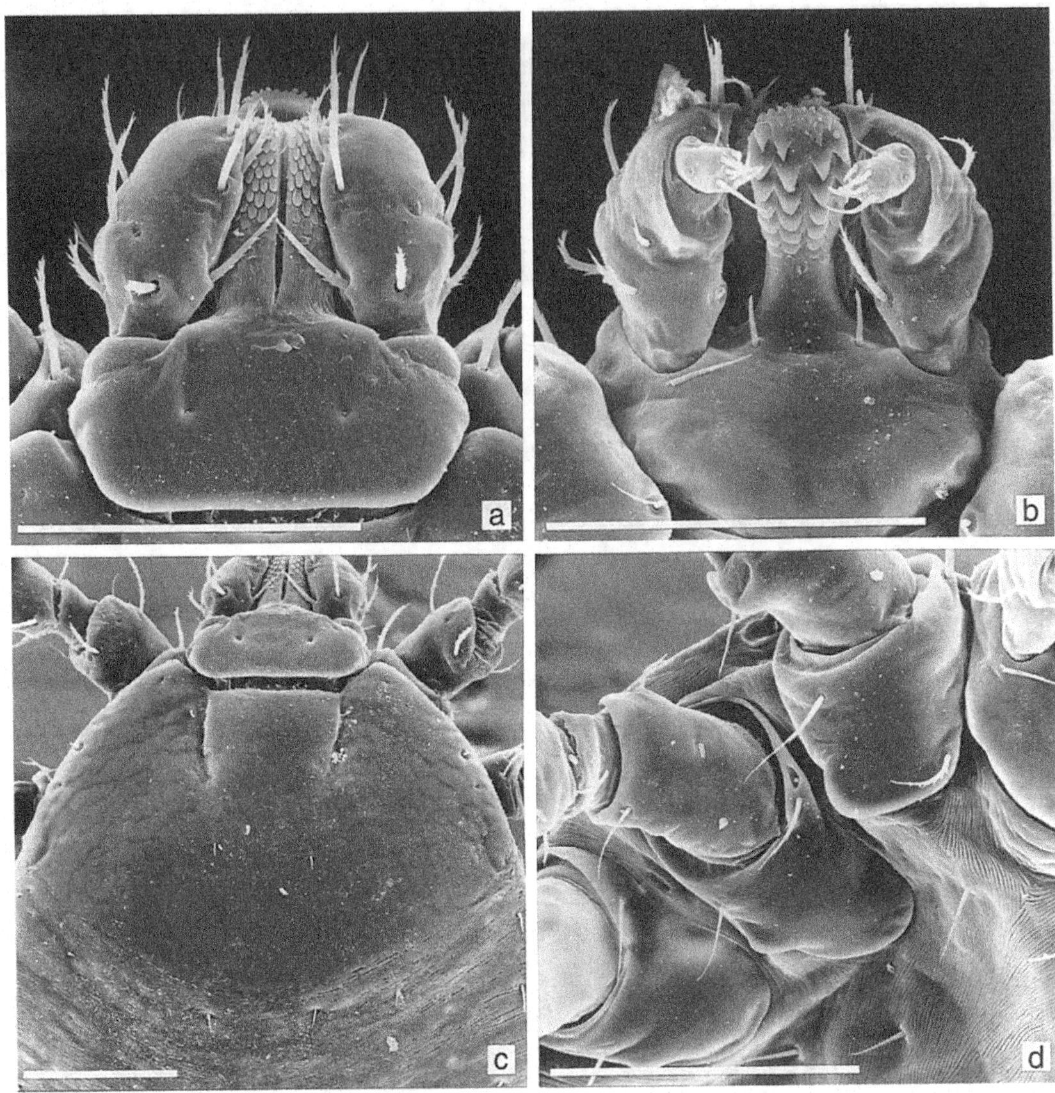

Figure 6. *Rhipicephalus appendiculatus* (B.S. 292/-, RML 66302, laboratory reared, original ♀ from a strain maintained at the Veterinary Research Laboratory, Kabete, Kenya, 1951). Larva: (a) capitulum, dorsal; (b) capitulum, ventral; (c) scutum; (d) coxae. Scale bars represent 0.10 mm. SEMs by M.D. Corwin.

from South-West Africa (now Namibia), and from hundreds of ♂♂ and ♀♀ taken from *Bos caffer* (i.e. *Syncerus caffer*) by [C.G.] Schillings (Berlin Museum). Lounsbury (pers. comm. referred to by Neumann, 1901) regarded this species as being very widely distributed in the Cape Colony.

Moritz & Fischer (1981, p. 343) recorded two syntype collections of *R. appendiculatus* in the Zoological Museum, Berlin: (1) ZMB 16890, 38 ♂♂, 23 ♀♀ from the Pangani River at the Mosimani Mountains of German East Africa [Tanzania] collected in September 1899 by [C.G.] Schillings, and (2) ZMB 17697, 7 ♂♂ on '*Bos caffer*', 1 ♂ as a microscope slide preparation (no. 10) by F. Zumpt from the Pangani River, German East Africa.

The Nuttall Tick Collection, deposited in The Natural History Museum, London, also contains syntypes of *R. appendiculatus* with the following data: Nuttall 2893 (Neumann No. 1143) 1 ♂, 1 ♀ from '*Bos caffer*, Tangani, German East Africa', which was received in 1900 from [C.G.] Schillings. Keirans (1985, p. 1207) confirmed the identity of these specimens but noted that there was no locality named Tangani in Tanzania, as indicated both in Nuttall's original handwritten catalogue and on the vial label. He suggested that the locality might be Tangeni, located at 06.56 S, 37.36 E. Judging by the information published by Moritz & Fischer (1981), though, it is apparent that the correct spelling of this locality is Pangani. In the circumstances we believe that it probably refers to the Pangani River near Masimani Mountain (as it is now spelt), inland at 04.13 S, 37.35 E, not to Pangani township on the Tanzanian coast.

We considered that it would be preferable to select a lectotype from amongst Neumann's Tanzanian syntypes, rather than those from either the Cape Colony or from South West Africa. This would avoid any possible confusion with either the closely-related species *R. nitens*, which occurs in the Eastern and Western Cape Provinces, South Africa, or with *R. zambeziensis*, which has been recorded in Namibia. We therefore designate the ♂ (Nuttall 2893; RML 111766) as lectotype and the ♀ (Nuttall 2893; RML 111766) as paralectotype, so mark them and correct the spelling of the type locality; these specimens are on deposit in The Natural History Museum, London.

Notes on identification

The measurements given above are those of specimens from a rather small laboratory-reared series of this tick. A greater range of sizes, associated with variations in the morphology of this species, has been recorded elsewhere (Nuttall, 1913; Walker, Norval & Corwin, 1981).

All stages of *R. appendiculatus* and *R. zambeziensis* are morphologically very similar in several respects. The primary feature distinguishing the adults is the density of the punctation pattern, which is usually much lighter in *R. appendiculatus* than it is in *R. zambeziensis* (compare Figs 3 and 220, pp. 60 and 471). The immature stages differ in the proportions of their basis capituli and shape of their palps (compare Figs 5(a) and 6(a) with 222(a) and 223(a), pp. 473 and 474). Even so it can be extremely difficult, or even impossible, to identify some individuals specifically because both species show such a wide range of morphological variation. Whenever possible it is better to examine series of these ticks rather than rely on small samples only, and to take the ecological conditions of the areas where they were obtained into consideration. Although their distributions undoubtedly sometimes overlap, their ecological preferences differ markedly and they commonly occur in separate areas.

The morphological differences between *R. appendiculatus*, *R. duttoni* and *R. nitens* can be seen by comparing Figs 3–6 (above), 50–53 (pp. 147–50) and 143–146 (pp. 317–20).

Hosts

A three-host species (Lounsbury, 1904). Cattle are the preferred domestic hosts of all stages of development (Yeoman & Walker, 1967; Walker, 1974; Norval, Walker & Colborne, 1982; Matthysse & Colbo, 1987) (Table 1). Many collections have, however, been taken from sheep and goats (Walker, 1974; Matthysse & Colbo, 1987). Ticks have also been collected from a surprisingly large number of domestic dogs as well as from various species of wild carnivores (Yeoman & Walker, 1967; Norval *et al.*, 1982; Horak *et al.*, 1987). The preferred wild hosts of all stages of development are the African buffalo, eland, various species of tragelaphine antelope and waterbuck (Norval *et al.*, 1982; Horak *et al.*, 1983, 1992; Horak, Boomker & Flamand, 1995). Smaller antelope species and hares are good hosts of the immature stages (Clifford, Flux & Hoogstraal, 1976; Norval *et al.*, 1982; Horak, 1982; Horak & Fourie, 1991). Collections consisting mainly of immature ticks have also been taken from numerous Burchell's zebra and warthog (Horak, De Vos & De Klerk, 1984; Horak *et al.*, 1988). The small number of records from

Table 1. *Host records of* Rhipicephalus appendiculatus

Hosts	Number of records
Domestic animals	
Cattle	Commonly parasitized (including immatures)
Water buffaloes	2
Sheep	245 (including immatures)
Goats	358 (including immatures)
Camels	5
Horses	31 (including immatures)
Donkeys	17 (including nymphs)
Pigs	1
Dogs	49 (including immatures)
Cats	1 (nymphs)
Cats (feral)	2 (including larvae)
Chickens	1 (1 ♂)
Wild animals	
Greater bushbaby (*Otolemur crassicaudatus*)	1 (including immatures)
Vervet monkey (*Chlorocebus aethiops*)	7 (including immatures)
Chacma baboon (*Papio ursinus*)	3 (including immatures)
Papio sp.	1 (immatures)
'Baboon'	1 (immatures)
Gorilla (*Gorilla gorilla*)	1
Side-striped jackal (*Canis adustus*)	3 (including immatures)
Black-backed jackal (*Canis mesomelas*)	7 (including immatures)
'Jackal' (*Canis* spp.)	9 (including immatures)
Hunting dog (*Lycaon pictus*)	2
Bat-eared fox (*Otocyon megalotis*)	2 (including immatures)
Cheetah (*Acinonyx jubatus*)	7 (including immatures)
African wild cat (*Felis lybica*)	2 (immatures)
Serval (*Leptailurus serval*)	3 (including 1 nymph)
Lion (*Panthera leo*)	21 (including immatures)
Leopard (*Panthera pardus*)	6 (including immatures)
Marsh mongoose (*Atilax paludinosus*)	1 (1 nymph)
Egyptian mongoose (*Herpestes ichneumon*)	2 (immatures)
White-tailed mongoose (*Ichneumia albicauda*)	26 (immatures, 1 with an adult)
Banded mongoose (*Mungos mungo*)	3 (immatures)
Meercat (*Suricata suricatta*)	2 (including larvae)
Spotted hyaena (*Crocuta crocuta*)	1 (1 nymph)
Striped hyaena (*Hyaena hyaena*)	1
Brown hyaena (*Parahyaena brunnea*)	12
'Hyaena'	1
Ratel (*Mellivora capensis*)	1 (including immatures)
Zorilla (*Ictonyx striatus*)	3 (immatures)
African civet (*Civettictis civetta*)	6 (including immatures)
Small-spotted genet (*Genetta genetta*)	2 (nymphs)
Rusty-spotted genet (*Genetta rubiginosa*)	1 (immatures)
Large-spotted genet (*Genetta tigrina*)	5 (including immatures)

Table 1. (*cont.*)

Hosts	Number of records
'Genet' (*Genetta* spp.)	3 (including nymphs)
African elephant (*Loxodonta africana*)	2
Burchell's zebra (*Equus burchellii*)	54 (including immatures)
Equus sp.	3 (including immatures)
White rhinoceros (*Ceratotherium simum*)	7 (including nymphs)
Black rhinoceros (*Diceros bicornis*)	4 (including immatures)
'Rhinoceros'	1 (immatures)
Yellow-spotted rock hyrax (*Heterohyrax brucei*)	2 (immatures)
Rock hyrax (*Procavia capensis*)	4 (including immatures)
Warthog (*Phacochoerus africanus*)	70 (including immatures)
Forest hog (*Hylochoerus meinertzhageni*)	1 (immatures)
Bushpig (*Potamochoerus larvatus*)	17 (including immatures)
Giraffe (*Giraffa camelopardalis*)	14 (including immatures)
Impala (*Aepyceros melampus*)	386 (including immatures)
Coke's hartebeest (*Alcelaphus buselaphus cokii*)	1 (including 1 nymph)
Alcelaphus sp.	1 (including immatures)
Blue wildebeest (*Connochaetes taurinus*)	49 (including immatures)
Tsessebe (*Damaliscus lunatus lunatus*)	2 (including immatures)
Blesbok (*Damaliscus pygargus phillipsi*)	5 (including immatures)
Lichtenstein's hartebeest (*Sigmoceros lichtensteinii*)	3 (including nymphs)
Springbok (*Antidorcas marsupialis*)	2 (including 1 nymph)
Grant's gazelle (*Gazella granti*)	4 (including immatures)
Thomson's gazelle (*Gazella thomsonii*)	2
Kirk's dik-dik (*Madoqua kirkii*)	7 (including nymphs)
Suni (*Neotragus moschatus*)	1
Klipspringer (*Oreotragus oreotragus*)	9 (including nymphs)
Oribi (*Ourebia ourebi*)	7 (including nymphs)
Steenbok (*Raphicerus campestris*)	6 (including nymphs)
African buffalo (*Syncerus caffer*)	78 (including immatures)
Eland (*Taurotragus oryx*)	53 (including immatures)
Nyala (*Tragelaphus angasii*)	90 (including immatures)
Bushbuck (*Tragelaphus scriptus*)	31 (including immatures)
Sitatunga (*Tragelaphus spekii*)	1
Greater kudu (*Tragelaphus strepsiceros*)	188 (including immatures)
Bay duiker (*Cephalophus dorsalis*)	1 (including immatures)
Blue duiker (*Cephalophus monticola*)	1 (immatures)
Red forest duiker (*Cephalophus natalensis*)	7 (including immatures)
Cephalophus sp.	2 (including immatures)
Common duiker (*Sylvicapra grimmia*)	37 (including immatures)
'Duiker'	1 (1 nymph)
Roan antelope (*Hippotragus equinus*)	5
Sable antelope (*Hippotragus niger*)	22 (including nymphs)
Waterbuck (*Kobus ellipsiprymnus*)	47 (including immatures)
Kob (*Kobus kob*)	7
Lechwe (*Kobus leche*)	2 (including immatures)
Kobus sp.	1

Table 1. (*cont.*)

Hosts	Number of records
Reedbuck (*Redunca arundinum*)	17 (including nymphs)
Mountain reedbuck (*Redunca fulvorufula*)	29 (including immatures)
Bohor reedbuck (*Redunca redunca*)	2 (including immatures)
Temminck's ground pangolin (*Manis temminckii*)	1 (nymphs)
Striped grass rat (*Lemniscomys striatus*)	2 (nymphs)
Brush-furred rat (*Lophuromys flavopunctatus*)	2 (immatures)
Swamp rat (*Otomys irroratus*)	5 (larvae)
Spring hare (*Pedetes capensis*)	3 (immatures)
South African porcupine (*Hystrix africaeaustralis*)	1 (including immatures)
Greater cane rat (*Thryonomys swinderianus*)	3 (including immatures)
Cape hare (*Lepus capensis*)	16 (immatures, 2 with adults)
Scrub hare (*Lepus saxatilis*)	139 (immatures, 3 with adults)
Savanna hare (*Lepus victoriae*)	21 (immatures, 1 with adults)
'Hare' (*Lepus* sp.)	3 (immatures, 2 with adults)
Birds	
Helmeted guineafowl (*Numida meleagris*)	11 (immatures, 2 with adults)
Long-crested helmet shrike (*Prionops plumata*)	1
Humans	18 (including immatures)

rodents confirms that these animals are not good hosts of any life stages of *R. appendiculatus*. The collections from birds are more a reflection of the abundance of ticks at a particular locality than of host preference.

Burdens of adult ticks can be very large and infestations exceeding 1000 ticks have been counted on cattle, African buffalo, eland, nyala, greater kudu, waterbuck and an old lion. Burdens of immature ticks, particularly larvae, often exceed several thousand.

The preferred sites of attachment of adult *R. appendiculatus* are the pinnae of the ears followed by the head. Nymphs are most commonly present on the ears, head, legs and feet, and larvae on the head, dewlap, legs and feet (Baker & Ducasse, 1967). According to Yeoman & Walker (1967) the prime site of attachment of adult ticks is the inside of the pinna, 'in particular the proximal third of the upper edge, where the fringe of long hair grows. With light infestations the vast majority of *R. appendiculatus* will be found here.' As infestations increase in size the ticks spread to other parts of the ear flap but usually avoid those parts of the integument lying directly over the cartilage bars. Most engorged larvae dropped from stalled cattle between 10:00 and 14:00 hours, most nymphs between 12:00 and 18:00 hours and most adults between 06:00 and 08:00 hours (Minshull, 1982).

In regions close to the equator more than one life cycle can be completed annually and no clear pattern of seasonal abundance is evident (McCulloch *et al.*, 1968; Matthysse & Colbo, 1987). Further south only one life cycle is completed annually and adult ticks are generally most abundant on hosts from mid to late summer (December to March), larvae from autumn to winter (April to August) and nymphs from winter to spring (July to October) (Wilson, 1946; Baker & Ducasse, 1967; Short & Norval, 1981a,b; Rechav, 1982). According to Short & Norval (1981a) the seasonal occurrence of the larvae and nymphs follows the pattern set by the adults. The seasonal occurrence of adult ticks on hosts may be regulated by a photoperiodically controlled diapause and these ticks can be present on the vegetation some months before they are found on animals (Rechav, 1981).

Zoogeography

The distribution of *R. appendiculatus* is confined to parts of eastern, central and south-eastern Africa (Map 2). Records of its presence outside this region are thought to be misidentifications, probably mostly of *R. duttoni, R. nitens* or *R. zambeziensis*. Within countries in nearly all these regions its distribution is patchy, being limited by various factors such as climate, vegetation and the availability of suitable hosts. Its distribution in relation to that of cattle and African buffalo, also other ticks of the *R. appendiculatus* group as well as climate and vegetation, has been plotted by Lessard *et al.* (1990).

Rhipicephalus appendiculatus is found at altitudes ranging from just above sea level to 2000 m, but there are some records of its presence above 2400 m. Rainfall for the regions in which it is most prevalent varies between 500 mm and 2000 mm annually. According to Theiler (1964) it does not occur in open grasslands without bush but requires tall grass interspersed with trees or bush. It has been recorded most frequently in East African bushland, woodland and coastal mosaic; evergreen and semi-evergreen bushland and thicket; wetter and drier Zambezian miombo woodland; and in undifferentiated woodland and Afromontane vegetation.

Disease relationships

Rhipicephalus appendiculatus is undoubtedly the most economically important tick of the genus *Rhipicephalus* in Africa. It owes this pre-eminence to the fact that it is the most efficient vector of *Theileria parva parva*, the causative organism of East Coast fever in cattle (Lounsbury, 1904, 1906; De Vos, 1981). It is also an efficient vector of *Theileria parva lawrencei* from African buffalo to cattle, causing Corridor or buffalo disease in the latter animals (Neitz, 1955; De Vos, 1981). *Theileria parva bovis* of cattle and *Theileria taurotragi* of eland and cattle are also transmitted by *R. appendiculatus* (Fivaz, Norval & Lawrence, 1989; Lawrence & MacKenzie, 1980), as is *Ehrlichia bovis* of cattle (Matson, 1967; Norval, 1979). It is a vector of the viruses causing Nairobi sheep disease and Kisenye sheep disease (Montgomery, 1917; Bugyaki, 1955), and of the virus responsible for louping ill in cattle, sheep and man (Alexander & Neitz, 1935). It is also a vector of *Rickettsia conori*, causing tick- bite fever in man (Yunker & Norval, 1992). Very large infestations with adult ticks may give rise to an immuno-suppressive toxicosis in cattle, resulting in the recurrence of tick-borne diseases to which the animals were previously immune (Thomas & Neitz, 1958; Van Rensburg, 1959). A similar toxicosis possibly occurs in wild antelope (Lightfoot & Norval, 1981). Very large infestations have also caused the death of eland calves as a result of both acute and chronic anaemia (Lewis, 1981).

REFERENCES

Alexander, R.A. & Neitz, W.O. (1935). The transmission of louping ill by ticks (*Rhipicephalus appendiculatus*). *Onderstepoort Journal of Veterinary Science and Animal Industry*, **5**, 15–33.

Baker, M.K. & Ducasse, F.B.W. (1967). Tick infestation of livestock in Natal. I. The predilection sites and seasonal variations of cattle ticks. *Journal of the South African Veterinary Medical Association*, **38**, 447–53.

Bugyaki, L. (1955). La 'maladie de Kisenyi' du mouton due à un virus filtrable et transmise par des tiques. *Bulletin Agricole du Congo Belge*, **46**, 1455–62.

Clifford, C.M., Flux, J.E. & Hoogstraal, H. (1976). Seasonal and regional abundance of ticks (Ixodidae) on hares (Leporidae) in Kenya. *Journal of Medical Entomology*, **13**, 40–7.

De Vos, A.J. (1981). *Rhipicephalus appendiculatus*: cause and vector of diseases in Africa. *Journal of the South African Veterinary Association*, **52**, 315–22.

Fivaz, B.H., Norval, R.A.I. & Lawrence, J.A. (1989). Transmission of *Theileria parva bovis* (Boleni strain) to cattle resistant to the brown ear tick *Rhipicephalus appendiculatus* (Neumann). *Tropical Animal Health and Production*, **21**, 129–34.

Horak, I.G. (1982). Parasites of domestic and wild animals in South Africa. XV. The seasonal prevalence of ectoparasites on impala and cattle in the northern Transvaal. *Onderstepoort Journal of Veterinary Research*, **49**, 85–93.

Horak, I.G., Boomker, J., De Vos, V. & Potgieter, F.T. (1988). Parasites of domestic and wild animals in South Africa. XXIII. Helminth and arthropod parasites of warthogs, *Phacochoerus aethiopicus*, in the eastern Transvaal Lowveld. *Onderstepoort Journal of Veterinary Research*, **55**, 145–52.

Horak, I.G., Boomker, J. & Flamand, J.R.B. (1995). Parasites of domestic and wild animals in South Africa. XXXIV. Arthropod parasites of nyalas in north-eastern KwaZulu-Natal. *Onderstepoort Journal of Veterinary Research*, **62**, 171–9.

Horak, I.G., Boomker, J., Spickett, A.M. & De Vos, V. (1992). Parasites of domestic and wild animals in South Africa. XXX. Ectoparasites of kudus in the eastern Transvaal Lowveld and the eastern Cape Province. *Onderstepoort Journal of Veterinary Research*, **59**, 259–73.

Horak, I.G., De Vos, V. & De Klerk, B.D. (1984). Parasites of domestic and wild animals in South Africa. XVII. Arthropod parasites of Burchell's zebra, *Equus burchelli*, in the eastern Transvaal Lowveld. *Onderstepoort Journal of Veterinary Research*, **51**, 145–54.

Horak, I.G. & Fourie, L.J. (1991). Parasites of domestic and wild animals in South Africa. XXIX. Ixodid ticks on hares in the Cape Province and on hares and red rock rabbits in the Orange Free State. *Onderstepoort Journal of Veterinary Research*, **58**, 261–71.

Horak, I.G., Jacot Guillarmod, A., Moolman, L.C. & De Vos, V. (1987). Parasites of domestic and wild animals in South Africa. XXII. Ixodid ticks on domestic dogs and on wild carnivores. *Onderstepoort Journal of Veterinary Research*, **54**, 573–80.

Horak, I.G., Potgieter, F.T., Walker, J.B., De Vos, V. & Boomker, J. (1983). The ixodid tick burdens of various large ruminant species in South African nature reserves. *Onderstepoort Journal of Veterinary Research*, **50**, 221–8.

Lawrence, J.A. & MacKenzie, P.K.I. (1980). Isolation of a non-pathogenic theileria of cattle transmitted by *Rhipicephalus appendiculatus*. *Zimbabwe Veterinary Journal*, **11**, 27–35.

Lessard, P., L'Eplattenier, R., Norval, R.A.I., Kundert, K., Dolan, T.T., Croze, H., Walker, J.B., Irvin, A.D. & Perry, B.D. (1990). Geographical information systems for studying the epidemiology of cattle diseases caused by *Theileria parva*. *Veterinary Record*, **126**, 255–62.

Lewis, A.R. (1981). The pathology of *Rhipicephalus appendiculatus* infestation of eland *Taurotragus oryx*. In *Proceedings of an International Conference on Tick Biology and Control*, ed. G.B. Whitehead & J.D. Gibson, pp. 15–20. Grahamstown: Rhodes University.

Lightfoot, C.J. & Norval, R.A.I. (1981). Tick problems in wildlife in Zimbabwe. I. The effects of tick parasitism on wild ungulates. *South African Journal of Wildlife Research*, **11**, 41–5.

Lounsbury, C.P. (1904). Transmission of African Coast Fever. *Agricultural Journal, Cape of Good Hope*, **24**, 428–32.

Lounsbury, C.P. (1906). Ticks and African Coast Fever. *Agricultural Journal, Cape of Good Hope*, **28**, 634–54.

Matson, B.A. (1967). Theileriosis in Rhodesia: I. A study of diagnostic specimens over two seasons. *Journal of the South African Veterinary Medical Association*, **38**, 93–102.

McCulloch, B., Kalaye, W.J., Tungaraza, R., Suda, B'Q.J. & Mbasha, E.M.S. (1968). A study of the life history of the tick *Rhipicephalus appendiculatus* – the main vector of East Coast fever – with reference to its behaviour under field conditions and with regard to its control in Sukumaland, Tanzania. *Bulletin of Epizootic Diseases of Africa*, **16**, 477–500.

Minshull, J.I. (1982). Drop-off rhythms of engorged *Rhipicephalus appendiculatus* (Acarina: Ixodidae). *Journal of Parasitology*, **68**, 484–9.

Montgomery, E. (1917). On a tick-borne gastro-enteritis of sheep and goats occurring in British East Africa. *Journal of Comparative Pathology and Therapeutics*, **30**, 28–57.

Neitz, W.O. (1955). Corridor disease: a fatal form of bovine theileriosis encountered in Zululand.

Bulletin of Epizootic Diseases of Africa, **3**, 121–3.

Neumann, L.G. (1901). Révision de la famille des Ixodidés. (4e Mémoire). *Mémoires de la Société Zoologique de France*, **14**, 249–372.

Norval, R.A.I. (1979). Tick infestations and tick-borne diseases in Zimbabwe Rhodesia. *Journal of the South African Veterinary Association*, **50**, 289–92.

Norval, R.A.I., Walker, J.B. & Colborne, J. (1982). The ecology of *Rhipicephalus zambeziensis* and *Rhipicephalus appendiculatus* (Acarina, Ixodidae) with particular reference to Zimbabwe. *Onderstepoort Journal of Veterinary Research*, **49**, 181–90.

Nuttall, G.H.F. (1913). Observations on the biology of Ixodidae. Part 1. *Parasitology*, **6**, 68–118.

Rechav, Y. (1981). Ecological factors affecting the seasonal activity of the brown ear tick *Rhipicephalus appendiculatus*. In *Proceedings of an International Conference on Tick Biology and Control*, ed. G.B. Whitehead & J.D. Gibson, pp. 187–91. Grahamstown: Rhodes University.

Rechav, Y. (1982). Dynamics of tick populations (Acari: Ixodidae) in the eastern Cape Province of South Africa. *Journal of Medical Entomology*, **19**, 679–700.

Short, N.J. & Norval, R.A.I. (1981a). Regulation of seasonal occurrence in the tick *Rhipicephalus appendiculatus* Neumann, 1901. *Tropical Animal Health and Production*, **13**, 19–25.

Short, N.J. & Norval, R.A.I. (1981b). The seasonal activity of *Rhipicephalus appendiculatus* Neumann, 1901 (Acarina: Ixodidae) in the highveld of Zimbabwe Rhodesia. *Journal of Parasitology*, **67**, 77–84.

Theiler, G. (1964). Ecogeographical aspects of tick distribution. In *Ecological Studies in Southern Africa*, ed. D.H.S. Davis, pp. 284–300. The Hague: Dr W. Junk Publishers.

Thomas, A.D. & Neitz, W.O. (1958). Rhipicephaline tick toxicosis in cattle: its possible aggravating effects on certain diseases. *Journal of the South African Veterinary Medical Association*, **29**, 39–50.

Van Rensburg, S.J. (1959). Haematological investigations into the rhipicephaline tick toxicosis syndrome. *Journal of the South African Veterinary Medical Association*, **30**, 75–95.

Walker, J.B., Norval, R.A.I. & Corwin, M.D. (1981). *Rhipicephalus zambeziensis* sp.nov., a new tick from eastern and southern Africa, together with a redescription of *Rhipicephalus appendiculatus* Neumann, 1901 (Acarina, Ixodidae). *Onderstepoort Journal of Veterinary Research*, **48**, 87–104.

Wilson, S.G. (1946). Seasonal occurrence of Ixodidae on cattle in Northern Province, Nyasaland. *Parasitology*, **37**, 118–25.

Yunker, C.E. & Norval, R.A.I. (1992). Observations on African tick typhus (tick-bite fever) in Zimbabwe. In *Tick Vector Biology, Medical and Veterinary Aspects*, ed. B. Fivaz, T. Petney & I. Horak, pp. 143–7. Berlin, Heidelberg: Springer-Verlag.

Also see the following Basic References (pp. 12–14): Keirans (1985); Matthysse & Colbo (1987); Moritz & Fischer (1981); Walker (1974); Yeoman & Walker (1967).

RHIPICEPHALUS AQUATILIS WALKER, KEIRANS & PEGRAM, 1993

The specific name *aquatilis*, from the Latin meaning 'living in or near water', refers to the fact that all the collections have been made from hosts in semi-aquatic habitats.

Diagnosis

A large dark reddish-brown to almost black species.

Male (Figs 7(a), 8(a) to (c))

Capitulum approximately as broad as long, length × breadth ranging from 0.66 mm × 0.66 mm to 0.85 mm × 0.82 mm. Lateral angles of basis capituli at about anterior third of its length, acute. Body length × breadth ranging from 2.55 mm × 1.66 mm to 3.20 mm × 2.16 mm; anterior process of coxae I not particularly prominent. In engorged specimens the body wall bulges slightly posterolaterally and posteriorly, forming a short blunt caudal process posteromedially. Conscutum with eyes marginal, flat, partially outlined by a few punctations. Cervical pits small, slightly convergent; internal cervical margins absent; external cervical margins marked by irregular rows of medium-sized punctations. Marginal lines deep, narrow, punctate, delimiting first festoons and extending anteriorly nearly to eyes. Posteromedian and posterolateral grooves well defined. Lateral borders, both anterior and posterior to eyes, smooth and shiny, with a few medium-sized setiferous punctations on scapulae and along external cervical margins; medially conscutum covered with uniformly small punctations. Legs increase very slightly in size from I to IV. Ventrally spiracles elongate, narrowing slightly as they curve towards the dorsal surface. Adanal plates usually sickle-shaped, but in some males more straight-sided; accessory adanal plates sharply pointed, well sclerotized.

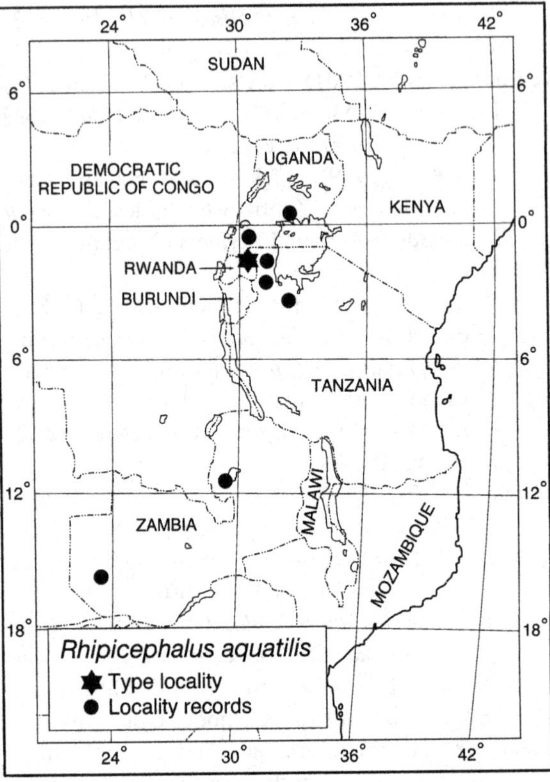

Map 3. *Rhipicephalus aquatilis*: distribution. (After Walker *et al.*, 1993.)

Female (Figs 7(b), 8(d) to (f))

Capitulum slightly broader than long, the length × breadth varying from 0.82 mm × 0.83 mm to 0.85 mm × 0.83 mm; porose areas small, subcircular. Scutum ovoid in shape, length × breadth ranging from 1.54 mm × 1.46 mm to 1.69 mm × 1.65 mm, broadest at eye level. Eyes marginal, flat to very slightly raised. Cervical pits short, deep, convergent; internal cervical margins shallow, initially converging then diverging; external cervical margins as slight declinations marked by irregular rows of medium-sized setiferous punctations; lateral borders slightly raised, generally smooth and shiny but with some setiferous punctations scattered on the scapulae and along the outer margins. Medially punctations numerous, small, uniformly distributed, but somewhat sparser in the anterior areas of the cervical fields. Ventrally genital aperture between coxae II, V-shaped with the sides of the opening as gentle sigmoid curves.

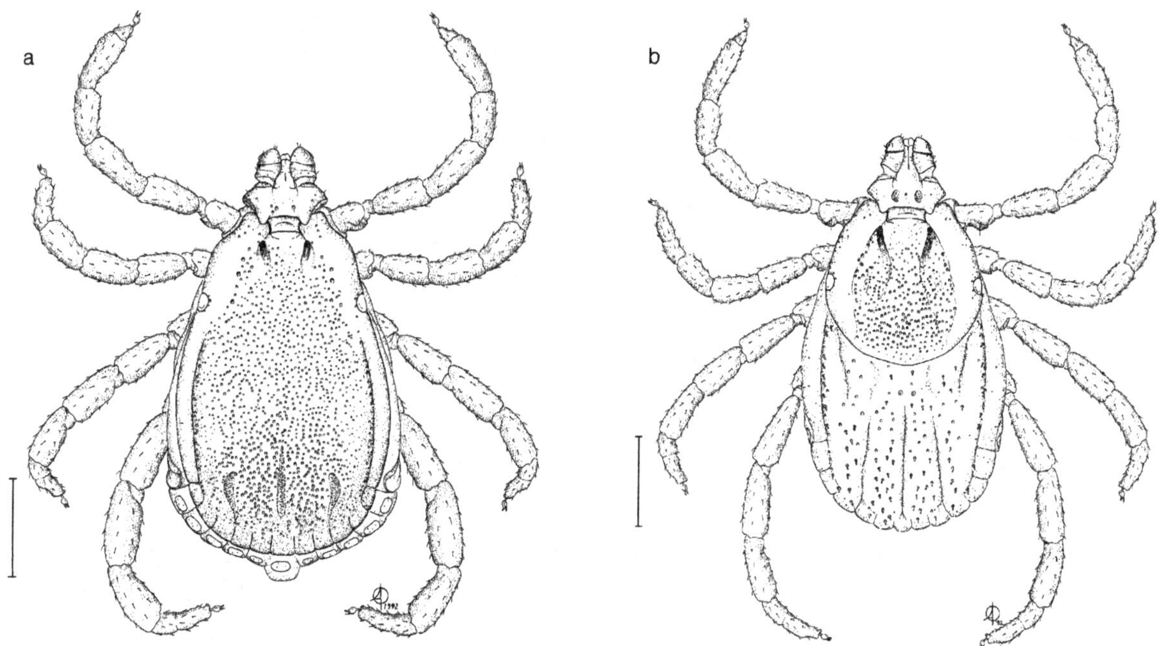

Figure 7. *Rhipicephalus aquatilis* [Onderstepoort Tick Collection 3143i, Tanzania Tick Collection WA/99, from sitatunga (*Tragelaphus spekii*), Kaisho, Karagwe, Tanzania, collected on 14 August 1959 by Mrs G. Tullock]. (a) Male, dorsal; (b) female, dorsal. Scale bars represent 1 mm. A. Olwage *del.* (From Walker *et al.*, 1993, figs 1 & 2, with kind permission from the Editor, *Onderstepoort Journal of Veterinary Research*).

Immature stages

Unknown.

Notes on identification

Rhipicephalus aquatilis was originally listed as *Rhipicephalus* sp. III by Yeoman & Walker (1967). Morphologically *R. aquatilis* somewhat resembles *R. hurti* (see p. 222). These two species, plus *R. jeanneli*, were included in their *R. hurti* group by Matthysse & Colbo (1987).

Hosts

Life cycle unknown. Only 15 collections of *R. aquatilis* are presently known (Table 2). The sitatunga appears to be the preferred host since the seven collections from this antelope comprised 41 ♂♂, 34 ♀♀ ticks whereas the eight collections from the other three host species comprised only 2 ♂♂, 8 ♀♀.

Table 2. *Host records of* Rhipicephalus aquatilis

Hosts	Number of records
Domestic animals	
Cattle	6
Wild animals	
Lion (*Panthera leo*)	1
Leopard (*Panthera pardus*)	1
Sitatunga (*Tragelaphus spekii*)	7

Zoogeography

This tick has been collected in Uganda, north-western Tanzania and Zambia from hosts living in various lakeside, swampy or seasonally inundated habitats (Map 3). Its main host, the sitatunga, is a semi-aquatic antelope that spends most of its life in swamps with dense papyrus, *Cyperus papyrus*, and reed beds, *Phragmites mauritianus* (Skinner & Smithers, 1990).

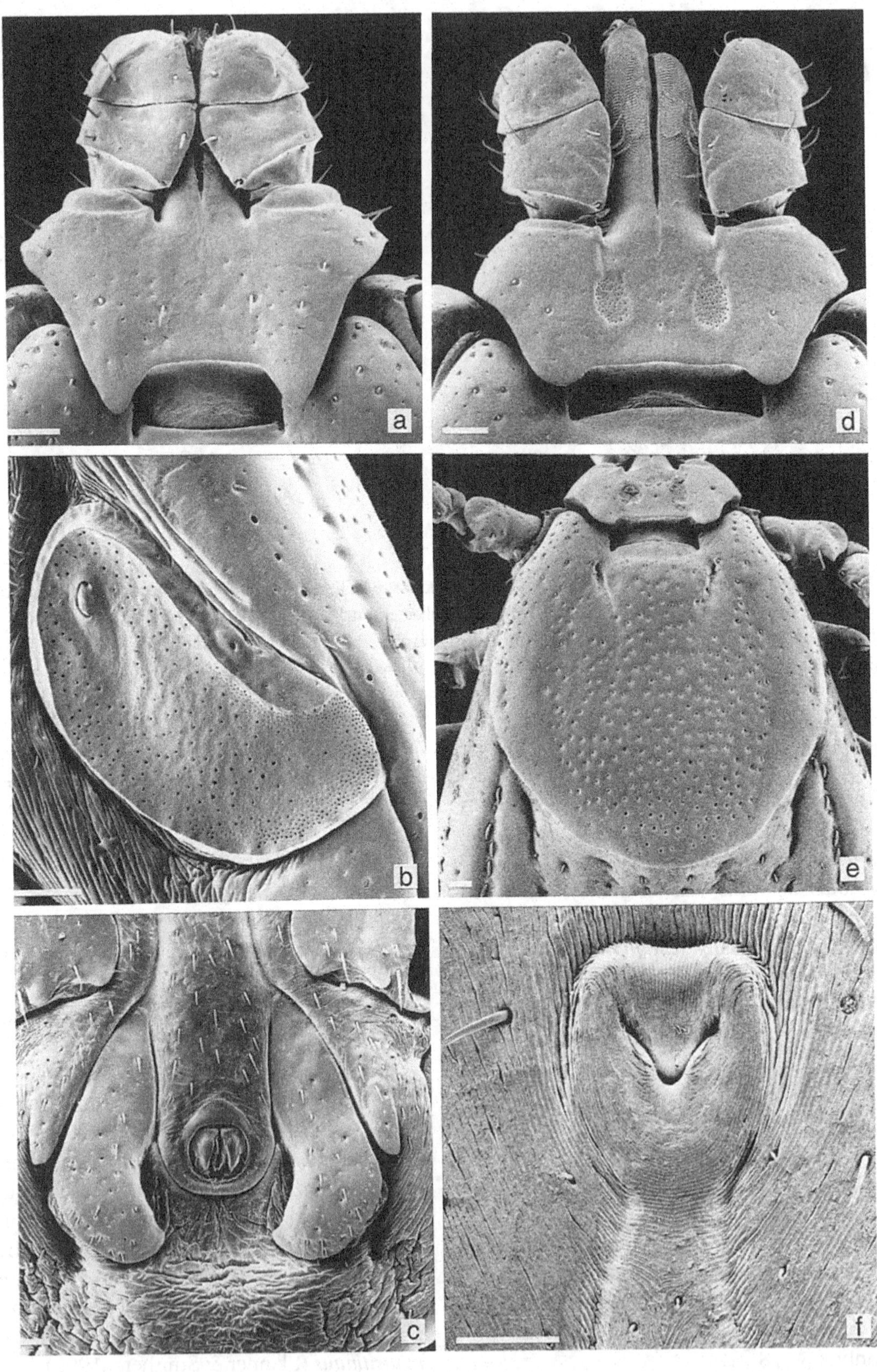
a
b
c
d
e
f

Disease relationships

Unknown.

REFERENCES

Walker, J.B., Keirans, J.E. & Pegram, R.G. (1993). *Rhipicephalus aquatilis* sp. nov. (Acari, Ixodidae), a new tick species parasitic mainly on the sitatunga, *Tragelaphus spekei* (*sic*), in East and Central Africa. *Onderstepoort Journal of Veterinary Research*, **60**, 205–10.

Also see the following Basic References (pp. 12–14): Matthysse & Colbo (1987); Skinner & Smithers (1990); Yeoman & Walker (1967).

Figure 8 (*opposite*). *Rhipicephalus aquatilis* [Onderstepoort Tick Collection 3143i, Tanzania Tick Collection WA/99, from sitatunga (*Tragelaphus spekii*), Kaisho, Karagwe, Tanzania, collected on 14 August 1959 by Mrs G. Tullock]. Male: (a) capitulum, dorsal; (b) spiracle; (c) adanal plates. Female: (d) capitulum, dorsal; (e) scutum; (f) genital aperture. Scale bars represent 0.10 mm. SEMs by J.F. Putterill. (From Walker *et al.*, 1993, figs 3–8, with kind permission from the Editor, *Onderstepoort Journal of Veterinary Research.*)

RHIPICEPHALUS ARMATUS POCOCK, 1900

The specific name *armatus*, from the Latin meaning 'armed', probably refers to the spines that become visible on the ventral body wall of engorged males, one under each adanal plate, as illustrated by Pocock (1900: pl. III, fig. 2c).

Diagnosis

A large glossy reddish-brown species.

Male (Figs 9(a), 10(a) to (c))

Capitulum broader than long, length × breadth ranging from 0.55 mm × 0.62 mm to 0.95 mm × 0.99 mm. Basis capituli with acute, sometimes forwardly directed, lateral angles at about mid-length. Palps tapering towards apices. Conscutum length × breadth ranging from 2.49 mm × 1.53 mm to 4.41 mm × 2.58 mm, narrower anteriorly and widening posterior to eyes; anterior process of coxae I rather inconspicuous, blunt. Eyes almost flat, sometimes edged dorsally by one or two large punctations. Cervical pits long, deep, convergent; external cervical margins indicated by irregular rows of very large, unevenly-shaped, deeply-sunken setiferous punctations. Marginal lines similarly punctate, extending from first festoons almost to eye level. Four broad, deeply-punctate longitudinal grooves characteristically present on the posterior half of the conscutum, comprising two short posteromedian grooves flanked on each side by a much longer, somewhat curved groove. A few slightly smaller, shallower punctations scattered medially on the conscutum. Ventrally spiracles elongatedly comma-shaped. Adanal plates broad, produced posterointernally into a long sub-bifid spine and posteroexternally into a shorter spine, with a concave posterior margin; in engorged males whose adanal plates have lifted away from the ventral body wall a short sharp spine is visible under each plate when these ticks are viewed from the ventral side; accessory adanal plates absent.

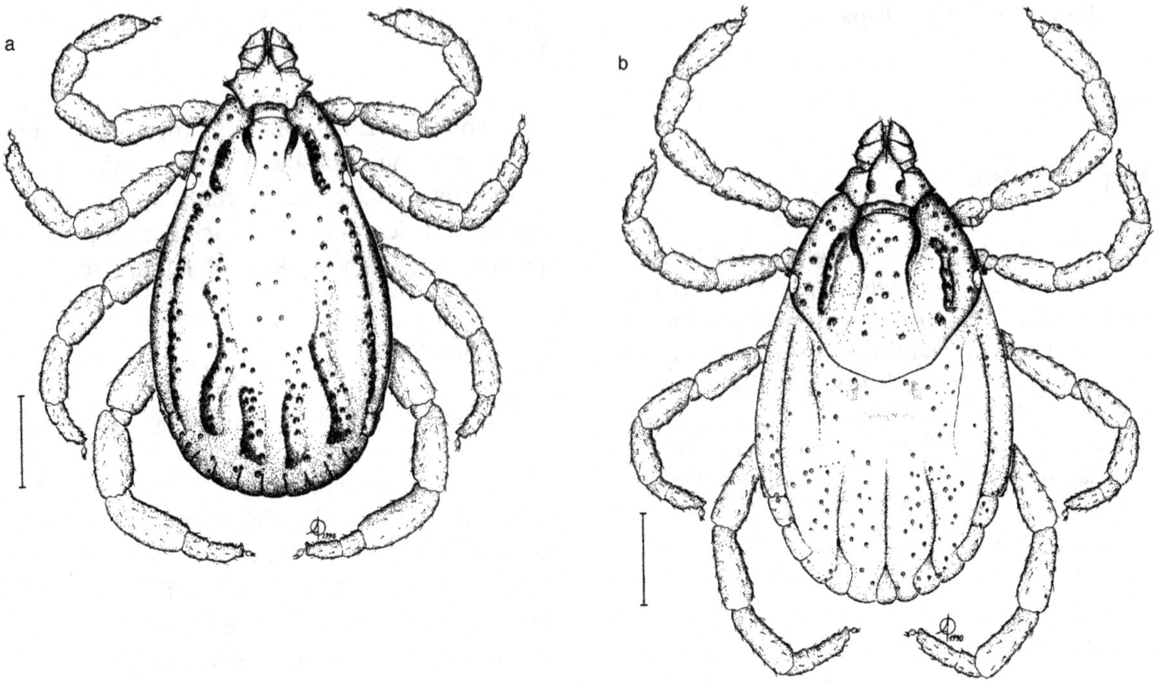

Figure 9. *Rhipicephalus armatus* [B.S. 908/-, RML 66303, laboratory reared, progeny of ♀ collected from zorilla (*Ictonyx striatus*), Nairange Tseikuru, Kenya, on 9 March 1960, by L.R. Rickman]. (a) Male, dorsal; (b) female, dorsal. Scale bars represent 1 mm. A. Olwage *del.*

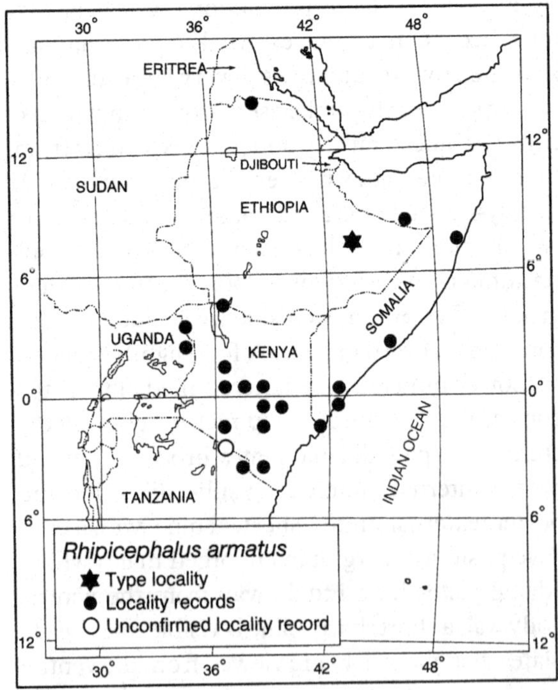

Map 4. *Rhipicephalus armatus* : distribution.

Female (Figs 9(b), 10(d) to (f))

Capitulum broader than long, length × breadth ranging from 0.68 mm × 0.79 mm to 1.07 mm × 1.16 mm. Basis capituli with acute lateral angles at about mid-length; porose areas large, not quite twice their own diameter apart. Palps tapering abruptly to narrowly rounded apices. Scutum broader than long, length × breadth ranging from 1.34 mm × 1.51 mm to 2.23 mm × 2.35 mm, posterior margin sinuous. Eyes almost flat, sometimes edged dorsally by one or two large punctations. Cervical pits long, convergent; cervical fields broad, slightly depressed, their outer margins delimited by irregular rows of very large, unevenly-shaped, setiferous punctations. A few slightly smaller setiferous punctations scattered medially and laterally on the scutum, interspersed with very fine punctations. Ventrally genital aperture broad, with the genital apron tucked under the straight posterior margin of the opening.

Table 3. *Host records of* Rhipicephalus armatus

Hosts	Number of records: Confirmed	Number of records: Unconfirmed
Domestic animals		
Sheep	1 (1 ♂ only)	
Donkey	1	
Dogs	1	1
Wild animals		
Four-toed hedgehog (*Atelerix albiventris*)	2	
Golden jackal (*Canis aureus*)	2	
*Black-backed jackal (*Canis mesomelas*)	2 (including nymphs)	1
'Jackal'	1	
Hunting dog (*Lycaon pictus*)	2	
*Bat-eared fox (*Otocyon megalotis*)	2 (nymphs)	1
Cheetah (*Acinonyx jubatus*)	3	
*Caracal (*Caracal caracal*)	1 (nymphs)	
African wild cat (*Felis lybica*)	1 (nymphs)	
Lion (*Panthera leo*)	3	2
Leopard (*Panthera pardus*)	1	
Spotted hyaena (*Crocuta crocuta*)	1	
Striped hyaena (*Hyaena hyaena*)	1	
'Hyaena'	1	
Zorilla (*Ictonyx striatus*)	1	
*Grant's gazelle (*Gazella granti*)	1 (1 nymph only)	2
Northern pygmy gerbil (*Gerbillus* sp.)	1 (1 ♂ only)	
Cape hare (*Lepus capensis*)	12 (immatures)	
Birds		
*Kori bustard (*Ardeotis kori*)	1 (1 nymph only)	
Humans	2 (1 ♂, 1 ♀ only)	

Note: *According to Morel (1980) the identification of the nymphs from these hosts was based on their differences from other *Rhipicephalus* sp. 'larvae' (*sic*) occurring in East Africa, especially Ethiopia; on the presence of *R. armatus* adults in the places where they were collected, and on the analogies between these nymphs and those of the related species *R. cuspidatus*.

Nymph (Fig. 11)

Capitulum broader than long, length × breadth ranging from 0.32 mm × 0.37 mm to 0.35 mm × 0.41 mm. Basis capituli over three times as broad as long, with acute lateral angles at about mid-length projecting over the scapulae; ventrally with posterior border broadly rounded. Palps long, almost parallel-sided for most of their length, broadly rounded apically, inclined inwards. Scutum broader than long, length × breadth ranging from 0.57 mm × 0.62 mm to 0.64 mm × 0.69 mm, posterior margin a deep smooth curve. Eyes at widest point, over halfway back, large, delimited dorsally by slight depressions. Cervical pits long, convergent, continuous with the long, divergent, depressed internal cervical margins; these depressions almost reach the posterolateral scutal margins and virtually divide the scutum into three. Ventrally coxae I each with a relatively narrow external spur and a much

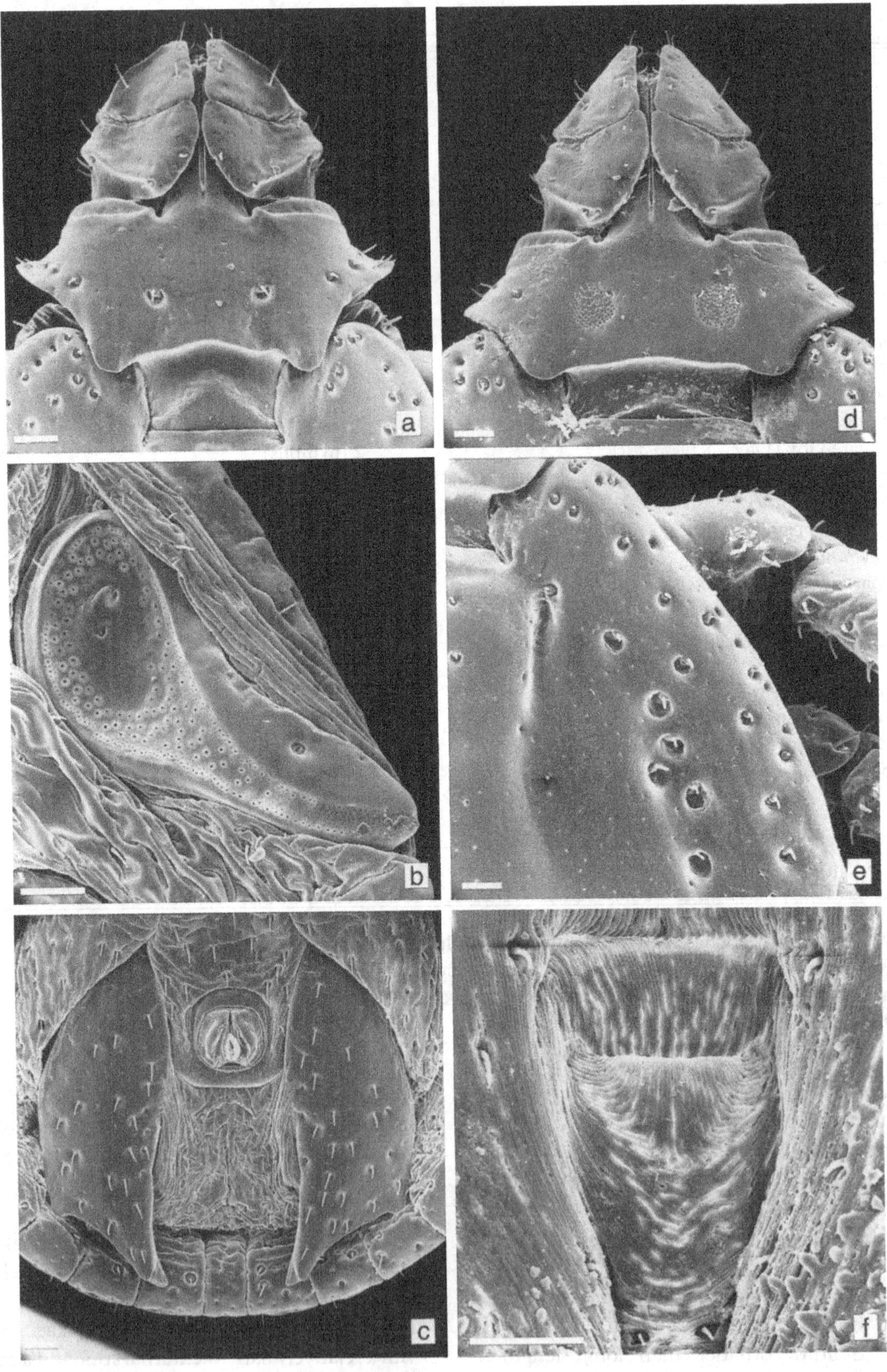
a
d
b
e
c
f

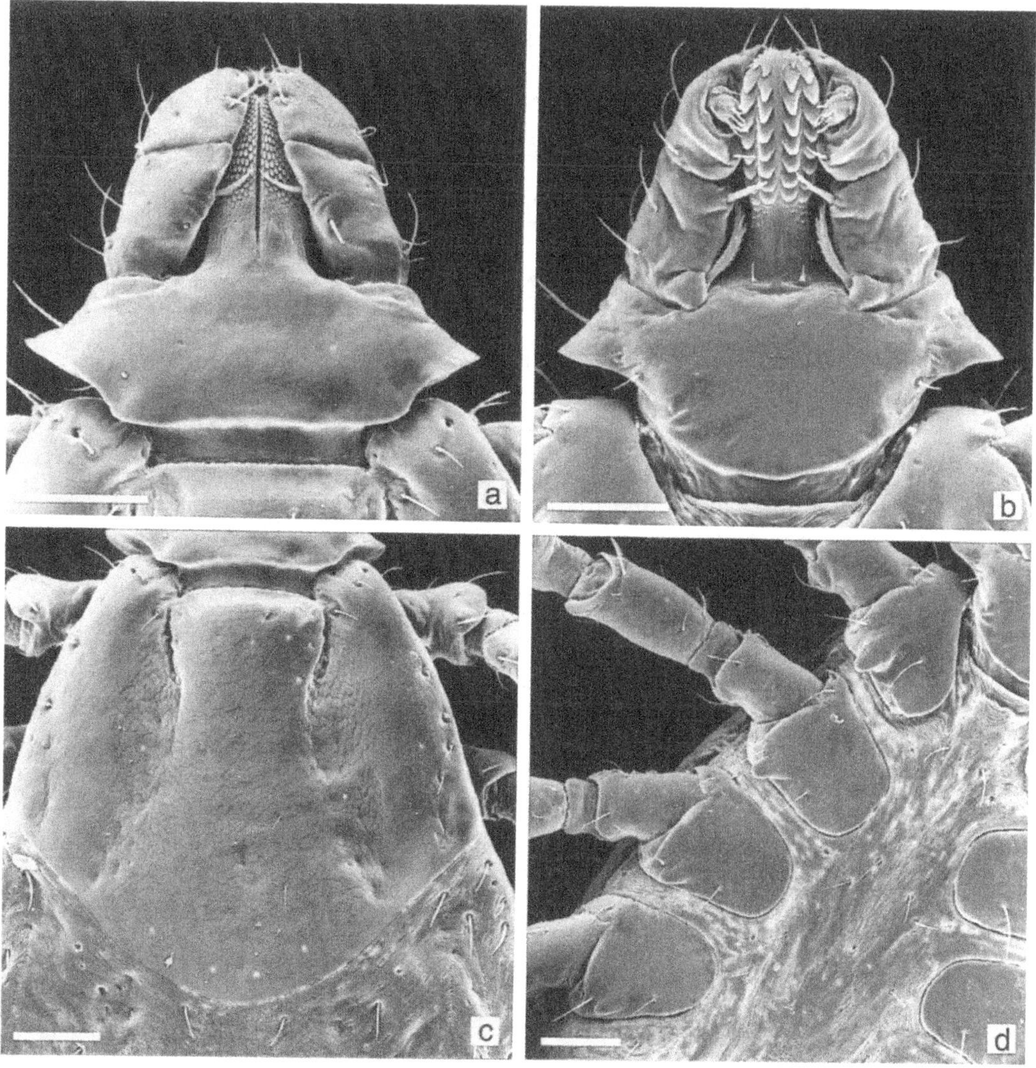

Figure 11 (*above*). *Rhipicephalus armatus* [B.S. 908/-, RML 66303, laboratory reared, progeny of ♀ collected from zorilla (*Ictonyx striatus*), Nairange Tseikuru, Kenya, on 9 March 1960, by L.R. Rickman]. Nymph: (a) capitulum, dorsal; (b) capitulum, ventral; (c) scutum; (d) coxae. Scale bars represent 0.10 mm. SEMs by M.D. Corwin.

Figure 10 (*opposite*). *Rhipicephalus armatus* [B.S. 908/-, RML 66303, laboratory reared, progeny of ♀ collected from zorilla (*Ictonyx striatus*), Nairange Tseikuru, Kenya, on 9 March 1960, by L.R. Rickman]. Male: (a) capitulum, dorsal; (b) spiracle; (c) adanal plates. Female: (d) capitulum, dorsal; (e) scapular area; (f) genital aperture. Scale bars represent 0.10 mm. SEMs by M.D. Corwin.

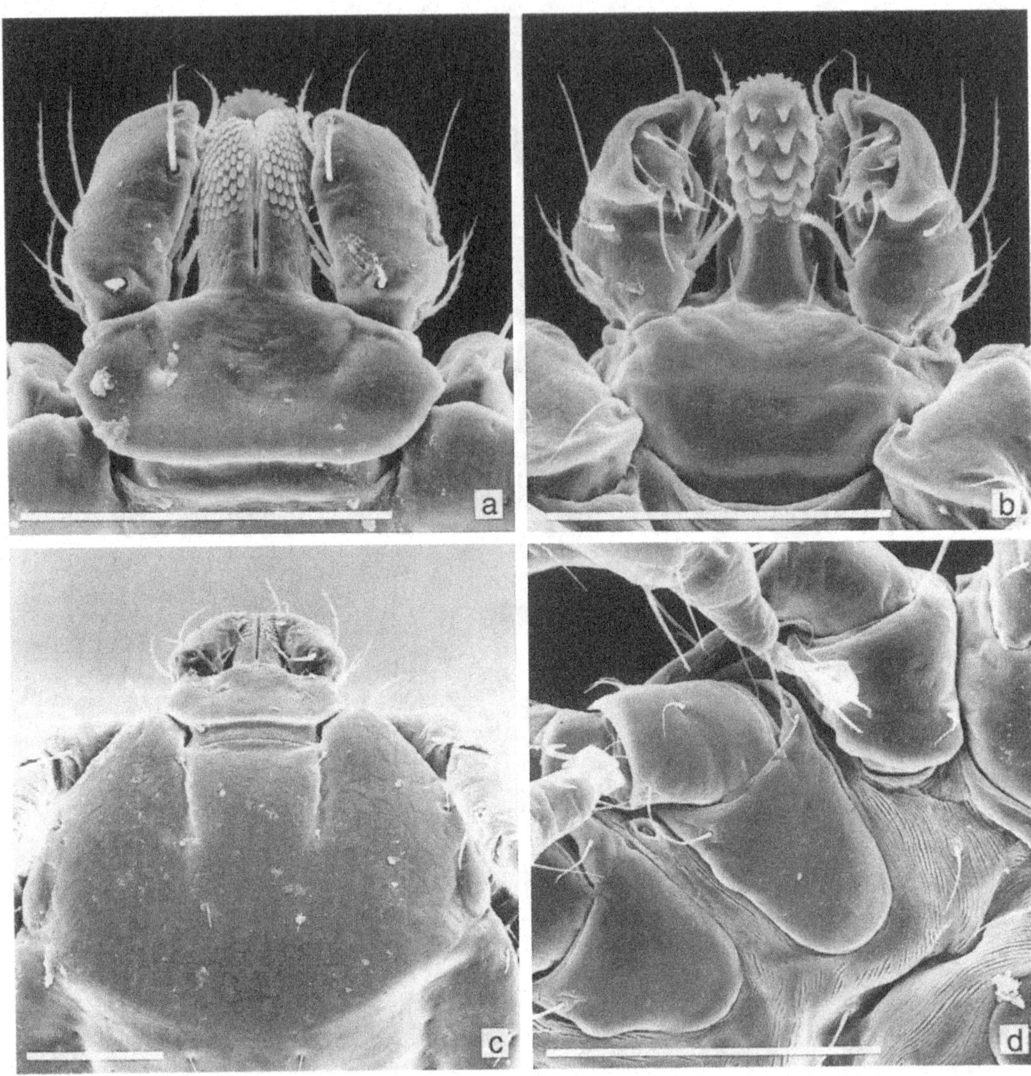

Figure 12. *Rhipicephalus armatus* [B.S. 908/-, RML 66303, laboratory reared, progeny of ♀ collected from zorilla (*Ictonyx striatus*), Nairange Tseikuru, Kenya, on 9 March 1960, by L.R. Rickman]. Larva: (a) capitulum, dorsal; (b) capitulum, ventral; (c) scutum; (d) coxae. Scale bars represent 0.10 mm. SEMs by M.D. Corwin.

broader internal spur; coxae II to IV each with a small external spur only.

Larva (Fig. 12)

Capitulum broader than long, length × breadth ranging from 0.138 mm × 0.159 mm to 0.151 mm × 0.170 mm. Basis capituli well over twice as broad as long with short blunt lateral angles, posterolaterally smoothly rounded. Palps slightly constricted proximally, otherwise broad, merely tapering slightly at their apices. Scutum much broader than long, length × breadth ranging from 0.256 mm × 0.423 mm to 0.283 mm × 0.455 mm; posterior margin a broad smooth curve. Eyes at widest part of scutum, slightly convex. Cervical grooves short, almost parallel. Ventrally coxae I each with a single broad spur, coxae II and III each with a small spur.

Notes on identification

This species has occasionally been misidentified. Adults in the Lewis Collection, The Natural History Museum, London, collected from a hunting dog (*Lycaon pictus*) at Benane, Kenya and listed by Lewis (1931) as *R. simus* var. *lunulatus*, were re-identified as *R. armatus* by Walker (1974). However, a female in this collection from a lion (*Panthera leo*) in the Narok area that was originally identified as *R. armatus* was initially re-identified as *R. simus sensu lato*; this is almost certainly *R. praetextatus*.

Morel (1980) suggested that the record of *R. cuspidatus* from Dire Dawa, Ethiopia, in Stella (1940) probably refers to *R. armatus*. He also stated that, according to Tonelli-Rondelli (1930), *R. armatus* had been collected from a bushpig (*Potamochoerus porcus*), at Sassabe (Mt. Ala), Eritrea; this is apparently a slip as we have been unable to find this record in Tonelli-Rondelli's paper.

Jooste (1969) recorded *R. armatus* in Zimbabwe on the basis of a single female. Norval (1985), who examined both this specimen and another female that he had collected, noted that they corresponded with the description given by Theiler (1947). Despite this he remained unconvinced that this species occurs in Zimbabwe. We have therefore omitted these records.

A comparison of the line drawing of the *R. armatus* male (Fig. 9(a)) with that of the *R. deltoideus* male (p. 137, Fig. 45(a)) shows that the general configuration of the capitulum and conscutum of these two species is similar in some respects. The individual punctations on *R. armatus* are, however, much larger than those on *R. deltoideus*.

Hosts

A three-host species (J.B.W., unpublished data, 1961). The majority of hosts recorded to date for this tick are carnivores, with a tendency for the adults to occur on the larger species and the immature stages on the smaller ones (Table 3). The largest collection from an individual animal, though, was 20 ♂♂, 4 ♀♀ taken from a four-toed hedgehog.

Cape hares sometimes act as hosts of the immature stages: 12 of the 245 hares examined in the Ololkisailie (= Olorgesailie)/Magadi area of Kenya were infested by a total of 70 nymphs and 1 larva, a mere 2.5% of the 2809 ticks of all stages collected from these animals (Clifford, Flux & Hoogstraal, 1976).

Zoogeography

Rhipicephalus armatus has been collected in north-eastern Uganda, northern and eastern Kenya, north-eastern Tanzania and at scattered points in Ethiopia and Somalia (Map 4). These are predominantly dry areas with mean annual rainfalls ranging from about 100 mm to 600 mm. The vegetation is mostly Somalia–Masai *Acacia*–*Commiphora* deciduous bushland and thicket to semi-desert grassland and shrubland.

Disease relationships

Unknown.

REFERENCES

Clifford, C.M., Flux, J.E. & Hoogstraal, H. (1976). Seasonal and regional abundance of ticks (Ixodidae) on hares (Leporidae) in Kenya. *Journal of Medical Entomology*, **13**, 40–7.

Jooste, K.F. (1969). The role of Rhodesia in ixodid tick distribution in central and southern Africa. In *Proceedings of a Symposium on the Biology and Control of Ticks in Southern Africa*, convenor G.B. Whitehead, pp. 37–42. Grahamstown, South Africa: Rhodes University.

Lewis, E.A. (1931). Report on tick survey in Kenya Colony. *Report of the Department of Agriculture, Kenya, for 1930*, pp. 151–62.

Norval, R.A.I. (1985). The ticks of Zimbabwe. XII. The lesser known *Rhipicephalus* species. *Zimbabwe Veterinary Journal*, **16**, 37–43.

Pocock, R.I. (1900). On a collection of insects and arachnids made in 1895–1897 in Somaliland, with description of new species. 9. Chilopoda and Arachnida. *Proceedings of the Zoological Society of London*, **Part 1**, 48–55, plate III, figs. 2–2f.

Stella, E. (1940). Nuovi dati sugli ixodidi dell'Africa Orientale Italiana. *Rivista di Biologia Coloniale*, 3, 431–5.

Tonelli-Rondelli, M. (1930). Ixodoidea del Museo di Milano. *Atti della Società Italiana di Scienzi Naturali, e del Museo Civile di Storia Naturale, Milano*, 59, 112–24.

Also see the following Basic References (pp. 12–14): Matthysse & Colbo (1987); Morel (1980); Scaramella (1988); Theiler (1947); Walker (1974); Yeoman & Walker (1967).

RHIPICEPHALUS ARNOLDI THEILER & ZUMPT, 1949

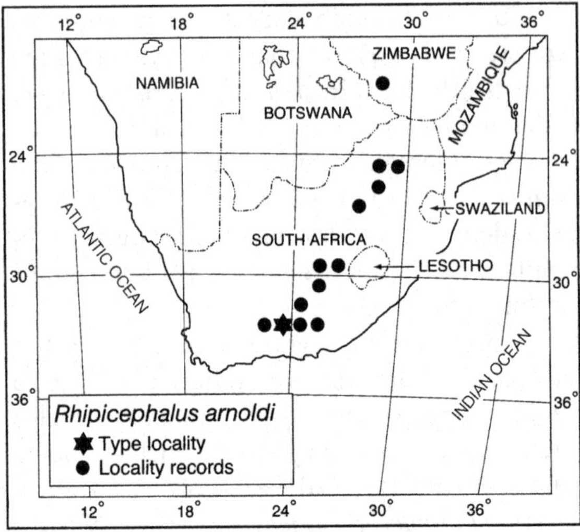

Map 5. *Rhipicephalus arnoldi*: distribution.

This species was named in honour of Mr Wolfgang Arnold of Johannesburg, who collected some of the specimens on which the original description was based.

Diagnosis

A small yellowish to reddish-brown pear-shaped tick.

Male (Figs 13(a), 14(a) to (c))

Capitulum much broader than long, length × breadth ranging from 0.40 mm × 0.53 mm to 0.49 mm × 0.62 mm. Basis capituli with long, sharp, forwardly tilted lateral angles at about anterior third of its length. Palps short, articles II and III tapering towards their rounded apices, inclined inwards. Conscutum length × breadth ranging from 1.89 mm × 1.22 mm to 2.24 mm × 1.40 mm, rather thinly sclerotized; anterior process of coxae I not particularly prominent. In engorged specimens the midgut caecae show through the conscutum. Eyes almost flat, edged dorsally by a few large setiferous punctations. Cervical pits deep, convergent, continuous with the long, divergent internal cervical margins; external cervical margins marked by rows of large setiferous punctations. Marginal lines short. Posterior grooves often inconspicuous; posteromedian groove long, narrow; posterolateral grooves shorter and broader. Large setiferous punctations present on the scapulae, scattered anteriorly in the shape of a female pseudoscutum and along the marginal lines, interspersed on the pseudoscutum with medium-sized punctations; posterior to the pseudoscutum the punctations medially on the conscutum are often sparser and finer. Ventrally spiracles elongatedly comma-shaped, with a long narrow dorsal prolongation. Adanal plates short, broad, almost pear-shaped; accessory adanal plates short, broad.

Female (Figs 13(b), 14(d) to (f))

Capitulum much broader than long, length × breadth ranging from 0.52 mm × 0.72 mm to 0.73 mm × 0.84 mm; from lateral angles forwards almost triangular in outline. Basis capituli with lateral angles well over halfway back, with particularly long, divergent anterolateral margins; porose areas medium-sized, about twice their own diameter apart. Palps tapering even more than in the male to narrowly rounded apices. Scutum longer than broad, length × breadth ranging from 1.07 mm × 0.99 mm to 1.33 mm × 1.28 mm, slightly sinuous posteriorly; in engorging specimens the midgut caecae show through the alloscutum. Eyes almost flat, edged dorsally by one or two large setiferous punctations. Cervical pits deep and convergent; cervical

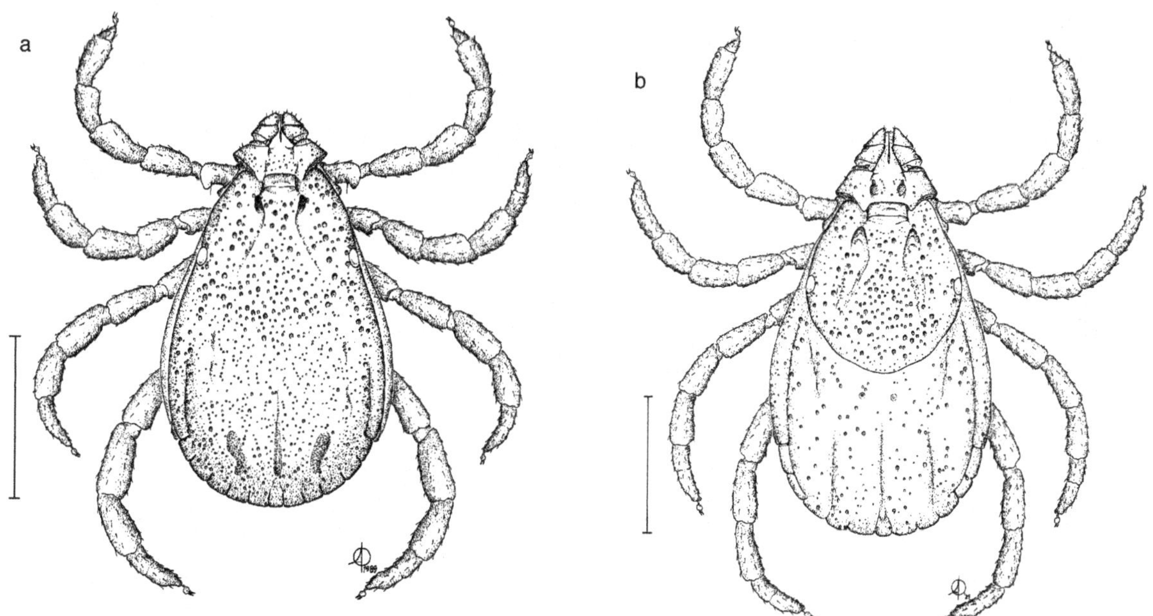

Figure 13. *Rhipicephalus arnoldi* [laboratory reared, progeny of ♀ collected from Smith's red rock rabbit (*Pronolagus rupestris*), Renosterberg, Eastern Cape Province, South Africa, on 24 August 1981, by Leonie Jordaan]. (a) Male, dorsal; (b) female, dorsal. Scale bars represent 1 mm. A. Olwage *del.*

Table 4. *Host records of* Rhipicephalus arnoldi

Hosts	Number of records
Wild animals	
Caracal (*Caracal caracal*)	2 (immatures)
Mountain zebra (*Equus zebra*)	1
Rock hyrax (*Procavia capensis*)	79 (immatures; only 1 with ♀♀)
Klipspringer (*Oreotragus oreotragus*)	1
Four-striped grass mouse (*Rhabdomys pumilio*)	1 (immatures)
'Field mouse'	1
Cape hare (*Lepus capensis*)	1
Scrub hare (*Lepus saxatilis*)	10 (including immatures)
'Wild hare'	2
Jameson's red rock rabbit (*Pronolagus randensis*)	12 (1 includes a nymph)
Smith's red rock rabbit (*Pronolagus rupestris*)	47 (including immatures)
Rock elephant shrew (*Elephantulus myurus*)	49 (immatures)

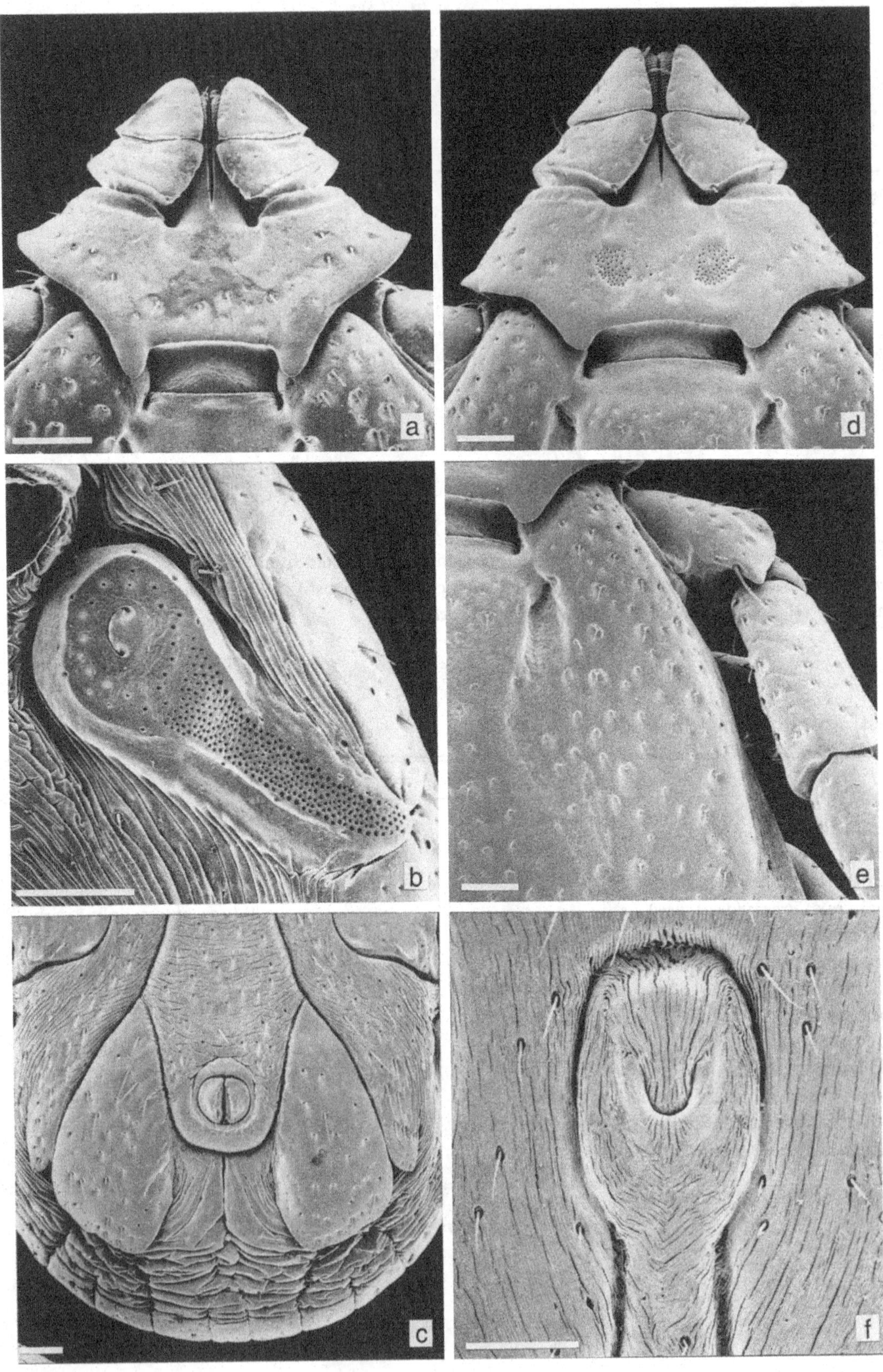
a
d
b
e
c
f

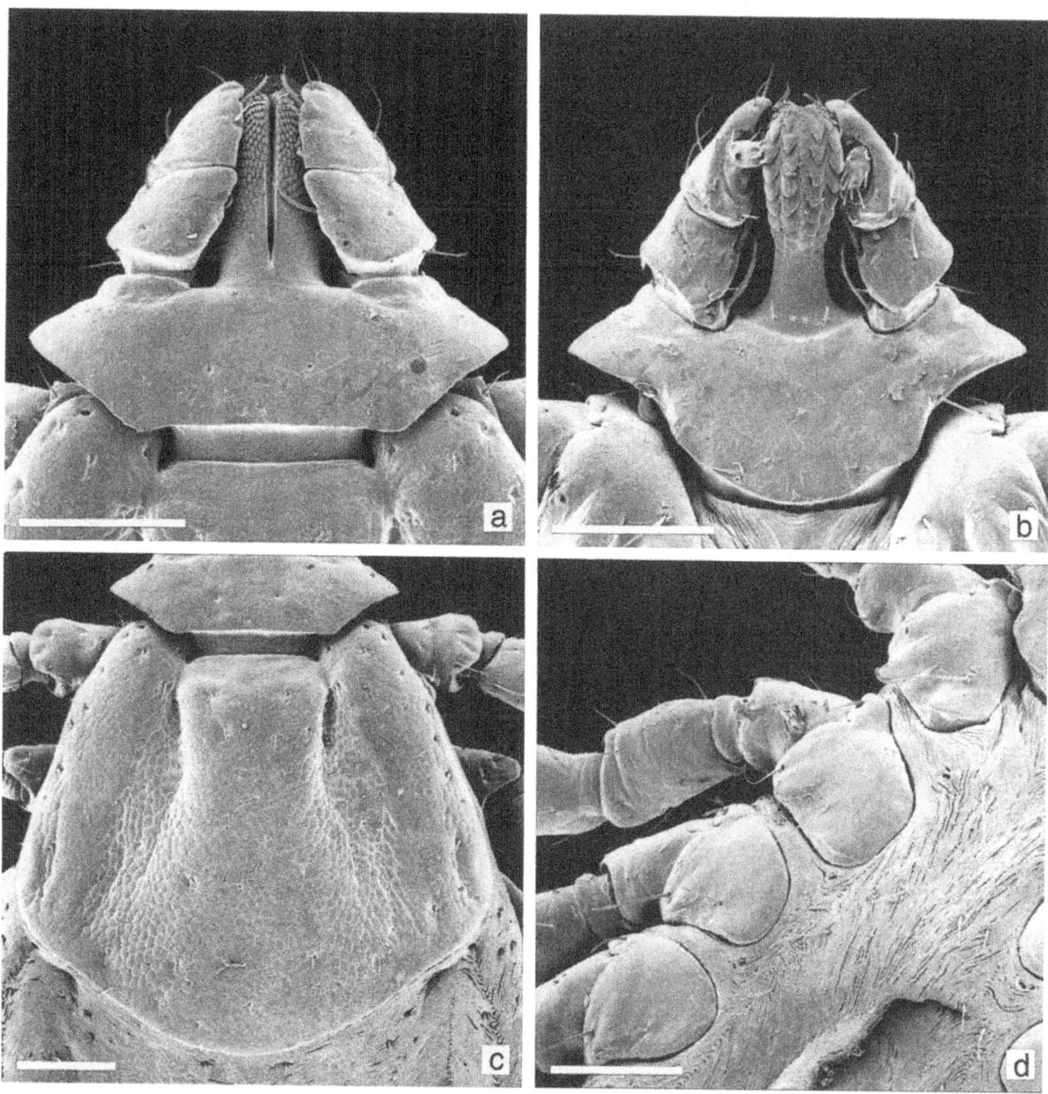

Figure 15 (*above*). *Rhipicephalus arnoldi* [laboratory reared, progeny of ♀ collected from Smith's red rock rabbit (*Pronolagus rupestris*), Renosterberg, Eastern Cape Province, South Africa, on 24 August 1981, by Leonie Jordaan]. Nymph: (a) capitulum, dorsal; (b) capitulum, ventral; (c) scutum; (d) coxae. Scale bars represent 0.10 mm. SEMs by J.F. Putterill.

Figure 14 (*opposite*). *Rhipicephalus arnoldi* [laboratory reared, progeny of ♀ collected from Smith's red rock rabbit (*Pronolagus rupestris*), Renosterberg, Eastern Cape Province, South Africa, on 24 August 1981, by Leonie Jordaan]. Male: (a) capitulum, dorsal; (b) spiracle; (c) adanal plates. Female: (d) capitulum, dorsal; (e) scapular area; (f) genital aperture. Scale bars represent 0.10 mm. SEMs by J.F. Putterill.

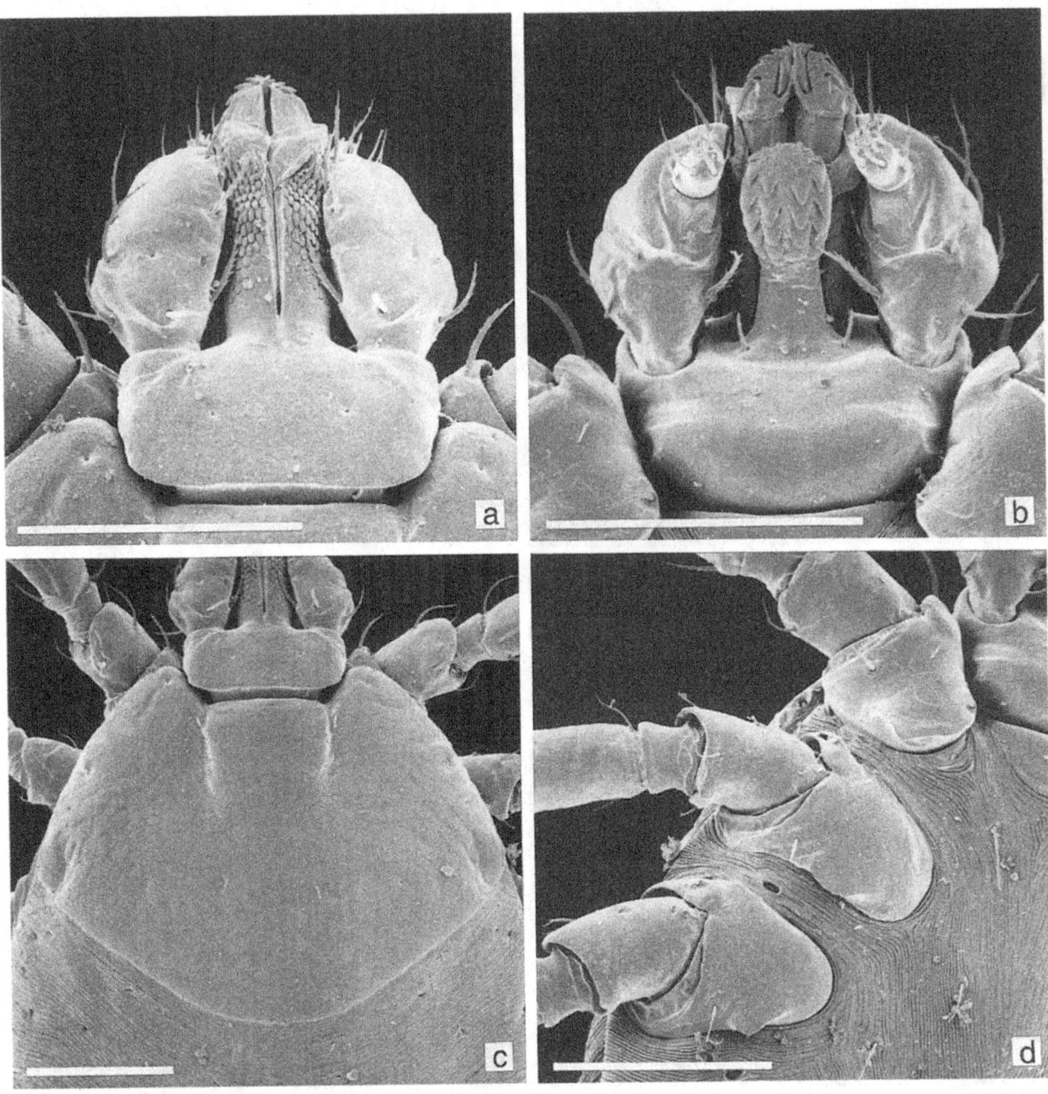

Figure 16. *Rhipicephalus arnoldi* [laboratory reared, progeny of ♀ collected from Smith's red rock rabbit (*Pronolagus rupestris*), Renosterberg, Eastern Cape Province, South Africa, on 24 August 1981, by Leonie Jordaan]. Larva: (a) capitulum, dorsal; (b) capitulum, ventral; (c) scutum; (d) coxae. Scale bars represent 0.10 mm. SEMs by J.F. Putterill.

fields long, tapering towards their posterior ends; external cervical margins marked by rows of large setiferous punctations. Large setiferous punctations scattered on the scapulae and medially on the scutum, where they are interspersed with medium-sized punctations; a few very fine punctations present anterior to the eyes. Ventrally genital aperture U-shaped.

Nymph (Fig. 15)

Capitulum much broader than long, length × breadth ranging from 0.22 mm × 0.33 mm to 0.27 mm × 0.37 mm. Basis capituli over three times as broad as long, with acute lateral angles in anterior half, overlapping the scapulae. Palps constricted proximally, then widening, tapering towards the rounded tips, inclined inwards. Scutum broader than long, length × breadth ranging from 0.49 mm × 0.54 mm to 0.50 mm × 0.56 mm; posterior margin broad, fairly deep. Eyes at widest point, over halfway back, mildly convex, delimited dorsally by slight depressions. Cervical pits deep and convergent ; cervical fields long, narrow, depressed, running parallel to anterolateral margins of scutum. Ventrally coxae I each with a long narrow external spur and short, broad internal spur, coxae II to IV each with mere indications of spurs.

Larva (Fig. 16)
Capitulum nearly as broad as long, length × breadth ranging from 0.143 mm × 0.141 mm to 0.165 mm × 0.156 mm. Basis capituli well over twice as broad as long, more or less rectangular in shape. Palps constricted proximally, then widening before they taper slightly to their tips; at their widest point extending laterally beyond the basis capituli. Scutum much broader than long, length × breadth ranging from 0.283 mm × 0.432 mm to 0.294 mm × 0.459 mm. Posterior margin broad, fairly deep. Eyes at widest point, about halfway back, slightly convex. Cervical grooves short, virtually parallel. Coxae each with an internal spur.

Notes on identification
The capituli of the adults of this small species, particularly of the female, are triangular in general appearance. Engorging females and nymphs are conspicuously pear-shaped.

Records of *R. arnoldi* from the Sudan (Hoogstraal, 1956) are now considered incorrect.

Hosts

A three-host species (J.B.W., unpublished data, 1981). Although scrub hares may be infested, the preferred hosts of the adults are Jameson's and Smith's red rock rabbits (Norval, 1985; Horak & Fourie, 1991). Rock hyraxes seldom carry adult ticks but, with rock elephant shrews and red rock rabbits, are the preferred hosts of the immature stages (Horak & Fourie, 1986; Fourie, Horak & Van den Heever, 1992) (Table 4). Infestations on other hosts should be regarded as accidental. There is no clear pattern of seasonal abundance for any of the life stages. This has been ascribed to the warmth and humidity provided by the preferred hosts while in their forms or refuges, thus creating ideal microhabitats for the year-round development of this tick (Horak *et al.*, 1991).

Zoogeography

Present indications are that *R. arnoldi* occurs only in South Africa and Zimbabwe (Map 5). In South Africa most collections have been made in the south-western Free State and in the Karoo regions of the Eastern and Western Cape Provinces. All records from Zimbabwe come from the Matopos region. The vegetation of these collection sites has been described as Highveld grassland and various types of undifferentiated woodland and of Karoo shrubland. However, the tick's distribution within any region is patchy, being restricted to the krantzes, rocky hillsides, boulder-strewn koppies and rocky ravines favoured by its preferred hosts. This association with a rocky habitat is emphasized by the fact that the common names of 6 of the 12 hosts listed in Table 4 include the words rock, klip (stone) or mountain.

Disease relationships

Unknown.

REFERENCES

Fourie, L.J., Horak, I.G. & Van den Heever, J.J. (1992). The relative host status of rock elephant shrews *Elephantulus myurus* and Namaqua rock mice *Aethomys namaquensis* for economically important ticks. *South African Journal of Zoology*, 27, 108–14.

Horak, I.G. & Fourie, L.J. (1986). Parasites of domestic and wild animals in South Africa. XIX.

Ixodid ticks and fleas on rock dassies (*Procavia capensis*) in the Mountain Zebra National Park. *Onderstepoort Journal of Veterinary Research*, **53**, 123–6.

Horak, I.G. & Fourie, L.J. (1991). Parasites of domestic and wild animals in South Africa. XXIX. Ixodid ticks on hares in the Cape Province and on hares and red rock rabbits in the Orange Free State. *Onderstepoort Journal of Veterinary Research*, **58**, 261–70.

Horak, I.G., Fourie, L.J., Novellie, P.A. & Williams, E.J. (1991). Parasites of domestic and wild animals in South Africa. XXVI. The mosaic of ixodid tick infestations on birds and mammals in the Mountain Zebra National Park. *Onderstepoort Journal of Veterinary Research*, **58**, 125–36.

Norval, R.A.I. (1985). The ticks of Zimbabwe. XII. The lesser known *Rhipicephalus* species. *Zimbabwe Veterinary Journal*, **16**, 37–43.

Theiler, G. & Zumpt, F. (1949). Description of new species. *Rhipicephalus* (s. str.) *arnoldi* Theiler and Zumpt, n. sp. In *Preliminary study to a revision of the genus* Rhipicephalus *Koch. Key to the adult ticks of the genus* Rhipicephalus *Koch and description of two new species,* author F. Zumpt, pp. 111–19. Published in: *Moçambique*, No. **60**, 57–123.

Also see the following Basic Reference (p. 13): Hoogstraal (1956).

RHIPICEPHALUS BEQUAERTI ZUMPT, 1949

Patronym in honour of Professor Joseph C. Bequaert, born in 1886. He was entomologist for the Belgian Sleeping Sickness Commission, Belgian Congo, 1910–1912, and in charge of exploration work, 1913. Later he became Alexander Agassiz Professor of Zoology, Museum of Comparative Zoology, Harvard University.

Diagnosis

A moderately large dark brown tick with a smooth lightly-punctate conscutum and scutum.

Male (Figs 17(a), 18(a) to (c))

Capitulum longer than broad, length × breadth ranging from 0.85 mm × 0.78 mm to 1.04 mm × 0.91 mm. Basis capituli with long, slightly-divergent lateral angles, not extending over the small, but definite, anterior processes of coxae I; posterior margin a concave arc between prominent cornua. Palps short, broad, flattened apically. Conscutum length × breadth ranging from 3.53 mm × 2.52 mm to 4.44 mm × 3.10 mm. In engorged males a single caudal process is seen. Eyes marginal, flat, may be edged dorsally with one or more punctations. Cervical fields inapparent, external cervical margins poorly defined with only a few shallow punctations. Marginal lines merely indicated by rows of shallow punctations or sometimes by punctations in a very shallow groove posteriorly near festoons. Posteromedian groove absent, posterolateral grooves absent or as shallow rounded depressions. Background punctations on conscutum fine with a few medium-sized punctations often appearing in a '*simus*' pattern. Without magnification the conscutum appears smooth and impunctate. Legs increase in size from I to IV. Ventrally spiracles narrow with apex of prolongation visible dorsally. Adanal plates roughly triangular with the internal margin scooped out around anus, and produced into large median cusps; accessory adanal plates elongate, pointed.

Female (Figs 17(b), 18(d) to (f))

Capitulum slightly longer than broad, length × breadth ranging from 0.81 mm × 0.80 mm to 1.01 mm × 0.95 mm. Basis capituli with moderately-curved lateral angles; porose areas large, nearly twice their own diameter apart. Palps short, broad, flattened apically. Scutum broader than long, length × breadth ranging from 1.51 mm × 1.62 mm to 1.87 mm × 2.05 mm. Eyes about halfway back, slightly bulging. Cervical fields broad, usually rugose anteriorly, almost reaching posterolateral scutal margin; internal cervical margins marked by a declination, external cervical margins marked by a declination anteriorly. Punctations small to medium-sized,

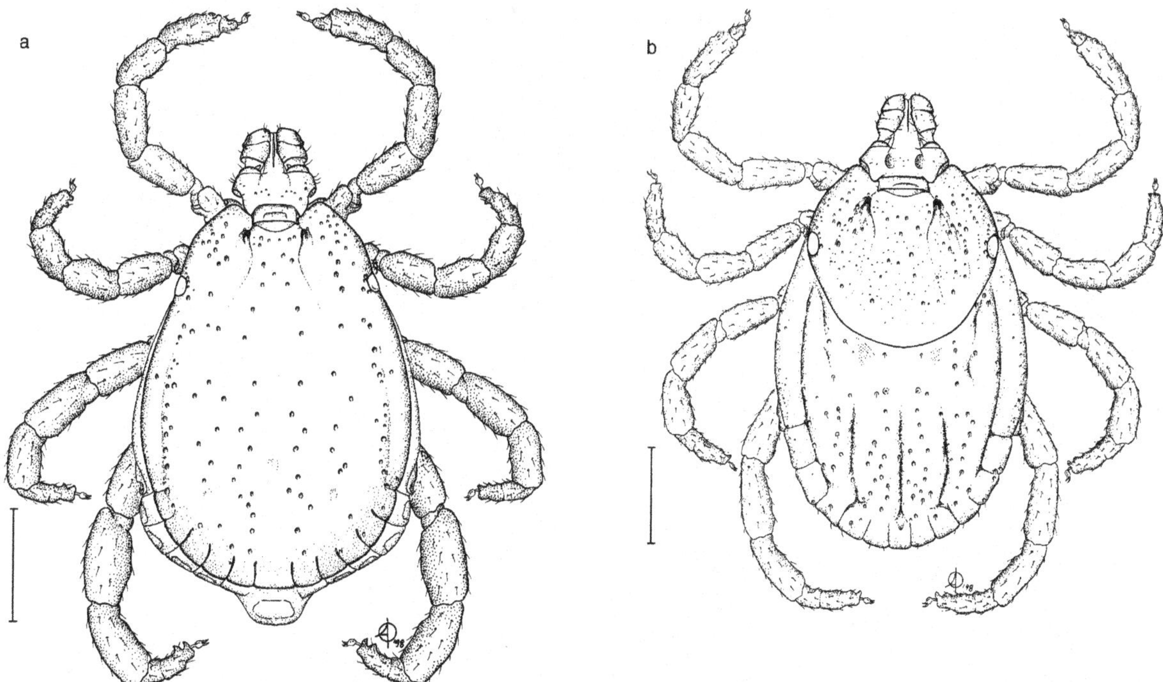

Figure 17. *Rhipicephalus bequaerti* [collected from African buffalo (*Syncerus caffer*), Icely's farm, Ol Arabel, Kenya on 29 July 1957 by S. F. Barnett.]. (a) Male, dorsal; (b) female, dorsal. Scale bars represent 1 mm. A. Olwage *del.*

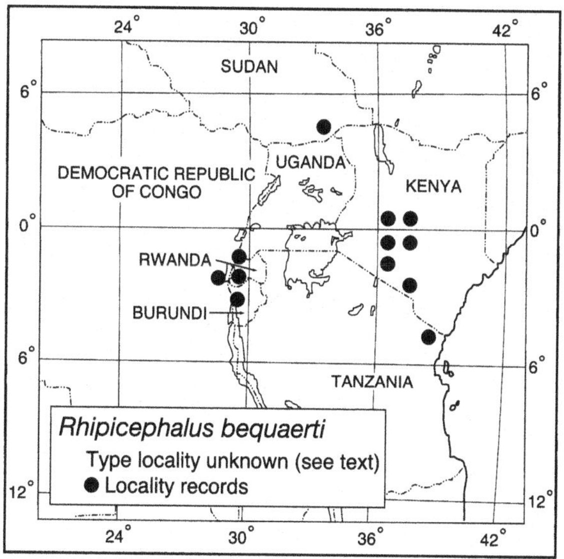

Map 6. *Rhipicephalus bequaerti*: distribution.

Table 5. *Host records of* Rhipicephalus bequaerti

Hosts	Number of records
Domestic animals	
Cattle	3
Goats	2
Wild animals	
Forest hog (*Hylochoerus meinertzhageni*)	1
Bushpig (*Potomochoerus larvatus*)	2
'Wild pig'	1
African buffalo (*Syncerus caffer*)	5
Humans	1

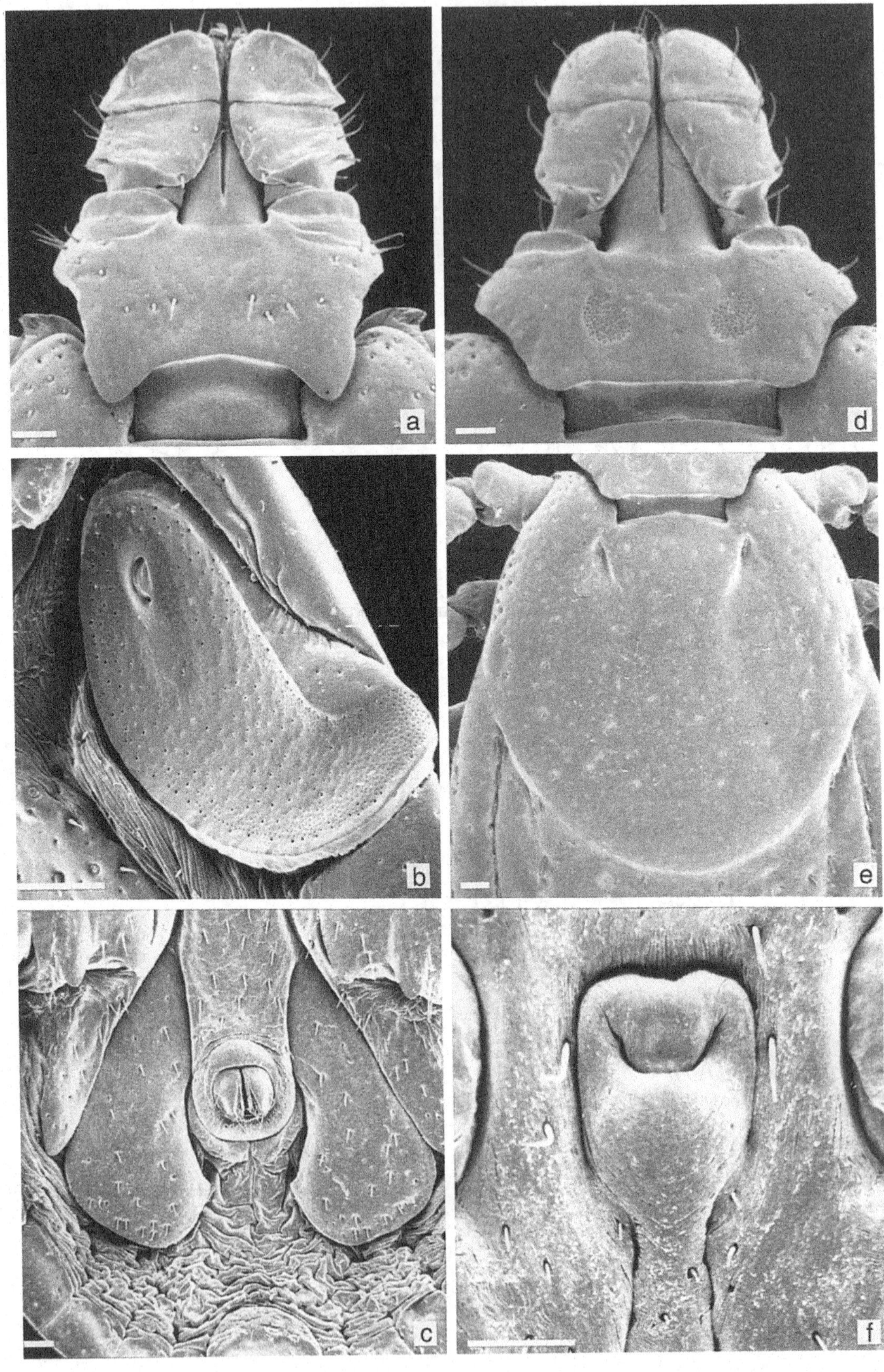
a
d
b
e
c
f

shallow, evenly distributed over the scutal surface. Ventrally genital aperture broadly U-shaped with apron bulging anteriorly, and posterior margin of the U also bulging.

Immature stages

Unknown.

Notes on identification

The type ♂ and ♀ of *R. bequaerti* were recorded from 'Lissenji', Central Africa. This has not been traced; Hoogstraal (1956) suggested that it might be a misspelling of Kisenyi (now called Gisenyi), Rwanda.

Matthysse & Colbo (1987) indicated that females of *R. bequaerti* may be confused with females of *R. appendiculatus* where the two species are sympatric. The bulging appearance of the genital aperture in the female of *R. bequaerti* helps to differentiate this species.

According to Morel (1980, p. 156) a female tick 'found free between the Amboni and Naremuru (*sic*) rivers (1800 m, Kenya)', listed as *R. simus planus* by Neumann (1913), proved on re-examination to be *R. bequaerti*.

Hosts

R. bequaerti has never been reared in the laboratory, but it is assumed to be a three-host species. It is a rare tick that has hardly ever been recorded in anything but very small numbers. Only a few specimens have been found on cattle and goats (Rousselot, 1951; Hoogstraal, 1956; Elbl & Anastos, 1966; Yeoman & Walker, 1967). Most collections have been obtained from African buffaloes, on which the ticks were attached on the head, ears and body, plus several collections from various wild suids (Elbl & Anastos, 1966; Walker, 1974). Elbl & Anastos (1966) also cited a large collection of 21 ♂♂, 3 ♀♀ from a human (Table 5).

Figure 18 (*opposite*). *Rhipicephalus bequaerti* [collected from African buffalo (*Syncerus caffer*), Icely's farm, Ol Arabel, Kenya, on 29 July 1957 by S.F. Barnett]. Male: (a) capitulum, dorsal; (b) spiracle; (c) adanal plates. Female: (d) capitulum, dorsal; (e) scutum; (f) genital aperture. Scale bars represent 0.10 mm. SEMs by J.F. Putterill.

Zoogeography

Present indications are that *R. bequaerti* has a discontinuous distribution from Nagichot, in the Didinga Mountains of south-eastern Sudan, southwards to northern Tanzania. Most records are from eastern Democratic Republic of Congo and central Kenya (Map 6). Matthysse & Colbo (1987) suggested that extensive collecting in the high altitude areas of Uganda might also reveal its presence there.

R. bequaerti has been found in high altitude forested areas from about 1800 m to 2500 m, with annual rainfalls ranging from *c.* 500 mm to *c.* 1500 mm.

Disease relationships

Unknown.

REFERENCES

Neumann, L.G. (1913). Ixodidae. In *Voyage de Ch. Alluaud et R. Jeannel en Afrique Orientale (1911–1912). Résultats Scientifiques. Arachnida*, II, 23–35. Paris: A. Schulz.

Rousselot, R. (1951). *Ixodes* de l'Afrique noire. *Bulletin de la Société de Pathologie Exotique*, **44**, 307–9.

Zumpt, F. (1949). Preliminary study to a revision of the genus *Rhipicephalus* Koch. Key to the adult ticks of the genus *Rhipicephalus* and description of two new species. *Moçambique*, **No. 60**, 57–123.

Also see the following Basic References (pp. 12–14): Elbl & Anastos (1966); Hoogstraal (1956); Matthysse & Colbo (1987); Morel (1980); Walker (1974); Yeoman & Walker (1967).

RHIPICEPHALUS BERGEONI MOREL & BALIS, 1976

This species was named after Dr P. Bergeon, who made a major contribution to our knowledge of Ethiopian ticks.

Diagnosis

A fairly large brown to reddish-brown tick.

Male (Figs 19(a), 20(a) to (c))
Capitulum slightly broader than long, length × breadth ranging from 0.56 mm × 0.57 mm to 0.63 mm × 0.64 mm. Basis capituli with short, broad lateral angles. Palps short, broad. Conscutum length × breadth ranging from 2.34 mm × 1.46 mm to 2.76 mm × 1.70 mm; anterior process of coxae I inconspicuous. In engorged specimens body wall expanded slightly laterally and posterolaterally and forming a short, blunt caudal process posteriorly. Eyes marginal, almost flat, edged dorsally by a few large punctations. Cervical fields inconspicuous. Marginal lines long, outlined by punctations. Posteromedian and posterolateral grooves well developed, their surfaces shagreened. A few large setiferous punctations scattered on the scapulae but largest and most conspicuous in irregular lines along the outer cervical margins; slightly smaller setiferous punctations scattered medially on the conscutum, interspersed with numerous finer punctations which become sparse to absent adjacent to the marginal lines. Legs increase slightly in size from I to IV. Ventrally spiracles short, broad, narrowing slightly only where they curve gently towards the dorsal surface. Adanal plates with their posteroexternal and posterointernal margins broadly rounded, their internal margins curving in towards each other posterior to the anus; accessory adanal plates short, pointed.

Female (Figs 19(b), 20(d) to (f))
Capitulum broader than long, length × breadth ranging from 0.64 mm × 0.71 mm to 0.79 mm × 0.89 mm. Basis capituli with broad lateral angles; porose areas large, slightly more than their own diameter apart. Palps broad, somewhat flattened apically. Scutum usually longer than broad, length × breadth ranging from 1.28 mm × 1.13 mm to 1.70 mm × 1.64 mm, posterior margin sinuous. Eyes about halfway back, almost flat, edged dorsally by a few large punctations. Cervical fields broad, slightly depressed, their outer margins indicated by a few large setiferous punctations. A few setiferous punctations also present on the scapulae and scattered medially on the scutum, interspersed with numerous fine punctations. Ventrally genital aperture with sides of opening converging to the broadly-rounded base, the genital apron depressed.

Immature stages
Unknown.

Notes on identification
According to Morel (1980), some specimens of *R. bergeoni* have in the past been confused with *R. appendiculatus*, *R. supertritus* and *R. turanicus*. He considered that this species belongs to the *R. sanguineus* group, but Pegram *et al.* (1987) regarded its position in this group as being somewhat equivocal. We feel that it will be necessary to study its immature stages in order to determine its real relationships within the genus.

Hosts

Life cycle unknown, but Morel (1980) suggested that it is probably a three-host species whose immature stages feed on rodents. The commonest hosts recorded for this species to date are cattle, and to a lesser extent sheep. Few collections have been made from other domestic animals, and even fewer from wild animals (Table 6). On cattle its predilection site is the ears, from which over 94% of the specimens in one collection were taken.

It is often associated with *Amblyomma variegatum* in the field (Pegram *et al.*, 1987; De Castro, 1994).

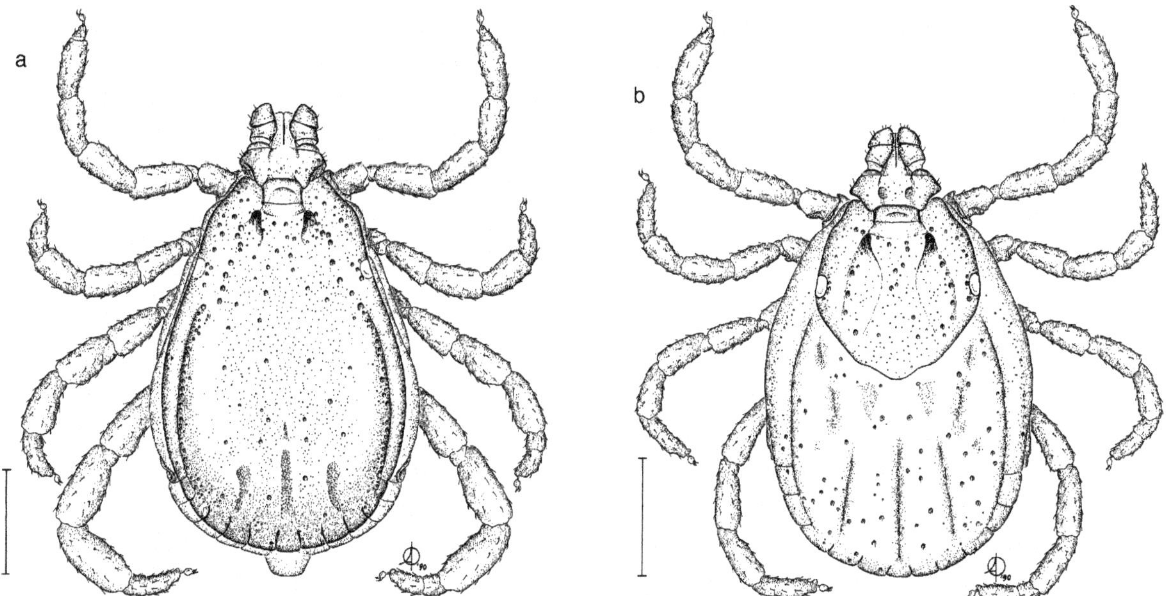

Figure 19. *Rhipicephalus bergeoni* (from bovine, Dodola, Bale, Ethiopia, collected in November 1971 and donated by P.C. Morel). (a) Male, dorsal; (b) female, dorsal. Scale bars represent 1 mm. A. Olwage *del.*

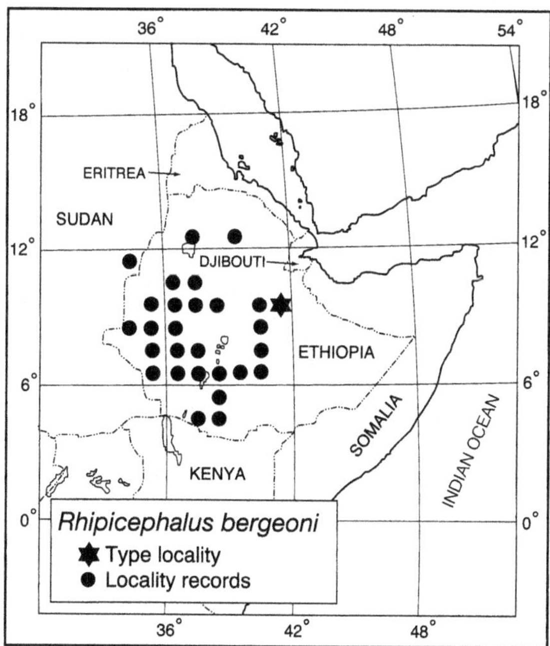

Map 7. *Rhipicephalus bergeoni*: distribution.

Table 6. *Host records of* Rhipicephalus bergeoni

Hosts	Number of records
Domestic animals	
Cattle	104
Sheep	33
Goats	6
Horses	5
Pigs	1
Wild animals	
Striped hyaena (*Hyaena hyaena*)	2
Mountain nyala (*Tragelaphus buxtoni*)	3
Bushbuck (*Tragelaphus scriptus*)	1
Greater kudu (*Tragelaphus strepsiceros*)	1

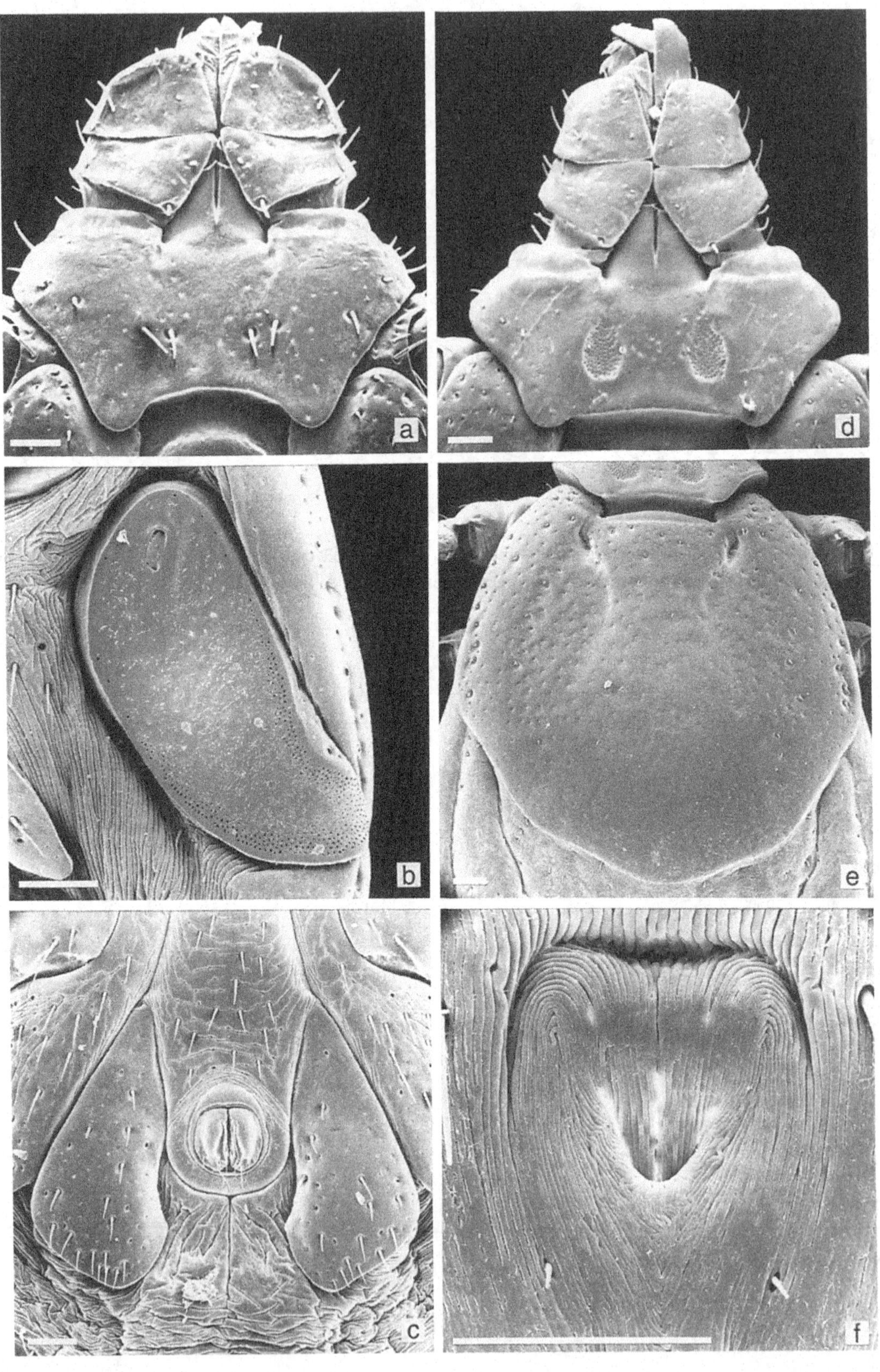
a
b
c
d
e
f

Of the 27 Ethiopian collections recorded prior to 1987, 26 were obtained between June and September, i.e. during the rainy season.

Zoogeography

Apart from one large collection from cattle at Damazein, eastern Sudan, *R. bergeoni* has been recorded only in Ethiopia (Map 7). There it occurs in the highlands and subhighlands at altitudes of up to 3000 m, with annual rainfalls ranging from about 1200 mm to over 2600 mm, mainly in highland forest communities. Occasionally it has been collected in montane grasslands near these forests (Pegram *et al.*, 1987; De Castro, 1994).

Disease relationships

Unknown.

Figure 20 (*opposite*). *Rhipicephalus bergeoni* [RML 105792 ((a), (d) and (e)), from bovine, Hubeta, Harrar, Ethiopia, collected on 26 June 1974 by P.C. Morel; RML 107629 ((b), (c) and (f)) also from bovine, Hubeta, collected on 24 June 1974 by P.C. Morel]. Male: (a) capitulum, dorsal; (b) spiracle; (c) adanal plates. Female: (d) capitulum, dorsal; (e) scutum; (f) genital aperture. Scale bars represent 0.10 mm. SEMs by Pat Hill ((a) and (d)) and M.D. Corwin ((b), (c), (e) and (f)).

REFERENCES

De Castro, J.J. (1994). *Tick survey – Ethiopia. A survey of the tick species in western Ethiopia, including previous findings and recommendations for further tick surveys in Ethiopia.* Rome: Food and Agriculture Organization of the United Nations, AG:DP/ETH/83/023, Technical Report, 83 pp.

Morel, P.C. & Balis, J. (1976). Description de *Rhipicephalus bergeoni* n. sp. (Acariens, Ixodida) des montagnes d'Ethiopie. *Revue d'Élevage et de Médecine Vétérinaire des Pays Tropicaux*, **29** (nouvelle série), 337–40.

Pegram, R.G., Keirans, J.E., Clifford, C.M. & Walker, J.B. (1987). Clarification of the *Rhipicephalus sanguineus* group (Acari, Ixodoidea, Ixodidae). II. *R. sanguineus* (Latreille, 1806) and related species. *Systematic Parasitology*, **10**, 27–44.

Also see the following Basic References (pp. 12–14): Morel, 1980; Pegram, 1979.

RHIPICEPHALUS BOUETI MOREL, 1957

This species was named in memory of Dr Bouet, who collected the type specimens in 1909.

Diagnosis

A small reddish-brown tick.

Male (Figs 21(a), 22(a) to (c))
Capitulum broader than long, length × breadth of the two specimens measured 0.51 mm × 0.59 mm and 0.57 mm × 0.60 mm respectively. Basis capituli with short acute lateral angles in the anterior third of its length. Palps short, broad, somewhat flattened apically. Conscutum length × breadth 2.60 mm × 1.70 mm and 2.66 mm × 1.75 mm respectively; anterior process on coxae I inconspicuous. Body wall of the specimen illustrated in the original description expanded posterolaterally and posteriorly with a single short bluntly rounded caudal process. Eyes almost flat, delimited dorsally by a few punctations. Cervical pits comma-shaped. Marginal lines short and shallow. Posteromedian groove relatively long and narrow, posterolateral grooves shorter and broader. Medium-sized setiferous punctations in two rows rather vaguely indicating the position of the external cervical margins and in the marginal lines, with another four rows in the '*simus*' pattern medially on the conscutum, all interspersed with numerous small punctations. Ventrally spiracles elongate with a long, gently-tapering dorsal prolongation that curves slightly only near its end. Adanal plates curved, almost banana-shaped, tapering to a point anteriorly and narrowly rounded posteriorly; accessory adanal plates conspicuous, strongly sclerotized, pointed.

Female (Figs 21(b), 22(d) to (f))
Capitulum broader than long, length × breadth of the two specimens measured 0.67 mm × 0.78 mm and 0.79 mm × 0.87 mm respectively. Basis capituli with acute lateral angles at about mid-length; porose areas medium sized, subcircular, about twice their own diameter apart. Palps with article III almost wedge-shaped, their apices narrowly rounded. Scutum longer than broad, length × breadth of the two specimens measured 1.34 mm × 1.27 mm and 1.57 mm × 1.37 mm respectively; posterior margin a deep slightly sinuous curve. Eyes at about mid-length, almost flat, delimited dorsally by a few punctations. Cervical pits convergent, continuous with short shallow divergent internal cervical margins. A few medium-sized setiferous punctations present on the scapulae, along the external margins of the cervical fields and medially on the scutum, interspersed with numerous fine punctations. Ventrally genital aperture a wide U, with its apron apparently not quite covering the opening.

Immature stages
Unknown.

Notes on identification
These descriptions are based on adults kindly donated to us by the late Dr P.C. Morel.

Hoogstraal (1954) recorded *R. ziemanni* 1 male, 1 female '. . . from behind ears of hyrax, Tinta-Atola, Mamfe, Cameroons . . .' but Morel (1957) suggested that these specimens might in fact be *R. boueti*.

Hosts

Life cycle unknown. Only two records of this species are known to us, both from the rock hyrax, *Procavia capensis* (syn. *Procavia latastei*, *P. ruficeps*). It is probably a specific parasite of these animals (Morel & Mouchet, 1965).

Zoogeography

Both records of *R. boueti* are from West Africa, the types from Agouagou (= Agouagon), Benin (formerly Dahomey), and the subsequent collection from Mboutwa (Maroua), Cameroon (Map 8). As the rock hyrax occurs in suitable habitats in a broad belt right across West Africa (Skinner & Smithers, 1990; see also p. 145, under

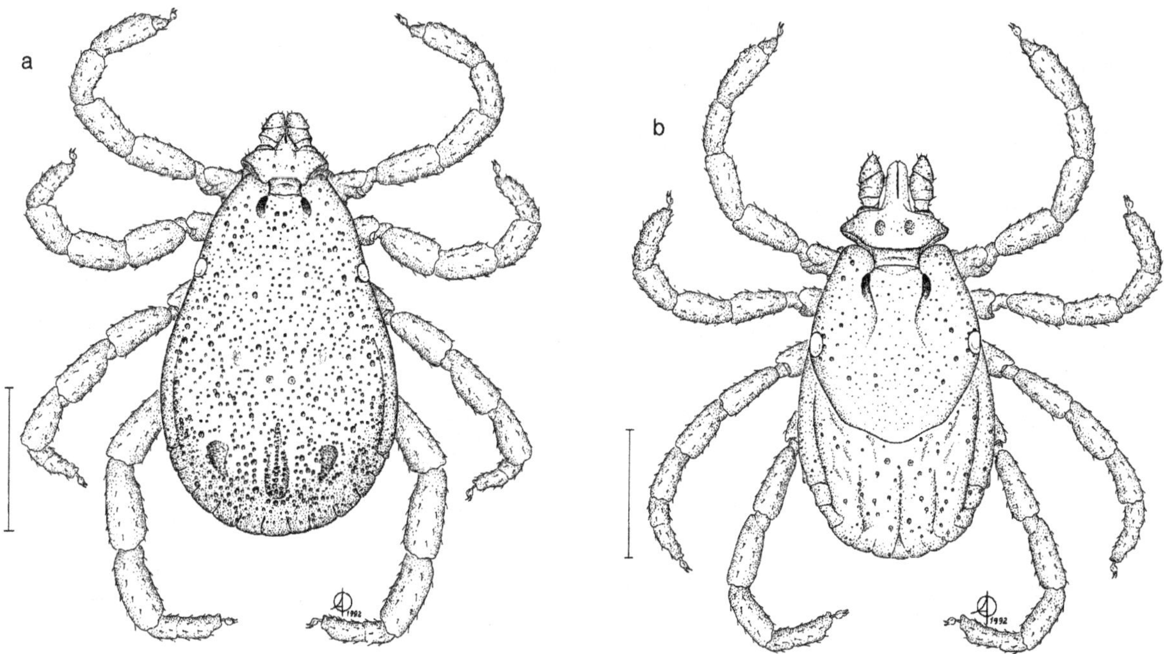

Figure 21. *Rhipicephalus boueti* [from rock hyrax (*Procavia capensis*), Maroua, Cameroon, collected in 1959 and donated by P.C. Morel]. (a) Male, dorsal; (b) female, dorsal. Scale bars represent 1 mm. A. Olwage *del.*

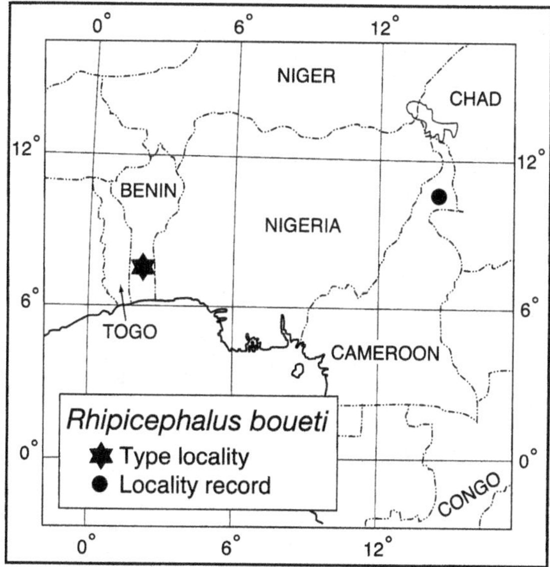

Map 8. *Rhipicephalus boueti*: distribution.

R. distinctus) this tick is probably also more widely distributed than these two records suggest.

Disease relationships

Unknown.

REFERENCES

Hoogstraal, H. (1954). Noteworthy African tick records in the British Museum (Natural History) collections. *Proceedings of the Entomological Society of Washington*, **56**, 273–9.

Morel, P.C. (1957). *Rhipicephalus boueti* n. sp. (Acarina, Ixodidae) parasites des damans du Dahomey. *Bulletin de la Société de Pathologie Exotique*, **50**, 696–700.

Morel, P.C. & Mouchet, J. (1965). Les tiques du Cameroun (Ixodidae et Argasidae) (2e note). *Annales de Parasitologie Humaine et Comparée*, **40**, 477–96.

Also see the following Basic Reference (p. 14): Skinner & Smithers (1990).

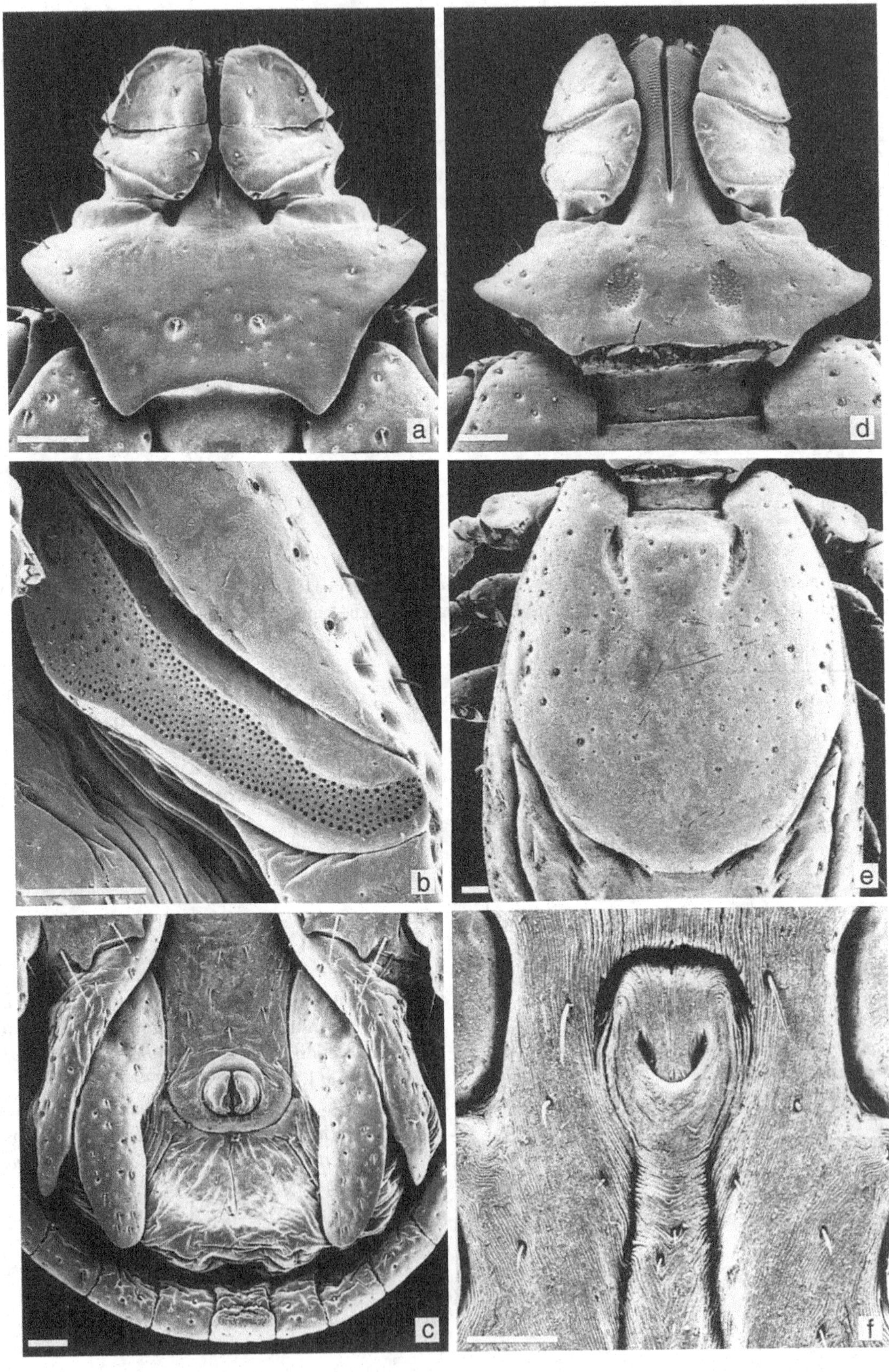
a
b
c
d
e
f

Figure 22 (*opposite*). *Rhipicephalus boueti* [from rock hyrax (*Procavia capensis*), Maroua, Cameroon, collected in 1959 and donated by P.C. Morel]. Male: (a) capitulum, dorsal; (b) spiracle; (c) adanal plates. Female: (d) capitulum, dorsal; (e) scutum; (f) genital aperture. Scale bars represent 0.10 mm. SEMs by J.F. Putterill.

RHIPICEPHALUS CAMICASI MOREL, MOUCHET & RODHAIN, 1976

This species was named after Dr J.-L. Camicas in recognition of his contributions to knowledge of the African *Haemaphysalis* species in the *H. leachi* group.

Diagnosis

A moderate-sized reddish-brown tick.

Male (Figs 23(a), 24(a) to (c))
Capitulum broader than long, length × breadth ranging from 0.43 mm × 0.48 mm to 0.52 mm × 0.57 mm. Basis capituli with acute lateral angles at about mid-length. Palps gently rounded apically. Conscutum length × breadth ranging from 1.72 mm × 1.07 mm to 2.20 mm × 1.31 mm; anterior process of coxae I bluntly rounded, inconspicuous. In engorged males body wall expanded laterally and posterolaterally, with a short bluntly rounded posteromedian protrusion. Eyes almost flat, edged dorsally by a few large setiferous punctations. Cervical pits comma-shaped, convergent; external cervical margins delimited by rows of large setiferous punctations. Marginal lines long but not always well defined. Posterior grooves broad, shallow. A few medium-sized setiferous punctations present on the scapulae, those medially on the conscutum larger and more distinct; interstitial punctations variable, usually light, becoming smaller and sparser laterally adjacent to the marginal lines. Ventrally spiracles relatively long and narrow, with the dorsal prolongation curving slightly just at its end. Adanal plates variable but in general also fairly narrow, with their internal margins scooped out posterior to the anus; accessory adanal plates small, pointed.

Female (Figs 23(b), 24(d) to (f))
Capitulum broader than long, length × breadth ranging from 0.54 mm × 0.64 mm to 0.68 mm × 0.76 mm. Basis capituli with acute lateral angles at about mid-length; porose areas quite small, about twice their own diameter apart. Palps gently rounded apically. Scutum slightly longer than broad, length × breadth ranging from 1.10 mm × 1.09 mm to 1.42 mm × 1.35 mm; posterior margin sinuous. Eyes almost flat, edged dorsally by a few large setiferous punctations. Cervical pits comma-shaped; cervical fields fairly long and narrow, their outer margins delimited by rather irregular rows of large setiferous punctations. A few medium-sized setiferous punctations scattered on the scapulae and others, somewhat larger, medially on the scutum, interspersed with numerous fine interstitial punctations. Ventrally genital aperture narrowly U-shaped, its internal flaps usually pigmented and parallel.

Nymph (Fig. 25)
Capitulum much broader than long, length × breadth ranging from 0.18 mm × 0.29 mm to 0.22 mm × 0.32 mm. Basis capituli about four times as broad as long, with sharp tapering lateral angles extending over the scapulae; ventrally with short spurs on the posterior margin. Palps tapering slightly to rounded apices, inclined inwards. Scutum broader than long, length × breadth ranging from 0.43 mm × 0.47 mm to 0.48 mm × 0.55 mm; posterior margin a smooth shallow curve. Eyes at widest point, well over halfway back, long and narrow, delimited dorsally by slight depressions. Cervical pits convergent; cervical fields long, narrow, slightly depressed, inconspicuous. Ventrally coxae I each with a relatively large external spur and a much smaller broader internal spur; coxae II and III with mere indications of a spur but none on IV.

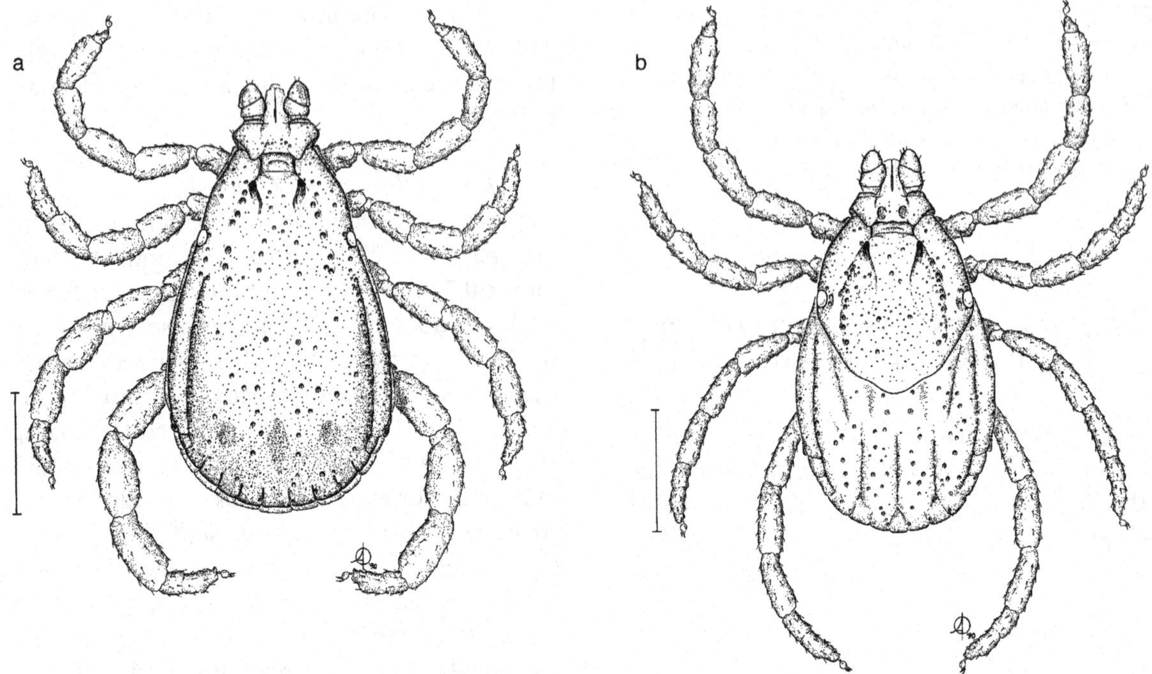

Figure 23. *Rhipicephalus camicasi* (Onderstepoort Tick Collection 3037v, collected from domestic sheep, Mersa Matruh, Western Desert, Egypt, on 9 August 1950 by H. Hoogstraal). (a) Male, dorsal; (b) female, dorsal. Scale bars represent 1 mm. A. Olwage *del.*

Table 7. *Host records of* Rhipicephalus camicasi

Hosts	Number of records
Domestic animals	
Cattle	13
Sheep	13
Goats	12
Camels	8
Donkeys	1
Wild animals	
Bat-eared fox (*Otocyon megalotis*)	1
Cheetah (*Acinonyx jubatus*)	1
Grevy's zebra (*Equus grevyi*)	2
'Zebra' (*Equus* sp.)	1
Aardvark (*Orycteropus afer*)	1
Warthog (*Phacochoerus africanus*)	1
Gerenuk (*Litocranius walleri*)	1
Eland (*Taurotragus oryx*)	2
Cape hare (*Lepus capensis*)	3

Figure 24 (*opposite*). *Rhipicephalus camicasi* (R.G. Pegram, series L66, laboratory reared, original ♀ collected from a donkey, Egypt). Male: (a) capitulum, dorsal; (b) spiracle; (c) adanal plates. Female: (d) capitulum, dorsal; (e) scutum; (f) genital aperture. Scale bars represent 0.10 mm. SEMs by M.D. Corwin. (Figs (b), (c), (e) & (f) from Pegram *et al.*, 1987, figs 22–24 & 26, with kind permission from Kluwer Academic Publishers.)

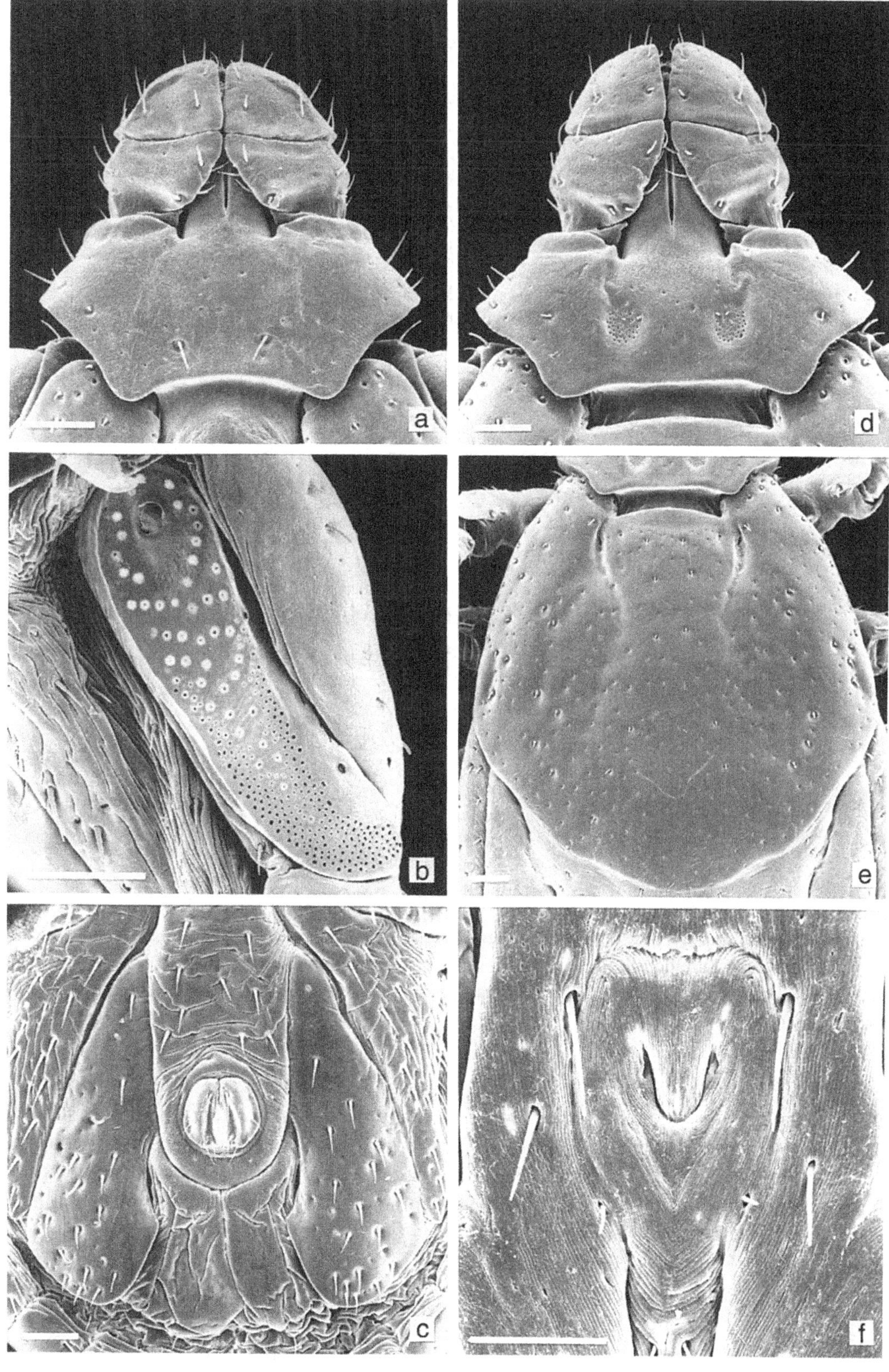
a
d
b
e
c
f

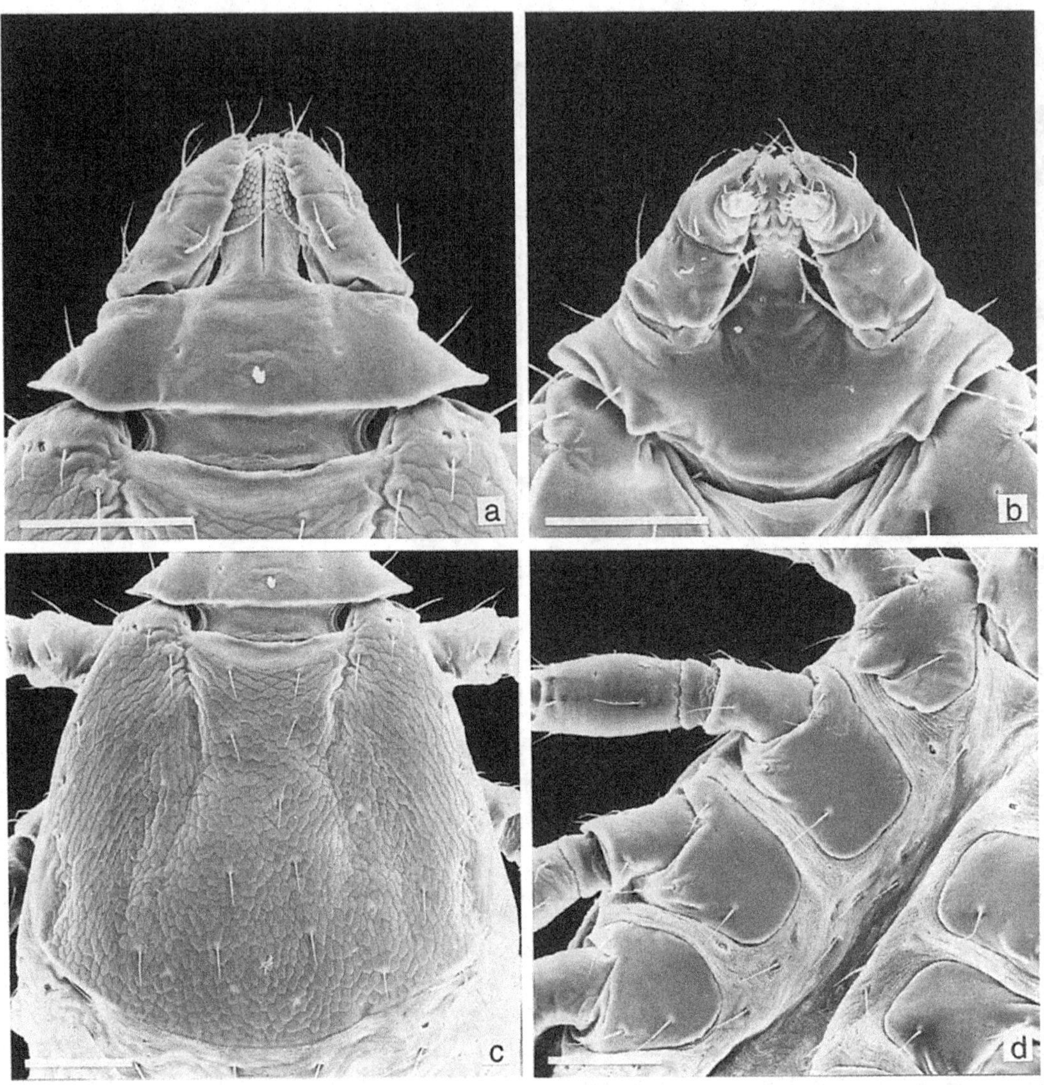

Figure 25. *Rhipicephalus camicasi* (R.G. Pegram, series L66, laboratory reared, original ♀ collected from a donkey, Egypt). Nymph: (a) capitulum, dorsal; (b) capitulum, ventral; (c) scutum; (d) coxae. Scale bars represent 0.10 mm. SEMs by M.D. Corwin. (Figs (a), (c) & (d) from Pegram *et al.*, 1987, figs 18–20, with kind permission from Kluwer Academic Publishers.)

Larva (Fig. 26)

Capitulum much broader than long, length × breadth ranging from 0.105 mm × 0.130 mm to 0.111 mm × 0.137 mm. Basis capituli about three times as broad as long with short blunt lateral angles. Palps broad, their external margins slightly convex, inclined inwards. Scutum much broader than long, length × breadth ranging from 0.204 mm × 0.330 mm to 0.223 mm × 0.362 mm; posterior margin a smooth shallow curve. Eyes at widest part of scutum, almost flat. Cervical grooves slightly convergent. Ventrally coxae I each with a broad salience on its posterior border; coxae II with a mere indication of a spur; coxae III lacks a spur.

Notes on identification

R. camicasi belongs to the *R. sanguineus* group (Morel *et al.*, 1976; Pegram *et al.*, 1987). Some records of this species were probably listed under

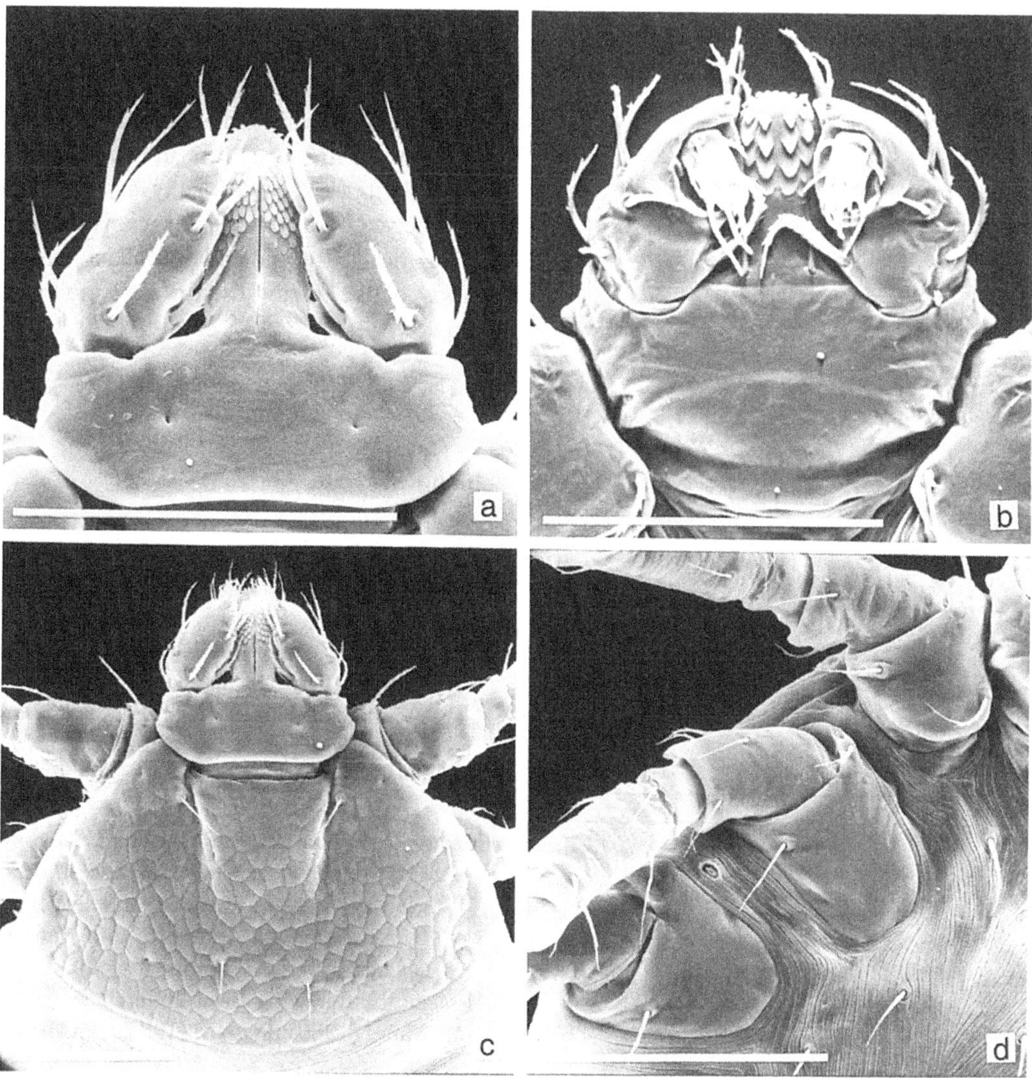

Figure 26. *Rhipicephalus camicasi* (R.G. Pegram, series L66, laboratory reared, original ♀ collected from a donkey, Egypt). Larva: (a) capitulum, dorsal; (b) capitulum, ventral; (c) scutum; (d) coxae. Scale bars represent 0.10 mm. SEMs by M.D. Corwin. (Figs (a), (c) & (d) from Pegram *et al.*, 1987, figs 15–17, with kind permission from Kluwer Academic Publishers.)

R. sanguineus sensu lato by Theiler (1962).

One feature that helps to distinguish *R. camicasi* from *R. sanguineus* is the structure of its mounted female genital aperture (see p. 387, Fig. 178).

Hosts

A three-host species (Pegram, 1984). Up to now domestic animals, particularly cattle, sheep, goats and camels, are the most commonly recorded hosts (Pegram *et al.*, 1987) (Table 7). In the Yemen Arab Republic it is the commonest rhipicephalid on domesticated animals (Pegram *et al.*, 1987). *Rhipicephalus camicasi* has, however, also been collected from a number of wild animals, of which zebras and Cape hares are the most numerous (Morel *et al.*, 1976; Morel, 1980; Pegram *et al.*, 1987). The hosts of the immature stages are unknown but are probably 'mouse-like

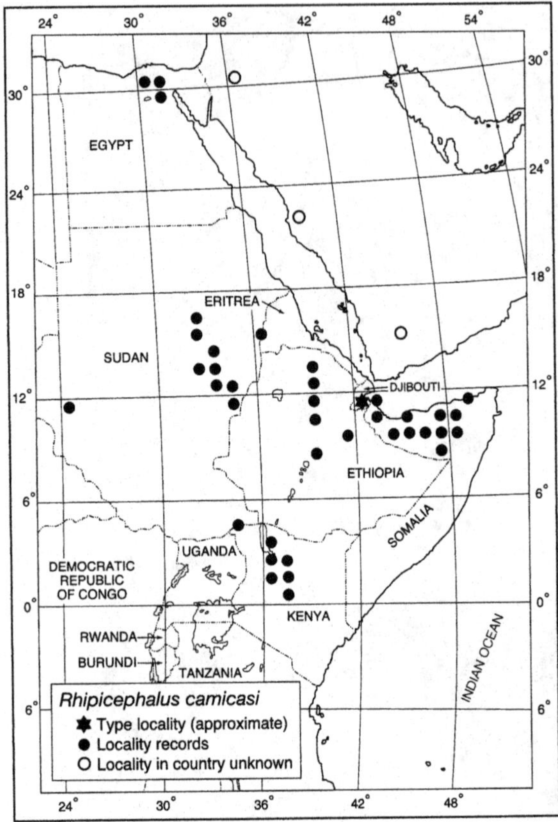

Map 9. *Rhipicephalus camicasi*: distribution.

rodents' (Morel, 1980). In both Ethiopia and Somalia *R. camicasi* appears to be most active during the dry season.

Zoogeography

Rhipicephalus camicasi occurs in the north-eastern corner of Africa. Its range extends from northern Kenya through Ethiopia to northern Somalia, Djibouti, western Sudan and northern Egypt (Map 9). Outside Africa its presence has been confirmed in the Yemen Arab Republic, Saudi Arabia, Jordan and Lebanon (Pegram *et al.*, 1987). It also occurs in Oman.

It is found in arid and semi-arid habitats generally receiving less than 250 mm of rainfall annually. The vegetation types of the regions in Africa in which it is present have been described as Somalia–Masai *Acacia–Commiphora* deciduous bushland and thicket, Sahel *Acacia* wooded grassland and deciduous bushland, and semi-desert grassland and shrubland. In Egypt in the vicinity of Cairo it occurs in stone and gravel desert regions, known locally as hamadas and regs, incised by dry valleys or wadis.

In both Ethiopia and Somalia it is apparently most active during the dry season (Pegram *et al.*, 1987).

Disease relationships

Unknown.

REFERENCES

Morel, P.C., Mouchet, J. & Rodhain, F. (1976). Description de *Rhipicephalus camicasi* n.sp. (Acariens, Ixodida) des steppes subdésertiques de la plaine afar. *Revue d'Élevage et de Médecine Vétérinaire des Pays Tropicaux*, **29** (nouvelle série), 337–40.

Pegram, R.G., Keirans, J.E., Clifford, C.M. & Walker, J.B. (1987). Clarification of the *Rhipicephalus sanguineus* group (Acari, Ixodoidea, Ixodidae). II. *R. sanguineus* (Latreille, 1806) and related species. *Systematic Parasitology*, **10**, 27–44.

Also see the following Basic References (p. 12–14): Morel (1980); Pegram (1984); Theiler (1962).

RHIPICEPHALUS CAPENSIS KOCH, 1844

The specific name *capensis* is derived from 'Cape Colony', South Africa, the origin of the type ♂, plus the Latin adjectival suffix *-ensis*, meaning 'belonging to'.

Diagnosis

A large dark brown extremely heavily punctate tick.

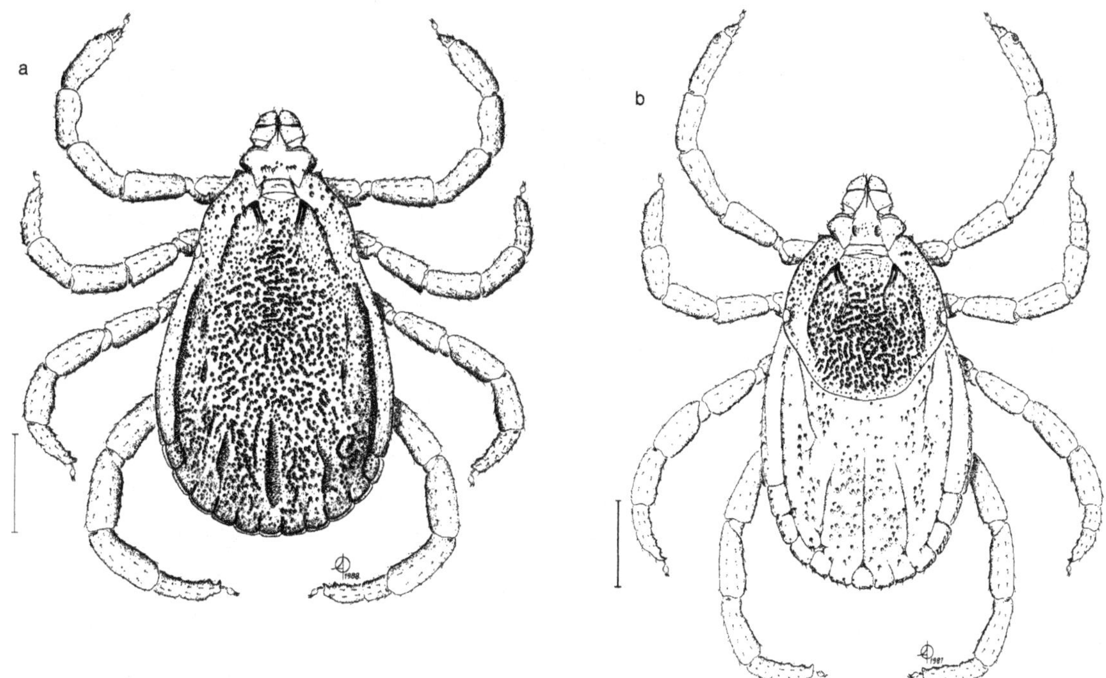

Figure 27. *Rhipicephalus capensis* [Onderstepoort Tick Collection 2826, RML 65719, laboratory reared, progeny of ♀ collected on 'Aties' (farm), Klawer, Gifberge, Western Cape Province, South Africa, probably in the mid-1940s]. (a) Male, dorsal; (b) female, dorsal. Scale bars represent 1 mm. A. Olwage *del.*

Table 8. *Host records of* Rhipicephalus capensis

Hosts	Number of records
Domestic animals	
Cattle	2
Horses	3
Wild animals	
Cape fox (*Vulpes chama*)	1
Mountain zebra (*Equus zebra*)	1
Bontebok (*Damaliscus pygargus dorcas*)	1
Eland (*Taurotragus oryx*)	3
Gemsbok (*Oryx gazella*)	2
Otomys sp.	2 (nymphs)

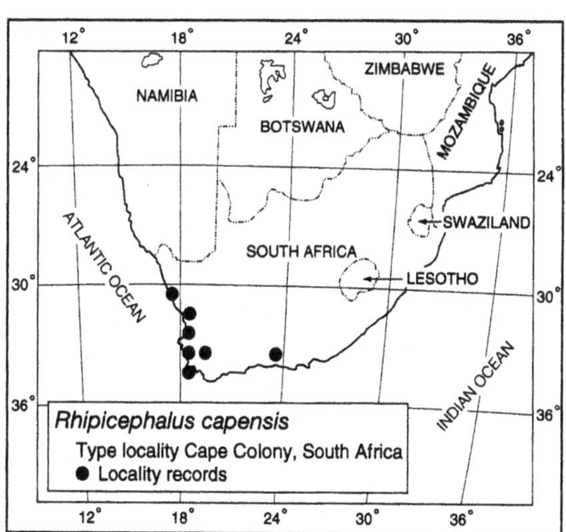

Map 10. *Rhipicephalus capensis*: distribution.

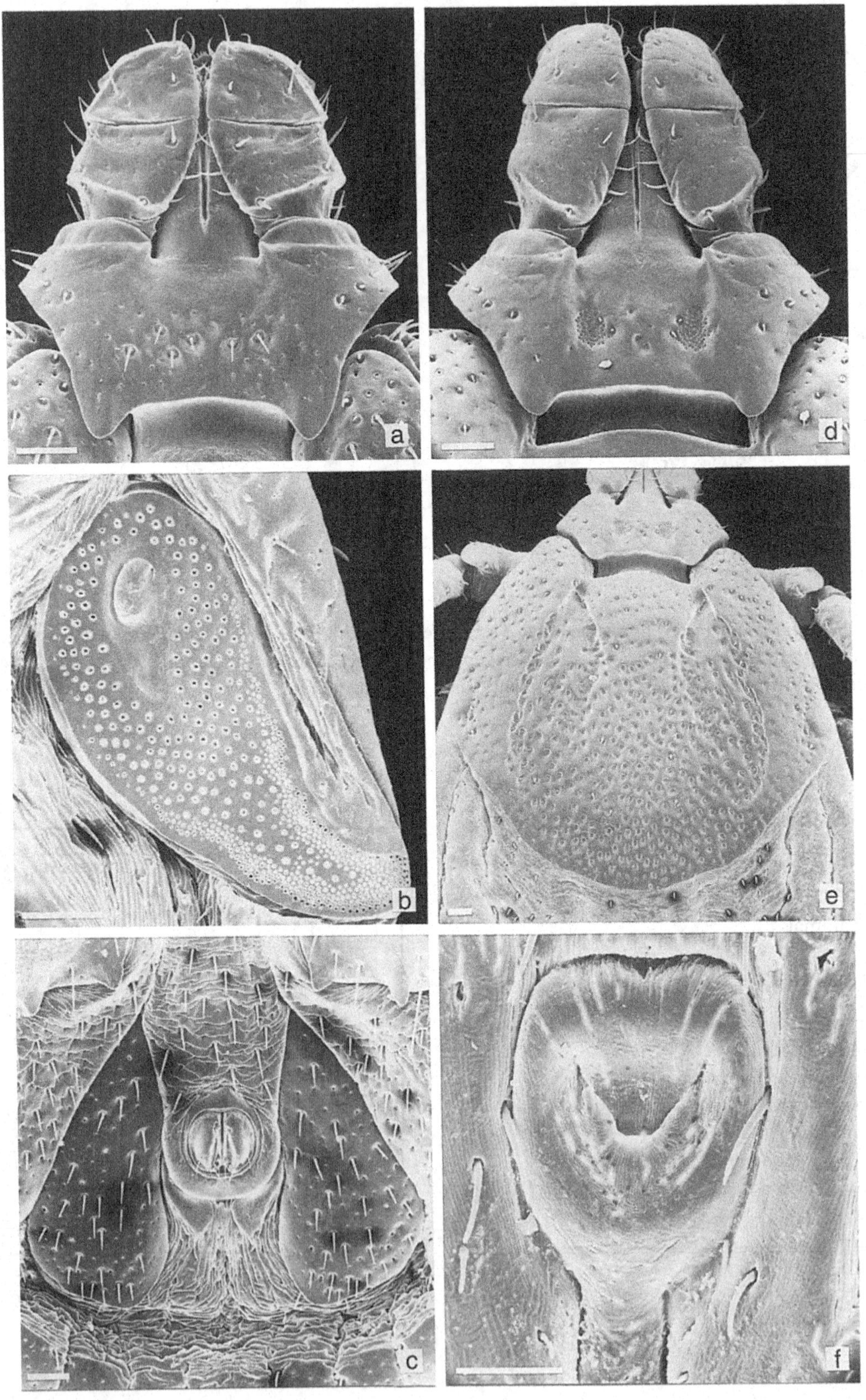
a
d
b
e
c
f

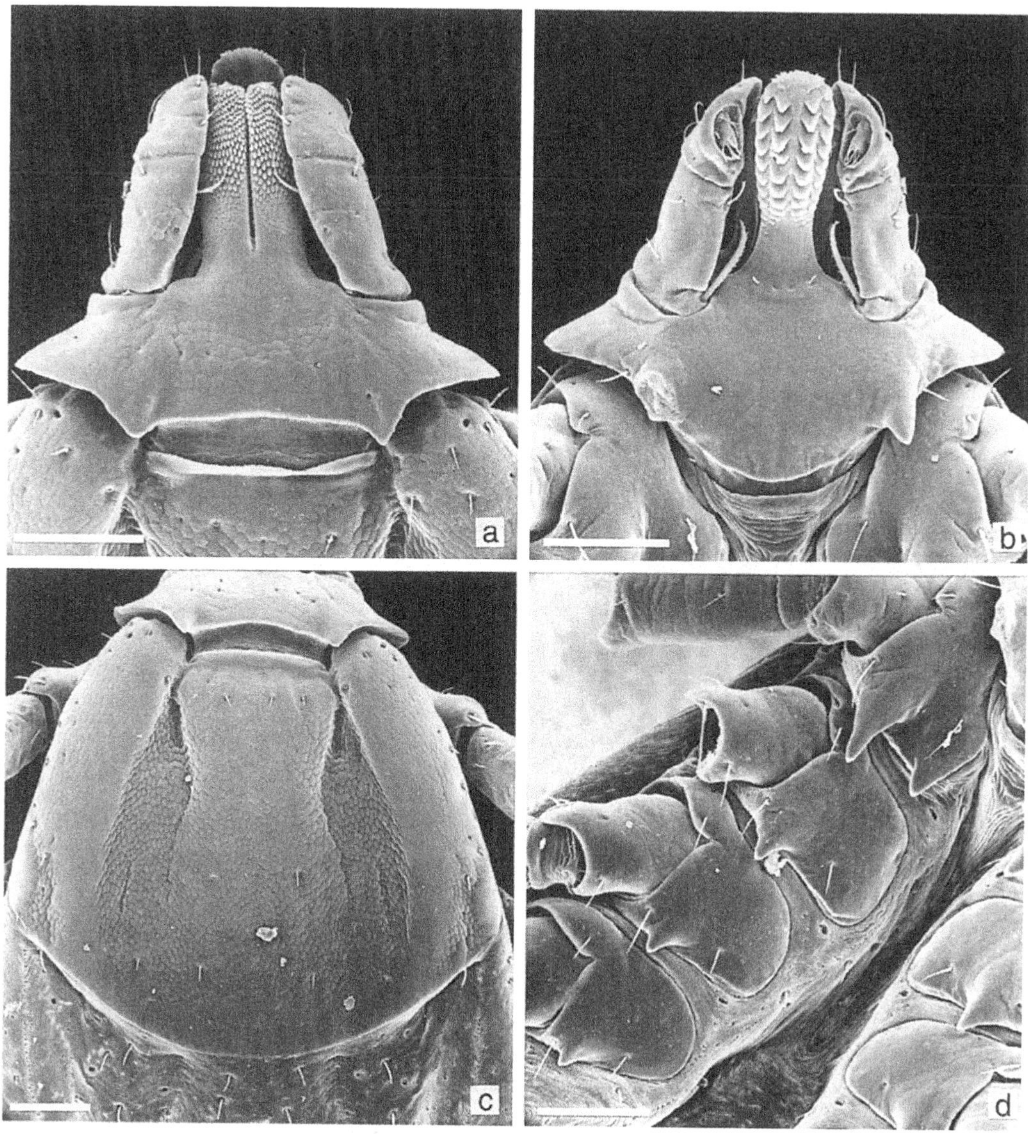

Figure 29 (*above*). *Rhipicephalus capensis* [Onderstepoort Tick Collection 2824, RML 65718, laboratory reared, progeny of ♀ collected on 'Aties' (farm), Klawer, Gifberge, Western Cape Province, South Africa, probably in the mid-1940s]. Nymph: (a) capitulum, dorsal; (b) capitulum, ventral; (c) scutum; (d) coxae. Scale bars represent 0.10 mm. SEMs by M.D. Corwin.

Figure 28 (*opposite*). *Rhipicephalus capensis* [Onderstepoort Tick Collection 2826, RML 65719, laboratory reared, progeny of ♀ collected on 'Aties' (farm), Klawer, Gifberge, Western Cape Province, South Africa, probably in the mid-1940s]. Male: (a) capitulum, dorsal; (b) spiracle; (c) adanal plates. Female: (d) capitulum, dorsal; (e) scutum; (f) genital aperture. Scale bars represent 0.10 mm. SEMs by M.D. Corwin.

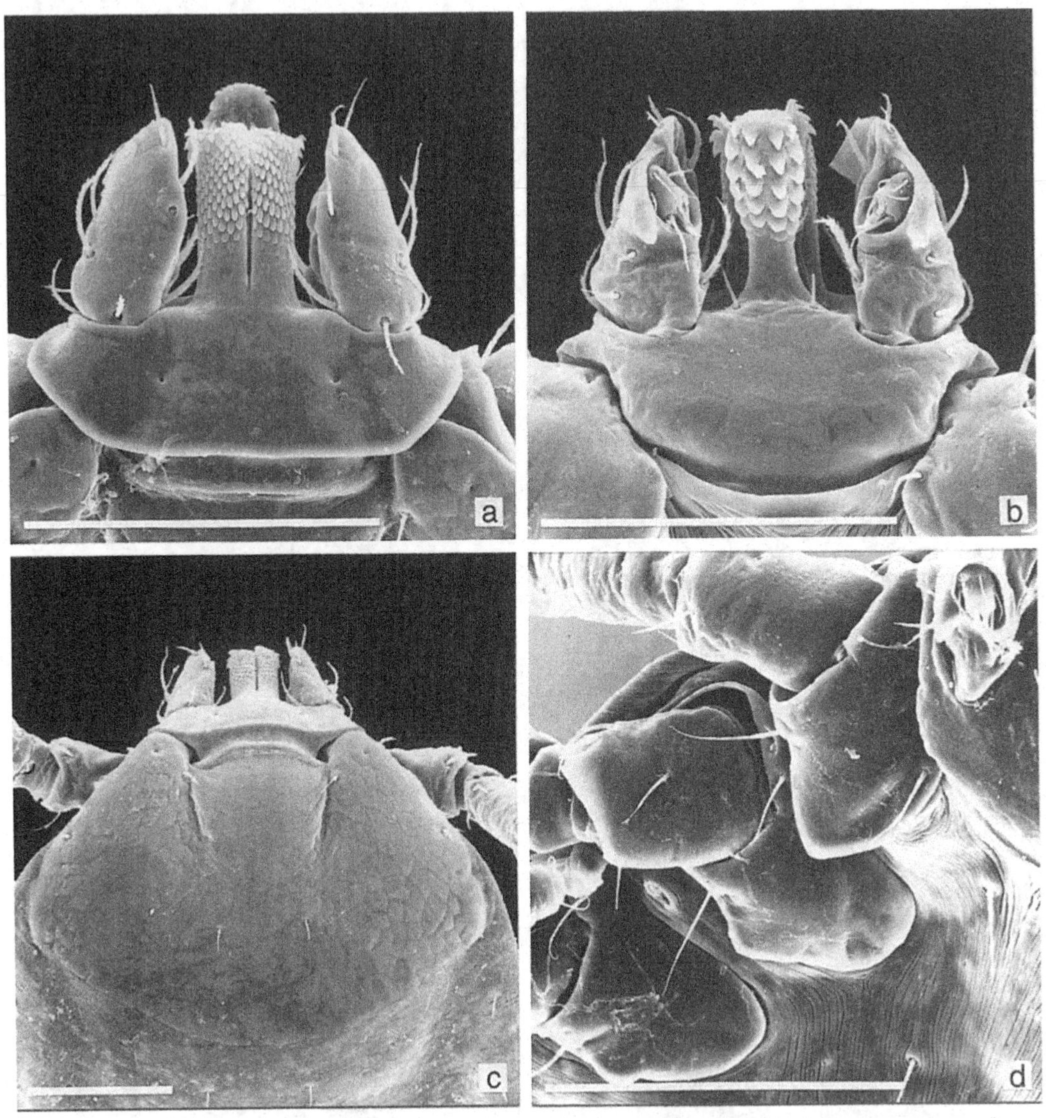

Figure 30. *Rhipicephalus capensis* [Onderstepoort Tick Collection 2824, RML 65718, laboratory reared, progeny of ♀ collected on 'Aties' (farm), Klawer, Gifberge, Western Cape Province, South Africa, probably in the mid-1940s]. Larva: (a) capitulum, dorsal; (b) capitulum, ventral; (c) scutum; (d) coxae. Scale bars represent 0.10 mm. SEMs by M.D. Corwin.

Male (Figs 27(a), 28(a) to (c))

Capitulum longer than broad, length × breadth ranging from 0.63 mm × 0.58 mm to 0.84 mm × 0.79 mm. Basis capituli with short acute lateral angles at about anterior third of its length. Palps short, broad, flattened apically. Conscutum length × breadth ranging from 2.31mm × 1.65 mm to 3.60 mm × 2.42 mm; anterior process of coxae I small. In engorged specimens body wall greatly expanded, forming two large bulges posterolaterally plus a single bulbous caudal process posteromedially. Eyes slightly convex, edged dorsally by a few punctations. Cervical pits comma-shaped; cervical fields long, narrow, depressed, their external margins sharply defined by large setiferous punctations. Marginal lines long, but not quite reaching eyes, deep and lined with numerous punctations. Posteromedian and posterolateral grooves distinct, rugose. Large punctations, some of them setiferous, scattered anteriorly on scapulae and forming a deep, extremely dense, partially confluent pattern medially on the conscutum. In many specimens the conscutum anterior to the posterior grooves is characteristically a lighter, more reddish brown in colour, visible to the naked eye. Ventrally spiracles comma-shaped, tapering gradually and curving gently towards the dorsal surface. Adanal plates almost pear-shaped, their internal margins slightly concave posterior to the anus, then widening posteriorly; accessory adanal plates small, narrow, sharply pointed.

Female (Figs 27(b), 28(d) to (f))

Capitulum as broad as long, length × breadth ranging from 0.75 mm × 0.79 mm to 0.91 mm × 0.88 mm. Basis capituli with short, acute lateral angles at anterior third of its length; porose areas small, round, twice their own diameter apart. Palps long. Scutum slightly broader than long, length × breadth ranging from 1.54 mm × 1.57 mm to 1.88 mm × 1.93 mm, posterior margin mildly sinuous. Eyes slightly convex, edged dorsally by a few punctations. Cervical pits comma-shaped; cervical fields broad, depressed, their external margins sharply defined by large contiguous setiferous punctations. Large punctations, some of them setiferous, scattered anteriorly on scapulae and forming a deep, dense, often confluent pattern medially on the scutum. Alloscutum of unfed females deeply folded. Ventrally genital aperture a wide V with a rounded base.

Nymph (Fig. 29)

Capitulum much broader than long, length × breadth ranging from 0.35 mm × 0.46 mm to 0.39 mm × 0.48 mm. Basis capituli well over three times as broad as long, with sharply tapering lateral angles projecting over the scapulae and also well-developed cornua; ventrally with sharp spurs on posterior border. Palps long, narrow, almost equal in width throughout their length, tapering slightly only at their tips. Scutum broader than long, length × breadth ranging from 0.67 mm × 0.78 mm to 0.72 mm × 0.81 mm; posterior margin a broad smooth curve. Eyes at widest point, well over halfway back, delimited dorsally by slight depressions. Cervical fields well defined, long, narrow, divergent. Ventrally coxae I each with a long sharp external spur and a shorter sharp internal spur; coxae II to IV each with a small, but sharp, external spur only.

Larva (Fig. 30)

Capitulum much broader than long, length × breadth ranging from 0.121 mm × 0.187 mm to 0.135 mm × 0.196 mm. Basis capituli over three times as broad as long, with short, bluntly-rounded lateral angles, posterior border straight. Palps broad proximally, tapering for distal third of their length to rounded apices. Scutum much broader than long, length × breadth ranging from 0.268 mm × 0.410 mm to 0.280 mm × 0.439 mm; posterior margin a wide smooth curve. Eyes at widest part of scutum, well over halfway back, mildly convex. Cervical grooves short, slightly convergent. Ventrally coxae I each with a large pointed spur; coxae II and III each with a short broad spur.

Notes on identification

Doss *et al.* (1974) list 189 references to *R. capensis* and its subspecies from various parts of the Afrotropical region. It has been possible to review only a small part of this mass of literature, and to re-examine just a few of the ticks concerned, many of which probably no longer exist. In general we endorse the comment by Theiler (1962), in her review of the distribution of *R. capensis* in South Africa, that this specific name 'appears to be a catchall for "*capensis*-like" ticks . . .' Two years earlier Feldman-Muhsam (1960) had contributed significantly to the solution of this problem with her redescription of *R. capensis*, based on Koch's holotype male from 'Capland' (Zoological Museum, Berlin, 1097), and her description of *R. gertrudae*. We think that three species whose adults have been confused, *R. capensis sensu stricto*, *R. follis sensu stricto* (p. 179), and *R.*

gertrudae (p. 193), are confined to southern Africa. The distributions of *R. capensis* and *R. gertrudae* overlap and both ticks may also be present on the same host. In the latter case it is often difficult to distinguish between the less heavily-punctate specimens of *R. capensis* and the more heavily-punctate specimens of *R. gertrudae.* However, the centre of the conscutum of *R. capensis* males, particularly of large, freshly collected or alcohol-preserved specimens, is usually reddish brown in colour compared to the adjacent dark brown surface. The conscutum of *R. gertrudae* males is uniformly dark brown in colour.

Hosts

A three-host species (Gertrud Theiler, unpublished data, 1942, for Onderstepoort Tick Collection Nos 2824–2826 ('Aties' strain)). Cattle and horses are the only domestic hosts from which we have seen *R. capensis* adults (Table 8). Large animals such as the eland and gemsbok appear to be the wild hosts most favoured by its adults. Two adult eland bulls examined in the Langebaan Nature Reserve, Western Cape Province, during February 1990 each harboured more than 700 *R. capensis* adults compared to less than 160 each on two adult gemsbok bulls, and only four and nil, respectively, on two adult bontebok rams examined in the reserve at the same time (I.G.H., unpublished data).

Although we have little concrete evidence of this at present it appears that, as in the case of the other two species referred to above, the immature stages prefer rodents. Neitz, Boughton & Walters (1972, citing G. Theiler, pers. comm., 1971) listed *R. theileri* nymphs from a swamp rat (*Otomys* sp.), Clanwilliam (Lambert's Bay), and from a 'bush otomys' (*Myotomys* sp., now regarded as a synonym of *Otomys*), Uniondale. These nymphs (Onderstepoort Tick Collection 3081i, iii) have been re-identified as *R. capensis* by I.G.H. and J.B.W.

Zoogeography

The distribution map labelled *R. capensis* published by Theiler (1950) includes under this name the distributions of *R. follis* and *R. gertrudae.* Unfortunately many of the collections on which Theiler based her paper no longer exist. The account given below is therefore based on specimens of *R. capensis* that we ourselves have seen.

The distribution of *R. capensis sensu stricto* as we see it is confined to the Western and Northern Cape Provinces, South Africa (Map 10). The tick has been collected from altitudes a few metres above sea level to 500 m, in regions with annual rainfall varying from 150 mm to 600 mm. All these collections have been taken in Cape shrubland (*fynbos*), succulent Karoo shrubland or bush Karoo-Namib shrubland.

Disease relationships

Unknown.

REFERENCES

Feldman-Muhsam, B. (1960). The South African ticks *Rhipicephalus capensis* Koch and *R. gertrudae* n. sp. *Journal of Parasitology*, **46**, 101–8.

Koch, C.L. (1844). Systematische Uebersicht über die ordung der Zecken. *Archiv für Naturgeschichte*, **10**, 217–39.

Neitz, W.O., Boughton, F. & Walters, H.S. (1972). Laboratory investigations on the life-cycle of *Rhipicephalus theileri* Bedford & Hewitt, 1925 (Ixodoidea: Ixodidae). *Onderstepoort Journal of Veterinary Research*, **39**, 117–23.

Theiler, G. (1950). Zoological Survey of the Union of South Africa. Tick Survey – Part IV. Distribution of *Rhipicephalus capensis*, the Cape brown tick. *Onderstepoort Journal of Veterinary Science and Animal Industry*, **24**, 7–32 + 1 map.

Also see the following Basic References (pp. 12–14): Doss *et al.* (1974); Theiler (1962); Walker (1990).

RHIPICEPHALUS CARNIVORALIS WALKER, 1966

The specific name *carnivoralis*, from the Latin *carnis* meaning 'flesh', refers to the fact that adults of this species usually feed on carnivores.

Diagnosis

A moderate-sized brown to reddish-brown tick that in some respects strongly resembles *R. appendiculatus*.

Male (Figs 31(a), 32(a) to (c))
Capitulum slightly broader than long, length × breadth ranging from 0.70 mm × 0.76 mm to 0.91 mm × 0.92 mm. Basis capituli with long sharp lateral angles. Palps short, broad. Conscutum length × breadth ranging from 2.40 mm × 1.50 mm to 3.66 mm × 2.36 mm; anterior process of coxae I large, strongly sclerotized. In engorged specimens body wall expanded slightly posterolaterally and forming a single caudal process posteriorly. Eyes marginal, very slightly bulging, edged with a few large punctations dorsally. Cervical fields broad, depressed, with finely reticulate surfaces. Marginal lines punctate, extending anteriorly almost to eye level. Posteromedian and posterolateral grooves well developed. Large setiferous punctations present on the scapulae, along the external margins of and between the cervical fields, and scattered medially on the conscutum, interspersed with fine punctations, though these are sparse or absent adjacent to the marginal lines. Legs increase markedly in size from I to IV. Ventrally spiracles elongate, comma-shaped. Adanal plates broad, tapering posterointernally to broadly rounded points, posterior margins almost straight; accessory adanal plates small to absent.

Female (Figs 31(b), 32(d) to (f))
Capitulum broader than long, length × breadth ranging from 0.71 mm × 0.83 mm to 0.87 mm × 0.99 mm. Basis capituli with long, sharp lateral angles overlapping the scapulae; porose areas large, nearly twice their own diameter apart. Palps rounded apically. Scutum longer than broad, length × breadth ranging from 1.10 mm × 1.05 mm to 1.74 mm × 1.62 mm. Eyes about halfway back, very slightly bulging, edged dorsally by a few punctations. Cervical fields well marked, depressed, with somewhat rugose, punctate surfaces. Large setiferous punctations present on the scapulae, along the external margins of the cervical fields and scattered centrally on the scutum, interspersed with fine punctations. Ventrally genital aperture broadly U-shaped.

Nymph (Fig. 33)
Capitulum much broader than long, length × breadth ranging from 0.25 mm × 0.34 mm to 0.29 mm × 0.38 mm. Basis capituli nearly four times as broad as long, lateral angles long, tapering, forwardly tilted and overlapping scapulae; ventrally small blunt spurs sometimes visible on posterior border. Palps narrower proximally than distally, inclined slightly inwards. Scutum much broader than long, length × breadth ranging from 0.45 mm × 0.55 mm to 0.52 mm × 0.65 mm; posterior margin a broad deep curve. Eyes at widest point, a little over halfway back, slightly convex. Cervical fields long, narrow, slightly depressed. Ventrally coxae I each with two large spurs; coxae II to IV each with an external spur only, decreasing progressively in size.

Larva (Fig. 34)
Capitulum broader than long, length × breadth ranging from 0.119 mm × 0.142 mm to 0.145 mm × 0.158 mm. Basis capituli over twice as broad as long, with short, blunt, forwardly-tilted lateral angles. Palps constricted proximally, then widening markedly, apices truncated. Scutum much broader than long, length × breadth ranging from 0.251 mm × 0.401 mm to 0.261 mm × 0.410 mm; posterior margin a broad deep curve. Eyes at widest point, slightly convex. Cervical grooves short, slightly convergent. Ventrally coxae I each with a large triangular spur; coxae II and III each with a small triangular spur.

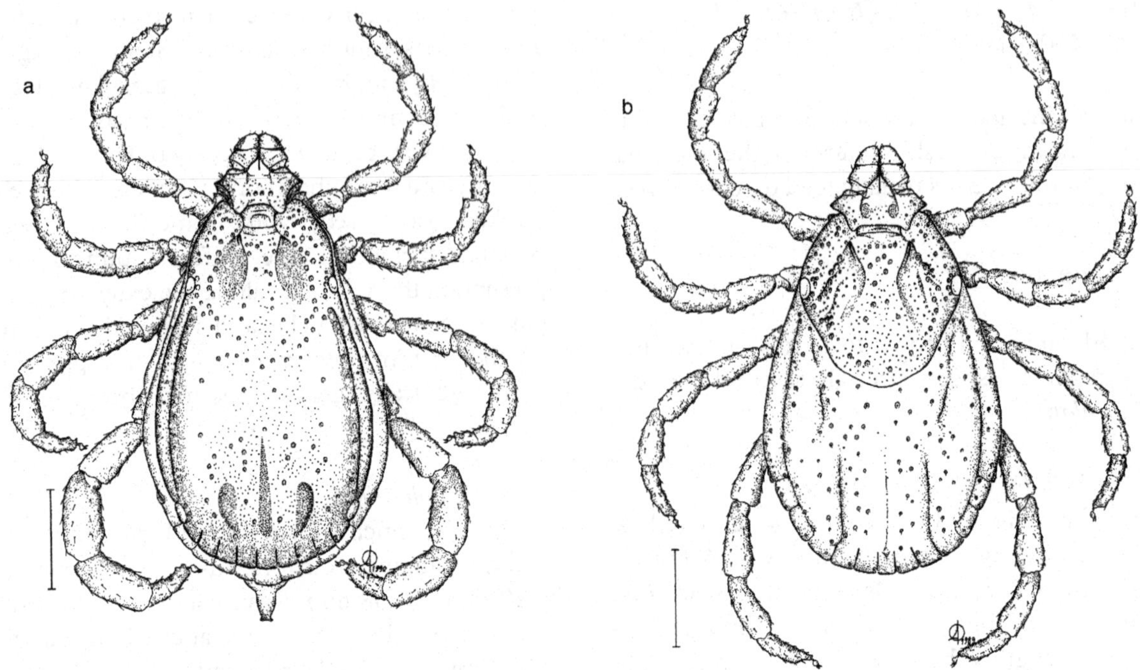

Figure 31. *Rhipicephalus carnivoralis* [B.S. 913/-, paratype series, laboratory reared, progeny of ♀ collected from lioness (*Panthera leo*), Muguga, Kiambu District, Kenya, on 8 October 1961, by J. B. Walker]. (a) Male, dorsal; (b) female, dorsal. Scale bars represent 1 mm. A. Olwage *del.*

Table 9. *Host records of* Rhipicephalus carnivoralis

Hosts	Number of records	
	Confirmed	Unconfirmed
Domestic animals		
Cattle	3	
Dogs	2	
Wild animals		
Cheetah (*Acinonyx jubatus*)	1	
Lion (*Panthera leo*)	31	
Leopard (*Panthera pardus*)	9	2
Spotted hyaena (*Crocuta crocuta*)	3	
Striped hyaena (*Hyaena hyaena*)	1	
'Hyaena'	1	
Yellow-spotted rock hyrax (*Heterohyrax brucei*)	2 (nymphs)	
Rock hyrax (*Procavia capensis*)	2 (nymphs)	
Humans	1	

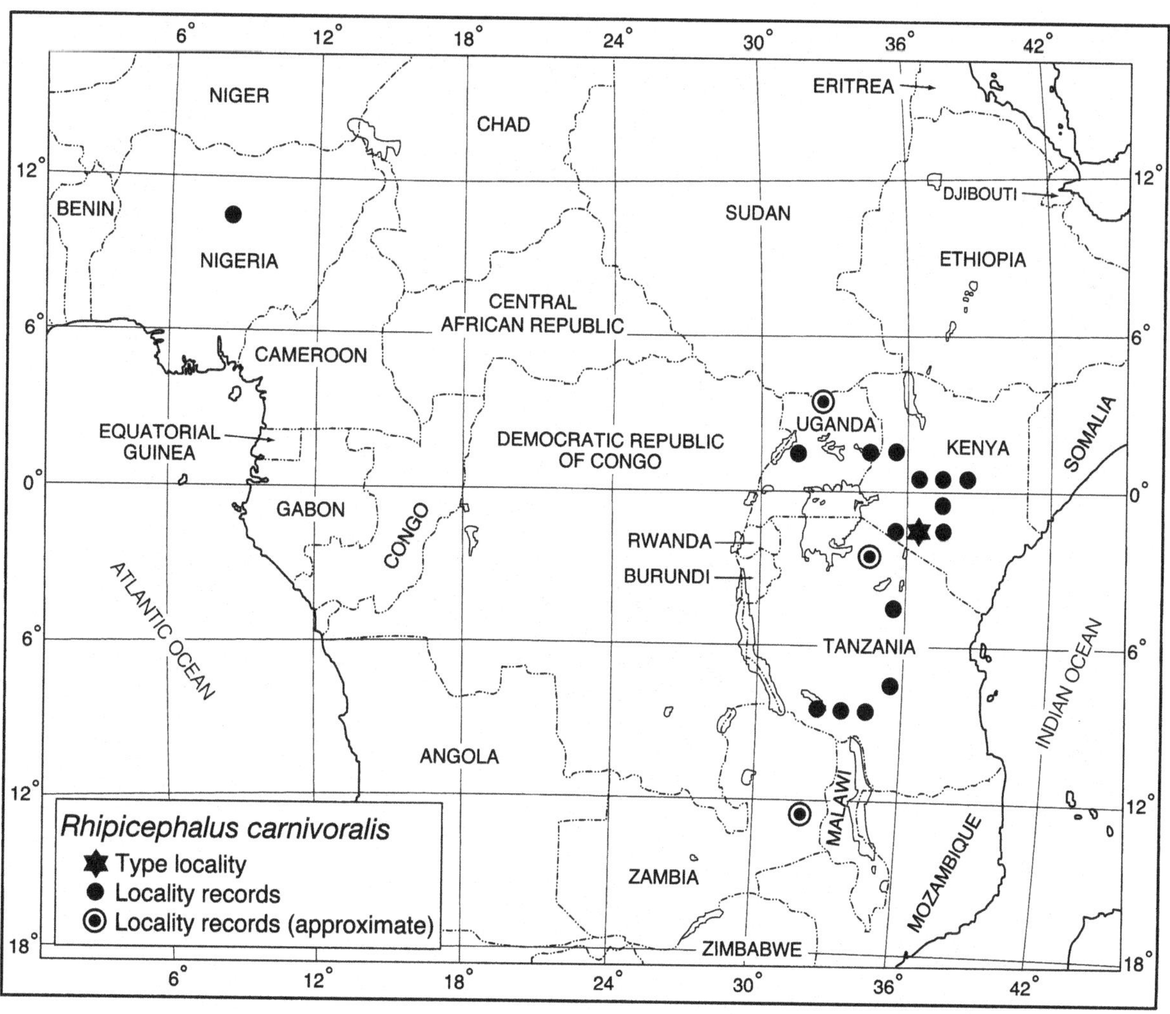

Map 11. *Rhipicephalus carnivoralis:* distribution.

Notes on identification

In the past *R. carnivoralis* has been confused with several other species in the genus. It is closest in general appearance to *R. appendiculatus* and *R. zambeziensis* but the male, female and nymph of *R. carnivoralis* are characterized by the shape of their basis capituli, which has longer, sharper lateral angles than either of the other two species. The *R. carnivoralis* larva also has a wider basis capituli than the *R. appendiculatus* larva, but it is difficult to distinguish it morphologically from the larva of *R. zambeziensis*.

From the description given by Hoogstraal (1956, pp. 637, 778) it seems likely that his *Rhipicephalus*? sp. larvae and nymphs collected from the yellow-spotted rock hyrax (*Heterohyrax brucei*) at Imatong and Imurok, southern Sudan, are *R. carnivoralis*, not *R. ? distinctus*. Although *R. carnivoralis* has not as yet been recorded from the Sudan it has been found in northern Uganda, and its nymphs have been collected from this hyrax elsewhere. It was described only in 1966 and thus would not have been known to Gertrud Theiler and Harry Hoogstraal 10 years earlier.

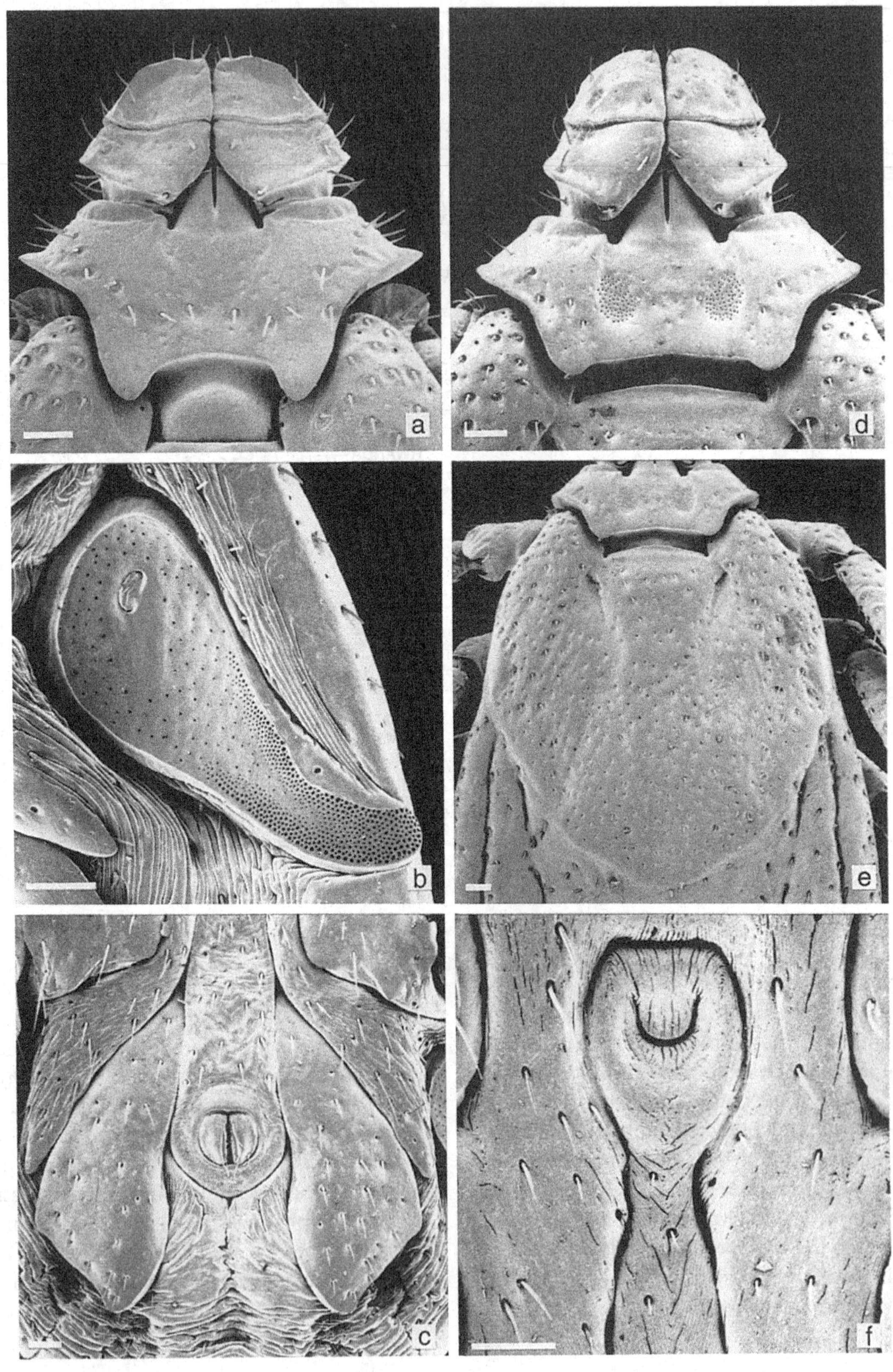
a
d
b
e
c
f

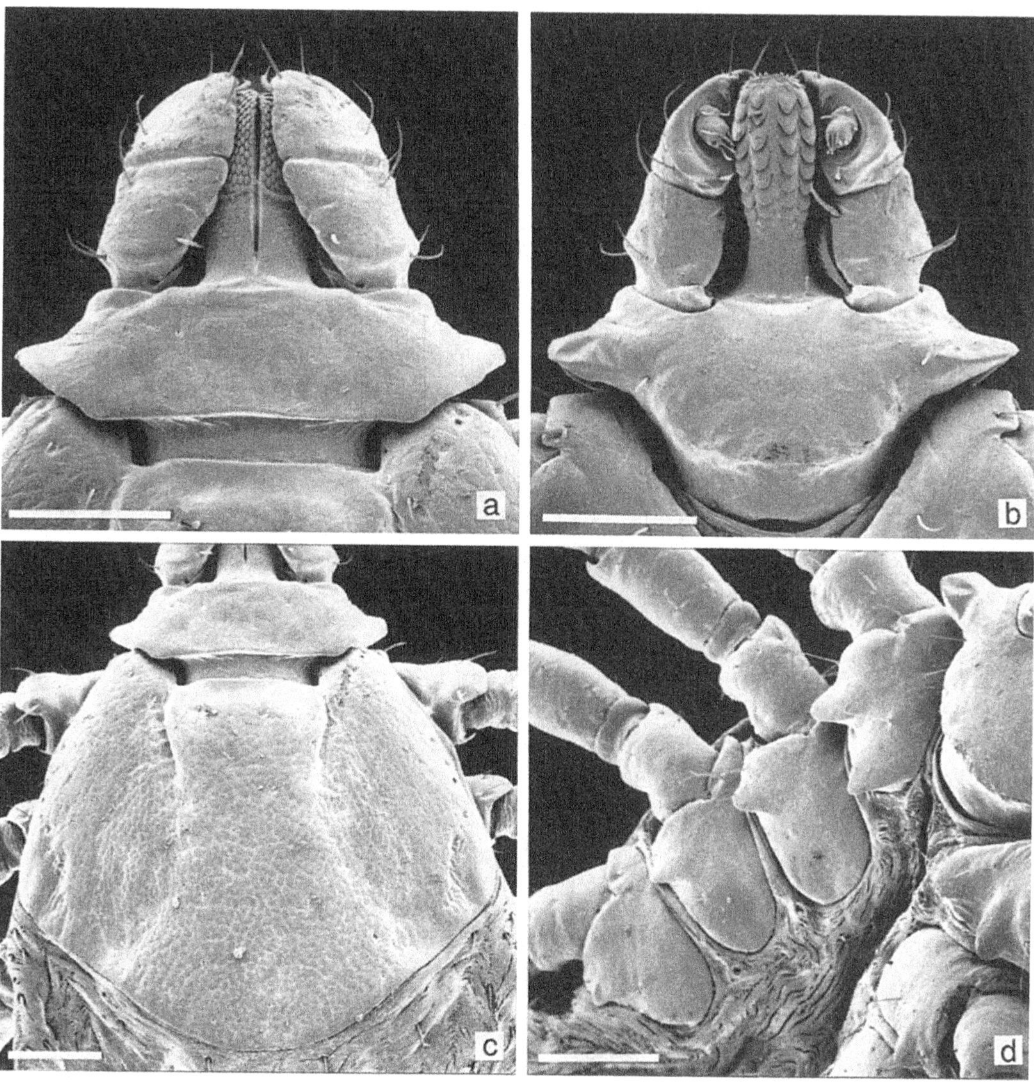

Figure 33 (*above*). *Rhipicephalus carnivoralis* [B.S. 913/-, paratype series, laboratory reared, progeny of ♀ collected from lioness (*Panthera leo*), Muguga, Kiambu District, Kenya, on 8 October 1961, by J.B. Walker]. Nymph: (a) capitulum, dorsal; (b) capitulum, ventral; (c) scutum; (d) coxae. Scale bars represent 0.10 mm. SEMs by J. F. Putterill.

Figure 32 (*opposite*). *Rhipicephalus carnivoralis* [B.S. 913/-, paratype series, laboratory reared, progeny of ♀ collected from lioness (*Panthera leo*), Muguga, Kiambu District, Kenya, on 8 October 1961, by J.B. Walker]. Male: (a) capitulum, dorsal; (b) spiracle; (c) adanal plates. Female: (d) capitulum, dorsal; (e) scutum; (f) genital aperture. Scale bars represent 0.10 mm. SEMs by J. F. Putterill.

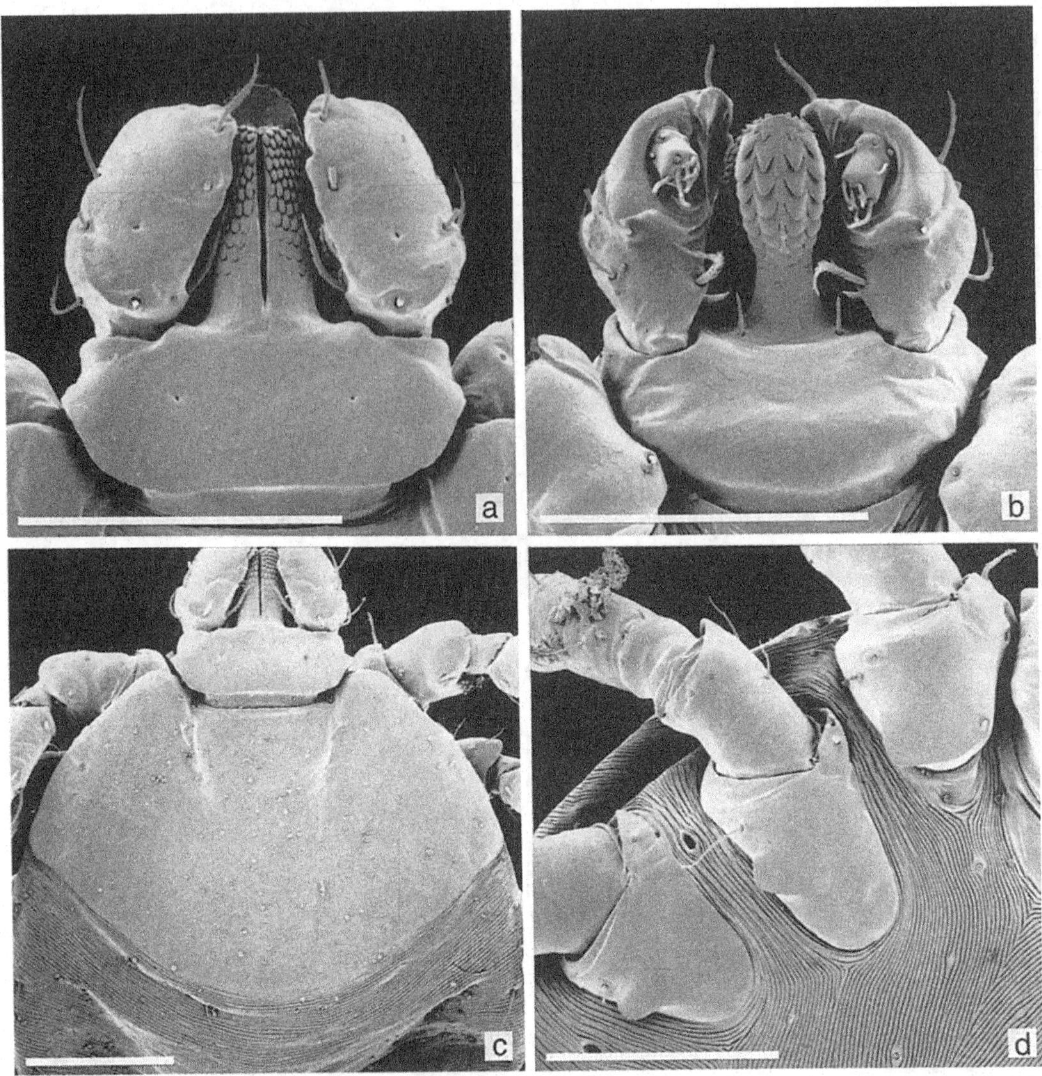

Figure 34. *Rhipicephalus carnivoralis* [B.S. 913/-, paratype series, laboratory reared, progeny of ♀ collected from lioness (*Panthera leo*), Muguga, Kiambu District, Kenya, on 8 October 1961, by J.B. Walker]. Larva: (a) capitulum, dorsal; (b) capitulum, ventral; (c) scutum; (d) coxae. Scale bars represent 0.10 mm. SEMs by J.F. Putterill.

Hosts

A three-host species (Walker, 1966). Almost all adults of this tick have been collected from carnivores (Table 9). Cattle are undoubtedly accidental hosts; since although many thousands of ticks have been collected from these animals (e.g. over 180 000 during the Tanzania tick survey alone (Yeoman & Walker, 1967)), only three of these, all males, were *R. carnivoralis*, each obtained from a different beast.

So far the only known hosts of the immature stages are two species of rock hyraxes. This host association probably reflects the fact that the large cats often lie up in the rocky areas that are also inhabited by these hyraxes.

Zoogeography

With only two exceptions these records of *R. carnivoralis* are from East Africa (Uganda,

Kenya and Tanzania: Map 11). Twenty-one of the 52 known collections of adults were obtained in the Serengeti National Park, Tanzania, nearly all from lions. In 1913 one collection (Nuttall 2548, originally identified as *R. appendiculatus*), was made from a leopard at Buji, on the Jos plateau in northern Nigeria, but it has not been recorded there since. A single male has also been collected from a lion in the Luangwa Valley, Zambia (MacLeod & Mwanaumo, 1978). Present indications are, though, that this tick does not occur throughout the ranges of its large carnivore hosts in the Afrotropical region.

These places lie in the various types of woodland, bushland and wooded grassland favoured by most of its hosts.

Disease relationships

Under laboratory conditions *R. carnivoralis* has been shown to be capable of transmitting *Theileria parva* (Brocklesby, Bailey & Vidler, 1966). Since it has so rarely been collected from cattle it probably never acts as a vector of this protozoan in the field.

REFERENCES

Brocklesby, D.W., Bailey, K.P. & Vidler, B.O. (1966). The transmission of *Theileria parva* (Theiler, 1904) by *Rhipicephalus carnivoralis* Walker, 1966. *Parasitology*, **56**, 13–14.

MacLeod, J. & Mwanaumo, B. (1978). Ecological studies of ixodid ticks (Acari: Ixodidae) in Zambia. IV. Some anomalous infestation patterns in the northern and eastern regions. *Bulletin of Entomological Research*, **68**, 409–29.

Walker, J.B. (1966). *Rhipicephalus carnivoralis* sp. nov. (Ixodoidea, Ixodidae). A new species of tick from East Africa. *Parasitology*, **56**, 1–12.

Also see the following Basic References (pp. 12–14): Hoogstraal (1956); Matthysse & Colbo (1987); Walker (1974); Yeoman & Walker (1967).

RHIPICEPHALUS COMPLANATUS NEUMANN, 1911

The specific name, derived from the Latin *complano* meaning 'to make level', refers to the characteristically slightly concave to flat scutum of the adults.

Synonym

planus complanatus.

Diagnosis

A large chestnut-brown tick with a slightly concave to flat moderately shiny scutum.

Male (Figs 35(a), 36(a) to (c))

Capitulum slightly longer than broad, length × breadth ranging from 0.73 mm × 0.70 mm to 1.02 mm × 0.97 mm. Basis capituli with acute somewhat recurved lateral angles in anterior third of its length. Palps short, broad. Conscutum length × breadth ranging from 3.02 mm × 2.22 mm to 4.32 mm × 3.19 mm; anterior process on coxae I small, rounded. In engorged specimens body wall expanded posterolaterally and posteromedially forming up to three smoothly rounded caudal processes, of which the middle one is the longest. Eyes flat. Cervical pits comma-shaped; cervical fields shallow and inconspicuous. Marginal lines well developed, usually almost reaching eye level, their inner margins rounded, lined by a few punctations. Posteromedian and posterolateral grooves, when present, shallow. Small additional depressions sometimes visible. Punctations subequal, in general fine, diffuse and very superficial. In many specimens, though, the punctations are virtually absent, in which case the surface of the conscutum has a uniformly dull shine. Adanal plates unique in shape, broad, curved, with their external margins convex and their internal margins concave; their posterior margins also concave, their junctions with the internal and external margins rounded; accessory adanal plates sharp, well sclerotized.

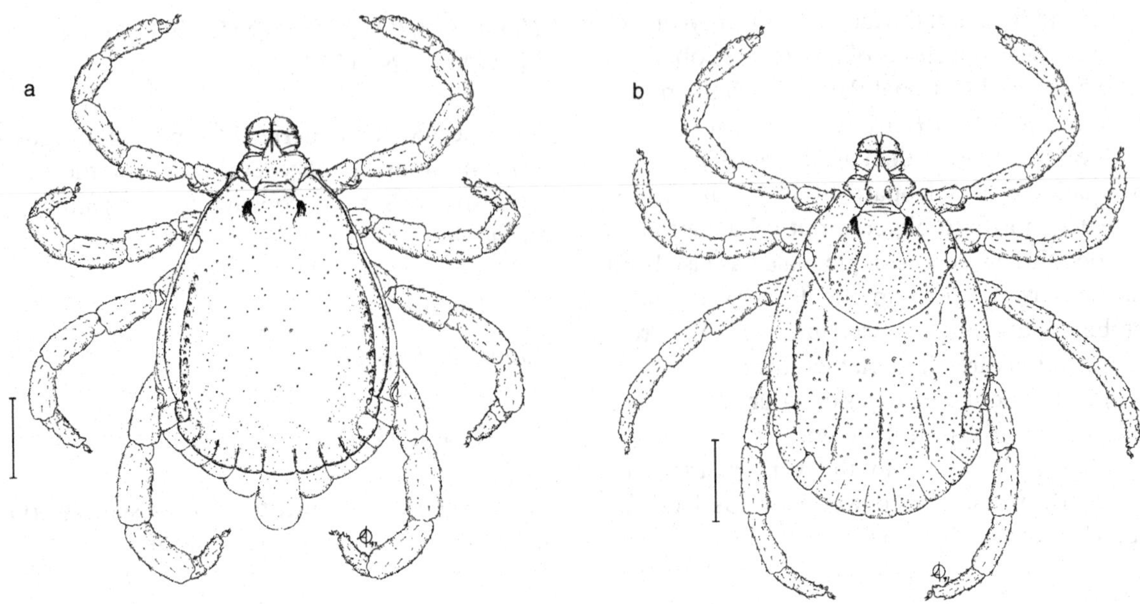

Figure 35. *Rhipicephalus complanatus* [Invertebrata, Acarina: Ixodidae, Ref. No. 70.016- 70.065, collected from red river hog (*Potamochoerus porcus*), Ibembo, Democratic Republic of Congo, on 6 July 1950 by R. Fr. Hutsebaut, by courtesy of the Musée Royal de l'Afrique Centrale, Tervuren, Belgium]. (a) Male, dorsal; (b) female, dorsal. Scale bars represent 1 mm. A. Olwage *del.*

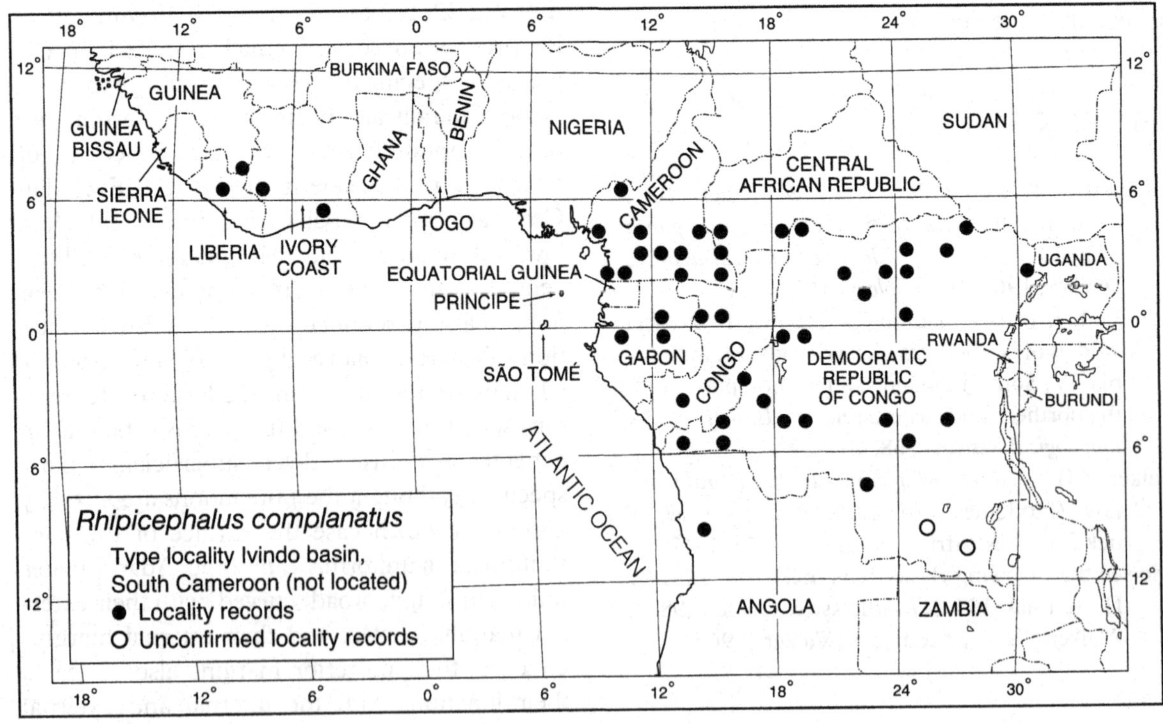

Map 12. *Rhipicephalus complanatus*: distribution. (Based on Morel, 1969).

Table 10. *Host records of* Rhipicephalus complanatus

Hosts	Number of records
Domestic animals	
Pigs	4
Wild animals	
Warthog (*Phacochoerus africanus*)	7
Forest hog (*Hylochoerus meinertzhageni*)	1
Bushpig (*Potamochoerus larvatus*)	1
Red river hog (*Potamochoerus porcus*)	35
'Wild pig'	4
African buffalo (*Syncerus caffer*)	5
Bushbuck (*Tragelaphus scriptus*)	1
Tullberg's soft-furred mouse (*Praomys tullbergi*)	1
Humans	1

Female (Figs 35(b), 36(d) to (f))

Capitulum slightly longer than broad, length × breadth ranging from 0.91 mm × 0.87 mm to 1.03 mm × 0.98 mm. Basis capituli with acute lateral angles a little anterior to mid-length; porose areas oval, just over their own diameter apart. Scutum broader than long, length × breadth ranging from 1.50 mm × 1.70 mm to 1.73 mm × 2.07 mm; in some specimens posterior margin broadly rounded, in others somewhat sinuous posterolaterally. Eyes flat. Cervical fields broad, their external margins well defined. Punctations generally somewhat more conspicuous than in the male but still subequal, fine and superficial.

Immature stages

Undescribed. Elbl & Anastos (1966) list, but do not describe, a nymph amongst the adults collected from *Potamochoerus* sp., listed as a red river hog in Table 10 (Musée Royal de l'Afrique Centrale, Collection No. T111386).

Notes on identification

This species was first described as *R. planus* by Neumann (1910). The following year Neumann (1911) changed its name to *R. complanatus* because he remembered that the name *planus* had already been given to a different entity, the subspecies *R. simus planus*, and was therefore preoccupied.

Elbl & Anastos (1966) comment that: 'The species is somewhat variable, primarily in the size of scutal punctations which range from medium to very large.'

Hosts

Life cycle unknown. *Rhipicephalus complanatus* adults parasitize domestic and wild pigs almost exclusively. Red river hogs appear to be its primary hosts, and to a lesser extent warthogs and other suids (Table 10). Other hosts are probably accidental (Morel & Mouchet, 1958; Morel, 1969).

Hoogstraal (1954) recorded a female from a Tullberg's soft-furred mouse collected at Mamfe, in south-west Cameroon near the Nigerian border. This mouse, which is by far the commonest small murid in natural rainforest in West Africa, is recorded as far east as western Cameroon (Rosevear, 1969; Happold, 1987). Morel & Mouchet (1958) thought that it is probably a host of the tick's immature stages. They suggested that adults only occur on it temporarily, merely using it to transport them to the outside world after moulting in its burrow.

This tick's attachment sites on its hosts have not been recorded. Apart from December, adults have been collected throughout the year.

Zoogeography

Rhipicephalus complanatus has been recorded mainly in the Democratic Republic of Congo and neighbouring countries just to its west. There are also a few records from further west, in the Ivory Coast and Liberia, and one to the east, in western Uganda (Map 12). It is largely

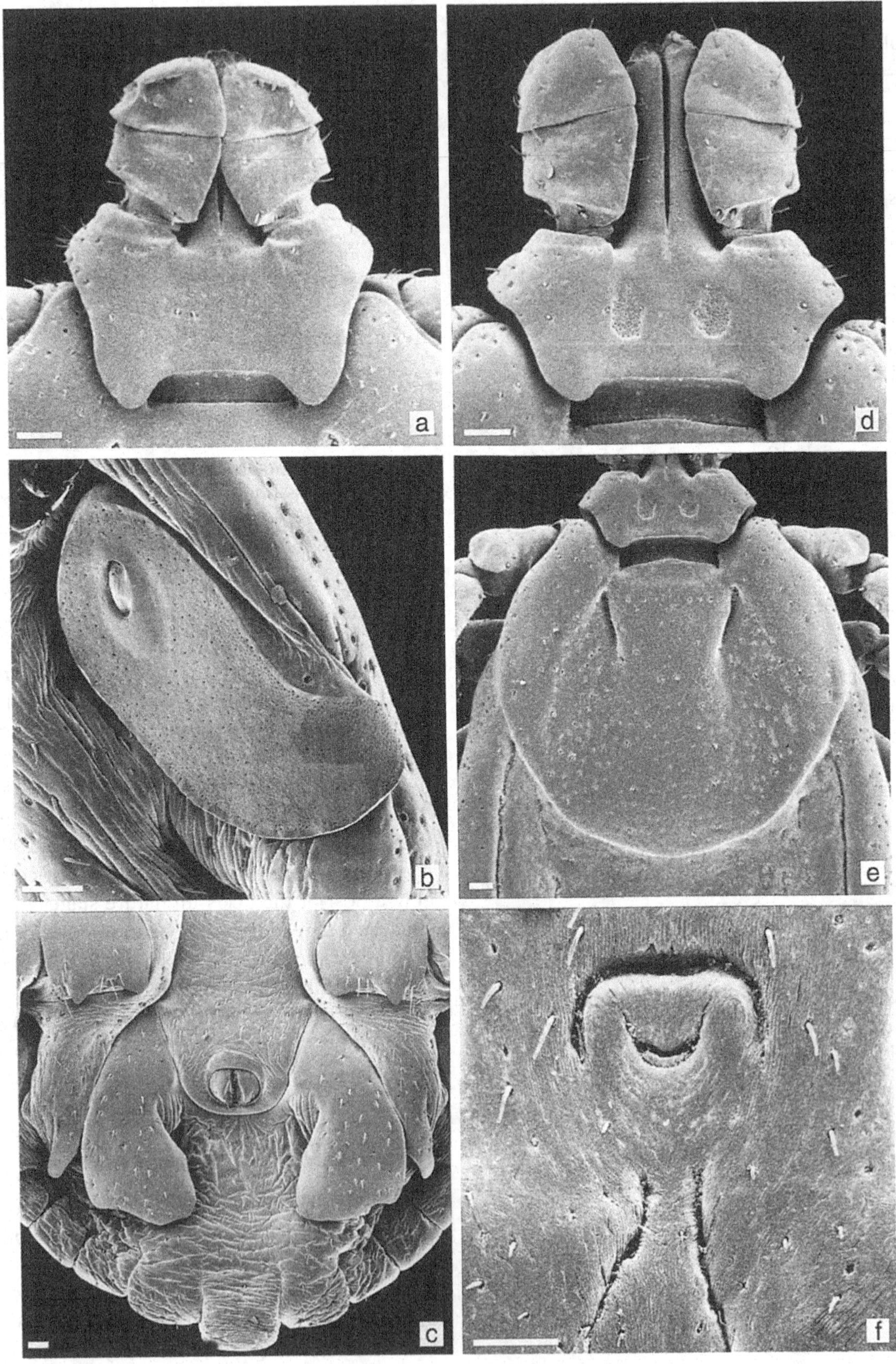
a
d
b
e
c
f

confined to dense humid tropical rainforest, secondary forest and riverine forest. It may also occur in the isolated patches of such forests that form part of the forest/savanna mozaic. These well-watered areas, which often include swamps and rivers whose banks become inundated at times, are the favoured habitat of its main host, the red river hog, and also of the forest hog. It has sometimes been collected from warthogs inhabiting savanna areas bordering these forests (Morel & Mouchet, 1958; Morel, 1969).

Disease relationships

Unknown.

REFERENCES

Happold, D.C.D. (1987). *The Mammals of Nigeria.* Oxford: Oxford University Press.

Hoogstraal, H. (1954). Noteworthy African tick records in the British Museum (Natural History) collections (Ixodoidea). *Proceedings of the Entomological Society of Washington*, **56**, 273–9.

Morel, P.C. & Mouchet, J. (1958) Les tiques du Cameroun (Ixodidae et Argasidae). *Annales de Parasitologie Humaine et Comparée*, **33**, 69–111.

Neumann, L.G. (1910). Sur quelques espéces d'Ixodidea nouvelles ou insuffisamment connues. *Annales des Sciences Naturelles, Zoologie*, 9e série, **12**, 161–76.

Neumann, L.G. (1911). Note rectificative à propos de deux espéces d'Ixodinae. *Archives de Parasitologie*, **14**, 415.

Rosevear, D.R. (1969). *The Rodents of West Africa.* London: Trustees of the British Museum (Natural History).

Figure 36 (*opposite*). *Rhipicephalus complanatus* [Invertebrata, Acarina: Ixodidae, Ref. No. 70.016-70.065, collected from red river hog (*Potamochoerus porcus*), Ibembo, Democratic Republic of Congo, on 6 July 1950 by R. Fr. Hutsebaut, by courtesy of the Musée Royal de l'Afrique Centrale, Tervuren, Belgium]. Male: (a) capitulum, dorsal; (b) spiracle; (c) adanal plates. Female: (d) capitulum, dorsal; (e) scutum; (f) genital aperture. Scale bars represent 0.10 mm. SEMs by J.F. Putterill.

Also see the following Basic References (pp. 12–14): Elbl & Anastos (1966); Matthysse & Colbo (1987); Morel (1969).

RHIPICEPHALUS COMPOSITUS NEUMANN, 1897

The specific name, from the Latin meaning 'put together, joined', probably refers to the fact that the numerous punctations on the scutum are often confluent.

Synonyms

ayrei, capensis compositus.

Diagnosis

A large dark brown to black heavily and evenly-punctate species.

Male (Figs 37(a), 38(a) to (c))

Capitulum slightly broader than long, length × breadth ranging from 0.67 mm × 0.69 mm to 1.02 mm × 1.05 mm. Basis capituli with acute lateral angles in anterior third of its length. Palps short, broad, flattened apically. Conscutum length × breadth ranging from 3.06 mm × 1.99 mm to 4.87 mm × 3.25 mm; anterior process of coxae I small. In engorged specimens body wall expanded posterolaterally and posteriorly, with a single short broadly-rounded caudal process posteromedially. Eyes slightly convex, edged dorsally by a few punctations. Cervical pits deep, convergent. Marginal lines long, but not quite reaching eye level, punctate. Posteromedian and posterolateral grooves rather small, their floors rugose. A few large setiferous punctations anteriorly on the scapulae and along the external margins of the cervical fields. Otherwise, apart from the smooth areas surrounding the eyes, adjacent to the marginal lines and the festoons, the conscutum is covered with a dense, even pattern of medium-sized punctations that are often confluent. Ventrally spiracles comma-shaped, narrowing and curving gently towards

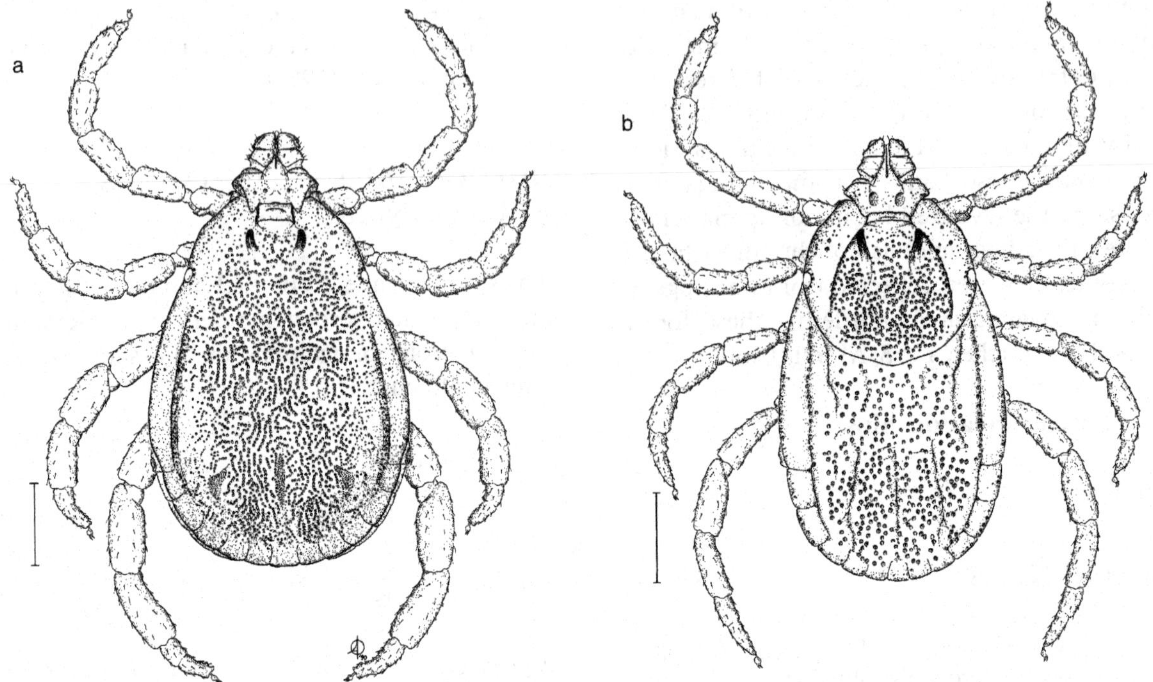

Figure 37. *Rhipicephalus compositus* [B.S.334/-, RML 66305, laboratory reared, original ♀ collected from African buffalo (*Syncerus caffer*), Thomson's Falls, Kenya, in *c.* 1951 by S.G. Wilson]. (a) Male, dorsal; (b) female, dorsal. Scale bars represent 1 mm. A. Olwage *del.*

the dorsal surface. Adanal plates large, broad, their external margins almost straight, their internal margins becoming concave posteriorly and joining the smoothly rounded posterior margins in sharp points; accessory adanal plates small, bluntly rounded, well sclerotized.

Female (Figs 37(b), 38(d) to (f))

Capitulum broader than long, length × breadth ranging from 0.56 mm × 0.64 mm to 0.99 mm × 1.02 mm. Basis capituli with broad lateral angles in anterior third of its length; porose areas large, oval, about twice their own diameter apart. Palps broad, slightly rounded apically. Scutum broader than long, length × breadth ranging from 1.10 mm × 1.18 mm to 1.88 mm × 2.06 mm; posterior margin broadly rounded. Eyes nearly halfway back, slightly convex, edged dorsally by a few punctations. Cervical fields broad, sharply demarcated laterally. A few large setiferous punctations present anteriorly on the scapulae, otherwise the lateral borders smooth and shiny. Medially the scutum is densely covered with medium-sized punctations that are often confluent. Alloscutum of unfed females deeply folded. Ventrally genital aperture a wide V, with the sides of the opening curving outwards from its narrowly-rounded base.

Nymph (Fig. 39)

Capitulum much broader than long, length × breadth ranging from 0.27 mm × 0.37 mm to 0.32 mm × 0.42 mm. Basis capituli nearly three times as broad as long, with tapering lateral angles extending over the scapulae and well-developed cornua; ventrally with bluntly-rounded spurs on posterior border. Palps long, narrow, equal in width for much of their length, tapering to pointed apices. Scutum as broad as long to slightly broader than long, length × breadth ranging from 0.56 mm × 0.56 mm to 0.60 mm × 0.65 mm; posterior margin a broad, smooth curve. Eyes at widest point, well over halfway back, delimited dorsally by slight de-

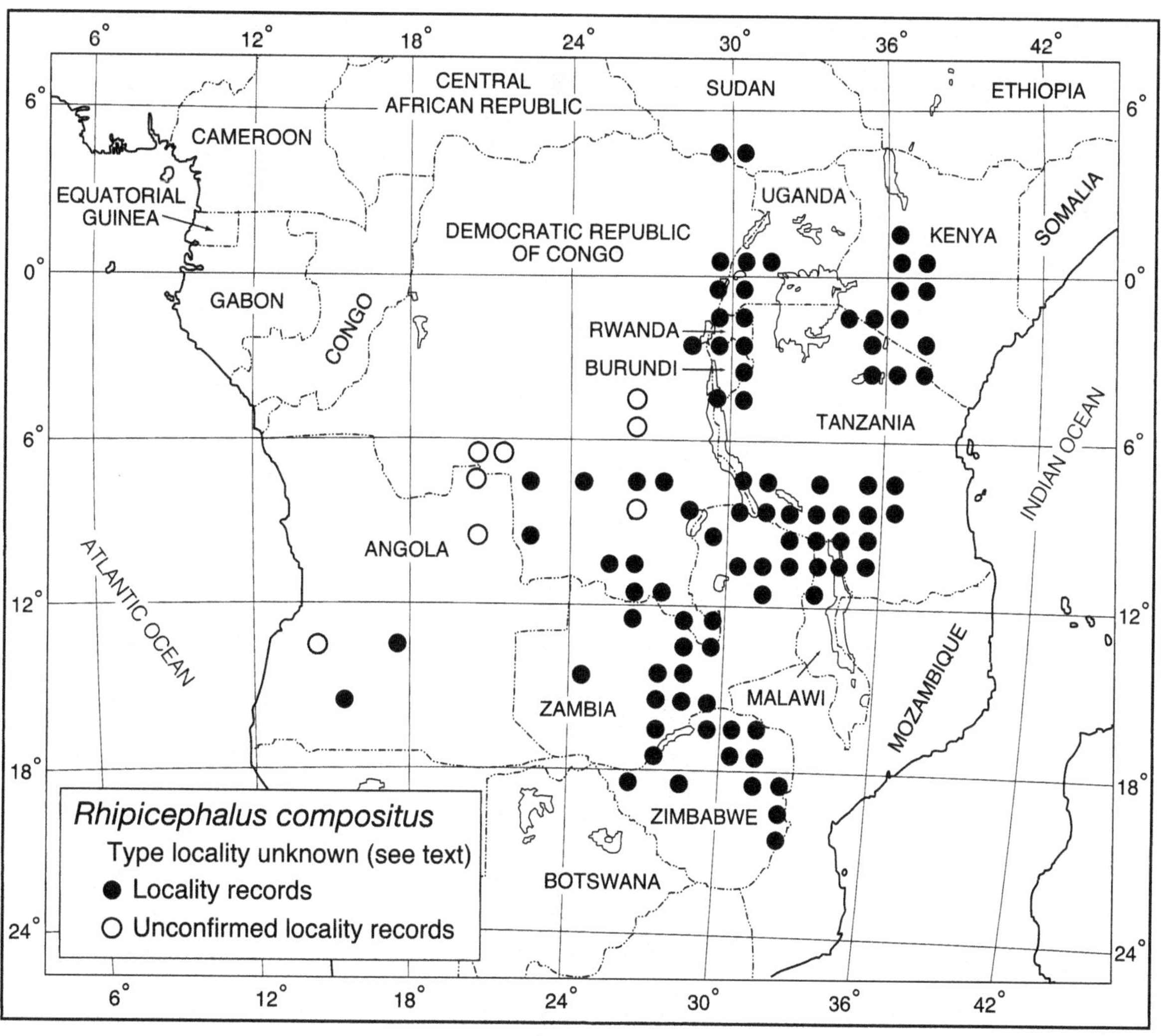

Map 13. *Rhipicephalus compositus*: distribution.

pressions. Cervical fields well defined, long, narrow, divergent. Ventrally coxae I each with an almost equal-sized external and internal spur; coxae II to IV each with a small blunt external spur only.

Larva (Fig. 40)

Capitulum much broader than long, length × breadth ranging from 0.116 mm × 0.174 mm to 0.143 mm × 0.184 mm. Basis capituli well over three times as broad as long, with tapering bluntly-rounded lateral angles, posterior border slightly concave; ventrally with blunt spurs on posterior border. Palps broad, tapering to fairly narrowly-rounded apices, inclined inwards. Scutum much broader than long, length × breadth ranging from 0.238 mm × 0.369 mm to 0.256 mm × 0.385 mm; posterior margin a wide smooth curve. Eyes at widest part of scutum, well over halfway back, slightly convex. Cervical grooves fairly short, slightly convergent. Ventrally coxae I each with a large pointed spur; coxae II and III each with an indication only of salience on its posterior border.

Notes on identification

The holotype male of *R. compositus* is said to have been taken from an unknown host at Khartoum,

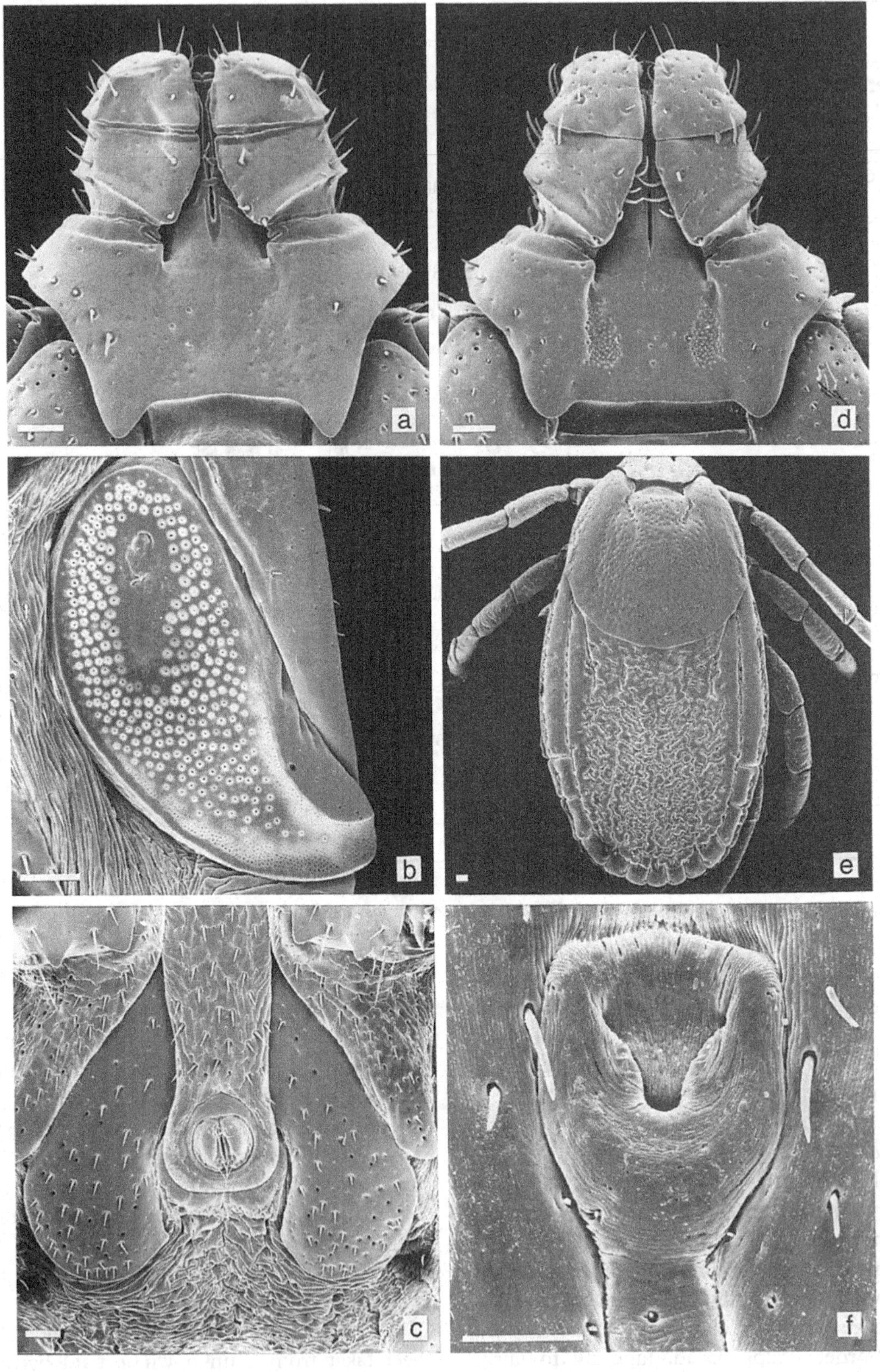
a
d
b
e
c
f

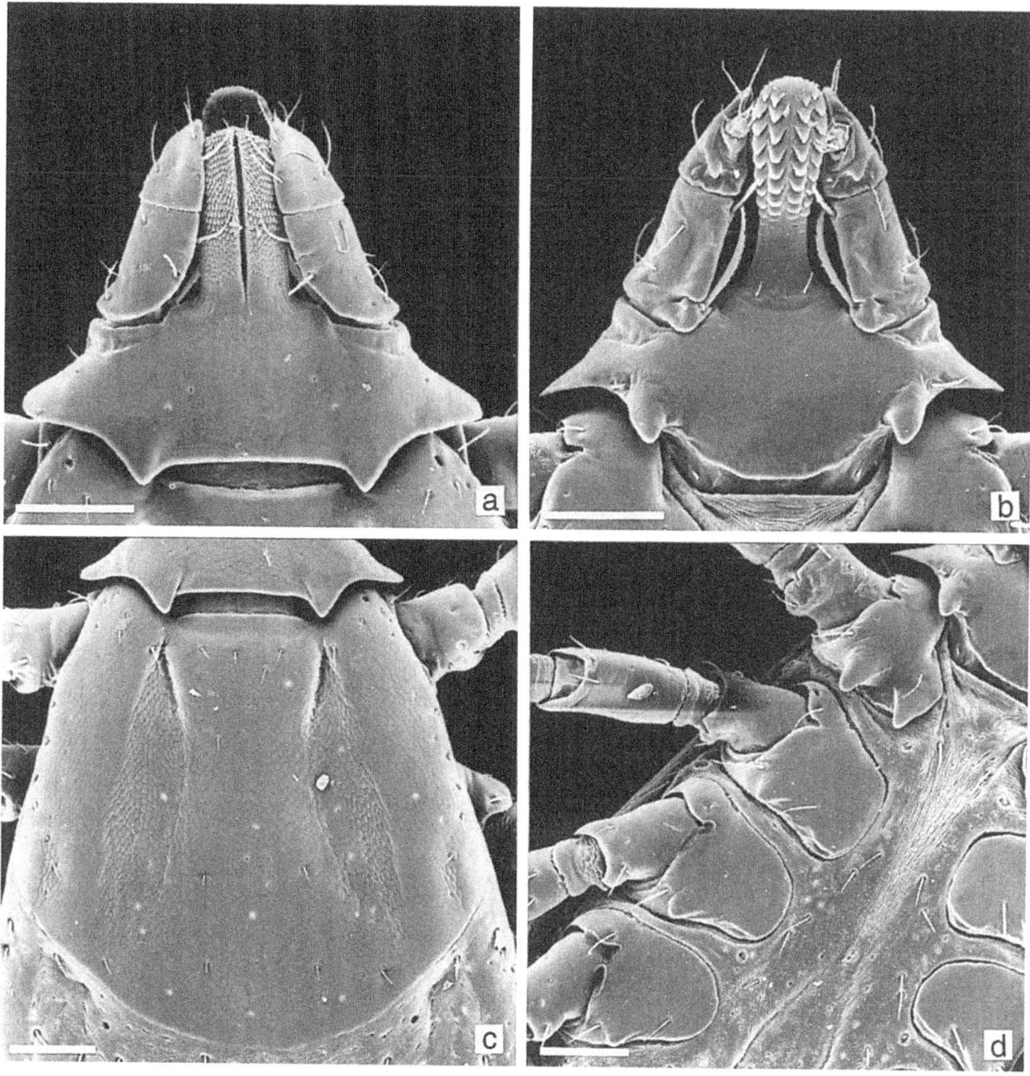

Figure 39 (*above*). *Rhipicephalus compositus* [B.S. 334/-, RML 66305, laboratory reared, original ♀ collected from African buffalo (*Syncerus caffer*), Thomson's Falls, Kenya, in *c.* 1951 by S.G. Wilson]. Nymph: (a) capitulum, dorsal; (b) capitulum, ventral; (c) scutum; (d) coxae. Scale bars represent 0.10 mm. SEMs by M.D. Corwin.

Figure 38 (*opposite*). *Rhipicephalus compositus* [B.S. 334/-, RML 66305, laboratory reared, original ♀ collected from African buffalo (*Syncerus caffer*), Thomson's Falls, Kenya, in *c.* 1951 by S.G. Wilson]. Male: (a) capitulum, dorsal; (b) spiracle; (c) adanal plates. Female: (d) capitulum, dorsal; (e) scutum and alloscutum, dorsal; (f) genital aperture. Scale bars represent 0.10 mm. SEMs by M.D. Corwin.

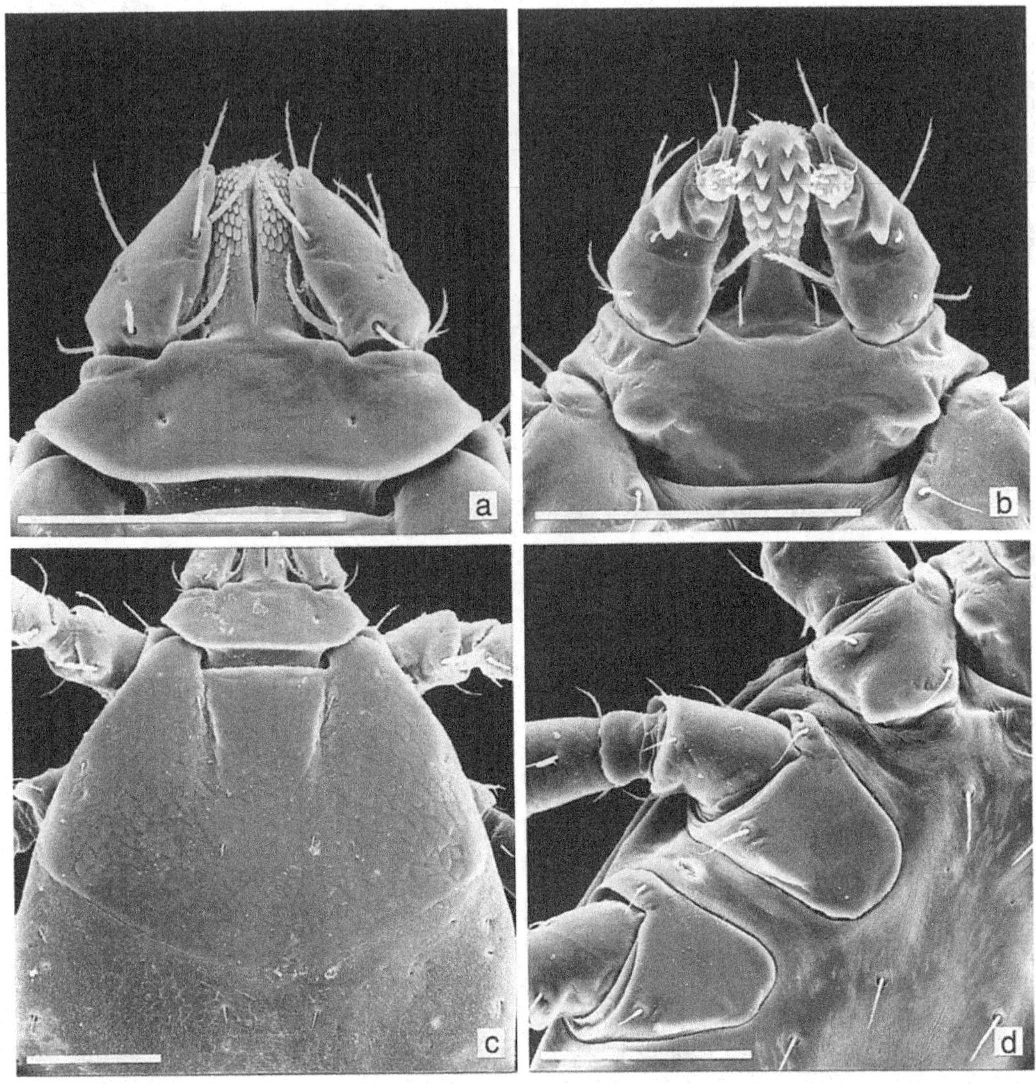

Figure 40. *Rhipicephalus compositus* [B.S. 334/-, RML 66305, laboratory reared, original ♀ collected from African buffalo (*Syncerus caffer*), Thomson's Falls, Kenya, in *c.* 1951 by S.G. Wilson]. Larva: (a) capitulum, dorsal; (b) capitulum, ventral; (c) scutum; (d) coxae. Scale bars represent 0.10 mm. SEMs by M.D. Corwin.

Sudan. Like Hoogstraal (1956) we believe that this is incorrect because it is a desert area that is not only ecologically unsuitable for this species but also far from any other places where it is known to occur. It was described as *Rhipicephalus ayrei* by Lewis (1933) and identified thus by A.J. Wiley (unpublished data, 1930s–1950s) and S.G. Wilson (unpublished data, 1951). This species was subsequently synonymized with *R. compositus* by Hoogstraal (1956).

Bergeon & Balis (1974) recorded *R. compositus* 1♂, 2♀♀ from cattle at Arjo, Wollega, Ethiopia, but Morel (1980) re-identified these ticks as atypically heavily-punctate *R. simus*. What they really are is somewhat open to question; specimens from Ethiopia identified as *R. simus* in the past are usually referred to as *R. praetextatus* now, but this is not a heavily-punctate species.

Walker (1974) listed two batches of ticks from the E.A. Lewis Collection, The Natural

Table 11. *Host records of* Rhipicephalus compositus *(provisional list)*

Hosts	Number of records
Domestic animals	
Cattle	149
Sheep	3
Goats	3
Donkeys	1
Pigs	1
Dogs	13
Wild animals	
Cheetah (*Acinonyx jubatus*)	1
Lion (*Panthera leo*)	10
Leopard (*Panthera pardus*)	1
African civet (*Civettictis civetta*)	1
Black rhinoceros (*Diceros bicornis*)	4
Warthog (*Phacochoerus africanus*)	7
Forest hog (*Hylochoerus meinertzhageni*)	1
Bushpig (*Potamochoerus larvatus*)	11
Giraffe (*Giraffa camelopardalis*)	1
Lichtenstein's hartebeest (*Sigmoceros lichtensteinii*)	1
African buffalo (*Syncerus caffer*)	72
Eland (*Taurotragus oryx*)	10
Bushbuck (*Tragelaphus scriptus*)	4
Sitatunga (*Tragelaphus spekii*)	2
Common duiker (*Sylvicapra grimmia*)	1
Roan antelope (*Hippotragus equinus*)	2
Waterbuck (*Kobus ellipsiprymnus*)	1
Puku (*Kobus vardonii*)	1
Nile rat (*Arvicanthis niloticus*)	1 (nymph)
Arvicanthis sp.[1]	1 (nymph)
Greater creek rat (*Pelomys fallax*)	1 (nymph)
'Creek rat' (*Pelomys* sp.)	1 (nymph)
Jackson's soft-furred rat (*Praomys jacksoni*)[2]	1 (larva)
Humans	1

Note: [1]Listed as an unstriped grass mouse (*Arvicanthis abyssinicus*) in Yeoman & Walker (1967). This species is now regarded as an Ethiopian endemic whose distribution range does not extend to Tanzania.
[2]Listed as Jackson's soft-furred mouse in Walker (1974).

History Museum, London, one from a rhinoceros, Mt. Kenya, 1930, labelled *R. sculptus*?, and the other from a buffalo, Kiago, labelled *R. protusus* (a *nomen nudem*), that she had re-identified as *R. compositus*.

Morel (1969, map 42) obviously accepted only some of the records from the Democratic Republic of Congo (formerly Zaire) listed as *R. compositus* by Elbl & Anastos (1966), apparently those from buffaloes in parts of the east and south of the country in Orientale, Kivu, southern Kasai and Katanga Provinces, which he plotted as '*Rhipicephalus compositus* (buffle)'. He entered their records from the western Democratic Republic of Congo as '*Rh. cliffordi* présumé'. We have followed his interpretation of their findings on the distribution of these two species (see *R. pseudolongus*, Map 48, p.361, and Map 13, p. 123).

Regarding his '*Rh. compositus* présumé' Morel stated: 'Dans la distribution figurée sur carte, à côté des symboles de *Rh. compositus* indubitable, ont été placés, sous un autre signe, des *Rhipicephalus* représentant vraisemblablement *Rh. compositus* du fait de leur distribution et de leur hôtes, et rapportés dans la littérature comme *Rh. capensis*'. We have mapped and designated these as 'Unconfirmed locality records'. Matthysse & Colbo (1987), who re-examined specimens from the Democratic Republic of Congo near the western border of Uganda, labelled *R. capensis* by Elbl & Anastos, also thought they were *R. compositus*.

To avoid possible confusion we have omitted the hosts listed for *R. compositus* in Elbl & Anastos (1966) from our provisional host list (see Table 11).

Nuttall Collection 1609, a male from *Potamochoerus larvatus* (listed as *P. porcus*) at Kaporo, Malawi, collected on 6 January 1909 by Dr J.B. Davey and identified as *R. gertrudae* by Keirans (1985), has been re-identified as *R. compositus*.

Matthysse & Colbo (1987) discussed the morphological differences between the adults of *R. pseudolongus* (listed as *R. cliffordi*), *R. compositus* and *R. longus* (see *R. pseudolongus*, p.

360). These species are sometimes extremely difficult to separate.

Hosts

A three-host species (Theiler, Walker & Wiley, 1956, as *R. ayrei*). Cattle are the most commonly recorded hosts of *R. compositus* adults, which are usually attached on the escutcheon and udder or scrotum. Apart from dogs, from which a number of collections have been made, other domestic animals appear to be of little consequence as hosts (Table 11). Amongst wild animals the large black African buffalo, as distinct from the smaller reddish forest buffalo, is by far its most important host, from which much greater numbers of adults have been recorded on individual animals than they have on other host species. They were attached on the ears, sternum, body, axillae, scrotum or teats, and tail. Collections have also been obtained from a few wild carnivores, pigs and various antelopes.

The few immature specimens identified thus far have been collected from rodents. Pegram *et al.* (1986) noted that their highest numbers of *R. compositus* adults were obtained from cattle in a herd kept permanently on or adjacent to the Kafue flood plains in Zambia. They continued: 'The few recorded hosts of immature stages of *R. compositus* include creek rats (*Pelomys fallax*) (Walker, 1974) which are common on the Kafue flats, and this may contribute to the greater abundance of adults of this species of tick'. It is therefore interesting that south-western Uganda, the only part of the country where *R. compositus* has been collected, is the only area where Delaney (1975) recorded *P. fallax*. He also recorded the Nile rat (*Arvicanthis niloticus*) quite commonly in Toro and Ankole Districts.

In both Zambia and Zimbabwe *R. compositus* adults are apparently most active in the hot dry season before the rains (September to October or November: MacLeod *et al.*, 1977; MacLeod & Mwanaumo, 1978; Norval & Tebele, 1984; Pegram *et al.*, 1986). MacLeod (1970) noted that in Northern Province, Zambia, where it occurs on cattle throughout the year, numbers start to increase in July, peak in September, and decline in January. In Uganda it has no seasonal activity pattern.

Zoogeography

Rhipicephalus compositus has been recorded primarily in parts of East and Central Africa, with extensions further south into Zimbabwe and west into Angola (Map 13).

It occurs under a wide range of ecological conditions at medium to high altitudes of approximately 1200 m to 1800 m, usually with mean annual rainfalls of at least 700 mm to over 1600 mm, in various types of forest, woodland, bushland and wooded and/or bushed grassland. It is sometimes associated with ticks in the *R. hurti/jeanneli* complex and with *R. punctatus*.

Disease relationships

According to Wilson (1953) *R. compositus* (listed as *R. ayrei*) can transmit *Theileria parva*, the causative agent of East Coast fever of cattle. This observation is thought to be based on laboratory experiments (Chief Field Zoologist, Kenya, unpublished report for 1951). Considering the hosts of this tick's immature stages it seems extremely unlikely that it would ever act as a field vector of this protozoan.

REFERENCES

Bergeon, P. & Balis, J. (1974). Contribution à l'étude de la répartition des tiques en Ethiopie (enquête effectuée de 1965 à 1969). *Revue d'Élevage et de Médecine Vétérinaire des Pays Tropicaux*, 27 (nouvelle série), 285–99.

Delaney, M.J. (1975). *The Rodents of Uganda*. London: Trustees of the British Museum (Natural History).

Lewis, E.A. (1933). *Rhipicephalus ayrei* n.sp. (a tick) from Kenya Colony. *Parasitology*, **25**, 269–72.

MacLeod, J. (1970). Tick infestation patterns in the Southern Province of Zambia. *Bulletin of Entomological Research*, **60**, 253–74.

MacLeod, J., Colbo, M.H., Madbouly, M.H. & Mwanaumo, B. (1977). Ecological studies of ixodid ticks (Acari: Ixodidae) in Zambia. III. Seasonal activity and attachment sites on cattle, with notes on other hosts. *Bulletin of Entomological Research*, **67**, 161–73.

MacLeod, J. & Mwanaumo, B. (1978). Ecological studies of ixodid ticks (Acari: Ixodidae) in Zambia. IV. Some anomalous infestation patterns in the northern and eastern regions. *Bulletin of Entomological Research*, **68**, 409–29.

Neumann, L.G. (1897). Révision de la famille des Ixodidés. (2e Mémoire). *Mémoires de la Société Zoologique de France*, **10**, 324–420.

Norval, R.A.I. & Tebele, N. (1984). The ticks of Zimbabwe. VIII. *Rhipicephalus compositus*. *Zimbabwe Veterinary Journal*, **15**, 3–8.

Pegram, R.G., Perry, B.D., Musisi, F.L. & Mwanaumo, B. (1986). Ecology and phenology of ticks in Zambia: seasonal dynamics on cattle. *Experimental and Applied Acarology*, **2**, 25–45.

Theiler, G., Walker, J.B. & Wiley, A.J. (1956). Ticks in the South African Zoological Survey Collection. Part VIII. Two East African ticks. *Onderstepoort Journal of Veterinary Research*, **27**, 83–99.

Wilson, S.G. (1953). A survey of the distribution of tick vectors of East Coast fever in East and Central Africa. *Proceedings of the 15th International Veterinary Congress, Stockholm*, **1**, 187–90.

Also see the following Basic References (pp. 12–14): Elbl & Anastos (1966); Hoogstraal (1956); Keirans (1985); Matthysse & Colbo (1987); Morel (1969, 1980); Walker (1974); Yeoman & Walker (1967).

RHIPICEPHALUS CUSPIDATUS NEUMANN, 1906

The specific name, from the Latin *cuspis* meaning 'point' or 'apex', refers to the characteristic shape of the elongated posterointernal angles of the male adanal plates.

Diagnosis

A moderate-sized reddish-brown tick.

Male (Figs 41(a), 42(a) to (c))

Capitulum slightly longer than broad, length × breadth ranging from 0.75 mm × 0.72 mm to 0.94 mm × 0.88 mm. Basis capituli with short blunt lateral angles in anterior quarter of its length. Palps short, broad, convex medially and apically. Conscutum flat, length × breadth ranging from 2.98 mm × 2.20 mm to 3.75 mm × 2.65 mm; anterior process of coxae I rounded, heavily sclerotized. Eyes flat. Cervical pits comma-shaped, convergent. Marginal lines deep. Posteromedian and posterolateral grooves either absent or only faintly indicated. Punctation pattern sparse. A few very large irregularly-shaped setiferous punctations along the external cervical margins link up posterior to the eyes with those in the marginal lines. Similar large punctations are scattered on the posterior half of the conscutum, interspersed laterally and anterior to the festoons with diffuse areas sprinkled with smaller punctations. Several smaller punctations also present on the scapulae and anteromedially on the conscutum. Ventrally spiracles long, with a broad curved dorsal prolongation. Adanal plates unique in shape, broadly triangular anteriorly with their posteroexternal margins rounded and their posterointernal margins produced into long convergent cusps that overlap the festoons and give this species its name; accessory adanal plates virtually absent.

Female (Figs 41(b), 42(d) to (f))

Capitulum broader than long, length × breadth ranging from 0.98 mm × 1.00 mm to 1.00 mm × 1.10 mm in the two specimens measured. Basis capituli with sharp, somewhat recurved lateral angles at about mid-length; porose areas large, round, about twice their own diameter apart. Palps short, broad, convex medially and apically. Scutum flat, broader than long, length × breadth ranging from 1.79 mm × 2.14 mm to 2.24 mm × 2.46 mm in the two specimens measured; posterolateral margins straight to slightly sinuous. Eyes just anterior to broadest part of scutum, flat. Cervical pits long, deep, convergent; cervical fields broad, slightly depressed, their external margins delimited by a few

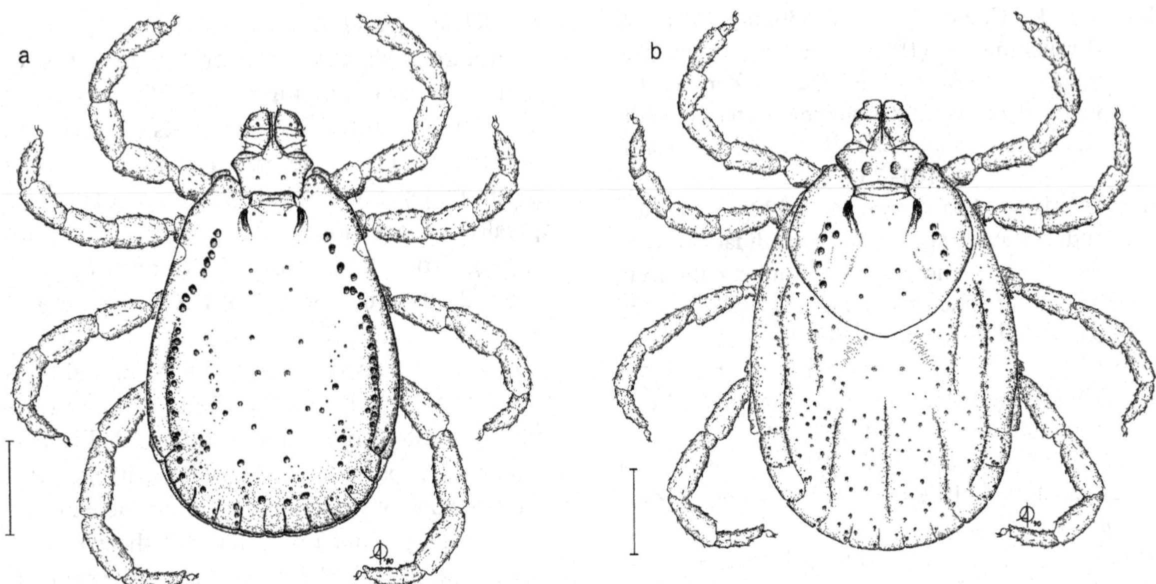

Figure 41. *Rhipicephalus cuspidatus* (Onderstepoort Tick Collection 2967i, collected in Galual-Nyang forest, Gogrial Toni, Bahr el Ghazal, Sudan on 24 February 1953 by H. Hoogstraal). (a) Male, dorsal; (b) female, dorsal. Scale bars represent 1 mm. A. Olwage *del.*

Table 12. *Host records of* Rhipicephalus cuspidatus

Hosts	Number of records
Wild animals	
Golden jackal (*Canis aureus*)	1 (immatures)
'Jackal'	1
Serval (*Leptailurus serval*)	1 (1 nymph)
Leopard (*Panthera pardus*)	1 (immatures)
Egyptian mongoose (*Herpestes ichneumon*)	1 (immatures)
White-tailed mongoose (*Ichneumia albicauda*)	1 (immatures)
Banded mongoose (*Mungos mungo*)	2 (nymphs)
Spotted hyaena (*Crocuta crocuta*)	2 (including nymphs)
Striped hyaena (*Hyaena hyaena*)	2 (including immatures)
'Hyaena'	1
Aardvark (*Orycteropus afer*)	5
Warthog (*Phacochoerus africanus*)	36 (including immatures)
Red river hog (*Potamochoerus porcus*)	1
Red-fronted gazelle (*Gazella rufifrons*)	1 (immatures)
Oribi (*Ourebia ourebi*)	1 (immatures)
Crested porcupine (*Hystrix cristata*)	6 (including immatures)
Greater cane rat (*Thryonomys swinderianus*)	1 (including immatures)
Birds	
Double-spurred francolin (*Francolinus bicalcaratus*)	1 (immatures)

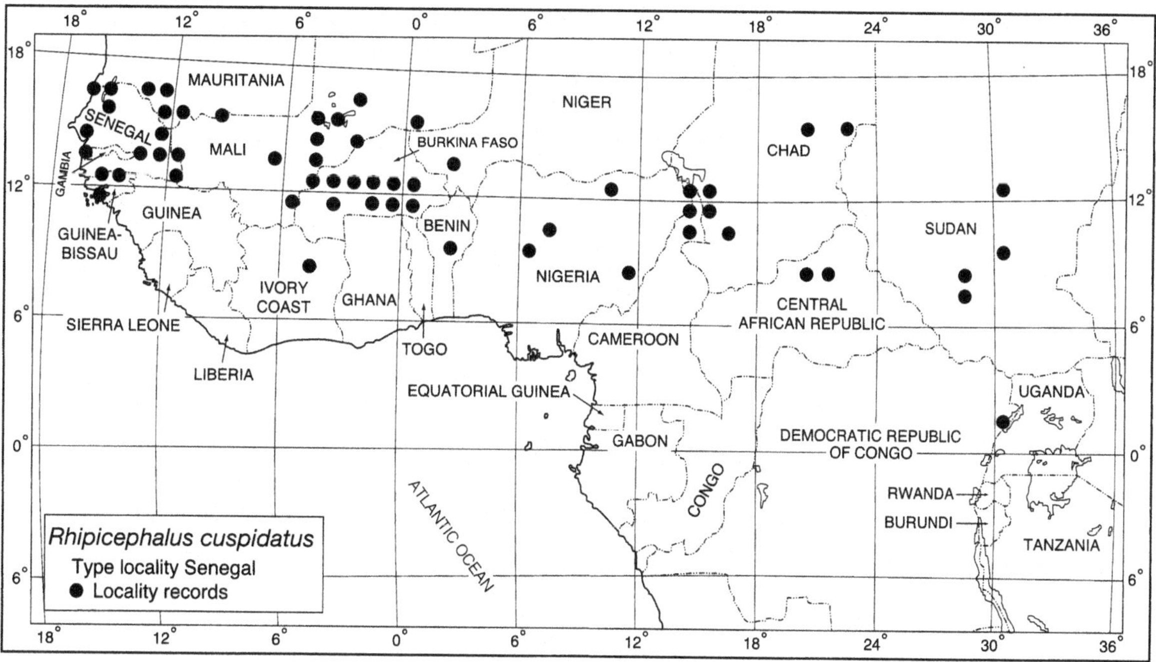

Map 14. *Rhipicephalus cuspidatus:* distribution. (Based in part on Morel, 1969).

very large setiferous punctations. Several small punctations present on the scapulae and a few others, slightly larger, between the cervical fields. Ventrally genital aperture very broad, crescentic.

Nymph (Fig. 43)

Capitulum slightly broader than long, length × breadth 0.36 mm × 0.37 mm in the only specimen measured. Basis capituli well over twice as broad as long, with short anteriorly-placed lateral angles. Palps short, broad, truncated apically. Scutum much broader than long, length × breadth 0.65 mm × 0.81 mm in the only specimen measured; posterior margin smoothly curved medially, slightly sinuous laterally. Eyes at widest point, flat. Cervical pits long, convergent, continuing as long, divergent cervical grooves that extend to the posterolateral margins, thus dividing the scutum into a median and two lateral areas. Ventrally coxae I each with two short stout spurs, almost equal in length; coxae II to IV each with an external spur only.

Larva (Fig. 44)

Capitulum slightly longer than broad, length × breadth ranging from 0.202 mm × 0.201 mm to 0.234 mm × 0.228 mm. Basis capituli twice as broad as long, very slightly convex laterally, posterior border straight. Palps broad, slightly constricted proximally, otherwise virtually equal in width throughout their length, truncated apically. Scutum much broader than long, length × breadth ranging from 0.333 mm × 0.519 mm to 0.352 mm × 0.545 mm; posterior margin smoothly curved medially, straight laterally. Eyes at widest part of scutum, halfway back, flat. Cervical grooves almost reaching eye level, slightly convergent. Ventrally coxae I each with a large broadly-rounded spur; coxae II and III each with a smaller spur.

Notes on identification

A detailed description of the adults of *R. cuspidatus*, with numerous measurements and illustrations of the variations in the structure of the male's adanal plates and adjacent structures, has been given by Tendeiro (1951).

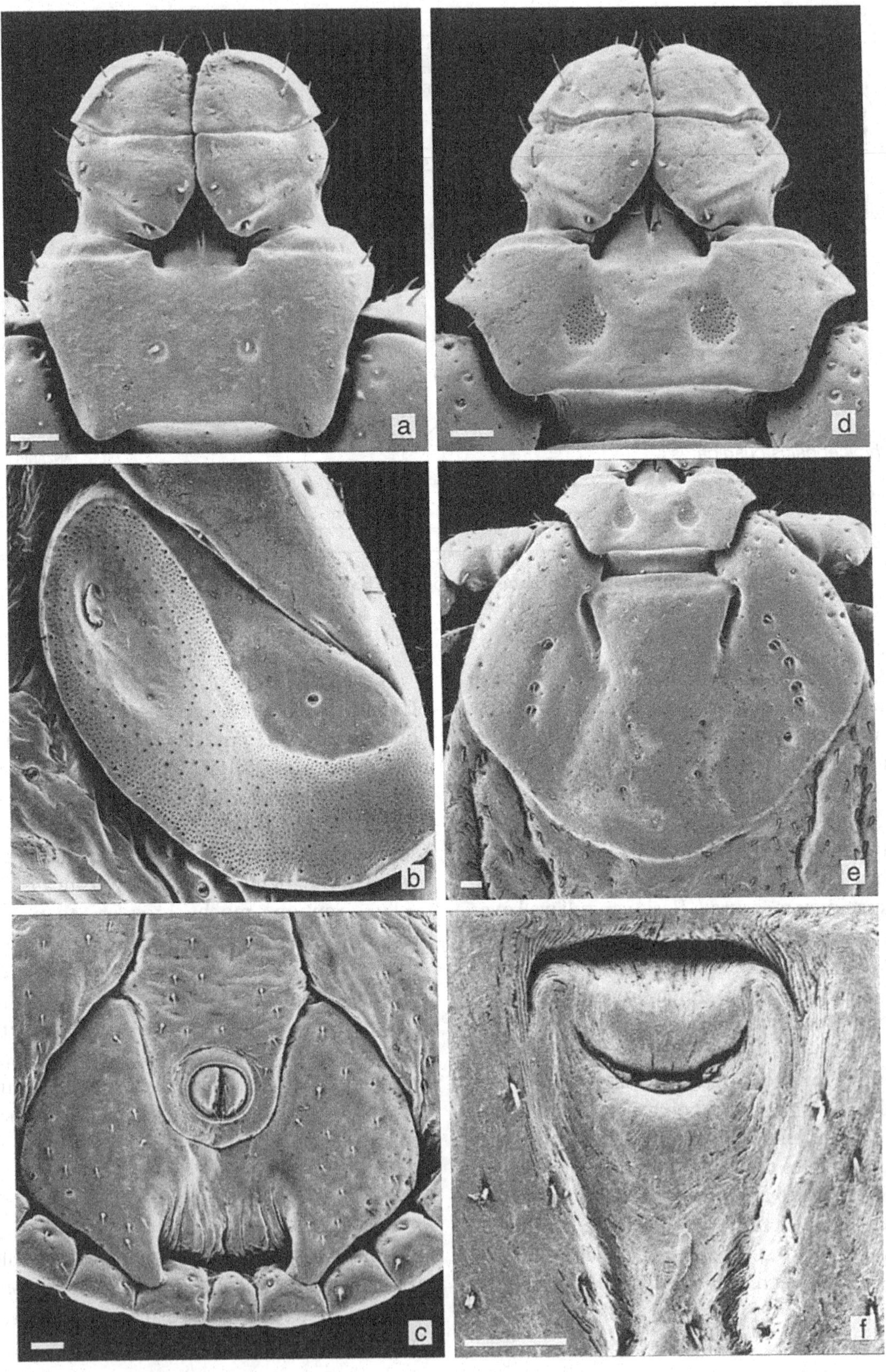
a
d
b
e
c
f

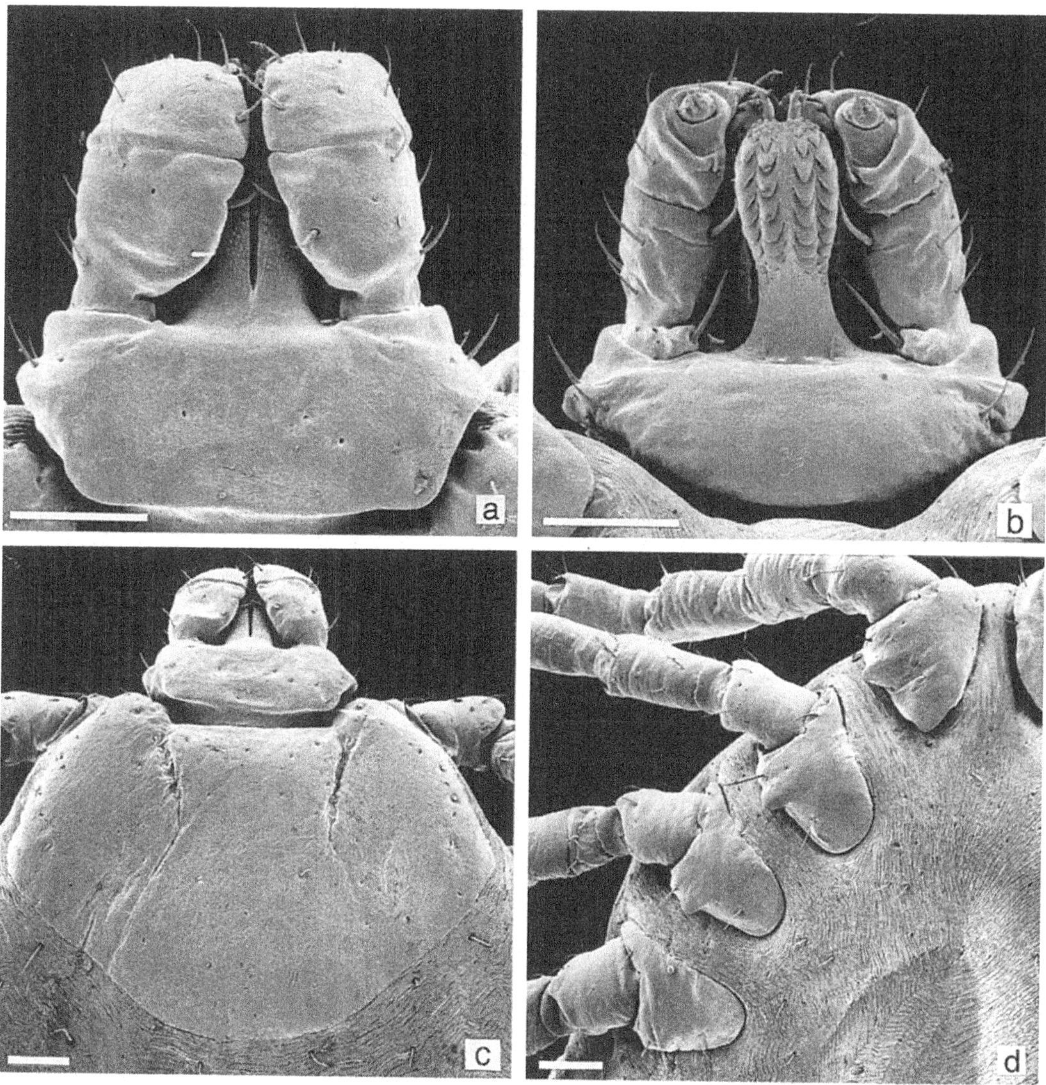

Figure 43 (*above*). *Rhipicephalus cuspidatus* [collected from warthog (*Phacochoerus africanus*), Njamena, Chad, on 11 November 1971 and donated by P.C. Morel]. Nymph: (a) capitulum, dorsal; (b) capitulum, ventral; (c) scutum; (d) coxae. Scale bars represent 0.10 mm. SEMs by J.F. Putterill.

Figure 42 (*opposite*). *Rhipicephalus cuspidatus* [collected from warthog (*Phacochoerus africanus*), Njamena, Chad, on 11 November 1971 and donated by P.C. Morel]. Male: (a) capitulum, dorsal; (b) spiracle; (c) adanal plates. Female: (d) capitulum, dorsal; (e) scutum; (f) genital aperture. Scale bars represent 0.10 mm. SEMs by J.F. Putterill.

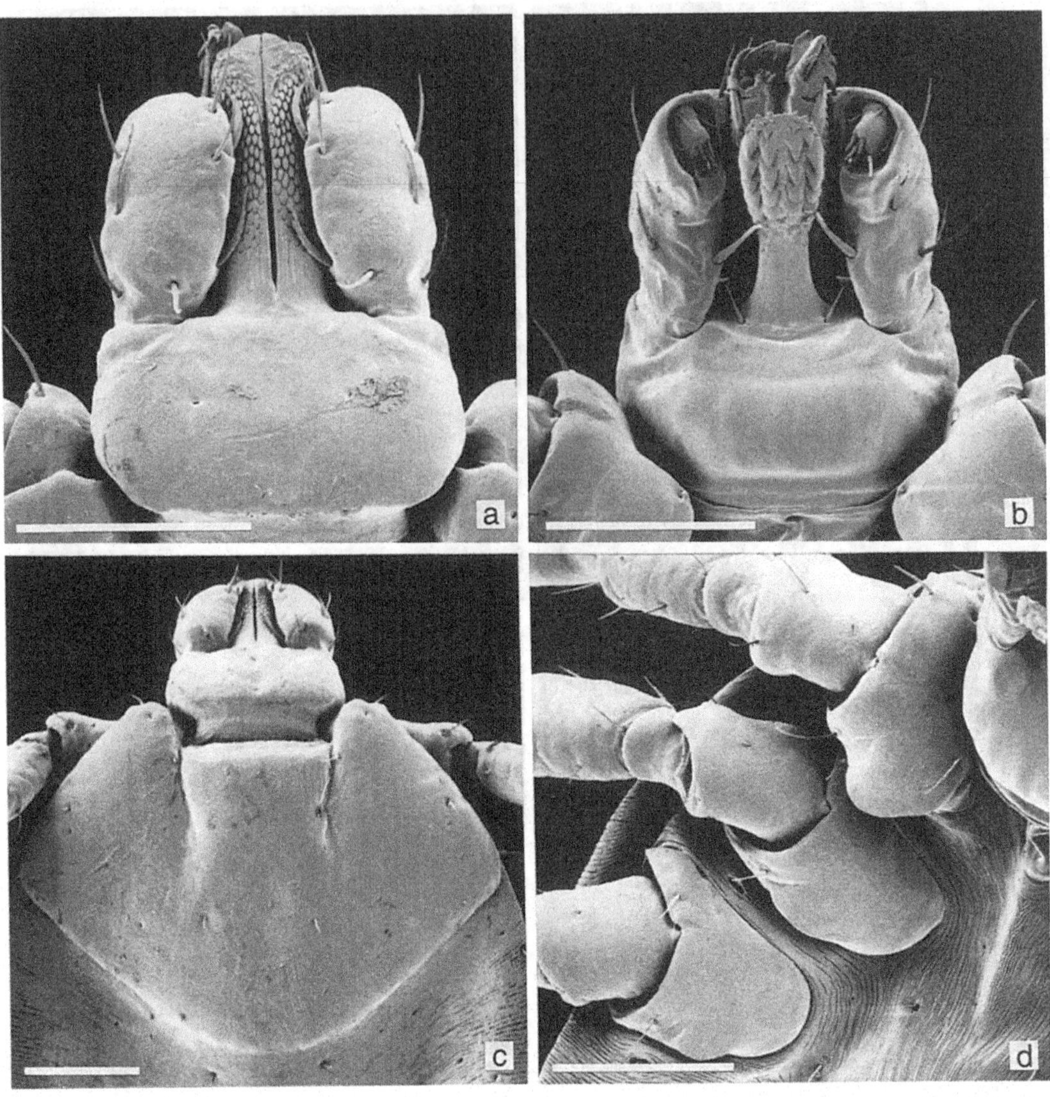

Figure 44. *Rhipicephalus cuspidatus* [progeny of ♀ collected from warthog (*Phacochoerus africanus*), St Louis, Senegal, on 27 January 1964 and donated by P.C. Morel]. Larva: (a) capitulum, dorsal; (b) capitulum, ventral; (c) scutum; (d) coxae. Scale bars represent 0.10 mm. SEMs by J.F. Putterill.

According to Hoogstraal (1956) this species was listed from the Harar area, Ethiopia, by Stella (1940), but he considered this record doubtful and it has not been included here. We know of no other records from Ethiopia.

Hosts

A three-host species (Morel & Mouchet, 1958). It has apparently not been reared in the laboratory but larvae were obtained from eggs laid by an engorged female. Nymphs were identified by collecting engorged specimens in the field and keeping them alive until they moulted into adults (Morel, 1956).

Morel (1969) regards the aardvark as the primary host of this tick, with other animals that share its burrows, especially warthogs, acting as secondary hosts. Porcupines and warthogs appear to be the usual hosts of the immature stages (Table 12). Overall, though, there are far more records from warthogs than aardvarks. This is probably because warthogs, besides being commoner animals, are active diurnally and are often seen. Aardvarks, though very widely distributed, are far less common. They are active almost exclusively nocturnally and consequently are rarely encountered. Some collections of *R. cuspidatus* adults have been made just inside the entrances of burrows inhabited by their hosts. The immature stages may also be found there during the dry season (Morel & Mouchet, 1958; Morel, 1969). However, Morel (1978) also listed numerous other hosts of this tick that do not inhabit burrows.

Zoogeography

Rhipicephalus cuspidatus has been recorded mainly in a band stretching nearly three-quarters of the way across Africa between latitudes 8 °N and 17 °N, from Senegal on the western seaboard of the continent to the southern Sudan (Map 14) (Morel, 1956, 1961; Morel & Finelle, 1961; Morel & Graber, 1961; Morel & Mouchet, 1958, 1965; Tendeiro, 1951). As yet there are few records outside this area, and the tick apparently does not follow its principal hosts throughout their range further eastwards and southwards in Africa.

Most records of *R. cuspidatus* come from places at less than 1000 m in altitude with mean annual rainfalls between 400 mm and 1400 mm, occasionally less, in Sudanian woodland or Sahel *Acacia* wooded grassland and deciduous bushland. It has sometimes been found in areas that experience either higher or lower rainfalls. The conditions there are undoubtedly moderated by the microhabitat in the hosts' burrows where it spends much of its time (Morel, 1969).

Disease relationships

According to Tendeiro (1954) *Coxiella burneti* has been isolated from this tick.

REFERENCES

Morel, P.C. (1956). Le Parc National du Niokola-Koba. (Premier Fascicule). XV. Tiques d'animaux sauvages. *Mémoires de l'Institut Français d'Afrique Noire*, **No. 48**, 229–32.

Morel, P.C. (1961). Le Parc National du Niokola-Koba. (Deuxième Fascicule). V. Tiques (Acarina, Ixodoidea). (Deuxième note). *Mémoires de l'Institut Français d'Afrique Noire*, **No. 62**, 83–90.

Morel, P.C. (1978). Tiques d'animaux sauvages en Haute-Volta. *Revue d'Élevage et de Médecine Vétérinaire des Pays Tropicaux*, **31** (nouvelle série), 69–78.

Morel, P.C. & Finelle, P. (1961). Les tiques des animaux domestiques du Centrafrique. *Revue d'Élevage et de Médecine Vétérinaire des Pays Tropicaux*, **14** (nouvelle série), 191–7.

Morel, P.C. & Graber, M. (1961). Les tiques des animaux domestiques du Tchad. *Revue d'Élevage et de Médecine Vétérinaire des Pays Tropicaux*, **14** (nouvelle série), 199–203.

Morel, P.C. & Mouchet, J. (1958). Les tiques du Cameroun (Ixodidae et Argasidae). *Annales de Parasitologie Humaine et Comparée*, **33**, 69–111.

Morel, P.C. & Mouchet, J. (1965). Les tiques du Cameroun. (2e note). *Annales de Parasitologie Humaine et Comparée*, **40**, 477–96.

Neumann, L.G. (1906). Notes sur les Ixodidés. IV. *Archives de Parasitologie*, **10**, 195–219.

Stella, E. (1940). Nuovi dati sugli ixodidi dell'Africa Orientale Italiana. *Revista di Biologia Coloniale*, **3**, 431–5.

Tendeiro, J. (1951). Ixodídeos da Guiné Portuguesa. Nota sobre duas carraças do género *Rhipicephalus*. *Boletim Cultural da Guiné Portuguesa*, **No. 24**, 909–28.

Tendeiro, J. (1954). Posição actual do problema da febre Q. *Revista de Ciencias Veterinarias*, **49** (350), 283–311.

Also see the following Basic References (pp. 12–14): Aeschlimann (1967); Elbl & Anastos (1966); Hoogstraal (1956); Morel (1969); Theiler (1947).

RHIPICEPHALUS DELTOIDEUS NEUMANN, 1910

The specific name *deltoideus*, based on the Greek letter *delta*, Δ, refers to the subtriangular shape of the male.

Diagnosis

A large shiny tick, dark brown posteriorly, lighter and more reddish anteromedially.

Male (Fig. 45(a) to (c))
Total length (including the capitulum) × breadth 4.2 mm × 2.4 mm. Capitulum 0.7 mm long. Basis capituli much broader than long (3.5:2), with long sharp lateral angles in the anterior third of its length and large cornua. Palps short and broad. Conscutum subtriangular in shape; anterior process on coxae I inconspicuous. Body wall expanded slightly laterally and posteriorly. Eyes flat, delimited dorsally by punctations. Cervical fields depressed, short, narrow and somewhat divergent. Marginal lines long, punctate, almost reaching eye level anteriorly. Anteromedially the slightly raised outline of a female pseudoscutum is visible, from which two mildly convergent grooves extend posteriorly. Towards the posterior end of the conscutum each of these grooves divides and their branches, which are filled with confluent punctations, bound three almost smooth elevated areas, each with a small round central pit. (These little hollows appear to represent the more usual elongated posteromedian and posterolateral grooves). Large punctations present on the scapulae, laterally and anteromedially on the pseudoscutum, and along the sides of the conscutum outside the marginal lines. Elsewhere punctations very fine. Ventrally spiracles comma-shaped. Adanal plates also comma-shaped, their anterointernal margins slightly concave, broadly rounded posteriorly; accessory adanal plates well developed.

Female (Fig. 45(d))
Capitulum broader than long. Basis capituli over three times as broad as long, length × breadth of the only existing specimen 0.51 mm × 1.68 mm; lateral angles at about mid-length, acute, cornua large; porose areas large, oval, deep, slightly more than their own diameter apart. Palps short, their apices broadly rounded. Scutum longer than broad, length × breadth of the specimen measured 1.97 mm × 1.77 mm; posterior margin sinuous. Eyes at about mid-length of scutum, flat, edged dorsally by a few large setiferous punctations. Cervical fields broad, slightly divergent, depressed, their external margins delimited by irregular rows of large setiferous punctations. Somewhat smaller setiferous punctations present on the scapulae and medially on the scutum, interspersed with numerous fine punctations. Ventrally genital aperture very broadly U-shaped with short, slightly diverging lateral arms.

Immature stages
Unknown.

Notes on identification
Neumann's original description was based on one male and three partly-engorged females. Of these only one cotype female, Nuttall Collection 2894, donated by Neumann to G.H.F. Nuttall, is now known to exist, and the above description and illustration have been based on this specimen, by courtesy of The Natural History Museum, London. The description and illustrations of the male are based on those originally published by Neumann.

It is interesting that in two respects the male of *R. deltoideus* is apparently similar to that of *R. armatus*. In both species the lateral angles of the basis capituli are anteriorly placed and elongated, and both have two longitudinal grooves on the conscutum that bifurcate posteriorly and surround three elevated areas in front of the festoons. The females of these two species, however, are not alike morphologically.

We have seen only the cotype female of *R. deltoideus*. Theiler (1947) listed several collections, all but one provisionally, but almost all

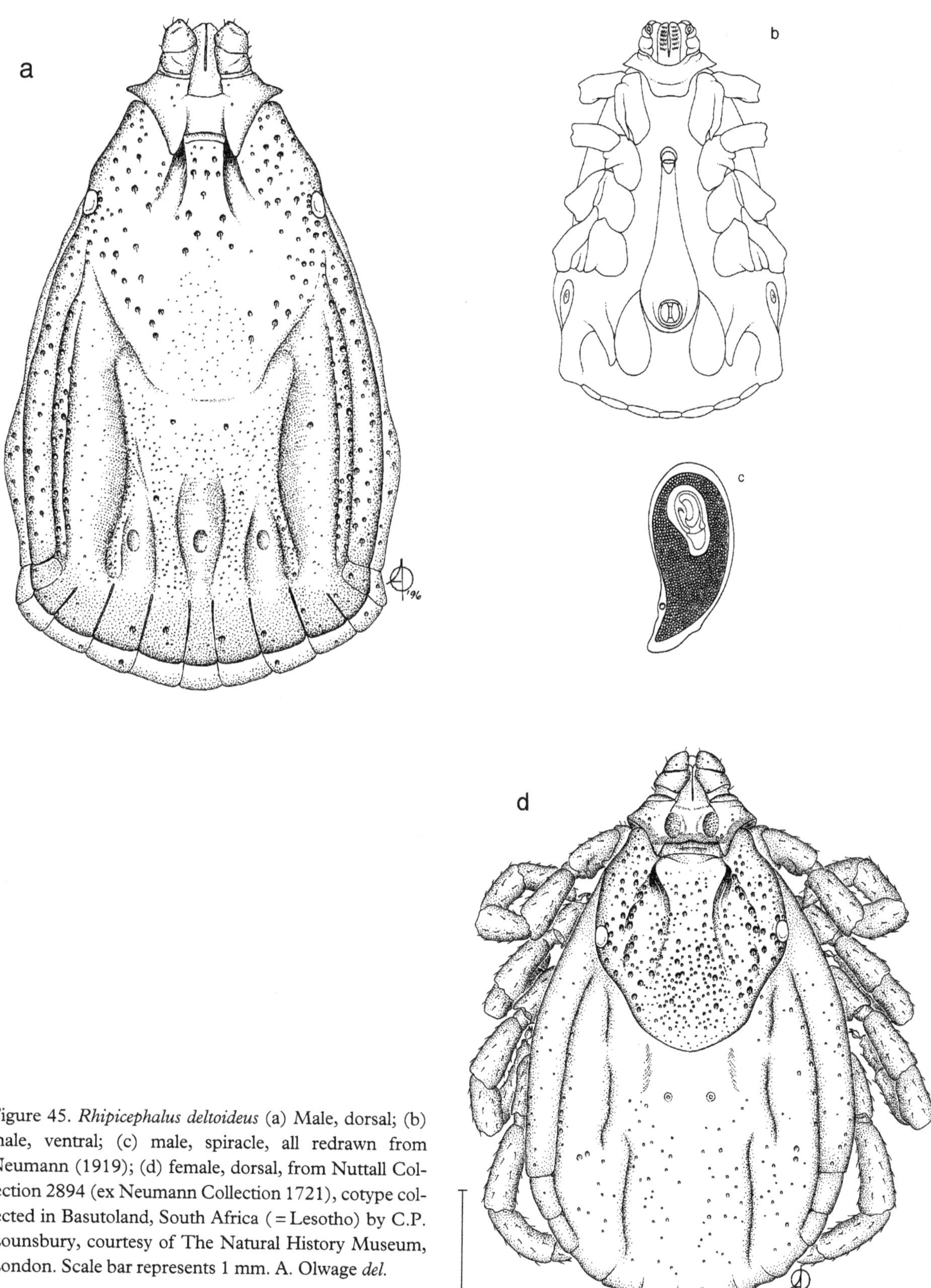

Figure 45. *Rhipicephalus deltoideus* (a) Male, dorsal; (b) male, ventral; (c) male, spiracle, all redrawn from Neumann (1919); (d) female, dorsal, from Nuttall Collection 2894 (ex Neumann Collection 1721), cotype collected in Basutoland, South Africa (= Lesotho) by C.P. Lounsbury, courtesy of The Natural History Museum, London. Scale bar represents 1 mm. A. Olwage *del.*

these ticks have subsequently been re-identified. The two females from a wild hare at Palmietfontein, near Richmond, Northern Cape Province, are *R. arnoldi* (OP 2583i), while the specimens from various hosts in Uganda, collected by T.W. Chorley, are *R. turanicus* (OP 2693i, 2731i & ii). Our attempt to borrow the collection from wild hares, Ishasa River, north of Rutshuru, Democratic Republic of Congo (formerly the Belgian Congo), on which the record in Bequaert (1931) was based, was unsuccessful. It was thought to be in the collections of the Museum of Comparative Zoology, Harvard University, but could not be found.

Hosts

Unknown.

Zoogeography

The original collection was made in Lesotho (formerly Basutoland), at an altitude of 2135 m, but the precise locality was not recorded by the collector, C.P. Lounsbury.

Disease relationships

Unknown.

REFERENCES

Bequaert, J.C. (1931). Synopsis des tiques du Congo Belge. *Revue de Zoologie et de Botanique Africaine*, **20**, 209–51.

Neumann, L.G. (1910). Description de deux nouvelles espèces d'Ixodinae. *Tijdschrift voor Entomologie*, **53**, 11–17 + 1 plate.

Also see the following Basic Reference (p. 14): Theiler (1947).

RHIPICEPHALUS DISTINCTUS BEDFORD, 1932

The specific name *distinctus*, from the Latin meaning 'distinguish', probably refers to the initial confusion regarding the naming of this tick. It was originally described as *R. punctatus* by Bedford (1929). In 1932, though, he proposed that, as this specific name was preoccupied by *R. neavei* var. *punctatus* Warburton, 1912, it should be renamed *distinctus*.

Synonym

punctatus Bedford, 1929 (*nec* Warburton, 1912).

Diagnosis

A small glossy dark brown tick.

Male (Figs 46(a), 47(a) to (c))

Capitulum broader than long, length × breadth ranging from 0.42 mm × 0.47 mm to 0.52 mm × 0.56 mm. Basis capituli with short acute lateral angles at anterior third of its length, narrowing conspicuously to straight posterior margin. Palps short, broad. Conscutum length × breadth ranging from 2.05 mm × 1.35 mm to 2.57 mm × 1.73 mm; anterior process on coxae I small. Eyes almost flat, edged dorsally by a few large setiferous punctations set in slight depressions. Cervical pits small; external margins of cervical fields merely indicated by rows of large setiferous punctations. Marginal lines short, outlined by large setiferous punctations that continue forwards to eye level. Posterior grooves only faintly indicated. In general conscutum smooth and glossy. Medium-sized setiferous punctations present on the scapulae and in a '*simus*' pattern anteromedially on the conscutum, becoming progressively larger and more numerous posteriorly. Ventrally spiracles long, narrowing markedly at about mid-length and curving gently towards the dorsal surface. Adanal plates more or less equal in width thoughout their length, their internal margins slightly concave

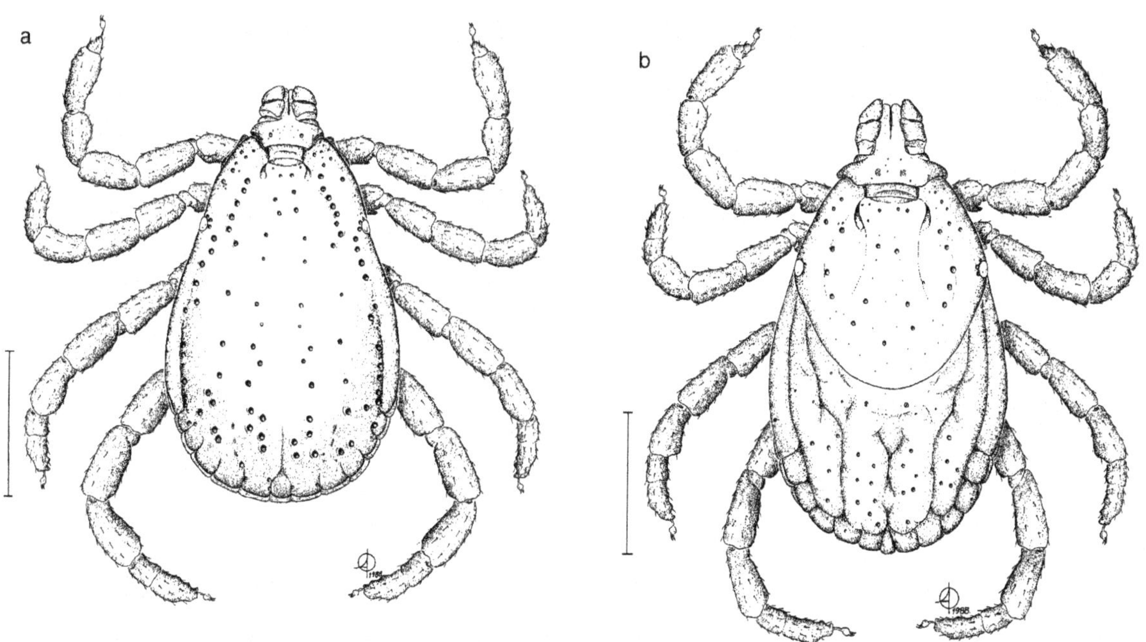

Figure 46. *Rhipicephalus distinctus* (Onderstepoort Tick Collection 2794v, RML 65726, collected from an unknown host, Kimberley District, South Africa, on 20 January 1963 by R.C. Bigalke). (a) Male, dorsal; (b) female, dorsal. Scale bars represent 1 mm. A. Olwage *del.* (Reprinted from Walker, 1990, figs 50 & 51, with kind permission from the Editor, *Onderstepoort Journal of Veterinary Research*).

Table 13. *Host records of* Rhipicephalus distinctus

Hosts	Number of records
Wild animals	
Caracal (*Caracal caracal*)	4 (immatures)
African wild cat (*Felis lybica*)	1
Yellow mongoose (*Cynictis penicillata*)	1 (nymphs)
Tree hyrax (*Dendrohyrax arboreus*)	1 (including immatures)
Yellow-spotted rock hyrax (*Heterohyrax brucei*)	10 (including immatures)
Rock hyrax (*Procavia capensis*)	153 (including immatures)
Procavia sp.	15 (including nymphs)
'Dassie'	12 (including nymphs)
Namaqua rock mouse (*Aethomys namaquensis*)	10 (immatures)
Swamp rat (*Otomys irroratus*)	1 (larva)
Bush Karoo rat (*Otomys unisulcatus*)	1 (larva)
'Field mouse'	1
Spring hare (*Pedetes capensis*)	1 (larva)
Scrub hare (*Lepus saxatilis*)	9 (immatures)
Jameson's red rock rabbit (*Pronolagus randensis*)	2 (including a larva)
Smith's red rock rabbit (*Pronolagus rupestris*)	4 (larvae)
Rock elephant shrew (*Elephantulus myurus*)	7 (immatures)
Humans	1 (including a nymph)

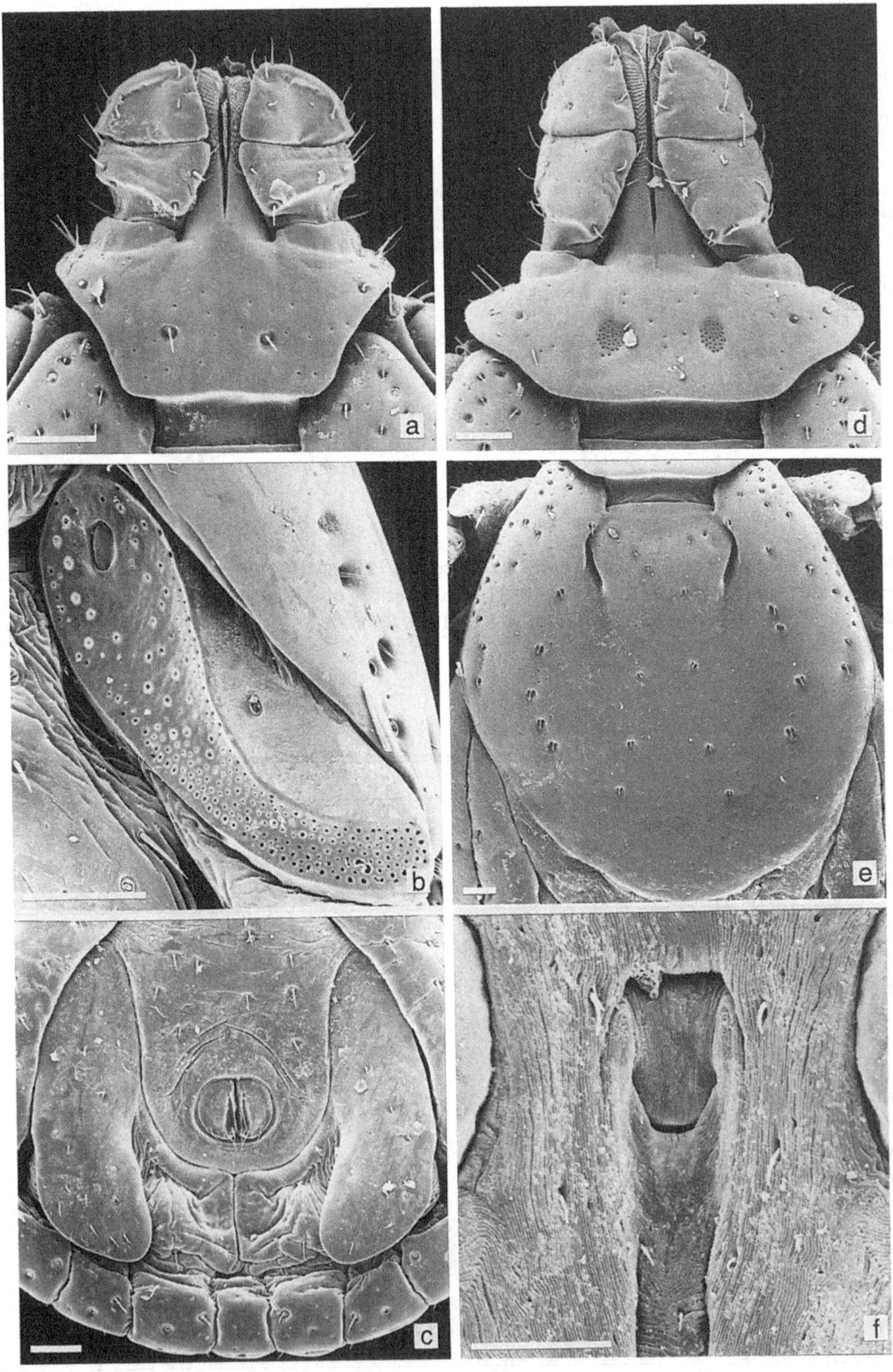
a
b
c
d
e
f

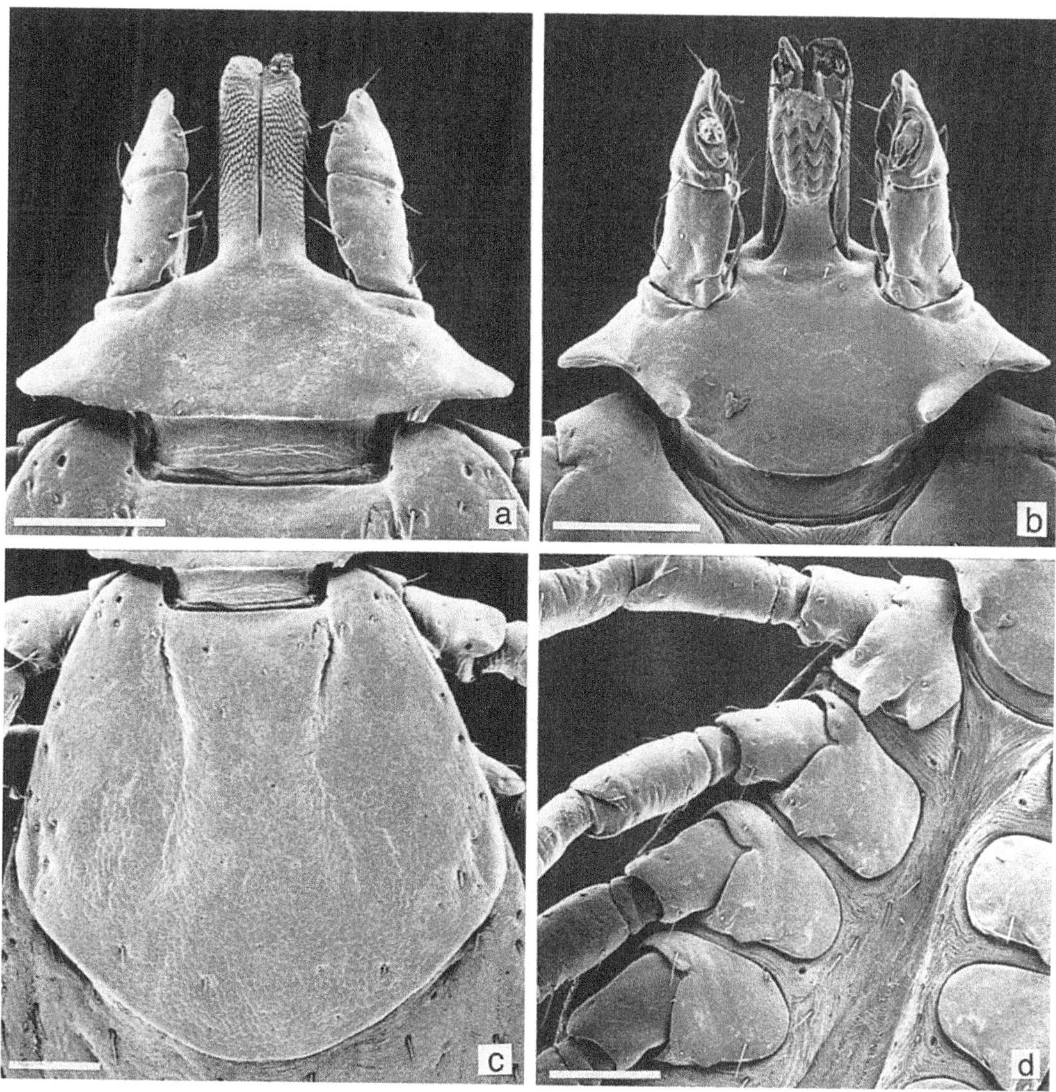

Figure 48 (*above*). *Rhipicephalus distinctus* [collected from rock hyrax (*Procavia capensis*), Mountain Zebra National Park, Eastern Cape Province, South Africa, in January 1981 by L.J. Fourie]. Nymph: (a) capitulum, dorsal; (b) capitulum, ventral; (c) scutum; (d) coxae. Scale bars represent 0.10 mm. SEMs by J.F. Putterill.

Figure 47 (*opposite*). *Rhipicephalus distinctus* (Onderstepoort Tick Collection 2794v, RML 65726, collected from an unknown host, Kimberley District, South Africa, on 20 January 1963 by R.C. Bigalke). Male: (a) capitulum, dorsal; (b) spiracle; (c) adanal plates. Female: (d) capitulum, dorsal; (e) scutum; (f) genital aperture. Scale bars represent 0.10 mm. SEMs by M.D. Corwin.

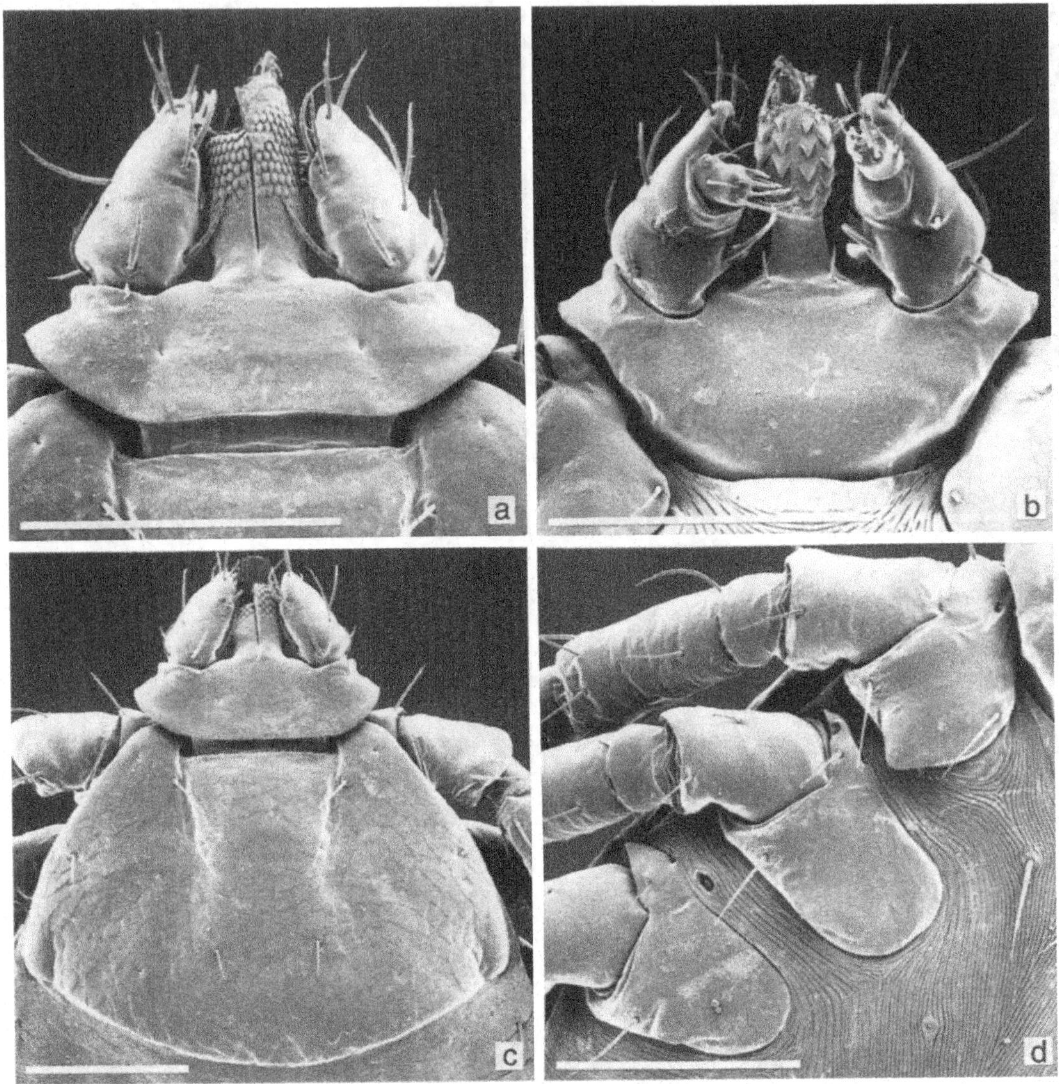

Figure 49. *Rhipicephalus distinctus* [collected from rock hyrax (*Procavia capensis*), Mountain Zebra National Park, Eastern Cape Province, South Africa, in January 1981 by L.J. Fourie]. Larva: (a) capitulum, dorsal; (b) capitulum, ventral; (c) scutum; (d) coxae. Scale bars represent 0.10 mm. SEMs by J.F. Putterill.

posterior to the anus, smoothly rounded posterointernally and posteroexternally; accessory adanal plates long sclerotized points.

Female (Figs 46(b), 47(d) to (f))

Capitulum broader than long, length × breadth ranging from 0.67 mm × 0.76 mm to 0.74 mm × 0.83 mm. Basis capituli well over twice as broad as long, with broadly-rounded lateral angles overlapping the scapulae; porose areas small, round, about three times their own diameter apart. Palps broad, long. Scutum slightly longer than broad, length × breadth ranging from 1.33 mm × 1.31 mm to 1.61 mm × 1.55 mm; posterior margin slightly sinuous to smoothly rounded. Eyes at widest part of scutum, at about mid-length, almost flat, delimited dorsally by a few setiferous punctations in slight depressions. Cervical pits comma-shaped, convergent; cervical fields broad but only faintly indicated. A few medium-sized setiferous punctations on scapulae; external cervical margins marked by slightly larger setiferous punctations, of which a few are also present medially on the scutum. In general scutum smooth and glossy. Ventrally genital aperture broadly V-shaped, its straight sides converging to the short straight posterior margin.

Nymph (Fig. 48)

Capitulum much broader than long, length × breadth ranging from 0.21 mm × 0.33 mm to 0.22 mm × 0.36 mm. Basis capituli four times as broad as long, with long gradually-tapering lateral angles overlapping the scapulae; ventrally with small sharp spurs on posterior margin. Palps slender, tapering to narrowly-rounded apices, inclined inwards. Scutum broader than long, length × breadth ranging from 0.47 mm × 0.51 mm to 0.51 mm × 0.53 mm; posterior margin a deep smooth curve. Eyes at widest point, well over halfway back, long and narrow, delimited dorsally by slight depressions. Cervical pits long, convergent, continuous with the long divergent internal cervical margins. Ventrally coxae I each with a long narrow external spur and a shorter broader internal spur; remaining coxae each with a small external spur only.

Larva (Fig. 49)

Capitulum much broader than long, length × breadth ranging from 0.102 mm × 0.165 mm to 0.119 mm × 0.168 mm. Basis capituli well over three times as broad as long, with broad acutely-pointed lateral angles at about mid-length, posterior margin slightly sinuous. Palps slightly constricted basally, almost immediately widening, then tapering to narrowly-rounded apices, inclined inwards. Scutum much broader than long, length × breadth ranging from 0.210 mm × 0.310 mm to 0.221 mm × 0.362 mm; posterior margin a broad shallow curve. Eyes at widest point, well over halfway back, large. Cervical grooves almost reaching eye level, slightly convergent. Ventrally coxae I each with a large broad spur; coxae II and III each with a mere indication of a spur.

The above descriptions are all based on field-collected specimens.

Notes on identification

The adults of *R. distinctus*, which feed almost exclusively on hyraxes (dassies), morphologically closely resemble those of *R. neumanni*, which have a predilection for the feet of sheep and goats. However, in the males there are marked differences in the appearance of the capitula and adanal plates of these two species, and in the females in the appearance of the capitula (Walker, 1990). The immature stages also differ morphologically, particularly in the shape of their capitula (see pp. 595–598).

A collection of *R. neumanni* from sheep at Victoria West, Northern Cape Province, South Africa, was recorded as *R. distinctus* by both Bedford (1932) and Theiler (1947). The record in Theiler (1962) from the bushpig, *Potamochoerus larvatus*, is based on an incorrect identification. The adults from a klipspringer (*Oreotragus oreotragus*), listed as *R. simpsoni* by Baker & Keep (1970) and as *R. distinctus* by Walker (1991), are now thought to be *R. oreotragi* sp. nov. (see p. 330).

Hoogstraal (1956, pp. 635–40) was doubtful about the specific status of a male tick which he listed as '*Rhipicephalus ? distinctus*', collected from a rock hyrax (*Procavia capensis*, syn. *Procavia habessinica slatini*) in the Sudan. So far as we know this problem has not been investigated since. We have, however, included this record because, according to Hoogstraal, this male differs from numerous specimens from Namibia (South West Africa) that he had seen merely in the shape of the posteroexternal angle of its adanal plates. We do not think that the immature specimens that he listed from the yellow-spotted rock hyrax (*Heterohyrax brucei*) (pp. 637, 778) belong to this species (see *R. carnivoralis*, p. 113).

Hosts

A three-host species. All developmental stages of *R. distinctus* are host-specific for hyraxes. They

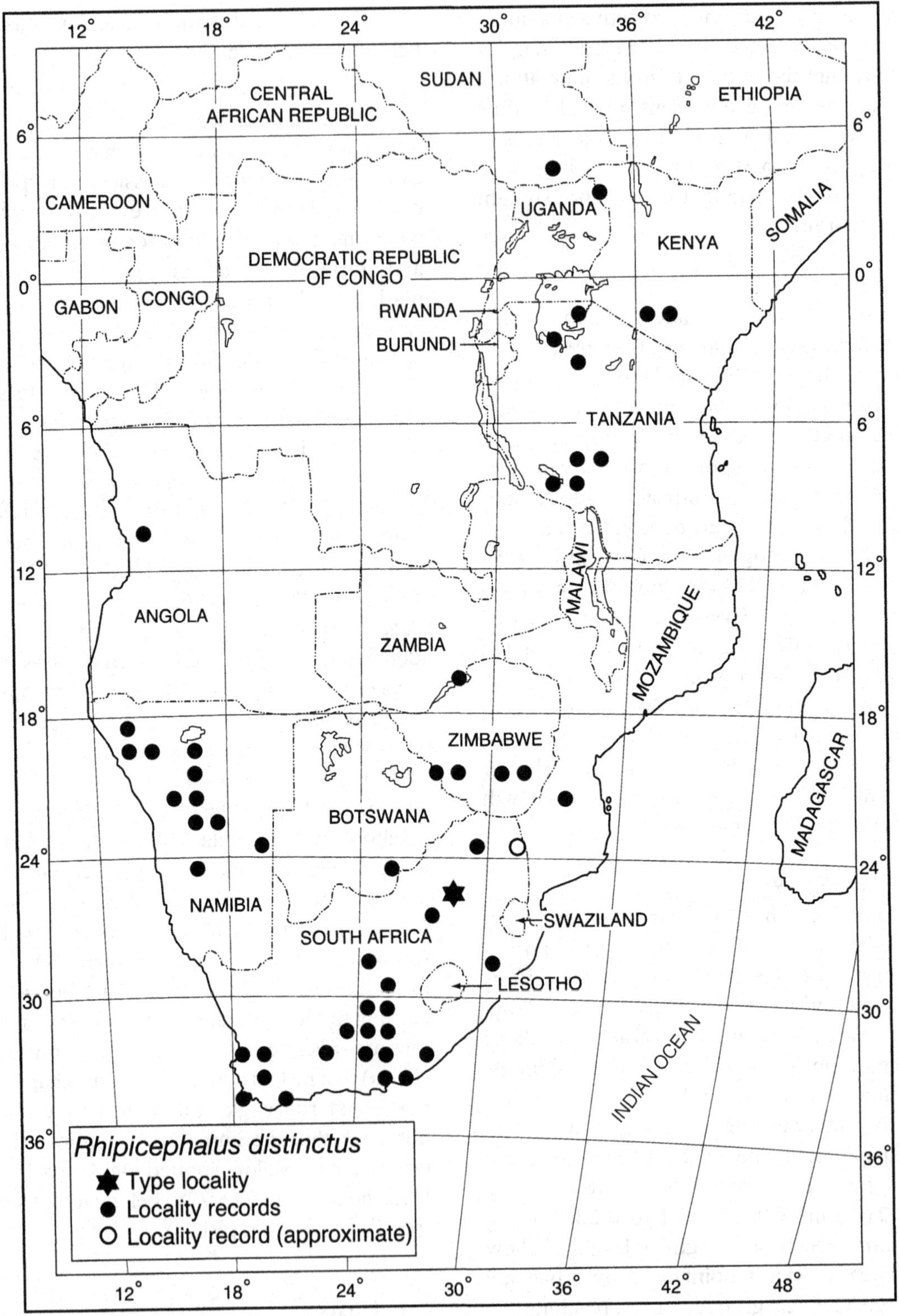

Map 15. *Rhipicephalus distinctus:* distribution.

are particularly prevalent on both species of rock hyrax, with a single record only from a tree hyrax (Table 13). The adults attach mainly to the face, lips, ears and neck. The former three sites are probably less prone to vigorous grooming with the claw-like nail of the inner toe of the hyrax's hindfoot.

Infestations on most hosts other than the various hyrax species are probably accidental. Immature ticks are, however, common on three animals that share the rocky outcrops favoured by rock hyraxes: the Namaqua rock mouse, red rock rabbit and rock elephant shrew (Horak *et al.*, 1991; Fourie, Horak & Van den Heever, 1992). In Zimbabwe Norval (1985) recorded a single *R. distinctus* male and a larva from two red rock rabbits identified as *Pronolagus crassicaudatus*. According to Skinner & Smithers (1990) this species does not occur in Zimbabwe whereas *Pronolagus randensis* does; we have therefore listed the latter as the host of these ticks.

The larvae of *R. distinctus* are most abundant on the rock hyrax in the Eastern Cape Province, South Africa from December to May, the nymphs from December to March and the adults from August to January (Horak & Fourie, 1986). Apparently, therefore, there is only one life cycle a year in this region.

Zoogeography

Rhipicephalus distinctus occurs in various parts of eastern, central and southern Africa in association with its specific hosts, the rock hyraxes (Map 15). Of these the most commonly parasitized species is *Procavia capensis*. Apparently, though, the tick is not present throughout the range of this hyrax, which is currently thought to have a wide, though patchy and discontinuous, distribution not only in Africa but also in various Middle Eastern countries where suitable habitat is present (Skinner & Smithers, 1990). Within their joint range the hyraxes and this tick generally occur where there are outcrops of rock in the form of cliffs, rocky hillsides or piles of loose boulders.

Ticks have been collected from hyraxes at altitudes from less than 100 m above sea level to about 2000 m; at localities with either winter, summer, or non-seasonal rainfall ranging from annual totals of less than 400 mm to over 1500 mm; and at places where snow may fall in winter, or where summer temperatures can reach 39 °C. It should be remembered, though, that this tick probably spends its non- parasitic phases in the rock refuges inhabited by the hyraxes where the temperature and humidity are relatively stable (Fourie, 1983).

Disease relationships

Unknown.

REFERENCES

Baker, M.K. & Keep, M.E. (1970). Checklist of the ticks found on the larger game animals in the Natal game reserves. *Lammergeyer*, **12**, 41–7.

Bedford, G.A.H. (1929). Notes on some South African ticks, with descriptions of three new species. *15th Annual Report of the Director of Veterinary Services, Union of South Africa*, 493–9.

Bedford, G.A.H. (1932). A synoptic check-list and host-list of the ectoparasites found on South African Mammalia, Aves, and Reptilia (2nd edn.). *18th Report of the Director of Veterinary Services and Animal Industry, Union of South Africa*, 223–523.

Fourie, L.J. (1983). *The population dynamics of the rock hyrax* Procavia capensis *(Pallas, 1766) in the Mountain Zebra National Park*. PhD thesis, Rhodes University, South Africa.

Fourie, L.J., Horak, I.G. & Van den Heever, J.J. (1992). The relative host status of rock elephant shrews *Elephantulus myurus* and Namaqua rock mice *Aethomys namaquensis* for economically important ticks. *South African Journal of Zoology*, **27**, 108–14.

Horak, I.G. & Fourie, L.J. (1986). Parasites of domestic and wild animals in South Africa. XIX. Ixodid ticks and fleas on rock dassies (*Procavia capensis*) in the Mountain Zebra National Park. *Onderstepoort Journal of Veterinary Research*, **53**, 123–6.

Horak, I.G., Fourie, L.J., Novellie, P.A. & Williams, E.J. (1991). Parasites of domestic and wild animals in South Africa. XXVI. The mosaic of ixodid tick infestations on birds and mammals in the Mountain Zebra National Park. *Onderstepoort Journal of Veterinary Research*, 58, 125–36.

Norval, R.A.I. (1985). The ticks of Zimbabwe. XII. The lesser known *Rhipicephalus* species. *Zimbabwe Veterinary Journal*, **16**, 37–43.

Walker, J.B. (1990). Two new species of ticks from southern Africa whose adults parasitize the feet of ungulates: *Rhipicephalus lounsburyi* n.sp. and *Rhipicephalus neumanni* n.sp. *Onderstepoort Journal of Veterinary Research*, 57, 57–75.

Walker, J.B. (1991). A review of the ixodid ticks (Acari, Ixodidae) occurring in southern Africa. *Onderstepoort Journal of Veterinary Research*, 58, 81–105.

Also see the following Basic References (pp. 12–14): Hoogstraal (1956); Matthysse & Colbo (1987); Skinner & Smithers (1990); Theiler (1947, 1962).

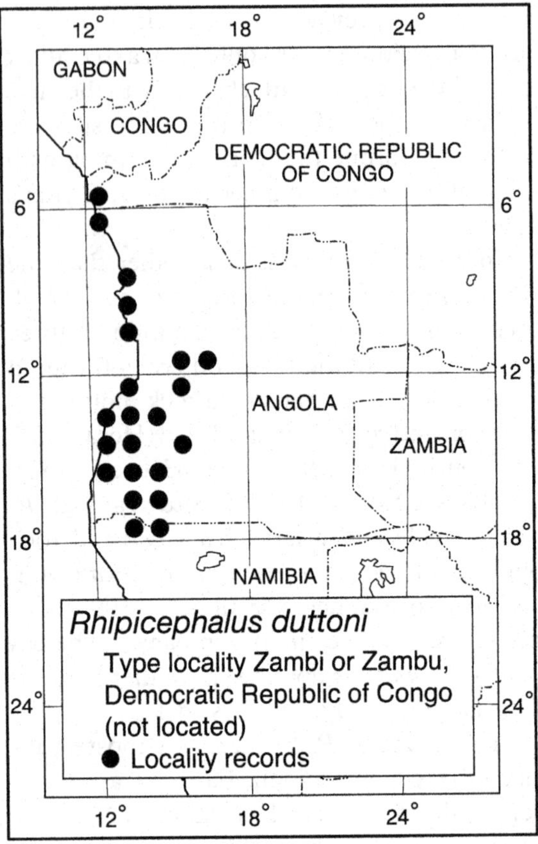

Map 16. *Rhipicephalus duttoni*: distribution.

RHIPICEPHALUS DUTTONI NEUMANN, 1907

This species was named in honour of the collector, Joseph Everett Dutton (1876–1905). This young doctor made important contributions to medical parasitology in Africa, including the first discovery of trypanosomes in human blood. He died, aged only 29, during an expedition to the Congo (Democratic Republic of Congo).

Diagnosis

A medium-sized dark reddish-brown tick.

Male (Figs 50(a), 51(a) to (c))

Capitulum longer than broad, length × breadth ranging from 0.53 mm × 0.51 mm to 0.76 mm × 0.69 mm. Basis capituli with lateral angles not quite halfway back, somewhat recurved. Palps smoothly rounded apically. Conscutum length × breadth ranging from 2.17 mm × 1.32 mm to 3.22 mm × 1.98 mm; large anterior process present on coxae I. A short median caudal process present on engorged specimens. Eyes mildly convex, edged dorsally by a few large punctations. Cervical fields broad and convergent anteriorly, becoming divergent and tapering posteriorly to beyond eye level, their external margins marked by large setiferous punctations and their surfaces shagreened. Marginal lines long, almost reaching eyes. Posteromedian and posterolateral grooves well developed, elongate. Large setiferous punctations present on the scapulae, also medially on the conscutum (where they may coalesce to form a roughened depressed patch just anterior to the posteromedian groove), and along the marginal lines, interspersed with finer punctations. Lateral areas of the conscutum adjacent to marginal lines almost impunctate but finely shagreened in places. Ventrally spiracles elongate, with a broad prolongation curving gently towards the dorsal surface. Adanal plates

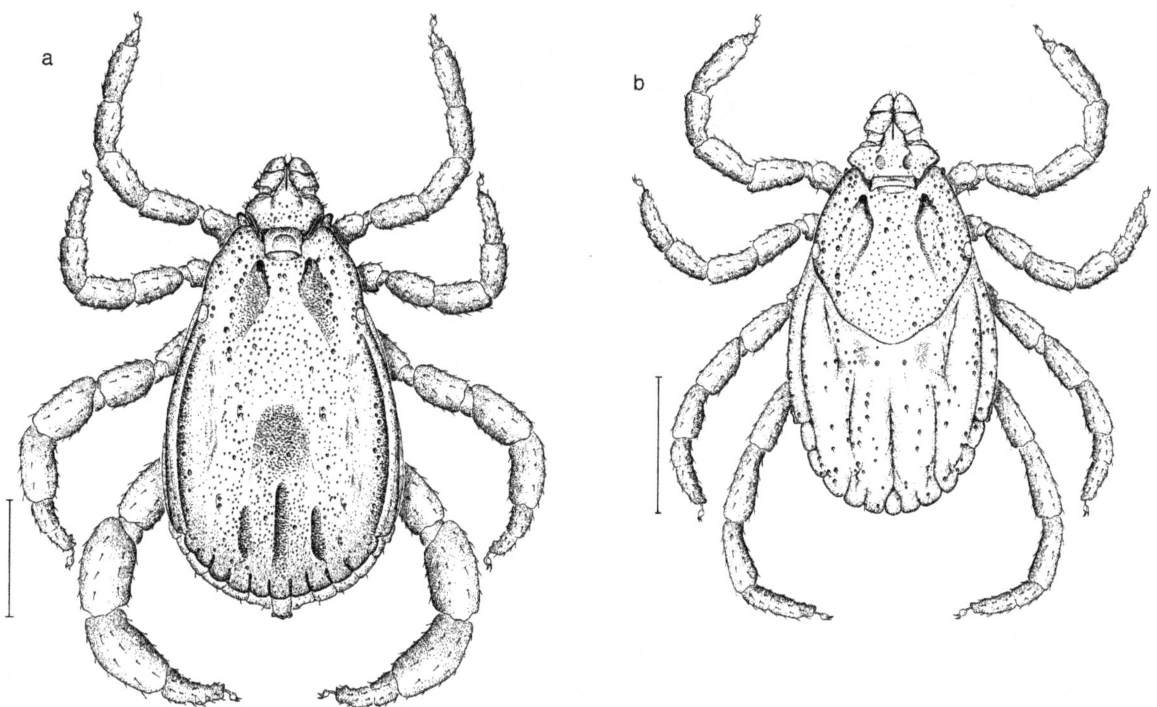

Figure 50. *Rhipicephalus duttoni* [Protozoology Section Tick Breeding Register, Onderstepoort, No. 3258, laboratory reared, progeny of ♀ collected at Huambo (Nova Lisboa), Angola, probably from a bovine, in November 1968 by F.M.H. Serrano]. (a) Male, dorsal; (b) female, dorsal. Scale bars represent 1 mm. A. Olwage *del.*

Table 14. *Host records of* Rhipicephalus duttoni

Hosts	Number of records
Domestic animals	
Cattle	Commonly parasitized (including immatures)
Sheep	2
Goats	1
Horses	1
Dogs	2
Wild animals	
African civet (*Civettictis civetta*)	1
Impala (*Aepyceros melampus*)	2 (including a nymph)
African buffalo (*Syncerus caffer*)	Commonly parasitized (including immatures)
Eland (*Taurotragus oryx*)	1
Common duiker (*Sylvicapra grimmia*)	1 (including nymphs)
Savanna hare (*Lepus victoriae*)	1 (including nymphs)
Birds	
Black-bellied bustard (*Eupadotis melanogaster*)	1

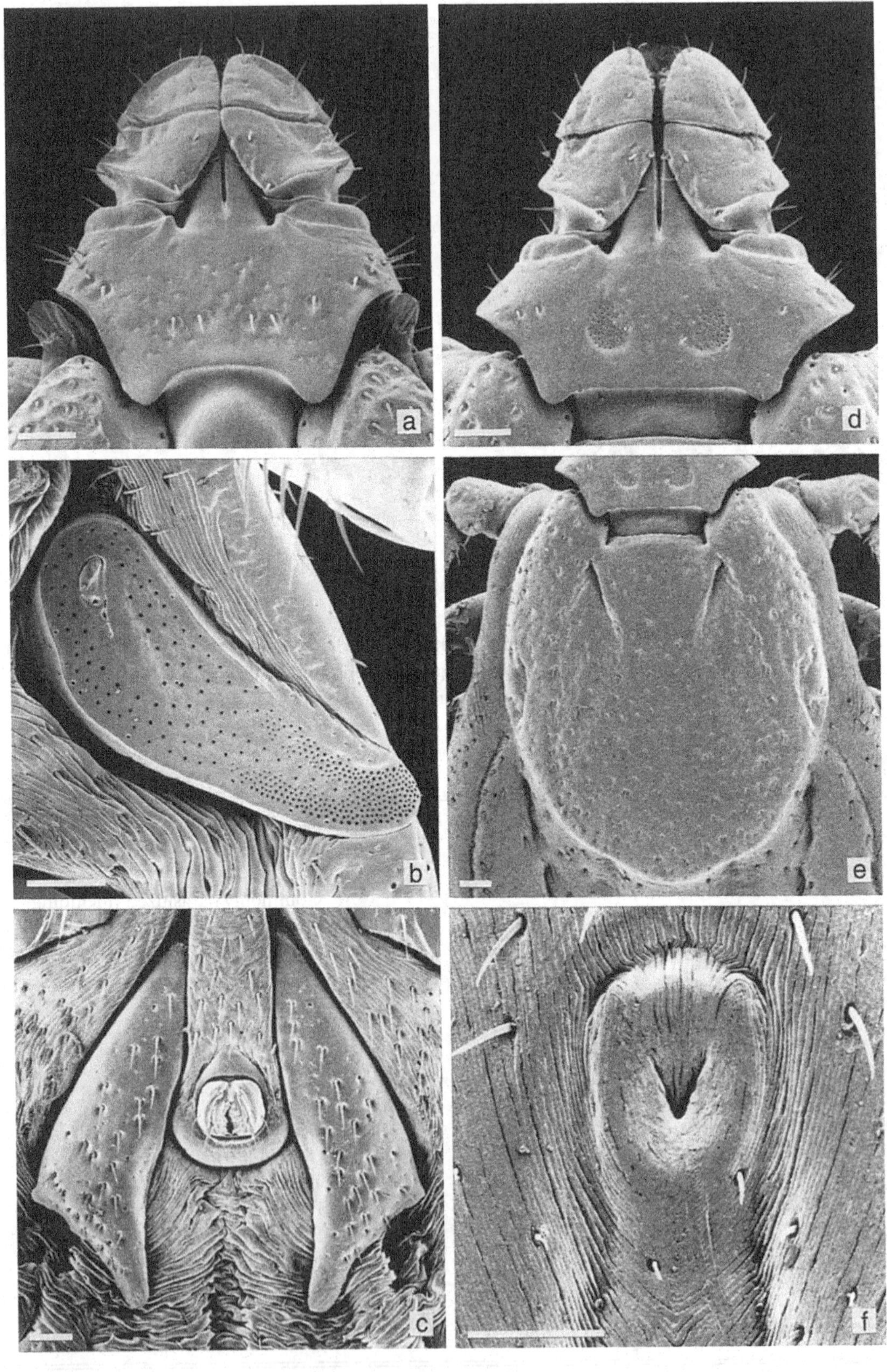
a
d
b
e
c
f

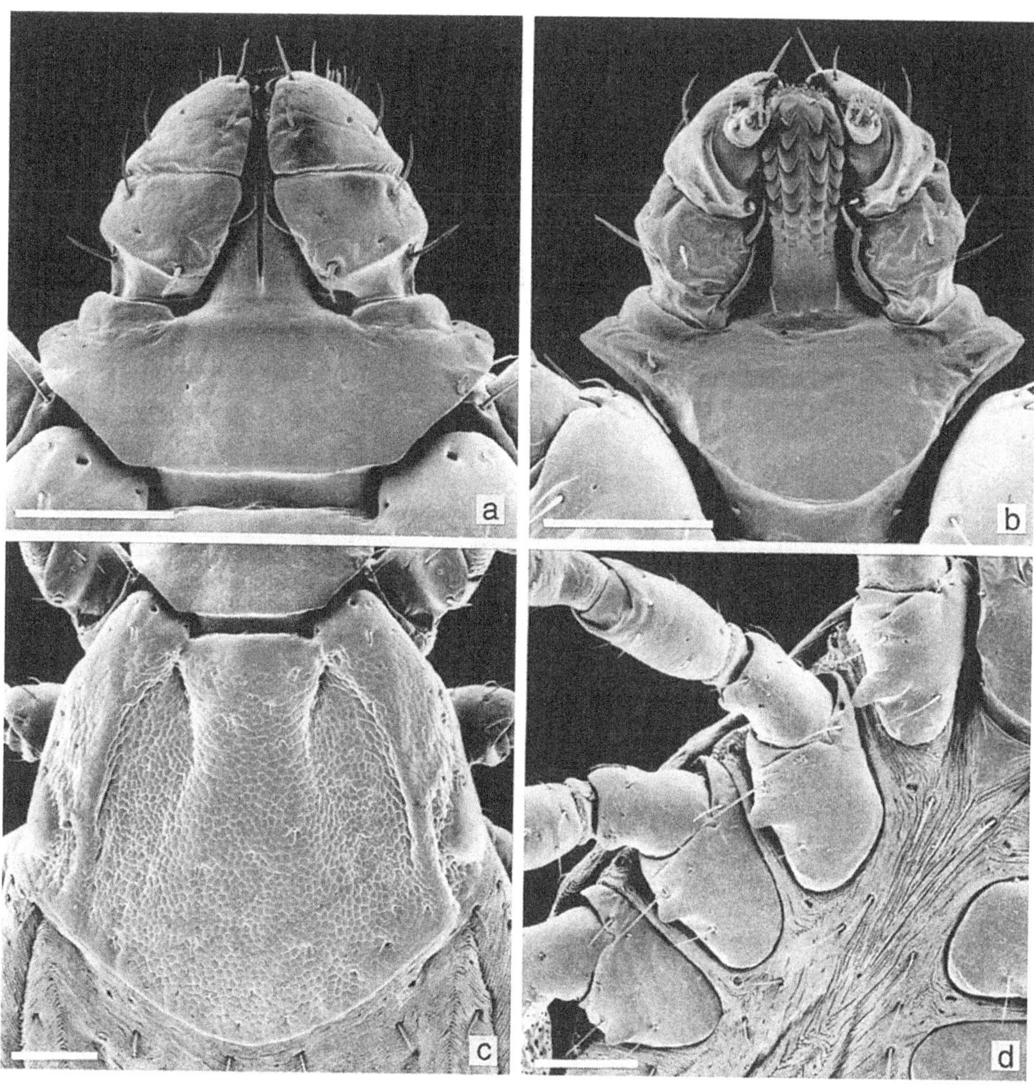

Figure 52 (*above*). *Rhipicephalus duttoni* [Protozoology Section Tick Breeding Register, Onderstepoort, No. 3258, laboratory reared, progeny of ♀ collected at Huambo (Nova Lisboa), Angola, probably from a bovine, in November 1968 by F.M.H. Serrano]. Nymph: (a) capitulum, dorsal; (b) capitulum, ventral; (c) scutum; (d) coxae. Scale bars represent 0.10 mm. SEMs by J.F. Putterill.

Figure 51 (*opposite*). *Rhipicephalus duttoni* [Protozoology Section Tick Breeding Register, Onderstepoort, No. 3258, laboratory reared, progeny of ♀ collected at Huambo (Nova Lisboa), Angola, probably from a bovine, in November 1968 by F.M.H. Serrano]. Male: (a) capitulum, dorsal; (b) spiracle; (c) adanal plates. Female: (d) capitulum, dorsal; (e) scutum; (f) genital aperture. Scale bars represent 0.10 mm. SEMs by J.F. Putterill.

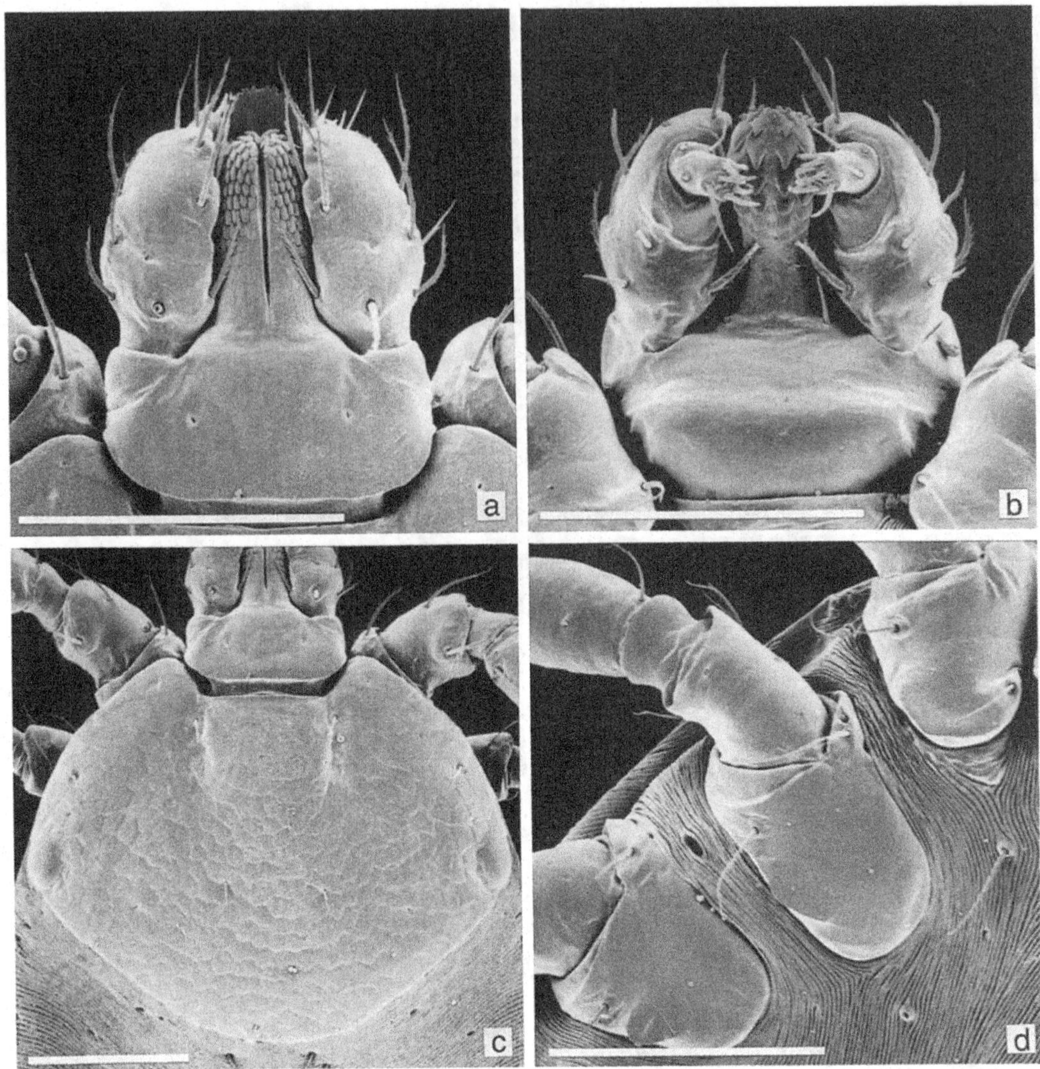

Figure 53. *Rhipicephalus duttoni* [Protozoology Section Tick Breeding Register, Onderstepoort, No. 3258, laboratory reared, progeny of ♀ collected at Huambo (Nova Lisboa), Angola, probably from a bovine, in November 1968 by F.M.H. Serrano]. Larva: (a) capitulum, dorsal; (b) capitulum, ventral; (c) scutum; (d) coxae. Scale bars represent 0.10 mm. SEMs by J.F. Putterill.

long, narrow, inclined inwards posteromedially, posterior margins long, slightly concave; accessory adanal plates absent.

Female (Figs 50(b), 51(d) to (f))

Capitulum broader than long, length × breadth ranging from 0.64 mm × 0.67 mm to 0.70 mm × 0.77 mm. Basis capituli with lateral angles at about mid-length, acute; porose areas large, slightly more than their own diameter apart. Palps a little longer than those of male, smoothly rounded apically. Scutum longer than broad, length × breadth ranging from 1.31 mm × 1.15 mm to 1.47 mm × 1.27 mm; posterior margin sinuous. Eyes mildly convex, edged dorsally by large punctations. Cervical fields long, broad; their surfaces slightly shagreened in places, especially along the internal margins; their external cervical margins marked by rows of large setiferous punctations. Large setiferous punctations present on the scapulae and scattered elsewhere on the scutum, interspersed with medium-sized to fine punctations. Ventrally genital aperture sharply V-shaped.

Nymph (Fig. 52)

Capitulum broader than long, length × breadth ranging from 0.19 mm × 0.24 mm to 0.25 mm × 0.28 mm. Basis capituli nearly three times as broad as long, with flange-like, forwardly-directed lateral angles. Palps constricted basally, then widening before they taper to broadly-rounded apices. Scutum broader than long, length × breadth ranging from 0.44 mm × 0.53 mm to 0.52 mm × 0.58 mm. Eyes at widest point, over halfway back, with a shallow furrow dorsally. Cervical fields long, broad, depressed, almost reaching posterolateral margins of scutum. Ventrally coxae I each with a large external and very small internal spur; coxae II to IV each with a small external spur only.

Larva (Fig. 53)

Capitulum broader than long, length × breadth ranging from 0.103 mm × 0.112 mm to 0.116 mm × 0.123 mm. Basis capituli over twice as broad as long, with very short lateral angles. Palps constricted proximally, then widening markedly, apices truncated. Scutum much broader than long, length × breadth ranging from 0.245 mm × 0.363 mm to 0.261 mm × 0.398 mm. Eyes at widest point, mildly convex. Cervical grooves short, almost parallel. Ventrally coxae I and II with broad spurs; coxae III with a smaller triangular spur.

Notes on identification

Neumann's original description of this species was based on a single male collected from a bovine at Zambie. This is thought to be in the Democratic Republic of Congo but has not been precisely located. According to Keirans (1985), Nuttall Collection 1614, labelled as '*Rhipicephalus duttoni* ♂, ♀, N's', and designated COTYPE, was obtained from 'cattle, Zambu, Africa, no date, Drs Dutton & Todd'. Nuttall noted: 'Compared with *R. duttoni* 1 ♂ type N. 1719 in Neumann coll., 18.VI.1914. Our ♂ more punctate and rougher, caudal process less protruding. We call it a co-type as it is from the same lot as the type and was rec'd from Liverpool'. Keirans also noted that this collection now contains 1♂, 1♀ and 2 nymphs that were found in a separate vial.

Doubts about the validity of *R. duttoni* as a species distinguishable from *R. appendiculatus* were expressed by Theiler (1947), though she did not reiterate these doubts in 1962. Other authors (e.g. Zumpt, 1949; Sousa Dias, 1950) regarded it as being closely related to *R. appendiculatus* but nevertheless distinct. In 1991 Gomes & Wouters, who carried out experiments with *R. appendiculatus*, *R. duttoni* and *R. zambeziensis*, concluded that: '*R. duttoni* is a distinct well-defined species in the *R. appendiculatus* group'.

Judging by morphology, we think that *R. duttoni* is most closely related to *R. nitens* and, were these two species to occur sympatrically, it would be extremely difficult to distinguish between them in field collections, especially the smaller specimens. Fortunately they occur in widely separated areas (see p. 146, Map 16, and *R. nitens* p. 316, Map 42). In addition there are a few distinct morphological differences between them. The larger males of *R. duttoni* have longer, more tapering, cervical fields than *R. nitens*, and the lateral areas of the conscutum do not bulge so conspicuously. In the nymphs there are slight differences in the shapes of the capitula. The nymph of *R. duttoni* has broader, more flange-like lateral angles to its basis capituli than does the *R.nitens* nymph. The larvae of these two species are virtually indistinguishable.

Two collections from the Musée Royal de l'Afrique Centrale, T8921/8927 and T8930, both from Zambi and labelled *R. appendiculatus* by Elbl & Anastos (1966), were re-identified as *R. duttoni* by J.B.W.

We ourselves have little personal experience of *R. duttoni*. Consequently we have had to base our account of its hosts and zoogeography almost exclusively on information in the literature. Unfortunately most references lack detail, e.g. of the number of hosts, such as cattle, from which collections were made.

Hosts

A three-host species (Protozoology Section Tick Breeding Register, Onderstepoort, No. 3258). Amongst domestic animals, cattle are probably its commonest hosts. In a recent survey conducted by Gomes, Pombal & Venturi (1994) it comprised 8.3% of the 3864 ticks collected from these animals. The preferred feeding site for the adults on cattle is their ears, on which they may be present in very large numbers (Gomes & Wouters, 1991). It is apparent, therefore, that in some parts of its range it is an important bovine parasite, though it is impossible to give an actual figure for the number of collections obtained from cattle. Other domestic animals have rarely been listed as hosts and we have no means of determining how often they really are parasitized by this tick (Table 14).

Only a few wild hosts of *R. duttoni* have been recorded in the literature (Fain, 1949; Santos Dias, 1956, 1957, 1983a,b). Many others probably remain to be discovered. The African buffalo is undoubtedly an important host because of its involvement in the transmission by this tick of *Theileria parva lawrencei* to cattle in Angola (Da Graça & Serrano, 1971).

Adults of *R. duttoni* are active mainly during the rainy season, though they can be found throughout the year.

Zoogeography

Rhipicephalus duttoni has been recorded most commonly in south-western Angola and, further north, in the vicinity of Luanda (Map 16). Its southernmost records to date are from Enyandi and Swartbooisdrif, on the north-western border of Namibia, and its northernmost from western Democratic Republic of Congo. It can apparently tolerate a wide range of ecological conditions. In the Luanda area the annual rainfall ranges from under 400 mm to about 600 mm and is associated with a relatively xerophytic type of North Zambezian undifferentiated woodland. Further south, in the hinterland away from the dry coastal zone, annual rainfalls ranging from 600 mm to well over 1200 mm may be obtained. The vegetation here is either wetter Zambezian miombo woodland, dominated by *Brachystegia*, *Julbernardia* and *Isoberlinia*, or *Colophospermum mopane* woodland.

Disease relationships

This tick is an efficient vector of Corridor disease of cattle, caused by *Theileria parva lawrencei*, in the field (Da Graça & Serrano, 1971).

REFERENCES

Da Graça, H.M. & Serrano, F.M.H. (1971). Contribuição para o estudo da Theileriose sincerina maligna dos bovinos, em Angola. *Acta Veterinária, Nova Lisboa*, 7, 1–8.

Fain, A. (1949). Contribution à l'étude des Arthropodes piquers dans le territoire de Banningville (Régions du Bas-Kwango et Bas-Kwilu). *Revue de Zoologie et de Botanique Africaines*, **XLII**, 176–82.

Gomes, A.F. & Wouters, G. (1991). Species identification of *Rhipicephalus duttoni* in relation to the *Rhipicephalus appendiculatus* group. *Journal of Medical Entomology*, **28**, 16–18.

Gomes, A.F., Pombal, A.M. & Venturi, L. (1994). Observations on cattle ticks in Huila Province (Angola). *Veterinary Parasitology*, **51**, 333–6.

Neumann, G. (1907). Description of two new species of African ticks. *Annals of Tropical Medicine and Parasitology*, **1**, 115–20.

Santos Dias, J.A.T. (1956). Notes sur quelques ixodidés d'Angola en collection dans le Laboratoire de Parasitologie de la Faculté de Médecine de Paris. *Bulletin de la Société de Pathologie Exotique*, **49**, 65–8.

Santos Dias, J.A.T. (1957). Notas sobre a Ixodofauna Angola. *Boletim da Sociedade de Estudios de Moçambique*, **No. 103**, 157–69.

Santos Dias, J.A.T. (1983a). Subsídios para o conhecimento da fauna ixodológica de Angola. *Garcia de Orta, Serie de Zoologia, Lisboa*, **11**, 57–68.

Santos Dias, J.A.T. (1983b). Alguns ixodídeos (Acarina-Ixodoidea-Ixodidae) coligidos em Angola pelo Dr. Crawford Cabral. *Garcia de Orta, Serie de Zoologia, Lisboa*, **11**, 69–76.

Also see the following Basic References (pp. 12–14): Elbl & Anastos (1966); Keirans (1985); Sousa Dias (1950); Theiler (1947, 1962); Zumpt (1949).

RHIPICEPHALUS DUX DÖNITZ, 1910

The Latin term *dux* means 'leader, ruler, guide' but the significance of this specific name for this tick is unknown.

Synonym

schwetzi.

Diagnosis

A large ornate rhipicephalid on which the ivory-coloured ornamentation on the adult's scutum characteristically consists of either one or two patches mid-dorsally in the male and a single patch posteromedially in the female.

Male (Figs 54(a), 55(a) to (c))
Capitulum longer than broad, length × breadth ranging from 0.83 mm × 0.81 mm to 1.25 mm × 1.07 mm. Basis capituli with sharp somewhat recurved lateral angles in anterior third of its length. Palps broadly rounded apically. Conscutum length × breadth ranging from 3.61 mm × 2.60 mm to 4.89 mm × 3.41 mm; anterior process on coxae I inconspicuous. In engorged males body wall bulges posterolaterally and forms a broadly-rounded process posteromedially. Eyes small, flat. Cervical fields flat, not particularly well delineated. Marginal lines not well developed, sometimes reaching eye level but sometimes shorter, then continued anteriorly merely by rows of punctations. Posteromedian groove long, posterolateral grooves more-or-less round. Large setiferous punctations present on scapulae, along external margins of cervical fields, and scattered medially on the conscutum and around the posterior grooves. These are interspersed with numerous smaller punctations that are quite prominent in some specimens but more superficial in others. An ivory-coloured pattern generally present mid-dorsally, often consisting of a relatively small patch anteriorly plus a larger one behind this that partially surrounds the anterior ends of the posterior grooves. These two patches may be joined, forming a single ivory-coloured area that usually extends from about eye-level to the anterior ends of the posterior grooves. Some specimens, however, lack the colour pattern. Ventrally spiracles with a short, broad, curved dorsal prolongation. Adanal plates short, broad, their posterior ends curving round smoothly towards each other; accessory adanal plates in the form of small sclerotized points.

Female (Figs 54(b), 55(d) to (f))
Capitulum often longer than broad but in some specimens broader than long, length × breadth ranging from 0.86 mm × 0.95 mm to 1.22 mm × 1.19 mm. Basis capituli with short acute lateral angles; porose areas large, deep set, more than their own diameter apart. Palps with a short narrow neck, otherwise broad, their apices truncated. Scutum broader than long, length × breadth ranging from 1.77 mm × 2.02 mm to 2.02 mm × 2.42 mm; posterior margin very slightly sinuous. Eyes small, flat. Cervical fields well defined, broad, depressed. A few large setiferous punctations present on the scapulae, along the cervical margins and scattered medially on the scutum, interspersed with numerous somewhat smaller punctations; in general the scutum is more densely punctate than that of the male. A single ivory-coloured patch, roughly triangular in shape, present posteromedially. Short white setae often scattered on the alloscutum, though they are easily rubbed off. Ventrally genital aperture U-shaped.

Immature stages
Unknown.

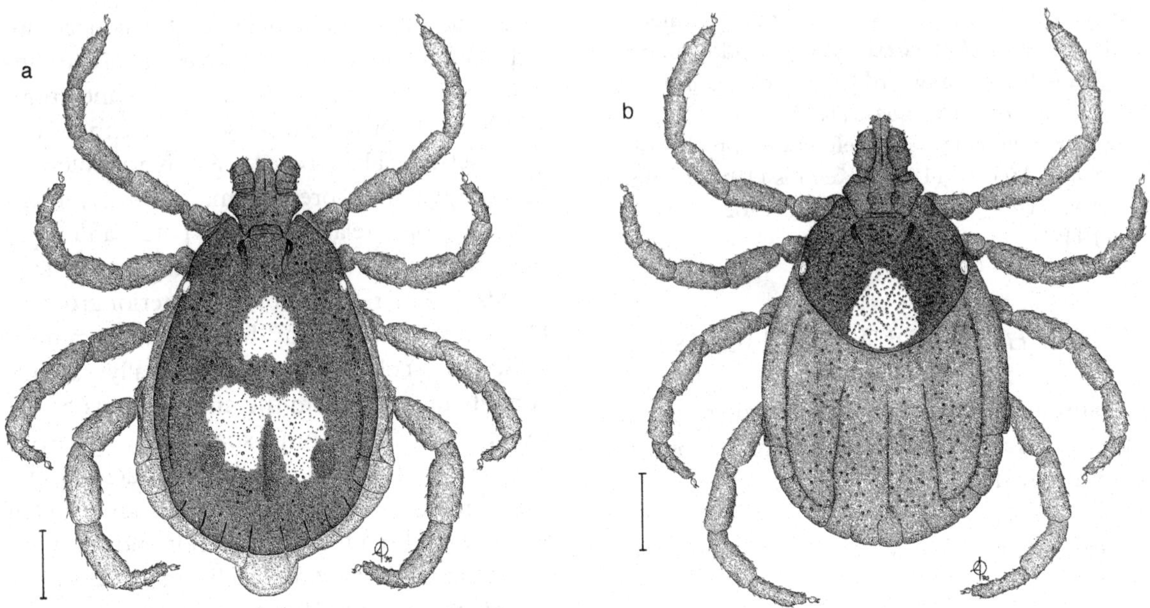

Figure 54. *Rhipicephalus dux* [Onderstepoort Tick Collection 2963i, collected from African buffalo (*Syncerus caffer*) at Angumu, Uélé, Democratic Republic of Congo, on 5 January 1953 by M. Wanson]. (a) Male, dorsal; (b) female, dorsal. Scale bars represent 1 mm. A. Olwage *del.*

Table 15. *Host records of* Rhipicephalus dux

Hosts	Number of records
Domestic animals	
Cattle	2
Pigs	2
Wild animals	
African elephant (*Loxodonta africana*)	2
Warthog (*Phacochoerus africanus*)	2
Forest hog (*Hylochoerus meinertzhageni*)	1
Bushpig (*Potamochoerus larvatus*)	2
African buffalo (*Syncerus caffer*)	21
Humans	1 (unattached)

Figure 55 (*opposite*). *Rhipicephalus dux* [Onderstepoort Tick Collection 2963i, collected from African buffalo (*Syncerus caffer*) at Angumu, Uélé, Democratic Republic of Congo, on 5 January 1953 by M. Wanson]. Male: (a) capitulum, dorsal; (b) spiracle; (c) adanal plates. Female: (d) capitulum, dorsal; (e) scutum; (f) genital aperture. Scale bars represent 0.10 mm. SEMs by J.F. Putterill.

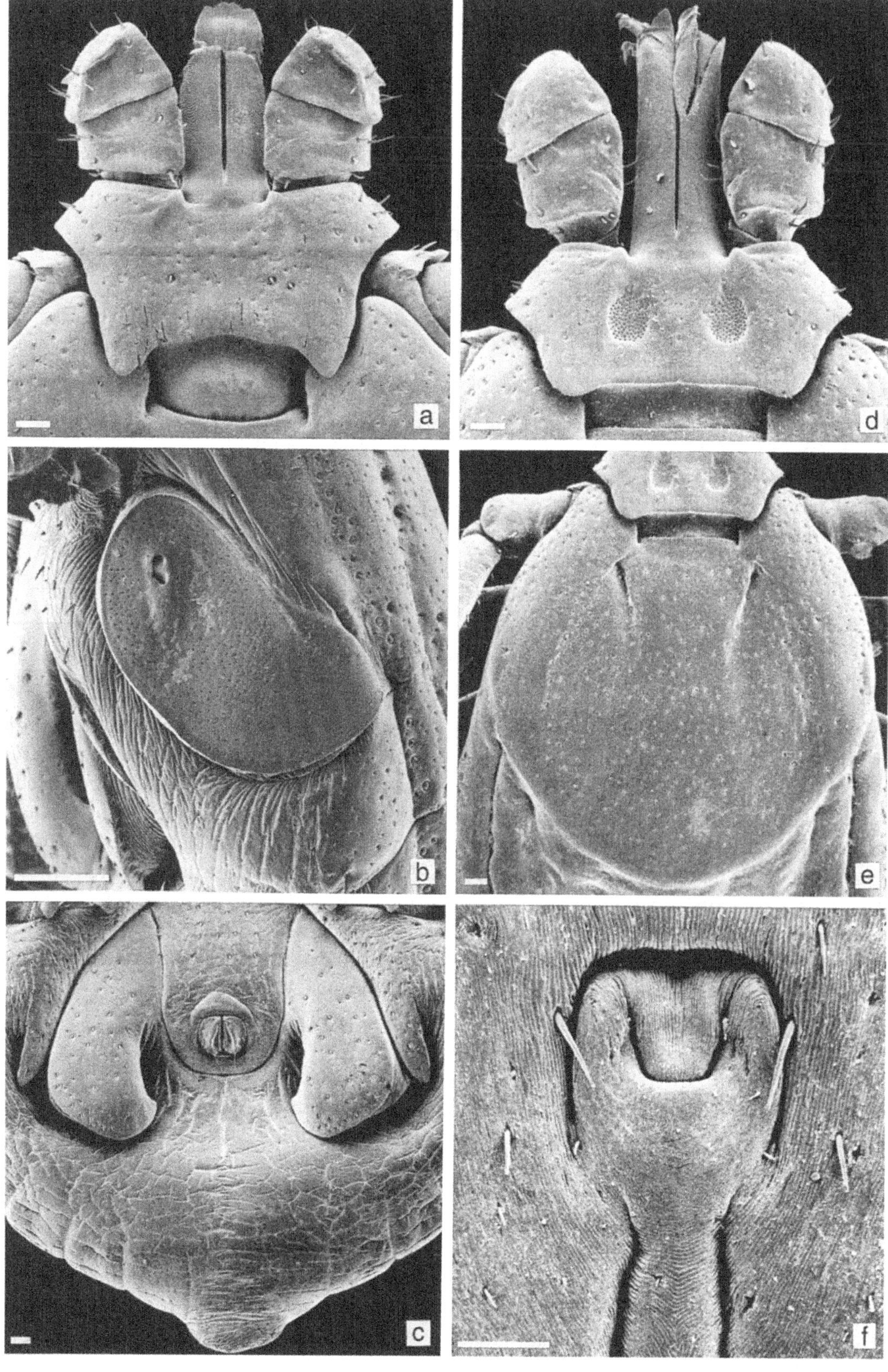
a
d
b
e
c
f

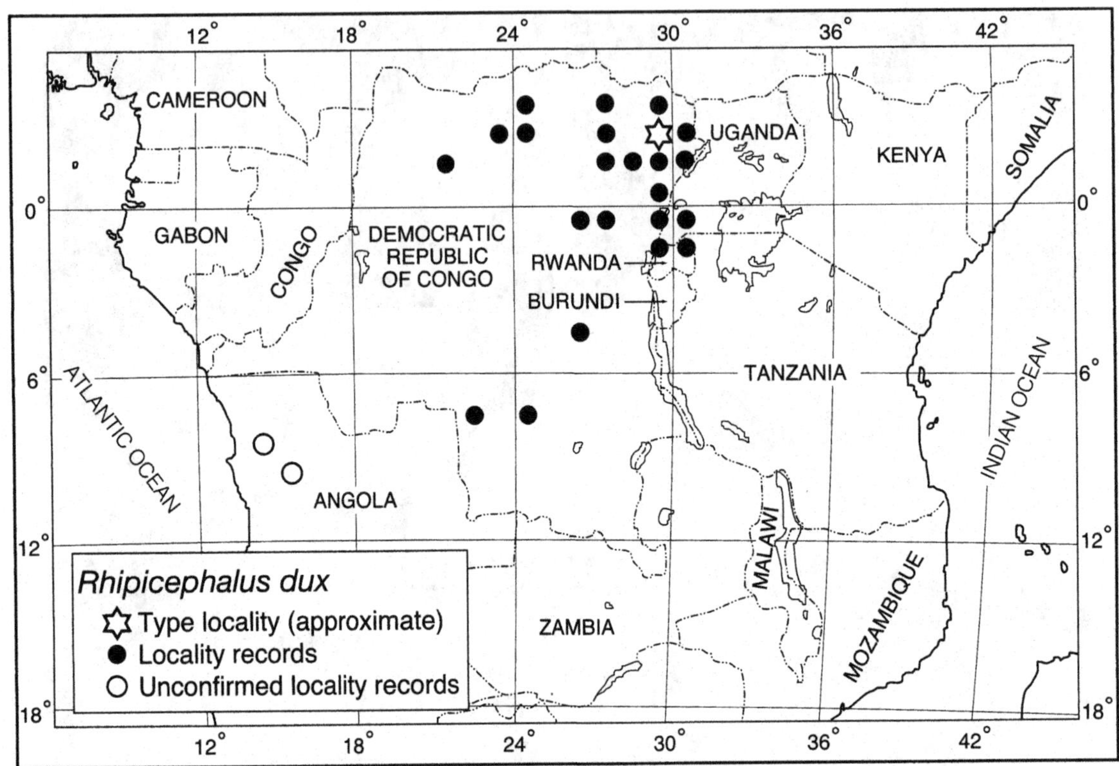

Map 17. *Rhipicephalus dux*: distribution. (Based largely on Morel, 1969).

Notes on identification

Zumpt (1942) first discussed the four ornate rhipicephalids, *R. dux, R. humeralis, R. maculatus* and *R. pulchellus*, as a group. Santos Dias (1962) then placed them together in his new subgenus *Tendeirodes* but we do not accept his definition of this subgenus based on ornamentation alone. Morel (1969) suggested that *R. dux* probably belongs to the *R. simus* group. Certainly its adults differ morphologically in several respects from the other three ornate species. As Morel noted, their affinities are with the *R. appendiculatus* group. In the *R. dux* male the basis capituli is somewhat broader in relation to its length than it is in the other three species; the anterior process of its coxae I is inconspicuous, not large and strongly sclerotized; the posterior grooves on its conscutum are better developed, and its caudal process is short and broadly rounded, not tail-like. In the *R. dux* female the scutum is broader in relation to its length than that of either *R. humeralis* or *R. pulchellus*, with only a small area posteromedially that is ivory-coloured. In the females of both *R. humeralis* and *R. pulchellus* almost the entire scutum is ivory-coloured. It is closest in general appearance to the *R. maculatus* female but its cervical fields are more definitely demarcated, and its scutum overall is far more punctate.

Only when its immature stages become available will it be possible to determine its real relationships within the genus.

Hosts

Life cycle unknown. There are few records of this species from domestic animals, which may simply reflect their relative scarcity in most areas where it has so far been collected rather than any aversion to them as hosts. It has been recorded most commonly from the African buffalo, and to a much lesser extent from three species of wild suids (Table 15). Morel (1969) regarded the

African elephant as its primary host but the information now available to us does not support this view.

Two adults were found on the genitalia of an African buffalo. Most collections have been made between May and October.

Zoogeography

Present indications are that *R. dux* occurs primarily in north-eastern Democratic Republic of Congo. It has also been found in a few places further south in that country as well as eastwards in south-western Uganda and northern Rwanda (Map 17). Apparently, therefore, it occurs most commonly in areas lying at an altitude of 500 m, but sometimes much higher, with high annual rainfalls ranging from about 1400 mm to 1800 mm or more, in Congolian lowland rainforest and mosaics of this forest and secondary grassland.

Morel (1969, map 51) recorded it in two areas in northern Angola. We also know of two Angolan records but cannot position either of them. Since together they comprise a total of only 4 ♂♂ and 3 ♀♀ (Santos Dias, 1964; RML 98961) we feel that its presence in that country requires confirmation.

Disease relationships

Unknown.

REFERENCES

Dönitz, W. (1910). Zwei neue Afrikanische Rhipicephalusarten (*R. dux*, *R. glyphis*). *Sitzungsberichte der Gesellschaft Naturforschende Freunde der Berlin*, **6**, 275–80.

Santos Dias, J.A.T. (1962). Contribuição ao estudo da systemática do género *Rhipicephalus* C.L. Koch, 1844 (Acarina, Ixodoidea). *Anais dos Serviços de Veterinária*, **No. 8 for 1960**, 1–13.

Santos Dias, J.A.T. (1964). Nova contribuição para o conhecimento da ixodofauna Angolana. Carraças colhidas por uma missão de estudo do Museu de Hamburgo. *Anais dos Serviços de Veterinária*, **No. 9 for 1961**, 79–98.

Zumpt, F. (1942). Die gefleckten *Rhipicephalus*-Arten. III. Vorstudie zu einer Revision der Gattung *Rhipicephalus*. *Zeitschrift für Parasitenkunde*, **12**, 433–43.

Also see the following Basic References (pp. 12–14): Elbl & Anastos (1966); Matthysse & Colbo (1987); Morel (1969); Theiler (1947).

RHIPICEPHALUS EVERTSI EVERTSI NEUMANN, 1897

This species was named after Dr J.G. Everts, who collected the type specimens in the province then called the Transvaal, South Africa.

Diagnosis

A medium-sized dark brown tick with reddish-orange legs.

Male (Figs 56(a), 57(a) to (c))

Capitulum as broad as long, length × breadth ranging from 0.63 mm × 0.63 mm to 0.75 mm × 0.76 mm. Basis capituli with short acute lateral angles in anterior third of its length. Palps short, broad, flattened apically. Conscutum length × breadth ranging from 2.81 mm × 1.78 mm to 3.30 mm × 2.10 mm; anterior process on coxae I sharp. Eyes beady, orbited. Cervical pits comma-shaped, continuous with the long divergent rugose internal cervical margins. Marginal lines long but not particularly conspicuous. Posteromedian and posterolateral grooves also shallow and inconspicuous. A few large setiferous punctations present anteriorly on scapulae, in irregular lines along external cervical margins and scattered elsewhere on the conscutum, interspersed with a dense pattern of medium-sized to fine punctations. Legs characteristically bright reddish-orange in colour, their colour contrasting strongly with the dark conscutum. Ventrally spiracles club-shaped, with a long, straight, narrow dorsal prolongation, surrounded by numerous long setae. Adanal plates almost drop-

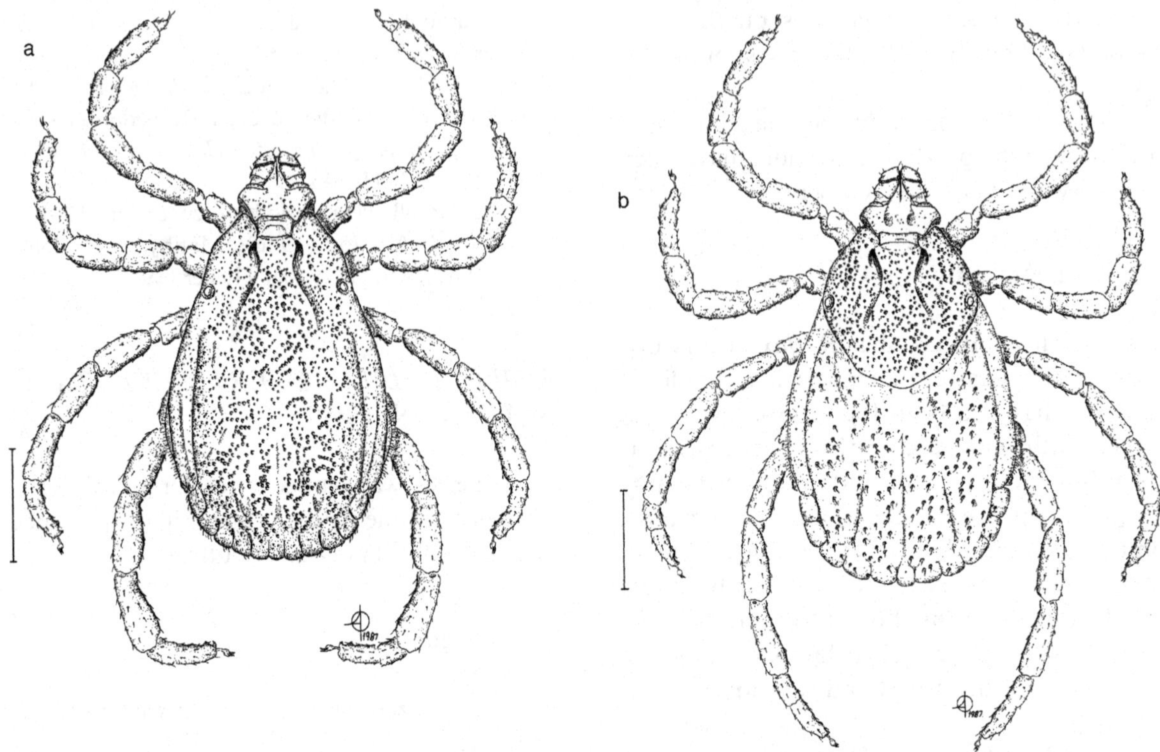

Figure 56. *Rhipicephalus evertsi evertsi* (Protozoology Section Tick Breeding Register, Onderstepoort, No. 3264, laboratory reared, origin of strain not recorded). (a) Male, dorsal; (b) female, dorsal. Scale bars represent 1 mm. A. Olwage *del.*

shaped, broadly rounded posteriorly; accessory adanal plates bluntly pointed.

Female (Figs 56(b), 57(d) to (f))

Capitulum broader than long, length × breadth ranging from 0.69 mm × 0.74 mm to 0.83 mm × 0.88 mm. Basis capituli with acute lateral angles just anterior to mid-length; porose areas large, oval, about 1.5 times their own diameter apart. Palps broad, flattened apically. Scutum longer than broad, length × breadth ranging from 1.47 mm × 1.33 mm to 1.80 mm × 1.57 mm; posterior margin sinuous. Eyes at widest point of scutum, beady, orbited. Cervical pits long, convergent, continuous with the long divergent rugose internal cervical margins. Large setiferous punctations most evident on scapulae and scattered between cervical fields, interspersed with numerous medium-sized to small punctations. Ventrally spiracles, as in the male, surrounded by numerous long setae. Genital aperture V-shaped.

Nymph (Fig. 58)

Capitulum slightly broader than long, length × breadth ranging from 0.23 mm × 0.24 mm to 0.26 mm × 0.27 mm. Basis capituli over twice as broad as long, mildly convex laterally. Palps almost parallel sided, their external margins nearly straight, their internal margins somewhat sinuous, broadly rounded apically, shorter than the hypostome. Scutum broader than long, length × breadth ranging from 0.46 mm × 0.55 mm to 0.53 mm × 0.61 mm; posterior margin concave laterally where the internal cervical grooves reach it. Eyes at widest point, well over halfway back, slightly bulging. Cervical grooves initially mildly convergent, becoming divergent and extending back to the posterolateral margins of the scutum, thus dividing the scutum into

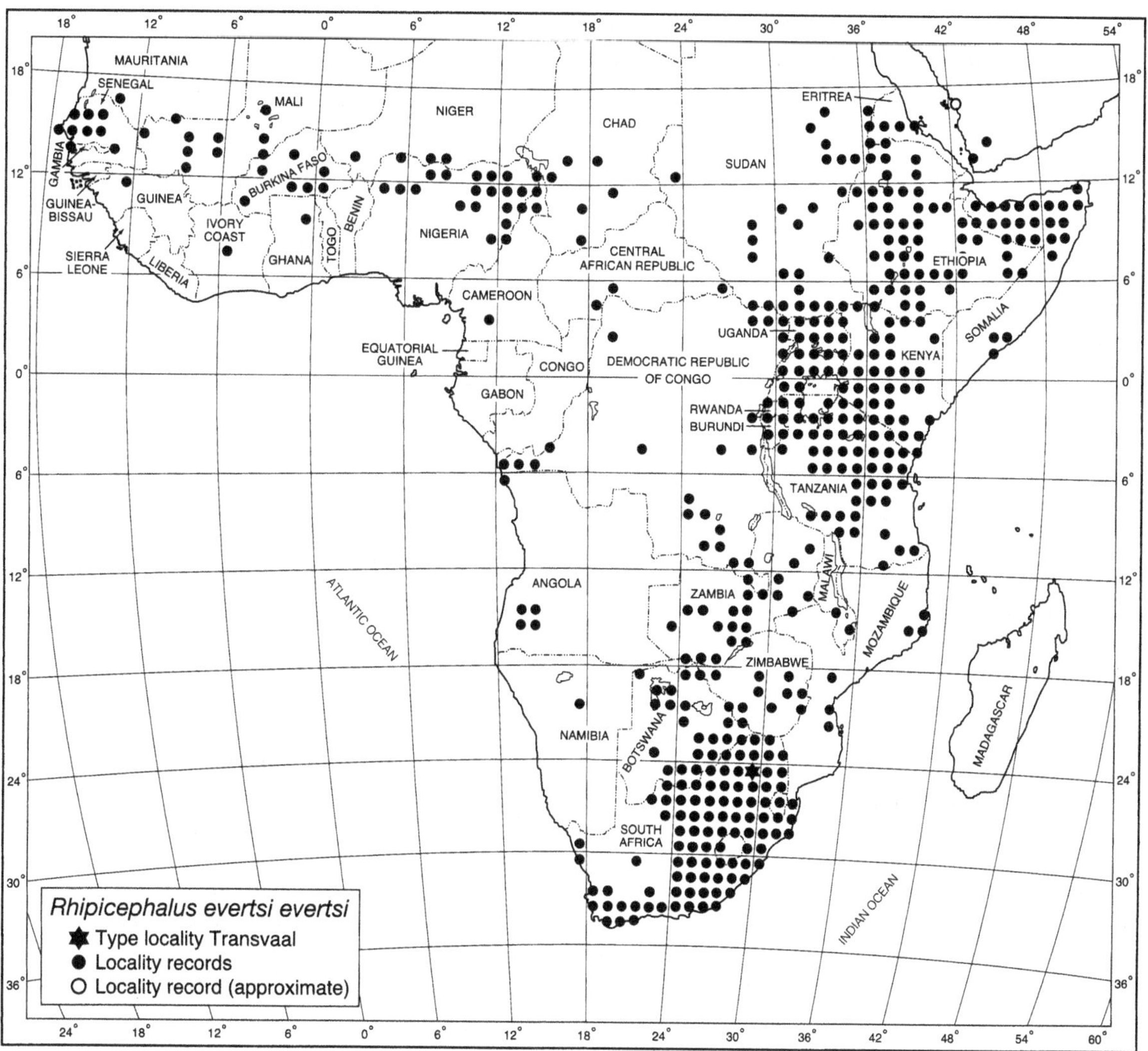

Map 18. *Rhipicephalus evertsi evertsi*: distribution.

three parts. Ventrally coxae I each with two short broad spurs; remaining coxae each with a bluntly-rounded external spur only.

Larva (Fig. 59)

Capitulum much longer than broad, length × breadth ranging from 0.173 mm × 0.137 mm to 0.188 mm × 0.155 mm. Basis capituli over twice as broad as long, almost rectangular in shape. Palps more or less sausage-shaped, broadly rounded apically. Scutum much broader than long, length × breadth ranging from 0.313 mm × 0.398 mm to 0.325 mm × 0.419 mm; posterior margin a smooth deep curve. Eyes at widest point of scutum, just over halfway back, slightly bulging. Cervical grooves short, almost parallel. Ventrally coxae I each with a short broad spur; remaining coxae each with a slight ridge-like salience on its posterior margin.

Notes on identification

The adults of this species are amongst the easiest rhipicephalids to identify, with their dark scutum and contrasting reddish-orange legs. In parts of the Eastern and Western Cape Provinces of South Africa, though, its distribution overlaps

Table 16. *Host records of* Rhipicephalus evertsi evertsi

Hosts	Number of records
Domestic animals	
Cattle	4819 (including immatures)
Sheep	667 (including immatures)
Goats	303 (including immatures)
Camels	18
Horses	104 (including immatures)
Donkeys	26 (including immatures)
Mules	3
Pigs	3 (including nymphs)
Dogs	30 (including immatures)
Cats	1 (1 nymph)
Rabbits	1
Wild animals	
Vervet monkey (*Chlorocebus aethiops*)	1 (1 larva)
Yellow baboon (*Papio cynocephalus*)	1
Chacma baboon (*Papio ursinus*)	2 (including immatures)
Papio sp.	1
'Baboon'	1
Colobus monkey (*Colobus* sp.)	1 (1 larva)
Side-striped jackal (*Canis adustus*)	2 (immatures)
Black-backed jackal (*Canis mesomelas*)	8 (immatures)
Canis sp.	3 (including nymphs)
'Jackal'	1
Hunting dog (*Lycaon pictus*)	1 (nymphs)
Cheetah (*Acinonyx jubatus*)	3 (including larvae)
Caracal (*Caracal caracal*)	23 (immatures)
African wild cat (*Felis lybica*)	2 (immatures)
Lion (*Panthera leo*)	7 (including larvae)
Leopard (*Panthera pardus*)	4
Marsh mongoose (*Atilax paludinosus*)	1 (immatures)
Slender mongoose (*Galerella sanguinea*)	1 (larvae)
Banded mongoose (*Mungos mungo*)	1 (immatures)
Meercat (*Suricata suricatta*)	1 (immatures)
Aardwolf (*Proteles cristatus*)	1 (larvae)
Ratel (*Mellivora capensis*)	1 (including immatures)
African civet (*Civettictis civetta*)	1 (larvae)
Genetta sp.	1 (larvae)
African elephant (*Loxodonta africana*)	2
Burchell's zebra (*Equus burchellii*)	170 (including immatures)
Grevy's zebra (*Equus grevyi*)	11 (including nymphs)
Mountain zebra (*Equus zebra*)	20 (including immatures)
'Zebra'	10 (including nymphs)
White rhinoceros (*Ceratotherium simum*)	1
Yellow-spotted rock hyrax (*Heterohyrax brucei*)	2 (including nymphs)
Rock hyrax (*Procavia capensis*)	5 (including immatures)
Warthog (*Phacochoerus africanus*)	19 (including immatures)

Table 16. (*cont.*)

Hosts	Number of records
Bushpig (*Potamochoerus larvatus*)	1
Giraffe (*Giraffa camelopardalis*)	46 (including immatures)
Impala (*Aepyceros melampus*)	472 (including immatures)
Red hartebeest (*Alcelaphus buselaphus caama*)	13 (including immatures)
Coke's hartebeest (*Alcelaphus buselaphus cokii*)	6 (including immatures)
Jackson's hartebeest (*Alcelaphus buselaphus jacksoni*)	4 (including larvae)
Black wildebeest (*Connochaetes gnou*)	25 (including immatures)
Blue wildebeest (*Connochaetes taurinus*)	172 (including immatures)
'Wildebeest'	1
Tsessebe (*Damaliscus lunatus lunatus*)	9 (including immatures)
Topi (*Damaliscus lunatus topi*)	5 (including nymphs)
Korrigum (*Damaliscus lunatus korrigum*)	1 (immatures)
Bontebok (*Damaliscus pygargus dorcas*)	9 (including immatures)
Blesbok (*Damaliscus pygargus phillipsi*)	14 (including immatures)
Lichtenstein's hartebeest (*Sigmoceros lichtensteinii*)	5 (including immatures)
Springbok (*Antidorcas marsupialis*)	33 (including immatures)
Grant's gazelle (*Gazella granti*)	16 (including immatures)
Thomson's gazelle (*Gazella thomsonii*)	36 (including immatures)
Gerenuk (*Litocranius walleri*)	4
Kirk's dik-dik (*Madoqua kirkii*)	4 (including nymphs)
'Dik-dik' (*Madoqua* sp.)	1
Klipspringer (*Oreotragus oreotragus*)	1 (immatures)
Oribi (*Ourebia ourebi*)	6 (including immatures)
Steenbok (*Raphicerus campestris*)	16 (including immatures)
Cape grysbok (*Raphicerus melanotis*)	1 (nymphs)
Sharpe's grysbok (*Raphicerus sharpei*)	1
African buffalo (*Syncerus caffer*)	181 (including immatures)
Eland (*Taurotragus oryx*)	80 (including immatures)
Nyala (*Tragelaphus angasii*)	32 (including immatures)
Lesser kudu (*Tragelaphus imberbis*)	1
Bushbuck (*Tragelaphus scriptus*)	28 (including immatures)
Greater kudu (*Tragelaphus strepsiceros*)	172 (including immatures)
'Kudu'	1
Blue duiker (*Cephalophus monticola*)	1 (immatures)
Red forest duiker (*Cephalophus natalensis*)	8 (including immatures)
Common duiker (*Sylvicapra grimmia*)	24 (including immatures)
'Duiker'	2
Roan antelope (*Hippotragus equinus*)	10
Sable antelope (*Hippotragus niger*)	37 (including immatures)
Gemsbok (*Oryx gazella*)	52 (including immatures)
Grey rhebok (*Pelea capreolus*)	6 (including immatures)
Waterbuck (*Kobus ellipsiprymnus*)	24 (including immatures)
Kob (*Kobus kob*)	3
Lechwe (*Kobus leche*)	8 (including immatures)
Puku (*Kobus vardonii*)	2
Kobus sp.	1

Table 16. (*cont.*)

Hosts	Number of records
Reedbuck (*Redunca arundinum*)	61 (including immatures)
Mountain reedbuck (*Redunca fulvorufula*)	25 (including immatures)
Redunca sp.	1
Smith's bush squirrel (*Paraxerus cepapi*)	1 (1 larva)
Huet's bush squirrel (*Paraxerus ochraceus*)	1
Unstriped ground squirrel (*Xerus rutilus*)	1
Short-tailed pouched mouse (*Saccostomus campestris*)	1 (1 larva)
Common fat mouse (*Steatomys pratensis*)	1 (immatures)
Red veld rat (*Aethomys chrysophilus*)	1 (immatures)
Namaqua rock mouse (*Aethomys namaquensis*)	1 (1 larva)
Arvicanthis sp.[1]	1 (1 larva)
Lemniscomys sp.	1 (immatures)
Spring hare (*Pedetes capensis*)	5 (immatures)
South African porcupine (*Hystrix africaeaustralis*)	1 (1 nymph)
Crested porcupine (*Hystrix cristata*)	1
Cape hare (*Lepus capensis*)[2]	136 (immatures)
Scrub hare (*Lepus saxatilis*)[2]	445 (immatures)
Savanna hare (*Lepus victoriae*)	1 (nymphs)
Lepus sp.	69 (immatures)
'Hare'	6 (including immatures)
Smith's red rock rabbit (*Pronolagus rupestris*)	9 (including immatures)
Pronolagus sp.	2 (immatures)
Short-snouted elephant shrew (*Elephantulus brachyrhynchus*)	1 (immatures)
Rufous elephant shrew (*Elephantulus rufescens*)	4 (immatures)
Short-eared elephant shrew (*Macroscelides proboscideus*)	1 (immatures)
Birds	
Reed cormorant (*Phalacrocorax africanus*)	2 (immatures)
Crested francolin (*Francolinus sephaena*)	1 (immatures)
Helmeted guineafowl (*Numida meleagris*)	7 (larvae)
White-backed mousebird (*Colius colius*)	1 (immatures)
Red-capped lark (*Calandrella cinerea*)	1 (including immatures)
Greater striped swallow (*Hirundo cucullata*)	1 (including immatures)
Brown-headed tchagra (*Tchagra australis*)	2 (immatures)
Arrow-marked babbler (*Turdoides jardineii*)	2 (immatures)
Red-billed quelea (*Quelea quelea*)	1 (immatures)
Reptiles	
Leopard tortoise (*Geochelone pardalis*)	1 (1 female)
Humans	3 (including 1 nymph)

Note: [1]Listed as an unstriped grass mouse by Yeoman & Walker (1967).
[2]A few hares harboured a single adult tick each.

with that of the related species *R. glabroscutatum* (see pp. 200–206) and this can cause some problems with the identification of field collections. Apart from their prominent eyes there is no morphological resemblance between the adults of these two species but in several respects their immature stages look alike and field-collected specimens are often difficult to separate. However, the basis capituli of an *R. evertsi evertsi* nymph is narrower and more rounded laterally than that of an *R. glabroscutatum* nymph, with its longer sharper lateral angles. The scuta of these nymphs are similar, with long cervical grooves that reach the posterolateral margins. Proportionally, though, the scutum of an *R. evertsi evertsi* nymph is narrower in relation to its length than that of an *R. glabroscutatum* nymph. The palps and basis capituli of an *R. evertsi evertsi* larva are more-or-less parallel sided, whereas the palps of an *R. glabroscutatum* larva are elongate oval in shape and its basis capituli narrows slightly posteriorly. The body of an engorged or semi-engorged *R. evertsi evertsi* larva is also nearly parallel sided, with a bluntly-rounded posterior end, compared with the egg-shaped *R. glabroscutatum* larva, which tapers towards its posterior end.

Hosts

A two-host species (Theiler, 1943; Rechav, Knight & Norval, 1977). The preferred hosts of all stages of development are large animals such as cattle, horses, zebras and eland (Yeoman & Walker, 1967; Walker, 1974; MacLeod *et al.*, 1977; Norval, 1981; Horak *et al.*, 1991b) (Table 16). Both Hoogstraal (1956) and Norval (1981) believe that domestic equids and wild zebras may be the pre-eminent hosts. It is the commonest species on donkeys in the Yemen (Pegram, Hoogstraal & Wassef, 1982). Many collections have also been taken from sheep, goats, impala, blue wildebeest, African buffalo and greater kudu (Walker, 1974; Carmichael, 1976; Norval, 1981; Horak *et al.*, 1992; Gallivan & Surgeoner, 1995). Sheep are good hosts of the adults only. Although they may harbour large numbers of larvae few of these develop to nymphs (Horak, Williams & Van Schalkwyk, 1991a). A similar situation pertains for several antelope species. Morel & Mouchet (1965) state that the absence of this species on wild ungulates in West Africa suggests that it was introduced on domestic livestock from East Africa. Scrub hares are good hosts of the immature stages. They seldom harbour large numbers but the translation on them of larvae into nymphs is excellent (Clifford, Flux & Hoogstraal, 1976; Rechav, Zeederberg & Zeller, 1987; Horak *et al.*, 1991b).

Although *R. evertsi evertsi* is widely distributed in both southern and eastern Africa, and parasitizes a great variety of hosts, adult ticks are seldom abundant. Individual collections exceeding 80 ticks from cattle, horses, zebras or eland must be considered large. Several collections exceeding 1000 immature ticks have been taken from the ears of zebras (Horak, De Vos & De Klerk, 1984; Horak *et al.*, 1991b).

The preferred site of attachment of adult ticks is the peri-anal area (Hoogstraal, 1956; Baker & Ducasse, 1967), but some may also be found on the inner thigh and groin region. The immature stages attach mainly in the external ear canals (Hoogstraal, 1956; Baker & Ducasse, 1967). Many larvae, both alive and dead, may be present in the fleece of woolled sheep (Horak *et al.*, 1991a).

Adult and immature *R. evertsi evertsi* are present on host animals throughout the year. In the southern regions of its distribution zone adult ticks are more abundant from spring to autumn and the immatures tend to be more abundant from autumn to spring (MacLeod *et al.*, 1977; Rechav, 1982; Colborne, 1988; Horak *et al.*, 1991a). In the north no clear pattern of seasonal abundance is evident (Kaiser *et al.*, 1988). Throughout its distribution range *R. evertsi evertsi* probably completes more than one life cycle annually. Field observations on the engorged females, eggs and larvae of this tick were carried out at Mazabuka, Zambia, by Gray (1951).

On the South African Highveld the synchronous emergence of large numbers of adults

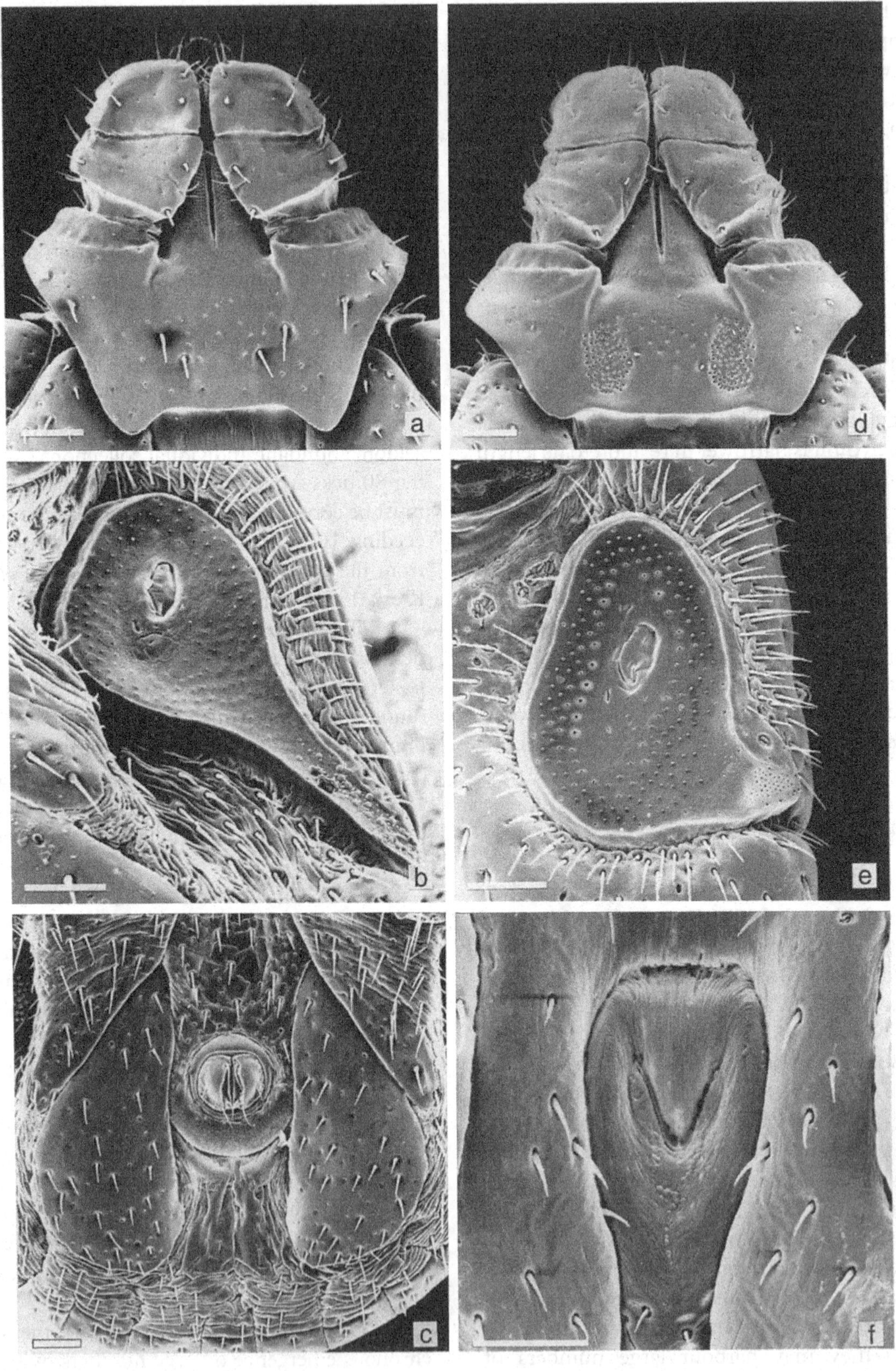
a
d
b
e
c
f

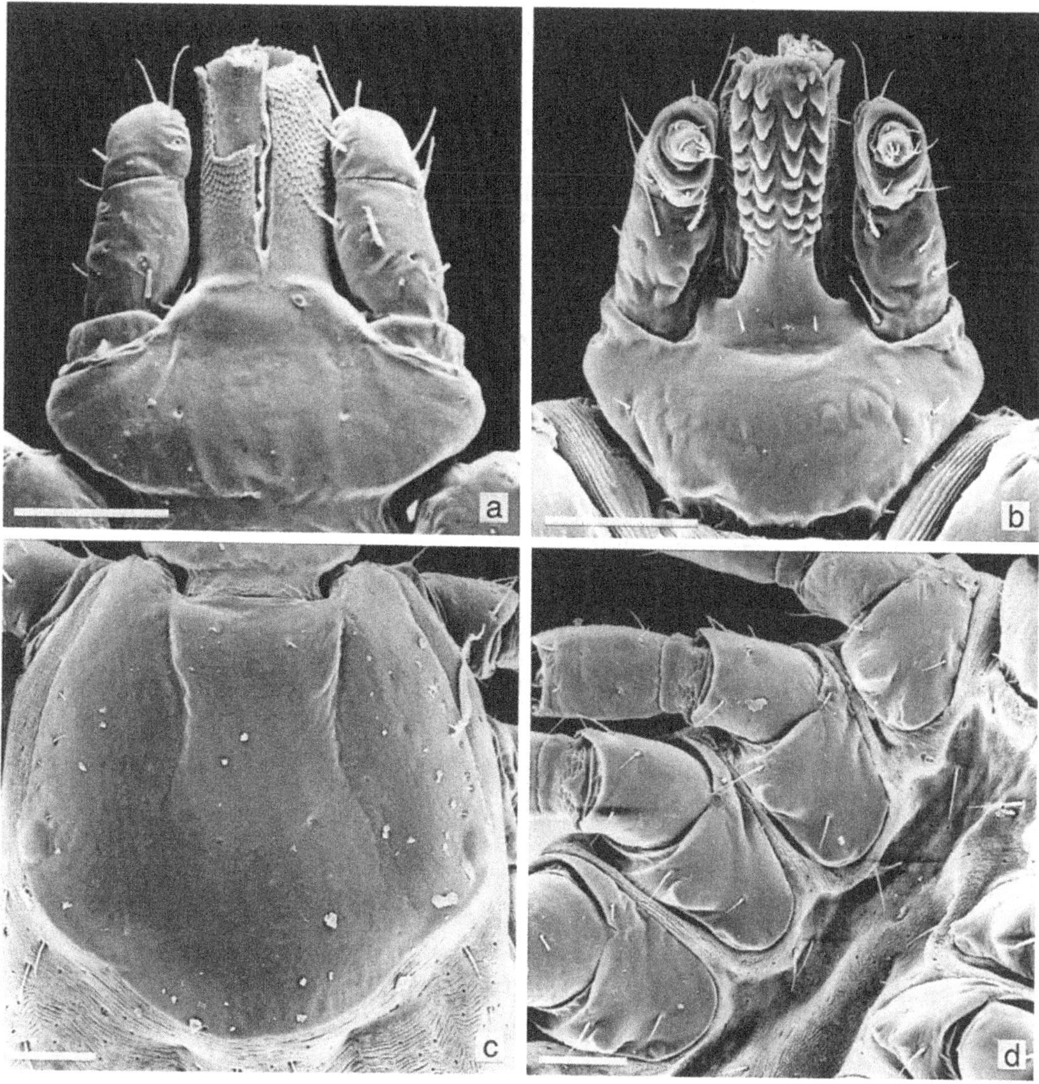

Figure 58 (*above*). *Rhipicephalus evertsi evertsi* (B.S. 250/-, RML 66306, laboratory reared, original ♀ collected from bovine, Kikuyu, Kenya, on 1 November 1950 by B. Gaithu). Nymph: (a) capitulum, dorsal; (b) capitulum, ventral; (c) scutum; (d) coxae. Scale bars represent 0.10 mm. SEMs by M.D. Corwin.

Figure 57 (*opposite*). *Rhipicephalus evertsi evertsi* (B.S. 250/-, RML 66306, laboratory reared, original ♀ collected from bovine, Kikuyu, Kenya, on 1 November 1950 by B. Gaithu). Male: (a) capitulum, dorsal; (b) spiracle; (c) adanal plates. Female: (d) capitulum, dorsal; (e) spiracle; (f) genital aperture. Scale bars represent 0.10 mm. SEMs by M.D. Corwin.

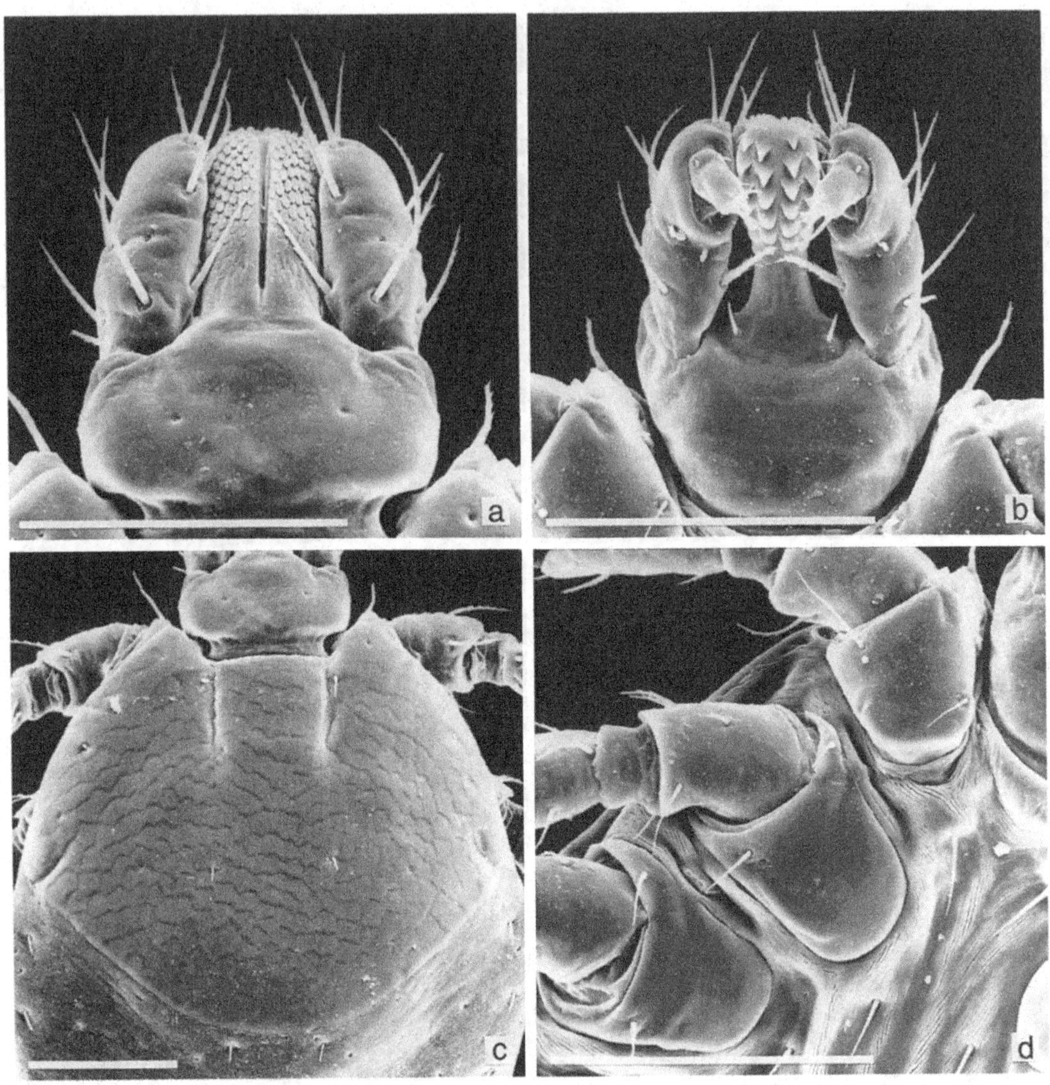

Figure 59. *Rhipicephalus evertsi evertsi* (B.S. 250/-, RML 66306, laboratory reared, original ♀ collected from bovine, Kikuyu, Kenya, on 1 November 1950 by B. Gaithu). Larva: (a) capitulum, dorsal; (b) capitulum, ventral; (c) scutum; (d) coxae. Scale bars represent 0.10 mm. SEMs by M.D. Corwin.

from overwintered nymphs can result in heavy burdens during spring (Horak *et al.*, 1991a). In young lambs this increase in infestation can result in paralysis (Hamel & Gothe, 1978).

Zoogeography

Rhipicephalus evertsi evertsi is the most widespread rhipicephalid in the Afrotropical region. It is prevalent in the eastern half of Africa from South Africa in the south to eastern Sudan in the north (Map 18). No collections have been made north of 18 °N. Within this region it is either absent or only marginally present in northern Zambia, Malawi, southern Tanzania, eastern and northern Kenya, southern Somalia and western Ethiopia. With the possible exception of low rainfall we do not know what constraints affect its distribution in these and other countries where it occurs. Its range in Mozambique is largely unknown because of a dearth of collections. In

South Africa, particularly in the western regions of that country, the tick is not necessarily as densely distributed as indicated by the degree square system of locality plotting that we have used. In the western half of the subcontinent it is sparsely distributed.

In West Africa it is present in a band extending roughly from 10 °N to 16 °N and 11 °W to 20 °E. As noted earlier Morel & Mouchet (1965) suggested that it was introduced into this region on domestic livestock from East Africa.

Doubtless *R. evertsi evertsi* was also introduced into the Yemen and Saudi Arabia. Pegram *et al.* (1982) found that it was common in the humid highlands and subhighlands in the Yemen but either rare or absent in the arid coastal plains. In Saudi Arabia, Al-Khalifa *et al.* (1987) recorded four collections of this subspecies from indigenous cattle and sheep in Gizan (Jazan) and Asir Provinces.

Rhipicephalus evertsi evertsi has been collected at altitudes varying from just above sea level to approximately 2500 m, and appears to be commonest in regions receiving between 400 mm and 1000 mm of rainfall annually. However, in Ethiopia, De Castro (1994) found that it was commonly present in habitats receiving between 1200 mm and 2600 mm of annual rainfall. Theiler (1950) suggested that increasing aridity limits its distribution and that the critical rainfall level lies between 250 mm and 375 mm annually. It occurs in woodland, bushland, wooded and bushed grassland, and grassland, and more rarely in forest and shrubland.

Disease relationships

Although Lounsbury (1906) showed that *R. evertsi evertsi* can transmit *Theileria parva parva* to cattle it is not regarded as an important vector. It is, however, an important vector of *Theileria separata* of sheep (Jansen & Neitz, 1956), and of *Babesia caballi* and *Babesia equi* of domestic equids (De Waal & Potgieter, 1987; De Waal, 1983–88, cited by De Waal & Van Heerden, 1994). It is less efficient as a vector of *Babesia bigemina* (Büscher, 1988). Intrastadial transmission of *Anaplasma marginale* to cattle by adult ticks and transstadial transmission of *Ehrlichia ovina* to sheep have been demonstrated experimentally (Neitz, 1952, cited by Neitz, 1956; Potgieter, 1981). *Rhipicephalus evertsi evertsi* is an important vector of *Borrelia theileri*, the causative agent of spirochaetosis in cattle, sheep, goats and domestic equids (Theiler, 1909). Engorging female ticks produce a toxin causing paralysis, particularly in lambs and possibly also in young calves (Clark, 1938; Hamel & Gothe, 1978). This tick may also play a role in the transmission of the virus causing Crimean-Congo haemorrhagic fever in humans (Swanepoel *et al.*, 1983), and of *Rickettsia conori*, the cause of tick-bite fever in humans (Gear, 1954).

REFERENCES

Al-Khalifa, M.S., Hussein, H.S., Al-Asgar, N.A. & Diab, F.M. (1987). Ticks (Acari: Ixodidae) infesting local domestic animals in western and southern Saudi Arabia. *Arab Gulf Journal of Scientific Research, Agriculture and Biology Science*, **B5**, 301–19.

Baker, M.K. & Ducasse, F.B.W. (1967). Tick infestation of livestock in Natal. I. The predilection sites and seasonal variations of cattle ticks. *Journal of the South African Veterinary Medical Association*, **38**, 447–53.

Büscher, G. (1988). The infection of various tick species with *Babesia bigemina*, its transmission and identification. *Parasitology Research*, **74**, 324–30.

Carmichael, I.H. (1976). Ticks from the African buffalo (*Syncerus caffer*) in Ngamiland, Botswana. *Onderstepoort Journal of Veterinary Research*, **43**, 27–30.

Clark, R. (1938). A note on paralysis in lambs caused apparently by *Rhipicephalus evertsi*. *Journal of the South African Veterinary Medical Association*, **9**, 143–5.

Clifford, C.M., Flux, J.E. & Hoogstraal, H. (1976). Seasonal and regional abundance of ticks (Ixodidae) on hares (Leporidae) in Kenya. *Journal of Medical Entomology*, **13**, 40–7.

Colborne, J.R.A. (1988). *The role of wild hosts in maintaining tick populations on cattle in the south-eastern Lowveld of Zimbabwe.* M. Phil. dissertation: University of Zimbabwe.

De Castro, J.J. (1994). *Tick Survey – Ethiopia. A survey of the tick species in western Ethiopia, including previous findings and recommendations for further tick surveys in Ethiopia.* Rome: Food and Agricultural Organisation of the United Nations, AG:DP/ETH/83/023, Technical Report, 83 pp.

De Waal, D.T. & Potgieter, F.T. (1987). The transstadial transmission of *Babesia caballi* by *Rhipicephalus evertsi evertsi. Onderstepoort Journal of Veterinary Research,* **54**, 655–6.

De Waal, D.T. & Van Heerden, J. (1994). Equine babesiosis. In *Infectious Diseases of Livestock with Special Reference to Southern Africa,* ed. J.A.W. Coetzer, G.R. Thomson & R.C. Tustin, pp. 295–304. Cape Town: Oxford University Press.

Gallivan, G.J. & Surgeoner, G.A. (1995). Ixodid ticks and other ectoparasites of wild ungulates in Swaziland: regional, host and seasonal patterns. *South African Journal of Zoology,* **30**, 169–77.

Gear, J. (1954). The rickettsial diseases of southern Africa. A review of recent studies. *South African Journal of Clinical Science,* **5**, 158–75.

Gray, W.J. (1951). *Rhipicephalus evertsi*: notes on free-living phases. *Bulletin of Epizootic Diseases of Africa,* **9**, 25–7.

Hamel, H.D. & Gothe, R. (1978). Influence of infestation rate on tick-paralysis in sheep induced by *Rhipicephalus evertsi evertsi* Neumann, 1897. *Veterinary Parasitology,* **4**, 183–91.

Horak, I.G., Boomker, J., Spickett, A.M. & De Vos, V. (1992). Parasites of domestic and wild animals in South Africa. XXX. Ectoparasites of kudus in the eastern Transvaal Lowveld and the eastern Cape Province. *Onderstepoort Journal of Veterinary Research,* **59**, 259–73.

Horak, I.G., De Vos, V. & De Klerk, B.D. (1984). Parasites of domestic and wild animals in South Africa. XVII. Arthropod parasites of Burchell's zebra, *Equus burchelli,* in the eastern Transvaal Lowveld. *Onderstepoort Journal of Veterinary Research,* **51**, 145–54.

Horak, I.G., Fourie, L.J., Novellie, P.A. & Williams, E.J. (1991b). Parasites of domestic and wild animals in South Africa. XXVI. The mosaic of ixodid tick infestations on birds and mammals in the Mountain Zebra National Park. *Onderstepoort Journal of Veterinary Research,* **58**, 125–36.

Horak, I.G., Williams, E.J. & Van Schalkwyk, P.C. (1991a). Parasites of domestic and wild animals in South Africa. XXV. Ixodid ticks on sheep in the north-eastern Orange Free State and in the eastern Cape Province. *Onderstepoort Journal of Veterinary Research,* **58**, 115–23.

Jansen, B.C. & Neitz, W.O. (1956). The experimental transmission of *Theileria ovis* by *Rhipicephalus evertsi. Onderstepoort Journal of Veterinary Research,* **27**, 3–6.

Kaiser, M.N., Sutherst, R.W., Bourne, A.S., Gorissen, L. & Floyd, R.B. (1988). Population dynamics of ticks on Ankole cattle in five ecological zones in Burundi and strategies for their control. *Preventive Veterinary Medicine,* **6**, 199–222.

Lounsbury, C.P. (1906). Ticks and African Coast fever. *Agricultural Journal, Cape of Good Hope,* **28**, 634–54.

MacLeod, J., Colbo, M.H., Madbouly, M.H. & Mwanaumo, B. (1977). Ecological studies of ixodid ticks (Acari: Ixodidae) in Zambia. III. Seasonal activity and attachment sites on cattle, with notes on other hosts. *Bulletin of Entomological Research,* **67**, 161–73.

Morel, P.C. & Mouchet, J. (1965). Les tiques du Cameroun (Ixodidae et Argasidae) (2e Note). *Annales de Parasitologie Humaine et Comparée,* **40**, 477–96.

Neitz, W.O. (1956). A consolidation of our knowledge of the transmission of tick-borne diseases. *Onderstepoort Journal of Veterinary Research,* **27**, 115–63.

Neumann, L.G. (1897). Révision de la famille des Ixodidés. (2e Mémoire). *Mémoires de la Société Zoologique de France,* **10**, 324–420.

Norval, R.A.I. (1981). The ticks of Zimbabwe. III. *Rhipicephalus evertsi evertsi. Zimbabwe Veterinary Journal,* **12**, 31–5.

Pegram, R.G., Hoogstraal, H. & Wassef, H.Y. (1982). Ticks (Acari: Ixodoidea) of the Yemen Arab Republic. 1. Species infesting livestock. *Bulletin of Entomological Research,* **72**, 215–27.

Potgieter, F.T. (1981). Tick transmission of anaplasmosis in South Africa. In *Proceedings of an International Conference on Tick Biology and Control,*

ed. G.B. Whitehead & J.D. Gibson, pp. 53–6. Grahamstown: Rhodes University.
Rechav, Y. (1982). Dynamics of tick populations (Acari: Ixodidae) in the eastern Cape Province. *Journal of Medical Entomology*, **19**, 679–700.
Rechav, Y., Knight, M.M. & Norval, R.A.I. (1977). Life cycle of the tick *Rhipicephalus evertsi evertsi* Neumann (Acarina: Ixodidae) under laboratory conditions. *Journal of Parasitology*, **63**, 575–9.
Rechav, Y., Zeederberg, M.E. & Zeller, D.A. (1987). Dynamics of African tick (Acari: Ixodoidea) populations in a natural Crimean-Congo hemorrhagic fever focus. *Journal of Medical Entomology*, **24**, 575–83.
Swanepoel, R., Struthers, J.K., Shepherd, A.J., McGillivray, G.M., Nel, H.J. & Jupp, P.G. (1983). Crimean-Congo hemorrhagic fever in South Africa. *American Journal of Tropical Medicine and Hygiene*, **32**, 1407–15.
Theiler, A. (1909). Transmission des spirelles et des piroplasmes par différentes espèces de tiques. *Bulletin de la Société de Pathologie Exotique*, **2**, 293–4.
Theiler, G. (1943). *Notes on the Ticks off Domestic Stock from Portuguese East Africa*. Estação Anti-Malárica de Lourenço Marques: Imprensa Nacional de Mocambique.
Theiler, G. (1950). Zoological Survey of the Union of South Africa. Tick Survey – Part V. Distribution of *Rhipicephalus evertsi*, the red tick. *Onderstepoort Journal of Veterinary Science and Animal Industry*, **24**, 33–6 + 1 map.
Also see the following Basic References (pp. 12–14): Hoogstraal (1956); Morel (1980); Scaramella (1988); Theiler (1962); Walker (1974); Walker, Mehlitz & Jones (1978); Yeoman & Walker (1967).

RHIPICEPHALUS EVERTSI MIMETICUS DÖNITZ, 1910

The subspecific name of this tick, from the Greek *mimetikos*, meaning 'imitation', is based on the fact that in various respects it mimics both the nominate subspecies *R. evertsi evertsi* and a *Hyalomma* sp.

Synonym

evertsi albigeniculatus.

Diagnosis

A moderate-sized dark brown tick with annulated legs (Fig. 60). Apart from the colour of their legs the adult and immature stages of this subspecies and those of *R. evertsi evertsi* appear to be morphologically identical.

Notes on identification

Attached adult ticks can easily be mistaken by the unwary for attached *Hyalomma* species, which also have banded legs, utilize the same peri-anal predilection attachment site and often occur in the same zoogeographical areas. An examination of their mouthparts readily separates these ticks.

Hosts

A two-host species. Following cattle, sheep and goats, most collections have been taken from domestic and wild equids and from greater kudu (Biggs & Langenhoven, 1984; Horak, Biggs & Reinecke, 1984; Horak *et al.*, 1992) (Table 17). Ticks on domestic dogs, lion, leopard and the chanting goshawk should be regarded as accidental infestations. Probably, as with *R. evertsi evertsi*, hares play a greater role as hosts of the immature stages than is indicated by the recorded host spectrum.

Adult ticks attach mainly in the peri-anal region while the majority of immatures are located in the external ear canals. The largest recorded burdens comprise 132 adults collected

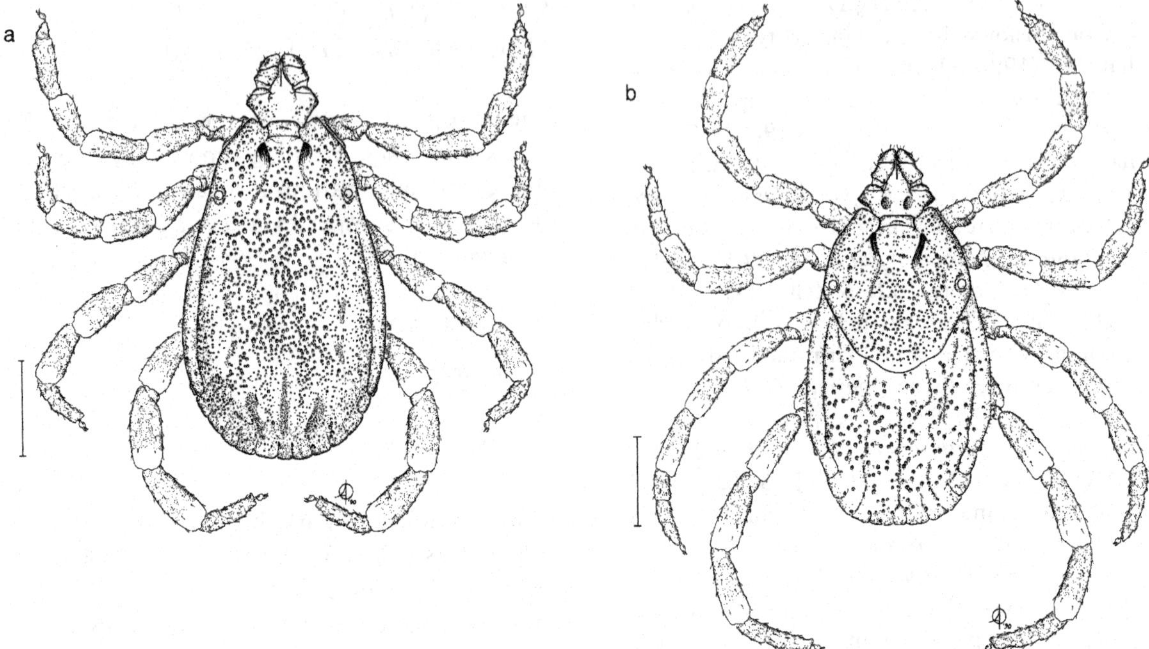

Figure 60. *Rhipicephalus evertsi mimeticus* (Protozoology Section Tick Breeding Register, Onderstepoort, No. 3349, laboratory reared, original ♀ collected from unknown host, Windhoek, Namibia in 1971). (a) Male, dorsal; (b) female, dorsal. Scale bars represent 1 mm. A. Olwage *del.*

from a horse and 5124 immatures from a greater kudu in Namibia (Horak *et al.*, 1984, 1992).

Adult and immature ticks are present throughout the year and more than one life cycle can probably be completed annually. Adults are most numerous on cattle in central Namibia from November to May and the immatures in February and March and from May to September (Biggs & Langenhoven, 1984).

Zoogeography

Rhipicephalus evertsi mimeticus has been recorded mainly in the drier regions of the sub-continent to the west of the areas occupied by *R. evertsi evertsi* (Map 19). Excluding the southern regions and the Namib desert, it is present throughout Namibia. It also occurs in western Democratic Republic of Congo, western and southern Angola, and western Botswana, where its distribution may overlap that of *R. evertsi evertsi* (Paine, 1982).

We question whether it ever became established in north-eastern Democratic Republic of Congo (Pierquin & Niemegeers, 1957) and have therefore omitted these records from Map 19. It appears to have been introduced more than once into South Africa, probably on livestock, but it remains to be seen whether it will succeed in establishing itself in that country.

It is found from just above sea level to 1500 m and at localities with rainfall varying from 100 mm to 400 mm annually, sometimes considerably more in the northern part of its range. The vegetation types coinciding with most of its habitat are Karoo-Namib shrubland, Kalahari *Acacia* wooded grassland and deciduous bushland, and *Colophospermum mopane* woodland and scrub woodland. In the northern parts of its distribution zone it has apparently been recorded in wetter Zambezian miombo woodland and in a lowland mosaic of rainforest and secondary grassland.

Table 17. *Host records of* Rhipicephalus evertsi mimeticus

Hosts	Number of records
Domestic animals	
Cattle	376 (including immatures)
Sheep	43 (including immatures)
Goats	50 (including immatures)
Camels	7
Horses	31 (including immatures)
Donkeys	13 (including immatures)
Pigs	2
Dogs	3
Wild animals	
Lion (*Panthera leo*)	2 (including immatures)
Leopard (*Panthera pardus*)	1
Burchell's zebra (*Equus burchellii*)	12 (including immatures)
Mountain zebra (*Equus zebra*)	24 (including immatures)
'Zebra'	15 (including immatures)
White rhinoceros (*Ceratotherium simum*)	1
Black rhinoceros (*Diceros bicornis*)	1
Warthog (*Phacochoerus africanus*)	8 (including immatures)
Giraffe (*Giraffa camelopardalis*)	12 (including immatures)
Impala (*Aepyceros melampus*)	5 (including immatures)
'Hartebeest'	1
Blue wildebeest (*Connochaetes taurinus*)	11 (including immatures)
Springbok (*Antidorcas marsupialis*)	5 (including larvae)
Kirk's dik-dik (*Madoqua kirkii*)	1 (including immatures)
Steenbok (*Raphicerus campestris*)	4 (including immatures)
African buffalo (*Syncerus caffer*)	2
Eland (*Taurotragus oryx*)	6
Greater kudu (*Tragelaphus strepsiceros*)	33 (including immatures)
Common duiker (*Sylvicapra grimmia*)	3 (including immatures)
Roan antelope (*Hippotragus equinus*)	4 (including nymphs)
Gemsbok (*Oryx gazella*)	13 (including nymphs)
'Hare'	1
Birds	
Pale chanting goshawk (*Melierax canorus*)	1

Disease relationships

Potgieter, De Waal & Posnett (1992) transmitted *Babesia equi* to horses by feeding immature *R. evertsi mimeticus* on an experimentally infected horse and the ensuing adults on susceptible animals. Neitz (1972) demonstrated that this tick can transmit *Theileria separata* (as *T. ovis*) to sheep. Large burdens of immature ticks in the ears of goat kids in Namibia are thought to have resulted in damage to the facial nerves of the kids, leading to accumulations of feed in the cheeks of affected animals (J.D. Bezuidenhout, 1972, unpublished data).

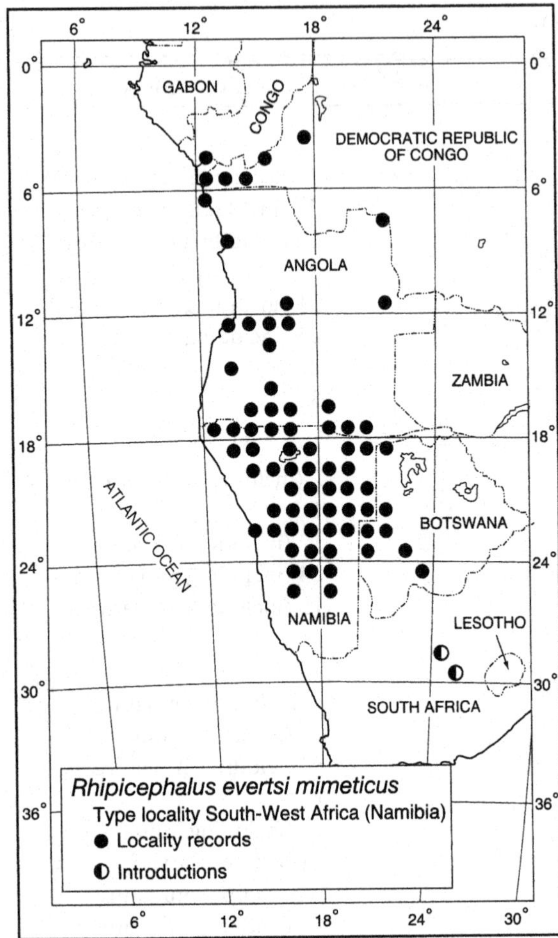

Map 19. *Rhipicephalus evertsi mimeticus*: distribution.

REFERENCES

Biggs, H.C. & Langenhoven, J.W. (1984). Seasonal prevalence of ixodid ticks on cattle in the Windhoek District of South West Africa/Namibia. *Onderstepoort Journal of Veterinary Research*, **51**, 175–82.

Dönitz, W. (1910). Die Zecken Südafrikas. In *Zoologische und Anthropologische Ergebnisse einer Forchungsreise im westlichen und zentralen Südafrika ausgefürt in den Jahren 1903–1905*, **4**, 3 Lieferung, ed. L. Schultze. *Denkschriften der Medizinisch-Naturwissenschaflichen Gesellschaft zu Jena*, **16**, 398–494, pls 15, 16a, b, & 17.

Horak, I.G., Anthonissen, M., Krecek, R.C. & Boomker, J. (1992). Arthropod parasites of springbok, gemsbok, kudus, giraffes and Burchell's and Hartmann's zebras in the Etosha and Hardap Nature Reserves, Namibia. *Onderstepoort Journal of Veterinary Research*, **59**, 253–7.

Horak, I.G., Biggs, H.C. & Reinecke, R.K. (1984). Arthropod parasites of Hartmann's mountain zebra, *Equus zebra hartmannae*, in South West Africa/Namibia. *Onderstepoort Journal of Veterinary Research*, **51**, 183–7.

Neitz, W.O. (1972). The experimental transmission of *Theileria ovis* by *Rhipicephalus evertsi mimeticus* and *R. bursa*. *Onderstepoort Journal of Veterinary Research*, **39**, 83–5.

Paine, G.D. (1982). Ticks (Acari: Ixodoidea) in Botswana. *Bulletin of Entomological Research*, **72**, 1–16.

Pierquin, L. & Niemegeers, K. (1957). Répertoire et distribution géographique des tiques au Congo Belge et au Ruanda-Urundi. *Bulletin Agricole du Congo Belge*, **48**, 1177–224.

Potgieter, F.T., De Waal, D.T. & Posnett, E.S. (1992). Transmission and diagnosis of equine babesiosis in South Africa. *Memórias do Instituto Oswaldo Cruz, Rio de Janeiro*, **87**, (Supplement III), 139–42.

Also see the following Basic References (pp. 12–14): Elbl & Anastos (1966); Morel (1980); Santos Dias (1983–84); Sousa Dias (1950); Theiler (1962); Walker, Mehlitz & Jones (1978).

RHIPICEPHALUS EXOPHTHALMOS KEIRANS & WALKER, 1993

The specific name *exophthalmos*, from the Greek meaning 'bulging eyes' refers to the shape of this species' eyes.

Diagnosis

A medium-sized light brown tick that in some respects resembles *R. oculatus*.

Male (Figs 61(a), 62 (a) to (c))

Capitulum slightly broader than long, length × breadth ranging from 0.53 mm × 0.56 mm to 0.74 mm × 0.76 mm. Basis capituli with short obtuse lateral angles at about anterior third

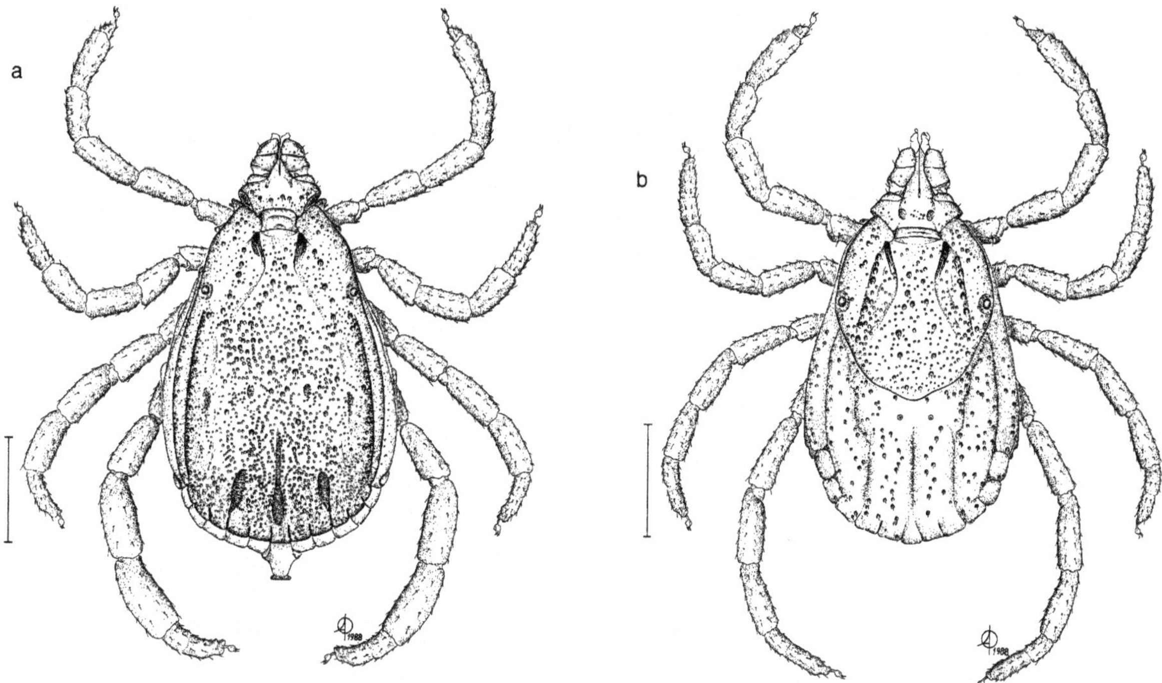

Figure 61. *Rhipicephalus exophthalmos* (Onderstepoort Tick Collection 3144, RML 65729, laboratory reared, progeny of ♀ collected on farm 'Karos II', Karas Region, Namibia, in March 1972). (a) Male, dorsal; (b) female, dorsal. Scale bars represent 1 mm. A. Olwage *del.* (From Keirans *et al.*, 1993, figs 1 & 2, with kind permission from the Editor, *Onderstepoort Journal of Veterinary Research*).

of its length. Palps broad, smoothly rounded apically. Conscutum length × breadth ranging from 2.25 mm × 1.44 mm to 3.35 mm × 2.07 mm; sharp anterior process present on coxae I. In engorged specimens body wall expanded laterally and posteriorly, with an elongate caudal process. Eyes submarginal, bulging, deeply orbited. Cervical pits deep; cervical fields shallow, tapering posteriorly beyond eye level. Marginal lines deep, punctate, extending anteriorly almost to eye level. Posteromedian and posterolateral grooves well developed. Large setiferous punctations present on scapulae, along external cervical margins and medially on conscutum, interspersed with numerous smaller punctations that are finer and sparser adjacent to marginal lines but sometimes denser posteriorly. Legs increase slightly in size from I to IV. Ventrally spiracles narrowly elongate with a long dorsal prolongation that can often be seen from the dorsal surface. Adanal plates broad, with a slightly convex

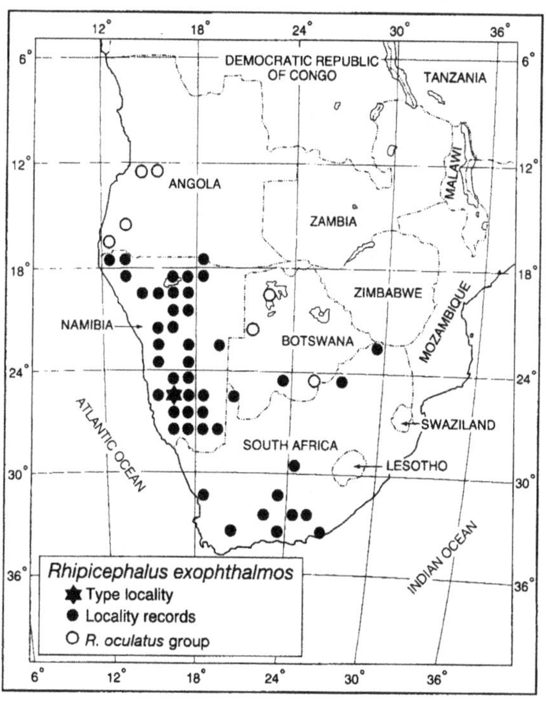

Map 20. *Rhipicephalus exophthalmos*: distribution. (After Keirans *et al.*, 1993).

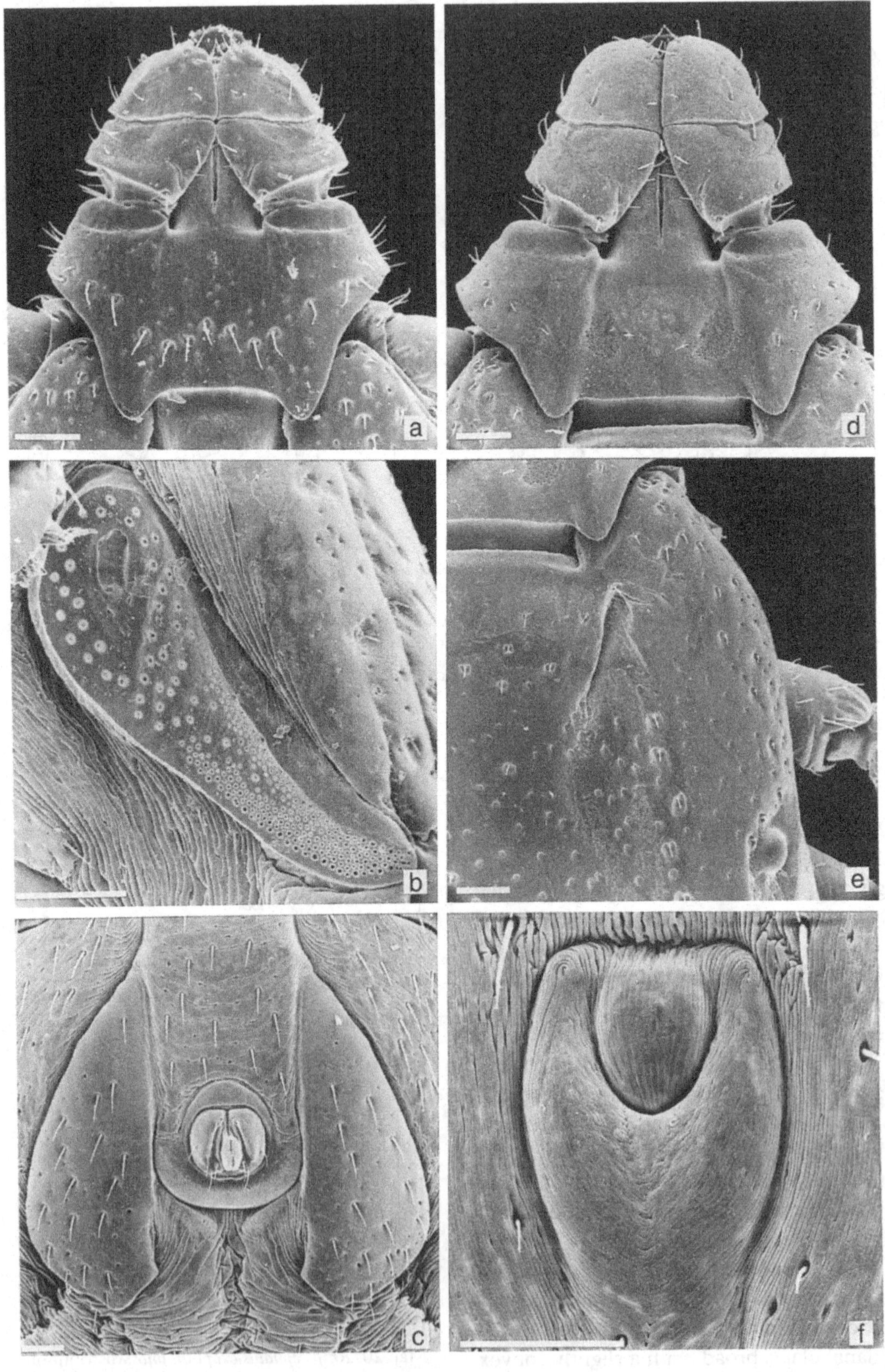
a
d
b
e
c
f

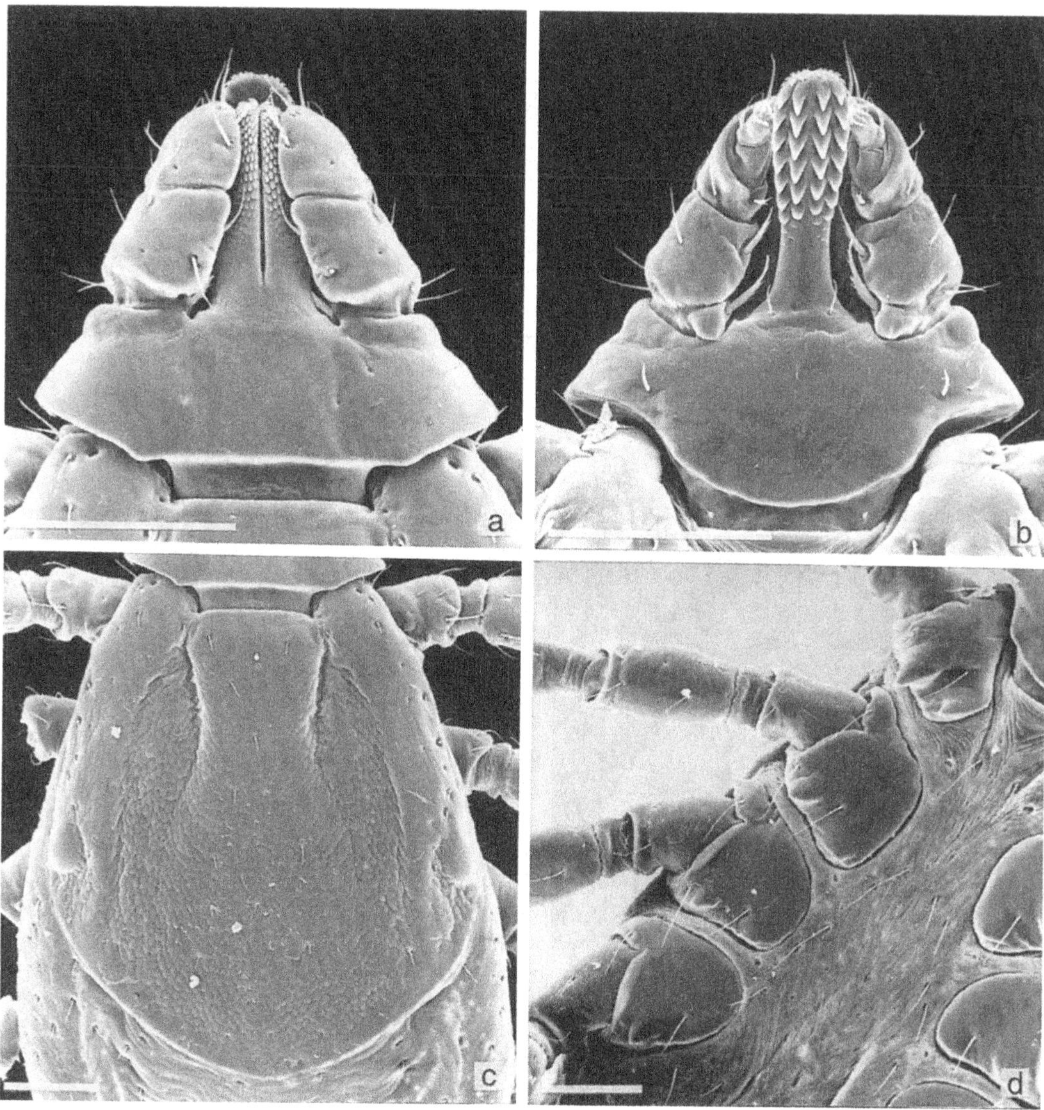

Figure 63. (*above*) *Rhipicephalus exophthalmos* (Onderstepoort Tick Collection 3144, RML 65729, laboratory reared, progeny of ♀ collected on farm 'Karos II', Karas Region, Namibia, in March 1972). Nymph: (a) capitulum, dorsal; (b) capitulum, ventral; (c) scutum; (d) coxae. Scale bars represent 0.10 mm. SEMs by M.D. Corwin. (From Keirans *et al.*, 1993, figs 9–12, with kind permission from the Editor, *Onderstepoort Journal of Veterinary Research*).

Figure 62 (*opposite*). *Rhipicephalus exophthalmos* (Onderstepoort Tick Collection 3144, RML 65729, laboratory reared, progeny of ♀ collected on farm 'Karos II', Karas Region, Namibia, in March 1972). Male: (a) capitulum, dorsal; (b) spiracle; (c) adanal plates. Female: (d) capitulum, dorsal; (e) scapular area; (f) genital aperture. Scale bars represent 0.10 mm. SEMs by M.D. Corwin. (From Keirans *et al.*, 1993, figs 3–8, with kind permission from the Editor, *Onderstepoort Journal of Veterinary Research*).

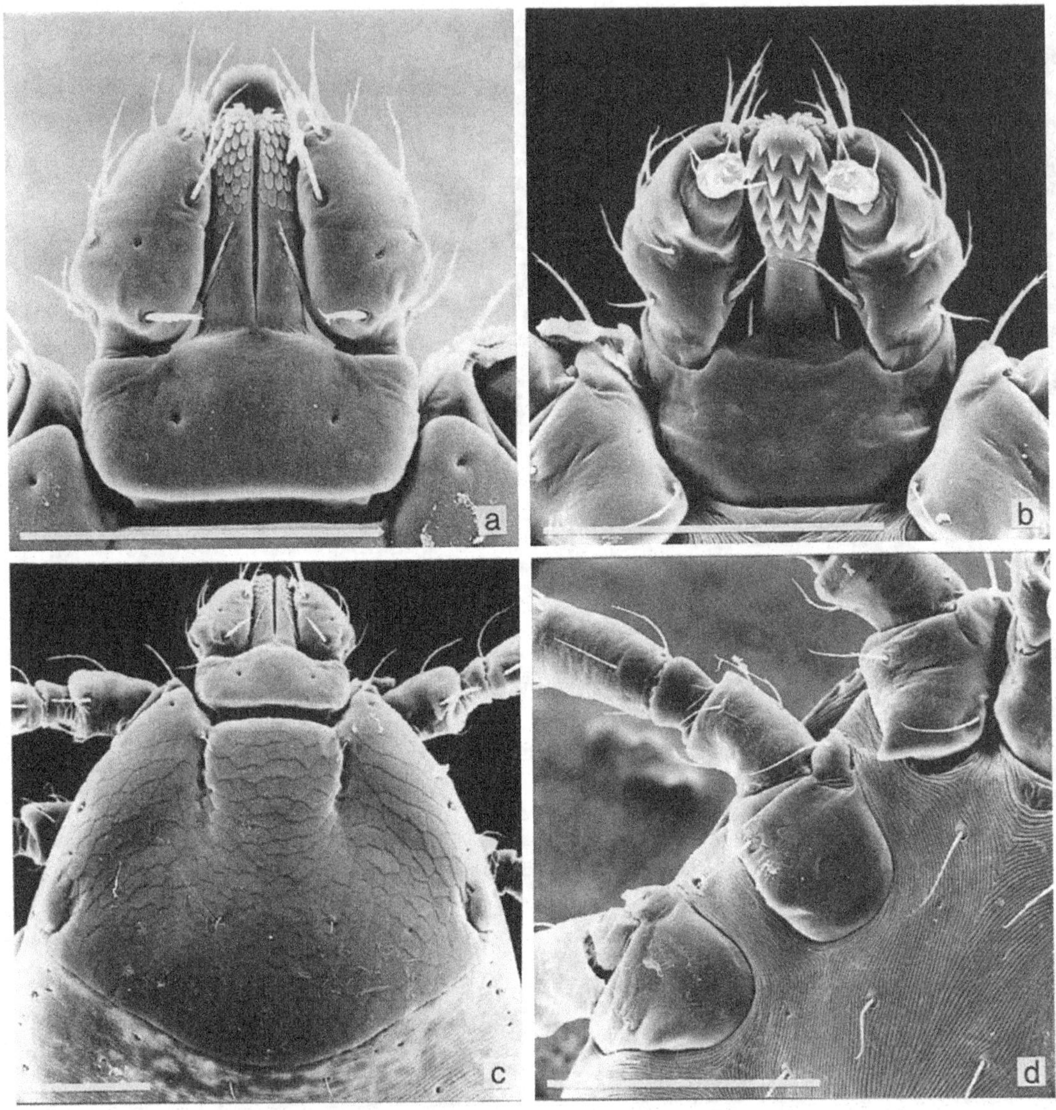

Figure 64. *Rhipicephalus exophthalmos* (Onderstepoort Tick Collection 3144, RML 65729, laboratory reared, progeny of ♀ collected on farm 'Karos II', Karas Region, Namibia, in March 1972). Larva: (a) capitulum, dorsal; (b) capitulum, ventral; (c) scutum; (d) coxae. Scale bars represent 0.10 mm. SEMs by M.D. Corwin. (From Keirans *et al.*, 1993, figs 13–16, with kind permission from the Editor, *Onderstepoort Journal of Veterinary Research*).

posterior margin leading posterointernally to a small medially-directed point; accessory adanal plates either absent or barely indicated.

Female (Figs 61(b), 62(d) to (f))

Capitulum broader than long, length × breadth ranging from 0.67 mm × 0.78 mm to 0.75 mm × 0.83 mm. Basis capituli with broad, slightly forwardly-directed lateral angles at about anterior third of its length; porose areas large, subcircular, about twice their own diameter apart. Palps broad, smoothly rounded apically. Scutum slightly longer than broad, length × breadth ranging from 1.41 mm × 1.33 mm to 1.53 mm × 1.48 mm; posterior margin slightly sinuous. Eyes submarginal, bulging, deeply orbited. Cervical pits deep; cervical fields tapering to beyond eye level, their surfaces often irregular-

Table 18. *Host records of* Rhipicephalus exophthalmos

Hosts	Number of records
Domestic animals	
Cattle	39
Sheep	43
Goats	16
Horses	2
Donkeys	1
Pigs	1
Dogs	2
Wild animals	
Lion (*Panthera leo*)	1
Burchell's zebra (*Equus burchellii*)	1
Warthog (*Phacochoerus africanus*)	9
Impala (*Aepyceros melampus*)	2
Springbok (*Antidorcas marsupialis*)	21
Steenbok (*Raphicerus campestris*)	5
Eland (*Taurotragus oryx*)	4
Greater kudu (*Tragelaphus strepsiceros*)	31
Gemsbok (*Oryx gazella*)	22
Grey rhebok (*Pelea capreolus*)	1
Mountain reedbuck (*Redunca fulvorufula*)	2
Namaqua rock mouse (*Aethomys namaquensis*)	1
Cape hare (*Lepus capensis*)	1
Scrub hare (*Lepus saxatilis*)	32 (including immatures)
'Hare'	20
Cape elephant shrew (*Elephantulus edwardii*)	3 (immatures)
Smith's rock elephant shrew (*Elephantulus rupestris*)	5 (immatures)
Short-eared elephant shrew (*Macroscelides proboscideus*)	1 (immatures)

ly ridged but sometimes smooth. Large setiferous punctations present on the scapulae, along the sharply delineated external cervical margins and scattered medially on the scutum, interspersed with numerous finer punctations. Ventrally genital aperture a broadly rounded U in shape, the area within the opening bulging.

Nymph (Fig. 63)

Capitulum much broader than long, length × breadth ranging from 0.20 mm × 0.29 mm to 0.26 mm × 0.32 mm. Basis capituli over three times as broad as long; lateral angles short, acute, posterior to mid-length. Palps narrow proximally, then widening before they taper to broadly rounded apices. Scutum slightly longer than broad, length × breadth ranging from 0.50 mm × 0.48 mm to 0.58 mm × 0.55 mm; posterior margin a deep smooth curve. Eyes bulging, partially orbited, on scutal margins immediately anterior to posterolateral angles. Internal cervical margins much shorter than the external margins, which almost reach the posterolateral borders of the scutum; cervical fields slightly depressed. Ventrally coxae I each with a long triangular external spur and a short triangular internal spur; coxae II to IV each with a small triangular external spur only.

Larva (Fig. 64)
Capitulum broader than long to slightly longer than broad, length × breadth ranging from 0.109 mm × 0.121 mm to 0.129 mm × 0.125 mm. Basis capituli about twice as broad as long; lateral margins mildly convex, curving smoothly to join the straight posterior margin. Palps constricted proximally, then becoming somewhat bulbous before tapering gently to bluntly-rounded apices. Scutum much broader than long, length × breadth ranging from 0.102 mm × 0.293 mm to 0.109 mm × 0.308 mm; posterior margin a smooth, fairly shallow curve. Eyes at widest point, bulging, rounded, slightly orbited. Cervical grooves short, slightly convergent. Ventrally coxae I each with a sharply-pointed triangular spur; coxae II each with a broadly-rounded spur; coxae III each with a small triangular spur.

Notes on identification
Rhipicephalus exophthalmos was described from a reared series that originated in southern Namibia by Keirans & Walker in Keirans *et al.* (1993). Previously various authors, including Zumpt (1949), Theiler & Robinson (1953), Theiler (1962) and Rechav & Knight (1983), and probably Sousa Dias (1950), had confused it with *R. oculatus* (see pp. 323–329). In publications by Horak and various co-workers records of *R. exophthalmos* on warthogs (*Phacochoerus africanus*) in northern Namibia were listed as *R. oculatus* and on several other host species in South Africa as *Rhipicephalus* sp. (near *R. oculatus*). A comparison of the descriptions and illustrations of these two species, though, shows that distinct morphological differences do exist between them. In the males the capitulum, punctation pattern, adanal plates, and caudal appendage differ, as do the capitulum, punctation pattern and genital aperture of the females. In the nymphs the capitulum, especially the position of the lateral angles on the basis capituli, the scutum and the coxal spurs should be examined, and in the larvae the capitulum, the scutum, especially the eyes, and the coxal spurs.

The record in Horak, Boomker & Flamand (1991) of *R. exophthalmos* [identified as *Rhipicephalus* sp. (near *R. oculatus*)] from a red forest duiker (*Cephalophus natalensis*) in north-eastern KwaZulu-Natal has, upon re-examination, proved to be *R. evertsi evertsi.*

Hosts

A three-host species (Theiler & Robinson, 1953; Rechav & Knight, 1983, both as *R. oculatus*). Domestic and wild ruminants as well as warthogs and hares are the preferred hosts of the adults (Horak *et al.*, 1983, 1992a, b) (Table 18). Few immature ticks have been collected; the scrub hare and elephant shrews appear to be their preferred hosts (Horak & Fourie, 1991; I.G.H., unpublished data). Although adult burdens seldom exceed 10 ticks, two yearling cattle examined in the Eastern Cape Province, South Africa and two gemsbok examined in southern Namibia each harboured more than 80 ticks (I.G.H., unpublished data).

Adult ticks were most abundant on cattle in the Eastern Cape Province during January (I.G.H., unpublished data) and on gemsbok in southern Namibia during November (Horak *et al.*, 1992a).

Zoogeography

Rhipicephalus exophthalmos has been recorded most frequently in Namibia, particularly in the southern parts of the country (Map 20). The majority of our records in South Africa come from the south-eastern regions, but the tick is probably commoner in the Northern Cape Province than the current records indicate. In Botswana there is one confirmed record only, from Sekoma Pan in the south. In south-western Angola, Sousa Dias (1950) recorded what is probably this species from Bocoio, in the Lobito area. Other collections from Botswana and Angola can be designated only as belonging to the *R. oculatus* group.

Rhipicephalus exophthalmos is present at altitudes ranging from about 200 m to 1500 m in regions where the climate can generally be described as semi-arid or arid with annual rainfall varying between 100 mm and 500 mm. The vegetation in many of these areas is semi-desert, bushy Karoo–Namib shrubland or dry wooded grassland and bushland. In south-eastern South Africa it is common in evergreen and semi-evergreen bushland and thicket referred to locally as Fish River bush.

Disease relationships

W.O. Neitz (cited by Keirans *et al.*, 1993) reported that a rabbit on which *R. exophthalmos* adults were fed became completely paralysed while the female ticks were feeding, but subsequently recovered. Two rabbits on which F_2 generation adults of the same strain of this species were fed also became paralysed. One of these rabbits recovered but the other died.

REFERENCES

Horak, I.G., Anthonissen, M., Krecek, R.C. & Boomker, J. (1992a). Arthropod parasites of springbok, gemsbok, kudus, giraffes and Burchell's and Hartmann's zebras in the Etosha and Hardap Nature Reserves, Namibia. *Onderstepoort Journal of Veterinary Research*, **59**, 253–7.

Horak, I.G., Biggs, H.C., Hanssen, T.S. & Hanssen, R.E. (1983). The prevalence of helminth and arthropod parasites of warthog *Phacochoerus aethiopicus*, in South West Africa/Namibia. *Onderstepoort Journal of Veterinary Research*, **50**, 145–8.

Horak, I.G., Boomker, J. & Flamand, J.R.B. (1991). Ixodid ticks and lice infesting red duikers and bushpigs in north-eastern Natal. *Onderstepoort Journal of Veterinary Research*, **58**, 281–4.

Horak, I.G., Boomker, J., Spickett, A.M. & De Vos, V. (1992b). Parasites of domestic and wild animals in South Africa. XXX. Ectoparasites of kudus in the eastern Transvaal Lowveld and the eastern Cape Province. *Onderstepoort Journal of Veterinary Research*, **59**, 259–73.

Horak, I.G. & Fourie, L.J. (1991). Parasites of domestic and wild animals in South Africa. XXIX. Ixodid ticks on hares in the Cape Province and on hares and red rock rabbits in the Orange Free State. *Onderstepoort Journal of Veterinary Research*, **58**, 261–70.

Keirans, J.E., Walker, J.B., Horak, I.G. & Heyne H. (1993). *Rhipicephalus exophthalmos* sp. nov., a new tick species from southern Africa, and redescription of *Rhipicephalus oculatus* Neumann, 1901, with which it has hitherto been confused (Acari: Ixodida: Ixodidae). *Onderstepoort Journal of Veterinary Research*, **60**, 229–46.

Rechav, Y. & Knight, M.M. (1983). Life cycle of *Rhipicephalus oculatus* (Acari: Ixodidae) in the laboratory. *Annals of the Entomological Society of America*, **76**, 470–2.

Theiler, G. & Robinson, B.N. (1953). Ticks in the South African Zoological Survey Collection. Part VII. Six lesser known African rhipicephalids. *Onderstepoort Journal of Veterinary Research*, **26**, 93–136 + 1 map.

Also see the following Basic References (pp. 12–14): Sousa Dias (1950); Theiler (1962); Zumpt (1949).

RHIPICEPHALUS FOLLIS DÖNITZ, 1910

According to Dönitz the specific name *follis* (Latin) means 'Lederbeutel' in German, i.e. a leather bag or pouch.

Diagnosis

A large, reddish-brown species.

Male (Figs 65(a), 66(a) to (c))

Capitulum as broad, or nearly as broad, as long in large specimens, but in the smallest broader than long, the length × breadth ranging from 0.46 mm × 0.53 mm to 0.90 mm × 0.88 mm. Basis capituli with short lateral angles at anterior third of its length. Palps somewhat flattened apically. Conscutum length × breadth ranging from 2.08 mm × 1.33 mm to 4.08 mm × 2.90 mm; anterior process of coxae I not very conspicuous. In engorged specimens a short, broad caudal process is present. Eyes slightly convex, edged dorsally by a

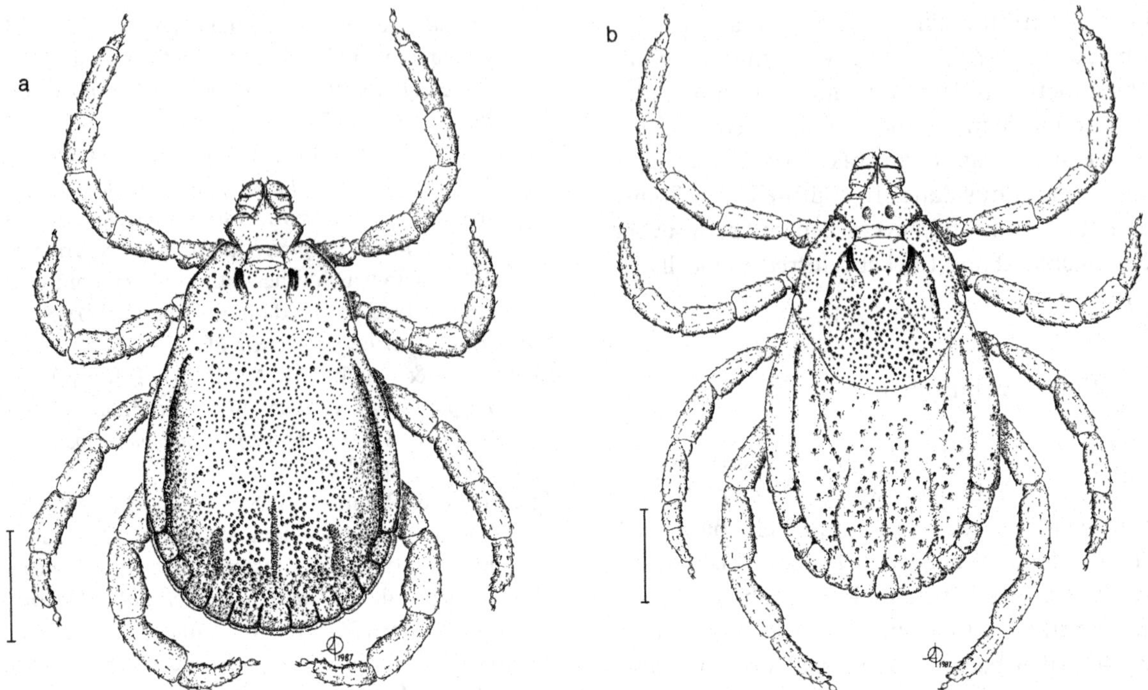

Figure 65. *Rhipicephalus follis* (Protozoology Section Tick Breeding Register, Onderstepoort, 3416, RML 65720, laboratory reared, progeny of ♀ collected from bovine, Lilystone Farm, East London, Eastern Cape Province, South Africa, in August 1972 by J.A.F. Baker). (a) Male, dorsal; (b) female, dorsal. Scale bars represent 1 mm. A. Olwage *del.*

Table 19. *Host records of* Rhipicephalus follis

Hosts	Number of records
Domestic animals	
Cattle	36
Sheep	12
Horses	6
Dogs	2
Wild animals	
Cheetah (*Acinonyx jubatus*)	1
Caracal (*Caracal caracal*)	3 (larvae)
Mountain zebra (*Equus zebra*)	7
Bushpig (*Potamochoerus larvatus*)	1
Black wildebeest (*Connochaetes gnou*)	2
Springbok (*Antidorcas marsupialis*)	1
African buffalo (*Syncerus caffer*)	2
Eland (*Taurotragus oryx*)	34
Bushbuck (*Tragelaphus scriptus*)	3
Greater kudu (*Tragelaphus strepsiceros*)	1
Gemsbok (*Oryx gazella*)	6
Four-striped grass mouse (*Rhabdomys pumilio*)	65 (immatures)
Swamp rat (*Otomys irroratus*)	1 (nymphs)

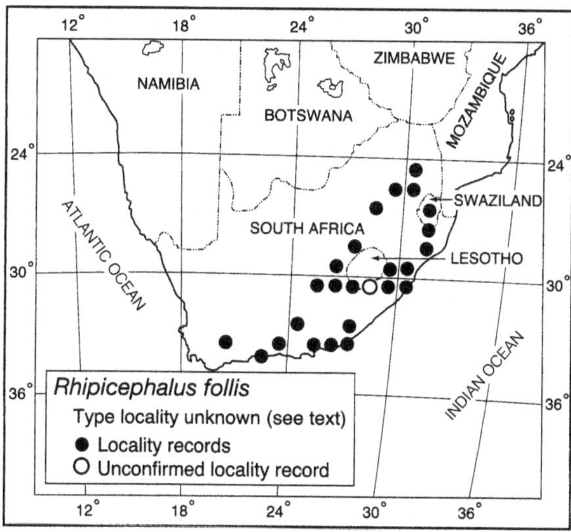

Map 21. *Rhipicephalus follis*: distribution.

few large punctations. Cervical pits short, convergent; internal cervical margins often short and indistinct, divergent; external cervical margins indicated by a few large setiferous punctations; cervical fields usually slightly depressed. Marginal lines long, almost reaching eye level, punctate. Posteromedian and posterolateral grooves usually present but often shallow and indistinct. A few large setiferous punctations present on the scapulae and scattered over the conscutum, interspersed with numerous finer punctations which increase in size posteriorly; only very fine, sparse punctations present adjacent to marginal lines; the overall impression is of a finely but densely punctate tick. Ventrally spiracles broadly comma-shaped. Adanal plates broad, their internal margins concave medially, smoothly rounded posteroexternally and posterointernally; accessory adanal plates short and broad.

Female (Figs 65(b), 66(d) to (f))
Capitulum broader than long, length × breadth ranging from 0.55 mm × 0.64 mm to 0.88 mm × 0.96 mm. Basis capituli with lateral angles at about anterior third of its length, broad; porose areas oval, about 1.5 times their own diameter apart. Palps longer than those of the male, broadly rounded apically. Scutum broader than long, length × breadth ranging from 0.96 mm × 1.11 mm to 1.69 mm × 1.98 mm, posterior margin somewhat sinuous. Eyes at about mid-length, slightly convex, edged dorsally by shallow punctate grooves. Cervical pits deep, convergent; internal cervical margins convergent initially, then diverging, becoming progressively shallower and almost reaching posterior margin of scutum; external cervical margins outlined by large setiferous punctations, almost reaching posterolateral margins of scutum; cervical fields slightly depressed. A few large setiferous punctations present on the scapulae and scattered medially on the scutum, interspersed with numerous smaller punctations. The latter vary, even among the progeny of one female, from a dense pattern of almost confluent, slightly angular pits to scattered discrete pinpoints. Ventrally genital aperture broadly V-shaped.

Nymph (Fig. 67)
Capitulum much broader than long, length × breadth ranging from 0.24 mm × 0.42 mm to 0.28 mm × 0.44 mm. Basis capituli well over three times as broad as long, with long tapering lateral angles overlapping the scapulae; ventrally with short stout posterolateral spurs. Palps broadest about mid-length, tapering to narrowly-rounded apices, inclined inwards. Scutum broader than long, length × breadth ranging from 0.58 mm × 0.64 mm to 0.62 mm × 0.67 mm; posterior margin a broad smooth curve. Eyes at widest point, over halfway back, slightly convex, edged dorsally by slight depressions. Cervical pits deep, convergent; cervical fields depressed, more-or-less parallel sided, almost reaching posterolateral margins of scutum. Ventrally coxae I with a long, sharp external spur and a shorter, blunter internal spur; coxae II to IV each with a small blunt external spur.

Larva (Fig. 68)
Capitulum approximately 1.5 times as broad as long, length × breadth ranging from 0.124 mm × 0.181 mm to 0.132 mm × 0.193 mm. Basis capituli over three times as broad as long, with short rounded forwardly directed lateral

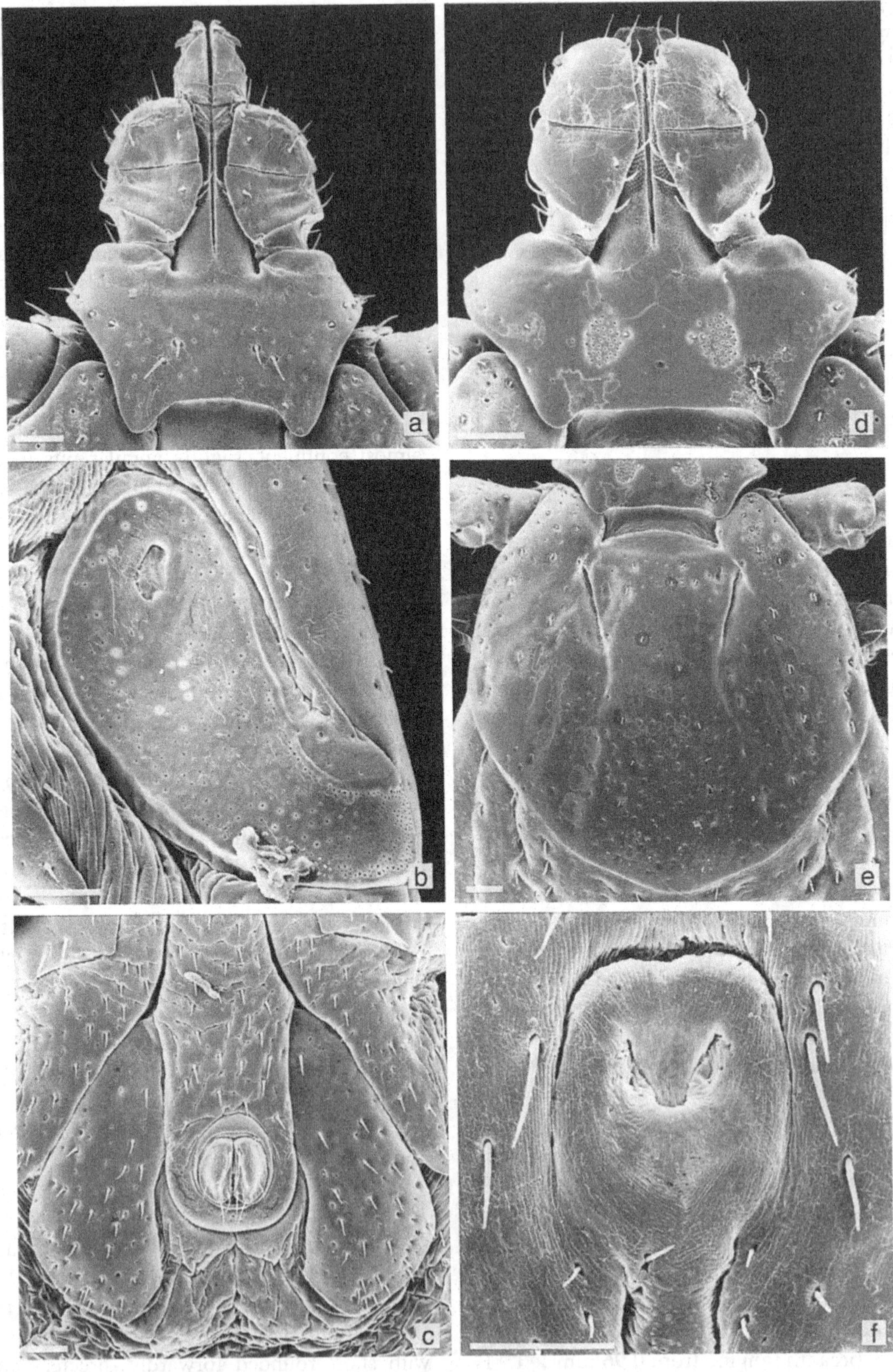
a
d
b
e
c
f

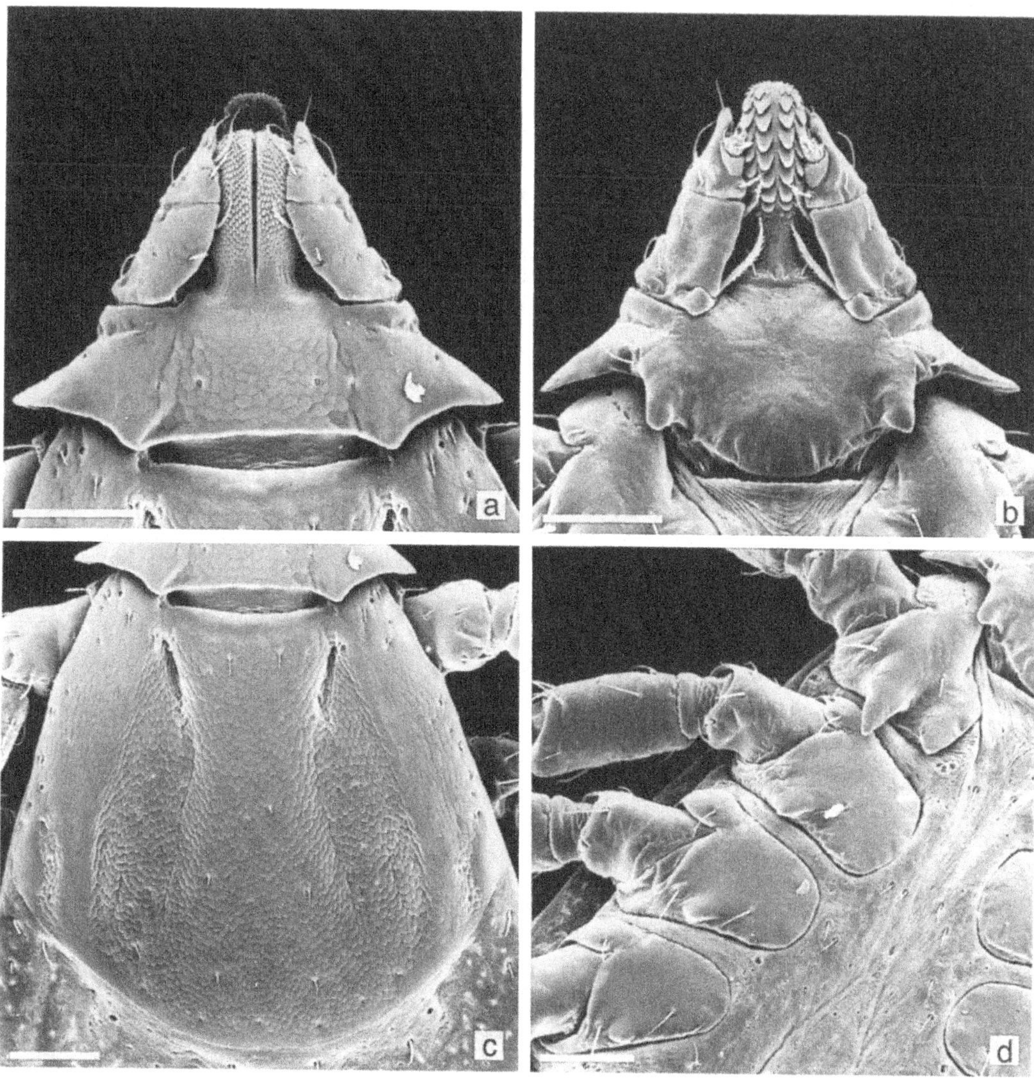

Figure 67 (*above*). *Rhipicephalus follis* (Protozoology Section Tick Breeding Register, Onderstepoort, 3417, RML 65721, laboratory reared, progeny of ♀ collected from bovine, Lilystone Farm, East London, Eastern Cape Province, South Africa, in August 1972 by J.A.F. Baker). Nymph: (a) capitulum, dorsal; (b) capitulum, ventral; (c) scutum; (d) coxae. Scale bars represent 0.10 mm. SEMs by M.D. Corwin.

Figure 66 (*opposite*). *Rhipicephalus follis* (Protozoology Section Tick Breeding Register, Onderstepoort, 3417, RML 65721, laboratory reared, progeny of ♀ collected from bovine, Lilystone Farm, East London, Eastern Cape Province, South Africa, in August 1972 by J.A.F. Baker). Male: (a) capitulum, dorsal; (b) spiracle; (c) adanal plates. Female: (d) capitulum, dorsal; (e) scutum; (f) genital aperture. Scale bars represent 0.10 mm. SEMs by M.D. Corwin.

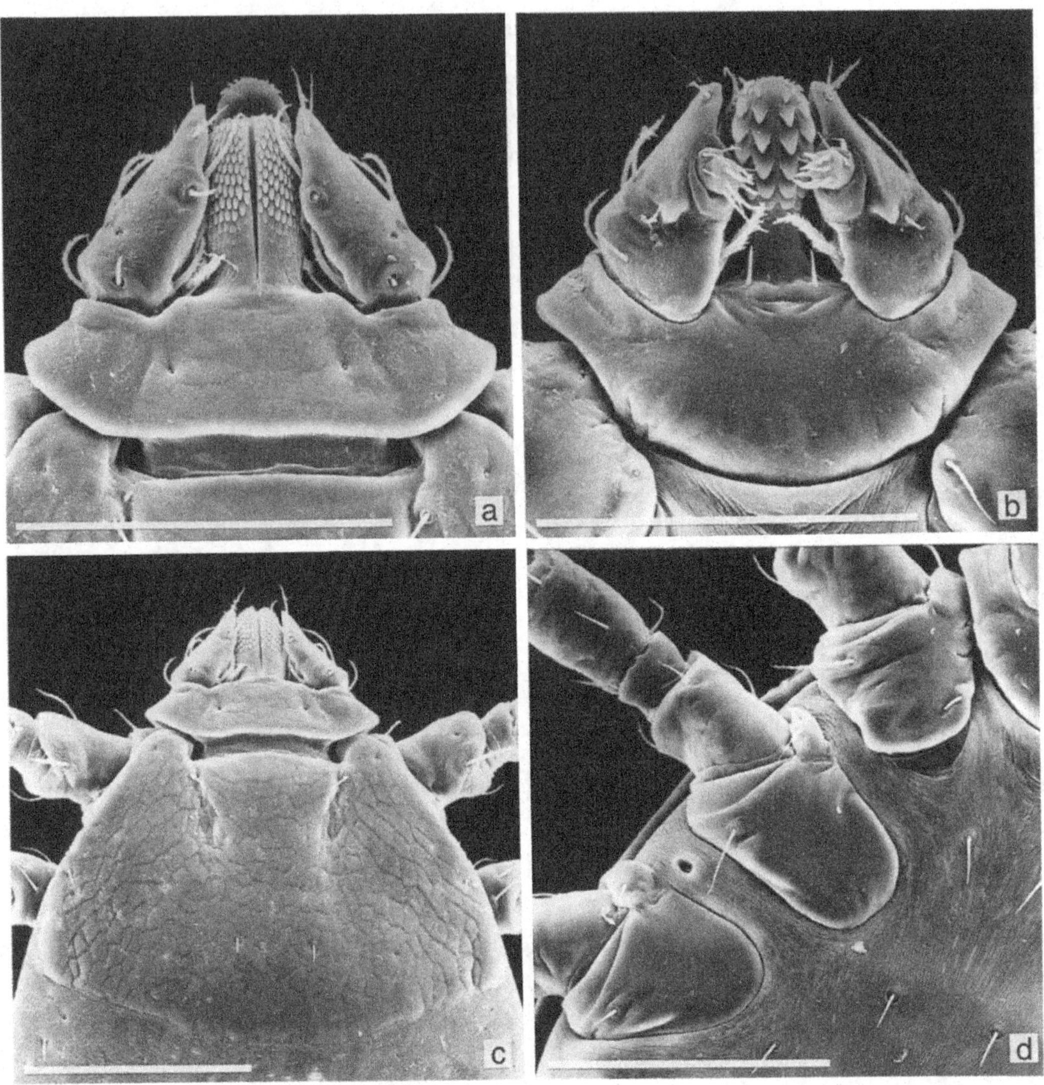

Figure 68. *Rhipicephalus follis* (Protozoology Section Tick Breeding Register, Onderstepoort, 3417, RML 65721, laboratory reared, progeny of ♀ collected from bovine, Lilystone Farm, East London, Eastern Cape Province, South Africa, in August 1972 by J.A.F. Baker). Larva: (a) capitulum, dorsal; (b) capitulum, ventral; (c) scutum; (d) coxae. Scale bars represent 0.10 mm. SEMs by M.D. Corwin.

angles extending over the scapulae. Palps broad for about two-thirds of their length, then tapering to narrowly rounded apices, their external margins almost straight, inclined inwards. Scutum over 1.5 times as broad as long; length × breadth ranging from 0.245 mm × 0.380 mm to 0.255 mm × 0.394 mm; posterior margin a wide shallow curve. Eyes at widest part of the scutum, well over halfway back. Cervical grooves short, convergent. Ventrally coxae I each with a broad spur; coxae II and III each with a broad salient ridge along its posterior border.

Notes on identification

Dönitz (1910) based his original description of *R. follis* on two syntype males, which he subsequently sent to G.H.F. Nuttall (Keirans, 1985: Nuttall Collection 2110; RML 111463). According to the labels accompanying them these males were collected off an undetermined host in

German S.W. Africa (D.S.W. Afrika, now Namibia) by P. Knuth. In his description of the species, though, Dönitz stated: 'Vaderland: Südafrika'. Presumeably Professor Knuth collected these ticks during his extensive study tour in 1906–1907, when he visited many places in both South Africa and Namibia (Knuth, 1938a,b). Although it is now impossible to determine from these conflicting data which is the country of origin of these syntypes we ourselves think, from what we know of this species' distribution, that it was South Africa.

The *R. follis* syntypes, which are now in alcohol, were formerly pinned. Keirans (1985) pointed out that the male without a pinhole in the same vial is probably not a syntype. He suggested that it 'may have been added inadvertently by an investigator comparing his specimen with the types'.

We have based our redescription of the male of *R. follis* on the two syntypes of this species plus laboratory-reared males (Protozoology Section Tick Breeding Register, Onderstepoort, 3416 (RML 65720), and 3417 (RML 65721)). These series are the progeny of two females from Lilystone Farm, East London (32°00′ S, 27°55′ E), Eastern Cape Province, South Africa, collected in August 1972 by J.A.F. Baker and reared by W.O. Neitz. Our redescriptions of the female, nymph and larva are also based on specimens from these reared series.

The re-examination of specimens in the Onderstepoort collection labelled *R. capensis* showed that some, but not all, resemble the syntypes of *R. follis*. Gertrud Theiler herself was well aware that in the past more than one species had been included under the name *R. capensis*. In 1962 she commented: '. . . "*R. capensis*" appears to be a catchall for "*capensis*-like" ticks . . .'. Our findings support this statement.

The species from Dordrecht, Eastern Cape Province, described as *R. follis* by Theiler & Robinson (1953) has since been described as *R. lounsburyi* (see p. 265). Theiler & Robinson had compared these ticks with the original description of *R. follis* but had not seen Dönitz's types. Theiler had assumed that these types were amongst the ticks collected by Professor L. Schultze and also described by Dönitz (1910). In 1962 she wrote: 'In so far as Schultze's travels were confined to South West Africa, the possibility exists that the tick described by Theiler and Robinson is not the true *R. follis*, but a species closely resembling Dönitz's species'. However, in the list of species given by Dönitz (1910, p. 404) there is no asterisk against the name *follis*; this indicates that the types were not collected by Schultze since Dönitz states (p. 403): '. . . die von L. SCHULTZE mitgebrachten haben ein Sternchen erhalten.'

In our experience heavily punctate, field-collected adult specimens of *R. follis* are difficult to separate from those of *R. gertrudae*. The immature stages closely resemble those of *R. simus*, with which they may occur sympatrically on four-striped grass mice (see p. 424). In addition, Neitz, Boughton & Walters (1972, citing G. Theiler, pers. comm., 1971) listed *R. theileri* nymphs from a swamp rat (*Otomys irroratus*), Port Alfred. These nymphs (Onderstepoort Tick Collection 3081 ii) have been re-identified as *R. follis* by I.G.H. and J.B.W.

Hosts

A three-host species (Protozoology Section Tick Breeding Register, Onderstepoort, Nos. 3416, 3417, as *R. capensis* group). The adults prefer large ruminant hosts such as cattle and eland but horses and Cape mountain zebras also appear to be good hosts (Table 19). The mean burdens of 11 eland, two horses and 14 Cape mountain zebras examined in the Mountain Zebra National Park, Cradock, Eastern Cape Province, were 129, 115 and two ticks respectively (Horak, Knight & De Vos, 1986; Horak *et al.*, 1991). The immature stages prefer four-striped grass mice as hosts (I.G.H., unpublished data).

In KwaZulu-Natal *R. follis* adults have been collected from cattle from August to April (Baker *et al.*, 1989). In the Cradock region of the Eastern Cape Province the tick appears to be present throughout the year. However, the largest numbers are generally collected from August to March (Horak *et al.*, 1991).

Zoogeography

We have recorded *R. follis* only in South Africa and Swaziland (Map 21). As discussed earlier we do not believe that the ticks on which Dönitz based his original descriptions came from Namibia.

In South Africa the majority of collections have been taken from animals in hilly or mountainous terrain and it has been collected at altitudes ranging from just above sea level to *c.* 1900 m. Annual rainfall in these regions varies from *c.* 200 mm in the more arid western parts of the tick's distribution to *c.* 1000 mm in the eastern coastal areas. It has been found in nearly every vegetation type in the southern, south-eastern, eastern, central and northern regions.

Disease relationships

Unknown.

REFERENCES

Baker, M.K., Ducasse, F.B.W., Sutherst, R.W. & Maywald, G.F. (1989). The seasonal tick populations on traditional and commercial cattle grazed at four altitudes in Natal. *Journal of the South African Veterinary Association*, **60**, 95–101.

Dönitz, W. (1910). Die Zecken Südafrikas. In *Zoologische und Anthropologische Ergebnisse einer Forschungsreise im westlichen und zentralen Südafrika ausgeführt in den Jahren 1903–1905*, 3 Lieferung, ed. L. Schultze. *Denkschriften der Medizinisch-naturwissenschaftlichen Gesellschaft zu Jena*, **16**, 398–494, pls. 15, 16a, b, & 17.

Horak, I.G., Fourie, L.J., Novellie, P.A. & Williams, E.J. (1991). Parasites of domestic and wild animals in South Africa. XXVI. The mosaic of ixodid tick infestations on birds and mammals in the Mountain Zebra National Park. *Onderstepoort Journal of Veterinary Research*, **58**, 125–36.

Horak, I.G., Knight, M.M. & De Vos, V. (1986). Parasites of domestic and wild animals in South Africa. XX. Arthropod parasites of the Cape mountain zebra (*Equus zebra zebra*). *Onderstepoort Journal of Veterinary Research*, **53**, 127–32.

Knuth, P. (1938a). Ueber meine Studienreise nach Afrika in den Jahren 1906 und 1907. Part 1. *Berliner Tierärztliche Wochenschrift*, **(1)**, 14–16.

Knuth, P. (1938b). Ueber meine Studienreise nach Afrika in den Jahren 1906 und 1907. Part 2. *Berliner Tierärztliche Wochenschrift*, **(2)**, 30–2.

Neitz, W.O., Boughton, F. & Walters, H.S. (1972). Laboratory investigations on the life-cycle of *Rhipicephalus theileri* Bedford & Hewitt, 1925 (Ixodoidea: Ixodidae). *Onderstepoort Journal of Veterinary Research*, **39**, 117–23.

Theiler, G. & Robinson, B.N. (1953). Ticks in the South African Zoological Survey Collection. Part VII. Six lesser known African rhipicephalids. *Onderstepoort Journal of Veterinary Research*, **26**, 93–136 + 1 map.

Also see the following Basic References (pp. 12–14): Keirans (1985); Theiler (1962).

RHIPICEPHALUS FULVUS NEUMANN, 1913

The specific name *fulvus*, from the Latin meaning 'reddish', refers to the general colour of this tick.

Diagnosis

A large glossy reddish-brown tick whose males have a unique strongly-sclerotized anterior process extending forwards from each scapula.

Male (Figs 69(a), 70(a) to (c))

Length × breadth overall ranging from 3.33 mm × 1.80 mm to 5.20 mm × 3.10 mm (Colas-Belcour, 1932). Capitulum longer than broad. Basis capituli with short somewhat rounded lateral angles at anterior third of its length. Palps short, broad. Conscutum with a large strongly-sclerotized anterior process extending forwards from each scapula on either side of the basis capituli to the level of its lateral angles. Eyes

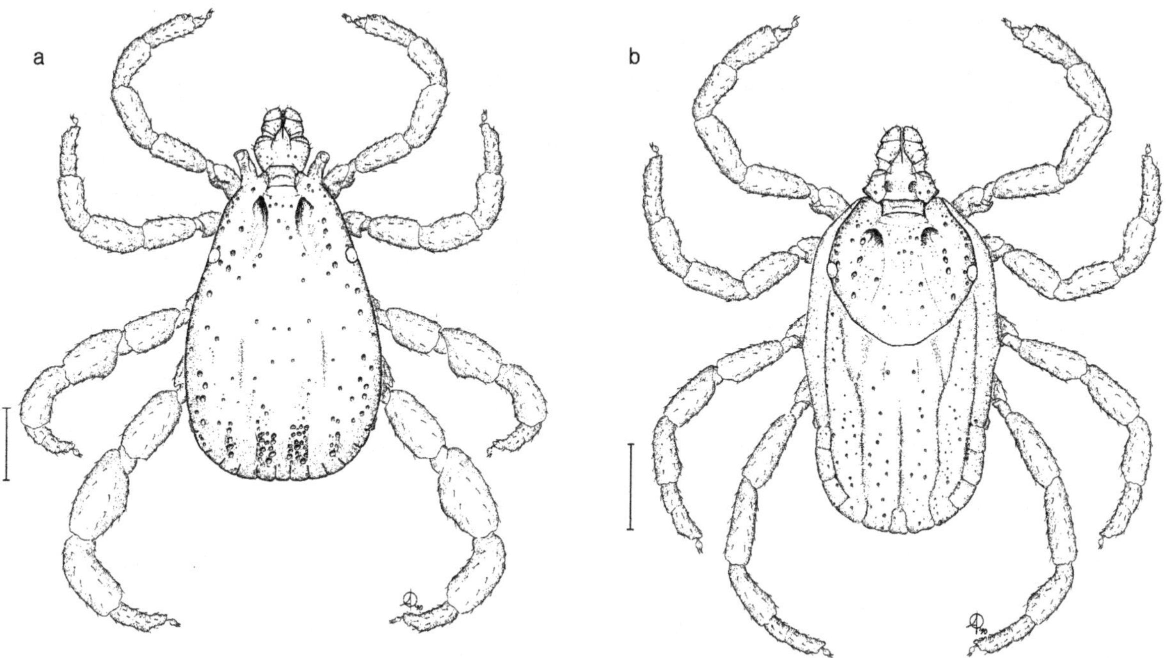

Figure 69. *Rhipicephalus fulvus* [RML 102106, HH 28405, from gundi (*Ctenodactylus gundi*), Gabes, Matmata, Tunisia, collected by P.C. Morel]. (a) Male, dorsal; (b) female, dorsal. Scale bars represent 1 mm. A. Olwage *del.*

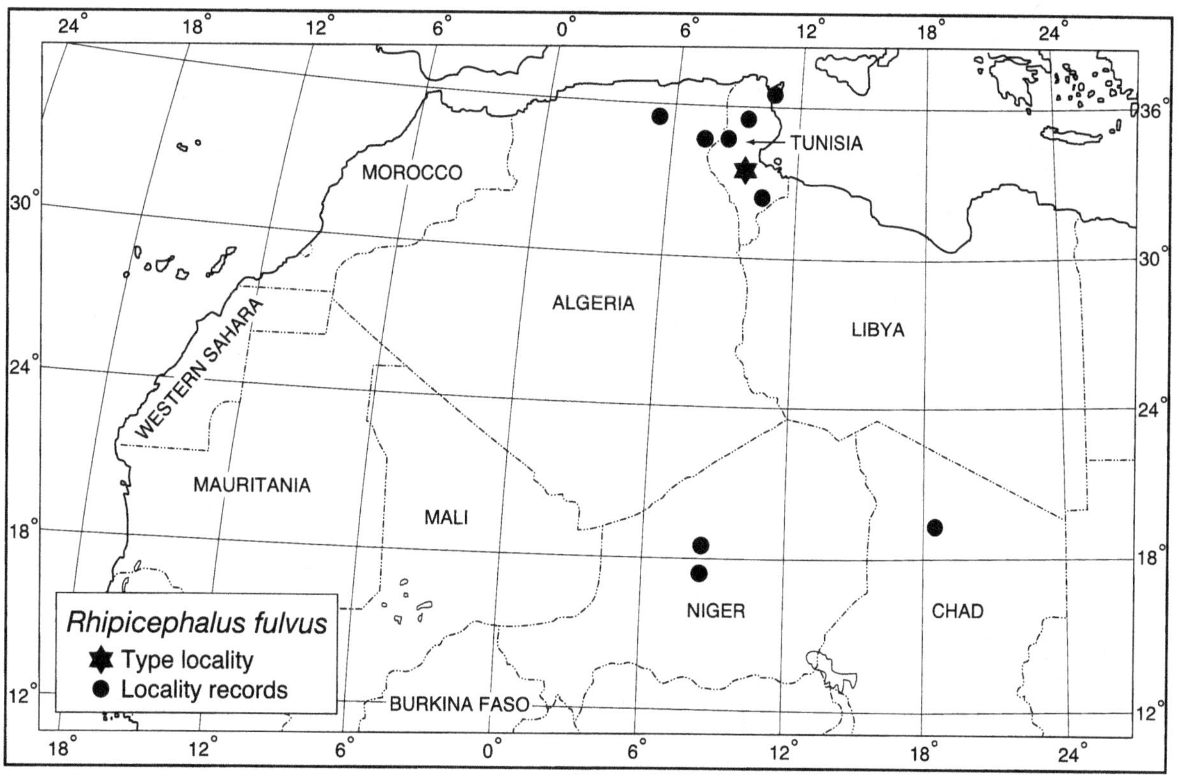

Map 22. *Rhipicephalus fulvus*: distribution.

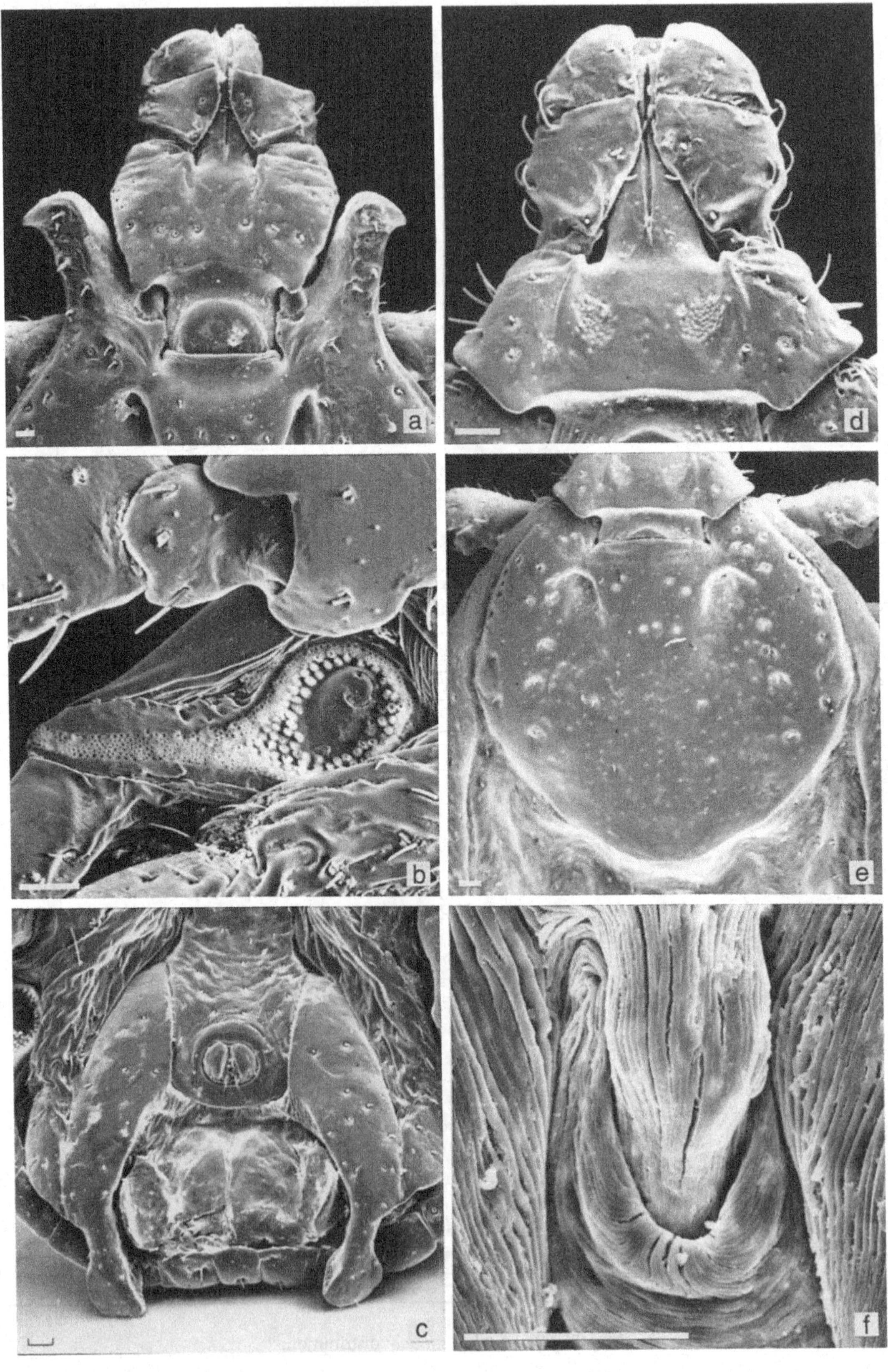
a
d
b
e
c
f

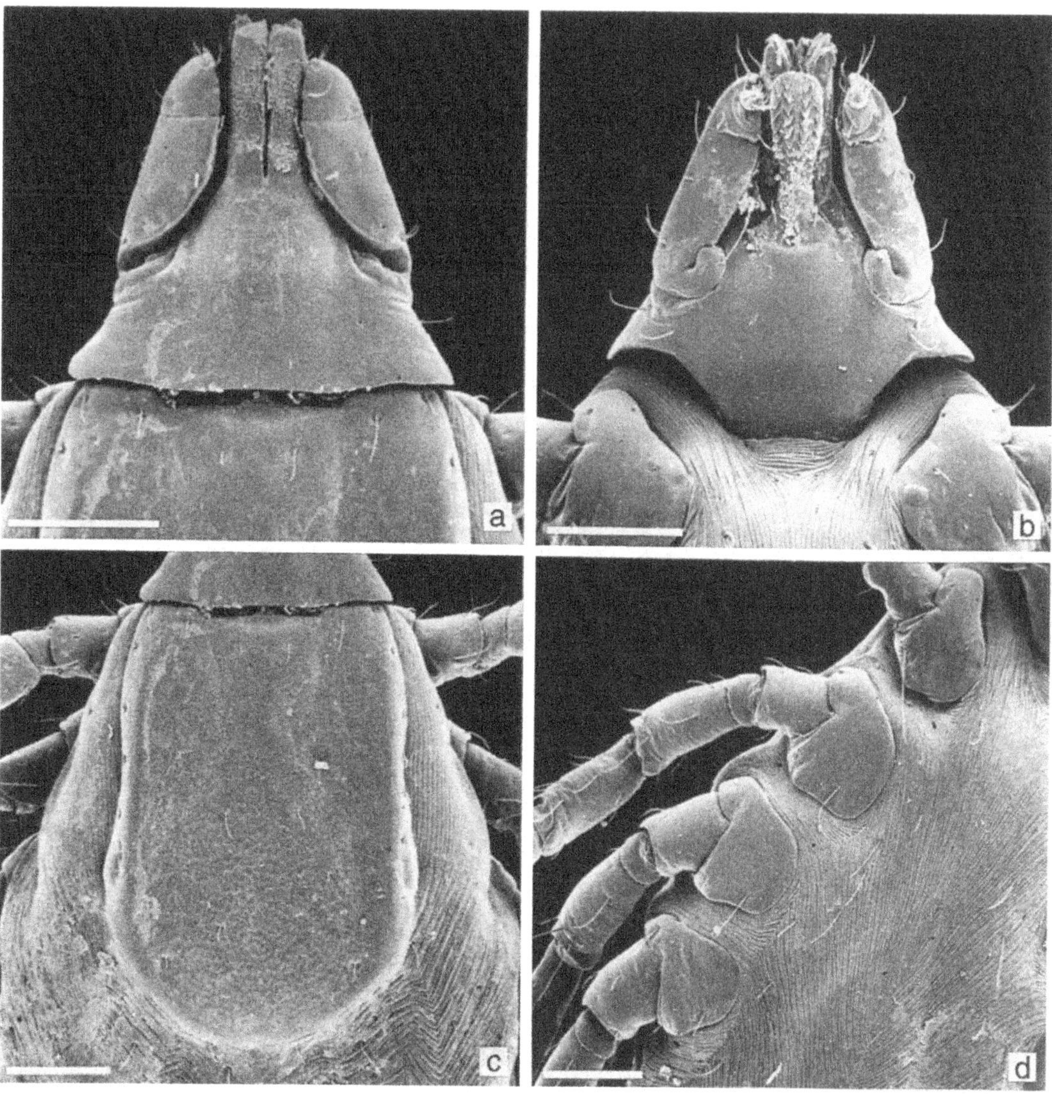

Figure 71 (*above*). *Rhipicephalus fulvus* [RML 111030, Nuttall Collection 2158, from gundi (*Ctenodactylus gundi*), El Kantara, Algeria, collected on 14 April 1913 by P.A. Buxton, by courtesy of The Natural History Museum, London]. Nymph: (a) capitulum, dorsal; (b) capitulum, ventral; (c) scutum; (d) coxae. Scale bars represent 0.10 mm. SEMs by J.F. Putterill.

Figure 70 (*opposite*). *Rhipicephalus fulvus* [RML 102106, HH 28405, from gundi (*Ctenodactylus gundi*), Gabes, Matmata, Tunisia, collected by P.C. Morel]. Male: (a) capitulum, dorsal; (b) spiracle; (c) adanal plates. Female: (d) capitulum, dorsal; (e) scutum; (f) genital aperture. Scale bars represent 0.10 mm. SEMs by Pat Hill.

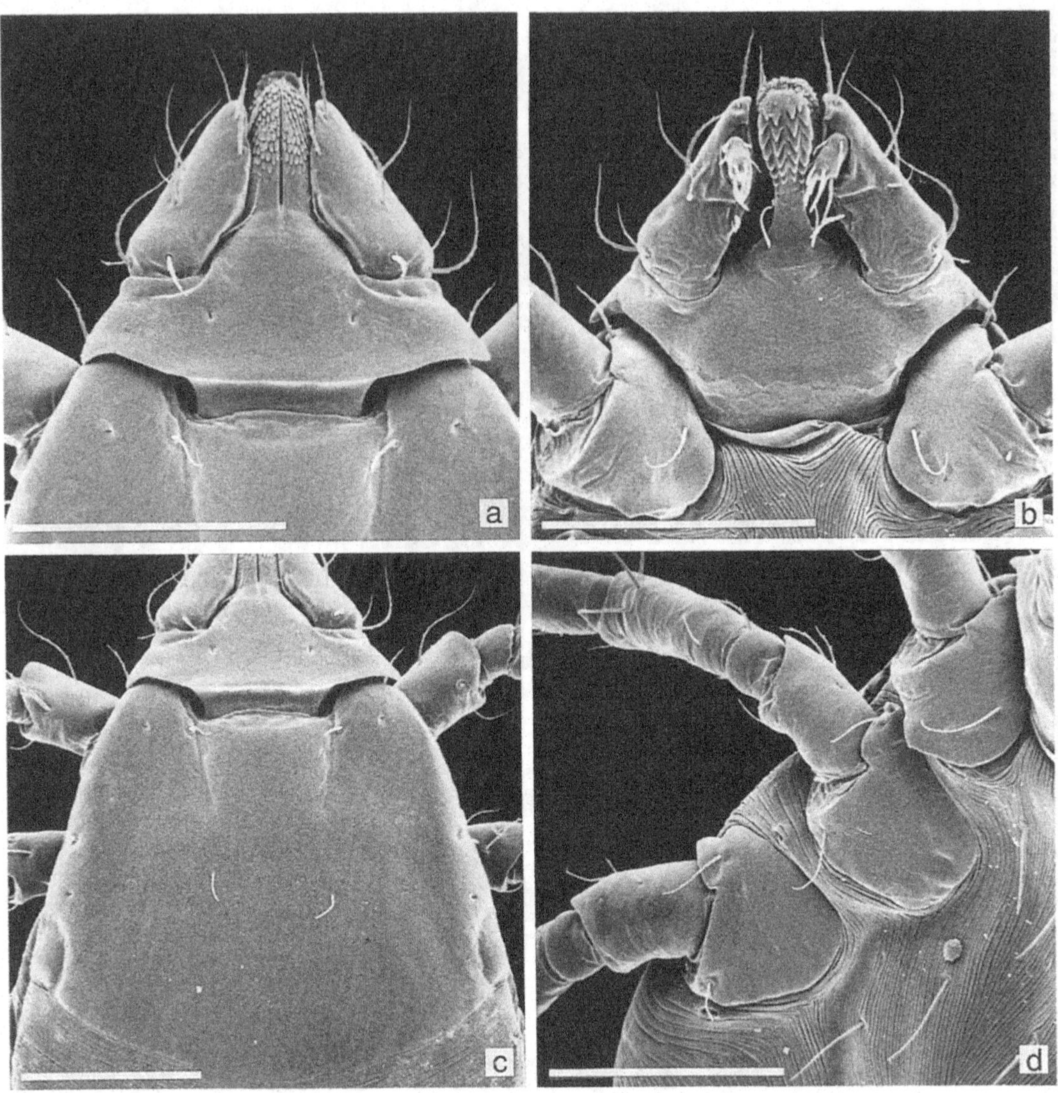

Figure 72. *Rhipicephalus fulvus* [RML 119964, HH 84132, from gundi (*Ctenodactylus gundi*), Medenine, Djebel Tameghza, Tunisia, collected on 29 October 1978 by N. Haffoudi]. Larva: (a) capitulum, dorsal; (b) capitulum, ventral; (c) scutum; (d) coxae. Scale bars represent 0.10 mm. SEMs by J.F. Putterill.

large, slightly convex. Cervical fields in the form of curved depressions, very deep anteriorly, becoming gradually shallower posteriorly and extending to just beyond eye level. Both marginal lines and posterior grooves absent. A few large setiferous punctations present along the external margins of the cervical fields and scattered laterally on the conscutum behind the eyes. Posteriorly four groups of conspicuous irregularly-shaped punctations present, consisting of a relatively large group on either side of the scutal midline, flanked by a much smaller group on each side. A few smaller punctations scattered on the scapulae and medially on the conscutum. In general appearance the conscutum is smooth and shiny. Legs increase markedly in size from I to IV. Ventrally spiracles with a long narrow almost straight dorsal prolongation. Adanal plates unique in shape, long, widest anteriorly, then narrowing abruptly just anterior to the festoons before finally expanding again into small knobs that project just beyond the festoons and are sometimes visible from the dorsal surface; accessory adanal plates oblong, terminating just in front of the festoons.

Female (Figs 69(b), 70 (d) to (e))

Length × breadth overall ranging from 4.00 mm × 2.00 mm to 5.00 mm × 2.80 mm (Colas-Belcour, 1932). Capitulum broader than long. Basis capituli with lateral angles just posterior to mid-length, directed slightly forwards; porose areas oval, nearly twice their own diameter apart. Palps longer than those of the male, broad. Scutum slightly broader than long, posterolateral margins slightly sinuous. Eyes large, slightly convex, delimited dorsally by one or two large punctations. Cervical pits broadly rounded anteriorly; cervical fields broad, slightly depressed, their outer margins delimited by a few large setiferous punctations. A few slightly smaller punctations scattered on the scapulae and medially on the scutum, interspersed by numerous extremely fine interstitial punctations. In general appearance the scutum is smooth and shiny. Ventrally genital aperture a wide V-shape.

Nymph (Fig. 71)

Length × breadth overall 1.4 mm × 0.6 mm (Colas-Belcour, 1932). Capitulum broader than long, almost triangular in general shape. Basis capituli nearly three times as broad as long, with its sharply-pointed lateral angles almost in line with the slightly sinuous posterior margin and overlapping the scapulae. Palps long, their external margins practically straight, their internal margins convex, with article II about twice as long as article III, tapering at each end and inclined inwards. Scutum elongate, narrowing slightly just anterior to mid-length, its posterior margin a deep smooth curve. Eyes posterior to mid-length, long, narrow, somewhat convex. Cervical fields broad, slightly depressed, their internal margins relatively short, their external margins more conspicuous, virtually parallel to the sides of the scutum and extending posteriorly past the eyes. Ventrally coxae each with an external spur, that on coxae I broadly triangular, those on coxae II to IV decreasing progressively in size.

Larva (Fig. 72)

Length × breadth overall 0.744 mm × 0.496 mm (Colas-Belcour, 1932). Capitulum broader than long, almost triangular in general shape. Basis capituli about four times as broad as long, with its sharp lateral angles recurved over the scapulae. Palps long, their external margins practically straight, their internal margins somewhat sinuous, tapering to rounded apices, inclined inwards. Scutum broader than long, its posterior margin a smooth shallow curve. Eyes well over halfway back, somewhat convex, conspicuous. Cervical grooves short, slightly convergent. Ventrally coxae I each with a large triangular spur, coxae II and III each with a small projection on its posterior border.

Notes on identification

The above descriptions are based in part on those given by Colas-Belcour (1932). Neumann (1913) originally described this species as a member of the genus *Rhipicephalus* and placed it in a new subgenus, *Pterygodes* (derived from the Greek *pterex*, meaning a 'wing', doubtless a reference to the unique anterior extensions of the scapulae of the male). Morel (1969), however, regarded *R. fulvus* as being intermediate morphologically between the genera *Rhipicephalus* and *Hyalomma* and raised *Pterygodes* to generic rank. He suggested that it constitutes a relic in the Sahara from a rhipicephaline line that had not developed further. He felt that it probably could not be regarded as either an 'Ethiopian' (i.e. an Afrotropical) or a Palaearctic species. Perhaps more conservatively than our late colleague we have decided to retain *R. fulvus* as a member of the genus *Rhipicephalus* (subgenus *Pterygodes*) but further studies may yet prove us wrong.

The unique morphological features of the male of this interesting species make it one of the easiest rhipicephalids to identify. So far as we know it has never been confused with any others in the genus.

Hosts

Its type of life cycle has not been stated by previous authors who have studied this tick. *R. fulvus* adults parasitize sheep, goats and camels and also the wild Barbary sheep (*Ammotragus lervia*). (Morel (1958, 1969) and Morel & Graber (1961) referred to the mouflon as the natural host of this tick but they apparently used this common name erroneously for *A. lervia*. The mouflon is the wild form of *Ovis aries* (syn. *O. orientalis*), which is not recorded from North Africa.) Adults have also been found on the gundi (*Ctenodactylus gundi*), although this rodent is regarded primarily as the host of the immature stages. A male was found once feeding on a human. There is also one record, stage not stated, from a northern pygmy gerbil (*Gerbillus campestris*).

Colas-Belcour (1932) stated that the larvae hatch in spring, and that they and recently moulted nymphs occur on gundis at the end of March, followed by engorged nymphs in April and early May (spring). He said that on goats young adults occur in May, and engorged females in November (winter). Subsequently Colas-Belcour & Rageau (1951) recorded both the immature stages and adults from gundis in January (winter) and June (summer). Small numbers of larvae and nymphs have also been collected from four gundis in October (autumn) (N. Haffoudi, unpublished data, 1978; Hoogstraal Collection nos. 84 132 – 84 135). Thus, although relatively few collections of this tick are known, they suggest that all stages may be active throughout much of the year.

Zoogeography

Rhipicephalus fulvus has been collected most commonly in Tunisia, with a few records only from further south in Niger and Chad (Map 22).

Our present somewhat scanty information indicates that this species can survive under extremely harsh conditions. As Morel (1969) commented, *R. fulvus* appears to be the only tick that is endemic to the Saharan mountain massifs of Aïr and Tibesti at altitudes of 1500 m to 3000 m in what he refers to as Saharo-mediterranean steppe. In this desert environment, where the annual rainfall is often less than 50 mm, suitable microclimates can be found only near temporary watercourses, along caravan routes or in areas frequented by domestic or wild herbivores. There the tick's immature stages can be found in the burrows and among the rocks where the gundis live. In Tunisia *R. fulvus* occurs in similar terrain but at the somewhat lower altitudes of 500 m to 1000 m.

Disease relationships

Unknown.

REFERENCES

Colas-Belcour, J. (1932). Contribution à l'étude de *Rhipicephalus (Pterygodes) fulvus* Neumann et sa biologie. *Archives de l'Institut Pasteur de Tunis*, 20, 430–43.

Colas-Belcour, J. & Rageau, J. (1951). Tiques de Tunisie – Ixodines. *Archives de l'Institut Pasteur du Maroc*, 4, 360–7.

Morel, P.C. (1958). Les tiques des animaux domestiques de l'Afrique occidentale française. *Revue d'Élevage et de Médecine Vétérinaire des Pays Tropicaux*, 11 (nouvelle série), 153–89.

Morel, P.C. & Graber, M. (1961). Les tiques des animaux domestiques du Tchad. *Revue d'Élevage et de Médecine Vétérinaire des Pays Tropicaux*, 14 (nouvelle série), 199–203.

Neumann, L.G. (1913). Un nouveau sous-genre et deux nouvelles espéces d'Ixodinés. *Bulletin de la Société Zoologique de France*, 38, 147–51.

Also see the following Basic Reference (p. 14): Morel (1969).

RHIPICEPHALUS GERTRUDAE FELDMAN-MUHSAM, 1960

This species was named in honour of Dr Gertrud Theiler (1897–1986), of the Onderstepoort Veterinary Institute. In her day she was the doyenne of African tick workers. She is remembered with respect and gratitude, not only for her contributions to our understanding of these parasites but also for her readiness to share her knowledge and experience with others.

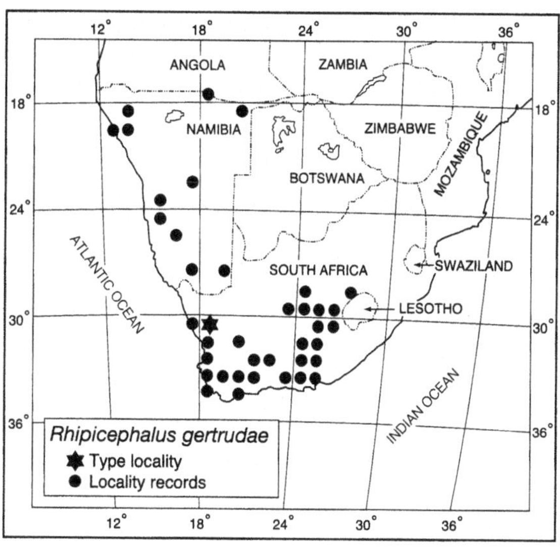

Map 23. *Rhipicephalus gertrudae*: distribution.

Diagnosis

A moderate-sized heavily punctate reddish-brown tick.

Male (Figs 73(a), 74(a) to (c))

Capitulum longer than broad, length × breadth ranging from 0.64 mm × 0.60 mm to 0.83 mm × 0.72 mm. Basis capituli with short, acute lateral angles in anterior third of its length. Palps broadly rounded apically. Conscutum length × breadth ranging from 2.94 mm × 1.91 mm to 3.67 mm × 2.48 mm; anterior process on coxae I small. In engorged specimens a short, bulbous caudal process present. Eyes slightly convex, edged dorsally by shallow depressions with one

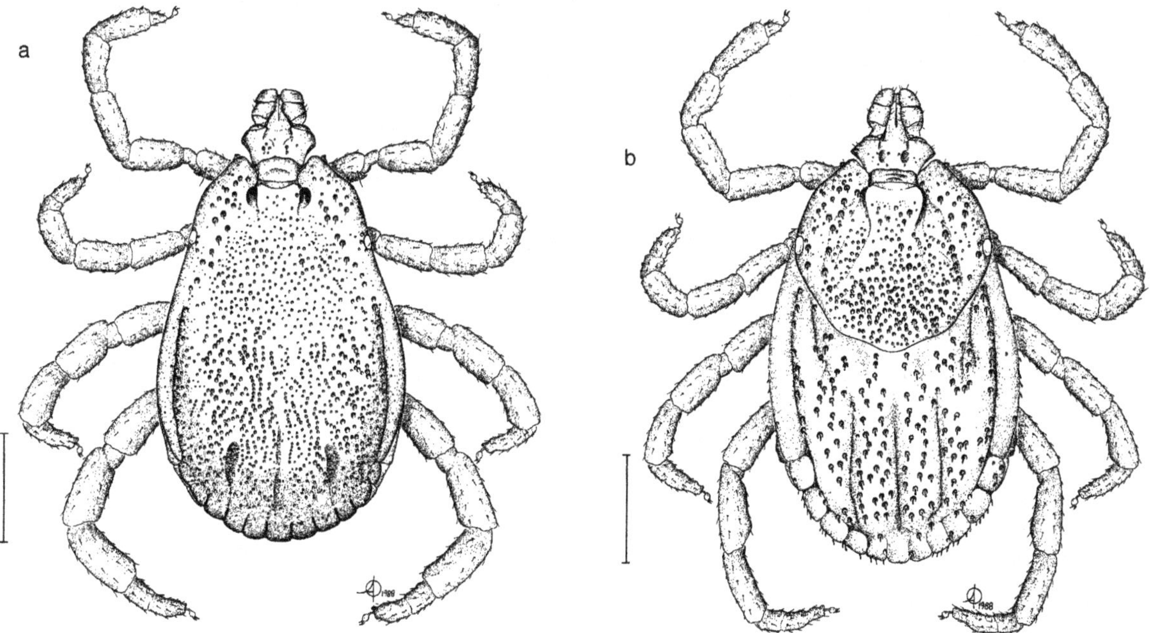

Figure 73. *Rhipicephalus gertrudae* [Onderstepoort Tick Collection 3094ii, original ♀ collected from bovine, 'Schoongesig' (farm), Schoombee, Eastern Cape Province, South Africa, on 12 February 1962]. (a) Male, dorsal; (b) female, dorsal. Scale bars represent 1 mm. A. Olwage *del.*

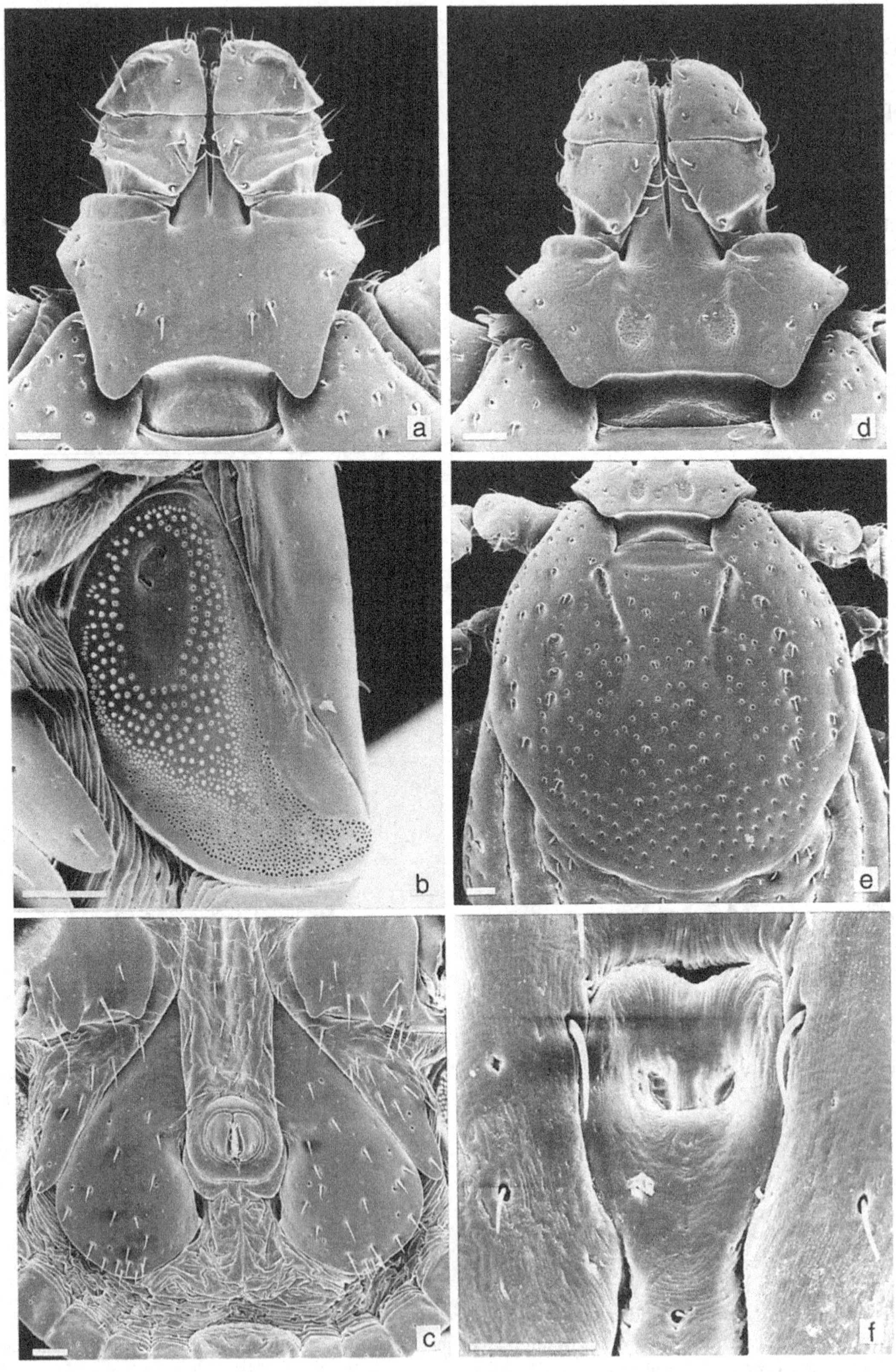
a
d
b
e
c
f

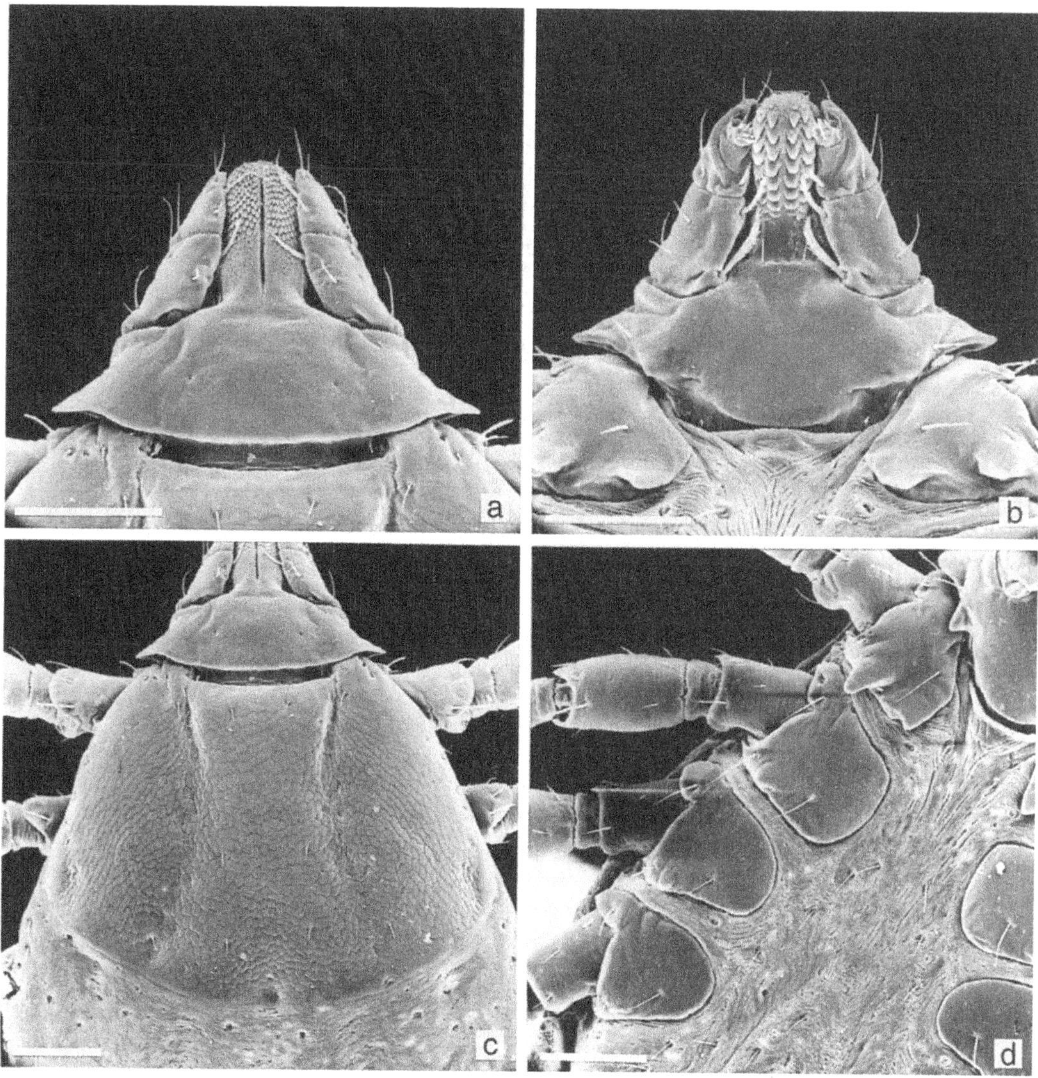

Figure 75 (*above*). *Rhipicephalus gertrudae* [Onderstepoort Tick Collection 3094ii, original ♀ collected from bovine, 'Schoongesig' (farm), Schoombee, Eastern Cape Province, South Africa, on 12 February 1962]. Nymph: (a) capitulum, dorsal; (b) capitulum, ventral; (c) scutum; (d) coxae. Scale bars represent 0.10 mm. SEMs by M.D. Corwin.

Figure 74 (*opposite*). *Rhipicephalus gertrudae* [Onderstepoort Tick Collection 3094ii, ♀ collected from bovine, 'Schoongesig' (farm), Schoombee, Eastern Cape Province, South Africa, on 12 February 1962]. Male: (a) capitulum, dorsal; (b) spiracle; (c) adanal plates. Female: (d) capitulum, dorsal; (e) scutum; (f) genital aperture. Scale bars represent 0.10 mm. SEMs by M.D. Corwin.

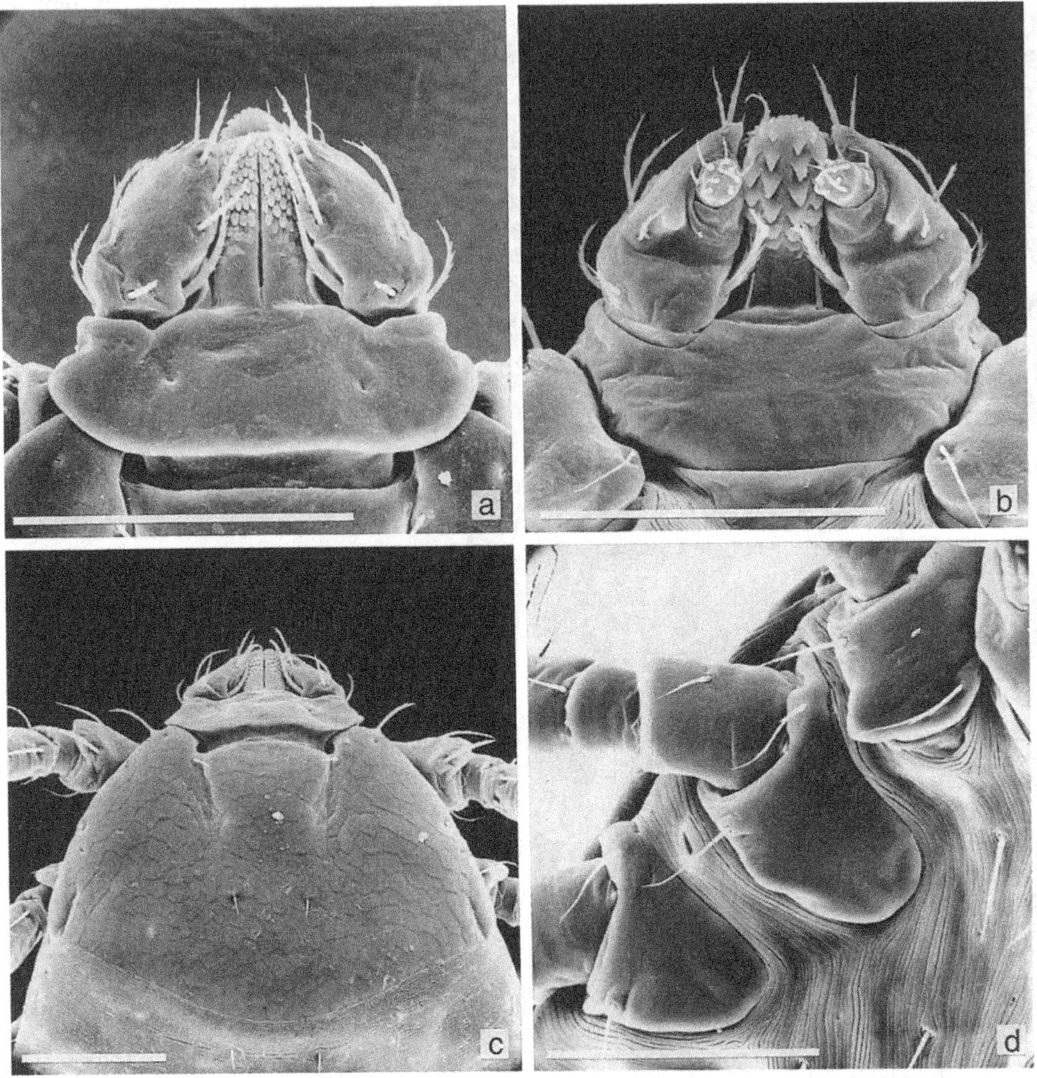

Figure 76. *Rhipicephalus gertrudae* [Onderstepoort Tick Collection 3094ii, original ♀ collected from bovine, 'Schoongesig' (farm), Schoombee, Eastern Cape Province, South Africa, on 12 February 1962]. Larva: (a) capitulum, dorsal; (b) capitulum, ventral; (c) scutum; (d) coxae. Scale bars represent 0.10 mm. SEMs by M.D. Corwin.

or two setiferous punctations. Cervical pits short, comma-shaped. Marginal lines deep, outlined by large punctations posteriorly, continued anteriorly by irregular lines of large setiferous punctations. Posteromedian groove narrow, superficial; posterolateral grooves kidney-shaped, sometimes also inconspicuous. Large setiferous punctations present on the scapulae and along the external cervical margins, also scattered anteromedially on the conscutum, becoming denser posterolaterally. These are interspersed with numerous somewhat smaller punctations that often run together in short lines, especially posteriorly. Ventrally spiracles broad, with a short abruptly-curved dorsal prolongation. Adanal plates broadly rounded posteriorly, tapering towards the anterior end; accessory adanal plates sharply pointed, well sclerotized.

Table 20. *Host records of* Rhipicephalus gertrudae

Hosts	Number of records
Domestic animals	
Cattle	58
Sheep	48
Goats	8
Horses	9
Donkeys	1
Dogs	3
Wild animals	
Chacma baboon (*Papio ursinus*)	3
Black-backed jackal (*Canis mesomelas*)	1
Bat-eared fox (*Otocyon megalotis*)	1
Cape fox (*Vulpes chama*)	4
Caracal (*Caracal caracal*)	3
Leopard (*Panthera pardus*)	1
Mountain zebra (*Equus zebra*)	4
Black-faced impala (*Aepyceros melampus petersi*)	1
Black wildebeest (*Connochaetes gnou*)	2
Bontebok (*Damaliscus pygargus dorcas*)	1
Springbok (*Antidorcas marsupialis*)	2
African buffalo (*Syncerus caffer*)	1
Eland (*Taurotragus oryx*)	7
Greater kudu (*Tragelaphus strepsiceros*)	1
Gemsbok (*Oryx gazella*)	11
Grey rhebok (*Pelea capreolus*)	2
Namaqua rock mouse (*Aethomys namaquensis*)	8 (immatures)
Four-striped grass mouse (*Rhabdomys pumilio*)	18 (immatures)
Bush Karoo rat (*Otomys unisulcatus*)	1 (nymph)
South African porcupine (*Hystrix africaeaustralis*)	2
Cape hare (*Lepus capensis*)	1 (nymph)
Scrub hare (*Lepus saxatilis*)	4 (nymphs)
Humans	3

Female (Figs 73(b), 74(d) to (f))

Capitulum as broad as long, length × breadth ranging from 0.79 mm × 0.79 mm to 0.85 mm × 0.85 mm. Basis capituli with acute lateral angles slightly anterior to mid-length; porose areas about 1.5 times their own diameter apart. Palps with article III virtually wedge-shaped. Scutum broader than long, length × breadth ranging from 1.71 mm × 1.73 mm to 1.82 mm × 2.02 mm, posterior margin mildly sinuous. Eyes slightly convex, edged dorsally by shallow depressions with one or two setiferous punctations. Cervical pits convergent; cervical fields slightly depressed, their external margins delimited by a few large setiferous punctations. Similar large punctations scattered on the scapulae and medially on the scutum, where they are interspersed with a dense pattern of somewhat smaller punctations. Ventrally genital aperture tripartite in appearance, with a vase-shaped central area flanked on either side by narrow, tapering depressed flaps.

Nymph (Fig. 75)
Capitulum much broader than long, length × breadth 0.19 mm × 0.35 mm on the single specimen available for measurement. Basis capituli over three times as broad as long, with sharp lateral angles projecting over the scapulae, posterior border a long smooth curve. Palps tapering to narrowly-rounded apices. Scutum much broader than long, length × breadth 0.52 mm × 0.66 mm; posterior margin a wide shallow curve. Eyes at widest point, well over halfway back, delimited dorsally by slight depressions. Cervical pits convergent, continuing as shallow divergent grooves that reach the posterior margin of the scutum and divide it into three. Ventrally coxae I each with a long, narrow external spur and a shorter internal spur; coxae II to IV each with a small external spur only.

Larva (Fig. 76)
Capitulum much broader than long, length × breadth ranging from 0.109 mm × 0.160 mm to 0.116 mm × 0.177 mm. Basis capituli just over three times as broad as long, with short bluntly-rounded slightly forwardly-directed lateral angles, posterior border straight. Palps slightly constricted basally, then widening before they taper to rounded apices, inclined inwards. Scutum much broader than long, length × breadth ranging from 0.213 mm × 0.365 mm to 0.217 mm × 0.386 mm; posterior margin a wide shallow curve. Cervical grooves short, slightly convergent. Ventrally coxae I each with a large broadly-rounded spur; coxae II with a broad, but rather indistinct, spur; coxae III with a mere indication of a spur.

Notes on identification
In the introduction to her descriptions of *R. gertrudae* and *R. capensis* Feldman-Muhsam (1960) states: 'It appears, in fact, that if certain individual characters are considered separately, specimens determined generally as *R. capensis* can be arranged in a continuous series of gradual variations. But if a complex of such variable characters is considered as a whole, the differentiation of types is feasible.' When one is confronted with large numbers of specimens it is these variations that make accurate diagnosis difficult. In the south- western regions of its distribution heavily-punctate specimens of *R. gertrudae* must be separated from *R. capensis*, with which it may occur on the same host. In the southern and eastern regions of its distribution lightly-punctate specimens must be distinguished from *R. follis*, a tick which has frequently been misidentified as *R. capensis* (Theiler, 1962; Walker, 1991). The female tick portrayed in Fig. 74(e) is a lightly-punctate specimen. The immature stages of these three ticks are not only fairly similar in appearance (see Figs 269–271, pp. 597–599) but they all prefer rodents as hosts.

Hosts

A three-host species. Cattle and sheep are the preferred domestic hosts of adult *R. gertrudae*. Large antelopes such as eland and gemsbok appear to be the favoured wild hosts of this stage of development (Table 20). Both horses and free-ranging zebras are good hosts, and domestic dogs and various wild carnivores may also be infested.

In addition the adults seem to have an affinity for primates. A troop of Chacma baboons living in an arid environment in Namibia were particularly heavily infested (Brain & Bohrmann, 1992). Two adult male baboons each harboured more than 400 ticks, of which the majority were attached to their ears. An infant carried 70 ticks, mostly on its muzzle, hands and feet. Three cases of human infestation have also been encountered in South Africa, one on the Cape Flats, Western Cape Province, and the others near Bloemfontein in the Free State.

The immature stages, which do not feed on the same hosts as the adults, prefer small rodents such as the Namaqua rock mouse and four-striped grass mouse (Fourie, Horak & Van den Heever, 1992; I.G.H., unpublished data). They may also be found on both Cape and scrub hares (Horak & Fourie, 1991).

In central Namibia and the south-western Free State, South Africa, adult ticks prefer the warmer months from September or October to February or March (Biggs & Langenhoven, 1984; Fourie, Kok & Heyne, 1996). In the Northern Cape and Western Cape Provinces, South Africa, they seemed to prefer the cooler months and were most abundant on Dorper sheep from May to October (Horak & Fourie, 1992).

Zoogeography

Thus far *R. gertrudae* has been recorded in Namibia and South Africa (Map 23). It is found at altitudes ranging from just above sea level to *c.*1800 m, in regions with winter, summer or non-seasonal rainfall and with annual precipitation varying from 200 mm to 600 mm. In the eastern parts of its distribution range *R. gertrudae* occurs in undifferentiated Afromontane vegetation and in Highveld grassland; in the central region of its distribution it is present in evergreen and semi-evergreen bushland and thicket in the south, Karoo grassy shrubland and its transition to Highveld as well as in Kalahari *Acacia* wooded grassland and deciduous bushland in the centre, and in a mosaic of dry deciduous forest and secondary grassland in the north; in the western section of its distribution it occurs in various types of Karoo shrubland as well as in undifferentiated and scrub woodland, deciduous bushland and wooded grassland, and in the extreme west in the Namib desert.

Disease relationships

More than half of the 18 recorded infant mortalities in a troop of Chacma baboons living in an arid environment in Namibia appeared to be related to heavy infestations with *R. gertrudae* (Brain & Bohrmann, 1992). Tick infestation of an infants' muzzle had caused acute inflammation of its nose and mouth and inhibited it from suckling, resulting in its death.

REFERENCES

Biggs, H.C. & Langenhoven, J.W. (1984). Seasonal prevalence of ixodid ticks on cattle in the Windhoek District of South West Africa/Namibia. *Onderstepoort Journal of Veterinary Research*, **51**, 175–82.

Brain, C. & Bohrmann, R. (1992). Tick infestation of baboons (*Papio ursinus*) in the Namib desert. *Journal of Wildlife Diseases*, **28**, 188–91.

Feldman-Muhsam, B. (1960). The South African ticks *Rhipicephalus capensis* Koch and *R. gertrudae* n. sp. *Journal of Parasitology*, **46**, 101–8.

Fourie, L.J., Horak, I.G. & Van den Heever, J.J. (1992). The relative host status of rock elephant shrews *Elephantulus myurus* and Namaqua rock mice *Aethomys namaquensis* for economically important ticks. *South African Journal of Zoology*, **27**, 108–14.

Fourie, L.J., Kok, D.J & Heyne, H. (1996). Adult ixodid ticks on two cattle breeds in the south-western Free State, and their seasonal dynamics. *Onderstepoort Journal of Veterinary Research*, **63**, 19–23.

Horak, I.G. & Fourie, L.J. (1991). Parasites of domestic and wild animals in South Africa. XXIX. Ixodid ticks on hares in the Cape Province and on hares and red rock rabbits in the Orange Free State. *Onderstepoort Journal of Veterinary Research*, **58**, 216–70.

Horak, I.G. & Fourie, L.J. (1992). Parasites of domestic and wild animals in South Africa. XXXI. Adult ixodid ticks on sheep in the Cape Province and in the Orange Free State. *Onderstepoort Journal of Veterinary Research*, **59**, 275–83.

Also see the following Basic References (pp. 12–14): Theiler (1962); Walker (1991).

RHIPICEPHALUS GLABROSCUTATUM DU TOIT, 1941

The specific name *glabroscutatum*, from the Latin *glaber* meaning 'smooth', plus *scutum*, refers to the smooth shiny scuta of the adults.

Diagnosis

A small shiny chestnut-brown species.

Male (Figs 77(a), 78(a) to (c))
Capitulum about as broad as long, length × breadth ranging from 0.47 mm × 0.53 mm to 0.72 mm × 0.70 mm. Basis capituli with lateral angles at about anterior third of its length, slightly recurved in shape. Palps bluntly rounded apically. Conscutum length × breadth ranging from 2.17 mm × 1.34 mm to 2.96 mm × 2.00 mm, smooth and shiny in general appearance; anterior process of coxae I large. Eyes beady, orbited. Cervical pits deep, convergent. Marginal lines long, finely punctate, not reaching eyes anteriorly. Posteromedian and posterolateral grooves present but shallow and indistinct in some specimens. Large setiferous punctations most numerous anterior to the eyes on the scap-

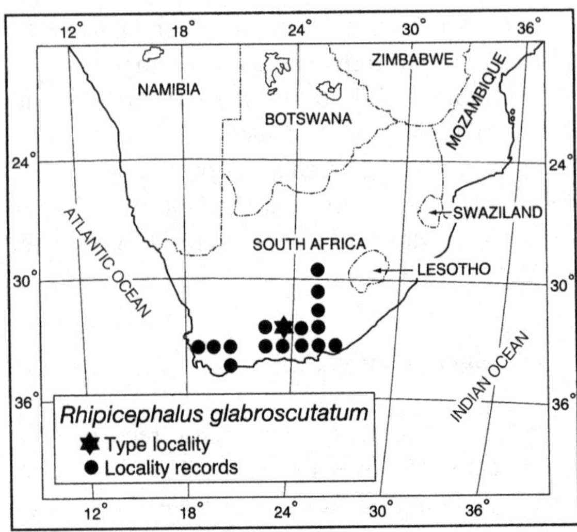

Map 24. *Rhipicephalus glabroscutatum*: distribution.

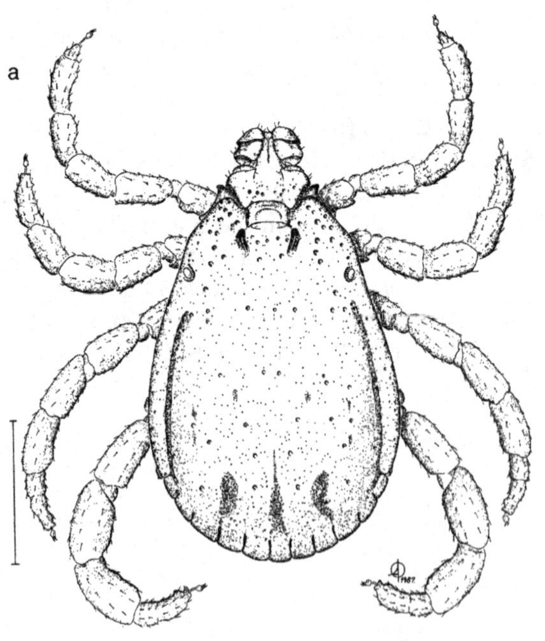

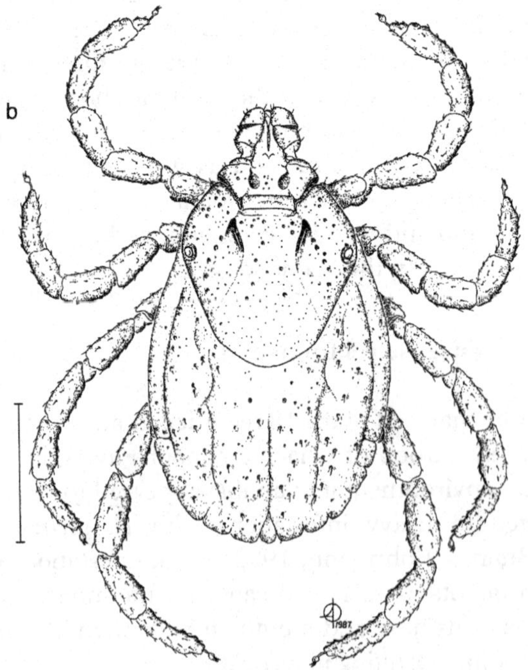

Figure 77. *Rhipicephalus glabroscutatum* [laboratory reared, original ♀ from greater kudu (*Tragelaphus strepsiceros*), 'Bucklands' (farm), near Grahamstown, Eastern Cape Province, South Africa, collected in 1979 by Y. Rechav]. (a) Male, dorsal; (b) female, dorsal. Scale bars represent 1 mm. A. Olwage *del.*

Table 21. *Host records of* Rhipicephalus glabroscutatum

Hosts	Number of records
Domestic animals	
Cattle	49 (including immatures)
Sheep	66 (including immatures)
Goats	614 (including immatures)
Horses	2 (including immatures)
Wild animals	
Caracal (*Caracal caracal*)	16 (immatures)
Mountain zebra (*Equus zebra*)	16 (including immatures)
Rock hyrax (*Procavia capensis*)	6 (immatures, 1 ♂)
Red hartebeest (*Alcelaphus buselaphus caama*)	3 (including immatures)
Black wildebeest (*Connochaetes gnou*)	9 (including immatures)
Bontebok (*Damaliscus pygargus dorcas*)	41 (including immatures)
Blesbok (*Damaliscus pygargus phillipsi*)	2 (including nymphs)
Springbok (*Antidorcas marsupialis*)	17 (including immatures)
Steenbok (*Raphicerus campestris*)	2 (including immatures)
African buffalo (*Syncerus caffer*)	2 (including nymphs)
Eland (*Taurotragus oryx*)	16 (including immatures)
Greater kudu (*Tragelaphus strepsiceros*)	60 (including immatures)
Common duiker (*Sylvicapra grimmia*)	11 (including immatures)
Gemsbok (*Oryx gazella*)	6 (including immatures)
Grey rhebok (*Pelea capreolus*)	44 (including immatures)
Mountain reedbuck (*Redunca fulvorufula*)	21 (including immatures)
Four-striped grass mouse (*Rhabdomys pumilio*)	1 (larva)
Spring hare (*Pedetes capensis*)	2 (immatures)
Scrub hare (*Lepus saxatilis*)	68 (immatures)
Smith's red rock rabbit (*Pronolagus rupestris*)	6 (immatures, 1 ♂)
Birds	
Helmeted guineafowl (*Numida meleagris*)	8 (immatures)

ulae, sparse elsewhere on the scutum, interspersed with numerous very fine punctations that are sometimes almost invisible. Ventrally spiracles comma-shaped, with a short broad dorsal prolongation. Adanal plates broad, smoothly rounded posterointernally and posteroexternally; accessory adanal plates well developed.

Female (Figs 77(b), 78(d) to (f))

Capitulum slightly broader than long, length × breadth ranging from 0.74 mm × 0.75 mm to 0.85 mm × 0.91 mm. Basis capituli with lateral angles at about anterior third of its length, slightly hunched in appearance, acute; porose areas large, twice their own diameter apart. Palps more elongate than those of male, with pedicel of article I easily visible dorsally and much narrower than article II, giving the palps a stalked appearance; bluntly rounded apically. Scutum longer than broad, length × breadth ranging from 1.32 mm × 1.10 mm to 1.62 mm × 1.42 mm; posterior margin slightly sinuous. Eyes far forward, beady, orbited. Internal cervical margins deep and convergent anteriorly, becoming increasingly shallow and divergent. Large setiferous punctations mostly present on the scapulae and along the external cervical margins, plus a few medially on the scutum,

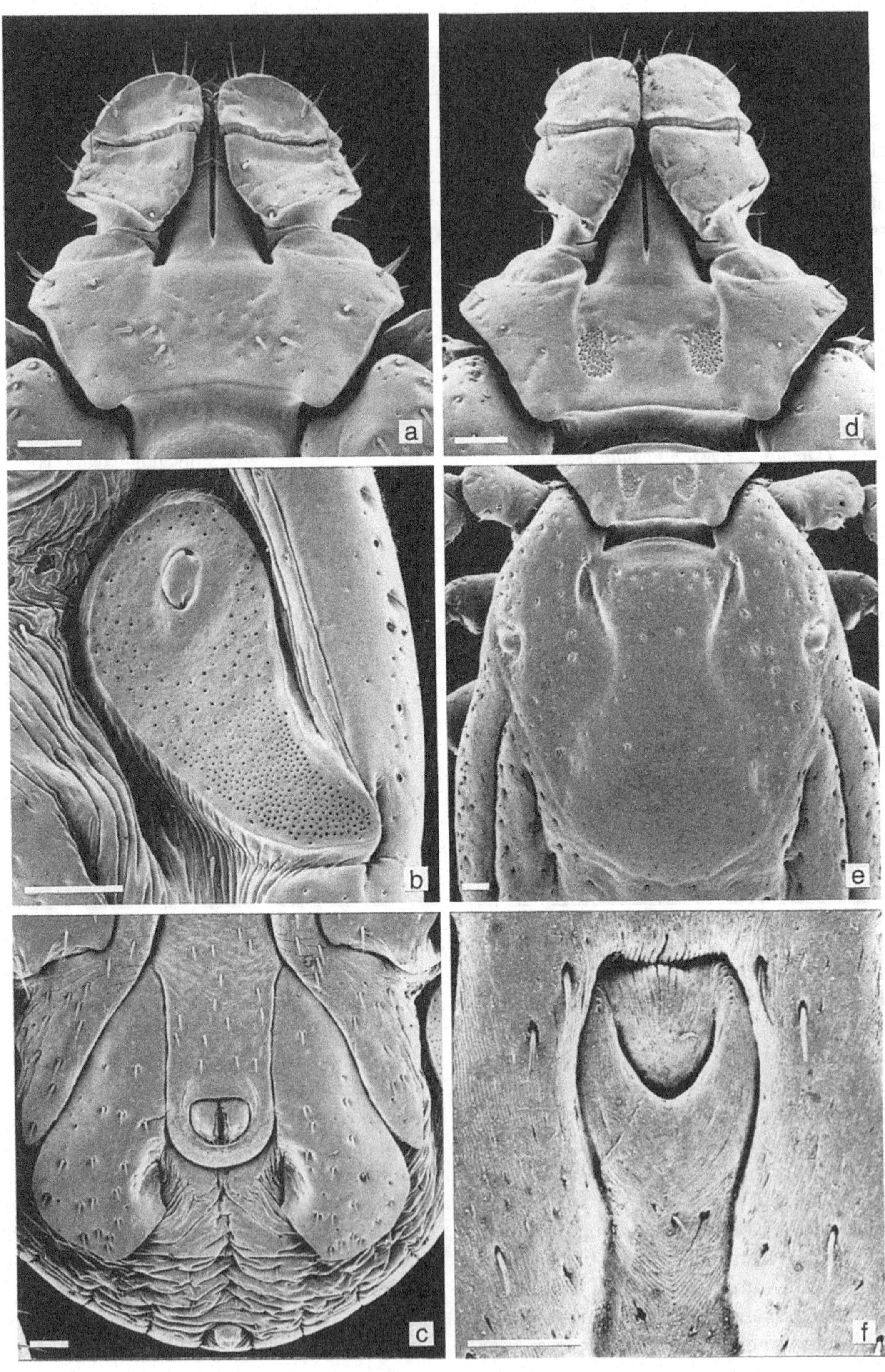
a
d
b
e
c
f

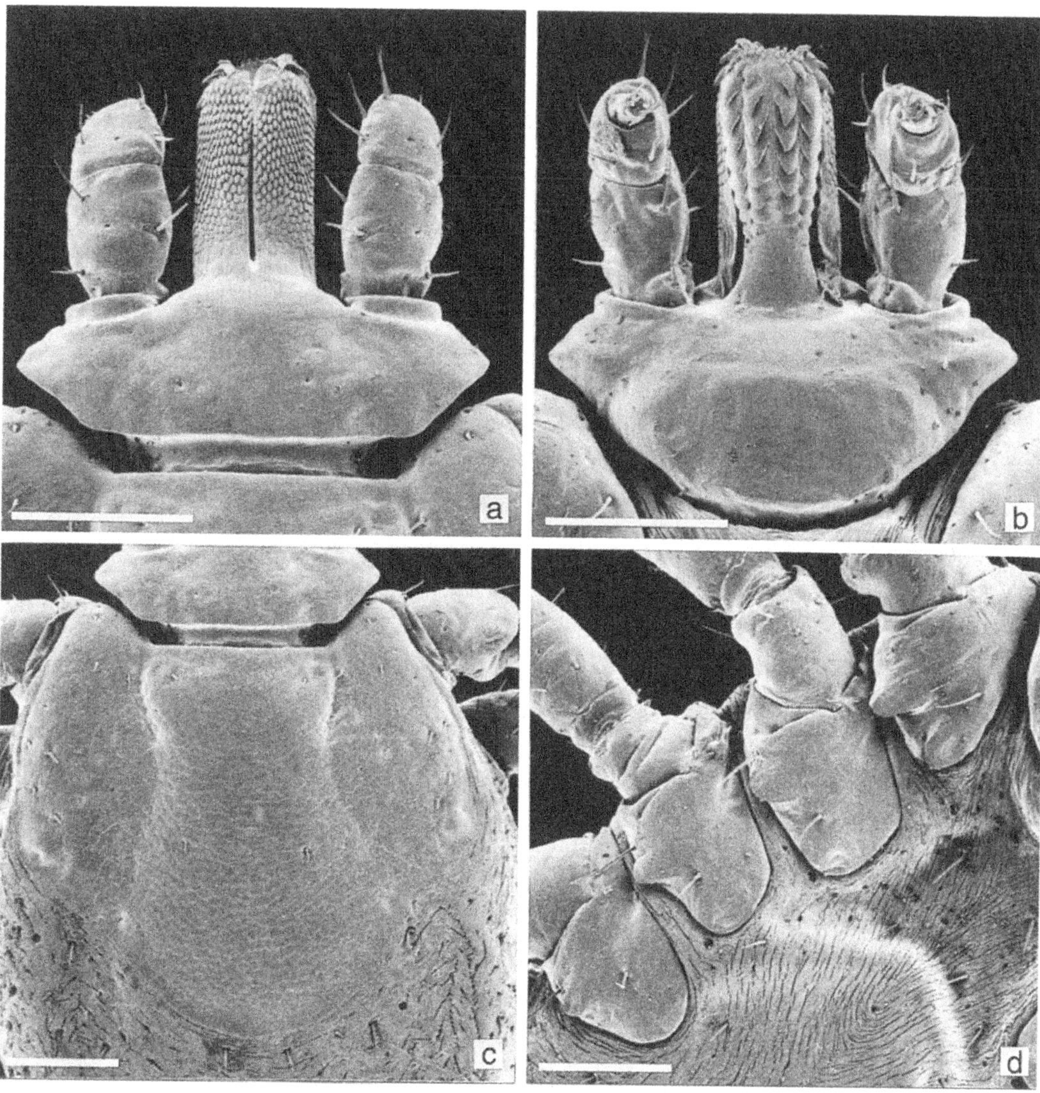

Figure 79 (*above*). *Rhipicephalus glabroscutatum* [laboratory reared, original ♀ from greater kudu (*Tragelaphus strepsiceros*), 'Bucklands' (farm), near Grahamstown, Eastern Cape Province, South Africa, collected in 1979 by Y. Rechav]. Nymph: (a) capitulum, dorsal; (b) capitulum, ventral; (c) scutum; (d) coxae. Scale bars represent 0.10 mm. SEMs by J.F. Putterill.

Figure 78 (*opposite*). *Rhipicephalus glabroscutatum* [laboratory reared, original ♀ from greater kudu (*Tragelaphus strepsiceros*), 'Bucklands' (farm), near Grahamstown, Eastern Cape Province, South Africa, collected in 1979 by Y. Rechav]. Male: (a) capitulum, dorsal; (b) spiracle; (c) adanal plates. Female: (d) capitulum, dorsal; (e) scutum; (f) genital aperture. Scale bars represent 0.10 mm. SEMs by J.F. Putterill.

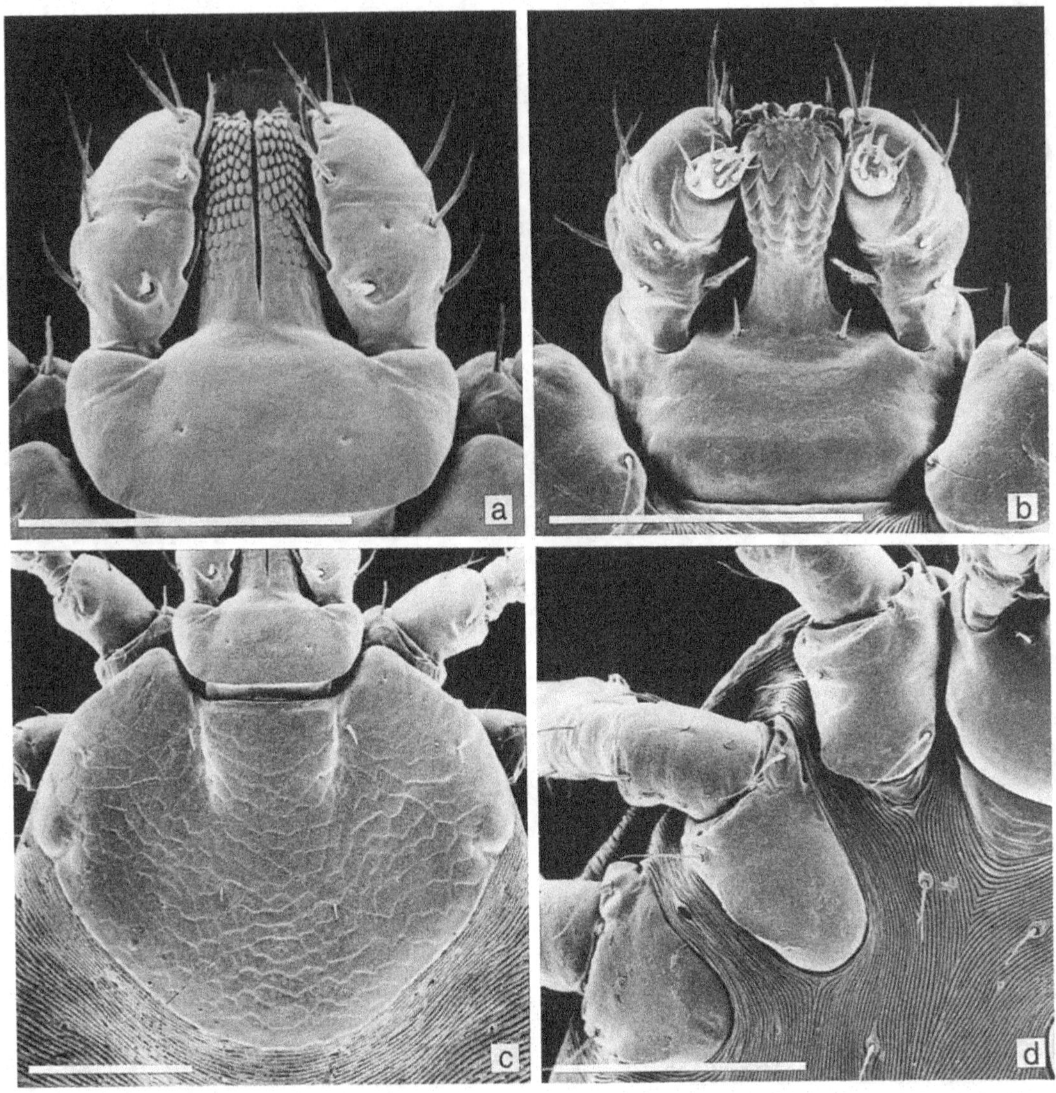

Figure 80. *Rhipicephalus glabroscutatum* [laboratory reared, original ♀ from greater kudu (*Tragelaphus strepsiceros*), 'Bucklands' (farm), near Grahamstown, Eastern Cape Province, South Africa, collected in 1979 by Y. Rechav]. Larva: (a) capitulum, dorsal; (b) capitulum, ventral; (c) scutum; (d) coxae. Scale bars represent 0.10 mm. SEMs by J.F. Putterill.

interspersed with very fine punctations that are sometimes virtually invisible. Ventrally genital aperture broadly V-shaped.

Nymph (Fig. 79)

Capitulum much broader than long, length × breadth ranging from 0.24 mm × 0.30 mm to 0.27 mm × 0.33 mm. Basis capituli over three times as broad as long, with acute lateral angles at about mid-length, overlapping the scapulae. Palps very slightly constricted proximally, more or less sausage-shaped. Chelicerae and hypostome project slightly beyond the palps. Scutum broader than long, length × breadth ranging from 0.45 mm × 0.52 mm to 0.51 mm × 0.55 mm; posterior margin a deep curve, indented posterolaterally. Eyes at widest point, at about mid-length, beady, orbited. Cervical grooves deep and convergent initially, becoming shallower and divergent and extending back to the slightly concave posterolateral margins of the scutum, thus dividing the scutum into three parts. Ventrally coxae I each with two broad spurs, coxae II to IV each with an external spur, decreasing in size from II to IV.

Larva (Fig. 80)
Capitulum approximately as broad as long, length × breadth ranging from 0.147 mm × 0.147 mm to 0.164 mm × 0.159 mm. Basis capituli over twice as broad as long, narrowing posteriorly. Palps constricted proximally, then widening before they taper to their apices. Scutum much broader than long, length × breadth ranging from 0.301 mm × 0.412 mm to 0.310 mm × 0.422 mm; posterior margin a deep smooth curve. Eyes at widest point, at about mid-length, slightly convex. Cervical grooves short, mildly convergent. Ventrally coxae I each with a broad spur, coxae II and III each with a broad salient ridge along its posterior border.

Notes on identification
Rhipicephalus glabroscutatum is closely related to *R. evertsi evertsi*, with which its distribution in South Africa overlaps. Apart from their beady eyes the adults of these two species differ considerably in their morphology. Their immature stages, though, are very similar in appearance and may be difficult to separate when they are present together in field collections (see p.157).

Hosts

A two-host species (Du Toit, 1941). All stages of development prefer ungulates: the majority of collections come from goats, followed by sheep and greater kudu (MacIvor, 1985; Horak & Knight, 1986). With the exception of scrub hares, which are good hosts of the immature stages, infestation of non-ungulate hosts appears to be accidental (Horak *et al.*, 1991) (Table 21). More than 80% of immature and adult ticks attach on the lower legs and between and/or around the hooves (Horak *et al.*, 1992). Larvae and nymphs are most numerous from March to August and adults from September to February (Horak *et al.*, 1992). One life cycle only is completed annually.

Total burdens, including all developmental stages, exceeding 8000 ticks have been recorded on a mountain reedbuck, 6000 ticks on a greater kudu and more than 4000 on a gemsbok (I.G.H., unpublished data).

Zoogeography

Rhipicephalus glabroscutatum occurs in South Africa where, with two minor exceptions, it has been recorded only in the Eastern and Western Cape Provinces (Map 24). These exceptions are a single male and a single female tick collected from sheep in the south-western and southern Free State respectively. They probably originated from ticks brought in with small stock translocated from the Eastern Cape Province. The vast majority of collection sites are at altitudes ranging from almost sea level to approximately 1900 m. Rainfall in these areas varies from approximately 200 mm to 600 mm annually, falling predominantly during summer in the east and during winter in the west of this tick's habitat. They are in various types of shrubland or in evergreen and semi-evergreen bushland and thicket (MacIvor, 1985).

The largest numbers of this tick on individual animals have been found in the extreme eastern and the extreme western parts of its range. In other areas where it occurs present indications are that the infestations are smaller.

Disease relationships

The presence of adult ticks, often in conjunction with adult *Amblyomma hebraeum*, is significantly associated with the occurrence of foot abscesses, and consequently lameness, in goats (MacIvor & Horak, 1987). Because the distribution of *R. glabroscutatum* overlaps the regions in which much of South Africa's mohair is produced this is an economically important condition. Sheep are frequently similarly affected. Despite large numbers of adult ticks around the hooves of antelopes these animals do not appear to develop foot abscesses.

REFERENCES

Du Toit, R. (1941). Description of a tick *Rhipicephalus glabroscutatum*, sp. nov., (Ixodidae) from the Karroo areas of the Union of South Africa. *Onderstepoort Journal of Veterinary Science and Animal Industry*, **16**, 115–18.

Horak, I.G., Boomker, J., Spickett, A.M. & De Vos, V. (1992). Parasites of domestic and wild animals in South Africa. XXX. Ectoparasites of kudus in the eastern Transvaal Lowveld and the eastern Cape Province. *Onderstepoort Journal of Veterinary Research*, **59**, 259–73.

Horak, I.G., Fourie, L.J., Novellie, P.A. & Williams, E.J. (1991). Parasites of domestic and wild animals in South Africa. XXVI. The mosaic of ixodid tick infestations on birds and mammals in the Mountain Zebra National Park. *Onderstepoort Journal of Veterinary Research*, **58**, 125–36.

Horak, I.G. & Knight, M.M. (1986). A comparison of the tick burdens of wild animals in a nature reserve and on an adjacent farm where tick control is practised. *Journal of the South African Veterinary Association*, **57**, 199–203.

MacIvor, K.M. [de F.] (1985). The distribution and hosts of *Rhipicephalus glabroscutatum*. *Onderstepoort Journal of Veterinary Research*, **52**, 43–6.

MacIvor, K.M. de F. & Horak, I.G. (1987). Foot abscess in goats in relation to the seasonal abundance of adult *Amblyomma hebraeum* and adult *Rhipicephalus glabroscutatum* (Acari: Ixodidae). *Journal of the South African Veterinary Association*, **58**, 113–18.

RHIPICEPHALUS GUILHONI MOREL & VASSILIADES, 1963

This species was named in honour of Professor M. Guilhon of the National Veterinary School, Maison Alfort, Paris, who was regarded by P.C. Morel as his first mentor.

Diagnosis

A moderate-sized fairly heavily punctate West African member of the *R. sanguineus* group.

Male (Figs 81(a), 82(a) to (c))

Capitulum broader than long, length × breadth ranging from 0.54 mm × 0.60 mm to 0.65 mm × 0.71 mm. Basis capituli with short acute lateral angles at about anterior third of its length. Palps short, broad, somewhat truncated apically. Conscutum length × breadth ranging from 2.42 mm × 1.53 mm to 2.73 mm × 1.71 mm; anterior process of coxae I inconspicuous. In engorged specimens body wall expanded posterolaterally and posteriorly, with a small additional bulge posteromedially. Eyes slightly raised, edged dorsally by a few large punctations that may be set in a groove. Cervical fields broad, somewhat depressed, their external margins delimited by large setiferous punctations. Marginal lines long, clearly defined, outlined by large punctations. Posteromedian groove short, tapering anteriorly; posterolateral grooves small, oval. A few large setiferous punctations scattered on the scapulae and more-or-less in a '*simus*' pattern medially on the conscutum, interspersed with numerous fine punctations that are larger and more conspicuous in some specimens than in others. Ventrally spiracles large, broad, their dorsal prolongations tapering and curving slightly just at their ends. Adanal plates broad, their internal margins barely indented posterior to the anus, their posterolateral margins smoothly curved; accessory adanal plates large, pointed, well sclerotized.

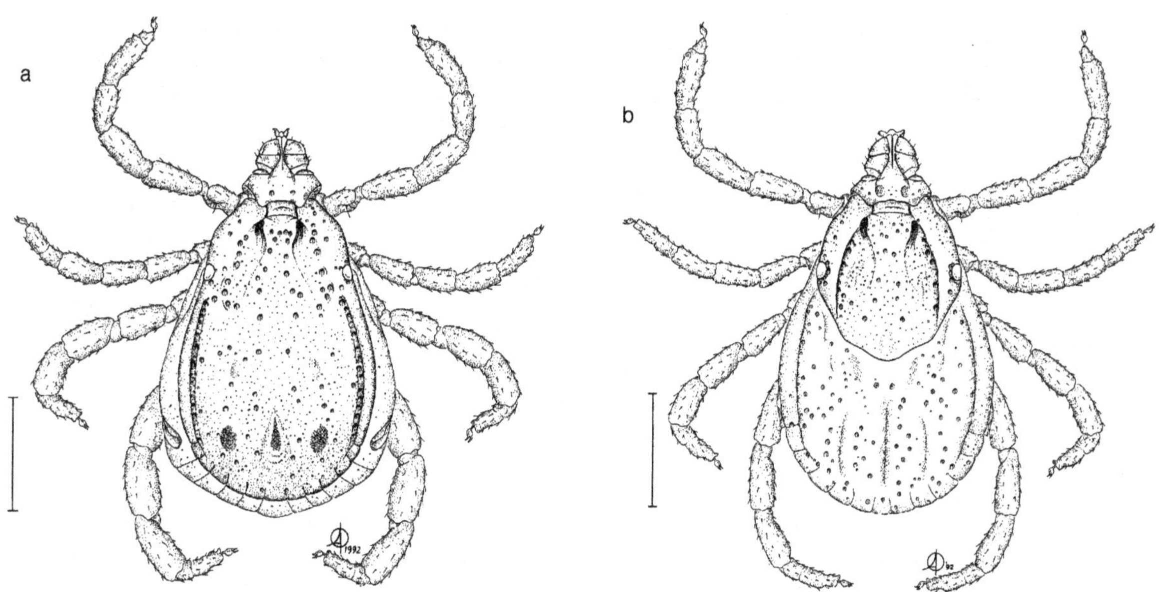

Figure 81. *Rhipicephalus guilhoni* (from sheep, Nioro, Mali, date of collection unknown, donated by P.C. Morel). (a) Male, dorsal; (b) female, dorsal. Scales represent 1 mm. A. Olwage *del.*

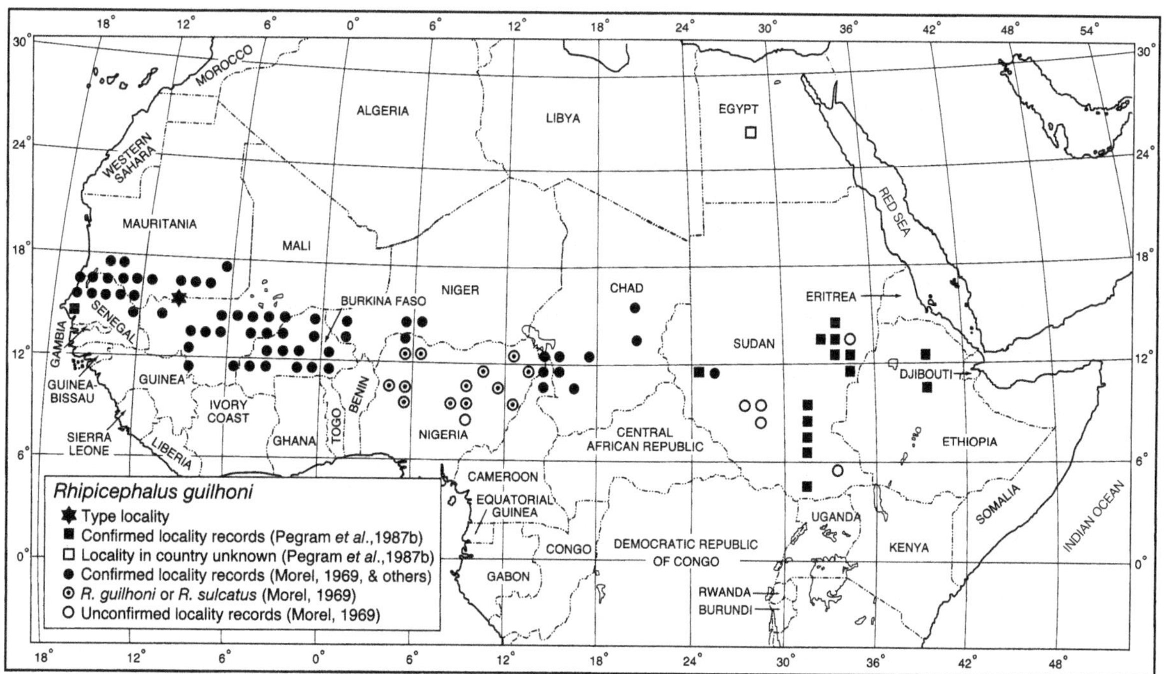

Map 25. *Rhipicephalus guilhoni*: distribution. (Based largely on Morel, 1969 and Pegram *et al.*, 1987b).

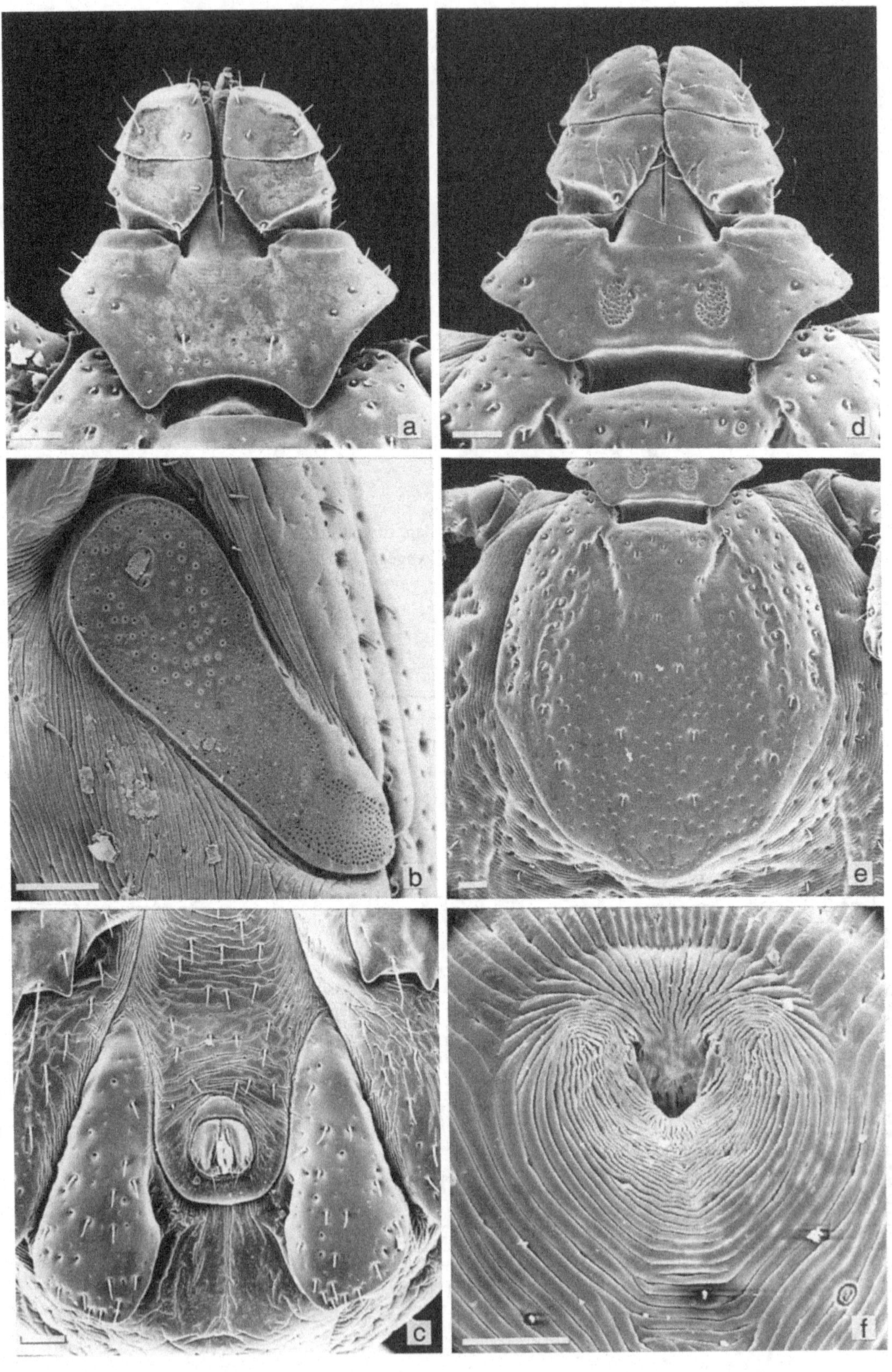
a
b
c
d
e
f

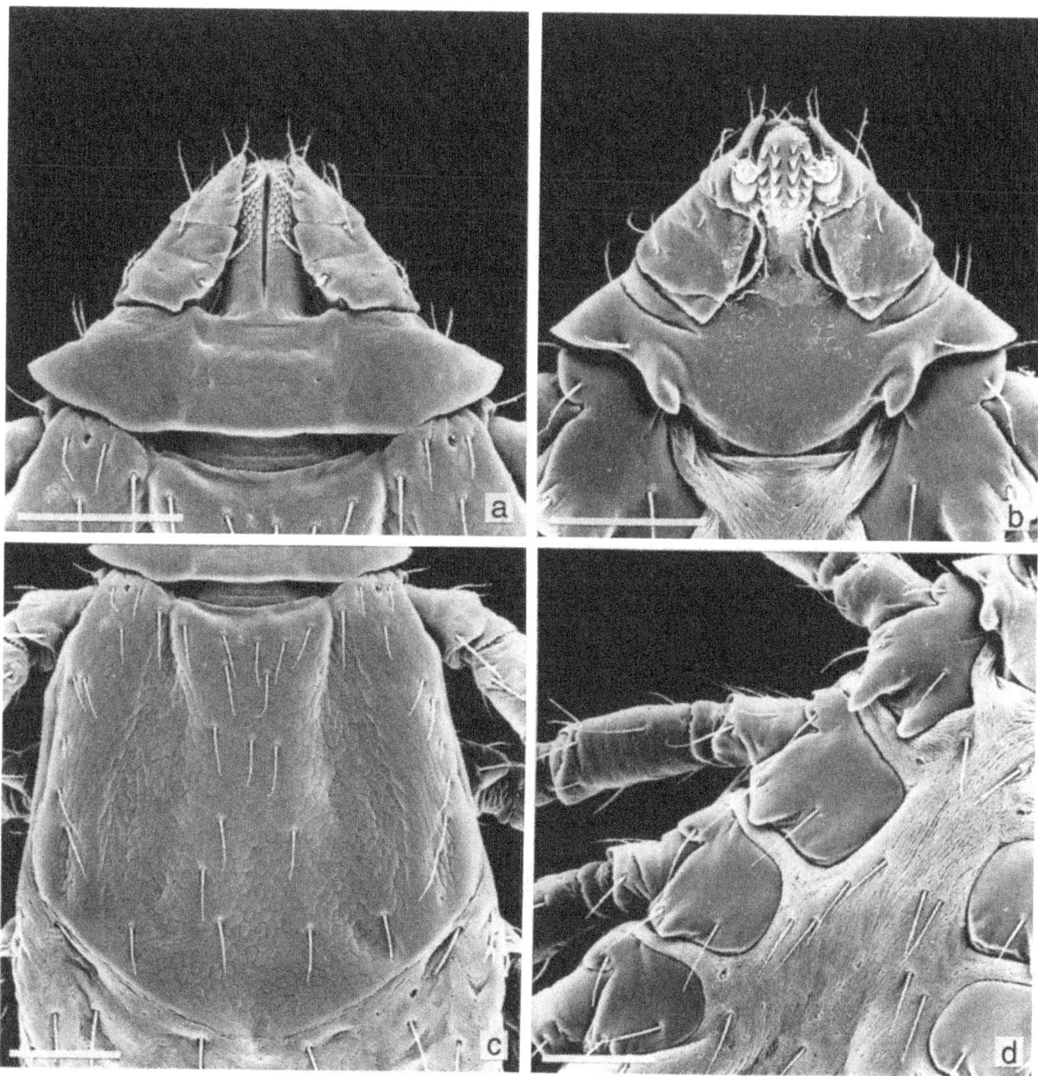

Figure 83 (*above*). *Rhipicephalus guilhoni* (RML 001330, provenance unknown). Nymph: (a) capitulum, dorsal; (b) capitulum, ventral; (c) scutum; (d) coxae. Scale bars represent 0.10 mm. SEMs by M.D. Corwin.

Figure 82 (*opposite*). *Rhipicephalus guilhoni* (RML 001330, male, provenance unknown; RML 49641, female collected from sheep, northern Nigeria in May 1967). Male: (a) capitulum, dorsal; (b) spiracle; (c) adanal plates. Female: (d) capitulum, dorsal; (e) scutum; (f) genital aperture. Scale bars represent 0.10 mm. SEMs by M.D. Corwin.

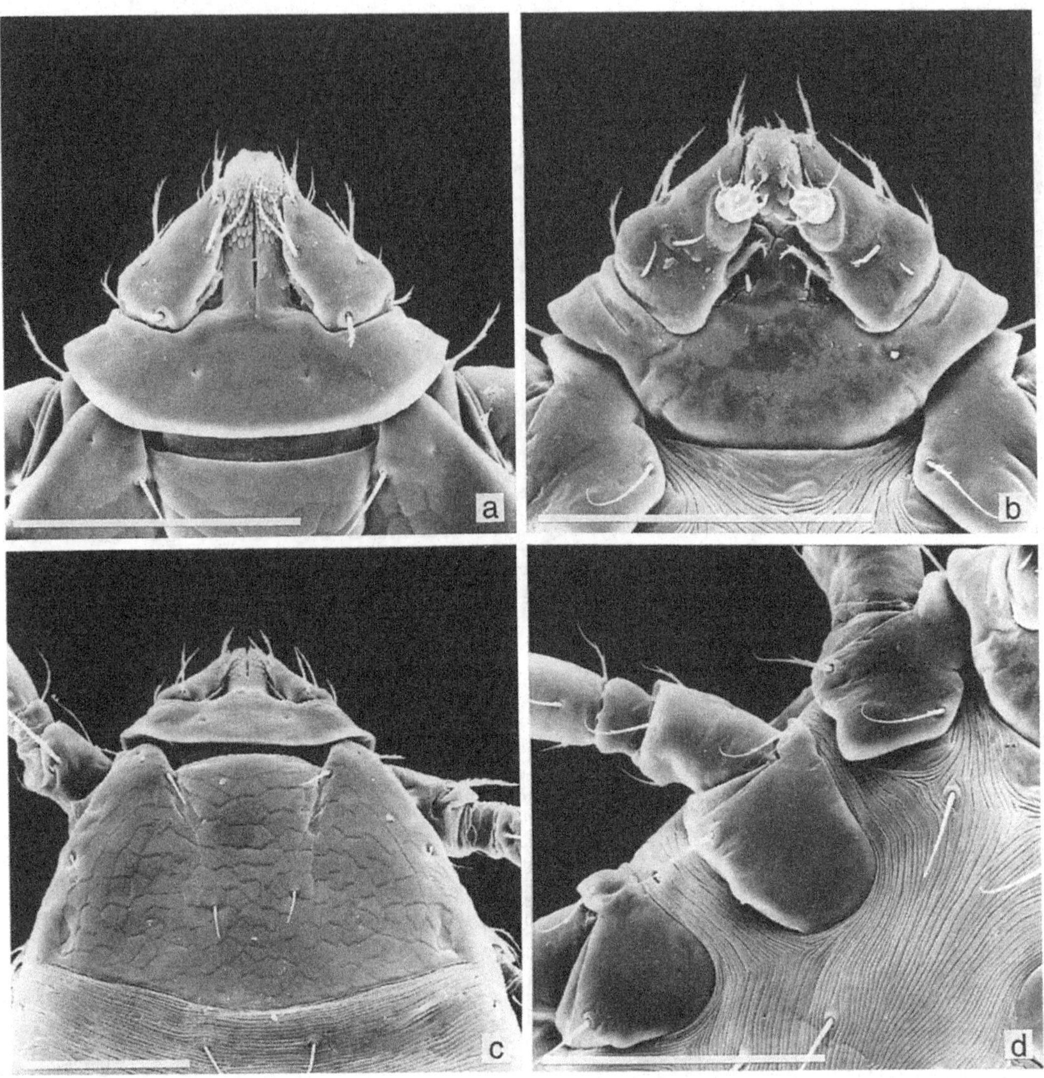

Figure 84. *Rhipicephalus guilhoni* (RML 001330, provenance unknown). Larva: (a) capitulum, dorsal; (b) capitulum, ventral; (c) scutum; (d) coxae. Scale bars represent 0.10 mm. SEMs by M.D. Corwin.

Female (Figs 81(b), 82(d) to (f))

Capitulum broader than long, length × breadth ranging from 0.60 mm × 0.71 mm to 0.75 mm × 0.82 mm. Basis capituli with broad lateral angles at about mid-length; porose areas large, oval, nearly twice their own diameter apart. Palps broad, their apices rounded. Scutum slightly longer than broad, length × breadth ranging from 1.31 mm × 1.27 mm to 1.59 mm × 1.55 mm; posterior margin sinuous. Eyes at about mid-length, slightly raised, edged dorsally by a few large punctations that may be set in a groove. Cervical fields broad, somewhat depressed, their external margins sharply defined, reaching posterolateral margins of scutum and outlined by large setiferous punctations that sometimes become confluent. A few large setiferous punctations scattered on scapulae and medially on scutum, interspersed with numerous fine interstitial punctations. Ventrally genital aperture broadly V-shaped.

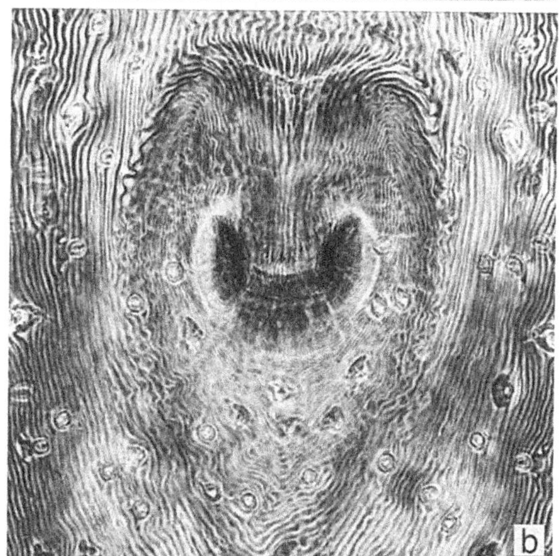

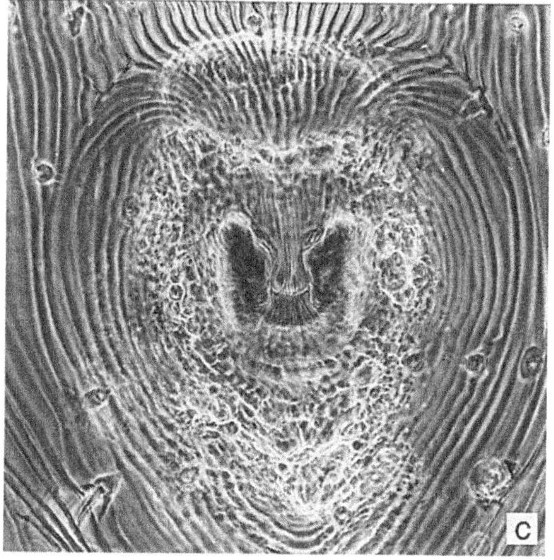

Nymph (Fig. 83)

Capitulum much broader than long, length × breadth ranging from 0.15 mm × 0.29 mm to 0.16 mm × 0.30 mm. Basis capituli nearly four times as broad as long, with broad tapering lateral angles overlapping the scapulae; ventrally with short, somewhat inwardly-directed spurs on posterior margin. Palps broad proximally, tapering to narrowly-rounded apices, inclined inwards. Scutum length × breadth ranging from 0.40 mm × 0.45 mm to 0.42 mm × 0.47 mm; posterior margin smoothly curved. Eyes at widest point, well over halfway back, large, slightly convex. Cervical fields broad, slightly depressed, divergent, with their external margins reaching the posterolateral margins of the scutum. Numerous long setae present medially on the scutum, between the cervical fields, and a few on the scapulae. Ventrally coxae I each with a relatively long external spur and a shorter internal spur; coxae II to IV each with an external spur only, decreasing progressively in size.

Larva (Fig. 84)

Capitulum broader than long, length × breadth ranging from 0.09 mm × 0.10 mm to 0.10 mm × 0.11 mm. Basis capituli about 3.5 times as broad as long, with short broad lateral angles. Palps broad proximally, tapering to narrowly rounded apices, inclined inwards. Scutum much broader than long, length × breadth ranging from 0.20 mm × 0.30 mm to 0.21 mm × 0.31 mm; posterior margin a broad extremely shallow curve. Eyes at widest part of scutum, far back, slightly convex. Cervical grooves short, slightly convergent. Ventrally coxae I each with a broad spur; coxae II and III each with an indication only of a spur on its posterior margin.

The measurements of the nymph and larva given above are quoted from Morel & Vassiliades (1963).

Figure 85 (*opposite*). Female genital apertures (mounted): (a) *Rhipicephalus guilhoni*; (b) *Rhipicephalus sulcatus*; (c) *Rhipicephalus turanicus*. (From Pegram *et al.*, 1987a), figs 39 & 40; (1987b), fig. 40, with kind permission from Kluwer Academic Publishers).

Table 22. *Host records of* Rhipicephalus guilhoni

Hosts	Number of records
Domestic animals	
Cattle	20
Sheep	14
Goats	4
Camels	2
Horses	4
Donkeys	4
Dogs	4
Cats	3
Wild animals	
Four-toed hedgehog (*Atelerix albiventris*)	3
'Jackal' (*Canis* sp.)	6
Pale fox (*Vulpes pallida*)	1
Caracal (*Caracal caracal*)	1
African wild cat (*Felis lybica*)	2
Serval (*Leptailurus serval*)	2
Spotted hyaena (*Crocuta crocuta*)	4
Zorilla (*Ictonyx striatus*)	2
African civet (*Civettictis civetta*)	1
Warthog (*Phacochoerus africanus*)	6
Giraffe (*Giraffa camelopardalis*)	1
Korrigum (*Damaliscus lunatus korrigum*)	1
Red-fronted gazelle (*Gazella rufifrons*)	2
Roan antelope (*Hippotragus equinus*)	2
Waterbuck (*Kobus ellipsiprymnus*)	1
Geoffroy's ground squirrel (*Xerus erythropus*)	1 (nymph)
Nile rat (*Arvicanthis* sp.)	4 (nymphs)
'Hare'	4
Birds	
Ostrich (*Struthio camelus*)	2
Saddle-bill stork (*Ephippiorhynchus senegalensis*)	1
Marabou stork (*Leptoptilos crumeniferus*)	1
African white-backed vulture (*Gyps africanus*)	1
Martial eagle (*Hieraaetus bellicosus*)	1
Denham's bustard (*Neotis denhami*)	2
Arabian bustard (*Ardeotis arabs*)	2

Notes on identification (Fig. 85)

Pegram *et al.* (1987a) noted that the morphological structure of the female genital aperture of *R. guilhoni* is such that 'its specific validity is unlikely to be disputed' (Fig. 85(a)). Pegram *et al.* (1987b) added: 'The larva and nymph of *R. guilhoni* were described by Morel & Vassiliades (1963), but Filippova (1981) believed that there is some confusion in their descriptions between this species and *R. turanicus*.'

Earlier workers on West African ticks included records of *R. guilhoni* under *R. sanguineus sensu lato* (Morel & Vassiliades, 1963; Morel, 1969; Mohammed, 1977).

This is not a species with which we are well acquainted ourselves. Our accounts of its hosts and distribution are based largely, but not exclusively, on those by Morel & Vassiliades (1963) and Morel (1969).

Hosts

A three-host species (Morel, 1969). Amongst domestic animals *R. guilhoni* adults have been recorded most commonly from cattle, also sheep, but less frequently from other animals. They have also been collected from a wide range of wild animals and various species of the larger non-passerine birds (Table 22). The immature stages apparently feed on sciuromorph (squirrel-like) and myomorph (rat-like) rodents. The hosts' holes, in which the immature stages develop, shelter them from the variations of humidity and temperature in the general environment.

Rhipicephalus guilhoni apparently develops through one generation annually. The adults become active during the rainy season, beginning in May to June, and persist until the cooler season (December to January), when they become rare. In northern Nigeria small numbers of adults were collected from the hooves of cattle from June to September (Mohammed, 1977).

Zoogeography

Rhipicephalus guilhoni has been recorded from Mauritania and Senegal eastwards across Africa to the Sudan and occasionally Ethiopia (Map 25). There is also one record from an unspecified locality in Egypt. It can, therefore, exist under relatively dry conditions. It is widely distributed in areas with 500 mm to 750 mm annual rainfall, and occurs in some places with only 250 mm to 500 mm in the southern part of the Sahel. The vegetation in which it characteristically occurs is Sudanian woodland, and to some extent Sahel *Acacia* wooded grassland and deciduous bushland.

In 1980, Matthysse regarded *R. guilhoni* as probably the commonest and most widely distributed tick on livestock in Mali. However, during their tick survey on sheep and goats in the Sahelian area of Senegal, Gueye *et al.* (1987) listed *R. guilhoni* as one of the species they had not found although it had apparently occurred there before the prevailing drought. Earlier they had noted that, during the previous 10 years or so, ecological changes had been caused by persistent drought, deforestation and increasing cultivation (Gueye *et al.*, 1986). These in turn had affected tick distributions and/or variations in population numbers.

Disease relationships

Unknown.

REFERENCES

Filippova, N.A. (1981). [On diagnosis of species of the genus *Rhipicephalus* Koch (Ixodoidea, Ixodidae) from the fauna of the USSR and adjoining territories by nymphal instar]. *Parazitologischeskii Sbornik*, **30**, 47–68. [In Russian; English summary].

Gueye, A., Camicas, J.L., Diouf, A. & Mbengue, Mb. (1987). Tiques et hémoparasitoses du bétail au Sénégal. II. La zone sahélienne. *Revue d'Élevage et de Médecine Vétérinaire des Pays Tropicaux*, **40** (nouvelle série), 119–25.

Gueye, A., Mbengue, Mb., Diouf, A. & Seye, M. (1986). Tiques et hémoparasitoses du bétail au Sénégal. I. La région des Niayes. *Revue d'Élevage et de Médecine Vétérinaire des Pays Tropicaux*, **39** (nouvelle série), 381–93.

Matthysse, J.G. (1980). *Research and training on vector-borne hemoparasites of livestock and their vectors in Mali. Phase I, objective 7: Survey to determine species diversity and distribution of ticks attacking cattle in Mali.* Technical Assistance Contract between the United States Agency for International Development and the International Division of Texas A. & M. University. Project AID/afr-c-1262.

Mohammed, A.N. (1977). The seasonal incidence of ixodid ticks of cattle in northern Nigeria. *Bulletin of Animal Health and Production in Africa*, 25, 273–93.

Morel, P.C. & Vassiliades, G. (1963). Les *Rhipicephalus* du groupe *sanguineus*: espèces africaines (Acariens: Ixodoidea). *Revue d'Élevage et de Médecine Vétérinaire des Pays Tropicaux*, **15 for 1962** (nouvelle série), 343–86.

Pegram, R.G., Clifford, C.M., Walker, J.B. & Keirans, J.E. (1987a). Clarification of the *Rhipicephalus sanguineus* group (Acari, Ixodoidea, Ixodidae). I. *R. sulcatus* Neumann, 1908 and *R. turanicus* Pomerantsev, 1936. *Systematic Parasitology*, **10**, 3–26.

Pegram, R.G., Keirans, J.E., Clifford, C.M. & Walker, J.B. (1987b). Clarification of the *Rhipicephalus sanguineus* group (Acari, Ixodoidea, Ixodidae). II. *R. sanguineus* (Latreille, 1806) and related species. *Systematic Parasitology*, **10**, 27–44.

Also see the following Basic Reference (p. 14): Morel (1969).

RHIPICEPHALUS HUMERALIS RONDELLI, 1926

The specific name *humeralis*, from the Latin *humerus* meaning 'the shoulder' or 'upper bone of the arm', refers to the light-coloured areas on the scapulae of the male.

Synonym

pulchellus humeralis.

Diagnosis

A moderate-sized ornate species.

Male (Figs 86(a), 87(a) to (c))
Capitulum much longer than broad, length × breadth ranging from 0.79 mm × 0.64 mm to 1.04 mm × 0.80 mm. Basis capituli with very short, bluntly-rounded lateral angles at about anterior quarter of its length. Palps short, broad.

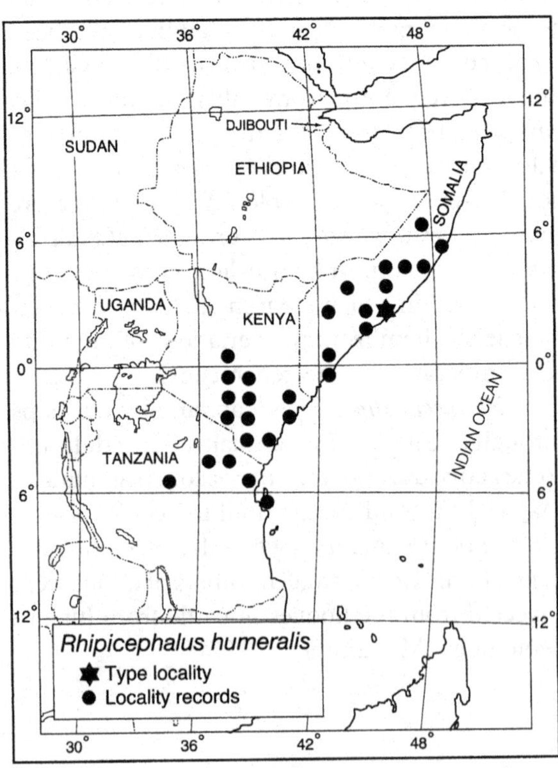

Map 26. *Rhipicephalus humeralis*: distribution.

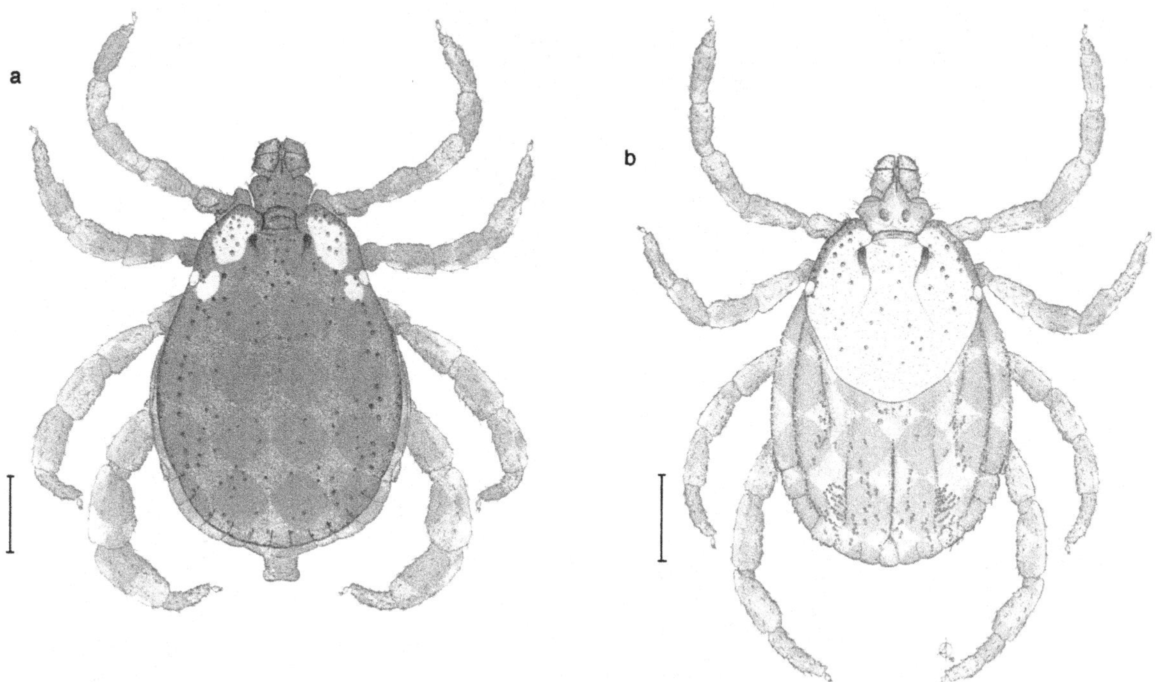

Figure 86. *Rhipicephalus humeralis*: [B.S. 274/-, RML 66307, laboratory reared, progeny of ♀ collected from African elephant (*Loxodonta africana*), Mackinnon Road, Kenya, on 3 October 1951 by D.L.W. Sheldrick]. (a) Male, dorsal; (b) female, dorsal. Scale bars represent 1 mm. A. Olwage *del.*

Conscutum length × breadth ranging from 3.12 mm × 2.18 mm to 4.13 mm × 2.86 mm; large heavily-sclerotized anterior process present on coxae I. In engorged specimens body wall expanded posterolaterally and forming a short broad caudal process. Eyes marginal, almost flat, delimited by slight depressions and a few punctations dorsally. Cervical pits deep, comma-shaped, continuous posteriorly with the short indistinct internal cervical margins. Marginal lines merely indicated by large setiferous punctations. Posterior grooves absent. Large setiferous punctations also present on the scapulae, and in small clusters in the positions of the posterolateral grooves; smaller punctations scattered medially on the conscutum. The conscutum is predominantly very dark brown in colour with a pattern of light smoky-brown patches anteriorly. A light-coloured patch always appears to be present on each scapula and often medially between the cervical pits. An isolated patch may also be present behind each eye or, in the most ornate specimens, these may extend forwards to join the adjacent scapular patches. Legs increase markedly in size from I to IV, with light-coloured mottling dorsally, especially on legs III and IV. Ventrally spiracles more-or-less comma-shaped, tapering gradually towards the dorsal surface. Adanal plates up to three times as long as broad, narrowly rounded posterointernally; accessory adanal plates absent.

Female (Figs 86(b), 87(d) to (f))

Capitulum slightly longer than broad, length × breadth ranging from 0.78 mm × 0.74 mm to 0.95 mm × 0.93 mm. Basis capituli with fairly short broad lateral angles at about anterior third of its length; porose areas round, slightly more than their own diameter apart. Palps long, truncated apically. Scutum longer than broad, length × breadth ranging from 1.70 mm × 1.62 mm to 2.21 mm × 2.06 mm. Eyes marginal,

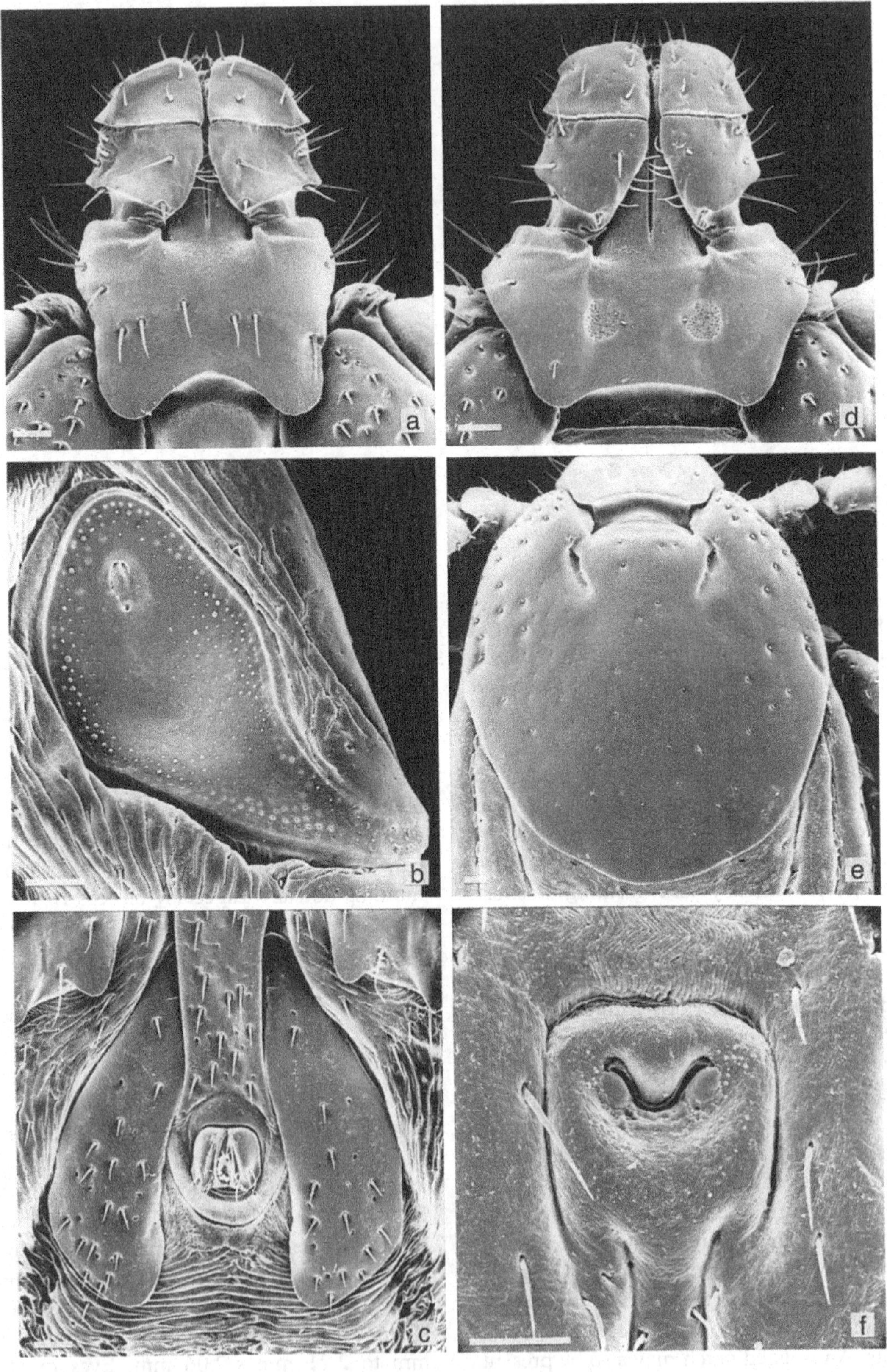
a
d
b
e
c
f

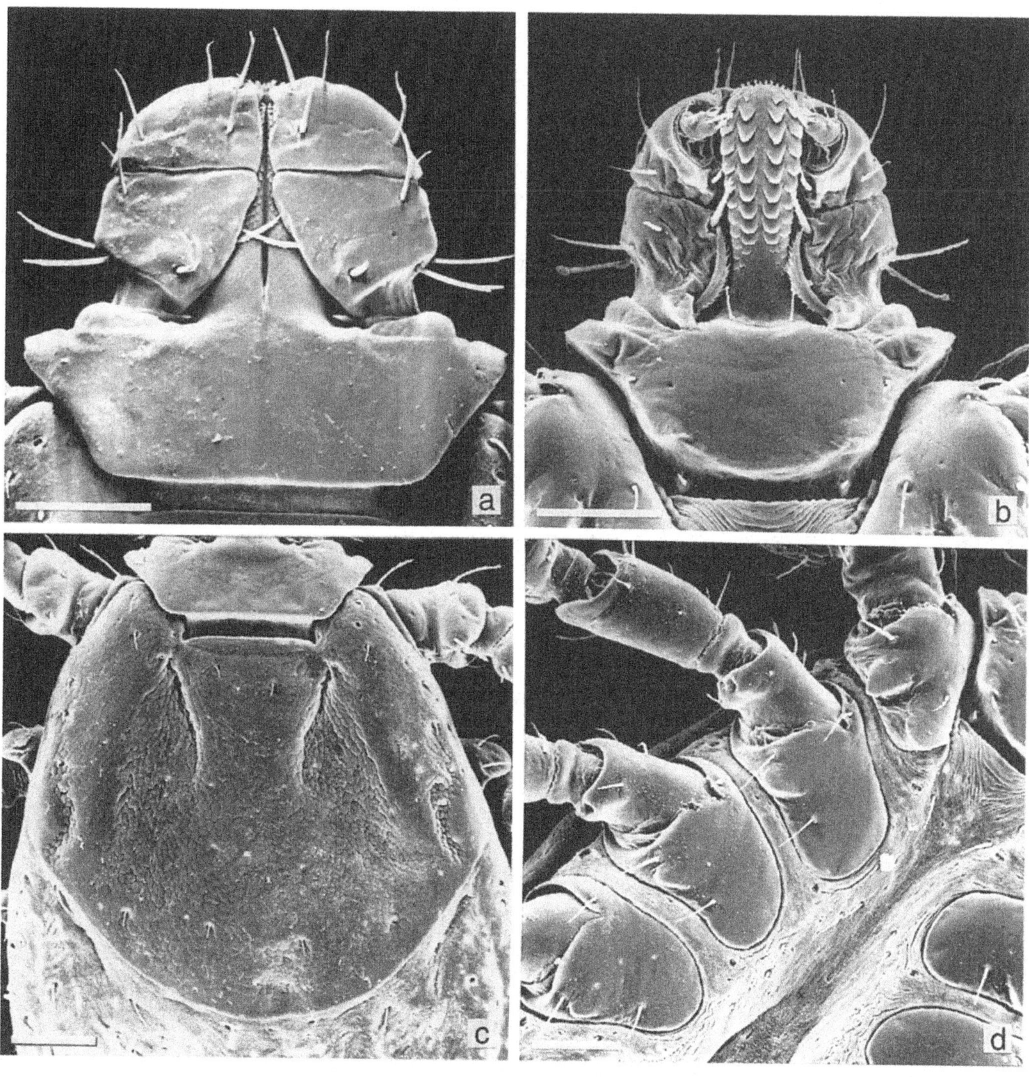

Figure 88 (*above*). *Rhipicephalus humeralis*: [B.S. 274/-, RML 66307, laboratory reared, progeny of ♀ collected from African elephant (*Loxodonta africana*), Mackinnon Road, Kenya, on 3 October 1951 by D.L.W. Sheldrick]. Nymph: (a) capitulum, dorsal; (b) capitulum, ventral; (c) scutum; (d) coxae. Scale bars represent 0.10 mm. SEMs by M.D. Corwin.

Figure 87 (*opposite*). *Rhipicephalus humeralis*: [B.S. 274/-, RML 66307, laboratory reared, progeny of ♀ collected from African elephant (*Loxodonta africana*), Mackinnon Road, Kenya, on 3 October 1951 by D.L.W. Sheldrick]. Male: (a) capitulum, dorsal; (b) spiracle; (c) adanal plates. Female: (d) capitulum, dorsal; (e) scutum; (f) genital aperture. Scale bars represent 0.10 mm. SEMs by M.D. Corwin.

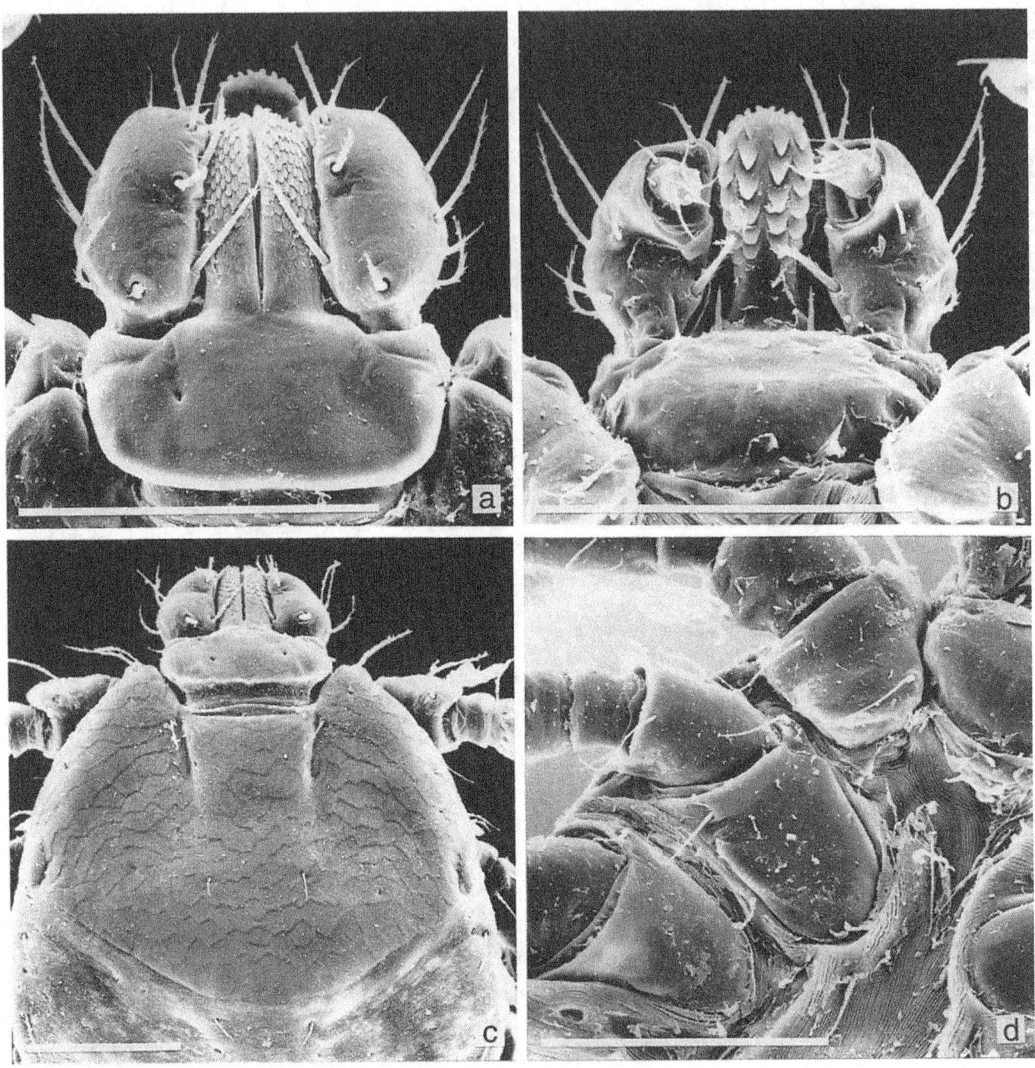

Figure 89. *Rhipicephalus humeralis*: [B.S. 274/-, RML 66307, laboratory reared, progeny of ♀ collected from African elephant (*Loxodonta africana*), Mackinnon Road, Kenya, on 3 October 1951 by D.L.W. Sheldrick]. Larva: (a) capitulum, dorsal; (b) capitulum, ventral; (c) scutum; (d) coxae. Scale bars represent 0.10 mm. SEMs by M.D. Corwin.

almost flat, delimited by slight depressions and a few punctations dorsally. Cervical pits convergent; internal cervical margins sometimes indicated by slight depressions. Largest setiferous punctations present on scapulae; smaller punctations scattered medially, sometimes interspersed with minute punctations. The scutum is predominantly brownish-cream with darker smoky-brown colouration round the eyes; this often extends anteriorly along the outer margins of the scapulae, and sometimes medially between the cervical pits. Alloscutum dark brown with a few stout white setae, especially posterolaterally. Legs with light-coloured mottling dorsally. Ventrally genital aperture sigmoid-shaped with large lateral plaques.

Nymph (Fig. 88)

Capitulum broader than long, length × breadth ranging from 0.27 mm × 0.31 mm to 0.32

Table 23. *Host records of* Rhipicephalus humeralis

Hosts	Number of records
Domestic animals	
Cattle	32
Sheep	5
Goats	5
Camels	21
Donkeys	2
Dogs	1
Wild animals	
Greater bushbaby (*Otolemus crassicaudatus*)	1
Hamadryas baboon (*Papio hamadryas*)	1
Black-backed jackal (*Canis mesomelas*)	1
Bat-eared fox (*Otocyon megalotis*)	1
Sand fox (*Vulpes rueppelli*)	1
Lion (*Panthera leo*)	2
African elephant (*Loxodonta africana*)	28
Black rhinoceros (*Diceros bicornis*)	10
Somali warthog (*Phacochoerus aethiopicus*)	1
Warthog (*Phacochoerus* sp.)	2
Bushpig (*Potamochoerus larvatus*)	1
Coke's hartebeest (*Alcelaphus buselaphus cokii*)	1
Speke's gazelle (*Gazella spekei*)	1
African buffalo (*Syncerus caffer*)	2
Eland (*Taurotragus oryx*)	2
Bushbuck (*Tragelaphus scriptus*)	2
Waterbuck (*Kobus ellipsiprymnus*)	1
Unstriped ground squirrel (*Xerus rutilus*)	1
Birds	
Great sparrow hawk (*Accipiter melanoleucus*)	1
Humans	1

mm × 0.35 mm. Basis capituli nearly three times as broad as long, lateral angles in anterior third of its length, short, acute; ventrally with two short broad spurs on posterior border. Palps broad, rounded apically. Scutum broader than long, length × breadth ranging from 0.56 mm × 0.59 mm to 0.63 mm × 0.66 mm; posterior margin a broad smooth curve. Eyes at widest point, large. Cervical fields broad, slightly depressed, their external margins almost reaching posterolateral margins of scutum. Ventrally coxae I each with a large external spur and a shorter, broader internal spur; coxae II to IV each with a rounded external spur, decreasing progressively in size.

Larva (Fig. 89)

Capitulum broader than long, length × breadth ranging from 0.131 mm × 0.141 mm to 0.135 mm × 0.154 mm. Basis capituli over twice as broad as long, very slightly convex anterolaterally. Palps broad, blunt apically. Scutum much broader than long, broadest over halfway back, at eye level, length × breadth ranging from 0.243 mm × 0.368 mm to 0.273 mm × 0.379

mm; posterior margin a wide curve, very slightly sinuous posterolaterally. Eyes large. Cervical grooves slightly convergent initially, becoming mildly divergent and almost reaching eye level. Ventrally coxae I each with a triangular spur; coxae II and III each with a broad salient ridge on its posterior border.

Notes on identification

In the past the systematic status of *R. humeralis* has been confused. It was originally described by Rondelli (1926) as a subspecies of *R. pulchellus* but was raised to specific status by Zumpt (1949).

Warburton (1933) referred to 'a variety of *R. pulchellus*' that he had received from the former Italian Somaliland 'in which all the white markings were obsolete except patches on the scapulae', ornamentation that is typical of *R. humeralis*. He maintained that intermediate forms in the same consignment linked these ticks with 'full-patterned specimens' of *R. pulchellus* but this observation has not been supported by those of other workers (Cunliffe, 1913; Walker, 1957).

Rhipicephalus humeralis has also been misidentified as *R. ecinctus* by several authors although the latter species is actually a synonym of *R. maculatus* (Walker, 1957; Morel, 1980).

The males of *R. humeralis* and *R. pulchellus* can easily be identified specifically (compare Figs 86(a) and 168(a)). It is much more difficult to distinguish the females and nymphs, and virtually impossible to separate the larvae (Walker, 1957). The scutum of *R. humeralis* females is predominantly brownish-cream in colour, with darker smoky-brown ornamentation, and its surface is slightly more uneven and pitted with rather deeper punctations than that of *R. pulchellus*, which is smoother in texture and ivory coloured with dark grey to black ornamentation. *Rhipicephalus humeralis* females also have fewer white setae on their alloscutum than *R. pulchellus* females (compare Figs 86(b) and 168(c)). The basis capituli of the *R. humeralis* nymph is somewhat broader than that of *R. pulchellus*, and its lateral angles project sideways rather than forwards as they do in the latter species.

Since the females and immatures of *R. humeralis* and *R. pulchellus* are so similar it would be unwise to identify these stages specifically in the absence of males with which they can be associated.

Hosts

A three-host species (Walker, 1957). Its most commonly recorded domestic animal hosts are cattle and camels (Table 23). Amongst wild animals it apparently has a predilection for elephants and black rhinos. It was present in every collection made from 10 elephants examined in two areas in Kenya. On these animals it was usually found on the ears, between the front legs and on the tail, and once each on the trunk and the scrotum. On the buffalo it was found on the rump.

The hosts of the immature stages have not been recorded, but they probably feed on the same animals as the adults.

Zoogeography

Rhipicephalus humeralis has been recorded from various parts of southern Somalia (Uilenberg, 1978), eastern Kenya and northern Tanzania (Map 26). The areas in which it occurs range from sea level to 1500 m in altitude, usually with annual rainfalls from about 800 mm to less than 400 mm. These localities are primarily in Somalia–Masai *Acacia–Commiphora* deciduous bushland and thicket. Its distribution generally overlaps that of *R. pulchellus* but it does not extend into the driest areas inhabited by the latter species, e.g. northern Kenya and northern Somalia.

Disease relationships

Unknown.

REFERENCES

Cunliffe, C. (1913). The variability of *Rhipicephalus pulchellus* (Gerstäcker, 1873), together with its geographical distribution. *Parasitology*, **6**, 204–16.

Rondelli, M.T. (1926). Alcuni ixodidi della Somalia Italiana. *Res Biologicae*, **1**, 33–43.

Uilenberg, G. (1978). *Report on a consultancy to Somalia: SOM/73/006.* Strengthening of Veterinary Laboratory Services, 22 November to 9 December 1977. Mimeographed, 18 pp.

Walker, J.B. (1957). *Rhipicephalus humeralis* Rondelli, 1926. *Parasitology*, **47**, 145–52.

Warburton, C. (1933). On five new species of ticks (Arachnida Ixodoidea). *Ixodes petauristae, I. ampullaceus, Dermacentor imitans, Amblyomma laticaudae* and *Aponomma draconis*, with notes on three previously described species, *Ornithodorus franchinii* Tonelli-Rondelli, *Haemaphysalis cooleyi* Bedford and *Rhipicephalus maculatus* Neumann. *Parasitology*, **24**, 558–73.

Also see the following Basic References (pp. 12–14): Morel (1980); Scaramello (1988); Walker (1974); Yeoman & Walker (1967); Zumpt (1949).

RHIPICEPHALUS HURTI WILSON, 1954

This species was named after Lieut.-Col. R.A.F. Hurt in gratitude for the numerous collections of ticks that he had sent to the author.

Diagnosis

A medium-sized dark brown tick.

Male (Figs 90(a), 91(a) to (c))

Capitulum slightly broader than long to slightly longer than broad, length × breadth ranging from 0.58 mm × 0.63 mm to 0.77 mm × 0.76 mm. Basis capituli with acute lateral angles at anterior third of its length. Palps short, broadly rounded apically. Conscutum length × breadth ranging from 2.49 mm × 1.63 mm to 3.40

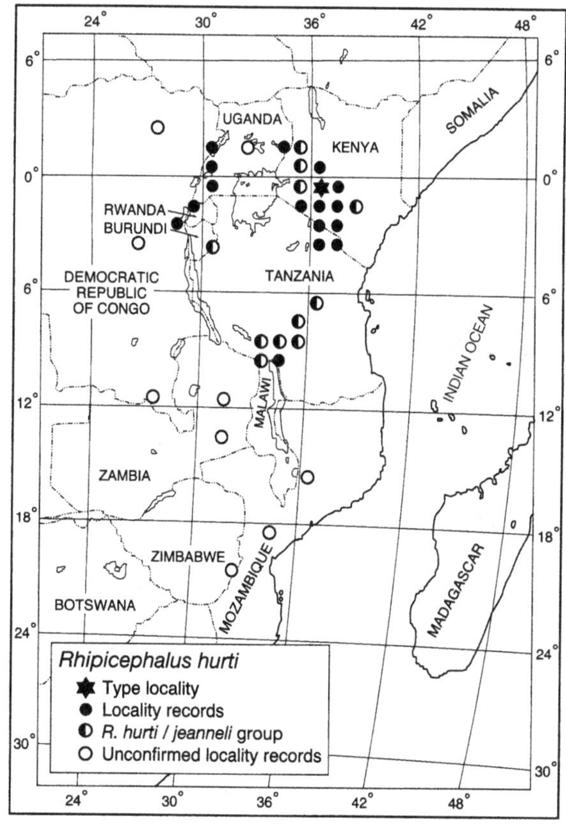

Map 27. *Rhipicephalus hurti*: distribution.

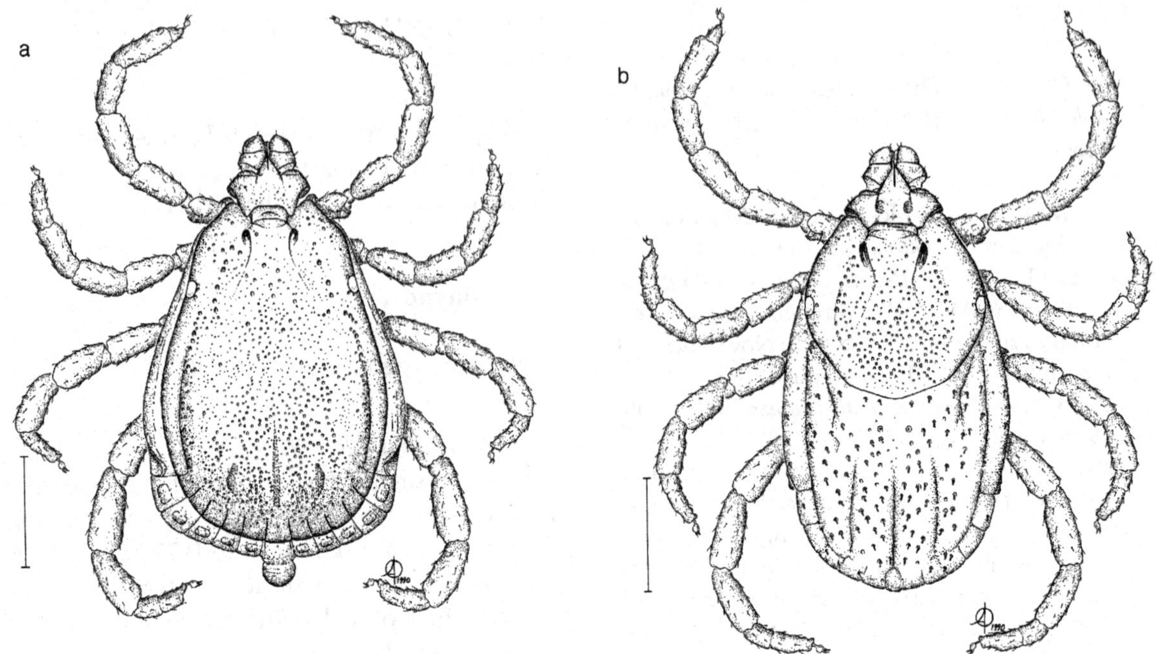

Figure 90. *Rhipicephalus hurti* [collected from eland (*Taurotragus oryx*) near Kijabe Hill, Kenya, on 30 April 1960 by D.W. Brocklesby]. (a) Male, dorsal; (b) female, dorsal. Scale bars represent 1 mm. A. Olwage *del.*

mm × 2.30 mm; anterior process of coxae I inconspicuous. In engorged specimens body wall expanded posterolaterally and posteriorly, with a short broad caudal process. Eyes flat to very slightly convex, edged dorsally by a few punctations. Cervical pits deep, convergent, continuous with shallow internal cervical margins. Marginal lines well defined posteriorly, becoming shallower anteriorly and finally indicated merely by punctations. Posterior grooves well defined, rugose; posteromedian groove long and narrow, posterolateral grooves semicircular to bow-shaped. Large to medium-sized setiferous punctations present anteriorly on the scapulae, along the external cervical margins and marginal lines, and scattered sparsely medially on the conscutum, especially between the cervical fields, interspersed with numerous somewhat smaller punctations that sometimes coalesce, especially near the posterior grooves. Anterior to the eyes and adjacent to the marginal lines the conscutum is usually almost free of punctations. Ventrally spiracles broad, narrowing and curving slightly towards the dorsal surface just at the end. Adanal plates broadly triangular, their external margins almost straight, their internal margins often more concave than in the specimen illustrated (Fig. 91(c)) and joining the rounded posterior margins in blunt points; accessory adanal plates represented by lightly-sclerotized points.

Female (Figs 90(b), 91(d) to (f))

Capitulum broader than long, length × breadth ranging from 0.55 mm × 0.63 mm to 0.78 mm × 0.80 mm. Basis capituli with broad lateral angles in anterior third of its length; porose areas oval, nearly twice their own diameter apart. Palps broad, rounded apically. Scutum ranging from broader than long to longer than broad, length × breadth ranging from 1.11 mm × 1.22 mm to 1.59 mm × 1.48 mm; posterior margin slightly sinuous. Eyes flat to slightly convex, edged dorsally by a few punctations. Cervical fields slightly depressed. A few large setiferous punctations present anteriorly on the scapulae, along the external margins of the cervical fields and scattered medially on the scutum among the numerous medium-sized punctations; the lateral

Table 24. *Host records of* Rhipicephalus hurti

Hosts	Number of records
Domestic animals	
Cattle	37
Dogs	8
Wild animals	
'Jackal' (*Canis* sp.)	1
Lion (*Panthera leo*)	3
Leopard (*Panthera pardus*)	1
'Genet' (*Genetta* sp.)	1
Black rhinoceros (*Diceros bicornis*)	3
Warthog (*Phacochoerus africanus*)	3
Bushpig (*Potamochoerus larvatus*)	1
'Wild pig'	1
Giraffe (*Giraffa camelopardalis*)	1
African buffalo (*Syncerus caffer*)	30
Eland (*Taurotragus oryx*)	4
Bushbuck (*Tragelaphus scriptus*)	9
Tragelaphus sp.	1
Common duiker (*Sylvicapra grimmia*)	1
Roan antelope (*Hippotragus equinus*)	1
'Antelope'	1
Angoni swamp rat (*Otomys angoniensis*)	1 (larva)
Humans	2

areas of the scutum generally smooth and glossy. Ventrally genital aperture broad, its posterior border straight, its sides somewhat angular.

Nymph (Fig. 92)

Capitulum much broader than long, length × breadth ranging from 0.21 mm × 0.35 mm to 0.25 mm × 0.41 mm. Basis capituli four times as broad as long, lateral angles long, tapering gradually to sharp points; ventrally small spurs present on posterior border. Palps slender, tapering to narrowly rounded apices, inclined inwards. Scutum much broader than long, length × breadth ranging from 0.46 mm × 0.53 mm to 0.52 mm × 0.65 mm; posterior margin a smooth shallow curve. Eyes at widest point, well over halfway back, very slightly convex. Cervical pits long, narrow, convergent, cervical fields inconspicuous. Ventrally coxae I each with a long sharp external spur and a short sharp internal spur; coxae II to IV each with a small external spur only, decreasing progressively in size.

Larva (Fig. 93)

Capitulum much broader than long, length × breadth (measured to two places of decimals only) ranging from 0.11 mm × 0.18 mm to 0.12 mm × 0.19 mm. Basis capituli well over four times as broad as long, with tapering bluntly rounded lateral angles; ventrally with blunt spurs on posterior border. Palps tapering to narrowly rounded apices, inclined inwards. Scutum much broader than long, length × breadth ranging from 0.22 mm × 0.35 mm to 0.23 mm × 0.36 mm; posterior margin a smooth shallow curve. Eyes at widest point, a little over halfway back, slightly convex. Cervical grooves convergent. Ventrally coxae I each with a relatively narrow

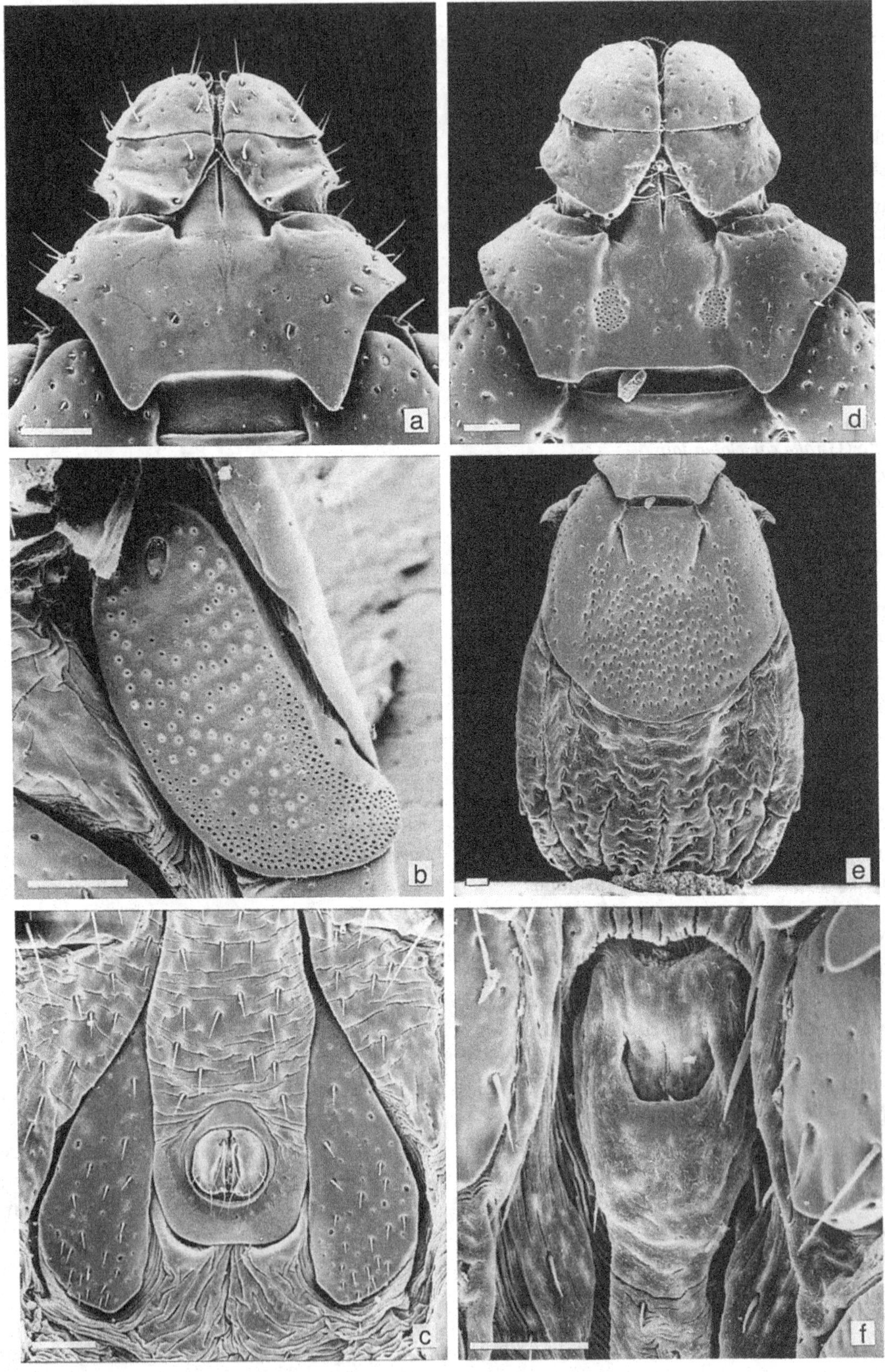
a
d
b
e
c
f

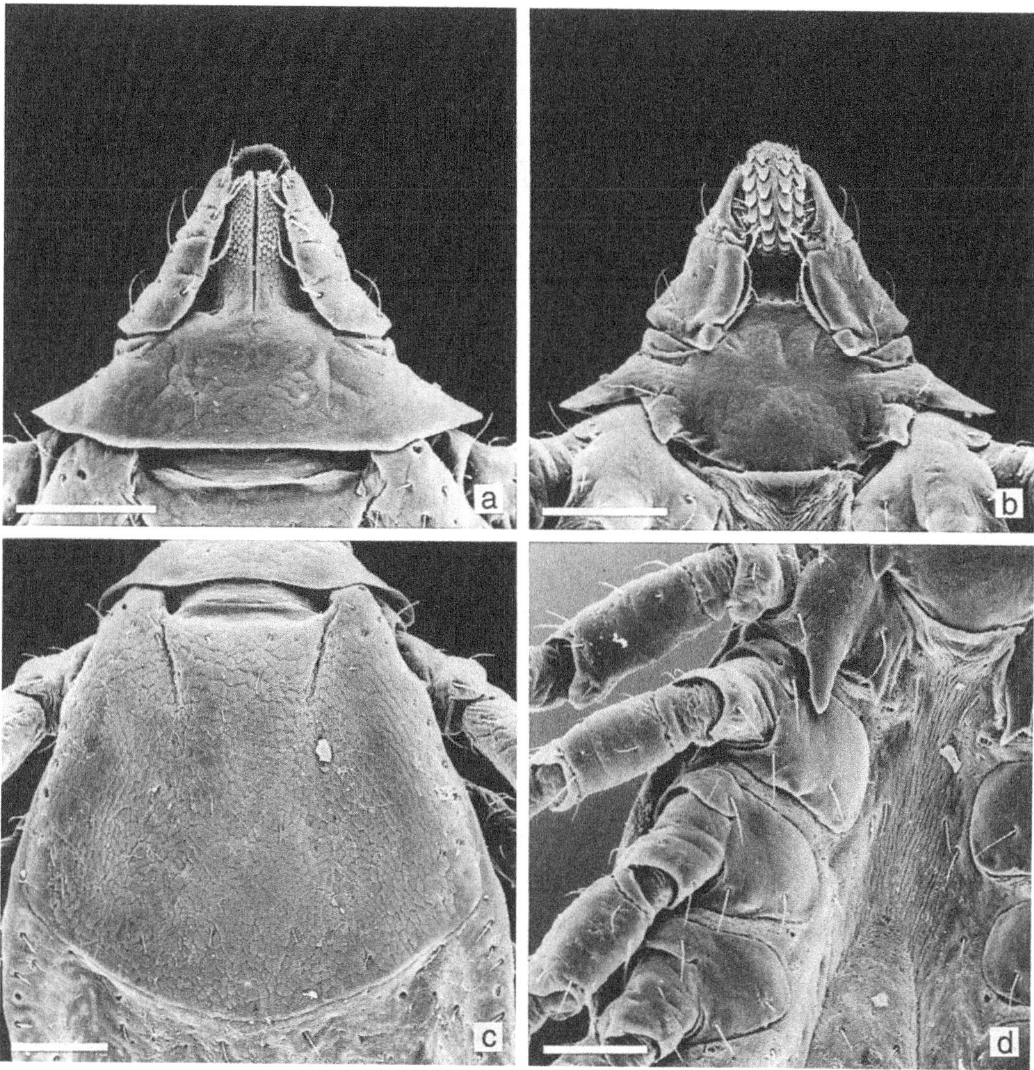

Figure 92 (*above*). *Rhipicephalus hurti* [B.S. 716/-, RML 66308; laboratory reared, original ♀ collected from bovine, Ngorongar, near Arusha, Tanzania, on 31 July 1956 by G.M. Kohls and F.J.W. Hampshire]. Nymph: (a) capitulum, dorsal; (b) capitulum, ventral; (c) scutum; (d) coxae. Scale bars represent 0.10 mm. SEMs by M.D. Corwin.

Figure 91 (*opposite*). *Rhipicephalus hurti* [B.S. 716/-, RML 66308; laboratory reared, original ♀ collected from bovine, Ngorongar, near Arusha, Tanzania, on 31 July 1956 by G.M. Kohls and F.J.W. Hampshire]. Male: (a) capitulum, dorsal; (b) spiracle; (c) adanal plates. Female: (d) capitulum, dorsal; (e) scutum and alloscutum, dorsal; (f) genital aperture. Scale bars represent 0.10 mm. SEMs by M.D. Corwin.

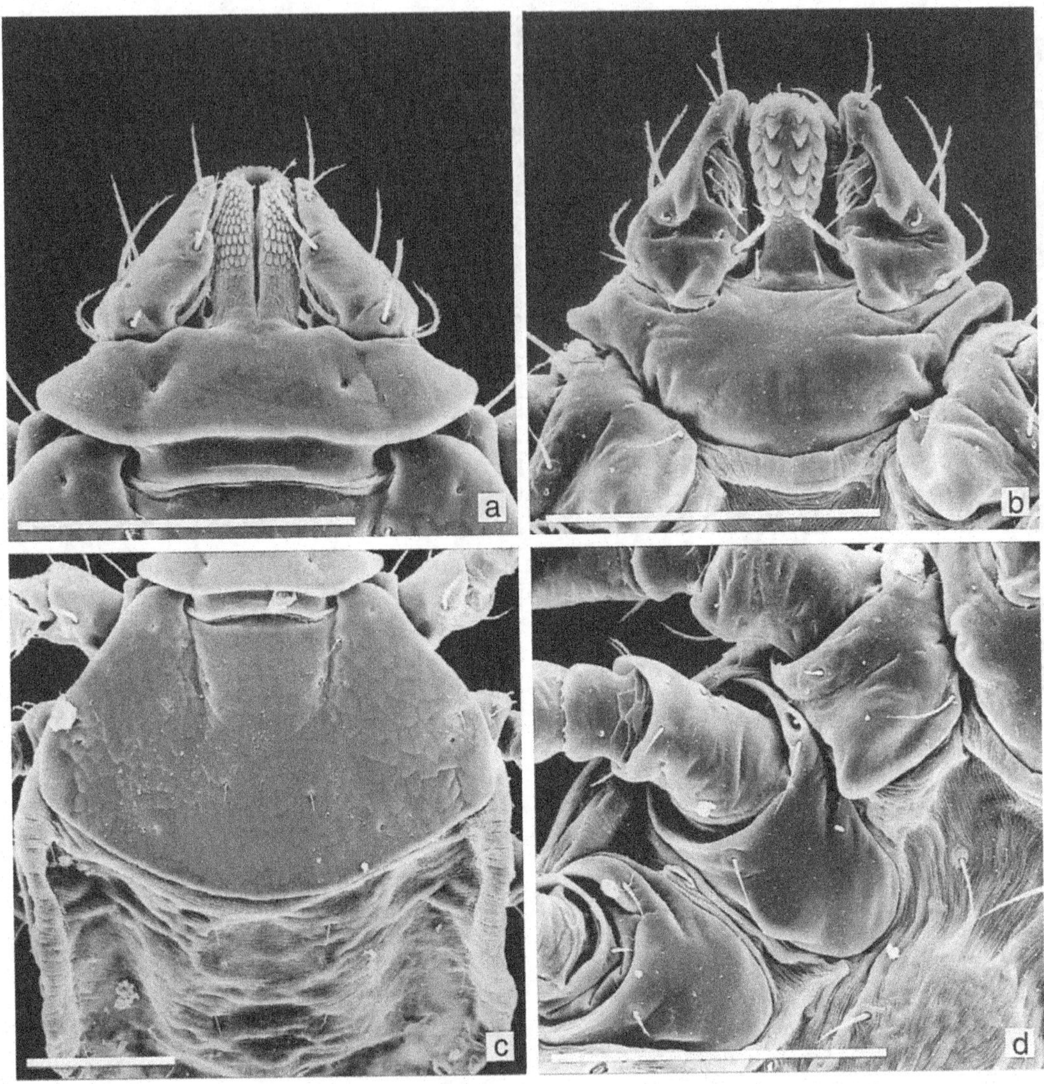

Figure 93. *Rhipicephalus hurti* [B.S. 716/-, RML 66308; laboratory reared, original ♀ collected from bovine, Ngorongar, near Arusha, Tanzania, on 31 July 1956 by G.M. Kohls and F.J.W. Hampshire]. Larva: (a) capitulum, dorsal; (b) capitulum, ventral; (c) scutum; (d) coxae. Scale bars represent 0.10 mm. SEMs by M.D. Corwin.

spur; coxae II and III each with a much broader, but shorter, spur.

Notes on identification

The taxonomic problem that exists in the *R. hurti/jeanneli* group has been discussed by several authors (Yeoman & Walker, 1967; Morel, 1969; Walker, 1974; Matthysse & Colbo, 1987). Typical adults of both *R. hurti* and *R. jeanneli* are reasonably easy to distinguish, and the immature stages of these two species differ quite markedly in appearance (compare Figs 90–93 above with those of *R. jeanneli*, pp. 232–238, Figs 96–99). The behaviour of the two in the laboratory also differs in that *R. hurti* is very difficult to rear and *R. jeanneli* very easy. Nevertheless it often remains virtually impossible, in our experience, to determine where the boundary lies between their adults in mixed collections that include morphologically intermediate forms. This group therefore remains in need of further detailed study.

The re-examination of ticks from Kenya in the E.A. Lewis Collection, The Natural History Museum, London, that were originally identified as *R. bursa*, *R. capensis* and *R. kochi* showed that they belong to the *R. hurti/jeanneli* group.

Another species whose appearance can be confused with that of *R. hurti* is *R. aquatilis* (*Rhipicephalus* sp. III of Yeoman & Walker, 1967) (see pp. 72–75). But, although *R. aquatilis* has been collected from cattle several times, it apparently has a predilection for the sitatunga (*Tragelaphus spekii*), consequently these two species generally occur in different habitats. Unfortunately Matthysse & Colbo (1987) included their records of *R. aquatilis* (listed as *Rhipicephalus* sp. III) with those of their *R. hurti* group.

Yeoman & Walker (1967) suspected that, in the past, ticks in the *R. hurti/jeanneli* group might also have been identified as *R. appendiculatus*, a mistake that could bedevil attempts to control the latter species.

Hosts

A three-host species (J.B. Walker, unpublished data, 1960). The only domestic animals from which *R. hurti* has so far been collected are cattle and a few dogs. Amongst wild animals the African buffalo is the most commonly recorded host, and to a lesser extent other wild bovids, mostly the larger species. Collections have also been made from various wild suids and carnivores (Table 24). The adults appear to have a predilection for their hosts' ears. They have also been collected from the neck and hind legs of a buffalo.

The immature stages probably feed on rodents, though the actual field evidence for this remains a single larva collected from an Angoni swamp rat.

Zoogeography

Rhipicephalus hurti has been recorded in East Africa, including Rwanda and Burundi (Map 27), usually in association with *R. jeanneli* (see pp. 232–238). We regard the few records that exist from further west and south in Africa with reserve: sometimes they are based on very small numbers of ticks, and also appear to be in ecologically unsuitable places (Elbl & Anastos, 1966; MacLeod & Mwanaumo, 1978; Santos Dias, 1987). We feel that this tick's presence in these areas requires confirmation.

This species occurs at high altitudes, from about 1500 m to over 2000 m, where there is a well-distributed mean annual rainfall between about 600 mm and 1500 mm in various types of Afromontane vegetation (forest, woodland, bushland and wooded or bushed grassland).

Disease relationships

Unknown.

REFERENCES

MacLeod, J. & Mwanaumo, B. (1978). Ecological studies of ixodid ticks (Acari: Ixodidae) in Zambia. IV. Some anomalous infestation patterns in the northern and eastern regions. *Bulletin of Entomological Research*, **68**, 409–29.

Santos Dias, J.A.T. (1987). Algumas observações sobre a fauna ixodológica (Acarina, Ixodoidea) de Moçambique, com a descrição de uma nova espécie do género *Boophilus* Curtice, 1891. *Garcia de Orta, Sér. Zool., Lisboa*, **14**, 17–26.

Wilson, S.G. (1954). *Rhipicephalus hurti* n.sp. (Ixodoidea) from Kenya game and domestic animals. *Parasitology*, **44**, 277–84.

Also see the following Basic References (pp. 12–14): Elbl & Anastos (1966); Matthysse & Colbo (1987); Morel (1969); Santos Dias (1960); Walker (1974); Yeoman & Walker (1967).

RHIPICEPHALUS INTERVENTUS WALKER, PEGRAM & KEIRANS, 1995

This specific name, a Latin term meaning 'to come between', refers to the fact that in various respects the adults of this tick are morphologically intermediate in appearance between those of *R. tricuspis* and *R. lunulatus*.

Diagnosis

A moderate-sized brown tick with bicuspid adanal and elongate accessory adanal plates giving a tricuspid appearance.

Male (Figs 94(a), 95(a) to (c))

Capitulum slightly broader than long, length × breadth ranging from 0.48 mm × 0.51 mm to 0.52 mm × 0.54 mm. Basis capituli with short, acute lateral angles in anterior third of its length. Palps short, broad, flattened apically. Conscutum length × breadth ranging from 2.17 mm × 1.40 mm to 2.36 mm × 1.58 mm; anterior process of coxae I small. In engorged specimens body wall expanded posterolaterally and posteriorly. Eyes marginal, flat, edged dorsally by a few medium-sized setiferous punctations. Cervical pits comma-shaped, convergent. Marginal lines shallow, extending forwards to just behind eyes. Posteromedian and posterolateral grooves either poorly developed or absent; when present posteromedian groove short, narrow; posterolaterals oval to circular. A few other slight depressions sometimes present anterior to posterolaterals. Moderately large setiferous punctations present on the scapulae, along the outer cervical margins and marginal lines, in a '*simus*' pattern medially on the conscutum and round the posterior grooves. Otherwise the conscutum is smooth and virtually impunctate. Ventrally spiracles broadly comma-shaped with a rather wide gently curving dorsal prolongation. Adanal plates elongate, internal margins slightly indented posterior to the anus, posterointernal margins broadly rounded, posteroexternal margins usually

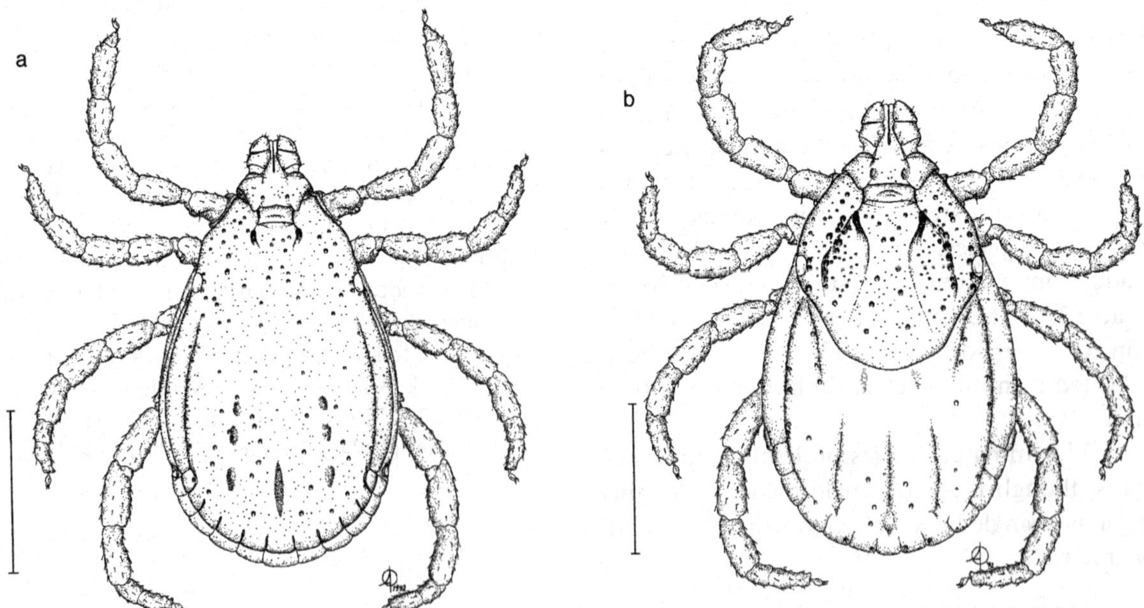

Figure 94. *Rhipicephalus interventus* (RML 53849, AHRC 67/2854, from cattle, Kawoko- Masaka, Masaka District, Uganda, collected on 2 March 1967 by J.G. Matthysse). (a) Male, dorsal; (b) female, dorsal. Scale bars represent 1 mm. A. Olwage *del.* (Reprinted from Walker *et al.*, 1995, figs 1 & 2, with kind permission from the Editor, *Onderstepoort Journal of Veterinary Research*).

Table 25. *Host records of* Rhipicephalus interventus

Hosts	Number of records
Domestic animals	
Cattle	25
Sheep	1
Dogs	2
Wild animals	
Topi (*Damaliscus lunatus topi*)	1
Grant's gazelle (*Gazella granti*)	1
Oribi (*Ourebia ourebi*)	1
Cape grysbok (*Raphicerus melanotis*)	1
Bushbuck (*Tragelaphus scriptus*)	1
'Antelope'	1

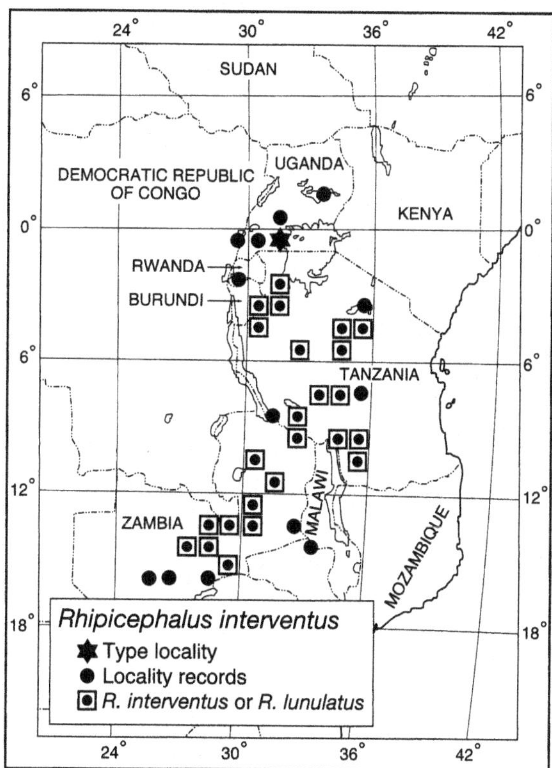

Map 28. *Rhipicephalus interventus:* distribution. (Based on Walker *et al.*, 1988, 1995).

extended into a long narrow cusp; accessory adanal plates long, narrow, well sclerotized.

Female (Figs 94(b), 95(d) to (f))

Capitulum broader than long, length × breadth ranging from 0.55 mm × 0.63 mm to 0.71 mm × 0.74 mm. Basis capituli with long sharp lateral angles just anterior to mid-length; porose areas large, more-or-less oval, about 1.5 times their own diameter apart. Palps with pedicels of articles I fairly long and narrow, giving them a slightly stalked appearance, palpal apices broadly rounded. Scutum length × breadth ranging from 1.25 mm × 1.22 mm to 1.43 mm × 1.46 mm, posterior margin broadly rounded. Eyes at widest point, flat to very slightly raised, edged dorsally with a few medium-sized setiferous punctations. Cervical pits comma-shaped, convergent; cervical fields long, broad, slightly depressed, their outer margins delimited by a few medium-sized setiferous punctations. Similar punctations scattered sparsely on the scapulae, interspersed with numerous very fine punctations. Ventrally genital aperture broadly U-shaped with smooth hyaline flaps visible laterally.

Immature stages

Unknown.

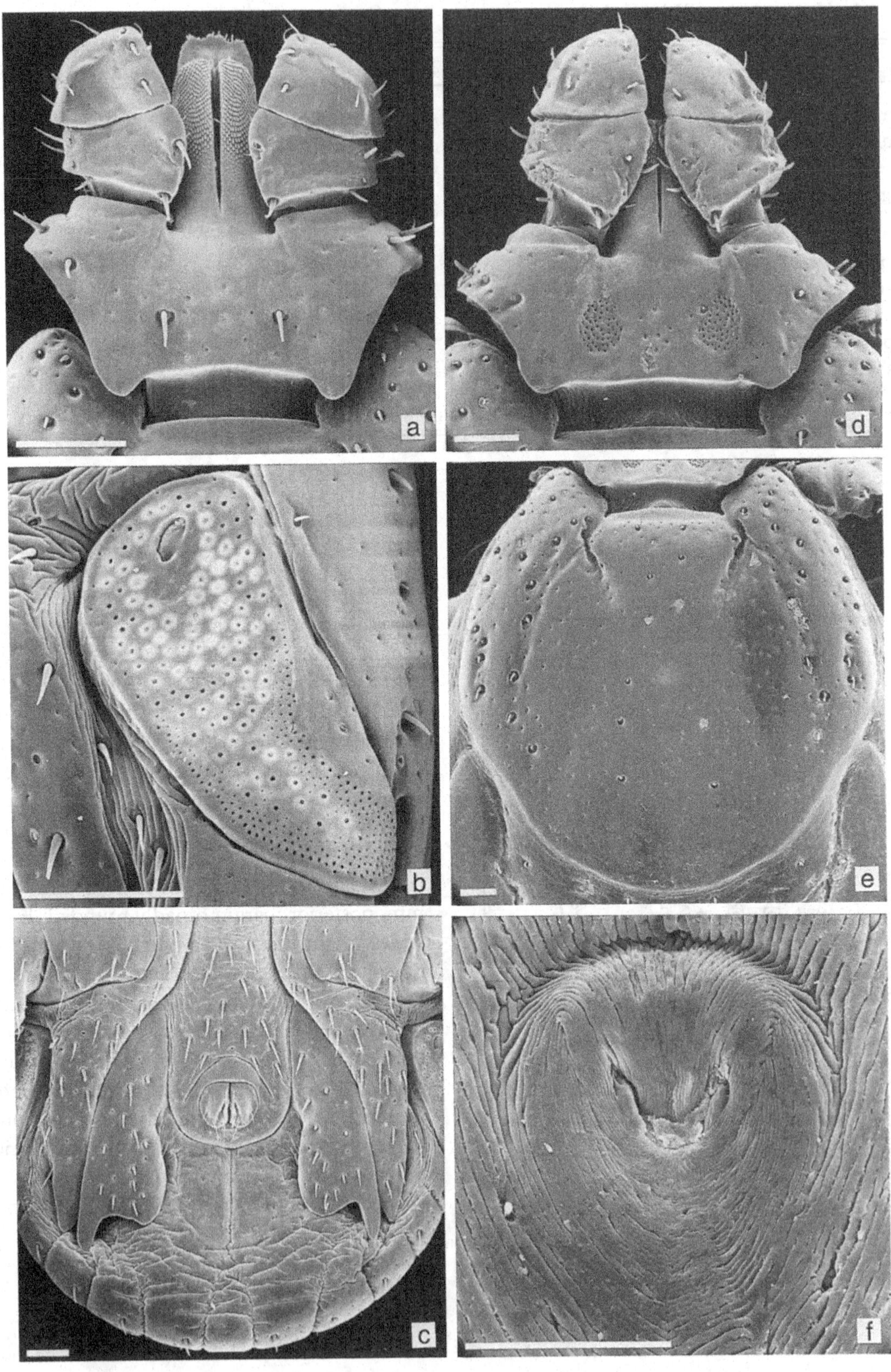
a
d
b
e
c
f

Notes on identification

The male of *R. interventus* resembles that of *R. tricuspis* in the general shape of its capitulum but its adanal plates, with their long posteroexternal cusps and broadly rounded posterointernal margins, are more like those of *R. lunulatus*. Its spiracular plates, with their relatively short, broad dorsal prolongations, differ in shape from those of both *R. tricuspis* and *R. lunulatus* with their longer narrower dorsal prolongations.

Although the female of *R. interventus* and that of *R. tricuspis* have similar capitula and genital apertures the small interstitial punctations on the scutum of *R. interventus* are generally sparser than they are on *R. tricuspis*. The *R. interventus* female can easily be distinguished from that of *R. lunulatus* because the latter has longer narrower palps and a broader shallower genital aperture.

It has been impossible to re-examine all of the collections featuring in the literature as *R. tricuspis sensu lato*, some of which probably no longer exist (Walker *et al.*, 1988, 1995). We have recorded here only collections that we have confirmed as *R. interventus*.

Hosts

Life cycle unknown. Thirty-four collections of *R. interventus* from various hosts exist at present (Table 25), plus one taken from pasture. The majority are from cattle, and most of the remainder are from various medium-sized to small antelopes. It is apparently not an abundant tick: the two largest collections, both from an unspecified number of cattle, contain only 11 adults. Many collections consist of a single adult only.

Figure 95 (*opposite*). *Rhipicephalus interventus* (RML 117418, AHRC 67/3058, from cattle, W. Mako, Masaka, Masaka District, Uganda, collected on 12 May 1967 by J.G. Matthysse). Male: (a) capitulum, dorsal; (b) spiracle; (c) adanal plates. Female: (d) capitulum, dorsal; (e) scutum; (f) genital aperture. Scale bars represent 0.10 mm. SEMs by R.G. Robbins. (Figs (a) & (c) reprinted from Matthyse & Colbo (1987). *Ixodid Ticks of Uganda*, plate 141, figs 2 & 3 as *R. tricuspis*, published by the Entomological Society of America; figs (b), (d), (e) & (f) reprinted from Walker *et al.*, 1995, figs 2b, d, e & f, with kind permission from the Editor, *Onderstepoort Journal of Veterinary Research*).

Zoogeography

Rhipicephalus interventus has been recorded from scattered localities in East and Central Africa, from Uganda southwards to Malawi and Zambia (Map 28). It is sometimes sympatric with *R. lunulatus*. Present indications are that it occurs mainly in various types of woodland (miombo), wooded grassland and sometimes grassland. In Uganda, where it was referred to by comparison with *R. lunulatus* as 'the smaller upland tick of Masaka, Ankole and Kigezi districts' (Matthysse & Colbo, 1987), it has been collected most commonly in dry wooded grassland dominated by *Acacia*.

Disease relationships

Unknown.

REFERENCES

Walker, J.B., Keirans, J.E., Pegram, R.G. & Clifford, C.M. (1988). Clarification of the status of *Rhipicephalus tricuspis* Dönitz, 1906 and *Rhipicephalus lunulatus* Neumann, 1907 (Ixodoidea, Ixodidae). *Systematic Parasitology*, **12**, 159–86.

Walker, J.B., Pegram, R.G. & Keirans, J.E. (1995). *Rhipicephalus interventus* sp. nov. (Acari: Ixodidae), a new tick species closely related to *Rhipicephalus tricuspis* Dönitz, 1906 and *Rhipicephalus lunulatus* Neumann, 1907, from East and Central Africa. *Onderstepoort Journal of Veterinary Research*, **62**, 89–95.

Also see the following Basic Reference (p. 14): Matthysse & Colbo (1987).

RHIPICEPHALUS JEANNELI NEUMANN, 1913

This species was named in honour of the collector, the distinguished French entomologist R. Jeannel (1879–1965). The types were obtained during an expedition that he made to East Africa with Ch. Alluaud in 1911–12.

Diagnosis

A medium-sized reddish-brown tick.

Male (Figs 96(a), 97(a) to (c))
Capitulum longer than broad, length × breadth from 0.71 mm × 0.67 mm to 0.83 mm × 0.79 mm. Basis capituli with short, somewhat recurved lateral angles in anterior half of its length. Palps broad, flattened apically. Conscutum length × breadth ranging from 2.98 mm × 2.13 mm to 3.78 mm × 2.61 mm; anterior process of coxae I small. In engorged males body wall expanded slightly laterally and posterolaterally and forming a short broadly-rounded caudal process posteromedially. Eyes slightly convex, edged dorsally by a few setiferous punctations set in slight depressions. Cervical pits deep. Marginal lines short, shallow, continued anteriorly by lines of medium-sized setiferous punctations. Posteromedian and posterolateral grooves, when present, small and indistinct. A few medium-sized setiferous punctations present anteriorly on the scapulae, along the external cervical margins and in a '*simus*' pattern medially on the conscutum and anterior to the festoons, interspersed with numerous fine punctations. Ventrally spiracles broad, narrowing rather abruptly and curving slightly towards the dorsal surface. Adanal plates large, broad, their external margins almost straight, their internal margins scooped out adjacent to the anus, then meeting the broadly-rounded posterior margins in sharp points; accessory adanal plates large, pointed, well sclerotized.

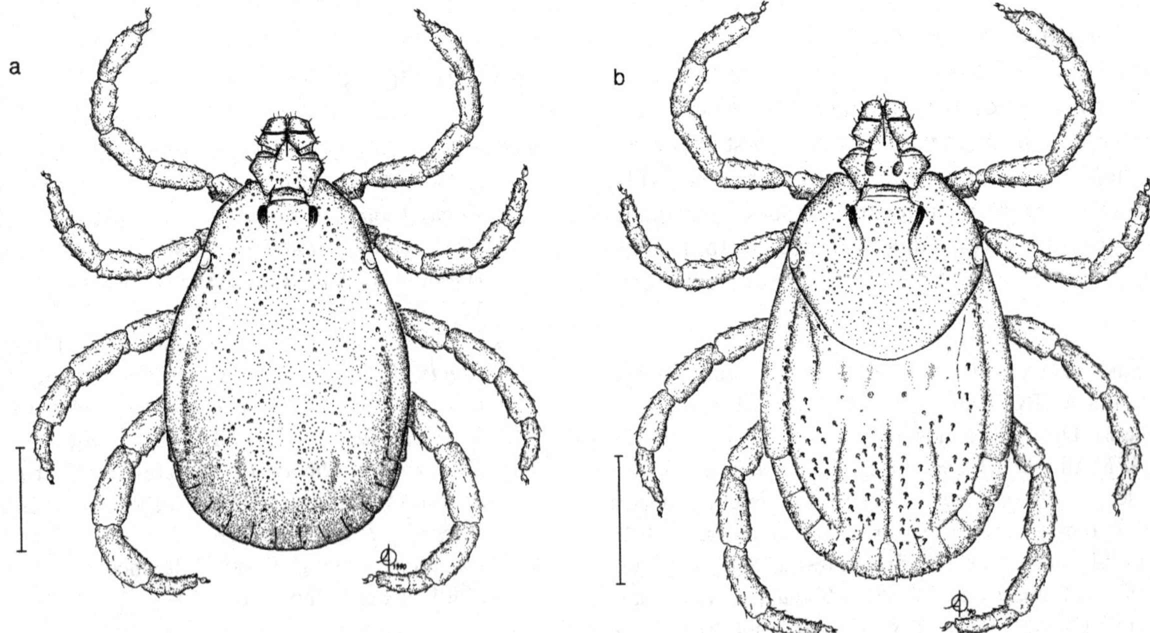

Figure 96. *Rhipicephalus jeanneli* (B.S. 556/-, RML 66309, laboratory reared, original ♀ collected from bovine near Nanyuki, Kenya, in October 1954 by J.A. Thorburn). (a) Male, dorsal; (b) female, dorsal. Scale bars represent 1 mm. A. Olwage *del.*

Table 26. *Host records of* Rhipicephalus jeanneli

Hosts	Number of records
Domestic animals	
Cattle	77
Sheep	1
Goats	1
Horses	1
Pigs	1
Dogs	1
Wild animals	
Lion (*Panthera leo*)	1
Burchell's zebra (*Equus burchellii*)	1
Black rhinoceros (*Diceros bicornis*)	4
Warthog (*Phacochoerus africanus*)	10
Forest hog (*Hylochoerus meinertzhageni*)	2
Bushpig (*Potamochoerus larvatus*)	6
'Wild pig'	1
African buffalo (*Syncerus caffer*)	20
Eland (*Taurotragus oryx*)	4
Bongo (*Tragelaphus eurycerus*)	2
Bushbuck (*Tragelaphus scriptus*)	5
Greater kudu (*Tragelaphus strepsiceros*)	1
Common duiker (*Sylvicapra grimmia*)	1
'Duiker'	1
Brush-furred rat (*Lophuromys flavopunctatus*)	1 (larva)
Swamp rat (*Otomys* sp.)	1 (nymph)
East African mole-rat (*Tachyoryctes splendens*)	1 (larva)
Birds	
'Spurfowl'	1
Humans	2

Female (Figs 96(b), 97(d) to (f))

Capitulum broader than long, length × breadth ranging from 0.71 mm × 0.78 mm to 0.92 mm × 0.95 mm. Basis capituli with acute lateral angles at about mid-length; porose areas round to oval, about 1.5 times their own diameter apart. Palps broad, their apices somewhat flattened. Scutum broader than long, length × breadth ranging from 1.37 mm × 1.55 mm to 1.87 mm × 2.00 mm; posterior margin slightly sinuous. Eyes slightly convex, edged dorsally by a few setiferous punctations set in shallow depressions. Cervical pits long, convergent; cervical fields broad, slightly depressed, their external margins delimited by irregular rows of medium-sized setiferous punctations. A few similar punctations scattered anteriorly on the scapulae and between the cervical fields, interspersed with numerous very fine punctations. Ventrally genital aperture small, broadly V-shaped.

Nymph (Fig. 98)

Capitulum much broader than long, length × breadth ranging from 0.26 mm × 0.37 mm to 0.32 mm × 0.43 mm. Basis capituli over three times as broad as long, with long tapering

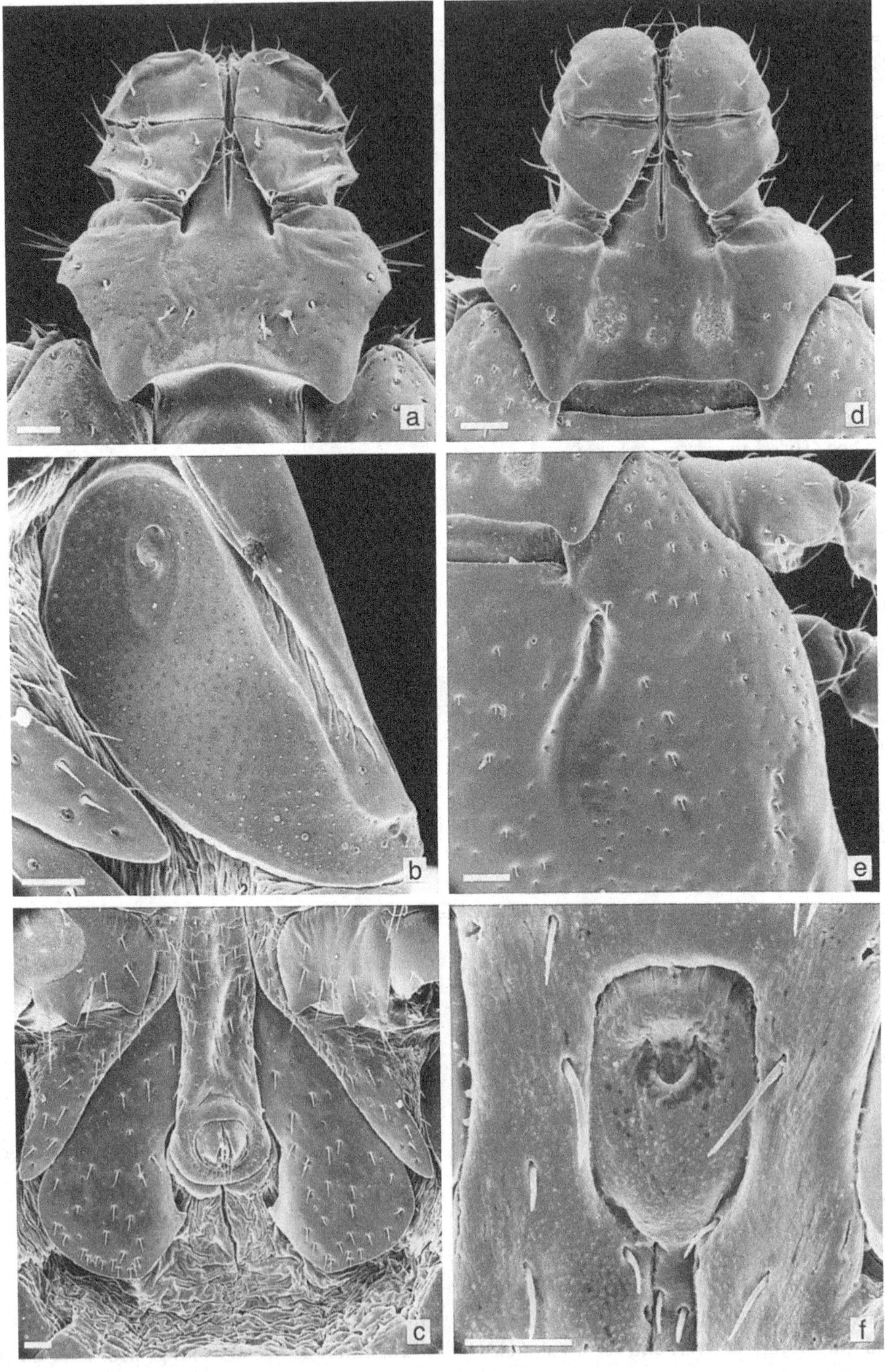
a
d
b
e
c
f

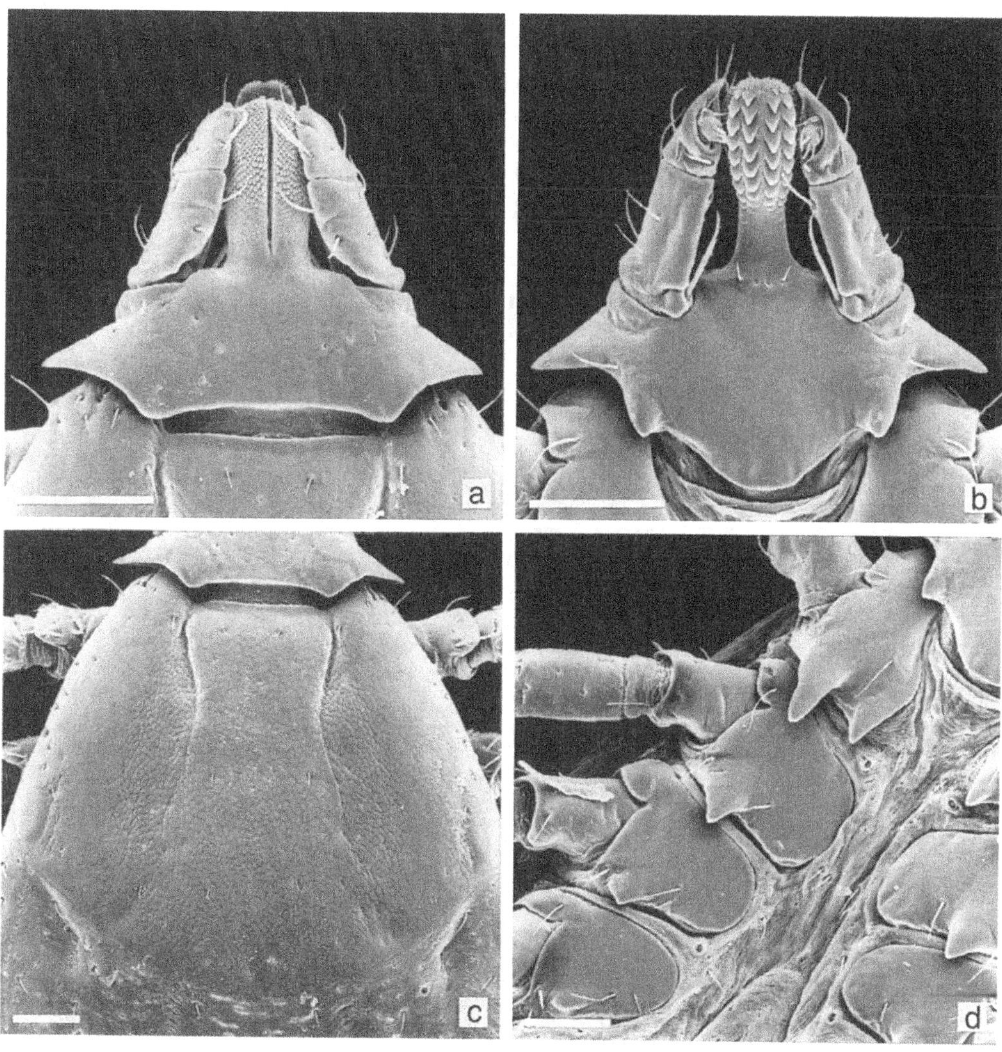

Figure 98 (*above*). *Rhipicephalus jeanneli* (B.S. 556/-, RML 66309, laboratory reared, original ♀ collected from bovine near Nanyuki, Kenya, in October 1954 by J.A. Thorburn). Nymph: (a) capitulum, dorsal; (b) capitulum, ventral; (c) scutum; (d) coxae. Scale bars represent 0.10 mm. SEMs by M.D. Corwin.

Figure 97 (*opposite*). *Rhipicephalus jeanneli* (B.S. 556/-, RML 66309, laboratory reared, original ♀ collected from bovine near Nanyuki, Kenya, in October 1954 by J.A. Thorburn). Male: (a) capitulum, dorsal; (b) spiracle; (c) adanal plates. Female: (d) capitulum, dorsal; (e) scapular area; (f) genital aperture. Scale bars represent 0.10 mm. SEMs by J.F. Putterill.

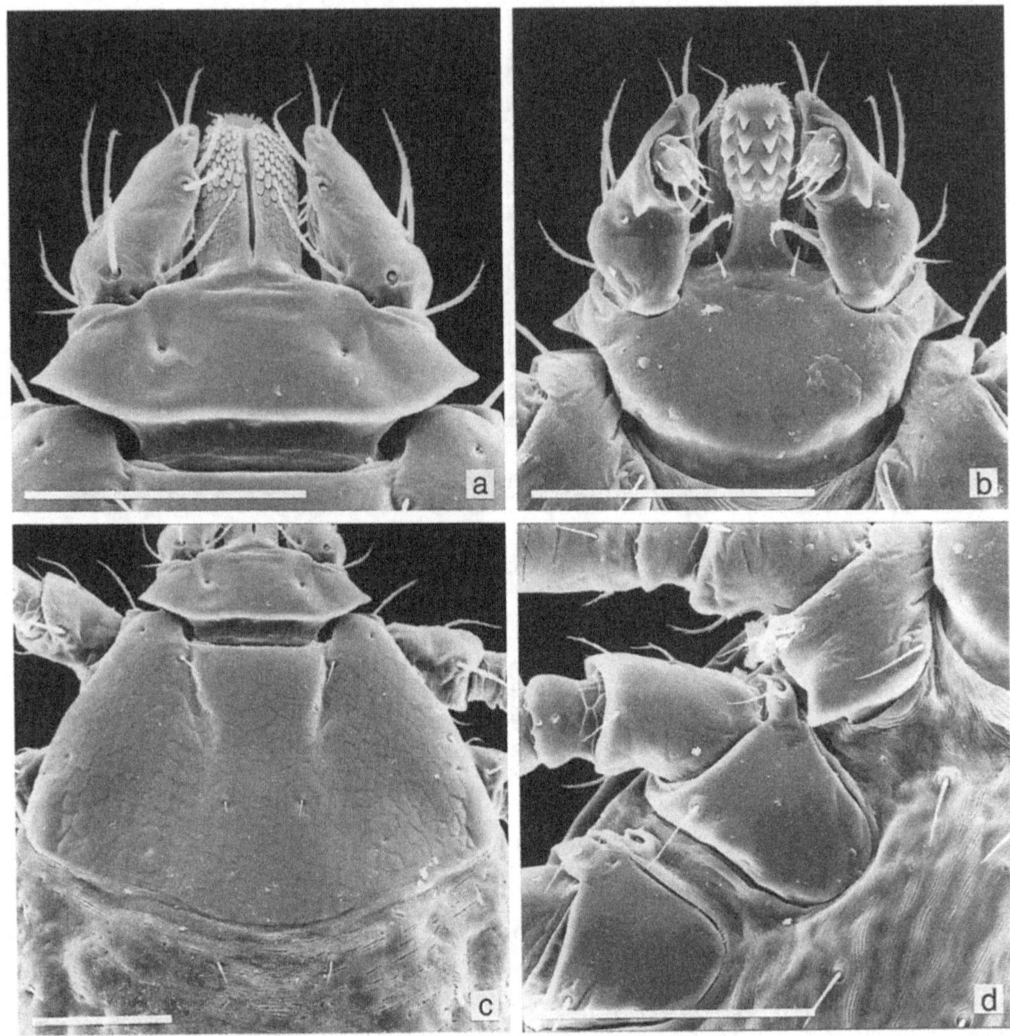

Figure 99. *Rhipicephalus jeanneli* (B.S. 556/-, RML 66309, laboratory reared, original ♀ collected from bovine near Nanyuki, Kenya, in October 1954 by J.A. Thorburn). Larva: (a) capitulum, dorsal; (b) capitulum, ventral; (c) scutum; (d) coxae. Scale bars represent 0.10 mm. SEMs by M.D. Corwin.

sharply-pointed lateral angles just posterior to mid-length and small broadly-rounded cornua; ventrally with small narrowly-rounded spurs on the posterior margin. Palps narrow, more-or-less equal in width for much of their length, tapering distally to rounded apices, inclined inwards. Scutum broader than long, length × breadth ranging from 0.58 mm × 0.69 mm to 0.65 mm × 0.74 mm; posterior margin a broad smooth curve. Eyes at widest point, well over halfway back, delimited dorsally by slight depressions. Cervical fields long, narrow, divergent, their internal margins more sharply defined than their external margins. Ventrally coxae I each with a long sharp external spur and a shorter sharp internal spur; coxae II to IV each with a sharp external spur only, decreasing progressively in size.

Larva (Fig. 99)

Capitulum much broader than long, length × breadth ranging from 0.136 mm × 0.172 mm to 0.139 mm × 0.187 mm. Basis capituli over three times as broad as long, with short sharply-

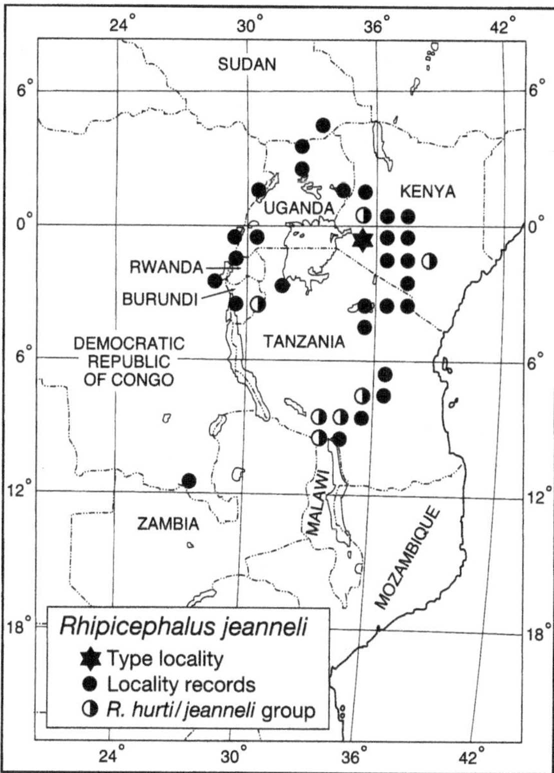

Map 29. *Rhipicephalus jeanneli*: distribution.

pointed lateral angles, posterior margin very slightly concave. Palps constricted proximally, almost immediately widening, then tapering gradually to narrowly-rounded apices, inclined inwards. Scutum much broader than long, length × breadth ranging from 0.257 mm × 0.360 mm to 0.279 mm × 0.376 mm; posterior margin broad, shallow, slightly sinuous. Eyes at widest point, far back, very slightly convex. Cervical grooves short, slightly convergent. Ventrally coxae I each with a sharp spur; coxae II and III each with a slight thickening only on its posterior border.

Notes on identification

In the past the taxonomy of *R. jeanneli* has been confused. The main problem arose when Zumpt (1943) noted that he was linking this species provisionally with *R. kochi* until such time as it could be shown that both species occurred at Sadani, one of the latter's type localities on the Tanzanian coast. In 1949 he finally synonymized these two species, although he gave no evidence to support his decision. Unfortunately his verdict was initially accepted uncritically by other tick workers and for about 17 years thereafter *R. jeanneli* was identified as *R. kochi*. In 1962, however, Clifford & Anastos expressed doubts regarding this synonymy. For one thing they pointed out that the female genital apertures of ticks then being identified as *R. kochi* were not the same as that of the type female of this species as illustrated by Feldman-Muhsam (1956). They therefore thought that the matter should be re-examined. In 1967, Yeoman & Walker formally resurrected *R. jeanneli* as a valid species because they considered it was very unlikely that this highland species would also occur in the coastal areas where the types of *R. kochi* had been collected. In addition they noted that the shape of both the female scutum and the genital aperture differ in these two species.

A further difficulty is that, although typical adults of *R. jeanneli* and *R. hurti* can be identified reasonably easily, field collections often include specimens that in some respects are morphologically intermediate between them. This problem requires further study (see p. 226, under *R. hurti*).

We have omitted the two records from Equateur and Bandundu (formerly Léopoldville) Provinces, Democratic Republic of Congo, quoted by Elbl & Anastos (1966). Not only is each based on a single male specimen but these areas are far outside the normal habitat of this tick and are ecologically unsuitable for it.

Yeoman & Walker (1967) believed that ticks in the *R. hurti/jeanneli* group have sometimes been misidentified as *R. appendiculatus*. This may have led to incorrect assessments regarding the requirements for East Coast fever control in parts of Tanzania.

Hosts

A three-host species (J.B.W., unpublished data, 1955). The only domestic animals of any consequence as hosts of *R. jeanneli* adults are cattle

(Table 26). Their most commonly recorded wild host is the African buffalo. The wild suids are also favoured. Adults have a predilection for the ears and tail brush of cattle and buffaloes. On buffaloes they have also been found round the eyes and on the neck, chest, shoulder, back, axillae, genitalia and legs. On a goat they attached under the tail.

Only three collections of immatures have been recorded, two larvae and a nymph, all from rodents.

Zoogeography

Rhipicephalus jeanneli was originally described from four collections, two from Kenya, one at Molo and the other from the Burguret River, Mt Kenya, and two from Tanzania, both from Kilimanjaro. Of these only the first is shown as the type locality (Map 29).

This species occurs in highland areas in eastern Africa, often in association with *R. hurti*. It has been recorded from the southernmost part of Sudan southwards to southern Tanzania and westwards through Rwanda and Burundi to adjacent areas in the Democratic Republic of Congo. It is common at higher altitudes in Rwanda (Schoenaers, 1951a,b, as *R. kochi*).

There is one isolated record from Lubumbashi in southern Democratic Republic of Congo; this requires confirmation as it is an ecologically atypical area for this species.

Rhipicephalus jeanneli usually occurs at altitudes between 1500 m and 2500 m, sometimes even higher, with mean annual rainfalls ranging from at least 500 mm to about 1500 mm. Characteristically these are areas with Afromontane vegetation, comprising various types of forest, woodland, bushland, wooded and/or bushed grassland and grassland.

Disease relationships

According to Wilson (1953) *R. jeanneli* will transmit *Theileria parva*, the causative agent of East Coast fever of cattle. This observation is thought to be based on experiments with laboratory-bred ticks described in the unpublished report by the Chief Field Zoologist, Kenya, for 1951. So far as we know, though, the immature stages of this species have never been collected from cattle under natural conditions. Consequently we do not regard it as a field vector of this disease.

REFERENCES

Feldman-Muhsam, B. (1956).The value of the female genital aperture and the peristigmal hairs for specific diagnosis in the genus *Rhipicephalus*. *Bulletin of the Research Council of Israel*, **5B**, 300–6 + 3 plates.

Neumann, L.G. (1913). Ixodidae. In *Voyage de Ch. Alluaud et R. Jeannel en Afrique Orientale (1911–1912). Résultats Scientifiques. Arachnida, II*, 23–35. Paris: A. Schulz.

Schoenaers, F. (1951a). Liste des tiques récoltées au cours d'un voyage d'études au Congo Belge. *Bulletin Agricole du Congo Belge*, **42**, 117–22.

Schoenaers, F. (1951b). Essai sur la répartition de la theileriose bovine et des tiques vectrices, au Ruanda-Urundi, en fonction de l'altitude. *Annales de la Société Belge de Médecine Tropicale*, **31**, 371–5.

Wilson, S.G. (1953). A survey of the distribution of tick vectors of East Coast fever in East and Central Africa. *Proceedings of the 15th International Veterinary Congress, Stockholm*, **1**, 187–90.

Zumpt, F. (1943). *Rhipicephalus aurantiacus* Neumann und ähnliche Arten. VIII. Vorstudie zu einer Revision der Gattung *Rhipicephalus* Koch. *Zeitschrift für Parasitenkunde*, **13**, 102–17.

Also see the following Basic References (pp. 12–14): Clifford & Anastos (1962); Elbl & Anastos (1966); Matthysse & Colbo (1987); Walker (1974); Yeoman & Walker (1967); Zumpt (1949).

RHIPICEPHALUS KOCHI DÖNITZ, 1905

This species was named in honour of Robert Koch (1843–1910), the great German bacteriologist. He began his career as a country doctor and carried out his original studies on the aetiology of anthrax at home. Later he made many other notable contributions to bacteriology. He ultimately became Director of the Institute for Infectious Diseases, which was founded for him in Berlin.

Synonym

neavei; piresi.

Diagnosis

A medium-sized brown to reddish-brown tick.

Male (Figs 100(a), 101(a) to (c))
Capitulum broader than long, length × breadth ranging from 0.50 mm × 0.55 mm to 0.70 mm × 0.79 mm. Basis capituli with acute lateral angles in anterior third of its length. Palps short, broad. Conscutum length × breadth ranging from 2.00 mm × 1.20 mm to 3.15 mm × 2.10 mm; anterior process on coxae I large. In engorged specimens body wall expanded posterolaterally and forming a single caudal process posteromedially. Eyes marginal, nearly flat, edged with a few medium-sized punctations dorsally. Cervical pits deep, convergent; cervical fields narrow, tapering posteriorly to just behind eyes, shallow. Marginal lines well developed, extending forwards almost to eye level. Posteromedian and posterolateral grooves well developed. Medium-sized setiferous punctations present on the scapulae, along the outer margins of the cervical fields and scattered medially on the conscutum, where they are interspersed with numerous fine punctations; lateral areas immediately anterior and mesial to the eyes almost devoid of punctations, smooth and shiny, and those adjacent to the marginal lines with only a few very fine punctations. Ventrally spiracles elongate, broadly comma-shaped with a short

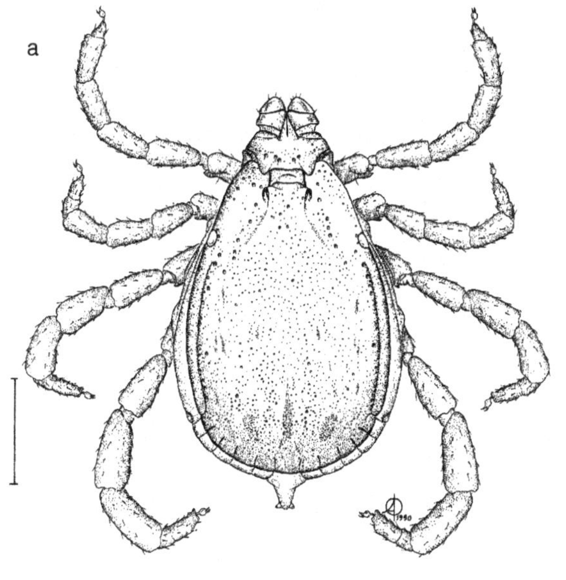

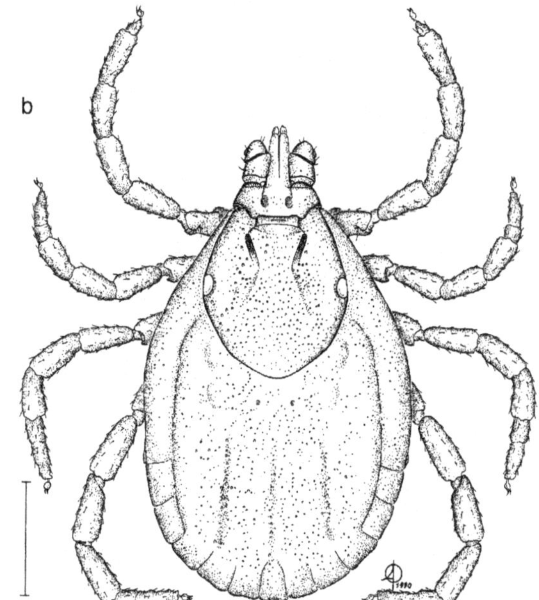

Figure 100. *Rhipicephalus kochi* [Collection No. TA 1, from nyala (*Tragelaphus angasii*), Pafuri, Kruger National Park, South Africa, on 6 October 1981 by I.G. Horak]. (a) Male, dorsal; (b) female, dorsal. Scale bars represent 1 mm. A. Olwage *del.*

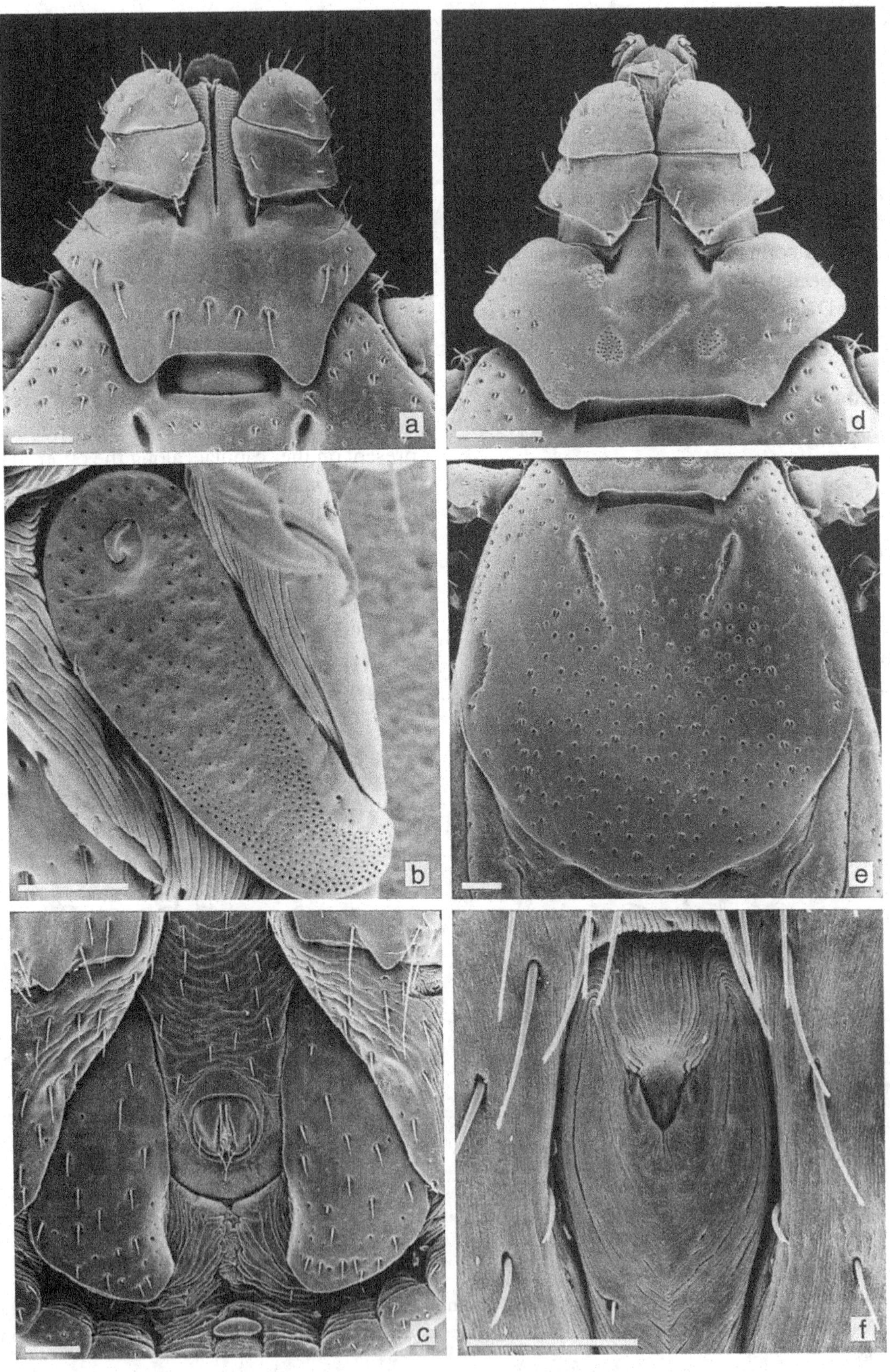
a
d
b
e
c
f

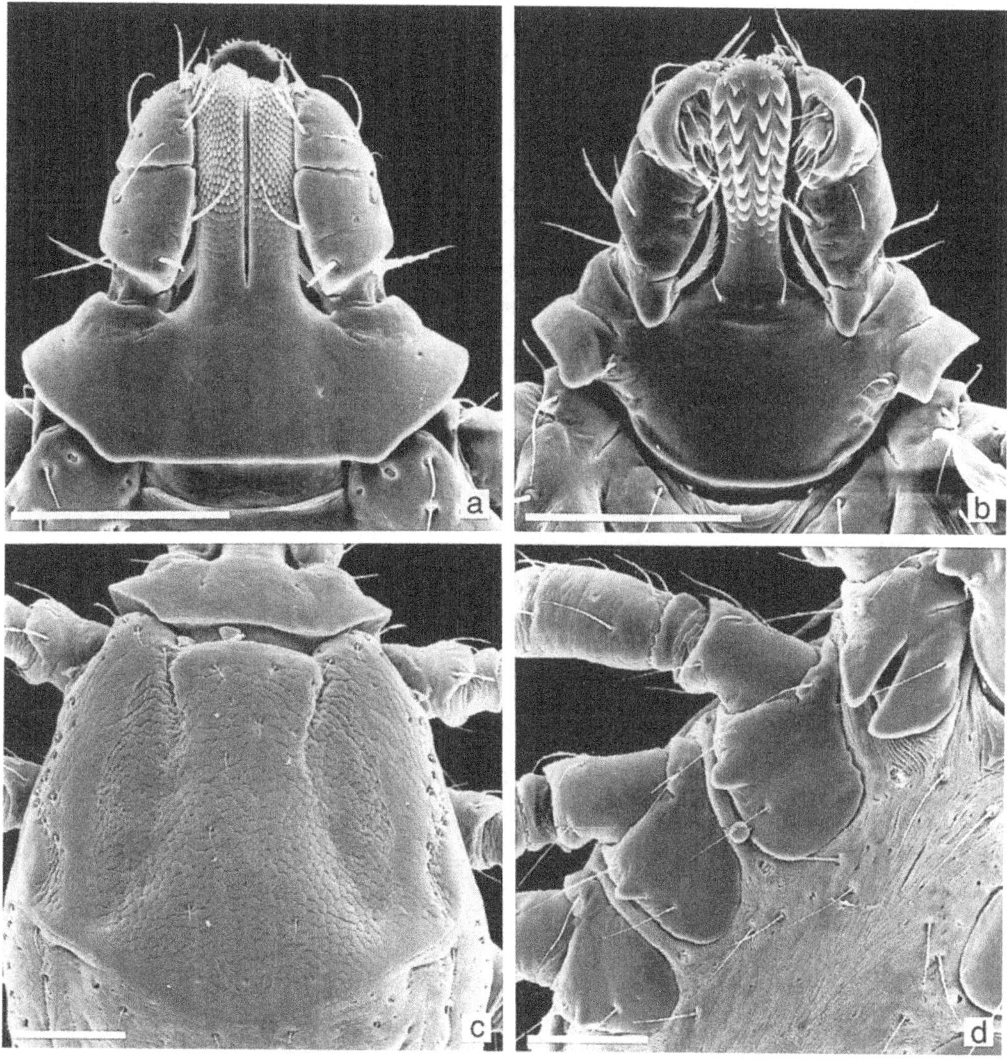

Figure 102 (*above*). *Rhipicephalus kochi* (RML 116144, laboratory reared, original ♀ collected from bovine at Z.A.D.L., Solwezi, Zambia, in 1981 by R.G. Pegram). Nymph: (a) capitulum, dorsal; (b) capitulum, ventral; (c) scutum; (d) coxae. Scale bars represent 0.10 mm. SEMs by M.D. Corwin. (From Clifford *et al.*, 1983, figs 18, 19, 21 & 22, with kind permission from the Editor, *Onderstepoort Journal of Veterinary Research*).

Figure 101 (*opposite*). *Rhipicephalus kochi* (RML 65686, collected from cattle, Tabora, Tanzania, in March 1975 by R. J. Tatchell). Male: (a) capitulum, dorsal; (b) spiracle; (c) adanal plates. Female: (d) capitulum, dorsal; (e) scutum; (f) genital aperture. Scale bars represent 0.10 mm. SEMs by M. D. Corwin. (From Clifford *et al.*, 1983, figs 4, 7, 10, 12, 16 & 17, with kind permission from the Editor, *Onderstepoort Journal of Veterinary Research*).

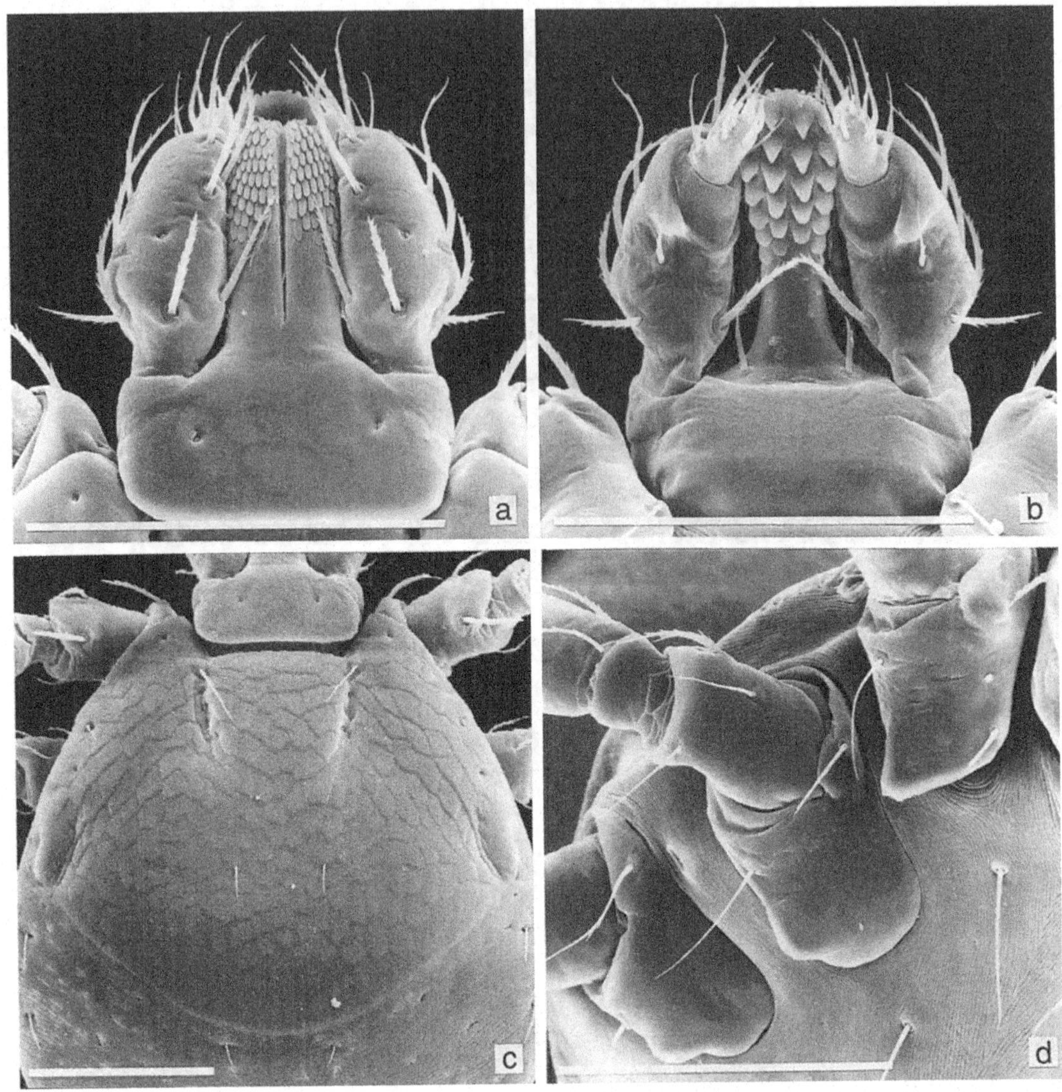

Figure 103. *Rhipicephalus kochi* (RML 116144, laboratory reared, original ♀ collected from bovine at Z.A.D.L., Solwezi, Zambia, in 1981 by R.G. Pegram). Larva: (a) capitulum, dorsal; (b) capitulum, ventral; (c) scutum; (d) coxae. Scale bars represent 0.10 mm. SEMs by M.D. Corwin. (From Clifford *et al.*, 1983, figs 24, 25, 27 & 28, with kind permission from the Editor, *Onderstepoort Journal of Veterinary Research*).

prolongation curving towards the dorsal surface. Adanal plates broad, with internal margins hollowed out posterior to anus, posterior margins as smooth shallow curves; accessory adanal plates represented by small sclerotized points.

Female (Figs 100(b), 101(d) to (f))

Capitulum broader than long, length × breadth ranging from 0.68 mm × 0.78 mm to 0.83 mm × 0.93 mm. Basis capituli with broad lateral angles a little anterior to mid-length; porose areas small, almost round, well over twice their own diameter apart. Palps with article I easily visible dorsally, broad and smoothly rounded apically. Scutum longer than broad, length × breadth ranging from 1.33 mm × 1.25 mm to 1.63 mm × 1.45 mm, posterior margin sinuous. Eyes about halfway back, flat, edged

Table 27. *Host records of* Rhipicephalus kochi

Hosts	Number of records
Domestic animals	
Cattle	199
Sheep	2
Goats	9
Camels	1
Horses	1
Donkeys	1
Dogs	1
Wild animals	
Lion (*Panthera leo*)	1
Leopard (*Panthera pardus*)	1
African civet (*Civettictis civetta*)	1 (nymph)
African elephant (*Loxodonta africana*)	1
Burchell's zebra (*Equus burchellii*)	3
Black rhinoceros (*Diceros bicornis*)	1
Warthog (*Phacochoerus africanus*)	9
Bushpig (*Potamochoerus larvatus*)	27
Giraffe (*Giraffa camelopardalis*)	2
Impala (*Aepyceros melampus*)	24 (including immatures)
Blue wildebeest (*Connochaetes taurinus*)	1
Lichtenstein's hartebeest (*Sigmoceros lichtensteinii*)	3
Kirk's dik-dik (*Madoqua kirkii*)	2
Suni (*Neotragus moschatus*)	3
Klipspringer (*Oreotragus oreotragus*)	6
Oribi (*Ourebia ourebi*)	1
Steenbok (*Raphicerus campestris*)	1
Cape grysbok (*Raphicerus melanotis*)	3
African buffalo (*Syncerus caffer*)	17
Eland (*Taurotragus oryx*)	6
Nyala (*Tragelaphus angasii*)	8 (including nymphs)
Bushbuck (*Tragelaphus scriptus*)	17 (including nymphs)
Greater kudu (*Tragelaphus strepsiceros*)	21 (including nymphs)
'Kudu'	3
Red forest duiker (*Cephalophus natalensis*)	3
Yellow-backed duiker (*Cephalophus silvicultor*)	1
Common duiker (*Sylvicapra grimmia*)	7
Roan antelope (*Hippotragus equinus*)	6
Sable antelope (*Hippotragus niger*)	12
Waterbuck (*Kobus ellipsiprymnus*)	1
Reedbuck (*Redunca arundinum*)	1
South African porcupine (*Hystrix africaeaustralis*)	1
Scrub hare (*Lepus saxatilis*)	20 (including immatures)
Savanna hare (*Lepus victoriae*)	7
Four-toed elephant shrew (*Petrodromus tetradactylus*)	2 (including immatures)
Birds	
Black-bellied bustard (*Eupodotis melanogaster*)	1

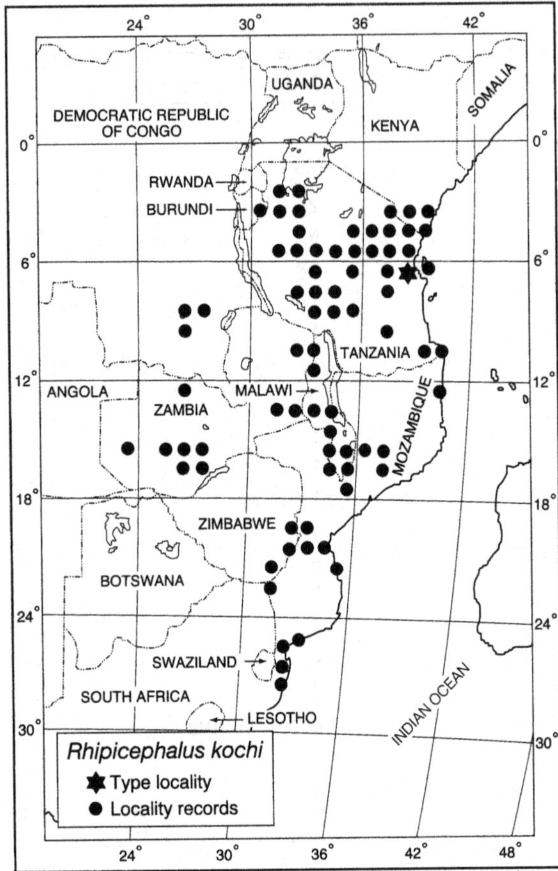

Map 30. *Rhipicephalus kochi*: distribution. (Based largely on Clifford, Walker & Keirans, 1983, fig. 35).

dorsally by shallow punctate grooves. Cervical pits convergent; cervical fields narrow, tapering posteriorly to well beyond eye level, shallow. Medium-sized setiferous punctations present on the scapulae and along the outer margins of the cervical fields; medially the scutum evenly covered with slightly smaller punctations; lateral areas anterior and mesial to the eyes smooth and shiny. Ventrally genital aperture V-shaped; area anterior to opening bulging.

Nymph (Fig. 102)

Capitulum much broader than long, length × breadth ranging from 0.23 mm × 0.29 mm to 0.24 mm × 0.30 mm. Basis capituli nearly three times as broad as long, lateral angles broad, overlapping scapulae; ventrally lateral angles unique in shape, rectangular. Palps narrow proximally, then widening; article III wedge-shaped. Scutum about as broad as long, length × breadth ranging from 0.48 mm × 0.48 mm to 0.50 mm × 0.51 mm. Eyes at widest point, well over halfway back, oval, slightly convex, edged dorsally by shallow furrows. Cervical fields long, with both their inner and outer margins gently convex, slightly depressed. Ventrally coxae I each deeply cleft into a relatively narrow external spur and a broader, more rounded internal spur; coxae II to IV each with an external spur, decreasing progressively in size.

Larva (Fig. 103)

Capitulum slightly longer than broad, length × breadth ranging from 0.112 mm × 0.108 mm to 0.116 mm × 0.112 mm. Basis capituli over twice as broad as long, lateral margins nearly straight, curving gently posteriorly to meet the straight posterior margin. Palps constricted proximally, then widening markedly before tapering to rounded apices. Scutum much broader than long, length × breadth ranging from 0.250 mm × 0.320 mm to 0.264 mm × 0.340 mm. Eyes at widest point, over halfway back, oval, slightly convex and edged dorsally by shallow grooves. Cervical grooves short, slightly convergent. Ventrally coxae I each with a large triangular spur; coxae II and III each with smaller rounded spurs.

Notes on identification

The status of *R. kochi*, including the confusion that surrounded the use of this name for many years, was reviewed in detail by Clifford, Walker & Keirans (1983).

This species was originally described by Dönitz from 1 ♂, 5 ♀♀ taken from cattle at Sadani and 3 ♀♀, also from cattle, at Lindi, on the Tanzanian coast. Of these specimens only the syntype female from Sadani (Zoological Museum, Berlin 8490) and the male from this batch (Nuttall Collection 2109) still exist.During his revision of the genus *Rhipicephalus* Zumpt (1943) saw the syntype female of *R. kochi*, which apparently puzzled him, and he linked it with the highland species *R. jeanneli* (see pp. 232–238). He suggested that, if *R. jeanneli* were later to be found at Sadani, these two species should be synonymized. In 1949, without presenting any further evidence or reasons for his decision, he finally did synonymize these two species. Consequently the name *R. kochi* was used for the highland tick by many authors for some 17 years thereafter.

Earlier Theiler (1947), who treated *R. kochi* and *R. jeanneli* as separate species, had already remarked on the many similarities between *R. kochi* and *R. neavei*. Unfortunately her observations, which subsequently proved to be correct, were overlooked. In 1956 Walker confused the situation further by synonymizing both *R. neavei* and *R. neavei punctatus* with *R. pravus*, a decision that was reversed by Yeoman & Walker (1967). At the same time these authors resurrected *R. jeanneli* as the valid name for the highland species.

Meanwhile studies by Feldman-Muhsam (1956), followed by Clifford & Anastos (1962), led Matthysse & Colbo to compare the types of *R. kochi* and *R. neavei* in the late 1960s. As Theiler had earlier suspected, they found these species to be conspecific but as a result of the turmoil in Uganda, where they had been working, they were able to publish their finding formally only in 1987.

Hosts

A three-host species (R.G. Pegram, unpublished data, 1982). Amongst its domestic hosts *R. kochi* adults feed most commonly on cattle (Clifford *et al.*, 1983) (Table 27). Although adult ticks have been recorded on a large variety of wild hosts most collections have been taken from wild suids, impala, African buffalo, tragelaphine antelopes, sable antelope and hares. They have also been collected from a bird, the black-bellied bustard (Clifford *et al.*, 1983). The immature stages probably infest many of the same species as the adults do: they have been found on impala, tragelaphine antelopes, scrub hares and the four-toed elephant shrew (Clifford *et al.*, 1983; Horak *et al.*, 1983, 1995). In most cases animals harbour only small numbers of adult ticks, but two collections each exceeding 270 ticks have been taken from bushpigs.

Wilson (1950) found that adult *R. kochi* attach most commonly to the udders and flanks of cattle. In Malawi he collected engorged females from September to March when the atmospheric humidity was low. In South Africa larvae were present in peak numbers on impala during May, nymphs during August and adults from March to May. In Zambia and Tanzania the adults were active mainly during the wet season (Pegram *et al.*, 1986; Tatchell & Easton, 1986).

Zoogeography

Rhipicephalus kochi occurs south of the equator in parts of eastern, central and southern Africa (Map 30). Records from Ghana and Uganda almost certainly represent other species and have been omitted. It is widespread in Tanzania, Zambia, Malawi and Mozambique, but has so far been found only in the south-eastern regions of Kenya, in the Democratic Republic of Congo and Zimbabwe and in north-eastern South Africa.

Rhipicephalus kochi is found in areas that range in altitude from a few metres above sea level to about 1800 m, with annual rainfalls between approximately 500 mm and 1300 mm. It is primarily an inhabitant of the East African coastal mosaic as well as miombo, scrub and undifferentiated woodland, and deciduous bushland and thicket.

Disease relationships

Rhipicephalus neavei has been identified as a vector of *Theileria parva*, the causative agent of East Coast fever in cattle (Lewis, Piercy & Wiley, 1946). This finding almost certainly refers to *R. pravus* and not to *R. kochi*. The strain of ticks used in these transmission experiments came from the

Machakos District of Kenya where *R. pravus* is common; *R. kochi* has not been collected there.

REFERENCES

Clifford, C.M., Walker, J.B. & Keirans, J.E. (1983). Clarification of the status of *Rhipicephalus kochi* Dönitz, 1905 (Ixodoidea, Ixodidae). *Onderstepoort Journal of Veterinary Research*, **50**, 77–89.

Dönitz, W. (1905). Die Zecken des Rindes als Krankheitsüberträger. *Sitzungsbericht der Gesellschaft naturforschender Freunde zu Berlin*, **No. 4**, 105–34 + 1 plate.

Feldman-Muhsam, B. (1956). The value of the female genital aperture and the peristigmal hairs for specific diagnosis in the genus *Rhipicephalus*. *Bulletin of the Research Council of Israel*, **5B**, 300–6 + 3 plates.

Horak, I.G., Potgieter, F.T., Walker, J.B., De Vos, V. & Boomker, J. (1983). The ixodid tick burdens of various large ruminant species in South African nature reserves. *Onderstepoort Journal of Veterinary Research*, **50**, 221–8.

Horak, I.G., Spickett, A.M., Braack, L.E.O., Penzhorn, B.L., Bagnall, R.J. & Uys, A.C. (1995). Parasites of domestic and wild animals in South Africa. XXXIII. Ixodid ticks on scrub hares in the north-eastern regions of Northern and Eastern Transvaal and of KwaZulu-Natal. *Onderstepoort Journal of Veterinary Research*, **62**, 123–31.

Lewis, E.A., Piercy, S.E. & Wiley, A.J. (1946). *Rhipicephalus neavei* Warburton, 1912 as a vector of East Coast fever. *Parasitology*, **37**, 60–4.

Pegram, R.G., Perry, B.D., Musisi, F.L. & Mwanaumo, B. (1986). Biology and phenology of ticks in Zambia: seasonal dynamics on cattle. *Experimental and Applied Acarology*, **2**, 25–45.

Tatchell, R.J. & Easton, E. (1986). Tick (Acari: Ixodidae) ecological studies in Tanzania. *Bulletin of Entomological Research*, **76**, 229–46.

Walker, J.B. (1956). *Rhipicephalus pravus* Dönitz, 1910. *Parasitology*, **46**, 243–60.

Wilson, S.G. (1950). A check-list and host-list of Ixodoidea found in Nyasaland, with descriptions and biological notes on some of the rhipicephalids. *Bulletin of Entomological Research*, **41**, 415–28.

Zumpt, F. (1943). *Rhipicephalus aurantiacus* Neumann und ähnliche Arten. VIII. Vorstudie zu einer Revision der Gattung *Rhipicephalus* Koch. *Zeitschrift für Parasitenkunde*, **13**, 102–17.

Also see the following Basic References (pp. 12–14): Clifford & Anastos (1962); Matthysse & Colbo (1987); Theiler (1947); Yeoman & Walker (1967); Zumpt (1949).

RHIPICEPHALUS LONGICEPS WARBURTON, 1912

The specific name *longiceps*, from the Latin *longus* meaning 'long' plus New Latin *ceps* meaning 'head', doubtless refers to the female capitulum, which Warburton described as being 'remarkably long'.

Diagnosis

A moderate-sized reddish-brown tick.

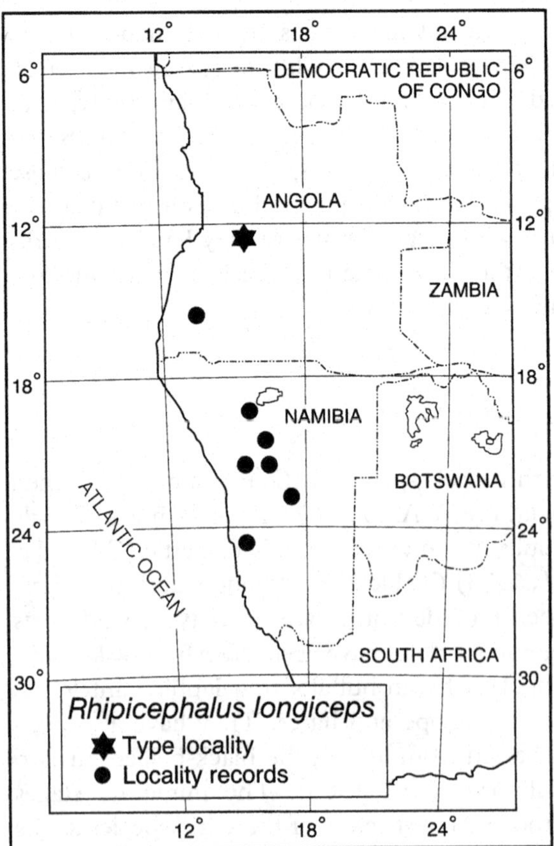

Map 31. *Rhipicephalus longiceps*: distribution.

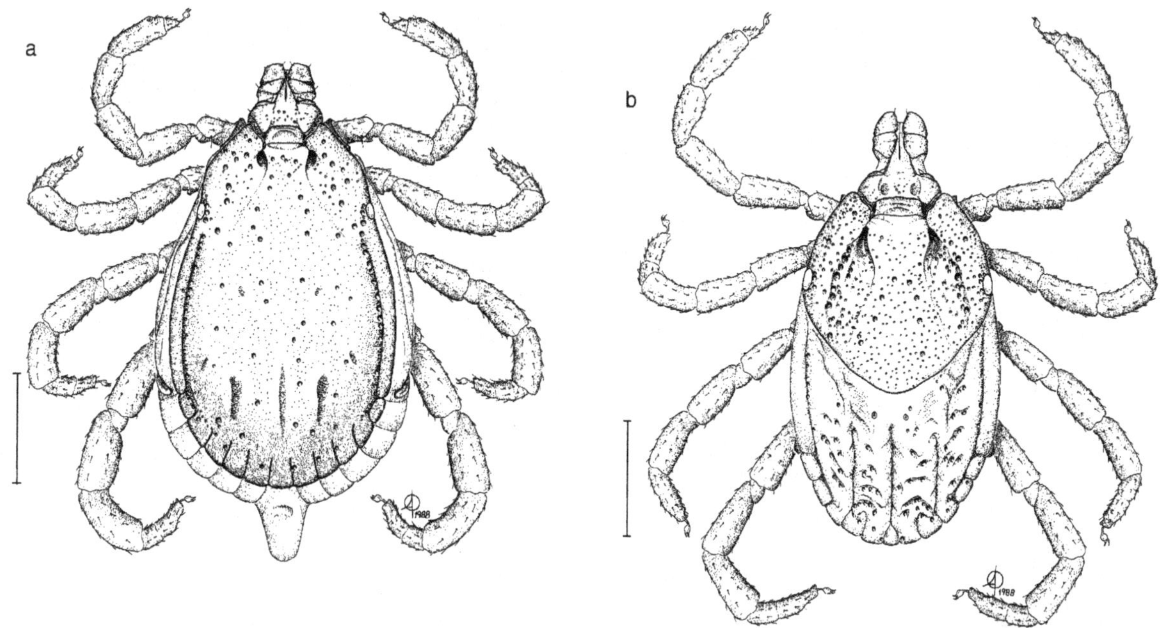

Figure 104. *Rhipicephalus longiceps* (RML 65711, collected on farm 'Lichtenstein Sud', near Windhoek, Namibia, in 1974 by J. Langenhoven). (a) Male, dorsal; (b) female, dorsal. Scale bars represent 1 mm. A. Olwage *del.*

Table 28. *Host records of* Rhipicephalus longiceps

Hosts	Number of records
Domestic animals	
Cattle	2
Pigs	1
Wild animals	
Warthog (*Phacochoerus africanus*)	4
Giraffe (*Giraffa camelopardalis*)	1
Klipspringer (*Oreotragus oreotragus*)	1
Gemsbok (*Oryx gazella*)	2

Male (Figs 104(a), 105(a) to (c))

Capitulum longer than broad, length × breadth ranging from 0.60 mm × 0.53 mm to 0.77 mm × 0.72 mm. Basis capituli with short obtuse lateral angles at about anterior third of its length. Palps somewhat elongated, with articles II and III broad. Conscutum length × breadth ranging from 2.64 mm × 1.61 mm to 3.50 mm × 2.30 mm; a small, sharp anterior process present on coxae I. In engorged specimens the body wall is expanded laterally and posterolaterally, then extends posteriorly as a broad smoothly-tapering caudal process. Eyes slightly convex, edged by a few large punctations dorsally. Cervical pits convergent, continuous with short depressed internal cervical margins. Marginal lines long, almost reaching eyes, deeply punctate. Posteromedian groove long and narrow, posterolateral grooves shorter and broader. Punctation pattern variable. In some specimens large setiferous punctations

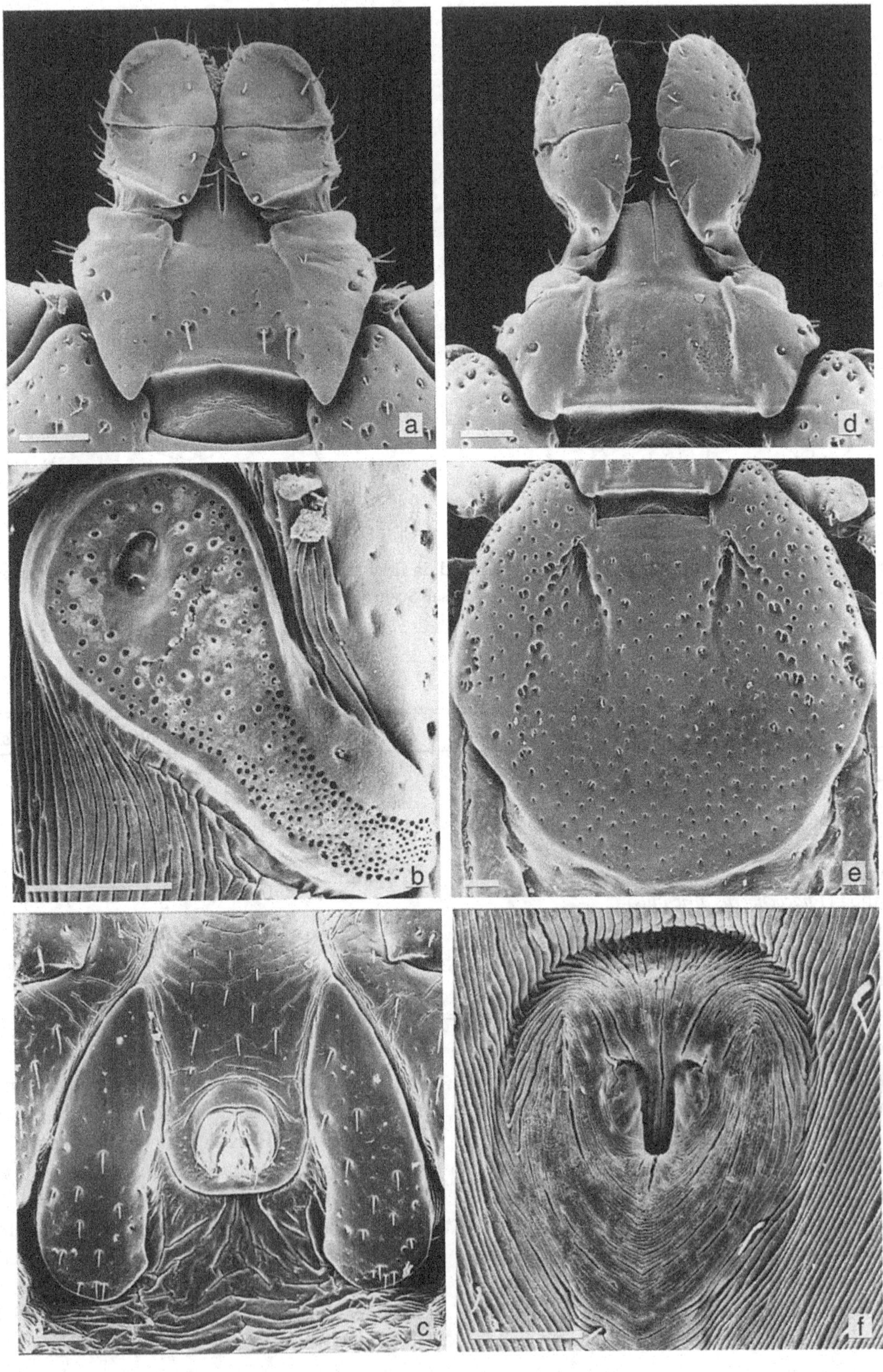
a
b
c
d
e
f

scattered all over the conscutum, especially anterior to the eyes, interspersed with numerous finer punctations. In other specimens the punctation pattern is more even. Legs increase slightly in size from I to IV. Ventrally spiracle comma-shaped, narrowing towards the dorsal surface. Adanal plates broadening posterior to the anus (sometimes more markedly than in specimen illustrated in Fig. 105(c)), smoothly rounded posteriorly; accessory adanal plates as sharp sclerotized points.

Female (Figs 104(b), 105(d) to (f))

Capitulum longer than broad, length × breadth ranging from 0.81 mm × 0.70 mm to 0.93 mm × 0.76 mm. Basis capituli with short hunched lateral angles at about mid-length; porose areas large, approximately twice their own diameter apart. Palps unique in shape, long; article I long and narrow; articles II and III much broader, with article II tapering to a point posteriorly and article III broadly rounded apically. Scutum about as broad as long, length × breadth ranging from 1.63 mm × 1.62 mm to 1.79 mm × 1.78 mm. Eyes about halfway back, edged dorsally by large deep punctations. Cervical fields broad, shallow, their external margins strongly marked by large setiferous punctations. A few large punctations scattered elsewhere on the scutum, interspersed with numerous medium-sized to fine punctations. Ventrally genital aperture a long narrow U-shape.

Immature stages

Unknown.

Figure 105 (*opposite*). *Rhipicephalus longiceps* (RML 65711, collected on farm 'Lichtenstein Sud', near Windhoek, Namibia, in 1974 by J. Langenhoven). Male: (a) capitulum, dorsal; (b) spiracle; (c) adanal plates. Female: (d) capitulum, dorsal; (e) scutum; (f) genital aperture. Scale bars represent 0.10 mm. SEMs by M.D. Corwin.

Notes on identification

One species with which *R. longiceps* could perhaps be confused initially is *R. lunulatus*, whose adults also have elongated capitula (see pp. 269–277). However, the palps of these two species, especially article II, differ in shape. In addition the males differ markedly in the way their body walls expand when they engorge and in the shape of their adanal plates. In the females the genital aperture of *R. longiceps* is reminiscent of a cow's teat in shape whereas that of *R. lunulatus* is a wide shallow curve.

Hosts

Life cycle unknown. This is a rare tick. The preferred hosts of the adults appear to be suids, with four collections taken from warthogs (Horak *et al.*, 1983) and one from a domestic pig (Table 28). The immature stages and their hosts are unknown. Adult ticks have been present in collections made from January to July.

Zoogeography

All records of *R. longiceps* are from Angola and Namibia (Map 31). The places in which it occurs range in altitude from 500 m to 1500 m with annual rainfalls from *c.* 100 mm to 400 mm or more. The vegetation in these areas varies from miombo, mopane and scrub woodland to deciduous bushland and wooded grassland and the Kalahari/Karoo–Namib transition in the west.

Disease relationships

Unknown.

REFERENCES

Horak, I.G., Biggs, H.C., Hanssen, T.S. & Hanssen, R.E. (1983). The prevalence of helminth and arthropod parasites of warthog, *Phacochoerus aethiopicus*, in South West Africa/Namibia. *Onderstepoort Journal of Veterinary Research*, **50**, 145–8.

Warburton, C. (1912). Notes on the genus *Rhipicephalus*, with the description of new species, and the consideration of some species hitherto described. *Parasitology*, **5**, 1–20.

Also see the following Basic Reference (p. 14): Theiler (1947).

RHIPICEPHALUS LONGICOXATUS NEUMANN, 1905

This specific name is based on the fact that the coxae of the first pair of legs of the adults are exceptionally long.

Synonym

camelopardalis.

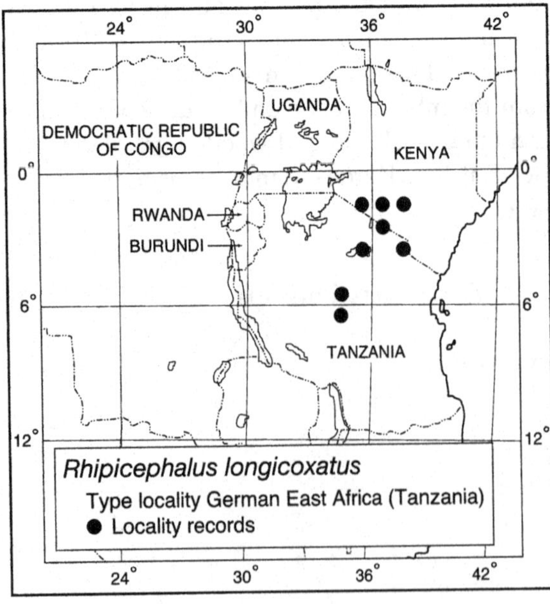

Map 32. *Rhipicephalus longicoxatus:* distribution.

Diagnosis

A large reddish-brown tick.

Male (Figs 106(a), 107(a) to (c))
Capitulum, in the three specimens measured, longer than broad, length × breadth ranging from 1.35 mm × 1.18 mm to 1.48 mm × 1.24 mm. Basis capituli with short blunt lateral angles at about anterior third of its length. Palps almost wedge-shaped apically. Conscutum length × breadth ranging from 4.62 mm × 2.68 mm to 4.92 mm × 2.75 mm, usually smooth and shiny in general appearance and almost rectangular in outline; anterior process of coxae I conspicuous. In engorged specimens body wall greatly expanded laterally, and rather less posteriorly, so that the posterior part of the body assumes a characteristic smoothly-rounded shape. Eyes large, sometimes slightly bulging. Cervical pits more-or-less rounded; cervical fields slightly depressed, their surfaces smooth. Marginal lines, when present, picked out by lines of large punctations that continue forwards nearly to eye level. Posterior grooves short, the posteromedian narrow, the posterolaterals broader. Punctations all shallow. A few medium-sized punctations present on the scapulae, with larger ones scattered along the length of the conscutum, especially on a small area anteromedially where they are interspersed amongst numerous smaller elements. Many very fine inconspicuous punctations present elsewhere on the conscutum apart from the cervical areas. Ventrally coxae I remarkably large, elongate, deeply divided. Spiracles also unique in shape, oval, with a short dorsal projection from one long side pointing almost straight towards the dorsal surface; the surrounding body wall setose. Adanal plates very broad posteriorly, their posterointernal and posteroexternal margins smoothly rounded, tapering anteriorly; accessory adanal plates absent.

Female (Figs 106(b), 107(d) to (f))
Capitulum slightly longer than broad, length × breadth of the two specimens measured 1.31 mm × 1.24 mm and 1.38 mm × 1.34 mm

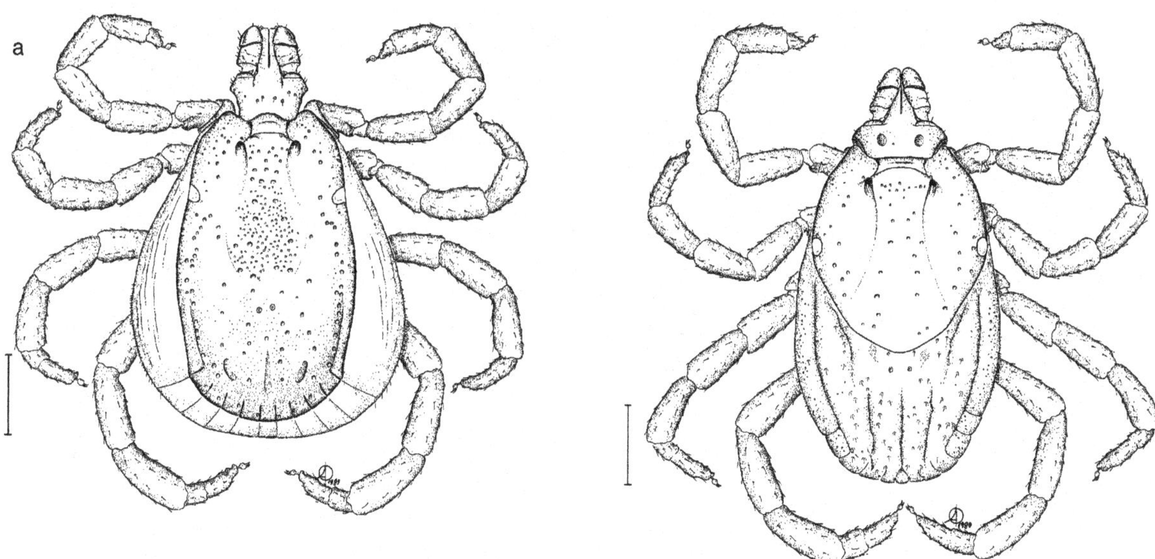

Figure 106. *Rhipicephalus longicoxatus* [collected from giraffe (*Giraffa camelopardalis*), Talek River, Cis-Mara area of western Maasailand, Kenya, in October 1955 by E.W. Temple-Boreham, originally designated as paratypes of *Rhipicephalus camelopardalis*]. (a) Male, dorsal; (b) female, dorsal. Scale bars represent 1 mm. A. Olwage *del.*

respectively. Basis capituli with short blunt lateral angles at about anterior third of its length; porose areas small, oval, two to three times their own diameter apart. Palps rather long, rounded apically. Scutum longer than broad, length × breadth of the two specimens measured 2.61 mm × 2.33 mm and 2.84 mm × 2.54 mm respectively; usually smooth, shiny and almost egg-shaped in outline. Eyes halfway back, sometimes bulging slightly. Cervical pits oval; cervical fields long, slightly depressed, their surfaces smooth. Punctations shallow. A few medium-sized punctations grouped on the scapulae and between the cervical pits; elsewhere scattered and interspersed with very fine punctations that may be virtually invisible in some specimens. Alloscutum with scattered fine white setae that easily become detached. Ventrally coxae I remarkably large, elongate, deeply divided. Genital aperture U-shaped, with a short anterolateral extension from each side.

Nymph

Unknown.

Larva (Fig. 108)

Capitulum about as long as broad, length × breadth ranging from 0.128 mm × 0.129 mm to 0.134 mm × 0.138 mm. Basis capituli about three times as broad as long, rounded laterally, posterior margin almost straight. Palps cone-shaped, tapering to narrowly-rounded apices, inclined slightly inwards. Scutum much broader than long, length × breadth ranging from 0.254 mm × 0.370 mm to 0.295 mm × 0.408 mm; posterior margin a broad fairly shallow curve. Eyes at widest point, over halfway back, slightly convex. Cervical grooves straight, slightly convergent. Ventrally coxae I each with a sharp spur; coxae II with a shallow broadly-rounded spur; coxae III with a mere indication of a spur.

Notes on identification

The type locality of *R. longicoxatus* is given merely as 'Afrique orientale allemande' (i.e. Tanzania) by Neumann (1905). His original description is not illustrated. It was doubtless this omission that led Hoogstraal (1956) to identify entirely different ticks as *R. longicoxatus* and his interpretation of this name, supported

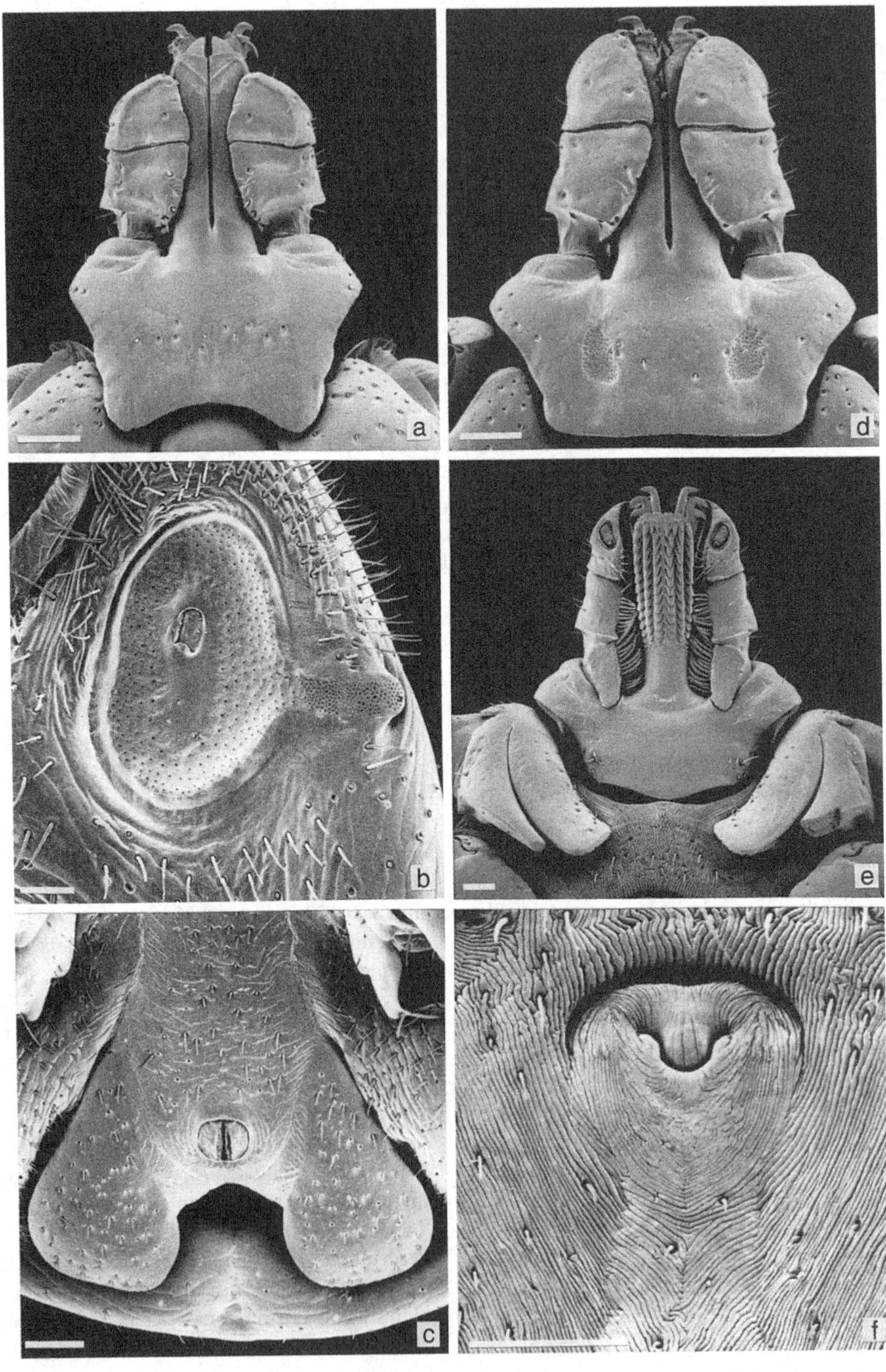
a
b
c
d
e
f

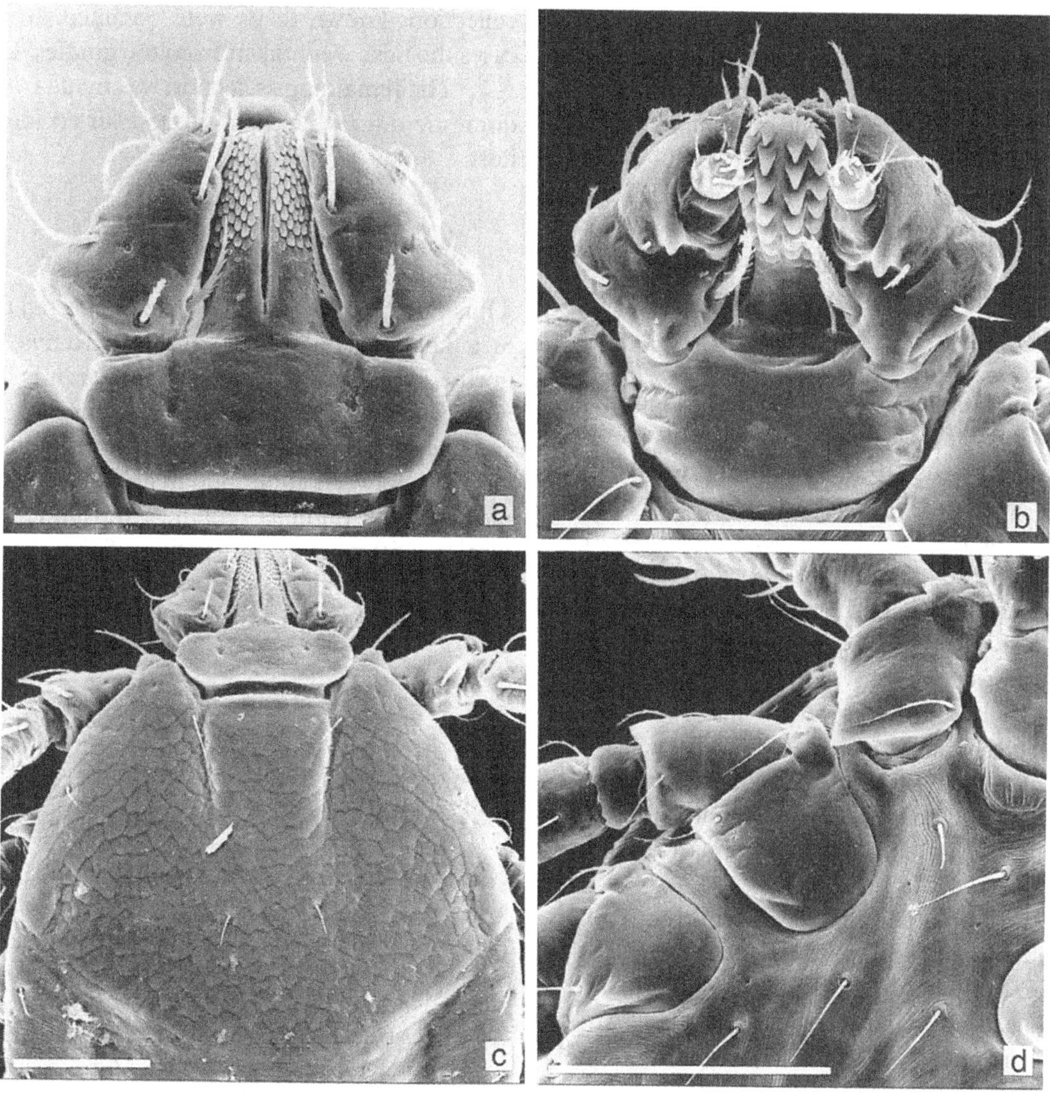

Figure 108 (*above*). *Rhipicephalus longicoxatus* [B.S. 912/-, RML 66304, progeny of ♀ collected from giraffe (*Giraffa camelopardalis*) at Athi River, Kenya, on 18 August 1961 by M.C. Round]. Larva: (a) capitulum, dorsal; (b) capitulum, ventral; (c) scutum; (d) coxae. Scale bars represent 0.10 mm. SEMs by J.F. Putterill.

Figure 107 (*opposite*). *Rhipicephalus longicoxatus* [Nuttall Collection 2577d, from giraffe (*Giraffa camelopardalis*), Manyoni, Tanzania, dated Kilimatinde, 14 July 1913 by Dr W. Bartels, donated by The Natural History Museum, London]. Male: (a) capitulum, dorsal; (b) spiracle; (c) adanal plates. Female: (d) capitulum, dorsal; (e) capitulum and coxae I, ventral; (f) genital aperture. Scale bars represent 0.10 mm. SEMs by J.F. Putterill.

by his well-illustrated, easily available description, has naturally been followed by other workers since. Recently, though, we compared the remaining paratype female of *R. longicoxatus* (Zoological Museum, Berlin, ZMB 17 740) with a paratype female of *R. camelopardalis* and it was immediately obvious that these two entities are the same. The name *R. camelopardalis* therefore falls as a junior synonym of *R. longicoxatus* (NEW SYNONYMY).

It has been possible to re-examine only six very small samples of ticks that were identified as *R. longicoxatus* according to Hoogstraal's description between 1956 and the present. Of these 1 ♀ from *Syncerus caffer*, Galual-Nyang Forest, Bahr El Ghazal Province, Sudan (RML 92536, H.H. 28438) appears to belong to the *R. simus* group. Several other collections of small-sized ticks from northern Somalia, plus one from Lali, eastern Kenya, may be *R. lunulatus*. However, they are atypical in that the adanal plates of the males are extremely broad with practically no indication of the usual cusps on their posterior margins (Hoogstraal, 1956, p. 660, figure 274). Further collections would be needed to solve this problem.

G.H.F. Nuttall and C. Warburton accurately identified the largest collection of *R. longicoxatus* known to us, 87 ♂♂, 49 ♀♀ from a giraffe, near Manyoni Ngogo, German East Africa (=Tanzania) (RML 66618, Nuttall Collection 2577d). Later, though, they misidentified as *R. longicoxatus* four specimens of *R. muehlensi*, a species unknown to them as it was described only in 1943 (Nuttall Collection 2824, 2825a; C. Clifford, pers. comm., 29 August 1963, to G. Theiler; Keirans, 1985).

Hosts

Life cycle unknown. Larvae hatched from eggs laid by an engorged female but died before an attempt could be made to feed them (J.B.W., unpublished data, 1961). This rare species is apparently a specific parasite of the giraffe (*Giraffa camelopardalis*), from which nine of the ten collections known to us were obtained. In two cases the ticks were taken from the giraffes' ears.

The remaining collection was made from a domestic dog, which is doubtless an accidental host.

Zoogeography

Thus far *R. longicoxatus* has been collected only in a fairly small area in southern Kenya and northern Tanzania (Map 32). It apparently does not occur throughout the giraffe's range in Africa. The places where it has been recorded lie at altitudes between about 300 m and 1800 m with mean annual rainfalls ranging from under 500 mm to 750 mm in the various forms of dry woodland, wooded or bushed grassland and bush/thicket favoured by giraffes.

Disease relationships

Unknown.

REFERENCES

Neumann, L.G. (1905). Notes sur les Ixodidés – III. *Archives de Parasitologie*, **9**, 225–41.

Walker, J.B. & Wiley, A.J. (1959). *Rhipicephalus camelopardalis* n.sp. (Ixodoidea, Ixodidae), a new species of tick from East African giraffes. *Parasitology*, **49**, 448–53.

Also see the following Basic References (pp. 12–14): Hoogstraal (1956); Keirans (1985); Walker (1974); Yeoman & Walker (1967).

RHIPICEPHALUS LONGUS NEUMANN, 1907

The specific name *longus* (Latin) means 'long' but the author gave no indication in his description as to why he gave the tick this name.

Synonyms

capensis longus; confusus; falcatus.

Diagnosis

A medium-sized dark brown tick whose males have sickle-shaped adanal plates.

Male (Figs 109(a), 110(a) to (c))
Capitulum longer than broad, length × breadth ranging from 0.79 mm × 0.73 mm to 0.90 mm × 0.83 mm. Basis capituli with short, sharply pointed lateral angles in anterior third of its length. Palps broad, gently rounded apically. Conscutum length × breadth ranging from 3.40 mm × 2.25 mm to 3.85 mm × 2.55 mm; anterior process of coxae I inconspicuous, rounded. In engorged specimens three caudal processes protrude posteromedially. Eyes almost flat, delimited dorsally by a few large setiferous punctations. Cervical pits deep. Marginal lines well developed, long, punctate. Posteromedian groove long, narrow, relatively inconspicuous; posterolateral grooves short, broad. Large setiferous punctations present along the external margins of the cervical fields, but those on the scapulae and in a '*simus*' pattern medially on the conscutum only medium-sized; the latter may be almost masked by the smaller interstitial punctations, which are numerous and dense in some specimens but sparser and shallower in others. Ventrally spiracles broadly comma-shaped with only a short gently-curved dorsal prolongation. Adanal plates broadly sickle-shaped; accessory adanal plates short, bluntly-rounded points.

Female (Figs 109(b), 110(d) to (f))
Capitulum slightly broader than long, length × breadth ranging from 0.81 mm × 0.84

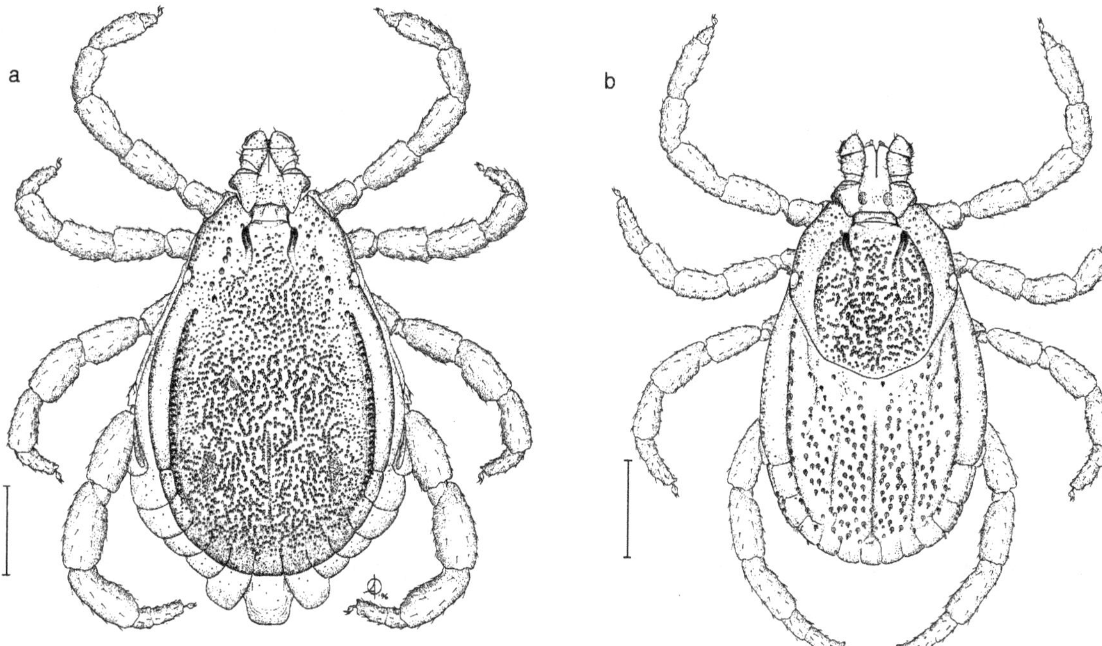

Figure 109. *Rhipicephalus longus* [collected from African buffalo (*Syncerus caffer*), Ankole District, Uganda in *c.* 1957/58 by W. Longhurst]. (a) Male, dorsal; (b) female, dorsal. Scale bars represent 1 mm. A. Olwage *del.*

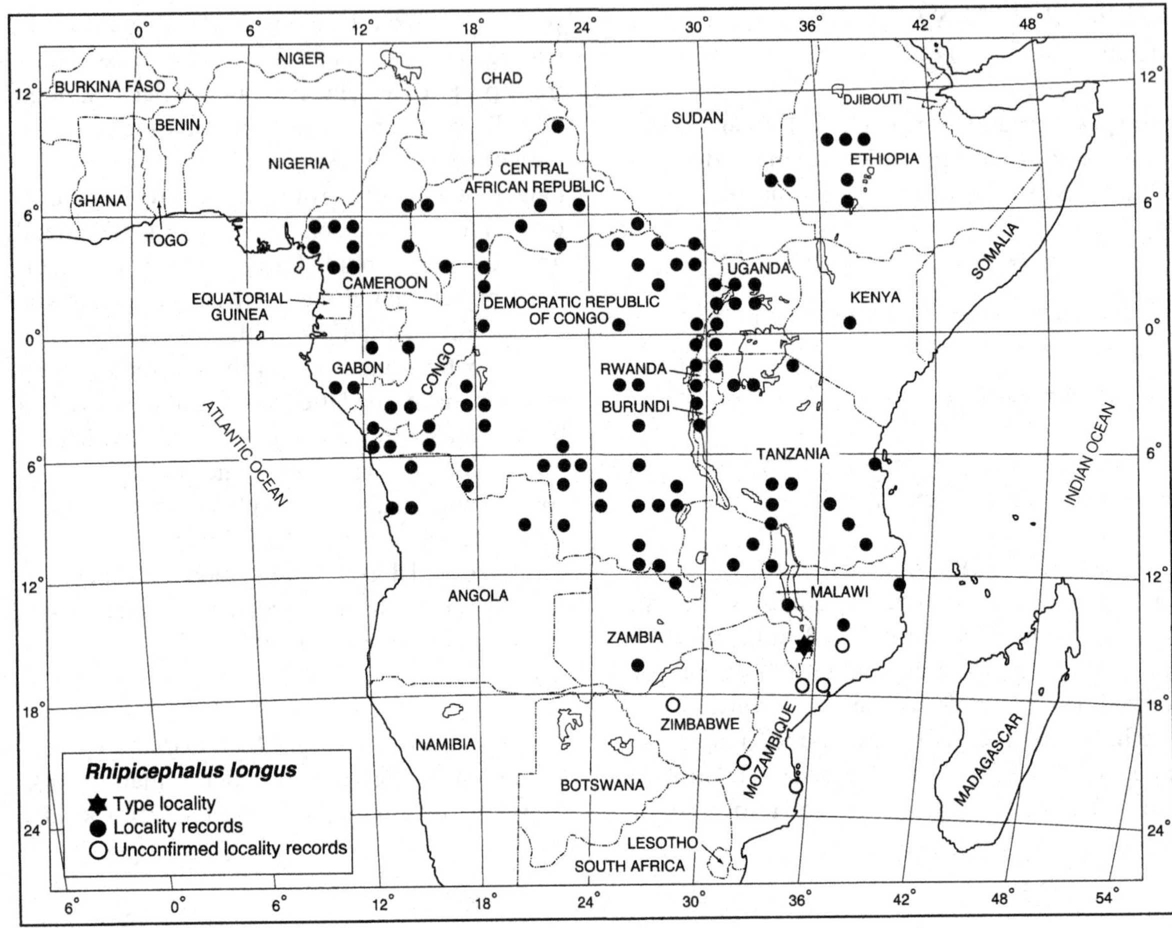

Map 33. *Rhipicephalus longus*: distribution. (Based partly on Morel, 1969).

mm to 0.95 mm × 0.96 mm. Basis capituli with broad lateral angles a little anterior to mid-length; porose areas round, somewhat more than their own diameter apart. Palps longer than those of the male, broadly rounded apically. Scutum about as long as broad, length × breadth ranging from 1.62 mm × 1.63 mm to 1.92 mm × 1.93 mm; posterior margin slightly sinuous. Eyes almost flat, delimited dorsally by a few medium-sized setiferous punctations. Cervical fields broad, slightly depressed, their outer margins delimited by rows of large and medium-sized setiferous punctations. A few medium-sized setiferous punctations scattered on the scapulae and medially on the scutum, where they may be almost masked by numerous interstitial punctations. Alloscutum of unfed females deeply convoluted. Ventrally genital aperture almost V-shaped in general outline, with a noticeable constriction towards its rounded posterior end.

Nymph (Fig. 111)

Capitulum much broader than long, length × breadth ranging from 0.23 mm × 0.31 mm to 0.26 mm × 0.33 mm. Basis capituli over three times as broad as long with tapering sharply-pointed lateral angles projecting over the scapulae, and its posterior margin almost straight; ventrally with broadly-rounded spurs. Palps quite narrow, more-or-less equal in width for most of their length, their apices broadly rounded, inclined inwards. Scutum broader than

Table 29. *Host records of* Rhipicephalus longus

Hosts	Number of records
Domestic animals	
Cattle	85
Sheep	2
Goats	1
Pigs	30
Dogs	10
Wild animals	
Side-striped jackal (*Canis adustus*)	1
Bat-eared fox (*Otocyon megalotis*)	1
Lion (*Panthera leo*)	1
Leopard (*Panthera pardus*)	1
Banded mongoose (*Mungos mungo*)	1
Spotted hyaena (*Crocuta crocuta*)	1
African elephant (*Loxodonta africana*)	1
Burchell's zebra (*Equus burchellii*)	2
Black rhinoceros (*Diceros bicornis*)	1
Aardvark (*Orycteropus afer*)	2
Warthog (*Phacochoerus africanus*)	44
Forest hog (*Hylochoerus meinertzhageni*)	3
Bushpig (*Potamochoerus larvatus*)	9
Red river hog (*Potamochoerus porcus*)	7
Bushpig/red river hog (*Potamochoerus* sp.)	2
'Wild pig'	1
Blue wildebeest (*Connochaetes taurinus*)	1
Lichtenstein's hartebeest (*Sigmoceros lichtensteinii*)	1
African buffalo (*Syncerus caffer*)	109
Eland (*Taurotragus oryx*)	3
Mountain nyala (*Tragelaphus buxtoni*)	1
Bongo (*Tragelaphus eurycerus*)	1
Roan antelope (*Hippotragus equinus*)	2
Sable antelope (*Hippotragus niger*)	4
Waterbuck (*Kobus ellipsiprymnus*)	3
Cape hare (*Lepus capensis*)	1
Humans	2

long, length × breadth ranging from 0.44 mm × 0.49 mm to 0.51 mm × 0.55 mm; posterior margin a broad smooth curve. Eyes at widest point, well over halfway back, delimited dorsally by slight depressions. Cervical fields long, narrow, slightly depressed, divergent. Ventrally coxae I each with two quite short broad subequal spurs; coxae II to IV each with a small external spur only, decreasing progressively in size with that on coxa IV being almost non-existent.

Larva (Fig. 112)

Capitulum much broader than long, length × breadth ranging from 0.110 mm × 0.150 mm to 0.117 mm × 0.162 mm. Basis capituli about

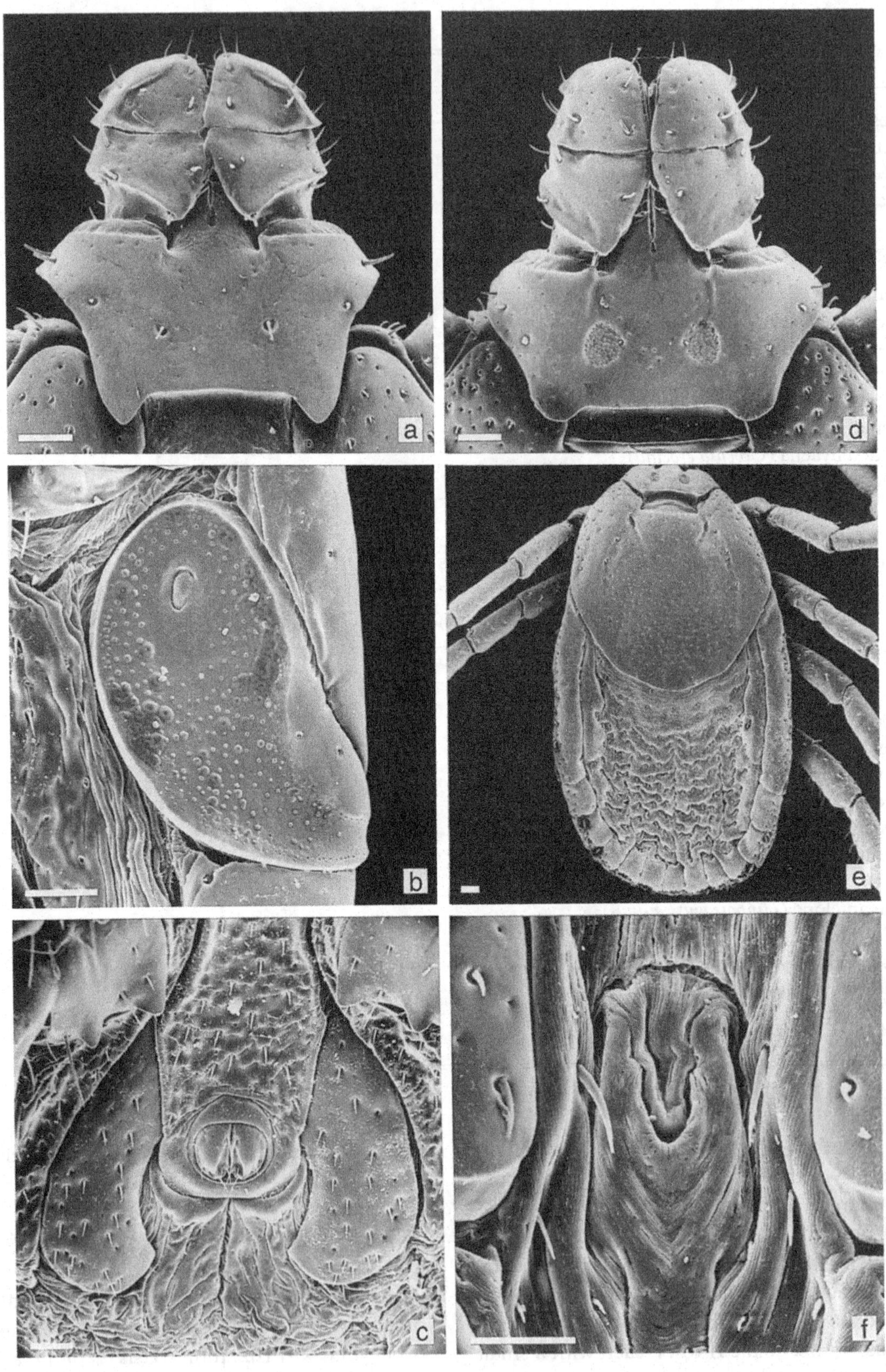
a
d
b
e
c
f

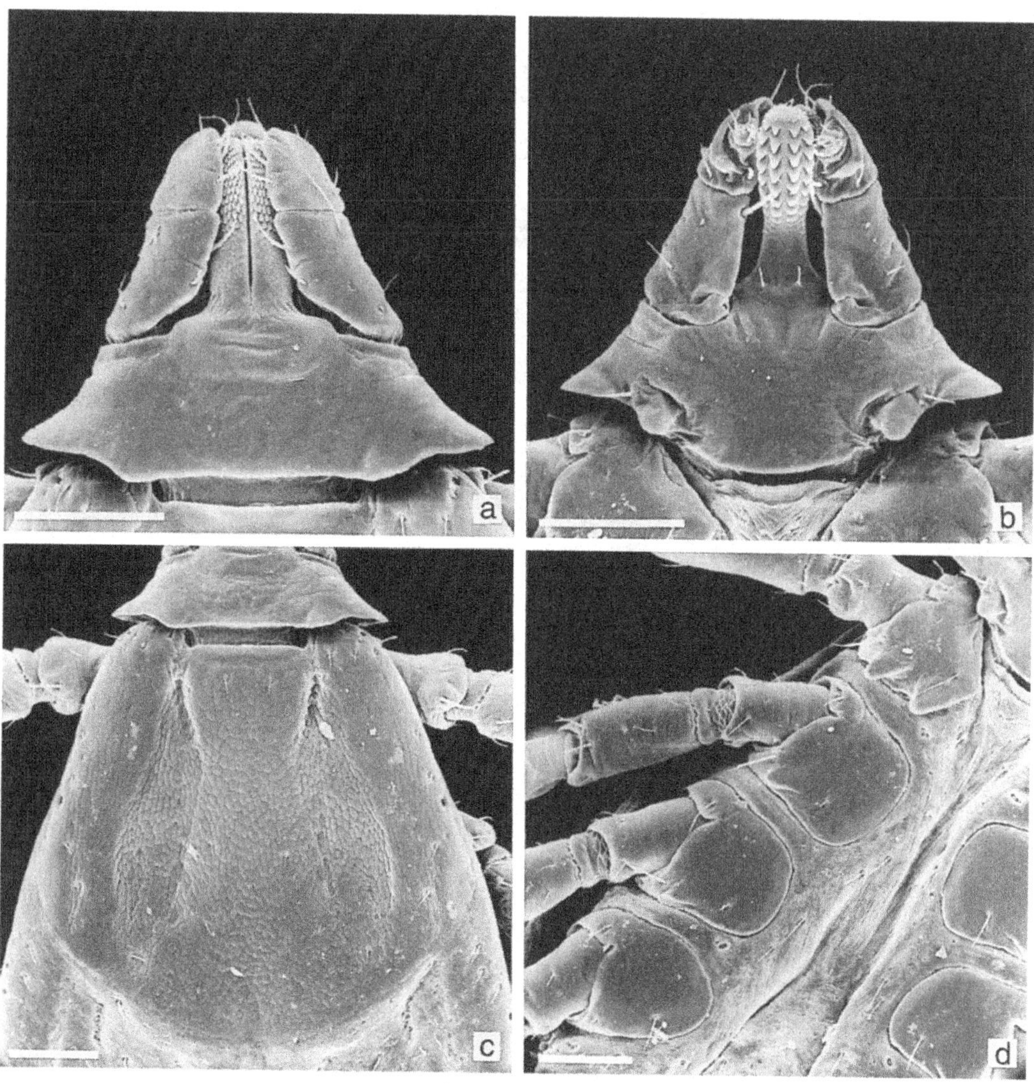

Figure 111 (*above*). *Rhipicephalus longus* [B.S. 701/-, RML 66310, laboratory reared, progeny of ♀ collected from warthog (*Phacochoerus africanus*), in Maruzi County, Lango, Uganda on 3 July 1956 by Eriasafu Okello]. Nymph: (a) capitulum, dorsal; (b) capitulum, ventral; (c) scutum; (d) coxae. Scale bars represent 0.10 mm. SEMs by M.D. Corwin.

Figure 110 (*opposite*). *Rhipicephalus longus* [B.S. 701/-, RML 66310, laboratory reared, progeny of ♀ collected from warthog (*Phacochoerus africanus*), in Maruzi County, Lango, Uganda on 3 July 1956 by Eriasafu Okello]. Male: (a) capitulum, dorsal; (b) spiracle; (c) adanal plates. Female: (d) capitulum, dorsal; (e) scutum and alloscutum, dorsal; (f) genital aperture. Scale bars represent 0.10 mm. SEMs by M.D. Corwin.

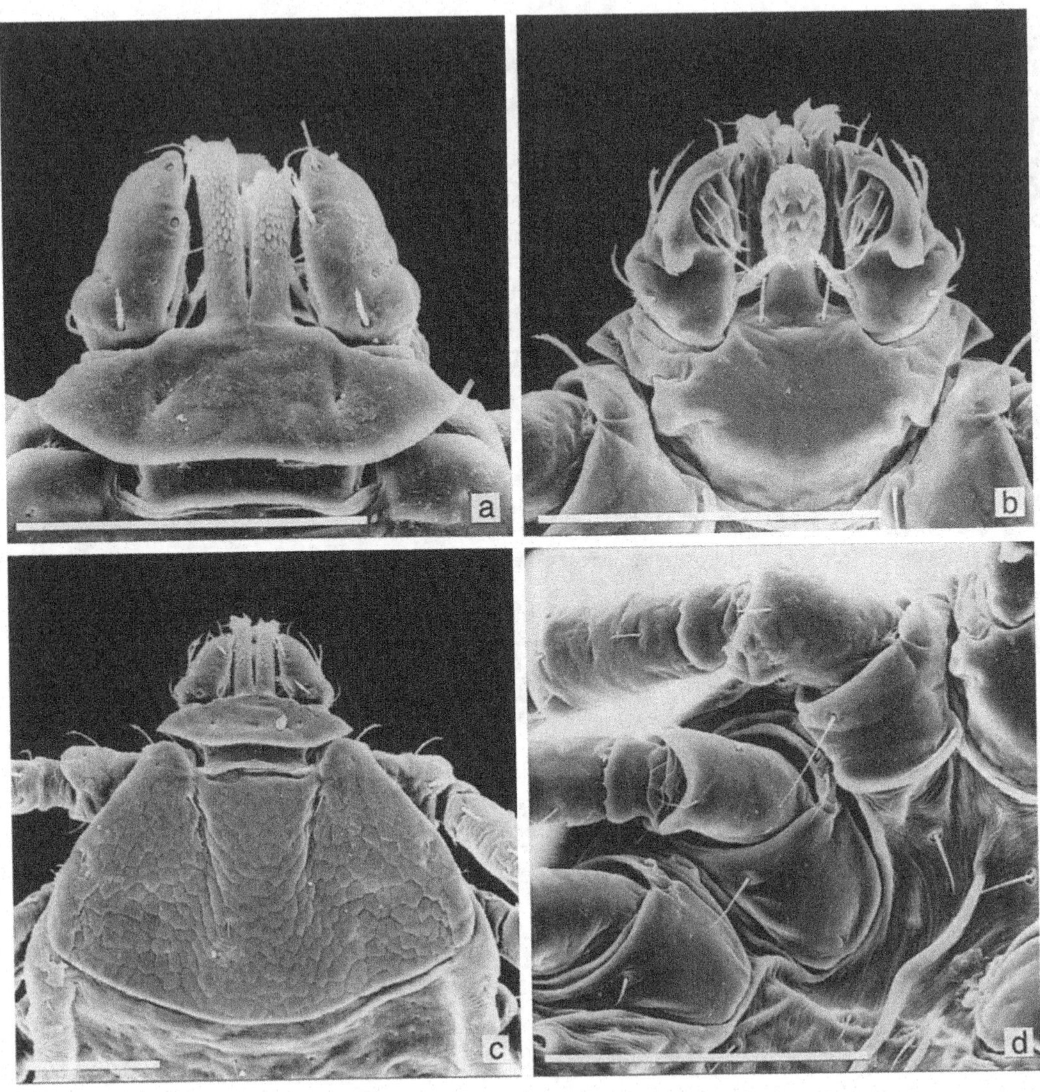

Figure 112. *Rhipicephalus longus* [B.S. 701/-, RML 66310, laboratory reared, progeny of ♀ collected from warthog (*Phacochoerus africanus*), in Maruzi County, Lango, Uganda on 3 July 1956 by Eriasafu Okello]. Larva: (a) capitulum, dorsal; (b) capitulum, ventral; (c) scutum; (d) coxae. Scale bars represent 0.10 mm. SEMs by M.D. Corwin.

3.5 times as broad as long, with bluntly pointed lateral angles, posterior margin slightly concave; ventrally with well-developed bluntly rounded spurs on its posterior margin. Palps with external margins sinuous, tapering quite abruptly to bluntly rounded apices, inclined inwards. Scutum much broader than long, length × breadth ranging from 0.219 mm × 0.330 mm to 0.239 mm × 0.356 mm; posterior margin a broad shallow smooth curve. Eyes at widest point, far back, almost flat. Cervical grooves quite long, slightly convergent. Ventrally coxae I each with a broad bluntly-rounded spur; coxae II and III each merely slightly salient along the posterior margin.

Notes on identification

In the past three of the *Rhipicephalus* species whose males all have sickle-shaped adanal plates, *R. pseudolongus*, *R. longus* and *R. senegalensis*, have often been confused. Their differentiation requires particular care, experience and attention

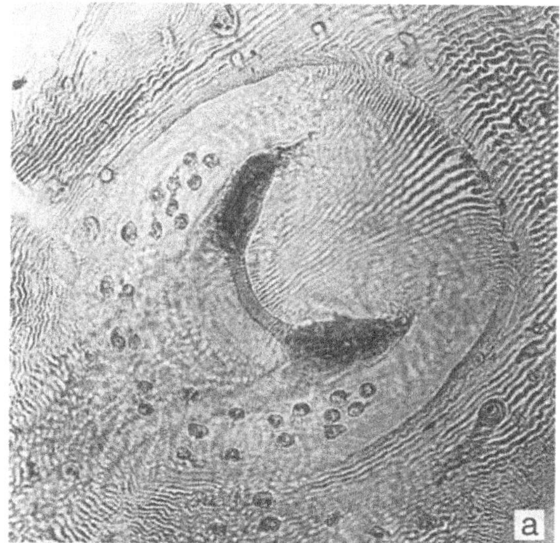

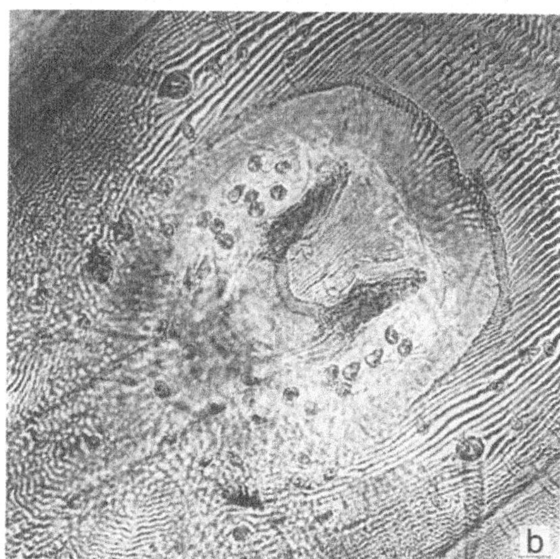

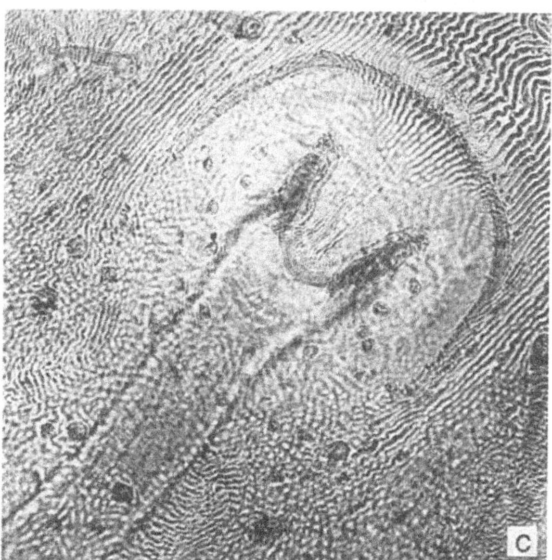

to detail, especially in the case of field collections from areas such as western Uganda where their distributions sometimes overlap (Matthysse & Colbo, 1987). Their diagnostic features, besides the sickle-shaped adanal plates of the males, were discussed in detail by Clifford & Anastos (1962), and can be summarized as follows:

1. *Rhipicephalus pseudolongus*. The male conscutum and female scutum are densely and evenly punctate medially, and a variable number of medium-sized to somewhat larger punctations are scattered on the scapulae. Large '*simus*'-like punctations are absent (p. 360, Fig. 165). The mounted female genital aperture is roughly V-shaped but is much shorter and wider than that of either *R. longus* or *R. senegalensis* (Fig. 113(a)). Nymphs of this species have definite cornua on their basis capituli.
2. *Rhipicephalus longus*. The male conscutum and female scutum are covered medially with numerous medium-sized to small punctations that may be interspersed with scattered larger '*simus*'-like punctations. The external margins of the cervical fields are indicated by irregular rows of conspicuous larger punctations, and several medium-sized to large punctations are scattered on the scapulae (Fig. 109). The mounted female genital aperture is more-or-less V-shaped (Fig. 113(b)). Nymphs of this species lack cornua on their basis capituli.
3. *Rhipicephalus senegalensis*. The male conscutum and female scutum are smooth and shiny in general appearance. Medially they have a '*simus*' pattern of large punctations interspersed with numerous minute interstitial punctations. The external margins of the cervical fields are indicated by irregular

Figure 113 (*opposite*). Female genital apertures (mounted): (a) *Rhipicephalus pseudolongus*; (b) *Rhipicephalus longus*; (c) *Rhipicephalus senegalensis*. (From Clifford & Anastos, 1962, figs 10, 2 & 8, respectively, published by l'Institut des Parcs Nationaux du Congo et du Rwanda).

rows of large punctations and several medium-sized punctations are scattered on the scapulae (p. 398, Fig. 183). The mounted female genital aperture is V-shaped and in general is rather similar to that of *R. longus*, but the aperture tends to be somewhat wider and its posterior edge more broadly rounded (Fig. 113(c)). Nymphs of this species lack cornua on their basis capituli.

The identification of field collections of these ticks is often far from easy, especially as a wide range of punctation patterns is currently accepted for *R. longus*. In the more heavily punctate specimens of this species the presence of the '*simus*' pattern of large punctations medially on the scutum together with the rows of large punctations along the external margins of the cervical fields are particularly important diagnostic features that serve to distinguish it from *R. pseudolongus*. Clifford & Anastos (1962) commented that both the *R. longus* type male and a cotype male of its synonym *R. falcatus* are relatively heavily punctate (Santos Dias, 1956, figures 1 & 2 respectively).

Rhipicephalus confusus was described as a new species from Mozambique by Santos Dias (1956, figures 3–5). Clifford & Anastos (1962) also regarded this tick as a synonym of *R. longus*, representing the more moderately-punctate end of the latter's range. We have examined a number of specimens identified as *R. confusus* by Santos Dias, including a paratype male from The Natural History Museum, London. For the present we support the above decision by Clifford & Anastos, though one of us (J.B.W.) has some reservations, primarily because so few specimens identified as *R. confusus* are available. Santos Dias himself was somewhat ambivalent about *R. confusus*. In 1960 he listed *R. longus* from Mozambique but did not mention *R. confusus*. In 1993, however, he omitted *R. longus* but reinstated *R. confusus* as a valid species.

At the lightly-punctate end of its range care must be taken to distinguish *R. longus* from *R. senegalensis*.

Hosts

A three-host species (J.B.W., unpublished data, 1956). On domestic animals *R. longus* has been recorded most commonly from cattle, and to a lesser extent from pigs and dogs (Table 29). It has similar host preferences among wild animals, the vast majority of collections having been obtained from the African buffalo and wild suids, plus some from the larger antelopes. Apparently other animals are only occasionally parasitized by this tick (Santos Dias, 1960, 1983–84; Morel & Finelle, 1961; Clifford & Anastos, 1962; Morel & Mouchet, 1965; Elbl & Anastos, 1966; Yeoman & Walker, 1967; MacLeod, 1970; Pegram, 1979; Matthysse & Colbo, 1987). In Uganda adults were collected throughout the year.

The immature stages have been found in rodent burrows (Morel, 1969).

Zoogeography

R. longus is presently thought to occur in a broad belt across Africa, from the Cameroons, Gabon, the Congo and Democratic Republic of Congo eastwards to western Ethiopia and thence southwards to Zambia, Malawi and Mozambique (Map 33). Norval (1985) listed one record from Zimbabwe but questioned whether it is really established in this country.

Morel (1969), who discussed its ecological preferences in detail, regarded *R. longus* essentially, though not exclusively, as a species occurring at medium altitudes of about 500 m to 1500 m with a hot humid climate and a well-distributed annual rainfall of over 1000 mm to 1250 mm. The vegetation in much of its habitat comprises the various types of lighter forest and woodland favoured by the large black African buffalo rather than the smaller reddish subspecies of this buffalo that characteristically inhabits dense forests.

Disease relationships

Unknown.

REFERENCES

MacLeod, J. (1970). Tick infestation patterns in the southern province of Zambia. *Bulletin of Entomological Research,* **60**, 253–74.

Morel, P.C. & Finelle, P. (1961). Les tiques des animaux domestiques du Centrafrique. *Revue d'Élevage et de Médecine Vétérinaire des Pays Tropicaux,* **14** (nouvelle série), 191–7.

Morel, P.C. & Mouchet, J. (1965). Les tiques du Cameroun (Ixodidae et Argasidae) (2e note). *Annales de Parasitologie Humaine et Comparée,* **40**, 477–96.

Neumann, L.G. (1907). Description of two new species of African ticks. *Annals of Tropical Medicine and Parasitology,* **1**, 115–20.

Norval, R.A.I. (1985). The ticks of Zimbabwe. XII. The lesser known *Rhipicephalus* species. *Zimbabwe Veterinary Journal,* **16**, 37–43.

Pegram, R.G. (1979). *Ticks (Ixodoidea) of Ethiopia with special reference to cattle and a critical review of the taxonomic status of species within the* Rhipicephalus sanguineus *group.* M. Phil. thesis, Brunel University, England, 169 pp., 18 tables, 40 figs., 123 plates.

Santos Dias, J.A.T. (1956). Sobre a verdadeira posição taxonomica de duos espécies ixodologicas da Africa Etiópeca. *Moçambique,* **No. 87**, 1–38 + 3 plates.

Santos Dias, J.A.T. (1983–84). Subsídios para o conhecimento da fauna ixodológica de Angola. *Garcia de Orta, Série de Zoologia, Lisboa,* **11**, 57–68.

Santos Dias, J.A.T. [1991 (1993)]. Some data concerning the ticks (Acarina-Ixodoidea) presently known in Mozambique. *Garcia de Orta, Série de Zoologia, Lisboa,* **18**, 27–48.

Also see the following Basic References (pp. 12–14): Clifford & Anastos (1962); Elbl & Anastos (1966); Hoogstraal (1956); Keirans (1985); Matthysse & Colbo (1987); Morel (1969); Santos Dias (1960); Walker (1974); Yeoman & Walker (1967).

RHIPICEPHALUS LOUNSBURYI WALKER, 1990

This species was named in honour of C.P. Lounsbury (1872–1955), the pioneer economic entomologist in South Africa from 1895–1927, for his remarkable contributions to our knowledge of ticks and tick-borne diseases of livestock.

Diagnosis

A relatively small shiny dark brown tick.

Male (Figs 114(a), 115(a) to (c))

Capitulum broader than long, length × breadth ranging from 0.41 mm × 0.45 mm to 0.67 mm × 0.71 mm. Basis capituli with lateral angles at about anterior third of its length, acute. Palps flattish apically. Conscutum length × breadth ranging from 1.75 mm × 1.21 mm to 2.58 mm × 1.71 mm; anterior process of coxae I not particularly prominent. Eyes about a quarter of the way back, slightly convex, edged dorsally by a few medium-sized punctations. Cervical pits deep, comma-shaped; external cervical margins delimited by medium-sized setiferous punctations. Marginal lines shallow, not quite reaching

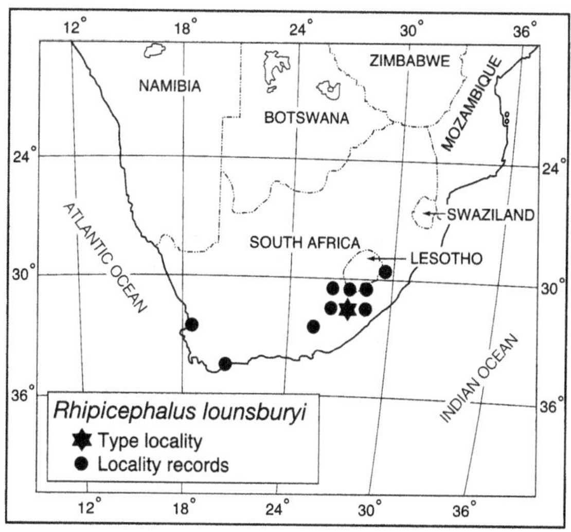

Map 34. *Rhipicephalus lounsburyi:* distribution.

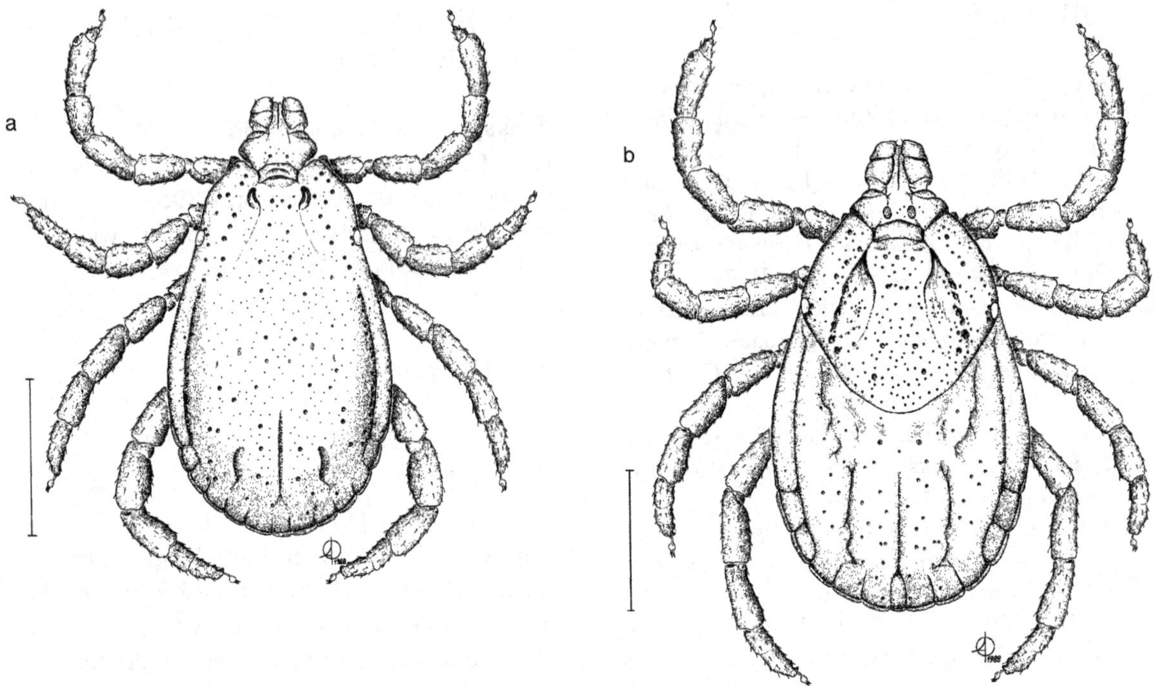

Figure 114. *Rhipicephalus lounsburyi* (Onderstepoort Tick Collection 2820, RML 105789, laboratory reared, progeny of ♀ collected at Dordrecht, Eastern Cape Province, South Africa, on 8 August 1945). (a) Male, dorsal; (b) female, dorsal. Scale bars represent 1 mm. A. Olwage *del.* (From Walker, 1990, figs 1 & 2, with kind permission from the Editor, *Onderstepoort Journal of Veterinary Research*).

Table 30. *Host records of* Rhipicephalus lounsburyi

Hosts	Number of records	
	Confirmed	Unconfirmed
Domestic animals		
Sheep	9	2
Wild animals		
Caracal (*Caracal caracal*)	1	
Black wildebeest (*Connochaetes gnou*)	1	
Bontebok (*Damaliscus pygargus dorcas*)	6	
Eland (*Taurotragus oryx*)	2	
Grey rhebok (*Pelea capreolus*)	9	
Mountain reedbuck (*Redunca fulvorufula*)	4	
Four-striped grass mouse (*Rhabdomys pumilio*)	1 (nymph)	

eyes anteriorly. Posteromedian and posterolateral grooves present but sometimes poorly defined. Punctation pattern rather inconspicuous, with a few medium-sized setiferous punctations scattered on the scapulae and medially on the conscutum, interspersed with numerous very fine punctations that are sometimes almost invisible. Ventrally spiracles long, narrowing at about two-thirds of their length and curving gently towards the dorsal surface. Adanal plates fairly broad, scooped out slightly posterior to anus, bluntly rounded posteriorly; accessory adanal plates sharp, well sclerotized.

Female (Figs 114(b), 115(d) to (f))
Capitulum much broader than long, length × breadth ranging from 0.47 mm × 0.56 mm to 0.64 mm × 0.78 mm. Basis capituli with lateral angles just anterior to mid-length, acute; porose areas medium-sized, twice their own diameter apart. Palps broadly rounded apically. Scutum longer than broad, length × breadth ranging from 0.91 mm × 0.85 mm to 1.46 mm × 1.40 mm, posterior margin slightly sinuous. Eyes slightly convex, demarcated dorsally by shallow grooves. Internal cervical margins deeply depressed and convergent initially, becoming shallower and divergent; cervical fields slightly depressed, their external margins indicated by rows of medium-sized setiferous punctations. A few similar setiferous punctations scattered on the scapulae and medially on the scutum, interspersed with numerous fine punctations. Ventrally genital aperture wide, deeply crescentic.

Nymph (Fig. 116)
Capitulum much broader than long, length × breadth ranging from 0.21 mm × 0.35 mm to 0.21 mm × 0.37 mm. Basis capituli four times as broad as long, with broad lateral angles projecting over the scapulae; ventrally with stout, broadly-rounded spurs on posterior border. Palps almost parallel sided for much of their length, then tapering to broadly-rounded apices, inclined slightly inwards. Scutum broader than long, length × breadth 0.49 mm × 0.58 mm to 0.57 mm × 0.60 mm; posterior margin a fairly deep curve. Eyes at widest part of scutum, edged dorsally by slight grooves. Cervical fields long, narrow, slightly depressed. Ventrally coxae I each with a long, sharp external spur overlapping coxae II and a short sharp internal spur; coxae II and III each with a sharp, but shorter, external spur; coxae IV each with a short blunt external spur.

Larva (Fig. 117)
Capitulum much broader than long, length × breadth ranging from 0.08 mm × 0.16 mm to 0.10 mm × 0.16 mm. Basis capituli over four times as broad as long, broadly rounded laterally. Palps broadest proximally, then tapering gradually to narrowly-rounded apices, inclined inwards. Scutum much broader than long, length × breadth *c.* 0.23 mm × 0.35 mm (length impossible to measure in most mounted specimens because the posterior margin of the scutum is obscured by faecal matter); posterior margin fairly deep, almost straight posterolaterally. Eyes at widest part of the scutum, edged dorsally by shallow grooves. Cervical grooves short, very slightly convergent. Ventrally coxae each with one spur, that on coxae I being the largest.

Notes on identification
This tick was originally described by Theiler & Robinson (1953) as *Rhipicephalus follis* (see p.185), an error that these authors would almost certainly not have made had they seen the two syntype males of the true *R. follis* (Nuttall Collection 2110). Gertrud Theiler herself later had doubts about their finding and in 1962 stated: '. . . the possibility exists that the tick described by Theiler & Robinson is not the true *R. follis* . . .' The collections made by Horak *et al.* (1986) were at that time described as a *Rhipicephalus* sp.

Hosts

A three-host species (Theiler & Robinson, 1953, as *R. follis*). The only domestic animals from which *R. lounsburyi* adults have been collected so

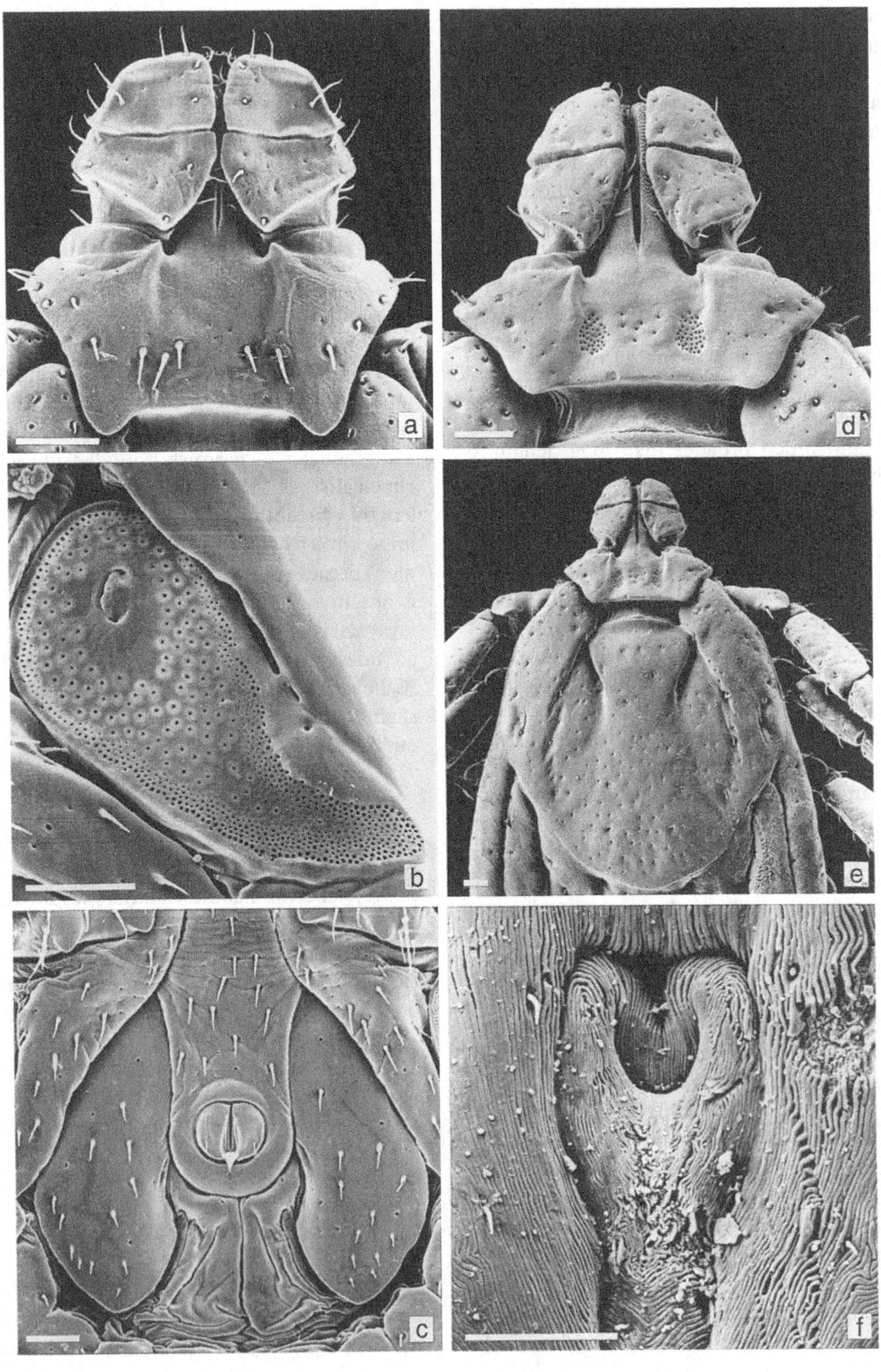
a
d
b
e
c
f

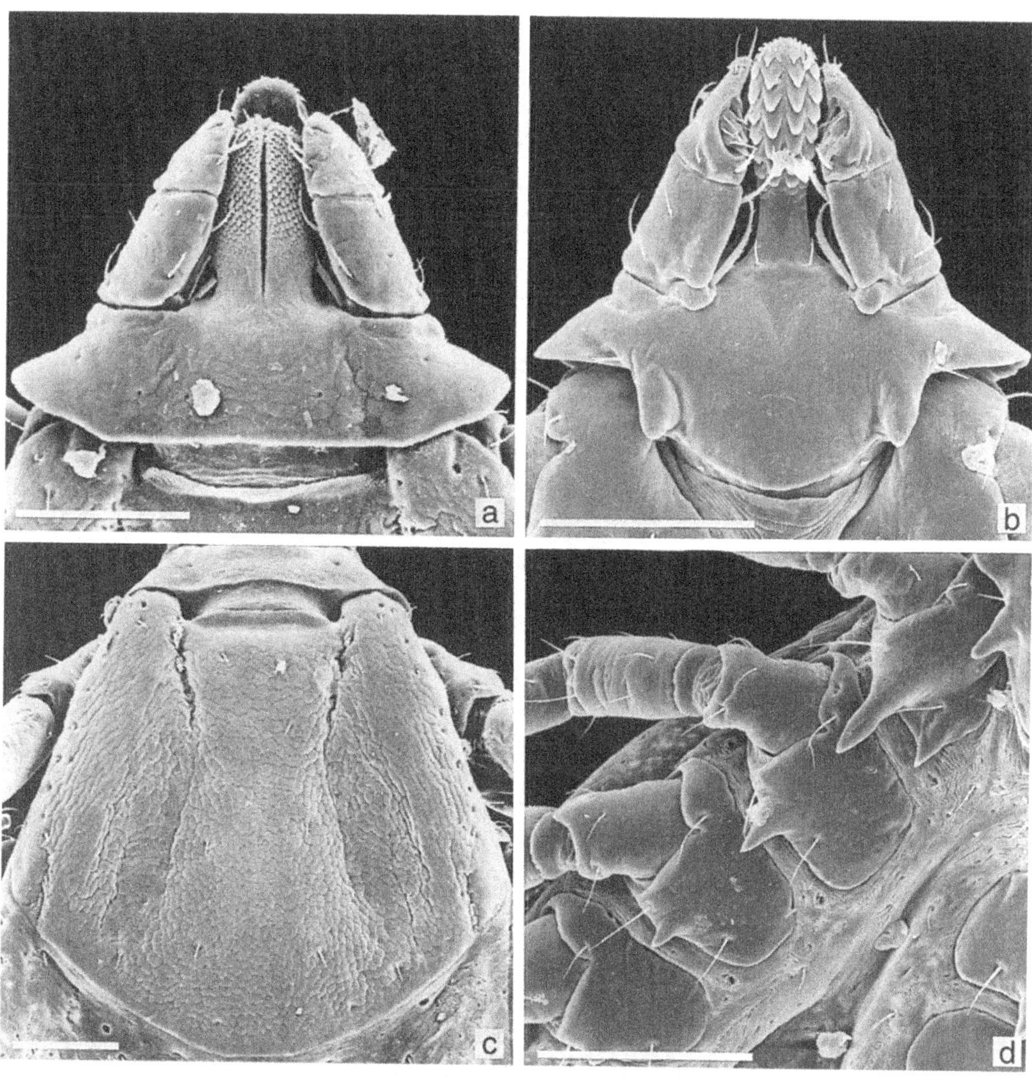

Figure 116 (*above*). *Rhipicephalus lounsburyi* (Onderstepoort Tick Collection 2820, RML 105789, laboratory reared, progeny of ♀ collected at Dordrecht, Eastern Cape Province, South Africa, on 8 August 1945). Nymph: (a) capitulum, dorsal; (b) capitulum, ventral; (c) scutum; (d) coxae. Scale bars represent 0.10 mm. SEMs (a) & (c) by M.D. Corwin; (b) & (d) by R.G. Robbins. (From Walker, 1990, figs 15, 16, 18 & 19, with kind permission from the Editor, *Onderstepoort Journal of Veterinary Research.*)

Figure 115 (*opposite*). *Rhipicephalus lounsburyi* (Onderstepoort Tick Collection 2820, RML 105789, laboratory reared, progeny of ♀ collected at Dordrecht, Eastern Cape Province, South Africa, on 8 August 1945). Male: (a) capitulum, dorsal; (b) spiracle; (c) adanal plates. Female: (d) capitulum, dorsal; (e) scutum; (f) genital aperture. Scale bars represent 0.10 mm. SEMs (a), (b) & (c) by M.D. Corwin; (d), (e) & (f) by J.F. Putterill. (From Walker, 1990, figs 3, 7–9, 11 & 13, with kind permission from the Editor, *Onderstepoort Journal of Veterinary Research.*)

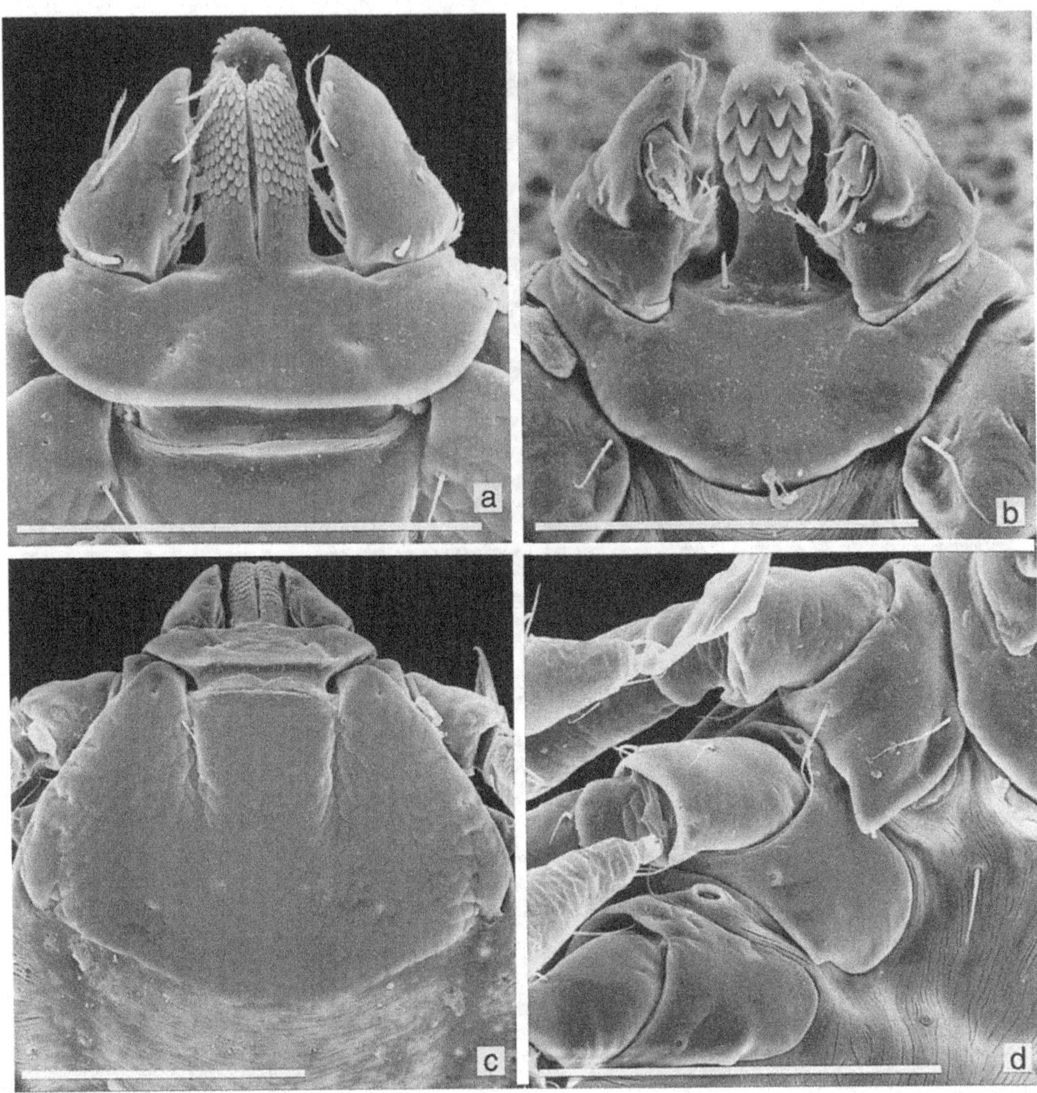

Figure 117. *Rhipicephalus lounsburyi* (Onderstepoort Tick Collection 2820, RML 105789, laboratory reared, progeny of ♀ collected at Dordrecht, Eastern Cape Province, South Africa, on 8 August 1945). Larva: (a) capitulum, dorsal; (b) capitulum, ventral; (c) scutum; (d) coxae. Scale bars represent 0.10 mm. SEMs (a), (c) & (d) by R. G. Robbins; (b) by M.D. Corwin. (From Walker, 1990, figs 21–24, with kind permission from the Editor, *Onderstepoort Journal of Veterinary Research.*)

far are sheep (Table 30), on which they attach on the feet, between the claws and on the heels. They have also been found on one wild feline, a caracal, and various antelopes. They are most active in autumn, winter and early spring (April to September) (Horak *et al.*, 1986, 1991).

The only known host of the immature stages is a four-striped grass mouse, from which a single nymph was recovered, at the same locality at which adult ticks were collected from bontebok and grey rhebok (I.G.H., unpublished data).

Zoogeography

This tick has been recorded only in South Africa (Map 34), most commonly in Eastern Cape

Province, especially in the mountainous areas of Barkly East, Dordrecht and Cradock Districts. In the Western Cape it has been found near Swellendam and, further north, between Clanwilliam and Graafwater. In Kwazulu-Natal there is an unconfirmed record from the Impendle area.

In the eastern part of its range, therefore, *R. lounsburyi* occurs mainly in mountainous and hilly areas with mean annual rainfalls of over 500 mm and various types of grassland vegetation. In the Western Cape, though, it has been found in somewhat lower areas with a winter rainfall regimen of approximately 500 mm annually and Cape shrubland (*fynbos*) vegetation.

Disease relationships

Rhipicephalus lounsburyi is not known to transmit any pathogenic organisms, but the adults themselves may have a deleterious effect on sheep. Gertrud Theiler (unpublished data, 1958) commented: 'Although sheep do not become lame the farmers contend they lose condition and in some cases sheep have even died as a consequence. Lambs also are attacked severely.'

REFERENCES

Horak, I.G., Fourie, L.J., Novellie, P.A. & Williams, E.J. (1991). Parasites of domestic and wild animals in South Africa. XXVI. The mosaic of ixodid tick infestations on birds and mammals in the Mountain Zebra National Park. *Onderstepoort Journal of Veterinary Research*, **58**, 125–36.

Horak, I.G., Sheppey, K., Knight, M.M. & Beuthin, C.L. (1986). Parasites of domestic and wild animals in South Africa. XXI. Arthropod parasites of vaal ribbok, bontebok and scrub hares in the western Cape Province. *Onderstepoort Journal of Veterinary Research*, **53**, 187–97.

Theiler, G. & Robinson, B.N. (1953). Ticks in the South African Zoological Survey Collection. Part VIII. Six lesser known African rhipicephalids. *Onderstepoort Journal of Veterinary Research*, **26**, 93–136 + map.

Walker, J.B. (1990). Two new species of ticks from southern Africa whose adults parasitize the feet of ungulates: *Rhipicephalus lounsburyi* n.sp. and *Rhipicephalus neumanni* n.sp. (Ixodoidea, Ixodidae). *Onderstepoort Journal of Veterinary Research*, 57, 57–75.

Also see the following Basic Reference (p. 214): Theiler (1962).

RHIPICEPHALUS LUNULATUS NEUMANN, 1907

The specific name *lunulatus*, from the Latin *lunatus* meaning 'crescent-shaped', refers to the crescentic posterior margins of the male's adanal plates.

Synonyms

attenuatus; glyphis; simus lunulatus.

Diagnosis

A medium-sized dark brown tick.

Male (Figs 118(a), 119(a) to (c))
Capitulum longer than broad, length × breadth ranging from 0.50 mm × 0.48 mm to 0.64 mm × 0.58 mm. Basis capituli with lateral angles at about anterior third of its length, acute. Palps elongate, with pedicel of article I visible dorsally and markedly narrower than article II. Conscutum length × breadth ranging from 2.36 mm × 1.61 mm to 3.07 mm × 2.07 mm; anterior process of coxae I well developed. In engorged specimens body wall expanded posterolaterally and posteriorly. Eyes marginal, flat, indistinct. Cervical pits deep, comma-shaped. Conspicuous setiferous punctations along external cervical margins, in '*simus*' pattern on central conscutum (see p. 416) and around posterolateral grooves; otherwise conscutum in general smooth and impunctate. Marginal lines shallow, demarcated by large punctations, almost reaching eyes. Posteromedian and posterolateral grooves, when present, merely indicated by slight inconspicuous depressions. Ventrally spiracles elongate, curving gently

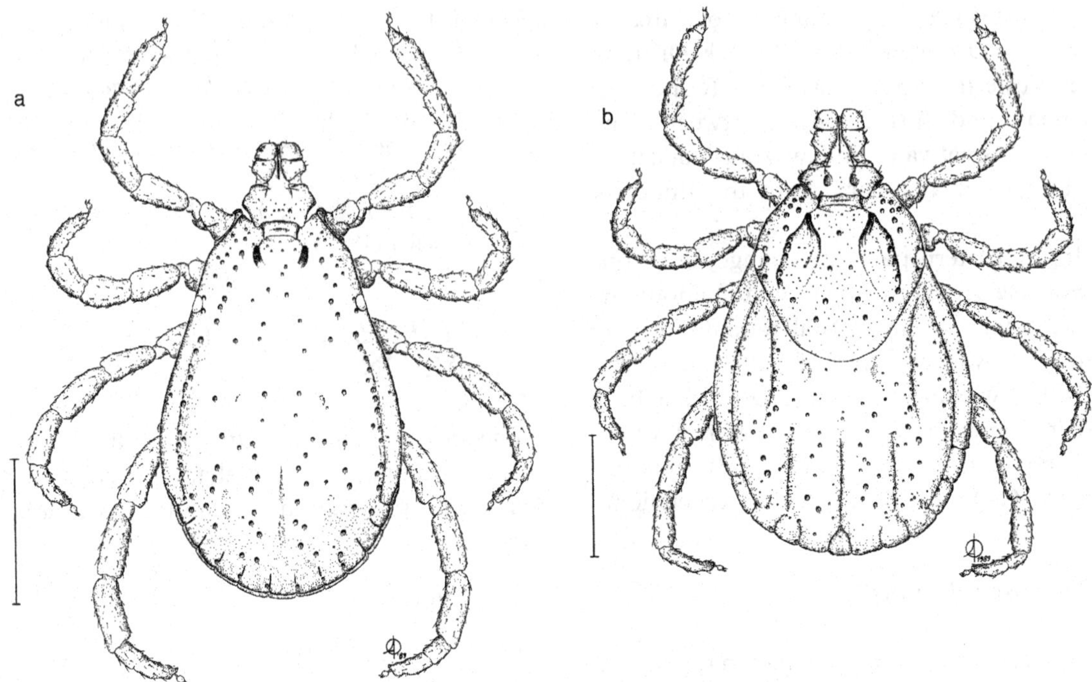

Figure 118. *Rhipicephalus lunulatus* (RML 116938, laboratory reared, progeny of ♀ collected from bovine, Ruware Ranch, Chiredzi, S. E. Zimbabwe, in the early 1980s by J. Colborne). (a) Male, dorsal; (b) female, dorsal. Scale bars represent 1 mm. A. Olwage *del.*

towards the dorsal surface. Adanal plates elongate, posterointernal margins broadly rounded, posteroexternal margins extended into narrowly elongated cusps; accessory adanal plates elongate triangular projections.

Female (Figs 118(b), 119(d) to (f))

Capitulum nearly as long as broad, length × breadth ranging from 0.59 mm × 0.66 mm to 0.79 mm × 0.79 mm. Basis capituli with lateral angles at about anterior third of its length; porose areas small, round, nearly twice their own diameter apart. Palps elongate, with pedicel of article I easily visible dorsally and markedly narrower than article II, giving the palps a stalked appearance. Scutum length × breadth ranging from 1.34 mm × 1.39 mm to 1.59 mm × 1.59 mm. Eyes at lateral angles, flat to slightly raised and edged dorsally by a few punctations. Cervical grooves deep and convergent anteriorly, becoming shallower and divergent. Large setiferous punctations along external cervical margins, on scapulae and scattered on central part of scutum, interspersed with very fine punctations. Ventrally genital aperture a very broad shallow U-shape.

Nymph (Fig. 120)

Capitulum triangular in outline dorsally, much broader than long, length × breadth ranging from 0.16 mm × 0.21 mm to 0.18 mm × 0.22 mm. Basis capituli well over three times as broad as long, with sharp lateral angles projecting over the scapulae, posterior border almost straight; ventrally with bluntly rounded spurs. Palps narrow, tapering towards apices and inclined inwards. Scutum length × breadth ranging from 0.48 mm × 0.32 mm to 0.50 mm × 0.36 mm; narrowly elongate in shape, posterior margin a deep curve. Eyes at lateral angles, long, narrow, slightly raised. Cervical grooves short, convergent; cervical fields long, narrow, indistinct. Ven-

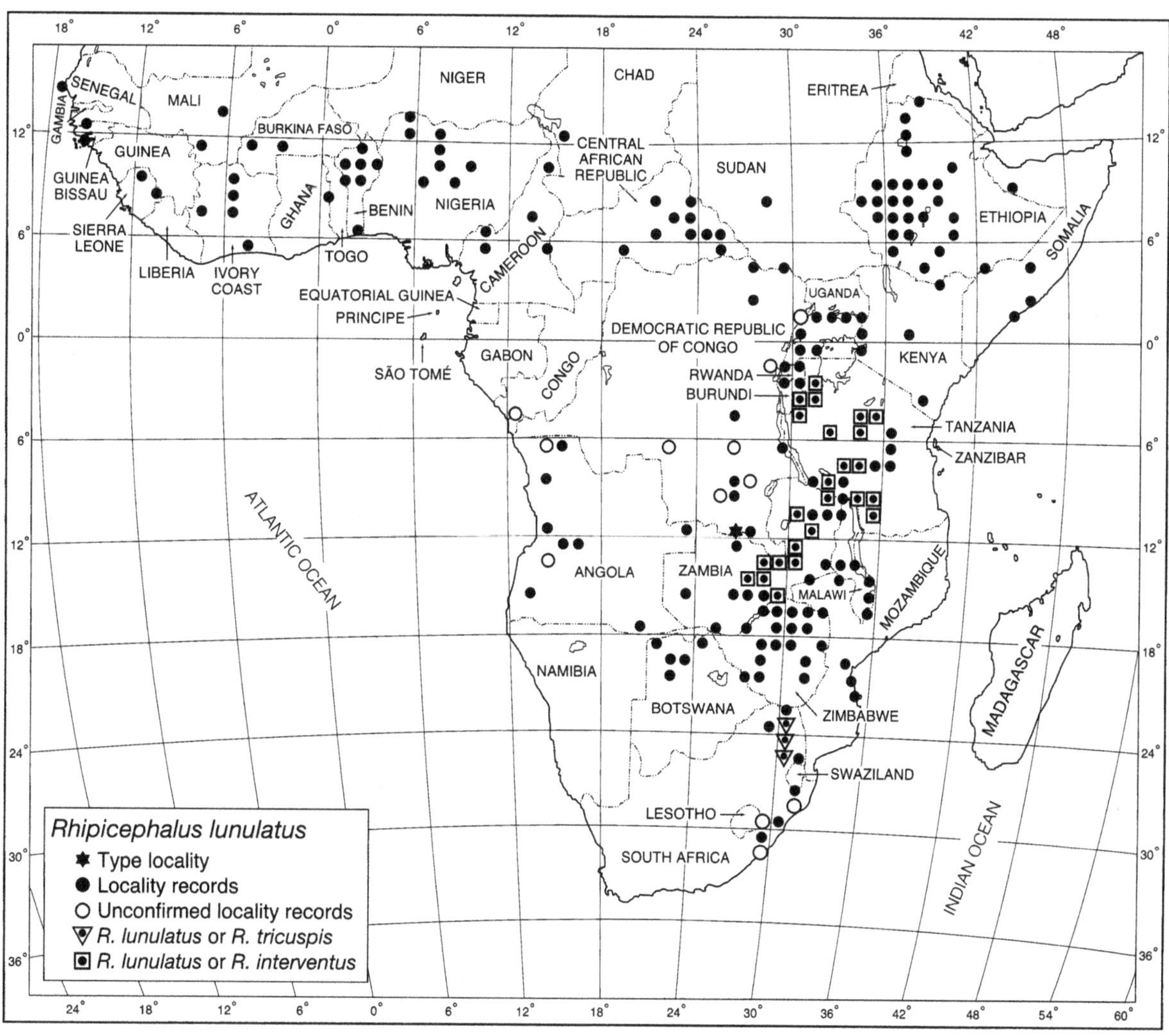

Map 35. *Rhipicephalus lunulatus*: distribution. (Based on Walker *et al.*, 1988).

trally coxae I each with a long bluntly-pointed external spur and a shorter bluntly-pointed internal spur; coxae II to IV each with a small external spur, decreasing in size from II to IV.

Larva (Fig. 121)

Capitulum much broader than long, length × breadth ranging from 0.114 mm × 0.178 mm to 0.134 mm × 0.202 mm. Basis capituli about four times as broad as long, with sharp lateral angles overlapping the scapulae; ventrally with short blunt spurs. Palps broad basally, tapering towards apices and inclined inwards. Scutum much broader than long, length × breadth ranging from 0.200 mm × 0.274 mm to 0.219 mm × 0.282 mm; posterior margin a wide shallow curve. Eyes at widest point, flat. Cervical grooves short, very slightly convergent. Ventrally coxae I each with a short, broad, bluntly-rounded internal spur; coxae II each with a slight salient ridge on posterior border; coxae III unarmed.

Notes on identification

Rhipicephalus lunulatus was first sunk as a synonym of *R. tricuspis* by Warburton (1912), then resurrected by Nuttall & Warburton (1916). It

Table 31. *Host records of* Rhipicephalus lunulatus

Hosts	Number of records: Confirmed or accepted as valid	Unconfirmed
Domestic animals		
Cattle	336	6
Sheep	60	
Goats	22	
Camels	2	
Horses	8	
Pigs	26	
Dogs	37	
Wild animals		
Bat ('flying fox')	1	
Side-striped jackal (*Canis adustus*)	3	
Black-backed jackal (*Canis mesomelas*)	1	
Jackal (*Canis* sp.)	2	
Caracal (*Caracal caracal*)	1	
Serval (*Leptailurus serval*)	1	2
Lion (*Panthera leo*)	6	
Leopard (*Panthera pardus*)	2	
African civet (*Civettictis civetta*)		1
Burchell's zebra (*Equus burchellii*)	2	
Black rhinoceros (*Diceros bicornis*)	1	
Aardvark (*Orycteropus afer*)	3	
Warthog (*Phacochoerus* sp., probably *P. africanus*)	21	3
Forest hog (*Hylochoerus meinertzhageni*)	1	
Bushpig (*Potamochoerus* sp.)	13	7
Giraffe (*Giraffa camelopardalis*)	1	
Impala (*Aepyceros melampus*)	12 (including 2 with larvae)	
Jackson's hartebeest (*Alcelaphus buselaphus jacksoni*)	1	
Lelwel hartebeest (*Alcelaphus buselaphus lelwel*)		1
Blue wildebeest (*Connochaetes taurinus*)	1	
Tiang (*Damaliscus lunatus tiang*)		1
Lichtenstein's hartebeest (*Sigmoceros lichtensteinii*)	2	
Oribi (*Ourebia ourebi*)	3	
African buffalo (*Syncerus caffer*)	41	
Eland (*Taurotragus oryx*)	4	
Mountain nyala (*Tragelaphus buxtoni*)	1	
Bushbuck (*Tragelaphus scriptus*)	6	
Greater kudu (*Tragelaphus strepsiceros*)	8	
Common duiker (*Sylvicapra grimmia*)	4	
Roan antelope (*Hippotragus equinus*)	2	1
Sable antelope (*Hippotragus niger*)	1	
Waterbuck (*Kobus ellipsiprymnus*)	6	

Table 31. (*cont.*)

Hosts	Number of records	
	Confirmed or accepted as valid	Unconfirmed
Wild animals (*cont.*)		
Reedbuck (*Redunca arundinum*)	1	
Reedbuck (*Redunca* sp.)	1	
'Antelope'	1	
Brush-furred rat (*Lophuromys flavopunctatus*)	1	
Multimammate mouse (*Mastomys* sp.)	3 (nymphs)	
Crested porcupine (*Hystrix cristata*)	1	
Scrub hare (*Lepus saxatilis*)	1 (nymph)	
Birds		
White-faced whistling duck (*Dendrocygna viduata*)	1	
Red-knobbed coot (*Fulica cristata*)	1	
Humans	5	

was synonymized with *R. tricuspis* for the second time by Theiler (1947). Her finding was accepted for nearly 30 years by some tick workers in Africa but others continued to insist that *R. tricuspis* and *R. lunulatus* are both distinct species. The issue was reviewed in detail, and *R. lunulatus* formally re-established as a full species, by Walker *et al.* (1988). Additional information that was not included in their review was published by Santos Dias (1983–84a, b, 1987), Merlin, Tsangueu & Rousvoal (1987) and Horak *et al.* (1989). Subsequently Walker, Pegram & Keirans (1995) described a third species in this group, *R. interventus* (p. 228).

Hosts

A three-host species (Saratsiotis, 1977; Colborne, 1985). *Rhipicephalus lunulatus* adults have been recorded from many species of domestic and wild animals (Table 31). Amongst domestic animals it has been collected most commonly from cattle, sheep, goats, pigs and also dogs. Its commonest wild hosts are the African buffalo, warthog and bushpig. It has also been found on many antelopes, especially the larger species, several carnivores, various other animals, two species of birds and occasionally on humans.

Nymphs have been collected from the multimammate mouse and a scrub hare (Zieger *et al.*, 1998).

On cattle *R. lunulatus* adults attach primarily on the legs, including the feet, and in the tail switch (Colborne, 1985). On bushbuck and common duikers adults attach on the lower legs and feet (Horak *et al.*, 1989). Throughout its range this tick appears to be most active during the rainy season. In south-eastern Zimbabwe adult ticks are most abundant on cattle during November and December (Colborne, 1985).

Zoogeography

Rhipicephalus lunulatus is widespread in the Afrotropical region (Map 35). It occurs in a variety of habitats, ranging from lowland rain forest (Ivory Coast) to woodland and bushland,

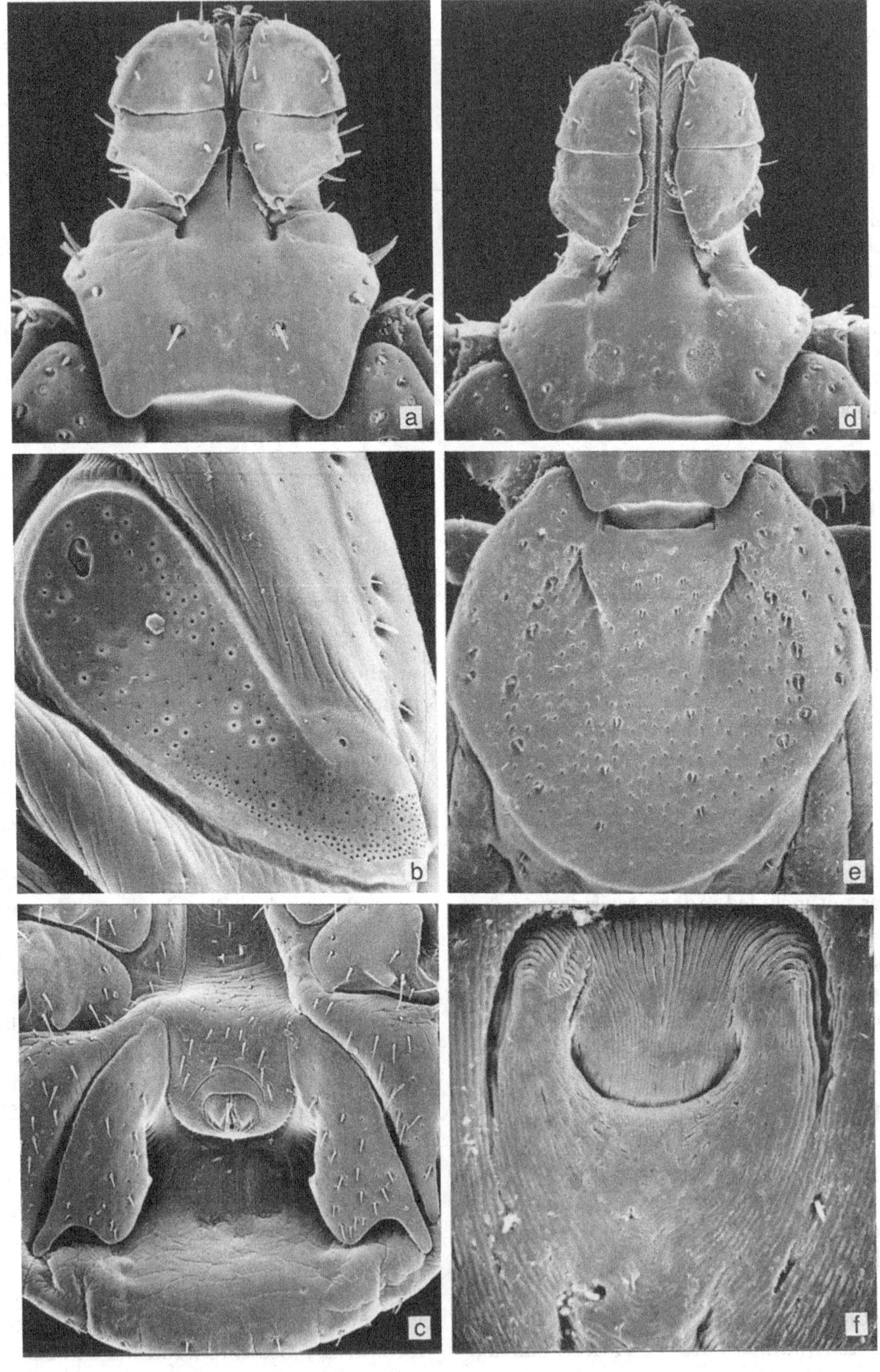
a
b
c
d
e
f

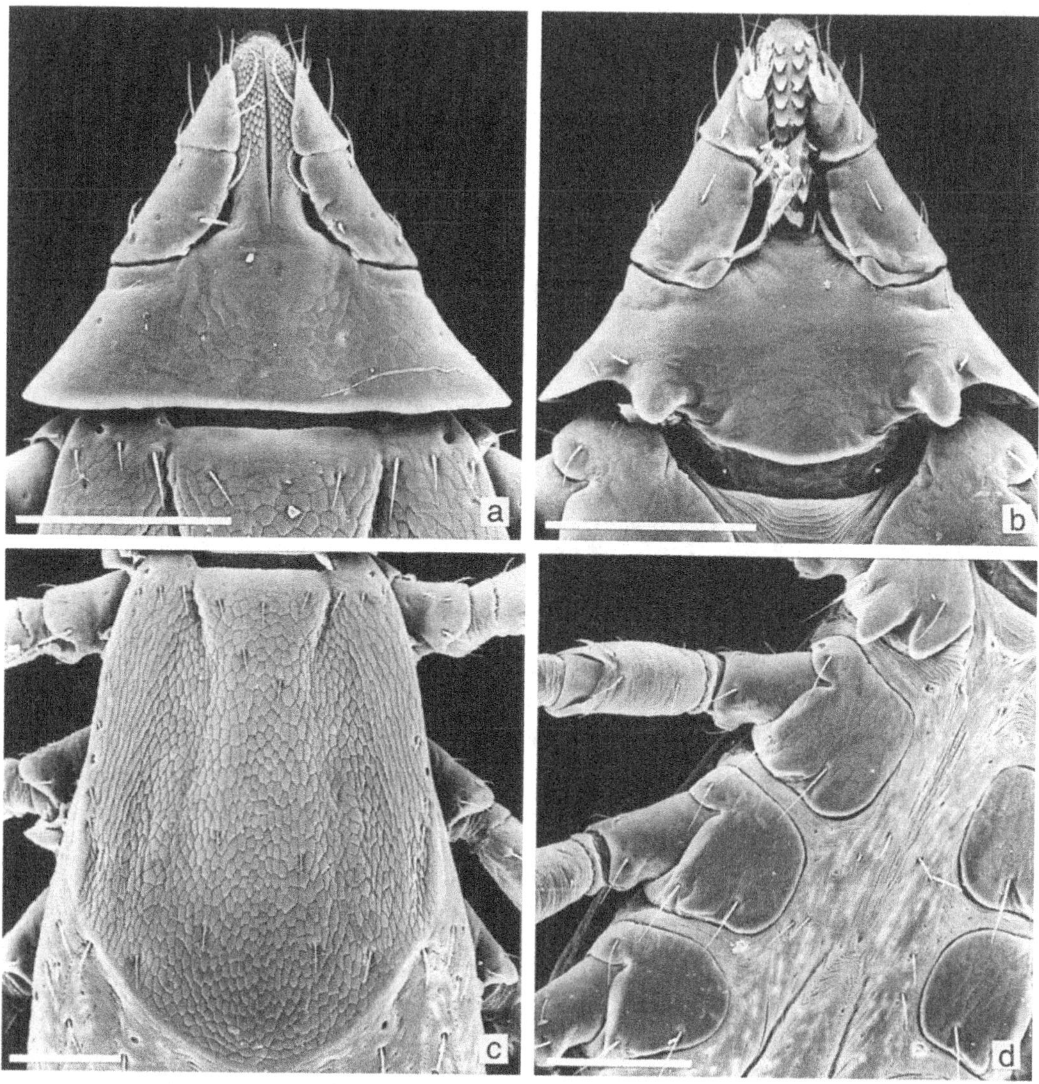

Figure 120 (*above*). *Rhipicephalus lunulatus* (RML 116938, laboratory reared, progeny of ♀ collected from bovine, Ruware Ranch, Chiredzi, S.E. Zimbabwe, in the early 1980s by J. Colborne). Nymph: (a) capitulum, dorsal; (b) capitulum, ventral; (c) scutum; (d) coxae. Scale bars represent 0.10 mm. SEMs by R.G. Robbins. (From Walker *et al.*, 1988, figs 36, 37, 39 & 40, with kind permission from Kluwer Academic Publishers.)

Figure 119 (*opposite*). *Rhipicephalus lunulatus* (Onderstepoort Tick Collection 2727ii, RML 105010, collected from cattle, Mazabuka, Zambia, 25 November 1952 by J.G. Matthysse). Male: (a) capitulum, dorsal; (b) spiracle; (c) adanal plates. Female: (d) capitulum, dorsal; (e) scutum; (f) genital aperture. (Scales not available). SEMs by M.D. Corwin. (From Walker *et al.*, 1988, figs 24, 28–30, 33 & 34, with kind permission from Kluwer Academic Publishers.)

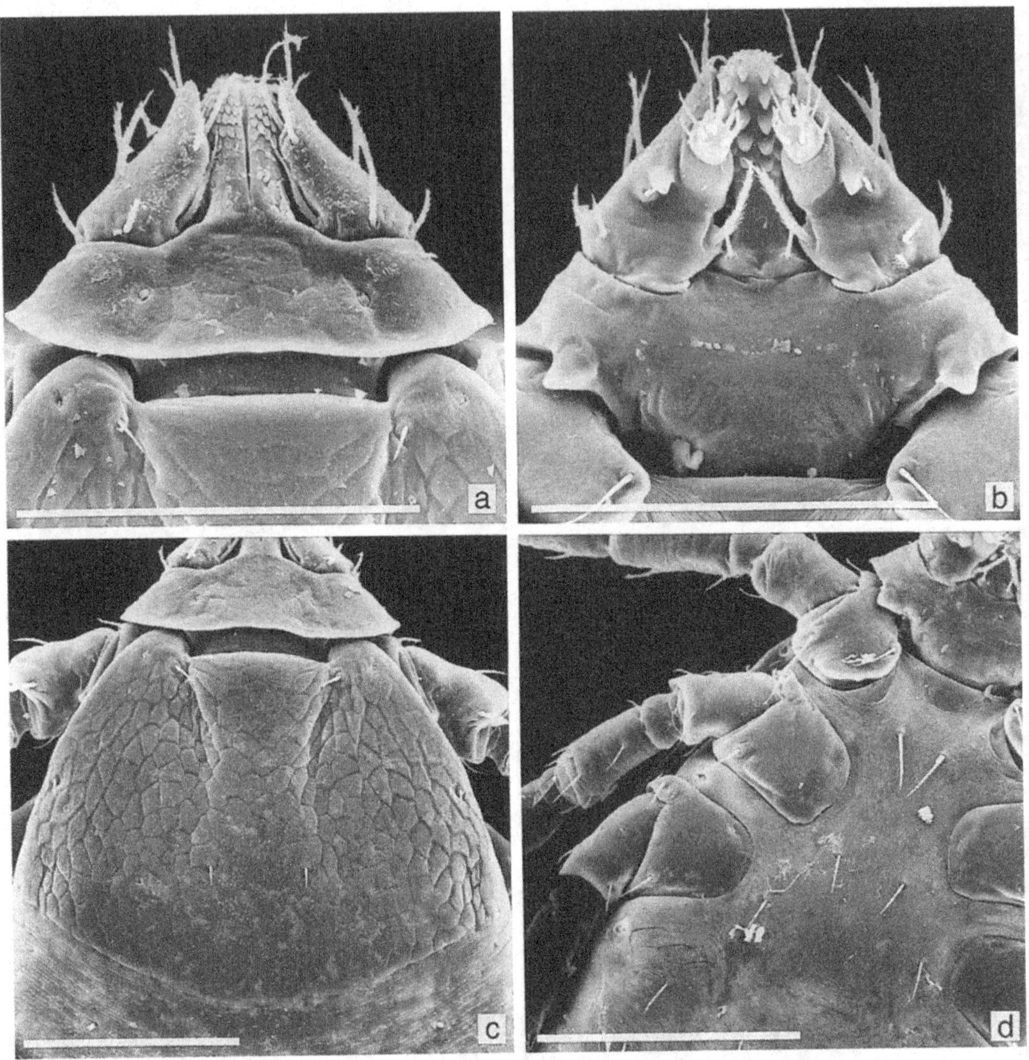

Figure 121. *Rhipicephalus lunulatus* (RML 116938, laboratory reared, progeny of ♀ collected from bovine, Ruware Ranch, Chiredzi, S.E. Zimbabwe, in the early 1980s by J. Colborne). Larva: (a) capitulum, dorsal; (b) capitulum, ventral; (c) scutum; (d) coxae. Scale bars represent 0.10 mm. SEMs by R.G. Robbins. (Figures (a), (b) & (d) from Walker *et al.*, 1988, figs 42, 43 & 45, with kind permission from Kluwer Academic Publishers.)

including dry Somalia–Masai *Acacia–Commiphora* deciduous bushland and thicket (Somalia and Tanzania). Most records, though, are from different types of woodland. These include the West African Sudanian woodland, in which *Isoberlinia* is dominant and, in central Africa, both wet and dry types of miombo with abundant *Brachystegia* and *Julbernardia*. Ticks have also been collected from animals in a habitat consisting of mixed mountain grassland, indigenous montane forest and plantations of exotic trees.

Disease relationships

Rhipicephalus lunulatus is suspected of being a vector of *Babesia trautmanni*, the cause of porcine piroplasmosis, in West Africa (Tendeiro, 1952).

In Zimbabwe this tick has been associated with tick paralysis in sheep and lambs in summer (Theiler, 1962, as *R. tricuspis*). This syndrome was also reported in calves by Lawrence & Norval (1979, also as *R. tricuspis*), who said that, although there were no published reports on this paralysis, it is well known to many local farmers and veterinarians. They observed: 'Affected animals show a typical flaccid paralysis and generally recover after the removal of the ticks. Verbal reports of tick paralysis from scattered localities throughout the country go back at least thirty years'.

REFERENCES

Colborne, J.R.A. (1985). The life cycle of *Rhipicephalus lunulatus* Neumann, 1907 (Acarina: Ixodidae) under laboratory conditions, with notes on its ecology in Zimbabwe. *Experimental and Applied Acarology*, **1**, 317–25.

Horak, I.G., Keep, M.E., Spickett, A.M. & Boomker, J. (1989). Parasites of domestic and wild animals in South Africa. XXIV. Arthropod parasites of bushbuck and common duiker in the Weza State Forest, Natal. *Onderstepoort Journal of Veterinary Research*, **56**, 63–6.

Lawrence, J.A. & Norval, R.A.I. (1979). A history of ticks and tick-borne diseases of cattle in Rhodesia. *Rhodesian Veterinary Journal*, **10**, 28–40.

Merlin, P., Tsangueu, P. & Rousvoal, D. (1987). Dynamique saisonnière de l'infestation des bovins par les tiques (Ixodoidea) dans les hauts plateaux de l'Ouest du Cameroun. II. Élevage extensif traditionnel. *Revue d'Élevage et de Médecine Vétérinaire des Pays Tropicaux*, **40** (nouvelle série), 133–40.

Neumann, L.G. (1907). Notes sur les Ixodidés – V. *Archives de Parasitologie*, **11**, 215–32.

Nuttall, G.H.F. & Warburton, C. (1916). Ticks of the Belgian Congo and the diseases they convey. *Bulletin of Entomological Research*, **6**, 313–52.

Santos Dias, J.A.T. (1983–84a). Subsídios para o conhecimento da fauna ixodológica de Angola. *Garcia de Orta, Sér. Zool., Lisboa*, **11**, 57–68.

Santos Dias, J.A.T. (1983–84b). Alguns ixodídeos (Acarina – Ixodoidea – Ixodidae) coligidos em Angola pelo Dr Crawford Cabral. *Garcia de Orta, Sér. Zool., Lisboa*, **11**, 69–76.

Santos Dias, J.A.T. (1987). Algumas observações sobre a fauna ixodológica (Acarina, Ixodoidea) de Moçambique, com a descrição de uma nova espécie do genero *Boophilus* Curtice, 1891. *Garcia de Orta, Sér. Zool., Lisboa*, **14**, 17–26.

Saratsiotis, A. (1977). Etudes sur les *Rhipicephalus* (Acariens, Ixodida) de l'Ouest africain. I. Réalisation du cycle évolutif des tiques *Rh. muhsamae* Morel et Vassiliades, 1965 et *Rh. lunulatus* Neumann, 1907. II. Etude comparative des stades préimaginaux dans le complexe *Rh. simus Rh. senegalensis* de l'Ouest africain. *Revue d'Élevage de Médecine Vétérinaire des Pays Tropicaux*, **30** (nouvelle série), 51–9.

Tendeiro, J. (1952). Infestação natural do porco da Guiné pela *Babesia trautmanni* (Knut e Du Toit). *Boletim Cultural da Guiné Portuguesa*, 7, 359–64.

Walker, J.B., Keirans, J.E., Pegram, R.G. & Clifford, C.M. (1988). Clarification of the status of *Rhipicephalus tricuspis* Dönitz, 1906 and *Rhipicephalus lunulatus* Neumann, 1907 (Ixodoidea, Ixodidae). *Systematic Parasitology*, **12**, 159–86.

Walker, J.B., Pegram, R.G. & Keirans, J.E. (1995). *Rhipicephalus interventus* sp. nov. (Acari: Ixodidae), a new tick species closely related to *Rhipicephalus tricuspis* Dönitz, 1906 and *Rhipicephalus lunulatus* Neumann, 1907, from East and Central Africa. *Onderstepoort Journal of Veterinary Research*, **62**, 89–95.

Zieger, U., Horak, I.G., Cauldwell, A.E. & Uys, A.C. (1998). Ixodid tick infestations of wild birds and mammals on a game ranch in Central Province, Zambia. *Onderstepoort Journal of Veterinary Research*, **65**, 113–24.

Also see the following Basic References (pp. 12–14): Theiler (1947, 1962); Warburton (1912).

RHIPICEPHALUS MACULATUS NEUMANN, 1901

The specific name *maculatus*, from the Latin meaning 'spotted', refers to the characteristic diffuse light-coloured pattern on the scutum of the adults.

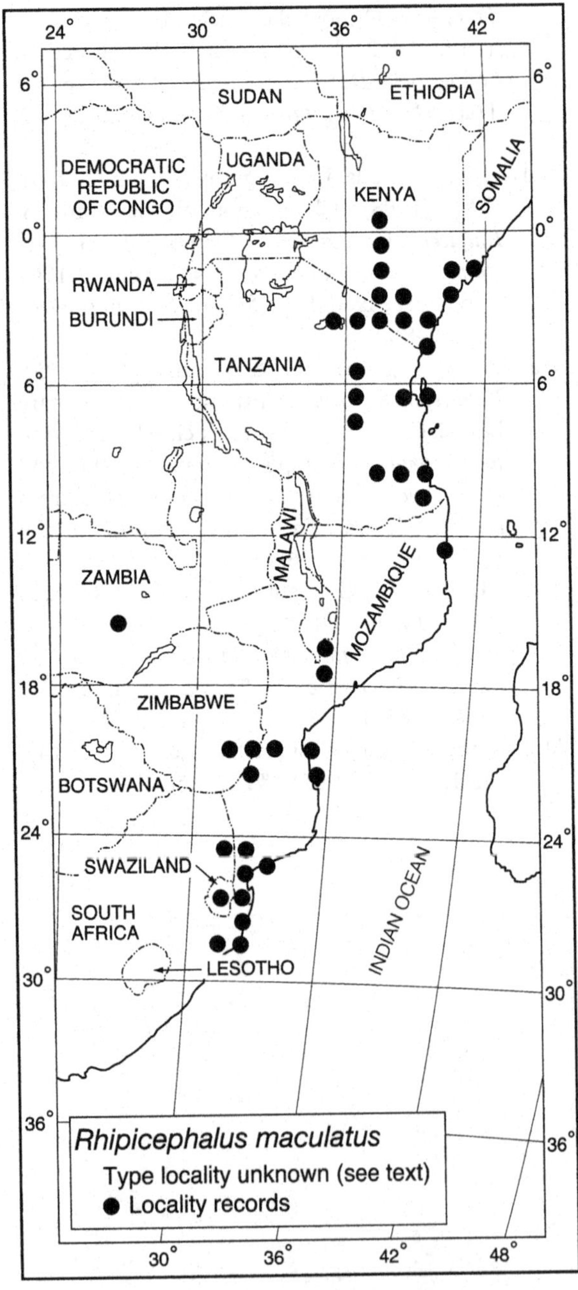

Map 36. *Rhipicephalus maculatus*: distribution.

Synonym

ecinctus (in part, male).

Diagnosis

A large ornate species with light-coloured patches commonly present on the shiny, predominantly dark brown, scutum of the adults.

Male (Figs 122(a) to (c), 123(a) to (c))
Capitulum much longer than broad, length × breadth ranging from 0.59 mm × 0.50 mm to 1.13 mm × 0.92 mm. Basis capituli with very short rounded lateral angles anteriorly, narrowing posteriorly. Palps broad, rounded apically. Conscutum length × breadth ranging from 2.38 mm × 1.90 mm to 4.54 mm × 3.42 mm, broadly ovate in shape; sharp anterior process present on coxae I. Eyes flat, virtually flush with surface of conscutum. Cervical pits large, deep; internal cervical margins sometimes, but not always, faintly indicated. Marginal lines absent. Posteromedian and posterolateral grooves usually absent but sometimes represented by faint shallow depressions. Punctations typically sparse; medium-sized setiferous punctations present on scapulae, in irregular rows along external cervical margins and marking the positions of marginal lines, also in small clusters laterally and posterolaterally, elsewhere scattered; sometimes interspersed with very fine punctations. A diffuse pattern of creamy-coloured patches, which is associated with some of the larger punctations and very variable in extent, is commonly, but not invariably, present. Legs increase in size from I to IV. Ventrally spiracles comma-shaped

Figure 122 (*opposite*). *Rhipicephalus maculatus* [collected from elephant (*Loxodonta africana*), Garsen, Lower Tana River, Kenya on 20 March 1960 by R.A.F. Hurt]. (a) Male, dorsal; (b) and (c) male, variations in colour pattern, after Zumpt (1942); (d) female, dorsal; (e) female, variation in colour pattern, after Zumpt (1942). Scale bars represent 1 mm for (a) and (d). A. Olwage *del.*

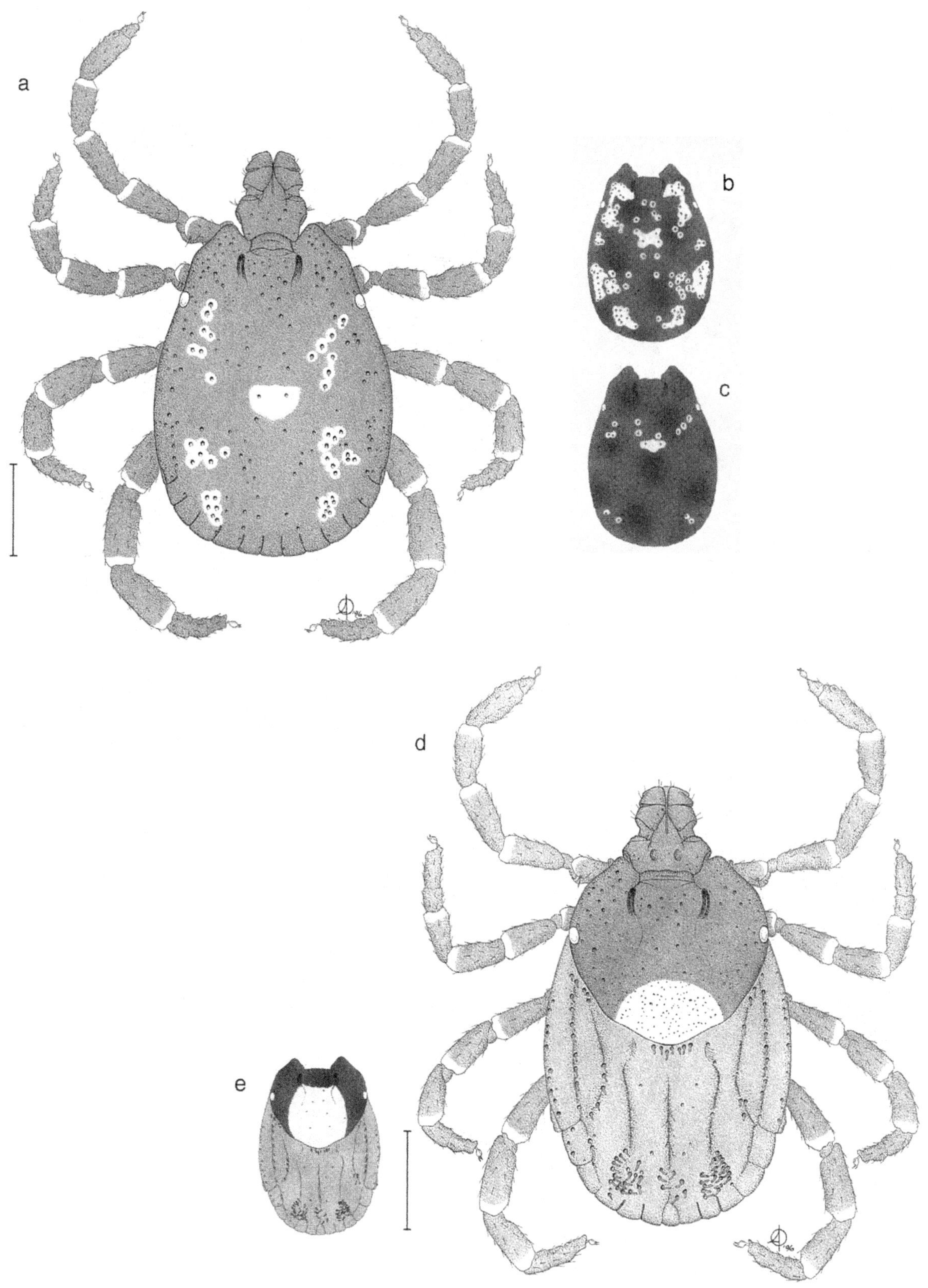
a
b
c
d
e

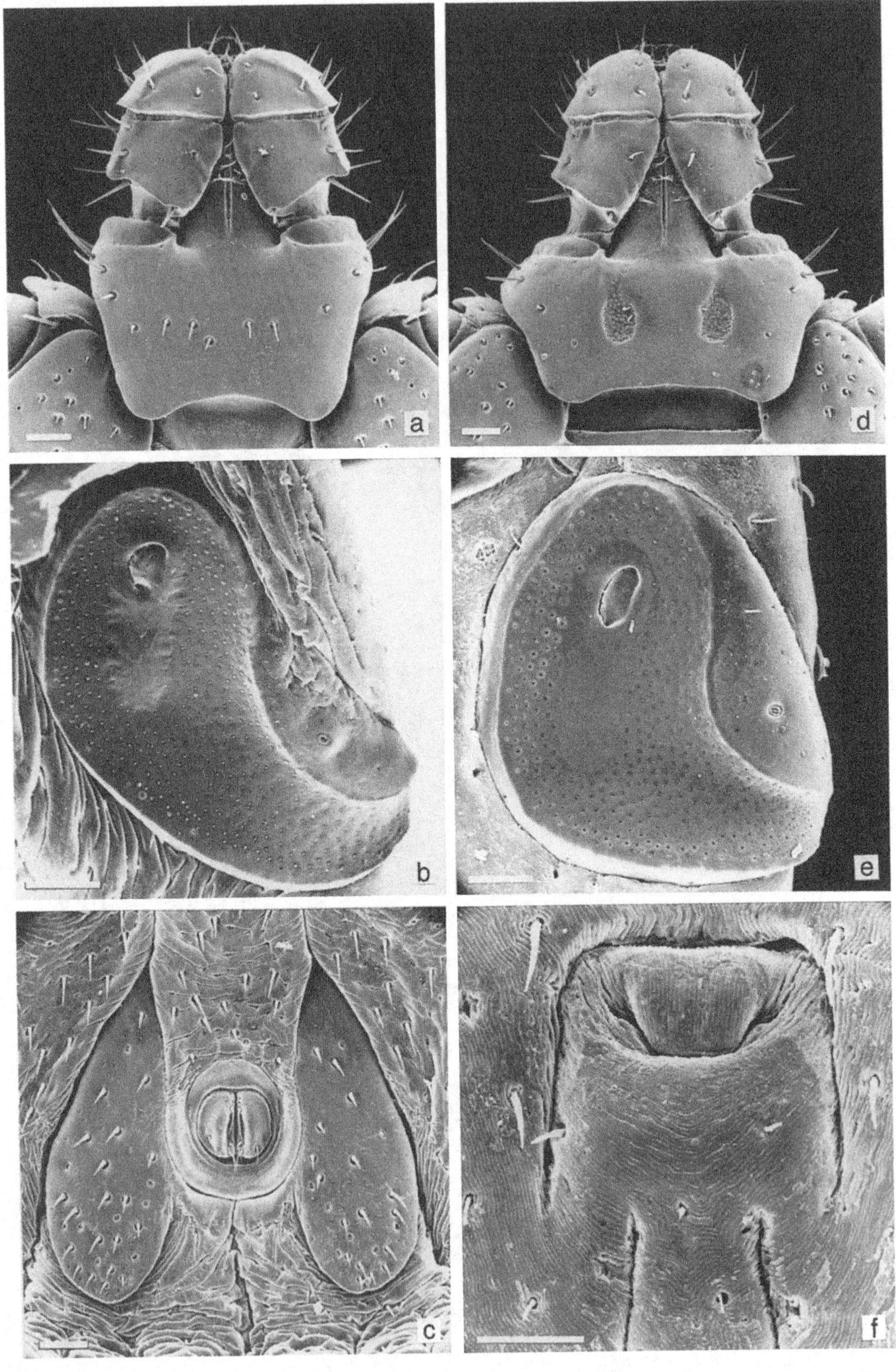
a
b
c
d
e
f

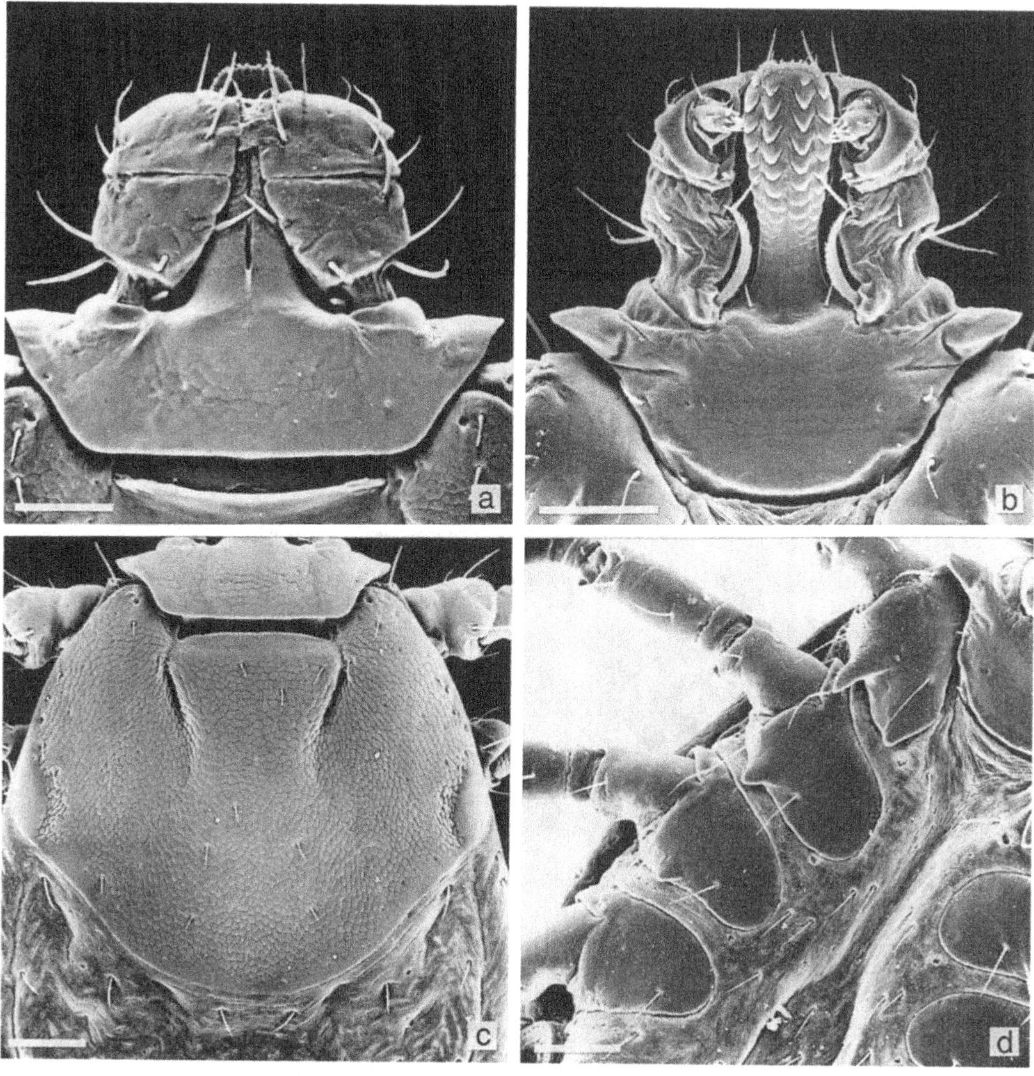

Figure 124 (*above*). *Rhipicephalus maculatus* [B.S. 703/-, RML 66311, laboratory reared, original ♀ collected from African elephant (*Loxodonta africana*), Maboyo, 12 miles west of Liwale, southern Tanzania, on 5 July 1956 by B.D. Nicholson]. Nymph: (a) capitulum, dorsal; (b) capitulum, ventral; (c) scutum; (d) coxae. Scale bars represent 0.10 mm. SEM (a) by P. Hill; (b), (c) & (d) by M.D. Corwin.

Figure 123 (*opposite*). *Rhipicephalus maculatus* [B.S. 703/-, RML 66311, laboratory reared, original ♀ collected from African elephant (*Loxodonta africana*), Maboyo, 12 miles west of Liwale, southern Tanzania, on 5 July 1956 by B.D. Nicholson]. Male: (a) capitulum, dorsal; (b) spiracle; (c) adanal plates. Female: (d) capitulum, dorsal; (e) spiracle; (f) genital aperture. Scale bars represent 0.10 mm. SEMs by M.D. Corwin.

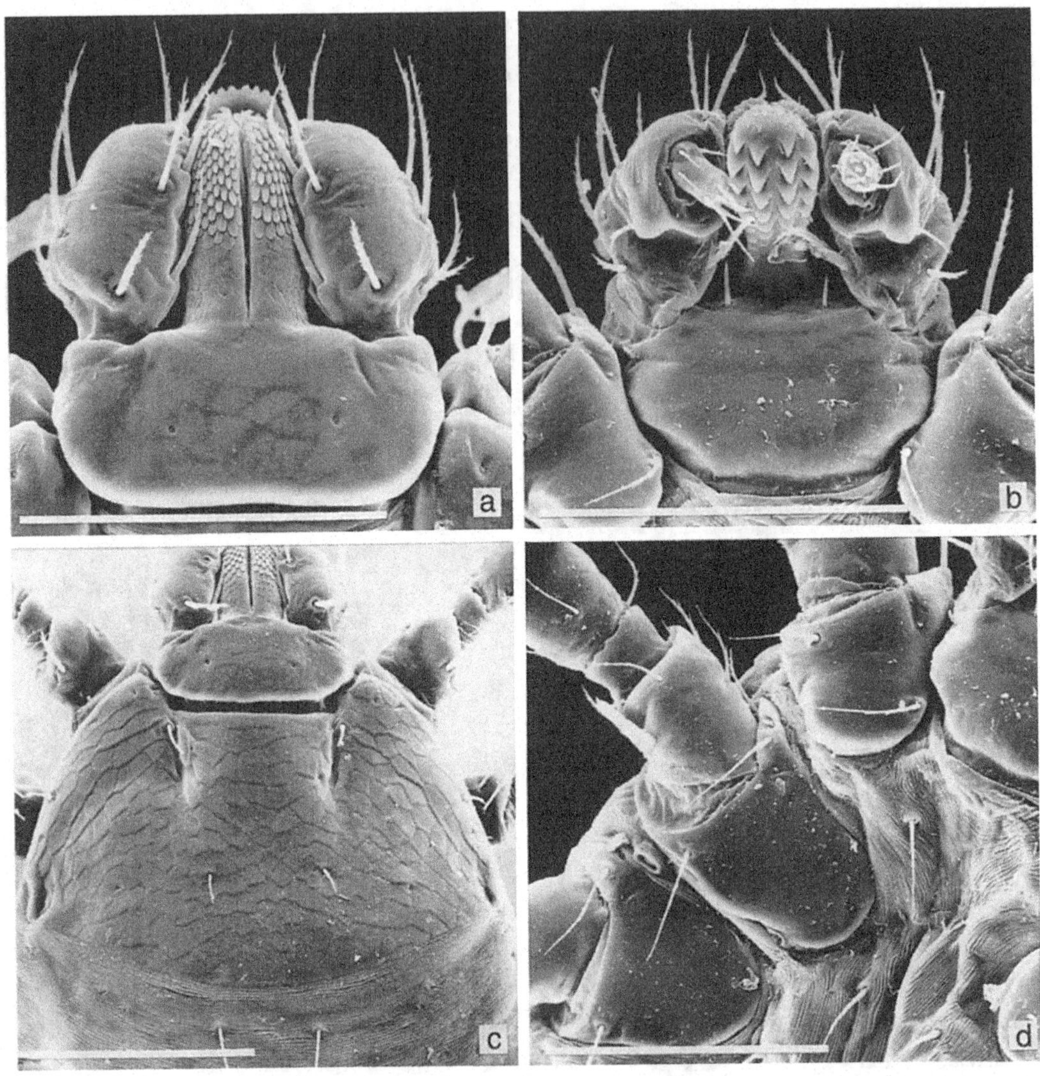

Figure 125. *Rhipicephalus maculatus* [B.S. 703/-, RML 66311, laboratory reared, original ♀ collected from African elephant (*Loxodonta africana*), Maboyo, 12 miles west of Liwale, southern Tanzania, on 5 July 1956 by B.D. Nicholson]. Larva: (a) capitulum, dorsal; (b) capitulum, ventral; (c) scutum; (d) coxae. Scale bars represent 0.10 mm. SEMs by M.D. Corwin.

with a short curved dorsal prolongation. Adanal plates elongate, almost teardrop-shaped, with slight concavities adjacent to the anus; accessory adanal plates absent.

Female (Figs 122(d,e), 123(d) to (f))

Capitulum slightly longer than broad, length × breadth ranging from 0.77 mm × 0.74 mm to 1.12 mm × 1.11 mm. Basis capituli with short, broad lateral angles in anterior third of its length; porose areas medium-sized, nearly twice their own diameter apart. Palps with article I long and narrow, giving them a somewhat stalked appearance. Scutum broader than long, length × breadth ranging from 1.39 mm × 1.55 mm to 2.18 mm × 2.49 mm. Eyes not quite halfway back, flat, virtually flush with the scutum. Cervical pits large, deep; internal cervical margins only faintly indicated. A few medium-sized setiferous punctations present, mainly on scapulae and between cervical pits, elsewhere sparsely scattered, interspersed with

Table 32. *Host records of* Rhipicephalus maculatus

Hosts	Number of records
Domestic animals	
Cattle	34 (including nymphs)
Sheep	1
Goats	1
Pigs	1
Wild animals	
Cheetah (*Acinonyx jubatus*)	2
Lion (*Panthera leo*)	1
Spotted hyaena (*Crocuta crocuta*)	2
African civet (*Civettictis civetta*)	2 (immatures)
African elephant (*Loxodonta africana*)	33
Burchell's zebra (*Equus burchellii*)	2
White rhinoceros (*Ceratotherium simum*)	4
Black rhinoceros (*Diceros bicornis*)	12 (including immatures)
'Rhinoceros'	5 (including immatures)
Warthog (*Phacochoerus africanus*)	22 (including immatures)
Bushpig (*Potamochoerus larvatus*)	20 (including immatures)
'Wild pig'	1
'Pig/hog'	3
Impala (*Aepyceros melampus*)	4 (immatures)
Blue wildebeest (*Connochaetes taurinus*)	2
Lichtenstein's hartebeest (*Sigmoceros lichtensteinii*)	1
Suni (*Neotragus moschatus*)	3 (immatures)
Oribi (*Ourebia ourebi*)	1 (nymphs)
African buffalo (*Syncerus caffer*)	23 (including immatures)
Eland (*Taurotragus oryx*)	3
Nyala (*Tragelaphus angasii*)	83 (including immatures)
Lesser kudu (*Tragelaphus imberbis*)	1 (nymph)
Bushbuck (*Tragelaphus scriptus*)	8 (including immatures)
Greater kudu (*Tragelaphus strepsiceros*)	6 (including immatures)
Red forest duiker (*Cephalophus natalensis*)	16 (immatures)
Common duiker (*Sylvicapra grimmia*)	6 (including immatures)
'Duiker'	1 (immatures)
Sable antelope (*Hippotragus niger*)	1
Waterbuck (*Kobus ellipsiprymnus*)	2
Reedbuck (*Redunca arundinum*)	7 (including immatures)
Scrub hare (*Lepus saxatilis*)	6 (nymphs)
Reptiles	
Water leguaan (*Varanus niloticus*)	1
Humans	7

very fine punctations. Colour pattern, when present, consists of a single large creamy-coloured patch posteromedianly; this may reach eye level (see Fig. 122(e)) but rarely further forward. Alloscutum bearing a few short stout white setae, especially along the marginal lines and in two small patches posteriorly. Ventrally genital aperture broad, but shallow, its straight posterior border somewhat tucked in.

Nymph (Fig. 124)

Capitulum much broader than long, length × breadth ranging from 0.24 mm × 0.37 mm to 0.32 mm × 0.42 mm. Basis capituli over three times as broad as long, widest anteriorly where the exceptionally long sharp lateral angles project sideways over the scapulae; ventrally small bulges represent the ventral spurs. Palps narrow proximally, widening markedly towards their broad, mildly-convex apices. Scutum much broader than long, length × breadth ranging from 0.53 mm × 0.71 mm to 0.64 mm × 0.75 mm, posterior margin a broad deep curve. Eyes at widest point, a little over halfway back, prominent. Cervical pits deep, convergent, continuous with the shallow, divergent internal cervical margins, but external cervical margins not delimited. Ventrally coxae I each with a long sharp external spur and a shorter broader internal spur; coxae II to IV each with an external spur only, decreasing progressively in size.

Larva (Fig. 125)

Capitulum broader than long, length × breadth ranging from 0.133 mm × 0.140 mm to 0.145 mm × 0.161 mm. Basis capituli well over twice as broad as long, rounded laterally and posterolaterally. Palps constricted proximally, then widening markedly, apices truncated. Scutum much broader than long, length × breadth ranging from 0.223 mm × 0.365 mm to 0.243 mm × 0.391 mm, posterior margin a wide shallow curve. Eyes at widest point, prominent. Cervical pits short, deep, slightly convergent. Ventrally coxae I each with a broadly-rounded spur; coxae II and III each with mere indications of saliences on their posterior borders.

Notes on identification

The ♂ and 2 ♀♀ on which the original description of *R. maculatus* was based were said to have been collected from *Psytalla horrida* (syn. *Platymeris horrida*) from Cameroon. This is an assassin bug (family Reduviidae) (C.D. Eardley, pers. comm., 1997), not a beetle, as suggested by Theiler (1962). This alleged host is undoubtedly incorrect. We know of no record of a tick feeding on an insect, nor has *R. maculatus* ever been collected in this part of Africa.

Zumpt (1942) illustrated a range of colour patterns as seen on the scuta of *R. maculatus* adults (Fig. 122).

As Gertrud Theiler (pers. comm. to J.B.W.) noted, the nymph that she had described as *R. masseyi* in 1947 (p. 281, fig. 31; p. 282, fig. 32) was in fact *R. maculatus*. These figures actually give a better impression of this nymph than do those in Theiler & Robinson (1953), in which the lateral angles of the basis capituli are too short. The larvae can be differentiated from those of *R. appendiculatus* and *R. muehlensi*, with which they are often present in mixed infestations, by their broad third palpal segments, the robust setae on their palps, and the wide shallow curve of the posterior margin of their scutum.

We have omitted records of *R. maculatus* immatures from the serval (*Leptailurus serval*) and korrigum (*Damaliscus korrigum*) (Theiler, 1962). We were unable to trace that from the serval, while that from the korrigum came from an area outside the known range of this tick so is probably a misidentification.

Hosts

A three-host species (Theiler & Robinson, 1953). Cattle are its preferred domestic hosts. Amongst wild animals the adults prefer large, thick-skinned species such as African elephants, black and white rhinoceroses, warthogs, bushpigs and African buffaloes (Baker & Keep, 1970; Walker, 1974; Horak *et al.*, 1983) (Table 32). Burdens exceeding several hundred adult ticks have been recorded on black rhinoceroses, bushpigs and African buffaloes (Horak *et al.*, 1983; Horak, Boomker & Flamand, 1991; I.G.H., unpublished data). With the possible exception of elephants, the immature stages also prefer these large hosts along with thicket and woodland dwelling antelopes such as nyalas, bushbuck and various duikers (Baker & Keep, 1970; Horak, Boomker & Flamand, 1991, 1995). The predilection site of attachment for adult ticks on the African buffalo appears to be their relatively hairless undersides where, in South Africa, *R. maculatus* is frequently encountered among large numbers of adult *Amblyomma hebraeum*.

In north-eastern KwaZulu-Natal, South Africa, larvae are most abundant from April to October and nymphs from June to December (Horak *et al.*, 1995). Adults appear to be most abundant from September to May.

Zoogeography

Rhipicephalus maculatus occurs mainly in the coastal regions of north-eastern KwaZulu- Natal, South Africa, and in Mozambique, Tanzania, eastern Kenya and southern Somalia (Santos Dias, 1960; Yeoman & Walker, 1967; Baker & Keep, 1970; Walker, 1974) (Map 36). It has been recorded less frequently further inland in Zambia, southern Malawi and Zimbabwe (Morel, 1969).

It is present at altitudes from sea level up to 1500 m in East Africa, and occasionally even higher, but seems to prefer the lower altitudes. Rainfall within its distribution range varies from 500 to 1250 mm annually. It is commonest in coastal mosaic vegetation and is also encountered in various types of woodland, deciduous bushland and thicket.

A small founding population of *R. maculatus* was probably introduced into the southern Kruger National Park, South Africa, from north-eastern KwaZulu-Natal either with translocated rhinoceroses or with 20 nyalas released along the Sabie River during 1980. Despite its apparently low numbers in this park *R. maculatus* must now be well established there because a male and female have been collected from an elephant and a nymph from the vegetation (Braack *et al.*, 1995). In this region the altitude is approximately 220 m, the annual rainfall is approximately 550 mm and the vegetation is undifferentiated woodland. Thus it does not differ substantially in these respects from the preferred habitat of this tick in KwaZulu-Natal and the southern regions of the Kruger National Park must now be included in its distribution range.

The distribution of *R. maculatus* and that of *R. muehlensi* overlap in much of their range.

Disease relationships

Unknown.

REFERENCES

Baker, M.K. & Keep, M.E. (1970). Checklist of the ticks found on the larger game animals in the Natal game reserves. *Lammergeyer*, **12**, 41–7.

Braack, L.E.O., Maggs, K.A.R., Zeller, D.A. & Horak, I.G. (1995). Exotic arthropods in the Kruger National Park, South Africa: modes of entry and population status. *African Entomology*, **3**, 39–48.

Horak, I.G., Boomker, J. & Flamand, J.R.B. (1991). Ixodid ticks and lice infesting red duikers and bushpigs in north-eastern Natal. *Onderstepoort Journal of Veterinary Research*, **58**, 281–4.

Horak, I.G., Boomker, J. & Flamand, J.R.B. (1995). Parasites of domestic and wild animals in South Africa. XXXIV. Arthropod parasites of nyalas in north-eastern KwaZulu-Natal. *Onderstepoort Journal of Veterinary Research*, **62**, 171–9.

Horak, I.G., Potgieter, F.T., Walker, J.B., De Vos, V. & Boomker, J. (1983). The ixodid tick burdens of various large ruminant species in South African nature reserves. *Onderstepoort Journal of Veterinary Research*, **50**, 221–8.

Neumann, L.G. (1901). Révision de la famille de Ixodidés (4e Mémoire). *Mémoires de la Société Zoologique de France*, **14**, 249–372.

Theiler, G. & Robinson, B.N. (1953). Ticks in the South African Zoological Survey Collection.

Part VII. Six lesser known African rhipicephalids. *Onderstepoort Journal of Veterinary Research*, **26**, 93–136.

Zumpt, F. (1942). Die gefleckten *Rhipicephalus*-Arten. III. Vorstudie zu einer Revision der Gattung *Rhipicephalus* Koch. *Zeitschrift für Parasitenkunde*, **12**, 433–43.

Also see the following Basic References (pp. 12–14): Morel (1969); Santos Dias (1960); Theiler (1947, 1962); Walker (1974); Yeoman & Walker (1967).

RHIPICEPHALUS MASSEYI NUTTALL & WARBURTON, 1908

This species was named in honour of the collector, Dr A. Yale Massey, of Tanganyika Concessions Ltd., Kansanshi, N.W. Rhodesia (now Zambia).

Synonym

tendeiroi.

Diagnosis

A moderate-sized reddish-brown tick.

Male (Figs 126(a), 127(a) to (c))

Capitulum longer than broad, length × breadth of the four specimens measured ranging from 0.81 mm × 0.76 mm to 0.98 mm × 0.94 mm. Basis capituli with short somewhat recurved lateral angles at about anterior third of its length. Palps short, broad. Conscutum length × breadth ranging from 3.16 mm × 2.15 mm to 4.13 mm × 2.88 mm; anterior process of coxae I bluntly rounded. In engorged specimens body wall bulging smoothly laterally and posterolaterally and forming a broad tapering caudal process. Eyes almost flat. Cervical pits comma-shaped, convergent; cervical fields shallow, divergent, slightly depressed. Marginal lines short, shallow, outlined by a few small punctations. Posteromedian groove long and narrow, posterolateral grooves small, rounded. Additional shallow rounded depressions often present

a

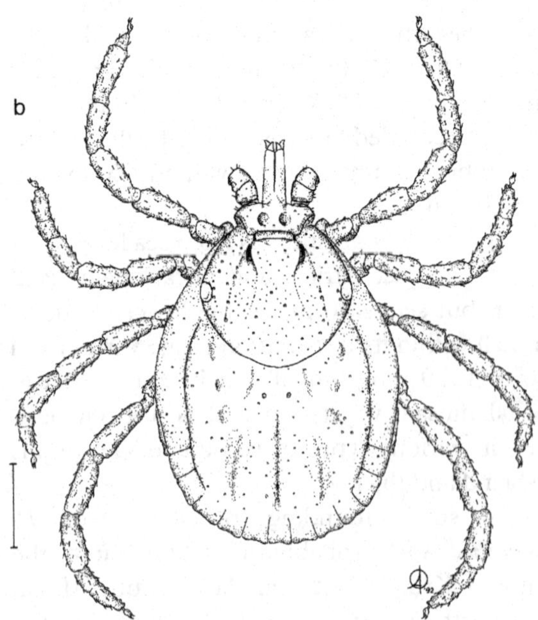

Figure 126. *Rhipicephalus masseyi* [Nuttall Collection 246, paralectotypes collected from African buffalo (*Syncerus caffer*) at Kansanshi, Zambia, in 1907 by A. Yale Massey, by courtesy of The Natural History Museum, London]. (a) Male, dorsal; (b) female, dorsal. Scale bars represent 1 mm. A. Olwage *del.*

Table 33. *Host records of* Rhipicephalus masseyi

Hosts	Number of records
Domestic animals	
Cattle	6
Sheep	2
Dogs	1
Wild animals	
Lion (*Panthera leo*)	5
Leopard (*Panthera pardus*)	2
Marsh mongoose (*Atilax paludinosus*)	1
Burchell's zebra (*Equus burchellii*)	2
Aardvark (*Orycteropus afer*)	1
Warthog (*Phacochoerus africanus*)	2
Bushpig (*Potamochoerus larvatus*)	11
African buffalo (*Syncerus caffer*)	4
Eland (*Taurotragus oryx*)	3
Nyala (*Tragelaphus angasii*)	1
Bushbuck (*Tragelaphus scriptus*)	3
Greater kudu (*Tragelaphus strepsiceros*)	1

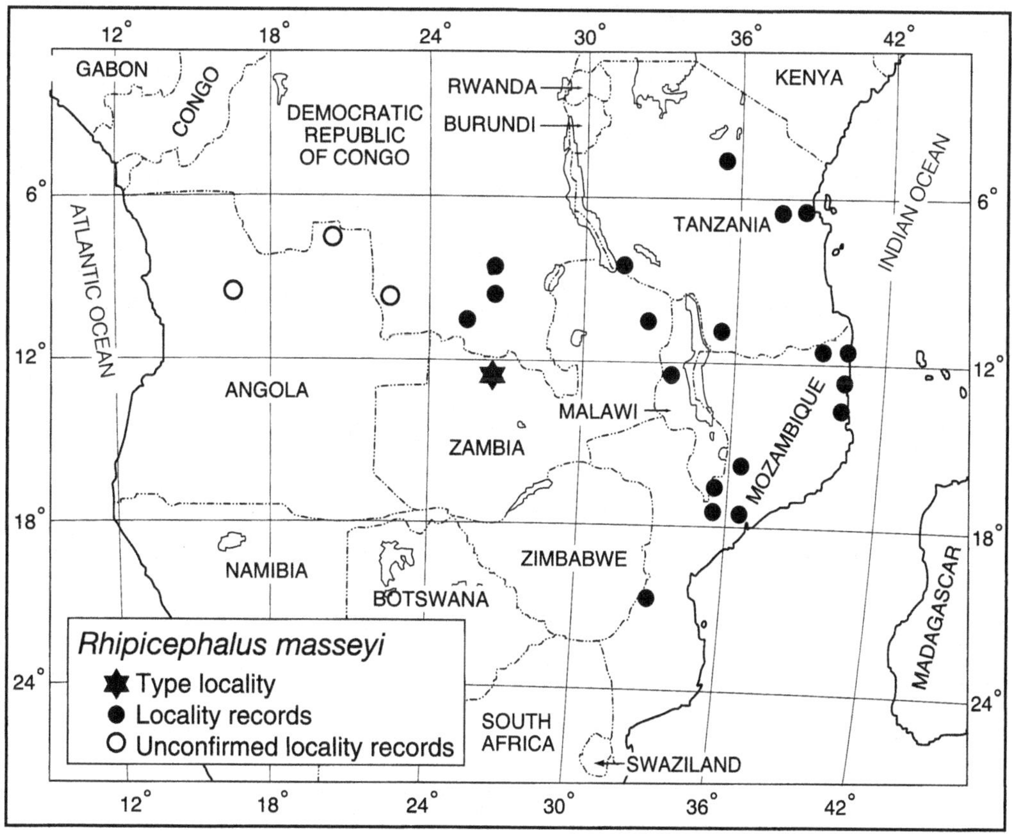

Map 37. *Rhipicephalus masseyi*: distribution.

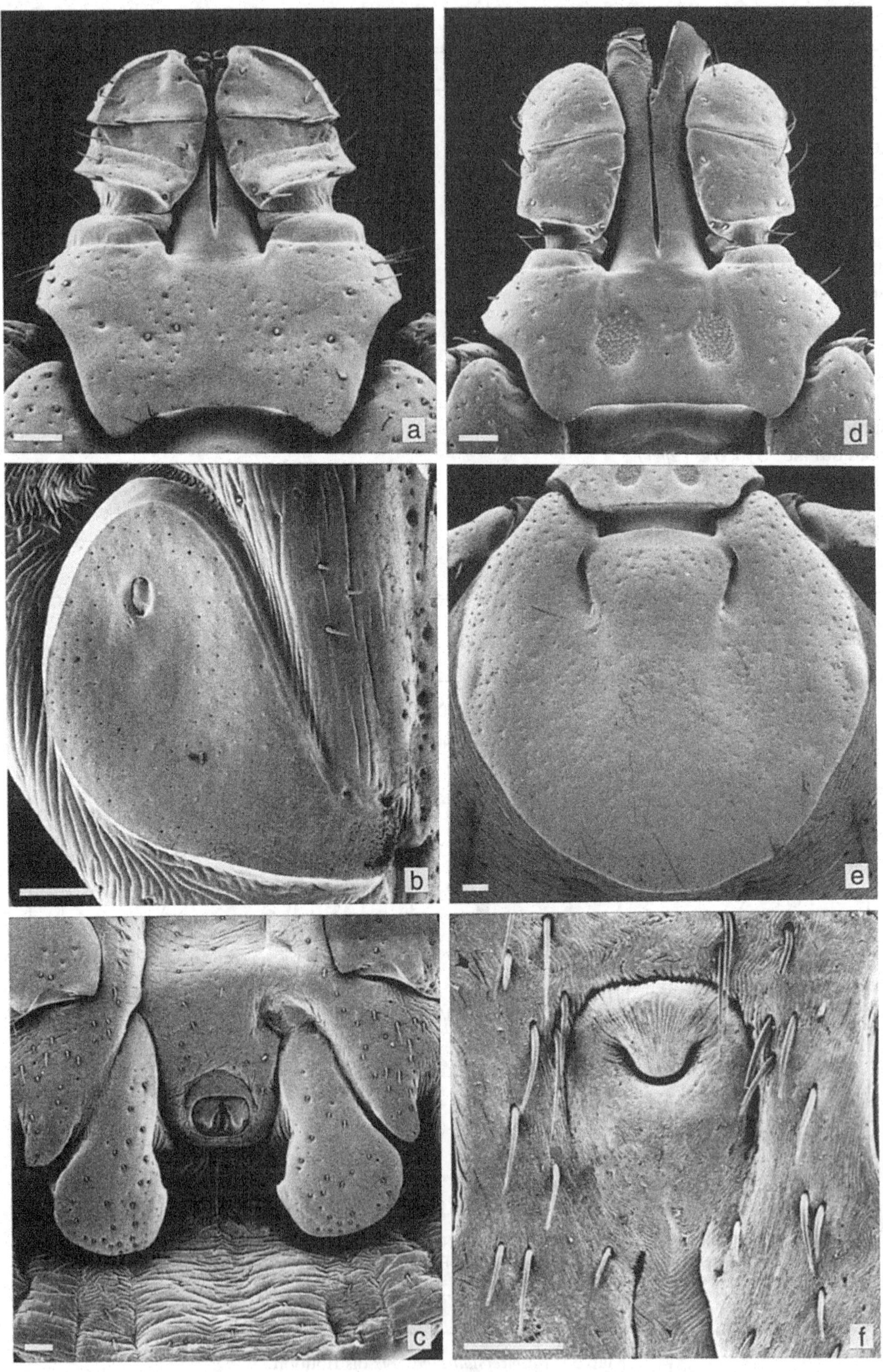
a
d
b
e
c
f

medially on the conscutum. Punctation pattern somewhat variable; on some specimens small superficial punctations are scattered over the conscutum, interspersed with very fine punctations, but on others the punctations are reduced and the conscutum is then smoother and shinier. Ventrally spiracles broadly comma-shaped, with only a short slightly curving dorsal prolongation. Adanal plates short, broad, almost pear-shaped, often with a small sharp, inwardly projecting point posterior to the anus; accessory adanal plates short, broad, bluntly pointed.

Female (Figs 126(b), 127(d) to (f))

Capitulum about as broad as long, length × breadth in the three specimens measured ranging from 0.83 mm × 0.82 mm to 0.95 mm × 0.95 mm. Basis capituli with short broad lateral angles just anterior to mid-length; porose areas large, round, slightly more than their own diameter apart. Palps with a short narrow neck, otherwise broad, smoothly rounded apically. Scutum ranging from about as long as broad to somewhat broader than long, length × breadth ranging from 1.57 mm × 1.57 mm to 1.84 mm × 1.91 mm; posterior margin broadly rounded. Eyes almost flat. Cervical pits comma-shaped, convergent; cervical fields depressed, indistinctly demarcated. A few small punctations scattered over the scutum, and sometimes along the external margins of the cervical fields, interspersed with fine punctations that are dense in some specimens but may be almost absent in others. Ventrally genital aperture a short wide U-shape.

Figure 127 (*opposite*). *Rhipicephalus masseyi* [Onderstepoort Tick Collection 3077i, collected from Mammal No. 1338/365–B: bushpig (*Potamochoerus larvatus*), Mabwe, Upemba National Park, Democratic Republic of Congo, on 14 January 1949 by Mission G.F. de Witte]. Male: (a) capitulum, dorsal; (b) spiracle; (c) adanal plates. Female: (d) capitulum, dorsal; (e) scutum; (f) genital aperture. Scale bars represent 0.10 mm. SEMs by J.F. Putterill.

Immature stages

Unknown.

Notes on identification

Rhipicephalus masseyi is not always easy to recognize. According to Zumpt (1943) two adults from Songea, southern Tanzania, were misidentified as *R. neavei* (= *R. kochi*) by Nuttall & Warburton. In 1947 Theiler confused the adults, in part, with those of *R. muehlensi*, and misidentified the nymph of *R. maculatus* as that of *R. masseyi*. She subsequently noted that in this paper figures 29 & 30 (p. 281) represent the male and female, respectively, of *R. muehlensi*, and figures 31 & 32 (pp. 281–2) the nymph of *R. maculatus* (G. Theiler, pers. comm. to J.B.W.). Her figure 28 (p. 281 in her paper) represents the holotype female of *R. attenuatus* Neumann, 1908, now regarded as a synonym of *R. lunulatus*. The specimens that she listed as *R. masseyi* in three collections from northern KwaZulu-Natal were later re-identified as *R. muehlensi* (Theiler, 1962). We have therefore omitted these records, which are apparently those quoted by Morel (1969: map 47), as we know of no others from South Africa. We have also omitted Morel's record from Namibia, which is believed to be incorrect, as is that from Angola (Santos Dias, 1964).

C.M. Clifford (unpublished data) has noted that it is sometimes difficult to distinguish the females of *R. masseyi* from those of *R. ziemanni* when these two species are sympatric.

Hosts

Life cycle unknown; probably three-host. *Rhipicephalus masseyi* has rarely been collected, and then in small numbers only, from domestic animals. It apparently has a predilection for the wild suids and some of the larger wild bovids (Table 33).

A sample from a lion was taken from one of its ears (MacLeod & Mwanaumo, 1978). Its attachment sites on other hosts have not been recorded. Judging from the collection dates the adults are active in summer, from August onwards.

Zoogeography

Thus far we have recorded *R. masseyi* only in parts of eastern and central Africa (Map 37). The northernmost record is from Kondoa, in northern Tanzania, and the southernmost from the Mossurize area in Mozambique, near the Zimbabwe border. Nearly a third of the collections were obtained in the Upemba National Park, south-eastern Democratic Republic of Congo, six of them from bushpigs (Clifford & Anastos, 1962).

These collection sites are at altitudes ranging from almost sea level to over 1000 m with mean annual rainfalls of approximately 800 mm to 1400 mm, often in Zambezian miombo woodland dominated by *Brachystegia* and *Julbernardia*, plus *Isoberlinia* in the wetter areas.

Disease relationships

Unknown.

REFERENCES

MacLeod, J. & Mwanaumo, B. (1978). Ecological studies of ixodid ticks (Acari: Ixodidae) in Zambia. IV. Some anomalous infestation patterns in the northern and eastern regions. *Bulletin of Entomological Research*, **68**, 409–29.

Nuttall, G.H.F. & Warburton, C. (1908). On a new genus of the Ixodoidea, together with a description of eleven new species of ticks. *Proceedings of the Cambridge Philosophical Society*, **14**, 392–416.

Santos Dias, J.A.T. (1964). Nova contribuição para o conhecimento da ixodofauna Angolana. Carraças colhidas por uma missão de estudo do Museu de Hamburgo. *Anais dos Serviços de Veterinária de Moçambique*, **No. 9 for 1961**, 79–98.

Zumpt, F. (1943). *Rhipicephalus aurantiacus* und ähnliche Arten. VIII. Vorstudie zu einer Revision der Gattung *Rhipicephalus* Koch. *Zeitschrift für Parasitenkunde*, **13**, 102–17.

Also see the following Basic References (pp. 12–14): Clifford & Anastos (1962); Elbl & Anastos (1966); Morel (1969); Pegram *et al.* (1986); Santos Dias (1960, 1993); Theiler (1947, 1962); Wilson (1950); Yeoman & Walker (1967).

RHIPICEPHALUS MOUCHETI MOREL, 1965

This species was named after J. Mouchet, a colleague of the author who collected the holotype male.

Diagnosis

A small dark reddish-brown heavily-punctate tick, the only species in the *R. sanguineus* group whose male has sickle-shaped adanal plates.

Male (Fig. 128(a))

Capitulum in the two males measured was slightly broader than long, length × breadth 0.53 mm × 0.57 mm and 0.60 mm × 0.61 mm, respectively. Basis capituli with acute lateral angles at about mid-length. Palps short, broad, rounded apically. Conscutum length × breadth in the two males measured 2.32 mm × 1.35 mm and 2.54 mm × 1.49 mm; anterior process on coxae I inconspicuous. Eyes almost flat, delimited dorsally by a few large punctations.

Table 34. *Host records of* Rhipicephalus moucheti

Hosts	Number of records
Domestic animals	
Cattle	1
Dogs	10
Wild animals	
Patas monkey (*Erythrocebus patas*)	1
African civet (*Civettictis civetta*)[1]	1

Note: [1]J.-L. Camicas (unpublished data, 1996).

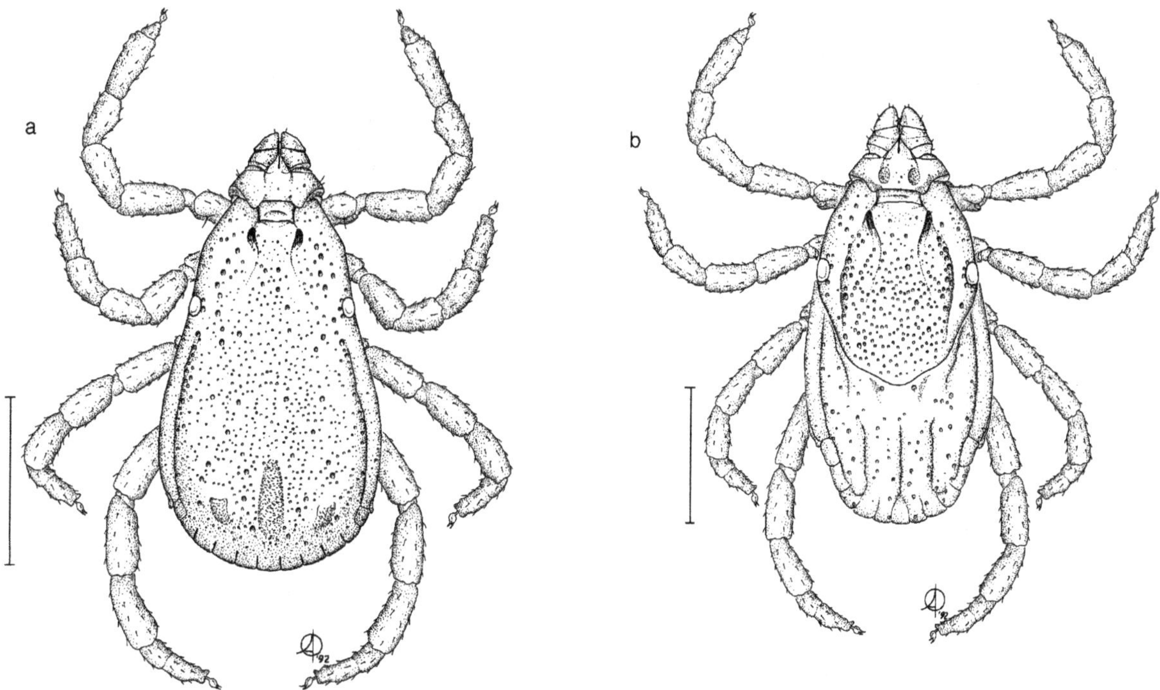

Figure 128. *Rhipicephalus moucheti* (Ref. no. GK8. Specimens collected from vegetation, Tabouna, Kindia area, Guinea, in April 1989 by O.K. Konstantinov; donated by J.-L. Camicas). (a) Male, dorsal; (b) female, dorsal. Scale bars represent 1 mm. A. Olwage *del.*

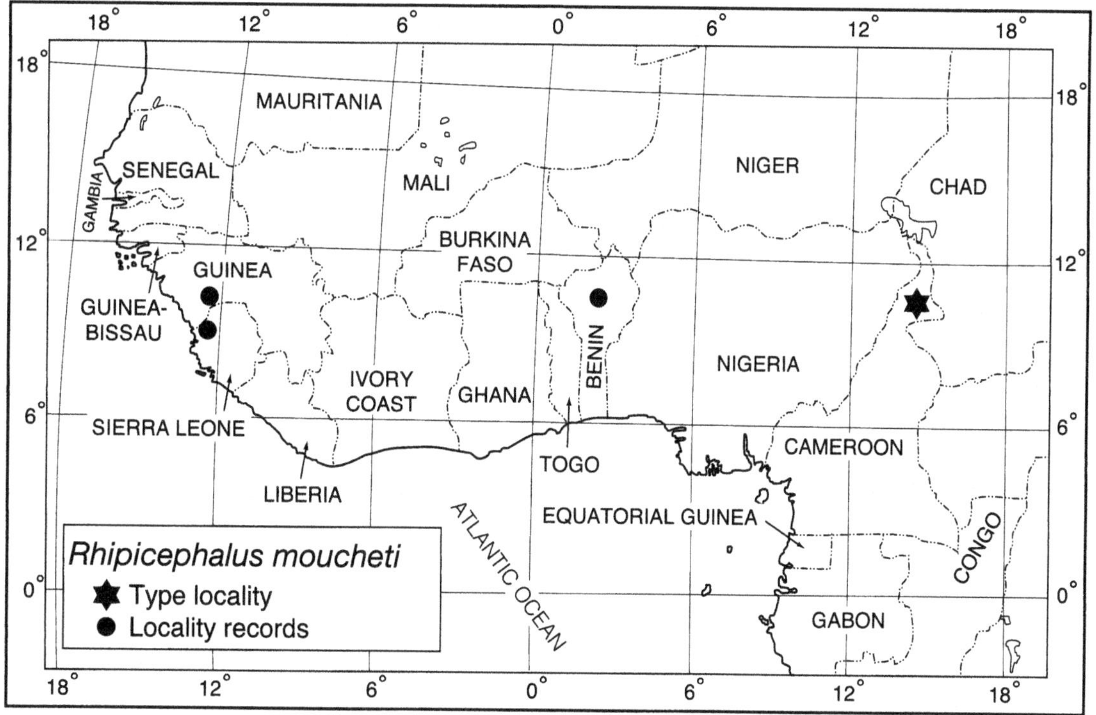

Map 38. *Rhipicephalus moucheti:* distribution.

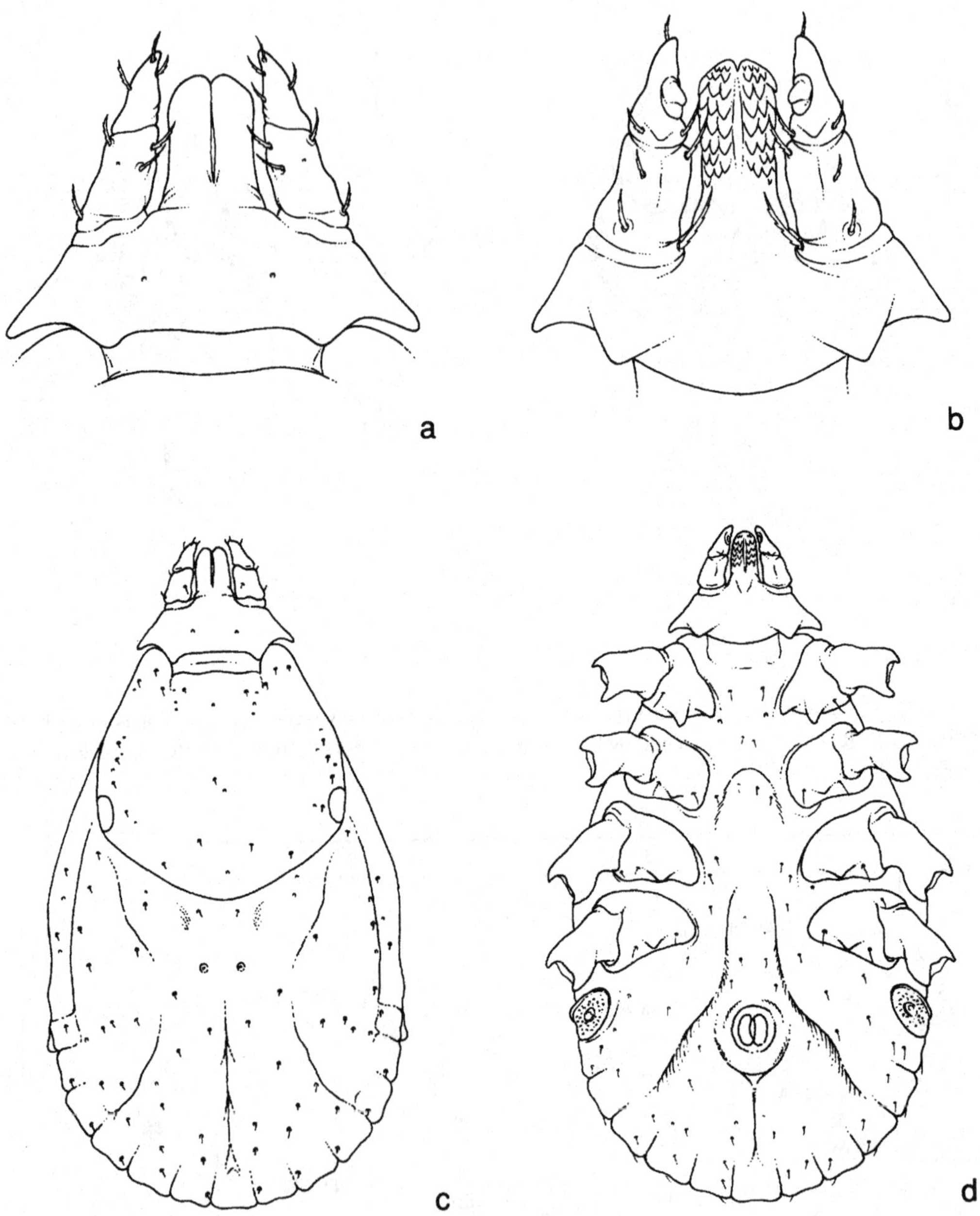

Figure 129. *Rhipicephalus moucheti*. Nymph: (a) capitulum, dorsal; (b) capitulum, ventral; (c) dorsal surface; (d) ventral surface. (Redrawn from Saratsiotis, 1981, plate 2, by A. Olwage, with kind permission from the Editor, *Acarologia*.)

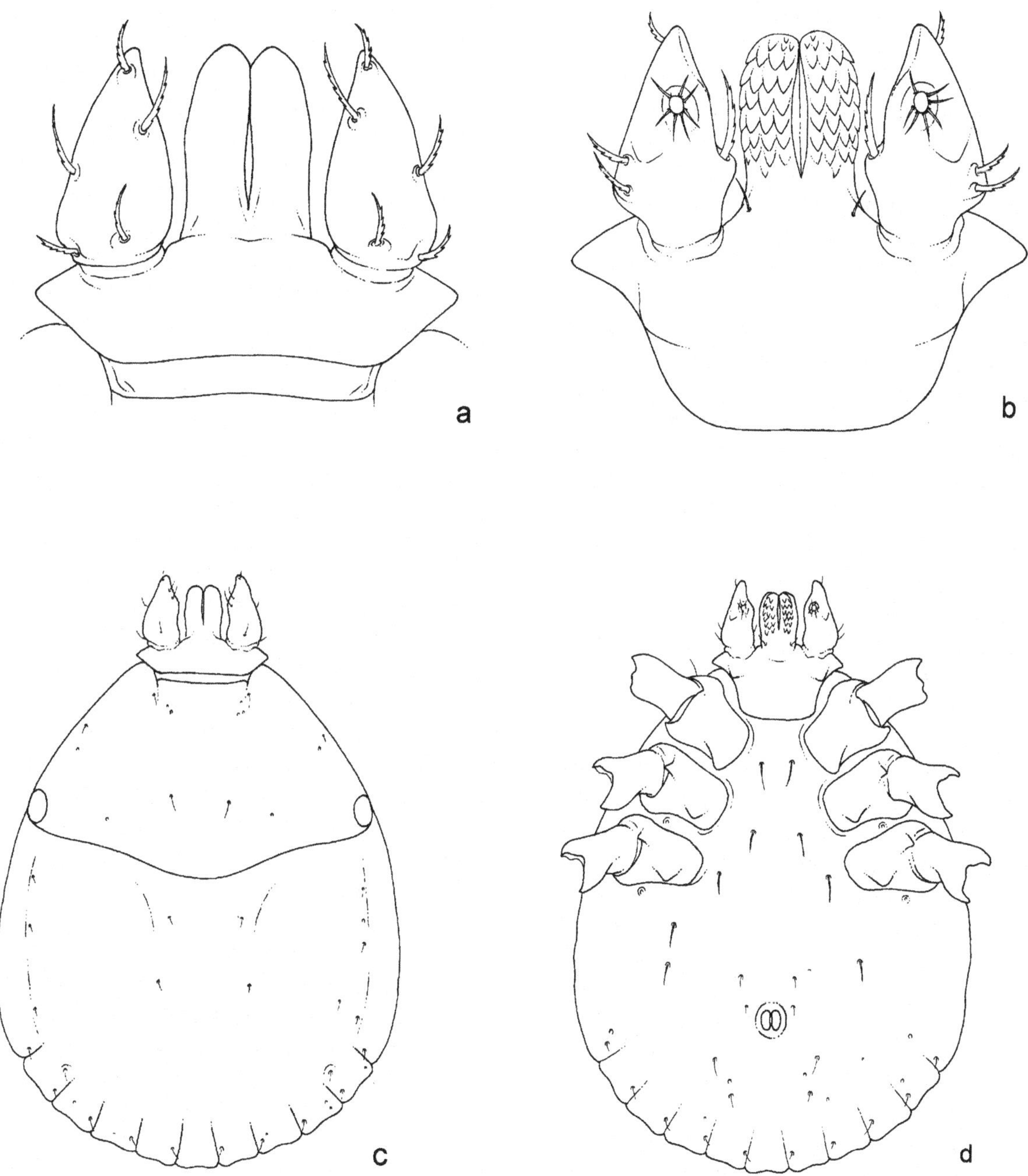

Figure 130. *Rhipicephalus moucheti.* Larva: (a) capitulum, dorsal; (b) capitulum, ventral; (c) dorsal surface; (d) ventral surface. (Redrawn from Saratsiotis, 1981, plate 1, by A. Olwage, with kind permission from the Editor, *Acarologia*.)

Cervical pits comma-shaped, convergent; cervical fields shallow, their external margins delimited by rows of large setiferous punctations. Marginal lines long, almost reaching eye level, containing numerous large, sometimes contiguous, punctations. Posteromedian groove long, broad, deep; posterolateral grooves much smaller but also deep. Punctation pattern dense apart from areas adjacent to the marginal lines where it is finer. Large setiferous punctations scattered medially on conscutum and around posterior grooves, interspersed with numerous medium-sized to fine punctations. Ventrally spiracles with a long broad dorsal prolongation that curves gently towards the dorsal surface just at the end. Adanal plates large in relation to the size of the tick, strongly sickle-shaped, their anterior ends tapered, otherwise almost equal in width throughout their length, their posterior ends rounded, curving towards each other and almost meeting mid-ventrally; accessory adanals long, narrow, pointed.

Female (Fig. 128(b))

Capitulum in the only female available to us slightly broader than long, length × breadth measuring 0.72 mm × 0.74 mm. Basis capituli with lateral angles at about mid-length; porose areas quite large, about their own diameter apart. Palps rounded apically. Scutum longer than broad, length × breadth measuring 1.52 mm × 1.24 mm; posterior margin sinuous. Eyes at about mid-length, almost flat, delimited dorsally by a few large punctations. Cervical fields long, narrow, shallow, their external margins strongly demarcated by numerous large, often contiguous, punctations. A few medium-sized punctations present on the scapulae, otherwise punctations on lateral areas of scutum fine. Medially scutum fairly heavily punctate, with the scattered larger punctations almost masked by the smaller ones. Ventrally genital aperture widely V-shaped.

Nymph (Fig. 129)

Capitulum much broader than long, length × breadth 0.17 mm × 0.28 mm. Basis capituli nearly three times as broad as long with tapering somewhat posteriorly-directed lateral angles; ventrally with prominent triangular spurs on posterior border. Palps tapering to narrowly-rounded apices, almost certainly inclined inwards when at rest. Scutum about as broad as long, length × breadth 0.50 mm × 0.51 mm; posterior margin a broad smooth curve. Eyes at widest point, well over halfway back, long and narrow. External margins of cervical fields said to be well developed, but not illustrated. Ventrally coxae I each with a narrow external spur and a somewhat broader internal spur; coxae II to IV each with an external spur only.

Larva (Fig. 130)

Capitulum 1.5 times broader than long, length × breadth 0.10 mm × 0.15 mm. Basis capituli with short acute lateral angles at about mid-length, posterior border very slightly concave. Palps with convex external margins, tapering to narrowly-rounded apices, almost certainly inclined inwards when at rest. Scutum much broader than long, length × breadth 0.20 mm × 0.36 mm; posterior margin concave posterolaterally. Eyes at widest part of scutum, about two-thirds of the way back, large. Ventrally coxae I each with a large pointed spur; coxae II to III each with a small spur.

Notes on identification

The only specimens of *R. moucheti* that we have seen are 2 ♂♂, 1 ♀ kindly donated to us by Dr J.-L. Camicas. Our descriptions and illustrations of the adults are based primarily on these specimens. The descriptions and illustrations of the nymph and larva are based on those given by Saratsiotis (1981), who described all stages from his laboratory-reared specimens.

The information on the adults given by Saratsiotis is somewhat puzzling. He reprinted Morel's original description of the male verbatim, and appears to have based his illustrations of the male's capitulum, its adanal and accessory adanal plates, and its spiracle on Morel's figures. His illustration of the male's dorsal surface, though, is quite unlike Morel's in that it shows a very lightly-punctate tick. His description and illustration of the female, on the other hand, show a tick with an overall pattern of large uniform punctations. This does not correspond with our specimen, which has punctations that are not uniform in size and do not cover the whole scutum evenly.

Both Morel (1965) and Saratsiotis (1981) regard this species as a member of the *R. sanguineus* group, and Saratsiotis discussed in detail the differentiation of its nymphs and larvae from those of other species in the group. We think, though, that it would be just as difficult, if not more so, to distinguish its immature stages from those of *R. muhsamae*, as described by Pegram *et al.* (1987). This species often occurs in the same areas as *R. moucheti* and its immature stages are also parasitic on rodents (see p. 303).

Hosts

A three-host species (Saratsiotis, 1981). The commonest hosts recorded thus far are domestic dogs (Table 34). Saratsiotis suggested that the hosts of its adults are probably some of the smaller animals, with only occasional infestations on the larger species. Rodents are thought to be the hosts of the immature stages.

Zoogeography

The localities where *R. moucheti* has been recorded are spread out across West Africa from central Guinea in the west to northern Cameroon in the east, all at about latitude 10 °N. (Map 38). It is apparently restricted to areas at an altitude of about 250 m, with moderate temperatures and mean annual rainfalls ranging from around 1000 mm to 3000 mm, in various types of humid tropical woodland. Its adults are most active during the second half of the rainy season, from about July to mid-October: of the 14 collections listed to date, including one from vegetation, nine were made in September, three in August, and one each in May and July.

Saratsiotis (1981) noted that a careful search made for this tick on dogs and other animals living in forested areas, rather than woodland, failed to reveal it. He regarded it as a tick with very specific ecological requirements.

Disease relationships

Unknown.

REFERENCES

Morel, P.C. (1965). Description de *Rhipicephalus moucheti* n. sp. (groupe de *Rh. sanguineus*; Acariens, Ixodoidea). *Revue d'Élevage et de Médecine Vétérinaire des Pays Tropicaux*, **17 for 1964** (nouvelle série), 615–17.

Pegram, R.G., Walker, J.B., Clifford, C.M. & Keirans, J.E. (1987). Comparison of populations of the *Rhipicephalus simus* group: *R. simus*, *R. praetextatus*, and *R. muhsamae*. *Journal of Medical Entomology*, **24**, 666–82.

Saratsiotis, A.G. (1981). Étude morphologique et biologique de *Rhipicephalus moucheti* Morel, 1964, groupe de *Rh. sanguineus* (Acariens; Ixodoidea), espèce Africaine. *Acarologia*, **22**, 15–24.

RHIPICEPHALUS MUEHLENSI ZUMPT, 1943

This species was named after Professor P. Mühlens, the Director of the Bernhard Nocht-Institut für Schiffs- und Tropenkrankheiten, Hamburg, where the author was working at the time.

Diagnosis

A small to medium-sized chestnut-brown tick.

Male (Figs 131(a), 132(a) to (c))
Capitulum much longer than broad, length × breadth ranging from 0.60 mm × 0.53 mm to 0.99 mm × 0.71 mm. Basis capituli long, with inconspicuous obtuse to rounded lateral angles that do not extend beyond the scapulae. Palps long, their outer margins almost in line with the sides of the basis capituli, thus accentuating the length of the capitulum. Conscutum length × breadth ranging from 1.91 mm × 1.26 mm to 3.29 mm × 2.23 mm; large anterior process present on coxae I. In engorged specimens body wall bulges out laterally and posteriorly into the characteristic shape illustrated; a separate caudal process is not formed. Eyes marginal, slightly bulging, sometimes delimited dorsally by shallow grooves and a few small punctations. Cervical pits deep; internal cervical margins indistinct; external cervical margins outlined by medium-sized setiferous punctations. Marginal lines short, indistinct, sometimes continued anteriorly by rows of punctations. Posteromedian groove long and narrow, posterolateral grooves oval to round, usually separated from the festoons. Punctations generally fine, usually most conspicuous in the area corresponding to the shape of the female scutum, sparse anterior to the eyes and adjacent to the marginal lines. Legs increase slightly in size from I to IV. Ventral surface, including the circumspiracular area and the adanal plates, covered with long white setae. Spiracles compact, with only a very short dorsal prolongation. Adanal plates broadly triangular,

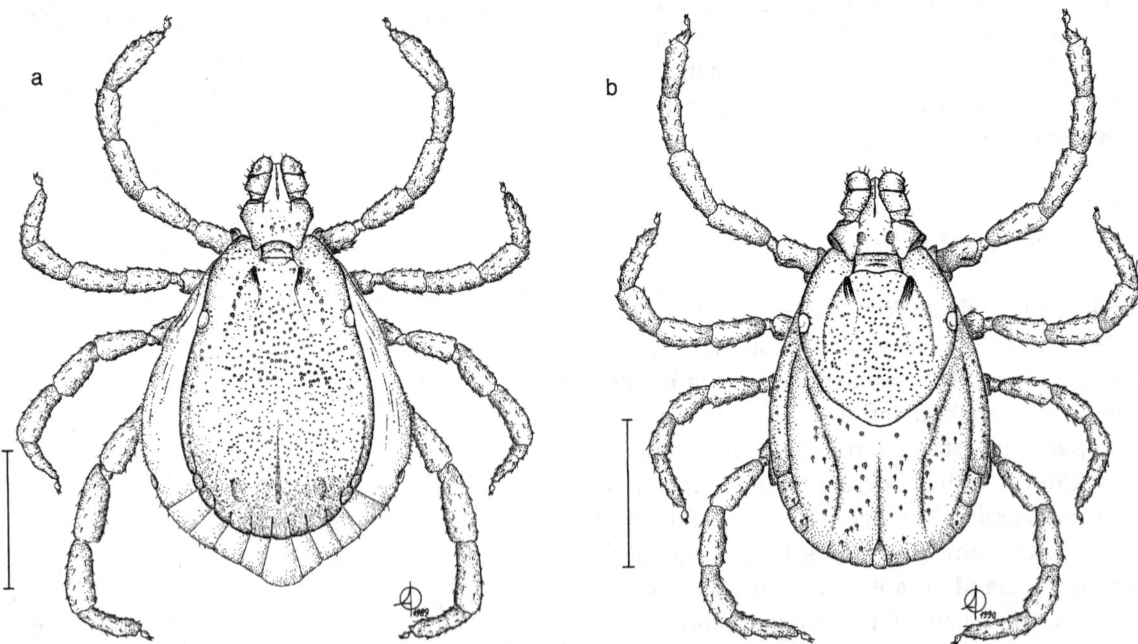

Figure 131. *Rhipicephalus muehlensi* [collected from impala (*Aepyceros melampus*), Hluhluwe Game Reserve, KwaZulu-Natal, South Africa, on 10 April 1970 by M.E. Keep]. (a) Male, dorsal; (b) female, dorsal. Scale bars represent 1 mm. A. Olwage *del.*

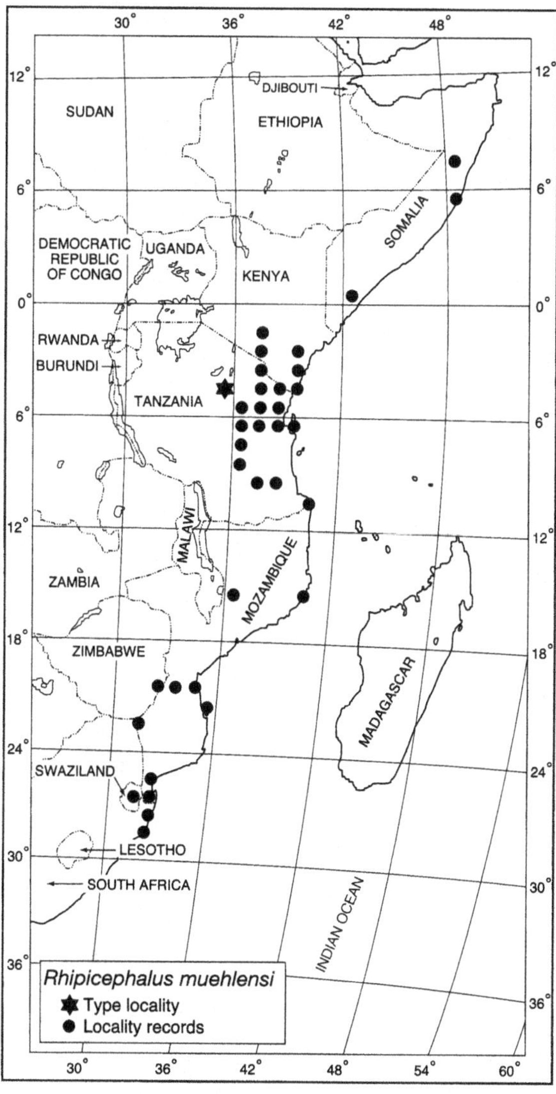

Map 39. *Rhipicephalus muehlensi*: distribution.

extending inwards somewhat behind the anus, their posterior angles smoothly curved; accessory adanal plates small.

Female (Figs 131(b), 132(d) to (f))
Capitulum about as long as broad, length × breadth ranging from 0.68 mm × 0.69 mm to 0.83 mm × 0.83 mm. Basis capituli with acute lateral angles overlapping the scapulae; porose areas small, about three times their own diameter apart. Palps long, broadly rounded apically. Scutum longer than broad, length × breadth ranging from 1.37 mm × 1.22 mm to 1.67 mm × 1.45 mm, posterior margin sinuous. Eyes not quite halfway back, very slightly bulging, sometimes edged dorsally by shallow grooves and a few punctations. Cervical pits deep; cervical fields shallow and indistinct, their external margins delimited by irregular rows of medium-sized setiferous punctations. A few setiferous punctations present on the scapulae and scattered medially on the scutum, interspersed with numerous small punctations except on the smooth lateral margins anterior to the eyes. Ventrally genital aperture long, U-shaped posteriorly, widening anteriorly.

Nymph (Fig. 133)
Capitulum much broader than long, length × breadth ranging from 0.23 mm × 0.31 mm to 0.27 mm × 0.32 mm. Basis capituli not quite three times as broad as long, with broad, slightly forwardly-directed lateral angles overlapping scapulae; ventrally small spurs present on posterior margin. Palps broad, inclined slightly inwards. Scutum broader than long, length × breadth ranging from 0.49 mm × 0.54 mm to 0.55 mm × 0.58 mm, posterior margin a deep smooth curve. Eyes at widest point, halfway back, slightly bulging and edged dorsally by grooves. Cervical fields fairly broad, slightly depressed, not quite reaching posterolateral margins of scutum. Ventrally coxae I each with a narrow external spur and a much broader internal spur; coxae II to IV each with an external spur only, decreasing progressively in size.

Larva (Fig. 134)
Capitulum slightly longer than broad, length × breadth ranging from 0.128 mm × 0.126 mm to 0.137 mm × 0.132 mm. Basis capituli well over twice as broad as long, with very short blunt lateral angles; ventrally with a mere indication of spurs on posterior margin. Palps constricted proximally, then widening markedly before tapering to narrowly-rounded apices. Scutum much broader than long, length × breadth ranging from 0.247 mm × 0.371 mm to 0.264 mm × 0.390 mm; posterior

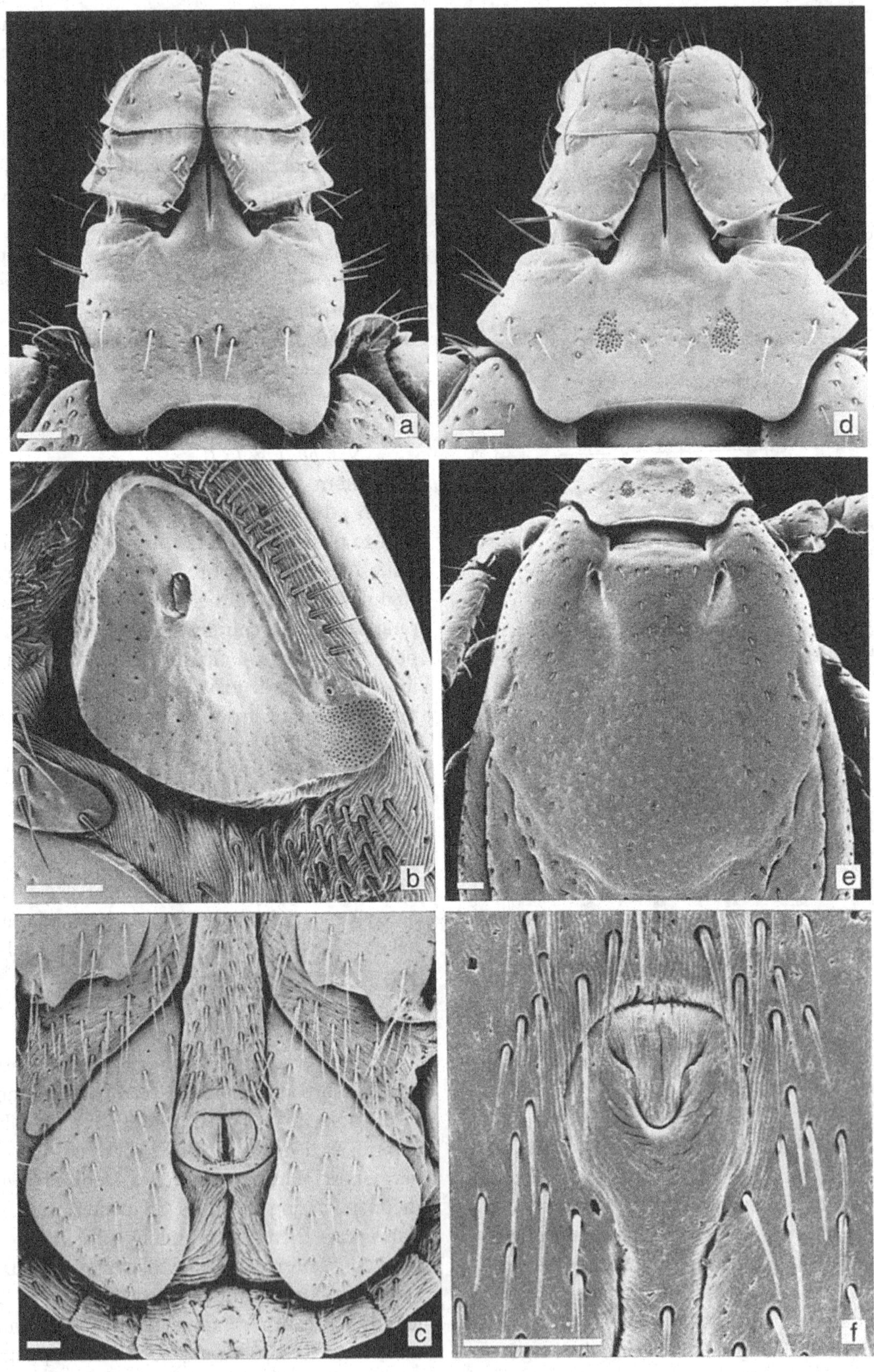
a
d
b
e
c
f

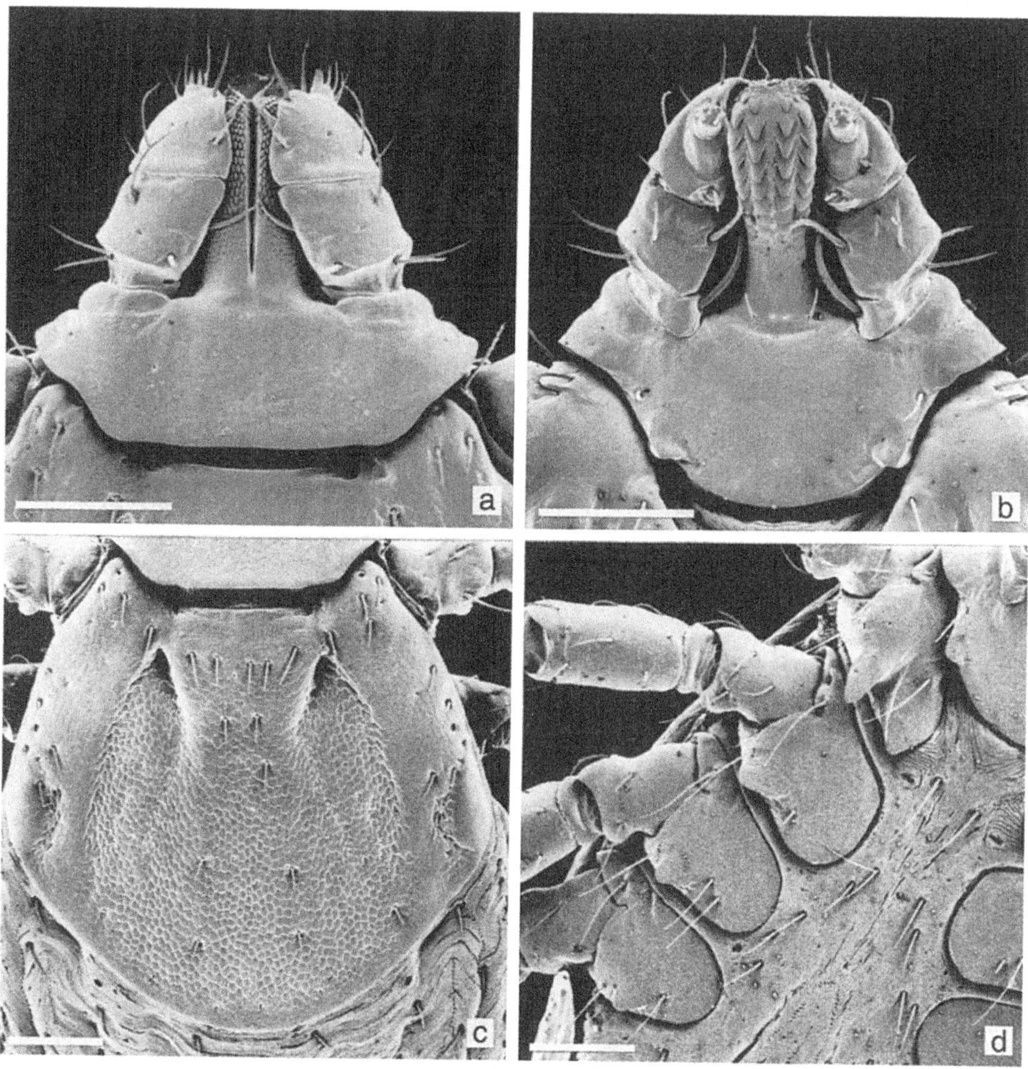

Figure 133 (*above*). *Rhipicephalus muehlensi* [Protozoology Section Tick Breeding Register, Onderstepoort, No. 4080, laboratory reared, original ♀ collected from nyala (*Tragelaphus angasii*), Mkuzi Game Reserve, KwaZulu-Natal, South Africa, on 9 March 1993 by I.G. Horak]. Nymph: (a) capitulum, dorsal; (b) capitulum, ventral; (c) scutum; (d) coxae. Scale bars represent 0.10 mm. SEMs by J.F. Putterill.

Figure 132 (*opposite*). *Rhipicephalus muehlensi* [Protozoology Section Tick Breeding Register, Onderstepoort, No. 4080, laboratory reared, original ♀ collected from nyala (*Tragelaphus angasii*), Mkuzi Game Reserve, KwaZulu-Natal, South Africa, on 9 March 1993 by I.G. Horak]. Male: (a) capitulum, dorsal; (b) spiracle; (c) adanal plates. Female: (d) capitulum, dorsal; (e) scutum; (f) genital aperture. Scale bars represent 0.10 mm. SEMs by J.F. Putterill.

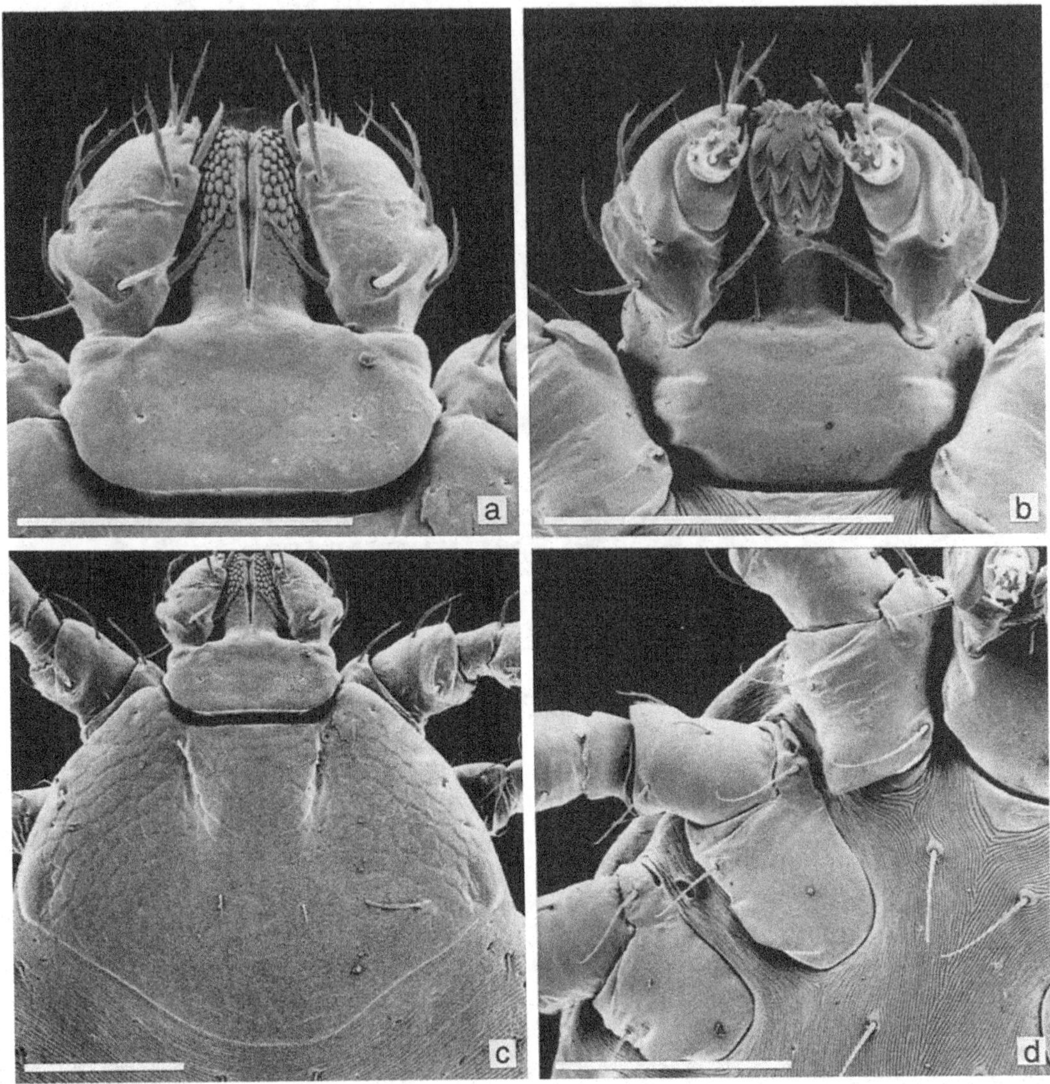

Figure 134. *Rhipicephalus muehlensi* [Protozoology Section Tick Breeding Register, Onderstepoort, No. 4080, laboratory reared, original ♀ collected from nyala (*Tragelaphus angasii*), Mkuzi Game Reserve, KwaZulu-Natal, South Africa, on 9 March 1993 by I.G. Horak]. Larva: (a) capitulum, dorsal; (b) capitulum, ventral; (c) scutum; (d) coxae. Scale bars represent 0.10 mm. SEMs by J.F. Putterill.

margin a broad smooth curve. Eyes at widest point, well over halfway back, mildly convex. Cervical grooves short, slightly convergent. Ventrally coxae I each with a large sharp triangular spur; coxae II and III each with a broad, bluntly rounded spur.

Notes on identification

Although *R. muehlensi* has been reported from the Sudan, the Democratic Republic of Congo and Rwanda, we do not now believe that it occurs in any of these countries. One of us (J.B.W.) has re-examined the female from a bovine at Yei on which Hoogstraal (1956) based his report of its occurrence in the Sudan, plus another male and female from a bovine at Kajo-Kaji, and regards these merely as small, atypical specimens of *R. appendiculatus*. Theiler & Robinson (1954) recorded *R. muehlensi* from several localities in Rwanda and Burundi but these specimens

Table 35. *Host records of* Rhipicephalus muehlensi

Hosts	Number of records
Domestic animals	
Cattle	53
Sheep	2
Goats	3
Dogs	1
Wild animals	
Black-backed jackal (*Canis mesomelas*)	4 (immatures)
Slender mongoose (*Galerella sanguinea*)	2 (immatures)
White-tailed mongoose (*Ichneumia albicauda*)	1 (including immatures)
'Lesser mongoose'	1 (nymphs)
'Hyaena'	1 (larvae)
African elephant (*Loxodonta africana*)	1
Burchell's zebra (*Equus burchellii*)	3 (including immatures)
Black rhinoceros (*Diceros bicornis*)	6 (including immatures)
'Rhinoceros'	1 (immatures)
Warthog (*Phacochoerus africanus*)	2 (including immatures)
Bushpig (*Potamochoerus larvatus*)	11 (including nymphs)
Giraffe (*Giraffa camelopardalis*)	6
Impala (*Aepyceros melampus*)	163 (including immatures)
Lichtenstein's hartebeest (*Sigmoceros lichtensteinii*)	1
Grant's gazelle (*Gazella granti*)	1
Gerenuk (*Litocranius walleri*)	2 (including immatures)
Suni (*Neotragus moschatus*)	1
Steenbok (*Raphicerus campestris*)	3
African buffalo (*Syncerus caffer*)	10 (including immatures)
Eland (*Taurotragus oryx*)	4
Nyala (*Tragelaphus angasii*)	107 (including immatures)
Lesser kudu (*Tragelaphus imberbis*)	5
Bushbuck (*Tragelaphus scriptus*)	15 (including immatures)
Greater kudu (*Tragelaphus strepsiceros*)	11 (including immatures)
Red forest duiker (*Cephalophus natalensis*)	28 (including immatures)
Common duiker (*Sylvicapra grimmia*)	6 (including immatures)
'Duiker'	7 (including immatures)
Roan antelope (*Hippotragus equinus*)	4
Sable antelope (*Hippotragus niger*)	1
Gemsbok (*Oryx gazella*)	1
Waterbuck (*Kobus ellipsiprymnus*)	2
Reedbuck (*Redunca arundinum*)	12 (including immatures)
Scrub hare (*Lepus saxatilis*)	29 (immatures)
Lepus sp.	1

apparently no longer exist. Elbl & Anastos (1966) recorded it from various places in the Democratic Republic of Congo but we question its occurrence there.

Nuttall and Warburton identified four specimens of *R. muehlensi*, a species unknown to them as it was described only in 1943, as *R. longicoxatus*. These specimens are in Nuttall Collection 2824, 1 ♂, 1 ♀ from giraffe, Negero, Africa, and Nuttall Collection 2825a, 2 ♂♂ from 'Africa' (C. Clifford, pers. comm., 29 August 1963, to G. Theiler; Keirans, 1985).

The record of immature *R. muehlensi* from an African civet quoted by Theiler (1962) is probably based on that listed by Salisbury (1959) from J.B.W. (unpublished data); the identification of this nymph was later corrected to *R. maculatus* by Walker (1974). We have been unable to trace Theiler's record from Smith's bush squirrel. We also question her record of immature ticks from *Procavia* sp.; this collection was made from an animal in the Sudan and *R. muehlensi* is not now thought to occur in that country. Unfortunately the specimens on which the last two records were based apparently no longer exist.

Hosts

A three-host species (Salisbury, 1959). A comparatively large number of collections have been taken from cattle, and a few from sheep and goats, but all these animals harboured only small numbers of ticks (Table 35). The preferred hosts of all stages of development are impala and tragelaphine antelopes, with nyala and bushbuck harbouring particularly large burdens (Horak *et al.*, 1988; Horak, Boomker & Flamand, 1995b; Gallivan & Surgeoner, 1995). All 79 of the nyala examined in three game reserves in north-eastern KwaZulu-Natal, South Africa, were infested and their mean burdens comprised 4078 larvae, 755 nymphs and 457 adult ticks. Several of the adult male animals each harboured considerably more than 1000 adult ticks. The preferred site of attachment of these ticks on the nyala was their ears (Horak *et al.*, 1995b).

Although many of both the larger and the smaller wild animals examined were infested with adult ticks they generally appeared to be better hosts of the immature stages, with the red forest duiker and other duikers being particularly favoured (Baker & Keep, 1970; Horak, Boomker & Flamand, 1991). Scrub hares must also be considered important hosts of immature *R. muehlensi* (Horak *et al.*, 1995a). Not only are a large proportion of these hares infested but individual animals may harbour as many as 200 ticks. In addition their distribution overlaps that of *R. muehlensi*.

No pattern of seasonal abundance was evident on nyala examined in north-eastern KwaZulu-Natal, South Africa, as all stages of development were present in large numbers on the animals throughout the year.

Zoogeography

This tick is present in all African countries with an eastern seaboard from Somalia to South Africa (Map 39). The majority of collection sites are at altitudes ranging from sea level to approximately 1500 m. With the exception of some parts of its range in Somalia, where the annual rainfall appears to be less than 200 mm, rainfall in the major portion of this tick's distribution zone varies from 500 mm to 1100 mm. Most sites are in East African coastal mosaic vegetation, miombo or undifferentiated woodland, and in *Acacia–Commiphora* deciduous bushland and thicket.

The preferred hosts of *R. muehlensi*, namely tragelaphine antelopes and various duikers, are browsers, while impala are mixed feeders. All these animals are found in habitats containing thickets or various types of woodland, so within its overall distribution range *R. muehlensi* is naturally commoner at localities encompassing these vegetation types.

Except in Somalia, where only *R. muehlensi* is present, and in Zambia and Zimbabwe, where only *R. maculatus* is present, the distribution of these two ticks often overlaps.

Disease relationships

Unknown.

REFERENCES

Baker, M.K. & Keep, M.E. (1970). Checklist of the ticks found on the larger game animals in the Natal game reserves. *Lammergeyer*, **12**, 41–7.

Gallivan, G.J. & Surgeoner, G.A. (1995). Ixodid ticks and other ectoparasites of wild ungulates in Swaziland: regional, host and seasonal patterns. *South African Journal of Zoology*, **30**, 169–77.

Horak, I.G., Boomker, J. & Flamand, J.R.B. (1991). Ixodid ticks and lice infesting red duikers and bushpigs in north-eastern Natal. *Onderstepoort Journal of Veterinary Research*, **58**, 281–4.

Horak, I.G., Boomker, J. & Flamand, J.R.B. (1995b). Parasites of domestic and wild animals in South Africa. XXXIV. Arthropod parasites of nyalas in north-eastern KwaZulu-Natal. *Onderstepoort Journal of Veterinary Research*, **62**, 171–79.

Horak, I.G., Keep, M.E., Flamand, J.R.B. & Boomker, J. (1988). Arthropod parasites of common reedbuck, *Redunca arundinum*, in Natal. *Onderstepoort Journal of Veterinary Research*, **55**, 19–22.

Horak, I.G., Spickett, A.M., Braack, L.E.O., Penzhorn, B.L., Bagnall, R.J. & Uys, A.C. (1995a). Parasites of domestic and wild animals in South Africa. XXXIII. Ixodid ticks on scrub hares in the north-eastern regions of Northern and Eastern Transvaal and of KwaZulu-Natal. *Onderstepoort Journal of Veterinary Research*, **62**, 123–31.

Salisbury, L.E. (1959). Ticks in the South African Zoological Survey Collection. Part X. *Rhipicephalus mühlensi*. *Onderstepoort Journal of Veterinary Research*, **28**, 125–32.

Theiler, G. & Robinson, B.N. (1954). Tick survey. VIII. Checklist of ticks recorded from the Belgian Congo and Ruanda Urundi, from Angola, and from Northern Rhodesia. *Onderstepoort Journal of Veterinary Research*, **26**, 447–61 + 3 maps.

Zumpt, F. (1943). *Rhipicephalus aurantiacus* Neumann und ähnliche Arten. VIII. Vorstudie zu einer Revision der Gattung *Rhipicephalus* Koch. *Zeitschrift für Parasitenkunde*, **13**, 102–17.

Also see the following Basic References (pp. 12–14): Elbl & Anastos (1966); Hoogstraal (1956); Keirans (1985); Santos Dias (1960); Theiler (1962); Walker (1974); Yeoman & Walker (1967).

RHIPICEPHALUS MUHSAMAE MOREL & VASSILIADES, 1965

This species was named after Dr Brouria Feldman-Muhsam, of the Department of Parasitology, the Hebrew University, Jerusalem, in recognition of her contributions to our knowledge of ixodid ticks, especially in the genera *Hyalomma* and *Rhipicephalus*.

Diagnosis

A large dark brown to blackish tick.

Male (Figs 135(a), 136(a) to (c))

Capitulum longer than broad, length × breadth ranging from 0.81 mm × 0.72 mm to 1.00 mm × 0.88 mm. Basis capituli with short acute lateral angles at about anterior third of its length. Palps broad, somewhat flattened apically. Conscutum length × breadth ranging from 3.47 mm × 2.20 mm to 4.14 mm × 2.81 mm; anterior process of coxae I rounded. Eyes flat, edged dorsally by a few medium-sized setiferous punctations. Cervical pits comma-shaped, discrete. Marginal lines long, usually enclosing two festoons posteriorly. Posteromedian and posterolateral grooves visible but superficial. Large setiferous punctations present along the external cervical margins and marginal lines; a few medium-sized punctations scattered on scapular apices and in the four irregular rows comprising the '*simus*' pattern medially on the conscutum, but interstitial punctation pattern light to absent. Ventrally spiracles variable, in general comma-shaped with a short broad prolongation curving

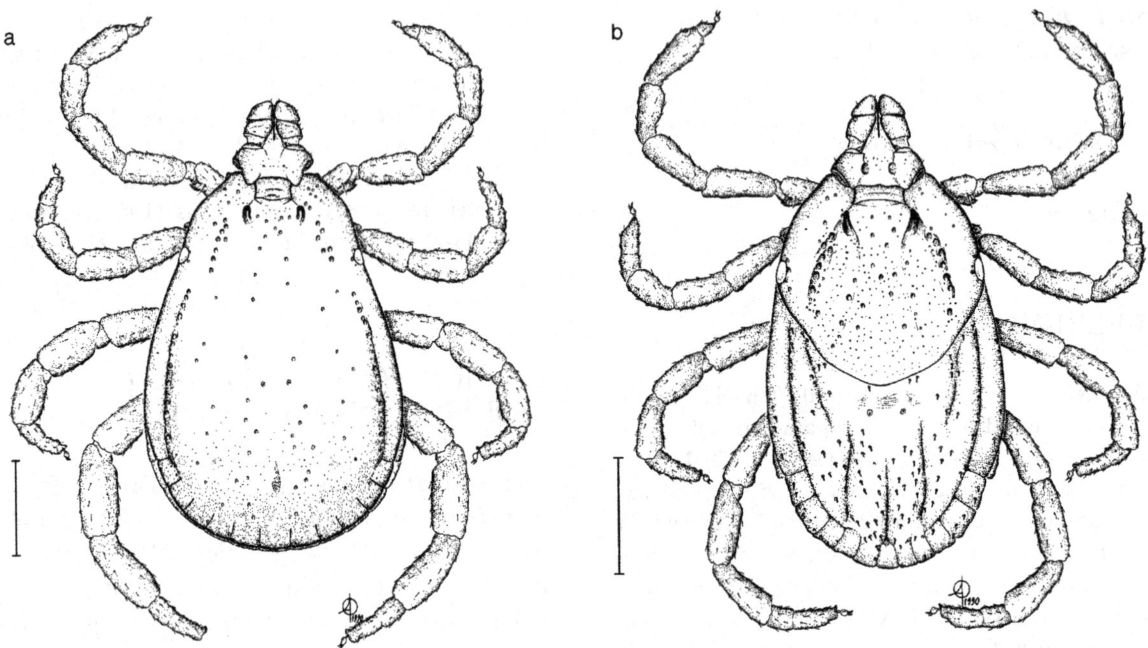

Figure 135. *Rhipicephalus muhsamae* [B.S.699/-, laboratory reared, progeny of ♀ collected from warthog (*Phacochoerus africanus*), Maruzi County, Lango, Uganda, on 3 July 1956 by Eriasafu Okello]. (a) Male, dorsal; (b) female, dorsal. Scale bars represent 1 mm. A. Olwage *del.*

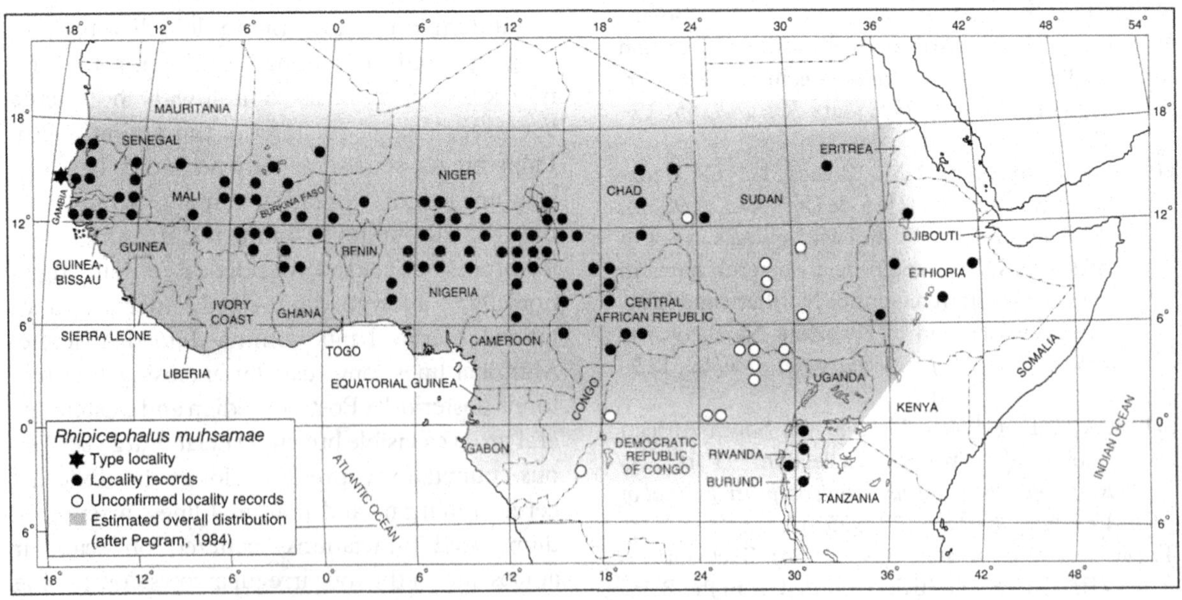

Map 40. *Rhipicephalus muhsamae*: distribution. (Based largely on Morel, 1969).

Table 36. *Host records of* Rhipicephalus muhsamae

Hosts	Number of records
Domestic animals	
Cattle	45
Sheep	2
Goats	1
Horses	2
Donkeys	1
Pigs	3
Dogs	5
Wild animals	
Side-striped jackal (*Canis adustus*)	3
'Jackal'	2
Lion (*Panthera leo*)	2
Leopard (*Panthera pardus*)	1
Spotted hyaena (*Crocuta crocuta*)	1
'Hyaena'	2
African civet (*Civettictis civetta*)	4
Warthog (*Phacochoerus africanus*)	11
Red river hog (*Potamochoerus porcus*)	1
'Wild pig'	1
Oribi (*Ourebia ourebi*)	2
African buffalo (*Syncerus caffer*)	2
Bushbuck (*Tragelaphus scriptus*)	1
Roan antelope (*Hippotragus equinus*)	1
Bohor reedbuck (*Redunca redunca*)	1
Geoffroy's ground squirrel (*Xerus erythropus*)	3 (nymphs; and adults in nest)
Nile rat (*Arvicanthis niloticus*)	2
Crested porcupine (*Hystrix cristata*)	2
'Rodent'	1 (nymphs)
Humans	3

towards the dorsal surface. Adanal plates broad, with internal margins posterior to the anus slightly concave; accessory adanal plates small, pointed.

Female (Figs 135(b), 136(d) to (f))

Capitulum longer than broad, length × breadth ranging from 0.86 mm × 0.79 mm to 1.06 mm × 0.96 mm. Basis capituli with acute lateral angles just anterior to mid-length; porose areas large, oval, about 1.5 times their own diameter apart. Palps with article I narrow relative to articles II and III and easily visible dorsally, apices broadly rounded. Scutum longer than broad, length × breadth ranging from 1.72 mm × 1.69 mm to 2.40 mm × 2.03 mm; posterior margin somewhat sinuous. Eyes just anterior to broadest part of scutum, almost flat, edged dorsally by a few medium-sized setiferous punctations. Cervical fields broad, slightly depressed, their external margins distinct, delimited by irregular rows of large setiferous punctations. Slightly smaller setiferous punctations present on the scapulae and scattered medially on the

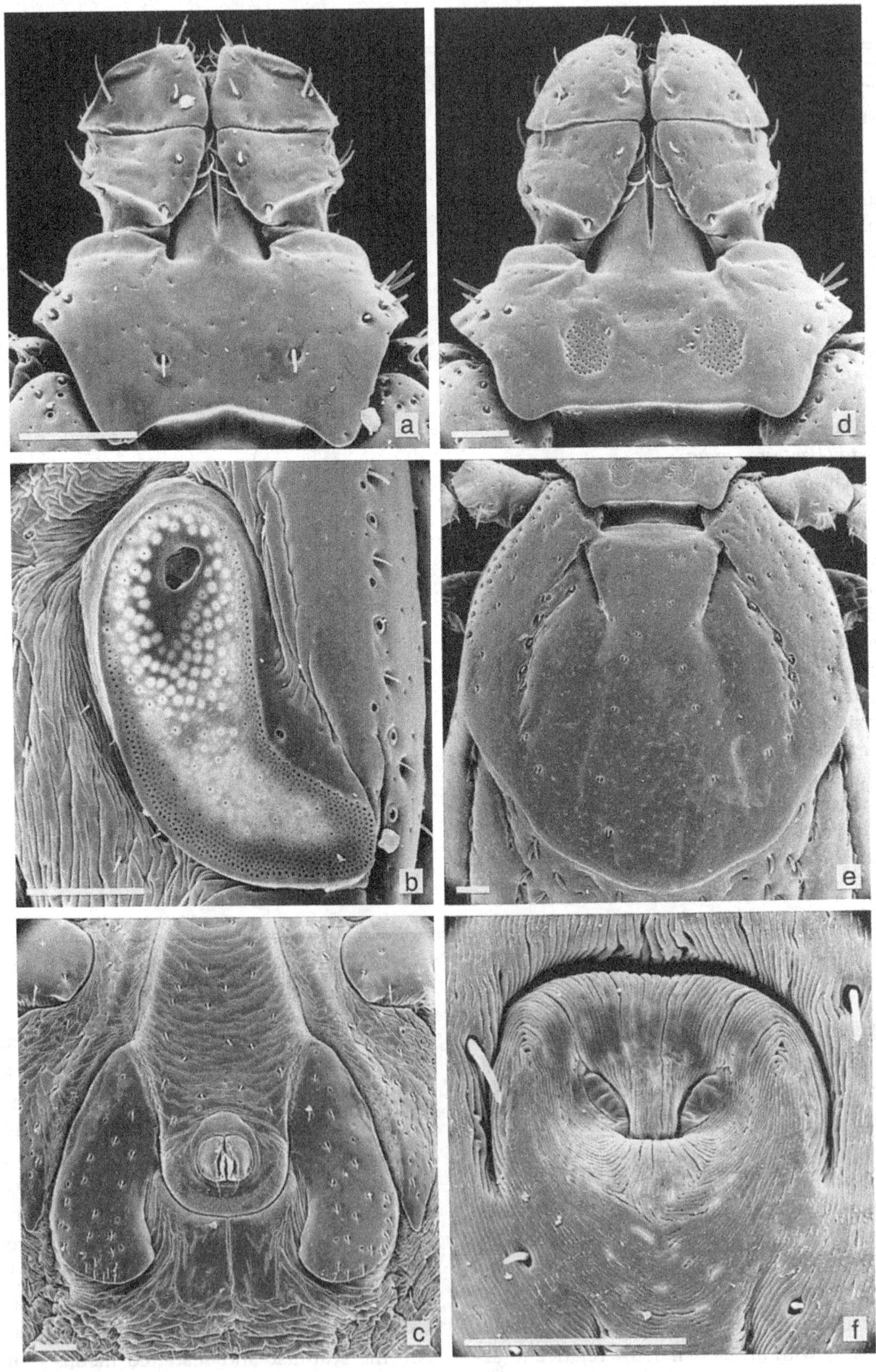
a
b
c
d
e
f

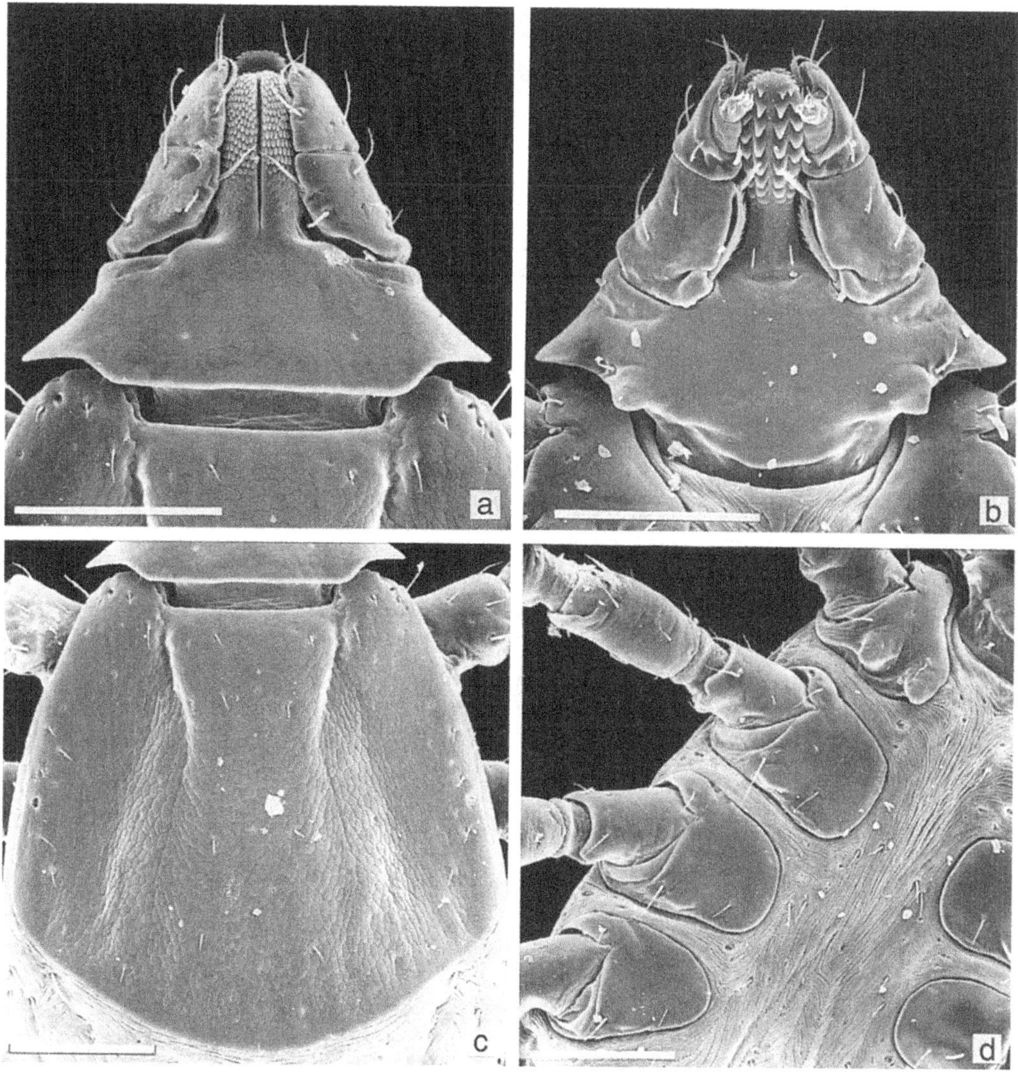

Figure 137 (*above*). *Rhipicephalus muhsamae* (L722, laboratory reared, progeny of ♀ originating from Senegal; host, date of collection and collector unknown). Nymph: (a) capitulum, dorsal; (b) capitulum, ventral; (c) scutum; (d) coxae. Scale bars represent 0.10 mm. SEMS by M.D. Corwin. (Figs (a), (c) & (d) from Pegram *et al.*, 1987, figs 28–30, with kind permission from the Entomological Society of America.)

Figure 136 (*opposite*). *Rhipicephalus muhsamae* (L722, laboratory reared, progeny of ♀ originating from Senegal; host, date of collection and collector unknown). Male: (a) capitulum, dorsal; (b) spiracle; (c) adanal plates. Female: (d) capitulum, dorsal; (e) scutum; (f) genital aperture. Scale bars represent 0.10 mm. SEMs by M.D. Corwin. (Figs (b), (c), (e) & (f) from Pegram *et al.*, 1987, figs 32–34 & 36, with kind permission from the Entomological Society of America.)

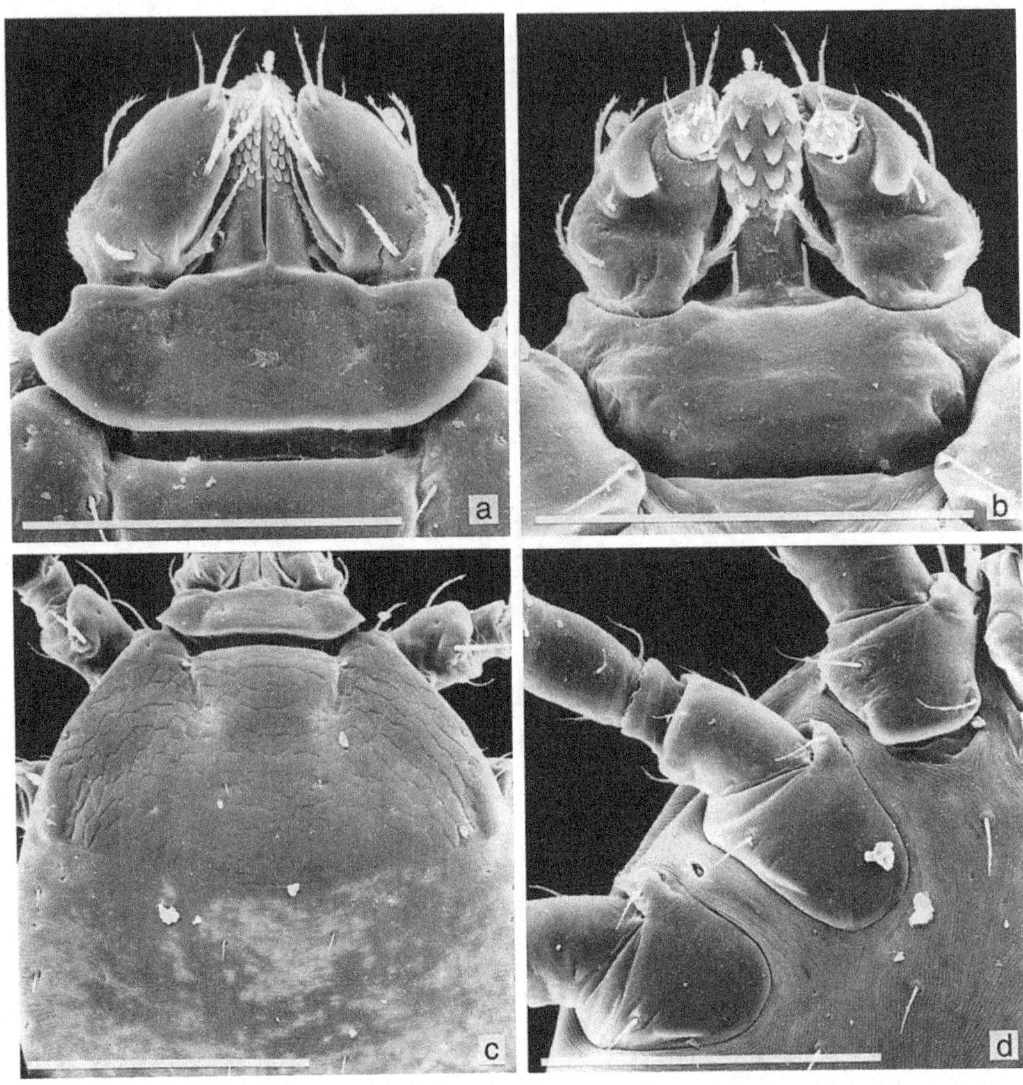

Figure 138. *Rhipicephalus muhsamae* (L722, laboratory reared, progeny of ♀ originating from Senegal; host, date of collection and collector unknown). Larva: (a) capitulum, dorsal; (b) capitulum, ventral; (c) scutum; (d) coxae. Scale bars represent 0.10 mm. SEMS by M.D. Corwin. (Figs (a), (c) & (d) from Pegram *et al.*, 1987, figs 25–27, with kind permission from the Entomological Society of America.)

scutum, interspersed with fine interstitial punctations that are numerous in some specimens but sparser, or even absent, in others. Ventrally genital aperture a narrow truncated U-shape, diverging markedly anteriorly; sclerotized margins usually wide and distinct.

Nymph (Fig. 137)

Capitulum much broader than long, length × breadth ranging from 0.26 mm × 0.37 mm to 0.27 mm × 0.38 mm. Basis capituli approaching four times as broad as long, with sharply-tapering lateral angles in posterior half, projecting over scapulae; ventrally with short blunt spurs on posterior margin. Palps broadest basally, tapering to narrowly-rounded apices. Scutum broader than long, length × breadth ranging from 0.50 mm × 0.57 mm to 0.57 mm × 0.63 mm; posterior margin a broad smooth curve. Eyes at widest point, well over halfway back, large, almost flat, delimited dorsally by faint depressions. Cervical fields long, narrow, slightly sunken, divergent. Ventrally coxae I each with external spur slightly longer and sharper than internal spur; coxae II to III each with small spurs; coxae IV each with indications only of spurs.

Larva (Fig. 138)

Capitulum much broader than long, length × breadth ranging from 0.123 mm × 0.175 mm to 0.135 mm × 0.192 mm. Basis capituli over three times as broad as long, with short broadly-rounded lateral angles, posterior border almost straight. Palps broad, tapering very slightly to bluntly-rounded apices, inclined inwards. Scutum much broader than long, length × breadth ranging from 0.240 mm × 0.369 mm to 0.251 mm × 0.394 mm; posterior margin a wide, very shallow curve. Eyes at widest point, well over halfway back, slightly convex. Cervical grooves short, slightly convergent. Ventrally coxae I each with a short broad spur; coxae II and III each with only a slight salience on their posterior margin.

Notes on identification

The nymphs and larvae of *R. muhsamae* differ slightly morphologically from those of *R. praetextatus* (p. 340, Figs 159, 160) and *R. simus* (p. 416, Figs 195, 196) but there is little to characterize its adults specifically apart from the structure of its mounted female genital aperture (p. 421, Fig. 197(c)) (Pegram, 1979; Morel, 1980; Pegram *et al.*, 1987).

According to Morel & Vassiliades (1965) *R. muhsamae* usually features in the earlier literature on West and Central African ticks as either *R. simus* or *R. s. simus*, and occasionally as *R. senegalensis*. The data on its hosts and distribution that we present here are largely based on this finding. It is virtually impossible now to either confirm or refute them because the present whereabouts of many of the specimens on which they were based, if they still exist, are unknown.

Some of the specimens from the Democratic Republic of Congo and Rwanda identified as *R. simus* by Elbl & Anastos (1966) may also prove to be *R. muhsamae*. So may the ticks listed by Yeoman & Walker (1967) as heavily-punctate *R. simus*, mainly in western and south-western Tanzania (Morel, 1980). Re-examination of these collections is, however, essential to clarify this situation (Pegram *et al.*, 1987). (See also *R. praetextatus*, p. 340, and *R. simus*, p. 416).

Hosts

A three-host species (Morel, 1980). We ourselves have little personal experience of this primarily West African tick. The accompanying list of its hosts (Table 36) has been compiled largely, but not exclusively, from the literature, in particular the publications quoted by Morel & Vassiliades (1965). (Records in those papers that they questioned have, obviously, been omitted). In 1980 Morel summarized its host relationships as follows: 'The preimagoes have been collected on Myomorpha and Sciuromorpha rodents and on hares. The host affinities of adults are exactly those of *Rh. simus* towards available ungulate and carnivore mammals, primates, aardvarks and hedgehogs.'

According to Mohammed (1977), on Fulani cattle the adults attach on the ears, neck, dewlap, genitalia and legs, including the hoof region. At Runka, in north-western Nigeria, adults were collected from cattle throughout most of the year. Further south, at Samaru, the adults were active both in January, during the dry season, and, like *R. senegalensis*, from May to September, during the wet season.

Zoogeography

Rhipicephalus muhsamae has been recorded across Africa from Senegal in the west to Ethiopia in the east, with an extension southwards into western Uganda, Rwanda and Burundi (Map 40). This distribution pattern is anomalous in some respects. In West Africa, where the vast majority of currently accepted records lie, it occurs mainly at altitudes of less than 1000 m in Sudanian woodland with mean annual rainfalls of 800 mm or less. At the other extreme, as Morel (1980) noted, it has been recorded in Ethiopia at high altitudes near highland forest communities. Pegram, Hoogstraal & Wassef (1981) collected it in the 'wetter western areas' of Ethiopia. In Uganda, Rwanda and Burundi it apparently also occurs at high altitudes.

Disease relationships

Morel (1980) quotes Tendeiro (1952) as stating that *R. muhsamae* can be spontaneously infected with *Coxiella burneti*.

REFERENCES

Mohammed, A.N. (1977). The seasonal incidence of ixodid ticks of cattle in northern Nigeria. *Bulletin of Animal Health and Production in Africa,* 25, 273–93.

Morel, P.C. & Vassiliades, G. (1965). Description de *Rhipicephalus muhsamae* n.sp. de l'Ouest-Africain (groupe de *Rh. simus*; Acariens, Ixodoidea). *Revue d'Élevage et de Médecine Vétérinaire des Pays Tropicaux,* **17 for 1964** (nouvelle série), 619–36.

Pegram, R.G., Hoogstraal, H. & Wassef, H.Y. (1981). Ticks (Acari: Ixodoidea) of Ethiopia. I. Distribution, ecology and host relationships of species infesting livestock. *Bulletin of Entomological Research,* 71, 339–59.

Pegram, R.G., Walker, J.B., Clifford, C.M. & Keirans, J.E. (1987). Comparison of populations of the *Rhipicephalus simus* group: *R. simus, R. praetextatus,* and *R. muhsamae* (Acari: Ixodidae). *Journal of Medical Entomology,* 24, 666–82.

Also see the following Basic References (pp. 12–14): Elbl & Anastos (1966); Matthysse & Colbo (1987); Morel (1969, 1980); Pegram (1979, 1984); Yeoman & Walker (1967).

RHIPICEPHALUS NEUMANNI WALKER, 1990

This species was named in honour of Professor L.G. Neumann (1846–1930), of the Toulouse Veterinary School in southern France. He described 19 species in this genus that are still regarded as valid and contributed greatly to our knowledge of ticks in general.

Diagnosis

A moderate-sized dark brown tick.

Male (Figs 139(a), 140(a) to (c))

Capitulum longer than broad, length × breadth ranging from 0.63 mm × 0.58 mm to 0.88 mm × 0.78 mm. Basis capituli with short blunt lateral angles anteriorly, at somewhat less than a third of its length. Palps broadly rounded apically. Conscutum length × breadth ranging from 2.77 mm × 1.78 mm to 3.52 mm × 2.21 mm; sharp anterior process present on coxae I. Eyes slightly bulging, edged dorsally by a few medium-sized punctations and faint depressions. Cervical pits comma-shaped, convergent. Marginal lines shallow, not reaching eye level. Posteromedian and posterolateral grooves, when

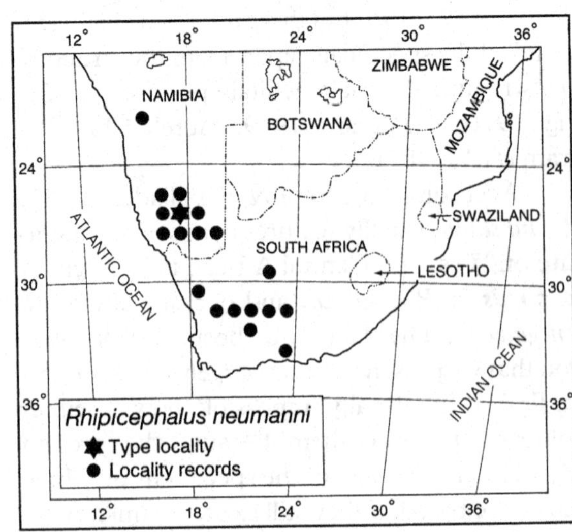

Map 41. *Rhipicephalus neumanni:* distribution.

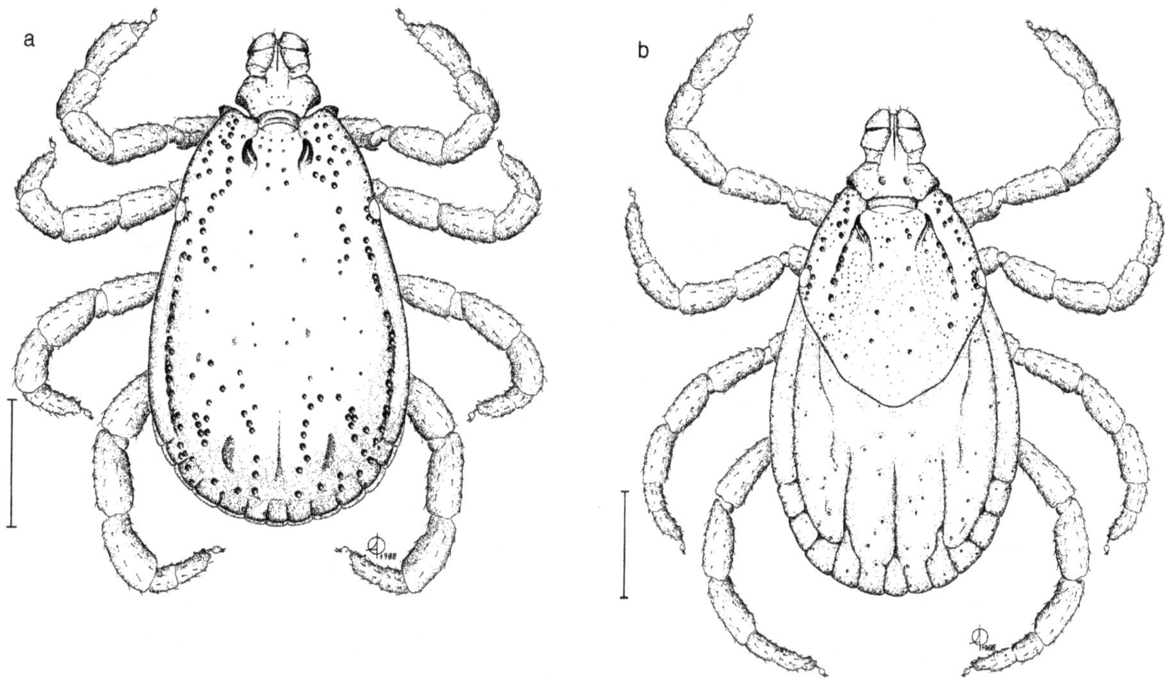

Figure 139. *Rhipicephalus neumanni* (Onderstepoort Tick Collection 3141iii; J.D. Bezuidenhout No. 5189; from sheep, Farm 'Wegkruip' No. 130, Karas Region, Namibia, November 1972, J.D. Bezuidenhout coll.) (a) Male, dorsal; (b) female, dorsal. Scale bars represent 1 mm. A. Olwage *del.* (From Walker, 1990, figs 48 & 49, with kind permission from the Editor, *Onderstepoort Journal of Veterinary Research.*)

Table 37. *Host records of* Rhipicephalus neumanni

Hosts	Number of records
Domestic animals	
Sheep	25
Goats	8
Horses	1
Wild animals	
Black wildebeest (*Connochaetes gnou*)	1
Springbok (*Antidorcas marsupialis*)	1
Gemsbok (*Oryx gazella*)	1
Grey rhebok (*Pelea capreolus*)	1
Mountain reedbuck (*Redunca fulvorufula*)	1
Namaqua rock mouse (*Aethomys namaquensis*)	1 (nymphs)

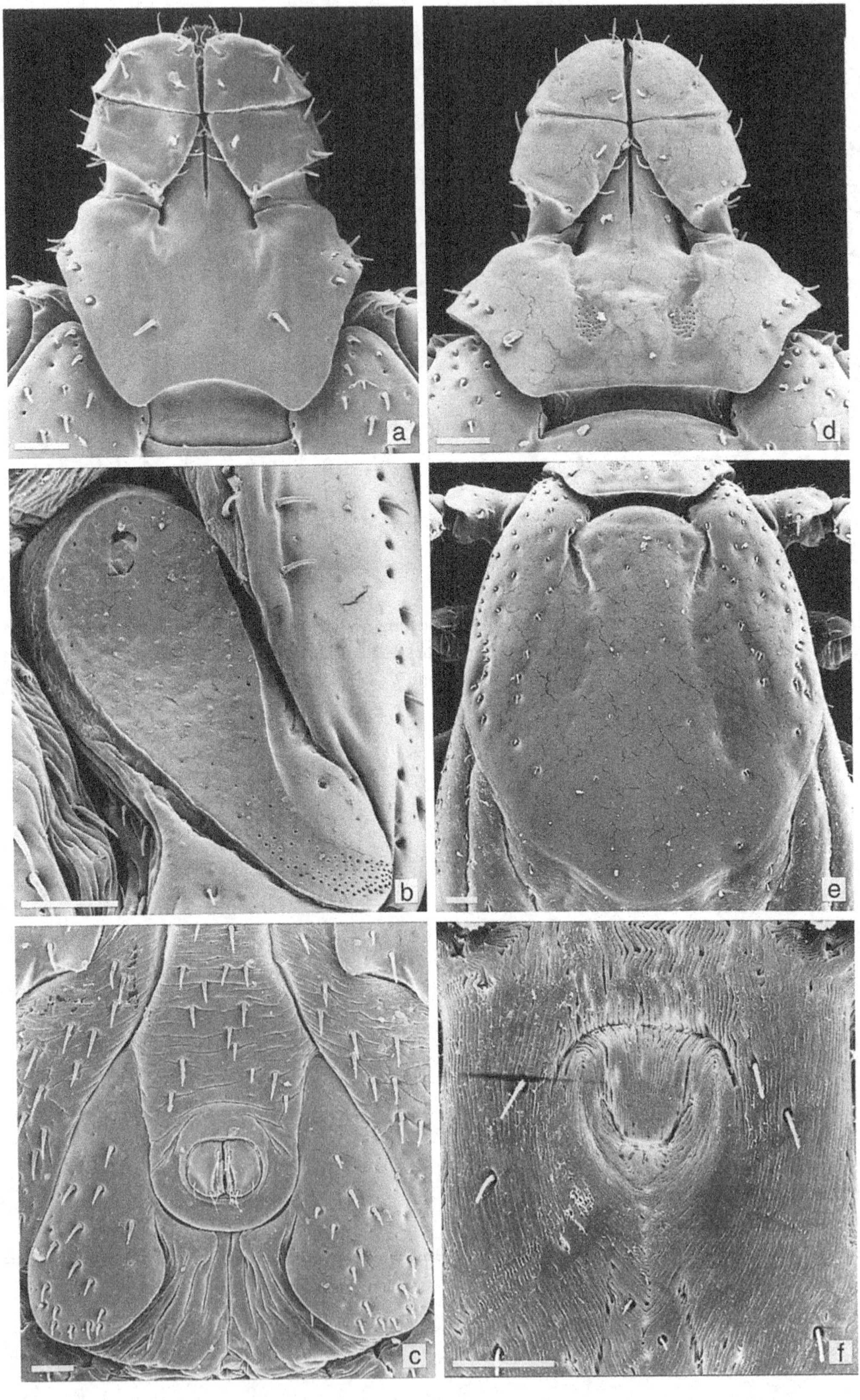
a
b
c
d
e
f

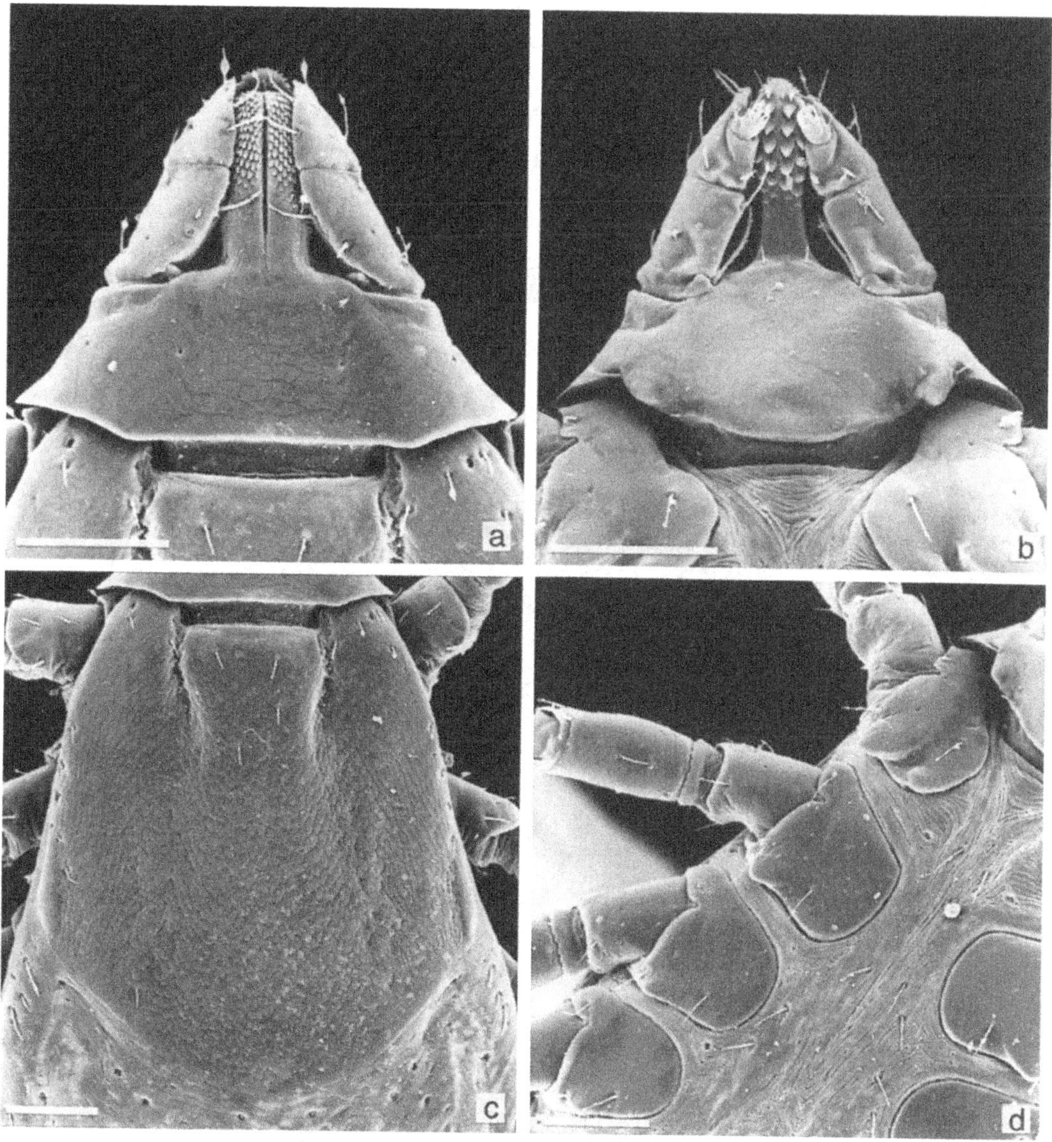

Figure 141 (*above*). *Rhipicephalus neumanni* (RML 65728; J.D. Bezuidenhout 5666, laboratory reared, original ♀ collected at Keetmanshoop, Karas Region, Namibia). Nymph: (a) capitulum, dorsal; (b) capitulum, ventral; (c) scutum; (d) coxae. Scale bars represent 0.10 mm. SEMs by M.D. Corwin. (From Walker, 1990, figs 38, 39, 41 & 42, with kind permission from the Editor, *Onderstepoort Journal of Veterinary Research.*)

Figure 140 (*opposite*). *Rhipicephalus neumanni* (RML 65728; J.D. Bezuidenhout 5666, laboratory reared, original ♀ collected at Keetmanshoop, Karas Region, Namibia). Male: (a) capitulum, dorsal; (b) spiracle; (c) adanal plates. Female: (d) capitulum, dorsal; (e) scutum; (f) genital aperture. Scale bars represent 0.10 mm. SEMs by M.D. Corwin. (From Walker, 1990, figs 26, 30–32, 35 & 36, with kind permission from the Editor, *Onderstepoort Journal of Veterinary Research.*)

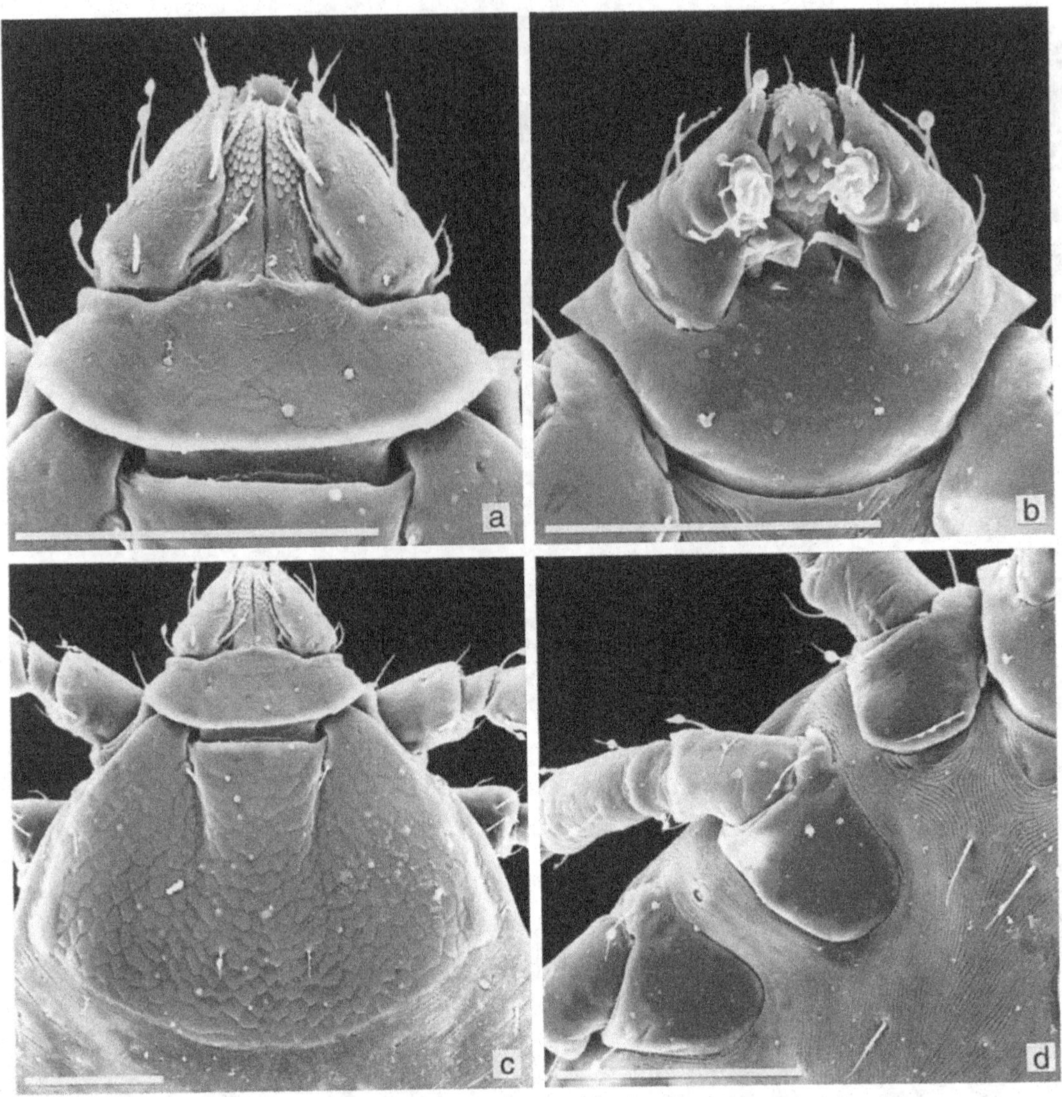

Figure 142. *Rhipicephalus neumanni* (RML 65728; J.D. Bezuidenhout 5666, laboratory reared, original ♀ collected at Keetmanshoop, Karas Region, Namibia). Larva: (a) capitulum, dorsal; (b) capitulum, ventral; (c) scutum; (d) coxae. Scale bars represent 0.10 mm. SEMs by M.D. Corwin. (From Walker, 1990, figs 44–47, with kind permission from the Editor, *Onderstepoort Journal of Veterinary Research*.)

present, shallow and inconspicuous. Large setiferous punctations present on the scapulae, along the external cervical margins and marginal lines, and encircling the posterolateral grooves. Slightly smaller setiferous punctations present medially on the conscutum, together with numerous minute interstitial punctations. Ventrally spiracles long, narrowing markedly at about two-thirds of their length and curving slightly towards the dorsal surface. Adanal plates broadly rounded posteriorly, tapering towards the anterior end; accessory adanal plates represented merely by small sclerotized points.

Female (Figs 139(b), 140(d) to (f))

Capitulum longer than broad, length × breadth ranging from 0.90 mm × 0.83 mm to 1.00 mm × 0.92 mm, but in some specimens slightly broader than long. Basis capituli with sharply pointed lateral angles at about mid-length; porose areas small, just over twice their own diameter apart. Palps with article III wedge-shaped, its outer margin smoothly curved, its inner and posterior margins almost straight. Scutum longer than broad, length × breadth ranging from 1.82 mm × 1.72 mm to 2.20 mm × 1.98 mm, posterior margin sinuous. Eyes just anterior to broadest part of scutum, slightly bulging, delimited dorsally by a few large, setiferous punctations. Cervical pits short, convergent; cervical fields slightly depressed, delimited along their external margins by irregular rows of large, setiferous punctations, and similar punctations scattered on the scapulae. A few slightly smaller setiferous punctations present medially on scutum and numerous minute interstitial punctations scattered all over scutum. Ventrally genital aperture quite wide, with the sides of the opening converging to join the straight posterior margin.

Nymph (Fig. 141)

Capitulum much broader than long, length × breadth *c.* 0.22 mm × 0.32 mm. Basis capituli three times as broad as long, with sharp lateral angles at about posterior two-thirds of its length, projecting over scapulae; ventrally with short, blunt spurs on posterior margin. Palps broadest at about mid-length, narrower proximally and tapering distally to sharp apices, inclined inwards. Scutum slightly longer than broad, length × breadth *c.* 0.48 mm × 0.46 mm; posterior margin a deep smooth curve. Eyes at widest point, well over halfway back, long and narrow, delimited dorsally by slight depressions. Cervical pits short, convergent; cervical fields long, narrow, divergent, inconspicuous. Ventrally coxae I each with a large external spur and a shorter internal spur; remaining coxae each with a small external spur only.

Larva (Fig. 142)

Capitulum much broader than long, length × breadth *c.* 0.09 mm × 0.13 mm. Basis capituli nearly three times as broad as long, with short bluntly rounded lateral angles, posterior border gently curved. Palps slightly constricted proximally, almost immediately widening then tapering gradually to narrowly-rounded apices, inclined inwards. Scutum much broader than long, length × breadth *c.* 0.21 mm × 0.34 mm; posterior margin a broad shallow curve. Eyes at widest part of scutum, slightly convex. Cervical grooves short, slightly convergent. Ventrally coxae I each with a broad protuberance on posterior border, coxae II each with only a slight convexity on posterior border, coxae III each with a straight posterior border.

Notes on identification

Rhipicephalus neumanni adults, which have a predilection for sheep and goats, closely resemble those of *R. distinctus*, a parasite of hyraxes (dassies) (see p. 138). Both Bedford (1932) and Theiler (1947) listed one collection of *R. neumanni* from sheep at Victoria West, Northern Cape Province, as *R. distinctus*.

The measurements of the nymph and larva quoted above were calculated from the scanning electron micrographs because the remaining unmounted specimens of this species had inadvertently been lost.

Hosts

A three-host species (Walker, 1990). The commonest recorded hosts of *R. neumanni* are sheep (Walker, 1990; Horak & Fourie, 1992), and to a lesser extent goats (Table 37), on both of which they usually attach on the feet between the claws. A single collection only exists from a horse. Single collections have also been taken from several species of antelopes. The adults are present from September to June, with most collections made from February to May.

The only known host of the nymphs is a Namaqua rock mouse.

Zoogeography

This species occurs commonly in southern Namibia (Walker, 1990). As yet a single record only (from a gemsbok) exists from further north in the country, at Omandumba, in the north-eastern Erongo Region. In South Africa it has been recorded in Northern, Western and Eastern Cape Provinces (Map 41).

It is present primarily in mountainous or hilly semi-desert areas, at altitudes from about 800 m to 1500 m, with 100 mm to 300 mm rainfall per annum in summer. Most collections have been made in bushy Karoo–Namib shrubland, though in South Africa a few come from dwarf Karoo shrubland (Walker, 1990). There is little doubt that this tick is more widespread in the Karoo than present records indicate.

Disease relationships

Farmers report that *R. neumanni* adults cause lameness and foot abscesses in sheep because they attach between the claws of the animals' hooves. Up to 30% of a flock may be affected (J.D. Bezuidenhout, pers. comm., 1989).

REFERENCES

Bedford, G.A.H. (1932). A synoptic check-list and host-list of the ectoparasites found on South African Mammalia, Aves, and Reptilia. (Second Edition). *18th Report of the Director of Veterinary Services and Animal Industry, Union of South Africa*, 223–523.

Horak, I.G. & Fourie, L.J. (1992). Parasites of domestic and wild animals in South Africa. XXXI. Adult ticks on sheep in the Cape Province and in the Orange Free State. *Onderstepoort Journal of Veterinary Research*, **59**, 275–83.

Walker, J.B. (1990). Two new species of ticks from southern Africa whose adults parasitize the feet of ungulates: *Rhipicephalus lounsburyi* n. sp. and *Rhipicephalus neumanni* n. sp. (Ixodoidea, Ixodidae). *Onderstepoort Journal of Veterinary Research*, 57, 57–75.

Also see the following Basic Reference (p. 14): Theiler (1947).

RHIPICEPHALUS NITENS NEUMANN, 1904

The specific name *nitens*, from the Latin *nitens* meaning 'shining', refers to the shiny scutum of the adults.

Diagnosis

A medium-sized shiny dark reddish-brown species.

Male (Figs 143(a), 144(a) to (c))

Capitulum about as broad as long, length × breadth ranging from 0.51 mm × 0.52 mm to 0.76 mm × 0.74 mm. Basis capituli with lateral angles at about anterior third of its length, short, acute. Palps short, bluntly rounded apically. Conscutum length × breadth ranging from 2.08 mm × 1.52 mm to 3.41 mm × 2.25 mm; large anterior process present on coxae I. A short, tail-like caudal process sometimes seen on engorged males. Eyes slightly convex, edged dorsally by a few punctations. Cervical fields broad, slightly depressed. Marginal lines long, outlined by a few large punctations, not quite reaching eyes anteriorly. Posteromedian and posterolateral

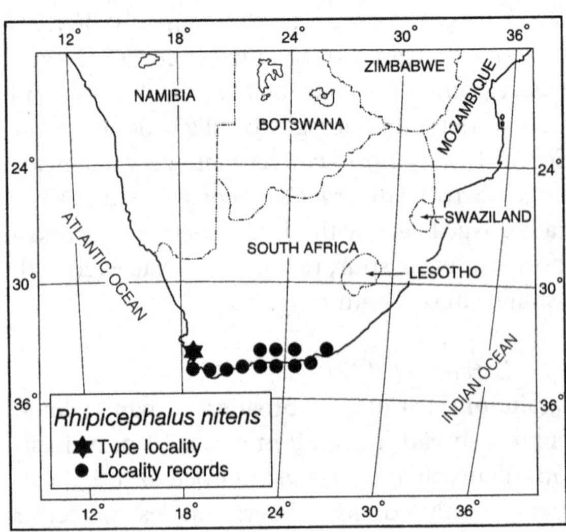

Map 42. *Rhipicephalus nitens*: distribution.

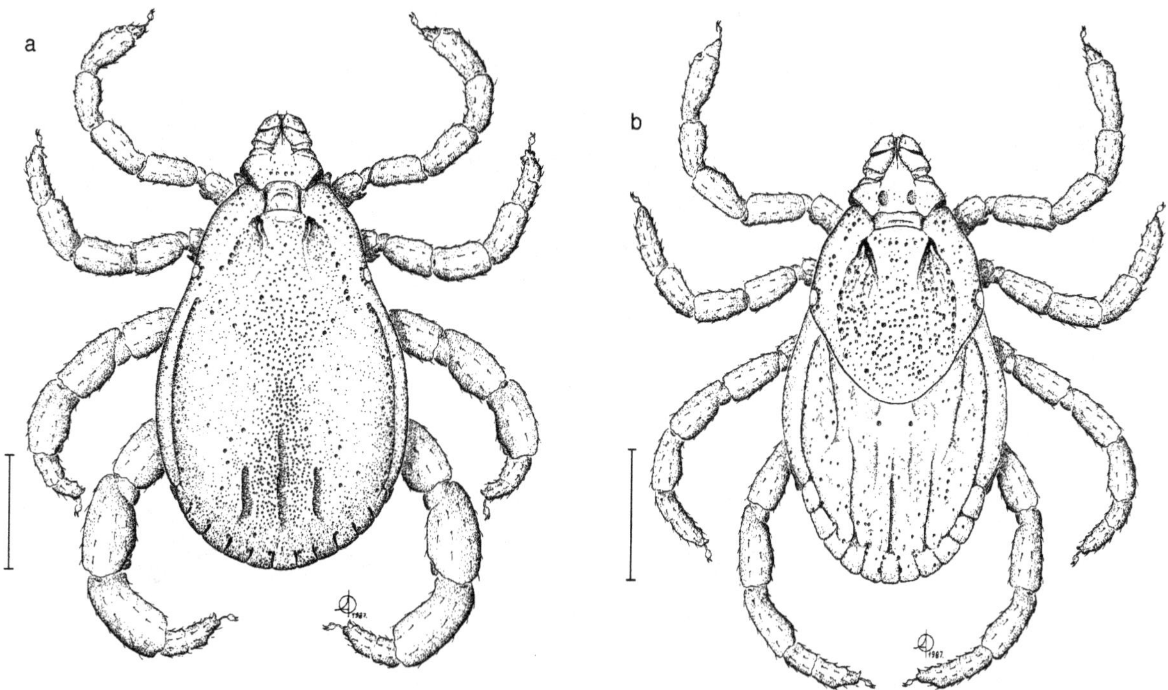

Figure 143. *Rhipicephalus nitens* [laboratory reared, original ♀ collected from grey rhebok (*Pelea capreolus*), Bontebok National Park, Swellendam, Western Cape Province, South Africa, in December 1979 by I.G. Horak & V. de Vos]. (a) Male, dorsal; (b) female, dorsal. Scale bars represent 1 mm. A. Olwage *del.*

Table 38. *Host records of* Rhipicephalus nitens

Hosts	Number of records
Domestic animals	
Cattle	2
Sheep	87 (including immatures)
Goats	2
Dogs	2
Wild animals	
Mountain zebra (*Equus zebra*)	1
Bontebok (*Damaliscus pygargus dorcas*)	55 (including immatures)
Springbok (*Antidorcas marsupialis*)	6 (including immatures)
Grey rhebok (*Pelea capreolus*)	62 (including immatures)
Scrub hare (*Lepus saxatilis*)	20 (including immatures)
Birds	
Helmeted guineafowl (*Numida meleagris*)	1 (larva)

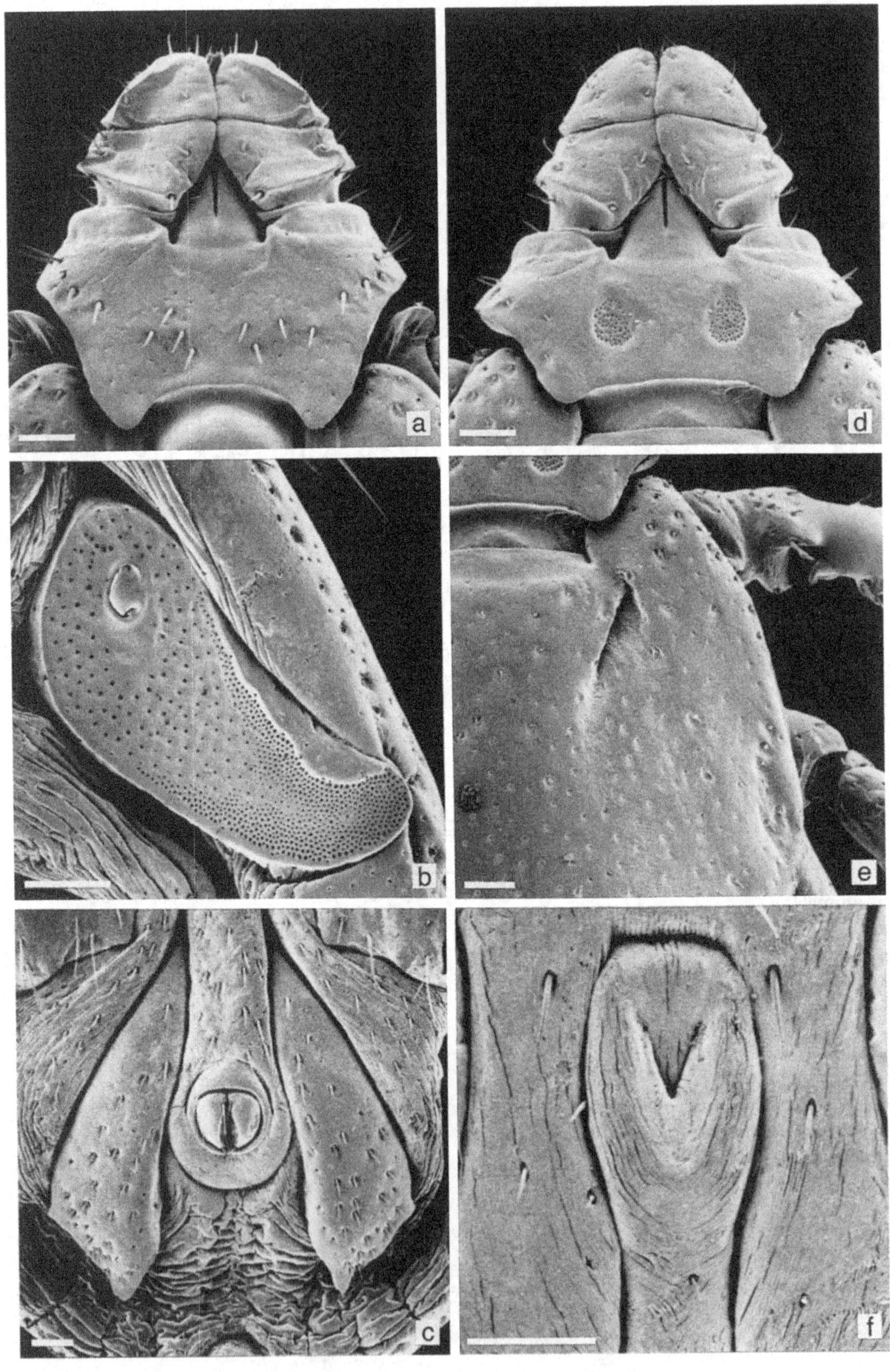
a
b
c
d
e
f

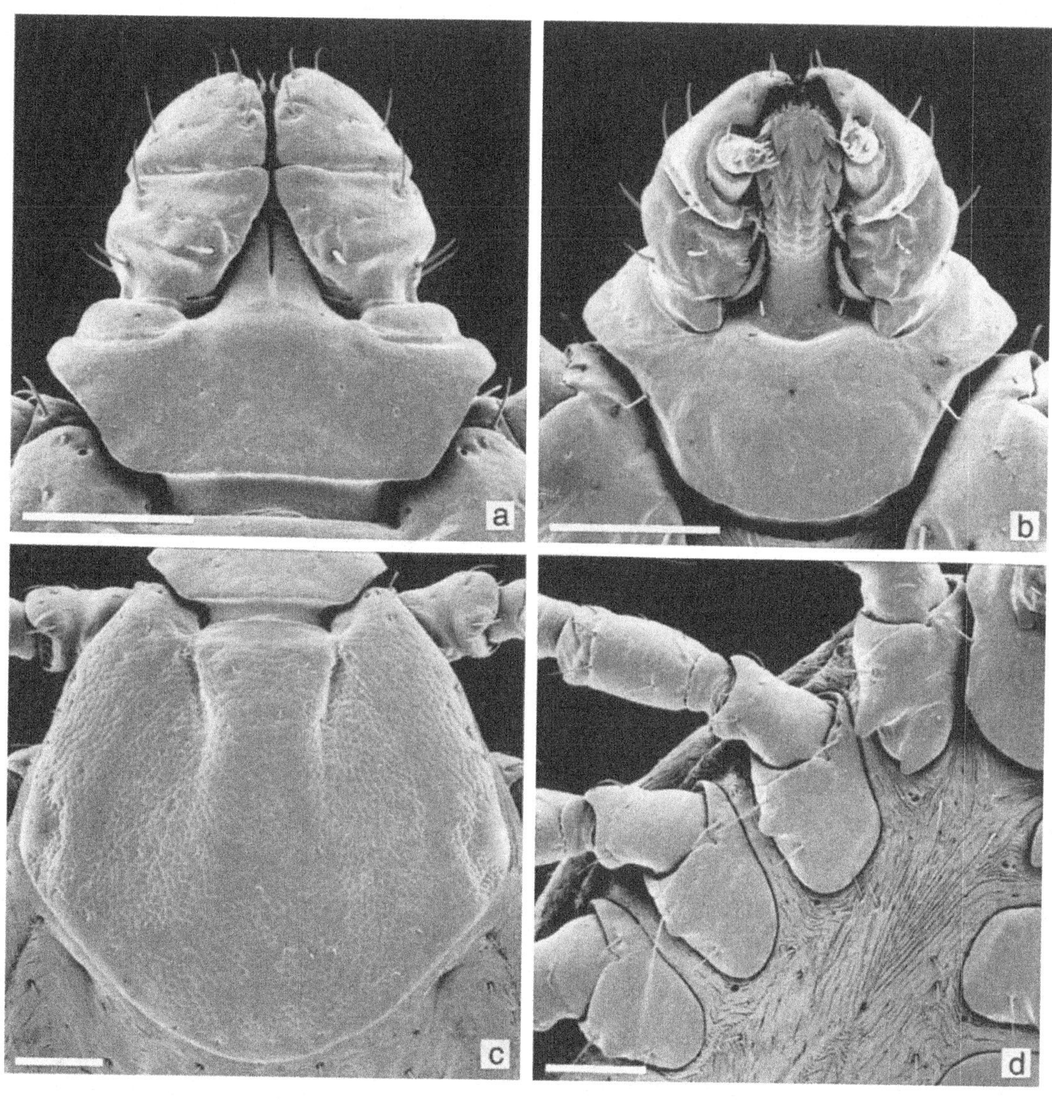

Figure 145 (*above*). *Rhipicephalus nitens* [Protozoology Section Tick Breeding Register, Onderstepoort, No. 3867, laboratory reared, original ♀ collected from grey rhebok (*Pelea capreolus*), Bontebok National Park, Swellendam, Western Cape Province, South Africa, on 9 February 1988 by I.G. Horak]. Nymph: (a) capitulum, dorsal; (b) capitulum, ventral; (c) scutum; (d) coxae. Scale bars represent 0.10 mm. SEMs by J.F. Putterill.

Figure 144 (*opposite*). *Rhipicephalus nitens* [Protozoology Section Tick Breeding Register, Onderstepoort, No. 3867, laboratory reared, original ♀ collected from grey rhebok (*Pelea capreolus*), Bontebok National Park, Swellendam, Western Cape Province, South Africa, on 9 February 1988 by I.G. Horak]. Male: (a) capitulum, dorsal; (b) spiracle; (c) adanal plates. Female: (d) capitulum, dorsal; (e) scapular area; (f) genital aperture. Scale bars represent 0.10 mm. SEMs by J.F. Putterill.

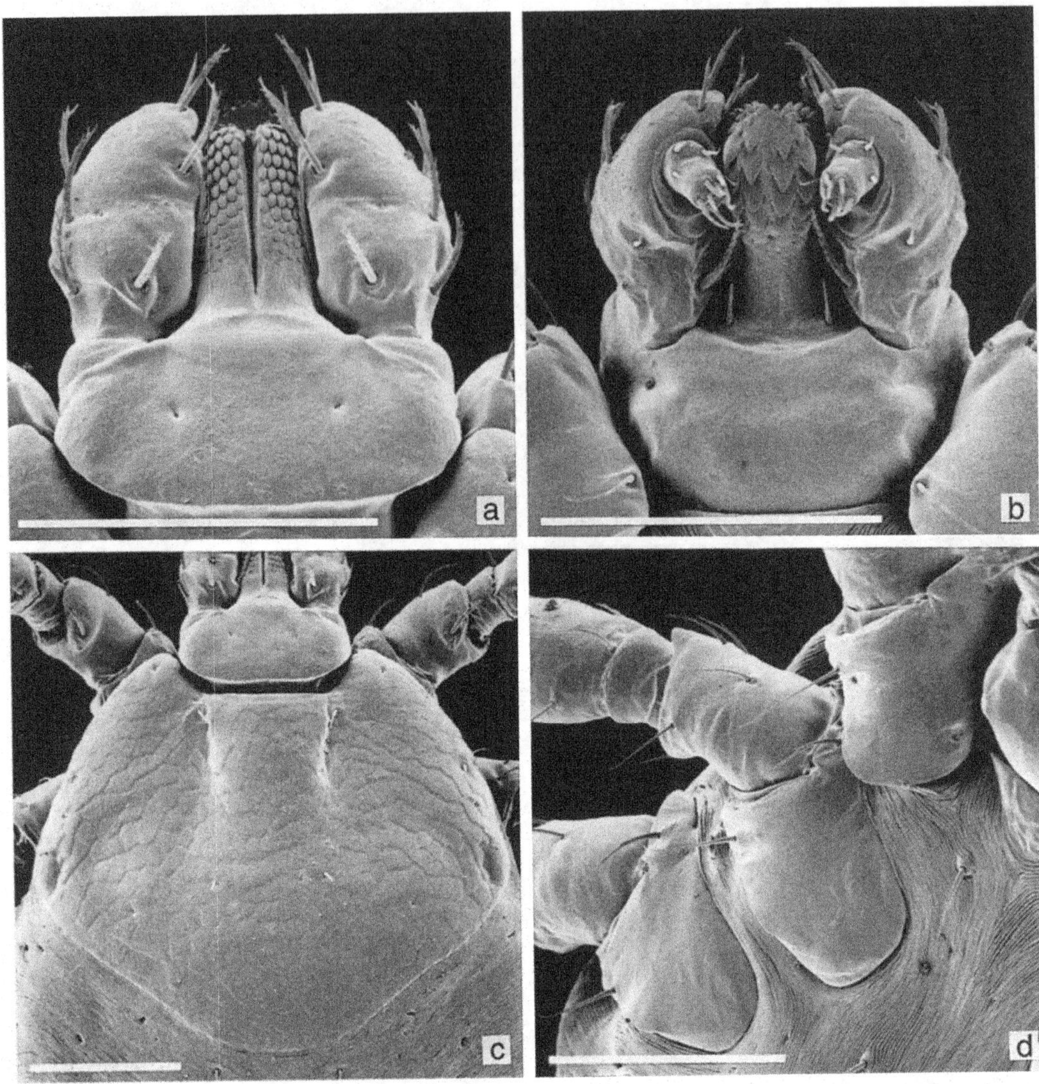

Figure 146. *Rhipicephalus nitens* [Protozoology Section Tick Breeding Register, Onderstepoort, No. 3867, laboratory reared, original ♀ collected from grey rhebok (*Pelea capreolus*), Bontebok National Park, Swellendam, Western Cape Province, South Africa, on 9 February 1988 by I.G. Horak]. Larva: (a) capitulum, dorsal; (b) capitulum, ventral; (c) scutum; (d) coxae. Scale bars represent 0.10 mm. SEMs by J.F. Putterill.

grooves well marked, long and narrow. A few large setiferous punctations present on the scapulae, along the external cervical margins and scattered over the conscutum. Medium-sized punctations present medially, sometimes dense (as in the male illustrated here), but on other specimens much sparser, separating the two broad, shiny, very finely-punctate lateral areas of the conscutum. In some males, especially engorged specimens, this median punctate area, including the posterior grooves, is deeply depressed between the two bulging lateral areas, a feature that can be easily seen with the naked eye. Numerous fine punctations scattered elsewhere on the conscutum. Ventrally spiracles elongate, comma-shaped, with a short curved dorsal prolongation. Adanal plates rather narrow, elongated posteromedially into rounded points, posterior margins almost straight; accessory adanal plates absent.

Female (Figs 143(b), 144(d) to (f))
Capitulum broader than long, length × breadth ranging from 0.65 mm × 0.68 mm to 0.77 mm × 0.82 mm. Basis capituli with lateral angles at about mid-length, acute; porose areas large, about twice their own diameter apart. Palps a little longer than those of male, bluntly rounded apically. Scutum sometimes, but not always, longer than broad, the length × breadth ranging from 1.21 mm × 1.19 mm to 1.51 mm × 1.48 mm, margin slightly sinuous posterolaterally. Eyes slightly convex, edged dorsally by a few punctations. Cervical fields broad, slightly depressed. Large setiferous punctations present on the scapulae, along the external cervical margins and scattered sparsely over the rest of the scutum, interspersed with minute punctations on the lateral margins and by more numerous medium-sized punctations medially. Ventrally genital aperture sharply V-shaped.

Nymph (Fig. 145)
Capitulum broader than long, length × breadth ranging from 0.25 mm × 0.29 mm to 0.28 mm × 0.32 mm. Basis capituli over twice as broad as long, its lateral angles anteriorly placed, short and broad with convex anterolateral margins. Palps constricted basally, otherwise broad, overlapping hypostome. Scutum broader than long, length × breadth ranging from 0.54 mm × 0.58 mm to 0.62 mm × 0.70 mm. Eyes at widest point, at about mid-length, slightly convex. Cervical grooves short, deep and convergent initially, becoming shallower and divergent; cervical fields broad, but shallow and indistinct. Ventrally coxae I each with a short broad internal spur and somewhat narrower external spur; coxae II to IV each with an external spur, decreasing in size from II to IV.

Larva (Fig. 146)
Capitulum usually slightly longer than broad, length × breadth ranging from 0.138 mm × 0.137 mm to 0.158 mm × 0.154 mm. Basis capituli over twice as broad as long, with mere indications of lateral angles. Palps slightly constricted proximally, then widening before they taper slightly towards the tips. Scutum much broader than long, length × breadth ranging from 0.303 mm × 0.430 mm to 0.319 mm × 0.448 mm, posterolateral margins very mildly sinuous. Eyes at widest point, about halfway back, very slightly convex. Cervical grooves short, almost parallel. Ventrally coxae I each with a broadly-rounded spur; coxae II and III each with broad ridge-like spurs.

Notes on identification
All stages of development are remarkably similar to those of *R. appendiculatus* (p. 59) and *R. duttoni* (p. 146). In fact on 20 June 1904 Neumann himself wrote personally to C.P. Lounsbury, who had sent the original specimens of *R. nitens* to him: 'Je vous envoie aujourd'hui deux travaux sur les Ixodidés. Dans les "Notes . . .", j'ai décrit *Rhipicephalus nitens* et je le regrette. Je supprime cette espèce dans mon travail d'ensemble pour "Das Tierreich" et je la fais synonyme de *Rh. appendiculatus*.' We have no record of any personal reply to this statement by Lounsbury but in 1906 Lounsbury himself observed: 'The species I know as *R. nitens* may be popularly called "Shiny Brown Tick". In a former report I included it under the name *R. appendiculatus*, but drew attention to the fact that there were constant differences between it and the species proper. It seems simpler to regard the two as distinct species!' We agree with Lounsbury.

Large males of *R. nitens* differ from those of *R. appendiculatus* in the shape of their shorter, broader basis capituli, the large bulging lateral scutal areas and the shape of their adanal plates. Small males of *R. nitens* have more clearly demarcated festoons than the small *R. appendiculatus* males. Females of *R. nitens* have narrow V-shaped genital apertures whereas those of *R. appendiculatus* are broadly U-shaped. The nymphs and larvae of these two species also differ somewhat in the shape of their basis capituli (see Figs 273 and 274, pp. 603, 604).

Both Theiler (1964) and later Howell, Walker & Nevill (1978) erroneously included the distribution zone of *R. nitens* in the Eastern and

Western Cape Provinces with that of *R. appendiculatus* on their maps.

However, in a footnote that was added to her paper in proof, Theiler stated: 'Present-day (1963) findings suggest that this Cape macchia tick is not the true *R. appendiculatus*.' We have therefore added her entries for '*R. appendiculatus*' from the area now known to be occupied by *R. nitens* to our Map 42. Unfortunately, though, most of the specimens on which her information from this area was based no longer exist.

The morphological differences between *R. nitens* and *R. duttoni* are discussed on p. 151.

Hosts

A three-host species (Protozoology Section Tick Breeding Register, Onderstepoort, No. 3867). Although a few adult ticks have been collected from dogs (Horak *et al.*, 1987), the majority of collections come from sheep, bontebok, grey rhebok and scrub hares (Table 38). All stages of development are present on these hosts (Horak *et al.*, 1986; Horak, Williams & Van Schalkwyk, 1991). The single larva collected from a helmeted guineafowl must be considered an accidental infestation. Adult ticks attach mainly around the heads of sheep and antelopes, particularly on the outer ear and the lower edge of the mandible. On hares they occur on the ears. The immature stages are predominantly present on the lower legs and around the feet of sheep and antelopes. The larvae are most abundant from February to June, nymphs from August to October, and adults from November to February (Horak *et al.*, 1986, 1991). It seems likely that there is only one life cycle per year.

With the exception of those from sheep, grey rhebok and bontebok, comparatively few collections of *R. nitens* have been made, particularly off the domestic livestock that are farmed intensively and extensively within this tick's distribution zone. Numerous ticks have, however, been seen on the ears and heads of cattle and eland (*Taurotragus oryx*) during the months of peak abundance of adult *R. nitens*. We suspect that these are *R. nitens* and think that this tick occurs far more frequently than current collection records would seem to indicate.

Zoogeography

Despite the fact that a number of authors have recorded this species in other parts of the Afrotropical region (Doss *et al.*, 1974) we believe that *R. nitens* occurs only in South Africa, where it is present in the Eastern and Western Cape Provinces (Map 42). It is found at altitudes varying from approximately 200 m to 800 m. Rainfall varies around 500 mm annually, falling predominantly during summer in the east and during winter in the west of this tick's habitat. All collection sites are in Cape shrubland (*fynbos*), or in the proximity of this vegetation type.

Disease relationships

Stoltsz (1994) has demonstrated experimentally that *Ehrlichia bovis* is acquired by the nymph and subsequently transmitted to cattle by the adults of *R. nitens*. According to Lounsbury (1906) this species can also transmit *Theileria parva* (syn. *Piroplasma parvum*). Large infestations of *R. nitens*, exceeding approximately 300 adult ticks, appear to have led to severe loss of condition and even deaths possibly due to toxicosis among springbok in the southern part of the Western Cape Province (I.G.H., unpublished data).

REFERENCES

Horak, I.G., Jacot Guillarmod, A., Moolman, L.C. & De Vos, V. (1987). Parasites of domestic and wild animals in South Africa. XXII. Ixodid ticks on domestic dogs and on wild carnivores. *Onderstepoort Journal of Veterinary Research*, 54, 573–80.

Horak, I.G., Sheppey, K., Knight, M.M. & Beuthin, C.L. (1986). Parasites of domestic and wild animals in South Africa. XXI. Arthropod parasites of vaal ribbok, bontebok and scrub hares in the western Cape Province. *Onderstepoort Journal of Veterinary Research*, **53**, 187–97.

Horak, I.G., Williams, E.J. & Van Schalkwyk, P.C. (1991). Parasites of domestic and wild animals in South Africa. XXV. Ixodid ticks on sheep in the north-eastern Orange Free State and in the eastern Cape Province. *Onderstepoort Journal of Veterinary Research*, **58**, 115–23.

Howell, C.J., Walker, J.B. & Nevill, E.M. (1978). *Ticks, mites and insects infesting domestic animals in South Africa. 1. Descriptions and biology*. [Pretoria]: Department of Agricultural Technical Services, Republic of South Africa (Science Bulletin, no. 393).

Lounsbury, C.P. (1906). Ticks and African Coast Fever. *Agricultural Journal, Cape Town*, **28**, 634–54.

Neumann, L.G. (1904). Notes sur les Ixodidés – II. *Archives de Parasitologie*, **8**, 444–64.

Stoltsz, W.H. (1994). Transmission of *Ehrlichia bovis* by *Rhipicephalus* spp. in South Africa. *Journal of the South African Veterinary Association*, **65**, 159.

Theiler, G. (1964). XXI. Ecogeographical aspects of tick distribution. In *Ecological Studies in Southern Africa, Monographiae Biologicae* XIV, ed. D.H.S. Davis, pp. 284–300. The Hague: Dr. W. Junk Publishers.

Also see the following Basic Reference (pp. 13): Doss *et al.* (1974).

RHIPICEPHALUS OCULATUS NEUMANN, 1901

The specific name *oculatus*, from the Latin *oculus* meaning 'eye', refers to this species' beady eyes.

Diagnosis

A large brown to dark brown tick that in some respects resembles *R. exophthalmos* (see pp. 172–179).

Male (Figs 147(a), 148(a) to (c))

Capitulum broader than long, length × breadth ranging from 0.70 mm × 0.81 mm to 0.87 mm × 0.99 mm. Basis capituli with acute lateral angles at about anterior third of its length. Palps short, broad, tapering towards their apices. Conscutum length × breadth ranging from 3.02 mm × 1.95 mm to 3.95 mm × 2.35 mm; large anterior process present on coxae I. In engorged specimens the body wall bulges considerably

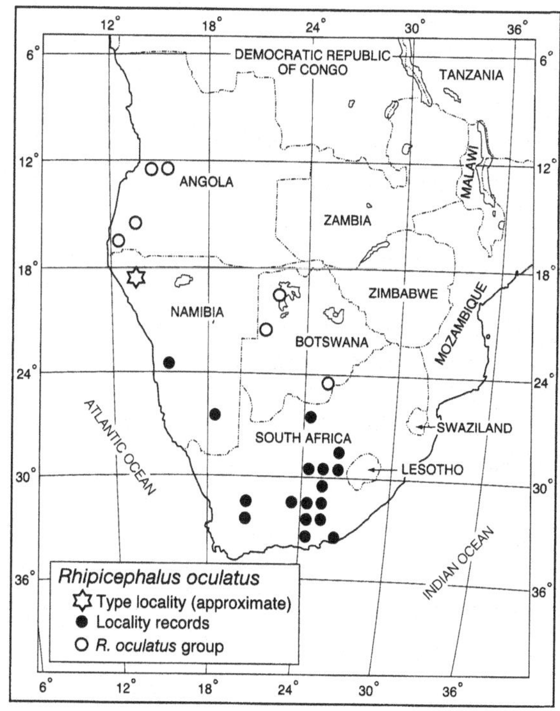

Map 43. *Rhipicephalus oculatus*: distribution. (After Keirans *et al.*, 1993, revised).

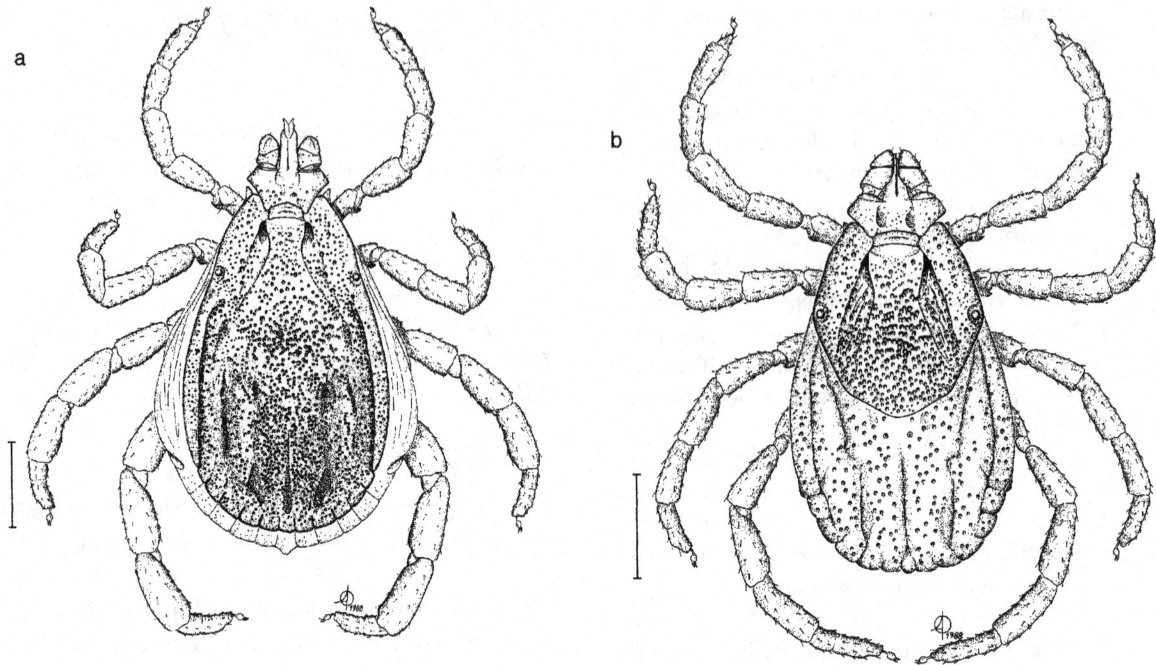

Figure 147. *Rhipicephalus oculatus* [collected from scrub hare No. 39 (*Lepus saxatilis*), Andries Vosloo Kudu Reserve, Grahamstown, South Africa, on 24 November 1986 by I.G. Horak]. (a) Male, dorsal; (b) female, dorsal. Scale bars represent 1 mm. A. Olwage *del.* (From Keirans *et al.*, 1993, figs 18 & 19, with kind permission from the Editor, *Onderstepoort Journal of Veterinary Research.*)

Table 39. *Host records of* Rhipicephalus oculatus

Hosts	Number of records
Wild animals	
'Duiker'	1
Mountain reedbuck (*Redunca fulvorufula*)	1 (larvae)
Springhare (*Pedetes capensis*)	2 (larvae)
Cape hare (*Lepus capensis*)	20 (including immatures)
Scrub hare (*Lepus saxatilis*)	102 (including immatures)
'Hare'	9 (including immatures)
Smith's red rock rabbit (*Pronolagus rupestris*)	1
Birds	
Helmeted guineafowl (*Numida meleagris*)	4 (immatures)

laterally and to a slightly lesser extent posteriorly, with only a very small pointed caudal process. Eyes submarginal, beady, deeply orbited. Cervical pits deep; cervical fields shallow, tapering posteriorly beyond eye level, sometimes inapparent. Marginal lines deep, punctate, extending anteriorly almost to eye level. Posteromedian and posterolateral grooves well developed, the scutal areas anterior to the posterolaterals often broadly indented. Punctation pattern dense; some large setiferous punctations scattered on the scapulae, a few along the external cervical margins and others scattered medially on the conscutum, interspersed with finer punctations that are particularly numerous medially and posteromedially but sparser in the indented areas of the conscutum parallel to the marginal lines. Legs increase slightly in size from I to IV. Ventrally spiracles narrowly elongate with a long dorsal prolongation. Adanal plates large, broadly triangular in shape.

Female (Figs 147(b), 148(d) to (f))

Capitulum broader than long, length × breadth ranging from 0.85 mm × 0.91 mm to 0.90 mm × 1.03 mm. Basis capituli with acute lateral angles at about mid-length; porose areas subcircular, about twice their own diameter apart. Palps broad, smoothly rounded apically. Scutum longer than broad, length × breadth ranging from 1.62 mm × 1.50 mm to 1.91 mm × 1.86 mm. Eyes submarginal, beady, deeply orbited. Cervical pits deep; cervical fields almost parallel sided, extending nearly to posterolateral margins of scutum. Large setiferous punctations present on the scapulae, often along the ridge-like external cervical margins and scattered medially among the dense slightly smaller punctations; lateral areas of scutum much more sparsely punctate. Ventrally genital aperture V-shaped, the genital apron depressed.

Nymph (Fig. 149)

Capitulum much broader than long, length × breadth ranging from 0.28 mm × 0.36 mm to 0.35 mm × 0.42 mm. Basis capituli about three times as broad as long, lateral angles short, acute, anterior to mid-length. Palps narrow proximally, then widening before they taper to rounded apices. Scutum generally broader than long, length × breadth ranging from 0.64 mm × 0.69 mm to 0.76 mm × 0.78 mm, posterior margin deeply curved. Eyes beady, orbited, on scutal margins immediately anterior to posterolateral angles. Internal cervical margins much shorter than the ridge-like external cervical margins, which almost reach the posterolateral margins of the scutum; cervical fields slightly depressed. Ventrally coxae I each with a long tapering external spur and a short triangular internal spur; coxae II to IV each with sharp tapering external spurs only.

Larva (Fig. 150)

Capitulum slightly longer than broad, length × breadth ranging from 0.158 mm × 0.157 mm to 0.181 mm × 0.168 mm. Basis capituli about twice as broad as long; lateral margins converging slightly posteriorly, curving smoothly to join the straight posterior margin. Palps constricted proximally, then becoming bulbous before tapering gently to their bluntly-rounded apices. Scutum much broader than long, length × breadth ranging from 0.334 mm × 0.440 mm to 0.379 mm × 0.484 mm; posterior margin a smooth, moderately deep curve. Eyes immediately anterior to posterolateral scutal angles, large, transversely ovoid with inner sides bulging, partially orbited. Cervical grooves short, parallel. Ventrally coxae I each with a moderately long, pointed internal spur; coxae II each broadly salient posteriorly; coxae III each with a moderately long pointed spur.

Notes on identification

Care must be taken to differentiate *R. oculatus* from *R. exophthalmos*, with which it was confused for many years (Keirans *et al.*, 1993), particularly as they may be found in mixed infestations on hares (see p. 329). Neumann (1901) described *R. oculatus* from 2 ♂♂, 2 ♀♀ collected in Damaraland, Namibia, and 1 ♀ from a bovine at Kilossa (= Kilosa, Tanzania). Of these only 1 ♂, 1 ♀ from Namibia (Zoological Museum, Berlin

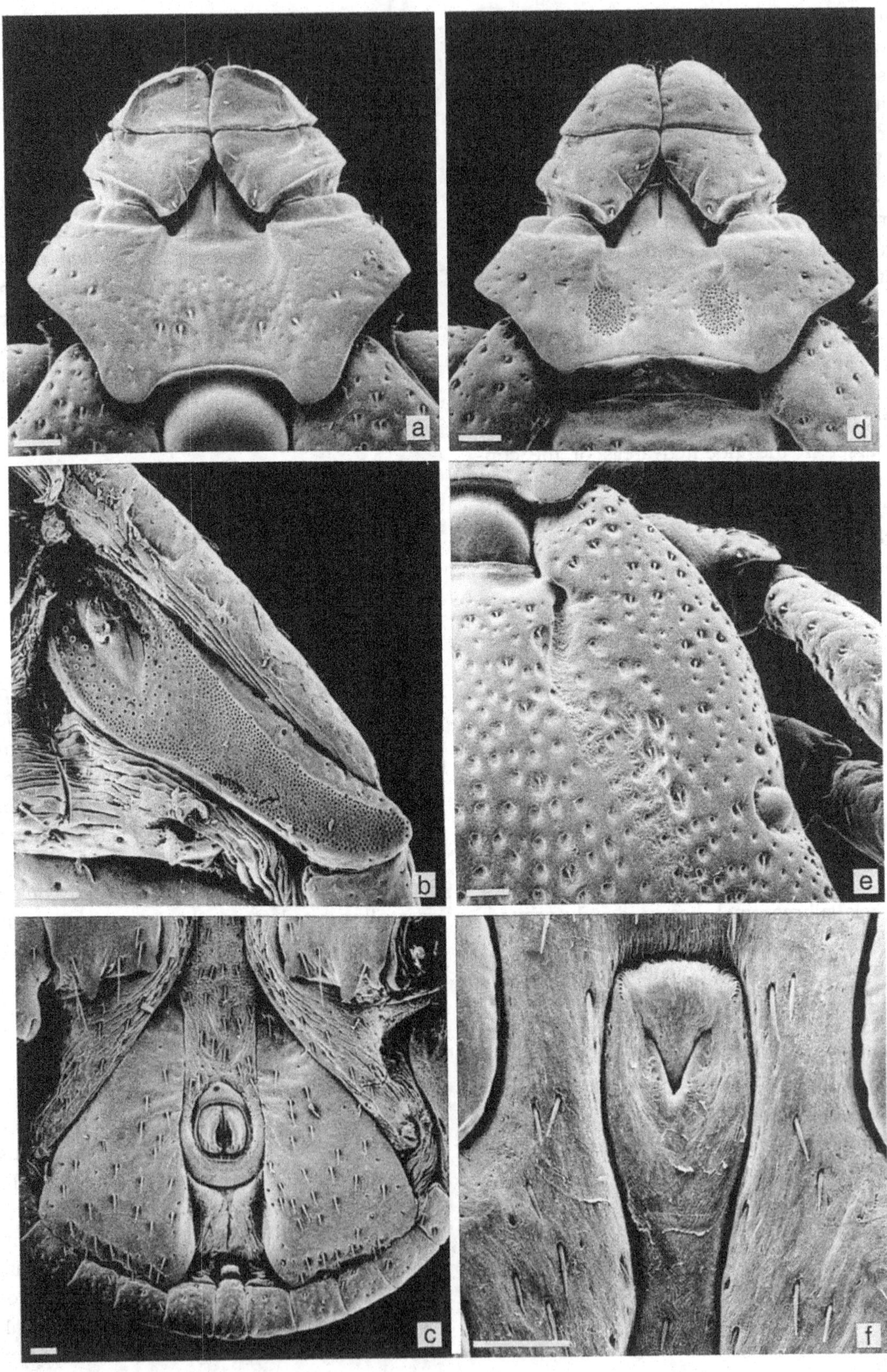
a
d
b
e
c
f

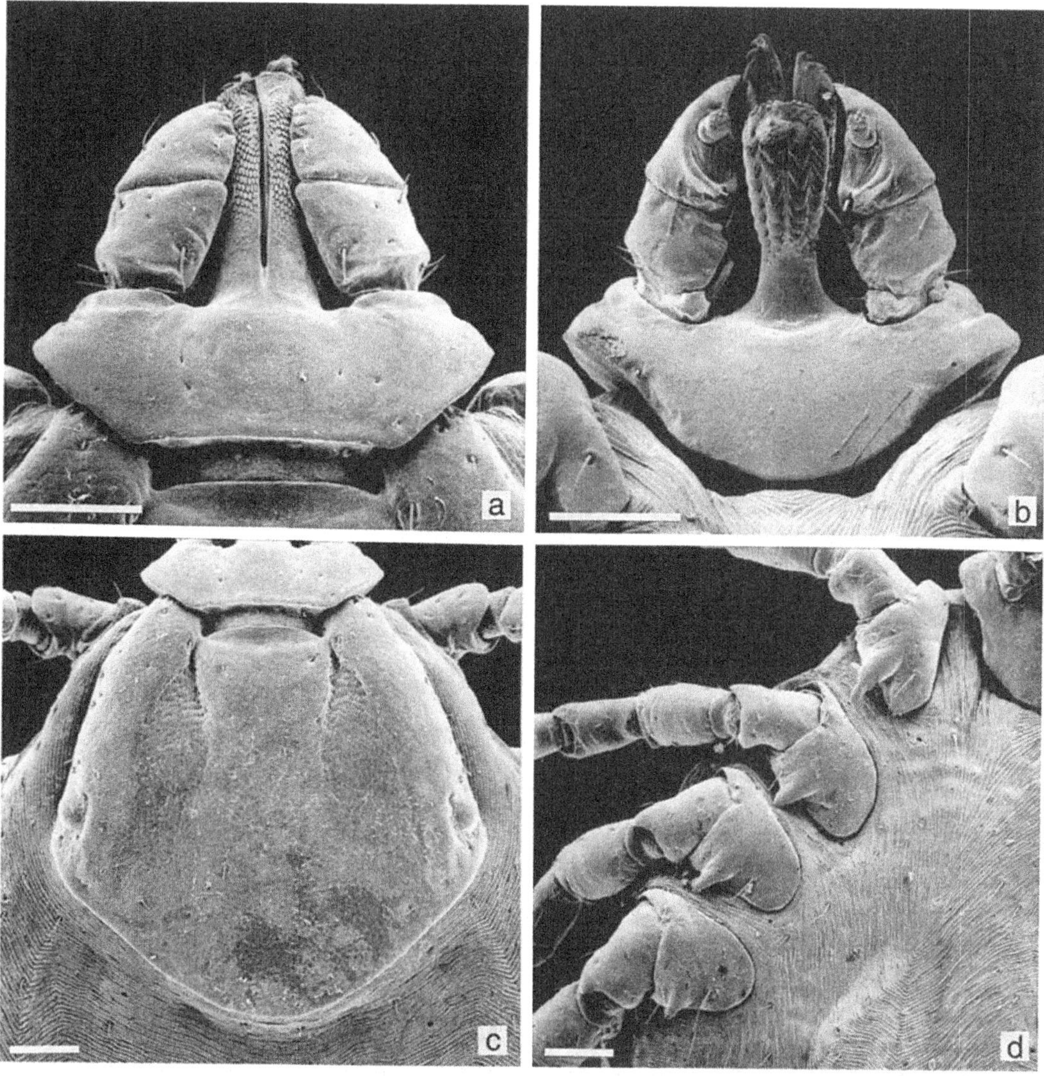

Figure 149 (*above*). *Rhipicephalus oculatus* [collected from scrub hare No. 28 (*Lepus saxatilis*), Andries Vosloo Kudu Reserve, Grahamstown, South Africa, on 19 June 1986 by I.G. Horak]. Nymph: (a) capitulum, dorsal; (b) capitulum, ventral; (c) scutum; (d) coxae. Scale bars represent 0.10 mm. SEMs by J.F. Putterill. (From Keirans *et al.*, 1993, figs 26–29, with kind permission from the Editor, *Onderstepoort Journal of Veterinary Research.*)

Figure 148 (*opposite*). *Rhipicephalus oculatus* [collected from scrub hare No. 12 (*Lepus saxatilis*), Andries Vosloo Kudu Reserve, Grahamstown, South Africa, on 18 August 1985 by I.G. Horak]. Male: (a) capitulum, dorsal; (b) spiracle; (c) adanal plates. Female: (d) capitulum, dorsal; (e) scapular area; (f) genital aperture. Scale bars represent 0.10 mm. SEMs by J.F. Putterill. (From Keirans *et al.*, 1993, figs 20–25, with kind permission from the Editor, *Onderstepoort Journal of Veterinary Research.*)

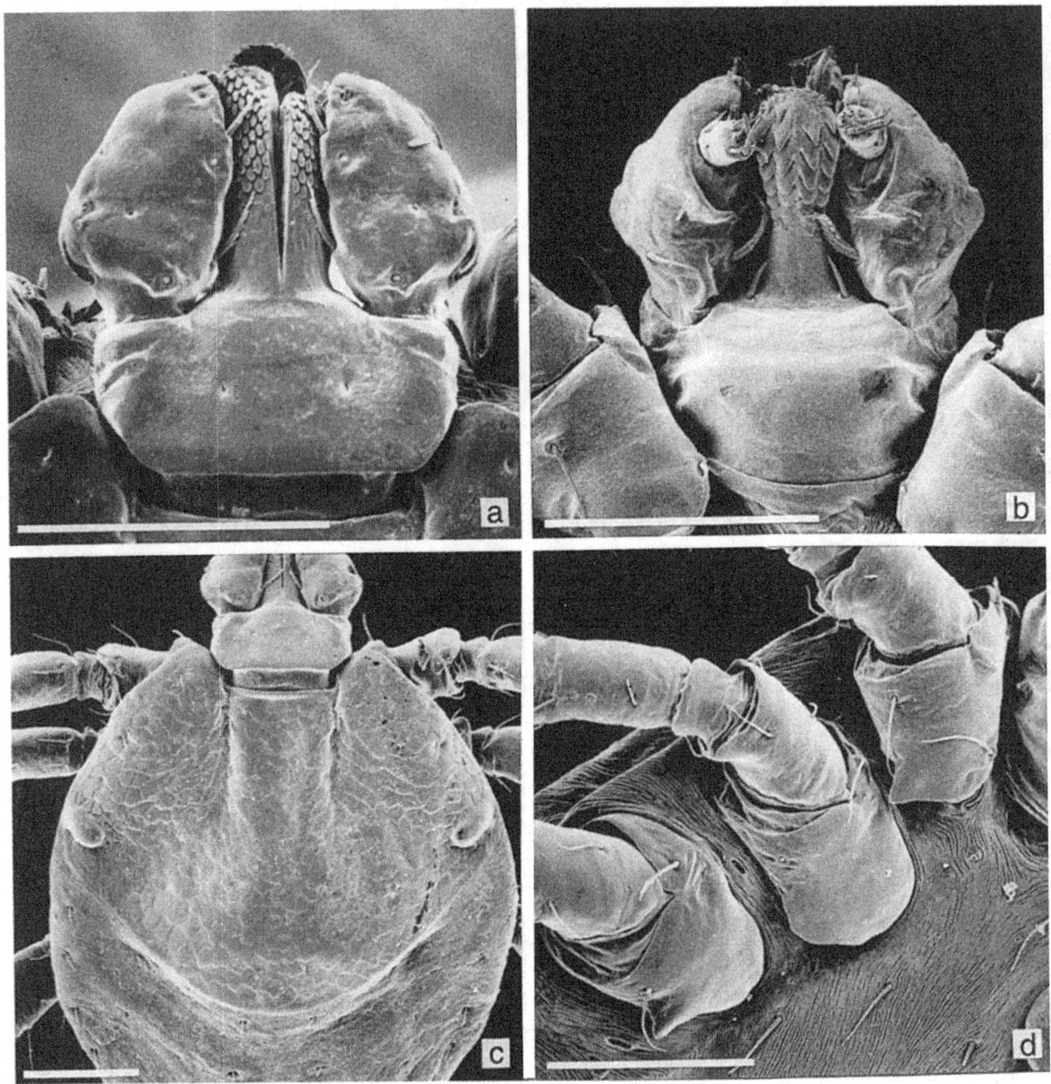

Figure 150. *Rhipicephalus oculatus* [collected from scrub hare No. 28 (*Lepus saxatilis*), Andries Vosloo Kudu Reserve, Grahamstown, South Africa, on 19 June 1986 by I.G. Horak]. Larva: (a) capitulum, dorsal; (b) capitulum, ventral; (c) scutum; (d) coxae. Scale bars represent 0.10 mm. SEMs by J.F. Putterill. (From Keirans *et al.*, 1993, figs 30–33, with kind permission from the Editor, *Onderstepoort Journal of Veterinary Research.*)

17613, 17614) still exist (Moritz & Fischer, 1981). The Kilosa specimen was probably *R. pravus*; we have no evidence that *R. oculatus* occurs in East Africa. Two reports of this species exist from western Zambia and one from Zimbabwe (Theiler & Robinson, 1953; Theiler, 1962) but unfortunately these specimens have apparently also been lost. Neither Norval (1985) nor we know of any other records of *R. oculatus* from these two countries.

The morphological differences between *R. oculatus* and *R. exophthalmos* are listed briefly on p. 178.

Hosts

This species has not as yet been reared in the laboratory. Its seasonal activity pattern indicates that it is a three-host species (Horak & Fourie,

1991). Cape hares and scrub hares are the preferred hosts of all stages of development (Table 39). There are marked differences in the prevalence of infestations on hares and Smith's red rock rabbits. Only one rock rabbit out of 28 examined in the Mountain Zebra National Park, South Africa, was infested, while 19 out of 26 scrub hares sampled at the same locality and at the same times harboured this species (Horak *et al.*, 1991). The hares' ears are the predilection site of attachment for adult ticks.

Only one life cycle per year seems probable as each of the parasitic life stages has a discrete period of seasonal occurrence. The larvae are most numerous from March to July, the nymphs from May to September and the adults from August to December (Horak & Fourie, 1991).

Zoogeography

There are confirmed records of *R. oculatus* from South Africa and Namibia plus unconfirmed records from Angola and Botswana (Map 43). Most collections have been made at altitudes ranging from 300 m to 800 m in semi-arid to arid areas with annual rainfalls of 200 mm to 500 mm. The vegetation in these areas ranges from evergreen and semi-evergreen bushland and thicket in the south-east through semi-desert types of montane, grassy, bushy and dwarf Karoo shrubland and the wooded grassland and deciduous bushland of the Kalahari to the Namib Desert in the north-west of the tick's distribution range.

Disease relationships

Unknown.

REFERENCES

Horak, I.G. & Fourie, L.J. (1991). Parasites of domestic and wild animals in South Africa. XXIX. Ixodid ticks on hares in the Cape Province and on hares and red rock rabbits in the Orange Free State. *Onderstepoort Journal of Veterinary Research*, **58**, 261–70.

Horak, I.G., Fourie, L.J., Novellie, P.A. & Williams, E.J. (1991). Parasites of domestic and wild animals in South Africa. XXVI. The mosaic of ixodid tick infestations on birds and mammals in the Mountain Zebra National Park. *Onderstepoort Journal of Veterinary Research*, **58**, 125–36.

Keirans, J.E., Walker, J.B., Horak, I.G. & Heyne, H. (1993). *Rhipicephalus exophthalmos* sp. nov., a new tick species from southern Africa, and a redescription of *Rhipicephalus oculatus* Neumann, 1901, with which it has hitherto been confused (Acari: Ixodida: Ixodidae). *Onderstepoort Journal of Veterinary Research*, **60**, 229–46.

Neumann, L.G. (1901). Révision de la famille de ixodidés. (4ᵉ Mémoire). *Mémoires de la Société Zoologique de France*, **14**, 249–372.

Norval, R.A.I. (1985). The ticks of Zimbabwe. XII. The lesser known *Rhipicephalus* species. *Zimbabwe Veterinary Journal*, **16**, 37–43.

Theiler, G. & Robinson, B.N. (1953). Ticks in the South African Zoological Survey Collection. Part VII. Six lesser known African rhipicephalids. *Onderstepoort Journal of Veterinary Research*, **26**, 93–136 + 1 map.

Also see the following Basic References (pp. 12–14): Moritz & Fischer (1981); Theiler (1962).

RHIPICEPHALUS OREOTRAGI WALKER & HORAK, *SP. NOV.*

This species is named after the only host from which it has been collected thus far, the klipspringer (*Oreotragus oreotragus*).

Diagnosis

A small lightly punctate yellowish-brown tick; females with a characteristically long narrow scutum. Punctations always discrete.

Male (Figs 151(a), 152(a) to (c))
Capitulum slightly broader than long, length × breadth ranging from 0.44 mm × 0.45 mm to 0.54 mm × 0.55 mm. Basis capituli with short acute lateral angles at anterior third of its length. Palps short, broad, almost flat apically. Conscutum length × breadth ranging from 2.02 mm × 1.25 mm to 2.61 mm × 1.62 mm; anterior process of coxae I fairly small but well sclerotized. Eyes slightly convex, delimited dorsally by a few medium-sized setiferous punctations. Cervical pits deep, discrete. External margins of cervical fields indicated by rows of large discrete setiferous punctations that continue past the eyes and mark the positions of the marginal lines. Posteromedian groove short, somewhat spindle-shaped; posterolateral grooves small, round. A few medium-sized setiferous punctations scattered on scapulae and anteromedially on the conscutum, becoming larger and more numerous towards the posterior end of the conscutum, where they encircle the posterior grooves. A few minute punctations scattered among the larger elements but in general the conscutum looks smooth and shiny. Ventrally spiracles broad, tapering rather abruptly to a short gently-curved dorsal prolongation. Adanal plates broad, their external margins slightly curved, their internal margins scooped out just posterior to the anus, their posterior margins smoothly rounded; accessory adanal plates large, pointed, well sclerotized.

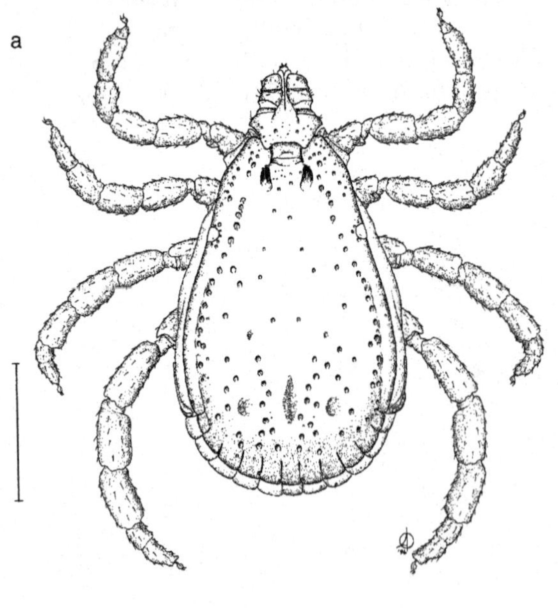

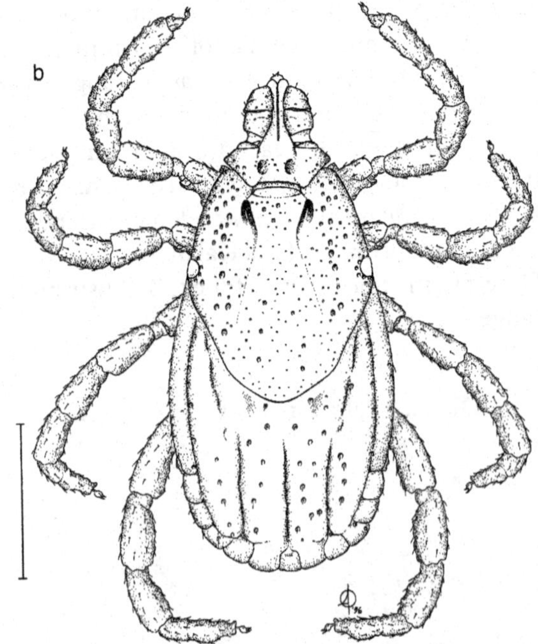

Figure 151. *Rhipicephalus oreotragi* Walker & Horak, *sp. nov.* [Onderstepoort Tick Collection 3145ii, paratypes, collected from klipspringer (*Oreotragus oreotragus*), Sentinel Ranch, 70 km W. of Beit Bridge, Zimbabwe, on 4 July 1992]. (a) Male, dorsal; (b) female, dorsal. Scale bars represent 1 mm. A. Olwage *del.*

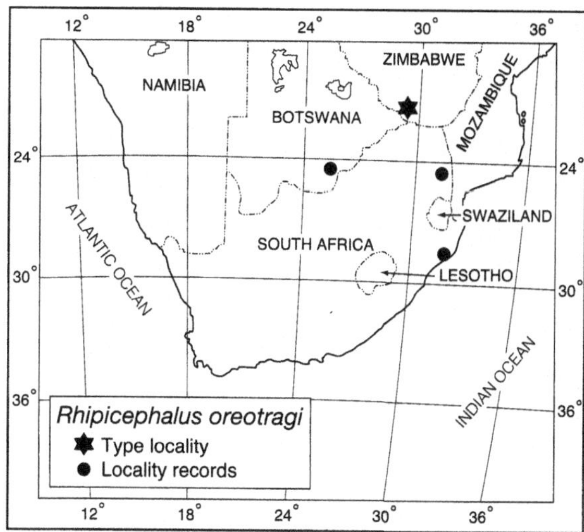

Map 44. *Rhipicephalus oreotragi* Walker & Horak, *sp. nov.*: distribution.

Female (Figs 151(b), 152(d) to (f))
Capitulum broader than long, length × breadth ranging from 0.50 mm × 0.56 mm to 0.67 mm × 0.73 mm. Basis capituli with long acutely-pointed lateral angles just anterior to mid-length; porose areas small, oval, three times their own diameter apart. Palps longer than those of the male, with article III more-or-less wedge-shaped. Scutum longer than broad, length × breadth ranging from 1.04 mm × 0.91 mm to 1.52 mm × 1.30 mm; posterior margin a deep slightly sinuous curve. Eyes at about mid-length, slightly convex, delimited dorsally by a few medium-sized setiferous punctations. Cervical pits long, convergent; cervical fields long, their internal margins inconspicuous, their external margins indicated by long rows of large discrete setiferous punctations. A few medium-sized setiferous punctations scattered on scapulae and medially on scutum, interspersed with numerous very fine punctations. Ventrally genital aperture broadly U-shaped.

Nymph and larva
Unknown.

Holotype
♂ collected from klipspringer (*Oreotragus oreotragus*), Sentinel Ranch, 70 km west of Beit Bridge, Zimbabwe, on 4 July 1992, deposited in Onderstepoort Tick Collection 3145i.

Allotype
♀, data as above.

Paratypes
♂♂, ♀♀ data as above, deposited in Onderstepoort Tick Collection 3145ii; 2 ♂♂, 2 ♀♀, data as above, deposited in the United States National Tick Collection, RML 122748; 2 ♂♂, 2 ♀♀, data as above, deposited in The Natural History Museum, London.

Notes on identification
After examining the ticks collected by Dr F. Zumpt during December 1957 from a klipspringer in Botswana, Gertrud Theiler (unpublished data) noted that these were '*R. simus* (with a difference)'. Later Baker & Keep (1970) identified ticks that they had taken from a klipspringer in KwaZulu-Natal as *R. simpsoni*. Walker (1991) thought the latter ticks could be *R. distinctus*, but we now think that they are *R. oreotragi* (see above under *R. distinctus*, p. 138).

Hosts

Life cycle undetermined but it is assumed to be a three-host species. The only animals from which adult ticks have been collected are klipspringer. Collections for which the dates were recorded were made during April, July, September and December.

Zoogeography

Collections have been made in Zimbabwe and Botswana as well as in the Mpumalanga and KwaZulu-Natal Provinces of South Africa (Map 44). As klipspringer appear to be the only hosts of adult *R. oreotragi* its distribution is likely to be the same as that of this animal. It will thus be

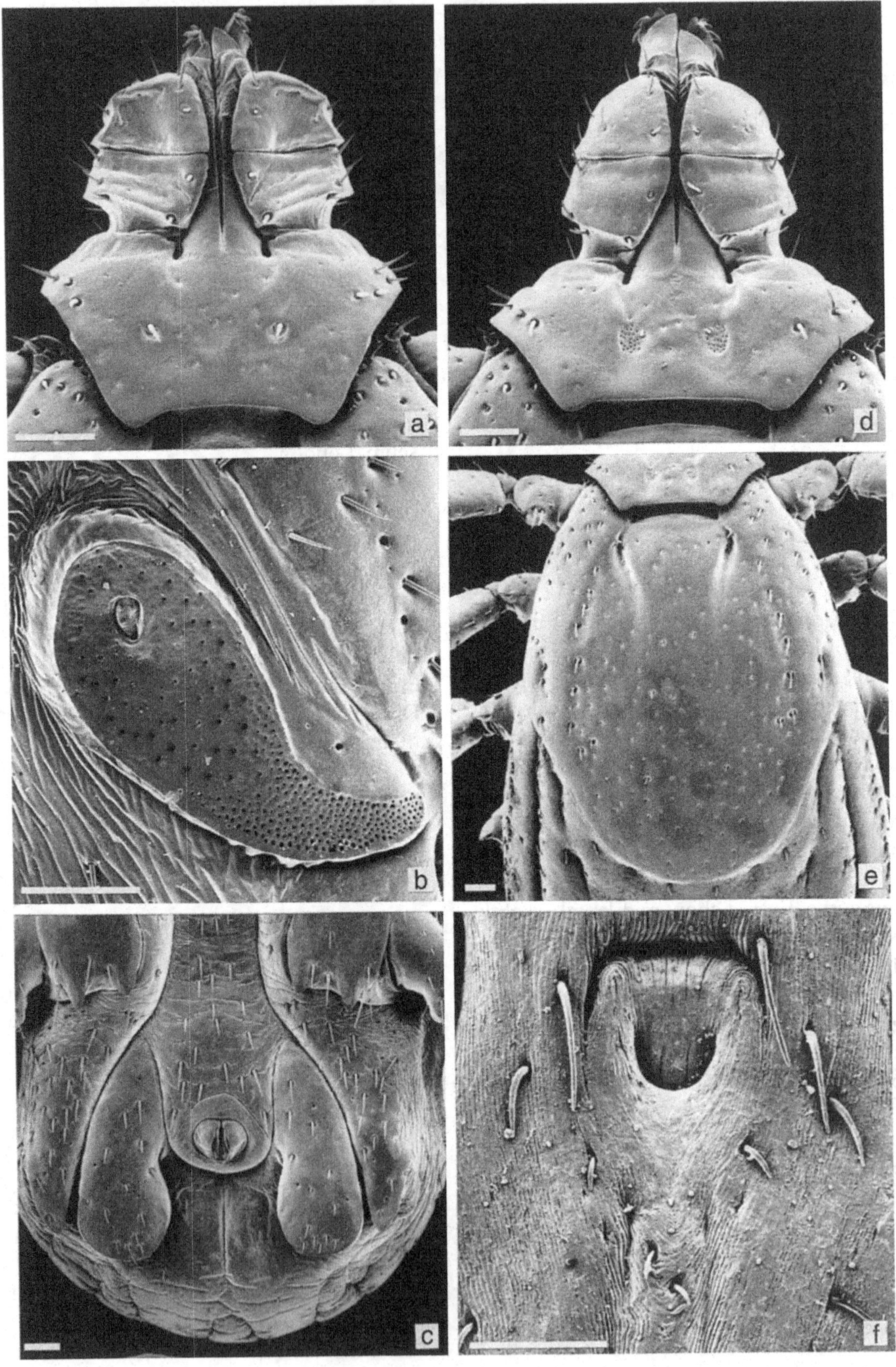
a
d
b
e
c
f

discontinuous and associated with rocky hills or outcrops, gorges with rocky sides or mountainous areas with krantzes, all habitats favoured by klipspringer (Skinner & Smithers, 1990).

Disease relationships

Unknown.

REFERENCES

Baker, M.K. & Keep, M.E. (1970). Checklist of the ticks found on the larger game animals in the Natal game reserves. *Lammergeyer*, **No. 12**, 41–7.

Walker, J.B. (1991). A review of the ixodid ticks (Acari, Ixodidae) occurring in southern Africa. *Onderstepoort Journal of Veterinary Research*, **58**, 81–105.

Also see the following Basic Reference (p. 14): Skinner & Smithers (1990).

Figure 152 (*opposite*). *Rhipicephalus oreotragi* Walker & Horak, *sp. nov.* [Onderstepoort Tick Collection 3145ii, collected from klipspringer (*Oreotragus oreotragus*), Sentinel Ranch, 70 km W. of Beit Bridge, Zimbabwe, on 4 July 1992]. Male: (a) capitulum, dorsal; (b) spiracle; (c) adanal plates. Female: (d) capitulum, dorsal; (e) scutum; (f) genital aperture. Scale bars represent 0.10 mm. SEMs by J.F. Putterill.

RHIPICEPHALUS PLANUS NEUMANN, 1907

The Latin word *planus*, meaning 'flat, level, even', doubtless refers to the appearance of the scutum, especially in the male.

Synonym

reichenowi; *simus planus*.

Diagnosis

A moderate-sized dark reddish-brown to almost black tick whose males have a flat to slightly concave conscutum.

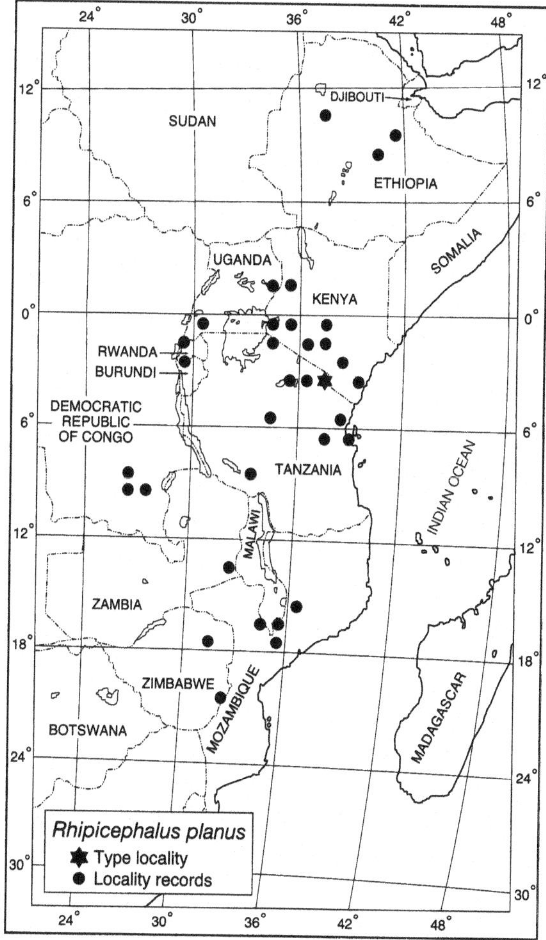

Map 45. *Rhipicephalus planus*: distribution.

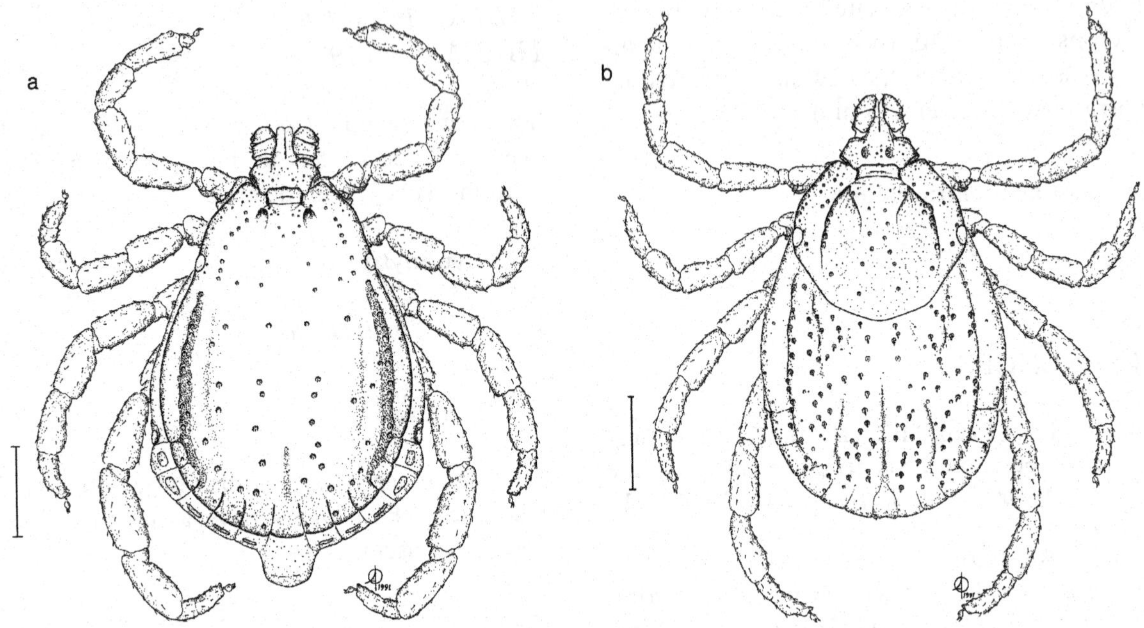

Figure 153. *Rhipicephalus planus* [collected from warthog (*Phacochoerus africanus*), Langata, Kenya, on 4 October 1955 by L.S.B. Leakey]. (a) Male, dorsal; (b) female, dorsal. Scale bars represent 1 mm. A. Olwage *del.*

Table 40. *Host records of* Rhipicephalus planus

Hosts	Number of records
Domestic animals	
Cattle	6
Wild animals	
Side-striped jackal (*Canis adustus*)	1
African elephant (*Loxodonta africana*)	1
Aardvark (*Orycteropus afer*)	1
Warthog (*Phacochoerus africanus*)	9
Bushpig (*Potamochoerus larvatus*)	15
'Wild pig'	1
Impala (*Aepyceros melampus*)	1
African buffalo (*Syncerus caffer*)	6
Common duiker (*Sylvicapra grimmia*)	1
Waterbuck (*Kobus ellipsiprymnus*)	1
Striped grass rat (*Lemniscomys striatus*)	1 (nymphs)
South African porcupine (*Hystrix africaeaustralis*)	2
Crested porcupine (*Hystrix cristata*)	1
'Porcupine'	2
Cape hare (*Lepus capensis*)	1 (larvae)
Humans	4

Male (Figs 153(a), 154(a) to (c))
Capitulum longer than broad, length × breadth ranging from 0.52 mm × 0.50 mm to 1.00 mm × 0.85 mm. Basis capituli with short sharp lateral angles at about anterior third of its length. Palps broadly rounded apically. Conscutum characteristically flat to slightly concave medially, length × breadth ranging from 2.68 mm × 1.84 mm to 3.65 mm × 2.60 mm; anterior process of coxae I inconspicuous. In engorged males body wall expanded posterolaterally and forming a short blunt caudal process posteromedially. Eyes almost flat, sometimes edged dorsally by a few moderate-sized setiferous punctations. Cervical pits deep, and only external margins of cervical fields marked by rows of punctations. Marginal lines long, shallow, conspicuous, their outer edges sharp, their inner edges rounded, picked out with large setiferous punctations. Posteromedian groove usually present but often rather indistinct; posterolateral grooves often absent, sometimes just indicated. Setiferous punctations on scapulae, along external cervical margins and anteromedially on conscutum moderate-sized, sharp-edged. Medially, on the flattened to sunken area of the conscutum, the setiferous punctations are larger and shallower, with more rounded edges. Minute interstitial punctations may be present all over the conscutum but are usually virtually invisible. Ventrally spiracles broad, with a short prolongation angled towards the dorsal surface. Adanal plates almost sickle-shaped, their internal margins concave, their posterior margins broadly rounded; accessory adanal plates small, pointed.

Female (Figs 153(b), 154(d) to (f))
Capitulum as broad as long, length × breadth ranging from 0.60 mm × 0.60 mm to 0.95 mm × 0.95 mm. Basis capituli with acute lateral angles at about mid-length; porose areas oval, sometimes over twice their own diameter apart, sometimes closer together. Palps broadly rounded apically. Scutum broader than long, length × breadth ranging from 1.10 mm × 1.25 mm to 1.70 mm × 2.00 mm; posterior margin sinuous. Eyes about halfway back, almost flat, sometimes edged dorsally by a few moderate-sized setiferous punctations. Cervical fields broad, depressed, their external margins usually sharp and delimited by rows of large setiferous punctations. Setiferous punctations on scapulae and anteromedially on the scutum moderate-sized, sharp-edged, becoming larger and shallower, with more rounded edges, further back. Minute interstitial punctations scattered all over the scutum, but sometimes barely visible. Ventrally genital aperture tripartite in appearance, with the narrow central area flanked on each side by a rounded depression.

Nymph (Fig. 155)
Capitulum much broader than long and almost triangular in general appearance, length × breadth ranging from 0.25 mm × 0.38 mm to 0.28 mm × 0.41 mm. Basis capituli with long tapering lateral angles posteriorly, and sometimes small cornua; ventrally with small sharp spurs on posterior border. Palps long, very slender, tapering to narrowly-rounded apices, inclined inwards. Scutum usually longer than broad, length × breadth ranging from 0.55 mm × 0.50 mm to 0.59 mm × 0.57 mm; posterior margin a broad smooth curve. Eyes at widest point, well over halfway back, long, narrow, delimited dorsally by shallow grooves. Cervical fields long, narrow, divergent, their internal margins deeper than their external margins. Ventrally coxae I each with a relatively long narrow external spur and a shorter broader internal spur; coxae II and III each with a small external spur but no spur on coxae IV.

Larva (Fig. 156)
Capitulum much broader than long, length × breadth ranging from 0.128 mm × 0.168 mm to 0.134 mm × 0.170 mm. Basis capituli nearly three times as broad as long, virtually hexagonal in general shape with broad pointed lateral angles over halfway back. Palps long, slender, tapering to narrowly-rounded apices, inclined inwards. Scutum much broader than long, length × breadth ranging from 0.237 mm × 0.325 mm to 0.254 mm × 0.354 mm; posterior margin a broad shallow curve. Eyes at widest point, well

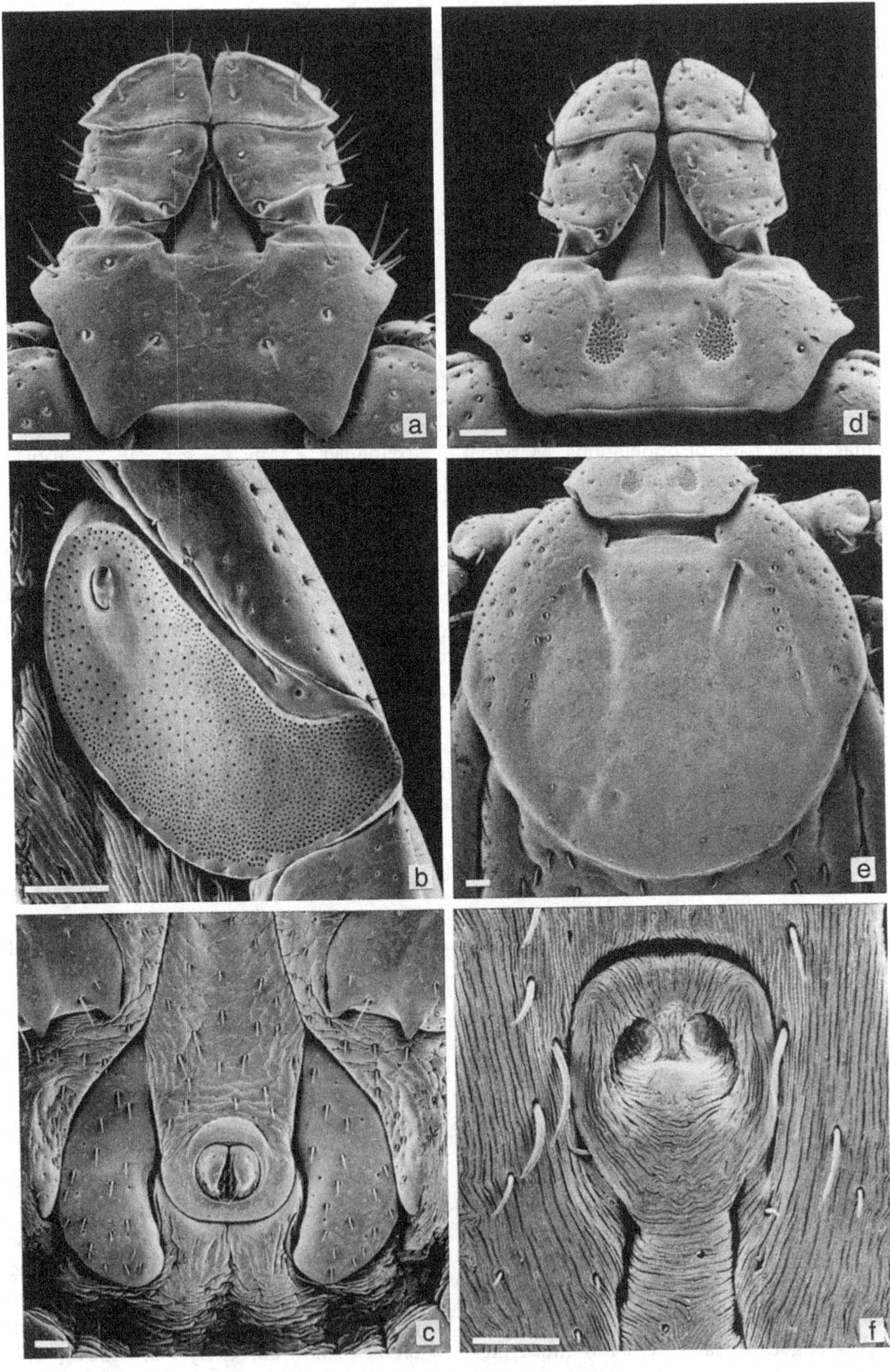
a
d
b
e
c
f

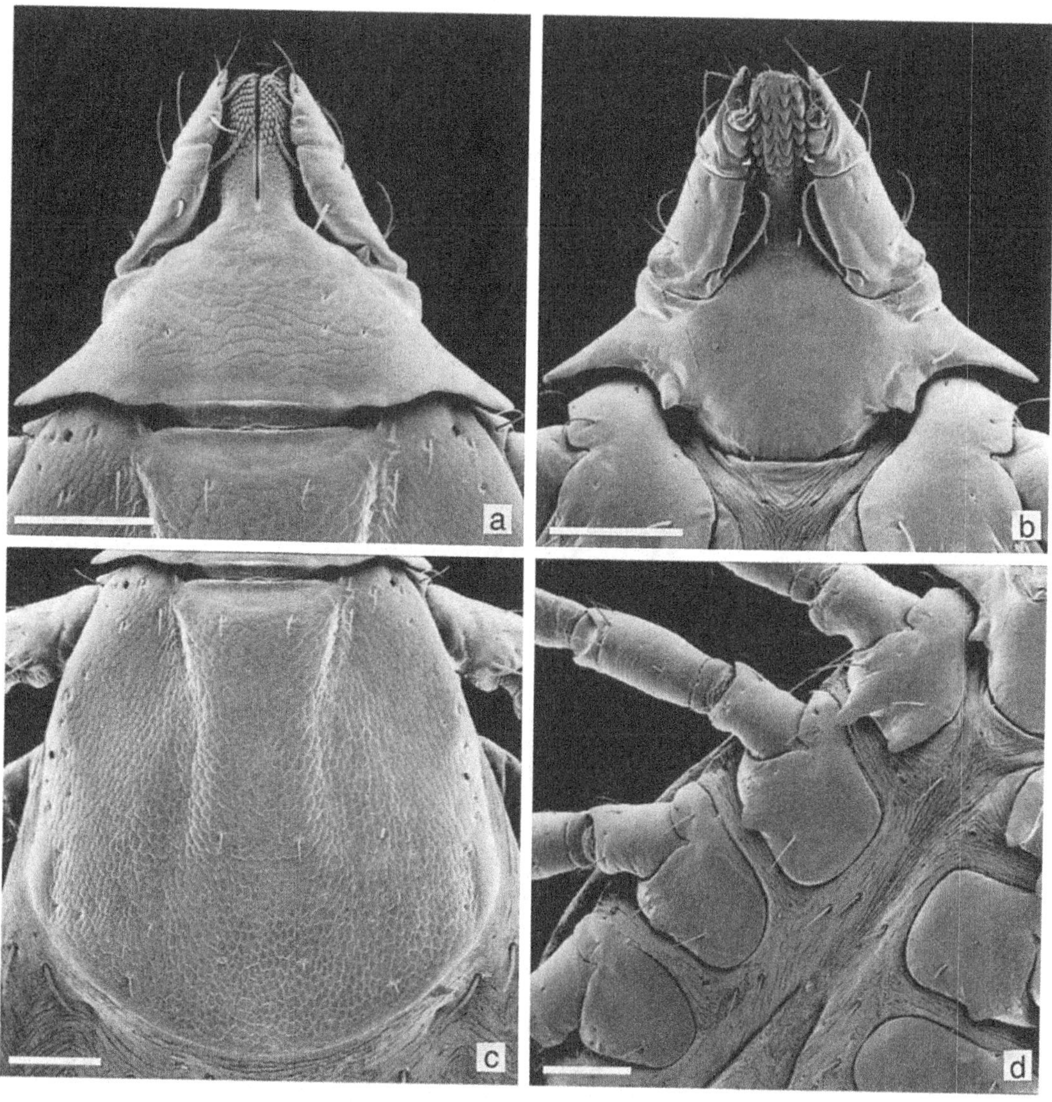

Figure 155 (*above*). *Rhipicephalus planus* (B.S. 720/-, laboratory reared, strain originating from ticks collected from humans and vegetation in forest on banks of Magumira River, north of Tengeru, near Arusha, Tanzania, in August 1956 by H. Hoogstraal, G.M. Kohls, G. Theiler & S. Gaber). Nymph: (a) capitulum, dorsal; (b) capitulum, ventral; (c) scutum; (d) coxae. Scale bars represent 0.10 mm. SEMs by J.F. Putterill.

Figure 154 (*opposite*). *Rhipicephalus planus* (B.S. 720/-, laboratory reared, strain originating from ticks collected from humans and vegetation in forest on banks of Magumira River, north of Tengeru, near Arusha, Tanzania, in August 1956 by H. Hoogstraal, G.M. Kohls, G. Theiler & S. Gaber). Male: (a) capitulum, dorsal; (b) spiracle; (c) adanal plates. Female: (d) capitulum, dorsal; (e) scutum; (f) genital aperture. Scale bars represent 0.10 mm. SEMs by J.F. Putterill.

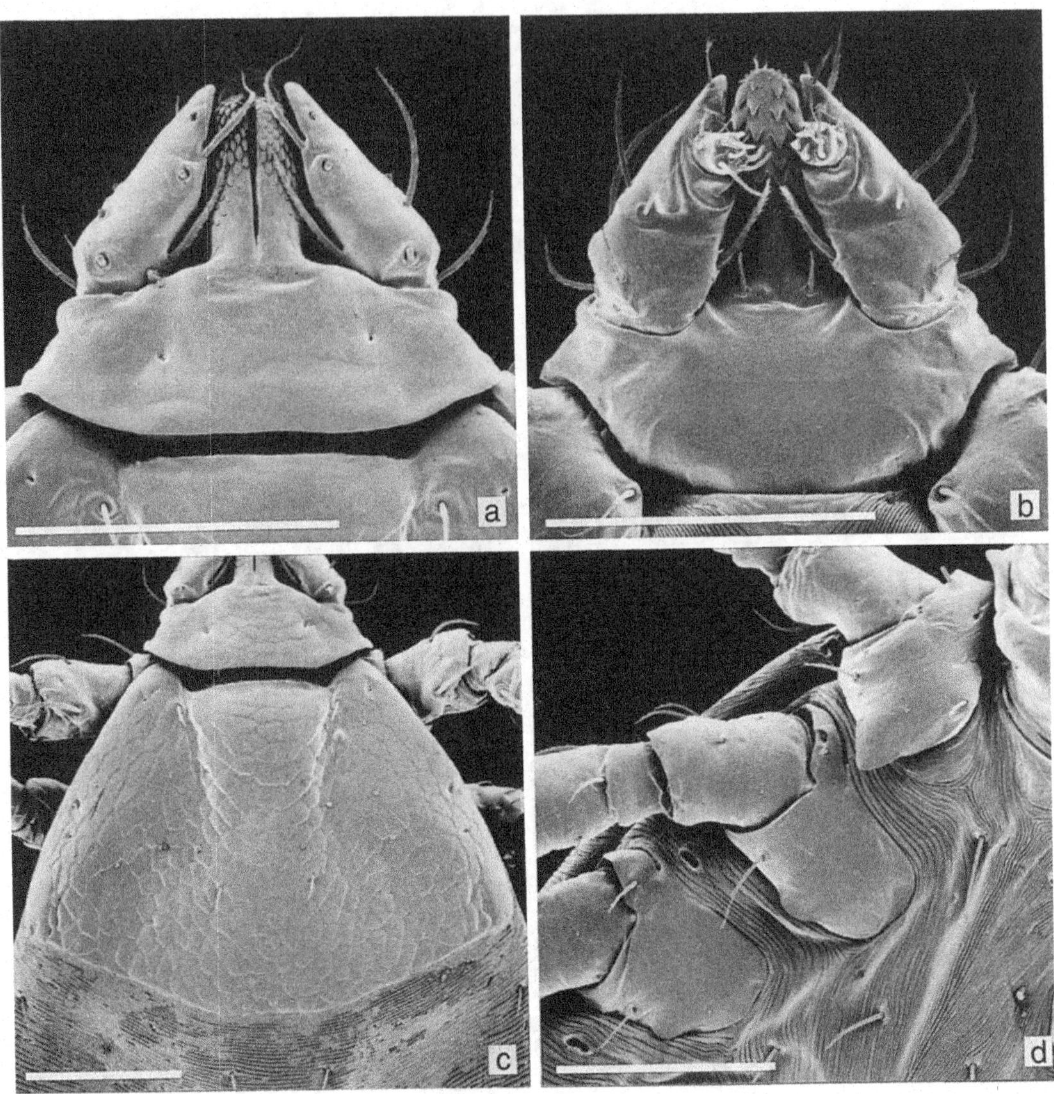

Figure 156. *Rhipicephalus planus* (B.S. 720/-, laboratory reared, strain originating from ticks collected from humans and vegetation in forest on banks of Magumira River, north of Tengeru, near Arusha, Tanzania, in August 1956 by H. Hoogstraal, G.M. Kohls, G. Theiler & S. Gaber). Larva: (a) capitulum, dorsal; (b) capitulum, ventral; (c) scutum; (d) coxae. Scale bars represent 0.10 mm. SEMs by J.F. Putterill.

over halfway back, large, almost flat. Cervical grooves shallow, extending back for about half of scutal length. Ventrally coxae I each with a large triangular spur; coxae II and III each with a small triangular spur.

Notes on identification

This species, which was originally described as *R. simus planus* by Neumann (1907), was raised to specific rank by Morel (1980). At the same time he sank *R. reichenowi* Zumpt, 1943 as a junior synonym of *R. planus*. We have examined a male tick labelled '*Rhipicephalus simus planus*, Kilimanjaro. Meru Sjöstedt 1905–06, cotype, auth. det.', deposited in the Muséum National d'Histoire Naturelle, Paris, and support Morel's findings regarding this species.

Morel (1980) also re-identified as *R. planus* two females collected from an African elephant, Guranni River, Kenya, that had been listed as *R.*

ecinctus by Neumann (1922). However, he re-identified as *R. bequaerti* a free-living female collected between the Amboni and Naromoru Rivers, Kenya, and originally recorded as *R. planus* by Neumann (1913).

Hosts

A three-host species (Walker, 1966, as *R. reichenowi*). The only domestic animals on which *R. planus* adults have been found thus far are cattle. Amongst wild animals adults have been recorded most commonly from bushpigs, warthogs, African buffaloes and also porcupines (Table 40). The source of the record given by Morel (1980) from hedgehogs is unknown. The record from zebras quoted by Walker (1966) and Morel is now thought to refer to *R. zumpti*. Its attachment sites on these hosts have not been recorded. Apart from June adults were collected throughout the year, especially in July and August.

The few records presently available indicate that the immature stages parasitize small animals. Besides the two nymphs found on a striped grass rat eight nymphs, plus a male, were collected from the refuse surrounding a nest, probably that of a squirrel, and one from an eagle's nest (Garnham, 1957, listed as members of the *R. simus* group). The latter may simply have fallen from a prey animal brought to its nest by the eagle. About 100 larvae were collected from a Cape hare (Clifford, Flux & Hoogstraal, 1976).

Zoogeography

Rhipicephalus planus has been recorded in eastern and central Africa from Ethiopia southwards to eastern Zimbabwe (Berggren, 1978; Norval, 1985) (Map 45). The ecological conditions in the different places where it has been found vary considerably (Morel, 1980). The altitudes range from about 300 m up to 2200 m, with rainfalls from some 500 mm to over 1000 mm annually, in montane grassland, wooded or bushed grassland, Somali–Masai bushland and thicket and various types of woodland.

Disease relationships

Unknown.

REFERENCES

Berggren, S.A. (1978). Cattle ticks in Malawi. *Veterinary Parasitology*, **4**, 289–97.

Clifford, C.M., Flux, J.E.C. & Hoogstraal, H. (1976). Seasonal and regional abundance of ticks (Ixodidae) on hares (Leporidae) in Kenya. *Journal of Medical Entomology*, **13**, 40–7.

Garnham, P.C.C. (1957). Trees, ticks and monkeys: further attempts to discover the invertebrate host of *Hepatocystis kochi. Zeitschrift für Tropenmedizin und Parasitologie*, **8**, 91–6.

Neumann, L.G. (1907). Ixodidae. In *Wissenschaftliche Ergebnisse der Schwedischen Zoologischen Expedition nach dem Kilimandjaro, dem Meru und den umgebenden Massaisteppen, Deutsch-Ostafrikas 1905–1906 (Sjöstedt)*, 3, *Abteilung 20: Arachnoidea*, (2), 17–30. Stockholm, Uppsala: Almqvist & Wiksells Boktryckeri-A-B.

Neumann, L.G. (1913). Ixodidae. In *Voyage de Ch. Alluaud et R. Jeannel en Afrique orientale (1911–12). Résultats Scientifiques. Arachnida*, II, 23–35. Paris: A. Schulz.

Neumann, L.G. (1922). Acariens: Ixodidae. In *Voyage de M. le Baron Maurice de Rothschild en Éthiopie et en Afrique Orientale Anglaise (1904–1905). Résultats Scientifiques, I*, 108–25. Paris: Imprimerie Nationale.

Norval, R.A.I. (1985). The ticks of Zimbabwe. XII. The lesser known *Rhipicephalus* species. *Zimbabwe Veterinary Journal*, **16**, 37–43.

Walker, J.B. (1966). *Rhipicephalus reichenowi* Zumpt, 1943: a re-description of the male and female and descriptions of the nymph and larva, together with an account of its known hosts and biology. *Parasitology*, **56**, 457–69.

Also see the following Basic References (pp. 12–14): Clifford & Anastos (1962); Elbl & Anastos (1966); Morel (1980); Santos Dias (1960); Walker (1974); Yeoman & Walker (1967).

RHIPICEPHALUS PRAETEXTATUS GERSTÄCKER, 1873

This specific name, a Latin term meaning 'bordered', is apparently based on Gerstäcker's statement in his original description that the type male has a narrow white margin posteriorly (. . . *retrorsum anguste albo-marginato* . . .). We believe that this is an artifact, visible only because this tick is a dry pinned specimen.

Diagnosis

A moderate-sized glossy brown to dark brown tick.

Male (Figs 157(a), 158(a) to (c))
Capitulum longer than broad, length × breadth ranging from 0.67 mm × 0.62 mm to 1.04 mm × 0.97 mm. Basis capituli with short acute lateral angles at about mid-length. Palps somewhat flattened apically. Conscutum length × breadth ranging from 2.96 mm × 2.00 mm to 4.67 mm × 3.19 mm; anterior process of coxae I small. In engorged specimens a single short blunt caudal process formed. Eyes flat, sometimes edged dorsally by a few medium-sized setiferous punctations. Cervical pits comma-shaped, discrete. Marginal lines shallow, outlined and continued anteriorly by a few medium-sized setiferous punctations. Posterior grooves, when present, inconspicuous. A few medium-sized punctations present on the scapulae, along the external cervical margins and in the '*simus*' pattern of four irregular rows medially on the conscutum; interstitial punctations either discrete and minute or absent. Ventrally spiracles large, with a broad prolongation curving gently towards the dorsal surface. Adanal plates broad, their inner margins slightly concave, their posterior margins smoothly rounded; accessory adanal plates pointed, well sclerotized.

Female (Figs 157(b), 158(d) to (f))
Capitulum slightly longer than broad, length × breadth ranging from 0.71 mm × 0.69

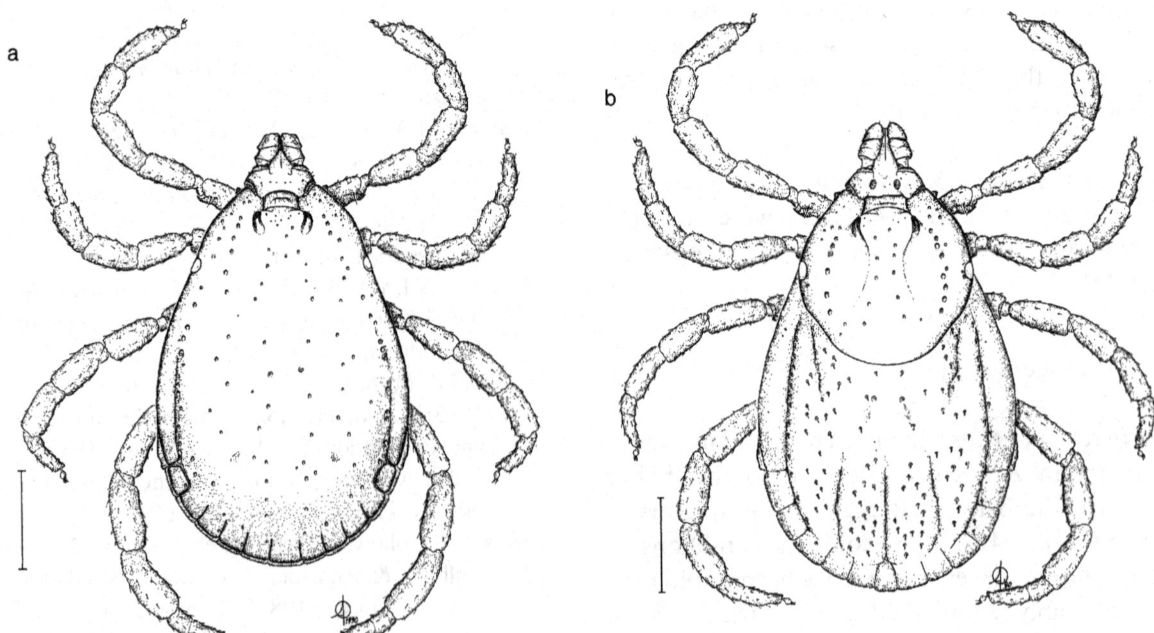

Figure 157. *Rhipicephalus praetextatus* (B.S.86/-, laboratory reared, original ♀ collected from calf, Kijabe, Kenya, on 17 May 1950 by R. Stevens). (a) Male, dorsal; (b) female, dorsal. Scale bars represent 1 mm. A. Olwage *del.*

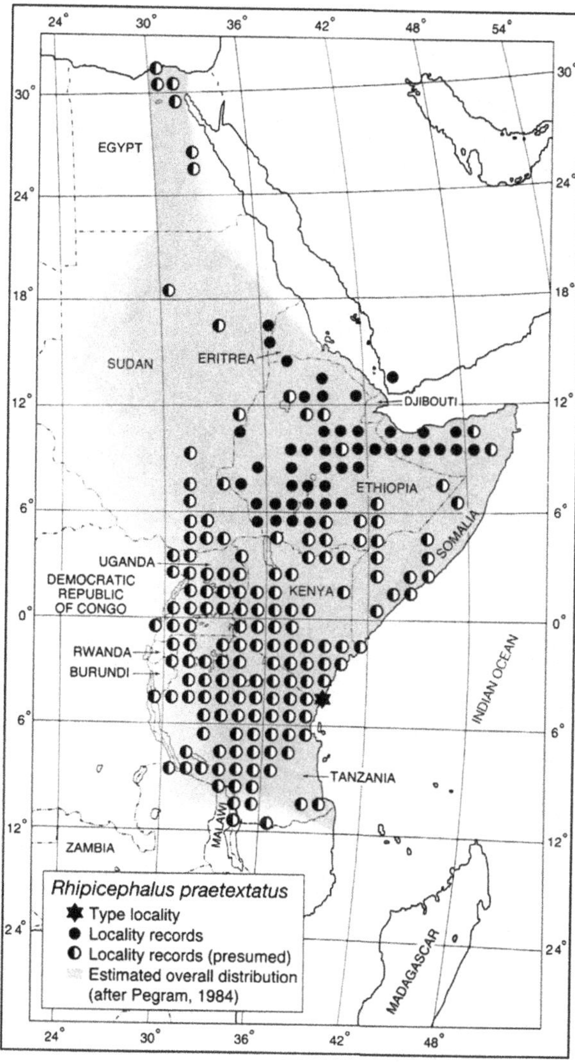

Map 46. *Rhipicephalus praetextatus*: distribution.

mm to 0.97 mm × 0.92 mm. Basis capituli with acute lateral angles at about mid-length; porose areas medium-sized, about 1.5 times their own diameter apart. Palps with article I long, easily visible from the dorsal surface. Scutum slightly broader than long, usually almost flat and glossy, length × breadth ranging from 1.46 mm × 1.51 mm to 1.95 mm × 1.99 mm; posterior margin a broad, fairly shallow curve. Eyes at mid-length, almost flat, edged dorsally by a few medium-sized setiferous punctations. Cervical grooves short; cervical fields broad, only slightly depressed, their external margins delimited by a few large shallow setiferous punctations. A few somewhat smaller setiferous punctations present on the scapulae and scattered medially on the scutum but interstitial punctations either very light or absent. Ventrally genital aperture a truncated U-shape, diverging slightly anteriorly; sclerotized margins absent.

Nymph (Fig. 159)

Capitulum much broader than long, length × breadth ranging from 0.24 mm × 0.34 mm to 0.25 mm × 0.37 mm. Basis capituli nearly four times as broad as long, with long tapering lateral angles extending over the scapulae; ventrally with sharp, rather slender, spurs on the posterior margin. Palps narrow, slightly curved, almost equal in width throughout their length, their apices broadly rounded, inclined inwards. Scutum broader than long, length × breadth ranging from 0.47 mm × 0.54 mm to 0.49 mm × 0.59 mm; posterior margin a broad smooth curve. Eyes at widest point, well over halfway back, long and narrow, delimited dorsally by slight depressions. Cervical fields long, narrow, divergent, inconspicuous. Ventrally coxae I each with two broad bluntly-rounded spurs, almost equal in length; coxae II to IV each with a broad shallow external spur, decreasing progressively in size.

Larva (Fig. 160)

Capitulum much broader than long, length × breadth ranging from 0.114 mm × 0.148 mm to 0.121 mm × 0.152 mm. Basis capituli over three times as broad as long, with short slightly forwardly-directed lateral angles, posterior margin almost straight. Palps broad, tapering slightly to their apices, external margins gently curved, inclined inwards. Scutum much broader than long, length × breadth ranging from 0.219 mm × 0.356 mm to 0.223 mm × 0.370 mm; posterior margin a broad shallow curve. Eyes at widest point, well over halfway back, delimited dorsally by faint depressions. Cervical grooves short, slightly convergent. Ventrally coxae I each with a small sharp spur; coxae II and III each with a small bluntly-rounded spur, that on coxae III being barely discernible.

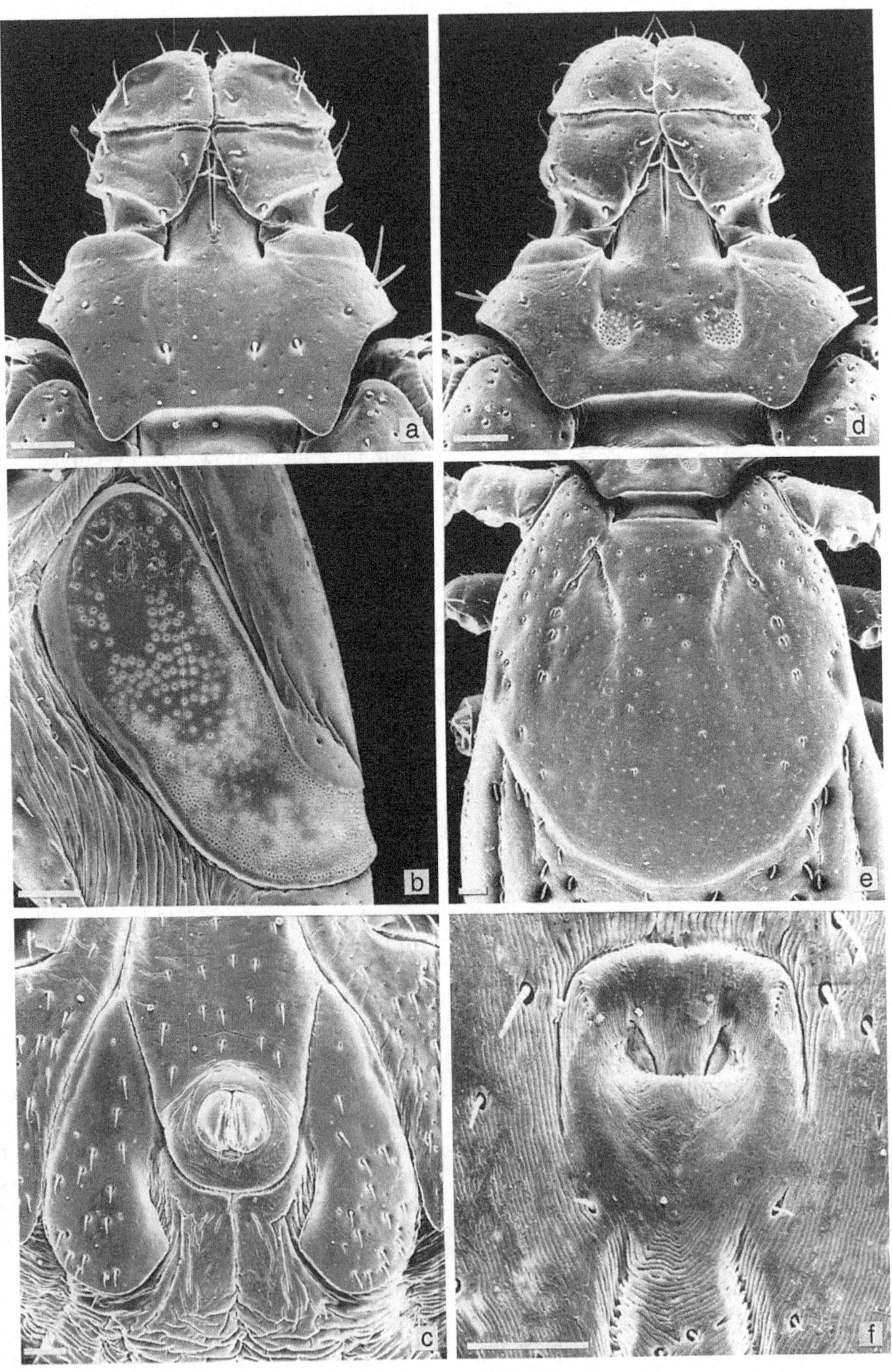
a
d
b
e
c
f

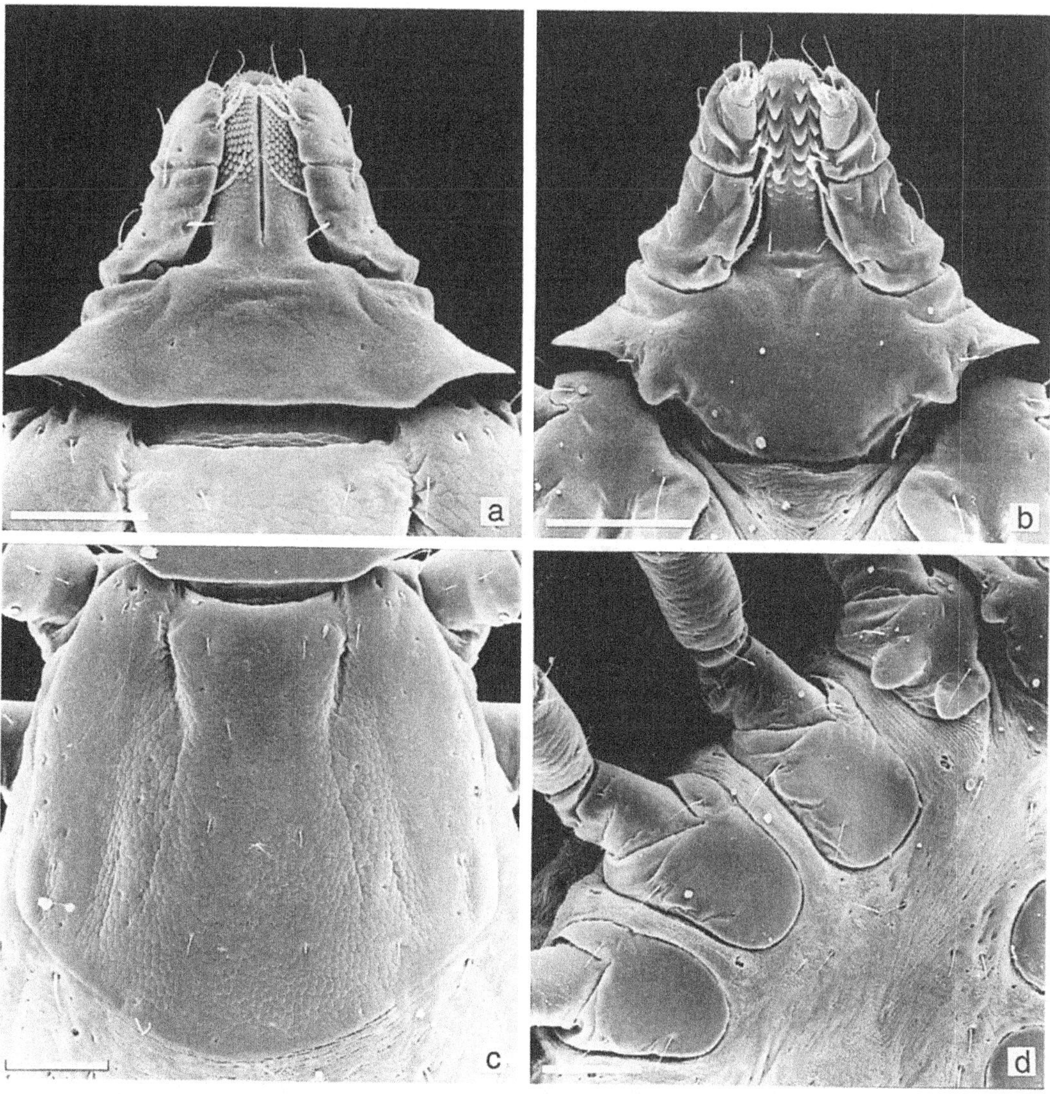

Figure 159 (*above*). *Rhipicephalus praetextatus* (L67, laboratory reared, original ♀ collected from sheep, Egypt, date of collection and collector unknown). Nymph: (a) capitulum, dorsal; (b) capitulum, ventral; (c) scutum; (d) coxae. Scale bars represent 0.10 mm. SEMs by M.D. Corwin. (Figs (a), (c) & (d) from Pegram *et al.*, 1987, figs 16–18, with kind permission from the Entomological Society of America).

Figure 158 (*opposite*). *Rhipicephalus praetextatus* (L67, laboratory reared, original ♀ collected from sheep, Egypt, date of collection and collector unknown). Male: (a) capitulum, dorsal; (b) spiracle; (c) adanal plates. Female: (d) capitulum, dorsal; (e) scutum; (f) genital aperture. Scale bars represent 0.10 mm. SEMs by M.D. Corwin. (Figure (c) from Pegram *et al.*, 1987, fig. 21, with kind permission from the Entomological Society of America).

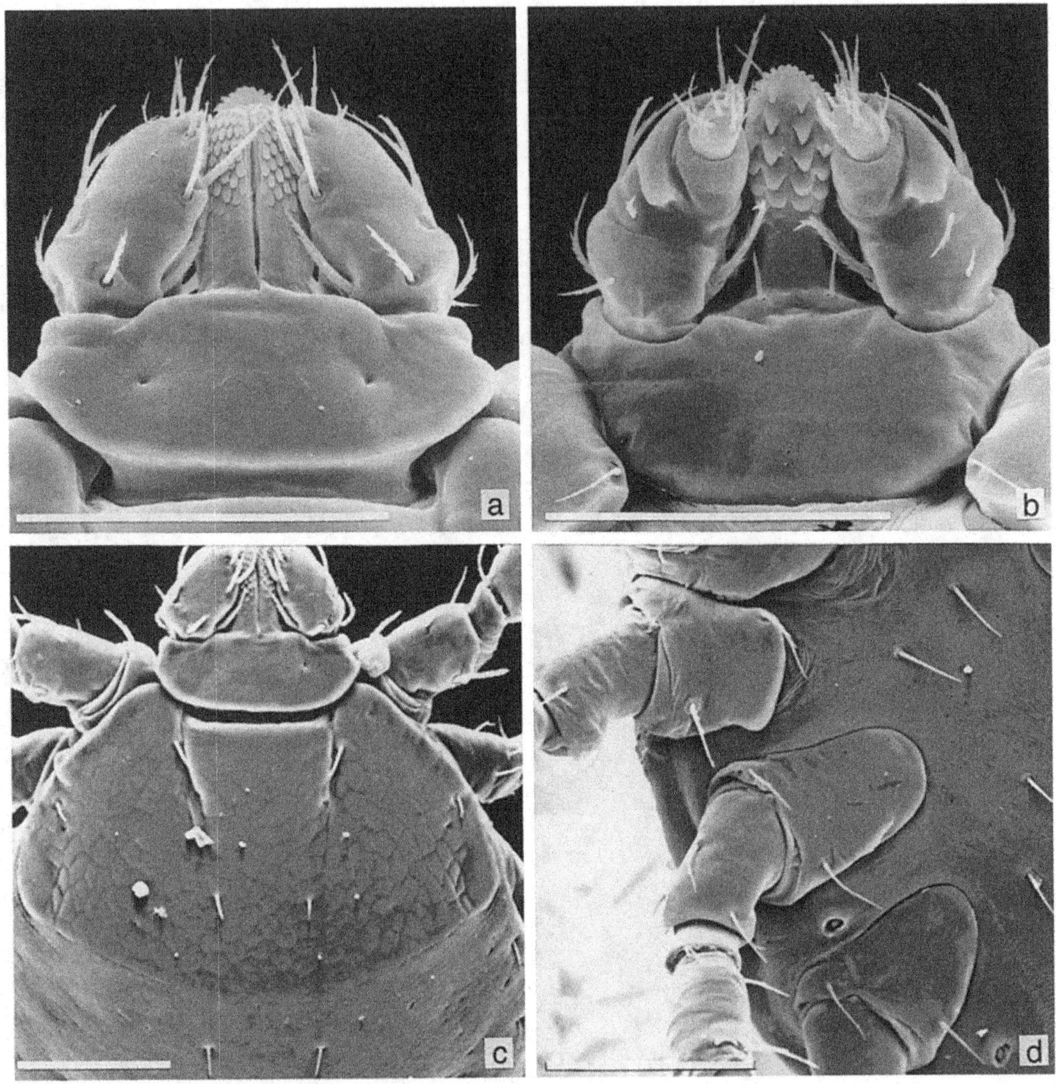

Figure 160. *Rhipicephalus praetextatus* (L67, laboratory reared, original ♀ collected from sheep, Egypt, date of collection and collector unknown). Larva: (a) capitulum, dorsal; (b) capitulum, ventral; (c) scutum; (d) coxae. Scale bars represent 0.10 mm. SEMs by M.D. Corwin.

Notes on identification

Neumann (1897) synonymized *R. praetextatus* with *R. simus* and, as a result, it features in much of the literature on African ticks under the latter name. It was redescribed as *R. simus* (eastern Africa) by Pegram (1984), and three years later the name *R. praetextatus* was formally resurrected for this taxon (Pegram *et al.*, 1987).

In the following account of the hosts and distribution of *R. praetextatus* we have included as much information as possible that we presently think refers to this species. Confirmed records comprise those from Ethiopia (Pegram, 1979, as *R. simus*; Pegram, Hoogstraal & Wassef, 1981, as *R. simus*; Pegram *et al.*, 1987; Morel, 1980, as *R. simus*); Somalia (Morel, 1969; Pegram, 1976, as *R. simus*); the Yemen (Pegram, Hoogstraal & Wassef, 1982) and Uganda (Matthysse & Colbo, 1987, though they note that their records could possibly include some of *R. muhsamae*). To these we have added unconfirmed records listed as *R. simus* from Kenya (Walker, 1974) and Tanzania (Yeoman & Walker, 1967, excluding as far as possible records of 'very heavily punctate' specimens from the western and south-western part of the country. See *R. muhsamae*, p. 303).

We have omitted from this analysis the records of *R. simus* listed by Hoogstraal (1956) from the Sudan, where both *R. muhsamae* and *R. praetextatus* occur, and by Elbl & Anastos (1966) from the Democratic Republic of Congo, where we feel that any of these three *R. simus* group species may be present in different areas.

Hosts

A three-host species (Lewis, 1932, as *R. simus*). Present indications are that *R. praetextatus* has a wide range of both domestic and wild hosts (Table 41). Its adults have been recorded from many wild carnivores, especially canids, and lions and leopards, and from numerous ungulates, particularly the wild suids and the African buffalo, also from porcupines. The immature stages feed primarily on the smaller burrowing and nest-inhabiting rodents, rarely on domestic livestock.

Pegram (1984, who referred to it as *R. simus* (eastern Africa)), noted that on cattle it commonly occurs 'in light to moderate infestations'. The two highest counts from individual cattle in Ethiopia were 84 and 69 ticks respectively, while in Tanzania the maximum single infestation on a bovine was only 28. In Egypt and the Yemen cattle and camels were evidently equally favoured as hosts: the highest single infestation on a Yemeni camel was 68 adult ticks. On cattle the tail brush and feet are the most commonly recorded attachment sites, but on Yemeni cattle it often attached on other parts of the body as well. Sheep and goats are less frequently infested, and only by very small numbers of ticks: in Tanzania the maximum single infestation on a sheep was only two adult ticks and on a goat four adults.

Zoogeography

Rhipicephalus praetextatus has been identified in collections made in Egypt, Ethiopia, Somalia and Uganda, and it is thought to be widespread in many parts of north-eastern Africa from Egypt to Tanzania. It has also been introduced into the Yemen (Map 46).

Existing information suggests that it is the only member of the closely related trio that includes *R. muhsamae* and *R. simus* occurring in Kenya, Somalia and the Yemen. In a few parts of the southern Sudan, Ethiopia and western Uganda, though, *R. muhsamae* has also been recorded and these two species can easily be confused. There is a further problem in Tanzania where, mainly in the extreme western and south-western parts of the country, Yeoman & Walker (1967) reported the presence in some collections of 'very heavily punctate' specimens of the ticks they were then calling '*R. simus*'. Morel (1980) postulated that these might also be *R. muhsamae* but Pegram (1984) considered that: 'This probably conforms to the southern African form', i.e. *R. simus sensu stricto*. Later Pegram *et al.* (1987) observed: 'We have seen typical *R. muhsamae* from Ankole Province, Uganda, which borders on both Rwanda and north-western Tanzania. We, therefore, agree with Morel's suggestion on zoogeographical grounds, but we emphasize that reexamination of the Tanzanian ticks is essential to settle the matter'. This problem has not as yet been resolved, nor has it been determined exactly how far westwards and southwards the distribution of *R. praetextatus* extends. It is worth noting that its absence from two large areas in western and south-eastern Tanzania may be apparent rather than real. Both are tsetse infested and few tick collections have been obtained from cattle there.

Rhipicephalus praetextatus occurs in a wide range of ecological conditions, from semi-arid habitats with a mean annual rainfall of *c.* 250 mm through tropical and subtropical savanna to wooded highland areas with a mean annual rainfall of *c.* 1500 mm. Wherever it occurs its adults are apparently most active during the rainy season.

Disease relationships

Rhipicephalus praetextatus can transmit Nairobi sheep disease virus, though it is not regarded as the most important vector (Lewis, 1949, as *R. simus*).

Table 41. *Host records of* Rhipicephalus praetextatus

Hosts	Number of records	
	Confirmed	Unconfirmed
Domestic animals		
Cattle	151	469
Sheep		52
Goats		27
Sheep & goats (pooled collections)	38	
Camels	37	21
Horses		3
Donkeys	3	1
Pigs		1
Dogs	4	54
Cats	1	
Wild animals		
Somali hedgehog (*Atelerix sclateri*)		1
Vervet monkey (*Chlorocebus aethiops*)		3
'Baboon' (*Papio* sp.)		2
Side-striped jackal (*Canis adustus*)		3
Golden jackal (*Canis aureus*)		2
Black-backed jackal (*Canis mesomelas*)		4
'Jackal' (*Canis* sp.)		11
Hunting dog (*Lycaon pictus*)		6
Bat-eared fox (*Otocyon megalotis*)		4
Cheetah (*Acinonyx jubatus*)		6
Caracal (*Caracal caracal*)		1
African wild cat (*Felis lybica*)		1
Serval (*Leptailurus serval*)		1
Lion (*Panthera leo*)	1	37
Leopard (*Panthera pardus*)		10
Marsh mongoose (*Atilax paludinosus*)		1
White-tailed mongoose (*Ichneumia albicauda*)		2
Spotted hyaena (*Crocuta crocuta*)		8
Striped hyaena (*Hyaena hyaena*)		5
'Hyaena'	2	6
Aardwolf (*Proteles cristatus*)		4
African civet (*Civettictis civetta*)		8
Small-spotted genet (*Genetta genetta*)		1
African elephant (*Loxodonta africana*)		12
Burchell's zebra (*Equus burchellii*)		25
Grevy's zebra (*Equus grevyi*)		2
Black rhinoceros (*Diceros bicornis*)		3
Aardvark (*Orycteropus afer*)		2
Warthog (*Phacochoerus africanus*)	1	42
Forest hog (*Hylochoerus meinertzhageni*)		2
Bushpig (*Potamochoerus larvatus*)		11
'Wild pig'	1	1

Table 41. (*cont.*)

Hosts	Number of records	
	Confirmed	Unconfirmed
Wild animals (*cont.*)		
Hippopotamus (*Hippopotamus amphibius*)	1	
Giraffe (*Giraffa camelopardalis*)		4
Coke's hartebeest (kongoni) (*Alcelaphus buselaphus cokii*)		1
Blue wildebeest (*Connochaetes taurinus*)		3
Topi (*Damaliscus lunatus topi*)		3
Dorcas gazelle (*Gazella dorcas*)		1
Grant's gazelle (*Gazella granti*)		1
Thomson's gazelle (*Gazella thomsonii*)		2
Steenbok (*Raphicerus campestris*)		1
African buffalo (*Syncerus caffer*)	4	52
Eland (*Taurotragus oryx*)		9
Mountain nyala (*Tragelaphus buxtoni*)		1
Bushbuck (*Tragelaphus scriptus*)		3
Greater kudu (*Tragelaphus strepsiceros*)		1
Red forest duiker (*Cephalophus natalensis*)		2
Common duiker (*Sylvicapra grimmia*)		1
Sable antelope (*Hippotragus niger*)		1
Gemsbok (*Oryx gazella*)		4
Bohor reedbuck (*Redunca redunca*)		1
Temminck's ground pangolin (*Manis temminckii*)		1
Unstriped ground squirrel (*Xerus rutilus*)		1
'Ground squirrel'		1
Black-tailed gerbil (*Tatera nigricauda*)		1 (nymphs)
'Gerbil'		1 (nymphs)
Nile rat (*Arvicanthis* sp.)		1 (immatures)
Striped grass rat (*Lemniscomys striatus*)		1 (nymphs)
Natal multimammate rat (*Mastomys natalensis*)		2 (immatures)
Black rat (*Rattus rattus*)		1
Four-striped grass mouse (*Rhabdomys pumilio*)		1
'Swamp rat' (*Otomys* sp.)		1 (immatures)
'Rat'		2 (immatures)
Crested porcupine (*Hystrix cristata*)		2
'Porcupine' (*Hystrix* sp.)		15
Cape hare (*Lepus capensis*)		2
'Hare'		1
Humans		
Ticks attached to host	1	4
Ticks crawling on host		9

In Tigre Province, Ethiopia, circumstantial evidence indicated that fairly heavy infestations of this tick were responsible for an outbreak of tick paralysis in cattle (Pegram *et al.*, 1981). It was thought to be one of the chief vectors of *Rickettsia conori*, causing tick typhus in humans, in the Nairobi area, Kenya (Heisch, McPhee & Rickman, 1957).

REFERENCES

Gerstäcker, A. (1873). II: Gliederthiere (Insekten, Arachniden, Myriopoden und Isopoden). In *Baron Carl Claus von der Decken's Reisen in Ost Afrika, in den Jahren 1859 bis 1861.* **III**: Wissenschaftliche Theil. Abteilung 2, ed. O. Kersten, pp. xvi + 542, 18 plates. (Ticks pp. 464–70, plate xviii). Leipsig & Heidelberg: C.J. Winter'sche Verlagshandlung.

Heisch, R.B., McPhee, R. & Rickman, L.R. (1957). The epidemiology of tick-typhus in Nairobi. *East African Medical Journal*, **34**, 459–77.

Lewis, E.A. (1932). Some tick investigations in Kenya Colony. *Parasitology*, **24**, 175–82.

Lewis, E.A. (1949). Nairobi sheep disease. *Report of the Veterinary Department, Kenya, for 1947*, pp. 45, 51.

Neumann, L.G. (1897). Révision de la famille des ixodidés. (2e Mémoire). Ixodinae. *Mémoires de la Société Zoologique de France*, **10**, 324–420.

Pegram, R.G. (1976). Ticks (Acarina, Ixodoidea) of the northern regions of the Somali Democratic Republic. *Bulletin of Entomological Research*, **66**, 345–63.

Pegram, R.G. (1979). *Ticks (Ixodoidea) of Ethiopia with special reference to cattle.* M. Phil. thesis, University of Brunel.

Pegram, R.G., Hoogstraal, H. & Wassef, H.Y. (1981). Ticks (Acari: Ixodoidea) of Ethiopia. I. Distribution, ecology and host relationships of species infesting livestock. *Bulletin of Entomological Research*, **71**, 339–59.

Pegram, R.G., Hoogstraal, H. & Wassef, H.Y. (1982). Ticks (Acari: Ixodoidea) of the Yemen Arab Republic. I. Species infesting livestock. *Bulletin of Entomological Research*, **72**, 215–27.

Pegram, R.G., Walker, J.B., Clifford, C.M. & Keirans, J.E. (1987). Comparison of populations of the *Rhipicephalus simus* group: *R. simus, R. praetextatus*, and *R. muhsamae* (Acari: Ixodidae). *Journal of Medical Entomology*, **24**, 666–82.

Also see the following Basic References (pp. 12–14): Elbl & Anastos (1966); Hoogstraal (1956); Matthysse & Colbo (1987); Morel (1969, 1980); Pegram (1984); Scaramella (1988); Walker (1974); Yeoman & Walker (1967).

RHIPICEPHALUS PRAVUS DÖNITZ, 1910 (INCLUDING *RHIPICEPHALUS* SP. NEAR *PRAVUS*)

The specific name *pravus* is derived from the Latin meaning 'crooked', 'irregular' or 'deformed'.

Diagnosis

A moderate-sized reddish-brown tick.

Male (Figs 161(a), 162(a) to (c))

Capitulum broader than long, length × breadth ranging from 0.58 mm × 0.63 mm to 0.82 mm × 0.84 mm. Basis capituli with acute lateral angles at about anterior third of its length. Palps broadly rounded apically. Conscutum length × breadth ranging from 2.45 mm × 1.38 mm to 3.29 mm × 1.89 mm; often rather narrow anteriorly; anterior process of coxae I large and heavily sclerotized. In engorged specimens body wall expanded laterally and forming a long tail-like caudal process posteromedially. Eyes convex, oval in shape and set obliquely, edged dorsally by medium-sized punctations that usually coalesce into a groove. Cervical pits deep, convergent; cervical fields long, narrow, slightly divergent, tapering posteriorly to points beyond eye level. Marginal lines well developed, long, punctate. Posteromedian groove long, relatively narrow; posterolateral grooves shorter and broader. Large setiferous punctations present anteriorly on the scapulae, along the external margins of the cervical fields and scattered medially on the conscutum, but absent laterally adjacent to the marginal lines. Numerous smaller punctations present medially on the conscutum. Punctations

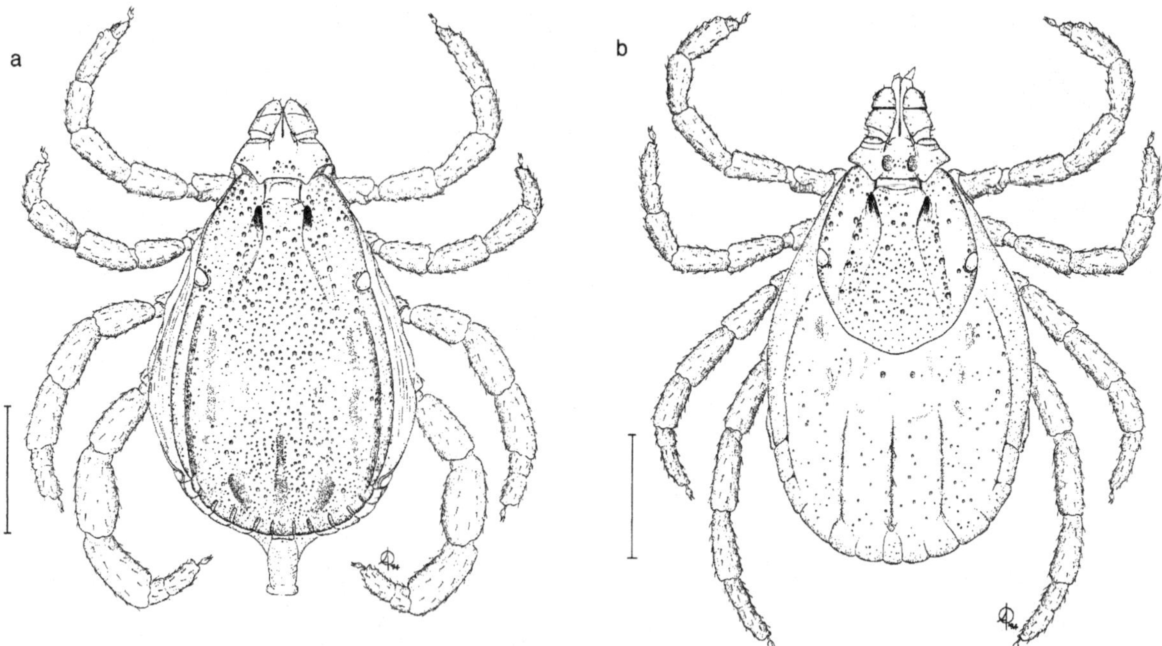

Figure 161. *Rhipicephalus pravus* [Nuttall Collection 1246, syntypes, collected from 'beisa oryx (*Oryx beisa*)', regarded herein as a synonym of the gemsbok (*Oryx gazella*), on the southern border of the Maasai Steppe, Tanzania, donated in April 1911 by W. Dönitz, by courtesy of The Natural History Museum, London]. (a) Male, dorsal; (b) female, dorsal. Scale bars represent 1 mm. A. Olwage *del.*

anterior to the eyes and next to the marginal lines minute, often almost invisible. Ventrally spiracles tapering very gently, only curving slightly at the end of the dorsal prolongation. Adanal plates broad, internal margin usually concave just posterior to the anus, anterointernal angle often sharply pointed, posterointernal angle acute to almost right-angled, posteroexternal angle broadly curved; accessory adanal plates, when present, small sclerotized points.

Female (Figs 161(b), 162(d) to (f))

Capitulum broader than long, length × breadth ranging from 0.70 mm × 0.79 mm to 0.80 mm × 0.88 mm. Basis capituli with acute lateral angles at about mid-length; porose areas medium-sized, oval, nearly twice their own diameter apart. Palps broadly rounded apically. Scutum longer than broad, length × breadth ranging from 1.43 mm × 1.25 mm to 1.60 mm × 1.44 mm; posterior margin sinuous. Eyes convex, oval, set obliquely, edged dorsally by medium-sized punctations that usually coalesce into a groove. Cervical pits convergent; cervical fields long, narrow, slightly divergent, tapering posteriorly and almost reaching posterolateral margins of scutum. Large setiferous punctations present on the scapulae, delimiting the external margins of the cervical fields and scattered medially on the scutum, where they are interspersed with numerous small to minute punctations. Lateral areas of the scutum surrounding the eyes minutely punctate in some specimens, smooth and shiny in others. Ventrally genital aperture broadly V-shaped.

Nymph (Fig. 163)

Capitulum much broader than long, length × breadth ranging from 0.18 mm × 0.28 mm to 0.22 mm × 0.30 mm. Basis capituli over three times as broad as long, with acute lateral angles projecting over the scapulae. Palps tapering gradually to smoothly-rounded apices, inclined inwards. Scutum longer than broad, length × breadth ranging from 0.49 mm × 0.44 mm to 0.53 mm × 0.47 mm; posterior margin a deep, slightly sinuous curve. Eyes immediately anterior

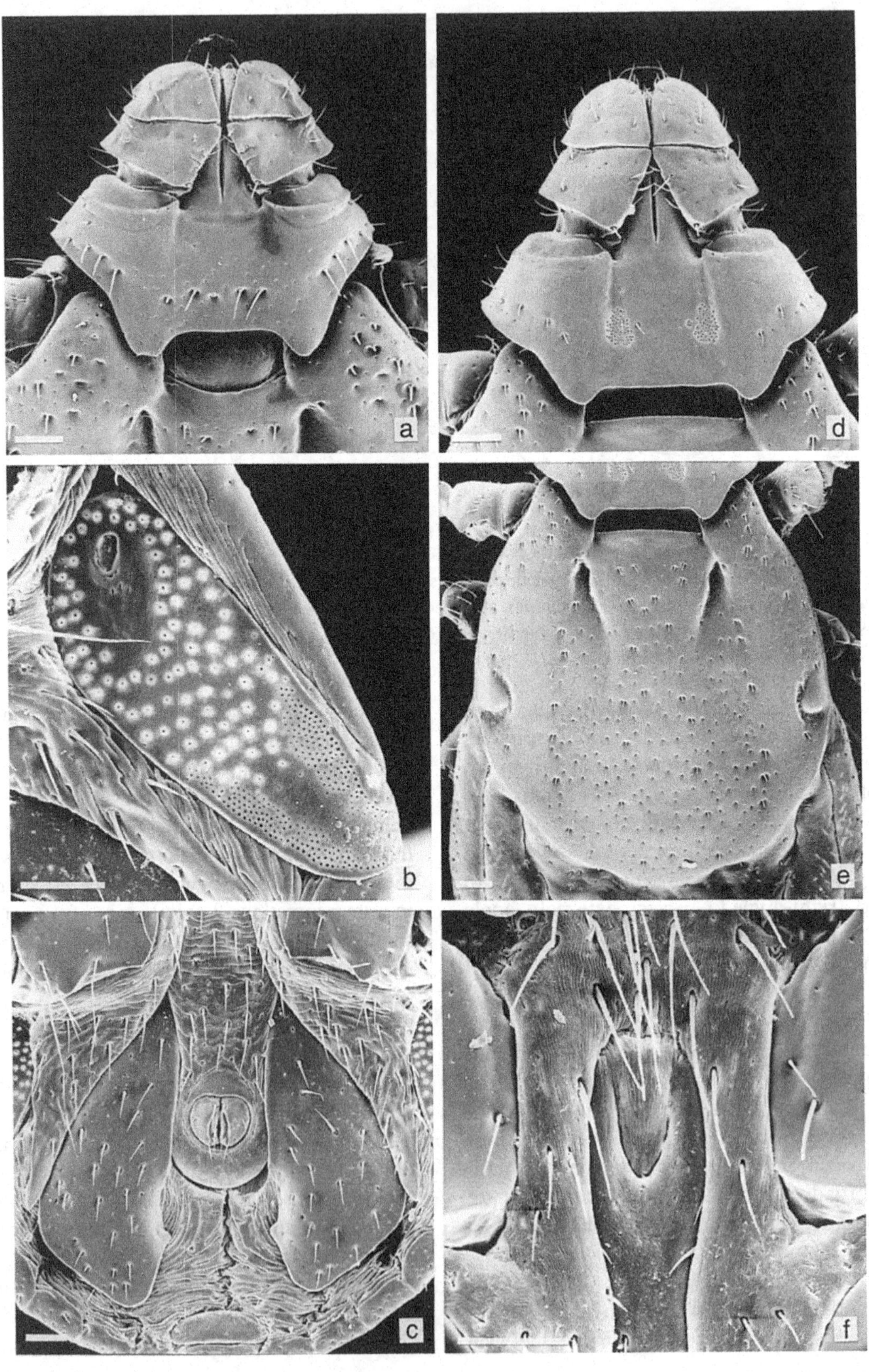
a
b
c
d
e
f

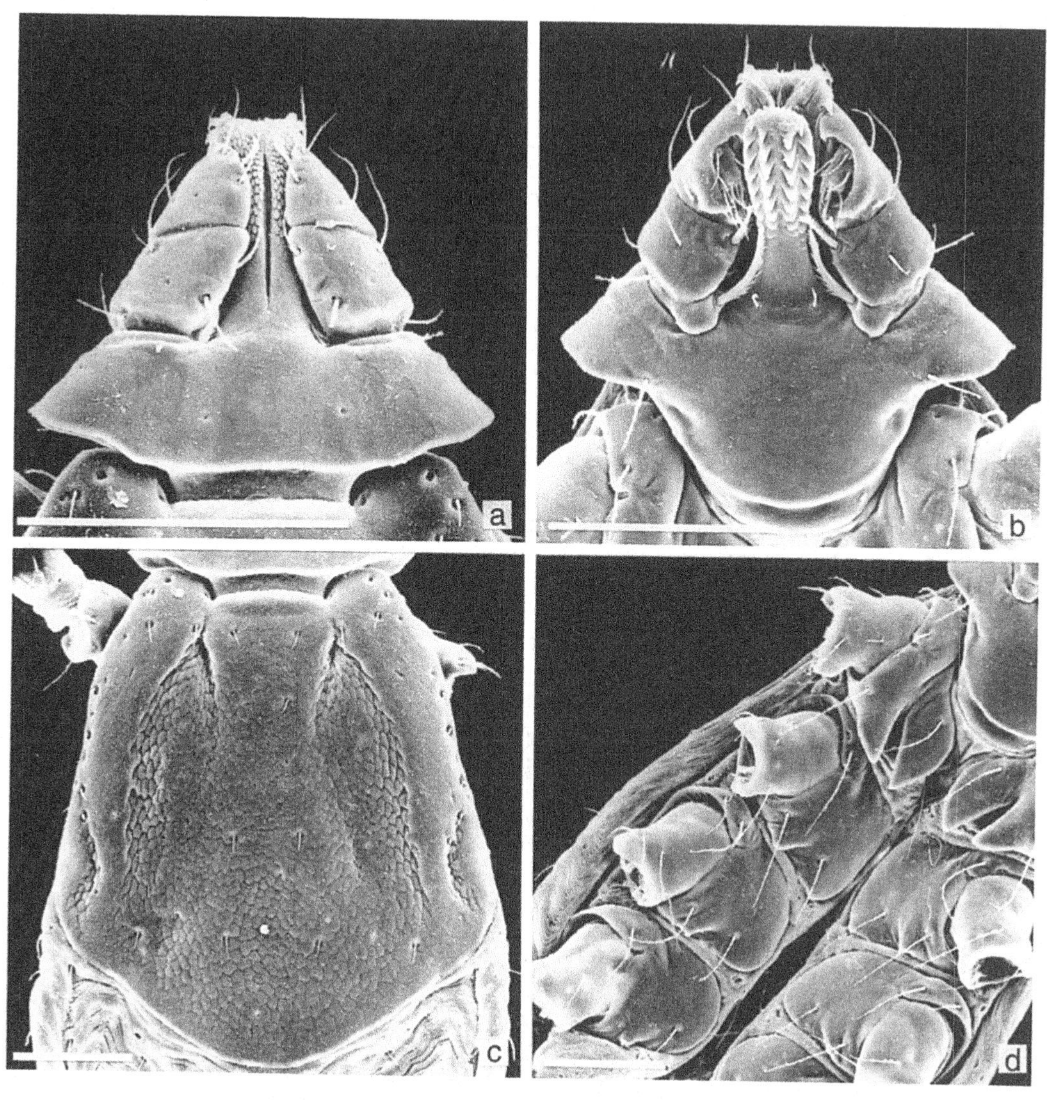

Figure 163 (*above*). *Rhipicephalus pravus* (B.S. 101/-, RML 66312, laboratory reared, progeny of ♀ collected from sheep, Marwa Sisal Estates, Kima, Kenya, in 1950 by W. Plowright). Nymph: (a) capitulum, dorsal; (b) capitulum, ventral; (c) scutum; (d) coxae. Scale bars represent 0.10 mm. SEMS by M.D. Corwin.

Figure 162 (*opposite*). *Rhipicephalus pravus* (B.S. 101/-, RML 66312, laboratory reared, progeny of ♀ collected from sheep, Marwa Sisal Estates, Kima, Kenya, in 1950 by W. Plowright). Male: (a) capitulum, dorsal; (b) spiracle; (c) adanal plates. Female: (d) capitulum, dorsal; (e) scutum; (f) genital aperture. Scale bars represent 0.10 mm. SEMs by M.D. Corwin.

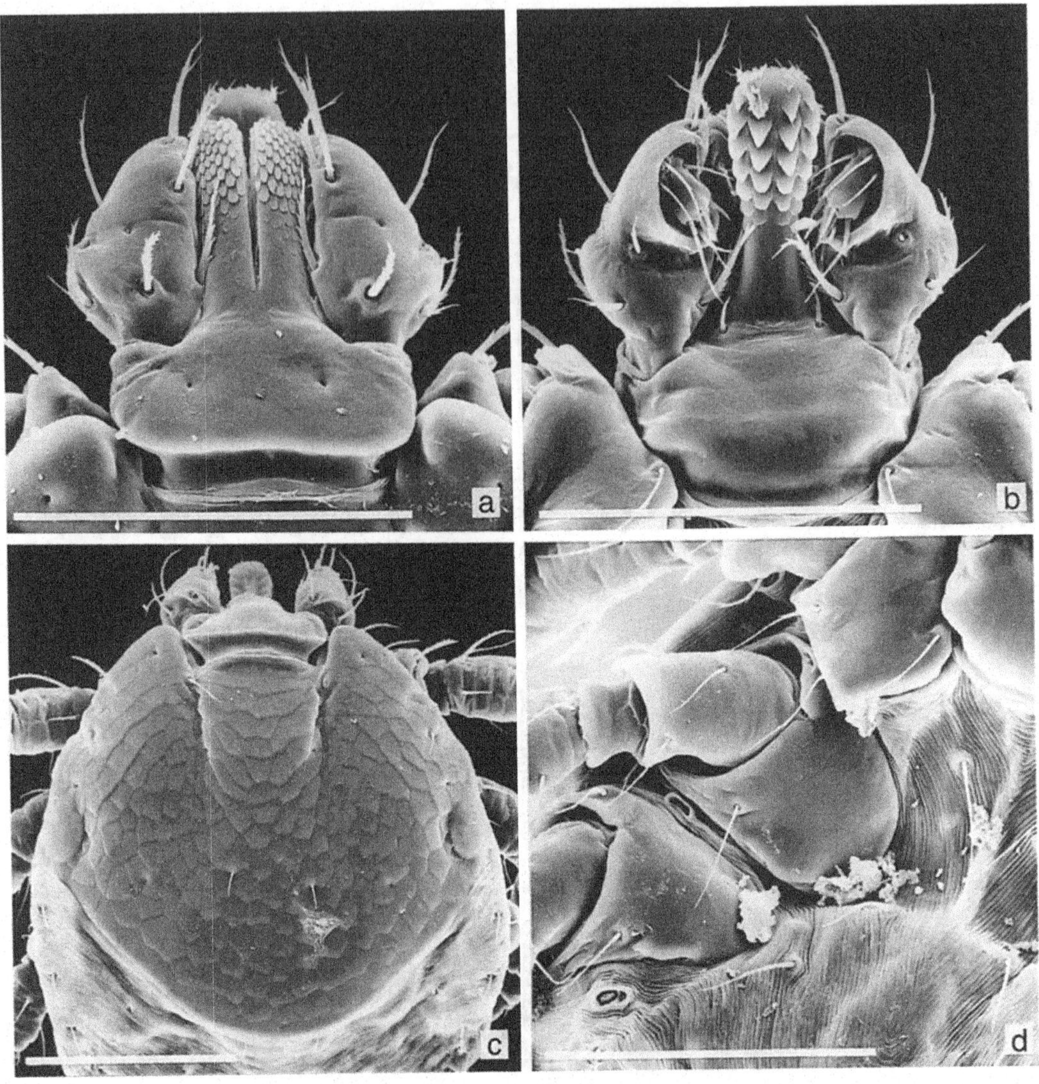

Figure 164. *Rhipicephalus pravus* (B.S. 101/-, RML 66312, laboratory reared, progeny of ♀ collected from sheep, Marwa Sisal Estates, Kima, Kenya, in 1950 by W. Plowright). Larva: (a) capitulum, dorsal; (b) capitulum, ventral; (c) scutum; (d) coxae. Scale bars represent 0.10 mm. SEMS by M.D. Corwin.

to widest point, well over halfway back, long, narrow, delimited dorsally by a marked depression. Cervical fields long, narrow, slightly divergent, their external margins curved outwards almost parallel to the sides of the scutum and extending posteriorly nearly to its posterolateral margins. Ventrally coxae I each with a large broad internal spur and a slightly smaller narrower external spur; remaining coxae each with a pointed external spur only, decreasing progressively in size.

Larva (Fig. 164)

Capitulum slightly broader than long, length × breadth ranging from 0.098 mm × 0.102 mm to 0.102 mm × 0.107 mm. Basis capituli over twice as broad as long, narrowly rectangular in shape with rounded corners. Palps somewhat bulbous proximally, then tapering rapidly to their rounded apices. Scutum broader than long, length × breadth ranging from 0.238 mm × 0.295 mm to 0.281 mm × 0.304 mm; posterior margin a deep smooth curve. Eyes approximately at mid-length, large, convex, edged dorsally by a groove. Cervical grooves slightly convergent. Ventrally coxae I each with a sharp spur; coxae II and III each with indications only of broad shallow spurs on their posterior border.

Notes on identification

The *R. pravus* group has presented us with one of our most intractable systematic problems and, although we feel that some progress has been made towards its resolution, various unanswered questions remain. In particular we are still unsure about the status of the tick closely resembling *R. pravus sensu stricto* that occurs in large parts of southern Africa.

The systematics of the East African species in the group have largely been clarified. Walker (1956) had confused the situation by sinking both *R. neavei* and *R. neavei punctatus* as synonyms of *R. pravus*. These findings were reversed by Yeoman & Walker (1967), who recognized three species in this group in Tanzania: *R. neavei, R. pravus* and *R. punctatus*. Of these *R. neavei* was later sunk again, this time as a synonym of *R. kochi*, by Matthysse & Colbo (1987).

Earlier Zumpt (1942) pointed out that Nuttall & Warburton had sometimes identified *R. pravus* as *R. neavei*. The re-examination of available specimens listed as *R. neavei* by Lewis (1934, 1939, and unpublished data); Wilson (1951, and unpublished data), and Wiley (1953, and unpublished data) has shown that all but one of their collections are also *R. pravus*, not *R. kochi*. Unfortunately many of their collections no longer exist (Walker, 1974); these are referred to below as unconfirmed records.

Like Morel (1969) we have omitted the records of *R. pravus* from the Democratic Republic of Congo, Rwanda and Burundi cited by Elbl & Anastos (1966). The ticks from Upemba Park that they documented under this name were earlier identified as *Rhipicephalus* sp. near *pravus* and discussed in detail by Clifford & Anastos (1962). One of us (J.B.W.) re-examined these specimens at that time and we have now accepted them as *R. kochi sensu stricto*. The remaining records listed by Elbl & Anastos (1966) are located either in rain-forest areas or in other regions that appear to be ecologically unsuitable for this dry-country tick. Matthysse & Colbo (1987) suggested that the records of *R. pravus* given by Elbl & Anastos (1966) in high-rainfall regions of the Democratic Republic of Congo and Rwanda close to western Uganda 'possibly refer to *R. kochi*'.

We have also omitted the records of *R. pravus* from Mozambique listed by Santos Dias (1960). Although we have not seen his specimens we believe that some belong to the *R. pravus*-like species occurring in southern Africa. (See below).

Hosts

A three-host species (Walker, 1956). Adults of *R. pravus* have often been collected from cattle, but less frequently from sheep, goats and camels, and rarely from other domestic animals (Table 42A). In general the adult's predilection sites on cattle resemble those of *R. appendiculatus*. The majority attach on the hosts' ears, especially inside the flap on the central area between the cartilage bars. There they characteristically feed 'standing on their heads' with their legs sticking out sideways. They apparently spread more readily than *R. appendiculatus* adults to other attachment sites, particularly the outside of the ear flap and its base, the horn base, eyelids, mouth commissure, neck and dewlap, brisket, abdomen, udder, escutcheon, perineum, groin and heels (Yeoman & Walker, 1967).

Rhipicephalus pravus adults have also been recorded from many wild animals, especially the giraffe, gemsbok and various smaller antelopes such as the impala, several species of gazelles and dik-diks (Yeoman & Walker, 1967; Walker, 1974; Pegram, 1976; Pegram, Hoogstraal & Wassef, 1981; Matthysse & Colbo, 1987). Many of the collections from the most commonly-recorded wild host of both the adults and immature stages of this tick, the Cape hare, and also the savanna hare, were obtained during a specific study of these leporids in East Africa (Clifford, Flux & Hoogstraal, 1976). Various species of

Table 42A. *Host records of* Rhipicephalus pravus

Hosts	Number of records	
	Confirmed	Unconfirmed
Domestic animals		
Cattle	506	200
Sheep	44	51
Goats	36	49
Camels	32	7
Horses	2	
Donkeys	5	
Pigs (feral)	1	
Dogs	8	1
Wild animals		
Four-toed hedgehog (*Atelerix albiventris*)	1	
Heart-nosed bat (*Cardioderma cor*)	1	
Banana bat (*Pipistrellus nanus*)	1	
Golden jackal (*Canis aureus*)	2	
Black-backed jackal (*Canis mesomelas*)	4	
'Jackal'	2	
Bat-eared fox (*Otocyon megalotis*)	1	
Lion (*Panthera leo*)	2	
'Genet' (*Genetta* sp.)	2 (including nymphs)	
African elephant (*Loxodonta africana*)	3	
Burchell's zebra (*Equus burchellii*)	3	
Black rhinoceros (*Diceros bicornis*)		1
Aardvark (*Orycteropus afer*)	1	
Warthog (*Phacochoerus africanus*)	5	3
Giraffe (*Giraffa camelopardalis*)	9	
Impala (*Aepyceros melampus*)	13	11
Coke's hartebeest (*Alcelaphus buselaphus cokii*)	1	10
Jackson's hartebeest (*Alcelaphus buselaphus jacksoni*)	1	
Blue wildebeest (*Connochaetes taurinus*)	1	1
Grant's gazelle (*Gazella granti*)	20	4
Soemmerring's gazelle (*Gazella soemmerringii*)	2	
Thomson's gazelle (*Gazella thomsonii*)	25	5
'Gazelle' (*Gazella* sp.)		1
Gerenuk (*Litocranius walleri*)	5	4
Günther's dik-dik (*Madoqua guentheri*)	6	
Kirk's dik-dik (*Madoqua kirkii*)	2	
Silver dik-dik (*Madoqua piacentinii*)	1	
'Dik-dik' (*Madoqua* sp.)	1	3
Klipspringer (*Oreotragus oreotragus*)	3	1
Oribi (*Ourebia ourebi*)	4 (one including a nymph)	
Steenbok (*Raphicerus campestris*)	5	3
African buffalo (*Syncerus caffer*)	4	3
Eland (*Taurotragus oryx*)	4	9
Lesser kudu (*Tragelaphus imberbis*)	4	4
Bushbuck (*Tragelaphus scriptus*)	2	4
Greater kudu (*Tragelaphus strepsiceros*)	4	
'Kudu' (*Tragelaphus* sp.)	1	

Table 42A. (*cont.*)

Hosts	Number of records	
	Confirmed	Unconfirmed
Wild animals (*cont.*)		
Common duiker (*Sylvicapra grimmia*)	3	
'Duiker'	1	
Gemsbok (*Oryx gazella*)	10 (one with nymphs only)	3
Waterbuck (*Kobus ellipsiprymnus*)	1	2
Bohor reedbuck (*Redunca redunca*)	1	
Unstriped ground squirrel (*Xerus rutilus*)	1	
Cape hare (*Lepus capensis*)	99 (including immatures)	2
Savanna hare (*Lepus victoriae*)	14 (including immatures)	1
'Hare' (*Lepus* sp.)	10 (including immatures)	2
Central African rabbit (*Poelagus marjorita*)	1	
Short-snouted elephant shrew (*Elephantulus brachyrhynchus*)	2 (nymphs)	
Uganda elephant shrew (*Elephantulus fuscipes*)	1 (nymphs)	
Rufous elephant shrew (*Elephantulus rufescens*)	42 (immatures)	2 (immatures)
'Elephant shrew' (*Elephantulus* sp.)	2 (immatures plus 1 ♀)	
Birds		
Ostrich (*Struthio camelus*)	1	
Great sparrow hawk (*Accipiter melanoleucus*)	1	
Secretary bird (*Sagittarius serpentarius*)	1	
Kori bustard (*Ardeotis kori*)	1 (nymphs)	
'Shrike'	1	
Reptiles		
'Tortoise'		1
Humans	9	

elephant shrews are also important hosts of the immature stages.

Zoogeography

Rhipicephalus pravus has been recorded in parts of eastern Africa from Ethiopia and Somalia southwards as far as northern Iringa District in Tanzania (Morel, 1969, 1980) (Map 47).

This tick species occurs throughout northern Somalia but is less common in the coastal areas (Pegram, 1976). It is probably more widely distributed in southern Somalia, eastern Ethiopia and north-eastern Kenya than present records indicate: access to these sparsely-inhabited areas is difficult (Walker, 1974; Pegram *et al.*, 1981). In Tanzania it is present throughout many important East Coast fever zones. It is particularly likely to be present in areas that are only marginally suitable for *R. appendiculatus*, where the small surviving populations of the latter species cause sporadic outbreaks of ECF (Yeoman & Walker, 1967).

In different parts of its range *R. pravus* has been recorded at altitudes varying from about 200 m to 2000 m with mean annual rainfalls of some 250 mm to over 1000 mm. A crucial factor with rainfall is not so much the annual total but

Table 42B. *Host records of* Rhipicephalus *sp. near* pravus

Hosts	Number of records	
	Confirmed	Unconfirmed
Domestic animals		
Cattle	21	
Goats	15	
Dogs	3	
Rabbits	1	
Wild animals		
Lion (*Panthera leo*)	2	
Burchell's zebra (*Equus burchellii*)	1	
Rock hyrax (*Procavia capensis*)	2	
Warthog (*Phacochoerus africanus*)	2	
Giraffe (*Giraffa camelopardalis*)	5	
Impala (*Aepyceros melampus*)	6	1
Tsessebe (*Damaliscus lunatus lunatus*)	2	
Suni (*Neotragus moschatus*)		1
Oribi (*Ourebia ourebi*)		2
Steenbok (*Raphicerus campestris*)	3	1
African buffalo (*Syncerus caffer*)		1
Eland (*Taurotragus oryx*)	4	1
Nyala (*Tragelaphus angasii*)	1	1
Bushbuck (*Tragelaphus scriptus*)		3
Greater kudu (*Tragelaphus strepsiceros*)	2	2
Red forest duiker (*Cephalophus natalensis*)		1
Common duiker (*Sylvicapra grimmia*)		2
Roan antelope (*Hippotragus equinus*)	2	
Sable antelope (*Hippotragus niger*)	1	1
Gemsbok (*Oryx gazella*)	1	
Waterbuck (*Kobus ellipsiprymnus*)		1
Greater cane rat (*Thryonomys swinderianus*)	1	
Cape hare (*Lepus capensis*)	1	
Scrub hare (*Lepus saxatilis*)	74 (including immatures)	3
'Hare' (*Lepus* sp.)	19	1
Short-snouted elephant shrew (*Elephantulus brachyrhynchus*)	5 (immatures)	
Rock elephant shrew (*Elephantulus myurus*)	10 (immatures)	
Elephantulus sp.	1 (immatures)	
Four-toed elephant shrew (*Petrodomus tetradactylus*)		1 (nymphs)
Humans	1	

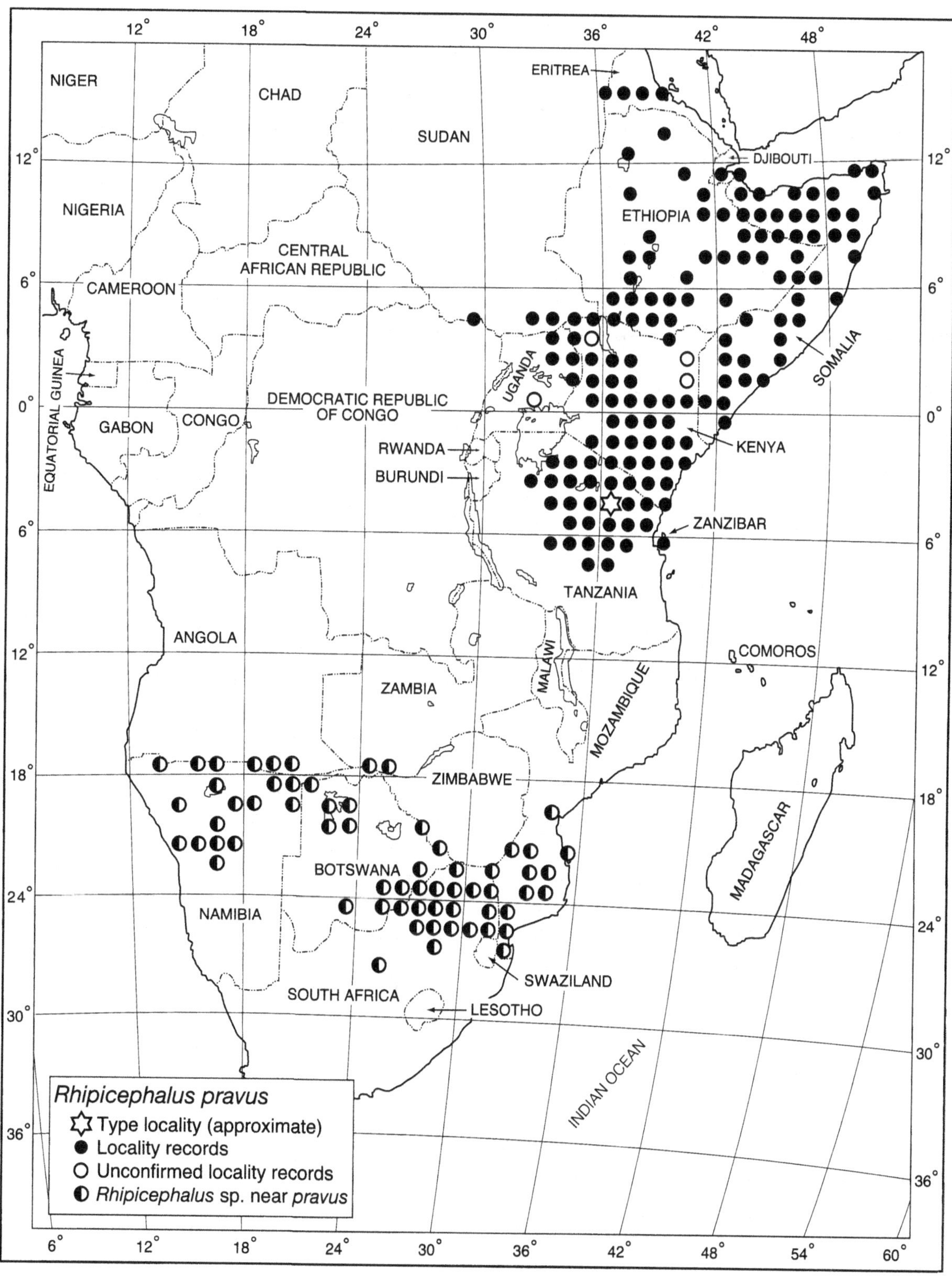

Map 47. *Rhipicephalus pravus* and *Rhipicephalus* sp. near *pravus*: distribution.

the length of the dry season. In some areas where this tick is common, such as parts of Karamoja District, Uganda, the dry season is 6 months long, with 50 mm rainfall or less per month, though the annual total may exceed 1000 mm. It is, however, absent from other parts of the country which may experience lower mean annual rainfalls but have only a 3-month dry season (Matthysse & Colbo, 1987). In Ethiopia most adults were collected from cattle during the rainy season, and in Somalia they were also slightly more prevalent during the rains (Pegram, 1976; Pegram *et al.*, 1981). In Uganda, and parts of Kenya and Tanzania, though, no seasonal trends in the adults' activity patterns on cattle were noted. During the 1967/68 survey on hares in Kenya the percentage of animals infested with the immature stages peaked during the dry seasons, followed slightly later by the adults. It is commonly, but not exclusively, found in Somalia–Masai *Acacia–Commiphora* deciduous bushland and thicket. In Tanzania it is noticeable that the southernmost point of the tick's range almost coincides with that of this vegetation type.

Disease relationships

In Kenya ticks identified as *R. neavei* transmitted *Theileria parva* under laboratory conditions (Lewis, Piercy & Wiley, 1946). These were almost certainly *R. pravus* as the strains used originated from Makueni and Kiboko where it is common. It is extremely unlikely that this species ever acts as a vector in the field because its immature stages are not known to feed on cattle.

Rhipicephalus sp. near *pravus*

Notes on identification

To us some of the ticks collected in the northern provinces of South Africa and in Namibia and Botswana strongly resemble *R. pravus sensu stricto.* Nevertheless we have decided to discuss them separately from the East African tick for two reasons. Firstly, they are widely removed geographically from *R. pravus* itself, which occurs in a discrete area in East Africa (Map 47). Secondly, even after the separation of *R. warburtoni sp. nov.* from within this group we believe that it may still encompass more than one species. Theiler & Robinson (1953) unwittingly highlighted this problem by including data on both this *Rhipicephalus* sp. near *pravus* and *R. warburtoni*, and possibly also the *R. punctatus*-like tick of Norval (1985), in their paper on *R. pravus.* Even their description of *R. pravus* was based on specimens collected at Edenburg in the central Free State, a locality at which only *R. warburtoni* occurs (p. 465, Map 61).

Hosts

A three-host species. The hosts of this *Rhipicephalus* sp. near *pravus* can be assigned to three groups: those on which only adult ticks occur, those on which only the immature stages are found, and those which harbour both adults and the immature stages (Table 42B). The preferred hosts of the adults are cattle, goats and various wild ruminants, particularly giraffe, impala and eland (Zumpt, 1958; Paine, 1982; Boomker, Horak & Ramsay, 1994). The hosts of the immature stages only are elephant shrews (I.G.H., unpublished data), and those of all stages of development are hares (Zumpt, 1958; Horak *et al.*, 1993, 1995).

Adult ticks have been collected from around the lips and from the cheeks and ears of eland, the forehead, cheeks and neck of giraffe and from the ears of scrub hares. Burdens of adult ticks seldom exceed 20, but 307 larvae and 47 nymphs have been taken from a single rock elephant shrew (I.G.H., unpublished data). No clear pattern of seasonal abundance for any of the life stages emerged during two surveys conducted on scrub hares in north-eastern South Africa (Horak *et al.*, 1993, 1995). Adult ticks were, however, present throughout the year.

Zoogeography

In southern Africa this *R.* sp. near *pravus* has been identified from the northern provinces of South Africa, south-eastern and north-western

Botswana and northern Namibia (see Map 47). Unconfirmed collections that we believe belong to this entity have been recorded in southern Mozambique (Santos Dias, 1960). It has been collected at altitudes varying between 100 m and 1650 m above sea level. The vegetation types in which it is present characterize regions with an annual rainfall of approximately 500 mm. It is found in miombo, *Colophospermum mopane* and undifferentiated woodland and is also present in deciduous bushland and wooded grassland.

Disease relationships

Unknown.

REFERENCES

Boomker, J., Horak, I.G. & Ramsay, K.A. (1994). Helminth and arthropod parasites of indigenous goats in the northern Transvaal. *Onderstepoort Journal of Veterinary Research*, **61**, 13–20.

Clifford, C.M., Flux, J.E.C. & Hoogstraal, H. (1976). Seasonal and regional abundance of ticks (Ixodidae) on hares (Leporidae) in Kenya. *Journal of Medical Entomology*, **13**, 40–7.

Dönitz, W. (1910). Die Zecken Südafrikas. In *Zoologische und Anthropologische Ergebnisse einer Forchungsreise in westliche und zentralen südafrika ausgeführt in den Jahren 1903–1905*, **4**, 3 Lieferung. L. Schultze. *Denkschriften der Medicinisch-Naturwissenschaftlichen Gesellschaft zu Jena*, **16**, 397–494, bls. 15, 16a, b & 17.

Horak, I.G., Spickett, A.M., Braack, L.E.O. & Penzhorn, B.L. (1993). Parasites of domestic and wild animals in South Africa. XXXII. Ixodid ticks on scrub hares in the Transvaal. *Onderstepoort Journal of Veterinary Research*, **60**, 163–74.

Horak, I.G., Spickett, A.M., Braack, L.E.O., Penzhorn, B.L., Bagnall, R.J. & Uys, A.C. (1995). Parasites of domestic and wild animals in South Africa. XXXIII. Ixodid ticks on scrub hares in the north-eastern regions of Northern and Eastern Transvaal and of KwaZulu-Natal. *Onderstepoort Journal of Veterinary Research*, **62**, 123–31.

Lewis, E.A. (1934). A study of ticks in Kenya Colony. The influence of natural conditions and other factors on their distribution and the incidence of tick-borne diseases. Part III. Investigations into the tick problem in the Masai Reserve. *Bulletin of the Department of Agriculture, Kenya*, **No. 7 of 1934**, 65 pp. + 3 maps.

Lewis, E.A. (1939). The ticks of East Africa. Part I. Species, distribution, influence of climate, habits and life histories. *Empire Journal of Experimental Agriculture*, 7, no. 27, 261–70.

Lewis, E.A., Piercy, S.E. & Wiley, A.J. (1946). *Rhipicephalus neavei* Warburton, 1912 as a vector of East Coast fever. *Parasitology*, **37**, 60–4.

Norval, R.A.I. (1985). The ticks of Zimbabwe. XII. The lesser known *Rhipicephalus* species. *Zimbabwe Veterinary Journal*, **16**, 37–43.

Paine, G.D. (1982). Ticks (Acari: Ixodoidea) in Botswana. *Bulletin of Entomological Research*, **72**, 1–16.

Pegram, R.G. (1976). Ticks (Acarina, Ixodoidea) of the northern regions of the Somali Democratic Republic. *Bulletin of Entomological Research*, **66**, 345–63.

Pegram, R.G., Hoogstraal, H. & Wassef, H.Y. (1981). Ticks (Acari: Ixodoidea) of Ethiopia. I. Distribution, ecology and host relationships of species infesting livestock. *Bulletin of Entomological Research*, **71**, 339–59.

Theiler, G. & Robinson, B.N. (1953). Ticks in the South African Zoological Survey Collection. Part VII. Six lesser known African rhipicephalids. *Onderstepoort Journal of Veterinary Research*, **26**, 93–136.

Walker, J.B. (1956). *Rhipicephalus pravus* Dönitz, 1910. *Parasitology*, **46**, 243–60.

Wiley, A.J. (1953). Notes on animal diseases. XXV. Common ticks of livestock in Kenya. *East African Agricultural Journal*, **19**, 1–6.

Wilson, S.G. (1951). Report by Chief Field Zoologist. Ticks and tick-borne diseases. Tick collections. *Report of the Department of Veterinary Services, Kenya, for 1951*, pp. 1–10. Mimeographed.

Zumpt, F. (1942). *Rhipicephalus appendiculatus* Neum. und verwandte Arten. VI. Vorstudie zu einer Revision der Gattung *Rhipicephalus* Koch. *Zeitschrift für Parasitenkunde*, **12**, 538–51.

Zumpt, F. (1958). A preliminary survey of the distribution and host-specificity of ticks (Ixodoidea) in the Bechuanaland Protectorate. *Bulletin of Entomological Research*, **49**, 201–23.

Also see the following Basic References (pp. 12–14): Clifford & Anastos (1962); Elbl & Anastos (1966); Hoogstraal (1956); Matthysse & Colbo (1987); Morel (1969, 1980); Santos Dias (1960); Scaramella (1988); Theiler (1962); Walker (1974); Walker, Mehlitz & Jones (1978); Yeoman & Walker (1967).

RHIPICEPHALUS PSEUDOLONGUS SANTOS DIAS, 1953

This specific name, from the Greek *pseudes* meaning 'false' plus *longus*, draws attention to the similarity between this species and *R. longus.*

Synonym

cliffordi.

Diagnosis

A large dark-brown to black heavily and evenly punctate species whose males have sickle-shaped adanal plates.

Table 43. *Host records of* Rhipicephalus pseudolongus

Hosts	Number of records
Domestic animals	
Cattle	5
Wild animals	
African buffalo (*Syncerus caffer*)	48
Waterbuck (*Kobus ellipsiprymnus*)	1
'Murid rodent'	1

Male (Figs 165(a), 166(a) to (c))

Capitulum longer than broad, length × breadth ranging from 0.80 mm × 0.76 mm to 1.10 mm × 1.00 mm. Basis capituli with acute lateral angles in anterior third of its length. Palps short, broad, slightly flattened apically. Conscutum length × breadth ranging from 3.30 mm × 2.00 mm to 4.38 mm × 2.79 mm; anterior process of coxae I small. In engorged specimens body wall ex-

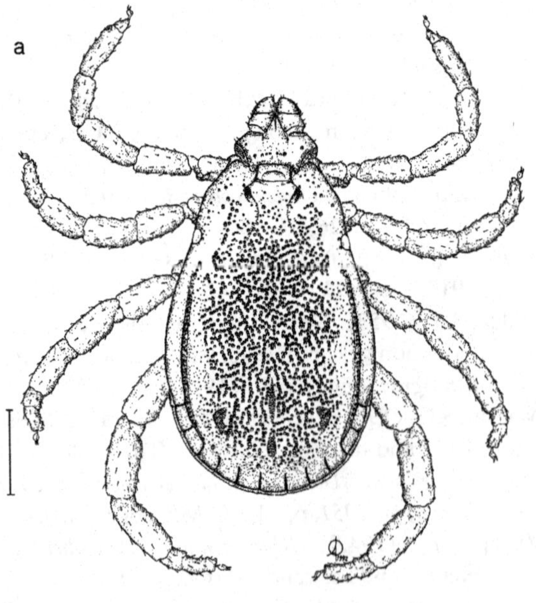

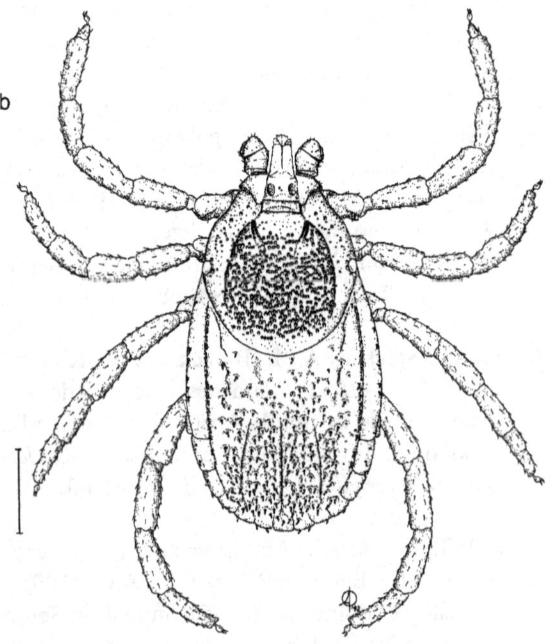

Figure 165. *Rhipicephalus pseudolongus* [collected from 'buffle', presumably African buffalo (*Syncerus caffer*), 'Est Centrafrique', in 1970, donated by P.C. Morel]. (a) Male, dorsal; (b) female, dorsal. Scale bars represent 1 mm. A. Olwage *del.*

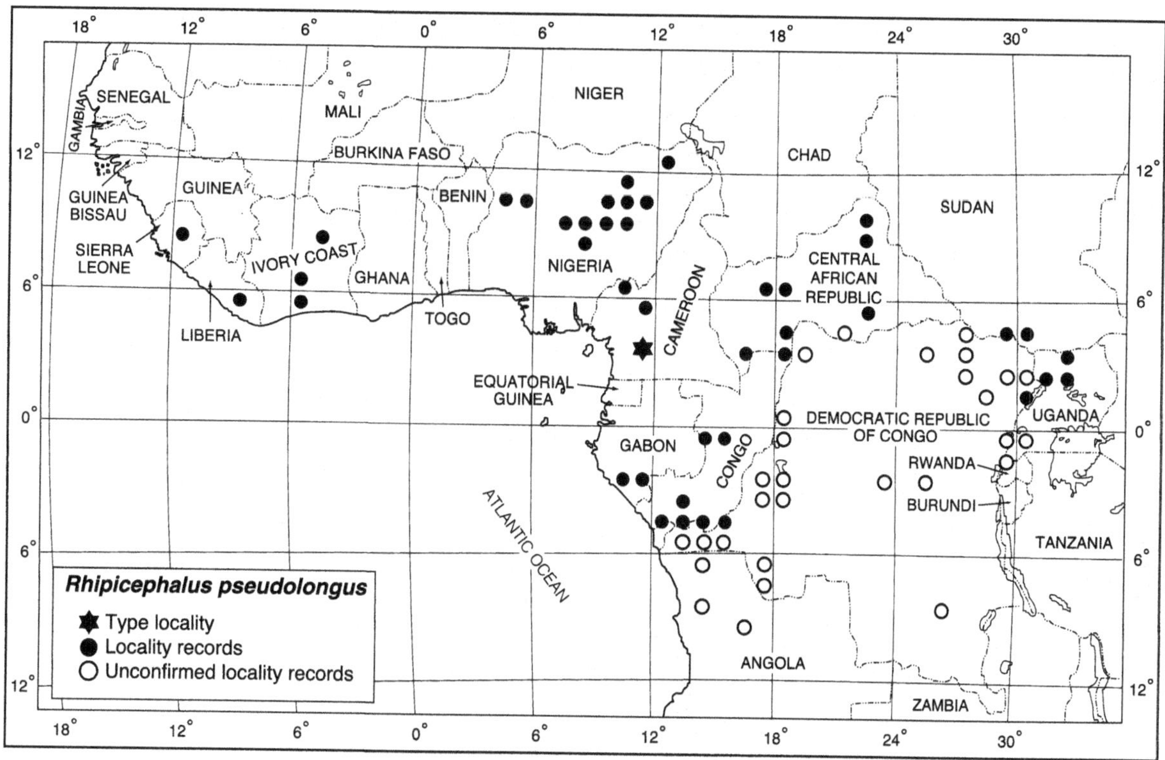

Map 48. *Rhipicephalus pseudolongus*: distribution. (Based largely on Morel, 1969, as *R. cliffordi*).

panded laterally and posteriorly, with three caudal processes protruding posteromedially. Eyes almost flat, delimited dorsally by shallow depressions. Cervical pits deep, convergent. Marginal lines long, but not quite reaching eye level, punctate. Posteromedian and posterolateral grooves quite small, their floors rugose. A few medium-sized setiferous punctations anteriorly on the scapulae.The areas surrounding the eyes, and adjacent to the marginal lines and festoons, almost smooth, otherwise the conscutum is virtually covered with a dense pattern of evenly sized punctations that often become confluent. Ventrally spiracles broad, with a short broad prolongation angled towards the dorsal surface. Adanal plates typically large, sickle-shaped, curving in towards each other posterior to the anus; accessory adanal plates quite large, pointed, well sclerotized.

Female (Figs 165(b), 166(d) to (f))

Capitulum varies from broader than long to longer than broad, length × breadth ranging from 0.64 mm × 0.73 mm to 1.12 mm × 1.03 mm. Basis capituli with broad lateral angles in anterior third of its length; porose areas large, oval, about 1.5 times their own diameter apart. Palps broad, almost flat apically. Scutum varies from about as broad as long to broader than long, length × breadth ranging from 1.34 mm × 1.33 mm to 2.14 mm × 2.39 mm; posterior margin broadly rounded. Eyes about halfway back, almost flat, delimited dorsally by shallow depressions. Cervical fields broad. A few medium-sized setiferous punctations interspersed with fine punctations scattered on scapulae. Medially the scutum is densely covered with medium-sized punctations that are often confluent. Alloscutum of unfed females deeply folded. Ventrally genital aperture broadly V-shaped.

Nymph (Fig. 167)

Capitulum much broader than long, length × breadth ranging from 0.25 mm × 0.31 mm to 0.32 mm × 0.39 mm. Basis capituli three times

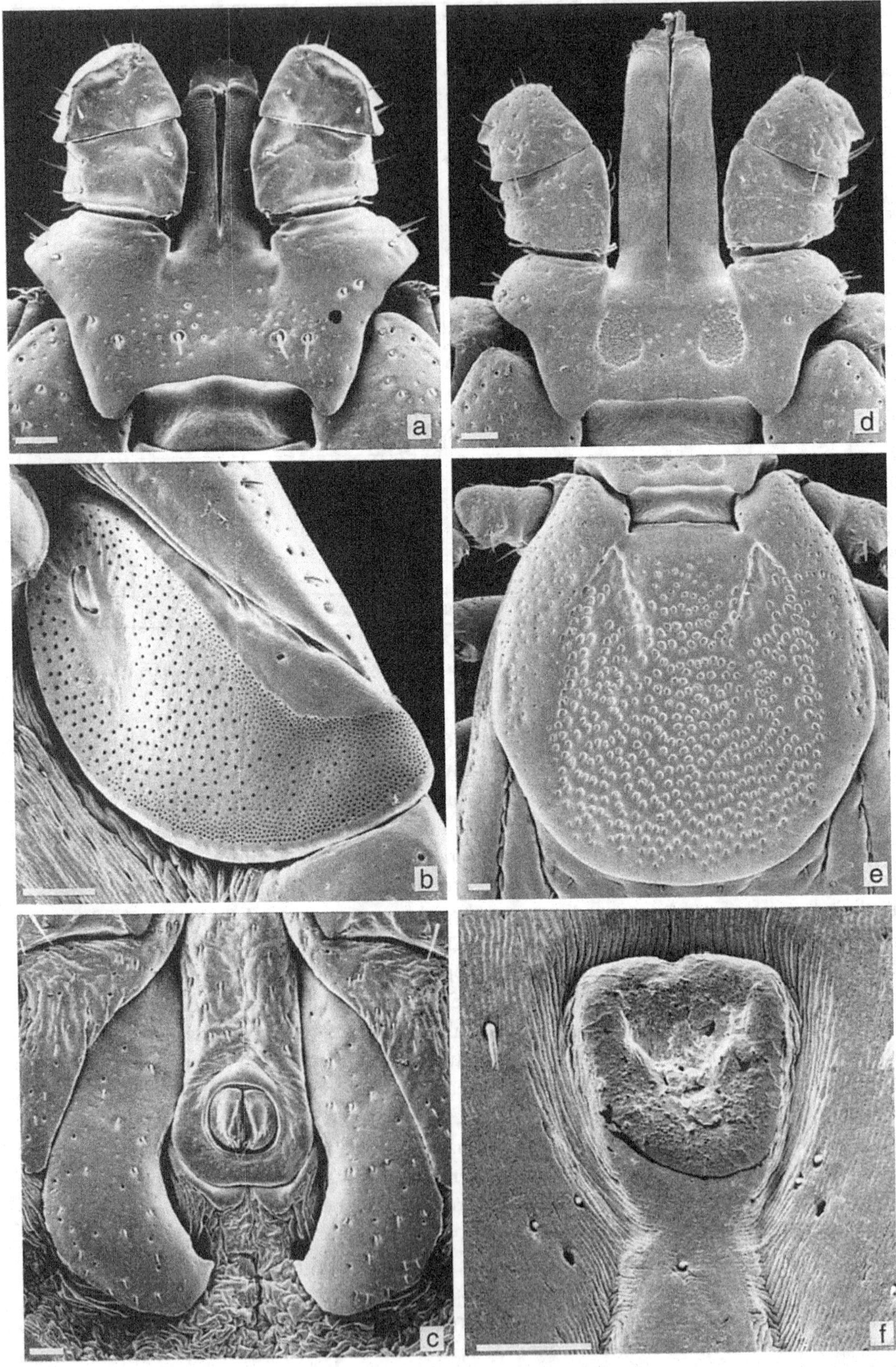
a
d
b
e
c
f

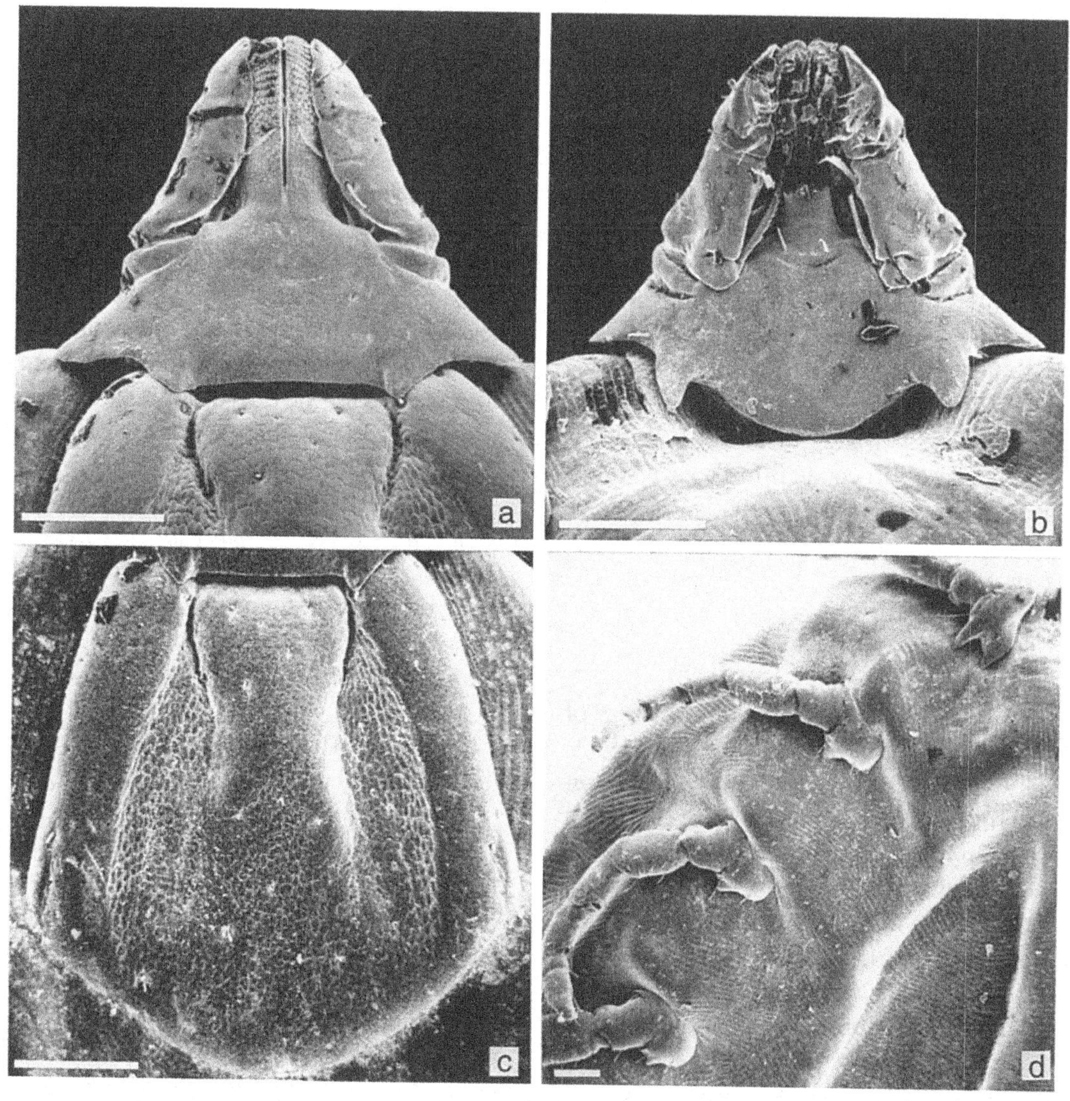

Figure 167 (*above*). *Rhipicephalus pseudolongus* (RML 36798, collected from rodent nest, Garamba Park, Democratic Republic of Congo, on 23 December 1950 by J. Verschuren). Nymph: (a) capitulum, dorsal; (b) capitulum, ventral; (c) scutum; (d) coxae. Scale bars represent 0.10 mm. SEMs by J.F. Putterill.

Figure 166 (*opposite*). *Rhipicephalus pseudolongus* [collected from 'buffle', presumably African buffalo (*Syncerus caffer*), 'Est Centrafrique', in 1970, donated by P.C. Morel]. Male: (a) capitulum, dorsal; (b) spiracle; (c) adanal plates. Female: (d) capitulum, dorsal; (e) scutum; (f) genital aperture. Scale bars represent 0.10 mm. SEMs by J.F. Putterill.

as broad as long, with broad tapering lateral angles extending over the scapulae and well-developed cornua; ventrally with sharp spurs on posterior margin. Palps equal in width for much of their length, tapering to rounded apices. Scutum virtually as broad as long to broader than long, length × breadth ranging from 0.47 mm × 0.51 mm to 0.55 mm × 0.56 mm; posterior margin smoothly curved. Eyes at widest point, in posterior third of the scutal length, long and narrow. Cervical fields well defined, long, narrow, divergent. Ventrally coxae I each with a relatively short triangular internal spur and a longer triangular external spur; coxae II to IV each with a small triangular external spur only.

Larva

Unknown.

Notes on identification

Pegram & Walker (1988) outlined the history of the name *R. pseudolongus*. Santos Dias (1953) originally described it as a subspecies of *R. capensis*, then raised it to specific rank in 1955, but finally synonymized it with *R. longus* in 1958. However, Clifford & Anastos (1962) re-examined the holotype male of *R. pseudolongus*, stated that it appeared to them to be entirely different from *R. longus* (*sensu* Neumann) and reinstated it as a valid species. Morel (1965) disagreed with this decision. He described a new species, *R. cliffordi*, with *R. pseudolongus sensu* Clifford & Anastos (1962) as its synonym, and sank *R. pseudolongus sensu* Santos Dias as a synonym of *R. longus*. He based his decisions on the hosts and ecology as well as the morphology of these ticks as he saw them but he did not re-examine the holotype of *R. pseudolongus*.

The situation was complicated further by Elbl & Anastos (1966), who were apparently unaware of Morel's findings the previous year. They synonymized *R. pseudolongus* with *R. compositus*, and there is no doubt that these two species are closely related. The major difference between them is that the larger *R. pseudolongus* males have markedly sickle-shaped adanal plates whereas *R. compositus* males do not, but runt males of the two, and the females whatever their size, are almost inseparable. When unmounted, even the female genital apertures of the two are virtually the same shape. Nevertheless we do not think that they are synonymous.

We feel that further study is required before final decisions regarding the nomenclature of these ticks can be made. This should include re-examination of the types of the various entities; we have not seen these types ourselves. For the present we have retained the name *R. pseudolongus* for the heavily-punctate ticks whose males have sickle-shaped adanal plates, as described by Clifford & Anastos (1962).

Besides the *R. pseudolongus* males those of *R. longus* and *R. senegalensis* also have sickle-shaped adanal plates. The differences between these three species, and the difficulties that can arise when trying to separate them, have been summarized earlier (see *R. longus*, p. 255).

Hosts

Life cycle unknown. The only domestic animals from which *R. pseudolongus* adults have been obtained are cattle. The overwhelming majority of collections have been made from the African buffalo (Table 43). Morel (1969) specified the smaller red forest buffalo (*Syncerus caffer nanus*) in particular but Matthysse & Colbo (1987) stated that all their collections were made from the larger black buffalo. Other possible hosts referred to by Morel are the bushpig, which in this context is probably the red river hog (*Potamochoerus porcus*), and various antelopes, including the waterbuck, living in the same habitat. (Matthysse & Colbo, 1987 also quote Morel, 1969 as having mentioned the warthog (*Phacochoerus africanus*) as a host but this is apparently in error).

The immature stages evidently feed on rodents, and possibly macroscelids. Clifford & Anastos (1964) list numerous collections from rodent nests in the Garamba National Park, one of them from a striped grass rat's nest (*Lemniscomys striatus*).

In Garamba adults were found on the vegetation throughout the year. Nymphs were collected from rodent nests from December to June, especially in January, though this could simply represent the period during which these nests were being excavated (Clifford & Anastos, 1964). The predilection sites of this tick's adults on its hosts have not been recorded.

Zoogeography

Rhipicephalus pseudolongus (referred to as *R. cliffordi*) has been recorded from Sierra Leone eastwards as far as western Uganda (Morel, 1969) (Map 48). He also included, as '*R. cliffordi* presumé', reports from the Democratic Republic of Congo, mostly in the north and west of the country, and seemingly extracted from the records of *R. compositus* listed by Elbl & Anastos (1966), plus three reports from northern Angola.

This picture does present some anomalies. Morel (1969) suggested that the distribution of *R. pseudolongus* is not influenced so much by altitude, since it has been found from near sea level to between 500 m and 1000 m, as by rainfall. He noted that the annual rainfall in the areas where it occurs is over 1250 mm, distributed during a period of over 6 months. While this is often true the records indicate that some places are drier. According to Morel the vegetation in the habitat favoured by its hosts is equatorial lowland rainforest, especially the dense gallery forest and thickets along the rivers and the edges of swampy areas, and various types of woodland. Matthysse & Colbo (1987), however, collected it only from the large black buffalo, which favours a somewhat more open habitat including forest mosaics and woodland.

Disease relationships

Unknown.

REFERENCES

Morel, P.C. (1965). Description de *Rhipicephalus cliffordi* n.sp. d'Afrique occidentale (groupe de *Rh. compositus*; Acariens, Ixodoidea). *Revue d'Élevage et de Médecine Vétérinaire des Pays Tropicaux*, **17 for 1964** (nouvelle série), 637–54.

Pegram, R.G. & Walker, J.B. (1988). Clarification of the biosystematics and vector status of some African *Rhipicephalus* species (Acarina: Ixodidae). In *Biosystematics of Haematophagous Insects*, ed. M.W. Service, pp. 61–76. Systematics Association Special Volume No. **37**. Oxford: Clarendon Press.

Santos Dias, J.A.T. (1953). Sobre uma nova subespécie de *Rhipicephalus* do 'Grupo *capensis*' Zumpt: *R. capensis pseudolongus* n. ssp. *Memórias e Estudos do Museu Zoológico da Universidade de Coimbra*, **No. 214**, 1–15.

Santos Dias, J.A.T. (1955). A propósito de uma colecção de carraças do Sudão Anglo-Egipcio. Algumas considerações sobre o *Rhipicephalus longus* Neumann, 1907. *Boletim da Sociedade de Estudos de Moçambique*, **No. 92**, 103–18.

Santos Dias, J.A.T. (1958). Notes on various ticks (Acarina-Ixodoidea) in collection at some entomological institutes in Paris and London. *Anais do Instituto de Medicina Tropical*, **15**, 459–563.

Also see the following Basic References (pp. 12–14): Clifford & Anastos (1962, 1964); Elbl & Anastos (1966); Matthysse & Colbo (1987); Morel (1969).

RHIPICEPHALUS PULCHELLUS (GERSTÄCKER, 1873)

The specific name *pulchellus*, a Latin term meaning 'beautiful', refers to the ornate scutal pattern of this species, especially the male.

Synonym

marmoreus.

Diagnosis

A medium-sized to large tick whose adults have a characteristic black-and-ivory pattern on their scuta.

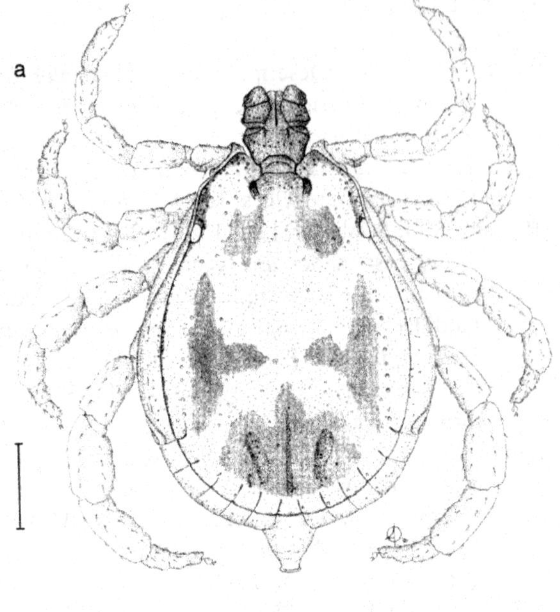

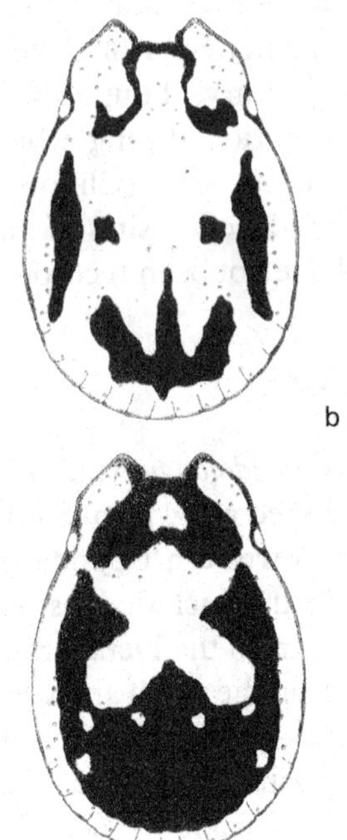

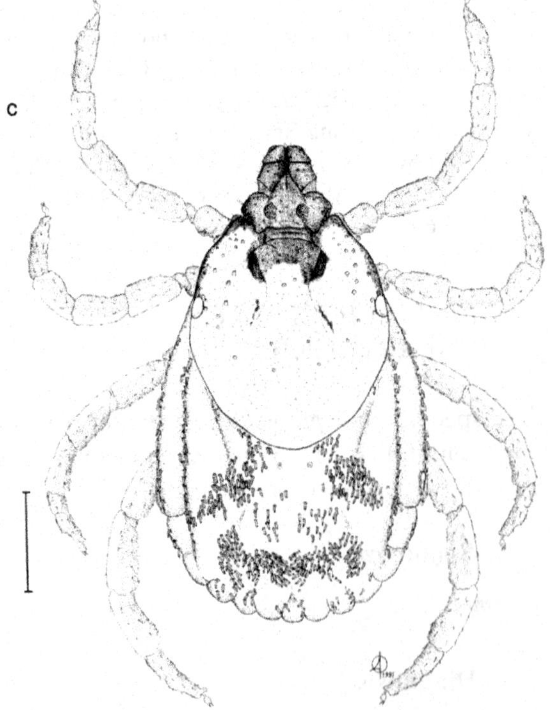

Figure 168. *Rhipicephalus pulchellus* [♂ collected from Burchell's zebra (*Equus burchellii*) at Mile 70 south of Arusha on Great North Road, Tanzania, in 1956 by W. Hilton; ♀ from B.S. 526/-, RML 66313, laboratory reared, original ♀ collected from an unknown host at Kiboko, Kenya, in *c.* 1950 by S.G. Wilson]. (a) Male, dorsal; (b) male, variations in colour pattern, after Cunliffe (1913); (c) female, dorsal. Scale bars represent 1 mm. A. Olwage *del.*

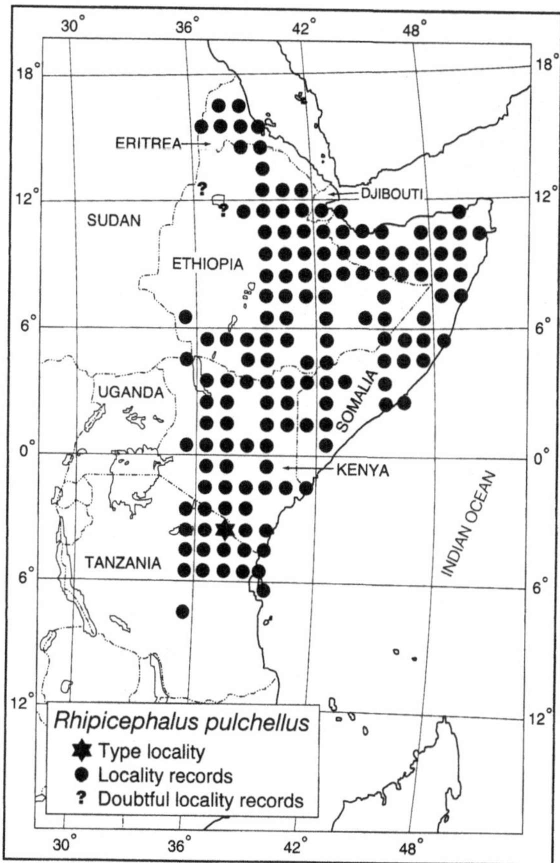

Map 49. *Rhipicephalus pulchellus*: distribution.

Male (Figs 168(a), 169(a) to (c))

Capitulum much longer than broad, length × breadth ranging from 0.84 mm × 0.69 mm to 1.07 mm × 0.83 mm. Basis capituli with very short, rounded lateral angles in anterior third of its length. Palps short, broad. Conscutum length × breadth ranging from 3.44 mm × 2.29 mm to 4.61 mm × 3.14 mm; large strongly-sclerotized anterior process present on coxae I. In engorged specimens body wall expanded laterally and forming a single stout caudal process. Eyes marginal, almost flat, delimited by slight depressions and a few punctations dorsally. Cervical pits deep, comma-shaped. Marginal lines represented merely by rows of large setiferous punctations. Posteromedian groove long and narrow; posterolateral grooves shorter and broader. A few large setiferous punctations present on the scapulae and scattered sparsely elsewhere on the conscutum, interspersed with numerous very fine punctations. The most conspicuous feature of this tick is its striking black pattern on an ivory-coloured background. The dark areas usually comprise two cervical patches that are continuous anterolaterally with the dark edging on the external margins of the scapulae; two lateral patches, and a single large posteromedian patch anterior to the festoons. Much less commonly the extent of these dark areas may be either increased or reduced. Legs increase markedly in size from I to IV, with light-coloured mottling dorsally. Ventrally spiracles comma-shaped, tapering gradually and curving gently towards the dorsal surface. Adanal plates broad, coming to a point posteroexternally, smoothly rounded posterointernally; accessory adanal plates absent.

Female (Figs 168(b), 169(d) to (f))

Capitulum slightly longer than broad, length × breadth ranging from 0.80 mm × 0.77 mm to 1.02 mm × 0.98 mm. Basis capituli with broad lateral angles at about anterior third of its length; porose areas large, round, slightly more than their own diameter apart. Palps long, truncated apically. Scutum longer than broad, length × breadth ranging from 1.82 mm × 1.68 mm to 2.50 mm × 2.17 mm. Eyes marginal, almost flat, delimited by slight depressions and a few punctations dorsally. Cervical pits convergent, continuous with shallow divergent internal cervical margins. Large setiferous punctations present on the scapulae and widely scattered medially on the scutum, interspersed with numerous very fine punctations. Scutum predominantly ivory coloured, with dark ornamentation between the cervical pits and extending anterolaterally along the external margins of the scapulae as far as the eyes. Alloscutum dark brown; in unfed specimens a dense pattern of stout white setae present, which may be shed as the tick engorges. Legs with light-coloured mottling dorsally. Ventrally genital aperture broadly U-shaped.

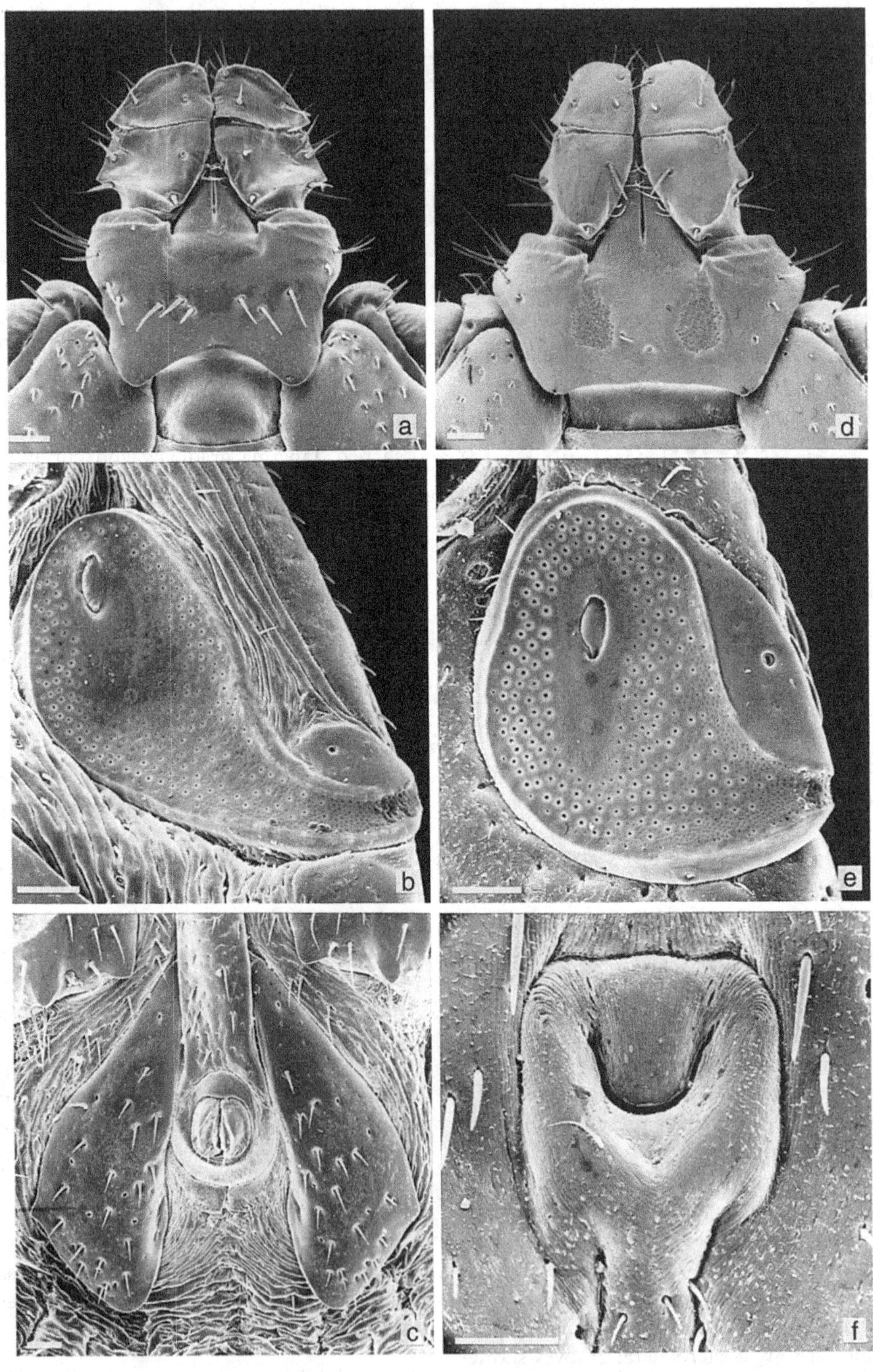
a
d
b
e
c
f

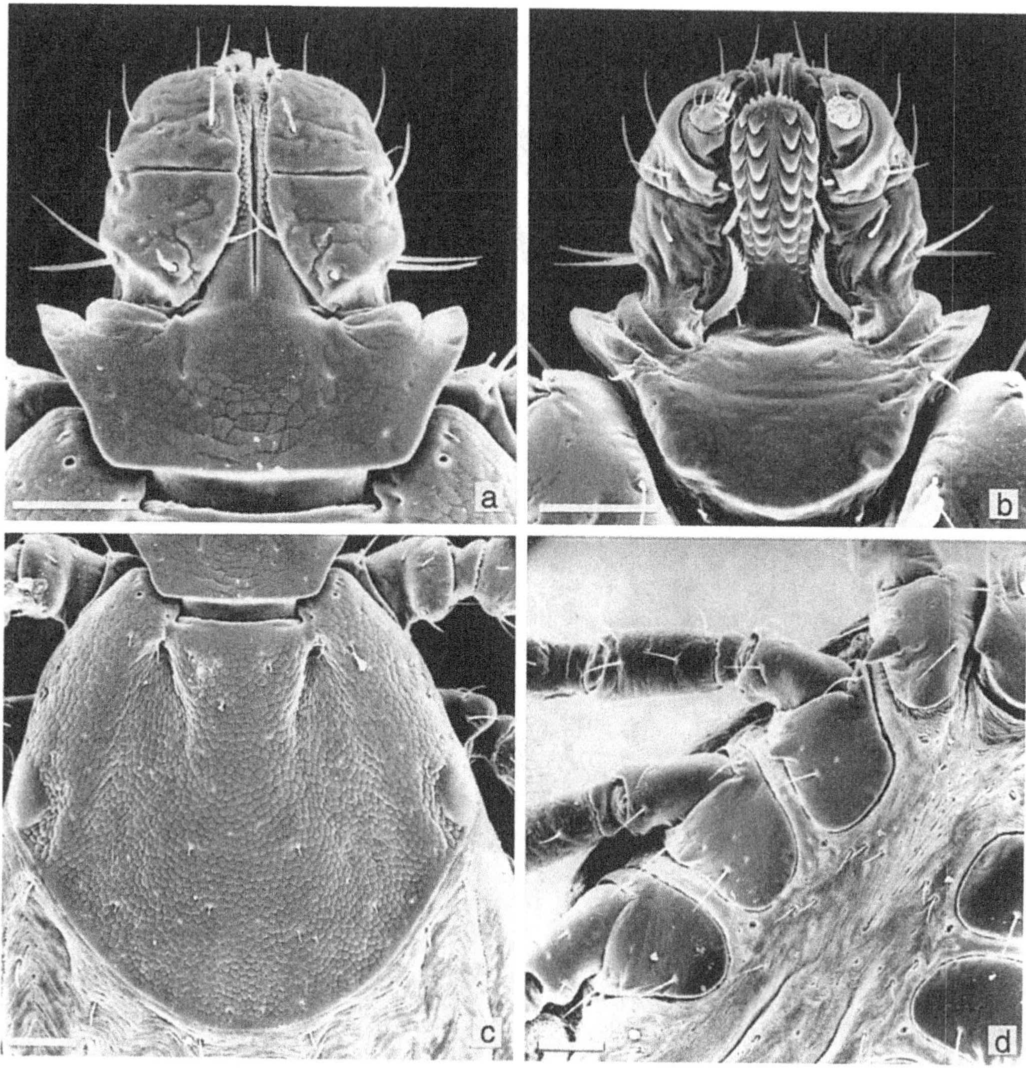

Figure 170 (*above*). *Rhipicephalus pulchellus* (B.S. 526/-, RML 66313, laboratory reared, original ♀ collected from an unknown host at Kiboko, Kenya, in *c.* 1950 by S.G. Wilson). Nymph: (a) capitulum, dorsal; (b) capitulum, ventral; (c) scutum; (d) coxae. Scale bars represent 0.10 mm. SEMs by M.D. Corwin.

Figure 169 (*opposite*). *Rhipicephalus pulchellus* (B.S. 526/-, RML 66313, laboratory reared, original ♀ collected from an unknown host at Kiboko, Kenya, in *c.* 1950 by S.G. Wilson). Male: (a) capitulum, dorsal; (b) spiracle; (c) adanal plates. Female: (d) capitulum, dorsal; (e) spiracle; (f) genital aperture. Scale bars represent 0.10 mm. SEMs by M.D. Corwin.

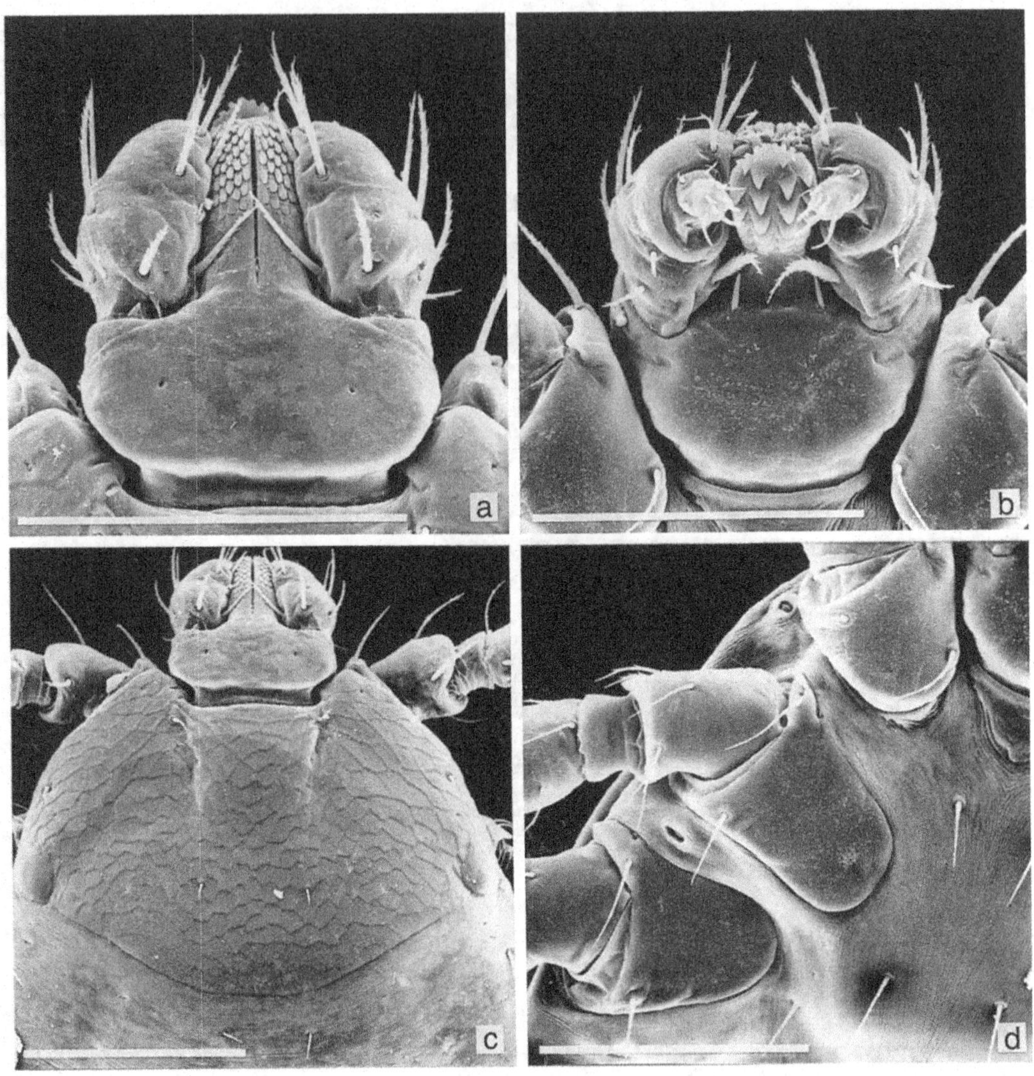

Figure 171. *Rhipicephalus pulchellus* (B.S. 526/-, RML 66313, laboratory reared, original ♀ collected from an unknown host at Kiboko, Kenya, in *c.* 1950 by S.G. Wilson). Larva: (a) capitulum, dorsal; (b) capitulum, ventral; (c) scutum; (d) coxae. Scale bars represent 0.10 mm. SEMs by M.D. Corwin.

Nymph (Fig. 170)

Capitulum broader than long, length × breadth ranging from 0.29 mm × 0.32 mm to 0.33 mm × 0.35 mm. Basis capituli just over 2.5 times as broad as long, lateral angles anteriorly placed, short and forwardly directed; ventrally with two short broad spurs on posterior margin. Palps broad, rounded apically. Scutum slightly broader than long, length × breadth ranging from 0.65 mm × 0.68 mm to 0.69 mm × 0.71 mm; posterior margin a deep smooth curve. Eyes at widest point, about halfway back, delimited dorsally by shallow depressions. Cervical fields broad, depressed, their external margins almost reaching posterolateral margins of scutum. Ventrally coxae I each with a long, narrow external spur and a shorter, broader internal spur; coxae II to IV each with a small external spur only, decreasing progressively in size.

Larva (Fig. 171)

Capitulum about as long as broad, length × breadth ranging from 0.134 mm × 0.142 mm to

0.151 mm × 0.151 mm. Basis capituli just over twice as broad as long, slightly convex laterally. Palps broad, flattened apically. Scutum much broader than long, broadest over halfway back at eye level, length × breadth ranging from 0.272 mm × 0.401 mm to 0.289 mm × 0.423 mm; posterior margin a smooth shallow curve. Eyes large. Cervical grooves almost parallel. Ventrally coxae I to III each with an indication only of a spur on its posterior margin.

Notes on identification

Rhipicephalus pulchellus was misidentified as *Dermacentor pulchellus* by Pavesi (1898). It was described as a new species, *R. marmoreus*, by Pocock (1900).

Rhipicephalus pulchellus males, with their unique colour pattern, are easy to identify but it is much more difficult to distinguish the females, nymphs and larvae from those of *R. humeralis*. This problem is discussed under *R. humeralis* (p. 214). The variability in the size, morphology and colour patterns of field-collected specimens of *R. pulchellus* was reviewed in detail by Cunliffe (1913).

Hosts

A three-host species (Walker, 1955). *Rhipicephalus pulchellus* has been recorded from an extremely wide range of both domestic and wild animals, especially ungulates (Table 44). In some parts of eastern Ethiopia and northern Somalia it is the commonest tick on domestic livestock (Pegram, 1976, 1979; Pegram, Hoogstraal & Wassef, 1981; Morel, 1980).

Approximate counts made *in situ* on individual cattle in Borana District, Sidamo, Ethiopia, revealed infestations of over 1000 adults per beast. In Harar Province it often constituted over 90% of the ticks obtained from cattle and camels. In Tanzania, though, collections of *R. pulchellus* adults from cattle were small, the maximum single infestation being only 28 ticks. On cattle, sheep and camels this tick attaches primarily on the ears and the underparts of the body, including the chest, belly, genital and anal areas, and also on the legs. It completely covered the ears and perianal areas of heavily-infested cattle examined in Ethiopia.

Amongst wild animals all stages of *R. pulchellus* parasitize various large ungulates. They are abundant on zebras: many of the Burchell's zebras and, in Kenya, all the Grevy's zebras examined were infested. The black rhinoceros is also a favourite host, as are the larger antelopes, for example the gemsbok, eland and various hartebeests. The smaller antelopes, such as the gazelles and Günther's dik-dik, may be parasitized by the immature stages only. This distinction also applies to the carnivores, with the larger species carrying all stages of this tick but the smaller ones only the immatures. Hares sometimes act as hosts of the immature stages. In a detailed survey carried out in the southern Rift Valley, Kenya, on the Cape hare, 84 out of 124 of these animals examined at Ololkisalie (= Olorgesailie) were infested by a total of 161 nymphs, 11 larvae of *R. pulchellus*. A little further north at Akira Ranch, though, only 2 out of 121 of these hares carried a single nymph each. Five nymphs, eight larvae were also recorded from two savanna hares obtained in nearby areas (Clifford, Flux & Hoogstraal, 1976).

Humans are undoubtedly far more frequently infested by *R. pulchellus* than our records suggest. Large numbers, particularly of the immature stages, sometimes attach on the legs of people walking through long grass in areas where it occurs.

Seasonal data on *R. pulchellus* from eastern Ethiopia and northern Somalia showed that it is most active during the rainy season. At Ololkisalie, though, the infestations of *R. pulchellus* immatures on hares apparently peaked during the dry seasons before and after the February to May rainy period.

Zoogeography

Rhipicephalus pulchellus is one of the commonest ticks occurring in the Horn of Africa, almost exclusively in and east of the Rift Valley, from Eritrea in the north as far south as north-eastern Tanzania (Map 49).

Table 44. *Host records of* Rhipicephalus pulchellus

Hosts	Number of records
Domestic animals	
Cattle	649 (including immatures)
Sheep	144 (including immatures)
Goats	133 (including immatures)
Camels	110
Horses	11
Donkeys	6
Mules	3
Pigs	5
Dogs	13
Wild animals	
Somali hedgehog (*Atelerix sclateri*)	1
Lesser mouse-tailed bat (*Rhinopoma hardwickei*)	1
Cape serotine (*Eptesicus capensis*)	1
Yellow baboon (*Papio cynocephalus*)	1
'Baboon' (*Papio* sp.)	1
Black-backed jackal (*Canis mesomelas*)	3 (2 include nymphs)
Hunting dog (*Lycaon pictus*)	2
Bat-eared fox (*Otocyon megalotis*)	2 (nymphs only)
Cheetah (*Acinonyx jubatus*)	1
African wild cat (*Felis lybica*)	2 (immatures)
Lion (*Panthera leo*)	7 (2 include immatures)
Leopard (*Panthera pardus*)	2 (includes nymphs)
Slender mongoose (*Galerella sanguinea*)	2
White-tailed mongoose (*Ichneumia albicauda*)	1
Striped hyaena (*Hyaena hyaena*)	4 (1 includes nymphs)
'Hyaena'	3
Aardwolf (*Proteles cristatus*)	2 (1 with nymph only)
Small-spotted genet (*Genetta genetta*)	1
Large-spotted genet (*Genetta tigrina*)	1
African elephant (*Loxodonta africana*)	3
Burchell's zebra (*Equus burchellii*)	41 (including immatures)
Grevy's zebra (*Equus grevyi*)	16 (including immatures)
'Zebra'	11
Black rhinoceros (*Diceros bicornis*)	34 (1 includes nymphs)
Aardvark (*Orycteropus afer*)	2
Warthog (*Phacochoerus* sp.)	19
Bushpig (*Potamochoerus larvatus*)	2
'Wild pig'	1
Giraffe (*Giraffa camelopardalis*)	8
Impala (*Aepyceros melampus*)	13 (including immatures)
Coke's hartebeest (*Alcelaphus buselaphus cokii*)	30 (2 include nymphs)
'Kenya hartebeest' (*A. b. cokii*/*A. b. jacksoni* intergrades)	3 (1 includes immatures)
'Hartebeest'	2
Topi (*Damaliscus lunatus topi*)	5
Dibatag (*Ammodorcas clarkei*)	1

Table 44. (*cont.*)

Hosts	Number of records
Wild animals (*cont.*)	
Grant's gazelle (*Gazella granti*)	15 (including nymphs)
Soemmerring's gazelle (*Gazella soemmerringii*)	7 (1 includes nymphs)
Speke's gazelle (*Gazella spekei*)	1
Thomson's gazelle (*Gazella thomsonii*)	12 (including immatures)
'Gazelle'	4
Gerenuk (*Litocranius walleri*)	10 (1 includes nymphs)
Günther's dik-dik (*Madoqua guentheri*)	2 (nymphs)
Silver dik-dik (*Madoqua piacentinii*)	1
African buffalo (*Syncerus caffer*)	20 (1 includes nymphs)
Eland (*Taurotragus oryx*)	33 (2 include nymphs)
Lesser kudu (*Tragelaphus imberbis*)	2
Bushbuck (*Tragelaphus scriptus*)	1
Greater kudu (*Tragelaphus strepsiceros*)	6
'Kudu'	2
Red forest duiker (*Cephalophus natalensis*)	3 (1 with nymph only)
Sable antelope (*Hippotragus niger*)	1
Gemsbok (*Oryx gazella*)	38 (including immatures)
Waterbuck (*Kobus ellipsiprymnus*)	6
Huet's bush squirrel (*Paraxerus ochraceus*)	1
Black-tailed gerbil (*Tatera nigricauda*)	1
Spiny mouse (*Acomys* sp.)	1
Black rat (*Rattus rattus*)	1
Cape hare (*Lepus capensis*)	48 (immatures, mostly nymphs)
Ethiopian hare (*Lepus fagani*)	1
Savanna hare (*Lepus victoriae*)	5 (immatures)
'Hare'	1 (nymphs only)
Birds	
Ostrich (*Struthio camelus*)	1
Yellow-necked francolin (*Francolinus leucoscepus*)	1
Humans	40

There is little doubt that its absence from large parts of south-eastern Ethiopia, southern Somalia and north-eastern Kenya merely reflects a lack of collections from these remote areas. However, the reasons for its apparent inability to extend its distribution to any significant extent west of the Rift Valley remain obscure. In northern Ethiopia Pegram *et al.* (1981) referred to two isolated records, indicated by question marks both on their map and ours, from west of the Rift Valley in Begemder and Gojam Provinces. They commented that this tick was absent from other extensive collections made from domestic livestock in Begemder. They suggested, though, that: 'At the northern end of the valley in Ethiopia, camel caravans passing from the east coast to western Ethiopia and the Sudan could easily carry numerous *R. pulchellus* west of the valley.' Further south, in Kenya, there are only occasional records of small numbers of this tick from west of the Rift Valley. It was suggested by Lewis (1934) that this might be because major movements of both livestock and wild animals were up and down the valley (i.e. northwards and southwards), not across it. A.J. Wilsmore (pers. comm. to R.G. Pegram) questioned this; he believes there is considerable seasonal movement of Maasai cattle across the Rift. It is noteworthy that *R. pulchellus* was not found on 12 Cape hares collected in the Lolgorien-Kilgoris area west of the valley in southern Kenya (Clifford *et al.*, 1976).

Broadly speaking *R. pulchellus* occurs at altitudes below 2000 m, only occasionally above this, with annual rainfall from 100 mm to 800 mm. It is associated primarily with either Somalia-Masai semi-desert vegetation or Somalia-Masai *Acacia-Commiphora* deciduous bushland and thicket, though it does not occur throughout the extensions of the latter vegetation type into north-western Kenya and the western Serengeti in Tanzania. It can apparently survive in harsher environmental conditions than the related species *R. humeralis*, with which it is sympatric in parts of its range (see p. 214).

Disease relationships

Rhipicephalus pulchellus has been associated with a wide variety of pathogenic organisms affecting both animals and man.

Theileria parva was transmitted by *R. pulchellus* in only 1 out of the 14 experiments attempted by Brocklesby (1965). Earlier experiments by Fotheringham & Lewis (1937) gave negative results. Since the areas where East Coast fever occurs lie outside the distribution zone of this tick it apparently does not act as a field vector of the disease.

Schizonts of a *Theileria* sp., possibly *T. taurotragi*, were found in the salivary glands of *R. pulchellus* taken from an eland (*Taurotragus oryx*). Adults from the same batch of ticks transmitted a *Theileria* sp. that was neither *T. parva* nor *T. mutans* to cattle, and produced only a transitory parasitaemia (Brocklesby, 1965).

Schizonts identified as those of *Babesia equi* were found in the salivary glands of *R. pulchellus* adults taken from zebras (Brocklesby, 1965). Naturally acquired infections of *B. equi* have been diagnosed in both Burchell's and Grevy's zebras (Dennig, 1965).

Trypanosoma theileri was found in the tissues and haemolymph of 19 out of 258 *R. pulchellus* adults from cattle near Negelli Borana, Sidamo Province, Ethiopia (Burgdorfer, Schmidt & Hoogstraal, 1973). Pegram (1975), who noted that *T. theileri* is usually regarded as being non-pathogenic, speculated that it might be involved in a chronic wasting disease, characterized by anaemia and known locally as 'Luta', seen in other cattle in the same area. Their symptoms were suggestive of trypanosomiasis but neither *T. vivax* nor *T. congolense* was detected in 'several hundred examinations' made. Pegram added that the large numbers of *R. pulchellus* on these cattle must themselves have contributed to the animals' anaemia and illthrift. (A mineral deficiency was also suspected.)

Nairobi sheep disease virus was first reported to have been transmitted experimentally by *R. pulchellus* by Lewis (1949). Trans-stadial and transovarial transmission of the virus through the tick was subsequently demonstrated by Pellegrini (1950). Epidemiological evidence linking *R. pulchellus* with outbreaks of the disease in northern Somalia has subsequently been documented (Edelsten, 1975; Pegram, 1975). The gastroenteritis that it causes is aggravated by pneumonia caused by *Pasteurella haemolytica*.

Evidence that *R. pulchellus* adults collected in Ethiopia were naturally infected with *Rickettsia conori*, one of the causative agents of human tick typhus, was presented by Philip *et al.* (1966). The bites alone of this tick may cause problems in humans: they cause irritating inflamed sores that, if left untreated, are liable to suppurate (Lewis, 1934).

Cowdry (1925) recorded a natural infection by a *Wolbachia* species in the Malpighian tubules of *R. pulchellus*. However, recent observations by Noda, Munderloh & Kurtti (1997), who studied 'endosymbiotic bacteria from four species of specific-pathogen-free ticks', suggest that the identity of this organism should be reconsidered. Noda *et al.* concluded that the organisms they found 'are closely related to bacterial pathogens transmitted by ticks', not to the *Wolbachia* spp. found in insects.

Crimean-Congo haemorrhagic fever virus of man was first reported to have been isolated from naturally infected *R. pulchellus* taken from a dead sheep in Kenya (Hoogstraal, 1975, *in litt.*, quoted by Morel, 1980). It was later isolated again from naturally infected ticks collected in Ethiopia (Wood *et al.*, 1978).

Dugbe virus was also isolated from naturally infected *R. pulchellus* collected in Ethiopia (Wood *et al.*, 1978).

REFERENCES

Brocklesby, D.W. (1965). Evidence that *Rhipicephalus pulchellus* (Gerstaecker, 1873) may be a vector of some piroplasms. *Bulletin of Epizootic Diseases of Africa*, **13**, 37–44.

Burgdorfer, W., Schmidt, M.L. & Hoogstraal, H. (1973). Detection of *Trypanosoma theileri* in Ethiopian cattle ticks. *Acta Tropica*, **30**, 340–6.

Clifford, C.M., Flux, J.E.C. & Hoogstraal, H. (1976). Seasonal and regional abundance of ticks (Ixodidae) on hares (Leporidae) in Kenya. *Journal of Medical Entomology*, **13**, 40–7.

Cowdry, E.V. (1925). A group of microorganisms transmitted hereditarily by ticks and apparently unassociated with disease. *Journal of Experimental Medicine*, **41**, 817–30.

Cunliffe, N. (1913). The variability of *Rhipicephalus pulchellus* (Gerstäcker, 1873). *Parasitology*, **6**, 204–16.

Dennig, H.K. (1965). The isolation of *Babesia* species from wild animals. *Proceedings of the 1st International Congress of Parasitology, Rome, September 21–26, 1964*, **1**, 262–3.

Edelsten, R.M. (1975). The distribution and prevalence of Nairobi sheep disease and other tick-borne infections of sheep and goats in Northern Somalia. *Tropical Animal Health and Production*, 7, 29–34.

Fotheringham, W. & Lewis, E.A. (1937). East Coast fever; its transmission by ticks in Kenya Colony. *Hyalomma impressum* near *planum* P.Sch. as a vector. *Parasitology*, **29**, 504–23.

Gerstäcker, A. (1873). II. Gliederthiere (Insekten, Arachniden, Myriopoden und Isopoden). In *Baron Carl Claus von der Decken's Reisen in Ost Afrika in den Jahren 1859 bis 1861*. **III**: Wissenschaftliche Theil. Abteilung 2, ed. O. Kersten, pp. xvi + 542, 18 plates. (Ticks pp. 464–70, plate xviii). Leipzig & Heidelberg: C.J. Winter'sche Verlagshandlung.

Lewis, E.A. (1934). A study of the ticks in Kenya Colony. The influence of natural conditions and other factors on their distribution and the incidence of tick-borne diseases. Part III. Investigations into the tick problem in the Masai Reserve. *Bulletin of the Department of Agriculture, Kenya*, **No. 7 of 1934**, 65 pp., 3 maps.

Lewis, E.A. (1949). Nairobi sheep disease. *Report of the Veterinary Department, Kenya, for 1947*, pp. 45, 51.

Noda, H., Munderloh, U.G. & Kurtti, T.J. (1997). Endosymbionts of ticks and their relationship to *Wolbachia* spp. and tick-borne pathogens of humans and animals. *Applied and Environmental Microbiology*, **63**, 3926–32.

Pavesi, P. (1898). Aracnidi raccolti nel paese del Somalia dall'ing. L. Bricchetti-Robecci. *Bolletino Scientifico, Pavia (1895–1898)*, an. 17, 5(2), 37–46. (Reference not seen by authors).

Pegram, R.G. (1975). *Ticks (Ixodoidea) of the northern regions of the Somali Democratic Republic*. Thesis submitted for Membership of the Institute of Biology, London, 119 pp., 29 figs.

Pegram, R.G. (1976). Ticks (Acarina, Ixodoidea) of the northern regions of the Somali Democratic Republic. *Bulletin of Entomological Research*, **66**, 345–63.

Pegram, R.G., Hoogstraal, H. & Wassef, H.Y. (1981). Ticks (Acari: Ixodoidea) of Ethiopia. I. Distribution, ecology and host relationships of species infesting livestock. *Bulletin of Entomological Research*, **71**, 339–59.

Pellegrini, D. (1950). La gastro-enterite emorragica delle pecore. Esperimenti di transmissione col '*Rhipicephalus pulchellus*'. *Bolletino della Società Italiana di Medicina e d' Igiene Tropicale*, **10**, 164–9.

Philip, C.B., Hoogstraal, H., Reiss-Gutfreund, R.J. & Clifford, C.M. (1966). Evidence of rickettsial disease agents in ticks from Ethiopian cattle. *Bulletin of the World Health Organisation*, **35**, 127–31.

Pocock, R.I. (1900). Chilopoda and Arachnida. *Proceedings of the Zoological Society of London*, **Part 1**, 48–55.

Walker, J.B. (1955). *Rhipicephalus pulchellus* Gerstäcker 1873: a description of the larva and the

nymph with notes on the adults and on its biology. *Parasitology*, **45**, 95–8.
Wood, O.L., Lee, V.H., Ash, J.S. & Casals, J. (1978). Crimean-Congo hemorragic fever, Thogoto, Dugbe and Jos viruses isolated from ixodid ticks in Ethiopia. *American Journal of Tropical Medicine and Hygiene*, **27**, 600–4.
Also see the following Basic References (pp. 12–14): Morel (1980); Pegram (1979); Walker (1974); Yeoman & Walker (1967).

RHIPICEPHALUS PUNCTATUS WARBURTON, 1912 (INCLUDING *RHIPICEPHALUS* SP. NEAR *PUNCTATUS*)

The specific name, a Latin term meaning 'pitted, punctate', refers to the numerous punctations on the male conscutum and female scutum.

Synonym

neavei punctatus

Diagnosis

A medium-sized reddish-brown heavily-punctate species.

Male (Figs 172(a), 173(a) to (c))
Capitulum ranging from longer than broad to broader than long, length × breadth [based on a paralectotype (Nuttall Collection 1411) and 2 other males only] ranging from 0.55 mm × 0.52 mm to 0.59 mm × 0.63 mm. Basis capituli with short acute lateral angles just anterior to mid-length. Palps short, broad, rounded apically. Conscutum length × breadth ranging from 2.39 mm × 1.32 mm to 2.49 mm × 1.54 mm; anterior process of coxae I large, sharp, well sclerotized. Eyes slightly convex, delimited dorsally by shallow punctate depressions. Cervical pits convergent, continuous with shallow divergent slightly shagreened internal cervical margins that extend posteriorly just beyond eye level. Marginal lines long, punctate. Posterior grooves well developed, posteromedian long and narrow, posterolaterals shorter and broader. Medium-

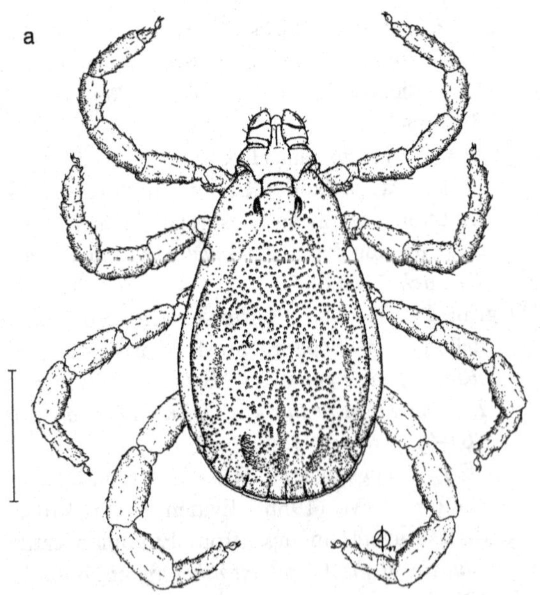

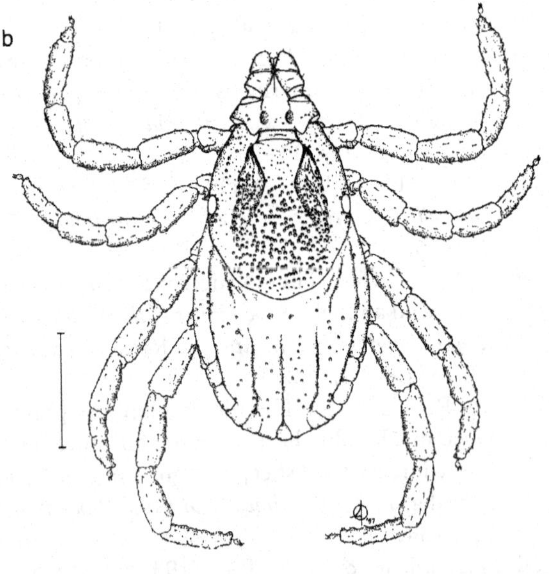

Figure 172. *Rhipicephalus punctatus* [Nuttall Collection 1411, RML 110224, paralectotypes, from greater kudu (*Tragelaphus strepsiceros*), near Fort Mlangeni, central Angoniland (= Ncheu District), Malawi, in May 1910 by S.A. Neave, by courtesy of The Natural History Museum, London]. (a) Male, dorsal; (b) female, dorsal. Scale bars represent 1 mm. A. Olwage *del.*

Table 45A. *Host records of* Rhipicephalus punctatus

Hosts	Number of records
Domestic animals	
Cattle	56
Wild animals	
Burchell's zebra (*Equus burchellii*)	2
Topi (*Damaliscus lunatus topi*)	1
Gerenuk (*Litocranius walleri*)	1
Kirk's dik-dik (*Madoqua kirkii*)	1
Klipspringer (*Oreotragus oreotragus*)	4
Oribi (*Ourebia ourebi*)	2
Greater kudu (*Tragelaphus strepsiceros*)	2
Sable antelope (*Hippotragus niger*)	1
Cape hare (*Lepus capensis*)	1
Savanna hare (*Lepus victoriae*)	8

sized setiferous punctations present on scapulae, along the external margins of the cervical fields and scattered medially on the conscutum, interspersed with numerous somewhat smaller punctations, creating an overall impression of a heavily-punctate tick. Ventrally spiracles broad, with the prolongation tapering gradually and curving slightly just at its end. Adanal plates large, broad, their external and posterior margins almost straight, their internal margins virtually parallel for much of their length, becoming divergent posterior to the anus and joining the posterior margins in broadly-rounded curves; accessory adanal plates relatively small sclerotized points.

Female (Figs 172(b), 173(d) to (f))

Capitulum broader than long, length × breadth [based on a paralectotype (Nuttall Collection 1411) and 2 other females only] ranging from 0.68 mm × 0.72 mm to 0.69 mm × 0.77 mm. Basis capituli with acute lateral angles just anterior to mid-length; porose areas large, oval, about twice their own diameter apart. Palps short, broad, rounded apically. Scutum longer than broad, length × breadth ranging from 1.43 mm × 1.26 mm to 1.47 mm × 1.31 mm; posterior margin slightly sinuous. Eyes slightly convex, delimited dorsally by shallow punctate depressions. Cervical pits deep, convergent; cervical fields long, slightly divergent. Medium-sized setiferous punctations present anteriorly on scapulae, along external margins of cervical fields and scattered medially on the scutum, interspersed with numerous somewhat smaller punctations, creating an overall impression of a heavily-punctate tick. Ventrally genital aperture more-or-less U-shaped, with the sides of the opening diverging slightly.

Immature stages

Unknown.

Notes on identification

Rhipicephalus punctatus was described originally as *R. neavei punctatus* by Warburton (1912), then raised to specific rank by Santos Dias (1951). In 1956 Walker sank it as a synonym of *R. pravus* but later realized that this was a mistake and it was reinstated as a full species by Yeoman & Walker (1967). In 1993 Santos Dias still listed it as a synonym of *R. pravus*.

It is the most heavily-punctate species in the *R. pravus* group. The relationship between this species and the somewhat more lightly-punctate entity that has been recorded further south in northern Mozambique, Zambia, Zimbabwe and Angola, referred to below as

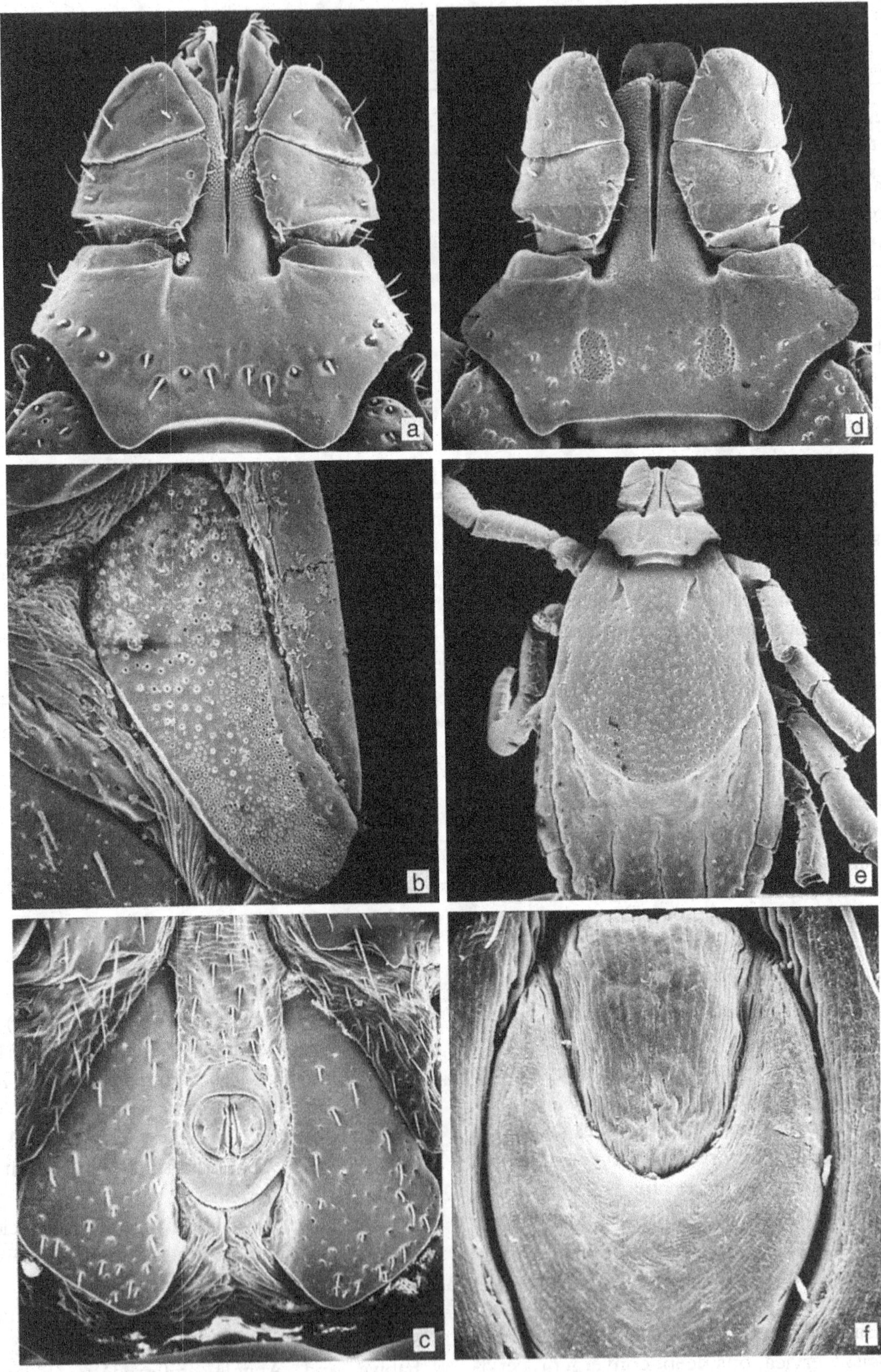
a
d
b
e
c
f

Rhipicephalus sp. near *punctatus*, requires further investigation.

Hosts

Life cycle unknown; since *R. punctatus* belongs to the *R. pravus* group it is almost certainly a three-host species. The only domestic animals from which its adults have been collected thus far, in very small numbers, are cattle. It has also been recorded from various wild ungulates and hares (Yeoman & Walker, 1967; Matthysse & Colbo, 1987) (Table 45A). The largest single collection (91 adults) was taken from a sable antelope, on which most of these ticks were attached in the axillae and on the tail, with a few in the mane and on the heels, but none in the ears. Almost all the other collections from ungulates yielded very few specimens of this tick. It occurs more commonly, and in somewhat larger numbers, on hares, particularly on their ears.

As noted above, the immature stages of *R. punctatus* have not been described. In Tanzania, though, *R. pravus* group nymphs were found on six hares in association with adults of this species (Yeoman & Walker, 1967). Probably, like the larvae and nymphs of other species in this group, they will feed on elephant shrews.

Zoogeography

Rhipicephalus punctatus, as envisaged here, has been recorded from south-western Uganda southwards through parts of western and south-western Tanzania to Malawi and Zambia (Yeoman & Walker, 1967; Keirans, 1985; Pegram *et al.*, 1986; Matthysse & Colbo, 1987) (Map 50).

Figure 173 (*opposite*). *Rhipicephalus punctatus* [Tanzania Tick Survey WA/80, RML 105012, collected from Burchell's zebra (*Equus burchellii*), Kitangule plains, Karagwe, Tanzania, on 5 April 1959 by Mrs G. Tullock]. Male: (a) capitulum, dorsal; (b) spiracle; (c) adanal plates. Female: (d) capitulum, dorsal; (e) dorsal surface; (f) genital aperture. (Scales not available). SEMs by M.D. Corwin.

There are also three isolated, and at present inexplicable, records in Tanzania from N. Maasai, Mpwapwa and Ndandawala, Kilwa District.

Most collections have been made at altitudes between about 1200 m to 1800 m, sometimes lower, with mean annual rainfalls ranging from approximately 500 mm to 1600 mm. It occurs in several vegetation types, from Afromontane vegetation to mosaics of lowland rainforest, secondary *Acacia* wooded grassland, evergreen bushland and secondary grassland.

Disease relationships

Unknown.

Rhipicephalus sp. near *punctatus*

Notes on identification

We have listed under this name ticks from Zimbabwe identified by Theiler & Robinson (1953) and Theiler (1962) as *R. pravus* and by Norval (1985) as *Rhipicephalus* sp. near *R. punctatus*. We have also included ticks from Zambia identified by MacLeod (1970), MacLeod *et al.* (1977) and MacLeod & Mwanaumo (1978) as belonging to the *R. pravus* group and containing both *R. punctatus* and *R. neavei* (now regarded as a synonym of *R. kochi*, which we have excluded from this account), as well as ticks recorded as *R. punctatus* by Pegram *et al.* (1986). The latter authors stated, though, that *R. punctatus sensu stricto* had been confirmed in only one collection from cattle at Solwezi, Northwestern Province. They commented: 'All other material referred to under this name is atypically lightly punctate, being intermediate between *R. punctatus sensu stricto* and the eastern African *Rhipicephalus pravus*'.

Hosts

Presumably a three-host species since it belongs to the *R. pravus* group. The preferred domestic hosts are cattle. The preferred wild hosts, while including greater kudu, are chiefly small antelopes such as steenbok, grysbok and common

Table 45B. *Host records of* Rhipicephalus *sp. near* punctatus

Hosts	Number of records
Domestic animals	
Cattle	128
Sheep	2
Goats	2
Horses	5
Donkeys	2
Dogs	2
Wild animals	
'Shrew' (*Crocidura* sp.)	3 (immatures)
Lion (*Panthera leo*)	1
Leopard (*Panthera pardus*)	2
Burchell's zebra (*Equus burchellii*)	2
Warthog (*Phacochoerus africanus*)	4
Bushpig (*Potamochoerus larvatus*)	3
Impala (*Aepyceros melampus*)	8
'Hartebeest'	1
Klipspringer (*Oreotragus oreotragus*)	1
Steenbok (*Raphicerus campestris*)	156 (including nymphs)
Sharpe's grysbok (*Raphicerus sharpei*)	28
African buffalo (*Syncerus caffer*)	8
Eland (*Taurotragus oryx*)	2
Nyala (*Tragelaphus angasii*)	1
Bushbuck (*Tragelaphus scriptus*)	3
Greater kudu (*Tragelaphus strepsiceros*)	34
Bushbuck and greater kudu (*Tragelaphus* spp.)	8
Common duiker (*Sylvicapra grimmia*)	20
Roan antelope (*Hippotragus equinus*)	1
Sable antelope (*Hippotragus niger*)	3
Waterbuck (*Kobus ellipsiprymnus*)	1
Natal multimammate rat (*Mastomys natalensis*)	1 (1 larva)
Spring hare (*Pedetes capensis*)	1
Cape hare (*Lepus capensis*)	1
'Hare'	8 (including 1 nymph)
Jameson's red rock rabbit (*Pronolagus randensis*)	1 (1 larva)

duiker (MacLeod, 1970; Norval, 1985) (Table 45B). Individual infestations of this tick always seem to be light. Adults have been collected from the chest, abdomen and around the anus of cattle (MacLeod *et al.*, 1977).

In Zimbabwe adults were present throughout the year (Norval, 1985). In Zambia they were active from December and January to July or August, with a peak in abundance from February to May (MacLeod, 1970; MacLeod *et al.*, 1977; Pegram *et al.*, 1986).

Zoogeography

With the exception of the south-western border area and the central region this *Rhipicephalus* sp. near *punctatus* has been recorded throughout Zimbabwe (see Map 50). Most collections have

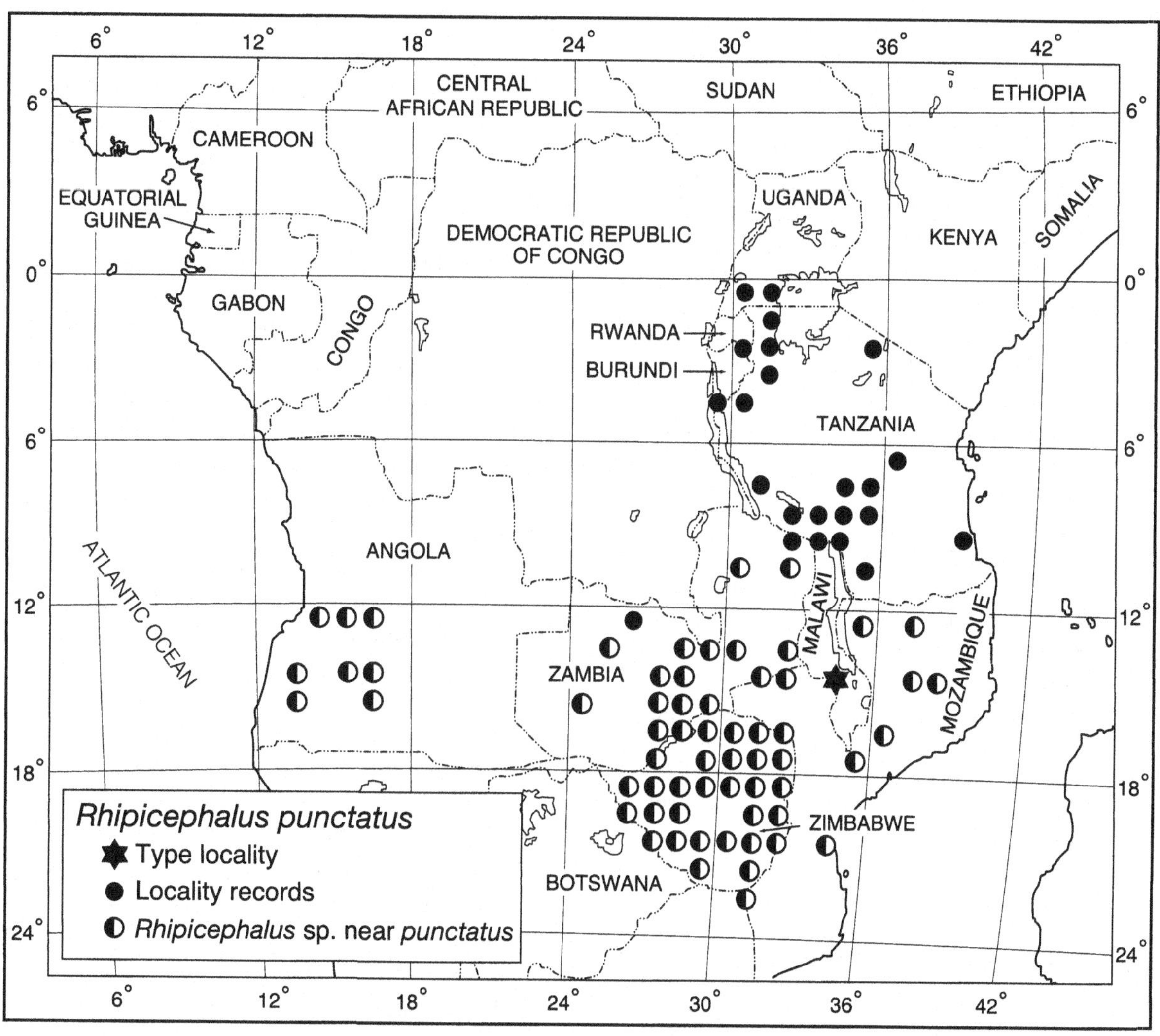

Map 50. *Rhipicephalus punctatus* and *Rhipicephalus* sp. near *punctatus*: distribution.

been made in the south-east of the country bordering Mozambique and in the west (Norval, 1985). In Zambia most collections come from the Southern and Central Provinces. It has been collected at localities between 400 m and 1400 m above sea level and receiving between 750 mm and 1000 mm of rainfall annually. It is found in both wetter and drier Zambezian miombo woodland, *Colophospermum mopane* woodland and undifferentiated northern Zambezian woodland.

Disease relationships

Unknown.

REFERENCES

MacLeod, J. (1970). Tick infestation patterns in the southern province of Zambia. *Bulletin of Entomological Research*, **60**, 253–74.

MacLeod, J., Colbo, M.H., Madbouly, M.H. & Mwanaumo, B. (1977). Ecological studies of ixodid ticks (Acari: Ixodidae) in Zambia. III. Seasonal activity and attachment sites on cattle, with notes on other hosts. *Bulletin of Entomological Research*, **67**, 161–73.

MacLeod, J. & Mwanaumo, B. (1978). Ecological studies of ixodid ticks (Acari: Ixodidae) in Zambia. IV. Some anomalous infestation patterns in the northern and eastern regions.

Bulletin of Entomological Research, **68**, 409–29.

Norval, R.A.I. (1985). The ticks of Zimbabwe. XII. The lesser known *Rhipicephalus* species. *Zimbabwe Veterinary Journal*, **16**, 37–43.

Pegram, R.G., Perry, B.D., Musisi, F.L. & Mwanaumo, B. (1986). Ecology and phenology of ticks in Zambia: seasonal dynamics on cattle. *Experimental and Applied Acarology*, **2**, 25–45.

Santos Dias, J.A.T. (1951). Mais um Ixodídeo do género *Rhipicephalus* s. str. Koch, 1844 para a fauna de Moçambique. *R. punctatus* Warburton, 1912. *Anais do Instituto de Medicina Tropical, Lisboa*, **8**, 373–90.

Theiler, G. & Robinson, B.N. (1953). Ticks in the South African Zoological Survey Collection. Part VII. Six lesser known African rhipicephalids. *Onderstepoort Journal of Veterinary Research*, **26**, 93–136.

Walker, J.B. (1956). *Rhipicephalus pravus* Dönitz, 1910. *Parasitology*, **46**, 243–60.

Warburton, C. (1912). Notes on the genus *Rhipicephalus* with the description of new species, and the consideration of some species hitherto described. *Parasitology*, **5**, 1–20.

Also see the following Basic References (pp. 12–14): Keirans (1985); Matthysse & Colbo (1987); Santos Dias (1993); Theiler (1962); Yeoman & Walker (1967).

RHIPICEPHALUS SANGUINEUS (LATREILLE, 1806)

This specific name from the Latin *sanguineus* meaning 'of blood' or 'bloody', refers to the blood-feeding habit of this tick.

Synonyms

becarii; bhamensis; breviceps; brevicollis; bursa americanus; carinatus; flavus; ?intermedius; limbatus; macropis; punctatissimus; rubicundus; rutilus; sanguineus brevicollis; sanguineus punctatissimus; siculus; stigmaticus; texanus.

Diagnosis

A medium-sized, pale yellowish-brown or reddish-brown tick with a '*simus*' pattern of punctations in the male, and scalpel-shaped cervical fields in the female.

Male (Figs 174(a), 175(a) to (c))

Capitulum broader than long, length x breadth ranging from 0.49 mm × 0.56 mm to 0.57 mm × 0.63 mm. Basis capituli with acutely-curved lateral angles, not overlapping scapular areas of the conscutum. Palps short, rounded apically. Conscutum slightly narrower anteriorly, broadening posterior to eyes, length × breadth ranging from 2.24 mm × 1.12 mm to 2.63 mm × 1.55 mm; anterior process of coxae I inconspicuous. In engorged specimens body wall expanded laterally and posteriorly. Eyes marginal, very slightly bulging, edged dorsally with a few punctations. Cervical pits short, deep. In some specimens the merest indication of cervical fields can be seen, but this condition is rare. Marginal lines deep and punctate, delimiting the first two festoons and extending anteriorly, ending posterior to eyes. Posteromedian groove narrowly elongate, posterolateral grooves subcircular. In many males examined the posteromedian and posterolateral grooves are larger and deeper than those illustrated in Fig. 174(a). Punctations range from scarce to numerous; usually the larger more robust male specimens are more densely punctate, but characteristically four more-or-less regular rows of widely-spaced punctations are visible. This '*simus*' pattern is composed of larger and somewhat deeper punctations than the background pattern of smaller interstitial punctations covering the conscutum. Ventrally spiracles elongate throughout, each with a narrow dorsal prolongation usually visible dorsally. Adanal plates variable but generally elongately subtriangular, distinctly broad in their posterior aspect, but occasionally truncated or rounded posteriorly; accessory adanal plates moderately distinct. Legs increase slightly in size from I to IV.

Female (Figs 174(b), 175(d) to (f), 178(a))

Capitulum broader than long, length × breadth ranging from 0.57 mm × 0.67 mm to 0.63 mm × 0.71 mm. Basis capituli with broad lateral angles; porose areas small, about twice their own diameter apart. Palps longer than those of male, narrowly rounded apically. Scutum longer than

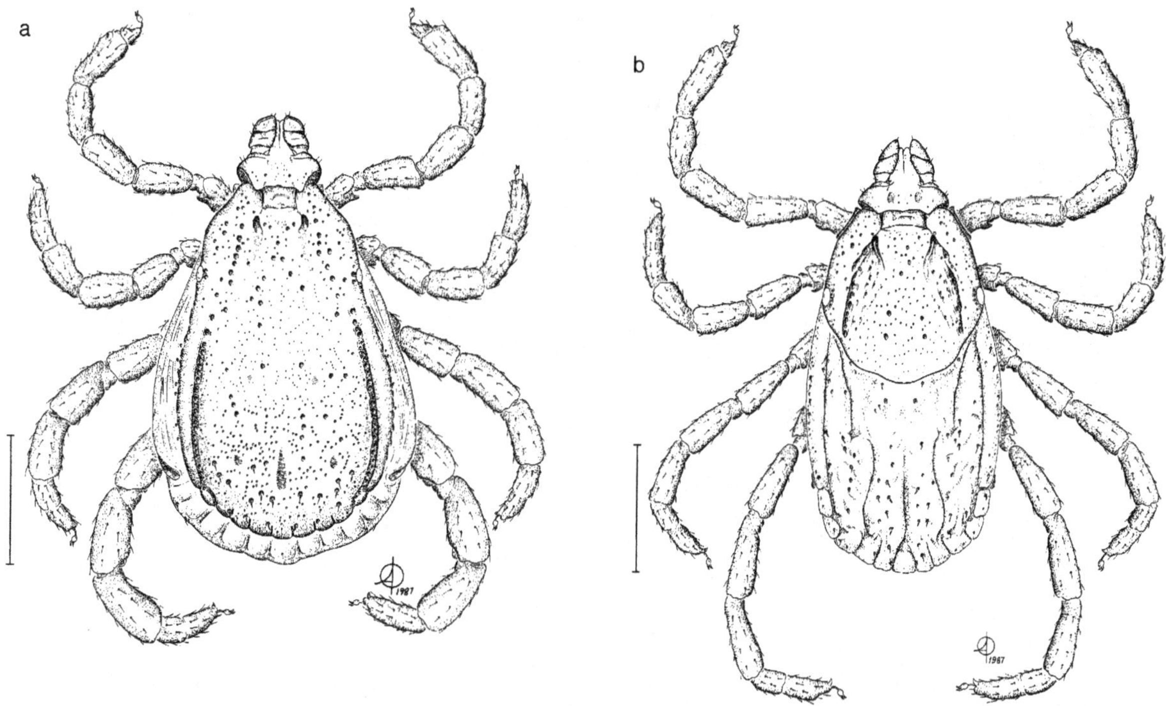

Figure 174. *Rhipicephalus sanguineus* (collected from dog, Pretoria North, Gauteng Province, South Africa, on 21 January 1980 by D. Roos). (a) Male, dorsal; (b) female, dorsal. Scale bars represent 1 mm. A. Olwage *del.*

Table 46. *Confirmed host records of* Rhipicephalus sanguineus *in the Afrotropical region*

Hosts	Number of records
Domestic animals	
Cattle	12
Goats	3
Horses	1
Dogs	547 (including immatures)
Cats	1
Wild animals	
'Jackal' (*Canis* spp.)	4
African wild cat (*Felis lybica*)	1
Serval (*Leptailurus serval*)	1
Brown hyaena (*Parahyaena brunnea*)	1
Zorilla (*Ictonyx striatus*)	1
African civet (*Civettictis civetta*)	2
Aardvark (*Orycteropus afer*)	1
Warthog (*Phacochoerus africanus*)	1
Giraffe (*Giraffa camelopardalis*)	1
Günther's dik-dik (*Madoqua guentheri*)	1
Scrub hare (*Lepus saxatilis*)	1
Humans	2

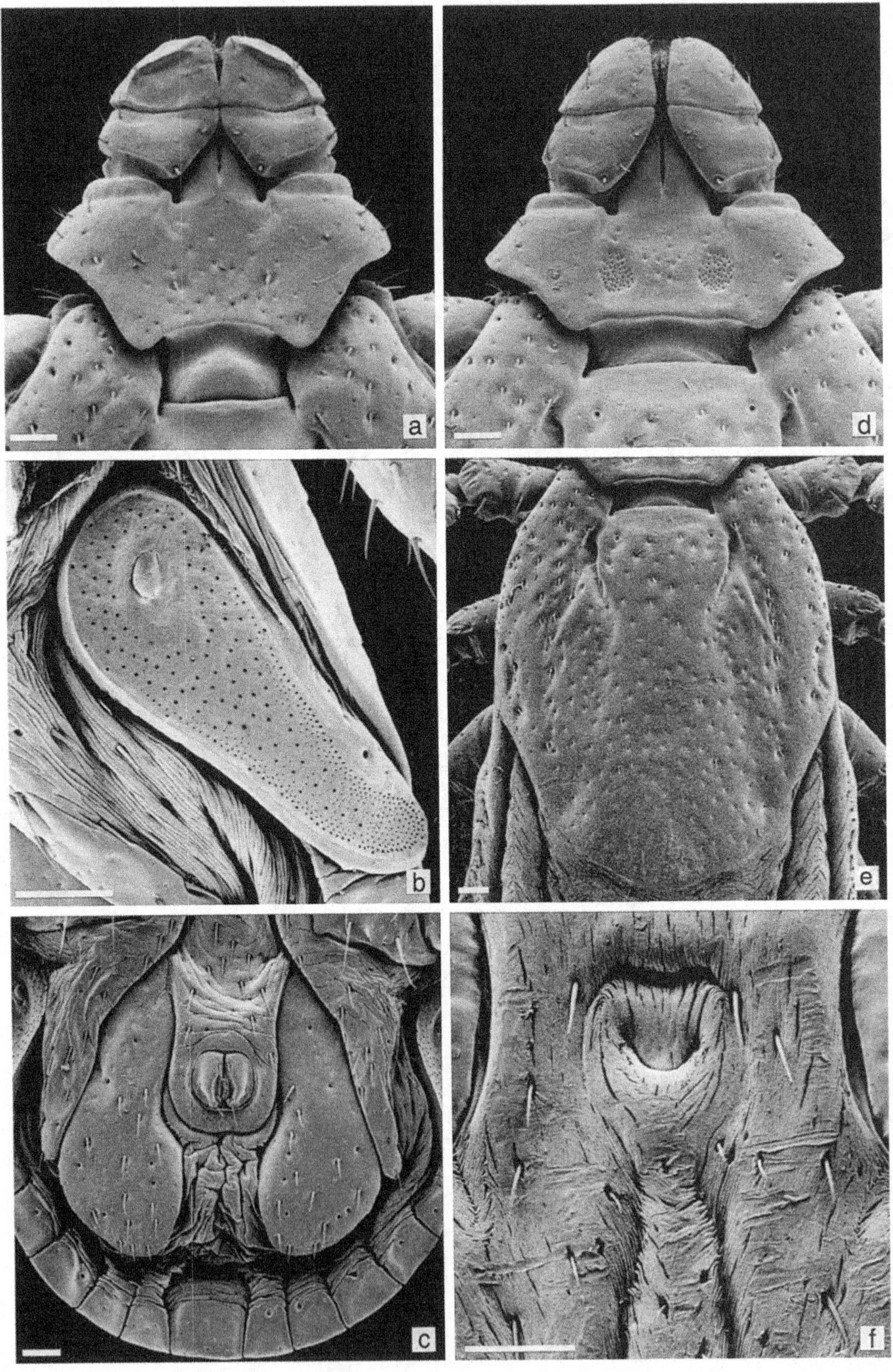
a
d
b
e
c
f

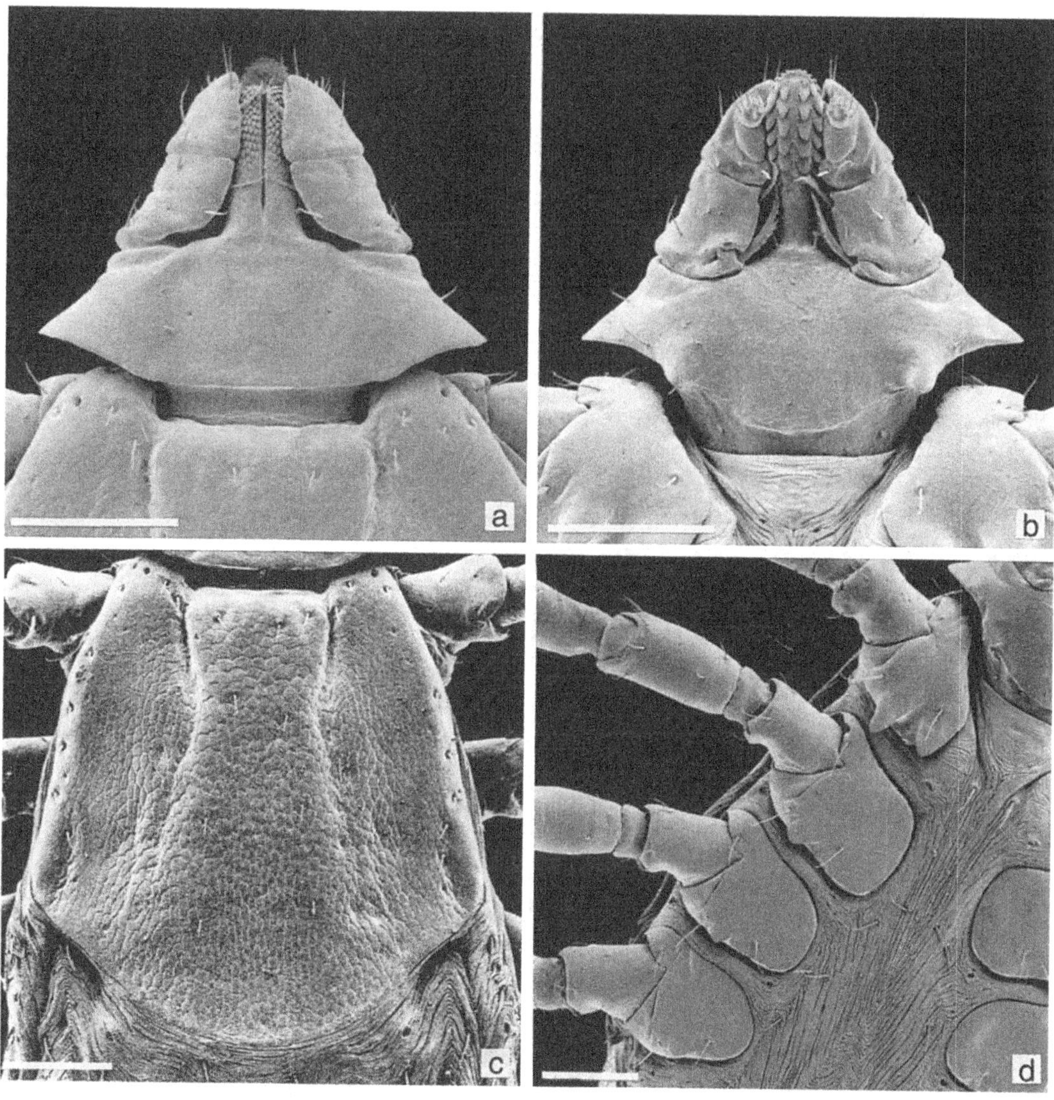

Figure 176 (*above*). *Rhipicephalus sanguineus* (Protozoology Section Tick Breeding Register, Onderstepoort, No. 3265, progeny of ♀ collected from dog, Fort Beaufort, Eastern Cape Province, South Africa, in April 1970). Nymph: (a) capitulum, dorsal; (b) capitulum, ventral; (c) scutum; (d) coxae. Scale bars represent 0.10 mm. SEMs by J.F. Putterill.

Figure 175 (*opposite*). *Rhipicephalus sanguineus* (Protozoology Section Tick Breeding Register, Onderstepoort, No. 3265, progeny of ♀ collected from dog, Fort Beaufort, Eastern Cape Province, South Africa, in April 1970). Male: (a) capitulum, dorsal; (b) spiracle; (c) adanal plates. Female: (d) capitulum, dorsal; (e) scutum; (f) genital aperture. Scale bars represent 0.10 mm. SEMs by J.F. Putterill.

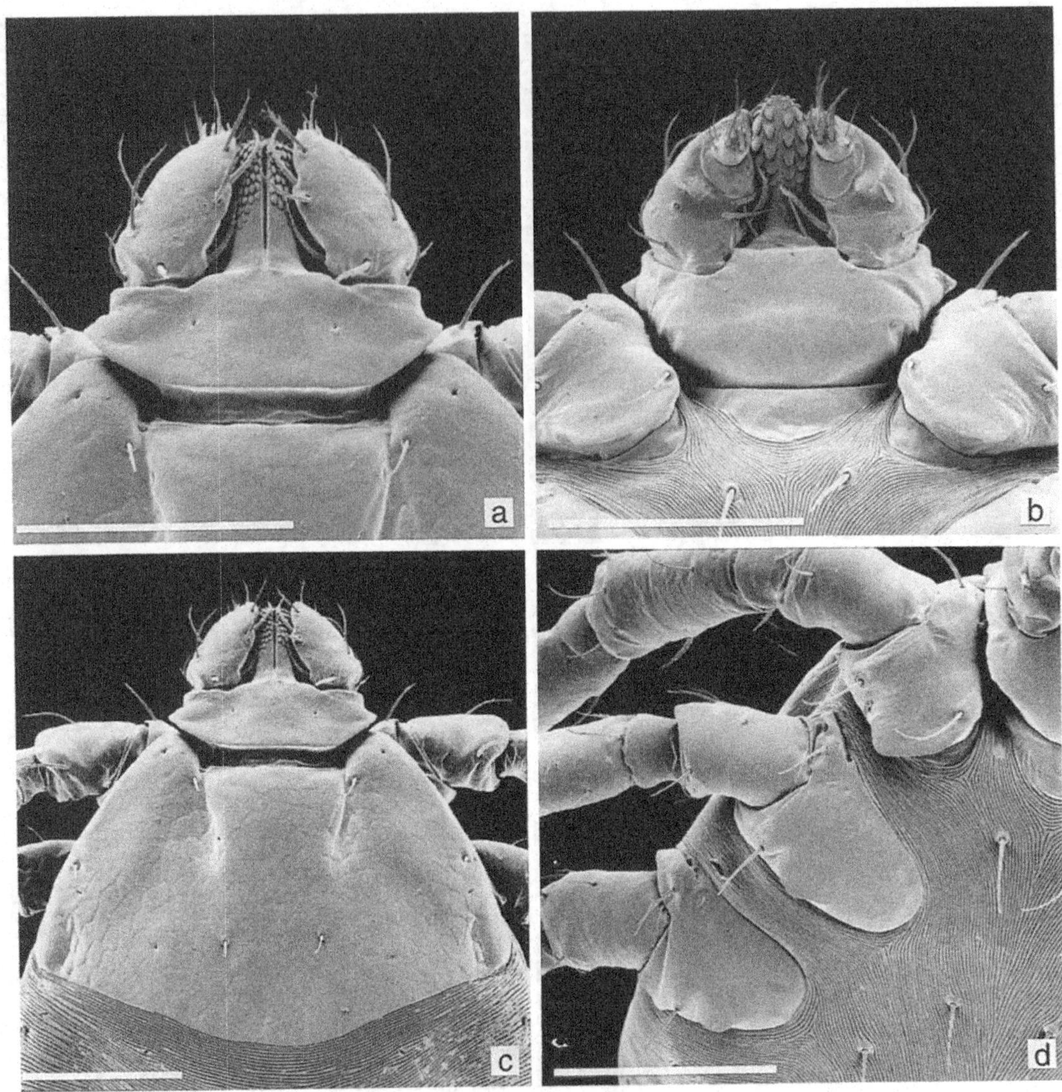

Figure 177. *Rhipicephalus sanguineus* (Protozoology Section Tick Breeding Register, Onderstepoort, No. 3265, progeny of ♀ collected from dog, Fort Beaufort, Eastern Cape Province, South Africa, in April 1970). Larva: (a) capitulum, dorsal; (b) capitulum, ventral; (c) scutum; (d) coxae. Scale bars represent 0.10 mm. SEMs by J.F. Putterill.

broad, length × breadth ranging from 1.28 mm × 1.14 mm to 1.36 mm × 1.23 mm; shield-shaped with posterior margin sinuous. Eyes at lateral angles, mildly convex, usually rimmed dorsally by a few punctations. Cervical fields slightly depressed, scalpel-shaped, usually beset with striations and some shagreening; external cervical margins marked by several larger punctations; a few larger punctations also present medially and in scapular areas. Ventrally genital aperture broadly U-shaped.

Nymph (Fig. 176)

Capitulum much broader than long, length × breadth ranging from 0.23 mm × 0.31 mm to 0.24 mm × 0.33 mm. Lateral angles shallowly curved. Palps short, apically acute. Scutum approximately as long as broad, length × breadth ranging from 0.49 mm × 0.51 mm to 0.54 mm × 0.54 mm. Lateral margins nearly straight, posterior margin broadly rounded. Eyes at widest point, slightly convex, marked dorsally by a few punctations. Cervical grooves short and

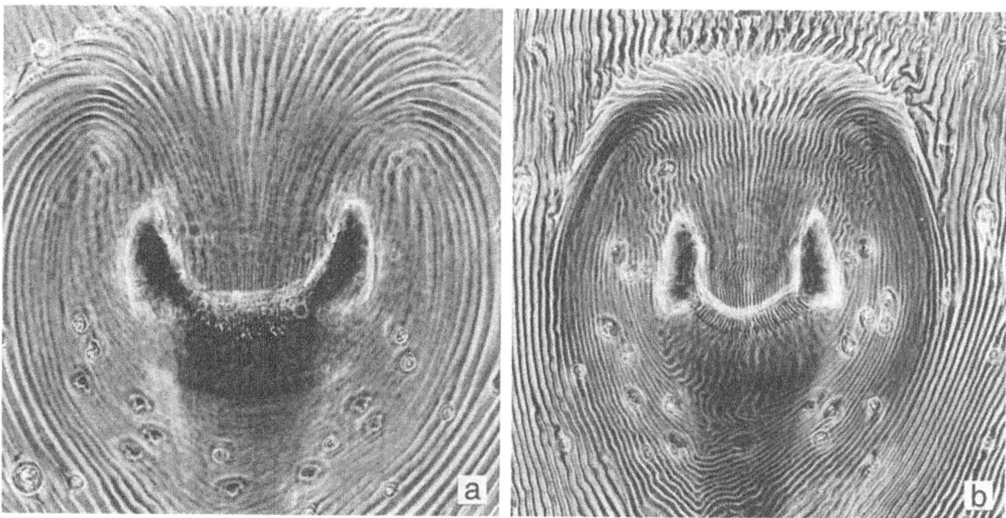

Figure 178. Female genital apertures (mounted): (a) *Rhipicephalus sanguineus*; (b) *Rhipicephalus camicasi*. (From Pegram *et al.*, 1987, figs 38 & 39, with kind permission from Kluwer Academic Publishers).

deep, scutal surface relatively impunctate. Ventrally coxae I each with two short, widely separated spurs, external longer than internal; a single rounded external spur, decreasing in size, on coxae II and III; coxae IV lacking spurs.

Larva (Fig. 177)

Capitulum much broader than long, length × breadth ranging from 0.098 mm × 0.144 mm to 0.113 mm × 0.156 mm. Basis capituli twice as broad as long, posterior margin broadly curved, lateral angles short, blunt. Palps short, external margin mildly convex, bluntly pointed apically. Scutum much broader than long, length × breadth ranging from 0.211 mm × 0.370 mm to 0.215 mm × 0.382 mm. A few very faint punctations present. Eyes at widest point, large, set well over halfway back. Cervical grooves shallow, moderately long, subparallel. Ventrally coxae I each with a broad salient ridge, coxae II each with a small spur; coxae III each with the slightest indication of a spur.

Notes on identification

The *R. sanguineus* group comprises several tick species. The biosystematic status of the majority of them has been confused, consequently they have often been misidentified. Pegram *et al.* (1987a, b) have critically reviewed the literature on this group. They illustrated the morphology and geographical distributions as well as supplying host lists of six of these ticks, including *R. sanguineus sensu stricto*. Two of these species, *R. camicasi* and *R. turanicus*, closely resemble *R. sanguineus* morphologically. In general, though, the scuta of male and female *R. camicasi* are less punctate than those of *R. sanguineus*, while those of *R. turanicus* are more punctate. In addition the genital apertures of female *R. camicasi* and *R. turanicus* are broadly V-shaped compared to the U-shape of that of *R. sanguineus*. Differences in the appearance of the mounted female genitalia of *R. sanguineus* and *R. camicasi* are illustrated in Fig. 178.

Another species in this group with which *R. sanguineus sensu stricto* may sometimes have been confused is *R. sulcatus*. The latter has, however, a much denser and more even punctation pattern than *R. sanguineus* (compare Fig. 198, p. 427, with Fig. 174). Its mounted female genital aperture also differs from that of the *R. sanguineus* female (compare Fig. 85(b), p. 211, with Fig. 178(a)).

In 1956, before the status of the *R. sanguineus* group of ticks had been clarified, Hoogstraal stated: 'A listing of each host reported by various authors would be of no practical value'. Pegram *et al.* (1987b), in their publication clarifying the status of *R. sanguineus*, wrote: 'Our results suggest that previously published records, which indicate a very wide host range for this species, require critical re-evaluation'. Bearing this in mind we have, in the Afrotropical region, accepted the published records of Norval, Daillecourt & Pegram (1983), Pegram *et al.* (1987b) and Horak (1982, 1995) as well as the unpublished records of H. Heyne, I.G.H. and J.B.W. as valid. All these workers were aware of the similarities between *R. sanguineus*, *R. sulcatus* and *R. turanicus* at the time and were able to differentiate between the three species. Their host records are listed in Table 46.

Hosts

A three-host species. Numerous workers have studied the life history and ovipositional habits of this tick (Hooker, Bishopp & Wood, 1912; Nuttall, 1913; Patton & Cragg, 1913; Sapre, 1944; Lombardini, 1950; Achan, 1961; Sweatman, 1967; Sardey & Rao, 1971 and others). Depending on altitude, laboratory hosts, temperature and relative humidity at which the studies were undertaken, larvae and nymphs feed for about 4 days and females for 8. The number of eggs deposited varies from 1164 to almost 5000. Nuttall's 1913 data showed that larvae can survive unfed for 253 days, nymphs for 97 and adults for as long as 568 days.

The preferred host for all stages of development of *R. sanguineus*, commonly known as the brown dog tick or the kennel tick, is the domesticated dog. In the Afrotropical region occasional collections have been taken from cattle, goats, some wild carnivores and other wild animals and also humans (Table 46). The latter hosts must, however, all be considered as accidental.

Preferred attachment sites on dogs of the immature stages are the legs, chest and belly with nymphs attaching also to the ears. Adults are found mainly on the ears, head and neck (Horak, 1982; Koch, 1982). In homes and apartments, this tick is often present around the bedding area of pets. Upon detaching many engorged nymphs and females migrate vertically providing the surface is suitable, before moulting or egg-laying commences in cracks or crevices, or between walls and the covering wallpaper.

In the Nearctic, *R. sanguineus* very rarely bites humans (Helm, 1952). However, Goddard (1989) recorded a focus of human parasitism at four United States Airforce installations in northern Texas and southern Oklahoma. These findings were subsequently confirmed by Carpenter, McMeans & McHugh (1990).

In studies conducted in both northern and southern Africa two generations of adult ticks have been noted in summer (Amin & Madbouly, 1973; Horak, 1982). The synchronous emergence of adults in spring from over-wintered nymphs gives rise to the first generation (Horak, 1982; Koch & Tuck, 1986). In humid tropical zones three generations may be possible annually (Aeschlimann, 1967).

Zoogeography

Rhipicephalus sanguineus is probably the most widely distributed tick in the world. Circumglobally it is found approximately between the latitudes of 50° N and 30° S, and its preference for dogs has facilitated its worldwide distribution (Leeson, 1951). In the tropics and subtropics *R. sanguineus* can bc found both indoors and out. In colder climates it is primarily found indoors, in homes, apartments, kennels and any other structures where dogs dwell. In 1946 Cooley listed the brown dog tick from 26 states in the U.S.A. It has now been recorded from all 50 states. Hoogstraal (1973) reported that *R. sanguineus* has never been recorded from Afghanistan; however, Le Riche *et al.* (1988) recorded it from 10 out of 105 stray dogs in Kabul. *Rhipicephalus sanguineus* is rare in Japan excluding Okinawa (Inokuma, Tamura & Onishi, 1995).

Disease relationships

Rhipicephalus sanguineus transmits the causative agents of canine tick fever or babesiosis infections in dogs, *Babesia canis* (Shortt, 1973) and *Babesia gibsoni* (Sen, 1933). The latter species occurs primarily in the Far East and North Africa, while *B. canis* is endemic in the U.S.A. and much, if not all, of Africa. Eaton (1934) first reported *B. canis* in greyhounds from Florida, U.S.A. It was thought that the disease was restricted to the southern states, but it is now known in the East, Middlewest and Southwest (Ewing, 1968). *Babesia canis* can be transmitted transovarially and transstadially (Shortt, 1973). The listing of *R. sanguineus* as a vector of *Babesia caballi* and *Babesia equi* of horses (Enigk, 1943) probably refers to *R. turanicus*.

Ehrlichia canis, the causative agent of canine ehrlichiosis (tropical canine pancytopenia) was described originally in Algeria (Donatien & Lestoquard, 1935). It was later recorded from the Middle East, the Orient, and numerous other areas of the globe (Ewing, 1972). The disease is transmitted by all parasitic stages of *R. sanguineus* (Groves *et al.*, 1975). It is the most commonly reported canine infectious disease in the U.S.A., and is present wherever the brown dog tick occurs. However, passage of an isolate of *E. canis* in cell culture apparently adversely affects its transmissibility by *R. sanguineus* (Mathew *et al.*, 1996).

Rhipicephalus sanguineus has been incriminated in the transmission of *Rickettsia rickettsi*, the cause of canine Rocky Mountain spotted fever (RMsf), but its significance as a vector is uncertain (Greene & Breitschwerdt, 1990). Pure breeds appear more likely to suffer clinical manifestations of the disease. In dogs the usual site of haemorrhage is in the mucous membranes rather than in the skin, the usual site in human RMsf. *Rickettsia rhipicephali*, another spotted fever group rickettsia, has also been recorded from the brown dog tick (Burgdorfer *et al.*, 1975, 1978).

Two minor canine diseases are also transmitted by *R. sanguineus*, *Haemobartonella canis*, an epierythrocytic rickettsial parasite with both transovarial and transstadial transmission occurring (Seneviratna, Weerasinghe & Ariyadasa, 1973), and *Hepatozoon canis* which is transmitted to dogs when they ingest or bite an infected tick (Nordgren & Craig, 1984).

The nymph to adult transmission of *Dipetalonema dracunculoides*, a filarial nematode of dogs, has been demonstrated experimentally (Olmeda-García, Rodríguez-Rodríguez & Roja-Vázquez, 1993). Viloria (1954) reported that an infestation consisting of several *R. sanguineus* caused tick paralysis in a dog in Venezuela, but this tick does not appear to be efficient in causing canine paralysis.

Various organisms pathogenic to humans have been found in the brown dog tick. *Rickettsia rickettsi*, the agent of Rocky Mountain spotted fever; *Rickettsia sibirica*, the agent of Siberian tick typhus, and *Rickettsia conori*, the agent of boutonneuse fever, have all been isolated from this tick (Hoogstraal, 1967). Parker (1933) reported larval to adult survival of *Francisella tularensis* in, and transmission of the causative organism of tularaemia by, *R. sanguineus*. However, the most recent review article on tularaemia (Hopla & Hopla, 1994) does not implicate *R. sanguineus* as a competent vector. In most regions of the world *R. sanguineus* does not feed on humans and is consequently not considered an important vector of the agents of RMsf, Siberian tick typhus or tularaemia.

Noda, Munderloh & Kurtti (1997) found symbionts in *R. sanguineus* whose 16S rDNA sequences 'were closely related to that for the Q-fever organism, *C[oxiella] burnetii*'.

In the Mediterranean Littoral, Crimea, northern and southern India, Burma, Malaysia, Thailand and Vietnam *R. sanguineus* is the primary vector of *R. conori*, the rickettsia of the fièvre boutonneuse group. This rickettsia is the causative agent of a condition variously known as South African, Kenyan, Crimean and Indian tick typhuses, Marseilles fever, Mediterranean spotted fever, as well as numerous other names. However, in Central and South Africa other ticks, endemic animals and local environments are involved (Philip & Burgdorfer, 1961; Hoogstraal, 1967).

Rickettsia conori was introduced into Switzerland in 1976 either from southern France or from Italy on a pet dog infested with *R. sanguineus* (Péter *et al.*, 1984). By the time control measures were initiated in 1981, four persons associated with the dog had been diagnosed with boutonneuse fever.

REFERENCES

Achan, P.D. (1961). Observation on the oviposition of *Rhipicephalus sanguineus* Latr. *Bulletin of Entomology*, **2**, 38–42.

Amin, O.M. & Madbouly, M.H. (1973). Distribution and seasonal dynamics of a tick, a louse fly, and a louse infesting dogs in the Nile Valley and Delta of Egypt. *Journal of Medical Entomology*, **10**, 295–8.

Burgdorfer, W., Sexton, D.J., Gerloff, R.K., Anacker, R.L., Philip, R.N. & Thomas, L.A. (1975). *Rhipicephalus sanguineus*: vector of a new spotted fever group rickettsia in the United States. *Infection and Immunity*, **12**, 205–10.

Burgdorfer, W., Brinton, L.P., Krinsky, W.L. & Philip, R.N. (1978). *Rickettsia rhipicephali*, a new spotted fever group rickettsia from the brown dog tick *Rhipicephalus sanguineus*. In *Rickettsiae and Rickettsial Diseases*, ed. J. Kazar, R.A. Ormsbee & N. Tarasevich, pp. 307–16. Bratislava, CSSR: VEDA.

Carpenter, T.L., McMeans, M.C. & McHugh, C.P. (1990). Additional instances of human parasitism by the brown dog tick (Acari: Ixodidae). *Journal of Medical Entomology*, **27**, 1056–66.

Cooley, R.A. (1946). The genera *Boophilus*, *Rhipicephalus*, and *Haemaphysalis* (Ixodidae) of the New World. *National Institutes of Health Bulletin*, **No. 187**, 54 pp.

Donatien, A. & Lestoquard, F. (1935). Existence en Algérie d'une *Rickettsia* du chien. *Bulletin de la Société de Pathologie Exotique*, **28**, 418–19.

Eaton, P. (1934). *Piroplasma canis* in Florida. *Journal of Parasitology*, **20**, 312–13.

Enigk, K. (1943). Die Überträger des Pferdepiroplasmose, ihre Verbreitung und Biologie. *Archiv für Wissenschaftliche und Praktische Tierheilkunde*, 78, 209–40.

Ewing, S.A. (1968). Differentiation of hematozoan parasites of dogs. *Southern Veterinarian*, **5**, 8–15.

Ewing, S.A. (1972). Geographic distribution and tick transmission of *Ehrlichia canis*. *Journal of Medical Entomology*, **9**, 597–8.

Goddard, J. (1989). Focus of human parasitism by the brown dog tick, *Rhipicephalus sanguineus* (Acari: Ixodidae). *Journal of Medical Entomology*, **26**, 628–9.

Greene, C.E. & Breitschwerdt, E.B. (1990). Rocky Mountain spotted fever and Q fever. In *Infectious Diseases of the Dog and Cat*, ed. C.E. Greene, pp. 419–33. Philadelphia : W.B. Saunders Company.

Groves, M.G., Dennis, G.L., Amyx, H.L. & Huxsoll, D.L. (1975). Transmission of *Ehrlichia canis* to dogs by ticks (*Rhipicephalus sanguineus*). *American Journal of Veterinary Research*, **36**, 937–40.

Helm, R.W. (1952). Report of the brown dog tick attacking humans. *Entomological News*, **63**, 214.

Hoogstraal, H. (1967). Ticks in relation to human diseases caused by *Rickettsia* species. *Annual Review of Entomology*, **12**, 377–420.

Hoogstraal, H. (1973). Biological patterns in the Afghanistan tick fauna. In *Proceedings of the 3rd International Congress of Acarology, Prague, August 31–September 6, 1971*, ed. M. Daniel & B. Rosický, pp. 511–14. Prague: Academia.

Hooker, W.A., Bishopp, F.C. & Wood, H.P. (1912). The life history and bionomics of some North American ticks. *Bulletin. Bureau of Entomology. United States Department of Agriculture*, **No. 106**, 239 pp.

Hopla, C.E. & Hopla, A.K. (1994). Tularemia. In *Handbook of Zoonoses*, 2nd edn, ed. G.W. Beren & J.H. Steele, pp. 113–25. Boca Raton, FL: CRC Press.

Horak, I.G. (1982). Parasites of domestic and wild animals in South Africa. XIV. The seasonal prevalence of *Rhipicephalus sanguineus* and *Ctenocephalides* spp. on kennelled dogs in Pretoria North. *Onderstepoort Journal of Veterinary Research*, **49**, 63–8.

Horak, I.G. (1995). Ixodid ticks collected at the Faculty of Veterinary Science, Onderstepoort, from dogs diagnosed with *Babesia canis* infection. *Journal of the South African Veterinary Association*, **66**, 170–1.

Inokuma, H., Tamura, K. & Onishi, T. (1995). Incidence of brown dog ticks, *Rhipicephalus san-*

guineus, at a kennel in Okayama Prefecture. *Journal of Veterinary Medical Science*, **57**, 567–8.

Koch, H.G. (1982). Seasonal incidence and attachment sites of ticks (Acari: Ixodidae) on domestic dogs in southeastern Oklahoma and northwestern Arkansas, USA. *Journal of Medical Entomology*, **19**, 293–8.

Koch, H.G. & Tuck, M.D. (1986). Molting and survival of the brown dog tick (Acari: Ixodidae) under different temperatures and humidities. *Annals of the Entomological Society of America*, **79**, 11–14.

Latreille, P.A. (1806). *Genera crustaceorum et insectorum secundum ordinem naturalem in familias disposita, iconibus exemplisque plurimis explicata*, **1**, 302 pp. Paris et Argentorati.

Le Riche, P.D., Soe, A.K., Alemzada, Q. & Sharifi, L. (1988). Parasites of dogs in Kabul, Afghanistan. *British Veterinary Journal*, **144**, 370–3.

Leeson, H.S. (1951). The recorded distribution of the tick *Rhipicephalus sanguineus* (Latreille). *Bulletin of Entomological Research*, **42**, 123–4.

Lombardini, G. (1950). Osservazioni biologiche ed anatomiche sul *Rhipicephalus sanguineus* Latr. (Acarina, Ixodidae). *Redia*, **35**, 173–83.

Mathew, J.S., Ewing, S.A., Barker, R.W., Fox, J.C., Dawson, J.E., Warner, C.K., Murphy, G.L. & Kocan, K.M. (1996). Attempted transmission of *Ehrlichia canis* by *Rhipicephalus sanguineus* after passage in cell culture. *American Journal of Veterinary Research*, **57**, 1594–8.

Noda, H., Munderloh, U.G. & Kurtti, T.J. (1997). Endosymbionts of ticks and their relationship to *Wolbachia* spp. and tick-borne pathogens of humans and animals. *Applied and Environmental Microbiology*, **63**, 3926–32.

Nordgren, R.M. & Craig, T.M. (1984). Experimental transmission of the Texas strain of *Hepatozoon canis*. *Veterinary Parasitology*, **16**, 207–14.

Norval, R.A.I., Daillecourt, T. & Pegram, R.G. (1983). The ticks of Zimbabwe. VI. The *Rhipicephalus sanguineus* group. *Zimbabwe Veterinary Journal*, **13**, 38–46.

Nuttall, G.H.F. (1913). Observations on the biology of Ixodidae. Part 1. *Parasitology*, **6**, 68–118.

Olmeda-García, A.S., Rodríguez-Rodríguez, J.A. & Rojo-Vázquez, F.A. (1993). Experimental transmission of *Dipetalonema dracunculoides* (Cobbold 1870) by *Rhipicephalus sanguineus* (Latreille 1806). *Veterinary Parasitology*, **47**, 339–42.

Parker, R.R. (1933). Recent studies of tick-borne diseases made at the United States Public Health Service Laboratory at Hamilton, Montana. *Proceedings of the 5th Pacific Science Congress*, **6**, 3367–74.

Patton, W.S. & Cragg, F.W. (1913). *A Textbook of Medical Entomology*. London, Madras and Calcutta: Christian Literature Society for India.

Pegram, R.G., Clifford, C.M., Walker, J.B. & Keirans, J.E. (1987a). Clarification of the *Rhipicephalus sanguineus* group (Acari, Ixodoidea, Ixodidae). I. *R. sulcatus* Neumann, 1908 and *R. turanicus* Pomerantsev, 1936. *Systematic Parasitology*, **10**, 3–26.

Pegram, R.G., Keirans, J.E., Clifford, C.M. & Walker, J.B. (1987b). Clarification of the *Rhipicephalus sanguineus* group (Acari, Ixodoidea, Ixodidae). II. *R. sanguineus* (Latreille, 1806) and related species. *Systematic Parasitology*, **10**, 27–44.

Péter, O., Burgdorfer, W., Aeschlimann, A. & Chatelanat, P. (1984). *Rickettsia conorii* isolated from *Rhipicephalus sanguineus* introduced into Switzerland on a pet dog. *Zeitschrift für Parasitenkunde*, **70**, 265–70.

Philip, C. B. & Burgdorfer, W. (1961). Arthropod vectors as reservoirs of microbial disease agents. *Annual Review of Entomology*, **6**, 391–412.

Sapre, S.N. (1944). Some observations on the life history of the dog tick *R. sanguineus* (Latr.) at Mukteswar. *Indian Journal of Veterinary Science*, **14**, 111–12.

Sardey, M.R. & Rao, S.R. (1971). Observations on the life-history and bionomics of *Rhipicephalus sanguineus* (Latreille, 1806) under different temperatures and humidities. *Indian Journal of Animal Science*, **41**, 500–3.

Sen, S.K. (1933). The vector of canine piroplasmosis due to *Piroplasma gibsoni*. *Indian Journal of Veterinary Science*, **3**, 356–63.

Seneviratna, P., Weerasinghe, N. & Ariyadasa, S. (1973). Transmission of *Haemobartonella canis* by the dog tick *Rhipicephalus sanguineus*. *Research in Veterinary Science*, **14**, 112–14.

Shortt, H.E. (1973). *Babesia canis*: the life cycle and laboratory maintenance in its arthropod and mammalian hosts. *International Journal for Parasitology*, **3**, 119–48.

Sweatman, G.K. (1967). Physical and biological factors affecting the longevity and oviposition of engorged *Rhipicephalus sanguineus*. *Journal of Parasitology*, **53**, 432–45.

Viloria, D. (1954). Paralisis por garrapatas en caninos. *Revista de Medicina Veterinaria y Parasitologia Maracay,* **13**, 67–70.

Also see the following Basic References (pp. 12–13): Aeschlimann (1967); Filippova (1997); Hoogstraal (1956).

RHIPICEPHALUS SCULPTUS WARBURTON, 1912

This specific name, from the Latin *sculptura* meaning 'carved', refers to the pattern on the dorsal surface of the male's conscutum.

Diagnosis

A large dark brown to blackish tick.

Male (Figs 179(a), 180(a) to (c))
Capitulum longer than broad, length × breadth ranging from 1.10 mm × 0.96 mm to 1.26 mm × 1.10 mm. Basis capituli with short blunt

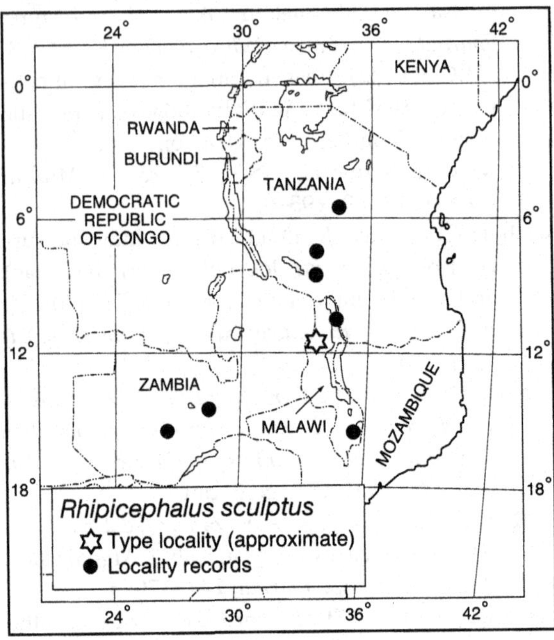

Map 51. *Rhipicephalus sculptus:* distribution.

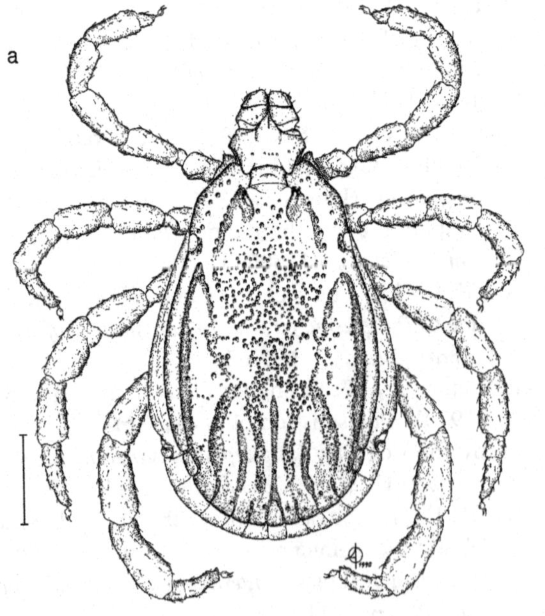

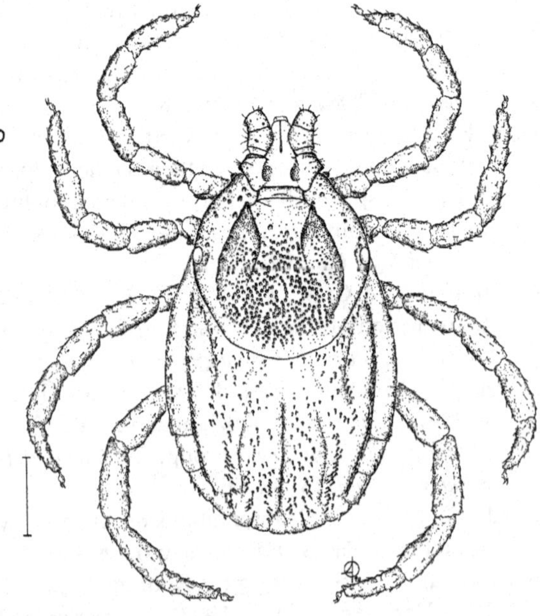

Figure 179. *Rhipicephalus sculptus* [Onderstepoort Tick Collection 2922i, collected from African buffalo (*Syncerus caffer*), Namwala, Zambia, 22 November 1951, by J.G. Matthysse]. (a) Male, dorsal; (b) female, dorsal. Scale bars represent 1 mm. A. Olwage *del.*

Table 47. *Host records of* Rhipicephalus sculptus

Hosts	Number of records
Wild animals	
Burchell's zebra (*Equus burchellii*)	5
Giraffe (*Giraffa camelopardalis*)	1
'Hartebeest'	1
African buffalo (*Syncerus caffer*)	1
Roan antelope (*Hippotragus equinus*)	1
Sable antelope (*Hippotragus niger*)	3
Waterbuck (*Kobus ellipsiprymnus*)	1

lateral angles at about anterior third of its length. Palps somewhat flattened apically. Conscutum length × breadth ranging from 4.29 mm × 2.98 mm to 5.49 mm × 3.89 mm; anterior process on coxae I conspicuous, strongly sclerotized. In engorged specimens body wall expanded laterally and posterolaterally, with a bluntly-rounded caudal process posteromedially. Eyes almost flat, delimited dorsally by deep punctate grooves. Cervical fields finely shagreened anteriorly and along their external margins. Marginal lines long, their outer margins sharply defined, their surfaces shagreened. Posteromedian and posterolateral grooves all long, narrow, tapering anteriorly, shagreened. Posterior to the eyes glossy ridges outline a distinct female pseudoscutum. Between this and the posterior grooves roughly shagreened tracts divide the conscutum into several separate raised areas. Large setiferous punctations present on the scapulae, along the external margins of the cervical fields and scattered amongst the more numerous smaller punctations on the raised medial areas of the conscutum. However, the lateral areas adjacent to the marginal lines and the surfaces surrounding the posterior grooves are glossy and almost free of punctations. The resulting sculptured pattern on the conscutum is very characteristic of this species. Ventrally spiracles with a long narrow gently-curved dorsal prolongation. Adanal plates unique in shape, narrow anteriorly, becoming progressively wider, their posterior margins convex and extending into short sharp points at their junctions with the internal and external margins; accessory adanal plates inconspicuous, sharply pointed.

Female (Figs 179(b), 180(d) to (f))

Capitulum slightly broader than long, length × breadth ranging from 1.09 mm × 1.15 mm to 1.18 mm × 1.23 mm. Basis capituli with blunt lateral angles in anterior third of its length; porose areas large, oval, not quite twice their own diameter apart. Palps broad, slightly flattened apically. Scutum usually slightly broader than long, though somewhat longer than broad in the specimen scanned, length × breadth ranging from 2.22 mm × 2.29 mm to 2.43 mm × 2.51 mm; posterior margin slightly sinuous. Eyes almost flat, delimited dorsally by deep punctate grooves. Cervical fields broad, finely shagreened anteriorly and along their external margins. A few large setiferous punctations present on the scapulae, along the external cervical margins and scattered among the numerous smaller punctations medially on the conscutum. The raised lateral borders and posterior marginal border of the scutum glossy, framing the punctate medial area. Alloscutum with numerous short white setae, especially along the marginal lines. Ventrally genital aperture almost tongue-shaped, smoothly rounded posteriorly.

Nymph (Fig. 181)

Capitulum broader than long, length × breadth ranging from 0.30 mm × 0.35 mm to 0.36 mm × 0.40 mm. Basis capituli about 2.5 times as broad as long with anteriorly-placed, broadly-

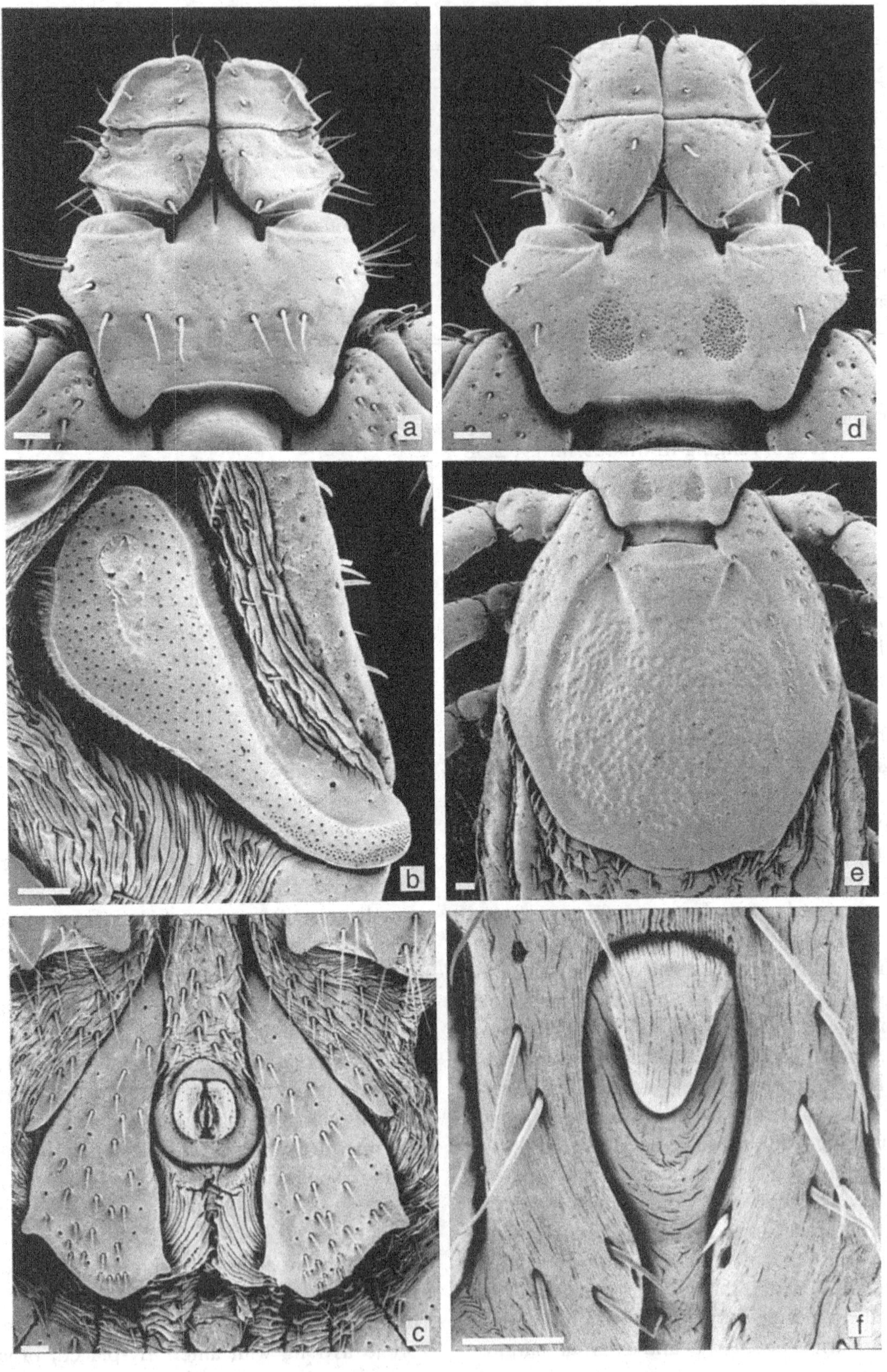
a
d
b
e
c
f

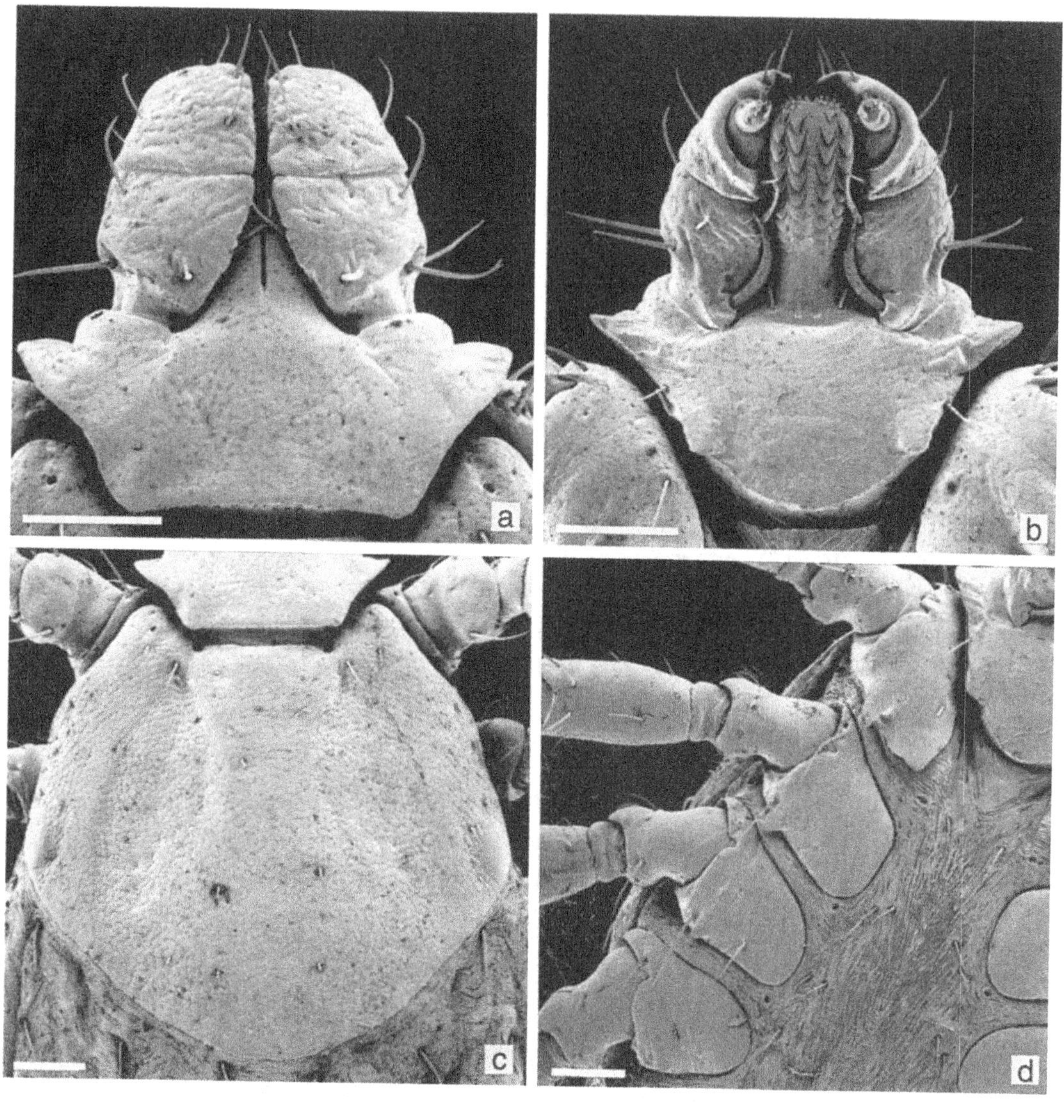

Figure 181 (*above*). *Rhipicephalus sculptus* [L 180. Reared specimens, progeny of ♀ collected from 'hartebeest', Kafue National Park, Zambia, in 1982 by R.G. Pegram]. Nymph: (a) capitulum, dorsal; (b) capitulum, ventral; (c) scutum; (d) coxae. Scale bars represent 0.10 mm. SEMs by J.F. Putterill.

Figure 180 (*opposite*). *Rhipicephalus sculptus* [L 180. Reared specimens, progeny of ♀ collected from 'hartebeest', Kafue National Park, Zambia, in 1982 by R.G. Pegram]. Male: (a) capitulum, dorsal; (b) spiracle; (c) adanal plates. Female: (d) capitulum, dorsal; (e) scutum; (f) genital aperture. Scale bars represent 0.10 mm. SEMs by J.F. Putterill.

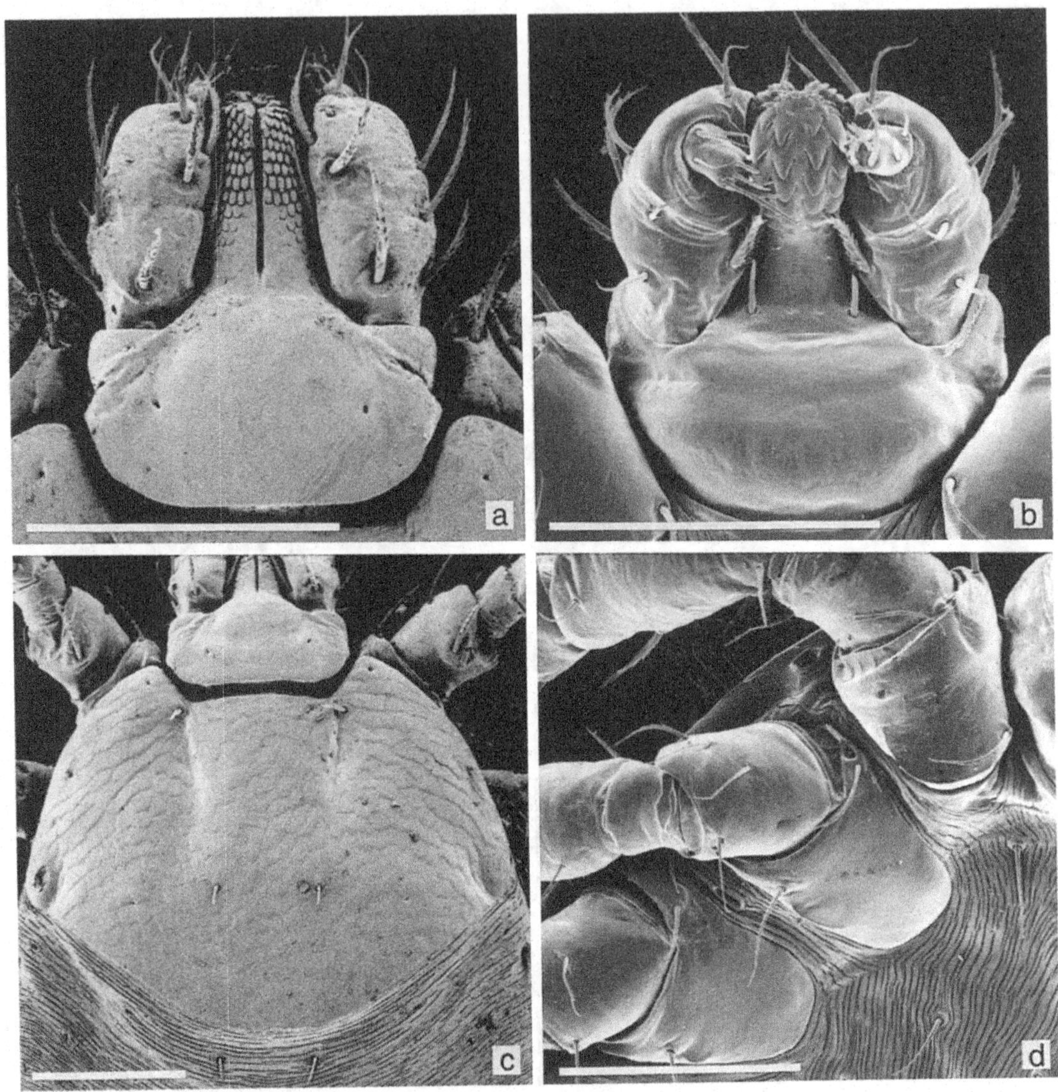

Figure 182. *Rhipicephalus sculptus* [L 180. Reared specimens, progeny of ♀ collected from 'hartebeest', Kafue National Park, Zambia, in 1982 by R.G. Pegram]. Larva: (a) capitulum, dorsal; (b) capitulum, ventral; (c) scutum; (d) coxae. Scale bars represent 0.10 mm. SEMs by J.F. Putterill.

pointed lateral angles projecting over the scapulae; ventrally with short blunt spurs on posterior margin. Palps broad, somewhat flattened apically. Scutum generally broader than long, length × breadth ranging from 0.62 mm × 0.62 mm to 0.69 mm × 0.77 mm; posterior margin a deep smooth curve. Eyes at widest point, a little over halfway back, slightly convex and delimited dorsally by shallow depressions. Cervical fields broad, slightly depressed. Ventrally coxae I each with a small narrow external spur and a larger broader internal spur; coxae II to IV each with progressively smaller external spurs only, that on IV being virtually obsolete.

Larva (Fig. 182)

Capitulum slightly broader than long, length × breadth ranging from 0.134 mm × 0.135 mm to 0.151 mm × 0.158 mm. Basis capituli approximately twice as broad as long with short bluntly-pointed lateral angles; junctions of the posterior and posterolateral sections of the margin smoothly rounded. Scutum much broader than long, length × breadth ranging from 0.257 mm × 0.367 mm to 0.272 mm × 0.379 mm; posterior margin a broad smooth curve. Eyes at widest part of scutum, slightly raised and delimited dorsally by shallow depressions. Cervical grooves short. Ventrally coxae I each with a broadly-rounded spur; coxae II to III each with progressively smaller spurs.

Notes on identification

Colbo (1973) referred to 'large *Rhipicephalus* sp. nymphs' that had been collected from 'Hartebeest' (presumably Lichtenstein's hartebeest, *Sigmoceros lichtensteinii*) and 'Grysbok' (presumably Sharpes' grysbok, *Raphicerus sharpei*) near Ngoma in the Kafue National Park, Zambia, and suggested that they might be either *R. supertritus* or *R. sculptus*. We have compared some of these nymphs from Lichtenstein's hartebeest with nymphs of *R. sculptus* that have since been reared but do not think they belong to this species. They may be *R. supertritus* (see p. 434).

Hosts

A three-host species (R.G. Pegram, unpublished data). The few available records of *R. sculptus* indicate that its preferred hosts are Burchell's zebras and the larger species of African bovids (Table 47). Its sites of attachment on its hosts are unknown.

Zoogeography

To date this species has been recorded from the Manyoni area in Tanzania southwards to parts of Malawi and Zambia (Wilson, 1950; Yeoman & Walker, 1967; MacLeod, 1970; MacLeod & Mwanaumo, 1978; Keirans, 1985) (Map 51). It would be reasonable to assume that it also occurs in adjacent areas of southern Democratic Republic of Congo and parts of northern Mozambique but it apparently does not follow its favoured hosts into all parts of their range, especially in southern Africa. Collections of adults have been made in summer (November to February), during the rainy season, and also in June. Its known collection sites lie at over 1000 m in altitude, with mean annual rainfalls of 800 mm to 1000 mm or more, in various types of woodland.

Disease relationships

Unknown.

REFERENCES

Colbo, M.H. (1973). Ticks of Zambian wild animals: a preliminary checklist. *Puku*, **No.** 7, 97–105.

MacLeod, J. (1970). Tick infestation patterns in the southern province of Zambia. *Bulletin of Entomological Research*, **60**, 253–74.

MacLeod, J. & Mwanaumo, B. (1978). Ecological studies of ixodid ticks (Acari: Ixodidae) in Zambia. IV. Some anomalous infestation patterns in the northern and eastern regions. *Bulletin of Entomological Research*, **68**, 409–29.

Warburton, C. (1912). Notes on the genus *Rhipicephalus*, with the description of new species, and the consideration of some species hitherto described. *Parasitology*, **5**, 1–20.

Wilson, S.G. (1950). A check-list and host-list of Ixodoidea found in Nyasaland, with descriptions and biological notes on some of the rhipicephalids. *Bulletin of Entomological Research*, **41**, 415–28.

Also see the following Basic References (pp. 12–14): Keirans (1985); Santos Dias (1960); Theiler (1947); Wilson (1950); Yeoman & Walker (1967).

RHIPICEPHALUS SENEGALENSIS KOCH, 1844

This specific name is derived from Senegal, where two of the type females apparently originated, plus the Latin adjectival suffix *-ensis* meaning 'belonging to'.

Synonyms

longoides; simus longoides; simus senegalensis.

Diagnosis

A large reddish-brown to dark brown lightly-punctate tick.

Male (Figs 183(a), 184(a) to (c))

Capitulum longer than broad, length × breadth ranging from 0.72 mm × 0.64 mm to 0.98 mm × 0.85 mm. Basis capituli with short acute lateral angles at anterior third of its length. Palps short, broad, somewhat flattened apically. Conscutum length × breadth ranging from 3.11 mm × 2.00 mm to 4.31 mm × 2.84 mm; anterior process of coxae I inconspicuous. In engorged specimens body wall expanded posterolaterally and forming three characteristic caudal processes posteromedially. Eyes almost flat, edged dorsally by a few medium-sized setiferous punctations. Cervical pits deep, discrete. Marginal lines long, punctate. Posterior grooves indistinct; posteromedian long and narrow, posterolaterals much shorter and more-or-less sickle-shaped. A few medium-sized setiferous punctations scattered anteriorly on scapulae; larger setiferous punctations present along external margins of cervical fields and in a '*simus*' pattern medially on the conscutum, becoming more numerous adjacent to the posterior grooves. Very fine punctations present all over the conscutum but these may be virtually invisible in some specimens, leaving the conscutum smooth and shiny. Ventrally spiracles broad with a short somewhat narrower dorsal prolongation extending almost at a right angle. Adanal plates

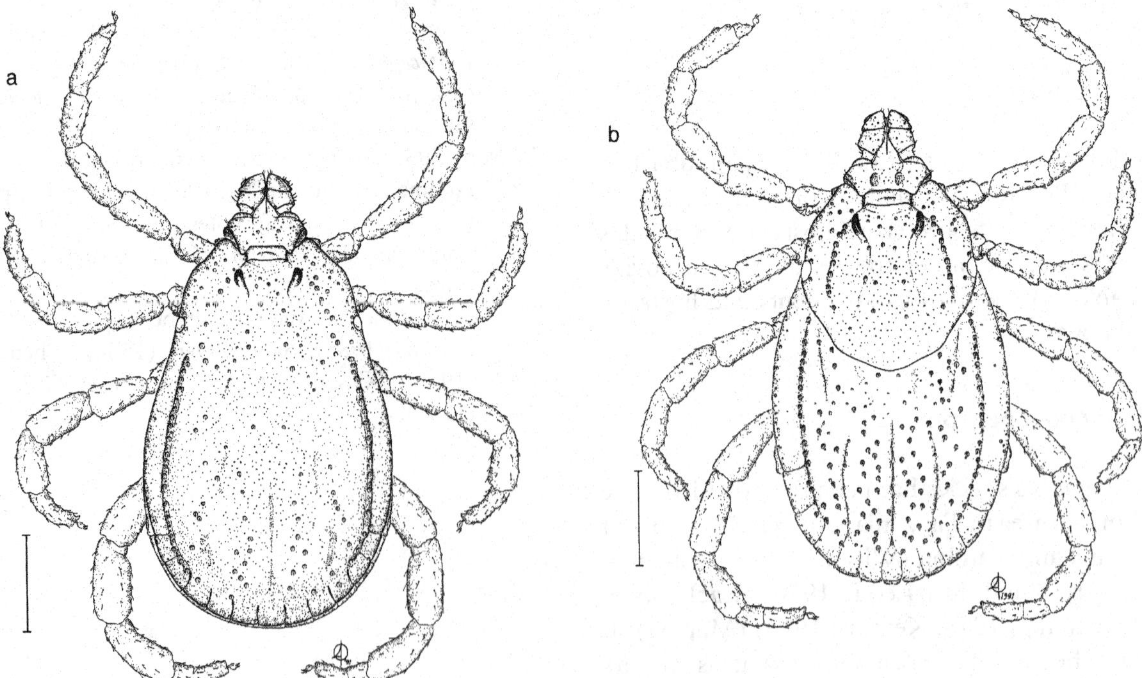

Figure 183. *Rhipicephalus senegalensis* (collected from roadside vegetation, Comoe National Park, Ivory Coast, on 22 to 23 September 1990 by R. Meiswinkel). (a) Male, dorsal; (b) female, dorsal. Scale bars represent 1 mm. A. Olwage *del.*

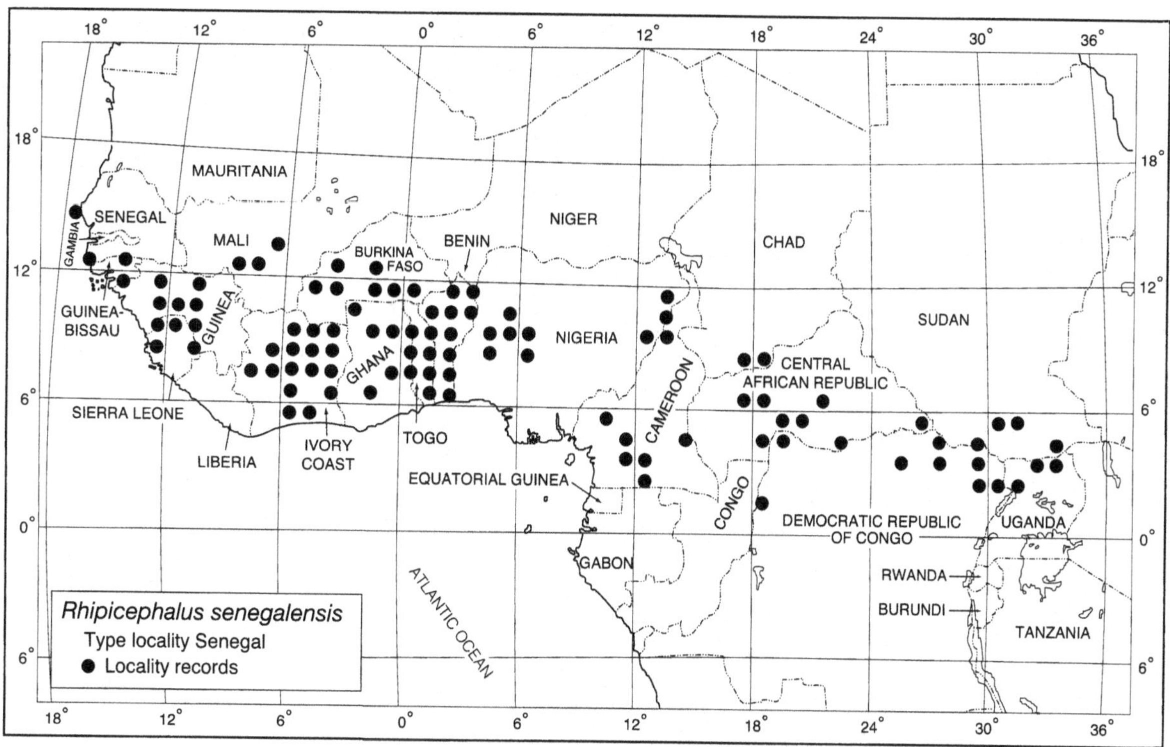

Map 52. *Rhipicephalus senegalensis*: distribution. (Based on Morel, 1969).

broad, sickle-shaped; accessory adanal plates large, pointed, well sclerotized.

Female (Figs 183(b), 184(d) to (f))

Capitulum longer than broad, length × breadth ranging from 0.72 mm × 0.69 mm to 0.95 mm × 0.90 mm. Basis capituli with acute lateral angles at mid-length; porose areas large, oval, about 1.5 times their own diameter apart. Palps broad, longer than those of male. Scutum about as broad as long, length × breadth ranging from 1.31 mm × 1.32 mm to 1.92 mm × 1.92 mm; posterior margin slightly sinuous. Eyes almost flat, edged dorsally by a few medium-sized setiferous punctations. Cervical pits deep, convergent; cervical fields broad, only slightly depressed, their external margins clearly delimited by rows of large setiferous punctations. A few medium-sized setiferous punctations scattered on scapulae and between the cervical fields, interspersed with numerous fine punctations, especially medially. Ventrally genital aperture with the sides of the opening curving outwards from its broadly-rounded posterior margin.

Nymph (Fig. 185)

Capitulum broader than long, length × breadth ranging from 0.28 mm × 0.33 mm to 0.31 × 0.36 mm. Basis capituli over three times as broad as long with tapering sharply-pointed lateral angles in posterior half of its length projecting over scapulae, posterior margin slightly concave; ventrally with long sharp spurs on posterior margin. Palps tapering to rounded apices, their lateral margins slightly concave, inclined inwards. Scutum longer than broad, length × breadth ranging from 0.51 mm × 0.46 mm to 0.62 mm × 0.57 mm; posterior margin an almost smooth curve. Eyes at widest point, well over halfway back, elongate and slightly raised. Cervical pits deep; cervical fields long, narrow, slightly divergent, with pronounced external margins running parallel to the lateral margins of the

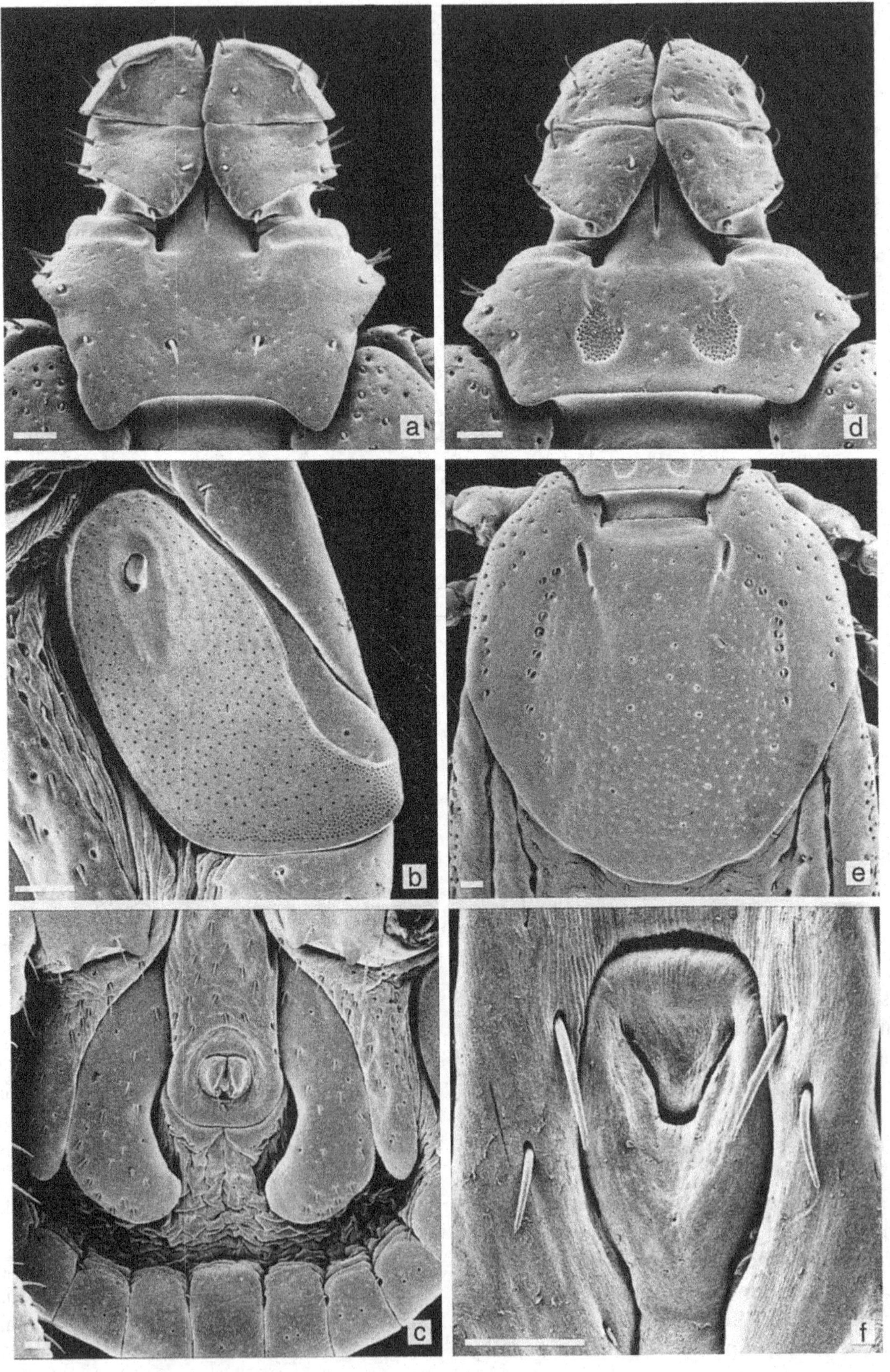
a
b
c
d
e
f

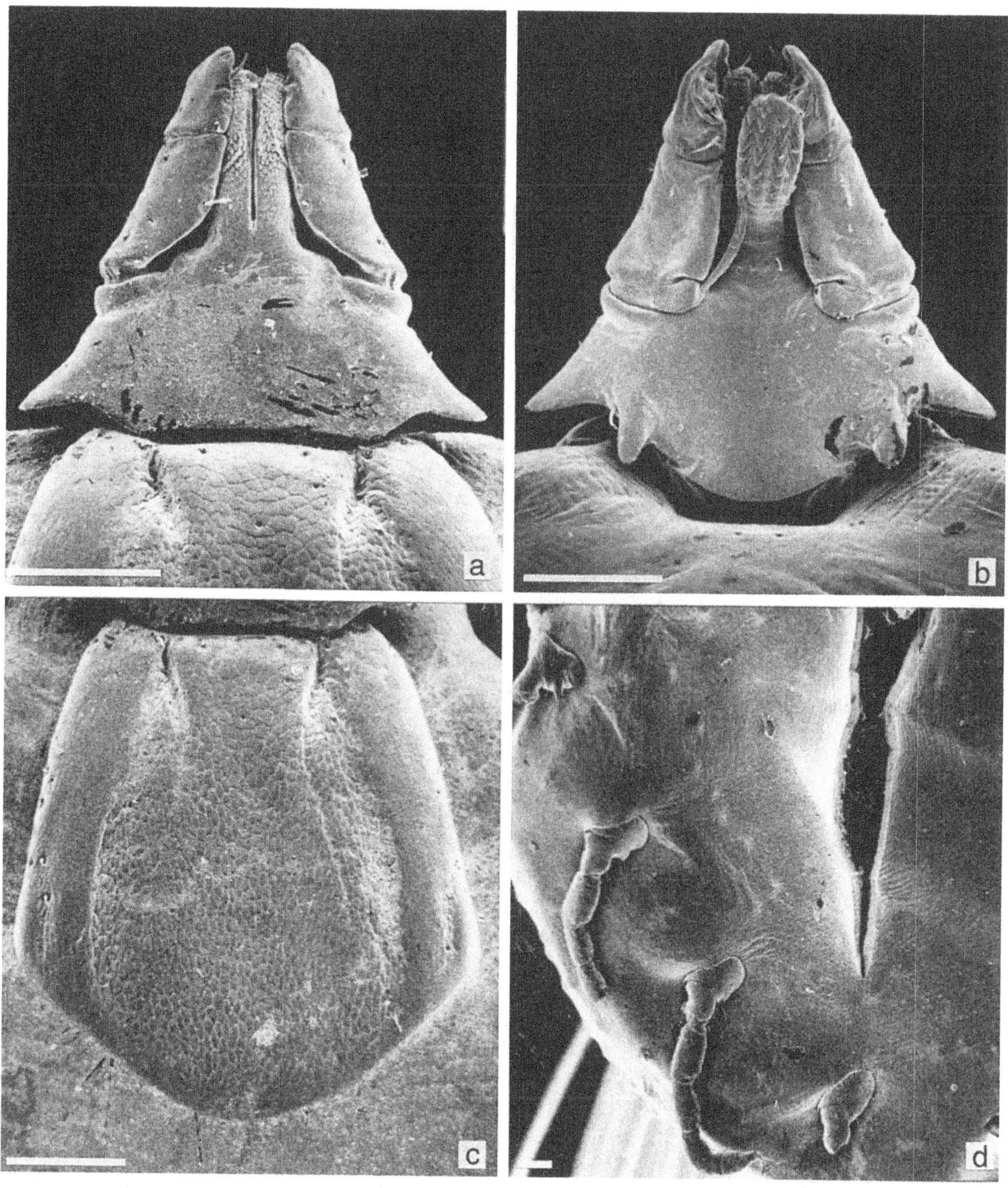

Figure 185 (*above*). *Rhipicephalus senegalensis* (RML 36780, collected from rodent nest, Garamba Park, Democratic Republic of Congo, on 1 August 1951 by J.V. Verschuren). Nymph: (a) capitulum, dorsal; (b) capitulum, ventral; (c) scutum; (d) coxae. Scale bars represent 0.10 mm. SEMs by J.F. Putterill.

Figure 184 (*opposite*). *Rhipicephalus senegalensis* (collected from roadside vegetation, Comoe National Park, Ivory Coast, on 22 to 23 September 1990 by R. Meiswinkel). Male: (a) capitulum, dorsal; (b) spiracle; (c) adanal plates. Female: (d) capitulum, dorsal; (e) scutum; (f) genital aperture. Scale bars represent 0.10 mm. SEMs by J.F. Putterill.

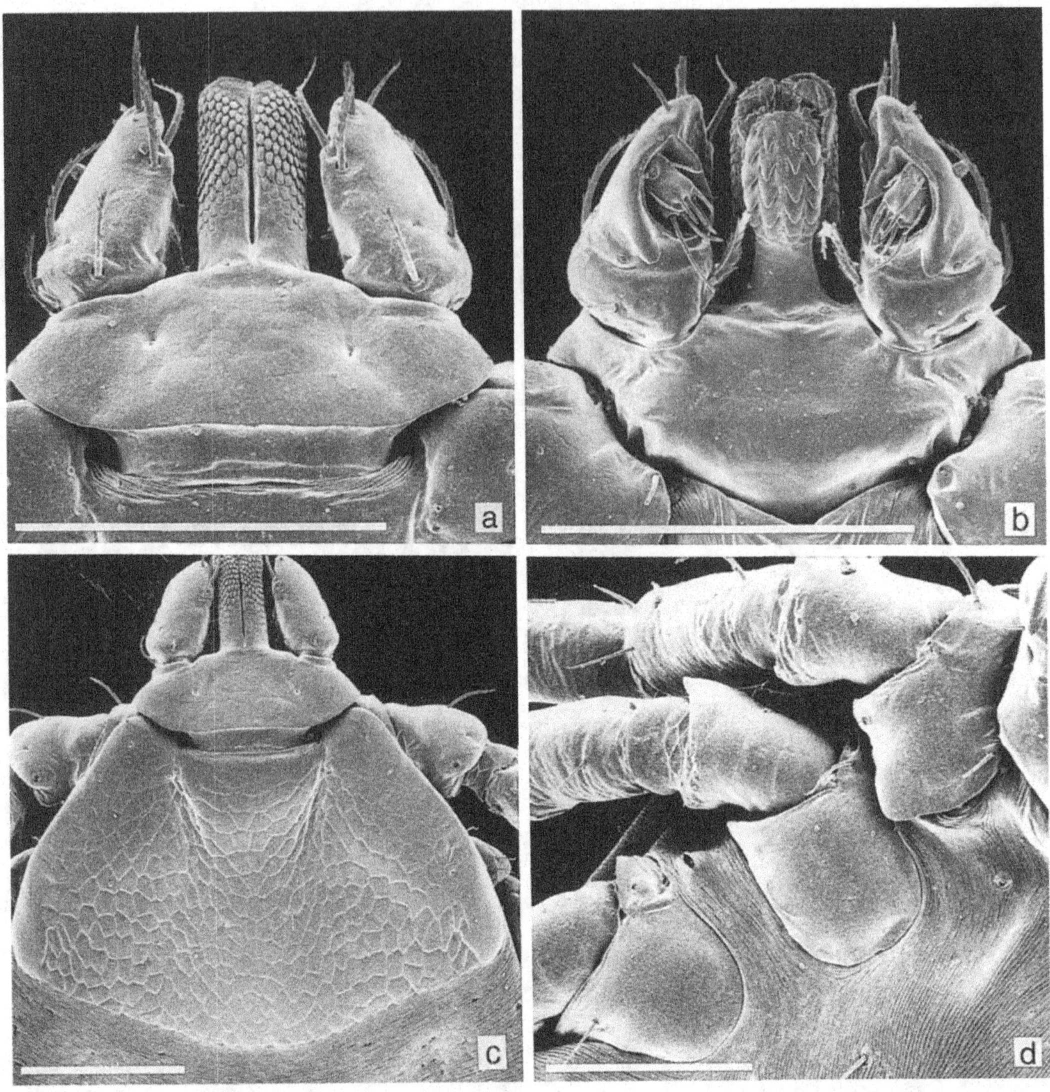

Figure 186. *Rhipicephalus senegalensis* (progeny of ♀ collected from bovine, Sangalkam, Senegal, in August 1963 and donated by P.C. Morel). Larva: (a) capitulum, dorsal; (b) capitulum, ventral; (c) scutum; (d) coxae. Scale bars represent 0.10 mm. SEMs by J.F. Putterill.

scutum. Ventrally coxae I each with two large triangular spurs; coxae II to IV each with a rounded external spur, decreasing progressively in size.

Larva (Fig. 186)

Capitulum much broader than long, length × breadth ranging from 0.081 mm × 0.145 mm to 0.108 mm × 0.157 mm. Basis capituli well over three times as broad as long with short broad bluntly-pointed lateral angles, and its long posterior margin very slightly concave. Palps broad, virtually equal in width for much of their length, tapering abruptly to broadly-rounded apices, inclined inwards. Scutum much broader than long, length × breadth ranging from 0.205 mm × 0.304 mm to 0.224 mm × 0.331 mm; posterior margin broad, shallow, very slightly concave posterolaterally. Eyes at widest part of scutum, very far back, large, flat. Cervical grooves short, convergent. Ventrally coxae I each with a broad bluntly-rounded spur; coxae II and III each with an indication only of a spur on its posterior margin.

Notes on identification

Vassiliades (1964) has given a detailed review of *R. senegalensis*, particularly its morphology and biology. Unfortunately, though, neither he nor any of us have re-examined the three syntype females of this species, one said to be from Egypt and two from Senegal (Zoological Museum, Berlin 1096), listed by Moritz & Fischer (1981). Probably no one now believes that *R. senegalensis* occurs in Egypt. Regarding this tick Hoogstraal (1956) noted: 'A female specimen of the *R. simus* group from Koch's time would be difficult to identify with any degree of certainty . . .'. He thought it had either been misidentified (page 756) or mislabelled (page 758). We agree and, like Vassiliades, have disregarded this record.

The synonymy of *R. senegalensis* is somewhat complicated (Clifford & Anastos (1962). Neumann (1897) synonymized this species with *R. simus* and his finding was accepted for some years by other authors. Later Zumpt (1943) described what he thought was a new subspecies *R. simus longoides*, though he mentioned the female from Senegal described by Koch (1844) and noted that it might be a synonym. In 1949, without explanation and apparently without having seen any of Koch's syntype females, Zumpt used the name *R. simus senegalensis*, with *R. simus longoides* as its synonym. These names appeared widely in the literature for some years. In 1955, though, Santos Dias stated that he regarded *R. senegalensis* as a valid species and this decision, which was supported by Tendeiro (1959) and Clifford & Anastos (1962), now prevails.

In more recent years increasing reliance has been placed on specific differences in the structure of the mounted genital aperture to separate the females of *R. pseudolongus*, *R. longus* and *R. senegalensis*, as discussed in detail by Clifford & Anastos (1962). (See p. 261, under *R. longus*, Fig. 113). Care must also be taken to distinguish *R. senegalensis* females from those of *R. praetextatus* (Pegram *et al.*, 1987). (See p. 421, under *R. simus*, Fig. 197).

We have taken a conservative view of the distribution of *R. senegalensis* and regard it as a species occurring north of the Equator from Senegal, its type locality in West Africa, eastwards to parts of the southern Sudan and northern Uganda (Clifford & Anastos, 1962; Morel, 1969). Further studies may show that this concept is too restricted. In the past it has also been recorded from further south in the Democratic Republic of Congo and from Kenya, Tanzania, Malawi and Mozambique (Hoogstraal, 1954; Santos Dias, 1960; Vassiliades, 1964; Elbl & Anastos, 1966; Morel, 1969; Keirans, 1985). It was not, however, listed from Mozambique by Santos Dias (1993).

Hosts

A three-host species (Vassiliades, 1964). *Rhipicephalus senegalensis* adults are essentially parasites of ungulates (Table 48). The most commonly recorded domestic hosts are cattle, and to a much lesser extent dogs. Numerous wild ungulates apparently act as hosts, especially the warthog and African buffalo. Morel (1958) also cited several earlier records of *R. simus* that probably refer to this species from various animals, including the African civet (*Civettictis civetta*) and aardvark (*Orycteropus afer*). We are not certain about this though, so these records do not qualify for inclusion in the table.

The immature stages parasitize small mammals, especially rodents. As well as several rodent species Clifford & Anastos (1964) listed all stages from rodent nests; the adults had doubtless moulted there from engorged nymphs that had fed on the rodents.

At Sangalkam, Senegal, the adults are most active from June to August, during the rainy season. In the Niayes region, along the north-western seaboard of Senegal, they are found only during the rains. They become extremely abundant on cattle then, particularly on their ears. When the greatest numbers are present, though, they may attach almost anywhere on the animal, including the eyelids, round the nostrils, on the

Table 48. *Host records of* Rhipicephalus senegalensis

Hosts	Number of records
Domestic animals	
Cattle	188
Sheep	7
Goats	4
Horses	5
Pigs	5
Dogs	23
Wild animals	
White-toothed shrew (*Crocidura* sp.)	1
Hunting dog (*Lycaon pictus*)	1
Lion (*Panthera leo*)	1
Leopard (*Panthera pardus*)	1
Banded mongoose (*Mungos mungo*)	1 (nymph)
Small-spotted genet (*Genetta genetta*)	1
African elephant (*Loxodonta africana*)	3
Black rhinoceros (*Diceros bicornis*)	1
Warthog (*Phacochoerus africanus*)	27
Red river hog (*Potamochoerus porcus*)	2
'Wild pig'	1
Giraffe (*Giraffa camelopardalis*)	1
Kanki (*Alcelaphus buselaphus major*)	1
Lichtenstein's hartebeest (*Sigmoceros lichtensteinii*)	1
Oribi (*Ourebia ourebi*)	1
African buffalo (*Syncerus caffer*)	23
Eland (*Taurotragus oryx*)	1
Bongo (*Tragelaphus eurycerus*)	1
Bushbuck (*Tragelaphus scriptus*)	1
Bay duiker (*Cephalophus dorsalis*)	1
Common duiker (*Sylvicapra grimmia*)	1
Roan antelope (*Hippotragus equinus*)	2
Bohor reedbuck (*Redunca redunca*)	1
Giant ground pangolin (*Manis gigantea*)	1
Cuvier's tree squirrel (*Funisciurus pyrrhopus*)	1 (1 nymph)
'Gerbil' (*Taterillus* sp.)	1 (immatures)
Striped grass rat (*Lemniscomys striatus*)	3 (1 ♂, 2 nymphs)
Crested porcupine (*Hystrix cristata*)	1
'Cane rat' (*Thryonomys* sp.)	1 (adults, 2 nymphs)
Cape hare (*Lepus capensis*)	1
'Elephant shrew' (*Elephantulus* sp.)	1 (nymph)
Birds	
White-faced whistling duck (*Dendrocygna viduata*)	1
Humans	2

lower jaw and on the shoulders, back, flanks, tail and feet (Vassiliades, 1964; Gueye *et al.*, 1986).

Zoogeography

We regard *R. senegalensis* primarily as a West African species whose distribution also extends eastwards through northern Democratic Republic of Congo, parts of northern Uganda and the southern Sudan (Hoogstraal, 1956; Clifford & Anastos, 1964; Morel, 1969; Matthyse & Colbo, 1987) (Map 52).

In West Africa it is prevalent at low altitudes, up to about 500 m, in lowland and Guineo–Congolian rainforest and mosaics of this forest and secondary grassland. It is usually abundant only in places receiving a minimum of 1000 mm rainfall annually, distributed over at least 5 to 6 months. It may, however, exist in some lower rainfall areas provided that the humidity in the microclimate is high enough.

East of the Cameroon-Adamawa mountains, on the periphery of the Congo basin, it is less abundant although the climate is analagous to that in West Africa. However it is somewhat higher in altitude, ranging from 500 m to 1000 m. In these areas its distribution overlaps that of *R. longus* (Morel, 1969).

Disease relationships

Specimens of *R. senegalensis* (syn. *R. simus senegalensis*) that were naturally infected with Q-fever (*Coxiella burneti*) were reported to have been found in Guinea-Bissau (formerly Portuguese Guinea) (Hoogstraal, 1956).

REFERENCES

Gueye, A., Mbengue, M., Diouf, A. & Seye, M. (1986). Tiques et hémoparasitoses du bétail au Sénégal. 1. La région des Niayes. *Revue d'Élevage et de Médecine Vétérinaire des Pays Tropicaux*, **39** (nouvelle série), 381–93.

Hoogstraal, H. (1954). Noteworthy African tick records in the British Museum (Natural History) collections. *Proceedings of the Entomological Society of Washington*, **56**, 273–9.

Koch, C.L. (1844). Systematische Uebersicht über die Ordung der Zecken. *Archiv für Naturgeschichte*, **10**, 217–39.

Morel, P.C. (1958). Les tiques des animaux domestiques de l'Afrique occidentale française. *Revue d'Élevage et de Médecine Vétérinaire des Pays Tropicaux*, **11** (nouvelle série), 153–89.

Neumann, L.G. (1897). Révision de la famille des ixodidés. (2ᵉ Mémoire). *Mémoires de la Société Zoologique de France*, **10**, 324–420.

Pegram, R.G., Walker, J.B., Clifford, C.M. & Keirans, J.E. (1987). Comparison of populations of the *Rhipicephalus simus* group: *R. simus, R. praetextatus*, and *R. muhsamae* (Acari: Ixodidae). *Journal of Medical Entomology*, **24**, 666–82.

Santos Dias, J.A.T. (1955). A propósito de uma coleção de carraças do Sudão Anglo-Egípcio. Algumas considerações sobre o *Rhipicephalus longus* Neumann, 1907. *Boletim da Sociedade de Estudos de Moçambique*, **No. 92**, 103–18.

Santo Dias, J.A.T. (1993). Some data concerning the ticks (Acarina – Ixodoidea) presently known in Mozambique. *Garcia de Orta, Serie de Zoologia, Lisboa*, **18 for 1991**, 27–48.

Tendeiro, J. (1959). Sur quelques ixodidés du Mozambique et de la Guinée Portugaise. 1. *Boletim Cultural da Guiné Portuguesa*, **14**(53), 21–95, figs 1–12.

Vassiliades, G. (1964). Contribution à la connaissance de la tique Africaine *Rhipicephalus senegalensis* Koch, 1844. *Annales de la Faculté des Sciences de l'Université de Dakar*, **14**, 71–104.

Zumpt, F. (1943). *Rhipicephalus simus* Koch und verwandte Arten. VII. Vorstudie zu einer Revision der Gattung *Rhipicephalus*. *Zeitschrift für Parasitenkunde*, **13**, 1–24.

Also see the following Basic References (pp. 12–14): Clifford & Anastos (1962, 1964); Elbl & Anastos (1966); Hoogstraal (1956); Keirans (1985); Matthysse & Colbo (1987); Morel (1969); Moritz & Fischer (1981); Santos Dias (1960); Zumpt (1949).

RHIPICEPHALUS SERRANOI SANTOS DIAS, 1950

This species was named after the collector of the type specimens, Dr António de Melo Serrano.

Diagnosis

A small reddish-brown tick.

Male (Figs 187(a), 188(a) to (c))
Capitulum as broad as long to slightly longer than broad, its length × breadth in the two specimens measured 0.42 mm × 0.40 mm and 0.44 mm × 0.44 mm, respectively. Basis capituli with short acute lateral angles in anterior half of its length. Palps short, their apices rounded. Conscutum length × breadth in the two specimens 1.96 mm × 1.26 mm and 2.03 mm × 1.43 mm respectively; anterior process of coxae I small, rounded. Even in partially engorged males no indication of a caudal process seen. Eyes flat, edged by a few small punctations. Cervical pits comma-shaped, convergent; external margins of cervical fields delimited by large setiferous punctations. Marginal lines long, deep, punctate. Posterior grooves in the form of two median punctate depressions flanked on either side by an irregularly shaped aggregation of large punctations. On the conscutum in general punctations medium-sized, numerous, tending to become larger, and sometimes more-or-less confluent, posteromedially. Ventrally spiracles with a long broad prolongation curving gently towards the

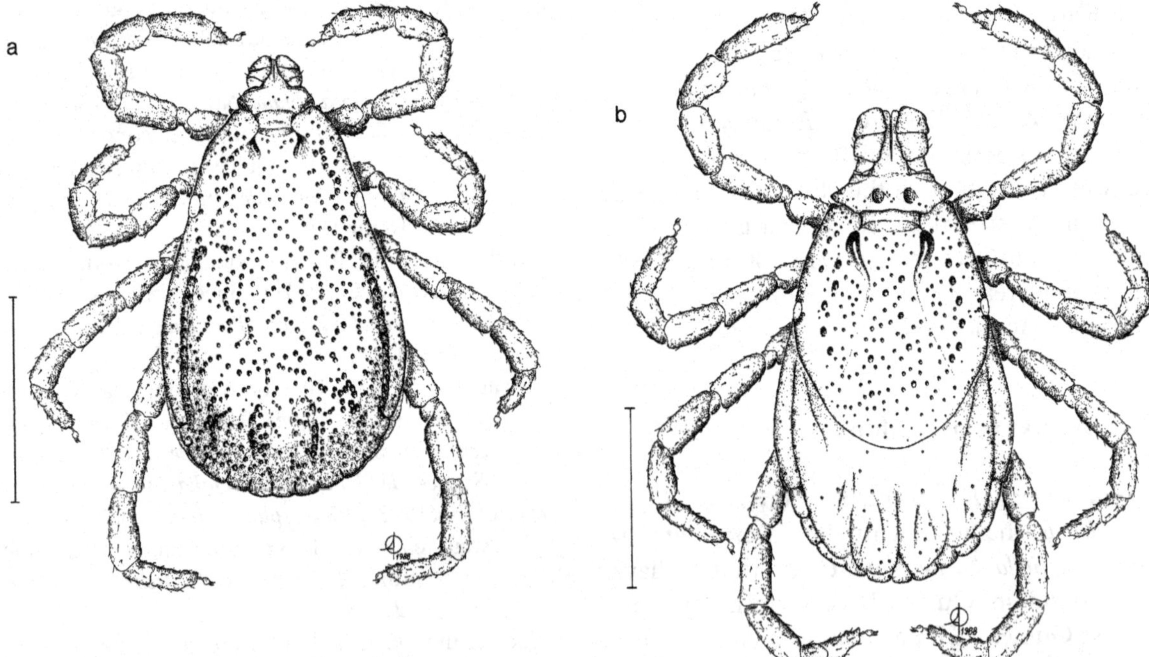

Figure 187 (*above*). *Rhipicephalus serranoi* [Zumpt's Ref. No. 142, collected from klipspringer (*Oreotragus oreotragus*), Chipangali, Zambia, on 28 March 1963 by F. Zumpt]. (a) Male, dorsal; (b) female, dorsal. Scale bars represent 1 mm. A. Olwage *del.*

Figure 188 (*opposite*). *Rhipicephalus serranoi* [Zumpt's Ref. No. 142, collected from klipspringer (*Oreotragus oreotragus*), Chipangali, Zambia, on 28 March 1963 by F. Zumpt]. Male: (a) capitulum, dorsal; (b) spiracle; (c) adanal plates. Female: (d) capitulum, dorsal; (e) scutum; (f) genital aperture. Scale bars represent 0.10 mm. SEMs by J.F. Putterill.

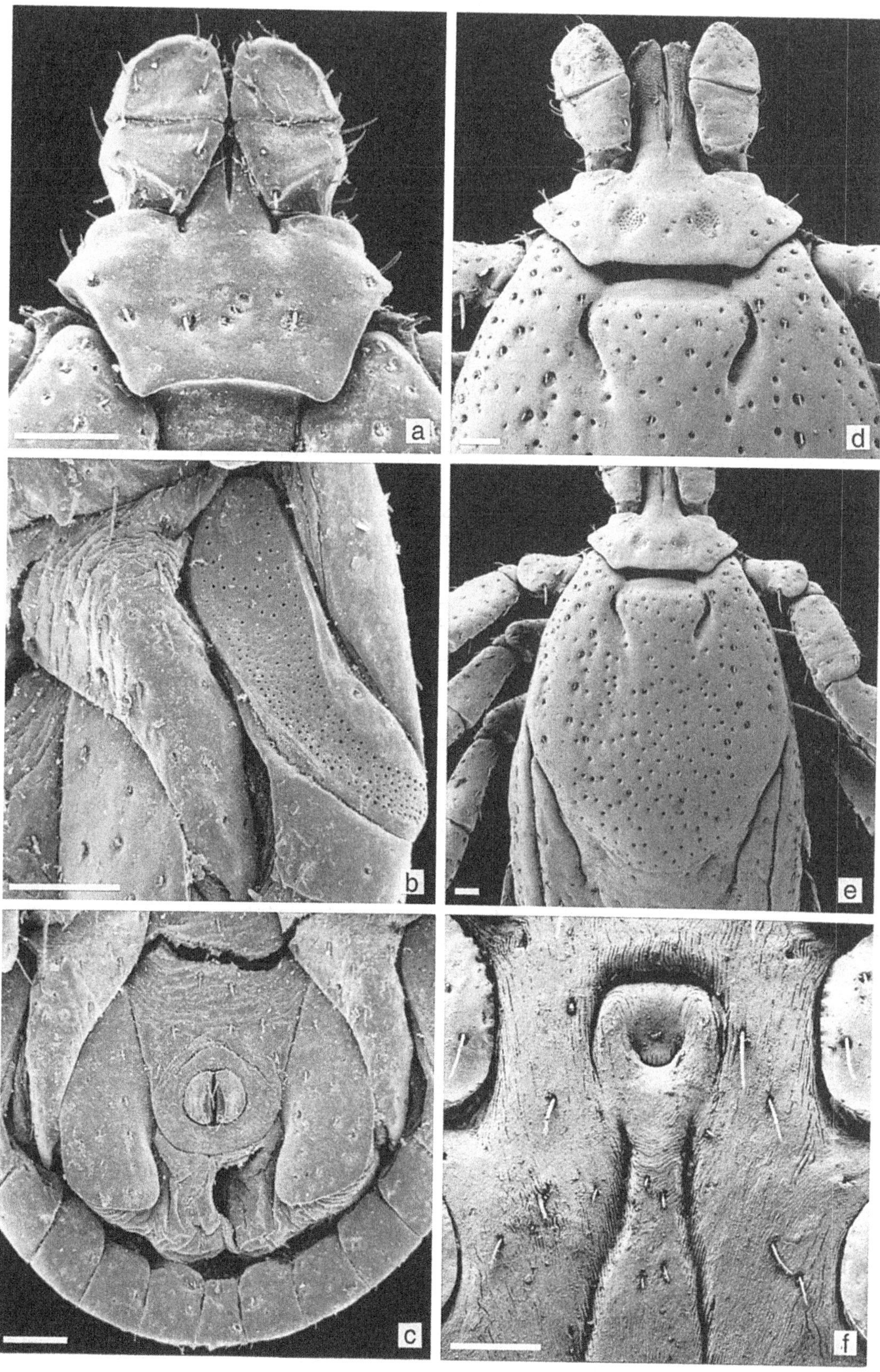
a
d
b
e
c
f

Table 49. *Host records of* Rhipicephalus serranoi

Hosts	Number of records
Wild animals	
Leopard (*Panthera pardus*)	1
Tree hyrax (*Dendrohyrax arboreus*)	1
Yellow-spotted rock hyrax (*Heterohyrax brucei*)	1
Rock hyrax (*Procavia capensis*)	1
Klipspringer (*Oreotragus oreotragus*)	2

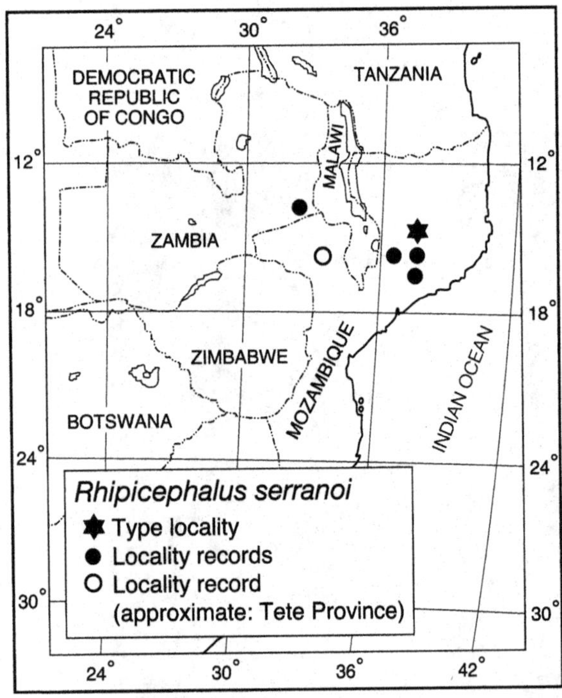

Map 53. *Rhipicephalus serranoi*: distribution.

dorsal surface. Adanal plates short, broad, their posteroexternal margins almost angular, their posterointernal margins smoothly rounded; accessory adanal plates sharply pointed.

Female (Figs 187(b), 188(d) to (f))

Capitulum broader than long, its length × breadth ranging from 0.61 mm × 0.63 mm to 0.64 mm × 0.70 mm in the five specimens measured. Basis capituli with acute lateral angles at about mid-length; porose areas rather small, almost twice their own diameter apart. Palps long, with article I easily visible. Scutum longer than broad, length × breadth ranging from 1.26 mm × 1.11 mm to 1.42 mm × 1.22 mm; posterior margin a broad deep curve, very slightly sinuous laterally. Eyes flat, edged dorsally by a few small punctations. Cervical pits long, convergent; cervical fields long, their external margins demarcated by large setiferous punctations. A few medium-sized setiferous punctations present on scapulae and scattered medially on the scutum, where they are interspersed with numerous evenly distributed small punctations. Ventrally genital aperture a broad shallow U-shape.

Immature stages

Unknown.

Notes on identification

We have not seen the type series of *R. serranoi* (2 XX, 5 YY). However, some of the specimens on which our descriptions of the adults were based were identified, and kindly donated to us, by Dr Santos Dias. In several important respects, though, our interpretation of this species differs markedly from his original description. For example, in his original illustration of the male's dorsal surface the tick's eyes are not flat but appear to bulge; the conscutum is more densely and evenly punctate than it is in our specimens, and he has indicated three evenly shaped posterior depressions on the conscutum (the usual arrangement in *Rhipicephalus* males), not the four irregularities seen in our specimens. In his illustration of the female he shows a more densely and evenly punctate tick than ours, with no clear indication of the rows of large punctations along the external cervical margins seen in our specimens. These anomalies should be remembered by anyone trying to identify this rare tick.

In our experience care must be taken not to confuse *R. serranoi* females with those of *R. oreotragi sp. nov.* (p. 330) as they are very similar morphologically. The males of these two species are easier to differentiate. Present indications are that these two species do not occur in the same areas.

Hosts

Life cycle unknown. Santos Dias (1993) regarded this rare rhipicephalid as a parasite of hyraxes (dassies) (Table 49). This may well be, but it could also prove to be primarily a klipspringer parasite. Four of its five known hosts, including the leopard, often inhabit the same rocky outcrops (Tendeiro, 1959). Leopards prey on both the hyraxes and klipspringers.

Zoogeography

To date *R. serranoi* has been recorded only in a small part of north-eastern Zambia and northern Mozambique, including Tete Province (Santos Dias, 1993) (Map 53). These localities are at altitudes between 500 m and 1500 m with mean annual rainfalls of ± 1000 mm in miombo woodland. It must be remembered, though, that the precise conditions under which the ticks actually live are probably moderated by the special habitat requrements of their hosts.

Disease relationships

Unknown.

REFERENCES

Santos Dias, J.A.T. (1950). Mais uma nova espécie de carraça para a fauna de Moçambique. *Moçambique*, **No. 63**, 143–51.

Santos Dias, J.A.T. (1993). Some data concerning the ticks (Acarina – Ixodoidea) presently known in Mozambique. *Garcia de Orta, Serie de Zoologia, Lisboa*, **18 for 1991**, 27–48.

Tendeiro, J. (1959). Sur quelques ixodidés du Mozambique et de la Guinée Portugaise. 1. *Boletim Cultural da Guiné Portuguesa*, **14** (53), 21–95, figs. 1–12.

RHIPICEPHALUS SIMPSONI NUTTALL, 1910

This species was named after the collector of the type specimens, J.J. Simpson, a British entomologist who worked for a number of years in West Africa.

Diagnosis

A medium-sized, glossy, often lightish-brown, tick.

Male (Figs 189(a), 190(a) to (c))

Capitulum broader than long to slightly longer than broad, length × breadth ranging from 0.43 mm × 0.49 mm to 0.81 mm × 0.78 mm. Basis capituli with sharply-pointed lateral angles in anterior third of its length. Palps short, tapering apically. Conscutum length × breadth ranging from 2.01 mm × 1.27 mm to 3.73 mm × 2.54 mm; anterior process of coxae I small, blunt. In engorged specimens a single short blunt caudal process present. Eyes flat, edged dorsally by a few punctations. Cervical pits short; cervical fields long, narrow, inconspicuous. Marginal lines long, though not reaching eye level, outlined by punctations. Posteromedian groove long and slender, posterolateral grooves shorter and broader. A few shallow medium-sized setiferous punctations present along the external margins of the cervical fields and in a '*simus*' pattern

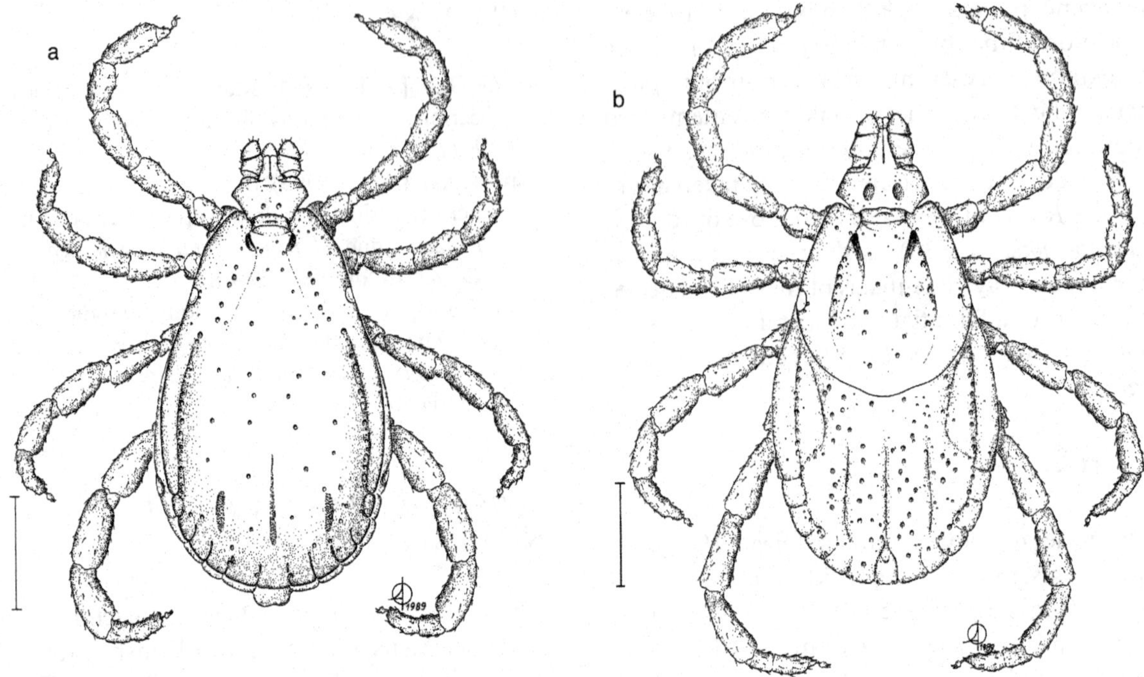

Figure 189. *Rhipicephalus simpsoni* [Onderstepoort Tick Collection 2490iii, collected from a greater cane rat (*Thryonomys swinderianus*), Yaoundé, Cameroon, on 28 July 1952 by J. Rageau]. (a) Male, dorsal; (b) female, dorsal. Scale bars represent 1 mm. A. Olwage *del.*

Table 50. *Host records of* Rhipicephalus simpsoni

Hosts	Number of records
Wild animals	
Sykes' monkey (*Cercopithecus albogularis*)	1
Royal antelope (*Neotragus pygmaeus*)	2
African buffalo (*Syncerus caffer*)	1
'Duiker'	2
Giant Gambian rat (*Cricetomys gambianus*)	3
'Marsh rat'	1
African brush-tailed porcupine (*Atherurus africanus*)	6 (nymphs)
Lesser cane rat (*Thryonomys gregorianus*)	3
Greater cane rat (*Thryonomys swinderianus*)	942 (including immatures)
Cane rat (*Thryonomys* sp.)	10
Birds	
Yellow warbler (*Chloropeta natalensis*)	1 (1 X)

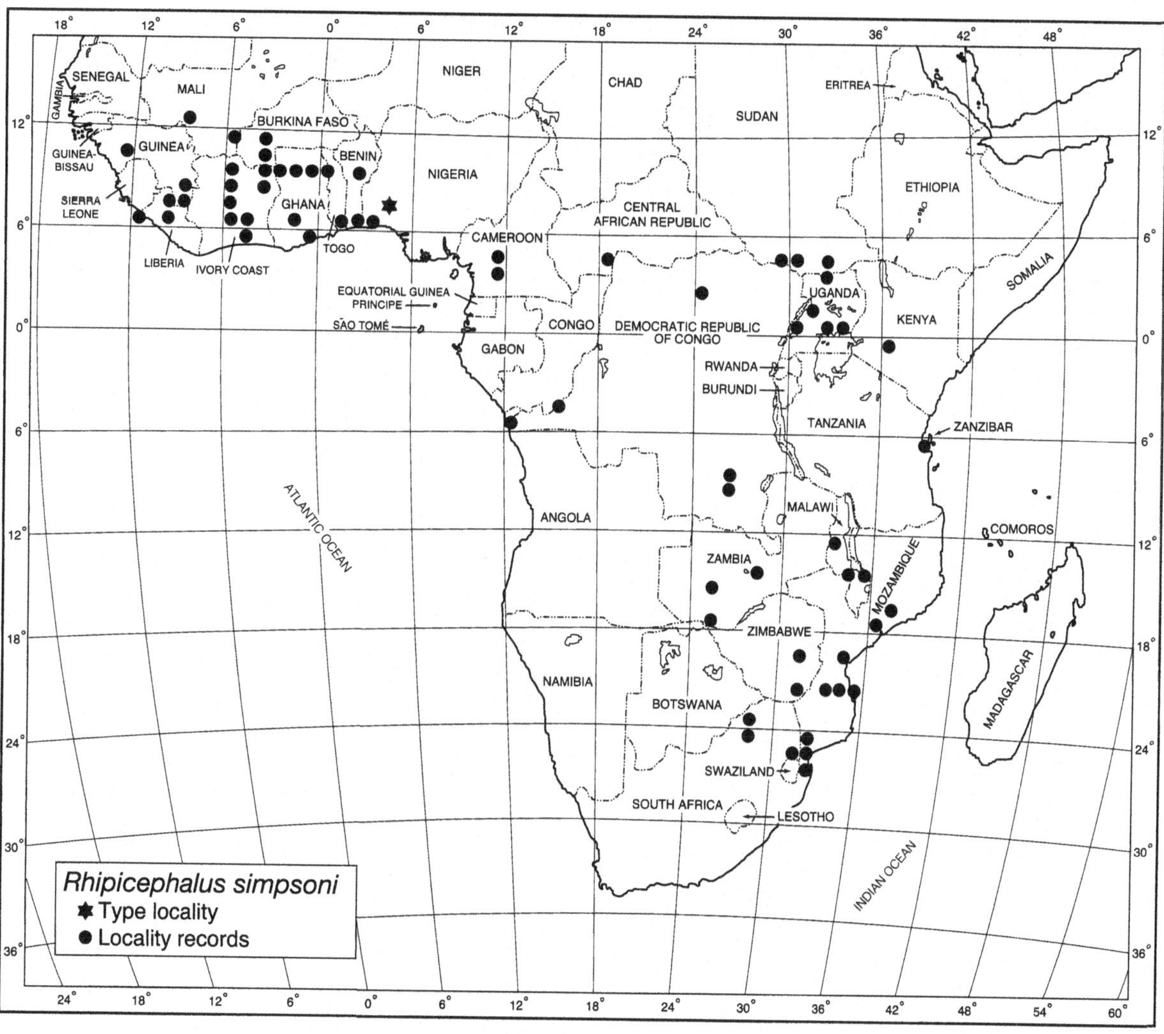

Map 54. *Rhipicephalus simpsoni:* distribution.

dorsally on the conscutum, sometimes interspersed with very fine superficial punctations, but the conscutum is usually smooth and shiny. Ventrally spiracles with a short broad curved dorsal prolongation. Adanal plates broadly sickle-shaped; accessory adanal plates long, pointed and well sclerotized.

Female (Figs 189(b), 190(d) to (f))

Capitulum broader than long to slightly longer than broad, length × breadth ranging from 0.58 mm × 0.64 mm to 0.88 mm × 0.87 mm. Basis capituli with sharply-pointed lateral angles at mid-length; porose areas rounded, not quite twice their own diameter apart. Palps quite long, narrowing apically. Scutum slightly longer than broad to broader than long, length × breadth ranging from 1.03 mm × 1.14 mm to 1.79 mm × 1.77 mm, posterior margin slightly sinuous. Eyes at mid-length, flat, edged dorsally by a few punctations. Cervical pits short, convergent; cervical fields long and narrow, their external margins delimited by steep ridges with a few medium-sized setiferous punctations. A few similar punctations scattered medially on the scutum, sometimes interspersed with numerous very fine superficial punctations, but the scutum is usually smooth and shiny. Ventrally genital

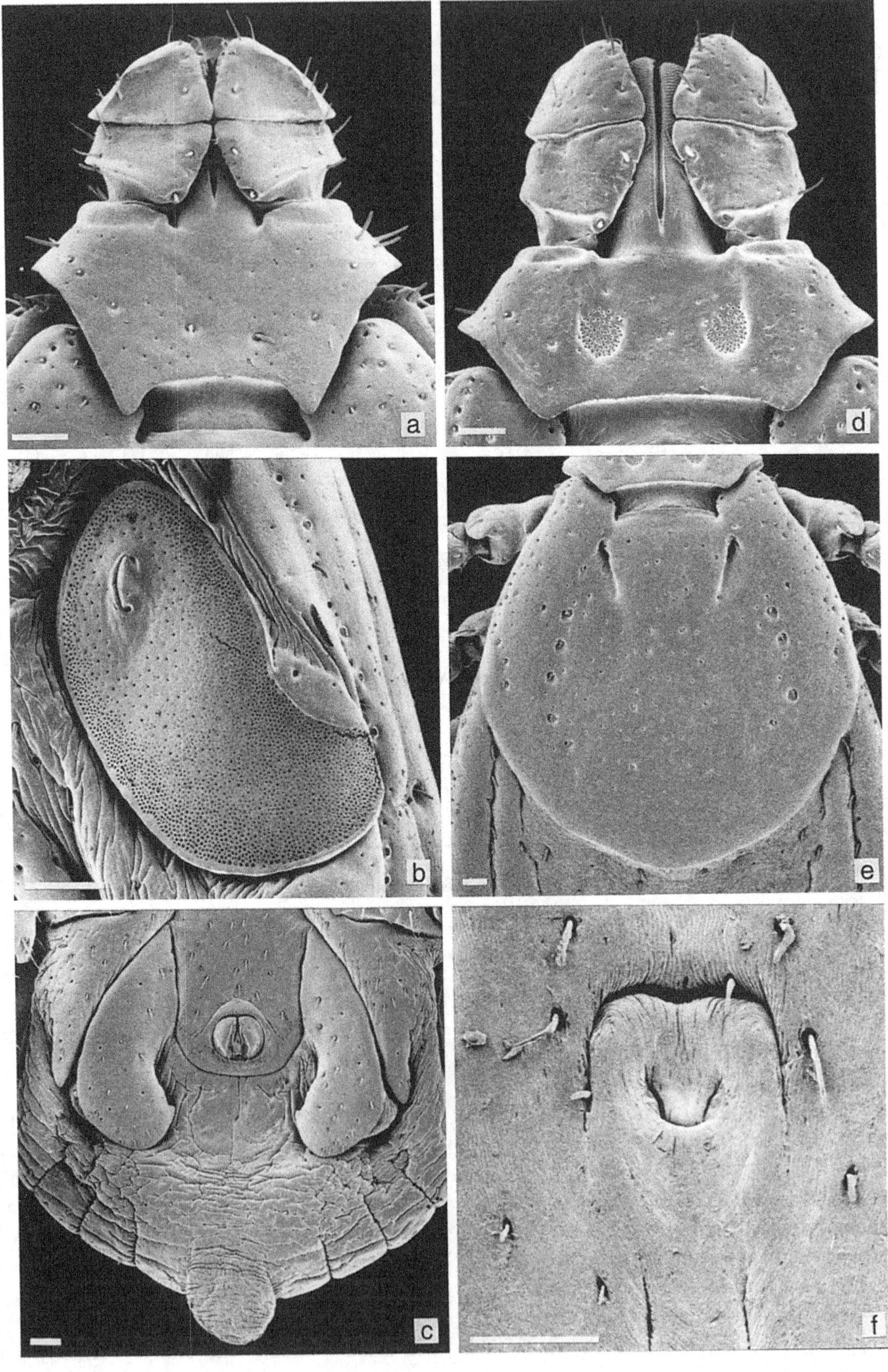
a
b
c
d
e
f

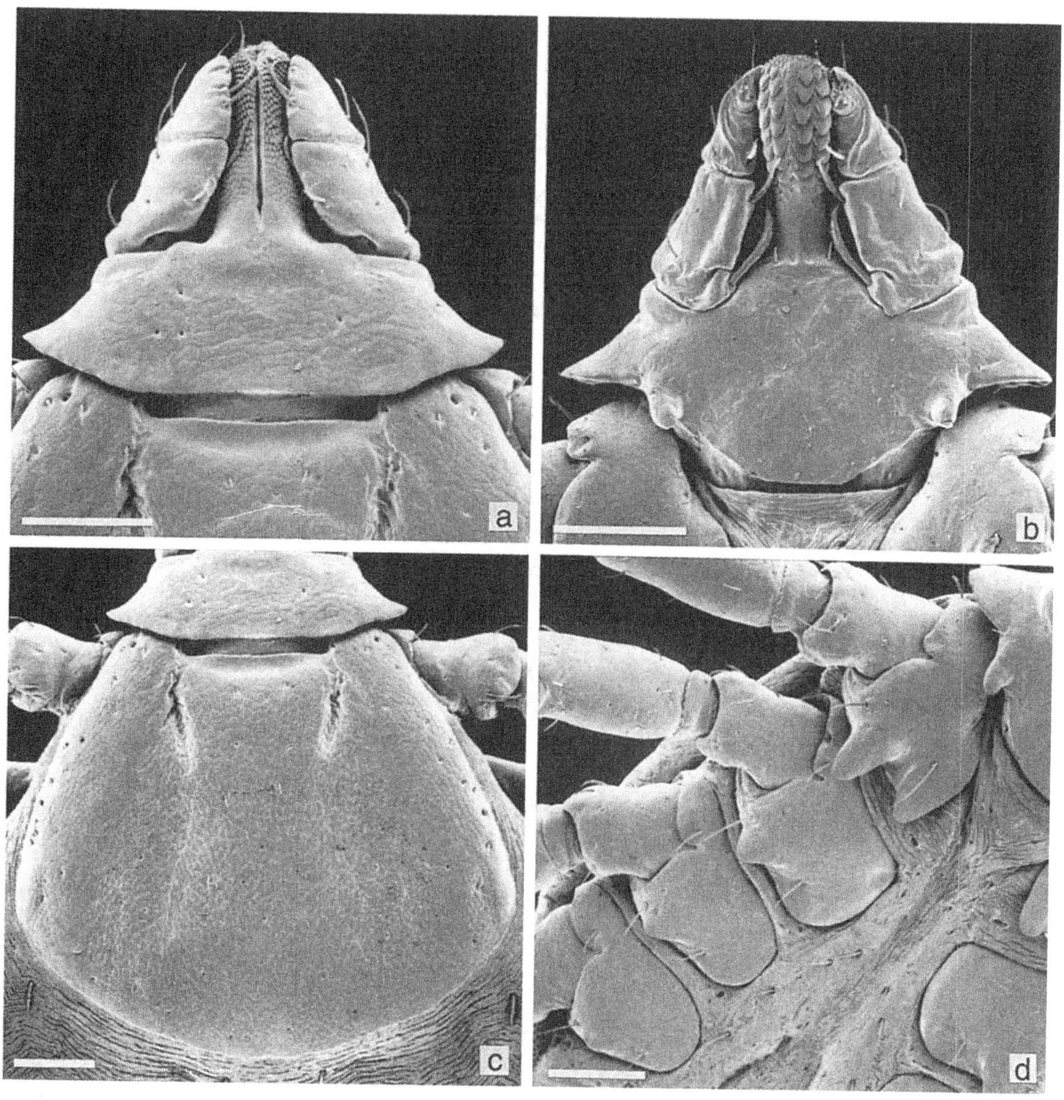

Figure 191 (*above*). *Rhipicephalus simpsoni* [Onderstepoort Tick Collection 2490viii, collected from a greater cane rat (*Thryonomys swinderianus*), Ndumu, northern KwaZulu-Natal, South Africa, on 22 December 1963 by J.E. Dixon]. Nymph: (a) capitulum, dorsal; (b) capitulum, ventral; (c) scutum; (d) coxae. Scale bars represent 0.10 mm. SEMs by J.F. Putterill.

Figure 190 (*opposite*). *Rhipicephalus simpsoni* [Onderstepoort Tick Collection 2490viii, collected from a greater cane rat (*Thryonomys swinderianus*), Ndumu, northern KwaZulu-Natal, South Africa, on 22 December 1963 by J.E. Dixon]. Male: (a) capitulum, dorsal; (b) spiracle; (c) adanal plates. Female: (d) capitulum, dorsal; (e) scutum; (f) genital aperture. Scale bars represent 0.10 mm. SEMs by J.F. Putterill.

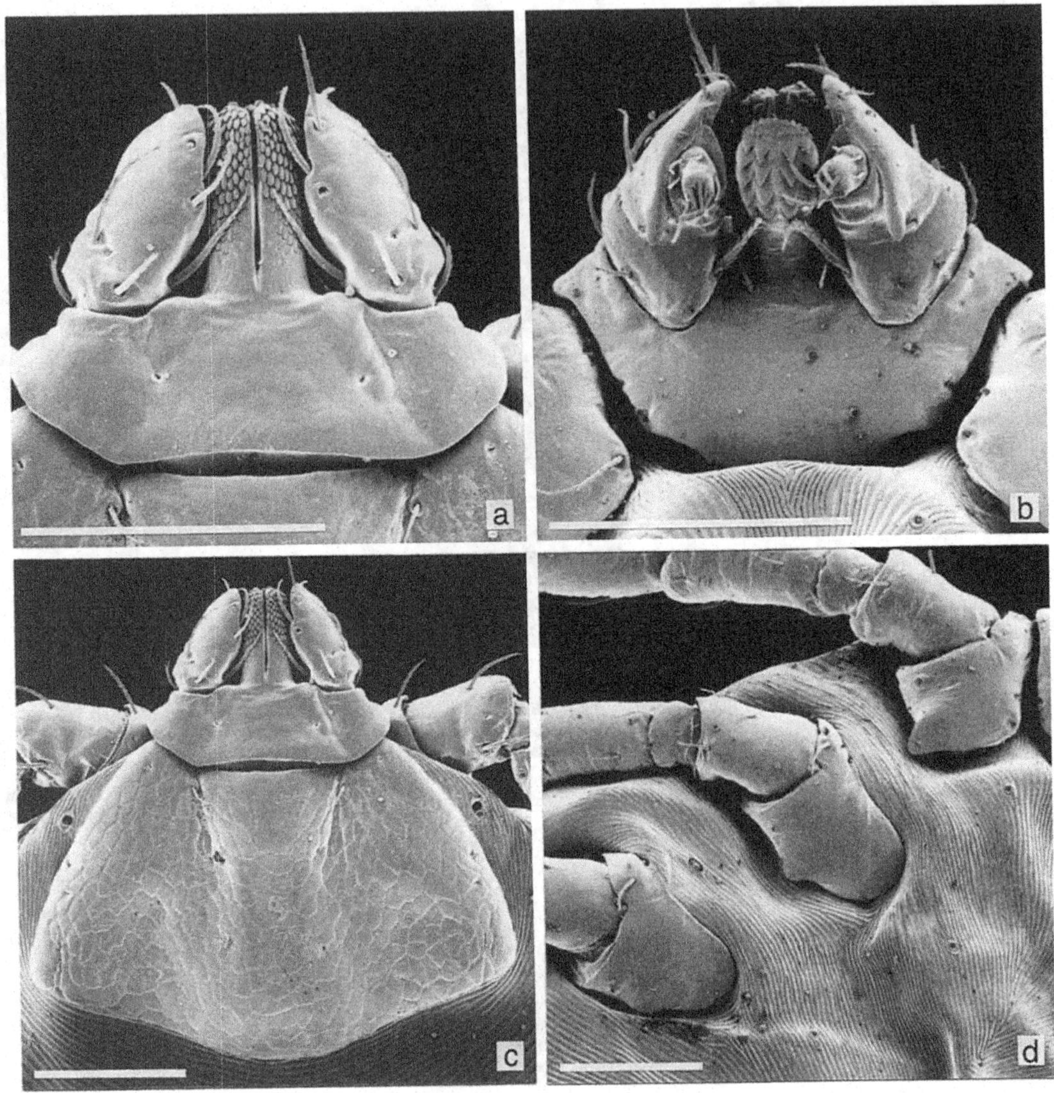

Figure 192. *Rhipicephalus simpsoni* [Onderstepoort Tick Collection 2490viii, collected from a greater cane rat (*Thryonomys swinderianus*), Ndumu, northern KwaZulu-Natal, South Africa, on 22 December 1963 by J.E. Dixon]. Larva: (a) capitulum, dorsal; (b) capitulum, ventral; (c) scutum; (d) coxae. Scale bars represent 0.10 mm. SEMs by J.F. Putterill.

aperture more or less U-shaped, with the sides of the opening converging slightly towards its evenly-curved posterior margin.

Nymph (Fig. 191)

Capitulum much broader than long, length × breadth ranging from 0.24 mm × 0.31 mm to 0.29 mm × 0.40 mm. Basis capituli well over three times as broad as long, its lateral angles long, sharp and slightly forwardly directed, its posterior margin almost straight; ventrally with short, blunt spurs on posterior margin. Palps tapering to quite narrowly-rounded apices, inclined inwards. Scutum broader than long, length × breadth ranging from 0.53 mm × 0.61 mm to 0.61 mm × 0.69 mm; posterior margin a broad smooth curve. Eyes at widest point, well over halfway back, flat. Cervical pits convergent; cervical fields long, shallow, divergent. Ventrally coxae I each with a long external spur and a shorter, somewhat broader, internal spur; coxae II to IV each with an external spur only, becoming progressively smaller in size.

Larva (Fig. 192)

Capitulum much broader than long, length × breadth of the only specimen measured 0.111 mm × 0.164 mm. Basis capituli over three times as broad as long, with broadly rounded lateral angles overlapping the scapulae, posterior margin mildly concave. Palps tapering to their apices, inclined inwards. Scutum much broader than long, length × breadth 0.209 mm × 0.327 mm; posterior margin broad, shallow and sinuous. Eyes at widest point of scutum, well over halfway back. Cervical grooves short, convergent. Ventrally coxae I to III each with a spur, decreasing progressively in size, those on coxae III being very small.

Notes on identification

The morphology of *R. simpsoni* was compared with that of *R. senegalensis, R. longus* and *R. pseudolongus* by Clifford & Anastos (1962). They considered that the mounted genital aperture of the *R. simpsoni* female, with its short thick lateral flaps, could be used to separate this species from the others.

Clifford & Anastos (1964) described the nymph of *R. simpsoni* from 'a single nymphal skin from which a fairly fully developed female of this species was removed.' Our nymphs, which were associated with adults collected in the field from a greater cane rat (*Thryonomys swinderianus*), correspond well with their description and illustrations of this stage. Our description of the larva is based on specimens from the same field collection as our nymphs.

The record in Baker & Keep (1970) from a klipspringer (*Oreotragus oreotragus*) was based on a misidentification (see above under *R. oreotragi sp. nov.*, p. 330).

Hosts

A three-host species (Ntiamoa-Baidu, 1987). Its preferred host, which can probably be regarded as its specific host, is the greater cane rat or grasscutter (Table 50). The numerous records from this rodent can be ascribed largely to an 8-month tick survey, involving some 2500 greater cane rats, conducted in Ghana (Campbell, Asibey & Ntiamoa-Baidu, 1978). In addition Aeschlimann (1967) drew attention to his records of nymphs collected from the African brush-tailed porcupine, which often occupies the same habitat as the greater cane rat. The lesser cane rat and the giant Gambian rat may also serve as hosts. Records of this tick from other hosts should be regarded as accidental infestations.

Peak numbers of adult *R. simpsoni* were collected from greater cane rats at Swedru, southern Ghana, during February and March. At Sunyani, in west central Ghana, very few ticks were collected in January but they rose to a peak during March, with a gradual decline thereafter until August when the survey ended (Campbell *et al.*, 1978).

Zoogeography

The distribution of *R. simpsoni* is dependent upon that of its preferred host, the greater cane rat (Map 54). This animal occurs south of the Sahara in a broad band stretching from Gambia in the west through Nigeria, the Central African Republic, southern Sudan and Uganda to western Kenya and thence southwards through Rwanda, Burundi and Tanzania. Further south it is found in another broad band stretching from Angola through the southern Democratic Republic of Congo, northern Botswana, Zambia, Malawi and Zimbabwe to Mozambique and thence through the eastern parts of South Africa as far as the eastern portion of the Eastern Cape Province.

According to Skinner & Smithers (1990): 'Greater canerats are specialized in their habitat requirements and are found in reedbeds or in areas of dense, tall grass of types with thick reed or cane-like stems.' In the southern regions of their distribution such associations occur in the vicinity of rivers, lakes and swamps, and greater cane rats are never found far from water. In West Africa they occur in the high forest zone but only where there are clearings within the forest, with a grassland invasion of thick-stemmed grass species. Wherever there are cultivated areas they will invade these to feed. Throughout their range they are absent from desert and semi-arid regions. Because of their specialized habitat requirements greater cane rats have a patchy and discontinuous distribution and there are large tracts of country in which they are not found. Norval (1985) noted that, in Zimbabwe, they are rarely examined for ticks. In general this is apparently true elsewhere and *R. simpsoni* is probably more widely distributed than present records indicate.

Disease relationships

Unknown.

REFERENCES

Baker, M.K. & Keep, M.E. (1970). Checklist of the ticks found on the larger game animals in the Natal game reserves. *Lammergeyer*, **No. 12**, 41–7.

Campbell, J.A., Asibey, E.A.O. & Ntiamoa-Baidu, Y. (1978). Rodent ticks in Ghana. In *Tick-borne Diseases and their Vectors*, ed. J.K.H. Wilde, pp. 68–74. Edinburgh: University of Edinburgh, Centre for Tropical Veterinary Medicine.

Norval, R.A.I. (1985). The ticks of Zimbabwe. XII. The lesser known *Rhipicephalus* species. *Zimbabwe Veterinary Journal*, **16**, 37–43.

Ntiamoa-Baidu, Y. (1987). *Rhipicephalus simpsoni* (Acari: Ixodidae) development under controlled conditions. *Journal of Medical Entomology*, **24**, 438–43.

Nuttall, G.H.F. (1910). New species of ticks (*Ixodes, Amblyomma, Rhipicephalus*). *Parasitology*, **3**, 408–16.

Also see the following Basic References (pp. 12–14): Aeschlimann (1967); Clifford & Anastos (1962, 1964); Matthysse & Colbo (1987); Skinner & Smithers (1990).

RHIPICEPHALUS SIMUS KOCH, 1844

The specific name, from the Latin meaning 'flat-nosed, snub-nosed', presumably refers to the shape of the adult's palps.

Synonyms

ecinctus (in part, female); *erlangeri*; *hilgerti*; *sanguineus simus*; *simus erlangeri*; *simus hilgerti*; *simus simus*.

Diagnosis

A large dark brown to blackish tick.

Male (Figs 193(a), 194(a) to (c))
Capitulum longer than broad, length × breadth ranging from 0.66 mm × 0.60 mm to 0.84 mm × 0.77 mm. Basis capituli with acute slightly-recurved lateral angles at about anterior third of its length. Palps broadly rounded apically. Conscutum length × breadth ranging from 2.71 mm × 1.79 mm to 3.53 mm × 2.45 mm; anterior process of coxae I inconspicuous. In engorged specimens a single blunt caudal process is formed. Eyes almost flat, edged dorsally by a few large setiferous punctations in shallow depressions. Cervical pits comma-shaped, discrete. Marginal lines long, usually enclosing one festoon posteriorly. Posteromedian and posterolateral grooves superficial to absent. Large setiferous punctations present along the external cervical margins and marginal lines; a few medium-sized setiferous punctations scattered on scapular apices and in four irregular rows, referred to as the '*simus*' pattern, medially on the conscutum, interspersed with numerous small to minute discrete interstitial punctations. Ventrally spiracles broadly comma-shaped with a short broad sharply-curved dorsal prolongation. Adanal plates large, robust, almost sickle-shaped; accessory adanal plates large, pointed, well sclerotized.

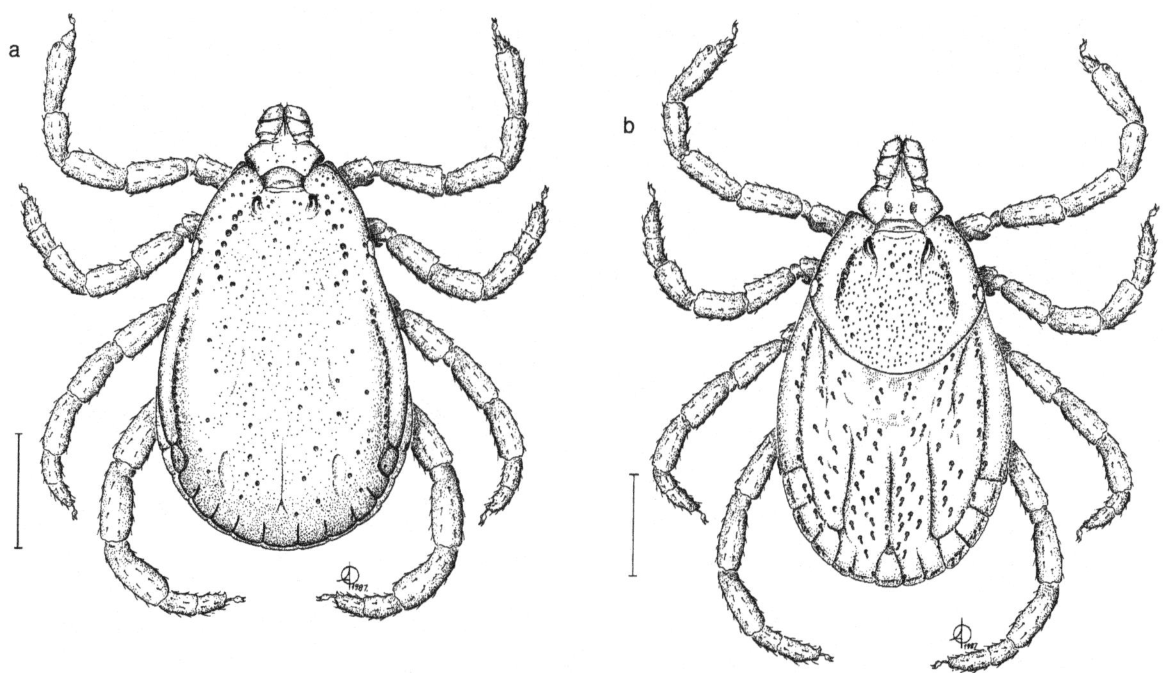

Figure 193. *Rhipicephalus simus* (L127, laboratory reared, original Y collected from bovine, Lutale, Zambia on 4 February 1982 by R.G. Pegram). (a) Male, dorsal; (b) female, dorsal. Scale bars represent 1 mm. A. Olwage *del.*

Female (Figs 193(b), 194(d) to (f), 197(a))

Capitulum slightly longer than broad, length × breadth ranging from 0.67 mm × 0.66 mm to 0.92 mm × 0.87 mm. Basis capituli with short lateral angles at about anterior third of its length; porose areas medium-sized, almost twice their own diameter apart. Palps with article I narrow relative to articles II and III, easily visible dorsally, apices broadly rounded. Scutum slightly broader than long, length × breadth ranging from 1.32 mm × 1.36 mm to 1.77 mm × 1.86 mm; posterior margin usually rounded. Eyes at broadest part of scutum, just anterior to mid-length, almost flat, edged dorsally by a few large setiferous punctations in faint depressions. Cervical pits short, convergent; cervical fields broad, slightly depressed, their external margins delimited by irregular rows of large setiferous punctations. A few medium-sized setiferous punctations scattered on scapular apices and medially on the scutum, where they may be masked by the numerous discrete interstitial punctations; the latter are usually more conspicuous than those of the male. Ventrally genital aperture a truncated U-shape, diverging anteriorly; narrow sclerotized margins to the genital aperture usually present.

Nymph (Fig. 195)

Capitulum much broader than long, length × breadth ranging from 0.22 mm × 0.32 mm to 0.25 mm × 0.35 mm. Basis capituli nearly four times as broad as long with acute lateral angles overlapping the scapulae; ventrally with sharp spurs on posterior border; (these are broken in the specimen shown in Fig. 195(b); see Fig. 195(d)). Palps narrow, slightly tapered apically, inclined inwards. Scutum broader than long, length × breadth ranging from 0.46 mm × 0.53 mm to 0.50 mm × 0.60 mm; posterior margin a wide smooth curve. Eyes at widest point, over halfway back, long, narrow, delimited dorsally by slight depressions. Cervical pits short, convergent; cervical fields divergent, long, narrow, slightly depressed. Ventrally coxae I each with a long narrow external spur and a shorter broader internal spur; coxae II and III

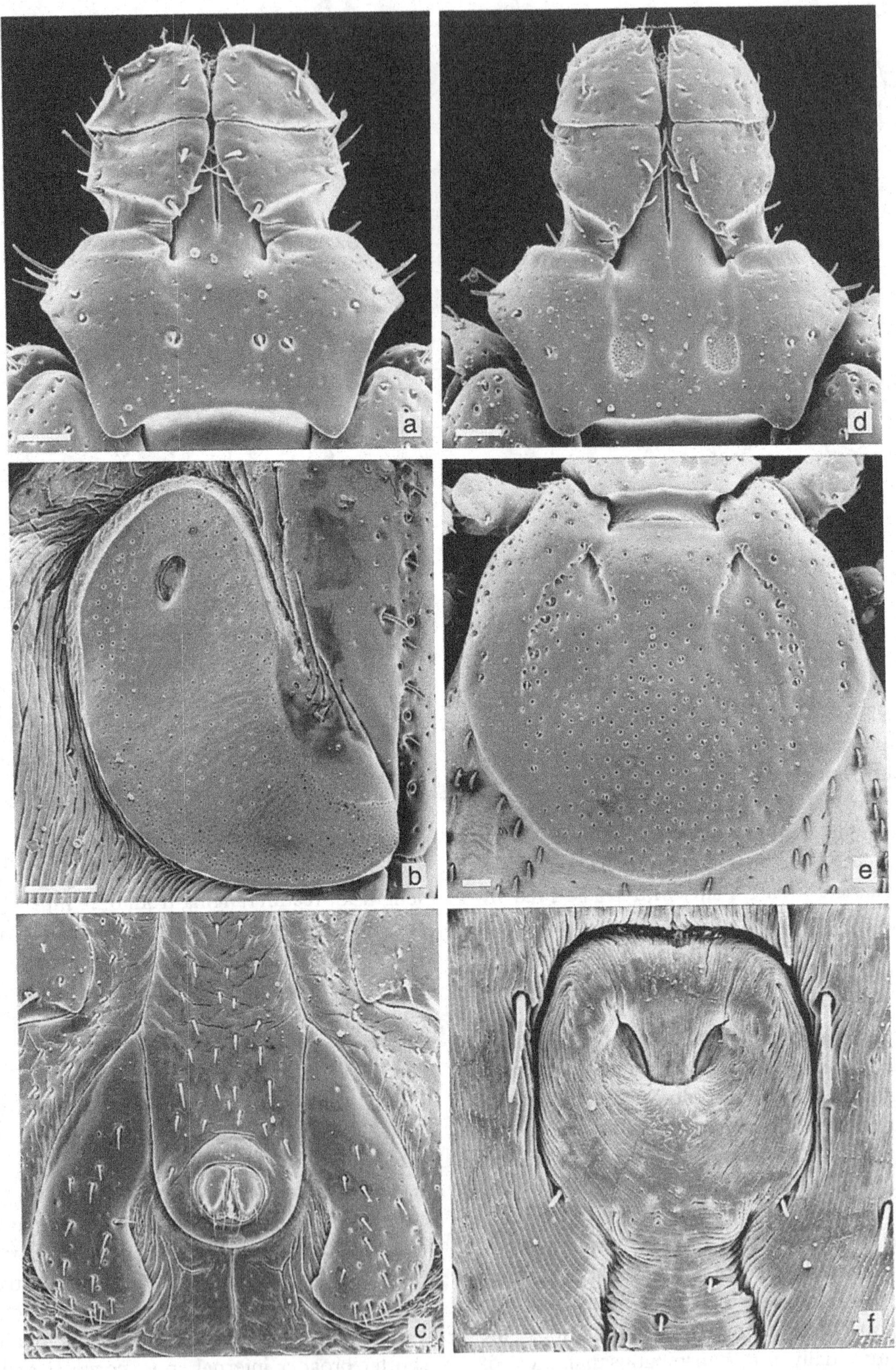
a
b
c
d
e
f

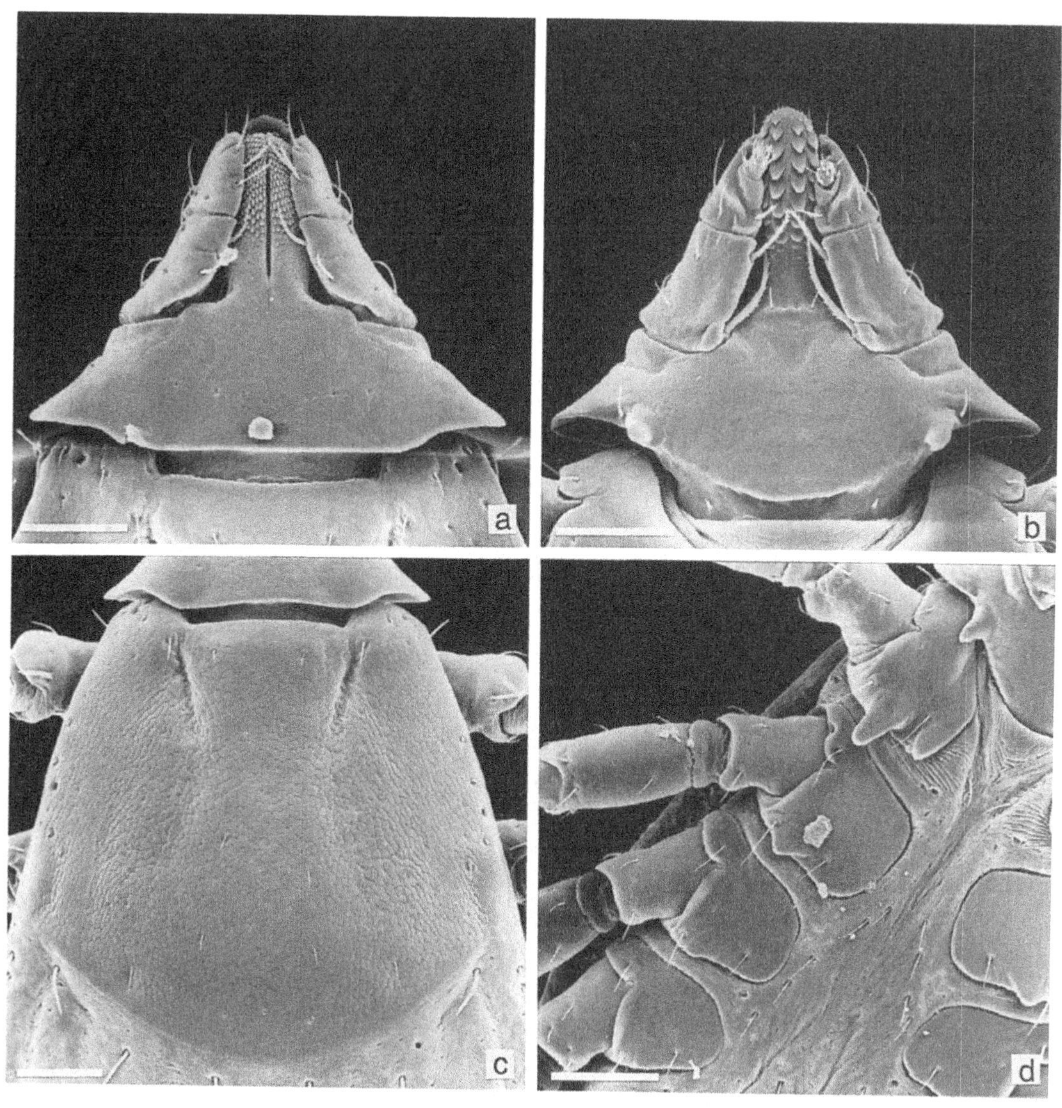

Figure 195 (*above*). *Rhipicephalus simus* (L42, laboratory reared, strain originating from ticks collected from pasture, Balmoral, Zambia, on 12 January 1981 by R.G. Pegram). Nymph: (a) capitulum, dorsal; (b) capitulum, ventral; (c) scutum (d) coxae. Scale bars represent 0.10 mm. SEMs by M.D. Corwin. (Figs (a) & (d) from Pegram *et al.*, 1987, figs 4 & 6, with kind permission from the Entomological Society of America.)

Figure 194 (*opposite*). *Rhipicephalus simus* (L42, laboratory reared, strain originating from ticks collected from pasture, Balmoral, Zambia, on 12 January 1981 by R.G. Pegram). Male: (a) capitulum, dorsal; (b) spiracle; (c) adanal plates. Female: (d) capitulum, dorsal; (e) scutum; (f) genital aperture. Scale bars represent 0.10 mm. SEMs by M.D. Corwin. (Figs (b), (c), (e) & (f) from Pegram *et al.*, 1987, figs 9–12, with kind permission from the Entomological Society of America.)

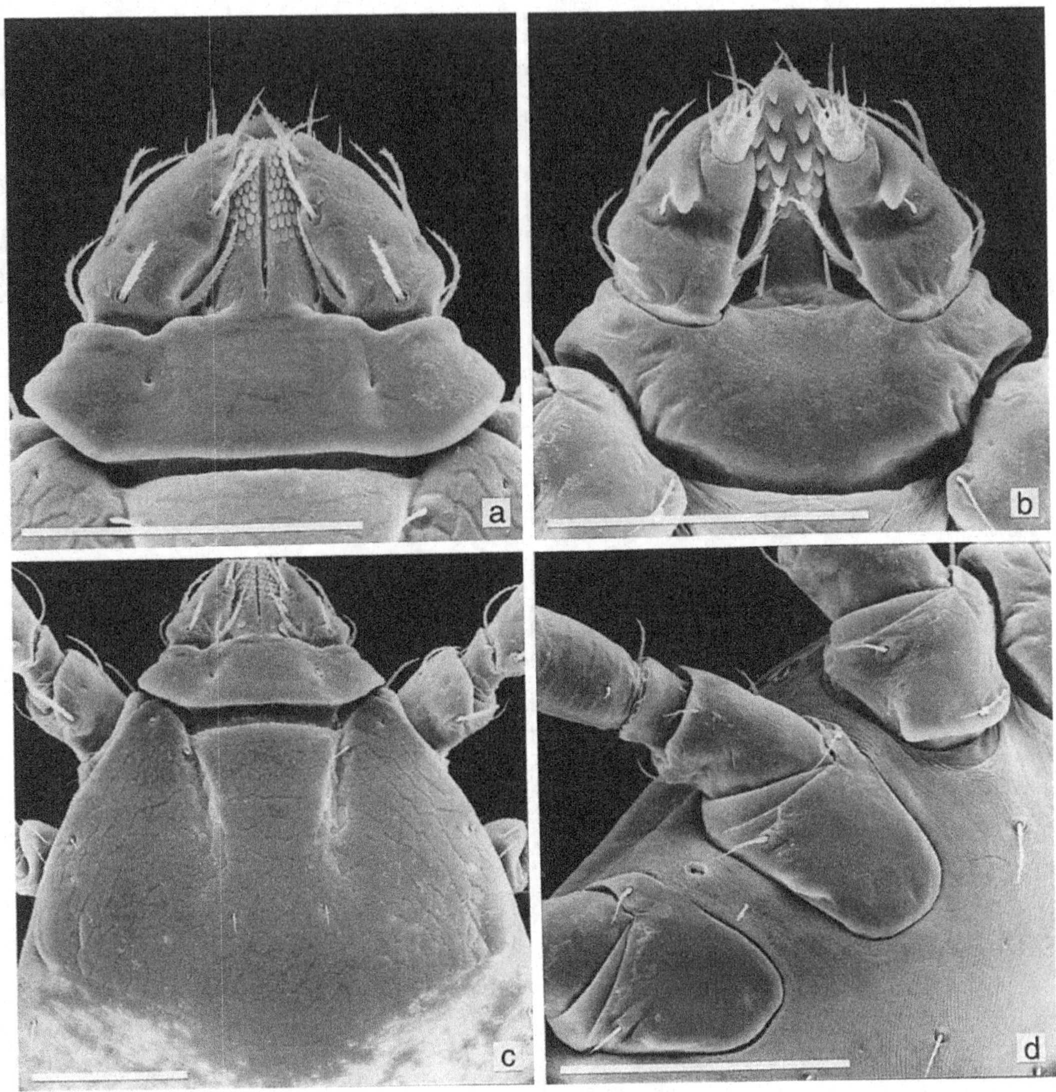

Figure 196. *Rhipicephalus simus* (L42, laboratory reared, strain originating from ticks collected from pasture, Balmoral, Zambia, on 12 January 1981 by R.G. Pegram). Larva: (a) capitulum, dorsal; (b) capitulum, ventral; (c) scutum; (d) coxae. Scale bars represent 0.10 mm. SEMs by M.D. Corwin. (Fig. (a) from Pegram *et al.*, 1987, fig. 1, with kind permission from the Entomological Society of America.)

each with a small sharp external spur; coxae IV each with a slight salience only on its posterior border.

Larva (Fig. 196)

Capitulum much broader than long, length × breadth ranging from 0.120 mm × 0.162 mm to 0.125 mm × 0.168 mm. Basis capituli well over three times as broad as long, with broad lateral angles, posterior margin long, slightly sinuous. Palps broad, tapering somewhat to rounded apices, inclined inwards. Scutum much broader than long, length × breadth ranging from 0.213 mm × 0.316 mm to 0.226 mm × 0.341 mm; posterior margin a wide shallow curve. Eyes at widest point, about three-quarters of the way back, long, narrow, delimited dorsally by slight depressions. Cervical grooves short, slightly

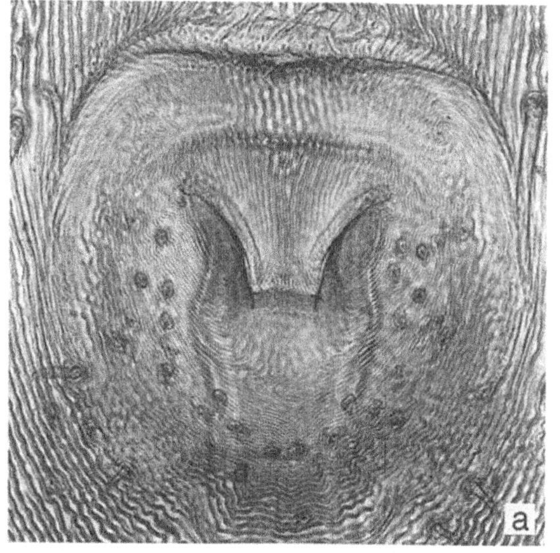

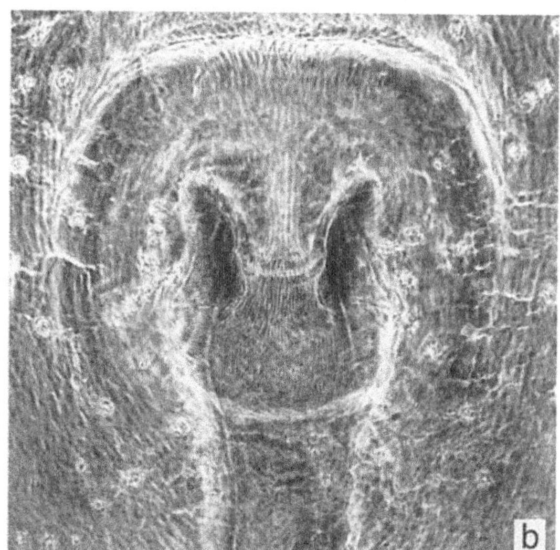

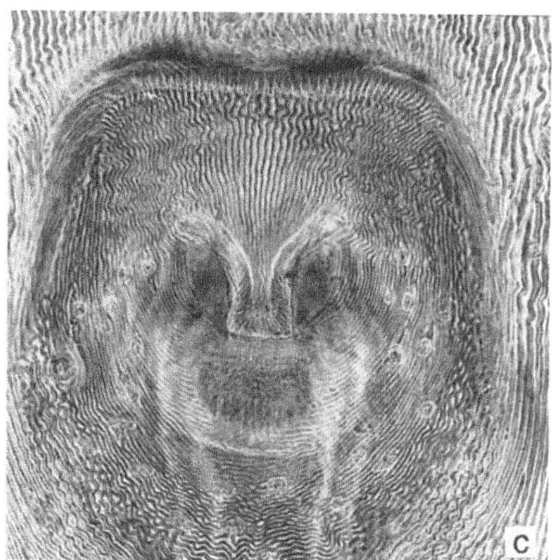

convergent. Ventrally coxae I each with a sharp spur; coxae II and III each with a broad ridge-like salience on its posterior border.

Notes on identification

The *R. simus* group, as we now understand it, comprises three species: *R. muhsamae*, a primarily western African tick that occurs across the continent from Senegal to Ethiopia; *R. praetextatus*, an eastern African tick whose distribution overlaps that of *R. muhsamae* in the Sudan, Ethiopia and western Uganda, and *R. simus* itself, a southern African tick whose distribution may overlap with that of *R. praetextatus* in the southern part of the latter's range.

In the earlier literature on African ticks many records of these three species were lumped together under the names *R. simus* or *R. simus simus*. In 1965 *R. muhsamae* was described by Morel & Vassiliades, who listed the publications that they thought referred to this tick. Records of *R. praetextatus* were, however, still included under the name *R. simus* until 1987, when Pegram *et al.* reviewed the literature on the group and differentiated these three species morphologically. Nevertheless, although these species have now been defined precisely, it remains difficult for many people, ourselves included, to identify them individually purely on morphological grounds because the differences between them at all stages of development are slight. One of the most important morphological features that can be used to differentiate the females is the structure of their mounted genital apertures (Fig. 197(a) to (c)).

Up to now it has also been impossible for anyone to re-examine and even try to re-identify

Figure 197 (*opposite*). Female genital apertures (mounted): (a) *Rhipicephalus simus*; (b) *Rhipicephalus praetextatus*; (c) *Rhipicephalus muhsamae*. (From Pegram *et al.*, 1987, figs 37, 38 & 39, with kind permission from the Entomological Society of America.)

Table 51. *Host records of* Rhipicephalus simus

Hosts	Number of records
Domestic animals	
Cattle	642
Sheep	14
Goats	15
Horses	31
Donkeys	2
Mules	1
Pigs	3
Pigs (feral)	2
Dogs	294
Cats	1
Cats (feral)	1
Wild animals	
South African hedgehog (*Atelerix frontalis*)	1
'Shrew'	1 (nymphs)
Side-striped jackal (*Canis adustus*)	1
Black-backed jackal (*Canis mesomelas*)	7
'Jackal'	10
Hunting dog (*Lycaon pictus*)	5
Cape fox (*Vulpes chama*)	1
Cheetah (*Acinonyx jubatus*)	7
Caracal (*Caracal caracal*)	5
African wild cat (*Felis lybica*)	2
Serval (*Leptailurus serval*)	1
Lion (*Panthera leo*)	29
Leopard (*Panthera pardus*)	6
Marsh mongoose (*Atilax paludinosus*)	1
White-tailed mongoose (*Ichneumia albicauda*)	1 (nymphs)
Banded mongoose (*Mungos mungo*)	1
Spotted hyaena (*Crocuta crocuta*)	9
Brown hyaena (*Parahyaena brunnea*)	13
'Hyaena'	2
Aardwolf (*Proteles cristatus*)	1
Ratel (*Mellivora capensis*)	1
African civet (*Civettictis civetta*)	15
'Genet'	1
African elephant (*Loxodonta africana*)	1
Burchell's zebra (*Equus burchellii*)	27
White rhinoceros (*Ceratotherium simum*)	9
Black rhinoceros (*Diceros bicornis*)	4
'Rhinoceros'	1
Aardvark (*Orycteropus afer*)	1
Warthog (*Phacochoerus africanus*)	124
Bushpig (*Potamochoerus larvatus*)	23
Giraffe (*Giraffa camelopardalis*)	8

Table 51. *(cont.)*

Hosts	Number of records
Wild animals *(cont.)*	
Impala (*Aepyceros melampus*)	7
Red hartebeest (*Alcelaphus buselaphus caama*)	1
Blue wildebeest (*Connochaetes taurinus*)	2
Klipspringer (*Oreotragus oreotragus*)	7
Steenbok (*Raphicerus campestris*)	2
African buffalo (*Syncerus caffer*)	56
Eland (*Taurotragus oryx*)	19
Nyala (*Tragelaphus angasii*)	3
Bushbuck (*Tragelaphus scriptus*)	3
Greater kudu (*Tragelaphus strepsiceros*)	18
Common duiker (*Sylvicapra grimmia*)	3
Sable antelope (*Hippotragus niger*)	4
Gemsbok (*Oryx gazella*)	1
Waterbuck (*Kobus ellipsiprymnus*)	1
Reedbuck (*Redunca arundinum*)	2
Temminck's ground pangolin (*Manis temminckii*)	5
Kuhl's tree squirrel (*Funisciurus congicus*)	1 (immatures)
Smith's bush squirrel (*Paraxerus cepapi*)	1 (nymph)
Gerbil (*Gerbillurus* sp.)	1 (immatures)
Bushveld gerbil (*Tatera leucogaster*)	2 (immatures)
Red veld rat (*Aethomys chrysophilus*)	37 (immatures)
Namaqua rock mouse (*Aethomys namaquensis*)	6 (immatures)
Single-striped mouse (*Lemniscomys griselda*)	2 (nymphs)
Natal multimammate rat (*Mastomys natalensis*)	10 (immatures)
Pigmy mouse (*Mus minutoides*)	1 (immatures)
Groove-toothed swamp rat (*Pelomys campanae*)	1 (nymph)
Four-striped grass mouse (*Rhabdomys pumilio*)	83 (immatures)
African swamp rat (*Otomys* sp.)	1 (nymph)
Karoo rat (*Parotomys* sp.)	1 (immatures)
'Murid rodents'	5 (immatures)
South African porcupine (*Hystrix africaeaustralis*)	7
Scrub hare (*Lepus saxatilis*)	49 (immatures)
Birds	
Reed cormorant (*Phalacrocorax africanus*)	1
Humans	27

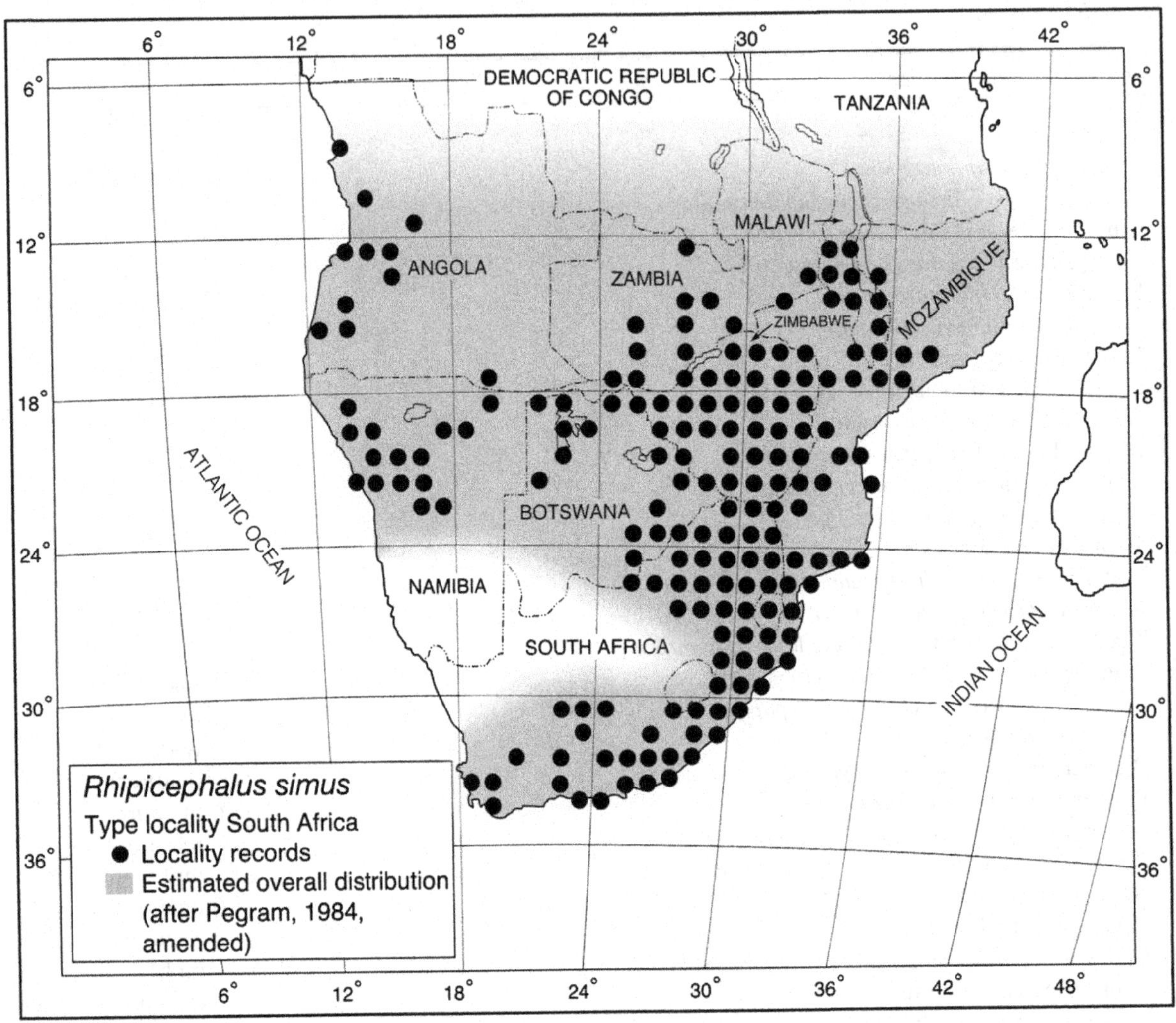

Map 55. *Rhipicephalus simus*: distribution.

the vast numbers of specimens labelled *R. simus sensu lato* deposited in various museum collections. At present, therefore, we can only list the hosts and map the distributions of these species provisionally. In Democratic Republic of Congo, especially, we do not know what the specimens identified by Elbl & Anastos (1966) as *R. simus* really are. That aside we feel that we should try to give readers an indication of the hosts and zoogeography of these ticks, imperfect as some of this information may later prove to be. For the purpose of this book we have designated southern Angola plus Zambia, Malawi and Mozambique as the northern limits of the distribution of *R. simus*.

In South Africa its distribution and that of *R. follis* overlap, particularly in the Eastern Cape Province and KwaZulu-Natal. In addition large numbers of the immature stages of these two species may occur together on the same host (I.G.H., unpublished data). We find it nearly impossible to separate these immature ticks, particularly in field collections.

Hosts

A three-host species (Norval & Mason, 1981). Its preferred hosts are large ruminants such as cattle and African buffalo as well as a variety of monogastric animals (MacLeod *et al.*, 1977; Norval & Mason, 1981) (Table 51). The latter include horses and dogs, large wild carnivores, Burchell's zebra, rhinoceroses, warthogs, bushpigs and also humans (Baker & Keep, 1970; Norval & Mason, 1981; Horak, De Vos & De Klerk, 1984; Horak *et al.*, 1988). The hosts of the immature stages are murid rodents, among which most collections have been taken from the red veld rat and the four-striped grass mouse (Rechav, 1982; Braack *et al.*, 1996). The scrub hare is also a good host of these stages (Horak *et al.*, 1993).

Although this species is widely distributed *R. simus* adults are never particularly abundant; burdens of more than 20 ticks per host are rare. However, collections comprising 713 and 521 adult ticks have been made from a sick lion and a sick leopard, respectively and 221 from a healthy warthog (I.G.H., unpublished data). Infestations with immature ticks can number several hundred on individual four-striped grass mice (I.G.H., unpublished data). The preferred sites of attachment of adults on cattle are the tail switch and feet; on sheep the feet; on dogs, wild carnivores and warthog the head and neck, and the tail of Burchell's zebra (Baker & Ducasse, 1967; I.G.H., unpublished data).

In South Africa and Zimbabwe the larvae of *R. simus* are most abundant from late summer to early winter (March to June) and the nymphs from early winter to spring (June to September) (Norval & Mason, 1981; Rechav, 1982; Braack *et al.*, 1996). Adults are most abundant from August to January or February in the southern regions of the tick's distribution (Horak *et al.*, 1987; Baker *et al.*, 1989). In the central regions this activity commences in October and extends to March, while in the north peak abundance extends from November to April (MacLeod *et al.*, 1977; Norval & Mason, 1981; Horak, 1982; Pegram *et al.*, 1986). Only one life cycle per year seems probable.

Zoogeography

Rhipicephalus simus has been recorded most frequently in southern, eastern and north-eastern South Africa; Swaziland; eastern and northern Botswana; throughout Zimbabwe apart from the western regions adjoining Botswana; central and northern Namibia; south-western Angola; Zambia, and throughout Malawi and southern Mozambique (Map 55). Within these countries it has been collected at altitudes between 100 m and 2000 m above sea level and in regions receiving between 450 mm and 1400 mm of rainfall annually. It is most commonly encountered in vegetation types variously described as undifferentiated woodland, scrub woodland, Zambezian miombo woodland and wooded grassland, also in East African coastal mosaic vegetation.

Disease relationships

Experimentally it has been shown that *R. simus* can transmit *Theileria parva parva, Theileria parva lawrencei, Anaplasma marginale* and *Anaplasma centrale* to cattle (Lounsbury, 1906; Neitz, 1962; Potgieter, 1981; Potgieter & Van Rensburg, 1987). Should natural transmission via this tick ever occur it is unlikely to be an important vector because its immature stages feed almost exclusively on rodents and hares. It has also been shown that *R. simus* can transmit *Babesia trautmanni* transovarially to splenectomized pigs (De Waal, López-Rebollar & Potgieter, 1992). Its listing by Lewis (1949) as a vector of the virus of Nairobi sheep disease probably refers to *R. praetextatus*. Adult ticks may produce a toxin that can cause paralysis in calves and lambs (Norval & Mason, 1981). It has been listed as a vector of *Rickettsia conori* to humans (Gear, 1992).

REFERENCES

Baker, M.K. & Ducasse, F.B.W. (1967). Tick infestation of livestock in Natal. I. The predilection sites and seasonal variations of cattle ticks. *Jour-*

nal of the South African Veterinary Medical Association, **38**, 447–53.

Baker, M.K., Ducasse, F.B.W., Sutherst, R.W. & Maywald, G.F. (1989). The seasonal tick populations on traditional and commercial cattle grazed at four altitudes in Natal. *Journal of the South African Veterinary Association*, **60**, 95–101.

Baker, M.K. & Keep, M.E. (1970). Checklist of the ticks found on the larger game animals in the Natal game reserves. *Lammergeyer*, **No. 12**, 41–7.

De Waal, D.T., López-Rebollar, L.M. & Potgieter, F.T. (1992). The transovarial transmission of *Babesia trautmanni* by *Rhipicephalus simus* to domestic pigs. *Onderstepoort Journal of Veterinary Research*, **59**, 219–21.

Braack, L.E.O., Horak, I.G., Jordaan, L.C., Segerman, J. & Louw, J.P. (1996). The comparative host status of red veld rats (*Aethomys chrysophilus*) and bushveld gerbils (*Tatera leucogaster*) for epifaunal arthropods in the southern Kruger National Park, South Africa. *Onderstepoort Journal of Veterinary Research*, **63**, 149–58.

Gear, J.H.S. (1992). Tick-bite fever (tick typhus) in southern Africa. In *Tick Vector Biology, Medical and Veterinary Aspects*, ed. B. Fivaz, T. Petney & I. Horak, pp. 135–42. Berlin, Heidelberg: Springer Verlag.

Horak, I.G. (1982). Parasites of domestic and wild animals in South Africa. XV. The seasonal prevalence of ectoparasites on impala and cattle in the northern Transvaal. *Onderstepoort Journal of Veterinary Research*, **49**, 85–93.

Horak, I.G., Boomker, J., De Vos, V. & Potgieter, F.T. (1988). Parasites of domestic and wild animals in South Africa. XXIII. Helminth and arthropod parasites of warthogs, *Phacochoerus aethiopicus*, in the eastern Transvaal Lowveld. *Onderstepoort Journal of Veterinary Research*, **55**, 145–52.

Horak, I.G., De Vos, V. & De Klerk, B.D. (1984). Parasites of domestic and wild animals in South Africa. XVII. Arthropod parasites of Burchell's zebra, *Equus burchelli*, in the eastern Transvaal Lowveld. *Onderstepoort Journal of Veterinary Research*, **51**, 145–54.

Horak, I.G., Jacot Guillarmod, A., Moolman, L.C. & De Vos, V. (1987). Parasites of domestic and wild animals in South Africa. XXII. Ixodid ticks on domestic dogs and on wild carnivores. *Onderstepoort Journal of Veterinary Research*, **54**, 573–80.

Horak, I.G., Spickett, A.M., Braack, L.E.O. & Penzhorn, B.L. (1993). Parasites of domestic and wild animals in South Africa. XXXII. Ixodid ticks on scrub hares in the Transvaal. *Onderstepoort Journal of Veterinary Research*, **60**, 163–74.

Koch, C.L. (1844). Systematische Uebersicht über die Ordung der Zecken. *Archiv für Naturgeschichte*, **10**, 217–39.

Lewis, E.A. (1949). Nairobi sheep disease. *Report of the Veterinary Department, Kenya, for 1947*, pp. 45, 51.

Lounsbury, C.P. (1906). Ticks and African Coast fever. *Agricultural Journal, Cape of Good Hope*, **28**, 634–54.

MacLeod, J., Colbo, M.H., Madbouly, M.H. & Mwanaumo, B. (1977). Ecological studies of ixodid ticks (Acari: Ixodidae) in Zambia. III. Seasonal activity and attachment sites on cattle, with notes on other hosts. *Bulletin of Entomological Research*, **67**, 161–73.

Morel, P.C. & Vassiliades, G. (1965). Description de *Rhipicephalus muhsamae* n. sp. de l'Ouest-Africain, (groupe de *Rh. simus*; Acariens, Ixodoidea). *Revue d'Élevage et de Médecine Vétérinaire des Pays Tropicaux*, **17 for 1964** (nouvelle série), 619–36.

Neitz, W.O. (1962). Review of recent developments in the protozoology of tick-borne diseases. In *Report of the Second Meeting of the FAO/OIE Expert Panel on Tick-borne Diseases of Livestock*, Appendix D, 34–5. Cairo: Food and Agricultural Organization of the United Nations.

Norval, R.A.I. & Mason, C.A. (1981). The ticks of Zimbabwe. II. The life cycle, distribution and hosts of *Rhipicephalus simus* Koch, 1844. *Zimbabwe Veterinary Journal*, **12**, 2–9.

Pegram, R.G., Perry, B.D., Musisi, F.L. & Mwanaumo, B. (1986). Ecology and phenology of ticks in Zambia: seasonal dynamics on cattle. *Experimental and Applied Acarology*, **2**, 25–45.

Pegram, R.G., Walker, J.B., Clifford, C.M. & Keirans, J.E. (1987). Comparison of populations of the *Rhipicephalus simus* group: *R. simus, R. praetextatus* and *R. muhsamae* (Acari: Ixodidae). *Journal of Medical Entomology*, **24**, 666–82.

Potgieter, F.T. (1981). Tick transmission of anaplasmosis in South Africa. In *Proceedings of an International Conference on Tick Biology and Control*, ed. G.B. Whitehead & J.D. Gibson, pp. 53–6. Grahamstown: Rhodes University.

Potgieter, F.T. & Van Rensburg, L. (1987). Tick transmission of *Anaplasma centrale*. *Onderstepoort Journal of Veterinary Research*, **54**, 5–7.

Rechav, Y. (1982). Dynamics of tick populations (Acari: Ixodidae) in the eastern Cape Province of South Africa. *Journal of Medical Entomology*, **19**, 679–700.

Also see the following Basic References (pp. 12–14): Elbl & Anastos (1966); Santos Dias (1960); Walker, Mehlitz & Jones (1978).

RHIPICEPHALUS SULCATUS NEUMANN, 1908

The specific name *sulcatus*, the Latin for 'furrowed', probably refers to the conspicuous marginal lines and posterior grooves on the male's conscutum.

Diagnosis

A reddish-brown species with a densely evenly-punctate scutum.

Male (Figs 198(a), 199(a) to (c))

Capitulum broader than long, length × breadth ranging from 0.47 mm × 0.51 mm to 0.57 mm × 0.62 mm. Basis capituli with acute lateral angles at anterior third of its length. Palps short, broadly rounded apically. Conscutum length × breadth ranging from 2.08 mm × 1.27 mm to 2.53 mm × 1.58 mm; anterior process of coxae I rounded, inconspicuous. Eyes flat, edged dorsally by a few large setiferous punctations. Cervical fields delimited laterally by rows of large setiferous punctations. Marginal lines long, outlined by numerous large setiferous punctations. Posteromedian groove narrow, tapering anteriorly; posterolateral grooves oval; all deep with reticulate surfaces. Punctations on the scapulae relatively sparsely distributed compared with the dense pattern elsewhere on the conscutum,

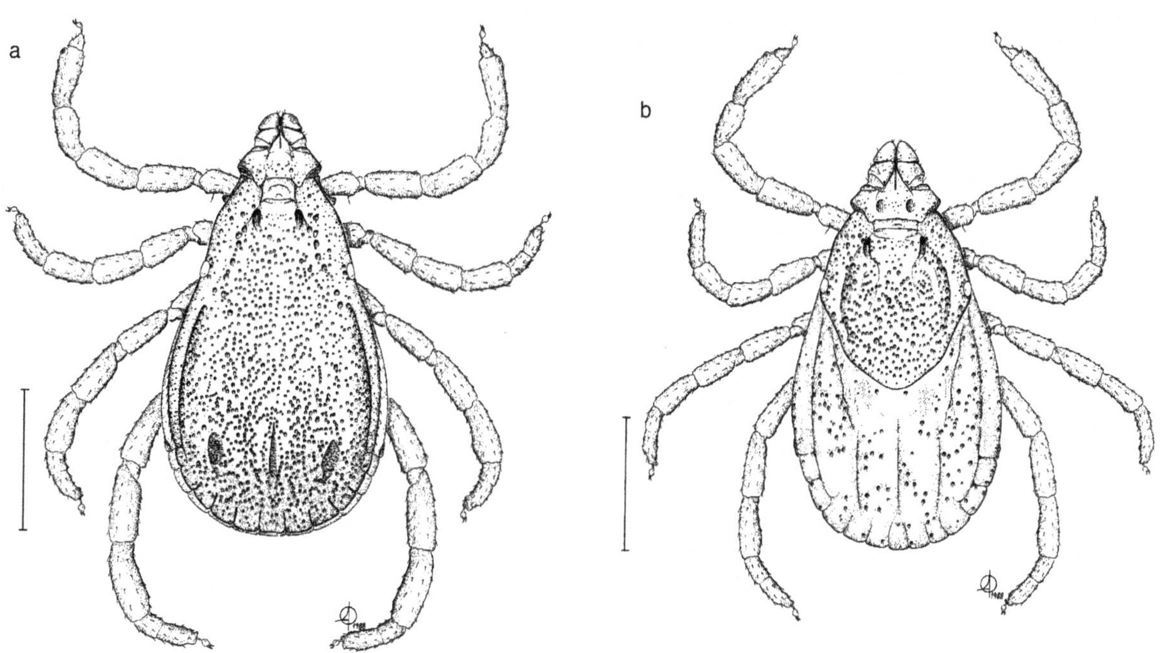

Figure 198. *Rhipicephalus sulcatus* (L40, laboratory reared, original ♀ collected from pasture, Balmoral, Zambia, on 12 January 1981 by R.G. Pegram). (a) Male, dorsal; (b) female, dorsal. Scale bars represent 1 mm. A. Olwage *del.*

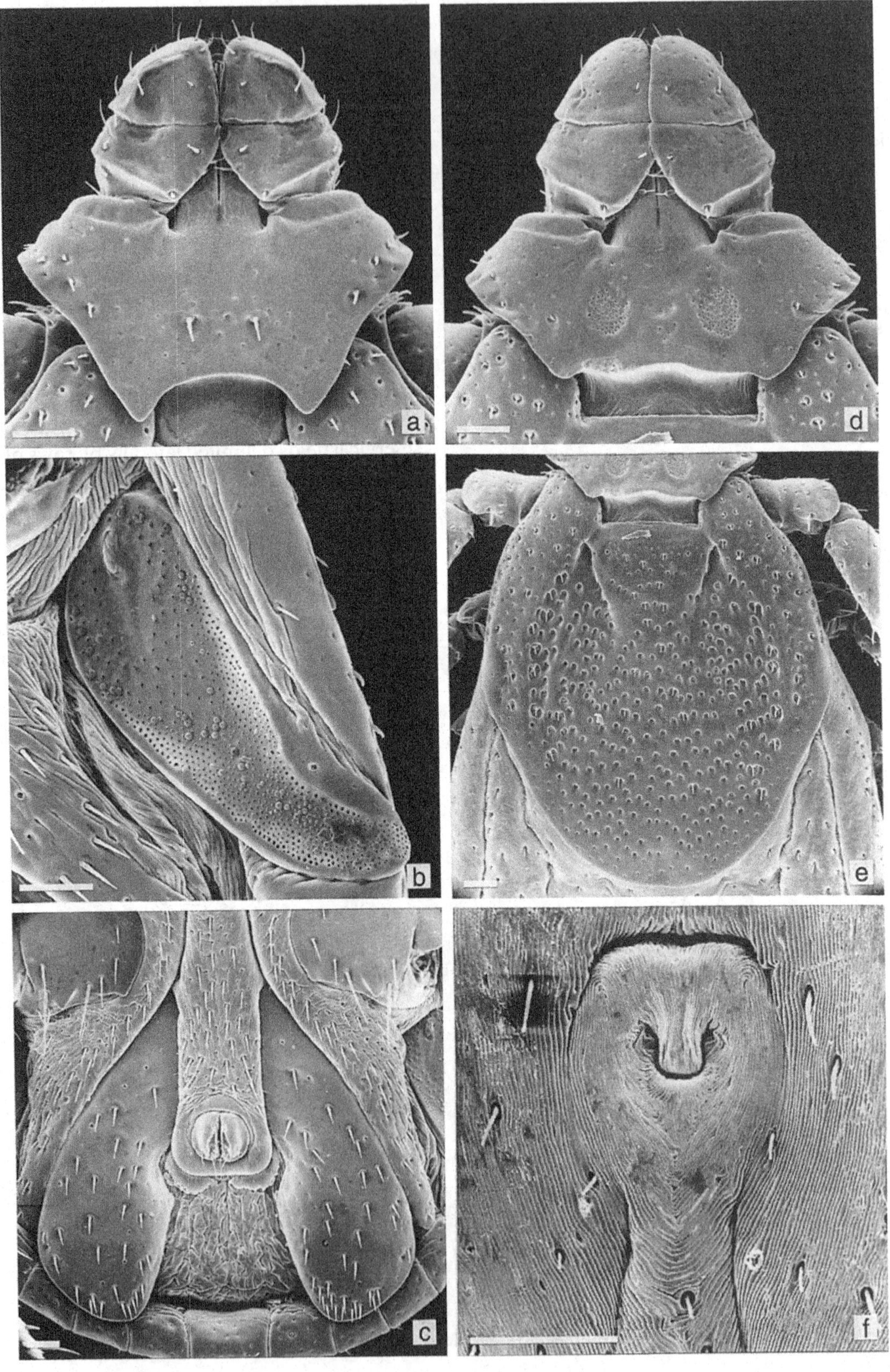
a
b
c
d
e
f

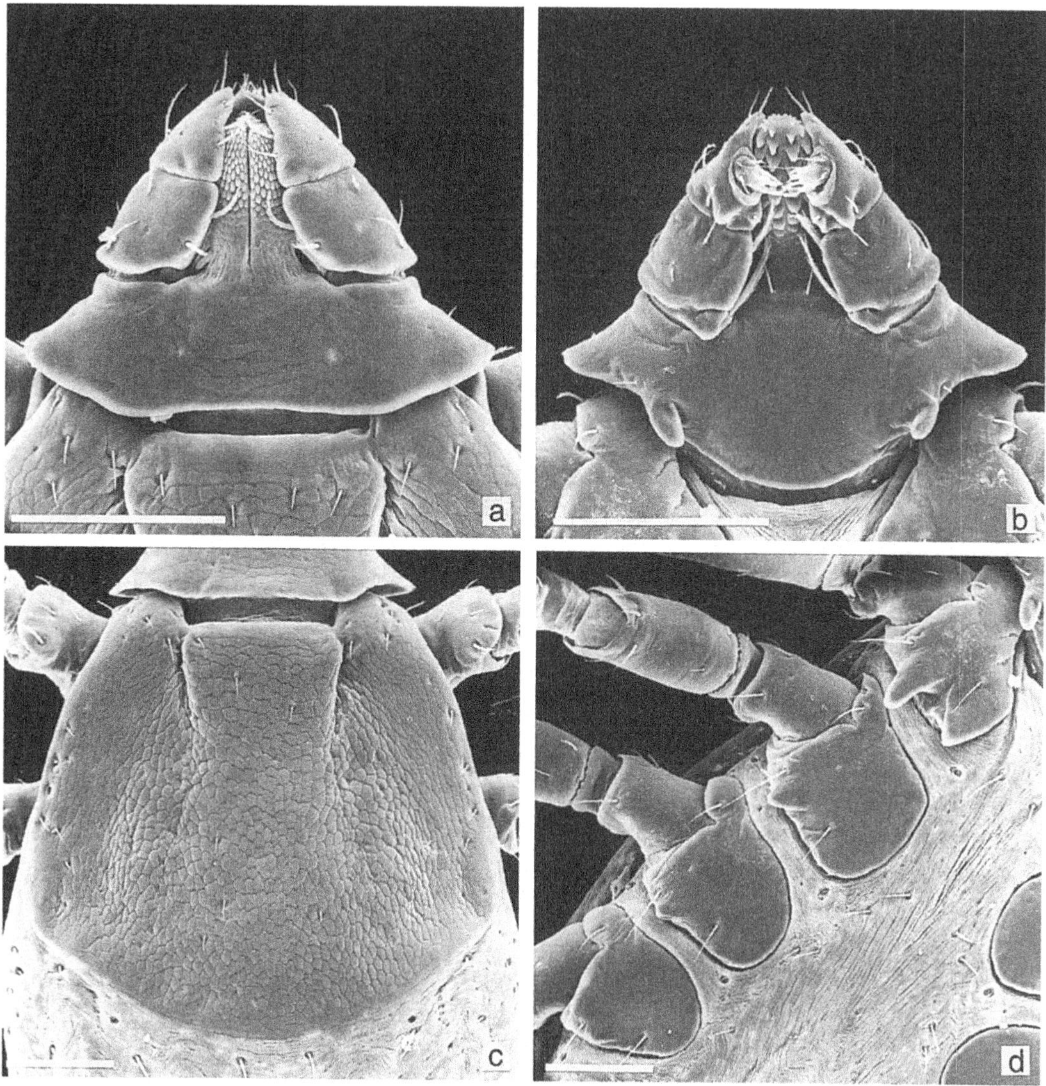

Figure 200 (*above*). *Rhipicephalus sulcatus* (L37, laboratory reared, original ♀ collected from hare, Balmoral, Zambia, on 5 January 1981 by R.G. Pegram). Nymph: (a) capitulum, dorsal; (b) capitulum, ventral; (c) scutum; (d) coxae. Scale bars represent 0.10 mm. SEMs by M.D. Corwin. (Figs (a), (c) & (d) from Pegram *et al.*, 1987, figs 5–7, with kind permission from Kluwer Academic Publishers).

Figure 199 (*opposite*). *Rhipicephalus sulcatus* (L37, laboratory reared, original ♀ collected from hare, Balmoral, Zambia, on 5 January 1981 by R.G. Pegram). Male: (a) capitulum, dorsal; (b) spiracle; (c) adanal plates. Female: (d) capitulum, dorsal; (e) scutum; (f) genital aperture. Scale bars represent 0.10 mm. SEMs by M.D. Corwin. (Figs (b), (c), (e) & (f) from Pegram *et al.*, 1987, figs 9–11 & 13, with kind permission from Kluwer Academic Publishers).

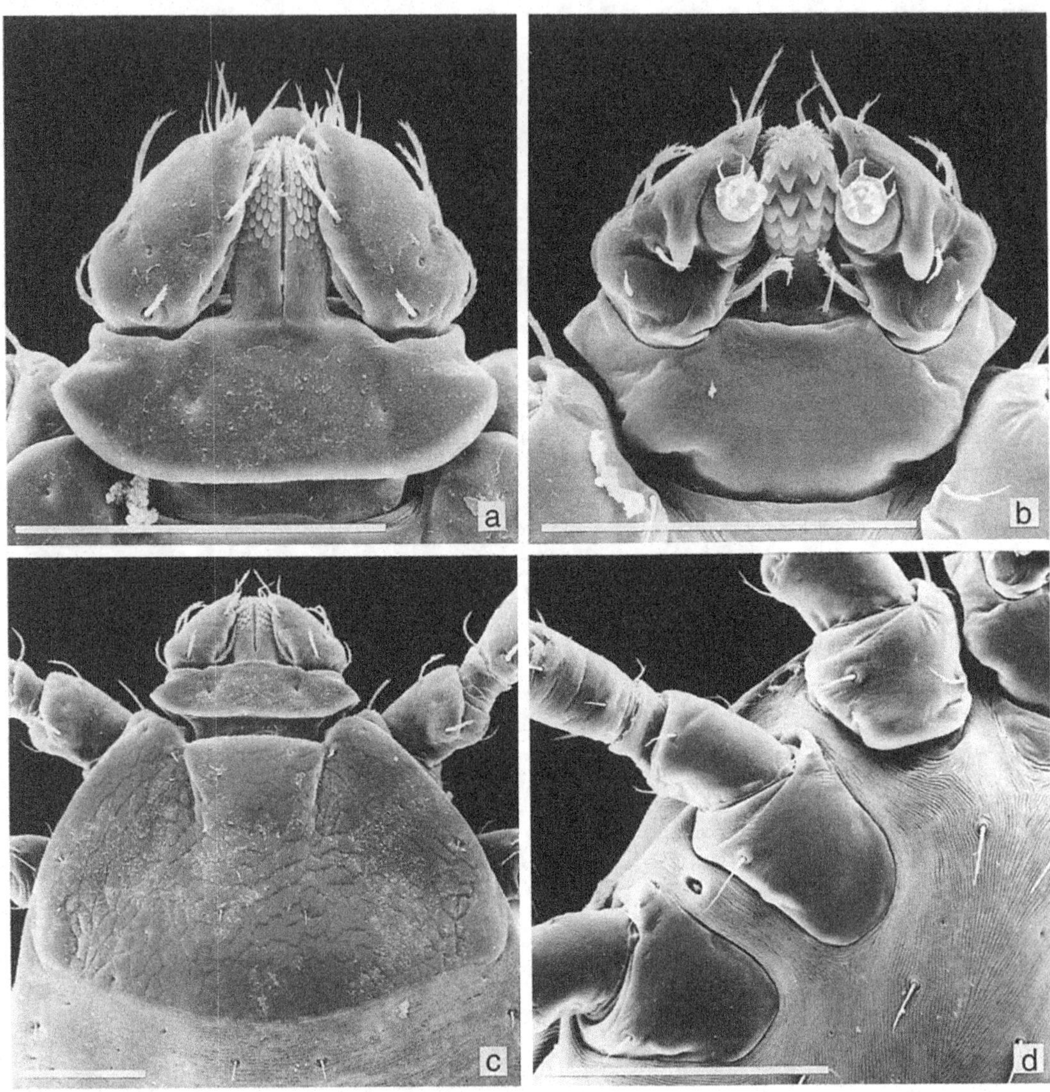

Figure 201. *Rhipicephalus sulcatus* (L37, laboratory reared, original ♀ collected from hare, Balmoral, Zambia, on 5 January 1981 by R.G. Pegram). Larva: (a) capitulum, dorsal; (b) capitulum, ventral; (c) scutum; (d) coxae. Scale bars represent 0.10 mm. SEMs by M.D. Corwin. (Figs (a), (c) & (d) from Pegram *et al.*, 1987, figs 2–4, with kind permission from Kluwer Academic Publishers).

where the setiferous punctations are often barely distinguishable from those around them. Ventrally spiracles long, gently curved towards dorsal surface and slightly tapered. Adanal plates elongate, their internal margins only slightly emarginate just posterior to the anus, their posterior margins smoothly rounded; accessory adanal plates small, pointed.

Female (Figs 198(b), 199(d) to (f), also 85(b))

Capitulum broader than long, length × breadth ranging from 0.47 mm × 0.58 mm to 0.66 mm × 0.77 mm. Basis capituli with acute lateral angles just anterior to mid-length; porose areas almost round, about 1.5 times their own diameter apart. Scutum longer than broad, length × breadth ranging from 1.05 mm × 1.01 mm to 1.41 mm × 1.27 mm; posterior margin slightly sinuous posterolaterally. Eyes at mid-

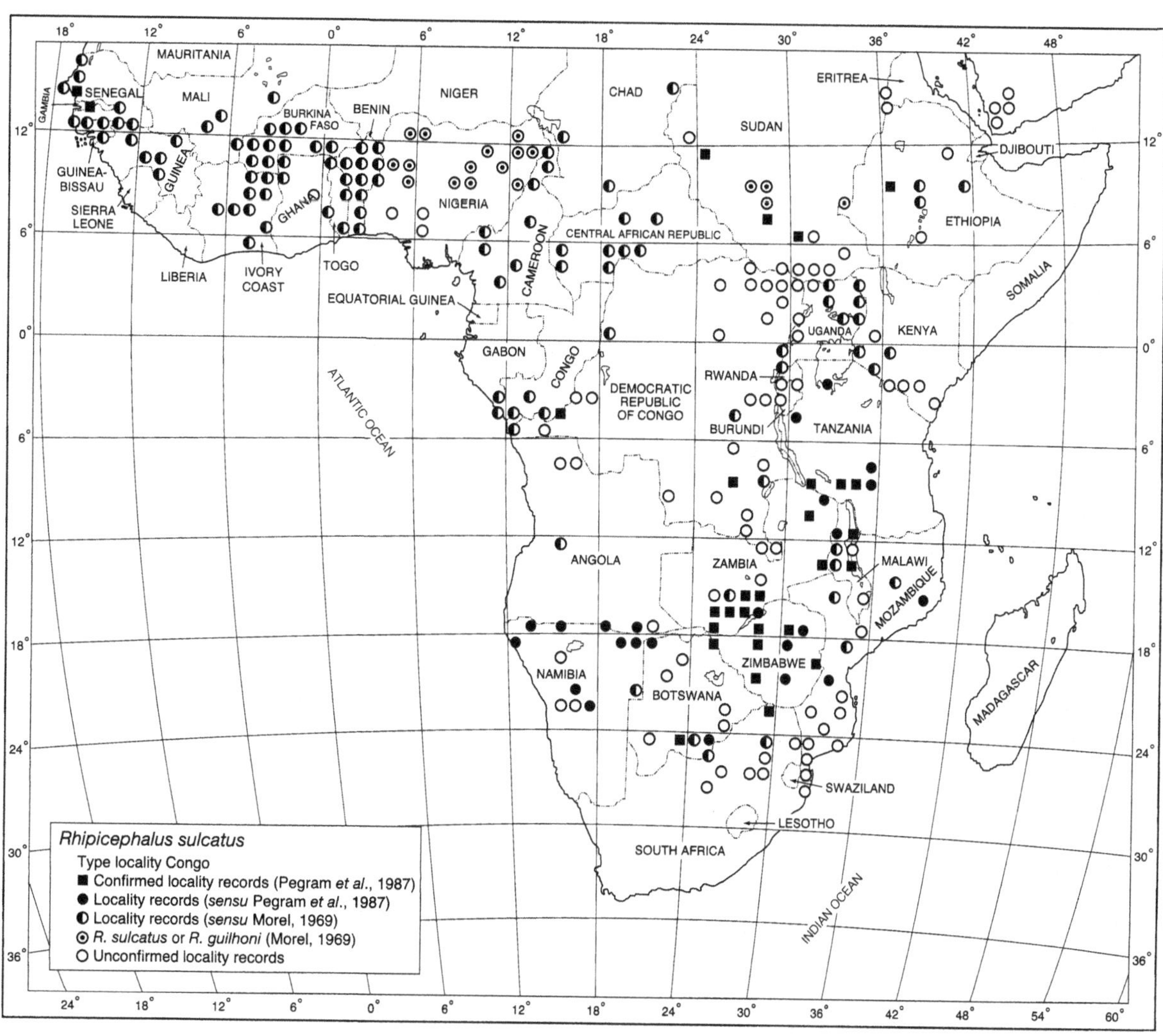

Map 56. *Rhipicephalus sulcatus*: distribution. (Based largely on Pegram *et al.*, 1987 and Morel, 1969).

length, flat, edged dorsally by a few large setiferous punctations. Cervical fields only slightly depressed, delimited laterally by rows of large setiferous punctations. Scapulae slightly raised, relatively less punctate compared with the dense pattern elsewhere on the scutum, where the setiferous punctations are often almost masked by those around them. Ventrally genital aperture broadly U-shaped. (Also see fig. 85(b), p.211).

Nymph (Fig. 200)

Capitulum much broader than long, length × breadth ranging from 0.21 mm × 0.30 mm to 0.21 mm × 0.32 mm. Basis capituli over three times as broad as long with broad tapering lateral angles at about mid-length extending over the scapulae; ventrally with small spurs on posterior margin. Palps tapering to slightly hunched apices, inclined inwards. Scutum broader than long, length × breadth ranging from 0.42 mm × 0.46 mm to 0.47 mm × 0.50 mm; posterior margin a broad smooth curve. Eyes at widest point, well over halfway back, long and narrow. Cervical fields long, narrow, slightly depressed, inconspicuous. Ventrally coxae I each with a longer narrower external spur and a shorter broader internal spur; coxae II to IV each with an external spur only, decreasing progressively in size.

Table 52. *Host records of* Rhipicephalus sulcatus

Hosts	Number of records	
	Confirmed	Unconfirmed
Domestic animals		
Cattle	14	5
Sheep	1	1
Goats	3	1
Dogs	33	1
Cats	1	2
Wild animals		
Greater bushbaby (*Otolemur crassicaudatus*)		1
Side-striped jackal (*Canis adustus*)		1
Golden jackal (*Canis aureus*)		1
'Jackal' (*Canis* spp.)	6	1
Cape fox (*Vulpes chama*)		1
African wild cat (*Felis lybica*)		1
Serval (*Leptailurus serval*)	1	
Lion (*Panthera leo*)	2	1
Leopard (*Panthera pardus*)	7	
Banded mongoose (*Mungos mungo*)		1
African civet (*Civettictis civetta*)	2	3
Rusty-spotted genet (*Genetta maculata*)		1
'Genet' (*Genetta* sp.)	1	
Aardvark (*Orycteropus afer*)		1
Impala (*Aepyceros melampus*)	1	
Lichtenstein's hartebeest (*Sigmoceros lichtensteinii*)	1	1
'Gazelle'		1
Oribi (*Ourebia ourebi*)	1	1
Cape grysbok (*Raphicerus melanotis*)	1	
African buffalo (*Syncerus caffer*)	1	
Greater kudu (*Tragelaphus strepsiceros*)	1	
Roan antelope (*Hippotragus equinus*)		1
Waterbuck (*Kobus ellipsiprymnus*)	1	
Common fat mouse (*Steatomys pratensis*)	1	
Spring hare (*Pedetes capensis*)	1	
Cape hare (*Lepus capensis*)	1	3
Scrub hare (*Lepus saxatilis*)	3	
Savanna hare (*Lepus victoriae*)		2
'Hare'	39	4
Central African rabbit (*Poelagus marjorita*)		1
Humans		2

Larva (Fig. 201)
Capitulum much broader than long, length × breadth ranging from 0.107 mm × 0.147 mm to 0.117 mm × 0.156 mm. Basis capituli about three times as broad as long, lateral angles broadly rounded, slightly forwardly directed, posterior margin a broad smooth curve. Palps with external margins slightly convex, apices pointed, inclined inwards. Scutum much broader than long, length × breadth ranging from 0.202 mm × 0.336 mm to 0.214 mm × 0.357 mm; posterior margin a smooth shallow curve. Eyes at widest part of scutum, flat. Cervical grooves slightly convergent. Ventrally coxae I each with a single spur; coxae II and III each with a mere indication of a spur on its posterior margin.

Notes on identification
Rhipicephalus sulcatus and *R. turanicus* are two species in the *R. sanguineus* group of ticks that have been confused both morphologically and ecologically (Pegram *et al.*, 1987). In West Africa in particular Morel & Vassiliades (1963) included both these ticks within their concept of *R. sulcatus*. This has resulted in erroneous conclusions regarding their host preferences and distribution.

While *R. sulcatus* can be confused with the more punctate forms of *R. turanicus* the less punctate forms of the latter tick resemble *R. sanguineus*. All three feed on dogs, but whereas all stages of development of *R. sanguineus* feed nearly exclusively on these animals, the adults of both *R. sulcatus* and *R. turanicus* feed on wild carnivores and hares as well as dogs. The hosts of the immature stages of *R. sulcatus* are unknown while those of *R. turanicus* feed on cats, hedgehogs, rodents and hares.

Hosts

A three-host species (Theiler & Robinson, 1953). The preferred domestic hosts of adult *R. sulcatus* are dogs and possibly cattle, while jackals, leopards and probably other carnivores as well as hares are the preferred wild hosts (MacLeod, 1970; Pegram *et al.*, 1987) (Table 52). *Rhipicephalus sulcatus* is never very abundant: one of the largest collections from a single host comprises 51 ticks from a jackal in Zimbabwe (Norval, Daillecourt & Pegram, 1983).

The hosts of the immature stages are unknown.

In Zimbabwe and Zambia the seasonal activity of adult *R. sulcatus* is confined to the main rainy season (November to April) (Norval *et al.*, 1983; Pegram *et al.*, 1987). In Ethiopia, in the northern hemisphere, the few collections that have been made also coincide with the months of highest rainfall (July to September) (Pegram *et al.*, 1987).

Zoogeography

Rhipicephalus sulcatus has been recorded in western, central and southern Africa, with most collections coming from Tanzania, Zambia, Zimbabwe and Namibia (Map 56). Pegram *et al.* (1987) suggested that it is restricted to ecological habitats receiving more than 500 mm of rainfall annually. Several collections from Namibia and Botswana indicate that its rainfall requirements may be lower. In West Africa *R. sulcatus* has been collected in drier rainforest; in East and Central Africa in coastal mosaic and Zambezian miombo woodland; and in Namibia and Botswana in woodland, wooded grassland, deciduous bushland and shrubland.

Disease relationships

Unknown.

REFERENCES

MacLeod, J. (1970). Tick infestation patterns in the southern province of Zambia. *Bulletin of Entomological Research*, **60**, 253–74.

Morel, P.C. & Vassiliades, G. (1963). Les *Rhipicephalus* du group *sanguineus*: espèces africaines (Acariens: Ixodoidea). *Revue d'Élevage et de Médecine Vétérinaire des Pays Tropicaux*, **15** (nouvelle série), 343–86.

Neumann, L.G. (1908). Description d'une nouvelle espèce d'Ixodiné. *Bulletin du Muséum d'Histoire Naturelle, Paris*, **14**, 352–5.

Norval, R.A.I., Daillecourt, T. & Pegram, R.G. (1983). The ticks of Zimbabwe. VI. The *Rhipicephalus sanguineus* group. *Zimbabwe Veterinary Journal*, **13**, 38–46.

Pegram, R.G., Clifford, C.M., Walker, J.B. & Keirans, J.E. (1987). Clarification of the *Rhipicephalus sanguineus* group (Acari, Ixodoidea, Ixodidae). I. *R. sulcatus* Neumann, 1908 and *R. turanicus* Pomerantsev, 1936. *Systematic Parasitology*, **10**, 3–26.

Theiler, G. & Robinson, B.N. (1953). Ticks in the South African Zoological Survey Collection. Part VII. Six lesser known African rhipicephalids. *Onderstepoort Journal of Veterinary Research*, **26**, 93–136 + 1 map.

Also see the following Basic References (pp. 12–14): Aeschlimann (1967); Elbl & Anastos (1966); Hoogstraal (1956); Morel (1969).

RHIPICEPHALUS SUPERTRITUS NEUMANN, 1907

The specific name *supertritus*, from the Latin *super* meaning 'above' plus the Greek *tritos* meaning 'one with two others', doubtless refers to the three finger-like caudal processes developed by engorged males. (Such a combination of Latin and Greek in a specific name is unusual, and is not recommended.)

Synonym

coriaceus.

Diagnosis

A large very dark brown tick with reddish-brown legs. It might initially be confused with heavily-punctate specimens of either *R. appendiculatus* or *R. zambeziensis*.

Male (Figs 202(a), 203(a) to (c))

Capitulum longer than broad, length × breadth ranging from 0.78 mm × 0.64 mm to 0.86 mm × 0.83 mm. Basis capituli with short pointed lateral angles at about anterior third of its length. Palps broadly rounded apically. Conscutum length × breadth ranging from 3.31 mm × 2.14 mm to 4.58 mm × 3.18 mm; anterior process on coxae I prominent, strongly sclerotized. Body wall of engorged specimens expanded laterally and posteriorly, exposing the dark sclerotized ventral plaques and forming three characteristic finger-like caudal processes posteromedially. Eyes small, flat, sometimes edged dorsally by a few punctations. Cervical fields broad, depressed, with finely-reticulate surfaces, their external margins more-or-less demarcated by variable numbers of large setiferous punctations. Marginal lines long, deep, clearly defined, with reticulate surfaces and only a few large punctations. A long narrow posteromedian groove plus two or more shorter broader posterolateral grooves or depressions present, all with finely-reticulate surfaces. Medium-sized to large punctations present on the scapulae and scattered medially on the conscutum, especially posteriorly, many of them containing unusually long fine white setae. They are interspersed with, and sometimes masked by, numerous smaller punctations and rugose areas. As in *R. appendiculatus* the lateral areas adjacent to the marginal lines are usually smooth or only very finely punctate. Legs increase in size from I to IV, markedly setose ventrally and usually reddish-brown. Ventrally spiracles comma-shaped with a short slightly-tapering dorsal prolongation curving gently towards the dorsal surface. Adanal plates elongate, their internal margins only mildly concave, their posterior margins drawn out and rounded to bluntly pointed; accessory adanal plates small, pointed.

Female (Figs 202(b), 203(d) to (f))

Capitulum about as broad as long, length × breadth ranging from 0.72 mm × 0.75 mm to 0.86 mm × 0.83 mm. Basis capituli with short broad lateral angles in anterior third of its length; porose areas medium-sized, subcircular, about twice their own diameter apart. Palps short, broad. Scutum slightly broader than long,

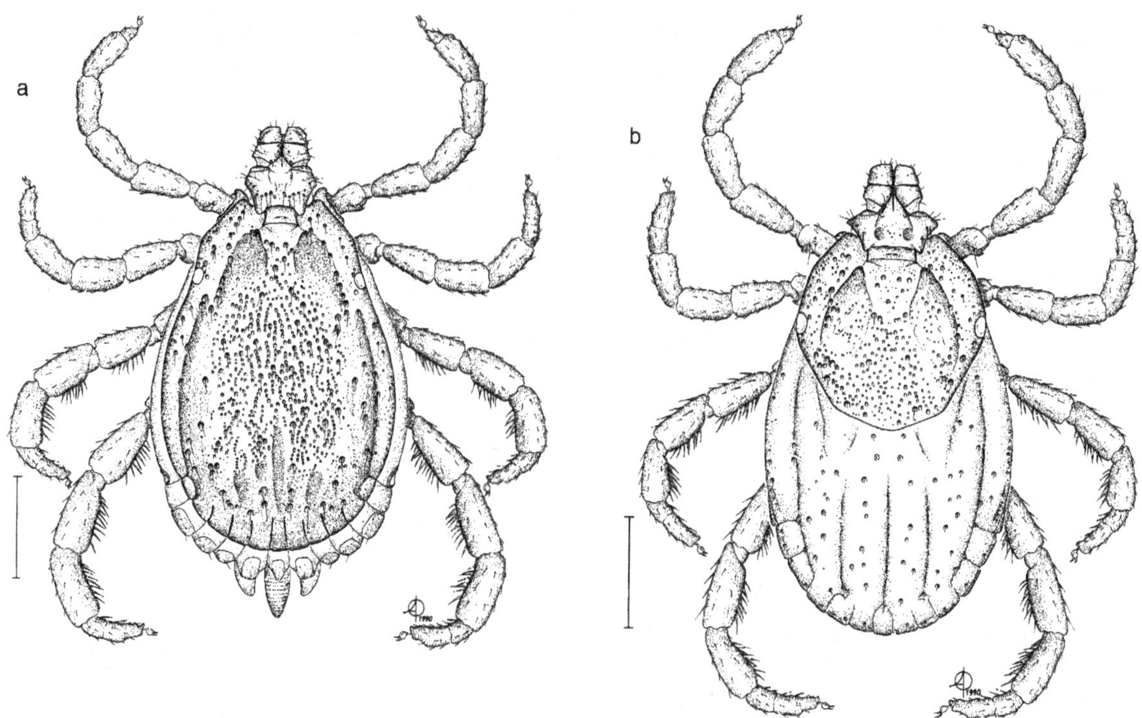

Figure 202. *Rhipicephalus supertritus* [Onderstepoort Tick Collection 2788iii, collected from African buffalo (*Syncerus caffer*), Juba, Sudan, in December 1950 by E.T.M. Reid]. (a) Male, dorsal; (b) female, dorsal. Scale bars represent 1 mm. A. Olwage *del.*

length × breadth ranging from 1.49 mm × 1.57 mm to 1.80 mm × 1.88 mm. Eyes at widest part of scutum, sometimes edged dorsally by a few punctations. Cervical fields very broad, divergent, somewhat depressed, their surfaces finely reticulate in places, especially adjacent to the external margins. Medium-sized to large punctations present on the scapulae, along the external cervical margins and medially on the scutum, often containing unusually long fine white setae. They are interspersed, especially medially, with numerous smaller punctations. Ventrally genital aperture a broad curve, with wide lateral flaps flanking the genital apron.

Immature stages

Unknown.

Notes on identification

Zumpt (1942) included *R. supertritus* in his *R. capensis* group but we think that its affinities lie with the *R. appendiculatus* group. Thus, although the immature stages of *R. supertritus* have not as yet been reared, we agree with Colbo (1973) that they will probably resemble those of *R. appendiculatus* (see p. 59), as do those of *R. sculptus* (see p. 392). We have examined the large engorged *appendiculatus*-like nymphs collected from a Lichtenstein's hartebeest to which Colbo refers and they do not appear to us to be *R. sculptus*. They may well be *R. supertritus*, but it will only be possible either to confirm or to correct this tentative identification once reared specimens of its immature stages become available.

We question whether *R. supertritus* occurs in Kenya. The ticks recorded by Lewis (1933) as *R. supertritus* were collected, together with the syntypes of *R. compositus* (syn. *R. ayrei*), from a rhinoceros (presumably *Diceros bicornis*), Mount Kenya on 29 October 1930 by A.F. Ayre. One of us (J.B.W.) has seen specimens from this rhinoceros labelled '*R. sculptus*?' by

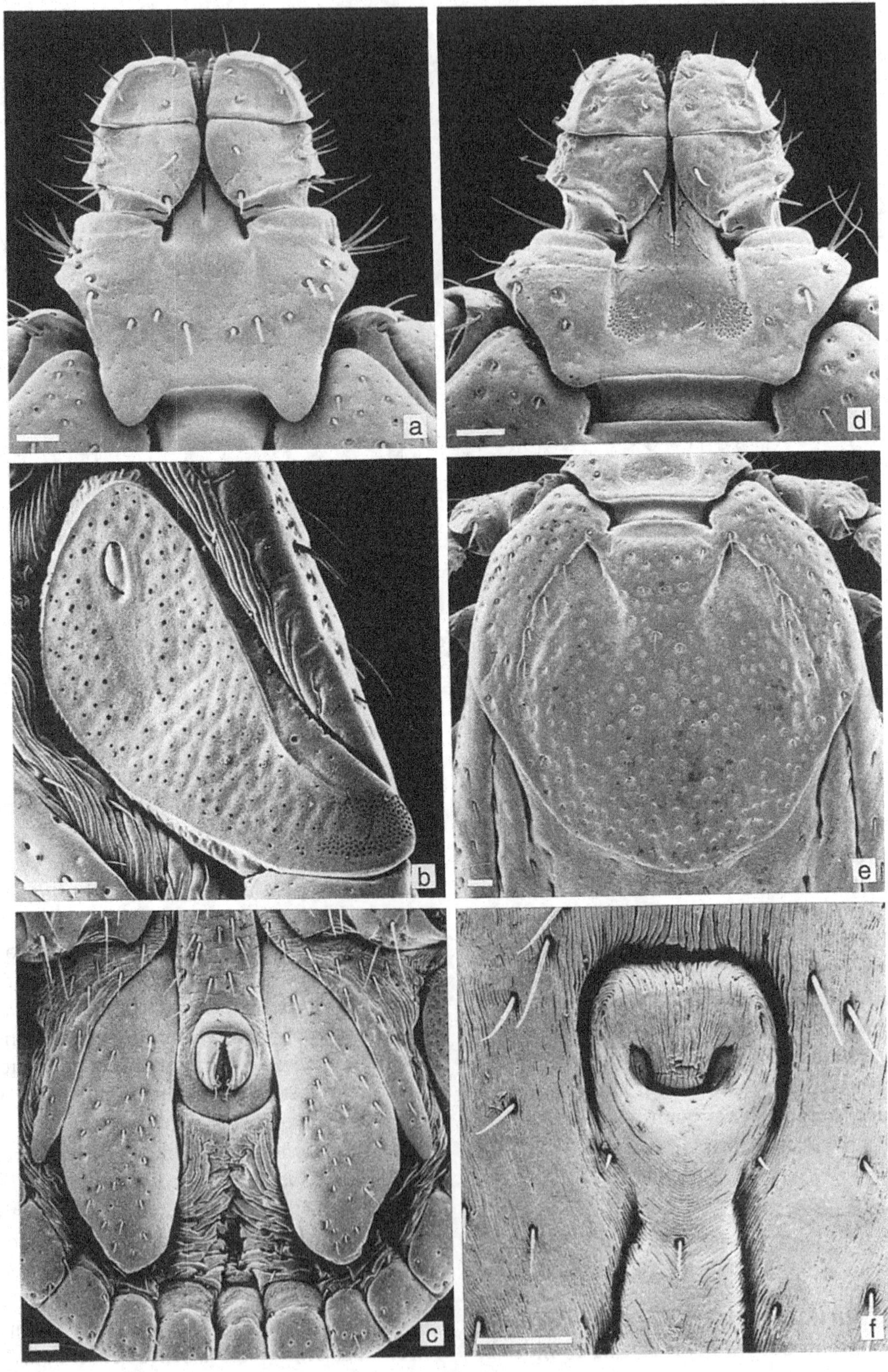
a
d
b
e
c
f

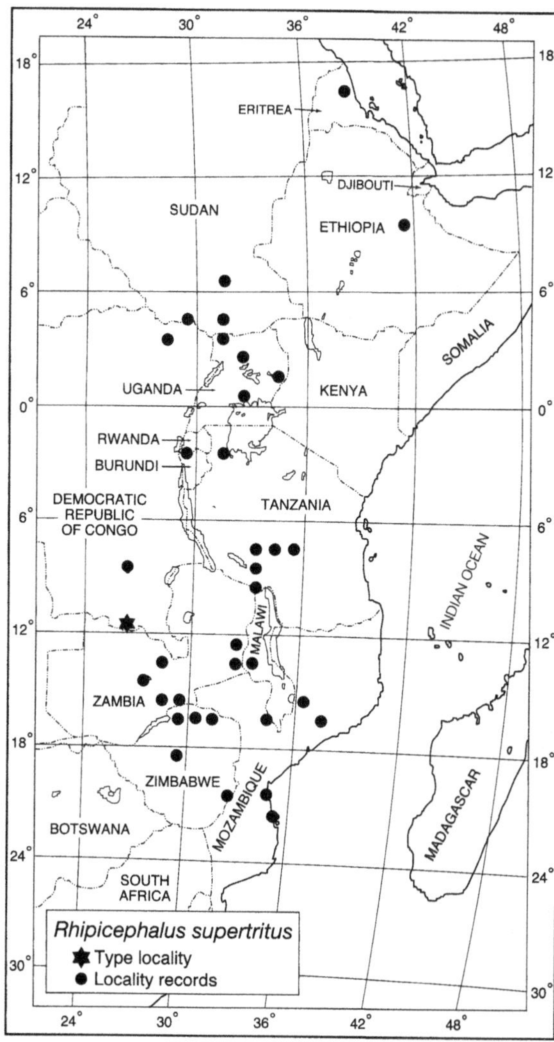

Map 57. *Rhipicephalus supertritus:* distribution.

Figure 203 (*opposite*). *Rhipicephalus supertritus* [Onderstepoort Tick Collection 2788iii, collected from African buffalo (*Syncerus caffer*), Juba, Sudan, in December 1950 by E.T.M. Reid]. Male: (a) capitulum, dorsal; (b) spiracle; (c) adanal plates. Female: (d) capitulum, dorsal; (e) scutum; (f) genital aperture. Scale bars represent 0.10 mm. SEMs by J.F. Putterill.

Lewis and re-identified them as *R. longus*, a species that has been confused with *R. supertritus* elsewhere. Hoogstraal (1956) referred to Lewis's 1933 record and it is doubtless the one quoted by Theiler (1962) and Zumpt (1964). As we know of no other collections of *R. supertritus* from the rhinoceros we have omitted this animal from our host list (see Table 53). The specimens recorded by Lewis (1933, 1934) from a lion, Sianna Plains, Maasailand, have been re-identified as *R. compositus* by J.B.W. The source of the record in Theiler (1962) from the Lambwe Valley, Nyanza, is unknown, as is another from Ololkisailie (= Olorgesaille) (Gertrud Theiler, unpublished data); as we have seen nothing to support these data they have been omitted from Map 57.

Hoogstraal (1954) identified Nuttall Collection 2394 from warthog (*Phacochoerus africanus*, listed as *P. aethiopicus*), Marimba, Malawi (formerly Nyasaland), July 1913, as *R. supertritus*. This collection was re-identified as *R. compositus* 1 ♂; *R. longus* 3 ♂♂, 2 ♀♀, and *R. simus* 1 ♂ by Keirans (1985).

There has been controversy regarding the identity of three collections from striped hyaenas, all from Alemaya, Harar, south-eastern Ethiopia (RML 97169: 6 ♂♂, 4 ♀♀; RML 97170: 8 ♂♂, 3 ♀♀, and RML 97171: 1 ♂). All three were originally identified by H. Hoogstraal and M. Kaiser as *R. supertritus*. The first two (14 ♂♂, 7 ♀♀) were re-identified as *R. bergeoni* by Morel & Rodhain (1973), a finding that was reiterated by Morel (1980). These specimens were, however, still listed as *R. supertritus* by Bergeon & Balis (1974), an opinion that was upheld by Pegram (1979) and that we also endorse.

Morel (1980) includes 'Zululand, Natal' in the range of this tick but we know of no records from South Africa. It may be based on a misinterpretation of the information in Zumpt (1964).

Hosts

Life cycle unknown. Amongst domestic animals the only hosts of *R. supertritus* of any conse-

Table 53. *Host records of* Rhipicephalus supertritus

Hosts	Number of records
Domestic animals	
Cattle	22
Goats	1
Horses	1
Dogs	2
Wild animals	
Hunting dog (*Lycaon pictus*)	1
Striped hyaena (*Hyaena hyaena*)	3
'Hyaena'	1
Burchell's zebra (*Equus burchellii*)	1 (1 ♀ only)
'Zebra'	2
Rock hyrax (*Procavia capensis*)	1 (1 ♂ only)
Warthog (*Phacochoerus africanus*)	1 (1 ♀ only)
Giraffe (*Giraffa camelopardalis*)	1
Coke's hartebeest (*Alcelaphus buselaphus cokii*)	1
Lichtenstein's hartebeest (*Sigmoceros lichtensteinii*)	1
'Hartebeest'	2
African buffalo (*Syncerus caffer*)	21
Giant eland (*Taurotragus derbianus*)	1 (1 ♂ only)
Eland (*Taurotragus oryx*)	8
Greater kudu (*Tragelaphus strepsiceros*)	5
Roan antelope (*Hippotragus equinus*)	1
Sable antelope (*Hippotragus niger*)	6
South African porcupine (*Hystrix africaeaustralis*)	3
Humans	1

quence appear to be cattle. It is apparently almost always present in very small numbers only, though it has been recorded as 'a significant pest of cattle' in a restricted part of the Chiota and Lupiya areas of Central Province, Zambia (MacLeod *et al.*, 1977). The comparatively few records from these animals that we have listed undoubtedly do not reflect their real status as hosts (Table 53).

Its most frequently recorded wild host is the African buffalo, followed by some of the larger antelopes such as the eland, greater kudu and sable antelope. Matthysse & Colbo (1987) listed a maximum of 64 adults collected from one buffalo. The collections from the Burchell's zebra, rock hyrax, warthog and giant eland consisted of a single adult only in each case.

On both cattle and the African buffalo its predilection site is the ears: of 21 specimens recorded from cattle by MacLeod *et al.* (1977), 18 were on the ears and three on the head. In Zambia the adults are active from November to January, during the rainy season.

Zoogeography

Rhipicephalus supertritus is apparently commonest in Central Africa, especially northern Zimbabwe (Norval, 1985), parts of Zambia, Malawi (Wilson, 1950), northern Mozambique and southern Tanzania. There are also scattered records from various parts of East Africa, the northernmost being that from Eritrea (Map 57).

It has seemingly been recorded in Angola on the basis of a single male only, collected from an unknown host in the Benguella area in 1907 and initially identified as *R. coriaceus*. Its presence there requires confirmation.

The arcas in which *R. supertritus* has been recorded range in altitude from about 750 m to 1500 m, with mean annual rainfalls around 800 mm to 1200 mm, occasionally less. Quite often the dry season extends for at least 3 to 5 months, especially in the more southerly part of its range. The vegetation is usually some type of woodland.

Disease relationships

Unknown.

REFERENCES

Bergeon, P. & Balis, J. (1974). Contribution à l'étude de la répartition des tiques en Éthiopie (enquête effectuée de 1965 à 1969). *Revue d'Élevage et de Médecine Vétérinaire des Pays Tropicaux*, **27** (nouvelle série), 285–99.

Colbo, M.H. (1973). Ticks of Zambian wild animals: a preliminary checklist. *Puku*, **No. 7**, 97–105.

Hoogstraal, H. (1954). Noteworthy African tick records in the British Museum (Natural History) collections. *Proceedings of the Entomological Society of Washington*, **56**, 273–9.

Lewis, E.A. (1933). *Rhipicephalus ayrei* n. sp. (a tick) from Kenya Colony. *Parasitology*, **25**, 269–72.

Lewis, E.A. (1934). A study of the ticks in Kenya Colony. The influence of natural conditions and other factors on their distribution and the incidence of tick-borne diseases. Part III. Investigations into the tick problem in the Masai Reserve. *Bulletin of the Department of Agriculture, Kenya*, **No. 7 of 1934**, 65 pp, 3 maps.

MacLeod, J., Colbo, M.H., Madbouly, M.H. & Mwanaumo, B. (1977). Ecological studies of ixodid ticks (Acari: Ixodidae) in Zambia. III. Seasonal activity and attachment sites on cattle, with notes on other hosts. *Bulletin of Entomological Research*, **67**, 161–73.

Morel, P.C. & Rodhain, F. (1973). Contribution à la connaissance des tiques (Ixodina) du sud de l'Éthiopie. Deuxième partie. *Bulletin de la Société de Pathologie Exotique*, **66**, 207–15.

Neumann, L.G. (1907). Notes sur les Ixodidés – V. *Archives de Parasitologie*, **11**, 215–32.

Norval, R.A.I. (1985). The ticks of Zimbabwe. XII. The lesser known *Rhipicephalus* species. *Zimbabwe Veterinary Journal*, **16**, 37–43.

Wilson, S.G. (1950). A check-list and host-list of Ixodoidea found in Nyasaland, with descriptions and biological notes on some of the rhipicephalids. *Bulletin of Entomological Research*, **41**, 415–28.

Zumpt, F. (1942). Zur kenntnis Afrikanischer Rhipicephalusarten. V. Vorstudie zu einer Revision der Gattung *Rhipicephalus* Koch. *Zeitschrift für Parasitenkunde*, **12**, 479–500.

Zumpt, F. (1964). Parasites of the white and the black rhinoceroses. *Lammergeyer*, **3**(1), 59–70.

Also see the following Basic References (pp. 12–14): Clifford & Anastos (1962, 1964); Elbl & Anastos (1966); Hoogstraal (1956); Keirans (1985); Matthysse & Colbo (1987); Morel (1980); Pegram (1979); Santos Dias (1960); Theiler (1947, 1962); Yeoman & Walker (1967).

RHIPICEPHALUS THEILERI BEDFORD & HEWITT, 1925

This species was named in honour of Sir Arnold Theiler (1867–1936), the founder of the Onderstepoort Veterinary Institute and first Dean of the Faculty of Veterinary Science, Onderstepoort. He became renowned for his research on animal diseases in South Africa, including several tickborne diseases of domestic animals.

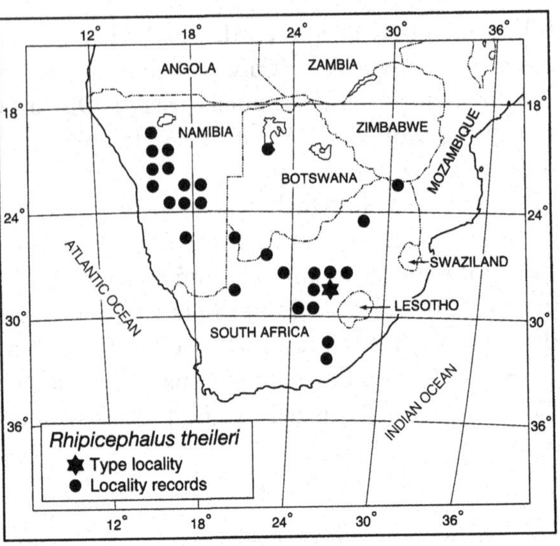

Map 58. *Rhipicephalus theileri*: distribution.

Diagnosis

A small broad yellowish to reddish-brown tick.

Male (Figs 204(a), 205(a) to (c))
Capitulum broader than long, length × breadth ranging from 0.50 mm × 0.55 mm to 0.66 mm × 0.72 mm. Basis capituli with lateral angles at about anterior third of its length. Palps short, broad. Conscutum length × breadth ranging from 2.35 mm × 1.51 mm to 3.28 mm × 2.22 mm; anterior process of coxae I small. Eyes flat. Cervical pits convergent; cervical fields with external margins delimited by irregular rows of large setiferous punctations but otherwise indistinct. Marginal lines punctate, long, almost

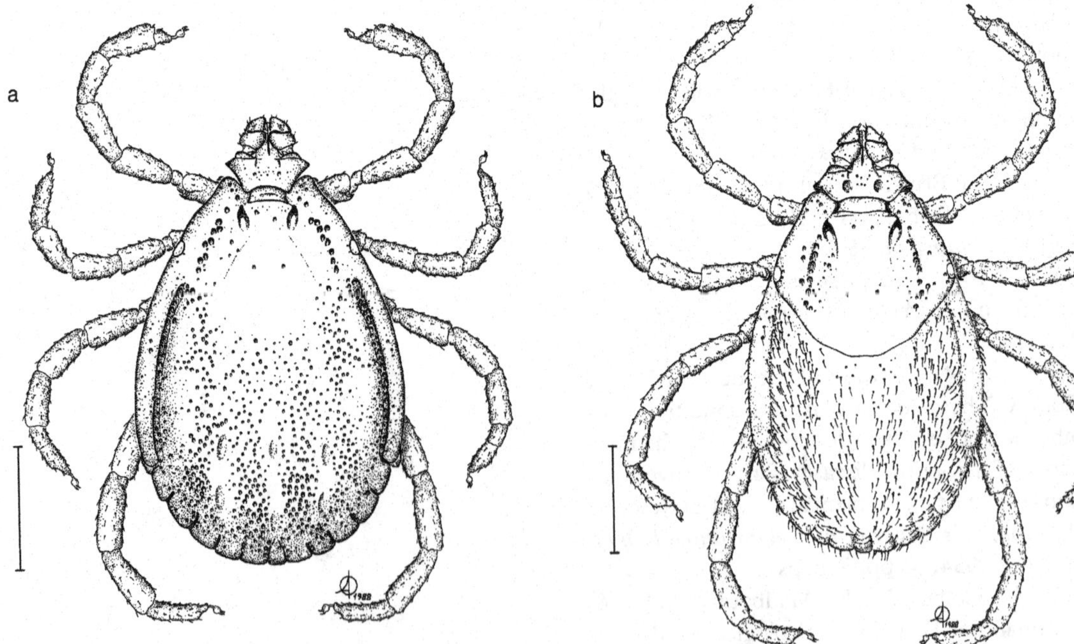

Figure 204. *Rhipicephalus theileri* [Protozoology Section Tick Breeding Register, Onderstepoort, 3106, laboratory reared, original ♀ collected from dog at Melville, Omaruru, Namibia on 8 February 1960 by J.S. Brown]. (a) Male, dorsal; (b) female, dorsal. Scale bars represent 1 mm. A. Olwage *del.*

Table 54. *Host records of* Rhipicephalus theileri

Hosts	Number of records
Domestic animals	
Cattle	3
Sheep	1
Dogs	2
Wild animals	
South African hedgehog (*Atelerix frontalis*)	1
Black-backed jackal (*Canis mesomelas*)	1
'Jackal'	1 (nymphs)
Cape fox (*Vulpes chama*)	1
Yellow mongoose (*Cynictis penicillata*)	28 (including immatures)
Meercat (*Suricata suricatta*)	8 (including immatures)
'Meercat'	1 (nymph)
Temminck's ground pangolin (*Manis temminckii*)	1
Smith's bush squirrel (*Paraxerus cepapi*)	16 (immatures)
Cape ground squirrel (*Xerus inauris*)	18 (including immatures)
Xerus sp.	8 (including immatures)
Namaqua rock mouse (*Aethomys namaquensis*)	1 (nymph, unconfirmed)

reaching external cervical margins. Posteromedian and posterolateral grooves may be indicated by small depressions but sometimes represented merely by smooth areas on the conscutum. Punctation pattern unique; anteriorly a distinct smooth shiny pseudoscutum present that is usually almost lacking in punctations apart from those along the external cervical margins; posterior to the pseudoscutum punctations numerous, small to medium-sized, finer and sparser adjacent to the marginal lines but increasing in size and density posteriorly. Ventrally spiracle comma-shaped, tapering rapidly to a long narrow dorsal prolongation. Adanal plates broad, tapering posterointernally to broadly-rounded points that may be visible from the dorsal surface of engorged specimens, their surfaces coarsely punctate; accessory adanal plates absent.

Female (Figs 204(b), 205(d) to (f))

Capitulum broader than long, length × breadth ranging from 0.64 mm × 0.74 mm to 0.81 mm × 0.97 mm. Basis capituli with short acute lateral angles at about mid-length; porose areas medium-sized, just over twice their own diameter apart. Palps short, broad, with article III wedge-shaped. Scutum broader than long, length × breadth ranging from 1.12 mm × 1.25 mm to 1.67 mm × 1.87 mm. Eyes about halfway back, flat. Cervical pits convergent; external cervical margins delimited by irregular rows of large setiferous punctations; cervical fields slightly depressed. A few medium-sized to small punctations scattered on the scutum, especially on the scapulae and cervical fields, but in general the scutum is smooth and shiny. Alloscutum with four broad conspicuous longitudinal bands of white setae dorsally. Ventrally a band of white setae fringes the posterior end of the alloscutum as far forward as the spiracles, elsewhere only a few short inconspicuous setae are present. Genital aperture a wide shallow curve.

Nymph (Fig. 206)

Capitulum much broader than long, length × breadth ranging from 0.21 mm × 0.32 mm to 0.28 mm × 0.36 mm. Basis capituli four times as broad as long, with short acute lateral

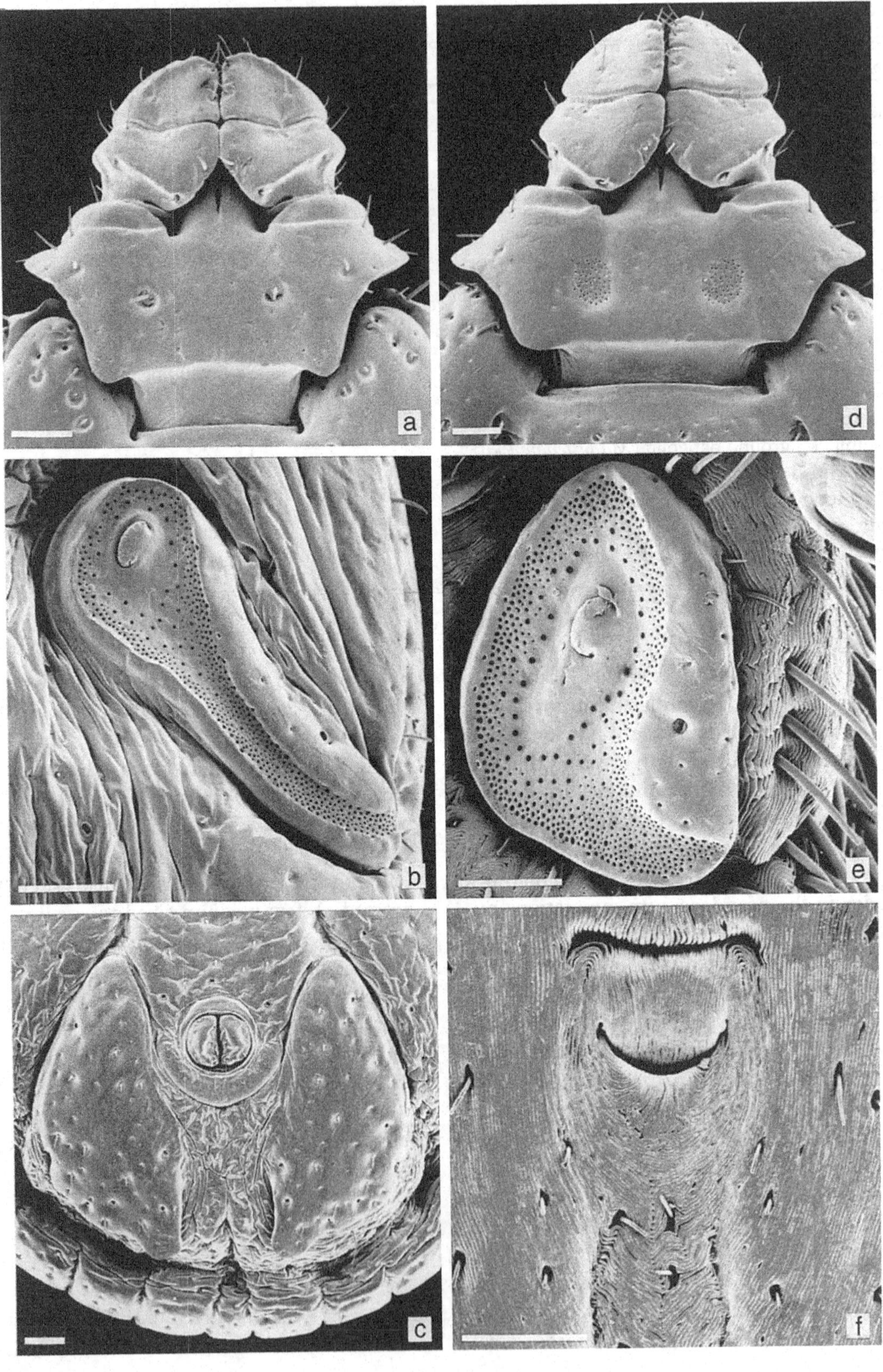
a
d
b
e
c
f

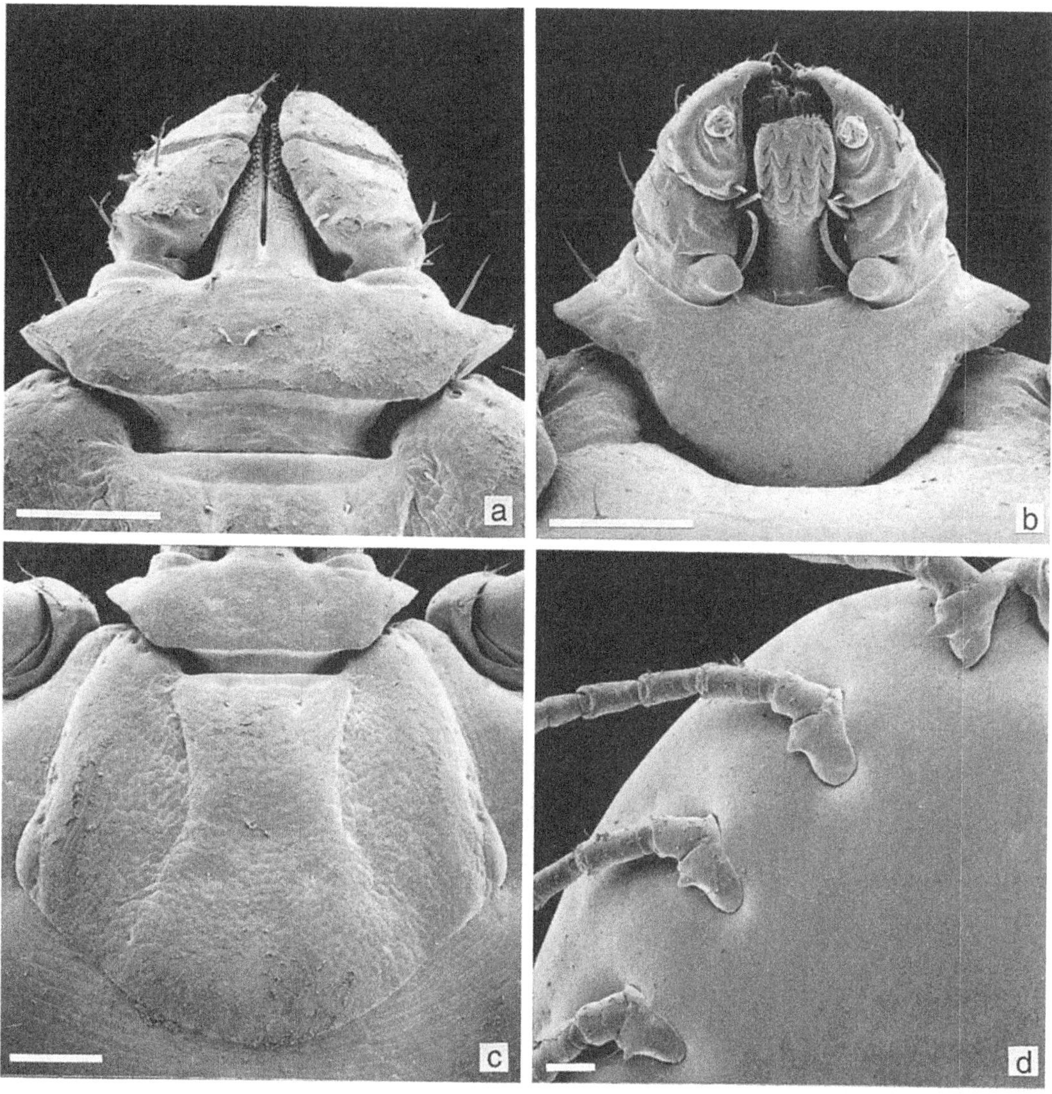

Figure 206 (*above*). *Rhipicephalus theileri* [Protozoology Section Tick Breeding Register, Onderstepoort, 3106, laboratory reared, original ♀ collected from dog at Melville, Omaruru, Namibia on 8 February 1960 by J.S. Brown]. Nymph: (a) capitulum, dorsal; (b) capitulum, ventral; (c) scutum; (d) coxae. Scale bars represent 0.10 mm. SEMs by J.F. Putterill.

Figure 205 (*opposite*). *Rhipicephalus theileri* [Protozoology Section Tick Breeding Register, Onderstepoort, 3106, laboratory reared, original ♀ collected from dog at Melville, Omaruru, Namibia on 8 February 1960 by J.S. Brown]. Male: (a) capitulum, dorsal; (b) spiracle; (c) adanal plates. Female: (d) capitulum, dorsal; (e) spiracle; (f) genital aperture. Scale bars represent 0.10 mm. SEMs by J.F. Putterill.

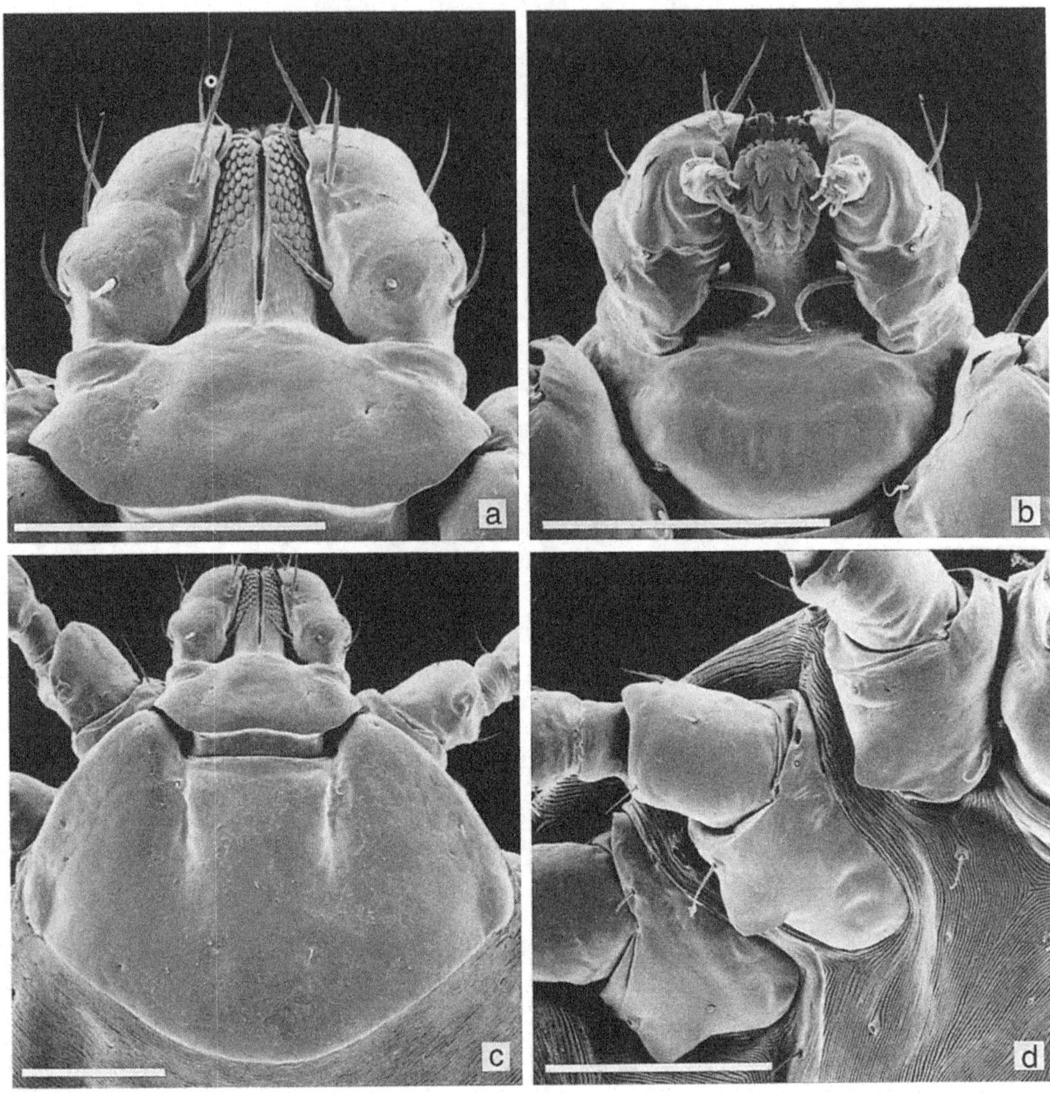

Figure 207. *Rhipicephalus theileri* [Protozoology Section Tick Breeding Register, Onderstepoort, 3106, laboratory reared, original ♀ collected from dog at Melville, Omaruru, Namibia on 8 February 1960 by J.S. Brown]. Larva: (a) capitulum, dorsal; (b) capitulum, ventral; (c) scutum; (d) coxae. Scale bars represent 0.10 mm. SEMs by J.F. Putterill.

angles at about mid-length. Palps tapering somewhat to broadly rounded apices. Scutum much broader than long, length × breadth ranging from 0.43 mm × 0.61 mm to 0.54 mm × 0.64 mm. Eyes at widest point, over halfway back, slightly bulging. Internal cervical margins convergent initially, becoming divergent and almost reaching posterolateral margins of scutum, but external cervical margins indistinct; cervical fields broad, slightly depressed. Ventrally coxae I each with a long narrow external spur and shorter broader internal spur; coxae II to IV each with an external spur only, decreasing progressively in size.

Larva (Fig. 207)

Capitulum broader than long, length × breadth ranging from 0.14 mm × 0.16 mm to 0.16 mm × 0.18 mm. Basis capituli three times as broad as long, with short acute lateral angles at about mid-length. Palps constricted proximally, then becoming somewhat bulbous before tapering towards their tips. Scutum much broader than long, length × breadth ranging from 0.24 mm × 0.45 mm to 0.27 mm × 0.47 mm, posterior margin shallow. Eyes at widest point, slightly convex. Cervical grooves short, almost parallel. Ventrally coxae I to III each with a large triangular spur.

Notes on identification

Theiler (1962) listed 'Chelonia-Tortoise' as a host of *R. theileri* but later she stated that this record was incorrect (G. Theiler, pers. comm., 1971, cited by Neitz, Boughton & Walters, 1972). In the same personal communication she apparently also listed the rodents *Otomys irroratus, Otomys* sp. and *Myotomys (Otomys)* sp. as hosts of the nymphs. However, when we re-examined these ticks we found them not to be *R. theileri* (see p. 110, *R. capensis* and p. 185, *R. follis*).

We have been unable to trace the immature specimens recorded from a Namaqua rock mouse (*Aethomys namaquensis*) at Okahandja, Namibia (Theiler, 1962; pers. comm., 1971), or those from Smith's bush squirrel (*Paraxerus cepapi*) at Gweru (= Gwelo), Zimbabwe (G. Theiler, pers. comm., 1971). Norval (1985), makes no mention of the latter collection nor does he list *R. theileri* as occurring in Zimbabwe.

Hosts

A three-host species (Neitz *et al.*, 1972). The preferred hosts of all stages of development are the yellow mongoose, meercat and Cape ground squirrel (Table 54). All three species inhabit burrows and can be found together, or in a combination of any two species, in the same burrow system. Each species may also occupy a burrow alone (Lynch, 1980). Only larvae and nymphs have as yet been collected from another favoured host, Smith's bush squirrel. Although these squirrels spend a great deal of their time foraging on the ground they make their nests in hollow trees (Skinner & Smithers, 1990). With the possible exception of some of the wild carnivores, ticks on other hosts are probably accidental infestations.

Although no single host species has been examined at regular intervals a pattern of seasonal abundance can be deduced from the collection dates of ticks from the various hosts. The majority of collections made between October and January contain adult ticks, while the majority made between April and September contain immatures.

Zoogeography

All our records of *R. theileri* are from South Africa, Botswana and Namibia (Map 58). The distribution of this species seems to be largely determined by the distribution of three of its preferred hosts, namely the yellow mongoose, meercat and Cape ground squirrel. These three small mammals are widespread but they all prefer the more arid western parts of the subregion (Lynch, 1980; Skinner & Smithers, 1990). Even within the distribution ranges of these animals the tick also seems to favour the arid western regions. None of the 38 yellow mongooses examined around Ermelo (26° 31′ S; 29° 59′ E), Mpumalanga, were infested, while 14 out of the 20 examined around Kuruman (27° 28′ S; 23° 26′ E), Northern Cape Province, harboured this tick (I.G.H., unpublished data).

In Zimbabwe the yellow mongoose occurs only in the southern part of Hwange National Park and in the Beit Bridge area. The meercat and Cape ground squirrel have not been recorded there (Skinner & Smithers, 1990).

The vegetation of the regions in which *R. theileri* has been collected ranges from Highveld and wooded grassland through various types of woodland, bushland and shrubland to Karoo grassy shrubland and the Kalahari/Karoo-Namib transition.

Disease relationships

Unknown.

REFERENCES

Bedford, G.A.H. & Hewitt, J. (1925). Descriptions and records of several new or little-known species of ticks from South Africa. *South African Journal of Natural History*, 5, 259–66.

Lynch, C.D. (1980). Ecology of the suricate, *Suricata suricatta* and yellow mongoose, *Cynictis penicillata* with special reference to their reproduction. *Memoirs of the National Museum, Bloemfontein*, **14**, vii + 145 pp.

Neitz, W.O., Boughton, F. & Walters, H.S. (1972). Laboratory investigations on the life-cycle of *Rhipicephalus theileri* Bedford & Hewitt, 1925 (Ixodoidea: Ixodidae). *Onderstepoort Journal of Veterinary Research*, **39**, 117–22.

Norval, R.A.I. (1985). The ticks of Zimbabwe. XII. The lesser known *Rhipicephalus* species. *Zimbabwe Veterinary Journal*, **16**, 37–43.

Also see the following Basic References (pp. 12–14): Skinner & Smithers (1990); Theiler (1947, 1962).

RHIPICEPHALUS TRICUSPIS DÖNITZ, 1906

The specific name *tricuspis*, from the Latin *tri* meaning 'three' plus *cuspis* meaning 'a point', refers to the presence ventrally in the male of two conspicuous cusps on the posterior border of each adanal plate plus a sharp point on each adjacent accessory adanal plate.

Synonym

simus tricuspis.

Diagnosis

A medium-sized reddish-brown species.

Male (Figs 208(a), 209(a) to (c))
Capitulum broader than long, length × breadth ranging from 0.50 mm × 0.54 mm to 0.59 mm × 0.65 mm. Basis capituli with lateral angles at about anterior third of its length, acute. Palps bluntly rounded apically. Conscutum length × breadth ranging from 2.31 mm × 1.60 mm to

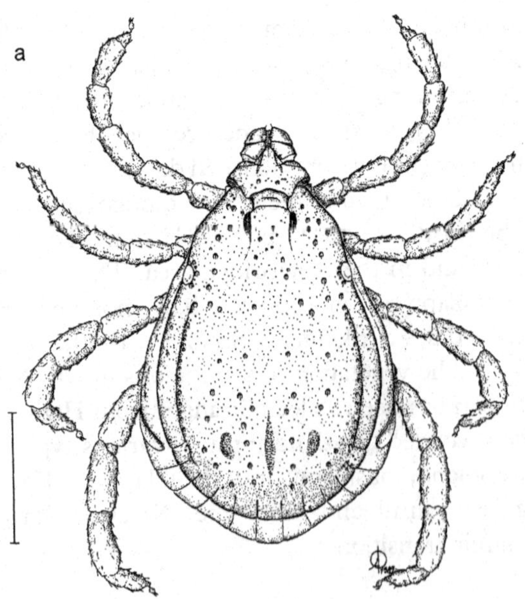

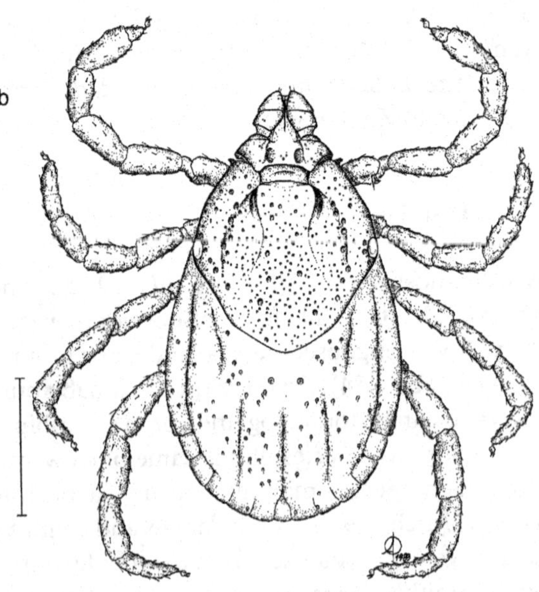

Figure 208. *Rhipicephalus tricuspis* [Onderstepoort Tick Collection 3022i, collected from a Cape fox (*Vulpes chama*), Debeete, Botswana on 20 January 1956 by F. Zumpt]. (a) Male, dorsal; (b) female, dorsal. Scale bars represent 1 mm. A. Olwage *del.*

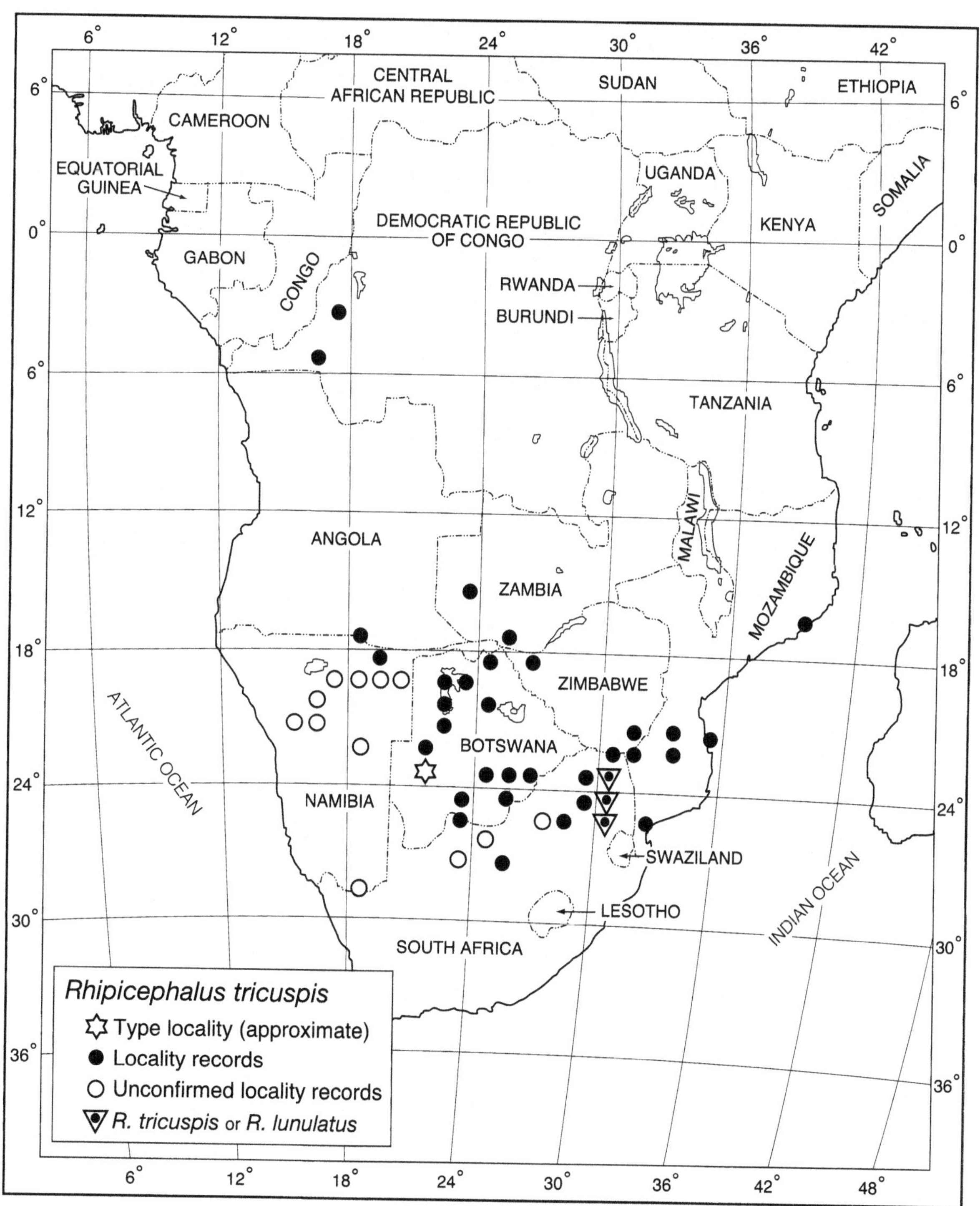

Map 59. *Rhipicephalus tricuspis:* distribution. (From Walker *et al.*, 1988, fig. 23, with kind permission from Kluwer Academic Publishers.)

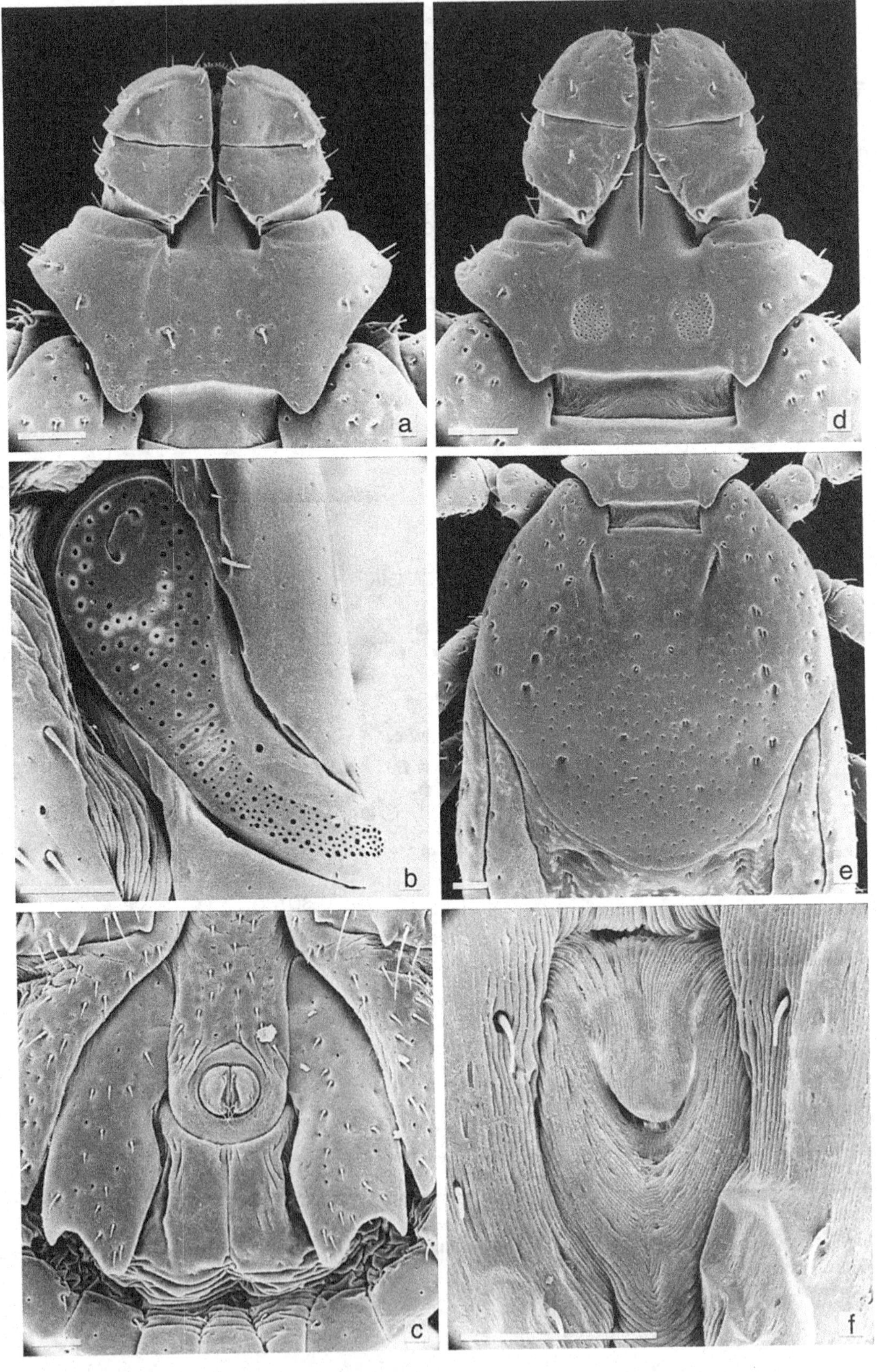
a
d
b
e
c
f

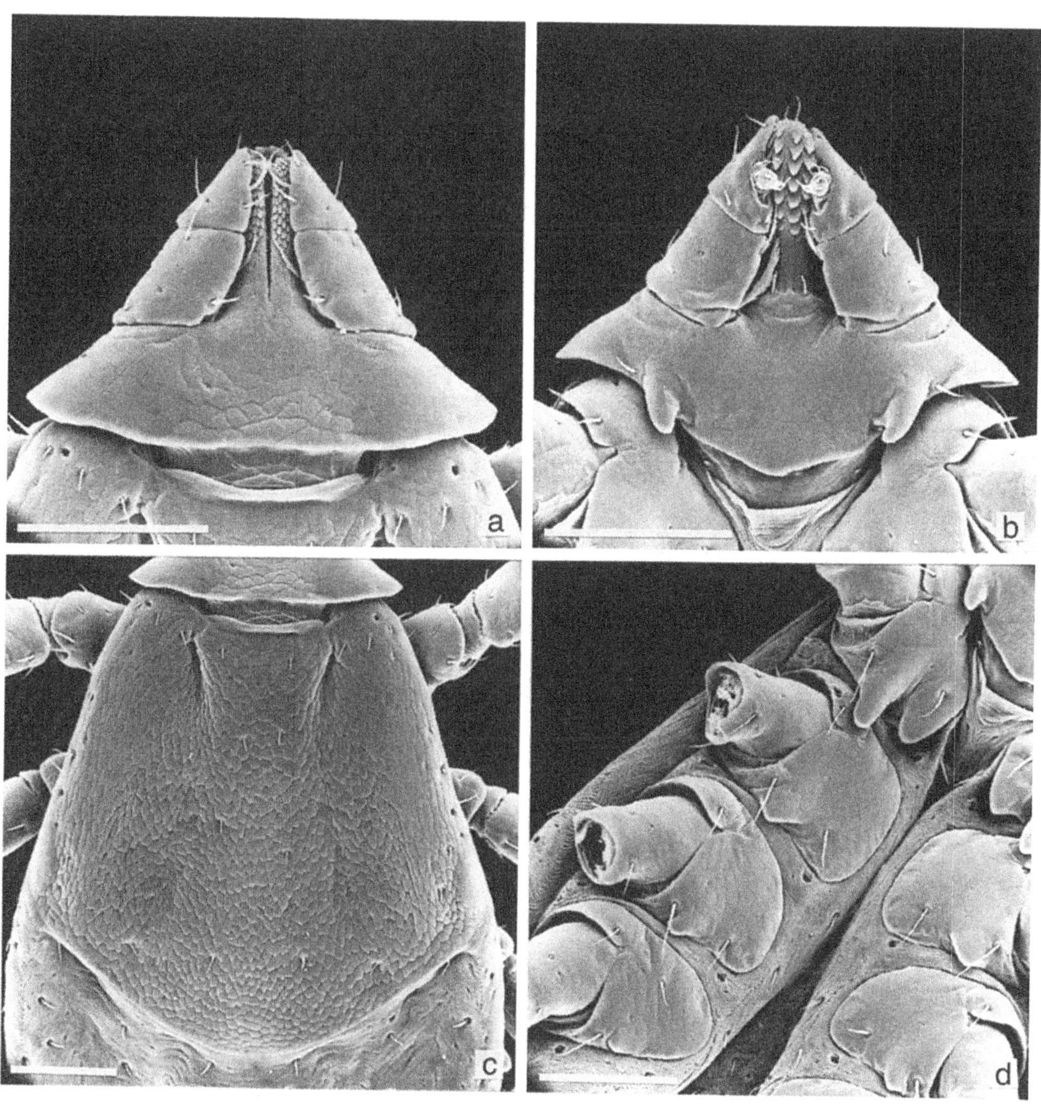

Figure 210 (*above*). *Rhipicephalus tricuspis* (Onderstepoort Tick Collection 2808, RML 109002, laboratory reared, progeny of ♀ collected from a sheep in Pretoria North, Gauteng Province, South Africa, on 26 January 1942 by G.E. Laurence). Nymph: (a) capitulum, dorsal; (b) capitulum, ventral; (c) scutum; (d) coxae. Scale bars represent 0.10 mm. SEMs by M.D. Corwin. (From Walker *et al.*, 1988, figs 13, 14, 16 & 17, with kind permission from Kluwer Academic Publishers.)

Figure 209 (*opposite*). *Rhipicephalus tricuspis* (Onderstepoort Tick Collection 2808, RML 109002, laboratory reared, progeny of ♀ collected from a sheep in Pretoria North, Gauteng Province, South Africa, on 26 January 1942 by G.E. Laurence). Male: (a) capitulum, dorsal; (b) spiracle; (c) adanal plates. Female: (d) capitulum, dorsal; (e) scutum; (f) genital aperture. Scale bars represent 0.10 mm. SEMs by M. D. Corwin. (From Walker *et al.*, 1988, figs 1, 5–7, 10 & 12, with kind permission from Kluwer Academic Publishers.)

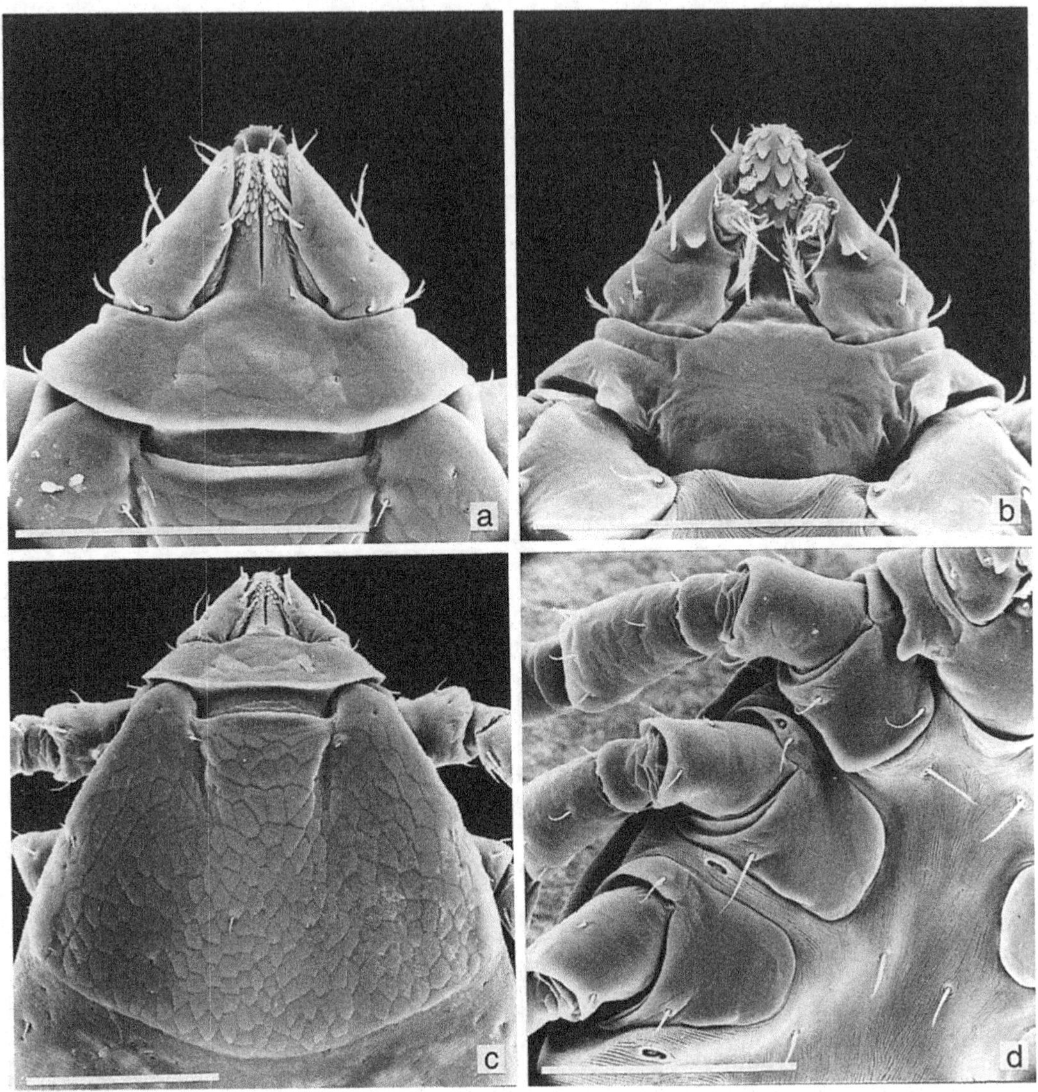

Figure 211. *Rhipicephalus tricuspis* (Onderstepoort Tick Collection 2808, RML 109002, laboratory reared, progeny of ♀ collected from a sheep in Pretoria North, Gauteng Province, South Africa, on 26 January 1942 by G.E. Laurence). Larva: (a) capitulum, dorsal; (b) capitulum, ventral; (c) scutum; (d) coxae. Scale bars represent 0.10 mm. SEMs by M.D. Corwin. (From Walker *et al.*, 1988, figs 19–22, with kind permission from Kluwer Academic Publishers.)

2.90 mm × 1.99 mm; anterior process on coxae I fairly small. In engorged specimens body wall expanded laterally and posteriorly. Eyes marginal, flat, edged dorsally with several medium-sized punctations. Cervical pits deep, comma-shaped; conspicuous setiferous punctations along external cervical margins and in '*simus*' pattern on central conscutum (see p. 416), interspersed with numerous fine punctations. Marginal lines as long punctate grooves delimiting first festoons and almost reaching eyes. Posteromedian and posterolateral grooves, when present, relatively shallow and inconspicuous. Ventrally spiracles comma-shaped, becoming narrowly elongated towards dorsal surface. Adanal plates unique in shape, bicuspid posteriorly; accessory adanal plates sharp, well sclerotized.

Table 55. *Host records of* Rhipicephalus tricuspis

Hosts	Number of records
Domestic animals	
Cattle	10
Goats	3
Dogs	3
Wild animals	
South African hedgehog (*Atelerix frontalis*)	1
Side-striped jackal (*Canis adustus*)	3
Black-backed jackal (*Canis mesomelas*)	1
'Jackal' (*Canis* sp.)	3
Bat-eared fox (*Otocyon megalotis*)	1
Cape fox (*Vulpes chama*)	1
Lion (*Panthera leo*)	1
Leopard (*Panthera pardus*)	1
Aardwolf (*Proteles cristatus*)	1
Aardvark (*Orycteropus afer*)	1
Warthog (*Phacochoerus africanus*)	3
Impala (*Aepyceros melampus*)	1
Steenbok (*Raphicerus campestris*)	17
Sharpe's grysbok (*Raphicerus sharpei*)	1
Greater kudu (*Tragelaphus strepsiceros*)	1
Red forest duiker (*Cephalophus natalensis*)	1
'Duiker'	1
Gemsbok (*Oryx gazella*)	1
Spring hare (*Pedetes capensis*)	11
Cape hare (*Lepus capensis*)	3
Scrub hare (*Lepus saxatilis*)	6
Lepus sp.	5
'Rabbit'	1

Female (Figs 208(b), 209(d) to (f))

Capitulum broader than long, length × breadth ranging from 0.57 mm × 0.67 mm to 0.71 mm × 0.79 mm. Basis capituli with lateral angles at about anterior third of its length, acute, tilting forwards slightly; porose areas large, more than their own diameter apart. Palps rounded apically. Scutum about as long as broad, length × breadth ranging from 1.26 mm × 1.27 mm to 1.50 mm × 1.47 mm; posterior margin sinuous. Eyes at lateral angles, almost flat, edged dorsally with several medium-sized punctations. Cervical fields slightly depressed. Large setiferous punctations along external cervical margins, scattered on raised lateral borders and on central area of scutum, interspersed with numerous fine punctations. Ventrally genital aperture a fairly broad U-shape.

Nymph (Fig. 210)

Capitulum much broader than long, length × breadth ranging from 0.13 mm × 0.21 mm to 0.15 mm × 0.22 mm. Basis capituli over three times as broad as long, with tapering lateral angles overlapping the scapulae; ventrally with well-developed spurs on posterior margin. Palps relatively short, tapering, inclined inwards and overlapping hypostome. Scutum about as broad as long, length × breadth ranging from 0.48 mm × 0.47 mm to 0.52 mm × 0.52 mm; posterior margin a broad smooth curve. Eyes at widest point, well over halfway back, marginal, flat. Cervical grooves short, convergent; cervical fields barely indicated. Ventrally coxae I each with a long bluntly pointed external spur and a shorter broader internal spur; coxae II to IV each with a short, broadly triangular external spur, decreasing in size from II to IV.

Larva (Fig. 211)
Capitulum much broader than long, length × breadth ranging from 0.091 mm × 0.156 mm to 0.096 mm × 0.170 mm. Basis capituli over three times as broad as long, with bluntly tapered lateral angles overlapping the scapulae. Palps broad basally, narrowing towards apices, inclined inwards. Scutum much broader than long, length × breadth ranging from 0.223 mm × 0.312 mm to 0.234 mm × 0.335 mm; posterior margin a broad shallow curve. Eyes at widest point, well over halfway back, flat. Cervical grooves short, slightly convergent. Ventrally coxae I each with a short broad bluntly rounded internal spur; coxae II and III each with an indication only of a broad salient ridge on its posterior margin.

Notes on identification
Since 1912 various authors have, at different times, applied the name *R. tricuspis* to three entities with similarly-shaped adanal plates. This situation was reviewed in detail by Walker *et al.* (1988), on whose findings our present account is based. They formally re-established *R. lunulatus* as a valid species (see p. 269) and referred briefly to a third, morphologically similar, species that has since been described as *R. interventus* (see p. 228).

Hosts

A three-host species (Gertrud Theiler, unpublished data, 1942–43). Adults have sometimes been recorded from cattle, goats and dogs but most collections have been made from relatively small wild animals (Table 55). These include various carnivores, particularly jackals and other wild canids, antelopes, mostly the smaller species, especially the steenbok, also the spring hare, and hares. In southern Africa the majority of collections were made during December to March (wet season).

The hosts of the immature stages are unknown; they are probably small mammals.

Zoogeography

Most records of *R. tricuspis* are from southern Africa, especially Botswana, the North-West and Northern Provinces of South Africa, and southern Mozambique (Map 59). Collections also exist from northern Namibia, Zimbabwe, western Zambia and western Democratic Republic of Congo. The vast majority of these collection sites are in various types of dry woodland.

Disease relationships

Statements that *R. tricuspis* may be involved in the transmission of porcine babesiosis, caused by *Babesia trautmanni*, refer to *R. lunulatus*.

REFERENCES

Dönitz, W. (1906). Über Afrikanische Zecken. *Sitzungsberichte der Gesellschaft Naturforschender Freunde der Berlin*, 5, 143–8 + 1 plate.

Walker, J.B., Keirans, J.E., Pegram, R.G. & Clifford, C.M. (1988). Clarification of the status of *Rhipicephalus tricuspis* Dönitz, 1906 and *Rhipicephalus lunulatus* Neumann, 1907 (Ixodoidea, Ixodidae). *Systematic Parasitology*, 12, 159–86.

RHIPICEPHALUS TURANICUS POMERANTSEV, 1936

This name probably refers to the Turan, which in Persian means Turkestan or, broadly speaking, the region north of the Amu-Darya (Oxus) River, east of the Caspian Sea on the borders of Uzbekistan and Turkmenistan; '*-icus*' is a Latin suffix meaning 'belonging to'.

Diagnosis

A moderate-sized heavily punctate reddish-brown tick belonging to the *R. sanguineus* group.

Male (Figs 212(a), 213(a) to (c))
Capitulum slightly broader than long, length × breadth ranging from 0.52 mm × 0.56 mm to 0.64 mm × 0.65 mm. Basis capituli with acute lateral angles at about anterior third of its length. Palps tapering somewhat to rounded apices. Conscutum length × breadth ranging from 2.34 mm × 1.49 mm to 2.90 mm × 1.95 mm; anterior process of coxae I inconspicuous. In engorged specimens body wall expanded considerably laterally but less so posteriorly, with a small rounded caudal process. Eyes flat, edged dorsally by a few large setiferous punctations. Cervical pits comma-shaped, convergent; cervical fields slightly depressed, their external margins delimited by large setiferous punctations. Marginal lines long, deep, outlined with numerous large punctations. Posteromedian and posterolateral grooves short, broad. A few large setiferous punctations present on scapulae and in four more-or-less distinct '*simus*' pattern rows medially on the conscutum, interspersed with numerous interstitial punctations. Although the punctation pattern varies considerably it is typically relatively deep and dense. Ventrally spiracles somewhat variable in shape but dorsal prolongation usually as wide as adjacent festoon and either gently curved or slightly angled. Adanal plates vary from broad and truncated to longer and more pointed posteriorly; accessory adanal plates well sclerotized, sharply pointed.

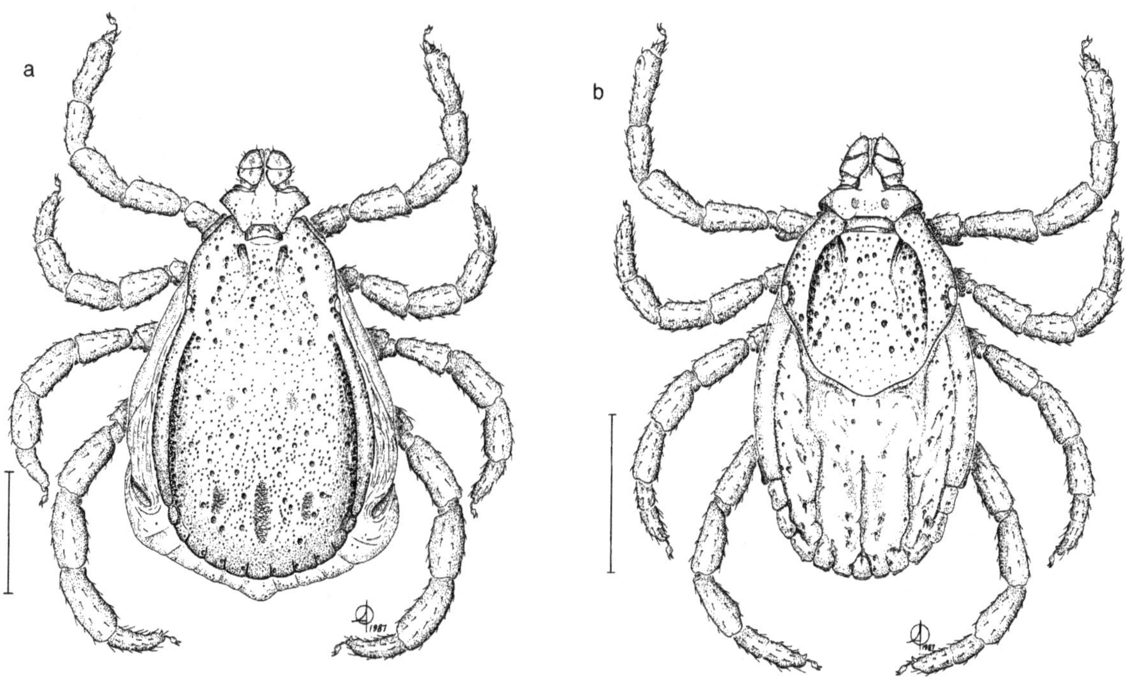

Figure 212. *Rhipicephalus turanicus* (Onderstepoort Tick Collection 2880viii, collected from sheep, Chirundu, Zambia, on 12 April 1952 by J.G. Matthysse). (a) Male, dorsal; (b) female, dorsal. Scale bars represent 1 mm. A. Olwage *del.*

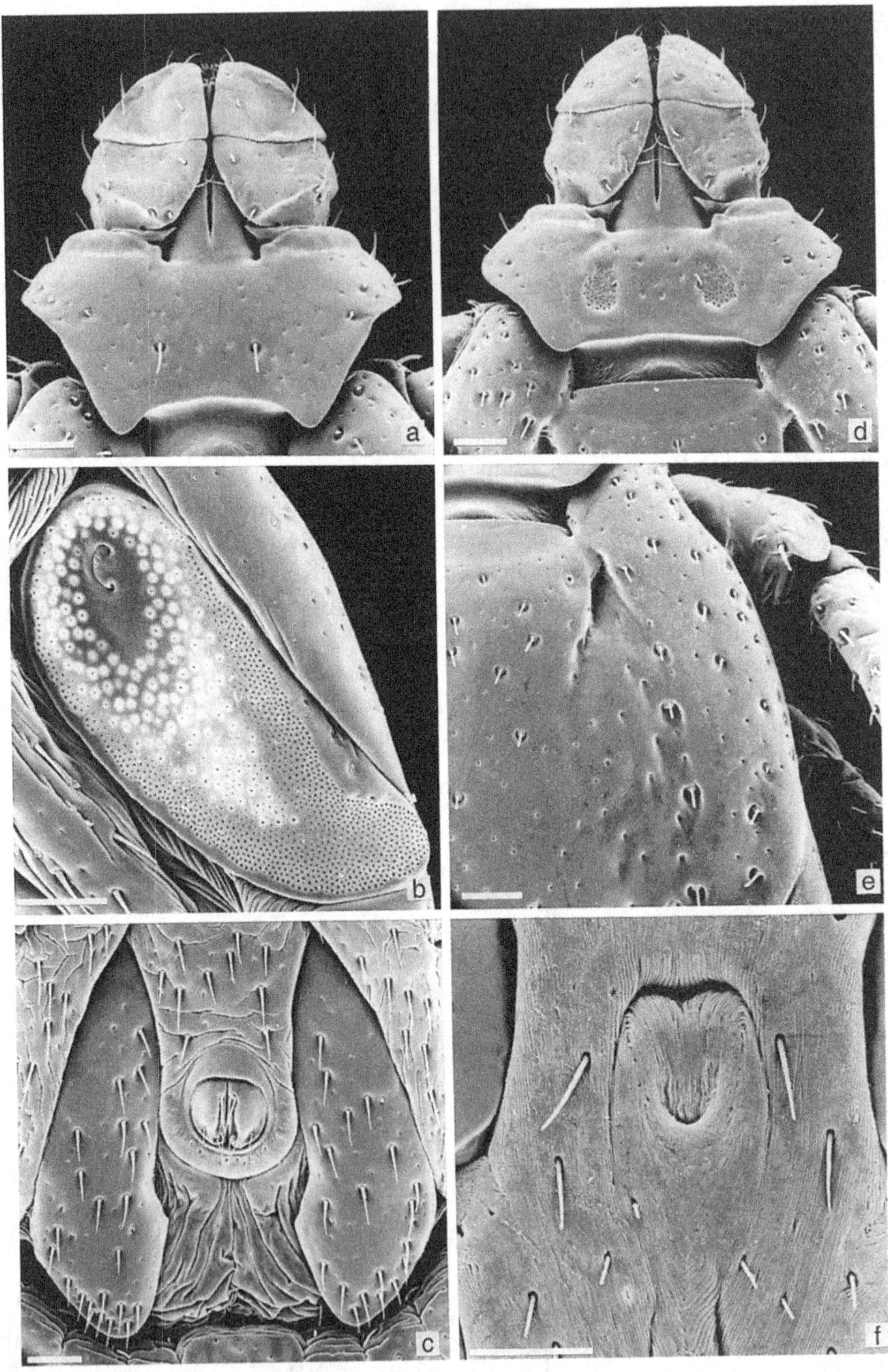

Figure 213. *Rhipicephalus turanicus* (L175, strain originating from Cyprus, ticks laboratory reared by R.G. Pegram). Male: (a) capitulum, dorsal; (b) spiracle; (c) adanal plates. Female: (d) capitulum, dorsal; (e) scapular area; (f) genital aperture. Scale bars represent 0.10 mm. SEMs by M.D. Corwin. (Figs (b), (c) & (f) from Pegram *et al.*, 1987, figs 31, 32 & 38, with kind permission from Kluwer Academic Publishers.)

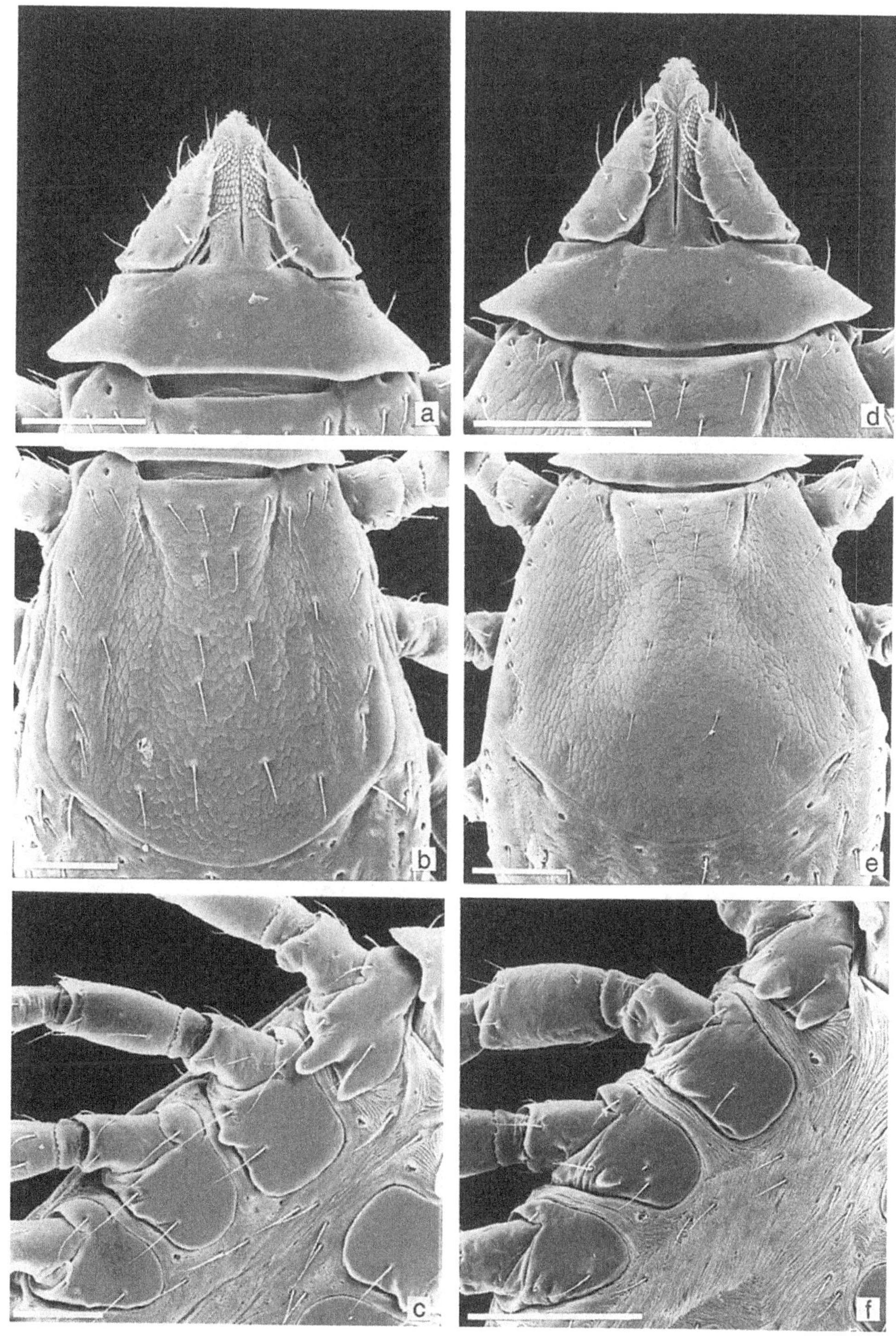

Figure 214. *Rhipicephalus turanicus* nymphs [figures (a), (b) & (c): L56, strain originating from Balmoral, Zambia; (d), (e) & (f): L175, strain originating from Cyprus; all ticks laboratory reared by R.G. Pegram]. Figures (a) & (d) capitulum, dorsal; (b) & (e) scutum; (c) & (f) coxae. Scale bars represent 0.10 mm. SEMs by M.D. Corwin. (From Pegram *et al.*, 1987, figs 21–26, with kind permission from Kluwer Academic Publishers.)

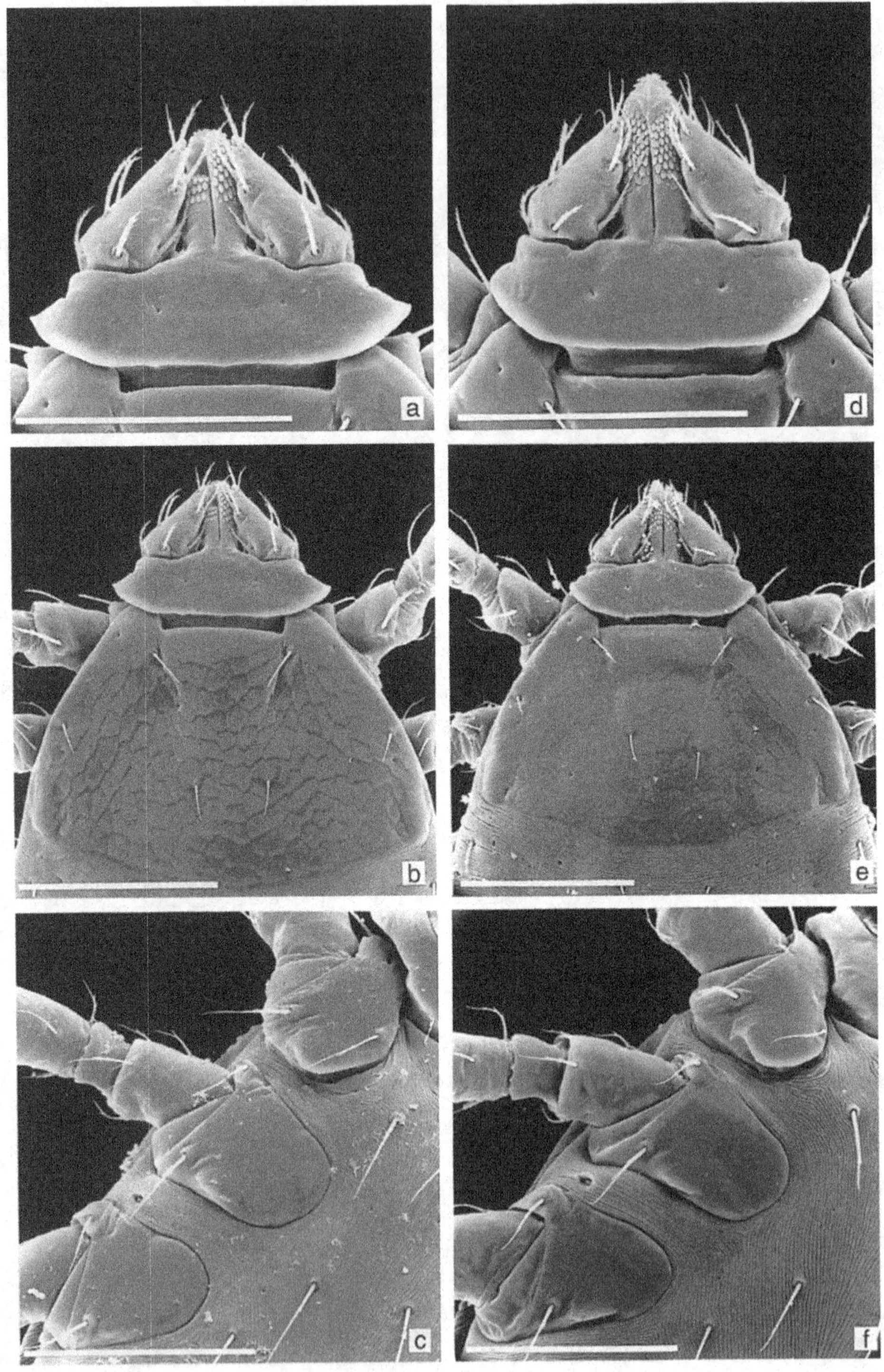

Figure 215. *Rhipicephalus turanicus* larvae [figures (a), (b) & (c): L56, strain originating from Balmoral, Zambia; (d), (e) & (f): L175, strain originating from Cyprus; all ticks laboratory reared by R.G. Pegram]. Figures (a) & (d) capitulum, dorsal; (b) & (e) scutum; (c) & (f) coxae. Scale bars represent 0.10 mm. SEMs by M.D. Corwin. (From Pegram *et al.*, 1987, figs 15–20, with kind permission from Kluwer Academic Publishers.)

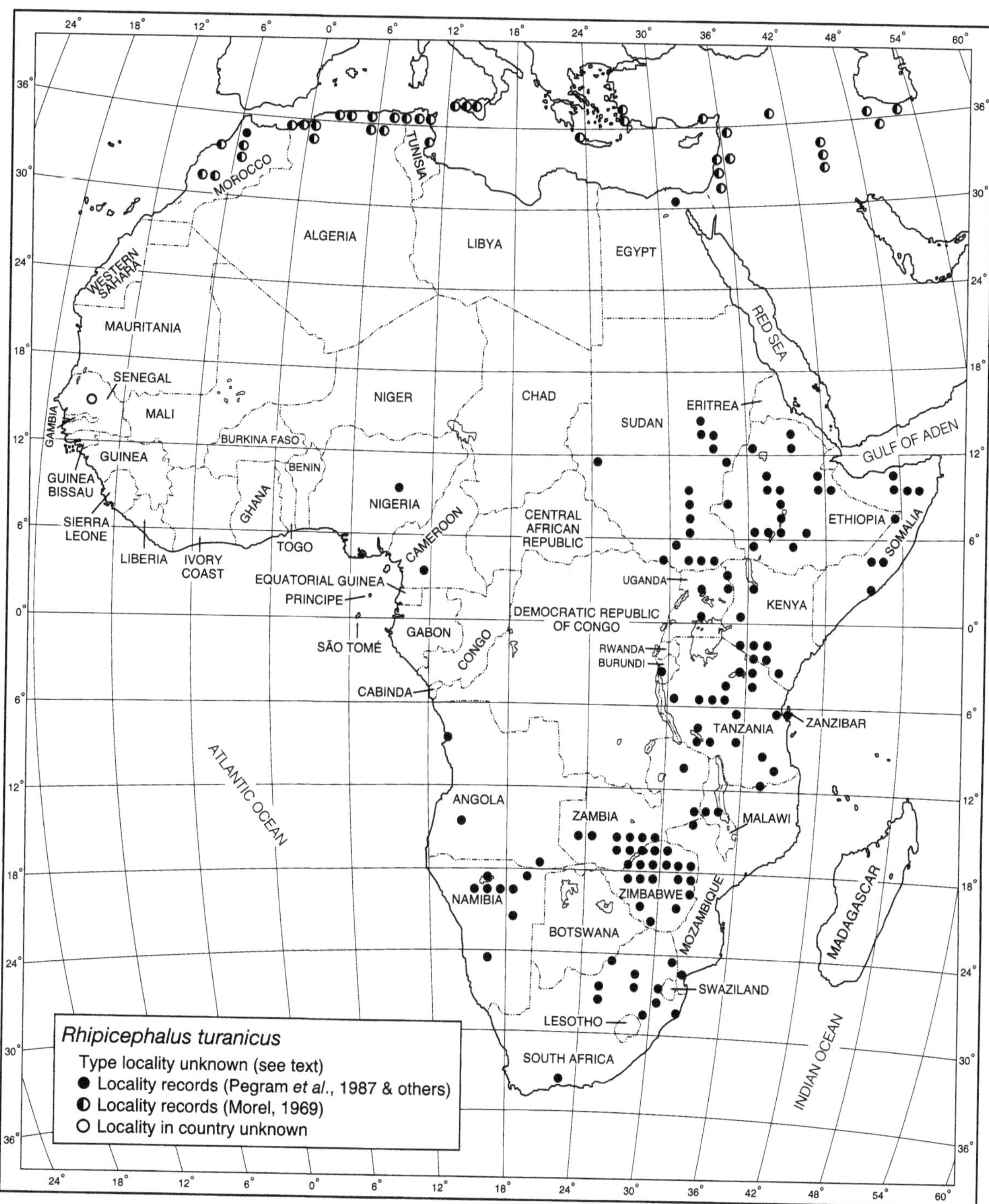

Map 60. *Rhipicephalus turanicus*: distribution in Africa, parts of southern Europe and the Middle East.

Female (Figs 212(b), 213(d) to (f), also 85(c))
Capitulum broader than long, length × breadth ranging from 0.56 mm × 0.67 mm to 0.72 mm × 0.78 mm. Basis capituli with blunt lateral angles a little anterior to mid-length; porose areas oval, about twice their own diameter apart. Palps tapering to quite narrowly-rounded apices. Scutum slightly longer than broad, length × breadth ranging from 1.20 mm × 1.16 mm to 1.53 mm × 1.45 mm; posterior margin markedly sinuous. Eyes flat, edged dorsally by a few large setiferous punctations. Cervical pits convergent; cervical fields slightly depressed, their external margins sharply defined, with numerous large deep setiferous punctations. A few large setiferous punctations also present on the scapulae and medially on the scutum, interspersed with numerous interstitial punctations. Punctation pattern variable but usually dense and conspicuous. Ventrally genital aperture small, U-shaped to broadly V-shaped. (Also see fig. 85(c), p. 211).

Nymph (Fig. 214)
Capitulum much broader than long, length × breadth ranging from 0.15 mm × 0.28 to 0.18 mm × 0.31 mm. Basis capituli over three times as broad as long, with tapering lateral angles overlapping the scapulae; ventrally with short spurs on posterior margin. Anterolateral margins of basis capituli approximately in line with external palpal margins. Palps tapering to narrowly-rounded apices, inclined inwards. Scutum longer than broad, length × breadth ranging from 0.39 mm × 0.33 mm to 0.46 mm × 0.38 mm; relatively shorter and broader in Mediterranean than in African strain; posterior margin a deep smooth curve. Eyes at widest point, well over halfway back, almost flat. Cervical fields long, narrow, slightly depressed, inconspicuous. Scutal setae longer and more obvious in African than in Mediterranean strain. Ventrally coxae I each with a slightly longer narrower external spur and a shorter broader internal spur; coxae II to IV each with an external spur only, decreasing progressively in size.

Larva (Fig. 215)
Capitulum much broader than long, length × breadth ranging from 0.097 mm × 0.146 mm to 0.103 mm × 0.160 mm. Basis capituli just over three times as broad as long, with lateral angles relatively sharp in African strain but more rounded in Mediterranean strain. Anterolateral margins of basis capituli approximately in line with external palpal margins. Palps tapering to narrowly-rounded apices, inclined inwards. Scutum much broader than long, length × breadth ranging from 0.201 mm × 0.275 mm to 0.215 mm × 0.299 mm; posterior margin a wide, very shallow curve. Eyes at widest point, almost flat. Cervical grooves convergent initially, becoming shallow and slightly divergent. Ventrally coxae I each with a large broadly-rounded spur; coxae II and III each with a slight salience only on the posterior margin.

Notes on identification
We have included illustrations of both an African and a Cypriot strain of *R. turanicus*. Until recently it was regarded as a Palaearctic species. However, research by Pegram (1984) showed that it is also widely distributed in the Afrotropical region. Pegram *et al.* (1987) found that there were considerable morphological differences, particularly in the immature stages, between the African and the Cypriot strains of this tick. Nevertheless these two strains were completely compatible cytoplasmically. They could therefore find no reason to continue to regard it as being of Palaearctic origin.

The earlier misconceptions regarding its distribution arose largely because it was confused with *R. sulcatus* by Morel & Vassiliades (1963). Up to now the collections of *R. sanguineus* group species on which they based their findings have not been re-examined and their systematic status re-evaluated. Consequently our knowledge of the precise distribution of *R. turanicus* and *R. sulcatus* as well as other species in this group in West Africa remains obscure.

In our review of this species in Africa we have included the ticks identified as *R. turanicus* in the surveys of Pegram (1976), Pegram, Hoogstraal & Wassef (1981, 1982) and Pegram *et al.*

(1987), in J.B.W.'s East African collections, in the Zimbabwean survey conducted by Norval, Daillecourt & Pegram (1983) (in which the ticks were referred to as a *Rhipicephalus* sp.), and more recently in surveys conducted in Namibia and South Africa.

Hosts

A three-host species (Pegram, 1984). In view of its wide distribution the host records for *R. turanicus* have been summarized in two tables. In the first its known hosts in Africa are listed and in the second its hosts outside Africa.

Hosts of African ticks

The domestic animals from which *R. turanicus* adults have been collected most frequently are cattle, sheep and dogs (Table 56A). Its favoured wild hosts include jackals, lions, leopards and hares and it has also been taken from a number of large ground-feeding birds. Apart from collections off scrub hares the immature stages have not been specifically identified from other hosts.

Cattle apparently harbour few ticks, since the largest numbers recorded from a single animal were 56 males and 38 females (Pegram *et al.*, 1987). Although no individual tick counts have been done on sheep they can be particularly heavily infested (Norval *et al.*, 1983). More than 450 adults were collected from an ostrich in Karamoja District, Uganda (Pegram *et al.*, 1987). The preferred site of attachment on sheep, goats and hares is their ears (Norval *et al.*, 1983; Horak *et al.*, 1995).

Most records of *R. turanicus* indicate that the adults are active mainly in the late rainy to early dry season. In Zambia 90% of collections were made between March and July (Pegram *et al.*, 1986, 1987), and in Zimbabwe 84% between November and April (Norval *et al.*, 1983). In South Africa most ticks were collected from the vegetation between February and May, from scrub hares between January and July, and from dogs between December and March (Horak *et al.*, 1995; I.G.H., unpublished data). In Morocco peak infestations were present on sheep during April and May, coinciding with the transition between wet and dry seasons (Ouhelli *et al.*, 1962), while collections in Ethiopia, northern Somalia and Tanzania were made during dry season periods (Pegram *et al.*, 1987).

Hosts of non-African ticks

The preferred domestic hosts of adult ticks are cattle, sheep, goats, dogs and cats (Table 56B). Most collections from wild hosts have been taken from hedgehogs, the golden jackal, red fox and hares. The immature stages have mainly been collected from domestic cats, hedgehogs, several species of rodents and the Indian hare.

Zoogeography

Although the majority of records of *R. turanicus* in Africa originate in the eastern parts of the continent, from Ethiopia, Sudan and the Somali Republic in the north to South Africa in the south, collections have also been made in Egypt and Morocco in north Africa; Senegal, Nigeria and Cameroon in the west, and in Angola and Namibia (Map 60). Within Africa *R. turanicus* is present at altitudes ranging from just above sea level to over 2000 m and in regions with annual rainfalls ranging from 100 mm to 1000 mm. It occurs in widely differing ecological habitats. It has been found in arid to semi-arid steppe and also in tropical and subtropical savanna and Afromontane scrub vegetation. Within an undifferentiated woodland habitat in South Africa it has been collected in greater numbers from the vegetation adjacent to gullies than from open grassland or more wooded subhabitats.

Outside Africa *R. turanicus* has been recorded in several of the Mediterranean countries and their immediate neighbours as well as in Afghanistan, China, India, Iran, Iraq, Nepal, Pakistan, Russia, Saudi Arabia, Sri Lanka, and Syria. The report of *R. turanicus* in the Tyrol of Austria by Sixl (1972) is most likely an introduction because the ticks were collected twice from the same dog.

Table 56A. *Host records of* Rhipicephalus turanicus *in Africa*

Hosts	Number of records
Domestic animals	
Cattle	81
Sheep	14
Goats	7
Camels	3
Horses	6
Dogs	51
Cats	4
Wild animals	
Hedgehog (*Atelerix* sp.)	2
Vervet monkey (*Chlorocebus aethiops*)	1
Side-striped jackal (*Canis adustus*)	2
Black-backed jackal (*Canis mesomelas*)	2
'Jackal' (*Canis* sp.)	3
Bat-eared fox (*Otocyon megalotis*)	1
Cheetah (*Acinonyx jubatus*)	3
Caracal (*Caracal caracal*)	1
African wild cat (*Felis lybica*)	5
Black-footed cat (*Felis nigripes*)	1
Serval (*Leptailurus serval*)	1
Lion (*Panthera leo*)	13
Leopard (*Panthera pardus*)	7
Egyptian mongoose (*Herpestes ichneumon*)	1
Brown hyaena (*Parahyaena brunnea*)	1
Aardwolf (*Proteles cristatus*)	1
African civet (*Civettictis civetta*)	6
'Genet'	4
Burchell's zebra (*Equus burchellii*)	4
Aardvark (*Orycteropus afer*)	1
Warthog (*Phacochoerus africanus*)	4
Grant's gazelle (*Gazella granti*)	3
Thomson's gazelle (*Gazella thomsonii*)	3
Eland (*Taurotragus oryx*)	1
'Eland' (*Taurotragus* sp.) (Equatoria, Sudan)	1
Greater kudu (*Tragelaphus strepsiceros*)	1
'Kudu' (*Tragelaphus* sp.)	1
Common duiker (*Sylvicapra grimmia*)	2
Roan antelope (*Hippotragus equinus*)	1
Gemsbok (*Oryx gazella*)	1
Lechwe (*Kobus leche*)	1
Cape hare (*Lepus capensis*)	1
Scrub hare (*Lepus saxatilis*)	34 (including immatures)
'Hare'	13
Birds	
Ostrich (*Struthio camelus*)	4
Marabou stork (*Leptoptilos crumeniferus*)	1

Table 56A. *(cont.)*

Hosts	Number of records
Birds *(cont.)*	
Black kite (*Milvus migrans*)	1
White-headed vulture (*Aegypius occipitalis*)	1
Wahlberg's eagle (*Hieraaetus wahlbergi*)	1
Secretary bird (*Sagittarius serpentarius*)	5
Denham's bustard (*Neotis denhami*)	2
Black bustard (*Eupodotis afra*)	1
Cape dikkop (*Burhinus capensis*)	1
'Dove' (*Streptopelia* sp.)	1
Reptiles	
Puff adder (*Bitis arietans*)	1
Humans	3

Table 56B. *Host records of* Rhipicephalus turanicus *outside Africa*

Hosts	Number of records
Domestic animals	
Cattle	150 +
Water buffaloes	3
Sheep	Numerous
Goats	Numerous
Camels	15
Donkeys	1
Pigs	1
Dogs	172
Cats	44 (including immatures)
Wild animals	
East European hedgehog (*Erinaceus concolor*)	5
West European hedgehog (*Erinaceus europaeus*)	4
Desert hedgehog (*Hemiechinus aethiopicus*)	3
Long-eared hedgehog (*Hemiechinus auritus*)	9 (including immatures)
Brandt's hedgehog (*Hemiechinus hypomelas*)	2
Indian hedgehog (*Hemiechinus micropus*)	7
Common European white-toothed shrew (*Crocidura russula*)	1 (immatures)
'White-toothed shrew' (*Crocidura zarudnyi*)	1 (immatures)
Golden jackal (*Canis aureus*)	7
Red fox (*Vulpes vulpes*)	13
Jungle cat (*Felis chaus*)	2
Eurasian badger (*Meles meles*)	2
Weasel (*Mustela nivalis*)	1 (immatures)
Argali (*Ovis ammon*)	3

Table 56B. (*cont.*)

Hosts	Number of records
Wild animals (*cont.*)	
Günther's vole (*Microtus guentheri*)	2 (immatures)
Grey hamster (*Cricetulus migratorius*)	7 (immatures)
Indian desert gerbil (*Meriones hurrinae*)	2
Libyan jird (*Meriones libycus*)	5 (immatures)
Persian jird (*Meriones persicus*)	2 (immatures)
Tristram's jird (*Meriones tristrami*)	1 (immatures)
Great gerbil (*Rhombomys opimus*)	1 (immatures)
Indian gerbil (*Tatera indica*)	1 (immatures)
Broad-toothed mouse (*Apodemus mystacinus*)	1 (immatures)
Lesser bandicoot rat (*Bandicota bengalensis*)	1 (immatures)
House mouse (*Mus musculus*)	3 (immatures)
Short-tailed bandicoot rat (*Nesokia indica*)	3 (immatures)
Black rat (*Rattus rattus*)	3 (immatures)
'Dormouse' (*Eliomys melanurus*)	1 (immatures)
Cape hare (*Lepus capensis*)	5
Indian hare (*Lepus nigricollis*)	8 (including immatures)
European rabbit (*Oryctolagus cuniculus*)	2
Humans	13

Disease relationships

Achuthan, Mahadevan & Lalitha (1980) have demonstrated that *R. turanicus* can transmit *Babesia canis* to splenectomized dogs. It has been listed as a vector of *Babesia equi* to horses by Friedhoff (1988), but four attempts by Potgieter, De Waal & Posnett (1992) to transmit this parasite with a South African strain of *R. turanicus* failed. As noted earlier, the listing of *R. sanguineus* as a vector of the other equine parasite, *Babesia caballi*, by Enigk (1943) probably also refers to *R. turanicus*. It has been shown that *R. turanicus* can transmit *Babesia trautmanni* transovarially to splenectomized pigs (López-Rebollar & De Waal, 1994). *Rhipicephalus turanicus* has also been implicated as a vector of the agents of Q fever and Siberian tick typhus (Balashov & Daiter, 1973; Berdyev, 1980).

REFERENCES

Achuthan, H.N., Mahadevan, S. & Lalitha, C.M. (1980). Studies on the developmental forms of *Babesia bigemina* and *Babesia canis* in ixodid ticks. *Indian Veterinary Journal*, 57, 181–4.

Balashov, Y.S. & Daiter, A.B. (1973). *Bloodsucking Arthropods and Rickettsiae*. Leningrad: Nauka. [In Russian].

Berdyev, A. (1980). *Ecology of Ixodid Ticks of Turkmenistan and their Importance in natural focal Diseases Epizootiology*. Ashkhabad, Turkmenistan: Ylym. [In Russian].

Enigk, K. (1943). Die Überträger des Pferdepiroplasmose, ihre Verbreitung und Biologie. *Archiv für Wissenschaftliche und Praktische Tierheilkunde*, 78, 209–40.

Friedhoff, K.T. (1988). Transmission of *Babesia*. In *Babesiosis in Domestic Animals and Man*, ed. M. Ristic, pp. 23–52. Boca Raton, FL: CRC Press.

Horak, I.G., Spickett, A.M., Braack, L.E.O., Penzhorn, B.L., Bagnall, R.J. & Uys, A.C. (1995). Parasites of domestic and wild animals in South Africa. XXXIII. Ixodid ticks on scrub hares in the north-eastern regions of Northern and Eastern Transvaal and of KwaZulu-Natal. *Onderstepoort Journal of Veterinary Research*, **62**, 123–31.

López-Rebollar, L.M. & De Waal, D.T. (1994). Tick vectors of *Babesia trautmanni* in domestic pigs in South Africa. *8th International Congress of Parasitology, October 10–14, Abstracts, vol. 1*, 231. 1994, Izmir: Turkey.

Morel, P.C. & Vassiliades, G. (1963). Les *Rhipicephalus* du groupe *sanguineus*: espèces africaines (Acariens: Ixodoidea). *Revue d'Élevage et de Médecine Vétérinaire des Pays Tropicaux*, **15** (nouvelle série), 343–86.

Norval, R.A.I., Daillecourt, T. & Pegram, R.G. (1983). The ticks of Zimbabwe. VI. The *Rhipicephalus sanguineus* group. *Zimbabwe Veterinary Journal*, **13**, 38–46.

Ouhelli, H., Pandey, V.S., Benzaouia, T. & Belkasmi, A. (1962). Seasonal prevalence of *Rhipicephalus turanicus* on sheep in Morocco. *Tropical Animal Health and Production*, **14**, 247–8.

Pegram, R.G. (1976). Ticks (Acarina, Ixodoidea) of the northern regions of the Somali Democratic Republic. *Bulletin of Entomological Research*, **66**, 345–63.

Pegram, R.G. (1984). *Biosystematic studies on the genus* Rhipicephalus: *the* R. sanguineus *and* R. simus *groups (Ixodoidea, Ixodidae)*. PhD. thesis, Brunel University, England. 160 pp.

Pegram, R.G., Clifford, C.M., Walker, J.B. & Keirans, J.E. (1987). Clarification of the *Rhipicephalus sanguineus* group (Acari, Ixodoidea, Ixodidae). I. *R. sulcatus* Neumann, 1908 and *R. turanicus* Pomerantsev, 1936. *Systematic Parasitology*, **10**, 3–26.

Pegram, R.G., Hoogstraal, H. & Wassef, H.Y. (1981). Ticks (Acari: Ixodoidea) of Ethiopia. I. Distribution, ecology and host relationships of species infesting livestock. *Bulletin of Entomological Research*, **71**, 339–59.

Pegram, R.G., Hoogstraal, H. & Wassef, H.Y. (1982). Ticks (Acari: Ixodoidea) of the Yemen Arab Republic. I. Species infesting livestock. *Bulletin of Entomological Research*, **72**, 215–27.

Pegram, R.G., Perry, B.D., Musisi, F.L. & Mwanaumo, B. (1986). Ecology and phenology of ticks in Zambia: seasonal dynamics on cattle. *Experimental and Applied Acarology*, **2**, 25–45.

Pomerantsev, B.I. (1936). [The morphology of the genus *Rhipicephalus* Koch in connection with the construction of a natural classification of Ixodoidea]. *Parasitologicheskii Sbornik*, **6**, 5–32. (In Russian; English summary).

Potgieter, F.T., De Waal, D.T. & Posnett, E.S. (1992). Transmission and diagnosis of equine babesiosis in South Africa. *Memórias do Instituto Oswaldo Cruz*, **87**, Supplement III, 139–42.

Sixl, W. (1972). Drie weitere Zeckenarten in Österreich. *Mitteilungen der Abteilung für Zoologischen des Landesmuseums Joanneum*, **1**, 51–2.

Also see the following Basic References (pp. 12–14): Filippova (1997); Hoogstraal (1956); Morel (1980); Theiler (1962); Walker (1974); Yeoman & Walker (1967).

RHIPICEPHALUS WARBURTONI WALKER & HORAK, *SP. NOV.*

This species is named in honour of Cecil Warburton (1854–1958), in recognition of his remarkable understanding of the difficult taxonomic problems in the genus *Rhipicephalus*. The analysis of morphological variations seen in individual species that he wrote in 1912 is still worthy of careful study by students of this genus today.

Diagnosis

A medium-sized dark brown species.

Male (Figs 216(a), 217(a) to (c))

Capitulum about as broad as long, length × breadth ranging from 0.59 mm × 0.61 mm to 0.79 mm × 0.78 mm. Basis capituli with short acute lateral angles at anterior third of its length. Palps short, broad, article III almost wedge-shaped. Conscutum length × breadth ranging from 2.59 mm × 1.52 mm to 3.46 mm × 2.13 mm; anterior process of coxae I prominent, well sclerotized. In engorged specimens body wall expanded somewhat laterally

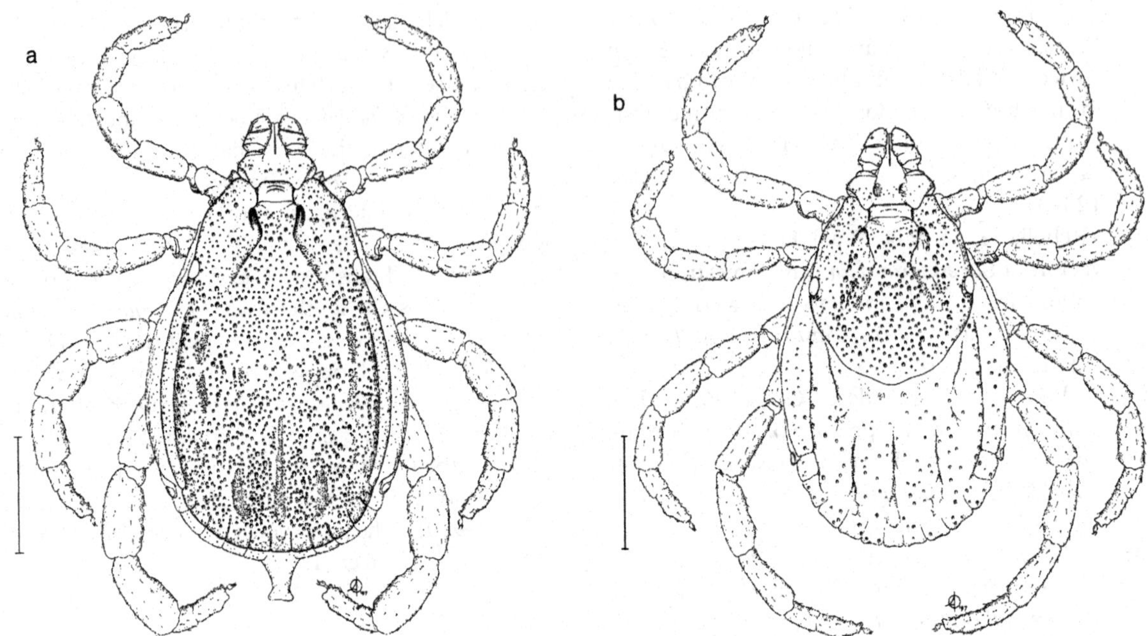

Figure 216. *Rhipicephalus warburtoni* Walker & Horak, *sp. nov.* [Onderstepoort Tick Collection 3146ii, paratypes, laboratory reared, originating from ♀ collected from Dorper sheep (D77), Preezfontein, Free State, South Africa, on 14 October 1993 by L.J. Fourie]. (a) Male, dorsal; (b) female, dorsal. Scale bars represent 1 mm. A. Olwage *del.*

Table 57. *Host records of* Rhipicephalus warburtoni

Hosts	Number of records
Domestic animals	
Cattle	26
Sheep	61
Goats	62
Dogs	2
Wild animals	
African wild cat (*Felis lybica*)	1
Gemsbok (*Oryx gazella*)	6
Namaqua rock mouse (*Aethomys namaquensis*)	69 (immatures)
Spring hare (*Pedetes capensis*)	14 (nymphs)
Cape hare (*Lepus capensis*)	12 (including immatures)
Scrub hare (*Lepus saxatilis*)	64 (including immatures)
Smith's red rock rabbit (*Pronolagus rupestris*)	7 (immatures)
Rock elephant shrew (*Elephantulus myurus*)	282 (immatures)

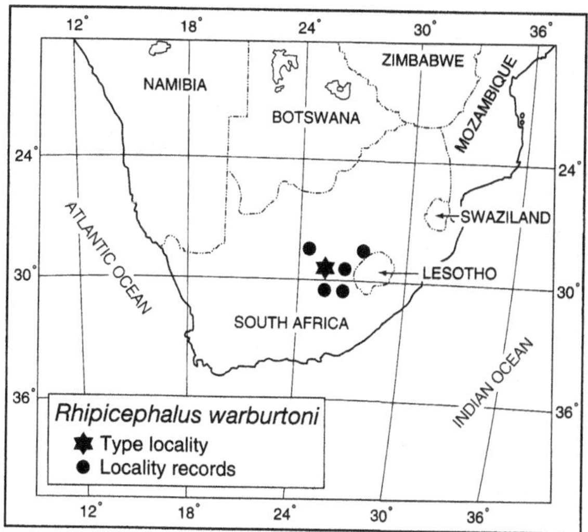

Map 61. *Rhipicephalus warburtoni* Walker & Horak, *sp. nov.*: distribution.

and posteriorly and forming a long tail-like caudal process posteromedially. Eyes convex, delimited dorsally by punctate grooves. Cervical pits deep, convergent; cervical fields divergent, tapering to points posterior to eyes, their external margins delimited by irregular rows of a few large discrete setiferous punctations. Marginal lines well developed, extending anteriorly almost to eye level, punctate. Posterior grooves conspicuous; posteromedian groove long and narrow, posterolaterals shorter and broader. A few large conspicuous setiferous punctations present on the scapulae and scattered medially on the conscutum, increasing in size and number towards the posterior end of the conscutum, interspersed with a dense pattern of discrete smaller punctations that also increase in size posteriorly. Small shagreened areas visible laterally on the conscutum, parallel to the marginal lines, and sometimes medially. Ventrally spiracles broad with a fairly short dorsal prolongation that curves slightly just at its tip. Adanal plates broad, their internal margins almost parallel anteriorly, becoming slightly divergent posterior to the anus; their posterointernal and posteroexternal margins smoothly rounded; accessory adanal plates indicated by small sclerotized points only.

Female (Figs 216(b), 217(d) to (f))

Capitulum broader than long, length × breadth ranging from 0.52 mm × 0.59 mm to 0.75 mm × 0.82 mm. Basis capituli with broad acute lateral angles just anterior to mid-length; porose areas small, oval, almost three times their own diameter apart. Palps longer and narrower than those of the male, their apices smoothly rounded. Scutum longer than broad, length × breadth ranging from 1.08 mm × 0.96 mm to 1.55 mm × 1.50 mm; posterior margin sinuous. Eyes at about mid-length, convex, delimited dorsally by punctate grooves. Cervical pits convergent; cervical fields broad, divergent, slightly depressed, their internal margins sometimes just reaching posterolateral margins of scutum, their external margins delimited by large setiferous punctations. A few medium-sized setiferous punctations scattered on scapulae and medially on scutum, interspersed with numerous discrete smaller punctations that are finest anterior to the eyes but larger between the cervical fields. Ventrally genital aperture broadly V-shaped.

Nymph (Fig. 218)

Capitulum much broader than long, length × breadth ranging from 0.21 mm × 0.27 mm to 0.22 mm × 0.29 mm. Basis capituli over three times as broad as long, lateral angles at about mid-length, very broad and overlapping scapulae, posterior margin concave. Palps narrow proximally, then widening, with article III tapering to narrowly-rounded apices, inclined inwards. Scutum slightly longer than broad, length × breadth ranging from 0.50 mm × 0.48 mm to 0.56 mm × 0.54 mm; posterior margin broadly curved. Eyes just anterior to widest point, well over halfway back, slightly convex, delimited dorsally by grooves. Cervical fields long, narrow, divergent, slightly depressed. Ventrally coxae I each with a short broad internal spur and a longer narrower external spur; coxae II to IV each with an external spur only, decreasing progressively in size.

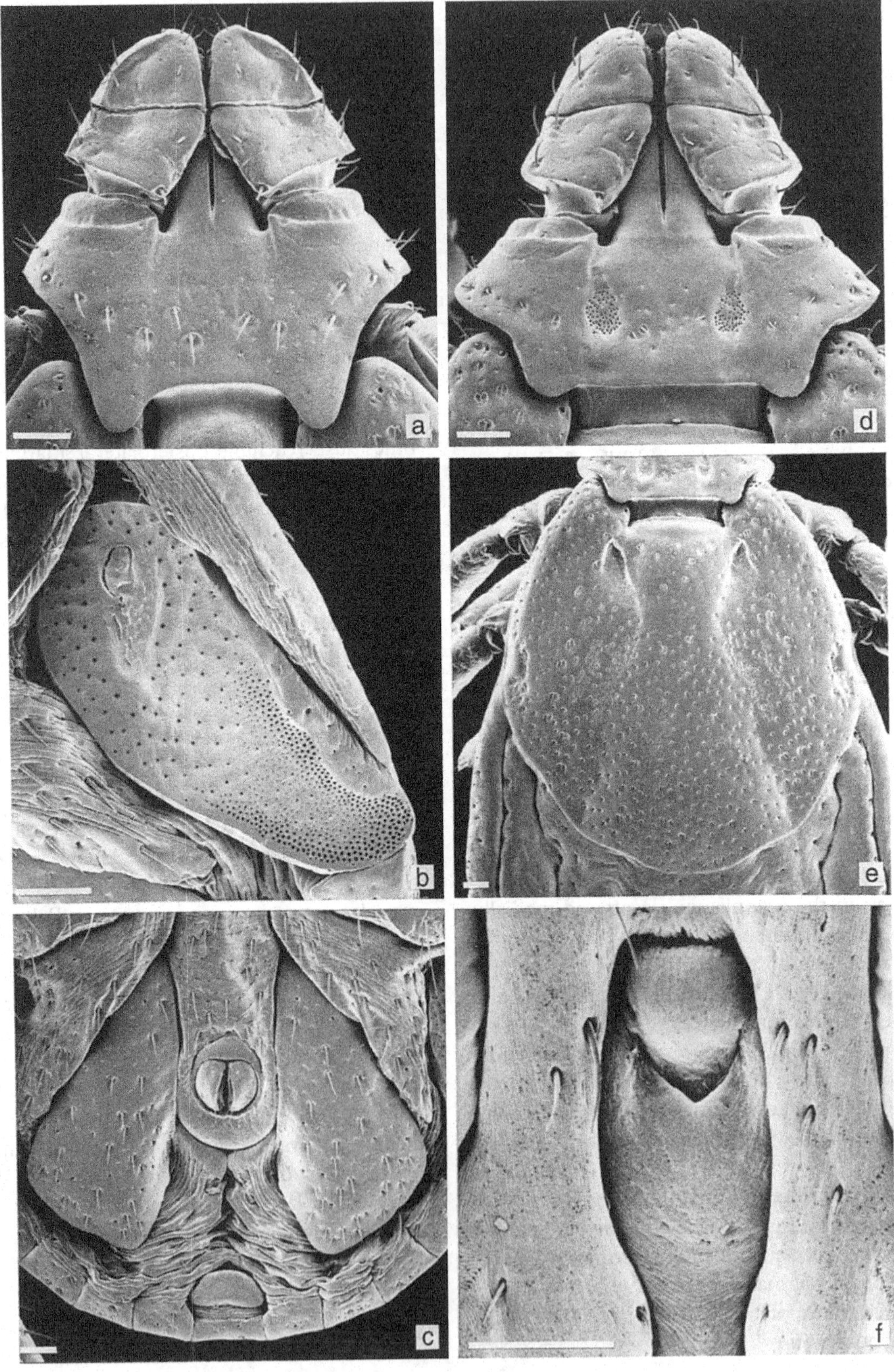
a
b
c
d
e
f

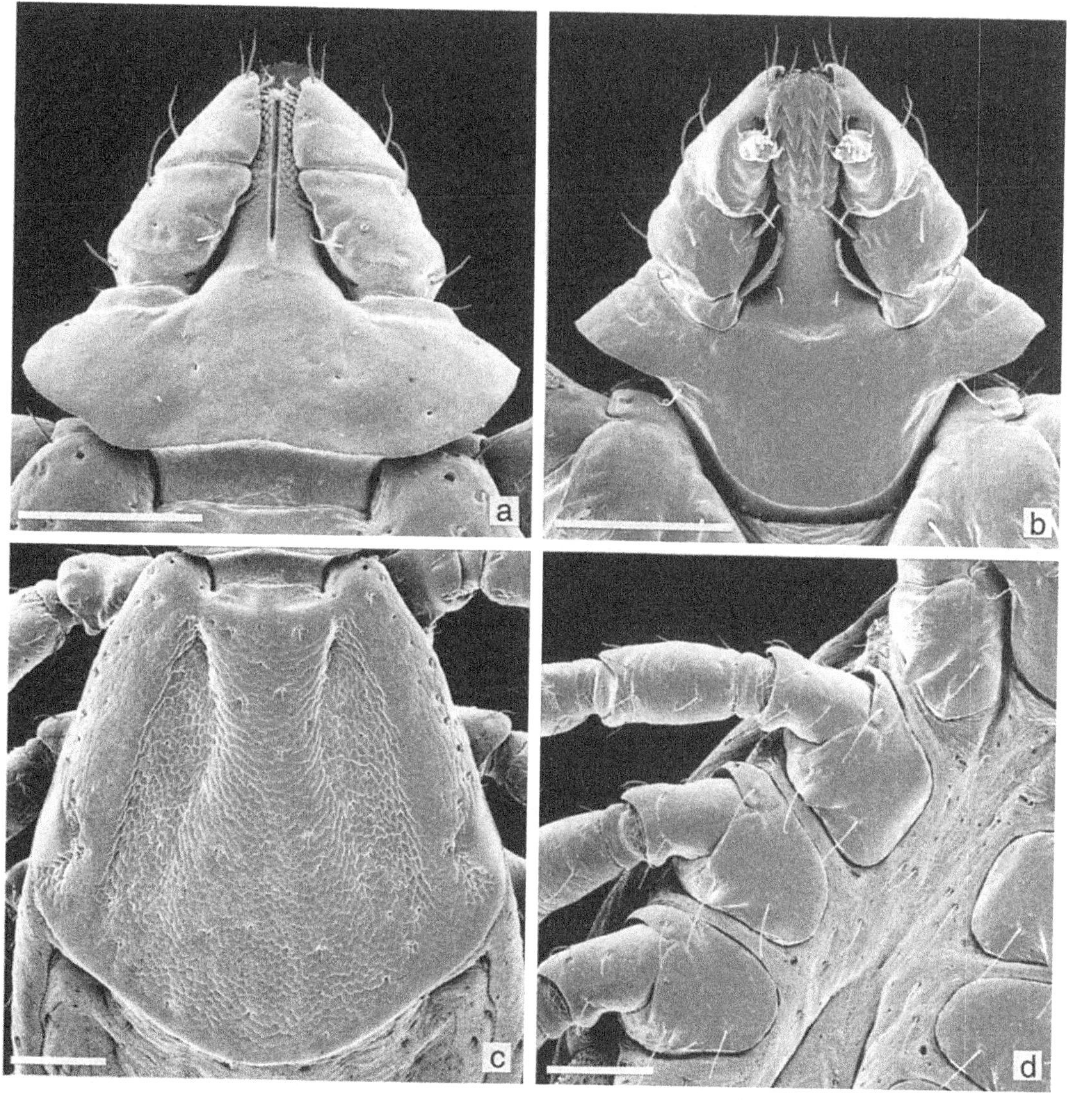

Figure 218 (*above*). *Rhipicephalus warburtoni* Walker & Horak, *sp. nov.* [laboratory reared, originating from ♀ collected from Dorper sheep (D77), Preezfontein, Free State, South Africa, on 14 October 1993 by L.J. Fourie]. Nymph: (a) capitulum, dorsal; (b) capitulum, ventral; (c) scutum; (d) coxae. Scale bars represent 0.10 mm. SEMs by J.F. Putterill.

Figure 217 (*opposite*). *Rhipicephalus warburtoni* Walker & Horak, *sp. nov.* [laboratory reared, originating from ♀ collected from Dorper sheep (D77), Preezfontein, Free State, South Africa, on 14 October 1993 by L.J. Fourie]. Male: (a) capitulum, dorsal; (b) spiracle; (c) adanal plates. Female: (d) capitulum, dorsal; (e) scutum; (f) genital aperture. Scale bars represent 0.10 mm. SEMs by J.F. Putterill.

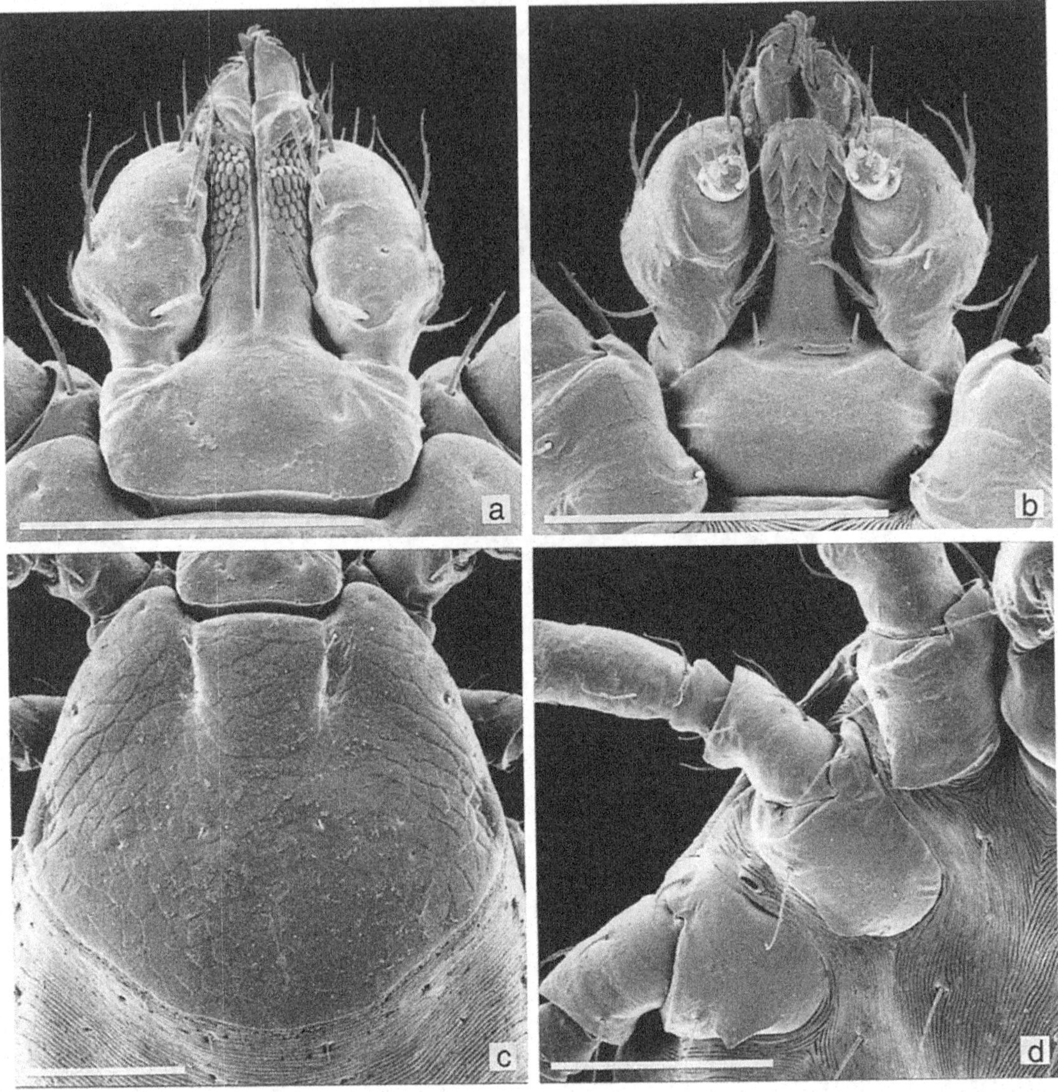

Figure 219. *Rhipicephalus warburtoni* Walker & Horak, *sp. nov.* [laboratory reared, originating from ♀ collected from Dorper sheep (D77), Preezfontein, Free State, South Africa, on 14 October 1993 by L.J. Fourie]. Larva: (a) capitulum, dorsal; (b) capitulum, ventral; (c) scutum; (d) coxae. Scale bars represent 0.10 mm. SEMs by J.F. Putterill.

Larva (Fig. 219)

Capitulum as long as broad, length × breadth ranging from 0.107 mm × 0.107 mm to 0.115 mm × 0.115 mm. Basis capituli nearly 2.5 times as broad as long, lateral margins almost straight, curving posteriorly to meet the slightly concave posterior margin. Palps constricted proximally, then widening markedly before tapering gently to rounded apices. Scutum much broader than long, length × breadth ranging from 0.276 mm × 0.356 mm to 0.281 mm × 0.372 mm; posterior margin broadly rounded. Eyes just anterior to widest point, about halfway back, slightly concave and delimited dorsally by grooves. Cervical grooves short, almost parallel. Ventrally coxae I each with a large sharp spur; coxae II and III each with more rounded spurs, that on III being quite small.

Holotype
♂, laboratory reared, progeny of ♀ collected from Dorper sheep (D 77), Preezfontein, Free State, South Africa, on 14 October 1993 by L.J. Fourie, deposited in Onderstepoort Tick Collection 3146i.

Allotype
♀, data as above.

Paratypes
♂♂, ♀♀, nymphs, larvae, data as above, deposited in Onderstepoort Tick Collection 3146ii; 3 ♂♂, 3 ♀♀, 10 nymphs, 10 larvae, data as above, deposited in the United States National Tick Collection, RML 122749; 3 ♂♂, 3 ♀♀, 10 nymphs, 10 larvae, data as above, deposited in The Natural History Museum, London.

Notes on identification
This tick is one of three closely-related species included under the name *R. pravus* by Theiler (1962). It also features in the South African literature under the following names: as an atypical strain of *R. appendiculatus* occurring in the Fauresmith region of the Free State (Theiler, 1949); as an *R. pravus*-like tick (Fourie, Horak & Marais, 1988a, b); as an *R. punctatus*-like tick (Fourie & Horak, 1990); and as *R. punctatus* in several publications on the ticks of the Free State by Fourie & Horak and their co-workers.

Hosts

A three-host species (L.J. Fourie, unpublished data, 1995). Although cattle, sheep and goats all serve as hosts of adult *R. warburtoni* (Table 57), its relative abundance on goats exceeds that on the other domestic animals (Fourie & Horak, 1991). The preferred wild host of all stages of development is the scrub hare. On two farms in the south-western Free State 46 out of 50 scrub hares examined were infested with a total of 1185 ticks. Twelve out of 34 Cape hares on the same farms carried 52 ticks, and seven out of 28 Smith's red rock rabbits harboured only 11 ticks (Horak & Fourie, 1991). The rock elephant shrew is the preferred host of the immature stages (Fourie, Horak & Van den Heever, 1992); 282 out of 287 elephant shrews examined in the Free State were infested, and 958 larvae and 22 nymphs were collected from a single animal.

The predilection sites of attachment of adult *R. warburtoni* on very young Angora goat kids are the head and ears (Fourie, Horak & Van Zyl, 1991). During the first few weeks of their lives these kids lie-up while their dams are feeding. This, coupled with the fact that the ticks actively quest for their hosts from the ground (Fourie *et al.*, 1993), could account for the large numbers attaching to their heads and ears. On older goats the ticks are found on the neck and brisket as well as the head and ears (Fourie *et al.*, 1991). On scrub hares the adults attach to the ears.

The larvae of *R. warburtoni* are most abundant on rock elephant shrews from December to July and the nymphs from April to October (Fourie *et al.*, 1992). Adults are most abundant on goats and cattle from September or October to February (Fourie & Horak, 1991; Fourie, Kok & Heyne, 1996).

Zoogeography

Rhipicephalus warburtoni has been found only in the Free State and Northern Cape Province, South Africa (Map 61). Here it has been collected at altitudes varying from 1200 m to 1600 m, with annual rainfall, which occurs mainly in summer, ranging from *c.* 200 mm to 600 mm. It occurs in Highveld grassland, Kalahari *Acacia* wooded grassland and in the transition from Karoo shrubland to Highveld. Within these regions the tick and its preferred hosts are most commonly associated with hilly habitats with vegetation cover.

Disease relationships

Paralysis has been reported in heavily-infested young Angora goat kids in the south-western

Free State during the period of peak adult tick abundance (Fourie *et al.*, 1988a). This paralysis is reversible provided the ticks are removed prior to the symptoms becoming too severe.

REFERENCES

Fourie, L.J. & Horak, I.G. (1990). Parasites of cattle in the south western Orange Free State. *Journal of the South African Veterinary Association*, **61**, 27–8.

Fourie, L.J. & Horak, I.G. (1991). The seasonal activity of adult ixodid ticks on Angora goats in the south western Orange Free State. *Journal of the South African Veterinary Association*, **62**, 104–6.

Fourie, L.J., Horak, I.G. & Marais, L. (1988a).An undescribed *Rhipicephalus* species associated with field paralysis of Angora goats. *Journal of the South African Veterinary Association*, **59**, 47–9.

Fourie, L.J., Horak, I.G. & Marais, L. (1988b). The seasonal abundance of adult ixodid ticks on Merino sheep in the south western Orange Free State. *Journal of the South African Veterinary Association*, **59**, 191–4.

Fourie, L.J., Horak, I.G. & Van den Heever, J.J. (1992).The relative host status of rock elephant shrews *Elephantulus myurus* and Namaqua rock mice *Aethomys namaquensis* for economically important ticks. *South African Journal of Zoology*, **27**, 108–14.

Fourie, L.J., Horak, I.G. & Van Zyl, J.M. (1991).Sites of attachment and intraspecific infestation densities of the brown paralysis tick (*Rhipicephalus punctatus*) on Angora goats. *Experimental and Applied Acarology*, **12**, 243–9.

Fourie, L.J., Kok, D.J. & Heyne, H. (1996). Species composition and seasonal dynamics of adult ixodid ticks on two cattle breeds in the south-western Free State. *Onderstepoort Journal of Veterinary Research*, **63**, 19–23.

Fourie, L.J., Snyman, A., Kok, D.J., Horak, I.G. & Van Zyl, J.M. (1993). The appetence behaviour of two South African paralysis-inducing ixodid ticks. *Experimental and Applied Acarology*, **17**, 921–30.

Horak, I.G. & Fourie, L.J. (1991). Parasites of domestic and wild animals in South Africa. XXIX. Ixodid ticks on hares in the Cape Province and on hares and red rock rabbits in the Orange Free State. *Onderstepoort Journal of Veterinary Research*, **58**, 261–70.

Theiler, G. (1949). Zoological Survey of the Union of South Africa: Tick Survey. Part III. Distribution of *Rhipicephalus appendiculatus*, the brown tick. *Onderstepoort Journal of Veterinary Science and Animal Industry*, **22**, 269–84 + 1 map.

Also see the following Basic Reference (p. 14): Theiler (1962).

RHIPICEPHALUS ZAMBEZIENSIS WALKER, NORVAL & CORWIN, 1981

This specific name is derived from the Zambezi, one of the major river valleys where this tick is commonly found, plus the Latin suffix *-ensis* meaning 'belonging to'.

Synonym

zambeziensis Lawrence & Norval, 1979 (*nomen nudum*).

Diagnosis

A moderate-sized dark brown tick whose adults closely resemble those of *R. appendiculatus* in general appearance but are more heavily punctate.

Male (Figs 220(a), 221(a) to (c))

Capitulum longer than broad, length × breadth ranging from 0.62 mm × 0.60 mm to 0.85 mm × 0.76 mm. Basis capituli with short sharp lateral angles at anterior third of its length. Palps short, broad, with slightly flattened to gently rounded apices. Conscutum length × breadth ranging from 2.65 mm × 1.70 mm to 3.60 mm × 2.40 mm; anterior process of coxae I large, heavily sclerotized. In engorged specimens body wall expanded posterolaterally and a tail-like caudal process formed posteromedially. Eyes marginal, almost flat, delimited by a shallow groove and a few large punctations dorsally. Cervical fields broad, depressed, with finely shag-

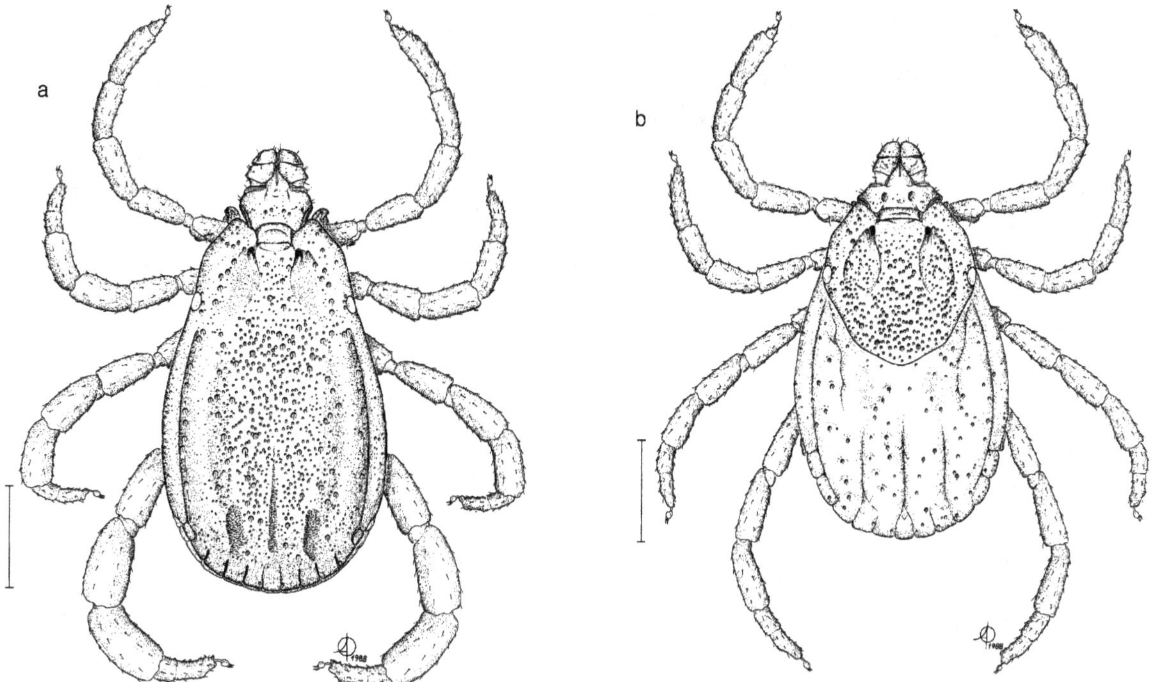

Figure 220. *Rhipicephalus zambeziensis* (Onderstepoort Tick Collection 3140ii, paratypes, laboratory reared, progeny of ♀ collected from bovine, Driehoek Ranch, near West Nicholson, Zimbabwe, in 1976 by R.A.I. Norval). (a) Male, dorsal; (b) female, dorsal. Scale bars represent 1 mm. A. Olwage *del.*

reened surfaces. Marginal lines well developed, extending anteriorly almost to eye level. Posteromedian groove long and narrow, posterolateral grooves shorter and broader, all with finely shagreened surfaces. Large setiferous punctations present on the scapulae, along the external margins of the cervical fields, in the marginal lines and scattered medially on the conscutum, where they are interspersed between numerous medium-sized punctations. Individual punctations usually discrete but sometimes so dense that they coalesce in places, giving the tick a somewhat rugose appearance. Minute pinpoint punctations scattered laterally on the conscutum, adjacent to the marginal lines. Legs increase markedly in size from I to IV. Ventrally spiracles elongate with a tapering slightly curved dorsal prolongation. Adanal plates long, narrow, tapering posterior to the anus to narrowly-rounded posterointernal angles; accessory adanal plates represented merely by very small sclerotized points.

Female (Figs 220(b), 221(d) to (f))

Capitulum broader than long, length × breadth ranging from 0.60 mm × 0.70 mm to 0.83 mm × 0.95 mm. Basis capituli with broad lateral angles in anterior third of its length; porose areas oval but sometimes slightly irregular in shape, about twice their own diameter apart. Palps broad, blunt apically. Scutum usually as broad as long, length × breadth ranging from 1.25 mm × 1.25 mm to 1.71 mm × 1.71 mm; posterior margin sinuous. Eyes about halfway back, marginal, almost flat, delimited dorsally by quite a deep groove with a few large punctations. Cervical fields broad, depressed, surfaces slightly shagreened in places, especially along the internal margins. Punctation pattern dense. Large setiferous punctations scattered on the scapulae, along the external margins of the cervical fields, where they sometimes coalesce, and scattered medially on the scutum amongst the numerous medium-sized punctations which may also coalesce in places. Fine pin-point

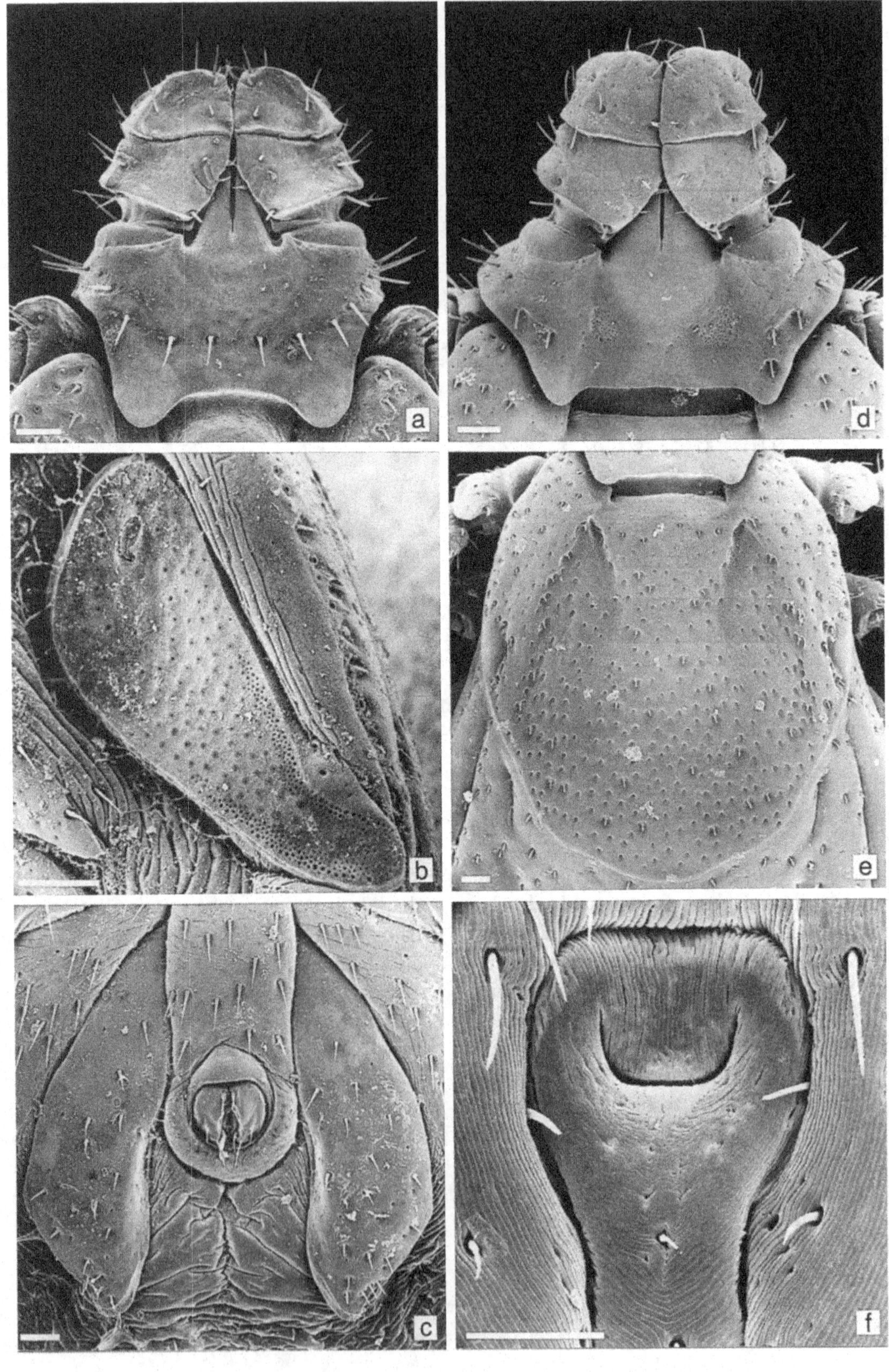

a
d
b
e
c
f

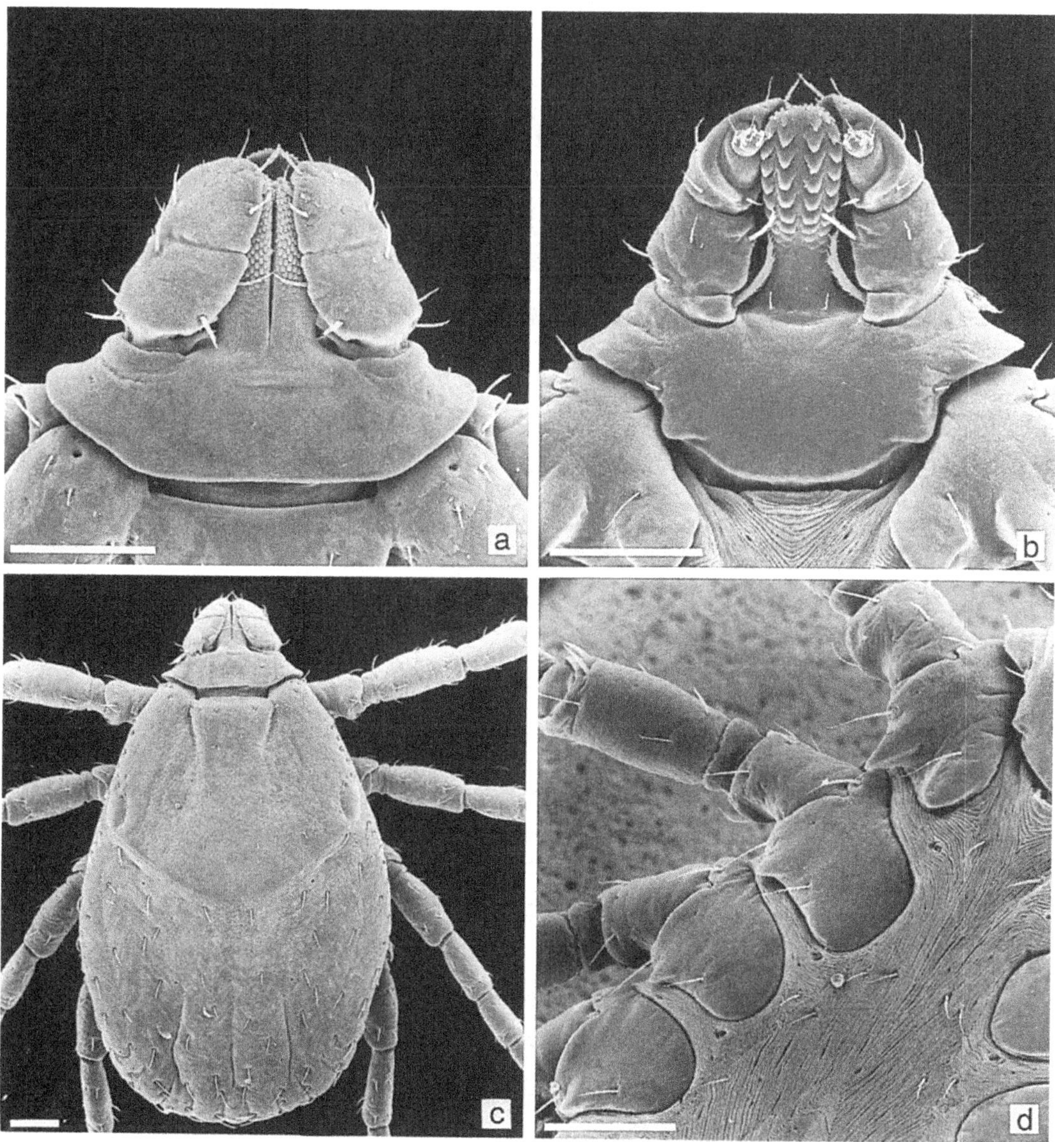

Figure 222 (*above*). *Rhipicephalus zambeziensis* (Onderstepoort Tick Collection 3140ii, laboratory reared, progeny of ♀ collected from bovine, Driehoek Ranch, near West Nicholson, Zimbabwe, in 1976 by R.A.I. Norval). Nymph: (a) capitulum, dorsal; (b) capitulum, ventral; (c) dorsal surface; (d) coxae. Scale bars represent 0.10 mm. SEMs by M.D. Corwin. (From Walker *et al.*, 1981, figs 21–23 & 25, with kind permission from the Editor, *Onderstepoort Journal of Veterinary Research*).

Figure 221 (*opposite*). *Rhipicephalus zambeziensis* (Onderstepoort Tick Collection 3140ii, laboratory reared, progeny of ♀ collected from bovine, Driehoek Ranch, near West Nicholson, Zimbabwe, in 1976 by R.A.I. Norval). Male: (a) capitulum, dorsal; (b) spiracle; (c) adanal plates. Female: (d) capitulum, dorsal; (e) scutum; (f) genital aperture. Scale bars represent 0.10 mm. SEMs by M.D. Corwin. (From Walker *et al.*, 1981, figs 5, 9, 10, 11, 13 & 15, with kind permission from the Editor, *Onderstepoort Journal of Veterinary Research*).

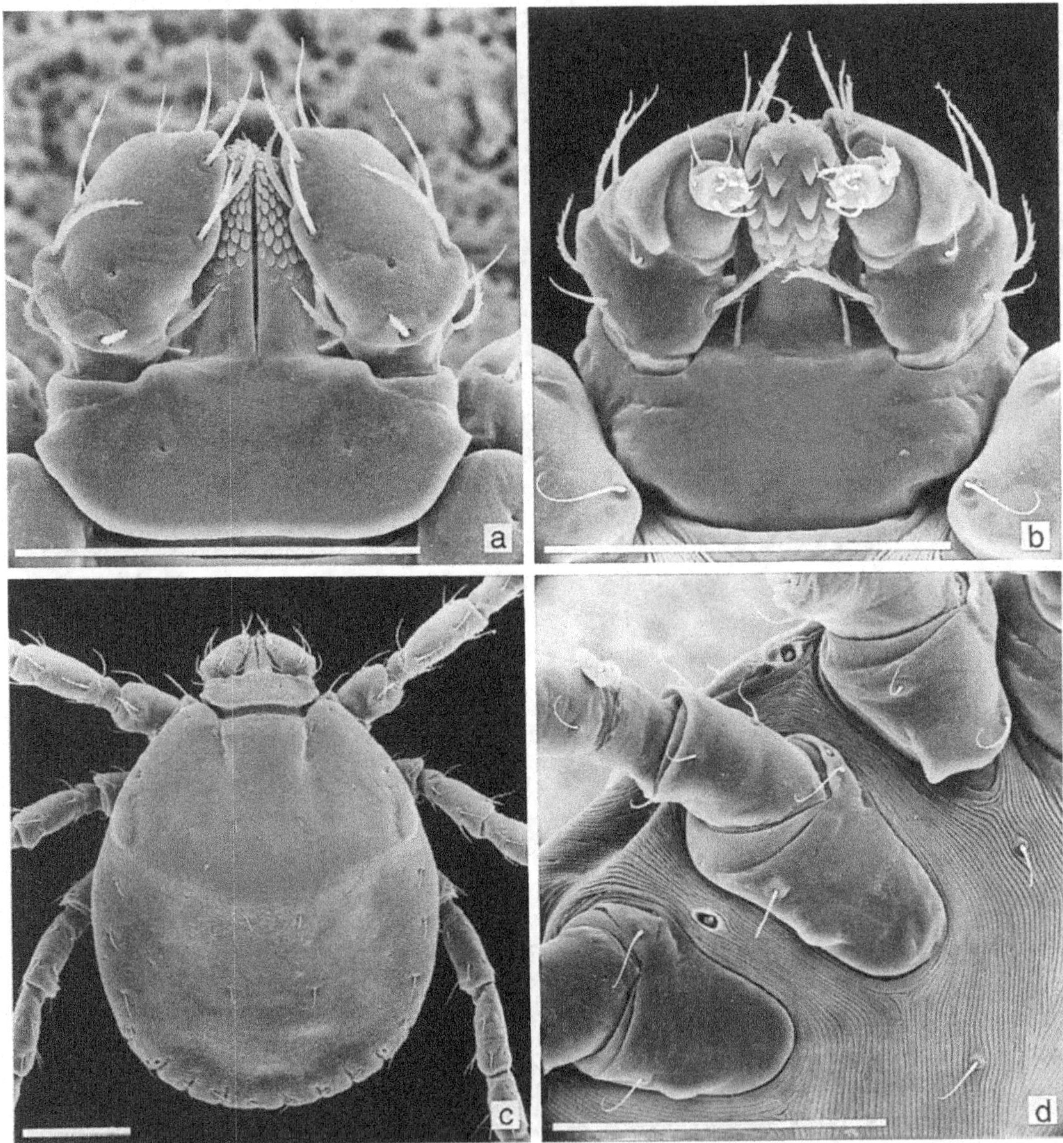

Figure 223. *Rhipicephalus zambeziensis* (Onderstepoort Tick Collection 3140ii, laboratory reared, progeny of ♀ collected from bovine, Driehoek Ranch, near West Nicholson, Zimbabwe, in 1976 by R.A.I. Norval). Larva: (a) capitulum, dorsal; (b) capitulum, ventral; (c) dorsal surface; (d) coxae. Scale bars represent 0.10 mm. SEMs by M.D. Corwin. (From Walker *et al.*, 1981, figs 27–29 & 31, with kind permission from the Editor, *Onderstepoort Journal of Veterinary Research*).

punctations present anterior to the eyes. Ventrally genital aperture wide, almost rectangular, its posterior margin straight, then curving forwards laterally.

Nymph (Fig. 222)

Capitulum much broader than long, length × breadth ranging from 0.21 mm × 0.31 mm to 0.23 mm × 0.32 mm. Basis capituli over three times as broad as long, lateral angles at about mid-length, long, tapering, sharply pointed; ventrally bluntly-rounded spurs present on posterior border. Palps broad, tapering to rounded apices. Scutum broader than long, length × breadth ranging from 0.47 mm × 0.56 mm to 0.53 mm × 0.63 mm; posterior margin a

Table 58. *Host records of* Rhipicephalus zambeziensis

Hosts	Number of records
Domestic animals	
Cattle	254 (including immatures)
Goats	2
Horses	6
Dogs	13
Cats	3 (immatures)
Wild animals	
Vervet monkey (*Chlorocebus aethiops*)	1 (nymph)
Black-backed jackal (*Canis mesomelas*)	2 (including immatures)
Hunting dog (*Lycaon pictus*)	3 (including immatures)
Cheetah (*Acinonyx jubatus*)	5 (including immatures)
Lion (*Panthera leo*)	21 (including immatures)
Leopard (*Panthera pardus*)	12 (including immatures)
Marsh mongoose (*Atilax paludinosus*)	1 (immatures)
Slender mongoose (*Galerella sanguinea*)	1 (nymph)
White-tailed mongoose (*Ichneumia albicauda*)	2 (immatures)
Banded mongoose (*Mungos mungo*)	2 (immatures)
Spotted hyaena (*Crocuta crocuta*)	3 (immatures)
Brown hyaena (*Parahyaena brunnea*)	1
African civet (*Civettictus civetta*)	3 (including immatures)
Burchell's zebra (*Equus burchellii*)	4 (including nymphs)
White rhinoceros (*Ceratotherium simum*)	1 (nymph)
Aardvark (*Orycteropus afer*)	1
Warthog (*Phacochoerus africanus*)	40 (including immatures)
Bushpig (*Potamochoerus larvatus*)	2
Giraffe (*Giraffa camelopardalis*)	3
Impala (*Aepyceros melampus*)	249 (including immatures)
Blue wildebeest (*Connochaetes taurinus*)	14 (including immatures)
Klipspringer (*Oreotragus oreotragus*)	1 (nymph)
Steenbok (*Raphicerus campestris*)	1 (nymph)
African buffalo (*Syncerus caffer*)	19 (including immatures)
Eland (*Taurotragus oryx*)	9 (including nymphs)
Nyala (*Tragelaphus angasii*)	2 (nymphs)
Bushbuck (*Tragelaphus scriptus*)	13 (including immatures)
Greater kudu (*Tragelaphus strepsiceros*)	117 (including immatures)
Common duiker (*Sylvicapra grimmia*)	3 (nymphs)
Roan antelope (*Hippotragus equinus*)	1
Sable antelope (*Hippotragus niger*)	7
Smith's bush squirrel (*Paraxerus cepapi*)	1 (immatures)
Spring hare (*Pedetes capensis*)	1 (nymph)
South African porcupine (*Hystrix africaeaustralis*)	1
Cape hare (*Lepus capensis*)	1 (nymph)
Scrub hare (*Lepus saxatilis*)	272 (immatures)
Birds	
Helmeted guineafowl (*Numida meleagris*)	15 (immatures)
Humans	1 (nymph)

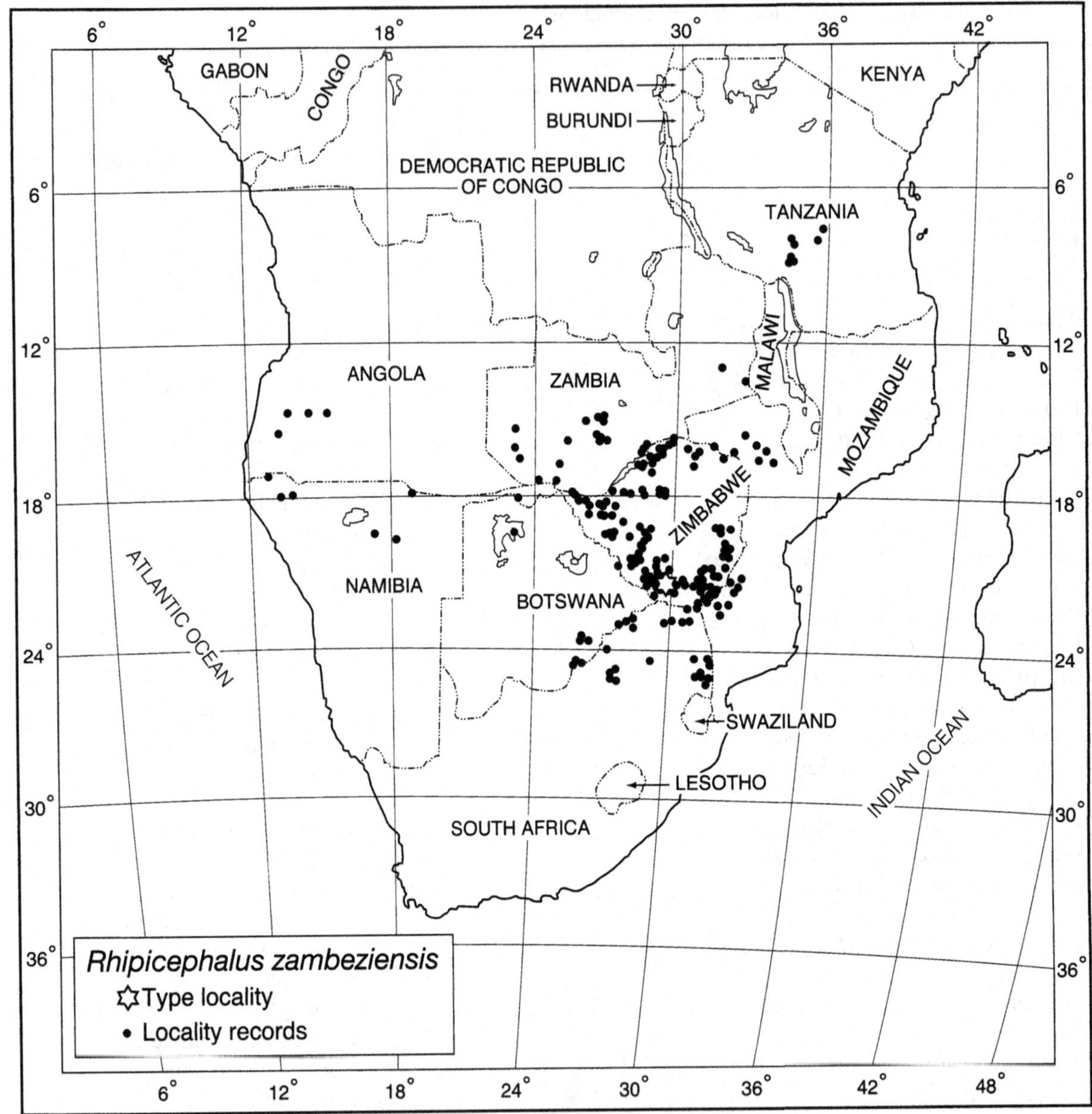

Map 62. *Rhipicephalus zambeziensis*: distribution (with acknowledgements to B.D. Perry & R. Kruska).

broad smooth curve. Eyes at widest point, slightly convex, edged dorsally by a shallow groove. Cervical fields long, narrow, depressed. Ventrally coxae I each with a long, tapering external spur and shorter, broader internal spur; coxae II to IV each with a small sharp external spur only.

Larva (Fig. 223)

Capitulum broader than long, length × breadth ranging from 0.115 mm × 0.133 mm to 0.119 mm × 0.141 mm. Basis capituli over twice as broad as long, lateral angles at about mid-length, short, sharp, slightly forwardly directed. Palps broad, truncated apically. Scutum much broader than long, length × breadth ranging from 0.230 mm × 0.353 mm to 0.242 mm × 0.372 mm; posterior margin a wide, smooth curve. Eyes at widest point, slightly convex, edged dorsally by a shallow groove. Cervical grooves short, slightly convergent. Ventrally coxae each with a single spur, those on coxae I being the largest and sharpest while those on coxae II and III are smaller and blunter.

Notes on identification

Prior to its description as a new species in 1981 *R. zambeziensis* was designated merely as *Rhipicephalus* sp. II by Yeoman & Walker (1967), MacLeod (1970), MacLeod *et al.* (1977), MacLeod & Mwanaumo (1978), and Walker, Mehlitz & Jones (1978).

The morphological differences between *R. zambeziensis* and *R. appendiculatus* are discussed on p. 65. Although typical specimens of these two species, especially the immature stages, are reasonably easy to differentiate, both can exhibit such wide ranges of morphological variation that it is virtually impossible to identify all the individuals present in field collections, especially in mixed infestations.

Hosts

A three-host species (Walker, Norval & Corwin, 1981). The preferred hosts of all stages of development are cattle, impala and greater kudu (Norval, Walker & Colborne, 1982; Horak *et al.*, 1992) (Table 58). Several carnivore species can be infested and lion in particular can harbour large burdens of adult ticks. The scrub hare is a preferred host of the immature stages (Horak *et al.*, 1993). Although warthog are frequently infested they usually harbour only fairly small numbers of immature ticks. The helmeted guineafowl must be regarded as an accidental host: the larvae and nymphs recovered from this bird reflect the abundance of ticks in the environment and not host preference (Horak *et al.*, 1991).

On cattle 68.9% of adult *R. zambeziensis* may attach to the ears, with 11.9% attaching to the remainder of the head (Colborne, 1988). On impala 79.9% attach to the muzzle, 4.8% to the head and only 0.8% to the ears (Matthee, 1996). On greater kudu the nymphs prefer the lower legs and feet, with 81.4% attaching here compared to 9.2% on the head and ears (Horak *et al.*, 1992).

In the southern Kruger National Park, South Africa, larvae are most abundant on the vegetation and on impala, greater kudu and the scrub hare during May to September, nymphs during August to October and adults on impala and greater kudu during February and March (Horak *et al.*, 1992, 1993; Spickett *et al.*, 1992). Adult ticks are also most abundant on cattle in Zimbabwe during the latter 2 months (Colborne, 1988).

Zoogeography

Existing records of *R. zambeziensis* are from Tanzania southwards to parts of Zambia, Zimbabwe, Angola, Namibia, Botswana, Mozambique and South Africa (Map 62). It is often found in the great river valleys and adjacent low-lying areas of these countries, including the Ruaha, Luangwa, Kafue, Zambezi, Cunene and Sabi/Limpopo systems. Most areas where this tick occurs therefore lie at altitudes below 900 m, but in Angola it is present above 1600 m. Mean annual rainfalls over much of its preferred habitat range between 400 and 700 mm, sometimes even less. In the southernmost regions of its distribution range *R. zambeziensis* is most frequently encountered in *Colophospermum mopane* woodland and scrub woodland. It is also present in Zambezian miombo woodland as well as various types of undifferentiated woodland. In Tanzania it occurs in Somalia-Masai *Acacia-Commiphora* deciduous bushland and thicket.

Disease relationships

It has been shown experimentally that *R. zambeziensis* can transmit *Theileria parva parva* to cattle from nymph to adult, and *Theileria parva lawrencei*, *Theileria parva bovis* and *Theileria taurotragi* from larva to nymph and nymph to adult (Lawrence, Norval & Uilenberg, 1983). In Zimbabwe the tick is believed to be a vector of *T. parva lawrencei* in the field (Lawrence *et al.*, 1983). *Rhipicephalus zambeziensis* can also experimentally transmit *Ehrlichia bovis* to cattle from larva to nymph and from nymph to adult; transovarial infection, however, failed (Stoltsz, 1994).

REFERENCES

Colborne, J.R.A. (1988). *The role of wild hosts in maintaining tick populations on cattle in the south-eastern Lowveld of Zimbabwe*. M.Phil. thesis, University of Zimbabwe.

Horak, I.G., Boomker, J., Spickett, A.M. & De Vos, V. (1992). Parasites of domestic and wild animals in South Africa. XXX. Ectoparasites of kudus in the eastern Transvaal Lowveld and the eastern Cape Province. *Onderstepoort Journal of Veterinary Research*, **59**, 259–73.

Horak, I.G., Spickett, A.M., Braack, L.E.O. & Penzhorn, B.L. (1993). Parasites of domestic and wild animals in South Africa. XXXII. Ixodid ticks on scrub hares in the Transvaal. *Onderstepoort Journal of Veterinary Research*, **60**, 163–74.

Horak, I.G., Spickett, A.M., Braack, L.E.O. & Williams, E.J. (1991). Parasites of domestic and wild animals in South Africa. XXVII. Ticks on helmeted guineafowls in the eastern Cape Province and eastern Transvaal Lowveld. *Onderstepoort Journal of Veterinary Research*, **58**, 137–48.

Lawrence, J.A. & Norval, R.A.I. (1979). A history of ticks and tick-borne diseases of cattle in Rhodesia. *Rhodesian Veterinary Journal*, **10**, 28–40.

Lawrence, J.A., Norval, R.A.I. & Uilenberg, G. (1983). *Rhipicephalus zambeziensis* as a vector of bovine Theileriae. *Tropical Animal Health and Production*, **15**, 39–42.

MacLeod, J. (1970). Tick infestation patterns in the southern province of Zambia. *Bulletin of Entomological Research*, **60**, 253–74.

MacLeod, J., Colbo, M.H., Madbouly, M.H. & Mwanaumo, B. (1977). Ecological studies of ixodid ticks (Acari: Ixodidae) in Zambia. III. Seasonal activity and attachment sites on cattle, with notes on other hosts. *Bulletin of Entomological Research*, **67**, 161–73.

MacLeod, J. & Mwanaumo, B. (1978). Ecological studies of ixodid ticks (Acari: Ixodidae) in Zambia. IV. Some anomalous infestation patterns in the northern and eastern regions. *Bulletin of Entomological Research*, **68**, 409–29.

Matthee, S. (1996). *The effectiveness of a live-sampling technique for estimating arthropod parasite populations on impala* (Aepyceros melampus). M.Sc. thesis, University of Pretoria, South Africa.

Norval, R.A.I., Walker, J.B. & Colborne, J. (1982). The ecology of *Rhipicephalus zambeziensis* and *Rhipicephalus appendiculatus* (Acarina, Ixodidae) with particular reference to Zimbabwe. *Onderstepoort Journal of Veterinary Research*, **49**, 181–90.

Spickett, A.M., Horak, I.G., Van Niekerk, A. & Braack, L.E.O. (1992). The effect of veldburning on the seasonal abundance of free-living ixodid ticks as determined by drag-sampling. *Onderstepoort Journal of Veterinary Research*, **59**, 285–92.

Stoltsz, W.H. (1994). Transmission of *Ehrlichia bovis* by *Rhipicephalus* spp. in South Africa. *Journal of the South African Veterinary Association*, **65**, 159.

Walker, J.B., Norval, R.A.I. & Corwin, M.D. (1981). *Rhipicephalus zambeziensis sp. nov.*, a new tick from eastern and southern Africa, together with a redescription of *Rhipicephalus appendiculatus* Neumann, 1901 (Acarina, Ixodidae). *Onderstepoort Journal of Veterinary Research*, **48**, 87–104.

Also see the following Basic References (pp. 12–14): Walker, Mehlitz & Jones (1978); Yeoman & Walker (1967).

RHIPICEPHALUS ZIEMANNI NEUMANN, 1904

This species was named after Dr Hans Ziemann (1865–1939), who collected the type specimens. He was a German medical doctor who worked in the German Colonial Service and was stationed in Cameroon for several years. He made a special study of malaria there.

Synonyms

aurantiacus; *brevicoxatus*; *cuneatus*; *ziemanni aurantiacus*.

Diagnosis

A large finely punctate reddish-brown tick.

Male (Figs 224(a), 225(a) to (c))
Capitulum slightly broader than long, length × breadth ranging from 0.62 mm × 0.63 mm to 0.90 mm × 0.93 mm. Basis capituli with short acute lateral angles just anterior to mid-length. Palps broad, relatively long, rounded apically. Conscutum length × breadth ranging from 2.36 mm × 1.63 mm to 3.50 mm × 2.44 mm; anterior process of coxae I inconspicuous. In engorged specimens body wall expanded somewhat posterolaterally and tapering to a broadly-rounded caudal process posteromedially. Eyes flat. Cervical pits slightly convergent, continuous with the faintly indicated internal cervical margins. Marginal lines ranging from short and inconspicuous to virtually absent and merely indicated by punctations. Posterior grooves shallow, the posteromedian long, the posterolaterals round to kidney-shaped. Punctations small, extremely numerous; those on the scapulae, on the lateral borders posterior to the eyes and on the festoons finer and sparser, those medially on the conscutum usually slightly larger, densely and evenly distributed. Ventrally spiracles broad, with a very short broad prolongation curving towards the dorsal surface. Adanal plates large, widening markedly posterior to the anus, their posterior margins almost straight; accessory adanal plates as small sclerotized points only.

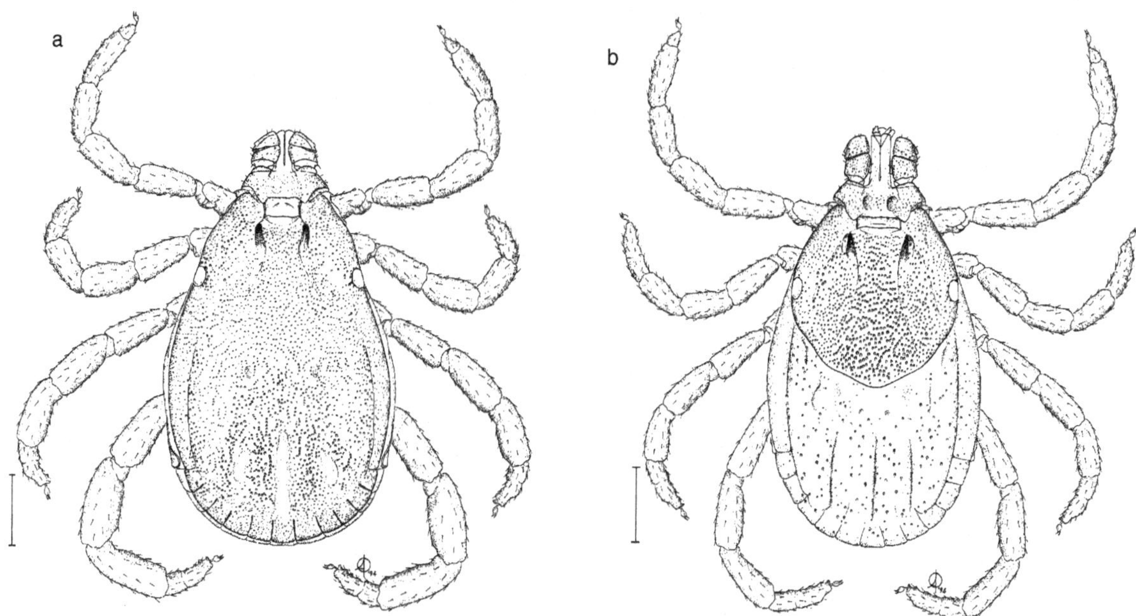

Figure 224. *Rhipicephalus ziemanni* [Onderstepoort Tick Collection 2889iv, collected from African buffalo (*Syncerus caffer*), Angumu, Uélé, Democratic Republic of Congo, received 5 January 1953 from Dr Wanson]. (a) Male, dorsal; (b) female, dorsal. Scale bars represent 1 mm. A. Olwage *del.*

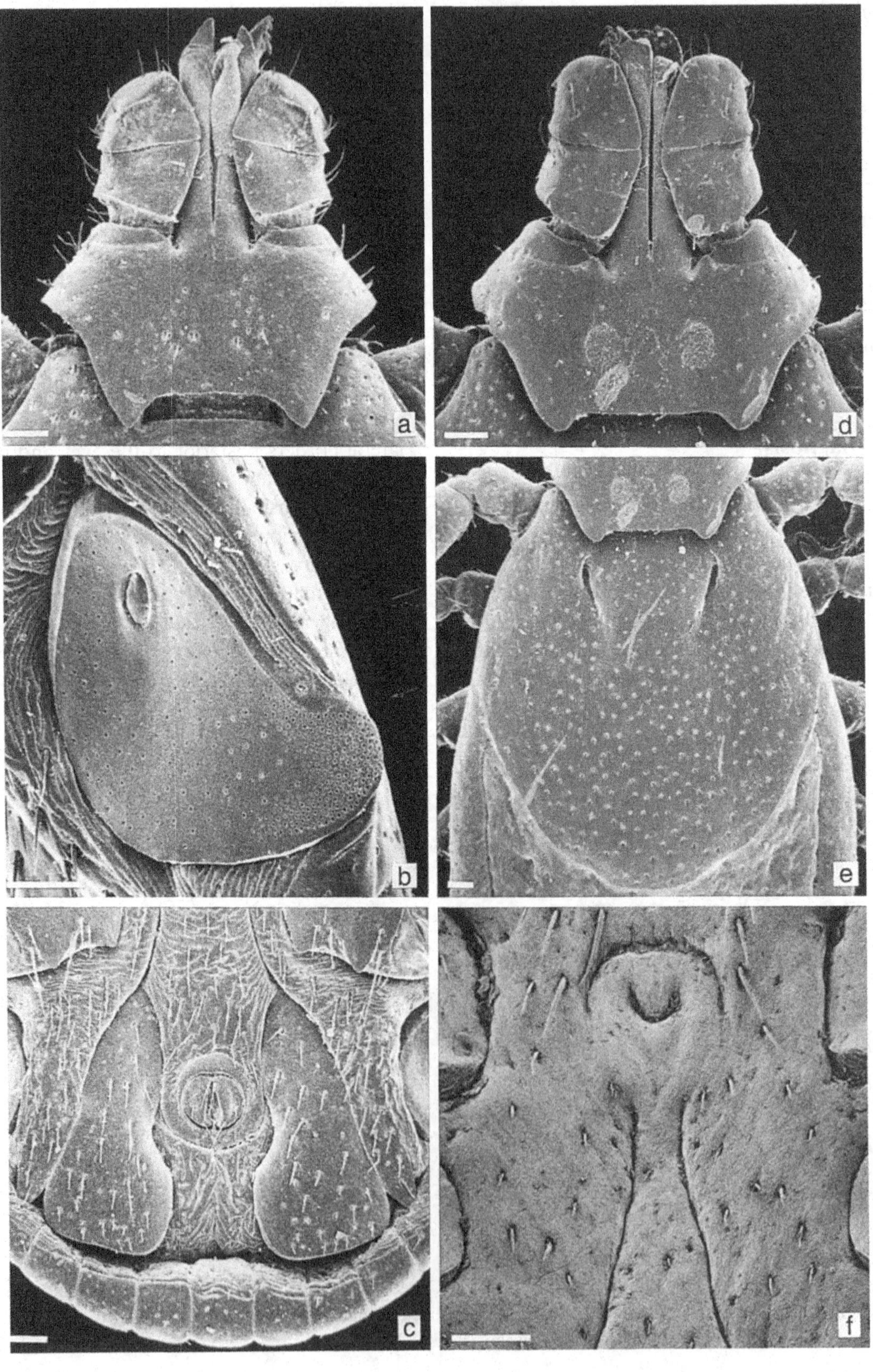
a
d
b
e
c
f

Table 59. *Host records of* Rhipicephalus ziemanni

Hosts	Number of records
Domestic animals	
Cattle	17
Sheep	1
Goats	4
Pigs	2
Dogs	8
Wild animals	
'Monkey'	1
Serval (*Leptailurus serval*)	1 (1 ♂ only)
Leopard (*Panthera pardus*)	4 (1 with 1 nymph (presumed))
'Otter'	3
African civet (*Civettictis civetta*)	2
African elephant (*Loxodonta africana*)	1 (1 ♀ only)
Warthog (*Phacochoerus africanus*)	3
Red river hog (*Potamochoerus porcus*)	8
'Wild pig'	1 (1 ♀ only)
Okapi (*Okapia johnstoni*)	2 (2 ♂♂, 2 ♀♀)
Royal antelope (*Neotragus pygmaeus*)	2
Oribi (*Ourebia ourebi*)	1
African buffalo (*Syncerus caffer*)	13
Bongo (*Tragelaphus eurycerus*)	2
Bushbuck (*Tragelaphus scriptus*)	14
Tragelaphus sp.	1
Bay duiker (*Cephalophus dorsalis*)	2
Maxwell's duiker (*Cephalophus maxwelli*)	1
Blue duiker (*Cephalophus monticola*)	1
Black duiker (*Cephalophus niger*)	3
Banded duiker (*Cephalophus zebra*)	1
Duiker (*Cephalophus* sp.)	1
African brush-tailed porcupine (*Atherurus africanus*)	2 (both including 1 nymph (presumed))
Humans	1

Figure 225 (*opposite*). *Rhipicephalus ziemanni* [Onderstepoort Tick Collection 2889iv, collected from African buffalo (*Syncerus caffer*), Angumu, Uélé, Democratic Republic of Congo, received 5 January 1953 from Dr Wanson]. Male: (a) capitulum, dorsal; (b) spiracle; (c) adanal plates. Female: (d) capitulum, dorsal; (e) scutum; (f) genital aperture. Scale bars represent 0.10 mm. SEMs by J.F. Putterill.

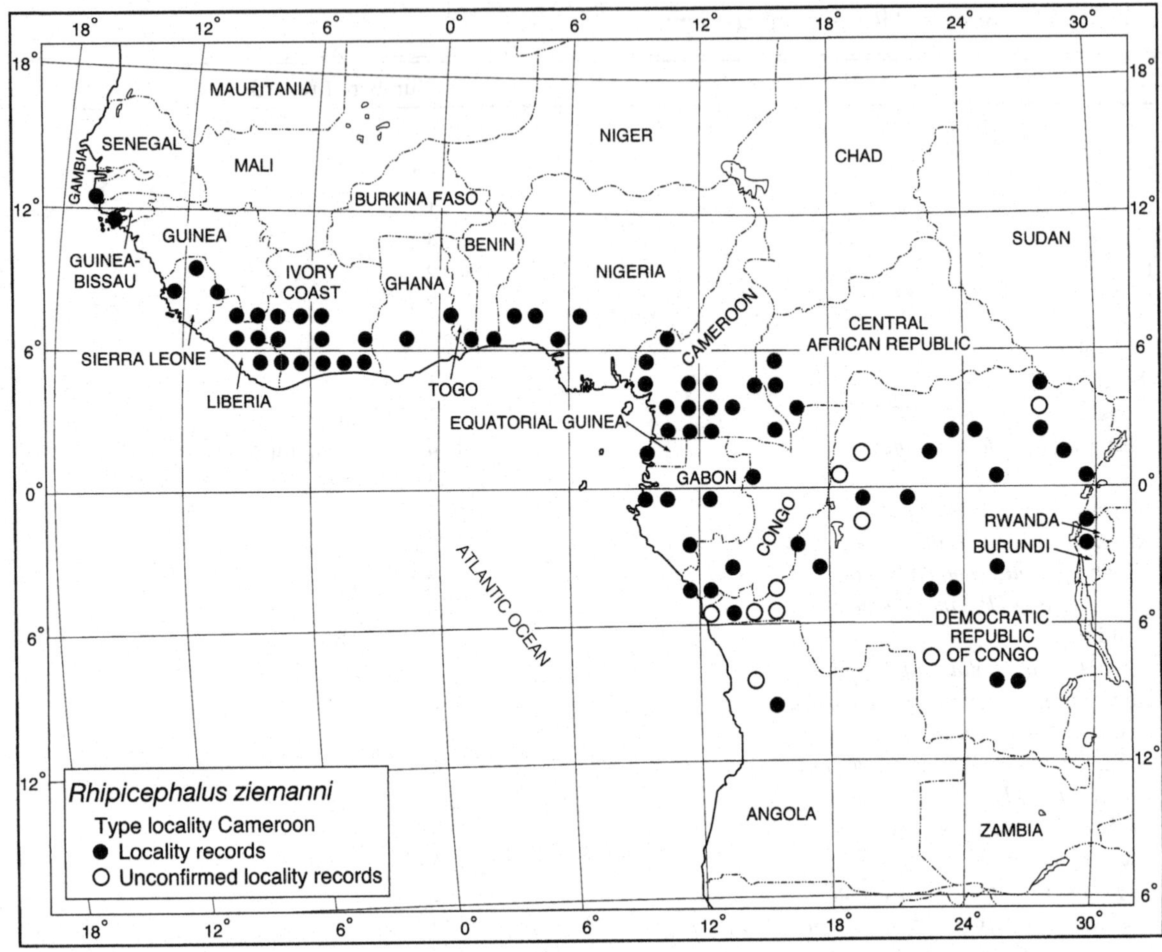

Map 63. *Rhipicephalus ziemanni*: distribution. (Based largely on Morel, 1969, including *R. aurantiacus*).

Female (Figs 224(b), 225(d) to (f))

Capitulum slightly broader than long, length × breadth ranging from 0.88 mm × 0.91 mm to 0.98 mm × 1.00 mm. Basis capituli with short broad lateral angles just anterior to mid-length; porose areas large, round, about 1.5 times their own diameter apart. Palps broad, long, rounded apically. Scutum length × breadth ranging from 1.59 mm × 1.59 mm to 1.80 mm × 1.70 mm; posterior margin slightly sinuous. Eyes flat. Cervical pits slightly convergent; cervical fields broad, but their margins very indistinct. Punctations small, densely distributed, finest on the scapulae and cervical fields, medially slightly larger. Ventrally genital aperture more-or-less U-shaped.

Immature stages

Undescribed.

Notes on identification

In his illustration of the *R. cuneatus* male, now regarded as a synonym of *R. ziemanni*, Neumann (1908) showed the scapulae as being markedly elongated. He apparently incorporated the anterior processes of the coxae as part of the tick's dorsum in his drawing (Clifford & Anastos, 1962). The scapulae of the *R. ziemanni* male are, as usual, merely rounded apically.

As noted earlier, Morel (1957) suggested that the two *R. ziemanni* adults listed from a hyrax by Hoogstraal (1954) might be *R. boueti* (see p. 96). In view of this uncertainty though, and in the absence of any other reports of *R. ziemanni* from hyraxes, we have omitted Hoogstraal's record from our host list.

The systematic position of *R. aurantiacus* is uncertain at present. Zumpt (1943) described it as a distinct entity, separate from *R. ziemanni*, though he noted that he could not distinguish the females of these two species. Theiler (1947) also treated *R. aurantiacus* as a separate species. Morel & Mouchet (1958), however, listed it as a 'form' of *R. ziemanni* (with *R. cuneatus* as its synonym), the others being their 'forme typique *ziemanni*' and their 'forme *brevicoxatus*'. In 1959 Tendeiro synonymized *R. aurantiacus* with *R. ziemanni*, as did Theiler (1962), but Morel (1963) referred to it as a separate species again, a position that he maintained in 1969, saying it was 'une espèce méconnue'.

Morel (1969) considered that *R. ziemanni* belongs to the *R. sanguineus* group, but did not amplify this statement. We question it.

We ourselves have little experience of *R. ziemanni*. For the present we have followed Theiler (1962) and included *R. aurantiacus* as its synonym. Further study, including examination of the types of *R. aurantiacus*, may later prove this decision to have been wrong. We have therefore noted below some of the distinctions that Morel (1969) made between these two species.

Hosts

Life cycle unknown. Existing records indicate that *R. ziemanni* adults will readily infest cattle, and also dogs, given the opportunity (Table 59). Aeschlimann (1967) regarded the larger forest antelopes, such as the bongo and bushbuck, as its favourite hosts. He felt that fewer of the numerous smaller antelopes he had examined, such as the royal antelope and various duikers, were parasitized by this tick. Morel (1969) stated that it would infest practically all ungulates and carnivores living in the forest and adjacent woodlands, and occasionally humans and monkeys. He noted that *R. aurantiacus* was found most frequently on the small forest buffalo, on ungulates such as the bongo and the 'potamochère' (doubtless in this case the red river hog) and sometimes on carnivores, but not on duikers. The attachment sites of *R. ziemanni* adults on their hosts have not been recorded.

Three nymphs have been listed, but not described, one from a leopard and two from African brush-tailed porcupines (Morel & Mouchet, 1958; Aeschlimann, 1967). We regard these identifications as provisional.

Zoogeography

Rhipicephalus ziemanni has been recorded from southern Senegal across the continent to the western slopes of mountains in the Kivu area, eastern Democratic Republic of Congo (Morel, 1969) (Map 63). The climate where it occurs is characteristically equatorial, with mean annual rainfalls of more than 1500 mm falling during a period of over 7 to 8 months. Morel regarded it as the most typical rhipicephalid occurring in the humid equatorial forests of West and Central Africa. Its distribution also extends into nearby patches of forest and mosaics of forest and woodland, and into the higher altitude woodlands and subtropical woodlands of the Katanga area in the Democratic Republic of Congo.

Morel emphasized that, judging by its hosts and their behaviour, it occurs in what he describes as the areas of firmer ground in dense forest, i.e. the drier parts. In his view this distinguishes *R. ziemanni* from *R. aurantiacus*, which he says is typically found on animals living in wetter places along rivers and in swamps.

Disease relationships

Unknown.

REFERENCES

Hoogstraal, H. (1954). Noteworthy African tick records in the British Museum (Natural History) collections. *Proceedings of the Entomological Society of Washington*, **56**, 273–9.

Morel, P.C. (1957). *Rhipicephalus boueti* n. sp. (Acarina Ixodidae) parasites des damans du Dahomey. *Bulletin de la Société de Pathologie Exotique*, **50**, 696–700.

Morel, P.C. (1963). La réserve naturelle intégral du Mont Nimba. II. Tiques (Acarina Ixodoidea). *Mémoires de l'Institut Français d'Afrique Noire*, **No. 66**, 33–40.

Morel, P.C. & Mouchet, J. (1958). Les tiques du Cameroun (Ixodidae et Argasidae). *Annales de Parasitologie Humaine et Comparée*, **33**, 69–111.

Neumann, L.G. (1904). Notes sur les ixodidés. II. *Archives de Parasitologie*, **8**, 444–64.

Neumann, L.G. (1908). Notes sur les ixodidés. VII. *Notes from the Leyden Museum*, **30**, 73–91.

Tendeiro, J. (1959). Sur quelques ixodidés du Mozambique et de la Guinée Portugaise. 1. *Boletim Cultural da Guiné Portuguesa*, **14** (53), 21–95 + 12 photographs.

Zumpt, F. (1943). *Rhipicephalus aurantiacus* Neumann und ähnlichen Arten. VIII. Vorstudie zu einer Revision der Gattung *Rhipicephalus* Koch. *Zeitschrift für Parasitenkunde*, **13**, 102–17.

Also see the following Basic References (pp. 12–14): Aeschlimann (1967); Clifford & Anastos (1962); Elbl & Anastos (1966); Morel (1969); Theiler (1947, 1962).

RHIPICEPHALUS ZUMPTI SANTOS DIAS, 1950

This species was named in honour of Dr F. Zumpt (1908–1985), in recognition of his contributions to our knowledge of the genus *Rhipicephalus*.

Diagnosis

A large dark-brown to black tick whose male conscutum is flat to slightly concave with a rather dull surface.

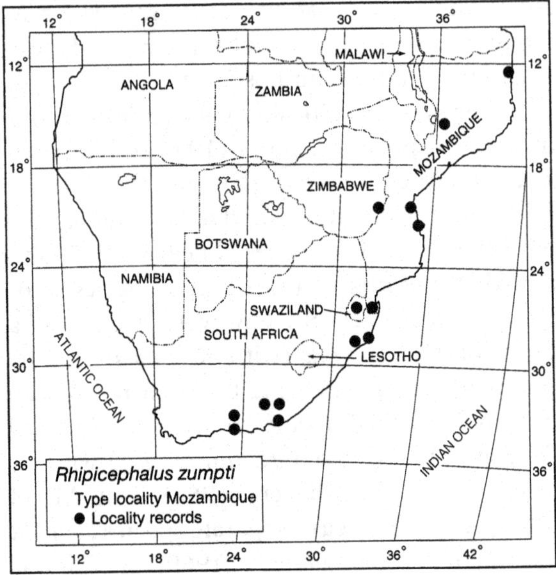

Map 64. *Rhipicephalus zumpti*: distribution.

Male (Figs 226(a), 227(a) to (c))

Capitulum about as broad as long, length × breadth ranging from 0.56 mm × 0.56 mm to 0.93 mm × 0.88 mm. Basis capituli with short lateral angles in anterior third of its length. Palps broadly rounded apically. Conscutum length × breadth ranging from 2.44 mm × 1.58 mm to 4.02 mm × 2.89 mm; anterior process of coxae I inconspicuous. In engorged males body wall expanded posterolaterally and forming a short broadly-rounded caudal process posteromedially. Eyes almost flat, edged dorsally by a few punctations. Cervical pits comma-shaped; cervical fields inconspicuous, their external margins delimited by an irregular pattern of medium-sized setiferous punctations. Marginal lines fairly short, shallow, outlined by large setiferous punctations. Posteromedian groove long, narrow; posterolateral grooves small, round; all shallow and inconspicuous. Small round shallow depressions sometimes present medially on scutum. Medium-sized setiferous punctations scattered on scapulae and anteriorly on scutum, becoming larger and more conspicuous posteriorly on the conscutum, especially around the posterior grooves. Interstitial puncta-

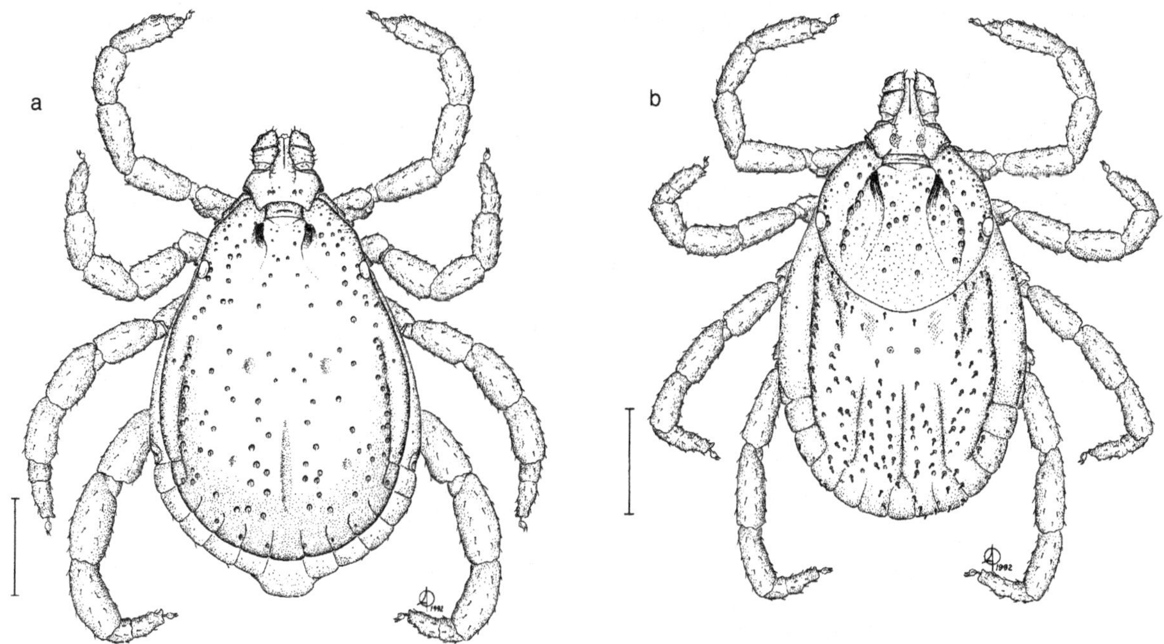

Figure 226. *Rhipicephalus zumpti* [collected from bushpig (*Potamochoerus larvatus*), Adelaide, Eastern Cape Province, South Africa, on 4 May 1972 by R. Joubert]. (a) Male, dorsal; (b) female, dorsal. Scale bars represent 1 mm. A. Olwage *del.*

tions virtually absent. Ventrally spiracles comma-shaped, with a short broad smoothly curved dorsal prolongation. Adanal plates short, broad, their internal margins concave posterior to the anus, their posterior margins smoothly rounded; accessory adanal plates as small sclerotized points.

Female (Figs 226(b), 227(d) to (f))

Capitulum slightly broader than long, length × breadth ranging from 0.70 mm × 0.72 mm to 0.96 mm × 1.01 mm. Basis capituli with short acute lateral angles at about mid-length; porose areas large, round, just more than their own diameter apart. Scutum broader than long, length × breadth ranging from 1.32 mm × 1.38 mm to 1.87 mm × 2.10 mm; posterior margin a broad smooth curve. Eyes almost flat, edged dorsally by a few punctations. Cervical fields broad, shallow, their external margins delimited by medium-sized to large setiferous punctations. A few medium-sized punctations scattered on the scapulae and between the cervical fields, interspersed with fine interstitial punctations that are sometimes more conspicuous than those in the male. Ventrally genital aperture tongue-shaped.

Nymph (Fig. 228)

Capitulum much broader than long, length × breadth ranging from 0.26 mm × 0.39 mm to 0.30 mm × 0.45 mm. Basis capituli over three times as broad as long with tapering lateral angles projecting posteriorly over the scapulae; ventrally with short spurs on posterior margin. Palps slender, tapering to narrowly-rounded apices, inclined inwards. Scutum broader than long, length × breadth ranging from 0.55 mm × 0.61 mm to 0.60 mm × 0.64 mm; posterior margin a broad smooth curve. Eyes at widest point, well over halfway back, long and very narrow. Cervical fields long, narrow, shallow, divergent. Ventrally coxae I each with a relatively long external spur and a shorter broader internal spur; coxae II to III each with a very small external spur; coxae IV each virtually without a spur.

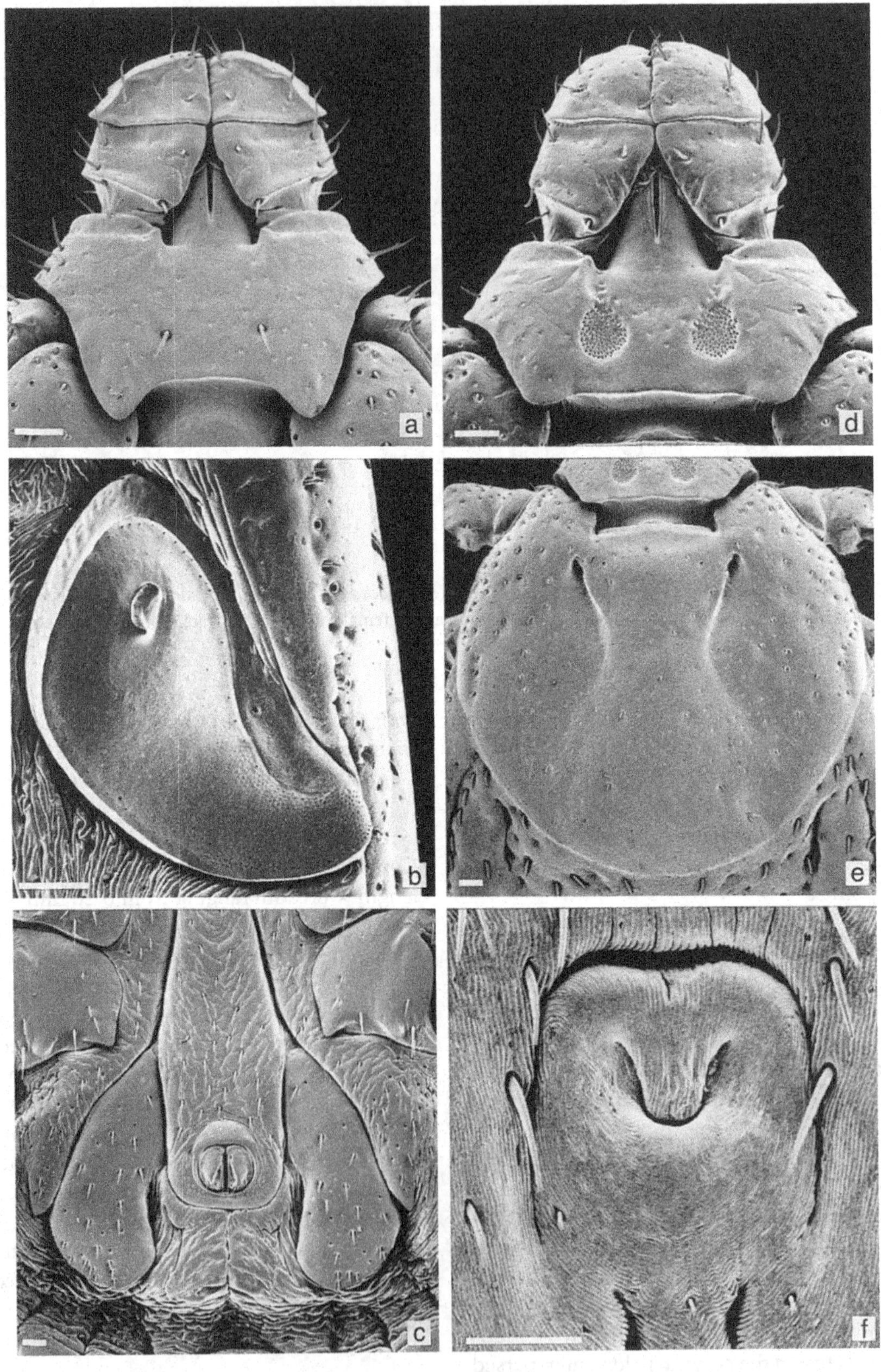
a
b
c
d
e
f

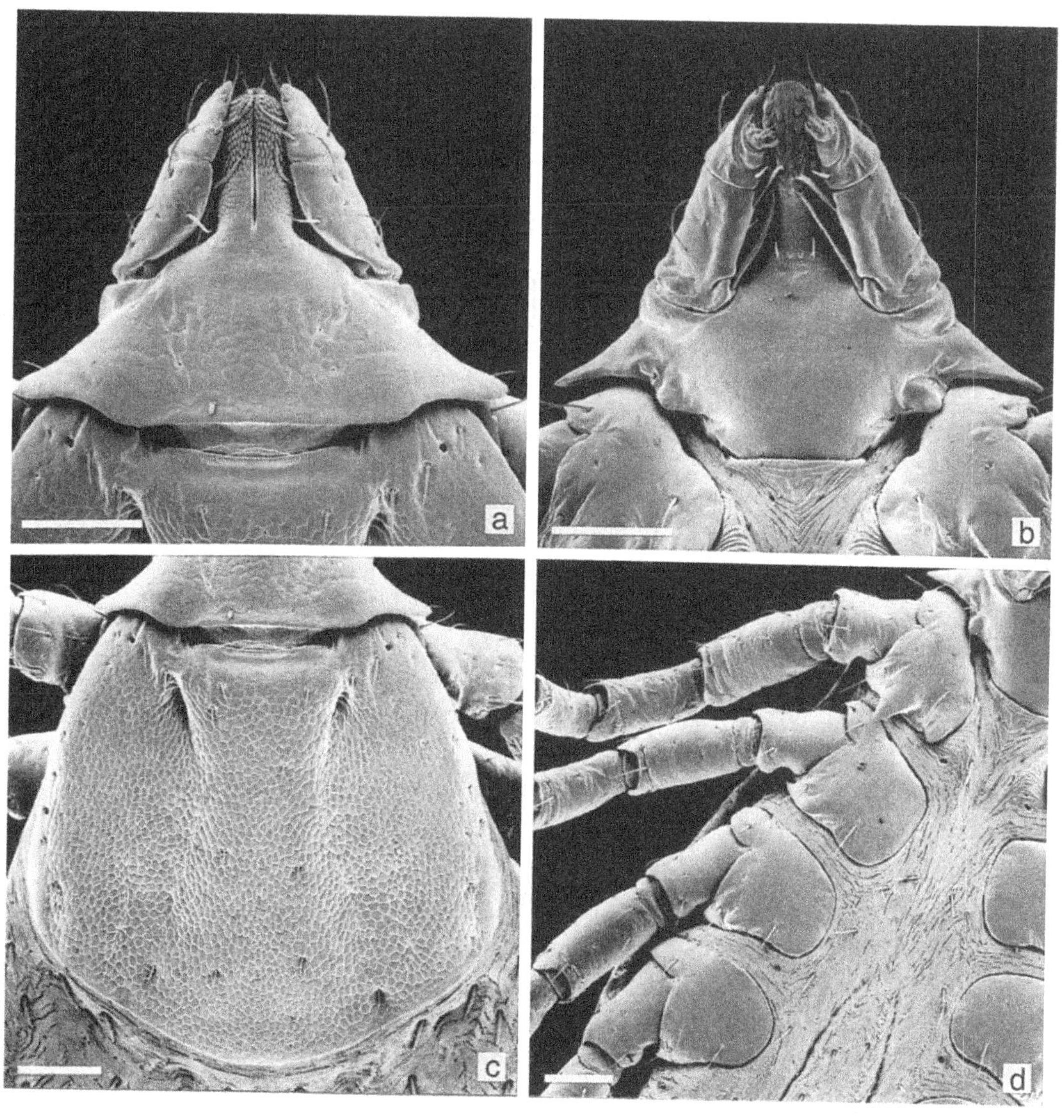

Figure 228 (*above*). *Rhipicephalus zumpti* (laboratory reared, progeny of adults collected from vegetation in Tsitsikama Forest, Eastern Cape Province, South Africa, in 1993 by I. McKay). Nymph: (a) capitulum, dorsal; (b) capitulum, ventral; (c) scutum; (d) coxae. Scale bars represent 0.10 mm. SEMs by J.F. Putterill.

Figure 227 (*opposite*). *Rhipicephalus zumpti* (laboratory reared, progeny of adults collected from vegetation in Tsitsikama Forest, Eastern Cape Province, South Africa, in 1993 by I. McKay). Male: (a) capitulum, dorsal; (b) spiracle; (c) adanal plates. Female: (d) capitulum, dorsal; (e) scutum; (f) genital aperture. Scale bars represent 0.10 mm. SEMs by J.F. Putterill.

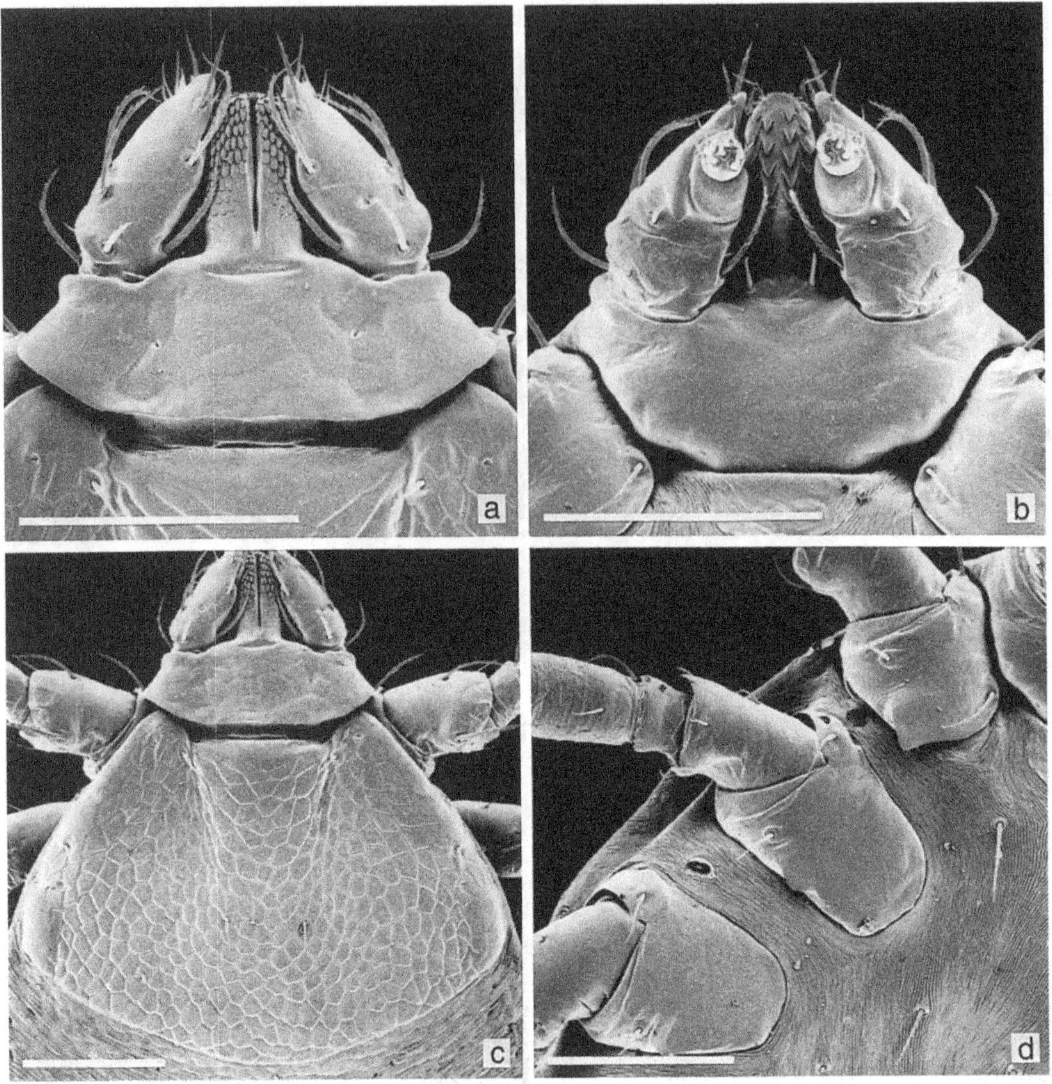

Figure 229. *Rhipicephalus zumpti* (laboratory reared, progeny of adults collected from vegetation in Tsitsikama Forest, Eastern Cape Province, South Africa, in 1993 by I. McKay). Larva: (a) capitulum, dorsal; (b) capitulum, ventral; (c) scutum; (d) coxae. Scale bars represent 0.10 mm. SEMs by J.F. Putterill.

Larva (Fig. 229)

Capitulum much broader than long, length × breadth ranging from 0.133 mm × 0.178 mm to 0.141 mm × 0.191 mm. Basis capituli almost hexagonal in shape, over three times as broad as long with short bluntly-rounded lateral angles. Palps narrow, their external margins very slightly convex, tapering to narrowly-rounded apices, inclined inwards. Scutum much broader than long, length × breadth ranging from 0.257 mm × 0.356 mm to 0.284 mm × 0.380 mm; posterior margin a smooth shallow curve. Eyes at widest part of scutum, well over halfway back, almost flat. Cervical grooves short, slightly convergent. Ventrally coxae I each with a sharp triangular spur; coxae II each with a small spur; coxae III each virtually without a spur.

Table 60. *Host records of* Rhipicephalus zumpti

Hosts	Number of records	
	Confirmed	Unconfirmed
Domestic animals		
Cattle		1
Dogs	5	
Wild animals		
Lion (*Panthera leo*)	1	
African civet (*Civettictis civetta*)	1	
Burchell's zebra (*Equus burchellii*)	1	
Black rhinoceros (*Diceros bicornis*)	1	
Warthog (*Phacochoerus africanus*)	3	
Bushpig (*Potamochoerus larvatus*)	9	2
Lichtenstein's hartebeest (*Sigmoceros lichtensteinii*)	1	
Oribi (*Ourebia ourebi*)	1	
Steenbok (*Raphicerus campestris*)	1	
African buffalo (*Syncerus caffer*)	1	1
Nyala (*Tragelaphus angasii*)	1	
Greater kudu (*Tragelaphus strepsiceros*)	1	
Common duiker (*Sylvicapra grimmia*)	1	
Waterbuck (*Kobus ellipsiprymnus*)	2	
South African porcupine (*Hystrix africaeaustralis*)	1	
Humans	1	1

Notes on identification

All stages of *R. zumpti* and *R. planus* are very much alike morphologically (see p. 333). The adults of *R. zumpti* also fairly closely resemble those of *R. simus*, with which they have been recorded sympatrically on dogs and bushpigs. *Rhipicephalus zumpti* can be differentiated by the dish-like shape of the large punctations on its conscutum or scutum and the flat to slightly concave conscutal surface that is particularly evident in large males.

Hosts

A three-host species (Protozoology Section Tick Breeding Register, Onderstepoort, No. 9039). Cattle and dogs are the only known domestic hosts of the adults (Table 60). Bushpigs and possibly also warthogs are their preferred wild hosts (Jooste, 1969; Horak, Boomker & Flamand, 1991; Gallivan & Surgeoner, 1995). Four out of the five infested dogs lived on rural properties on which bushpigs were also present (I.G.H., unpublished data). The comparatively large number of monogastric host species from which *R. zumpti* adults have been collected suggests that they have host preferences similar to those of adult *R. simus*.

The preferred sites of attachment of the tick's adults and their seasonal abundance are unknown. The hosts of the immature stages are also unknown but are probably murid rodents.

Zoogeography

Rhipicephalus zumpti has been recorded in Mozambique (Santos Dias, 1960), eastern Zimbabwe (as *R. reichenowi*) (Jooste, 1969; Norval, 1985) and the eastern lowveld of Swaziland (Gallivan & Surgeoner, 1995) as well as in eastern and southern South Africa (Baker & Keep, 1970; Horak *et al.*, 1991) (Map 64).

It occurs at altitudes from just above sea level up to *c.* 1500 m and in regions receiving between 500 mm and 1000 mm of rainfall annually. Several localities at which it has been collected, either from the vegetation or from host animals, are situated in forested regions or are associated with tracts of densely-wooded vegetation. These forested or wooded areas lie within vegetation types classified as coastal mosaic, evergreen and semi-evergreen bushland and thicket, miombo or *Colophospermum mopane* or undifferentiated woodland, or undifferentiated Afromontane vegetation.

Disease relationships

Unknown.

REFERENCES

Baker, M.K. & Keep, M.E. (1970). Checklist of the ticks found on the larger game animals in the Natal game reserves. *Lammergeyer*, **No. 12**, 41–7.

Gallivan, G.J. & Surgeoner, G.A. (1995). Ixodid ticks and other ectoparasites of wild ungulates in Swaziland: regional, host and seasonal patterns. *South African Journal of Zoology*, **30**, 169–77.

Horak, I.G., Boomker, J. & Flamand, J.R.B. (1991). Ixodid ticks and lice infesting red duikers and bushpigs in north-eastern Natal. *Onderstepoort Journal of Veterinary Research*, **58**, 281–4.

Jooste, K.F. (1969). The role of Rhodesia in ixodid tick distribution in central and southern Africa. In *Proceedings of a Symposium on the Biology and Control of Ticks in Southern Africa*, convenor G.B. Whitehead, pp. 37–42. Grahamstown, South Africa: Rhodes University.

Norval, R.A.I. (1985). The ticks of Zimbabwe. XII. The lesser known *Rhipicephalus* species. *Zimbabwe Veterinary Journal*, **16**, 37–43.

Santos Dias, J.A.T. (1950). Contribuição para o conhecimento da fauna ixodológica de Moçambique. *Moçambique*, **No. 61**, 113–70.

Also see the following Basic Reference (p. 14): Santos Dias (1960).

8

Host/parasite list for the Afrotropical *Rhipicephalus* species

Unless otherwise stated the records in the host/parasite checklist refer to adult ticks. The number in brackets appearing after a tick's name represents the number of collections made from the host species under which it is listed. It has been difficult to determine the exact number of records for *R. fulvus*. Consequently where the name of this tick appears in the host/parasite list it is followed by (+) which indicates that it has been collected from a particular host species.

When a tick's name appears in **bold type** this indicates that the animal under which it is listed is a preferred host of the adults. However, when the hosts of the immature stages are entirely unrelated to those of the adults the names of these ticks also appear in **bold type** under the names of the preferred hosts of their immature stages. These ticks are *R. follis, R. gertrudae* and *R. simus*. When more collections of their immature stages have been made we feel sure that the same will apply for *R. compositus, R. lounsburyi, R. lunulatus, R. muhsamae, R. neumanni, R. planus, R. praetextatus, R. senegalensis* and *R. zumpti*, and perhaps other species too.

Despite the fact that the immature stages of *R. pravus, R.* sp. near *pravus* and *R. warburtoni* are found on hares, on which their adults also occur, they undoubtedly have a preference for elephant shrews. The names of these ticks are also **typed in bold** under their respective elephant shrew hosts. We are confident that the same will apply for *R. punctatus* and *R.* sp. near *punctatus* once additional collections of their immature stages have been made.

Numerous surveys on the seasonal abundance of ixodid ticks have been conducted within the distribution range of the genus *Rhipicephalus*. In these surveys long series of various species of domestic and wild animals have frequently been examined. In addition large numbers of 'sought after' wild animal species have been shot by hunters and their carcasses made available for the collection of ticks. The many collections taken from animals from these two sources may make the number of records of certain tick species from particular hosts seem disproportionately large and thus give a false impression of host preference. Such collections have been made from domestic cattle, sheep, goats and dogs and from wild caracal, lion, brown hyaena, Burchell's zebra, mountain zebra, rock hyrax, warthog, bushpig, giraffe, impala, blue wildebeest, bontebok, springbok, Thomson's gazelle, steenbok, Sharpe's grysbok, African buffalo, eland, nyala, bushbuck, greater kudu, red forest duiker, common duiker, gemsbok, grey rhebok, reedbuck, mountain reedbuck, red veld rat, four-striped grass mouse, spring hare, greater cane rat, Cape hare, scrub hare, savanna hare, Smith's red rock rabbit and the rock elephant shrew.

The host groupings under which the tick collections are recorded in the host/parasite checklist follow the same sequence as those in the host record tables accompanying the individual tick species accounts, namely domestic animals, wild animals, birds, reptiles and humans.

DOMESTIC ANIMALS

Bos indicus/taurus **Cattle**

- ***R. appendiculatus*** (commonly parasitized, including immatures)
- *R. aquatilis* (6)
- *R. bequaerti* (3)
- ***R. bergeoni*** (104)
- ***R. camicasi*** (13)
- *R. capensis* (2)
- *R. carnivoralis* (3)
- ***R. compositus*** (149)
- ***R. duttoni*** (commonly parasitized, including immatures)
- *R. dux* (2)
- ***R. evertsi evertsi*** (4819, including immatures)
- ***R. evertsi mimeticus*** (376, including immatures)
- ***R. exophthalmos*** (39)
- ***R. follis*** (36)
- ***R. gertrudae*** (58)
- ***R. glabroscutatum*** (49, including immatures)
- *R. guilhoni* (20)
- ***R. humeralis*** (32)
- *R. hurti* (37)
- *R. interventus* (25)
- ***R. jeanneli*** (77)
- ***R. kochi*** (199)
- *R. longiceps* (2)
- ***R. longus*** (85)
- ***R. lunulatus*** (336; 6, unconfirmed)
- *R. maculatus* (34, including nymphs)
- *R. masseyi* (6)
- *R. moucheti* (1)
- *R. muehlensi* (53)
- ***R. muhsamae*** (45)
- *R. nitens* (2)
- *R. planus* (6)
- ***R. praetextatus*** (151; 469, unconfirmed)
- ***R. pravus*** (506; 200, unconfirmed)
- ***R. sp. near pravus*** (21)
- *R. pseudolongus* (5)
- ***R. pulchellus*** (649, including immatures)
- *R. punctatus* (56)
- ***R. sp. near punctatus*** (128)
- *R. sanguineus* (12)
- ***R. senegalensis*** (188)
- ***R. simus*** (642)
- *R. sulcatus* (14; 5, unconfirmed)
- *R. supertritus* (22)
- *R. theileri* (3)
- *R. tricuspis* (10)
- *R. turanicus* (81)
- *R. warburtoni* (26)
- ***R. zambeziensis*** (254, including immatures)
- *R. ziemanni* (17)
- *R. zumpti* (1, unconfirmed)

Bubalus bubalis **Water buffaloes**

- *R. appendiculatus* (2)

Ovis aries **Sheep**

- *R. appendiculatus* (245, including immatures)
- *R. armatus* (1)
- *R. bergeoni* (33)
- ***R. camicasi*** (13)
- *R. compositus* (3)
- *R. duttoni* (2)
- ***R. evertsi evertsi*** (667, including immatures)
- *R. evertsi mimeticus* (43, including immatures)
- *R. exophthalmos* (43)
- *R. follis* (12)
- *R. fulvus* (+)
- ***R. gertrudae*** (48)
- ***R. glabroscutatum*** (66, including immatures)
- *R. guilhoni* (14)
- *R. humeralis* (5)
- *R. interventus* (1)
- *R. jeanneli* (1)
- *R. kochi* (2)

Sheep (*cont.*)

R. longus (2)
R. lounsburyi (9; 2, unconfirmed)
R. lunulatus (60)
R. maculatus (1)
R. masseyi (2)
R. muehlensi (2)
R. muhsamae (2)
R. neumanni (25)
R. nitens (87, including immatures)
R. praetextatus (52, unconfirmed)
R. pravus (44; 51, unconfirmed)
R. pulchellus (144, including immatures)
R. sp. near *punctatus* (2)
R. senegalensis (7)
R. simus (14)
R. sulcatus (1; 1, unconfirmed)
R. theileri (1)
R. turanicus (14)
R. warburtoni (61)
R. ziemanni (1)

Capra hircus **Goats**

R. appendiculatus (358, including immatures)
R. bequaerti (2)
R. bergeoni (6)
R. camicasi (12)
R. compositus (3)
R. duttoni (1)
R. evertsi evertsi (303, including immatures)
R. evertsi mimeticus (50, including immatures)
R. exophthalmos (16)
R. fulvus (+)
R. gertrudae (8)
R. glabroscutatum (614, including immatures)
R. guilhoni (4)
R. humeralis (5)
R. jeanneli (1)
R. kochi (9)
R. longus (1)
R. lunulatus (22)
R. maculatus (1)
R. muehlensi (3)
R. muhsamae (1)
R. neumanni (8)
R. nitens (2)
R. praetextatus (27, unconfirmed)
R. pravus (36; 49, unconfirmed)
R.* sp. near *pravus (15)
R. pulchellus (133, including immatures)
R. sp. near *punctatus* (2)
R. sanguineus (3)
R. senegalensis (4)
R. simus (15)
R. sulcatus (3; 1, unconfirmed)
R. supertritus (1)
R. tricuspis (3)
R. turanicus (7)
R. warburtoni (62)
R. zambeziensis (2)
R. ziemanni (4)

Sheep and goats (pooled collections)

R. praetextatus (38)

Camelus dromedarius **Camels**

R. appendiculatus (5)
R. camicasi (8)
R. evertsi evertsi (18)
R. evertsi mimeticus (7)
R. fulvus (+)
R. guilhoni (2)
R. humeralis (21)
R. kochi (1)
R. lunulatus (2)
R. praetextatus (37; 21, unconfirmed)
R. pravus (32; 7, unconfirmed)
R. pulchellus (110)
R. turanicus (3)

Equus caballus **Horses**

R. appendiculatus (31, including immatures)
R. bergeoni (5)
R. capensis (3)
R. duttoni (1)
R. evertsi evertsi (104, including immatures)
R. evertsi mimeticus (31, including immatures)
R. exophthalmos (2)
R. follis (6)
R. gertrudae (9)

Horses (*cont.*)

R. glabroscutatum (2, including immatures)
R. guilhoni (4)
R. jeanneli (1)
R. kochi (1)
R. lunulatus (8)
R. muhsamae (2)
R. neumanni (1)
R. praetextatus (3, unconfirmed)
R. pravus (2)
R. pulchellus (11)
R. sp. near *punctatus* (5)
R. sanguineus (1)
R. senegalensis (5)
R. simus (31)
R. supertritus (1)
R. turanicus (6)
R. zambeziensis (6)

Equus asinus **Donkeys**

R. appendiculatus (17, including nymphs)
R. armatus (1)
R. camicasi (1)
R. compositus (1)
R. evertsi evertsi (26, including immatures)
R. evertsi mimeticus (13, including immatures)
R. exophthalmos (1)
R. gertrudae (1)
R. guilhoni (4)
R. humeralis (2)
R. kochi (1)
R. muhsamae (1)
R. praetextatus (3; 1, unconfirmed)
R. pravus (5)
R. pulchellus (6)
R. sp. near *punctatus* (2)
R. simus (2)

Equus caballus* × *E. asinus **Mules**

R. evertsi evertsi (3)
R. pulchellus (3)
R. simus (1)

Sus scrofa **Pigs**

R. appendiculatus (1)
R. bergeoni (1)
R. complanatus (4)
R. compositus (1)
R. dux (2)
R. evertsi evertsi (3, including nymphs)
R. evertsi mimeticus (2)
R. exophthalmos (1)
R. jeanneli (1)
R. longiceps (1)
R. longus (30)
R. lunulatus (26)
R. maculatus (1)
R. muhsamae (3)
R. praetextatus (1, unconfirmed)
R. pulchellus (5)
R. senegalensis (5)
R. simus (3)
R. ziemanni (2)

Pigs (feral)

R. pravus (1)
R. simus (2)

Canis familiaris **Dogs**

R. appendiculatus (49, including immatures)
R. armatus (1; 1, unconfirmed)
R. carnivoralis (2)
R. compositus (13)
R. duttoni (2)
R. evertsi evertsi (30, including immatures)
R. evertsi mimeticus (3)
R. exophthalmos (2)
R. follis (2)
R. gertrudae (3)
R. guilhoni (4)
R. humeralis (1)
R. hurti (8)
R. interventus (2)
R. jeanneli (1)
R. kochi (1)
R. longicoxatus (1)
R. longus (10)
R. lunulatus (37)
R. masseyi (1)
R. moucheti (10)
R. muehlensi (1)
R. muhsamae (5)
R. nitens (2)

Dogs (*cont.*)
R. praetextatus (4; 54, unconfirmed)
R. pravus (8; 1, unconfirmed)
R. sp. near *pravus* (3)
R. pulchellus (13)
R. sp. near *punctatus* (2)
R. sanguineus (547, including immatures)
R. senegalensis (23)
R. simus (294)
R. sulcatus (33; 1, unconfirmed)
R. supertritus (2)
R. theileri (2)
R. tricuspis (3)
R. turanicus (51)
R. warburtoni (2)
R. zambeziensis (13)
R. ziemanni (8)
R. zumpti (5)

Felis catus **Cats**
R. appendiculatus (1, nymphs)
R. evertsi evertsi (1, nymph)
R. guilhoni (3)
R. praetextatus (1)
R. sanguineus (1)
R. simus (1)
R. sulcatus (1; 2, unconfirmed)
R. turanicus (4)
R. zambeziensis (3, immatures)

Cats (feral)
R. appendiculatus (2, including larvae)
R. simus (1)

Rabbits
R. evertsi evertsi (1)
R. sp. near *pravus* (1)

Chickens
R. appendiculatus (1 ♂)

WILD ANIMALS

CLASS MAMMALIA

Order Insectivora

Family Erinaceidae

Atelerix albiventris **Four-toed hedgehog**
R. armatus (2)
R. guilhoni (3)
R. pravus (1)

Atelerix frontalis **South African hedgehog**
R. simus (1)
R. theileri (1)
R. tricuspis (1)

Atelerix sclateri **Somali hedgehog**
R. praetextatus (1, unconfirmed)
R. pulchellus (1)

***Atelerix* sp.** **Hedgehog**
R. turanicus (2)

Family Soricidae

***Crocidura* spp.** **White-toothed shrews**
R. sp. near *punctatus* (3, immatures)
R. senegalensis (1)

'Shrew'
R. simus (1, nymphs)

Order Chiroptera

Family Rhinopomatidae

Rhinopoma hardwickei **Lesser mouse-tailed bat**
R. pulchellus (1)

Family Megadermatidae

Cardioderma cor **Heart-nosed bat**
R. pravus (1)

Family Vespertilionidae

Eptesicus capensis **Cape serotine**
R. pulchellus (1)

Pipistrellus nanus **Banana bat**
R. pravus (1)

'Bat'
R. lunulatus (1)

Order Primates

Family Galagonidae

Otolemur crassicaudatus **Greater bushbaby**
R. appendiculatus (1, including immatures)
R. humeralis (1)
R. sulcatus (1, unconfirmed)

Family Cercopithecidae

Cercopithecus albogularis **Sykes' monkey**
R. simpsoni (1)

Chlorocebus aethiops **Vervet monkey**
R. appendiculatus (7, including immatures)
R. evertsi evertsi (1, larva)
R. praetextatus (3, unconfirmed)
R. turanicus (1)
R. zambeziensis (1, nymph)

'Monkey'
R. ziemanni (1)

Erythrocebus patas **Patas monkey**
R. moucheti (1)

Papio cynocephalus **Yellow baboon**
R. evertsi evertsi (1)
R. pulchellus (1)

Papio hamadryas **Hamadryas baboon**
R. humeralis (1)

Papio ursinus **Chacma baboon**
R. appendiculatus (3, including immatures)
R. evertsi evertsi (2, including immatures)
R. gertrudae (3)

Papio **sp.** **Baboon**
R. appendiculatus (1, immatures)
R. evertsi evertsi (1)
R. praetextatus (2, unconfirmed)
R. pulchellus (1)

'Baboon'
R. appendiculatus (1, immatures)
R. evertsi evertsi (1)

Colobus **sp.** **Colobus monkey**
R. evertsi evertsi (1, larva)

Family Hominidae

Gorilla gorilla **Gorilla**
R. appendiculatus (1)

Order Carnivora

Family Canidae

Canis adustus **Side-striped jackal**
R. appendiculatus (3, including immatures)
R. evertsi evertsi (2, immatures)
R. longus (1)
R. lunulatus (3)
R. muhsamae (3)
R. planus (1)
R. praetextatus (3, unconfirmed)
R. simus (1)
R. sulcatus (1, unconfirmed)
R. tricuspis (3)
R. turanicus (2)

Canis aureus **Golden jackal**
R. armatus (2)
R. cuspidatus (1, immatures)
R. praetextatus (2, unconfirmed)
R. pravus (2)
R. sulcatus (1, unconfirmed)

Canis mesomelas **Black-backed jackal**
R. appendiculatus (7, including immatures)
R. armatus (2, including nymphs; 1, unconfirmed)
R. evertsi evertsi (8, immatures)
R. gertrudae (1)
R. humeralis (1)
R. lunulatus (1)
R. muehlensi (4, immatures)
R. praetextatus (4, unconfirmed)
R. pravus (4)
R. pulchellus (3, including nymphs)
R. simus (7)
R. theileri (1)
R. tricuspis (1)
R. turanicus (2)
R. zambeziensis (2, including immmatures)

Canis **spp.**
R. appendiculatus (9, including immatures)
R. evertsi evertsi (3, including nymphs)
R. guilhoni (6)
R. hurti (1)
R. lunulatus (2)
R. praetextatus (11, unconfirmed)
R. sanguineus (4)
R. sulcatus (6; 1, unconfirmed)
R. tricuspis (3)
R. turanicus (3)

'Jackal'
R. armatus (1)
R. cuspidatus (1)
R. evertsi evertsi (1)
R. muhsamae (2)
R. pravus (2)
R. simus (10)
R. theileri (1, nymphs)

Lycaon pictus **Hunting dog**
R. appendiculatus (2)
R. armatus (2)
R. evertsi evertsi (1, nymphs)
R. praetextatus (6, unconfirmed)
R. pulchellus (2)
R. senegalensis (1)
R. simus (5)
R. supertritus (1)
R. zambeziensis (3, including immatures)

Otocyon megalotis **Bat-eared fox**
R. appendiculatus (2, including immatures)
R. armatus (2, nymphs; 1, unconfirmed)
R. camicasi (1)
R. gertrudae (1)
R. humeralis (1)
R. longus (1)
R. praetextatus (4, unconfirmed)
R. pravus (1)
R. pulchellus (2, nymphs)
R. tricuspis (1)
R. turanicus (1)

Vulpes chama **Cape fox**
R. capensis (1)
R. gertrudae (4)
R. simus (1)
R. sulcatus (1, unconfirmed)
R. theileri (1)
R. tricuspis (1)

Vulpes pallida **Pale fox**
R. guilhoni (1)

Vulpes rueppelli **Sand fox**
R. humeralis (1)

Family Felidae

Acinonyx jubatus **Cheetah**
R. appendiculatus (7, including immatures)
R. armatus (3)
R. camicasi (1)
R. carnivoralis (1)
R. compositus (1)
R. evertsi evertsi (3, including larvae)
R. follis (1)
R. maculatus (2)
R. praetextatus (6, unconfirmed)
R. pulchellus (1)
R. simus (7)
R. turanicus (3)
R. zambeziensis (5, including immatures)

Caracal caracal **Caracal**
R. armatus (1, nymphs)
R. arnoldi (2, immatures)
R. distinctus (4, immatures)
R. evertsi evertsi (23, immatures)
R. follis (3, larvae)
R. gertrudae (3)
R. glabroscutatum (16, immatures)
R. guilhoni (1)
R. lounsburyi (1)
R. lunulatus (1)
R. praetextatus (1, unconfirmed)
R. simus (5)
R. turanicus (1)

Felis lybica **African wild cat**
R. appendiculatus (2, immatures)
R. armatus (1, nymphs)
R. distinctus (1)
R. evertsi evertsi (2, immatures)
R. guilhoni (2)
R. praetextatus (1, unconfirmed)
R. pulchellus (2, immatures)

African wild cat (*cont.*)
R. sanguineus (1)
R. simus (2)
R. sulcatus (1, unconfirmed)
R. turanicus (5)
R. warburtoni (1)

Felis nigripes **Black-footed cat**
R. turanicus (1)

Leptailurus serval **Serval**
R. appendiculatus (3, including 1 nymph)
R. cuspidatus (1, nymph)
R. guilhoni (2)
R. lunulatus (1; 2, unconfirmed)
R. praetextatus (1, unconfirmed)
R. sanguineus (1)
R. simus (1)
R. sulcatus (1)
R. turanicus (1)
R. ziemanni (1, 1 ♂ only)

Panthera leo **Lion**
R. appendiculatus (21, including immatures)
R. aquatilis (1)
R. armatus (3; 2, unconfirmed)
R. carnivoralis (31)
R. compositus (10)
R. evertsi evertsi (7, including larvae)
R. evertsi mimeticus (2, including immatures)
R. exophthalmos (1)
R. humeralis (2)
R. hurti (3)
R. jeanneli (1)
R. kochi (1)
R. longus (1)
R. lunulatus (6)
R. maculatus (1)
R. masseyi (5)
R. muhsamae (2)
R. praetextatus (1; 37, unconfirmed)
R. pravus (2)
R. sp. near *pravus* (2)
R. pulchellus (7, 2 including immatures)
R. sp. near *punctatus* (1)
R. senegalensis (1)
R. simus (29)
R. sulcatus (2; 1, unconfirmed)
R. tricuspis (1)
R. turanicus (13)
R. zambeziensis (21, including immatures)
R. zumpti (1)

Panthera pardus **Leopard**
R. appendiculatus (6, including immatures)
R. aquatilis (1)
R. armatus (1)
R. carnivoralis (9; 2, unconfirmed)
R. compositus (1)
R. cuspidatus (1, immatures)
R. evertsi evertsi (4)
R. evertsi mimeticus (1)
R. gertrudae (1)
R. hurti (1)
R. kochi (1)
R. longus (1)
R. lunulatus (2)
R. masseyi (2)
R. muhsamae (1)
R. praetextatus (10, unconfirmed)
R. pulchellus (2, including nymphs)
R. sp. near *punctatus* (2)
R. senegalensis (1)
R. serranoi (1)
R. simus (6)
R. sulcatus (7)
R. tricuspis (1)
R. turanicus (7)
R. zambeziensis (12, including immatures)
R. ziemanni (4, one with 1 nymph (presumed))

Family Herpestidae

Atilax paludinosus **Marsh mongoose**
R. appendiculatus (1, nymph)
R. evertsi evertsi (1, immatures)
R. masseyi (1)
R. praetextatus (1, unconfirmed)
R. simus (1)
R. zambeziensis (1, immatures)

Cynictis penicillata **Yellow mongoose**
R. distinctus (1, nymphs)
R. theileri (28, including immatures)

Galerella sanguinea **Slender mongoose**
R. evertsi evertsi (1, larvae)
R. muehlensi (2, immatures)
R. pulchellus (2)
R. zambeziensis (1, nymph)

Herpestes ichneumon **Egyptian mongoose**
R. appendiculatus (2, immatures)
R. cuspidatus (1, immatures)
R. turanicus (1)

Ichneumia albicauda **White-tailed mongoose**
R. appendiculatus (26, immatures, only 1 with an adult)
R. cuspidatus (1, immatures)
R. muehlensi (1, including immatures)
R. praetextatus (2, unconfirmed)
R. pulchellus (1)
R. simus (1, nymphs)
R. zambeziensis (2, immatures)

Mungos mungo **Banded mongoose**
R. appendiculatus (3, immatures)
R. cuspidatus (2, nymphs)
R. evertsi evertsi (1, immatures)
R. longus (1)
R. senegalensis (1, nymph)
R. simus (1)
R. sulcatus (1, unconfirmed)
R. zambeziensis (2, immatures)

'Lesser mongoose'
R. muehlensi (1, nymphs)

Suricata suricatta **Meercat**
R. appendiculatus (2, including larvae)
R. evertsi evertsi (1, immatures)
R. theileri (8, including immatures)

'Meercat'
R. theileri (1, nymph)

Family Hyaenidae
Crocuta crocuta **Spotted hyaena**
R. appendiculatus (1, nymph)
R. armatus (1)
R. carnivoralis (3)
R. cuspidatus (2, including nymphs)
R. guilhoni (4)
R. longus (1)
R. maculatus (2)
R. muhsamae (1)
R. praetextatus (8, unconfirmed)
R. simus (9)
R. zambeziensis (3, immatures)

Hyaena hyaena **Striped hyaena**
R. appendiculatus (1)
R. armatus (1)
R. bergeoni (2)
R. carnivoralis (1)
R. cuspidatus (2, including immatures)
R. praetextatus (5, unconfirmed)
R. pulchellus (4, including nymphs)
R. supertritus (3)

Parahyaena brunnea **Brown hyaena**
R. appendiculatus (12)
R. sanguineus (1)
R. simus (13)
R. turanicus (1)
R. zambeziensis (1)

'Hyaena'
R. appendiculatus (1)
R. armatus (1)
R. carnivoralis (1)
R. cuspidatus (1)
R. muehlensi (1, larvae)
R. muhsamae (2)
R. praetextatus (2; 6, unconfirmed)
R. pulchellus (3)
R. simus (2)
R. supertritus (1)

Family Protelidae
Proteles cristatus **Aardwolf**
R. evertsi evertsi (1, larvae)
R. praetextatus (4, unconfirmed)
R. pulchellus (2, including a nymph)
R. simus (1)
R. tricuspis (1)
R. turanicus (1)

Family Mustelidae
'Otter'
R. ziemanni (3)

Mellivora capensis **Ratel**
R. appendiculatus (1, including immatures)
R. evertsi evertsi (1, including immatures)
R. simus (1)

Ictonyx striatus **Zorilla**
R. appendiculatus (3, immatures)
R. armatus (1)
R. guilhoni (2)
R. sanguineus (1)

Family Viverridae

Civettictis civetta **African civet**
R. appendiculatus (6, including immatures)
R. compositus (1)
R. duttoni (1)
R. evertsi evertsi (1, larvae)
R. guilhoni (1)
R. kochi (1, nymph)
R. lunulatus (1, unconfirmed)
R. maculatus (2, immatures)
R. moucheti (1)
R. muhsamae (4)
R. praetextatus (8, unconfirmed)
R. sanguineus (2)
R. simus (15)
R. sulcatus (2; 3, unconfirmed)
R. turanicus (6)
R. zambeziensis (3, including immatures)
R. ziemanni (2)
R. zumpti (1)

Genetta genetta **Small-spotted genet**
R. appendiculatus (2, nymphs)
R. praetextatus (1, unconfirmed)
R. pulchellus (1)
R. senegalensis (1)

Genetta rubiginosa **Rusty-spotted genet**
R. sulcatus (1, unconfirmed)

Genetta tigrina **Large-spotted genet**
R. appendiculatus (5, including immatures)
R. pulchellus (1)

***Genetta* spp.**
R. appendiculatus (3, including immatures)
R. evertsi evertsi (1, larvae)
R. hurti (1)
R. pravus (2, including nymphs)
R. sulcatus (1)

'Genet'
R. simus (1)
R. turanicus (4)

Order Proboscidea

Family Elephantidae

Loxodonta africana **African elephant**
R. appendiculatus (2)
R. dux (2)
R. evertsi evertsi (2)
R. humeralis (28)
R. kochi (1)
R. longus (1)
R. maculatus (33)
R. muehlensi (1)
R. planus (1)
R. praetextatus (12, unconfirmed)
R. pravus (3)
R. pulchellus (3)
R. senegalensis (3)
R. simus (1)
R. ziemanni (1, 1 ♀ only)

Order Perissodactyla

Family Equidae

Equus burchellii **Burchell's zebra**
R. appendiculatus (54, including immatures)
R. evertsi evertsi (170, including immatures)
R. evertsi mimeticus (12, including immatures)
R. exophthalmos (1)
R. jeanneli (1)
R. kochi (3)
R. longus (2)
R. lunulatus (2)
R. maculatus (2)
R. masseyi (2)
R. muehlensi (3, including immatures)
R. praetextatus (25, unconfirmed)
R. pravus (3)
R. sp. near *pravus* (1)

Burchell's zebra (*cont.*)
R. pulchellus (41, including immatures)
R. punctatus (2)
R. sp. near *punctatus* (2)
R. sculptus (5)
R. simus (27)
R. supertritus (1)
R. turanicus (4)
R. zambeziensis (4, including nymphs)
R. zumpti (1)

Equus grevyi **Grevy's zebra**
R. camicasi (2)
R. evertsi evertsi (11, including nymphs)
R. praetextatus (2, unconfirmed)
R. pulchellus (16, including immatures)

Equus zebra **Mountain zebra**
R. arnoldi (1)
R. capensis (1)
R. evertsi evertsi (20, including immatures)
R. evertsi mimeticus (24, including immatures)
R. follis (7)
R. gertrudae (4)
R. glabroscutatum (16, including immatures)
R. nitens (1)

Equus **spp.**
R. appendiculatus (3, including immatures)
R. camicasi (1)

'Zebra'
R. evertsi evertsi (10, including nymphs)
R. evertsi mimeticus (15, including immatures)
R. pulchellus (11)
R. supertritus (2)

Family Rhinocerotidae

Ceratotherium simum **White rhinoceros**
R. appendiculatus (7, including nymphs)
R. evertsi evertsi (1)
R. evertsi mimeticus (1)
R. maculatus (4)
R. simus (9)
R. zambeziensis (1, nymph)

Diceros bicornis **Black rhinoceros**
R. appendiculatus (4, including immatures)
R. compositus (4)
R. evertsi mimeticus (1)
R. humeralis (10)
R. hurti (3)
R. jeanneli (4)
R. kochi (1)
R. longus (1)
R. lunulatus (1)
R. maculatus (12, including immatures)
R. muehlensi (6, including immatures)
R. praetextatus (3, unconfirmed)
R. pravus (1; 1, unconfirmed)
R. pulchellus (34, 1 includes nymphs)
R. senegalensis (1)
R. simus (4)
R. zumpti (1)

'Rhinoceros'
R. appendiculatus (1, immatures)
R. maculatus (5, including immatures)
R. muehlensi (1, immatures)
R. simus (1)

Order Hyracoidea

Family Procaviidae *Hyraxes, dassies*

Dendrohyrax arboreus **Tree hyrax**
R. distinctus (1, including immatures)
R. serranoi (1)

Heterohyrax brucei **Yellow-spotted rock hyrax**
R. appendiculatus (2, immatures)
R. carnivoralis (2, nymphs)
R. distinctus (10, including immatures)
R. evertsi evertsi (2, including nymphs)
R. serranoi (1)

Procavia capensis **Rock hyrax**
R. appendiculatus (4, including immatures)
R. arnoldi (78, immatures; 1 with ♀♀)
R. boueti (2)
R. carnivoralis (2, nymphs)
R. distinctus (153, including immatures)
R. evertsi evertsi (5, including immatures)

Rock hyrax (*cont.*)
R. glabroscutatum (6, immatures, 1 ♂)
R. sp. near *pravus* (2)
R. serranoi (1)
R. supertritus (1)

***Procavia* sp.**
R. distinctus (15, including nymphs)

'Dassie'
R. distinctus (12, including nymphs)

Order Tubulidentata

Family Orycteropodidae

Orycteropus afer **Aardvark (Antbear)**
R. camicasi (1)
R. cuspidatus (5)
R. longus (2)
R. lunulatus (3)
R. masseyi (1)
R. planus (1)
R. praetextatus (2, unconfirmed)
R. pravus (1)
R. pulchellus (2)
R. sanguineus (1)
R. simus (1)
R. sulcatus (1, unconfirmed))
R. tricuspis (1)
R. turanicus (1)
R. zambeziensis (1)

Order Artiodactyla

Family Suidae

Phacochoerus aethiopicus **Somali warthog**
R. humeralis (1)

Phacochoerus africanus **Warthog**
R. appendiculatus (70, including immatures)
R. camicasi (1)
R. complanatus (7)
R. compositus (7)
R. cuspidatus (36, including immatures)
R. dux (2)
R. evertsi evertsi (19, including immatures)
R. evertsi mimeticus (8, including immatures)
R. exophthalmos (9)
R. guilhoni (6)
R. humeralis (2)
R. hurti (3)
R. jeanneli (10)
R. kochi (9)
R. longiceps (4)
R. longus (44)
R. lunulatus (21; 3, unconfirmed)
R. maculatus (22, including immatures)
R. masseyi (2)
R. muehlensi (2, including immatures)
R. muhsamae (11)
R. planus (9)
R. praetextatus (1; 42, unconfirmed)
R. pravus (5; 3, unconfirmed)
R. sp. near *pravus* (2)
R. pulchellus (19)
R. sp. near *punctatus* (4)
R. sanguineus (1)
R. senegalensis (27)
R. simus (124)
R. supertritus (1)
R. tricuspis (3)
R. turanicus (4)
R. zambeziensis (40, including immatures)
R. ziemanni (3)
R. zumpti (3)

Hylochoerus meinertzhageni **Forest hog**
R. appendiculatus (1, immatures)
R. bequaerti (1)
R. complanatus (1)
R. compositus (1)
R. dux (1)
R. jeanneli (2)
R. longus (3)
R. lunulatus (1)
R. praetextatus (2, unconfirmed)

Potamochoerus larvatus **Bushpig**
R. appendiculatus (17, including immatures)
R. bequaerti (2)
R. complanatus (1)
R. compositus (11)
R. dux (2)
R. evertsi evertsi (1)
R. follis (1)

Bushpig (*cont.*)
R. humeralis (1)
R. hurti (1)
R. jeanneli (6)
R. kochi (27)
R. longus (9)
R. lunulatus (13; 7, unconfirmed)
R. maculatus (20, including immatures)
R. masseyi (11)
R. muehlensi (11, including nymphs)
R. planus (15)
R. praetextatus (11, unconfirmed)
R. pulchellus (2)
R. sp. near *punctatus* (3)
R. simus (23)
R. zambeziensis (2)
R. zumpti (9; 2, unconfirmed)

Potamochoerus porcus **Red river hog**
R. complanatus (35)
R. cuspidatus (1)
R. longus (7)
R. muhsamae (1)
R. senegalensis (2)
R. ziemanni (8)

***Potamochoerus* sp.**
R. longus (2)

'Wild pig'
R. bequaerti (1)
R. complanatus (4)
R. hurti (1)
R. jeanneli (1)
R. longus (1)
R. maculatus (1)
R. muhsamae (1)
R. planus (1)
R. praetextatus (1; 1, unconfirmed)
R. pulchellus (1)
R. senegalensis (1)
R. ziemanni (1, 1 ♀ only)

'Pig/hog'
R. maculatus (3)

Family Hippopotamidae
Hippopotamus amphibius **Hippopotamus**
R. praetextatus (1)

Family Giraffidae
Giraffa camelopardalis **Giraffe**
R. appendiculatus (14, including immatures)
R. compositus (1)
R. evertsi evertsi (46, including immatures)
R. evertsi mimeticus (12, including immatures)
R. guilhoni (1)
R. hurti (1)
R. kochi (2)
R. longiceps (1)
R. longicoxatus (9)
R. lunulatus (1)
R. muehlensi (6)
R. praetextatus (4, unconfirmed)
R. pravus (9)
R.* sp. near *pravus (5)
R. pulchellus (8)
R. sanguineus (1)
R. sculptus (1)
R. senegalensis (1)
R. simus (8)
R. supertritus (1)
R. zambeziensis (3)

Okapia johnstoni **Okapi**
R. ziemanni (2)

Family Bovidae
Aepyceros melampus **Impala**
R. appendiculatus (386, including immatures)
R. duttoni (2, including a nymph)
R. evertsi evertsi (472, including immatures)
R. evertsi mimeticus (5, including immatures)
R. exophthalmos (2)
R. kochi (24, including immatures)
R. lunulatus (8, including 2 with larvae)
R. maculatus (4, immatures)
R. muehlensi (163, including immatures)
R. planus (1)
R. pravus (13; 11, unconfirmed)
R. sp. near *pravus* (6; 1, unconfirmed)
R. pulchellus (13, including immatures)

Impala (*cont.*)
R. sp. near *punctatus* (8)
R. simus (7)
R. sulcatus (1)
R. tricuspis (1)
R. zambeziensis (249, including immatures)

Aepyceros melampus petersi **Black-faced impala**
R. gertrudae (1)

Alcelaphus buselaphus caama **Red hartebeest**
R. evertsi evertsi (13, including immatures)
R. glabroscutatum (3, including immatures)
R. simus (1)

Alcelaphus buselaphus cokii **Coke's hartebeest (kongoni)**
R. appendiculatus (1, including 1 nymph)
R. evertsi evertsi (6, including immatures)
R. humeralis (1)
R. praetextatus (1, unconfirmed)
R. pravus (1; 10, unconfirmed)
R. pulchellus (30, 2 include nymphs)
R. supertritus (1)

Alcelaphus buselaphus jacksoni **Jackson's hartebeest**
R. evertsi evertsi (4, including larvae)
R. lunulatus (1)
R. pravus (1)

Alcelaphus buselaphus cokii × jacksoni **'Kenya hartebeest'**
R. pulchellus (3, including immatures)

Alcelaphus buselaphus lelwel **Lelwel hartebeest**
R. lunulatus (1, unconfirmed)

Alcelaphus buselaphus major **Kanki**
R. senegalensis (1)

***Alcelaphus* sp.**
R. appendiculatus (1, including immatures)

'Hartebeest'
R. evertsi mimeticus (1)
R. pulchellus (2)
R. sp. near *punctatus* (1)
R. sculptus (1)
R. supertritus (2)

Connochaetes gnou **Black wildebeest**
R. evertsi evertsi (25, including immatures)
R. follis (2)
R. gertrudae (2)
R. glabroscutatum (9, including immatures)
R. lounsburyi (1)
R. neumanni (1)

Connochaetes taurinus **Blue wildebeest**
R. appendiculatus (49, including immatures)
R. evertsi evertsi (172, including immatures)
R. evertsi mimeticus (11, including immatures)
R. kochi (1)
R. longus (1)
R. lunulatus (1)
R. maculatus (2)
R. praetextatus (3, unconfirmed)
R. pravus (1; 1, unconfirmed)
R. simus (2)
R. zambeziensis (14, including immatures)

'Wildebeest'
R. evertsi evertsi (1)

Damaliscus lunatus lunatus **Tsessebe**
R. appendiculatus (2, including immatures)
R. evertsi evertsi (9, including immatures)
R. sp. near *pravus* (2)

Damaliscus lunatus topi **Topi**
R. evertsi evertsi (5, including nymphs)
R. interventus (1)
R. praetextatus (3, unconfirmed)
R. pulchellus (5)
R. punctatus (1)

Damaliscus lunatus tiang **Tiang**
R. lunulatus (1, unconfirmed)

Damaliscus lunatus korrigum **Korrigum**
R. evertsi evertsi (1, immatures)
R. guilhoni (1)

Damaliscus pygargus dorcas **Bontebok**
R. capensis (1)
R. evertsi evertsi (9, including immatures)
R. gertrudae (1)

Bontebok (*cont.*)
- ***R. glabroscutatum*** (41, including immatures)
- *R. lounsburyi* (6)
- ***R. nitens*** (55, including immatures)

Damaliscus pygargus phillipsi **Blesbok**
- *R. appendiculatus* (5, including immatures)
- *R. evertsi evertsi* (14, including immatures)
- *R. glabroscutatum* (2, including nymphs)

Sigmoceros lichtensteinii **Lichtenstein's hartebeest**
- *R. appendiculatus* (3, including nymphs)
- *R. compositus* (1)
- *R. evertsi evertsi* (5, including immatures)
- *R. kochi* (3)
- *R. longus* (1)
- *R. lunulatus* (2)
- *R. maculatus* (1)
- *R. muehlensi* (1)
- *R. senegalensis* (1)
- *R. sulcatus* (1; 1, unconfirmed)
- *R. supertritus* (1)
- *R. zumpti* (1)

Ammodorcas clarkei **Dibatag**
- *R. pulchellus* (1)

Antidorcas marsupialis **Springbok**
- *R. appendiculatus* (2, including 1 nymph)
- *R. evertsi evertsi* (33, including immatures)
- *R. evertsi mimeticus* (5, including larvae)
- ***R. exophthalmos*** (21)
- *R. follis* (1)
- *R. gertrudae* (2)
- *R. glabroscutatum* (17, including immatures)
- *R. neumanni* (1)
- *R. nitens* (6, including immatures)

Gazella dorcas **Dorcas gazelle**
- *R. praetextatus* (1, unconfirmed)

Gazella granti **Grant's gazelle**
- *R. appendiculatus* (4, including immatures)
- *R. armatus* (1, nymph; 2, unconfirmed)
- *R. evertsi evertsi* (16, including immatures)
- *R. interventus* (1)
- *R. muehlensi* (1)
- *R. praetextatus* (1, unconfirmed)
- ***R. pravus*** (20; 4, unconfirmed)
- *R. pulchellus* (15, including nymphs)
- *R. turanicus* (3)

Gazella rufifrons **Red-fronted gazelle**
- *R. cuspidatus* (1, immatures)
- *R. guilhoni* (2)

Gazella soemmerringii **Soemmerring's gazelle**
- *R. pravus* (2)
- *R. pulchellus* (7, 1 includes nymphs)

Gazella spekei **Speke's gazelle**
- *R. humeralis* (1)
- *R. pulchellus* (1)

Gazella thomsonii **Thomson's gazelle**
- *R. appendiculatus* (2)
- *R. evertsi evertsi* (36, including immatures)
- *R. praetextatus* (2, unconfirmed)
- ***R. pravus*** (25; 5, unconfirmed)
- *R. pulchellus* (12, including immatures)
- *R. turanicus* (3)

'Gazelle'
- *R. pravus* (1, unconfirmed)
- *R. pulchellus* (4)
- *R. sulcatus* (1, unconfirmed)

Litocranius walleri **Gerenuk**
- *R. camicasi* (1)
- *R. evertsi evertsi* (4)
- *R. muehlensi* (2, including immatures)
- *R. pravus* (5; 4, unconfirmed)
- *R. pulchellus* (10, 1 includes nymphs)
- *R. punctatus* (1)

Madoqua guentheri **Günther's dik-dik**
- *R. pravus* (6)
- *R. pulchellus* (2, nymphs)
- *R. sanguineus* (1)

Madoqua kirkii **Kirk's dik-dik**
- *R. appendiculatus* (7, including nymphs)
- *R. evertsi evertsi* (4, including nymphs)
- *R. evertsi mimeticus* (1, including immatures)
- *R. kochi* (2)
- *R. pravus* (2)
- *R. punctatus* (1)

Madoqua piacentinii **Silver dik-dik**
R. pravus (1)
R. pulchellus (1)

Madoqua **spp.** **'Dik-dik'**
R. evertsi evertsi (1)
R. pravus (1; 3, unconfirmed)

Neotragus moschatus **Suni**
R. appendiculatus (1)
R. kochi (3)
R. maculatus (3, immatures)
R. muehlensi (1)
R. sp. near *pravus* (1, unconfirmed)

Neotragus pygmaeus **Royal antelope**
R. simpsoni (2)
R. ziemanni (2)

Oreotragus oreotragus **Klipspringer**
R. appendiculatus (9, including nymphs)
R. arnoldi (1)
R. evertsi evertsi (1, immatures)
R. kochi (6)
R. longiceps (1)
R. oreotragi (4)
R. pravus (3; 1, unconfirmed)
R. punctatus (4)
R. sp. near *punctatus* (1)
R. serranoi (2)
R. simus (7)
R. zambeziensis (1, nymph)

Ourebia ourebi **Oribi**
R. appendiculatus (7, including nymphs)
R. cuspidatus (1, immatures)
R. evertsi evertsi (6, including immatures)
R. interventus (1)
R. kochi (1)
R. lunulatus (3)
R. maculatus (1, nymphs)
R. muhsamae (2)
R. pravus (4, including 1 nymph)
R. sp. near *pravus* (2, unconfirmed)
R. punctatus (2)
R. senegalensis (1)
R. sulcatus (1; 1, unconfirmed)
R. ziemanni (1)
R. zumpti (1)

Raphicerus campestris **Steenbok**
R. appendiculatus (6, including nymphs)
R. evertsi evertsi (16, including immatures)
R. evertsi mimeticus (4, including immatures)
R. exophthalmos (5)
R. glabroscutatum (2, including immatures)
R. kochi (1)
R. muehlensi (3)
R. praetextatus (1, unconfirmed)
R. pravus (5; 3, unconfirmed)
R. sp. near *pravus* (3; 1, unconfirmed)
R.* sp. near *punctatus (156, including nymphs)
R. simus (2)
R. tricuspis (17)
R. zambeziensis (1, nymph)
R. zumpti (1)

Raphicerus melanotis **Cape grysbok**
R. evertsi evertsi (1, nymphs)
R. interventus (1)
R. kochi (3)
R. sulcatus (1)

Raphicerus sharpei **Sharpe's grysbok**
R. evertsi evertsi (1)
R.* sp. near *punctatus (28)
R. tricuspis (1)

Syncerus caffer **African buffalo**
R. appendiculatus (78, including immatures)
R. bequaerti (5)
R. complanatus (5)
R. compositus (72)
R. duttoni (commonly parasitized, including immatures)
R. dux (21)
R. evertsi evertsi (181, including immatures)
R. evertsi mimeticus (2)
R. follis (2)
R. gertrudae (1)
R. glabroscutatum (2, including nymphs)
R. humeralis (2)
R. hurti (30)

African buffalo (*cont.*)
- ***R. jeanneli*** (20)
- *R. kochi* (17)
- ***R. longus*** (109)
- ***R. lunulatus*** (41)
- ***R. maculatus*** (23, including immatures)
- *R. masseyi* (4)
- *R. muehlensi* (10, including immatures)
- *R. muhsamae* (2)
- *R. planus* (6)
- *R. praetextatus* (4; 52, unconfirmed)
- *R. pravus* (4; 3, unconfirmed)
- *R.* sp. near *pravus* (1, unconfirmed)
- ***R. pseudolongus*** (48)
- *R. pulchellus* (20, 1 includes nymphs)
- *R.* sp. near *punctatus* (8)
- *R. sculptus* (1)
- ***R. senegalensis*** (23)
- *R. simpsoni* (1)
- ***R. simus*** (56)
- *R. sulcatus* (1)
- ***R. supertritus*** (21)
- *R. zambeziensis* (19, including immatures)
- ***R. ziemanni*** (13)
- *R. zumpti* (1; 1, unconfirmed)

Taurotragus derbianus **Giant eland**
- *R. supertritus* (1)

Taurotragus oryx **Eland**
- ***R. appendiculatus*** (53, including immatures)
- *R. camicasi* (2)
- ***R. capensis*** (3)
- *R. compositus* (10)
- *R. duttoni* (1)
- ***R. evertsi evertsi*** (80, including immatures)
- *R. evertsi mimeticus* (6)
- *R. exophthalmos* (4)
- ***R. follis*** (34)
- *R. gertrudae* (7)
- ***R. glabroscutatum*** (16, including immatures)
- *R. humeralis* (2)
- *R. hurti* (4)
- *R. jeanneli* (4)
- *R. kochi* (6)
- *R. longus* (3)
- *R. lounsburyi* (2)
- *R. lunulatus* (4)
- *R. maculatus* (3)
- *R. masseyi* (3)
- *R. muehlensi* (4)
- *R. praetextatus* (9, unconfirmed)
- *R. pravus* (4; 9, unconfirmed)
- *R.* sp. near *pravus* (4; 1, unconfirmed)
- ***R. pulchellus*** (33, 2 include nymphs)
- *R.* sp. near *punctatus* (2)
- *R. senegalensis* (1)
- *R. simus* (19)
- ***R. supertritus*** (8)
- *R. turanicus* (1)
- *R. zambeziensis* (9, including nymphs)

***Taurotragus* sp.** **'Eland'** (Equatoria, Sudan)
- *R. turanicus* (1)

Tragelaphus angasii **Nyala**
- ***R. appendiculatus*** (90, including immatures)
- *R. evertsi evertsi* (32, including immatures)
- *R. kochi* (8, including nymphs)
- *R. maculatus* (83, including immatures)
- *R. masseyi* (1)
- ***R. muehlensi*** (107, including immatures)
- *R.* sp. near *pravus* (1; 1, unconfirmed)
- *R.* sp. near *punctatus* (1)
- *R. simus* (3)
- *R. zambeziensis* (2, nymphs)
- *R. zumpti* (1)

Tragelaphus buxtoni ***Mountain nyala***
- *R. bergeoni* (3)
- *R. longus* (1)
- *R. lunulatus* (1)
- *R. praetextatus* (1, unconfirmed)

Tragelaphus eurycerus **Bongo**
- *R. jeanneli* (2)
- *R. longus* (1)
- *R. senegalensis* (1)
- ***R. ziemanni*** (2)

Tragelaphus imberbis **Lesser kudu**
- *R. evertsi evertsi* (1)
- *R. maculatus* (1, nymph)
- *R. muehlensi* (5)

Lesser kudu (*cont.*)
- *R. pravus* (4; 4, unconfirmed)
- *R. pulchellus* (2)

Tragelaphus scriptus **Bushbuck**
- *R. appendiculatus* (31, including immatures)
- *R. bergeoni* (1)
- *R. complanatus* (1)
- *R. compositus* (4)
- *R. evertsi evertsi* (28, including immatures)
- *R. follis* (3)
- *R. humeralis* (2)
- *R. hurti* (9)
- *R. interventus* (1)
- *R. jeanneli* (5)
- ***R. kochi*** (17, including nymphs)
- *R. lunulatus* (6)
- *R. maculatus* (8, including immatures)
- *R. masseyi* (3)
- ***R. muehlensi*** (15, including immatures)
- *R. muhsamae* (1)
- *R. praetextatus* (3, unconfirmed)
- *R. pravus* (2; 4, unconfirmed)
- *R.* sp. near *pravus* (3, unconfirmed)
- *R. pulchellus* (1)
- *R.* sp. near *punctatus* (3)
- *R. senegalensis* (1)
- *R. simus* (3)
- *R. zambeziensis* (13, including immatures)
- ***R. ziemanni*** (14)

Tragelaphus spekii **Sitatunga**
- *R. appendiculatus* (1)
- ***R. aquatilis*** (7)
- *R. compositus* (2)

Tragelaphus strepsiceros **Greater kudu**
- ***R. appendiculatus*** (188, including immatures)
- *R. bergeoni* (1)
- *R. evertsi evertsi* (172, including immatures)
- *R. evertsi mimeticus* (33, including immatures)
- ***R. exophthalmos*** (31)
- *R. follis* (1)
- *R. gertrudae* (1)
- ***R. glabroscutatum*** (60, including immatures)
- *R. jeanneli* (1)
- ***R. kochi*** (21, including nymphs)
- *R. lunulatus* (8)
- *R. maculatus* (6, including immatures)
- *R. masseyi* (1)
- *R. muehlensi* (11, including immatures)
- *R. praetextatus* (1, unconfirmed)
- *R. pravus* (4)
- *R.* sp. near *pravus* (2; 2, unconfirmed)
- *R. pulchellus* (6)
- *R. punctatus* (2)
- *R.* sp. near *punctatus* (34)
- *R. simus* (18)
- *R. sulcatus* (1)
- *R. supertritus* (5)
- *R. tricuspis* (1)
- *R. turanicus* (1)
- ***R. zambeziensis*** (117, including immatures)
- *R. zumpti* (1)

'Kudu'
- *R. evertsi evertsi* (1)
- *R. kochi* (3)
- *R. pulchellus* (2)

Tragelaphus **spp.**
- *R. hurti* (1)
- *R. pravus* (1)
- *R.* sp. near *punctatus* (8)
- *R. turanicus* (1)
- *R. ziemanni* (1)

Ammotragus lervia **Barbary sheep**
- *R. fulvus* (+)

Cephalophus dorsalis **Bay duiker**
- *R. appendiculatus* (1, including immatures)
- *R. senegalensis* (1)
- *R. ziemanni* (2)

Cephalophus maxwelli **Maxwell's duiker**
- *R. ziemanni* (1)

Cephalophus monticola **Blue duiker**
- *R. appendiculatus* (1, immatures)
- *R. evertsi evertsi* (1, immatures)
- *R. ziemanni* (1)

Cephalophus natalensis **Red forest duiker**

- *R. appendiculatus* (7, including immatures)
- *R. evertsi evertsi* (8, including immatures)
- *R. kochi* (3)
- *R. maculatus* (16, immatures)
- *R. muehlensi* (28, including immatures)
- *R. praetextatus* (2, unconfirmed)
- *R.* sp. near *pravus* (1, unconfirmed)
- *R. pulchellus* (3, including a nymph)
- *R. tricuspis* (1)

Cephalophus niger **Black duiker**

- *R. ziemanni* (3)

Cephalophus silvicultor **Yellow-backed duiker**

- *R. kochi* (1)

Cephalophus zebra **Banded duiker**

- *R. ziemanni* (1)

Cephalophus **spp.**

- *R. appendiculatus* (2, including immatures)
- *R. ziemanni* (1)

Sylvicapra grimmia **Common duiker**

- *R. appendiculatus* (37, including immatures)
- *R. compositus* (1)
- *R. duttoni* (1, including nymphs)
- *R. evertsi evertsi* (24, including immatures)
- *R. evertsi mimeticus* (3, including immatures)
- ***R. glabroscutatum*** (11, including immatures)
- *R. hurti* (1)
- *R. jeanneli* (1)
- *R. kochi* (7)
- *R. lunulatus* (4)
- *R. maculatus* (6, including immatures)
- *R. muehlensi* (6, including immatures)
- *R. planus* (1)
- *R. praetextatus* (1, unconfirmed)
- *R. pravus* (3)
- *R.* sp. near *pravus* (2, unconfirmed)
- ***R.* sp. near *punctatus*** (20)
- *R. senegalensis* (1)
- *R. simus* (3)
- *R. turanicus* (2)
- *R. zambeziensis* (3, nymphs)
- *R. zumpti* (1)

'Duiker'

- *R. appendiculatus* (1, nymph)
- *R. evertsi evertsi* (2)
- *R. jeanneli* (1)
- *R. maculatus* (1, immatures)
- *R. muehlensi* (7, including immatures)
- *R. oculatus* (1)
- *R. pravus* (1)
- *R. simpsoni* (2)
- *R. tricuspis* (1)

Hippotragus equinus **Roan antelope**

- *R. appendiculatus* (5)
- *R. compositus* (2)
- *R. evertsi evertsi* (10)
- *R. evertsi mimeticus* (4, including nymphs)
- *R. guilhoni* (2)
- *R. hurti* (1)
- *R. kochi* (6)
- *R. longus* (2)
- *R. lunulatus* (2; 1, unconfirmed)
- *R. muehlensi* (4)
- *R. muhsamae* (1)
- *R.* sp. near *pravus* (2)
- *R.* sp. near *punctatus* (1)
- *R. sculptus* (1)
- *R. senegalensis* (2)
- *R. sulcatus* (1, unconfirmed)
- *R. supertritus* (1)
- *R. turanicus* (1)
- *R. zambeziensis* (1)

Hippotragus niger **Sable antelope**

- *R. appendiculatus* (22, including nymphs)
- *R. evertsi evertsi* (37, including immatures)
- *R. kochi* (12)
- *R. longus* (4)
- *R. lunulatus* (1)
- *R. maculatus* (1)
- *R. muehlensi* (1)
- *R. praetextatus* (1, unconfirmed)
- *R.* sp. near *pravus* (1; 1, unconfirmed)
- *R. pulchellus* (1)
- *R. punctatus* (1)
- *R.* sp. near *punctatus* (3)
- *R. sculptus* (3)
- *R. simus* (4)

Sable antelope (*cont.*)
R. supertritus (6)
R. zambeziensis (7)

Oryx gazella **Gemsbok**
R. capensis (2)
R. evertsi evertsi (52, including immatures)
R. evertsi mimeticus (13, including nymphs)
R. exophthalmos (22)
R. follis (6)
R. gertrudae (11)
R. glabroscutatum (6, including immatures)
R. longiceps (2)
R. muehlensi (1)
R. neumanni (1)
R. praetextatus (4, unconfirmed)
R. pravus (10, 1 with nymphs only; 3, unconfirmed)
R. sp. near *pravus* (1)
R. pulchellus (38, including immatures)
R. simus (1)
R. tricuspis (1)
R. turanicus (1)
R. warburtoni (6)

Pelea capreolus **Grey rhebok**
R. evertsi evertsi (6, including immatures)
R. exophthalmos (1)
R. gertrudae (2)
R. glabroscutatum (44, including immatures)
R. lounsburyi (9)
R. neumanni (1)
R. nitens (62, including immatures)

Kobus ellipsiprymnus **Waterbuck**
R. appendiculatus (47, including immatures)
R. compositus (1)
R. evertsi evertsi (24, including immatures)
R. guilhoni (1)
R. humeralis (1)
R. kochi (1)
R. longus (3)
R. lunulatus (6)
R. maculatus (2)
R. muehlensi (2)
R. planus (1)
R. pravus (1; 2, unconfirmed)
R. sp. near *pravus* (1, unconfirmed)
R. pseudolongus (1)
R. pulchellus (6)
R. sp. near *punctatus* (1)
R. sculptus (1)
R. simus (1)
R. sulcatus (1)
R. zumpti (2)

Kobus kob **Kob**
R. appendiculatus (7)
R. evertsi evertsi (3)

Kobus leche **Lechwe**
R. appendiculatus (2, including immatures)
R. evertsi evertsi (8, including immatures)
R. turanicus (1)

Kobus vardonii **Puku**
R. compositus (1)
R. evertsi evertsi (2)

Kobus **spp.**
R. appendiculatus (1)
R. evertsi evertsi (1)

Redunca arundinum **Reedbuck**
R. appendiculatus (17, including nymphs)
R. evertsi evertsi (61, including immatures)
R. kochi (1)
R. lunulatus (1)
R. maculatus (7, including immatures)
R. muehlensi (12, including immatures)
R. simus (2)

Redunca fulvorufula **Mountain reedbuck**
R. appendiculatus (29, including immatures)
R. evertsi evertsi (25, including immatures)
R. exophthalmos (2)
R. glabroscutatum (21, including immatures)
R. lounsburyi (4)
R. neumanni (1)
R. oculatus (1, larvae)

Redunca redunca **Bohor reedbuck**
R. appendiculatus (2, including immatures)
R. muhsamae (1)
R. praetextatus (1, unconfirmed)
R. pravus (1)
R. senegalensis (1)

Redunca **spp.**
R. evertsi evertsi (1)
R. lunulatus (1)

'Antelope'
R. hurti (1)
R. interventus (1)
R. lunulatus (1)

Order Pholidota

Family Manidae

Manis gigantea **Giant ground pangolin**
R. senegalensis (1)

Manis temminckii **Temminck's ground pangolin**
R. appendiculatus (1, nymphs)
R. praetextatus (1, unconfirmed)
R. simus (5)
R. theileri (1)

Order Rodentia

Family Sciuridae

Funisciuris congicus **Kuhl's tree squirrel**
R. simus (1, immatures)

Funisciuris pyrrhopus **Cuvier's tree squirrel**
R. senegalensis (1, 1 nymph)

Paraxerus cepapi **Smith's bush squirrel**
R. evertsi evertsi (1, larva)
R. simus (1, nymph)
R. theileri (16, immatures)
R. zambeziensis (1, immatures)

Paraxerus ochraceus **Huet's bush squirrel**
R. evertsi evertsi (1)
R. pulchellus (1)

Xerus erythropus **Geoffroy's ground squirrel**
R. guilhoni (1, nymph)
R. muhsamae (3, nymphs; adults in nest)

Xerus inauris **Cape ground squirrel**
R. theileri (18, including immatures)

Xerus rutilus **Unstriped ground squirrel**
R. evertsi evertsi (1)
R. humeralis (1)
R. praetextatus (1, unconfirmed)
R. pravus (1)

Xerus **sp.**
R. theileri (8, including immatures)

'Ground squirrel'
R. praetextatus (1)

Family Muridae

Cricetomys gambianus **Giant Gambian rat**
R. simpsoni (3)

Saccostomus campestris **Short-tailed pouched mouse**
R. evertsi evertsi (1, larva)

Steatomys pratensis **Common fat mouse**
R. evertsi evertsi (1, immatures)
R. sulcatus (1)

Gerbillurus **sp.** **Southern pygmy gerbils**
R. simus (1, immatures)

Gerbillus campestris **Northern pygmy gerbil**
R. fulvus (+)

Gerbillus **sp.** **Northern pygmy gerbils**
R. armatus (1, 1♂)

Tatera leucogaster **Bushveld gerbil**
R. simus (2, immatures)

Tatera nigricauda **Black-tailed gerbil**
R. praetextatus (1, nymphs, unconfirmed)
R. pulchellus (1)

Taterillus **sp.** **Taterilline gerbils**
R. senegalensis (1, immatures)

'Gerbil'
R. praetextatus (1, nymphs, unconfirmed)

***Acomys* sp.** **African spiny mice**
R. pulchellus (1)

Aethomys chrysophilus **Red veld rat**
R. evertsi evertsi (1, immatures)
R. simus (37, immatures)

Aethomys namaquensis **Namaqua rock mouse**
R. distinctus (10, immatures)
R. evertsi evertsi (1, larva)
R. exophthalmos (1)
R. gertrudae (8, immatures)
R. neumanni (1, nymphs)
R. simus (6, immatures)
R. theileri (1, nymph, unconfirmed)
R. warburtoni (69, immatures)

Arvicanthis niloticus **Nile rat**
R. compositus (1, nymph)
R. muhsamae (2)

***Arvicanthis* spp.** **Nile rats**
R. compositus (1, nymph)
R. evertsi evertsi (1, larva)
R. guilhoni (4, nymphs)
R. praetextatus (1, immatures, unconfirmed)

'Marsh rat'
R. simpsoni (1)

Lemniscomys griselda **Single-striped mouse**
R. simus (2, nymphs)

Lemniscomys striatus **Striped grass rat**
R. appendiculatus (2, nymphs)
R. planus (1, nymphs)
R. praetextatus (1, nymphs, unconfirmed)
R. senegalensis (3; 1 ♂, 2 nymphs)

***Lemniscomys* sp.**
R. evertsi evertsi (1, immatures)

Lophuromys flavopunctatus **Brush-furred rat**
R. appendiculatus (2, immatures)
R. jeanneli (1, larva)
R. lunulatus (1)

Mastomys natalensis **Natal multimammate rat**
R. praetextatus (2, immatures, unconfirmed)
R. sp. near *punctatus* (1, larva)
R. simus (10, immatures)

***Mastomys* sp.** **Multimammate mice**
R. lunulatus (3, nymphs)

Mus minutoides **Pigmy mouse**
R. simus (1, immatures)

Pelomys campanae **Groove-toothed swamp rat**
R. simus (1, nymph)

Pelomys fallax **Greater creek rat**
R. compositus (1, nymph)

***Pelomys* sp.** **Groove-toothed swamp rats**
R. compositus (1, nymph)

Praomys jacksonii **Jackson's soft-furred rat**
R. compositus (1, larva)

Praomys tullbergi **Tullberg's soft-furred mouse**
R. complanatus (1)

Rattus rattus **Black rat**
R. praetextatus (1, unconfirmed)
R. pulchellus 1

'Rat'
R. praetextatus (2, immatures, unconfirmed)

Rhabdomys pumilio **Four-striped grass mouse**
R. arnoldi (1, immatures)
R. follis (65, immatures)
R. gertrudae (18, immatures)
R. glabroscutatum (1, larva)
R. lounsburyi (1, nymph)
R. praetextatus (1, unconfirmed)
R. simus (83, immatures)

'Field mouse'
R. arnoldi (1)
R. distinctus (1)

Otomys angoniensis **Angoni swamp rat**
R. hurti (1, larva)

Otomys irroratus **Swamp rat**
R. appendiculatus (5, larvae)
R. distinctus (1, larva)
R. follis (1, nymphs)

Otomys unisulcatus **Bush Karoo rat**
R. distinctus (1, larva)
R. gertrudae (1, nymph)

***Otomys* spp.**
R. capensis (2, nymphs)
R. jeanneli (1, nymph)
R. praetextatus (1, immatures, unconfirmed)
R. simus (1, nymph)

***Paratomys* sp.**
R. simus (1, immatures)

Tachyoryctes splendens **East African mole-rat**
R. jeanneli (1, larva)

'Murid rodents'
R. pseudolongus (1)
R. simus (5, immatures)

Family Pedetidae

Pedetes capensis **Spring hare**
R. appendiculatus (3, immatures)
R. distinctus (1, larva)
R. evertsi evertsi (5, immatures)
R. glabroscutatum (2, immatures)
R. oculatus (2, larvae)
R. sp. near *punctatus* (1)
R. sulcatus (1)
R. tricuspis (11)
R. warburtoni (14, nymphs)
R. zambeziensis (1, nymph)

Family Ctenodactylidae

Ctenodactylus gundi **Gundi**
R. fulvus (+, including immatures)

Family Hystricidae

Atherurus africanus **African brush-tailed porcupine**
R. simpsoni (6, nymphs)
R. ziemanni [2, both including 1 nymph (presumed)]

Hystrix africaeaustralis **South African porcupine**
R. appendiculatus (1, including immatures)
R. evertsi evertsi (1, nymph)
R. gertrudae (2)
R. kochi (1)
R. planus (2)
R. simus (7)
R. supertritus (3)
R. zambeziensis (1)
R. zumpti (1)

Hystrix cristata **Crested porcupine**
R. cuspidatus (6, including immatures)
R. evertsi evertsi (1)
R. lunulatus (1)
R. muhsamae (2)
R. planus (1)
R. praetextatus (2, unconfirmed)
R. senegalensis (1)

***Hystrix* sp.**
R. praetextatus (15, unconfirmed)

'Porcupine'
R. planus (2)

Family Thryonomyidae

Thryonomys gregorianus **Lesser cane rat**
R. simpsoni (3)

Thryonomys swinderianus **Greater cane rat**
R. appendiculatus (3, including immatures)
R. cuspidatus (1, including immatures)
R. sp. near *pravus* (1)
R. simpsoni (942, including immatures)

***Thryonomys* spp.** **Cane rat**
R. senegalensis (1, including nymphs)
R. simpsoni (10)

'Rodent'

R. muhsamae (1, nymphs)

Order Lagomorpha

Family Leporidae

Lepus capensis **Cape hare**

R. appendiculatus (16, immatures, 2 with adults)
R. armatus (12, immatures)
R. arnoldi (1)
R. camicasi (3)
R. evertsi evertsi (136, a few with adults)
R. exophthalmos (1)
R. gertrudae (1, nymph)
R. longus (1)
R. oculatus (20, including immatures)
R. planus (1, larvae)
R. praetextatus (2, unconfirmed)
R. pravus (99, including immatures; 2, unconfirmed)
R. sp. near *pravus* (1)
R. pulchellus (48, immatures, mostly nymphs)
R. punctatus (1)
R. sp. near *punctatus* (1)
R. senegalensis (1)
R. sulcatus (1; 3, unconfirmed)
R. tricuspis (3)
R. turanicus (1)
R. warburtoni (12, including immatures)
R. zambeziensis (1, nymph)

Lepus fagani **Ethiopian hare**

R. pulchellus (1)

Lepus saxatilis **Scrub hare**

R. appendiculatus (139, immatures, 3 with adults)
R. arnoldi (10, including immatures)
R. distinctus (9, immatures)
R. evertsi evertsi (445, a few with adults)
R. exophthalmos (32, including immatures)
R. gertrudae (4, nymphs)
R. glabroscutatum (68, immatures)
R. kochi (20, including immatures)
R. lunulatus (1, nymph)
R. maculatus (6, nymphs)
R. muehlensi (29, immatures)
R. nitens (20, including immatures)
R. oculatus (102, including immatures)
R.* sp. near *pravus (74, including immatures; 3, unconfirmed)
R. sanguineus (1)
R. simus (49, immatures)
R. sulcatus (3)
R. tricuspis (6)
R. turanicus (34, including immatures)
R. warburtoni (64, including immatures)
R. zambeziensis (272, immatures)

Lepus victoriae **Savanna hare**

R. appendiculatus (21, immatures, 1 with adults)
R. duttoni (1, including nymphs)
R. evertsi evertsi (1, nymphs)
R. kochi (7)
R. pravus (14, including immatures; 1, unconfirmed)
R. pulchellus (5, immatures)
R. punctatus (8)
R. sulcatus (2, unconfirmed)

***Lepus* spp.**

R. appendiculatus (3, immatures, 2 with adults)
R. evertsi evertsi (69, immatures)
R. muehlensi (1)
R. pravus (10, including immatures; 2, unconfirmed)
R.* sp. near *pravus (19; 1, unconfirmed)
R. tricuspis (5)

'Hare'

R. arnoldi (2)
R. evertsi evertsi (6, including immatures)
R. evertsi mimeticus (1)
R. exophthalmos (20)
R. guilhoni (4)
R. oculatus (9, including immatures)
R. praetextatus (1, unconfirmed)
R. pulchellus (1, nymphs)
R. sp. near *punctatus* (8, including 1 nymph)

'Hare' (*cont.*)
- ***R. sulcatus*** (39; 4, unconfirmed)
- ***R. turanicus*** (13)

Poelagus marjorita **Central African rabbit**
- *R. pravus* (1)
- *R. sulcatus* (1, unconfirmed)

Pronolagus randensis **Jameson's red rock rabbit**
- ***R. arnoldi*** (12, including a nymph)
- *R. distinctus* (2, including a larva)
- *R.* sp. near *punctatus* (1, larva)

Pronolagus rupestris **Smith's red rock rabbit**
- ***R. arnoldi*** (47, including immatures)
- *R. distinctus* (4, larvae)
- *R. evertsi evertsi* (9, including immatures)
- *R. glabroscutatum* (6, immatures, 1 with 1 ♂)
- *R. oculatus* (1)
- *R. warburtoni* (7, immatures)

Pronolagus **sp.**
- *R. evertsi evertsi* (2, immatures)

'Rabbit'
- *R. tricuspis* (1)

Order Macroscelidea

Family Macroscelididae

Elephantulus brachyrhynchus **Short-snouted elephant shrew**
- *R. evertsi evertsi* (1, immatures)
- *R. pravus* (2, nymphs)
- ***R.* sp. near *pravus*** (5, immatures)

Elephantulus edwardii **Cape elephant shrew**
- *R. exophthalmos* (3, immatures)

Elephantulus fuscipes **Uganda elephant shrew**
- *R. pravus* (1, nymphs)

Elephantulus myurus **Rock elephant shrew**
- *R. arnoldi* (49, immatures)
- *R. distinctus* (7, immatures)
- ***R.* sp. near *pravus*** (10, immatures)
- ***R. warburtoni*** (282, immatures)

Elephantulus rufescens **Rufous elephant shrew**
- *R. evertsi evertsi* (4, immatures)
- ***R. pravus*** (42, immatures; 2, unconfirmed, immatures)

Elephantulus rupestris **Smith's rock elephant shrew**
- *R. exophthalmos* (5, immatures)

Elephantulus **spp.**
- *R. pravus* (2, immatures plus 1 ♀)
- *R.* sp. near *pravus* (1, immatures)
- *R. senegalensis* (1, nymph)

Macroscelides proboscideus **Short-eared elephant shrew**
- *R. evertsi evertsi* (1, immatures)
- *R. exophthalmos* (1, immatures)

Petrodromus tetradactylus **Four-toed elephant shrew**
- *R. kochi* (2, including immatures)
- *R.* sp. near *pravus* (1, nymphs, unconfirmed)

BIRDS

CLASS AVES

Order Struthioniformes

Family Struthionidae

Struthio camelus **Ostrich**
- *R. guilhoni* (2)
- *R. pravus* (1)
- *R. pulchellus* (1)
- *R. turanicus* (4)

Order Pelicaniformes

Family Phalacrocoracidae

Phalacrocorax africanus **Reed cormorant**
- *R. evertsi evertsi* (2, immatures)
- *R. simus* (1)

Order Ciconiiformes

Family Ciconiidae

Ephippiorhyncus senegalensis **Saddle-bill stork**
R. guilhoni (1)

Leptoptilos crumeniferus **Marabou stork**
R. guilhoni (1)
R. turanicus (1)

Order Anseriformes

Family Anatidae

Dendrocygna viduata **White-faced whistling duck**
R. lunulatus (1)
R. senegalensis (1)

Order Falconiformes

Family Accipitridae

Milvus migrans **Black kite**
R. turanicus (1)

Aegypius occipitalis **White-headed vulture**
R. turanicus (1)

Gyps africanus **African white-backed vulture**
R. guilhoni (1)

Melierax canorus **Pale chanting goshawk**
R. evertsi mimeticus (1)

Accipiter melanoleucus **Great sparrow hawk**
R. humeralis (1)
R. pravus (1)

Hieraaetus bellicosus **Martial eagle**
R. guilhoni (1)

Hieraaetus wahlbergi **Wahlberg's eagle**
R. turanicus (1)

Family Sagittariidae

Sagittarius serpentarius **Secretary bird**
R. pravus (1)
R. turanicus (5)

Order Galliformes

Family Phasianidae

Francolinus bicalcaratus **Double-spurred francolin**
R. cuspidatus (1, immatures)

Francolinus leucoscepus **Yellow-necked francolin**
R. pulchellus (1)

Francolinus sephaena **Crested francolin**
R. evertsi evertsi (1, immatures)

***Francolinus* sp.** **'Spurfowl'**
R. jeanneli (1)

Numida meleagris **Helmeted guineafowl**
R. appendiculatus (11, immatures, including 2 with adults)
R. evertsi evertsi (7, larvae)
R. glabroscutatum (8, immatures)
R. nitens (1, larva)
R. oculatus (4, immatures)
R. zambeziensis (15, immatures)

Order Gruiformes

Family Rallidae

Fulica cristata **Red-knobbed coot**
R. lunulatus (1)

Family Otidae

Neotis denhami **Denham's bustard**
R. guilhoni (2)
R. turanicus (2)

Ardeotis arabs **Arabian bustard**
R. guilhoni (2)

Ardeotis kori **Kori bustard**
R. armatus (1, 1 nymph)
R. pravus (1, nymphs)

Eupodotis afra **Black bustard**
R. turanicus (1)

Eupodotis melanogaster **Black-bellied bustard**
R. duttoni (1)
R. kochi (1)

Order Charadriiformes

Family Burhinidae

Burhinus capensis **Cape dikkop**

R. turanicus (1)

Order Columbiformes

Family Columbidae

Streptopelia **sp.** **'Dove'**

R. turanicus (1)

Order Coliiformes

Family Coliidae

Colius colius **White-backed mousebird**

R. evertsi evertsi (1, immatures)

Order Passeriformes

Family Alaudidae

Callandrella cinerea **Red-capped lark**

R. evertsi evertsi (1, including immatures)

Family Hirundinidae

Hirundo cucullata **Greater striped swallow**

R. evertsi evertsi (1, including immatures)

Family Laniidae

Prionops plumata **Long-crested helmet shrike**

R. appendiculatus (1)

'Shrike'

R. pravus (1)

Tchagra australis **Brown-headed tchagra**

R. evertsi evertsi (2, immatures)

Family Timaliidae

Turdoides jardineii **Arrow-marked babbler**

R. evertsi evertsi (2, immatures)

Family Sylviidae

Chloropeta natalensis **Yellow warbler**

R. simpsoni (1, 1 ♂)

Family Ploceidae

Quelea quelea **Red-billed quelea**

R. evertsi evertsi (1, immatures)

REPTILES

CLASS REPTILIA

Order Chelonii

Family Testudinidae

Geochelone pardalis **Leopard tortoise**

R. evertsi evertsi (1, female)

'Tortoise'

R. pravus (1, unconfirmed)

Order Squamata

Family Viperidae

Bitis arietans **Puff adder**

R. turanicus (1)

Family Varanidae

Varanus niloticus **Water leguaan**

R. maculatus (1)

HUMANS

Humans

R. appendiculatus (18, including immatures)
R. armatus (2)
R. bequaerti (1)
R. carnivoralis (1)
R. complanatus (1)
R. compositus (1)
R. distinctus (1, including a nymph)
R. dux (1, unattached)
R. evertsi evertsi (3, including 1 nymph)
R. fulvus (+)
R. gertrudae (3)
R. humeralis (1)
R. hurti (2)

Humans (*cont.*)

R. jeanneli (2)
R. longus (2)
R. lunulatus (5)
R. maculatus (7)
R. muhsamae (3)
R. praetextatus (1; 4, unconfirmed, attached to host; 9, unconfirmed, crawling on host)
R. pravus (9)
R. sp. near *pravus* (1)
R. pulchellus (40)
R. sanguineus (2)
R. senegalensis (2)
R. simus (27)
R. sulcatus (2, unconfirmed)
R. supertritus (1)
R. turanicus (3)
R. zambeziensis (1, nymph)
R. ziemanni (1)
R. zumpti (1; 1, unconfirmed)

9

Rhipicephalus species occurring outside the Afrotropical region

HISTORICAL REVIEW

The description of *Rhipicephalus sanguineus* by Pierre Latreille (1806) early in the nineteenth century was followed much later in the same century by that of *R. bursa*. This tick, which is common in the Mediterranean region, was collected from a wild boar and described by Professor Giovanni Canestrini and F. Fanzago in an 1878 monograph on Italian Acari. In 1897, also in Italy, Supino described two junior synonyms of *R. sanguineus (R. bhamensis* and *R. flavus*). He also described *R. haemaphysaloides*, an important and widespread ectoparasite of large mammals from Afghanistan and the Indian subcontinent to south-east Asia and Indonesia.

In 1911, Professor V.L. Yakimov and his wife Nina Kol-Yakimova of St. Petersburg, Russia, while working in the laboratory of Professor L.G. Neumann at Toulouse, France, described *R. rossicus* collected from dogs in the governmental district of Saratov. Two other Russian tick workers each described a single species of *Rhipicephalus*. Dr. N.O. Olenev, of the Museum of the Russian Academy of Sciences in St. Petersburg, described *R. schulzei* in 1929 but gave no details of the host or locality. In 1948, Dr. B.I. Pomerantsev of the Zoological Institute of the U.S.S.R. Academy of Sciences produced a description of *R. leporis*, published posthumously through the assistance of his widow, G. Serdyukova. These specimens were taken from a hare in Uzbekistan.

Paul Schulze worked alone in Rostock, Germany, for most of his career. He was essentially an 'armchair taxonomist', not visiting the major tick collections or collaborating with other tick taxonomists. Tick names authored by Schulze, including those at the generic, subgeneric, specific, subspecific and infraspecific levels, and combinations thereof, number well over 800. However, most of these names are now considered junior synonyms. He described four species of *Rhipicephalus*, two of which, both described in 1935, are still considered valid: *R. pilans* collected from an unknown host on Flores Island, Indonesia, and *R. pumilio* collected from an unknown host at Maralbashi (now Pach'u), Sinkiang Province (now Xinjiang), China.

The Spanish entomologist and museum conservator, J. Gil Collado described *R. pusillus* as a subspecies of *R. bursa* in 1936. The two female specimens were collected from a fox in Barcelona. Two years later he raised *R. pusillus* to specific rank.

Jaime Augusto Travassos Santos Dias, who was a native of Mozambique, spent most of his life there working on trypanosomiasis and many other parasitic diseases. Almost all of his tick work centred on species found in sub-Saharan Africa, but in the 1950s he made a tour outside Africa to several museums containing significant tick collections. At the Zoological Museum in

Hamburg, Germany he found and described *R. scalpturatus*, a very rare species found in northern India and Nepal but never yet recovered from a host animal.

In 1966, the Indian entomologist, Vijai Dhanda of the Virus Research Centre in Poona, described *R. ramachandrai*, a parasite of the Indian gerbil in Maharashtra State, India.

Two other highly respected tick workers have contributed greatly to our knowledge of the genus *Rhipicephalus*. Brouria Feldman-Muhsam at the Hebrew University in Jerusalem and Carleton M. Clifford at the Rocky Mountain Laboratory in Hamilton, Montana. In a series of articles, published initially with S. Adler and culminating in one published in 1956 on a number of *Rhipicephalus* spp., Feldman-Muhsam demonstrated the value of the female genital aperture for the identification of ixodid tick species (Adler & Feldman-Muhsam, 1946, 1948; Feldman-Muhsam, 1951, 1956). Clifford & Anastos (1962), concentrating on the genus *Rhipicephalus*, showed in a series of 23 illustrations of both mounted and unmounted preparations of female genital apertures that this technique is almost indispensable for the specific determination of many female African *Rhipicephalus*.

REFERENCES

Adler, S. & Feldman-Muhsam, B. (1946). The differentiation of ticks of the genus *Hyalomma* in Palestine. *Refuah Veterinarith*, 3, 91–4.

Adler, S. & Feldman-Muhsam, B. (1948). A note on the genus *Hyalomma* Koch in Palestine. *Parasitology*, 39, 95–101.

Feldman-Muhsam, B. (1951). The value of the female genital aperture for specific diagnosis in the genera *Amblyomma* and *Dermacentor*. *Bulletin of the Research Council of Israel*, 1, 164–5.

Feldman-Muhsam, B. (1956). The value of the female genital aperture and the peristigmal hairs for specific diagnosis of the genus *Rhipicephalus*. *Bulletin of the Research Council of Israel*, **5B**, 300–6.

Also see the following Basic Reference (p. 13): Clifford & Anastos (1962).

KEY TO THE *RHIPICEPHALUS* SPECIES MALES OCCURRING OUTSIDE THE AFROTROPICAL REGION

1a. Adanal plates sickle-shaped. 2

1b. Adanal plates variable but not sickle-shaped 3

2a. Conscutum smooth with conscutal setae small, inapparent. Primarily a parasite of domestic and wild ungulates from Pakistan eastward to Indonesia . . . . ***R. haemaphysaloides***

2b. Conscutum irregular with conscutal setae conspicuous . . . . ***R. pilans***

3a. A very large male with legs increasing markedly from I to IV; cervical fields deep, relatively narrow, curved troughs; posterior body margin indented at central festoon. Recorded from the lowland areas of Nepal and northern India . . . . . ***R. scalpturatus***

3b. Lacking this combination of male characters 4

4a. Entire conscutal area punctate, densely so in posterior one-third, less so centrally; marginal lines distinct, composed of large deep punctations extending anteriorly from festoon I to cervical pits; festoons poorly delineated. Primarily a parasite of the Indian gerbil in India and Nepal . . . . ***R. ramachandrai***

4b. Lacking this combination of characters 5

5a. Numerous circumspiracular setae surrounding each spiracle. . . . . . . ***R. bursa***

5b. Spiracles without numerous circumspiracular setae 6

6a. Length of conscutum less than 2.0 mm. Primarily a parasite of the European rabbit. . . . . ***R.pusillus***

6b. Length of conscutum greater than 2.0 mm. 7

7a. Adanal plates narrow posteriorly; their length about three times their breadth. . 8

7b. Adanal plates broad posteriorly; their length about two to two and a half times their breadth. 10

8a. Marginal lines shallow, quite indistinct and lightly punctate . . . . ***R. camicasi***
8b. Marginal lines deep, distinctly punctate . 9
9a. Dorsal prolongation of spiracles broad, equal to breadth of adjacent festoon . . . . ***R. turanicus***
9b. Dorsal prolongation of spiracles narrow, equal to about one-half the breadth of the adjacent festoon. Primarily a parasite of domestic dogs worldwide . . . . ***R. sanguineus***
10a. Spiracular plates with a broad dorsal prolongation . . . . ***R. rossicus***
10b. Spiracular plates with a narrow dorsal prolongation 11
11a. Adanal plates with small, medially-directed cusps. A parasite of a wide variety of mammals in southern Russia, Turkmenistan, Tajikistan, Kashmir and Mongolia . . . . ***R. pumilio***
11b. Adanal plates without small, medially-directed cusps 12
12a. Marginal lines delimit no, or only the first, festoons; spiracles small, each with a very long, narrow dorsal prolongation. Primarily a parasite of the Tolai hare in Kazakhstan, Turkmenistan, Afghanistan and Iran . . . . ***R. leporis***
12b. Marginal lines delimit first two festoons; spiracles narrowly elongate, dorsal prolongation slightly narrower than the macular area. Primarily parasitic on sousliks . . . . ***R. schulzei***

KEY TO THE *RHIPICEPHALUS* SPECIES FEMALES OCCURRING OUTSIDE THE AFROTROPICAL REGION

1a Spiracles surrounded by numerous circumspiracular setae . . . . ***R. bursa***
1b. Spiracles not surrounded by numerous circumspiracular setae 2
2a. Scutum less than 1.0 mm in length. Primarily a parasite of the Indian gerbil in India and Nepal . . . . ***R. ramachandrai***
2b. Scutum greater than 1.0 mm in length 3
3a. Genital aperture with lateral arms flaring outward anteriorly . . . . ***R. pilans***
3b. Genital aperture with lateral arms not flaring outward anteriorly 4
4a. Genital aperture broadly U-shaped with bulging genital apron. Recorded from the lowland areas of Nepal and northern India . . . . ***R. scalpturatus***
4b. Genital aperture broadly or narrowly U-shaped, but without a bulging genital apron 5
5a. Scutum with external cervical margins not marked by a series of larger punctations 6
5b. Scutum with external cervical margins marked by a series of larger punctations 7
6a. Genital aperture narrowly U-shaped. Primarily a parasite of wild and domestic ungulates from Pakistan eastward to Indonesia . . . . ***R. haemaphysaloides***
6b. Genital aperture broadly U-shaped. A parasite of a wide variety of mammals in southern Russia, Turkmenistan, Tajikistan, Kashmir and Mongolia . . . . . ***R. pumilio***
7a. Genital aperture narrowly U-shaped, with slight pigmented hyaline flaps laterally . . . . ***R. camicasi***
7b. Genital aperture narrowly or broadly U-shaped, but without slight pigmented hyaline flaps laterally 8
8a. Scutum rounded posteriorly 9
8b. Scutum sinuous posteriorly giving it a shield shape 11
9a. Scutum greater than 1.5 mm in length . . . . ***R. rossicus***
9b. Scutum less than 1.5 mm in length 10
10a. Palps pointed apically; cervical fields rugose; scapulae punctate. Primarily parasitic on sousliks . . . . ***R. schulzei***
10b. Palps rounded apically; cervical fields smooth, not rugose; scapulae impunctate. Primarily a parasite of the European rabbit . . . . ***R. pusillus***
11a. Genital aperture broadly U-shaped . . . 12

11b. Genital aperture narrowly U-shaped .... ***R. turanicus***

12a. Cervical fields with striations and shagreening; scapulae with a few large punctations. Primarily a parasite of domestic dogs worldwide........... ***R. sanguineus***

12b. Cervical fields and scapulae smooth, quite impunctate. Primarily a parasite of the Tolai hare in Kazakhstan, Turkmenistan, Afghanistan and Iran....... ***R. leporis***

10

Accounts of individual species occurring outside the Afrotropical region

RHIPICEPHALUS BURSA CANESTRINI & FANZAGO, 1878

The specific name *bursa*, from Middle Latin meaning a 'pouch' or a 'purse made of skin', refers to the bloated pouch-like appearance of an engorged female of this tick species.

Synonym

?*bilenus*.

Diagnosis

A large, light brown species.

Male (Figs 230(a), 231(a) to (c))
Capitulum usually slightly broader than long, length × breadth ranging from 0.66 mm × 0.71 mm to 0.88 mm × 0.88 mm. Basis capituli with short acute lateral angles in the anterior third of its length. Palps short, broad. Conscutum narrower anteriorly, broadening posterior to eyes, length × breadth ranging from 1.55 mm × 1.02 mm to 3.34 mm × 2.26 mm; the light tan body wall expanded laterally and posteriorly in engorged specimens. Eyes marginal, slightly bulging, not edged with punctations. Cervical pits deep, then extending as very shallow troughs just posterior to eyes. Punctations numerous, fine, evenly distributed over scutal surface, a few larger punctations in scapular areas. Marginal lines long and narrow, impunctate, delimiting first festoons, and almost reaching eyes. Posteromedian groove narrow, shallow and usually inconspicuous; posterolateral grooves broadly-oval shallow depressions. Ventrally spiracles elongate in anteroposterior axis, each with a short, narrow dorsal prolongation; circumspiracular setae present. Adanal plates large, pointed anteriorly, broadly rounded posteriorly with cusps or pointed prominences on their internal margins; accessory adanal plates absent or present as very small, light-brown, sclerotized points.

Female (Figs 230(b), 231(d) to (f))
Capitulum broader than long, length × breadth ranging from 0.70 mm × 0.77 mm to 0.94 mm × 1.05 mm. Basis capituli with broad lateral angles at about the anterior third of its length; porose areas large, about 1.5 times their own diameter apart. Palps longer than those of male, with pedicel of article I easily visible dorsally. Scutum broader than long, length × breadth ranging from 1.41 mm × 1.44 mm to 1.82 mm × 1.98 mm, posterior margin sinuous. Eyes at lateral angles, slightly bulging, not edged with punctations. Internal cervical margins short, deep, convergent; cervical fields narrow, slightly sunken; external cervical margins slightly raised, not marked by punctations. Punctations on scutal surface fine with a few larger punctations laterally. Ventrally spiracles with

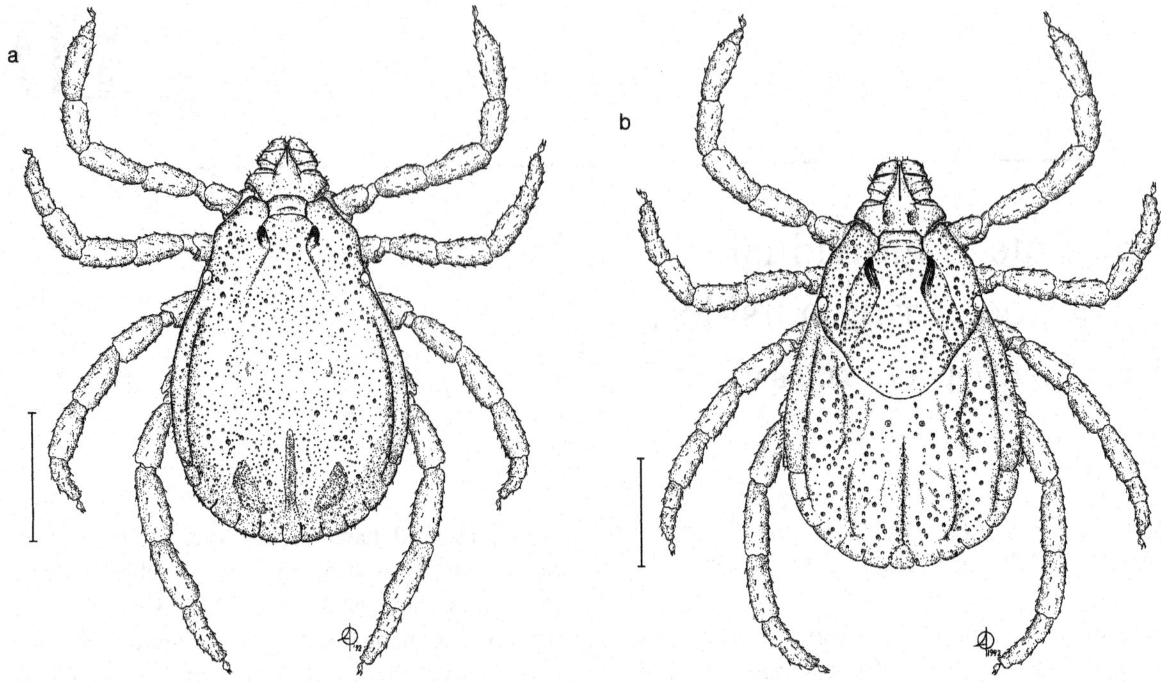

Figure 230. *Rhipicephalus bursa* [RML 120344; HH 85114, collected on domestic bovine, Los Barrios, Cadiz, Spain, May 26, 1978 by L. E. Hueli]. (a) Male, dorsal; (b) female, dorsal. Scale bars represent 1 mm. A. Olwage *del.*

Table 61. *Host records of* Rhipicephalus bursa

Hosts	Number of records
Domestic animals	
Cattle	40
Sheep	91
Goats	35 (including immatures)
Camels	1
Horses	10
Donkeys	1
Dogs	1
Wild animals	
Mountain gazelle (*Gazella gazella*)	1
Argali (*Ovis ammon*)	2
Wood mouse (*Apodemus sylvaticus*)	7 (immatures)
Garden dormouse (*Eliomys quercinus*)	1 (immatures)
European rabbit (*Oryctolagus cuniculus*)	3
Humans	2

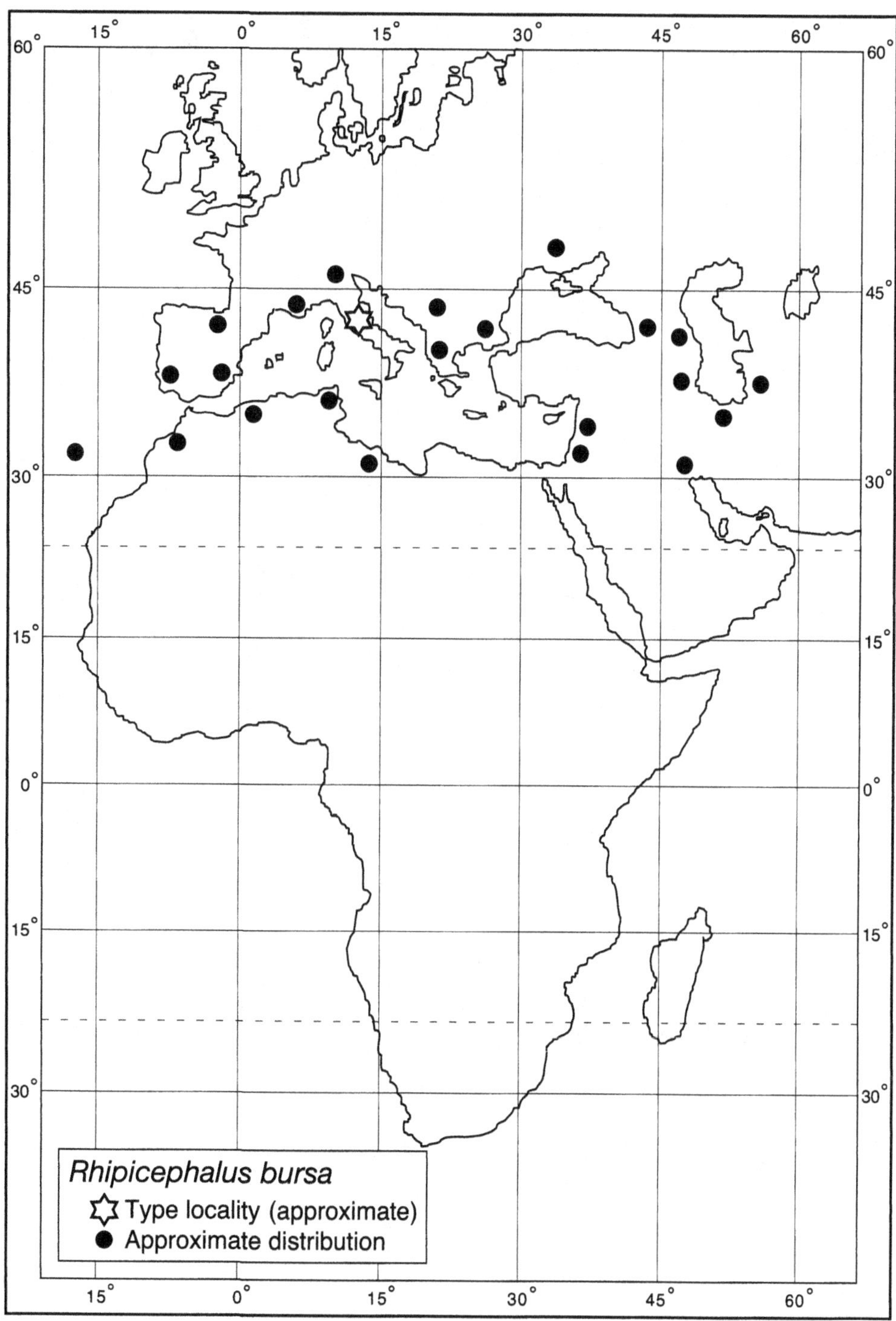

Map 65. *Rhipicephalus bursa*: distribution.

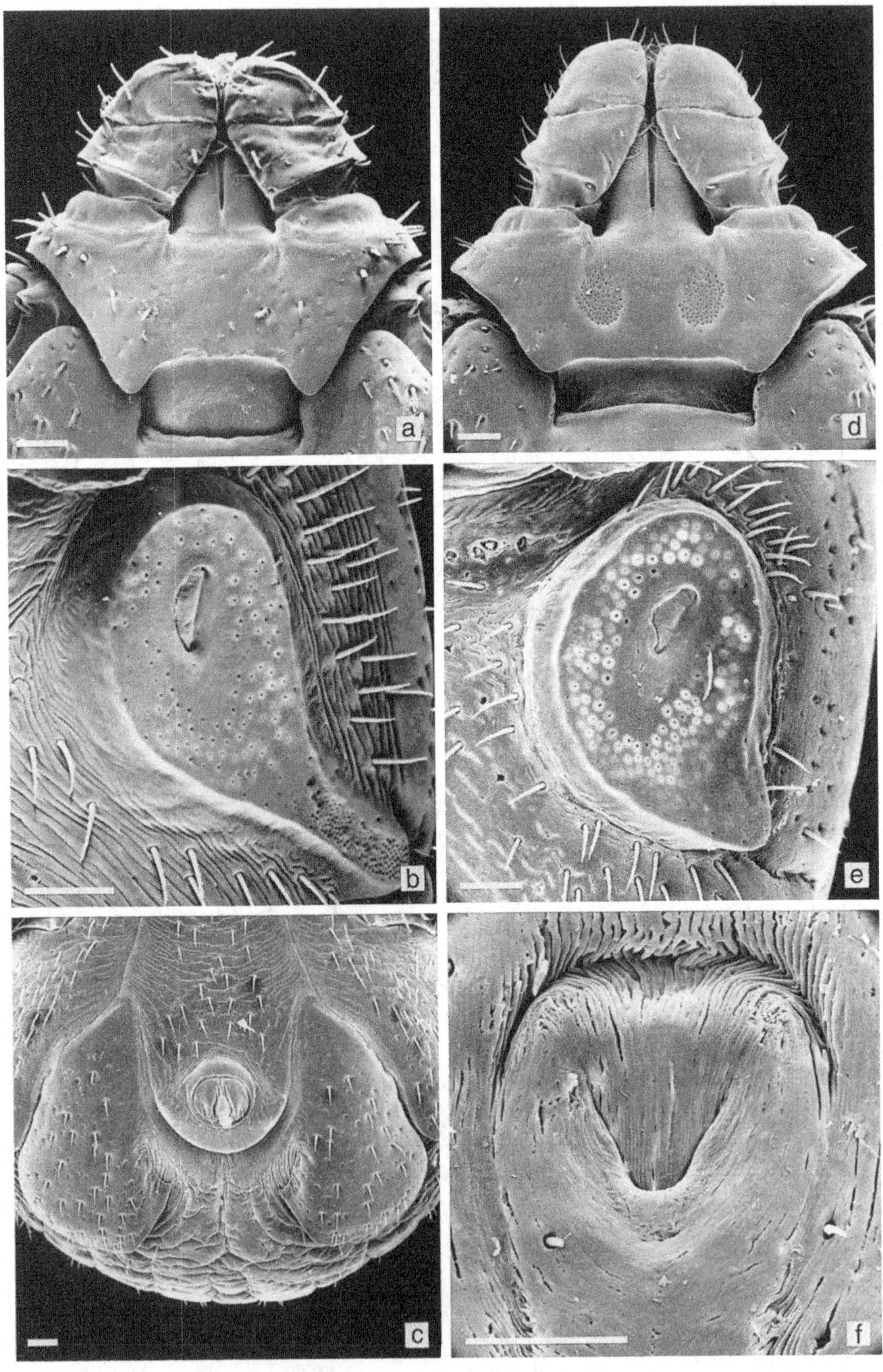
a
d
b
e
c
f

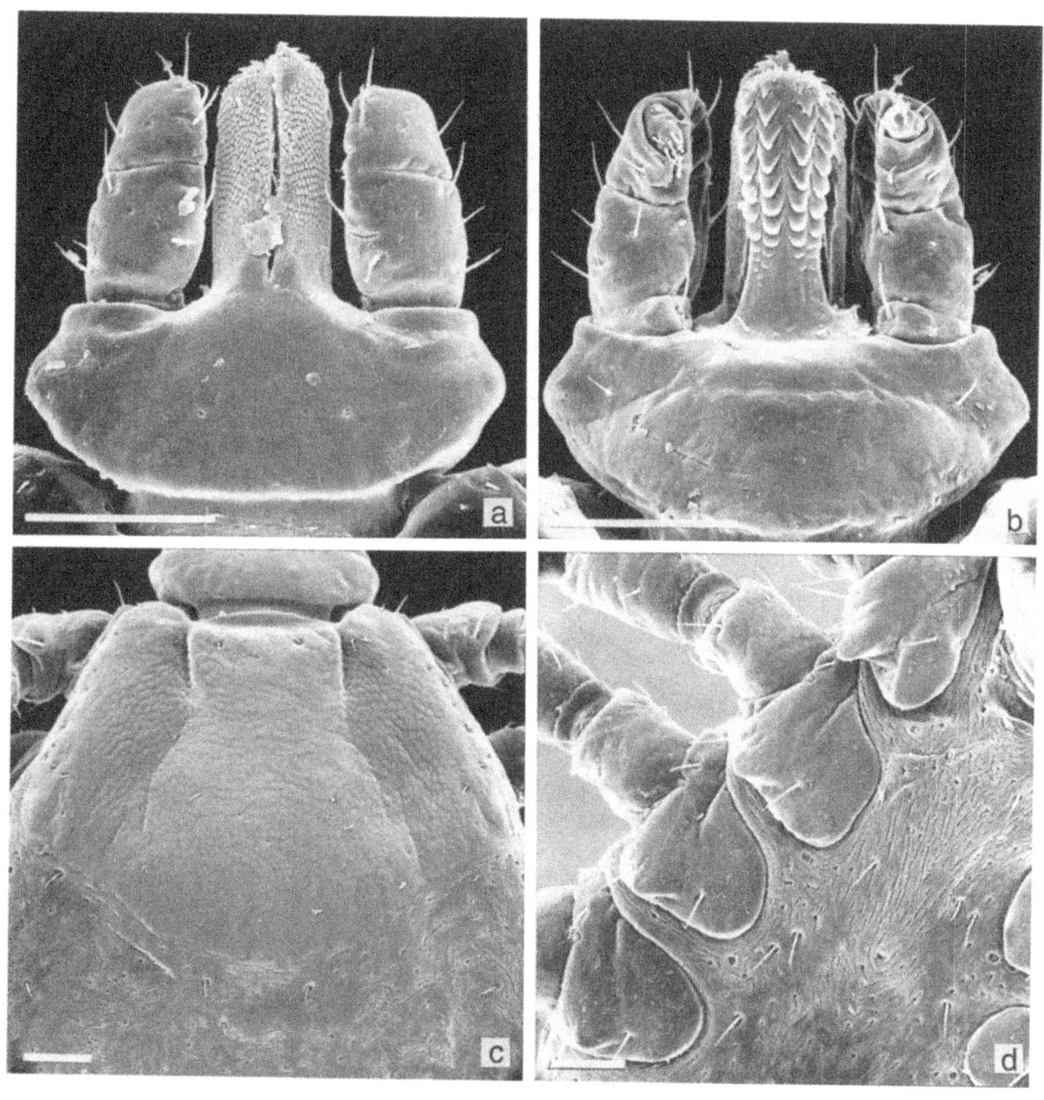

Figure 232. *Rhipicephalus bursa* [RML 85335; HH 81549, collected on domestic sheep, Karamlase, Mosul, Iraq, December 14, 1965 by J. Robson]. Nymph: (a) capitulum, dorsal; (b) capitulum, ventral; (c) scutum; (d) coxae. Scale bars represent 0.01 mm. SEMs by P. Hill.

Figure 231 (*opposite*). *Rhipicephalus bursa* [RML 120344; HH 85114, collected on *Bos taurus*, Los Barrios, Cadiz, Spain, May 26, 1978 by L. E. Hueli]. Male: (a) capitulum, dorsal; (b) spiracle; (c) adanal plates. Female: (d) capitulum, dorsal; (e) spiracle; (f) genital aperture. Scale bars represent 0.10 mm. SEMs by P. Hill.

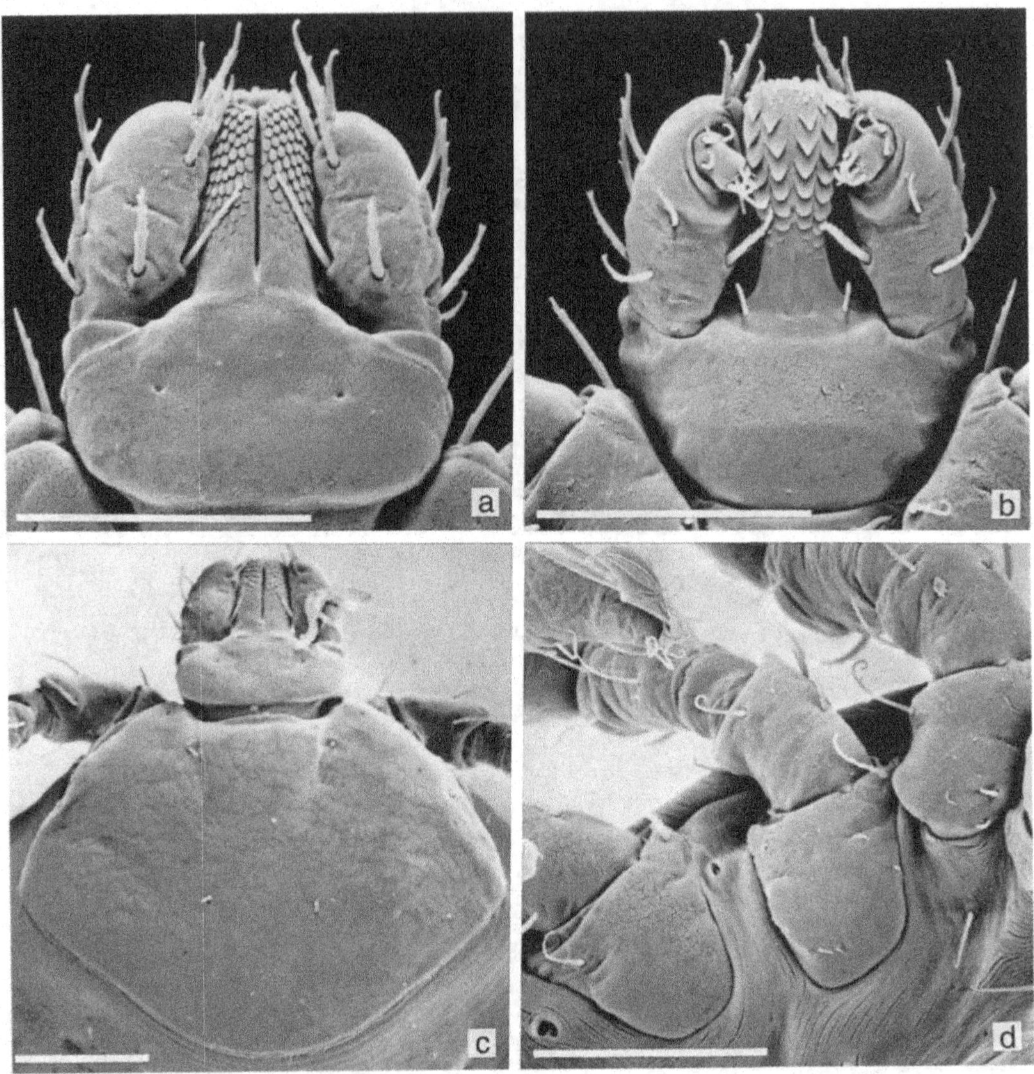

Figure 233. *Rhipicephalus bursa* [RML 85335; HH 81549, collected on domestic sheep, Karamlase, Mosul, Iraq, December 14, 1965 by J. Robson]. Larva: (a) capitulum, dorsal; (b) capitulum, ventral; (c) scutum; (d) coxae. Scale bars represent 0.10 mm. SEMs by P. Hill.

circumspiracular setae as in male. Genital aperture narrowly V-shaped with the base of the V rounded, not pointed.

Nymph (Fig. 232)

Capitulum broader than long, length × breadth ranging from 0.28 mm × 0.30 mm to 0.31 mm × 0.36 mm. Basis capituli with broadly-rounded lateral angles, posterior margin slightly convex, cornua absent. Palps as long as hypostome, cylindrical. Scutum much broader than long, length × breadth ranging from 0.51 mm × 0.66 mm to 0.59 mm × 0.72 mm, posterior margin a wide, fairly shallow curve. Eyes at widest point, well over halfway back, marginal, very slightly bulging. Cervical grooves deep initially, becoming shallower and diverging to the posterolateral margins of the scutum. Ventrally coxae I each with a long narrow external spur and a shorter, broader internal spur; coxae II to IV each with an external spur only.

Larva (Fig. 233)
Capitulum slightly broader than long, length × breadth ranging from 0.132 mm × 0.137 mm to 0.144 mm × 0.146 mm. Basis capituli twice as broad as long, with rounded lateral angles, posterior margin convex. Palps cylindrical, about the same breadth throughout their length. Scutum much broader than long, length × breadth ranging from 0.262 mm × 0.356 mm to 0.303 mm × 0.404 mm. Eyes at widest point, flat to very slightly raised. Cervical grooves short, parallel to slightly convergent. Ventrally coxae I each with a broad spur; coxae II and III each with a broad salient ridge on its posterior margin.

Notes on identification
A tick with a superficially but densely punctate appearance and circumspiracular setae obvious in both males and females. Zumpt (1942) found a great variability in the size and shape of the punctations, adanal plates and shape of the spiracles in the adult offspring from a single female. In addition, larvae from the same female can have scuta of greatly differing shapes (Feldman-Muhsam, 1953). Beati *et al.* (1998), using a phylogenetic tree based on mitochondrial 12SrDNA gene sequences, found that species of *Boophilus* clustered with the *R. bursa* group of ticks. They suggested that there may not be a justification for *Boophilus* being a separate tick genus from *Rhipicephalus.*

Hosts

A two-host species (Enigk, 1943) of which immatures and adults infest the same hosts. Cattle, sheep, goats and other domesticated animals are now the primary hosts of this tick (Table 61). Wild ungulates were the original hosts (Hoogstraal & Valdez, 1980). *Rhipicephalus bursa* will also feed on the wood mouse, rabbits and occasionally humans. It has a predilection for dropping from the host at night (Yeruham *et al.*, 1995b).

Adult ticks attach mainly on the inner surface of the ears of sheep but can also be present on the dorsal outer surface as well as in the perianal and inguinal regions of these animals. In northern Spain adults are present on sheep from July to September (Estrada-Peña, Dehesa & Sanchez, 1990), and in Turkey during June and July (Göksu, 1969). In Israel they are present from April to July and the immatures from October to February (Yeruham *et al.*, 1996). Only one life cycle a year is completed.

Zoogeography

The entire Mediterranean, Adriatic and Aegean basins, including their islands and the North African countries of Algeria, Morocco and Tunisia account for most records of this tick (Map 65). It has not been recorded from Egypt but quite recently *R. bursa* was found in Libya (Beesley & Gabaj, 1991). It is also present in Portugal including the Machico area of Madeira (Torres de Almeida, 1995) which is the westernmost extent of its range. Feldman-Muhsam (1953) found that *R. bursa*, although present in Israel, was not common. In Syria, Liebisch & Zukari (1978) encountered *R. bursa* infrequently and only in the south-western arid and semi-arid part of the country. It occurs in Switzerland, Bulgaria, Romania, northern Iraq, northern Iran, Azerbaijan, Georgia, Kazakhstan, Turkmenistan and Ukraine. Thus in the Palearctic region it is generally found between the latitudes of 31° to 45° North (Yeruham *et al.*, 1985). *Rhipicephalus bursa* prefers grassy slopes and low to medium altitude mountain slopes as well as certain modified steppe and semi-desert environments (Hoogstraal & Valdez, 1980).

Records of *R. bursa* from various countries in sub-Saharan Africa, Mexico, Venezuela, Uruguay, Haiti, Cuba, Timor and Xinjiang, China are either misidentifications or accidental importations.

Disease relationships

It has been demonstrated that *R. bursa* can transmit *Babesia bigemina, Babesia bovis* and

Anaplasma marginale to cattle (Brumpt, 1931; Sergent *et al.*, 1931); *Babesia caballi* and *Babesia equi* to horses (Markov, Kurchatov & Dzasokov, 1940, cited by Neitz, 1956; Enigk, 1943); and *Babesia motasi, Babesia ovis, Theileria separata, Anaplasma ovis* and possibly *Ehrlichia ovina* to sheep (Rastegaïeff, 1933; Donatien & Lestoquard, 1937; Markov & Abramov, 1970; Neitz, 1972). In a study on *B. ovis* in Israel, Yeruham *et al.* (1995a) showed that morbidity and mortality in sheep flocks occurred about 2 weeks after detection of adult *R. bursa* on the animals, and no clinical cases of babesiosis were found when the pre-imaginal stages of *R. bursa* were active. Crimean-Congo haemorrhagic fever virus has been isolated from this tick in Greece and Kirgizia (Antoniadis & Casals, 1982), and Thogoto virus in Sicily (Albanese *et al.*, 1972). However, Darwish & Hoogstraal (1981) consider that Thogoto isolates from Italy represent a separate virus in the Thogoto serogroup. It is also known to cause paralysis in sheep (Anonymous, 1984).

REFERENCES

Albanese, M., Bruno-Smiraglia, C., Di Cuonzo, G., Lavagnino, A. & Srihongse, S. (1972). Isolation of Thogoto virus from *Rhipicephalus bursa* ticks in western Sicily. *Acta Virologica*, **16**, 267. (Letter to the editor).

Anonymous (1984). *Ticks and Tick-borne Disease Control. A Practical Field Manual. Volume 1. Tick Control.* Rome: Food and Agriculture Organization of the United Nations.

Antoniadis, A. & Casals, J. (1982). Serological evidence of human infection with Congo-Crimean hemorrhagic fever virus in Greece. *American Journal of Tropical Medicine and Hygiene*, **31**, 1066–7. (Brief communication).

Beati, L., Tasmandjan, A., Keirans, J.E. & Raoult, D. (1998). Phylogenetic relationships of hard-tick species belonging to the genera *Rhipicephalus* and *Boophilus* based on mitochondrial 12SrDNA gene sequences. *Program and Abstracts, 47th Annual Meeting, American Journal of Tropical Medicine and Hygiene*, **59**, 210–11.

Beesley, W.N. & Gabaj, M.M. (1991). New records for *Rhipicephalus bursa, Boophilus microplus, B. decoloratus* and *Hypoderma lineatum* from Libya. *Medical and Veterinary Entomology*, **5**, 259–60.

Brumpt, E. (1931). Transmission d'*Anaplasma marginale* par *Rhipicephalus bursa* et par *Margaropus. Annales de Parasitologie Humaine et Comparée*, **9**, 4–9.

Canestrini, G. & Fanzago, F. (1878). Intorno agli acari italiani. *Atti della Reale Istituto Veneto di Scienze, Lettere ed Arte, ser.* 5, **4**, 69–208.

Darwish, M. & Hoogstraal, H. (1981). Arboviruses infecting humans and lower animals in Egypt: a review of thirty years of research. *Journal of the Egyptian Public Health Association*, **56**, 1–112.

Donatien, A. & Lestoquard, R. (1937). État actual des connaissance sur les rickettsioses animales. *Archives de l' Institut Pasteur d' Algérie*, **15**, 142–87, 6 pls.

Enigk, K. (1943). Die Überträger der Pferdepiroplasmose, ihre Verbreitung und Biologie. *Archiv für wissenschaftliche und praktische Tierheilkunde*, **78**, 209–40.

Estrada-Peña, A., Dehesa, V. & Sanchez, C. (1990). The seasonal dynamics of *Haemaphysalis punctata, Rhipicephalus bursa* and *Dermacentor marginatus* (Acari : Ixodidae) on sheep of Pais Vasco (Spain). *Acarologia*, **31**, 17–24.

Feldman-Muhsam, B. (1953). *Rhipicephalus bursa* in Israel. *Bulletin of the Research Council of Israel*, **3**, 201–6.

Göksu, K. (1969). Bio-ecological studies of *Rhipicephalus bursa* Canestrini and Fanzago, 1877 (*sic*) (Acarina: Ixodoidea) under field and laboratory conditions. *Ankara Üniversitesi Veteriner Fakültesi Dergisi*, **16**, 295-312. [In Turkish; English summary].

Hoogstraal, H. & Valdez, R. (1980). Ticks (Ixodoidea) from wild sheep and goats in Iran and medical and veterinary implications. *Fieldiana: Zoology, New Series*, **No. 6**, 1–16.

Liebisch, A. & Zukari, M. (1978). Biological and ecological studies on ticks of the genera *Boophilus, Rhipicephalus* and *Hyalomma* in Syria. In *Tick-borne Diseases and their Vectors*, ed. J.K.H. Wilde, pp. 150-62. Edinburgh: Edinburgh University Press.

Markov, A.A. & Abramov, I.V. (1970). Results of a twenty years' observation on repeated cycles of

Babesia ovis in forty-four generations of *Rhipicephalus bursa*. *Trudy Vsesoyuznogo Instituta Éksperimental'noy Veterinarii*, **38**, 5-14. [In Russian; translation 1636, NAMRU-3, Cairo].

Neitz, W.O. (1956). Classification, transmission and biology of piroplasms of domestic animals. *Annals of the New York Academy of Sciences*, **64**, 56–111.

Neitz, W.O. (1972). The experimental transmission of *Theileria ovis* by *Rhipicephalus evertsi mimeticus* and *R. bursa*. *Onderstepoort Journal of Veterinary Research*, **39**, 83–5.

Rastegaïeff, E.F. (1933). Zur Frage der Überträger der Schaf-piroplasmosen in Azerbaidschan (Transkaukasien). *Archiv für Wissenschaftliche und Praktische Tierheilkunde*, **67**, 176–86.

Sergent, E., Donatien, A., Parrot, L. & Lestoquard, F. (1931). Transmission héréditaire de *Piroplasma bigeminum* chez *Rhipicephalus bursa*. Persistence du parasite chez des tiques nourries sur des chevaux. *Bulletin de la Société de Pathologie Exotique*, **24**, 195–8.

Torres de Almeida, V. C. (1995). Sobre as áreas de ocorrência das espécies da família Ixodidae conhecidas na Ilha da Madeira. *Revista do Sindicato Nacional dos Médicos Veterinarios*, **5**, 36–40.

Yeruham, I., Hadani, A., Galker, F. (Kronthal), Mauer, E., Rubina, M. & Rosen, S. (1985). The geographical distribution and animal hosts of *Rhipicephalus bursa* (Canestrini and Fanzago, 1877) (*sic*), in Israel. *Revue d'Élevage et de Médecine Vétérinaire des Pays Tropicaux*, **38** (nouvelle série), 173–9.

Yeruham, I., Hadani, A., Galker, F. & Rosen, S. (1995a). A study of an enzootic focus of sheep babesiosis (*Babesia ovis*, Babes, 1892). *Veterinary Parasitology*, **60**, 349–54.

Yeruham, I., Hadani, A., Galker, F. & Rosen, S. (1996). The seasonal occurrence of ticks (Acari : Ixodidae) on sheep and in the field in the Judean area of Israel. *Experimental and Applied Acarology*, **20**, 47–56.

Yeruham, I., Hadani, A., Galker, F. (Kronthal), Rosen, S. & Gunders, A. (1995b). The daily distribution and circadian rhythm of detachment of engorged *Rhipicephalus bursa* ticks from lambs and rabbits. *Medical and Veterinary Entomology*, **9**, 445–7.

Zumpt, F. (1942). Die Variationsbreite der Nachkommen eines Weibchens von *R. bursa* Can. u. Fanz. IV. Vorstudie zu einer Revision der Gattung *Rhipicephalus*. *Zeitschrift für Parasitenkunde*, **12**, 444–50.

RHIPICEPHALUS HAEMAPHYSALOIDES (SUPINO, 1897)

The specific name *haemaphysaloides*, from the Greek genitive, *haimatos* meaning 'blood', plus the Greek genitive *physalidos* meaning 'a bladder', refers to the bloated, bladder-like, blood-engorged abdomen of females of this tick.

Synonyms

haemaphysaloides expeditus; haemaphysaloides niger; haemaphysaloides ruber; ruber.

Diagnosis

A large, brown species, with sickle-shaped adanal plates in the male.

Male (Figs 234(a), 235(a) to (c))

Capitulum slightly longer than broad, length × breadth ranging from 0.76 mm × 0.72 mm to 0.89 mm × 0.88 mm. Basis capituli with acute lateral angles at about anterior third of its length. Conscutum narrower anteriorly, broadening posterior to eyes, length × breadth ranging from 3.18 mm × 2.20 mm to 3.88 mm × 2.61 mm. Body wall slightly expanded laterally in engorged specimens; in partially engorged specimens the middle festoon protrudes as a caudal process; in fully engorged specimens, the first two festoons on either side protrude as do the middle three festoons. Eyes marginal, flat, not edged dorsally with punctations. Cervical pits deep, short. Scutal surface smooth with scattered, shallow, medium-sized punctations. Marginal lines appear as long, punctate grooves delimiting first two festoons, and ending well before eyes. Posteromedian groove long, broad and deep, ending in a bulbous depression posteriorly;

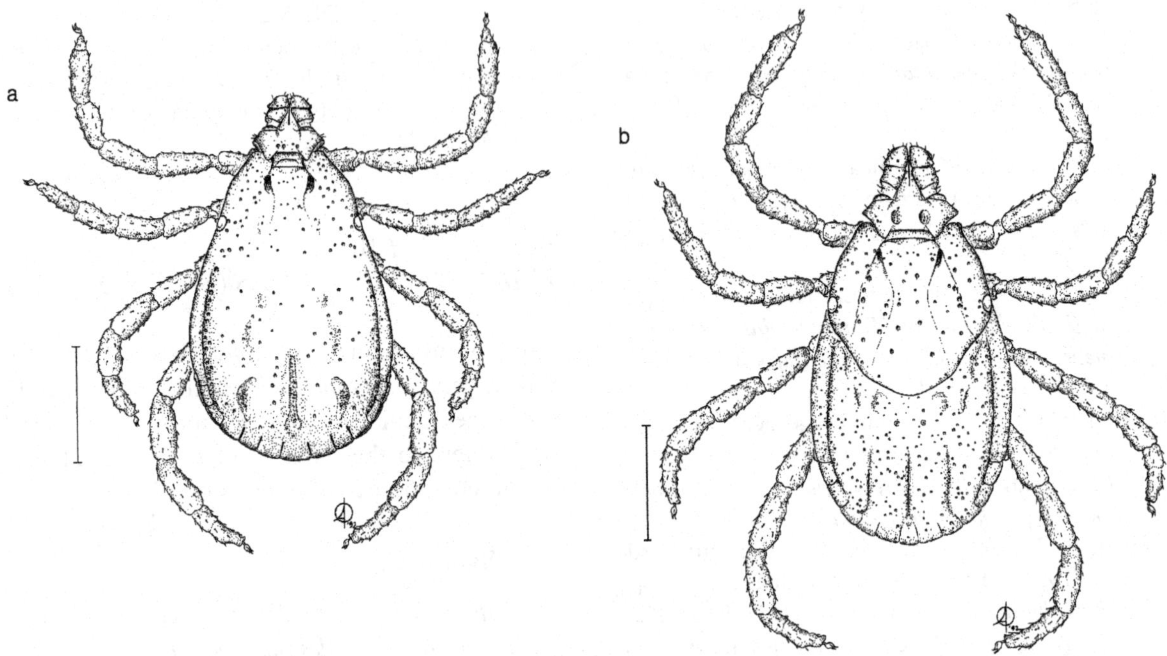

Figure 234. *Rhipicephalus haemaphysaloides* (RML 91868; HH 81556, collected on vegetation around corn fields and the Trisuli River bank, Syaburbensi, Rasuwa, Nepal, June 30, 1970 by R. M. Mitchell). (a) Male, dorsal; (b) female, dorsal. Scale bars represent 1 mm. A. Olwage *del.*

posterolateral grooves circular, comma, or tear-drop-shaped, shallower than posteromedian groove. Ventrally spiracles comma-shaped, broad throughout their length. Adanal plates sickle-shaped; accessory adanal plates weakly developed, small sclerotized points.

Female (Figs 234(b), 235(d) to (f))

Capitulum broader than long, length × breadth ranging from 0.86 mm × 0.90 mm to 0.96 mm × 1.05 mm. Basis capituli with lateral angles not markedly curved; porose areas small, subcircular, about 1.5 times their own diameter apart. Scutum about as long as broad, length × breadth ranging from 1.68 mm × 1.65 mm to 1.90 mm × 1.92 mm, posterior margin sinuous. Eyes at lateral angles, flat, yellowish, elongate, not edged with punctations. Internal cervical margins short, convergent; cervical fields slightly depressed, scalpel-shaped, a few irregular punctations along external cervical margins. Raised lateral borders and central area of scutum relatively impunctate. Ventrally genital aperture narrowly U-shaped.

Nymph (Fig. 236)

Capitulum much broader than long, length × breadth ranging from 0.22 mm × 0.32mm to 0.27mm × 0.36 mm. Basis capituli three times as broad as long with sharply pointed lateral angles; posterior margin straight, then becoming concave posterolaterally over scapulae; cornua present but small; ventrally spurs on posterior margin. Palps relatively short, tapering over hypostome. Scutum longer than broad, length × breadth ranging from 0.45 mm × 0.43 mm to 0.55 mm × 0.46 mm. Eyes at posterolateral angles, well over halfway back, marginal, flat. Cervical grooves short, convergent; cervical fields narrow, shagreened, barely indicated. Ventrally coxae I each with two sharply-pointed subequal spurs, the external narrower; coxae II to IV each with a single, small external spur.

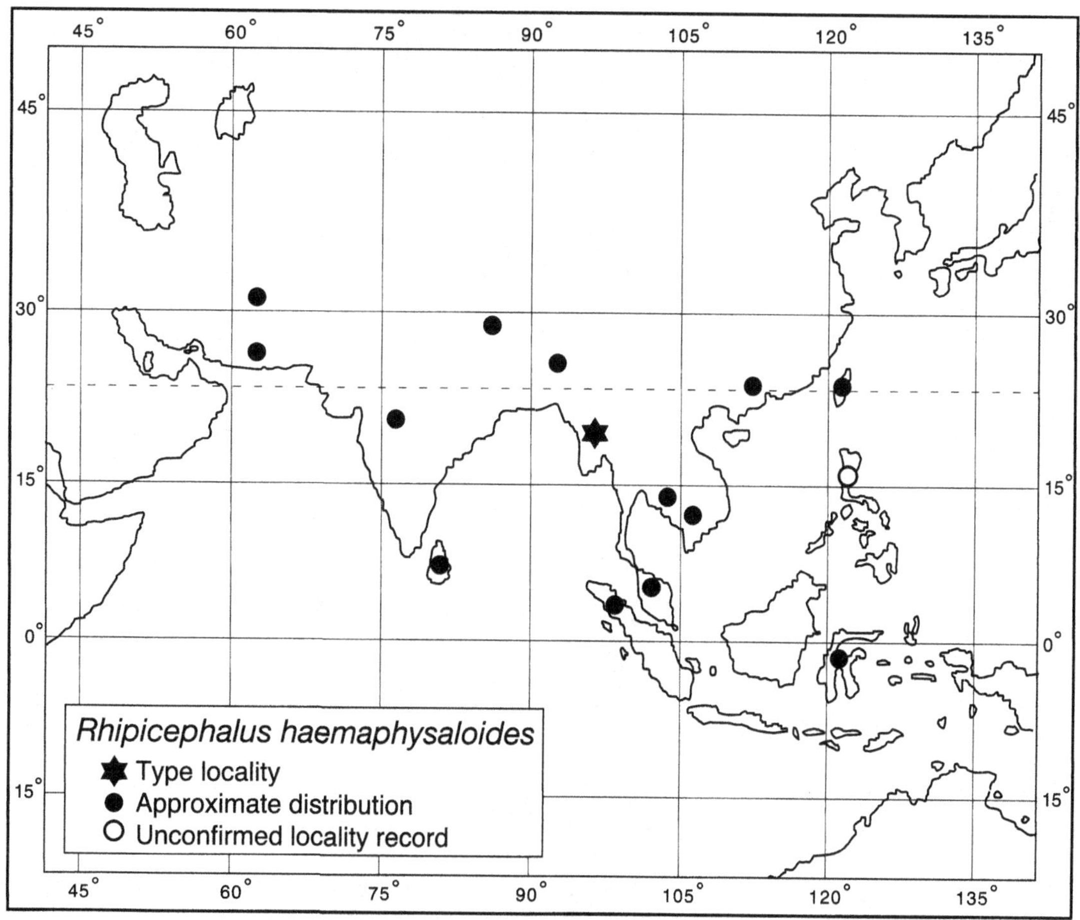

Map 66. *Rhipicephalus haemaphysaloides*: distribution.

Larva (Fig. 237)

Capitulum broader than long, length × breadth ranging from 0.122 mm × 0.158 mm to 0.160 mm × 0.177 mm. Basis capituli about 2.5 times as broad as long; posterior margin distinctly convex, lateral margins tapering slightly anteriorly towards palps. Palps short, exterior margins convex, about same length as hypostome. Scutum broader than long, length × breadth ranging from 0.283 mm × 0.346 mm to 0.328 mm × 0.373 mm. Eyes at widest point, approximately halfway back, flat to very slightly bulging. Cervical grooves very slight and shallow. Ventrally coxae I each with a broadly-rounded spur; coxae II and III each with a small rounded spur.

Notes on identification

Males of *R. haemaphysaloides* have sickle-shaped adanal plates and small, inconspicuous conscutal setae. Females of this species have a narrow U-shaped genital aperture and sparse alloscutal setae. This tick resembles *R. pilans* which has long scutal and alloscutal setae, and *R. sanguineus* which, in the male, does not have sickle-shaped adanal plates and, in the female has a broad U-shaped genital aperture. Kadarsan (1971) separated larval specimens of *R. haemaphysaloides* from *R. pilans* by noting that the former had palps that are bluntly pointed and a basis capituli as wide as the anterior margin of the scutum, while the latter species had acutely-pointed palps and a basis capituli wider than the anterior margin of the scutum.

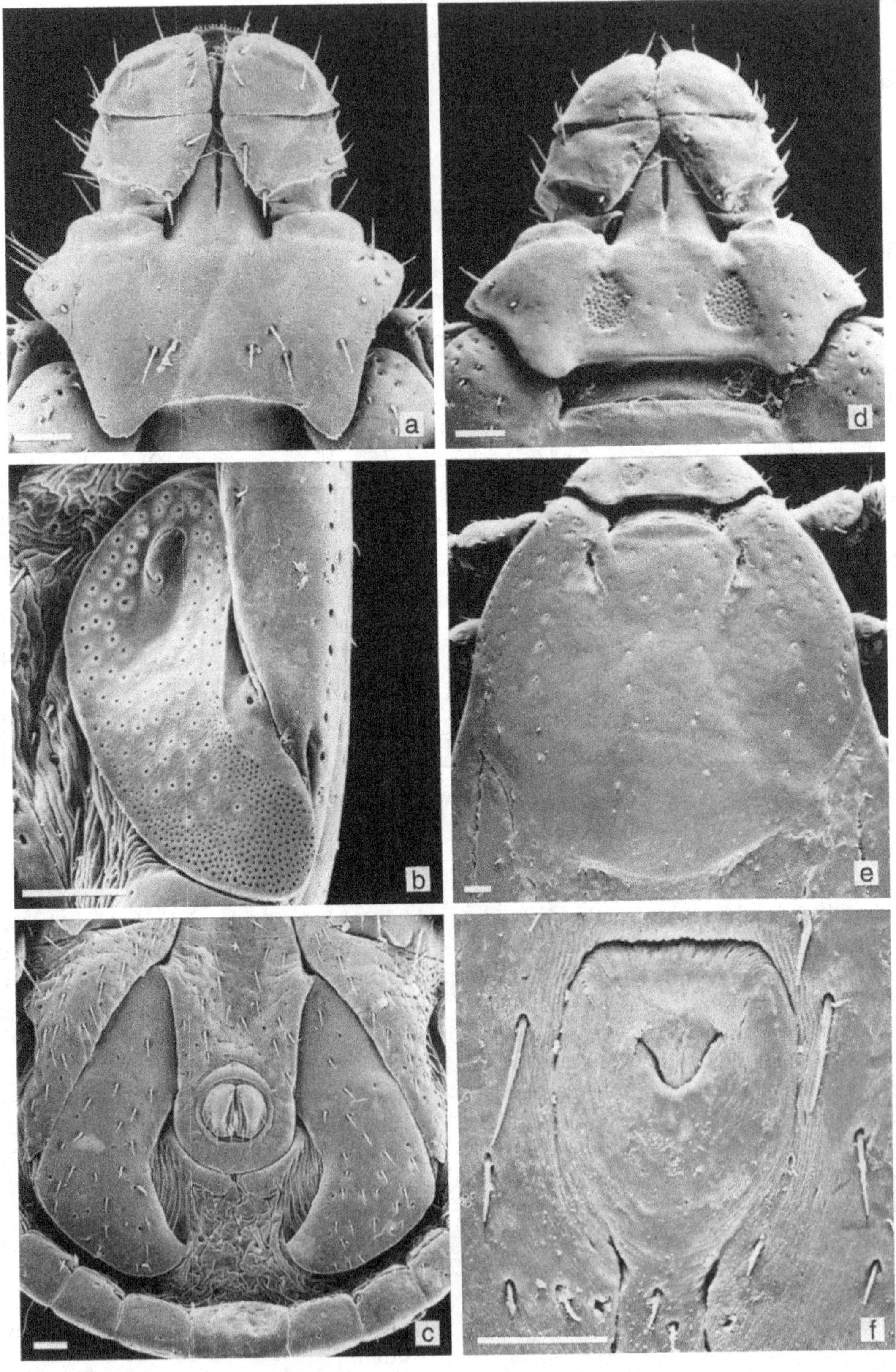
a
d
b
e
c
f

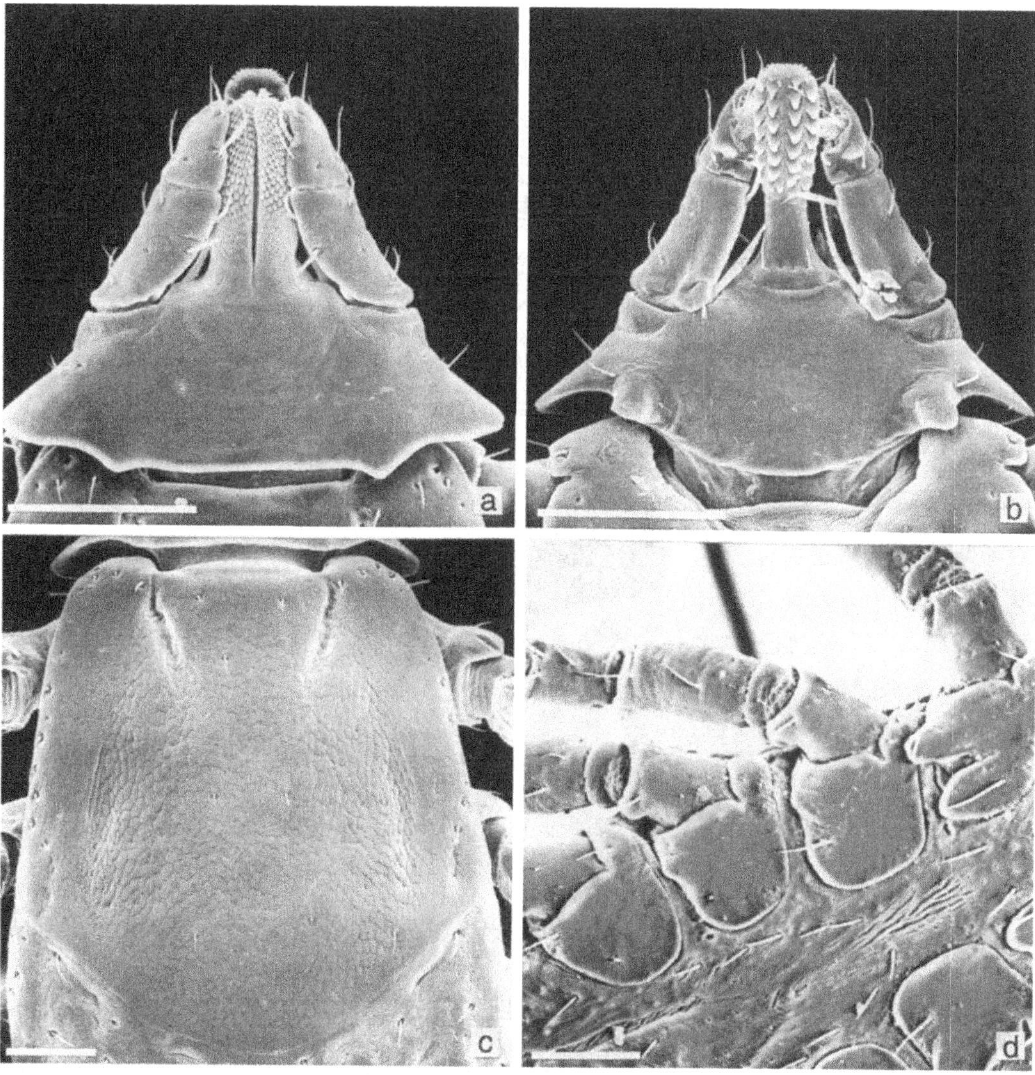

Figure 236 (*above*). *Rhipicephalus haemaphysaloides* (RML 84781; HH 82526, collected on greater bandicoot rat (*Bandicota indica*), University Hospital, Chingmai, Thailand, January 1972 by M. Kliks). Nymph: (a) capitulum, dorsal; (b) capitulum, ventral; (c) scutum; (d) coxae. Scale bars represent 0.10 mm. SEMs by P. Hill.

Figure 235 (*opposite*). *Rhipicephalus haemaphysaloides* (RML 91868; HH 81556, collected on vegetation around corn fields and the Trisuli River bank, Syaburbensi, Rasuwa, Nepal, June 30, 1970 by R. M. Mitchell). Male: (a) capitulum, dorsal; (b) spiracle; (c) adanal plates. Female: (d) capitulum, dorsal; (e) scutum; (f) genital aperture. Scale bars represent 0.10 mm. SEMs by P. Hill.

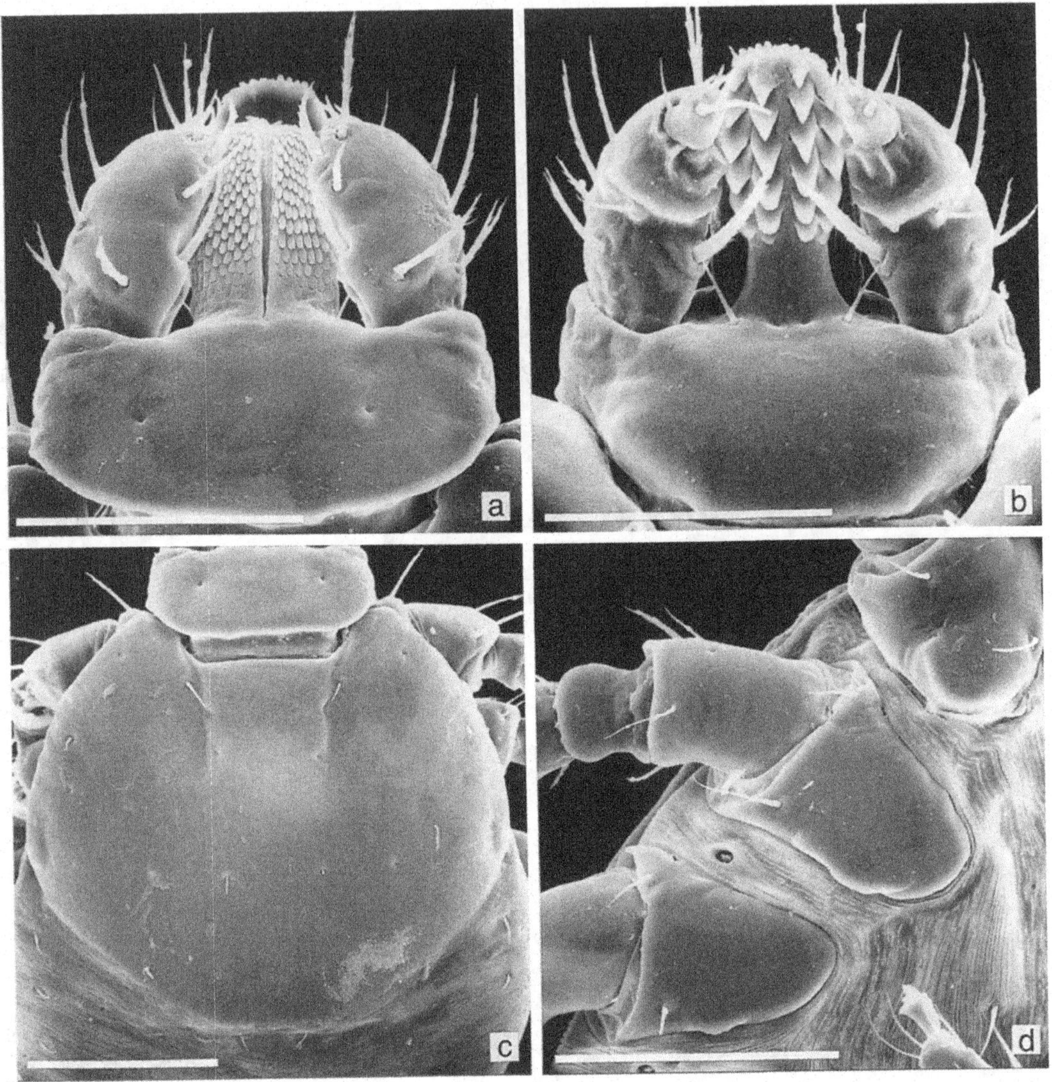

Figure 237. *Rhipicephalus haemaphysaloides* (RML 91868; HH 81556, collected on vegetation around corn fields and the Trisuli River bank, Syaburbensi, Rasuwa, Nepal, June 30, 1970 by R.M. Mitchell). Larva: (a) capitulum, dorsal; (b) capitulum, ventral; (c) scutum; (d) coxae. Scale bars represent 0.10 mm. SEMs by P. Hill.

Hosts

A three-host species (Jagannath, Alwar & Lalitha, 1972). In the first study of the life cycle of *R. haemaphysaloides*, Jagannath *et al.* (1972) found that the life cycle took from 77 to 112 days and that 10 females kept at 37 °C laid between 1102 and 4972 eggs. Gill & Bhattacharyulu (1981) found that the life cycle took from 72 to 107 days at 27 °C to 33 °C and 75% to 85% relative humidity (RH), and that females laid between 4448 and 6540 eggs. Ismail *et al.* (1982) found that the life cycle varied between 166 and 212 days at a temperature ranging from 23.9 °C to 28.9 °C, and that a single female deposited 3779 eggs over 19 days.

Ungulates account for most collections of this tick. It will feed on all domestic livestock, as well as wild sheep, goats and pigs (Table 62). In a study of ticks in Maharashtra State, India, Geevarghese & Dhanda (1995) recorded 591 adult *R. haemaphysaloides* on 1438 cattle, 154 on

Table 62. *Host records of* Rhipicephalus haemaphysaloides

Hosts	Number of records
Domestic animals	
Cattle	11
Water buffaloes	2
Sheep	5
Goats	8
Camels	2
Horses	4
Donkeys	2
Pigs	4
Dogs	6
Wild animals	
House shrew (*Suncus murinus*)	2 (immatures)
Golden jackal (*Canis aureus*)	3
Dhole (*Cuon alpinus*)	1
Asian golden cat (*Catopuma temminckii*)	1
Leopard (*Panthera pardus*)	2
Tiger (*Panthera tigris*)	1
Spotted deer (*Axis axis*)	1
Sambar (*Cervus unicolor*)	2
Indian muntjac (*Muntiacus muntjak*)	3
Nilgiri tahr (*Hemitragus hylocrius*)	1
Himalayan tahr (*Hemitragus jemlahicus*)	1
Common goral (*Naemorhedus sumatraensis*)	1
'Marmot' (*Marmota* sp.)	1
Indian gerbil (*Tatera indica*)	1 (immatures)
Greater bandicoot rat (*Bandicota indica*)	1 (nymph)
Indian bush rat (*Golunda ellioti*)	1 (immatures)
Little Indian field mouse (*Mus booduga*)	4 (immatures)
House mouse (*Mus musculus*)	15 (immatures)
Indian brown spiny mouse (*Mus platythrix*)	1 (immatures)
Mus saxicola	1 (immatures)
Blandford's rat (*Rattus blandfordi*)	3 (immatures)
Black rat (*Rattus rattus*)	8 (immatures)
Indian hare (*Lepus nigricollis*)	1
'Hare' (*Lepus* sp.)	1
Birds	
Crow pheasant (*Centropus sinensis*)	1 (immatures?)
Oriental yellow-eyed babbler (*Chrysomma sinense*)	1 (immatures?)
Humans	9

900 water buffaloes, 313 on 1065 sheep, and 406 on 1785 goats. Preferred attachment sites were the ears, but the axilla and inguinal areas were also utilized. It will also parasitize monkeys and humans. Rodents are the primary hosts for immatures, which will also feed on shrews. Geevarghese & Dhanda (1995) listed a total of 965 larvae and 300 nymphs of *R. haemaphysaloides* parasitizing the following hosts: the Indian gerbil, Indian bush rat, Blandford's rat, black rat, little Indian field mouse, Indian brown spiny mouse and *Mus saxicola,* as well as the following birds, the oriental yellow-eyed babbler and crow pheasant.

Mitchell (1979) listed 16 species of mammalian hosts for this tick in Nepal, and Robbins *et al.* (1997) recorded a single female *R. haemaphysaloides* from the endangered Asian golden cat.

Zoogeography

Rhipicephalus haemaphysaloides is distributed in the lowlands from Afghanistan, Pakistan, Nepal, India and Sri Lanka eastward throughout continental Southeast Asia to Taiwan and the Indonesian islands of Sumatra Utara and Sulawesi (Anastos, 1950; Kadarsan, 1971; Kolonin, 1992) (Map 66). This tick has also recently been recorded from the Lao People's Democratic Republic (Petney & Keirans, 1996; Robbins *et al.*, 1997). The single record from Luzon in the Philippines (Pippin, 1966) needs verification. For many years *R. haemaphysaloides* was considered to have a distribution in the northern portion of the Oriental region, with *R. pilans* occupying the southern portion of this region. However, this allopatric distribution has been questioned by Kadarsan (1971) with the finding of *R. haemaphysaloides* on Sulawesi where *R. pilans* is also known to occur.

Disease relationships

In the laboratory, *R. haemaphysaloides* has been shown to be a competent vector of Kyasanur Forest disease virus (Bhat *et al.*, 1978). However, it was unable to transmit *Theileria annulata*, a causative agent of theileriosis, transstadially from the larval stage to the nymph and to the adult (Bhattacharyulu, Chaudheri & Gill, 1975).

REFERENCES

Anastos, G. (1950). The scutate ticks, or Ixodidae, of Indonesia. *Entomologica Americana*, **30**(n.s.), 1–144.

Bhat, H.R., Naik, S.V., Ilkal, M.A. & Banerjee, K. (1978). Transmission of Kyasanur Forest disease virus by *Rhipicephalus haemaphysaloides* ticks. *Acta Virologica*, **22**, 241–4.

Bhattacharyulu, Y., Chaudheri, R.P. & Gill, B.S. (1975). Transstadial transmission of *Theileria annulata* through common ixodid ticks infesting Indian cattle. *Parasitology*, **71**, 1–7.

Geevarghese, G. & Dhanda, V. (1995). Ixodid ticks of Maharashtra state, India. *Acarologia*, **36**, 309–13.

Gill, H.S. & Bhattacharyulu, Y. (1981). Note on laboratory studies on the life history of *Rhipicephalus haemaphysaloides* Supino, 1897 (Acarina: Ixodidae). *Indian Journal of Animal Science*, **51**, 901–2.

Ismail, S., Nadchatram, M., Ho, T.M. & Rajamanickam, C. (1982). The life cycle of *Rhipicephalus h. haemaphysaloides* (Acarina: Ixodidae) under laboratory conditions. *Malayan Nature Journal*, **35**, 73–5.

Jagannath, M.S., Alwar, V.S. & Lalitha, C.M. (1972). Study on the life-history of *Rhipicephalus haemaphysaloides* Supino, 1897 (Acarina: Ixodidae). *Indian Journal of Animal Science*, **42**, 847–60.

Kadarsan, S. (1971). *Larval ixodid ticks of Indonesia (Acarina: Ixodidae).* PhD thesis: University of Maryland, College Park, 182 pp.

Kolonin, G.V. (1992). The fauna of ixodid ticks (Acarina, Ixodidae) of vertebrate animals in Vietnam, In *Zoological Research in Vietnam*, ed. V.E. Sokolov, pp. 242-76. Moscow: Science. [In Russian].

Mitchell, R.M. (1979). A list of ectoparasites from Nepalese mammals, collected during the Nepal ectoparasite program. *Journal of Medical Entomology*, **16**, 227–33.

Petney, T.N. & Keirans, J.E. (1996). Ticks of the genera *Boophilus, Dermacentor, Nosomma* and *Rhipicephalus* (Acari: Ixodidae) in South-east Asia. *Tropical Biomedicine*, **13**, 73–84.

Pippin, W.F. (1966). Scientific note. An addition to the tick fauna of the Philippines. *Philippine Journal of Science*, **94**, 395.

Robbins, R.G., Karesh, W.B., Rosenberg, S., Schonwalter, N. & Inthavong, C. (1997). Two noteworthy collections of ticks (Acari: Ixodida: Ixodidae) from endangered carnivores in the Lao People's Democratic Republic. *Entomological News*, **108**, 60–2.

Supino, F. (1897). Nuovi Ixodes della Birmania. (Nota preventiva). *Atti della Società Veneto-Trentia di Scienze Naturali in Padova, 2nd Ser*, **3**, 230–8.

RHIPICEPHALUS LEPORIS POMERANTSEV, 1946

The specific name *leporis* is derived from the Latin *lepus*, genitive *leporis*, meaning 'belonging to the hare'.

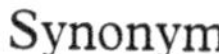

Synonym

?*pomeranzevi*.

Diagnosis

A small, light to reddish-brown species.

Male (Figs 238(a), 239(a) to (c))

Capitulum broader than long, length × breadth ranging from 0.55 mm × 0.62 mm to 0.62 mm × 0.69 mm. Basis capituli with short, acute lateral angles at the anterior third of its length. Palps short, broad. Conscutum narrow anteriorly, broadening posterior to eyes, length × breadth ranging from 2.53 mm × 1.34 mm to 2.80 mm × 1.74 mm. Body wall expanded laterally and posteriorly with festoon VI slightly protruded in engorged specimens. Eyes marginal, very slightly raised from conscutal surface, edged dorsally with two or three small punctations.

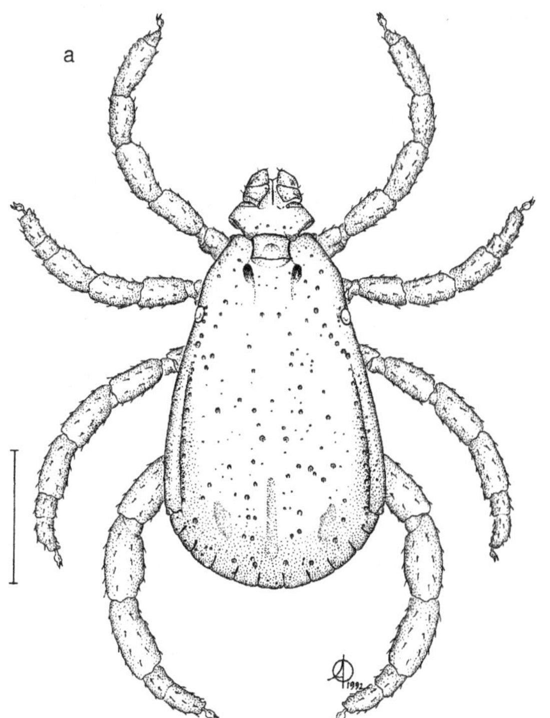

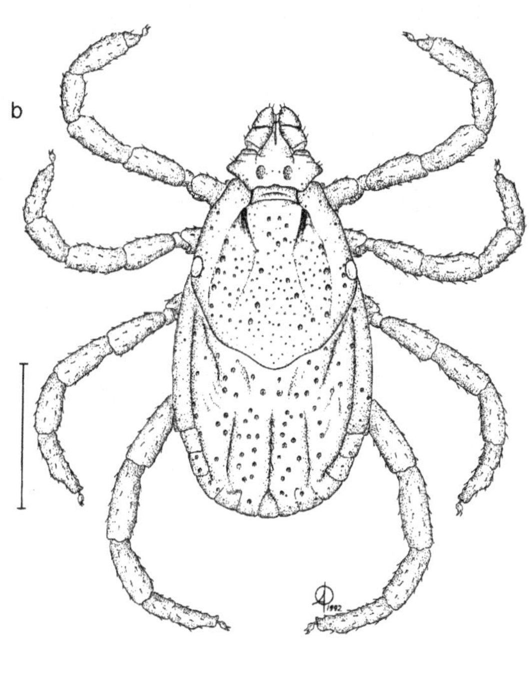

Figure 238. *Rhipicephalus leporis* (RML 74681; HH 30542, collected on red fox (*Vulpes vulpes*), 8–11 km west of Herat, Afghanistan, September 19, 1965 by the Street Expedition). (a) Male, dorsal; (b) female, dorsal. Scale bars represent 1 mm. A. Olwage *del.*

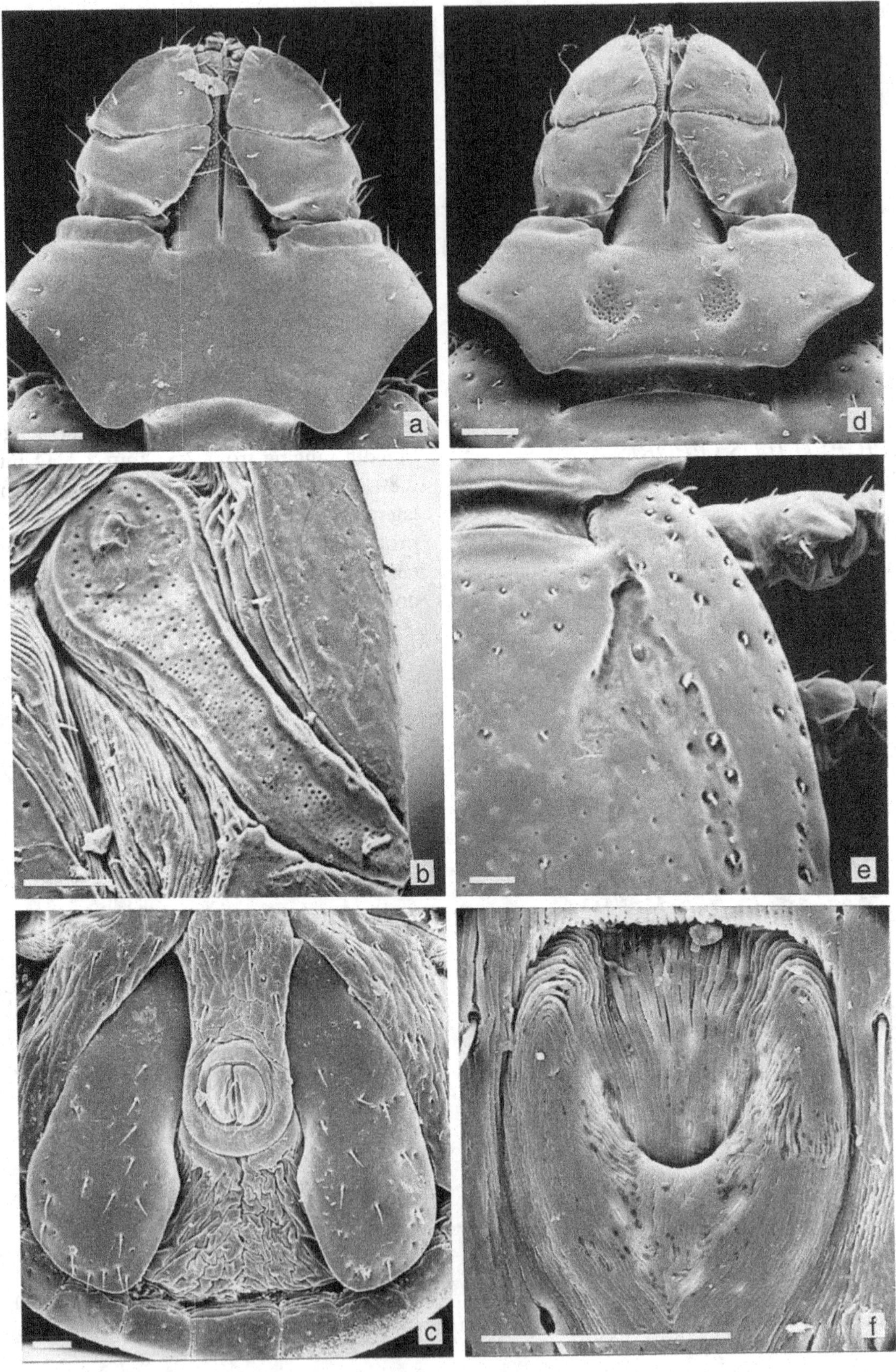
a
d
b
e
c
f

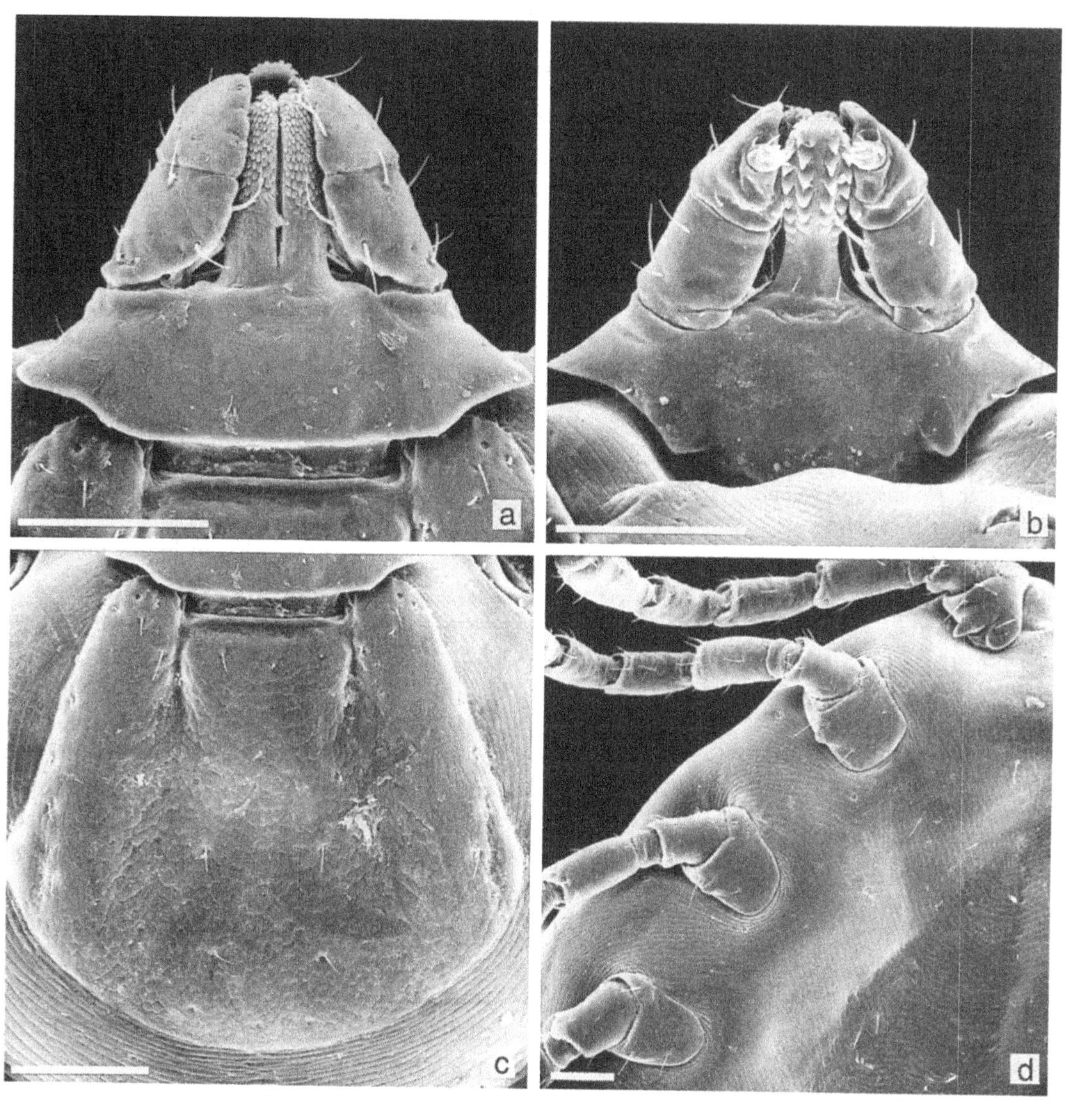

Figure 240 (*above*). *Rhipicephalus leporis* (RML 74681; HH 30542, collected on red fox (*Vulpes vulpes*), 8–11 km west of Herat, Afghanistan, September 19, 1965 by the Street Expedition). Nymph: (a) capitulum, dorsal; (b) capitulum, ventral; (c) scutum; (d) coxae. Scale bars represent 0.10 mm. SEMs by P. Hill.

Figure 239 (*opposite*). *Rhipicephalus leporis* (RML 74681; HH 30542, collected on red fox (*Vulpes vulpes*), 8–11 km west of Herat, Afghanistan, September 19, 1965 by the Street Expedition). Male: (a) capitulum, dorsal; (b) spiracle; (c) adanal plates. Female: (d) capitulum, dorsal; (e) scapular area; (f) genital aperture. Scale bars represent 0.10 mm. SEMs by P. Hill.

Table 63. *Host records of* Rhipicephalus leporis

Hosts	Number of records
Domestic animals	
Goats	1
Dogs	2
Wild animals	
Desert hedgehog (*Hemiechinus aethiopicus*)	2
Long-eared hedgehog (*Hemiechinus auritus*)	3
Brandt's hedgehog (*Hemiechinus hypomelas*)	2
Golden jackal (*Canis aureus*)	3
Red fox (*Vulpes vulpes*)	6 (including nymphs)
Steppe polecat (*Mustela eversmannii*)	1
Great gerbil (*Rhombomys opimus*)	1
Cape hare (*Lepus capensis*)	5
Tolai hare (*Lepus tolai*)	12 (including immatures)
'Hare'	3

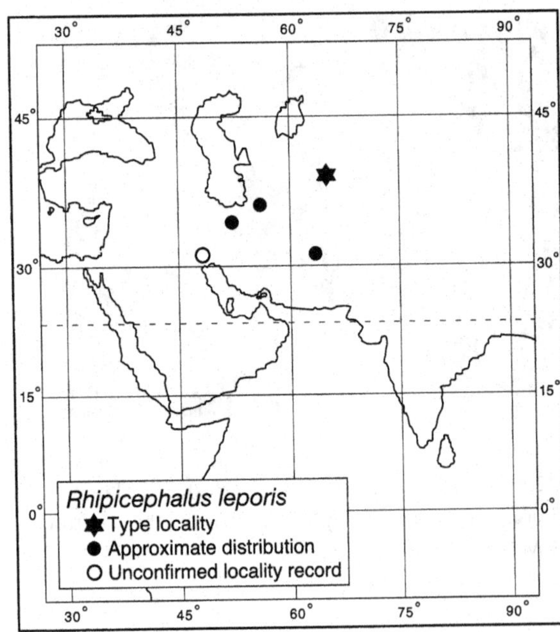

Map 67. *Rhipicephalus leporis*: distribution.

Cervical pits short, deep, comma-shaped to slightly convergent; external cervical margins indicated by a short row of medium-sized setiferous punctations; an indication of punctations in a '*simus*' pattern on central conscutum, interspersed with numerous fine punctations especially posteriorly. Marginal lines narrow, shallow, punctate, usually not delimiting first festoons and ending well before eyes. Posteromedian groove elongate, narrow; posterolateral grooves semicircular, not as deep as posteromedian groove. Ventrally spiracles small, compact each with a very long, narrow dorsal prolongation. Adanal plates tear-drop-shaped, slightly concave medially with rounded posterior margins; accessory adanal plates represented by small almost translucent hyaline points.

Female (Figs 238(b), 239(d) to (f))

Capitulum much broader than long, length × breadth ranging from 0.56 mm × 0.66 mm to 0.65 mm × 0.80 mm. Basis capituli with short bluntly-pointed lateral angles at about one half of its length; porose areas small, subcircular, almost twice their own diameter apart. Palps short, broad. Scutum longer than broad, length × breadth ranging from 1.26 mm × 1.17 mm to 1.51 mm × 1.43 mm, posterior margin sinuous giving the scutum a shield-like shape. Eyes at lateral angles, flat to very slightly raised, with or without edging punctations. Internal cervical margins convergent then divergent; cervical fields depressed, scalpel-shaped, external cervical margins relatively steep, edged with large punctations; raised lateral borders with a few punctations. Ventrally genital aperture broadly U-shaped.

Nymph (Fig. 240)

Only two specimens available for measurements. Capitulum broader than long, length × breadth ranging from 0.23 mm × 0.27 mm to 0.25 mm × 0.30 mm. Basis capituli more than three times as broad as long with sharply-pointed lateral angles; posterior margin straight, then becoming concave posterolaterally over scapulae, cornua absent; ventrally with small spurs on its posterior margin. Palps short, tapering over hypostome, external margins straight throughout most of their length. Scutum much broader than long, length × breadth ranging from 0.44 mm × 0.54 mm to 0.46 mm × 0.55 mm. Eyes at posterolateral angles, well over halfway back, marginal, slightly raised. Cervical grooves short, parallel to slightly convergent; cervical fields broad, only slightly indicated. Ventrally coxae I each with two very small pointed spurs; coxae II and III each with a minute external spur; coxae IV each without spurs.

Larva

Unknown.

Notes on identification

Males of *R. leporis* superficially resemble males of *R. pumilio*, but the small compact spiracles with their extremely narrow dorsal prolongations and the tear-drop shaped adanal plates separate them from *R. pumilio*. The smaller-sized scutum with large punctations on the external cervical margins separates females of *R. leporis* from those of *R. pumilio*.

Hosts

Presumed to be a three-host tick. The primary host of *R. leporis* is the Tolai hare, on which all stages will feed. Other hosts include hedgehogs, medium-sized carnivores and the Cape hare (Table 63). The U.S. National Tick Collection contains specimens from dogs, the golden jackal and Brandt's hedgehog.

In Turkmenistan *R. leporis* parasitizes the Tolai hare in all months of the year except December (Afanaseva & Sapozhenkov, 1965).

Zoogeography

Rhipicephalus leporis is known from Kazakhstan, Turkmenistan, Afghanistan and Iran (Hoogstraal, 1973; Arsen'eva & Neronov, 1980) (Map 67). Reports of this tick species in Iraq are considered tentative identifications (Shamsuddin & Mohammad, 1988).

Disease relationships

Unknown.

REFERENCES

Afanaseva, O.V. & Sapozhenkov, Yu.F. (1965). Data on the ixodid ticks of the hare *Lepus* in Turkmenia. *Izvestiya Akademii Nauk Turkmenskoy SSR*, 4, 91–2. [In Russian].

Arsen'eva, L.P. & Neronov, V.M. (1980). Ticks – ectoparasites of wild and domestic animals in Afghanistan. *Meditsinskaya Parazitologiya, Moskva*, **49**, 37–42. [In Russian, English translation T1553, NAMRU-3, Cairo].

Hoogstraal, H. (1973). Biological patterns in the Afghanistan tick fauna. In *Proceedings of the 3rd International Congress of Acarology, Prague, August 31 – September 6, 1971*, ed. M. Daniel & B. Roziky, pp. 511–14. Prague: Academia.

Pomerantsev, B.I. (1946). Les tiques (Ixodidae) de la faune de l'URSS et des pays limitrophes. *Opredeliteli po Faune SSSR, Izdavaemye Zoologicheskim Institutom Akademii Nauk SSSR*, (26), 1–28. [In Russian, French title].

Shamsuddin, M. & Mohammad, M. K. (1988). Incidence, distribution, and host relationships of some ticks (Ixodoidea) in Iraq. *Journal of the University of Kuwait (Sciences)*, **15**, 321–9.

Also see the following Basic Reference (p. 13): Filippova (1997).

RHIPICEPHALUS PILANS SCHULZE, 1935

The specific name, from the Latin, *pil, pilus* meaning 'hair', refers to the distinctive short white setae found on this species.

Synonyms

haemaphysaloides pilans; haemaphysaloides paulopunctata; paulopunctatus.

Diagnosis

A medium-sized, dark brown species.

Male (Figs 241(a), 242(a) to (c))
Capitulum longer than broad to slightly broader than long, length × breadth ranging from 0.69 mm × 0.65 mm to 0.87 mm × 0.88 mm. Basis

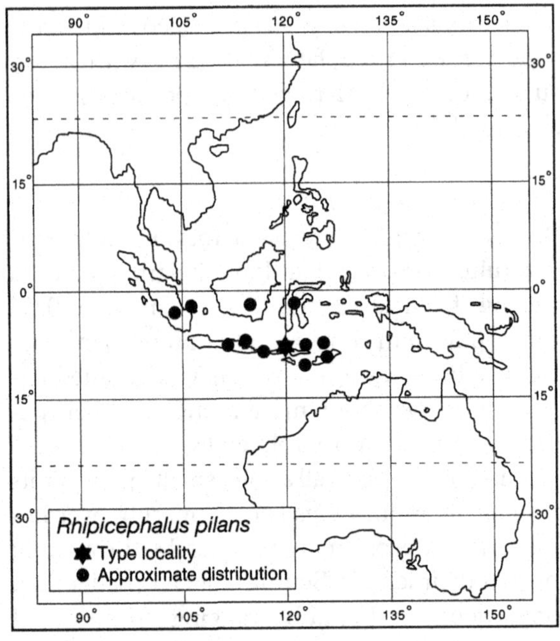

Map 68. *Rhipicephalus pilans*: distribution.

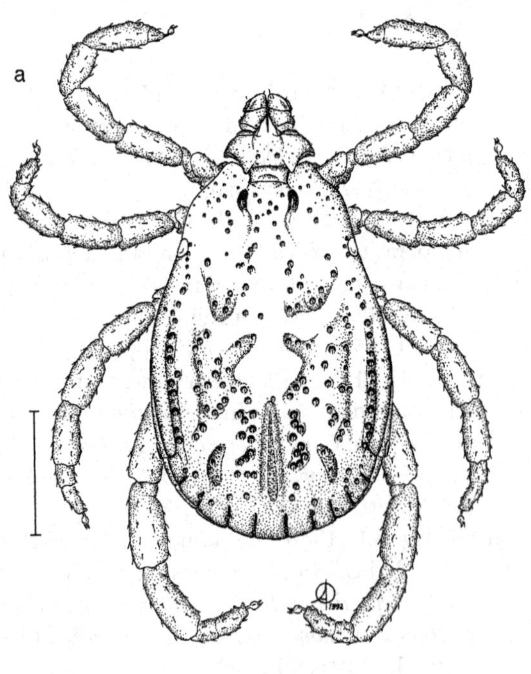

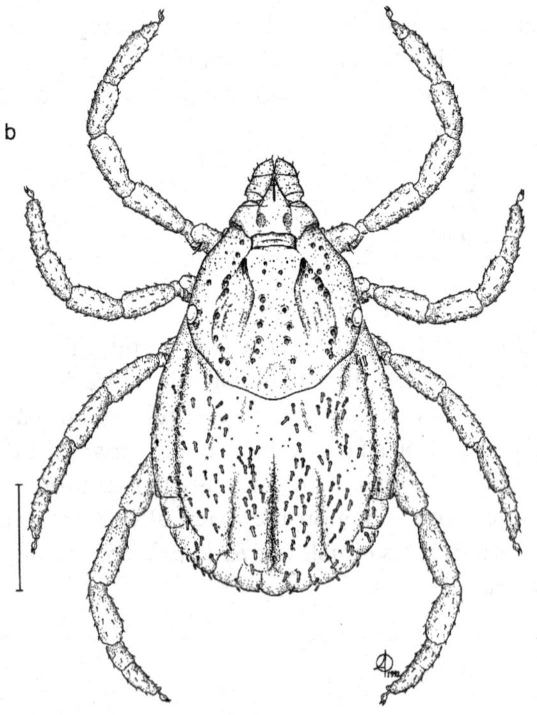

Figure 241. *Rhipicephalus pilans* ♂ (RML 94881; HH 81785, collected on common palm civet (*Paradoxurus hermaphroditus*), Lampung Selatan, Sumatra, Indonesia, June 11, 1971, by H. Hoogstraal, P.F.D. Van Peenan and S. Gaber). ♀ (RML 94875; HH 81778, collected on vegetation, Lampsing, Lampung Selatan, Sumatra, Indonesia, June 8, 1971, by H. Hoogstraal and S. Gaber). (a) Male, dorsal; (b) female dorsal. Scale bars represent 1 mm. A. Olwage *del.*

Table 64. *Host records of* Rhipicephalus pilans

Hosts	Number of records
Domestic animals	
Cattle	5
Water buffaloes	12
Horses	8 (including immatures)
Pigs	4
Dogs	6 (including immatures)
Wild animals	
Lesser moonrat (*Hylomys suillus*)	1 (immatures)
Sulawesi white-toothed shrew (*Crocidura nigripes*)	2 (immatures)
House shrew (*Suncus murinus*)	2
Leschenault's rousette (*Rousettus leschenaulti*)	1 (immatures)
Tiger (*Panthera tigris*)	1
Common palm civet (*Paradoxurus hermaphroditus*)	1
Timor deer (*Cervus timorensis*)	2 (including immatures)
Sambar (*Cervus unicolor*)	1
Greater bandicoot rat (*Bandicota indica*)	2 (immatures)
Sulawesi rat (*Bunomys penitus*)	1 (immatures)
Oriental spiny rat (*Maxomys bartelsii*)	2 (immatures)
Chestnut rat (*Niviventer fulvescens*)	1 (immatures)
Ricefield rat (*Rattus argentiventer*)	3 (immatures)
Polynesian rat (*Rattus exulans*)	24 (immatures)
Rattus tanezumi	3 (immatures)
Malaysian field rat (*Rattus tiomanicus*)	2 (immatures)
'Rat' (*Rattus* sp.)	2 (immatures)
Humans	2

capituli with short, acute lateral angles in the anterior third of its length. Palps short, broad. Conscutum narrower anteriorly, broadening posterior to eyes, length × breadth ranging from 2.63 mm × 1.63 mm to 3.50 mm × 2.46 mm. Body wall only slightly expanded laterally and posteriorly in engorged specimens with a small bulbous caudal protrusion. Eyes marginal, flat, usually edged dorsally with a few small punctations. Cervical pits deep, short, slightly converging to comma-shaped. Numerous large, deep, setiferous punctations scattered over conscutum giving the surface a pilose, irregularly roughened appearance. Marginal lines deep, punctate, delimiting first festoons and extending almost to eyes. Posteromedian and posterolateral grooves very conspicuous; posteromedian groove a deep trough, posterolateral grooves deep and comma-shaped. Ventrally spiracles elongate in the anteroposterior plane, each with a short broad dorsal prolongation. Adanal plates narrow anteriorly, broad posteriorly with relatively straight posterior margins, distinctly concave medially with small pointed cusps; accessory adanal plates sharply V-shaped sclerotized points.

Female (Figs 241(b), 242(d) to (f))

Capitulum broader than long, length × breadth ranging from 0.65 mm × 0.74 mm to 0.79 mm × 0.89 mm. Basis capituli with lateral angles at about the anterior third of its length; porose areas small, suboval, almost twice their own

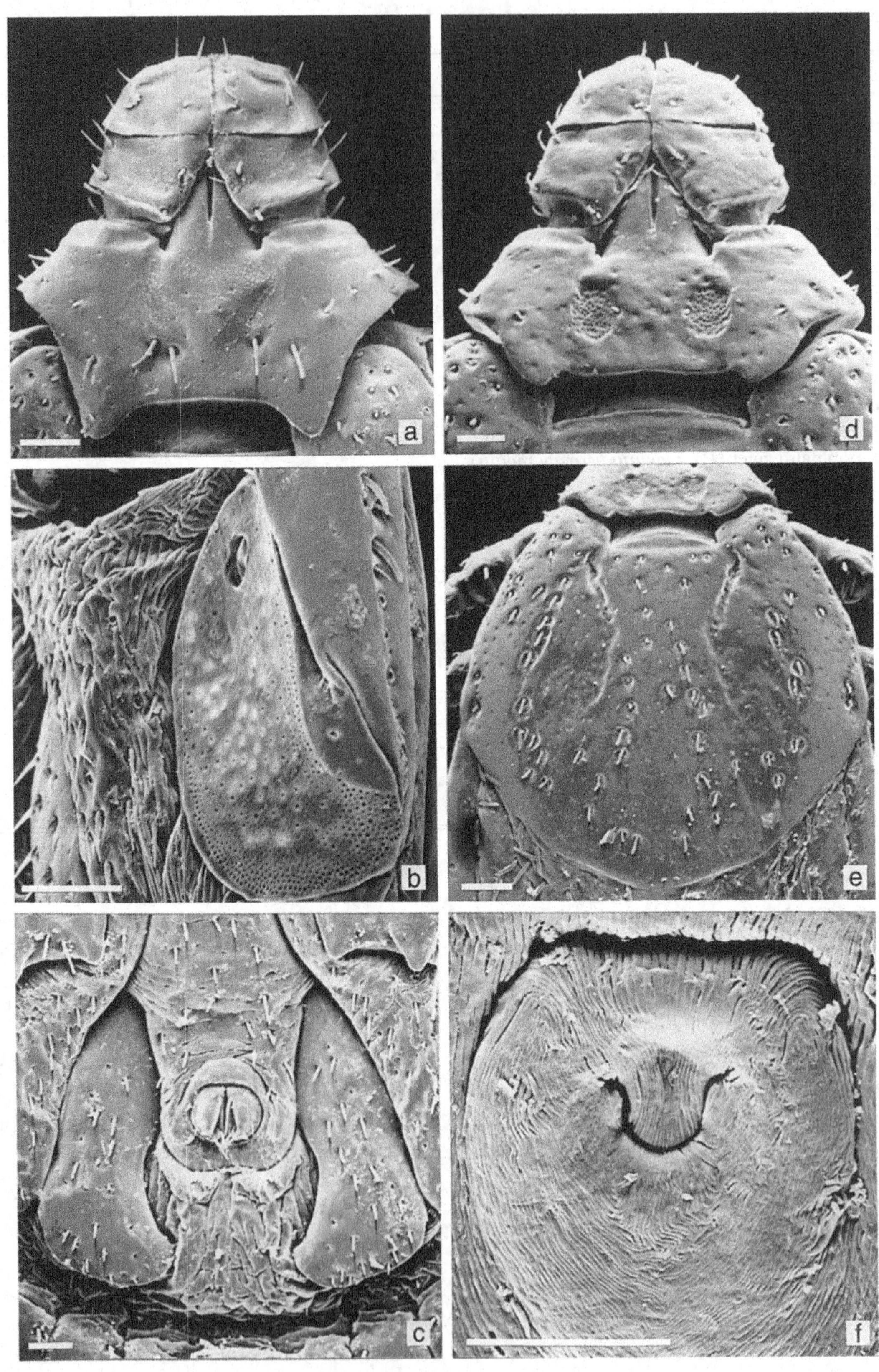
a
d
b
e
c
f

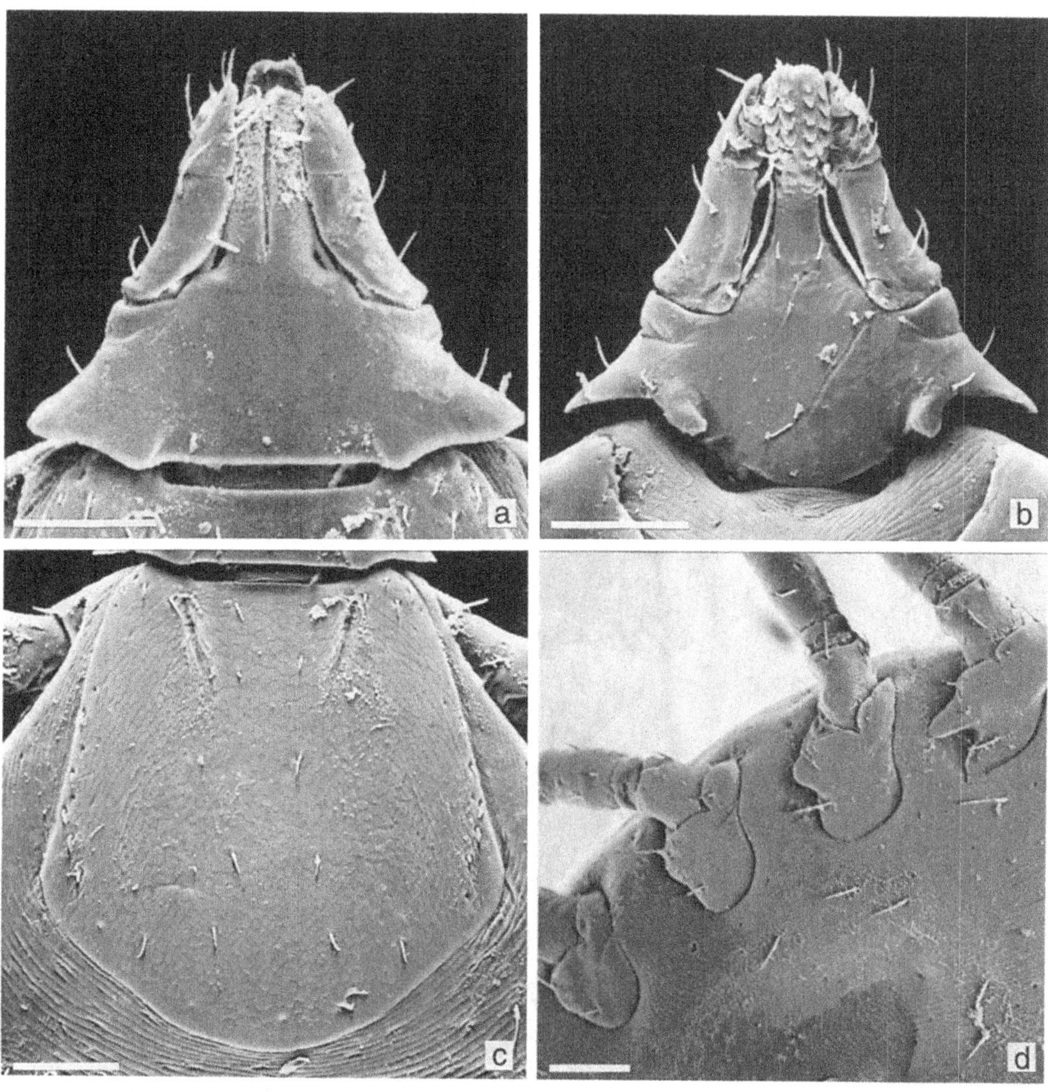

Figure 243 (*above*). *Rhipicephalus pilans* (RML 120345; HH 84011, collected on greater bandicoot rat (*Bandicota indica*), Sunter, Jakarta, Java, Indonesia, November 23, 1976, by J. D. M.). Nymph: (a) capitulum, dorsal; (b) capitulum, ventral; (c) scutum; (d) coxae. Scale bars represent 0.10 mm. SEMs by P. Hill.

Figure 242 (*opposite*). *Rhipicephalus pilans* ♂ (RML 120339; HH 81781, collected on vegetation in disturbed original forest, Lampung Tengah, Sumatra, Indonesia, June 10, 1971, by H. Hoogstraal and S. Gaber). Male: (a) capitulum, dorsal; (b) spiracle; (e) adanal plates; female: (RML 94873; HH 81572, collected on domestic pig, Loho Liang Valley, Komodo Island, Indonesia, March 27, 1970, by W. Auffenberg), (d) capitulum, dorsal; (e) scutum; (f) genital aperture. Scale bars represent 0.10 mm. SEMs by P. Hill.

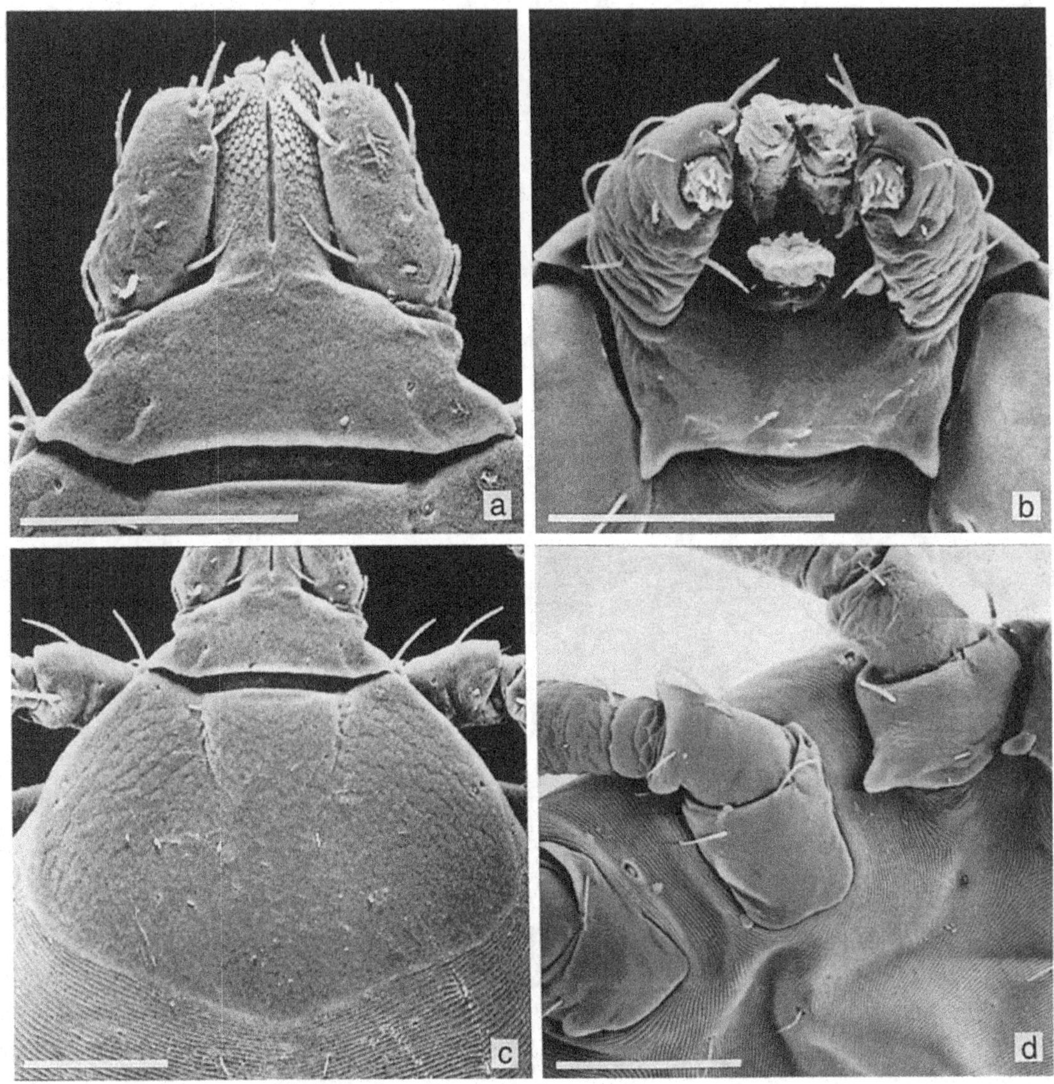

Figure 244. *Rhipicephalus pilans* (RML 94929; HH 82608, collected on Timor deer (*Cervus timorensis*), Uranga, Lake Lindu, Sulawesi, Indonesia, December 4, 1972). Larva: (a) capitulum, dorsal; (b) capitulum, ventral; (c) scutum; (d) coxae. Scale bars represent 0.10 mm. SEMs by P. Hill.

diameter apart. Scutum slightly longer than broad to slightly broader than long, length × breadth ranging from 1.40 mm × 1.35 mm to 1.62 mm × 1.67 mm, posterior margin sinuous. Eyes at lateral angles, flat to slightly raised, edged dorsally with a few punctations. Internal cervical margins relatively steep declinations, convergent then divergent; cervical fields sunken, roughly scalpel-shaped, with several medium-sized to large setiferous punctations along their external margins and on the central scutal area, fewer and smaller on raised lateral borders. Alloscutal surface pilose. Ventrally genital aperture U-shaped with lateral arms curving outward at the top.

Nymph (Fig. 243)

Capitulum much broader than long, length × breadth ranging from 0.23 mm × 0.31 mm to 0.30 mm × 0.35 mm. Basis capituli over 2.5 times as broad as long with tapering lateral angles overlapping the scapulae; cornua absent to very slightly indicated; ventrally with spurs on posterior margin. Palps elongate, tapering, external margin distinctly concave. Scutum longer than broad, length × breadth ranging from 0.43 mm × 0.41 mm to 0.53 mm × 0.43 mm. Eyes at widest point, well over halfway back, flat, outlined with a few small setiferous punctations. Cervical fields barely indicated. Ventrally coxae I each with two subequal sharply-pointed spurs, the external spur slightly longer; coxae II to IV each with a single small external spur.

Larva (Fig. 244)

Capitulum much broader than long, length × breadth ranging from 0.097 mm × 0.128 mm to 0.127 mm × 0.205 mm. Basis capituli over three times as broad as long, posterior margin slightly concave; lateral angles irregular, tapering. Scutum much broader than long, length × breadth ranging from 0.200 mm × 0.262 mm to 0.332 mm × 0.392 mm. Eyes at widest point, over halfway back, flat. Cervical grooves short, shallow, slightly convergent. Ventrally coxae I each with a broadly-rounded spur, coxae II and III each with a small spur close to the coxal margins.

Notes on identification

Although *R. pilans* was mentioned in a key by Schulze (1935) the full description of this species only appeared a year later (Schulze, 1936). *Rhipicephalus pilans* is closely related to, and has sometimes been considered to be a subspecies of, *R. haemaphysaloides* (Zumpt, 1940, 1943). However, we consider it a valid species. In addition to being a darker brown and somewhat larger tick than *R. haemaphysaloides, R. pilans* has numerous, deep scutal pits bearing thick white setae. Males of *R. pilans* have a furrowed and sculptured appearance because of irregular scutal indentations not found on males of *R. haemaphysaloides*.

Hosts

Presumed to be a three-host tick. Cattle, water buffaloes, horses, pigs, dogs, as well as deer are hosts of the adults, and Zumpt (1943) recorded it from a tiger (Table 64). Immatures have been collected from horses, dogs and rats (Kadarsan, 1971).

Zoogeography

Rhipicephalus pilans has been recorded from the Philippines and the Indonesian islands of Sumatra Selatan, Java, Bawean, Madura, Kalimantan, Sulawesi, Bali, Lombok, Sumbawa, Sumba, Komodo, Flores, Sawu, Roti, Timor and Alor (Anastos, 1950; Kadarsan, 1971) (Map 68).

Disease relationships

Unknown, but this tick was probably the species that was the basis for complaints of bites and distress of United States troops at the time of reoccupation of the Philippine island of Mindoro in early 1945 (Kohls, 1950).

REFERENCES

Anastos, G. (1950). The scutate ticks, or Ixodidae, of Indonesia. *Entomologica Americana*, **30** (n.s.), 1–144.

Kadarsan, S. (1971). *Larval ixodid ticks of Indonesia (Acarina: Ixodidae).* PhD thesis: University of Maryland, College Park. 182 pp.

Kohls, G.M. (1950). Ticks (Ixodoidea) of the Philippines. *National Institutes of Health Bulletin*, **No. 192**, 1–28.

Schulze, P. (1935). Acarina: Ixodoidea. *Wissenschaftliche Ergebnisse der Niederländischen Expeditionen in dem Karakorum (1922–1930)*, **1**, 178–86.

Schulze, P. (1936). Zwei neue *Rhipicephalus* und eine neue *Haemaphysalis* nebst Bemerkungen über Zeckenarten aus verschiedenen Gattungen. *Zeitschrift für Parasitenkunde*, **8**, 521–7.

Zumpt, F. (1940). Zur Kenntnis der ausserafrikanischen Rhipicephalusarten. II. Vorstudie zu einer Revision der Gattung *Rhipicephalus* Koch. *Zeitschrift für Parasitenkunde*, **11**, 669–78.

Zumpt, F. (1943). *Rhipicephalus simus* Koch und verwandte Arten. VII. Vorstudie zu einer Revision der Gattung *Rhipicephalus* Koch. *Zeitschrift für Parasitenkunde*, **13**, 1–24.

RHIPICEPHALUS PUMILIO SCHULZE, 1935

The specific name *pumilio*, from the Latin, *pumilius* meaning 'diminutive', 'dwarfish', refers to the small size of this species.

Diagnosis

A small to medium-sized light brown tick.

Male (Figs 245(a), 246(a) to (c))
Capitulum broader than long, length × breadth ranging from 0.57 mm × 0.58 mm to 0.65 mm × 0.75 mm. Basis capituli with acute lateral angles at about the anterior third of its length. Palps with gently-curved external margins leading to acute bluntly-tipped apices. Conscutum very slightly narrower anteriorly, thereafter about the same breadth throughout its length, length × breadth ranging from 2.30 mm × 1.21 mm to 3.05 mm × 1.45 mm. Body wall slightly expanded laterally in engorged specimens. Eyes

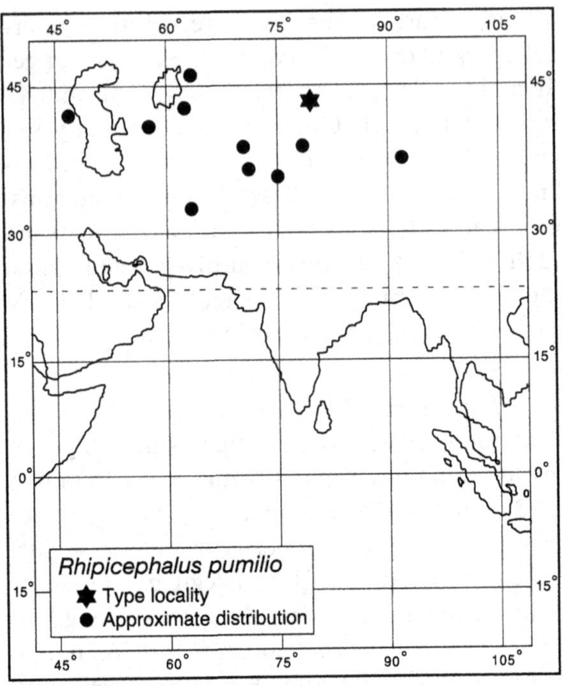

Map 69. *Rhipicephalus pumilio*: distribution.

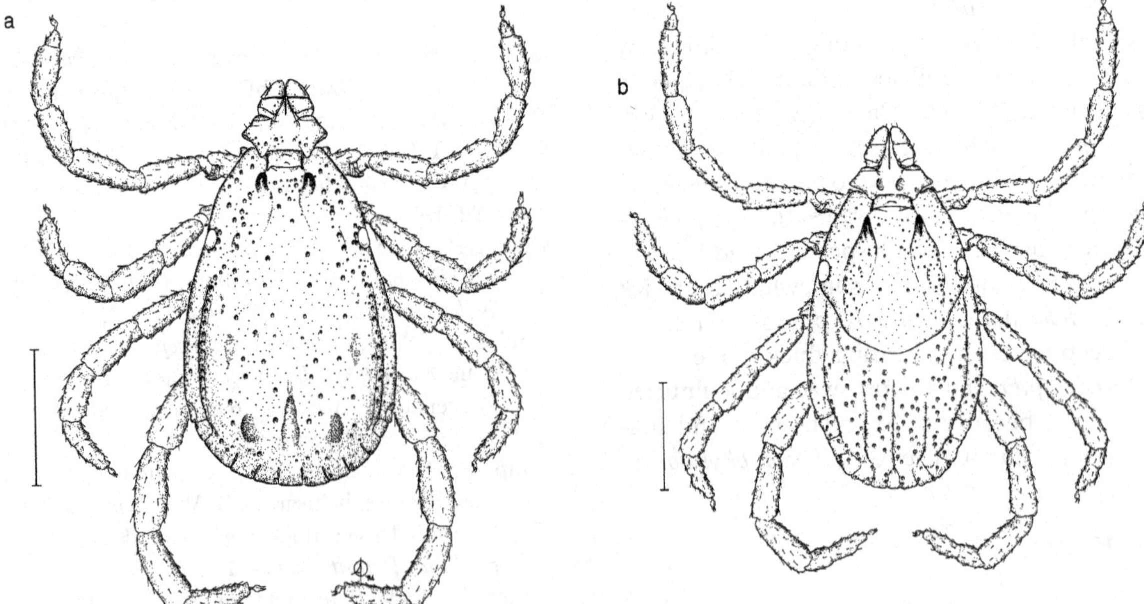

Figure 245. *Rhipicephalus pumilio* (RML 98604; HH 31277, collected on a rabbit, Bornova field, Izmir, Turkey, May 4, 1967 by Garrett). (a) Male, dorsal; (b) female, dorsal. Scale bars represent 1 mm. A. Olwage *del.*

marginal, flat, edged dorsally by a few punctations. Cervical pits deep anteriorly, then very shallow, wedge-shaped. Only the barest indication of cervical fields shown by slight depressions, not delimited by punctations along their internal margins, a few punctations usually present along external cervical margins; remainder of conscutum lightly punctate giving a smooth appearance. Marginal lines narrow, shallow, delimiting first festoons and ending well behind the eyes. Posteromedian and posterolateral grooves shallow and inconspicuous; posterolaterals subcircular. Ventrally spiracles narrow throughout, leading to a narrowly-elongate dorsal prolongation. Adanal plates large, subtriangular, with straight posterior margins and small internal cusps medially posterior to anus; accessory adanal plates prominent, elongate sclerotized points.

Female (Figs 245(b), 246(d) to (f))
Capitulum broader than long, length × breadth ranging from 0.67 mm × 0.76 mm to 0.72 mm × 0.82 mm. Basis capituli with pointed lateral angles; porose areas small, subcircular, slightly more than their own diameter apart. Scutum longer than broad, length × breadth ranging from 1.40 mm × 1.33 mm to 1.56 mm × 1.48 mm, posterior margin sinuous. Eyes at lateral angles, almost flat, edged dorsally with rather deep punctations. Internal cervical margins converging then diverging; cervical fields very slightly sunken internally, more deeply sunken externally giving external cervical margins a steeper declination; without punctations along external cervical margins. Small to medium-sized punctations scattered on central area of scutum, finer and less numerous on raised lateral borders. Ventrally genital aperture broadly U-shaped.

Nymph (Fig. 247)
Capitulum much broader than long, length × breadth ranging from 0.21 mm × 0.31 mm to 0.25 mm × 0.32 mm. Basis capituli over three times as broad as long with concave lateral angles not overlapping the scapulae; cornua absent. Palps short, tapering, external margins straight. Scutum broader than long, length × breadth ranging from 0.47 mm × 0.49 mm to 0.50 mm × 0.53 mm. Eyes marginal, large, slightly bulging, well over halfway back. Internal cervical margins converging then diverging; external cervical margins a slight declination; cervical fields narrowly scalpel-shaped. Ventrally coxae I each with a long pointed external spur and a shorter, more rounded internal spur; coxae II and III each with a single, short external spur; coxae IV each without spurs.

Larva (Fig. 248)
Capitulum much broader than long, length × breadth ranging from 0.114 mm × 0.154 mm to 0.122 mm × 0.167 mm. Basis capituli over three times as broad as long, posterior margin straight. Palps broad basally, narrowing slightly towards their apices. Scutum much broader than long, length × breadth ranging from 0.234 mm × 0.324 mm to 0.260 mm × 0.341 mm. Eyes at widest point, well over halfway back. Cervical grooves short, shallow, convergent.

Notes on identification
The presence of internal cusps on the adanal plates helps to separate males of this species from those of *R. schulzei*. The female capitula of the two species are also morphologically distinct. Zahler *et al.* (1997) found that *R. pumilio* and *R. rossicus* possessed identical DNA sequences in the target gene investigated, and suggested cross-breeding experiments to see whether these two species are conspecific.

Hosts

Presumed to be a three-host species with a developmental cycle of 1 year. The host range of *R. pumilio* is quite wide. Adults feed chiefly on hares, especially the Tolai hare, and hedgehogs, but also on larger-sized rodents, wolves, jackals, gazelle, domestic mammals and humans (Table 65). Nymphs and larvae prefer gerbils, hares and hedgehogs. Pomerantsev (1950) lists hosts in

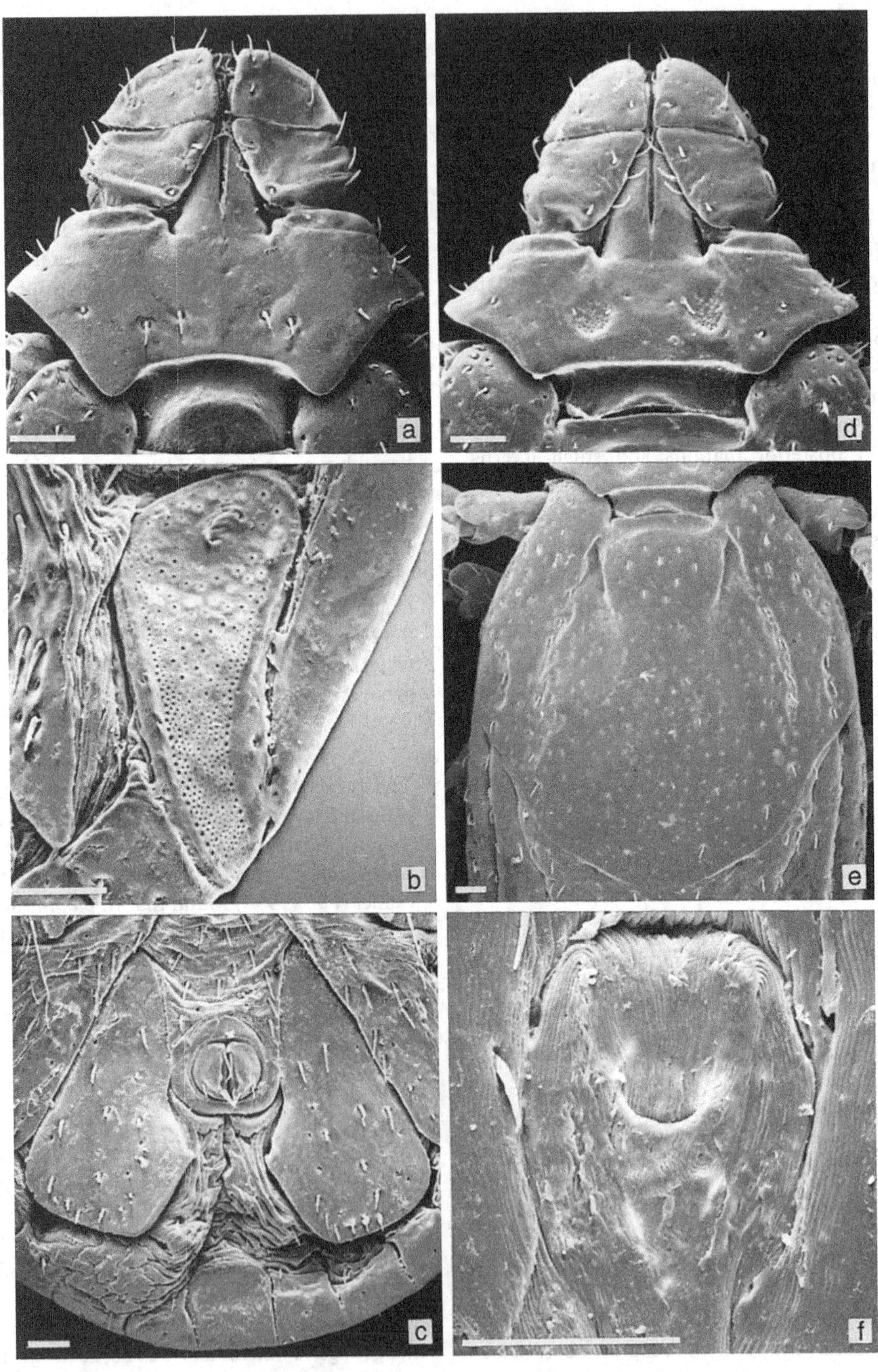
a
d
b
e
c
f

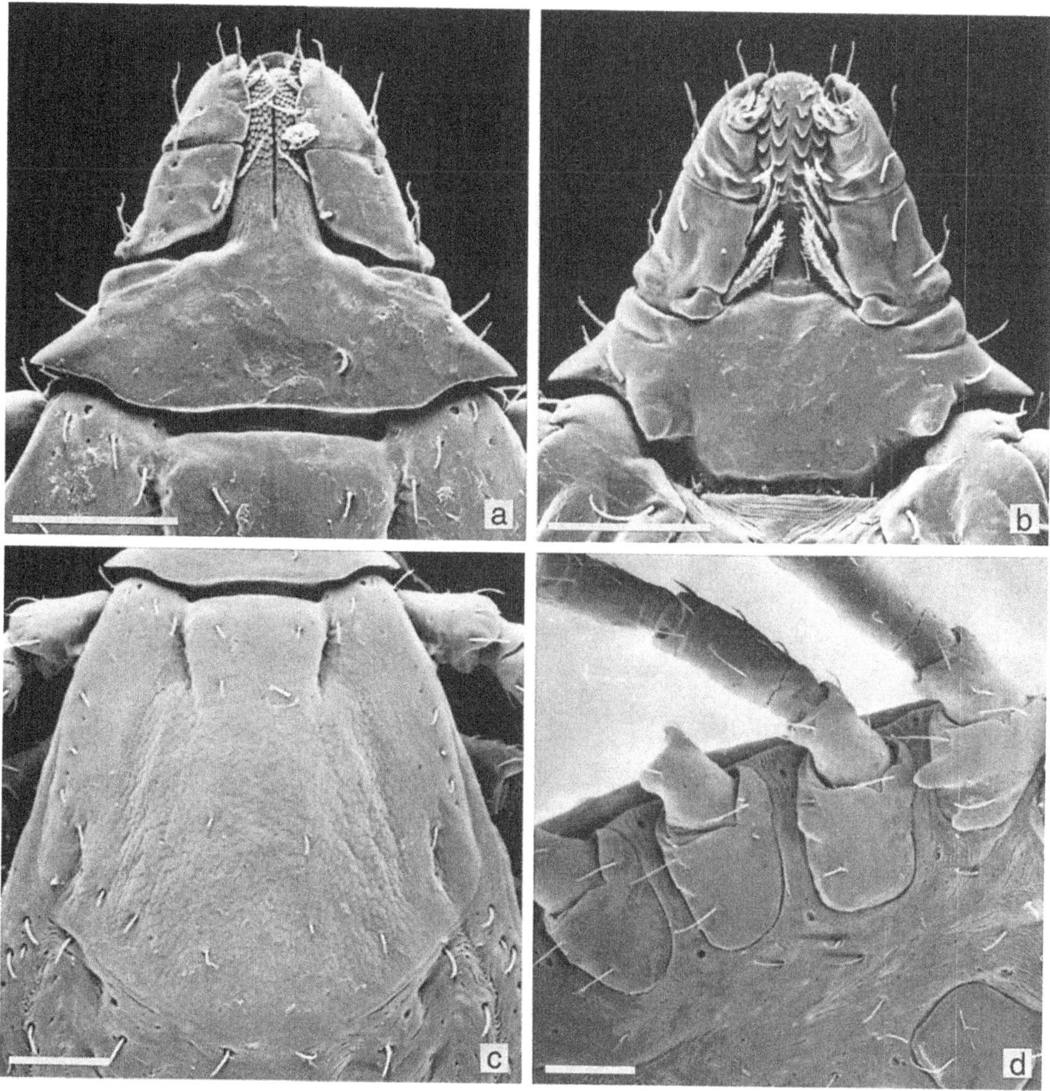

Figure 247 (*above*). *Rhipicephalus pumilio* (RML 98604; HH 31277, collected on a rabbit, Bornova field, Izmir, Turkey, May 4, 1967 by Garrett). Nymph: (a) capitulum, dorsal; (b) capitulum, ventral; (c) scutum; (d) coxae. Scale bars represent 0.10 mm. SEMs by P. Hill.

Figure 246 (*opposite*). *Rhipicephalus pumilio* (RML 98604; HH 31277, collected on a rabbit, Bornova field, Izmir, Turkey, May 4, 1967 by Garrett). Male: (a) capitulum, dorsal; (b) spiracle; (c) adanal plates. Female: (d) capitulum, dorsal; (e) scutum; (f) genital aperture. Scale bars represent 0.10 mm. SEMs by P. Hill.

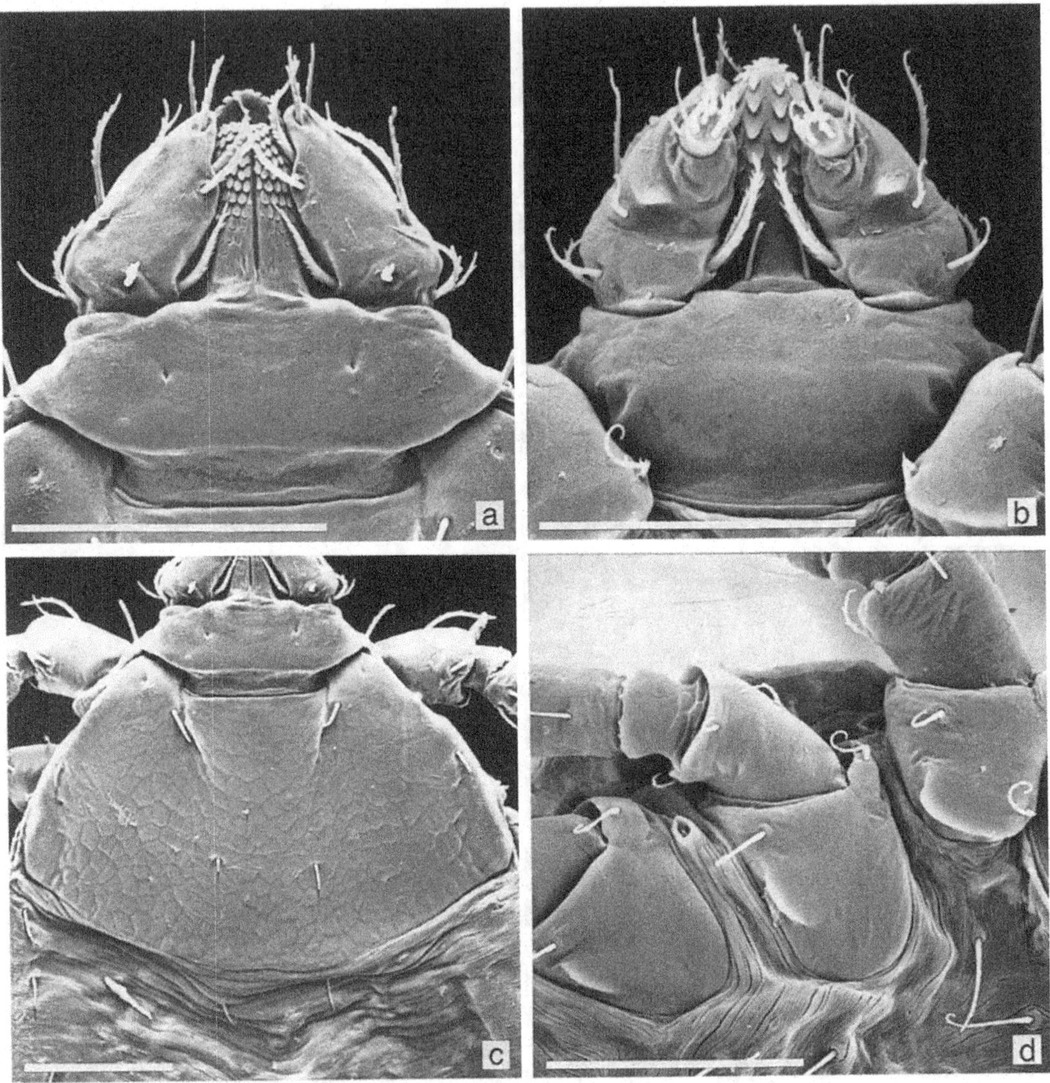

Figure 248. *Rhipicephalus pumilio* (RML 98604; HH 31277, collected on a rabbit, Bornova field, Izmir, Turkey, May 4, 1967 by Garrett). Larva: (a) capitulum, dorsal; (b) capitulum, ventral; (c) scutum; (d) coxae. Scale bars represent 0.10 mm. SEMs by P. Hill.

order of preference for adult *R. pumilio* as hares, hedgehogs, dogs, cattle, camels, humans, pigs, jackals, rabbits, goats, gazelle and horses.

In China adults are most active from April to June, and the immatures have two peaks of activity, the first in April and the second in July and August (Teng & Zaijie, 1991). In the former USSR adults are parasitic from April to the last days of August, nymphs from May to August, and larvae in July (Pomerantsev, 1950).

Zoogeography

This tick is found throughout southern Russia including Daghestan, and in Kazakhstan, Uzbekistan, Tajikistan (Tadzhiskistan) and Turkmenistan in regions with lakes and rivers in biotopes from desert plains to mid-mountainous territories (Voltsit, 1982) (Map 69). Balashov (1972) considered this species to favour humid habitats along river valleys, oases and foothills,

Table 65. *Host records of* Rhipicephalus pumilio

Hosts	Number of records
Domestic animals	
Cattle	2
Sheep	1
Goats	1
Camels	1
Horses	1
Donkeys	1
Pigs	1
Dogs	2
Wild animals	
Long-eared hedgehog (*Hemiechinus auritus*)	2 (immatures)
Brandt's hedgehog (*Hemiechinus hypomelas*)	1
Golden jackal (*Canis aureus*)	1
Wolf (*Canis lupus*)	1
Red fox (*Vulpes vulpes*)	1
Polecat (*Mustela putorius*)	1
Marbled polecat (*Vormela peregusna*)	1
Asian wild ass (*Equus hemionus*)	1
Goitred gazelle (*Gazella subgutturosa*)	1
Mid-day gerbil (*Meriones meridianus*)	1 (immatures)
Tamarisk gerbil (*Meriones tamariscinus*)	1 (immatures)
Great gerbil (*Rhombomys opimus*)	1
Tolai hare (*Lepus tolai*)	8 (including immatures)
Yarkand hare (*Lepus yarkandensis*)	1 (immatures)
'Rabbit'	1
Humans	2

but in contrast Pomerantsev (1950) considered it to be a desert species. It has been found on sheep and goats at an altitude of 2100 m to 2300 m in the Mountain-Badakhshan district of the West Pamirs (Voltsit, 1986). Beyond the borders of the former Soviet Union, it is known from the Karakorum Mountains of Kashmir, the Bachu area of Xinjiang Autonomous Region (Huang & Teng, 1980), and Inner Mongolia Autonomous Region (south-western Mongolia), China.

Disease relationships

The disease agents found in *R. pumilio*, including *Francisella tularensis*, *Coxiella burneti*, and the virus of Crimean-Congo haemorrhagic fever in Turkmenistan and Uzbekistan, have been reviewed by Hoogstraal (1979). In 1991 a new spotted fever group rickettsia causing Astrakhan fever was described (Tarasevich *et al.*, 1991). The rickettsia is similar if not identical to the agent of Israeli tick typhus and to *Rickettsia conori*, the agent of Mediterranean tick typhus, except that both the Israeli and Astrakhan agents lack the *tache noire* symptoms of *R. conori* infections (Eremeeva *et al.*, 1994). In Russia, *R. pumilio* has been incriminated as the vector of the Astrakhan agent (Eremeeva *et al.*, 1994; Galimzyanov *et al.*, 1996).

REFERENCES

Balashov, Yu. S. (1972). *Bloodsucking Ticks (Ixodoidea) – Vectors of Diseases of Man and Animals*. Leningrad Department, Leningrad: Nauka Publishers, 1967, (Published in 1968). [Translation T500, NAMRU-3, Cairo]. Published by the Entomological Society of America, 1972.

Eremeeva, M.E., Beati, L., Markova, V.A., Fetisova, N.F., Tarasevich, I.V., Balayeva, N.M. & Raoult, D. (1994). Astrakhan fever rickettsiae: antigenic and genotypic analysis of isolates obtained from human and *Rhipicephalus pumilio* ticks. *American Journal of Tropical Medicine and Hygiene*, **51**, 697–706.

Galimzyanov, K.M., Rasskazov, N.I., Altukhov, S.A. & Mesnyankin, A.P. (1996). Primary effect and course characteristic of new tick-borne rickettsiosis – Astrakhan fever. *Terapevticheskii Arkhiv*, **68**, 77-9. [In Russian].

Hoogstraal, H. (1979). The epidemiology of tick-borne Crimean-Congo hemorrhagic fever in Asia, Europe, and Africa. *Journal of Medical Entomology*, **15**, 307–417.

Huang, C.-A. & Teng, K.-F. (1980). Ticks from Hashi and neighbouring districts, Xinjiang. *Acta Entomologica Sinica*, **23**, 93–5. [In Chinese].

Pomerantsev, B.I. (1950). Ixodid ticks (Ixodidae). *Fauna SSSR, Paukoobraznye*, n.s. (41), 4(2), 224 pp. [In Russian, English translation by Elbl, A. and edited by Anastos, G. The American Institute of Biological Sciences, Washington, DC, 199 pp.]

Schulze, P. (1935). Acarina: Ixodoidea. *Wissenschaftliche Ergebnisse der Niederländischen Expeditionen in dem Karakorum*, **1**, 178–86.

Tarasevich, I.V., Makarova, V., Fetisova, N.F., Stepanov, A., Miskarova, A., Balayeva, E. & Raoult, D. (1991). Astrakhan fever: new spotted fever group rickettsiosis. *Lancet*, 337, 172–3.

Teng, K.-F. & Zaijie, J. (1991). *Economic Insect Fauna of China*, fasc. 39, Acari: Ixodidae. Beijing: Science Press. [In Chinese].

Voltsit, O.V. (1982). The first finding of the tick *Rhipicephalus pumilio* P. Sch., 1935 (Ixodidae) in the western Pamirs. *Parazitologiya, Leningrad*, **16**, 242–3. [In Russian, translation T1760, NAMRU-3, Cairo].

Voltsit, O.V. (1986). New data on the fauna of ixodid ticks from the West Pamirs. *Parazitologiya, Leningrad*, **20**, 483–4. [In Russian].

Zahler, M., Filippova, N.A., Morel, P.C., Gothe, R. & Rinders, H. (1997). Relationship between species of the *Rhipicephalus sanguineus* group: a molecular approach. *Journal of Parasitology*, **83**, 302–6.

Also see the following Basic Reference (p. 13): Filippova (1997).

RHIPICEPHALUS PUSILLUS GIL COLLADO, 1936

The specific name *pusillus*, from the Latin meaning 'very small', 'weak', refers to the small size of this tick species.

Synonym

bursa pusillus.

Diagnosis

A small light brown species of the *R. sanguineus* group lacking cervical fields in the male.

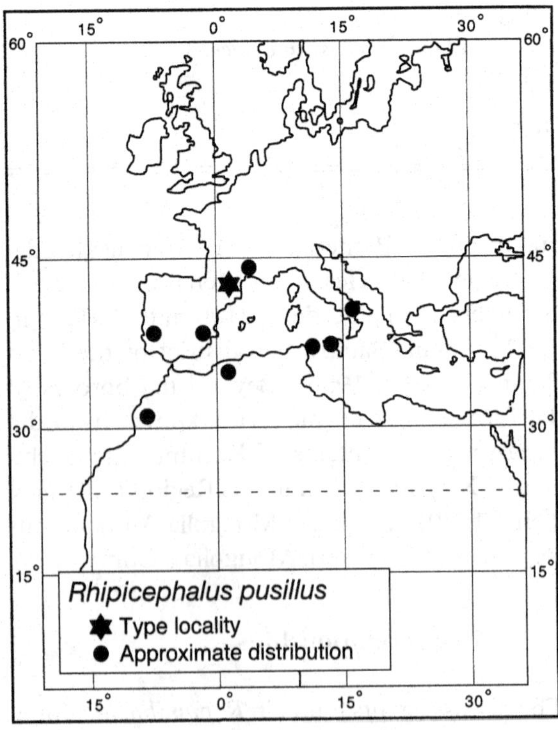

Map 70. *Rhipicephalus pusillus*: distribution.

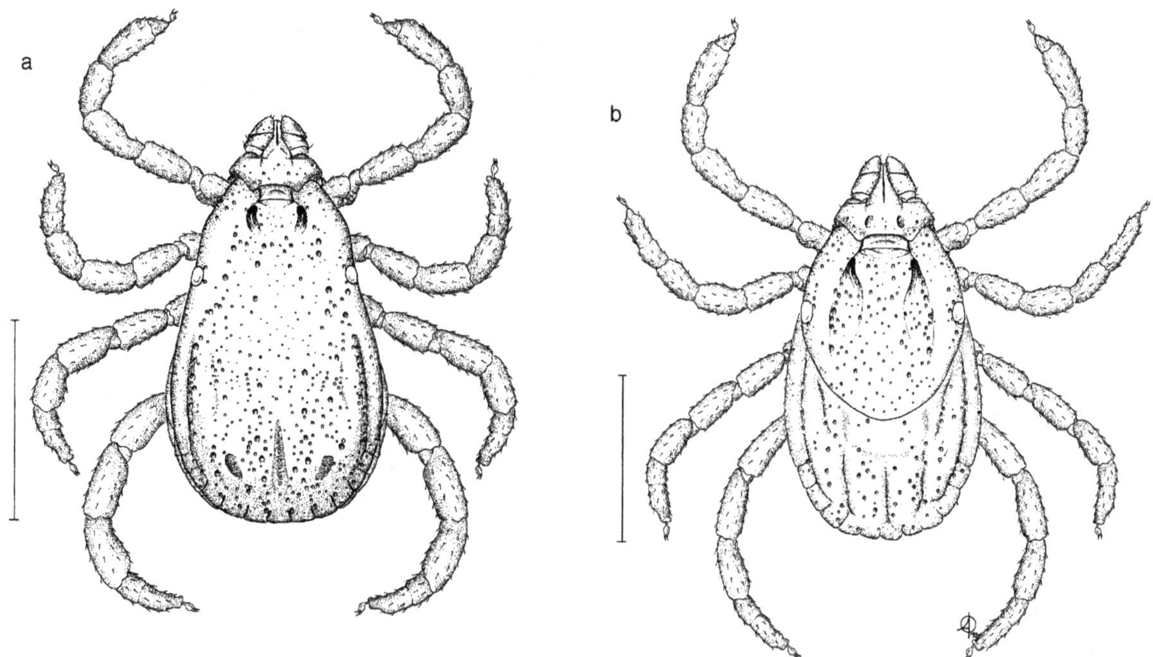

Figure 249. *Rhipicephalus pusillus* ♂ (RML 97206; HH 30360, collected on European rabbit (*Oryctolagus cuniculus*), Cota Donata, Spain, June 26, 1965 by Allen, received from the Hon. Miriam Rothschild; ♀ (RML86176; HH 82622, collected on vegetation, Bouches-du-Rhone, Salin-de-Bacon, France, June 29, 1954, received from Institut Pasteur). (a) Male, dorsal; (b) female, dorsal. Scale bars represent 1 mm. A. Olwage *del.*

Table 66. *Host records of* Rhipicephalus pusillus

Hosts	Number of records
Domestic animals	
Pigs (feral)	2
Dogs	3
Wild animals	
Algerian hedgehog (*Atelerix algirus*)	1
'Hedgehog'	1
Red fox (*Vulpes vulpes*)	5
Weasel (*Mustela nivalis*)	1
'Genet'	1
'Deer'	1
Muskrat (*Ondatra zibethicus*)	1
House mouse (*Mus musculus*)	1
'Orchard rat'	1
Dormouse (*Eliomys melanurus*)	1
European rabbit (*Oryctolagus cuniculus*)	Many (including immatures)

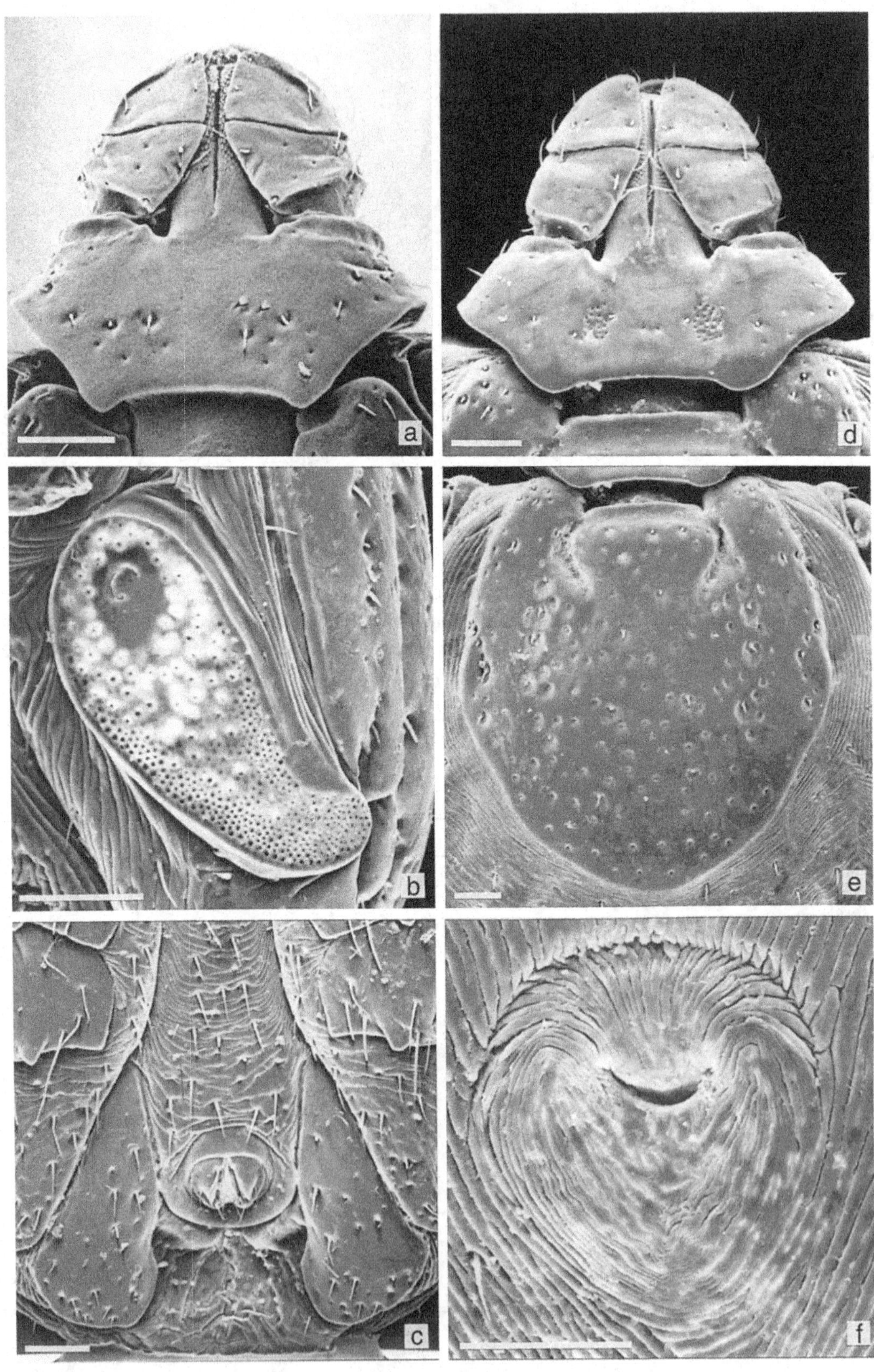
a
b
c
d
e
f

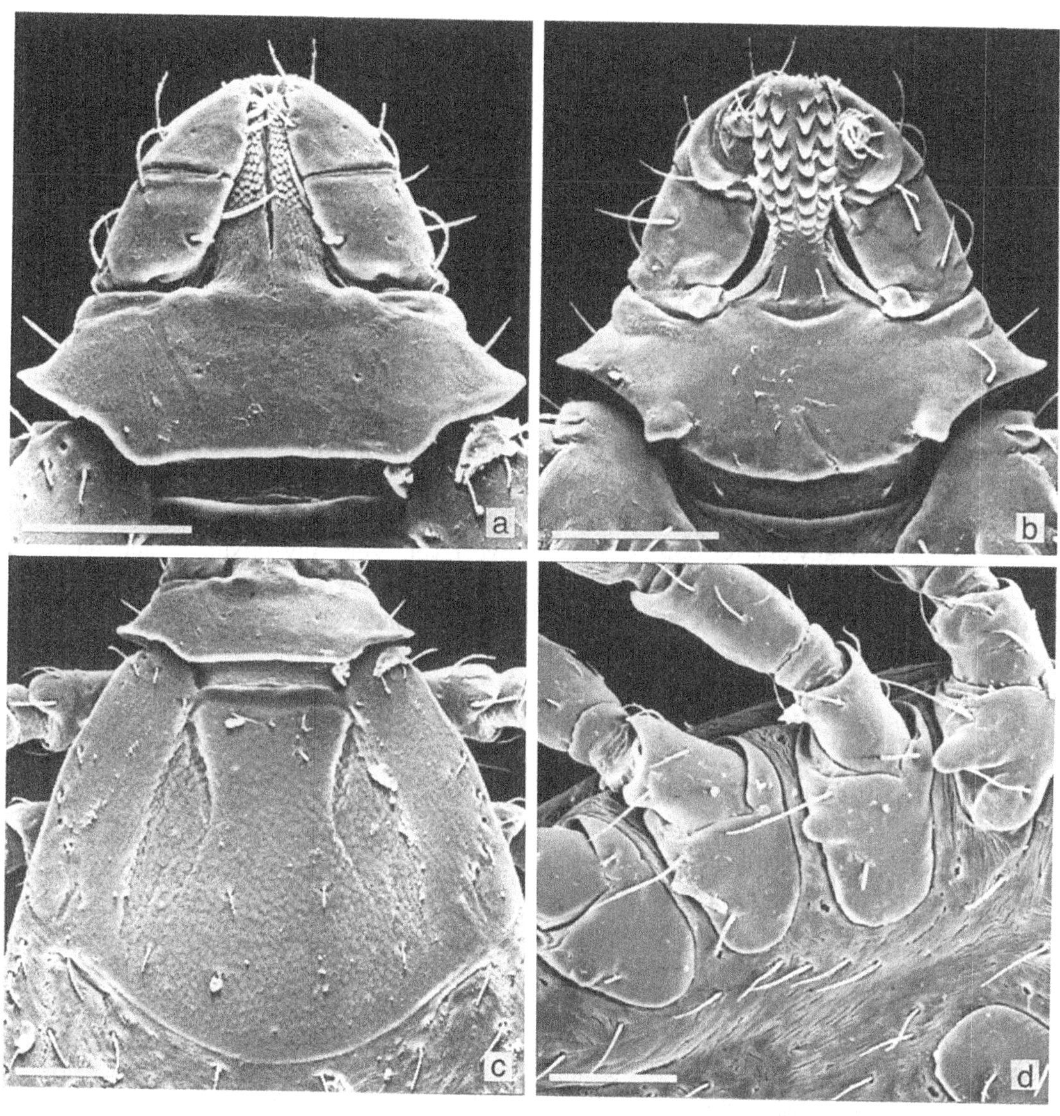

Figure 251 (*above*). *Rhipicephalus pusillus* (RML 97205; HH 30359, collected on European rabbit (*Oryctolagus cuniculus*), Cota Donata, Spain, June 25, 1965 by Allen, received from the Hon. Miriam Rothschild). Nymph: (a) capitulum, dorsal; (b) capitulum, ventral; (c) scutum; (d) coxae. Scale bars represent 0.10 mm. SEMs by P. Hill.

Figure 250 (*opposite*). *Rhipicephalus pusillus* ♂ (RML 97206; HH 30360, collected on European rabbit (*Oryctolagus cuniculus*), Cota Donata, Spain, June 26, 1965 by Allen, received from the Hon. Miriam Rothschild); ♀ (RML86176; HH 82622, collected on vegetation, Bouches-du-Rhone, Salin-de-Bacon, France, June 29, 1954, received from Institut Pasteur). Male: (a) capitulum, dorsal; (b) spiracle; (c) adanal plates; female: (d) capitulum, dorsal; (e) scutum; (f) genital aperture. Scale bars represent 0.10 mm. SEMs by P. Hill.

Male (Figs 249(a), 250(a) to (c))

Capitulum broader than long to slightly longer than broad, length × breadth ranging from 0.40 mm × 0.43 mm to 0.49 mm × 0.48 mm. Basis capituli broadly curved posterolaterally, but in some specimens very shallowly curved, not overlapping scapular areas of conscutum or the short anterior process of coxae I. Palps short, rounded apically. Conscutum much longer than broad, length × breadth ranging from 1.50 mm × 0.90 mm to 1.76 mm × 1.05 mm. Eyes flat to very mildly convex, usually not edged dorsally by punctations. Cervical fields absent. Marginal lines quite short, shallow and punctate, ending well posterior to eye level, delimiting one or two festoons, occasionally one festoon on one side and two festoons on the other in the same specimen. Scapular areas and areas lateral to marginal lines relatively free of punctations, otherwise with many medium-sized punctations scattered over entire conscutum. Posteromedian groove short, posterolateral grooves irregular depressions tending towards the subcircular. Ventrally spiracles elongate, narrowing only slightly from macula to dorsal prolongation, which is just visible from above. Adanal plates long, narrow, curved inwards posteromedially, posterior margin slightly concave; accessory adanal plates absent.

Female (Figs 249(b), 250(d) to (f))

Capitulum broader than long, length × breadth ranging from 0.53 mm × 0.60 mm to 0.57 mm × 0.64 mm. Basis capituli with lateral angles at about mid-length, acute; porose areas small, about twice their own diameter apart. Palps slightly longer than those of male, rounded apically. Scutum longer than broad, length × breadth ranging from 1.02 mm × 0.92 mm to 1.16 mm × 1.05 mm; in some specimens the posterior margin is sinuous. Eyes mildly convex, edged dorsally with a few punctations. Cervical fields scalpel-shaped, narrowly elongate but disappearing posterior to eye level; external cervical margins marked by several larger punctations. Scapular areas smooth, impunctate; numerous small punctations centrally and posteriorly on scutum. Ventrally genital aperture small and U-shaped.

Nymph (Fig. 251)

Capitulum broader than long, length × breadth ranging from 0.23 mm × 0.28 mm to 0.25 mm × 0.29 mm. Basis capituli about three times as broad as long; lateral angles sharp, extending over scapulae. Palps not constricted basally, tapering only at apex. Scutum longer than broad to as broad as long, length × breadth ranging from 0.39 mm × 0.35 mm to 0.42 mm × 0.42 mm. Eyes at widest point, well over halfway back, very slightly convex. Cervical fields long and narrow, depressed, reaching posteriorly to eye level. Ventrally coxae I each with a large internal and external spur; coxae II to IV each with a small external spur only.

Larva

Unavailable for study.

Notes on identification

A tick belonging to the *R. sanguineus* group. Using gas chromatography of cuticular hydrocarbons of the four *Rhipicephalus* species found in Spain, Estrada-Peña, Estrada-Peña & Peiró (1991) were able to differentiate *R. pusillus* from *R. bursa, R. sanguineus* and *R. turanicus*.

Hosts

Presumed to be a three-host tick. The European rabbit is the primary host of *R. pusillus*, but it is also found feeding on the red fox (Estrada-Peña *et al.*, 1992), and on dogs and feral pigs (Table 66). In Portugal, in addition to the above mentioned hosts, Caeiro & Simões (1991) have found this tick on deer, hedgehogs, genet, weasel, orchard rat and muskrat. In Morocco, Blanc, Delage & Ascione (1962) recorded *R. pusillus* (as *R. sanguineus*) from the Algerian hedgehog, and Morel & Vassiliades (1963) recorded it from the garden dormouse and the house mouse. The latter authors consider any animal that visits rabbit warrens to be a potential host for this tick.

Zoogeography

The original distribution of the European rabbit, including the Iberian peninsula and Morocco in north-west Africa, encompasses the tick's distribution. It is also found in southern France and in Italy, including the islands of Sicily and Ustica (Gallo, Rilli & Sobrero, 1977), and Tunisia (Island of Zembra) (Map 70). Recent collections in Italy have extended the range of this tick to San Martino, Mal di Ventre and the Island of Serpentara (Manilla, 1991).

Disease relationships

Antibodies to *Rickettsia conori* and *Rickettsia slovaca* were detected in 78.9% of European rabbits collected in Tuscany and parasitized by *R. pusillus* (Ciceroni *et al.*, 1988).

REFERENCES

Blanc, G., Delage, B. & Ascione, L. (1962). Etude épidémio-écologique de la forêt du Cherrat. *Archives de l'Institut Pasteur Maroc*, **6**, 223–92.

Caeiro, V.M.P. & Simões, A.L. (1991). Ixodoidea da fauna silvestre em Portugal continental. Interesse do seu conhecimento. *Revista Portuguesa de Ciências Veterinárias*, **86**, 20–30.

Ciceroni, L., Pinto, A., Rossi, C., Khoury, C., Rivosecchi, L., Stella, E. & Cacciapuoti, B. (1988). Rickettsiae of the spotted fever group associated with the host-parasite system *Oryctolagus cuniculus/Rhipicephalus pusillus*. *Zentrablatt für Bakteriologie und Hygiene, Series A*, **269**, 211–17.

Estrada-Peña, A., Estrada-Peña, R. & Peiró, J.M. (1991). Differentiation of *Rhipicephalus* ticks (Acari : Ixodidae) by gas chromatography of cuticular hydrocarbons. *Journal of Parasitology*, 78, 982–93.

Estrada-Peña, A., Osácar, J.J., Gortázar, C., Calvete, C. & Lucientes, J. (1992). An account of the ticks of the northeastern of Spain (Acarina: Ixodidae). *Annales de Parasitologie Humaine et Comparée*, **67**, 42–9.

Gallo, C., Rilli, S. & Sobrero, L. (1977). *Rhipicephalus pusillus* Gil Collado in Italia. *Rivista di Parassitologia*, **38**, 89–91.

Gil Collado, J. (1936). Acaros ixodoideos de Cataluña y Baleares. *Treballs del Museu de Ciències Naturals de Barcelona, Ser. Entomológica*, **11**, 1–8.

Manilla, G. (1991). Nuove osservazioni faunistiche e biologiche sulle zecche d'Italia (Acari: Ixodoidea) (Nata V). *Atti della Società Italiana di Scienze Naturali e del Museo Civico di Storia Naturale di Milano*, **131**, 433–52.

Morel, P.C. & Vassiliades, G. (1963). Les *Rhipicephalus* du groupe *sanguineus* espèces africaines (Acarines, Ixodoidea). *Revue d'Élevage et de Médecine Vétérinaire des Pays Tropicaux*, **15** (nouvelle série), 343–86.

RHIPICEPHALUS RAMACHANDRAI DHANDA, 1966

The specific name *ramachandrai* is a patronym for Dr T. Ramachandra Rao of the Virus Research Centre, Poona.

Synonym

arakeri.

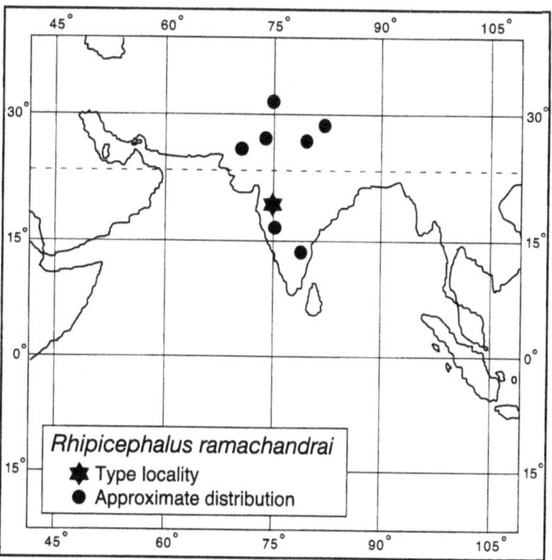

Map 71. *Rhipicephalus ramachandrai*: distribution.

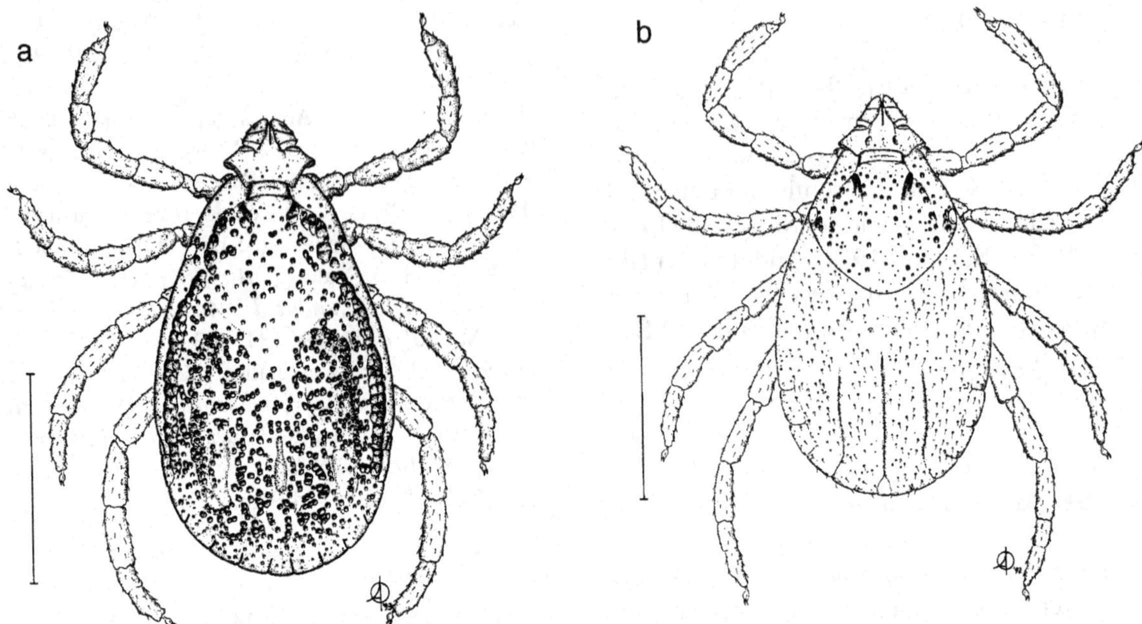

Figure 252. *Rhipicephalus ramachandrai* (RML 101124; HH 30307, collected on Indian gerbil (*Tatera indica*), Stud farm, Manjiri, Poona, Maharashtra, India, March 17, 1964, received from Virus Research Centre, Poona). (a) Male, dorsal; (b) female, dorsal. Scale bars represent 1 mm. A. Olwage *del.*

Table 67. *Host records of* Rhipicephalus ramachandrai

Hosts	Number of records
Wild animals	
Bengal fox (*Vulpes bengalensis*)	1
Jungle striped squirrel (*Funambulus tristriatus*)	1
Indian desert gerbil (*Meriones hurrianae*)	2 (immatures)
Indian gerbil (*Tatera indica*)	34 (including immatures)
Lesser bandicoot rat (*Bandicota bengalensis*)	2
Sand-coloured rat (*Millardia gleadowi*)	2 (immatures)
Soft-furred field rat (*Millardia meltada*)	2 (immatures)

Diagnosis

A small, light brown, heavily-punctate species with some darkening when preserved in alcohol.

Male (Figs 252(a), 253(a) to (c))

Capitulum broader than long, length × breadth ranging from 0.35 mm × 0.42 mm to 0.42 mm × 0.47 mm. Basis capituli with several punctations dorsally. Palps short, broad. Conscutum much longer than broad, length × breadth ranging from 1.74 mm × 1.01 mm to 2.05 mm × 1.26 mm; broadest at level of spiracles. Eyes marginal, small, flat, indistinct. Scapular area anterior to eyes raised due to sunken punctate cervical pits. Entire scutal area punctate, densely so in posterior one-third, less so centrally anterior to posteromedian and posterolateral grooves. Marginal lines distinct, composed of large deep punctations extending anteriorly from festoon I to cervical pits. Festoons poorly delineated. Legs do not increase in size from I to IV. Ventrally spiracles elongate, relatively broad throughout without a narrow dorsal prolongation. Adanal plates large, teardrop-shaped with inner margins slightly concave, posterior margins broadly rounded; accessory adanal plates small, bluntly rounded.

Female (Figs 252(b), 253(d) to (f))
Capitulum much broader than long, length × breadth ranging from 0.32 mm × 0.46 mm to 0.43 mm × 0.53 mm. Basis capituli with moderately long, sharp lateral angles extending over the anterior processes of coxae I; porose areas small, superficial, not depressed into the surface of the basis capituli. Palps short, rounded apically. Scutum approximately as broad as long, length × breadth ranging from 0.64 mm × 0.61 mm to 0.87 mm × 0.87 mm. Eyes at widest point, flat, edged with several punctations, some of which are deep. Cervical fields well marked externally by a declination and punctations. Several large setiferous punctations scattered over the scutum, interspersed with fine punctations. Ventrally genital aperture broadly U-shaped.

Nymph (Fig. 254)
Capitulum broader than long, length × breadth ranging from 0.25 mm × 0.30 mm to 0.35 mm × 0.37 mm. Lateral angles of capitulum shallowly curved, tips of scapulae usually visible; ventrally with prominent triangular spurs on posterior margin (damaged in Fig. 254(b)). Palps narrower distally than proximally, external margins straight to mildly concave. Scutum slightly longer than broad, length × breadth ranging from 0.51 mm × 0.49 mm to 0.57 mm × 0.54 mm; posterior margin broadly U-shaped. Eyes at widest point, set well over halfway back, flat. Cervical fields long, very narrow, the inner margins often difficult to see. Alloscutum with numerous long white setae dorsally. Ventrally coxae I each with two large spurs, external longer than internal; a single external spur decreasing in size on coxae II to IV.

Larva (Fig. 255)
Capitulum much broader than long, length × breadth ranging from 0.088 mm × 0.114 mm to 0.096 mm × 0.128 mm. Basis capituli twice as broad as long, posterior margin convex, lateral angles bluntly rounded. Palps broad, conical, blunt apically. Scutum much broader than long, length × breadth ranging from 0.180 mm × 0.263 mm to 0.217 mm × 0.292 mm. Eyes at widest point, flat, not easily discernible, set well over halfway back. Cervical grooves as slight, short indentations in the scutal surface. Ventrally coxae I each with a short triangular spur; coxae II and III each with short ridge-like spurs.

Notes on identification
The small size of both the male and female of *R. ramachandrai* and the heavily punctate appearance of the male scutum easily separates this species from all other *Rhipicephalus* species on the Indian subcontinent.

Hiregoudar (1975) described *Rhipicephalus arakeri* and mentioned that his new species 'resembles closely *R. ramachandrai* in most respects but differs from it in having many large punctations and the setiferous setae on the ventral aspect of palpal segment I, two in male and four in female instead of three in *R. ramachandrai*'. The holotype and allotype of his new species were deposited in the Zoological Survey of India. These specimens could not be located, but Dr Hiregoudar kindly sent one of us (J.E.K.) two slides of *R. arakeri* from his personal collection; one containing 2 ♀♀, the other, labelled 'paratypes', contained 1 ♂ and 1 nymph. These slide-mounted ticks had been over-cleared, but examination of the specimens has led us to the conclusion that *R. arakeri* is a junior subjective synonym of *R. ramachandrai*.

Hosts

From collection data presented by Dhanda (1966) this is undoubtedly a three-host tick.

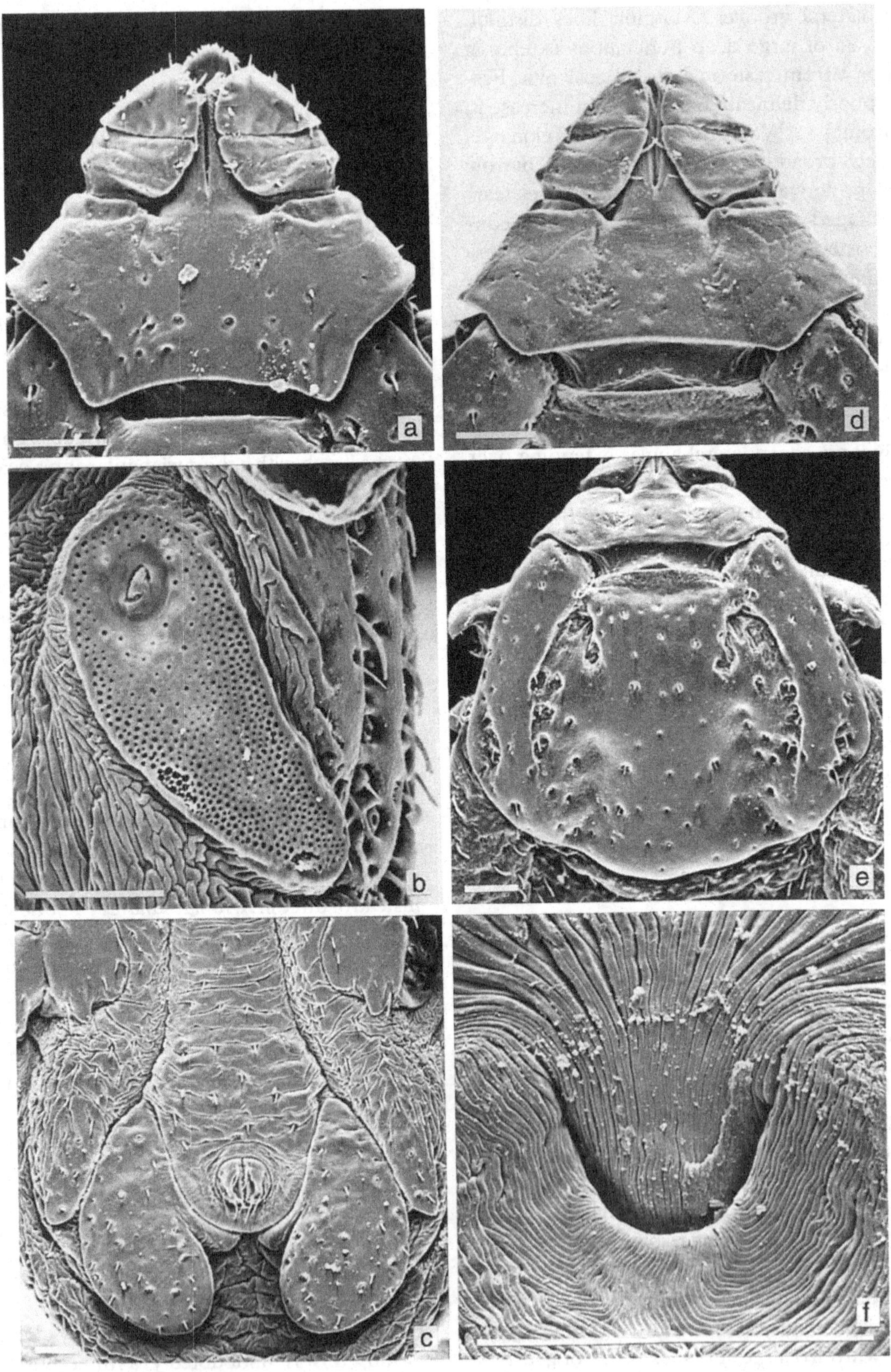
a
d
b
e
c
f

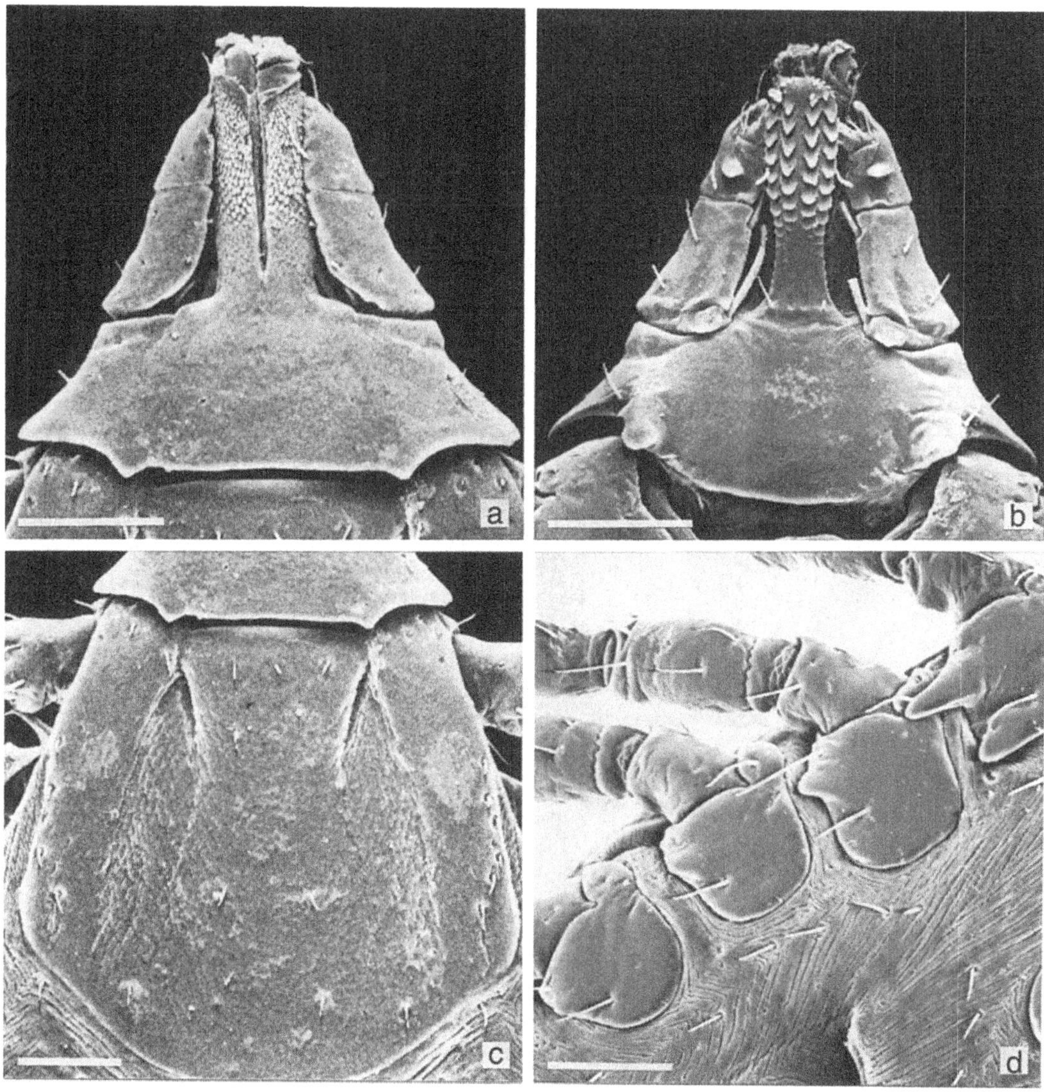

Figure 254 *(above)*. *Rhipicephalus ramachandrai* (RML 76745; HH 80172, collected on Indian gerbil (*Tatera indica*), Changa Manga Forest, Lahore, Pakistan, date unknown, R. Traub). Nymph: (a) capitulum, dorsal; (b) capitulum, ventral; (c) scutum; (d) coxae. Scale bars represent 0.10 mm. SEMs by P. Hill.

Figure 253 *(opposite)*. *Rhipicephalus ramachandrai* ♂ (RML 107134; HH 83834, collected on Indian gerbil (*Tatera indica*), Malir Cantonment, Karachi, Pakistan, October 21, 1976, S. Telford); ♀ (RML 76745; HH 80172, collected on Indian gerbil (*Tatera indica*), Changa Manga Forest, Lahore, Pakistan, April 1969, Z. B. Mirza). Male: (a) capitulum, dorsal; (b) spiracle; (c) adanal plates; female: (d) capitulum, dorsal; (e) scutum; (f) genital aperture. Scale bars represent 0.10 mm. SEMs by P. Hill.

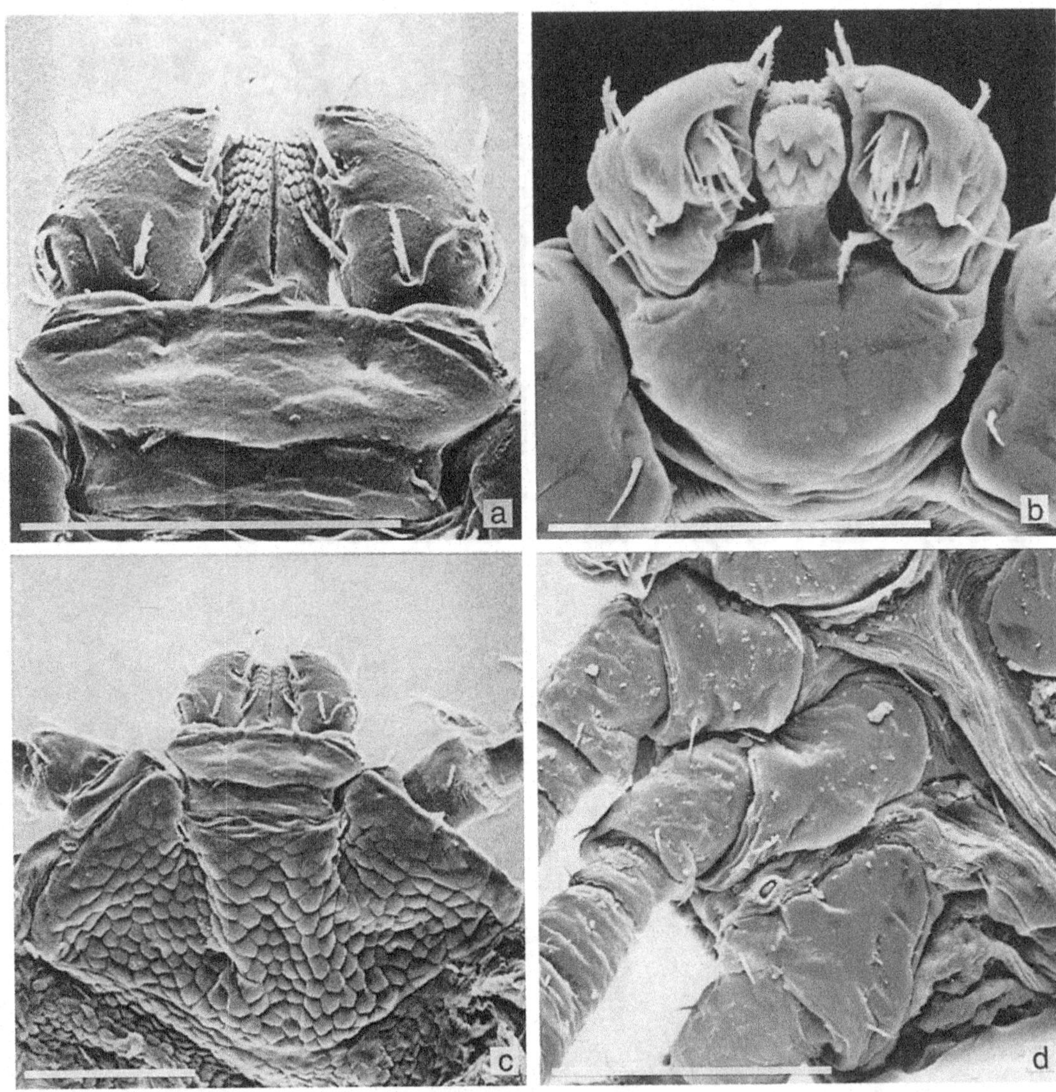

Figure 255. *Rhipicephalus ramachandrai* (RML 101124; HH 30307, collected on Indian gerbil (*Tatera indica*), Stud farm, Manjiri, Poona, Maharashtra, India, March 17, 1964, received from Virus Research Centre, Poona). Larva: (a) capitulum, dorsal; (b) capitulum, ventral; (c) scutum; (d) coxae. Scale bars represent 0.10 mm. SEMs by P. Hill.

Rhipicephalus ramachandrai was originally described from the Indian gerbil, and all stages of the tick can be found in the rodent's burrow; a highly unusual ecological relationship in the genus *Rhipicephalus*. In a survey of ticks in the state of Maharashta, India, Geevarghese & Dhanda (1995) found *R. ramachandrai* only on the Indian gerbil. Mitchell (1979) also collected it from the same host in Nepal. Additional published records are from the jungle striped squirrel (Miranpuri & Gill, 1983) and the soft-furred field rat (Kaul *et al.*, 1978) (Table 67). Unpublished U.S. National Tick Collection records show that it is also ectoparasitic on the Bengal fox, Indian desert gerbil, lesser bandicoot rat and sand-coloured rat.

Zoogeography

Rhipicephalus ramachandrai is found in Karnataka, Maharashtra, Punjab, Rajasthan and Uttar Pradesh, India; Sind, Pakistan; and Bahwanipur, Nepal (Map 71). The most recent collections of this tick are from the Chitoor District of Andhra Pradesh, India (Saxena, 1997).

Disease relationships

On the alluvial plain of the Indus river, 21 km east of Lahor, Pakistan, Manawa virus was isolated from vulture (*Gyps bengalensis*) nests infested with the argasid tick, *Argas abdussalami* (data summarized by Karabatsos, 1985). Eighteen of the 19 ticks tested contained the arbovirus. Hoogstraal (1973) reported that contrary to the characteristic pattern of viruses from ticks parasitizing birds in restricted areas, Manawa virus was also isolated from *R. ramachandrai* and *R. turanicus* near Lahore. As the pathogenicity of this virus has not been determined we have excluded it from the tables of animal and human diseases transmitted by ticks.

REFERENCES

Dhanda, V. (1966). *Rhipicephalus ramachandrai* sp. n. (Acarina: Ixodidae) from the Indian gerbil, *Tatera indica* (Hardwicke, 1807) (Rodentia: Muridae). *Journal of Parasitology*, **52**, 1025–31.

Geevarghese, G. & Dhanda, V. (1995). Ixodid ticks of Maharashtra state, India. *Acarologia*, **36**: 309–13.

Hiregoudar, L.S. (1975). On *Rhipicephalus* ticks of Gujarat, together with a description of a new species – *Rhipicephalus arakeri* from a common Indian rat. *Mysore Journal of Agricultural Science*, **9**, 473–9.

Hoogstraal, H. (1973). Viruses and ticks. In *Viruses and Invertebrates*, ed. A.J. Gibbs, pp. 349-90. Amsterdam and London: North Holland Publishing Co.

Karabatsos, N. (ed.) (1985). *International Catalogue of Arboviruses including certain other Viruses of Vertebrates*, 3rd edn. San Antonio, TX: American Society of Tropical Medicine and Hygiene.

Kaul, H.N., Mishra, A.C., Dhanda, V., Kulkarni, S.M. & Guttikar, S.N. (1978). Ectoparasitic arthropods of birds and mammals from Rajasthan State, India. *Indian Journal of Parasitology*, **2**, 19–25.

Miranpuri, G.S. & Gill, H.S. (1983). *Ticks of India*. Edinburgh: Lindsay & Macleod.

Mitchell, R.M. (1979). A list of ectoparasites from Nepalese mammals, collected during the Nepal ectoparasite program. *Journal of Medical Entomology*, **16**, 227–33.

Saxena, V.K. (1997). Ixodid ticks infesting rodents and sheep in diverse biotopes of southern India. *Journal of Parasitology*, **83**, 766–7.

RHIPICEPHALUS ROSSICUS YAKIMOV & KOL-YAKIMOVA, 1911

The origin of the specific name *rossicus is* unknown; perhaps from the Latin *russus* meaning 'reddish', or Italian *rosso* meaning 'red'.

Synonym

sanguineus rossicus.

Diagnosis

A moderate-sized, light to medium-brown tick of the *R. sanguineus* group.

Male (Figs 256(a), 257(a) to (c))

Capitulum broader than long, length × breadth ranging from 0.74 mm × 0.81 mm to 0.83 mm × 0.87 mm. Basis capituli with broad lateral angles projecting over the slightly protruding anterior process of coxae I; paired depressions in the basis may be seen in some specimens in the position of the porose areas of the female. Palps short, broad. Conscutum length × breadth ranging from 3.27 mm × 2.10 mm to 3.50 mm × 2.30 mm. In engorged specimens the body wall expands laterally posterior to the eyes; a short blunt caudal process may protrude. Eyes marginal, flat, may or may not be edged with a few small dorsal punctations. Cervical fields inapparent; cervical grooves as shallow depressions extending posteriorly to eye level.

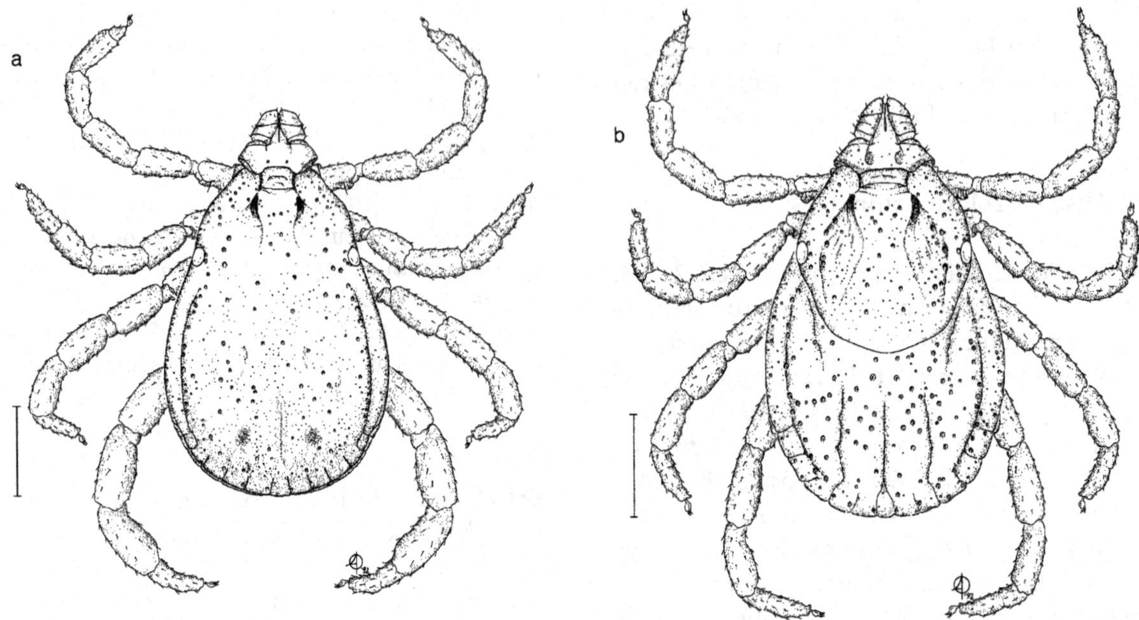

Figure 256. *Rhipicephalus rossicus* (RML 102337; HH 28603, collected from an unknown host in the former USSR, specific locality, date and collector unknown). (a) Male, dorsal; (b) female, dorsal. Scale bars represent 1 mm. A. Olwage *del.*

Table 68. *Host records of* Rhipicephalus rossicus

Hosts	Number of records
Domestic animals	
Cattle	8
Goats	1
Camels	1
Horses	2
Dogs	3
Wild animals	
East European hedgehog (*Erinaceus concolor*)	1
Serotine (*Eptesicus serotinus*)	1 (immatures ?)
European water vole (*Arvicola terrestris*)	1
'Rodents'	Several records (immatures)
Pallas's pika (*Ochotona pallasi*)	2
European rabbit (*Oryctolagus cuniculus*)	1
Humans	2

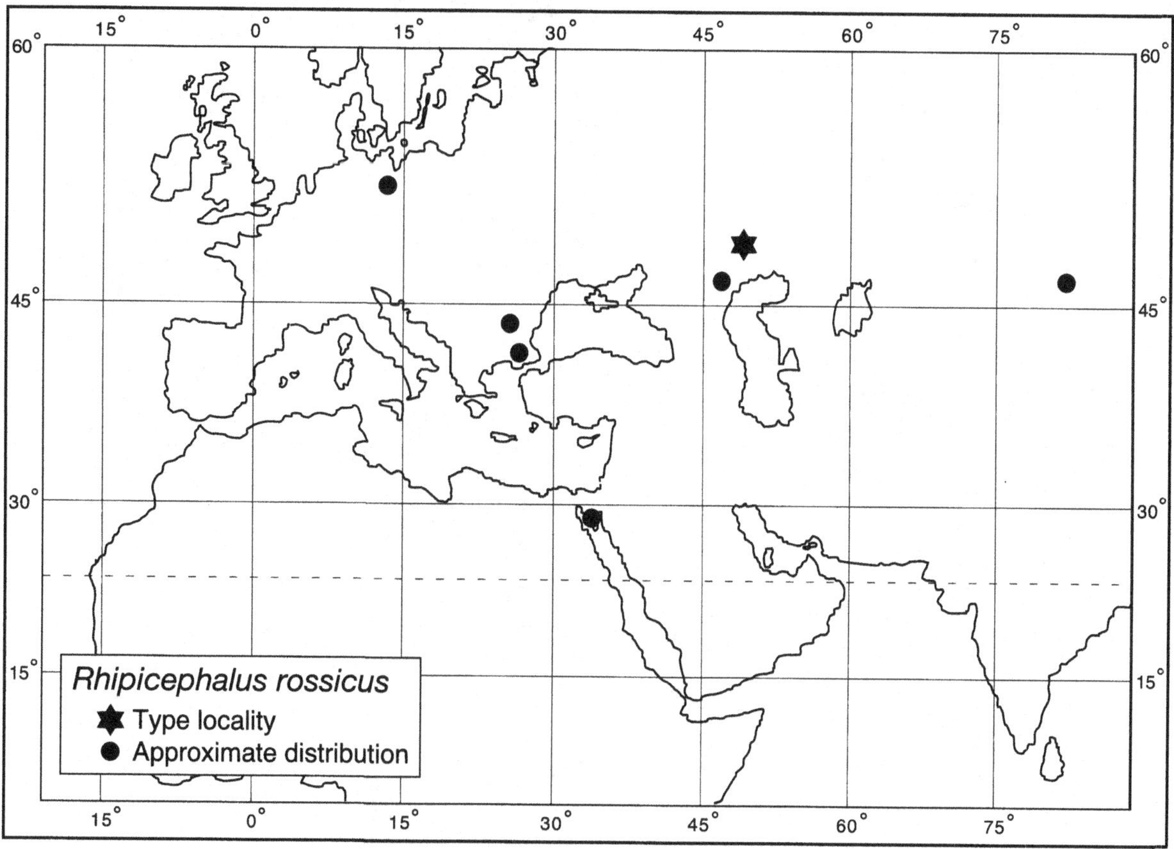

Map 72. *Rhipicephalus rossicus*: distribution.

Marginal lines shallow punctate troughs delimiting first festoons and becoming less obvious as they extend anteriorly to just posterior of eye level. Conscutum appears impunctate and smooth except for depressed posteromedian and posterolateral grooves; a few large punctations in cervical areas and a poorly-developed '*simus*' pattern of punctations present on a background of evenly-distributed fine punctations. Posteromedian groove broad, deep; posterolateral grooves deep, subcircular. Legs increase slightly in size from I to IV. Ventrally spiracles elongate, each with a broad dorsal prolongation; goblet cells uniformally fine with only a few larger cells near the macula. Adanal plates very large, broad posteriorly with medial projections; accessory adanal plates, elongate, narrowly V-shaped.

Female (Figs 256(b), 257(d) to (f))

Capitulum broader than long, length × breadth ranging from 0.79 mm × 0.86 mm to 0.99 mm × 1.10 mm. Basis capituli with broad lateral angles not overlapping the scapulae; porose areas circular, deep set, about as broad as their distance apart. Palps short, broad. Scutum slightly longer than broad to slightly broader than long, length × breadth ranging from 1.53 mm × 1.51 mm to 1.94 mm × 2.00 mm. Eyes about halfway back, flat to very slightly bulging, edged dorsally by a few punctations. Cervical fields scalpel shaped, with ridge-like elevations running anterior to posterior; slightly depressed with raised external cervical margins usually marked by a few larger punctations. Remainder of scutum not markedly punctate. Ventrally genital aperture broadly U-shaped.

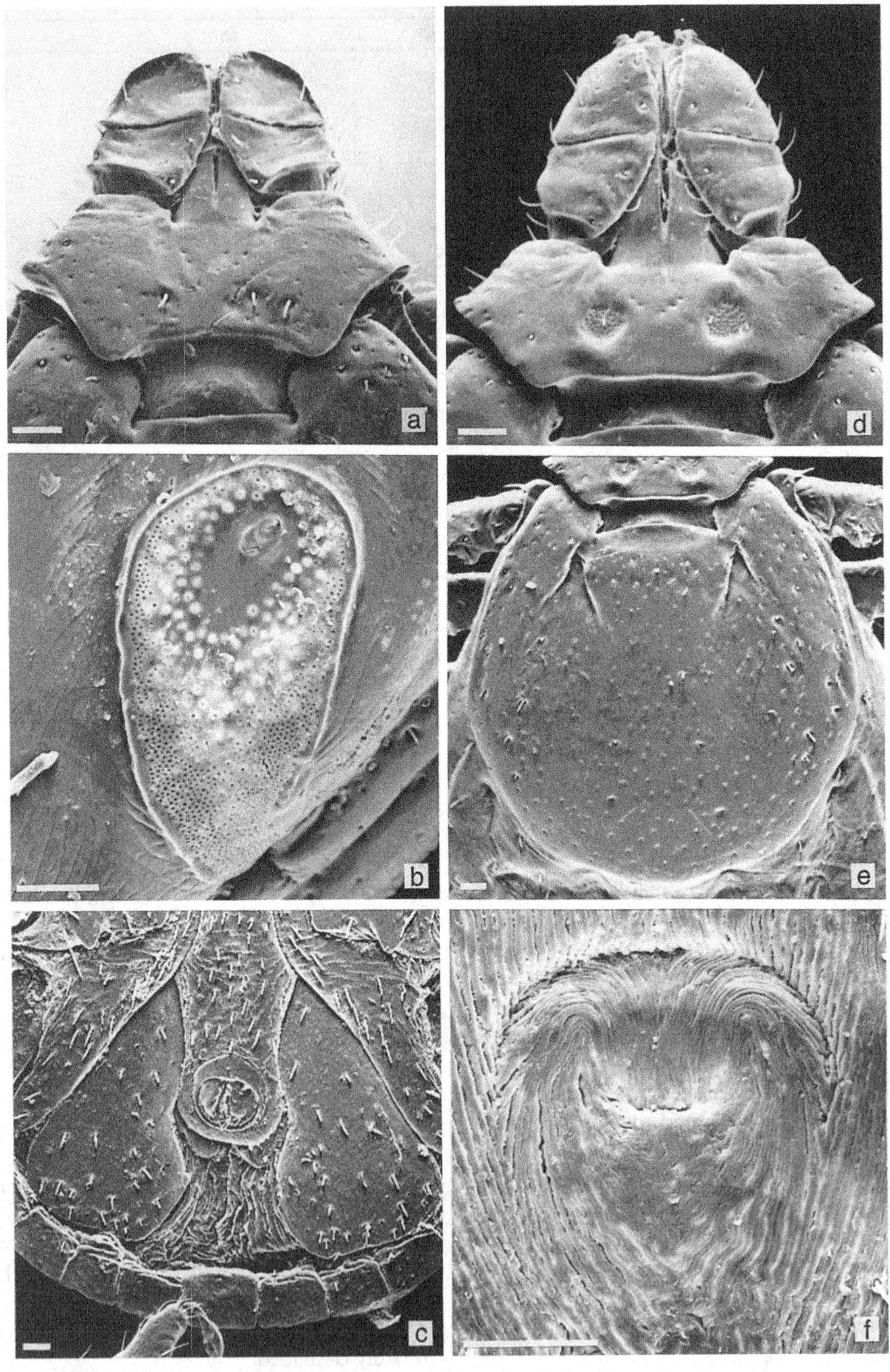
a
d
b
e
c
f

Figure 257 (*opposite*). *Rhipicephalus rossicus* (RML 102335; HH 28601, collected from Pallas's pika (*Ochotona pallasi*), Kazakhstan, June 28, 1953, C. Hoare). Male: (a) capitulum, dorsal; (b) spiracle; (c) adanal plates. Female: (d) capitulum, dorsal; (e) scutum; (f) genital aperture. Scale bars represent 0.10 mm. SEMs by P. Hill.

Nymph (Fig. 258)

Capitulum much broader than long, length × breadth 0.26 mm × 0.36 mm. Lateral angles of capitulum very acute, posterior to mid-length; cornua absent; tips of scapulae usually visible; posteriorly-directed spurs ventrally on capitulum. Palps roughly cylindrical, apices bluntly rounded. Scutum broader than long, length × breadth 0.50 mm × 0.55 mm; posterior margin very broadly U-shaped. Eyes at widest point, set in posterior third of scutum. Cervical fields long and narrow, internal margins often difficult to see. Numerous feathered setae on the dorsal and ventral surfaces of the alloscutum. Coxae I each with two moderate spurs, external spur longer than internal; a single external spur, decreasing in size, on coxae II to IV, coxal spur IV often absent.

Larva (Fig. 259)

Capitulum much broader than long, length × breadth ranging from 0.090 mm × 0.167 mm to 0.110 mm × 0.175 mm. Basis capituli three times as broad as long, with bluntly-rounded lateral angles, posterior margin slightly convex. Palps pointed apically, external margins straight. Scutum much broader than long, length × breadth ranging from 0.235 mm × 0.329 mm to 0.258 mm × 0.378 mm, posterior margin broadly curved. Eyes at widest point, flat. Cervical grooves short, slightly convergent. Ventrally coxae I each with a large, blunt triangular spur; coxae II each with a small spur; no obvious spur on coxae III.

Notes on identification

A member of the *R. sanguineus* group. *Rhipicephalus rossicus* has broadly-elongate spiracles, and relatively impunctate scuta in both the male and female, and small circular deep set porose areas in the female, all of which help to separate it from *R. sanguineus*. Yakimov (1923) presented a chart as an aid for separating *R. rossicus* from *R. sanguineus*. A neotype male for *R. rossicus* has been designated by Filippova (1996), who deposited it in the Zoological Institute of the Russian Academy of Sciences in St. Petersburg. However, because syntypes exist, this neotype has no validity. Zahler *et al.* (1997) found that *R. pumilio* and *R. rossicus* possessed identical DNA sequences in the target gene investigated, and suggested cross-breeding experiments to see if these two species are conspecific.

We had no nymphs available for study. The above description of this stage is based on those by Feider (1965) and Filippova (1981).

Hosts

A three-host tick, with a life cycle that takes 2 to 3 years to complete. Adults are parasitic on domestic animals, hedgehogs and occasionally humans (Table 68). Immatures feed on hedgehogs, rodents and hares in lowlands and mountain steppes. Adults are active in the warmer months with a maximum activity peak in May to June. Larvae appear at the same time as adults in April and also peak in May to June with a possible second peak in August. A nymphal peak on rodents occurs in June to July.

Zoogeography

Rhipicephalus rossicus is known from the District of Tomaszów Lubelski, Poland (Dutkiewicz & Siuda, 1969), and from Romania, Bulgaria, and the Sinai Peninsula of Egypt (Feldman-Muhsam, 1960), to western Kazakhstan, in lowland areas and on mountain steppes (Map 72). It has also been recorded recently in Xinjiang Province, China (Yu, Ye & Gong, 1997).

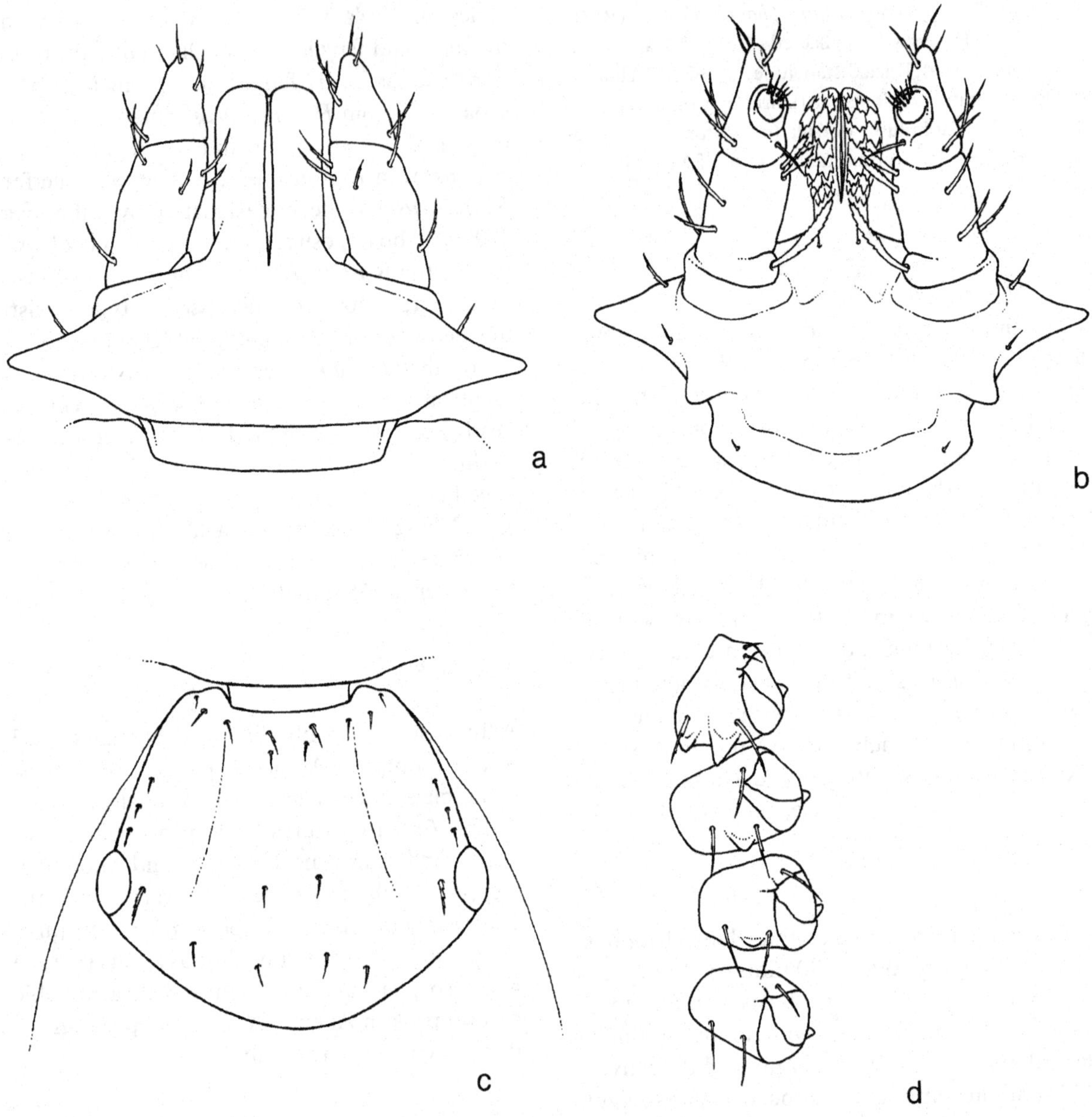

Figure 258. *Rhipicephalus rossicus.* Nymph: (a) capitulum, dorsal; (b) capitulum, ventral; (c) scutum; (d) coxae. (Redrawn from Filippova, 1981, figs 9–12, by A. Olwage, with kind permission from the author and from the Editor-in-Chief, *Parazitologicheskiy Sbornik*, of the Zoological Institute, Russian Academy of Sciences).

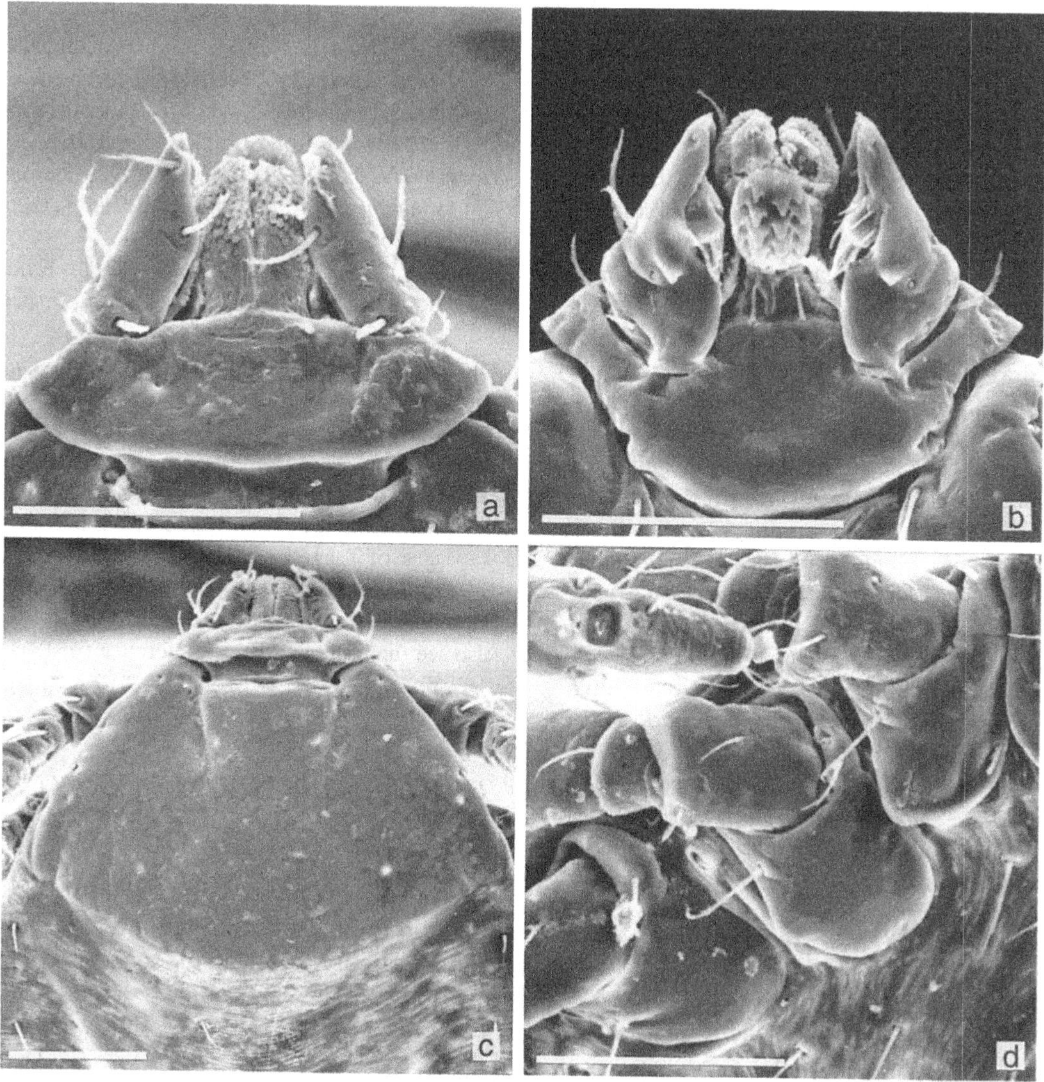

Figure 259. *Rhipicephalus rossicus* (RML 102338; HH 30766, collected from a horse, Bekhtar, Odessa, Ukraine, June 8, 1934, received from M.V. Pospelova-Shtrom). Larva: (a) capitulum, dorsal; (b) capitulum, ventral; (c) scutum; (d) coxae. Scale bars represent 0.10 mm. SEMs by P. Hill.

Disease relationships

Rhipicephalus rossicus is a vector of the virus of Crimean-Congo haemorrhagic fever (Kondratenko *et al.*, 1970; Watts *et al.*, 1988). It is also a well-known vector of *Francisella tularensis* (Borodin, Samsonova & Koroleva, 1958; Borodin *et al.*, 1965). *Coxiella burneti* has also been isolated from this tick species (Hoogstraal, 1979). West Nile virus, usually a mosquito-borne virus, survives experimentally in *R. rossicus* from the larval to the adult stage and is irregularly transmitted transovarially to the F_1 generation (Hoogstraal, 1979).

REFERENCES

Borodin, V.P., Samsonova, A.P. & Koroleva, A.P. (1958). Two cases of allergic reaction in persons inoculated against tularemia, resulting from bites by infected ticks of the species *Rhipicephalus rossicus*. *Journal of Microbiology, London*, **29**, 1827–9.

Borodin, V.P., Spitsin, N.A., Samsonova, A.P., Koroleva, A.P., Ermolova, N.D. & Chunikhin, V.P. (1965). Contribution to further study of natural ravine-steppe type tularemia focus in Volgograd Oblast. In *Materialy Nauchno Prakticheskoy Konferentsii Tularemia*, ed. G.V. Kornilova, pp. 14–16. Omsk: Ministerstvo Zdravookhraneniya RSFSR .

Dutkiewicz, J. & Siuda, K. (1969). *Rhipicephalus rossicus* Yakimov and Kol-Yakimova – a genus and species of tick new to the fauna of Poland. *Fragmenta Faunistica*, **15**, 99-105. [In Polish with a Russian and German summary].

Feider, Z. (1965). *Fauna Republicii Populare Romane. Arachnida, vol. 5, fasc. 2. Acaromorpha, Suprafamilia Ixodoidea*. Bucharesti: Editura Academiei Republicii Populare Romane.

Feldman-Muhsam, B. (1960). The ticks of Sinai. *Bulletin of the Research Council of Israel, Section B: Zoology*, **9B**, 57–64.

Filippova, N.A. (1981). Diagnosis of species of the genus *Rhipicephalus* Koch (Ixodoidea, Ixodidae) of the USSR fauna and neighboring countries according to the nymphal instar. *Parazitologicheskii Sbornik Zoologicheskogo Muzeya Akademii Nauk SSSR*, **30**, 47–68. [In Russian; NIH Library translation NIH-87-349].

Filippova, N.A. (1996). Designation of the neotypes for two species of ticks family Ixodidae. *Parasitologiya*, **30**, 404–9. [In Russian].

Hoogstraal, H. (1979). The epidemiology of tick-borne Crimean-Congo hemorrhagic fever in Asia, Europe, and Africa. *Journal of Medical Entomology*, **15**, 307–417.

Kondratenko, V.F., Blagoveshchenskaya, N.M., Butenko, A.M., Vishnivetskaya, L.K., Zarubina, L.V., Milyutin, V.N., Kuchin, V.V., Novikova, E.N., Rabinovich, V.D., Shevchenko, S.F. & Chumakov, M.P. (1970). Results of virological investigation of ixodid ticks in Crimean hemorrhagic fever focus in Rostov Oblast. In *Crimean Hemorrhagic Fever*, ed. M.P. Chumakov, pp. 29–35. Materialy III Oblastnoy Nauchno-Prakticheskoy Konferentsii, Rostov-na-Donu, May 1970. [In Russian, Translation T524, NAMRU-3, Cairo].

Watts, D.M., Ksiazek, T.G., Linthicum, K.J. & Hoogstraal, H. (1988). Crimean-Congo hemorrhagic fever. In *The Arboviruses: Epidemiology and Ecology*, vol. 2, ed. T.P. Monath, pp. 177-222. Boca Raton, FL: CRC Press, Inc.

Yakimov, V. L. (1923). A propos du *Rhipicephalus sanguineus* et *Rhipicephalus rossicus*. *Parasitology*, **15**, 256–7.

Yakimov, L. & Kol-Yakimova, N. (1911). Étude des Ixodidés de Russie. *Archives de Parasitologie*, **14**, 416–23.

Yu, X., Ye, R.-Y. & Gong, Z.-D. (1997). *The Tick Fauna of Xinjiang*. Urumqi, China: Xinjiang Scientific, Technological and Medical Publishing House.

Zahler, M., Filippova, N.A., Morel, P.C., Gothe, R. & Rinders, H. (1997). Relationship between species of the *Rhipicephalus sanguineus* group: a molecular approach. *Journal of Parasitology*, **83**, 302–6.

Also see the following Basic Reference (p. 13): Filippova (1997).

RHIPICEPHALUS SCALPTURATUS SANTOS DIAS, 1959

The specific name *scalpturatus*, from the Latin *scalptura* meaning 'a cutting' or 'engraving', 'an engraved figure', refers to the sculptured appearance of the male conscutum.

Diagnosis

A large, light brown tick with the male conscutum distinctly grooved.

Male (Figs 260(a), 261(a) to (c))

Only a single male specimen was available for measurement. Capitulum slightly broader than long, length × breadth 0.79 mm × 0.82 mm. Basis capituli with long slightly curving lateral angles not concealing the large apical process of coxae I; posterior margin of basis short and straight between long triangular cornua. Palps short, broad. Conscutum length × breadth 3.45 mm × 1.95 mm. Eyes marginal, slightly bulging. Cervical fields in the form of deep relatively

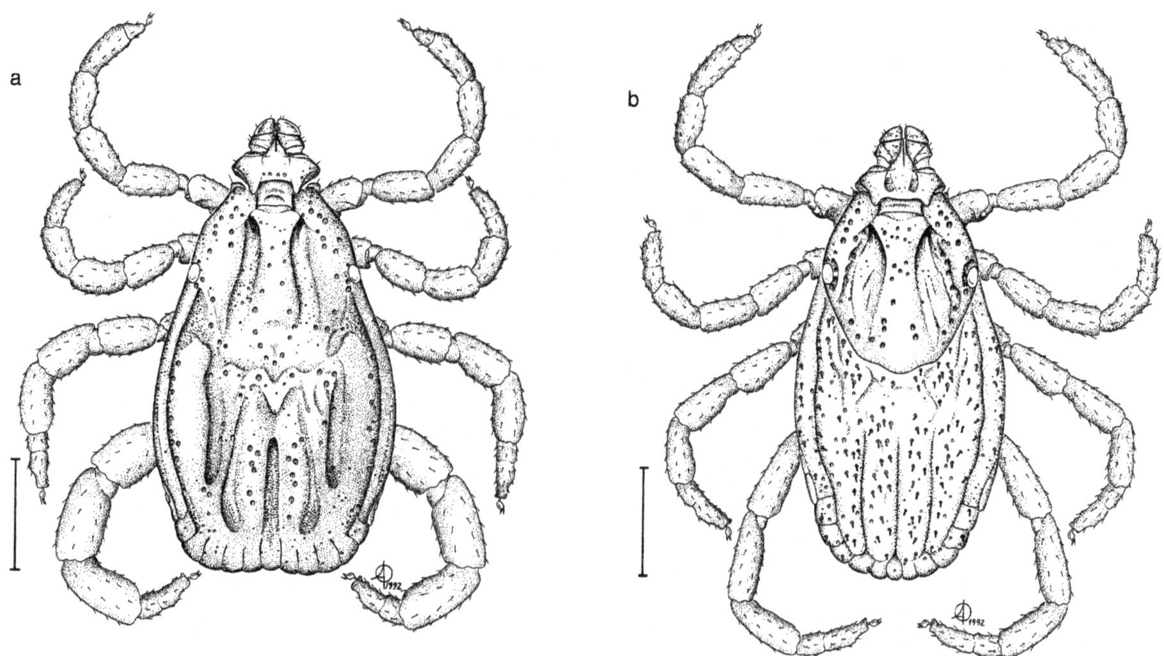

Figure 260. *Rhipicephalus scalpturatus* (RML 107652; HH 83652, collected by sweeping long dense grass vegetation near Lake Rover, Sukla Phanta Reserve, Kanchanpur, Nepal, November 11, 1974, C. Dietrich Schaaf). (a) Male, dorsal; (b) female, dorsal. Scale bars represent 1 mm. A. Olwage *del.*

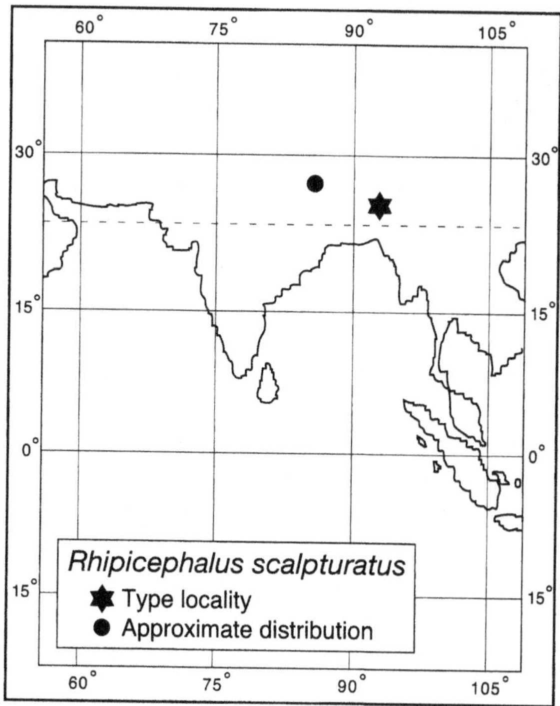

Map 73. *Rhipicephalus scalpturatus*: distribution.

narrow curved troughs; surface between external cervical margins and eyes raised. Marginal lines deep, extending anteriorly to eye level. A few moderately-large punctations scattered over scutal surface. Conscutum broadens posterior to eyes, indented at central festoon. Posteromedian groove broad, deep and long, extending to central festoon; posterolateral grooves as irregular depressions on either side of posteromedian groove, not well defined. Legs increase markedly in size from I to IV. Ventrally spiracles very large and broad throughout their length, extending from just posterior to coxae IV to festoons I. Adanal plates elongate, triangular, extending slightly medially posterior to anus but without posterointernal points, posterior margins almost straight; accessory adanal plates elongate rods.

Female (Figs 260(b), 261(d) to (f))

Only a single specimen was available for measurement. Capitulum broader than long, length × breadth 0.77 mm × 0.84 mm. Basis capituli with broad lateral angles not overlapping

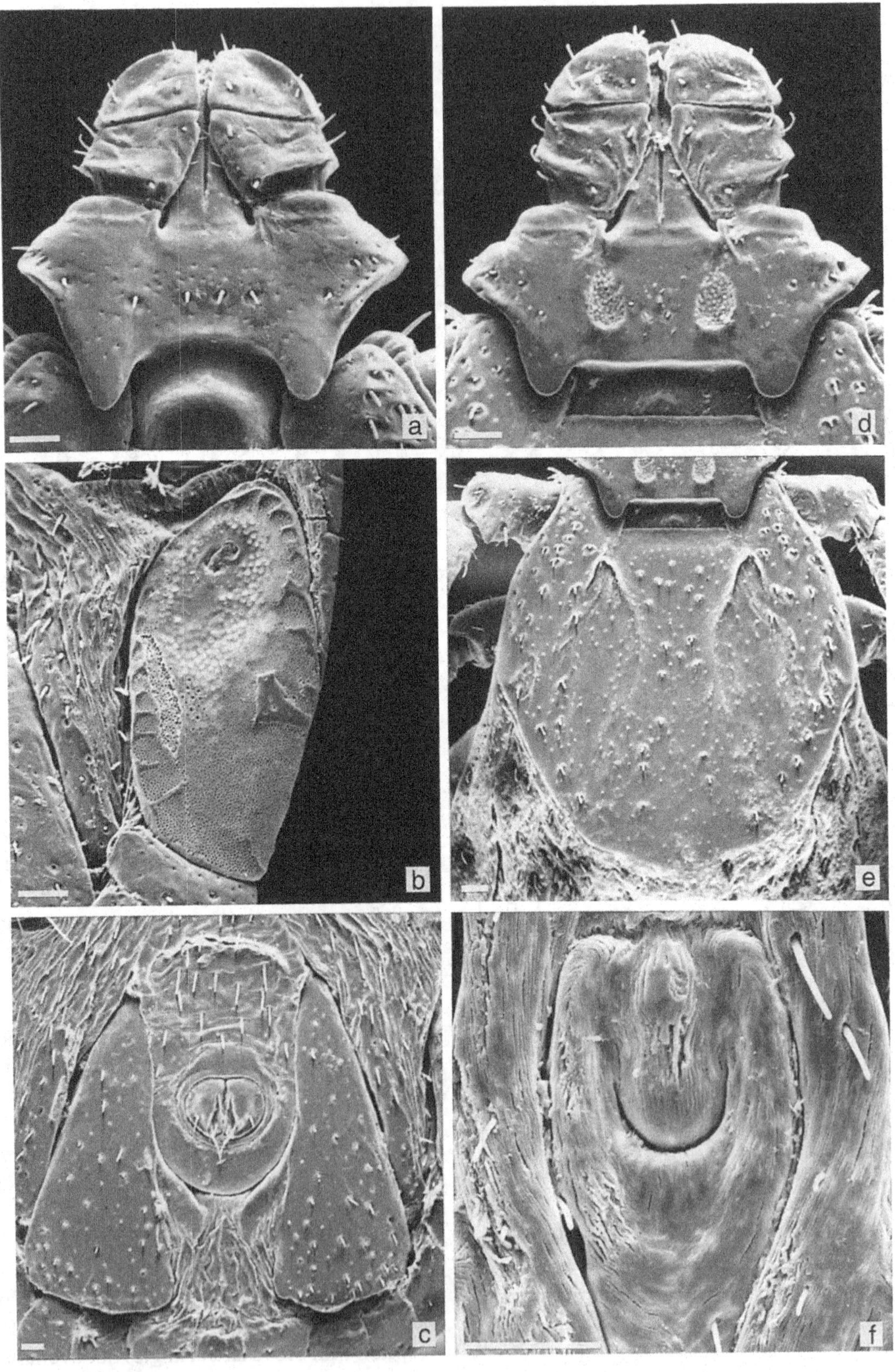
a
b
c
d
e
f

scapulae; porose areas large, longer in the anteroposterior axis. Palps short, broad. Scutum longer than broad, length × breadth 1.57 mm × 1.43 mm. Eyes about halfway back, very slightly bulging, edged dorsally with a few small punctations. Cervical fields broad, elongate, depressed between raised internal and external cervical margins, surface striated. Punctations relatively few, shallow, giving the scutum an impunctate appearance. Ventrally genital aperture broadly U-shaped with a bulging genital apron.

Nymph and Larva
Unknown.

Notes on identification
Males of *R. scalpturatus* are very large ticks with a conspicuously grooved conscutum that is indented at the central festoon, and legs increasing markedly in size from I to IV. Females are also very large ticks with a broadly U-shaped genital aperture with a bulging genital apron.

Hosts

Presumed to be a three-host tick. *Rhipicephalus scalpturatus* was originally described from museum specimens in the Zoological Museum, Hamburg, Germany by Santos Dias (1959), who erected the monotypic subgenus *Pomerantzevia* for this species. The only collection data available was that the host was unknown and the collection locality was given as the Khasi Hills, Umsan, Assam. (Umsan was a typographical error for Umsaw). This collection was cited by Hoogstraal & Rack (1967), and Miranpuri & Gill (1983). One additional collection in the U.S. National Tick Collection is from sweeping vegetation in long dense grass near Lake Rover, Sukla Phanta Reserve, Kanchanpur, Nepal. Immatures are unknown and adults have never been collected from an animal.

Zoogeography

Rhipicephalus scalpturatus is known only from grassy hill environments in Assam, India and the Nepal terai (Map 73).

Disease relationships

Unknown.

REFERENCES

Hoogstraal, H. & Rack, G. (1967). Ticks (Ixodidae) collected by Deutsche Indien-expedition, 1955–1958. *Journal of Medical Entomology*, **4**, 284–8.
Miranpuri, G.S. & Gill, H.S. (1983). *Ticks of India*. Edinburgh: Lindsay & Macleod.
Santos Dias, J.A.T. (1959). Notas Ixodológicas. VI. Descrição de um novo subgénero e de uma nova espécie de *Rhipicephalus* (Acarina, Ixodoidea) da Região Oriental. *Memórias e Estudos do Museu Zoológico da Universidade de Coimbra*, (**256**): 1–6.

Figure 261 (*opposite*). *Rhipicephalus scalpturatus* (RML 107652; HH 83652, collected by sweeping long dense grass vegetation near Lake Rover, Sukla Phanta Reserve, Kanchanpur, Nepal, November 11, 1974, C. Dietrich Schaaf). Male: (a) capitulum, dorsal; (b) spiracle; (c) adanal plates. Female: (d) capitulum, dorsal; (e) scutum; (f) genital aperture. Scale bars represent 0.10 mm. SEMs by P. Hill.

RHIPICEPHALUS SCHULZEI OLENEV, 1929

The specific name *schulzei* is a patronym in honour of Professor Dr Leopold Ernst Paul Schulze (1887–1949), German tick researcher, Professor at the University of Rostock, and among the founders and publishers of the journal *Zeitschrift für Parasitenkunde*. He died of a heart attack while at his desk writing a monograph summarizing all of his tick research.

Synonym

sanguineus schulzei.

Diagnosis

A small, light brown, lightly-punctate tick.

Male (Figs 262(a), 263(a) to (c))
Capitulum broader than long, length × breadth ranging from 0.47 mm × 0.55 mm to 0.54

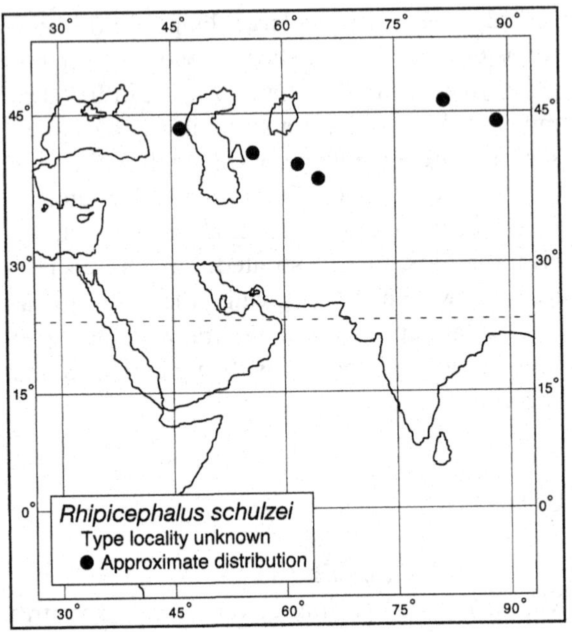

Map 74. *Rhipicephalus schulzei*: distribution.

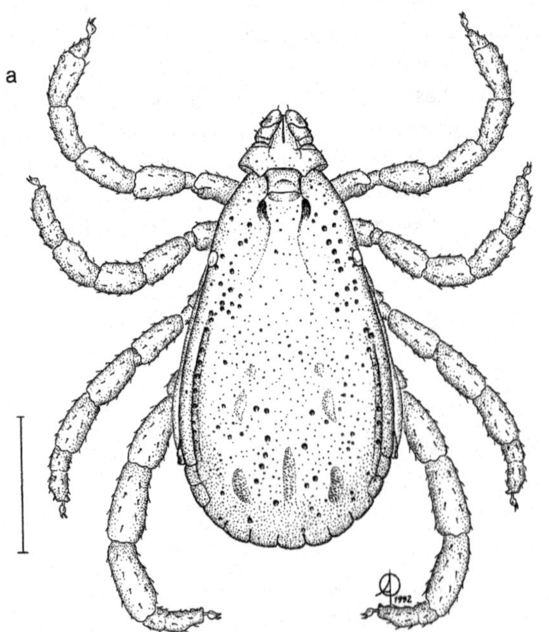

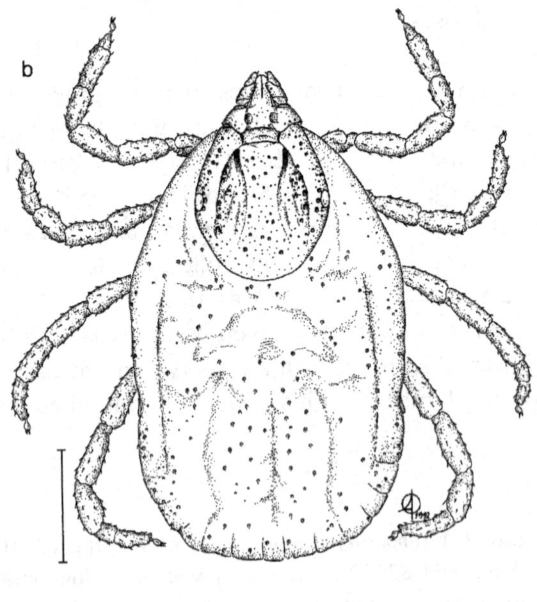

Figure 262. *Rhipicephalus schulzei* ♂ (RML 49397, collected from an unknown host and locality in the former USSR, 1932, W.W. Sceknew); ♀ (RML 102322; HH 29327, collected on large-toothed souslik (*Spermophilus fulvus*), near the Caspian Plains, Kazakhstan, June 20, 1951, by Galuzo and received from C.A. Hoare). (a) Male,dorsal; (b) female, dorsal. Scale bars represent 1 mm. A. Olwage *del.*

Table 69. *Host records of* Rhipicephalus schulzei

Hosts	Number of records
Domestic animals	
Cattle	1
Sheep	1
Goats	1
Camels	1
Dogs	1
Wild animals	
Steppe polecat (*Mustela eversmannii*)	2
Red-cheeked souslik (*Spermophilus erythrogenys*)	1
Large-toothed souslik (*Spermophilus fulvus*)	3
Little souslik (*Spermophilus pygmaeus*)	7 (including immatures)
Tien Shan souslik (*Spermophilus relictus*)	4
Long-tailed souslik (*Spermophilus undulatus*)	1
Great jerboa (*Allactaga major*)	1
Thick-tailed jerboa (*Stylodipus telum*)	1
Social vole (*Microtus socialis*)	1
Eversmann's hamster (*Allocricetulus eversmanni*)	1
Common hamster (*Cricetus cricetus*)	1
Mid-day gerbil (*Meriones meridianus*)	1
Tamarisk gerbil (*Meriones tamariscinus*)	1
Great gerbil (*Rhombomys opimus*)	2
House mouse (*Mus musculus*)	1
Humans	1

mm × 0.61 mm. Basis capituli with short broad lateral angles extending over the slightly protruding anterior process of coxae I. Palps short, slender, external margins slightly curved. Conscutum length × breadth ranging from 2.20 mm × 1.26 mm to 2.45 mm × 1.52 mm. Eyes marginal, flat, may or may not be edged dorsally with a few small punctations. Cervical fields only sketchily indicated and not depressed, a few punctations present in the area of the external cervical margins. Marginal lines punctate, delimiting first two festoons, ending well posterior to eye level. Posteromedian groove short and narrow, posterolateral grooves subcircular; all grooves well developed. Conscutum relatively impunctate, punctations most obvious anterolaterally. Legs increase in size only very slightly from I to IV. Ventrally spiracles narrowly elongate, dorsal prolongation slightly narrower than macular area of spiracle. Adanal plates rounded posteriorly, tapering posteromedially to a broad curve without cusps; accessory adanal plates elongate, narrowly V-shaped.

Female (Figs 262(b), 263(d) to (f))

Capitulum broader than long, length × breadth ranging from 0.58 mm × 0.66 mm to 0.64 mm × 0.76 mm. Basis capituli with long sharp lateral angles overlapping the scapulae; porose areas small to moderate in size, less than twice their own diameter apart. Palps pointed apically, external margins straight. Scutum slightly longer than broad, length × breadth ranging from 1.12 mm × 1.09 mm to 1.36 mm × 1.31 mm. Eyes about halfway back, flat, edged dorsally with a few fine punctations. Cervical fields scalpel-shaped,

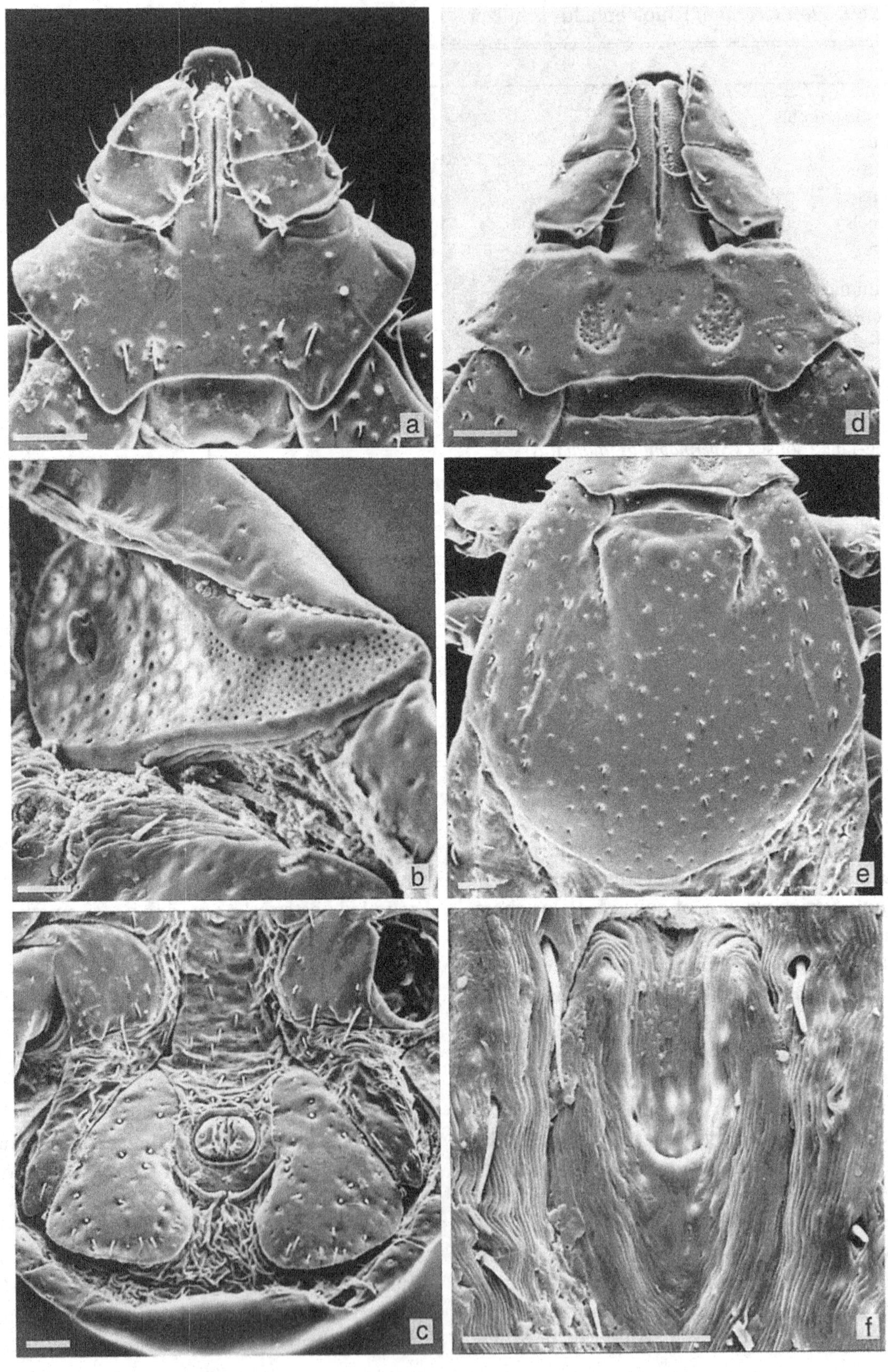
a
d
b
e
c
f

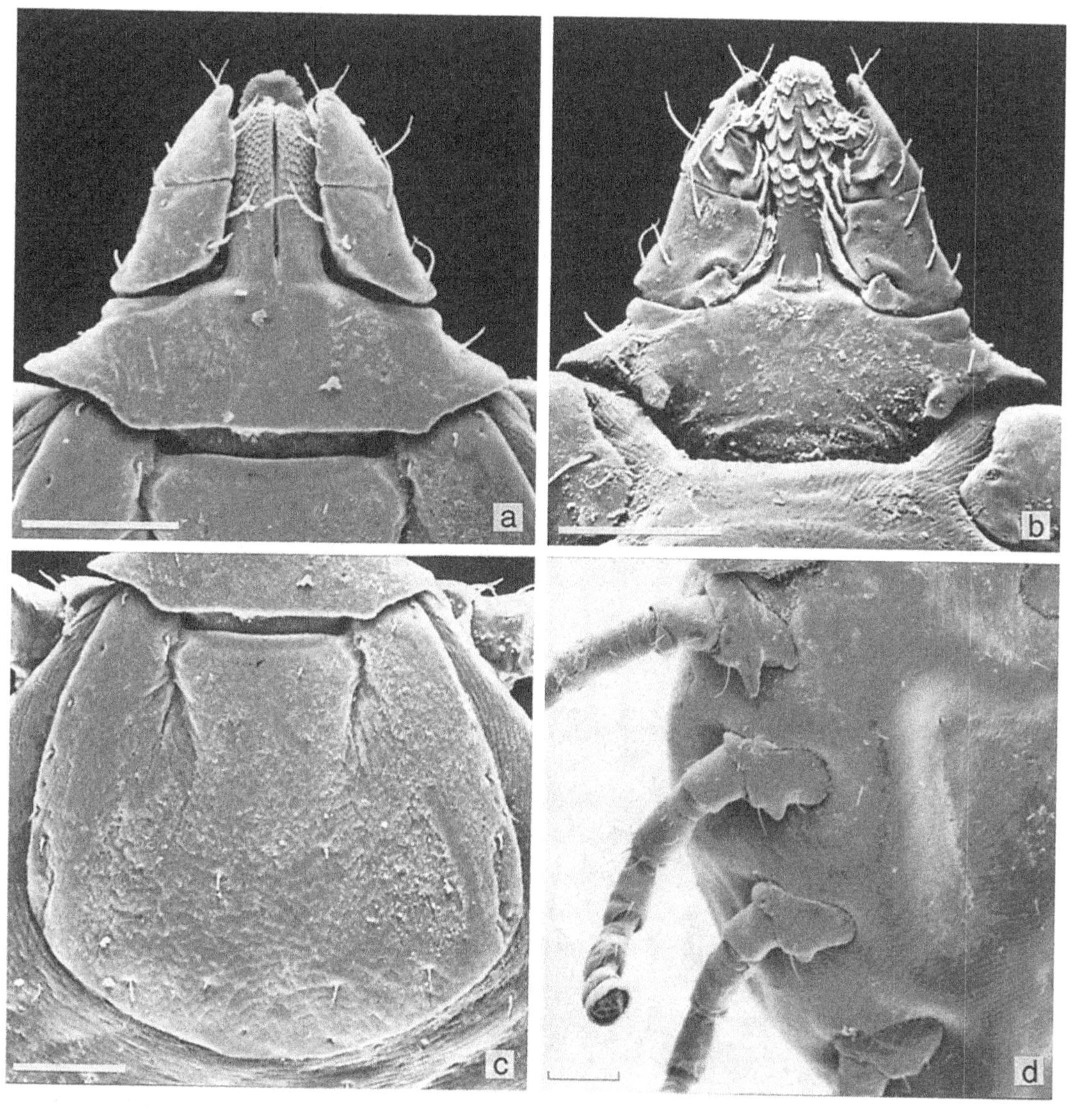

Figure 264 (*above*). *Rhipicephalus schulzei* (RML 118120, collected on little souslik (*Spermophilus pygmaeus*), southern shore of Aral Sea, Russia, date and collector unknown, received from G.V. Kolonin). Nymph: (a) capitulum, dorsal; (b) capitulum, ventral; (c) scutum; (d) coxae. Scale bars represent 0.10 mm. SEMs by P. Hill.

Figure 263 (*opposite*). *Rhipicephalus schulzei* (RML 102324; HH 29419, collected on great gerbil (*Rhombomys opimus*), Gur-Yev, Kazakhstan, date and collector unknown, received from M. Pospelova-Shtrom). Male: (a) capitulum, dorsal; (b) spiracle; (c) adanal plates. Female: (d) capitulum, dorsal; (e) scutum; (f) genital aperture. Scale bars represent 0.10 mm. SEMs by P. Hill.

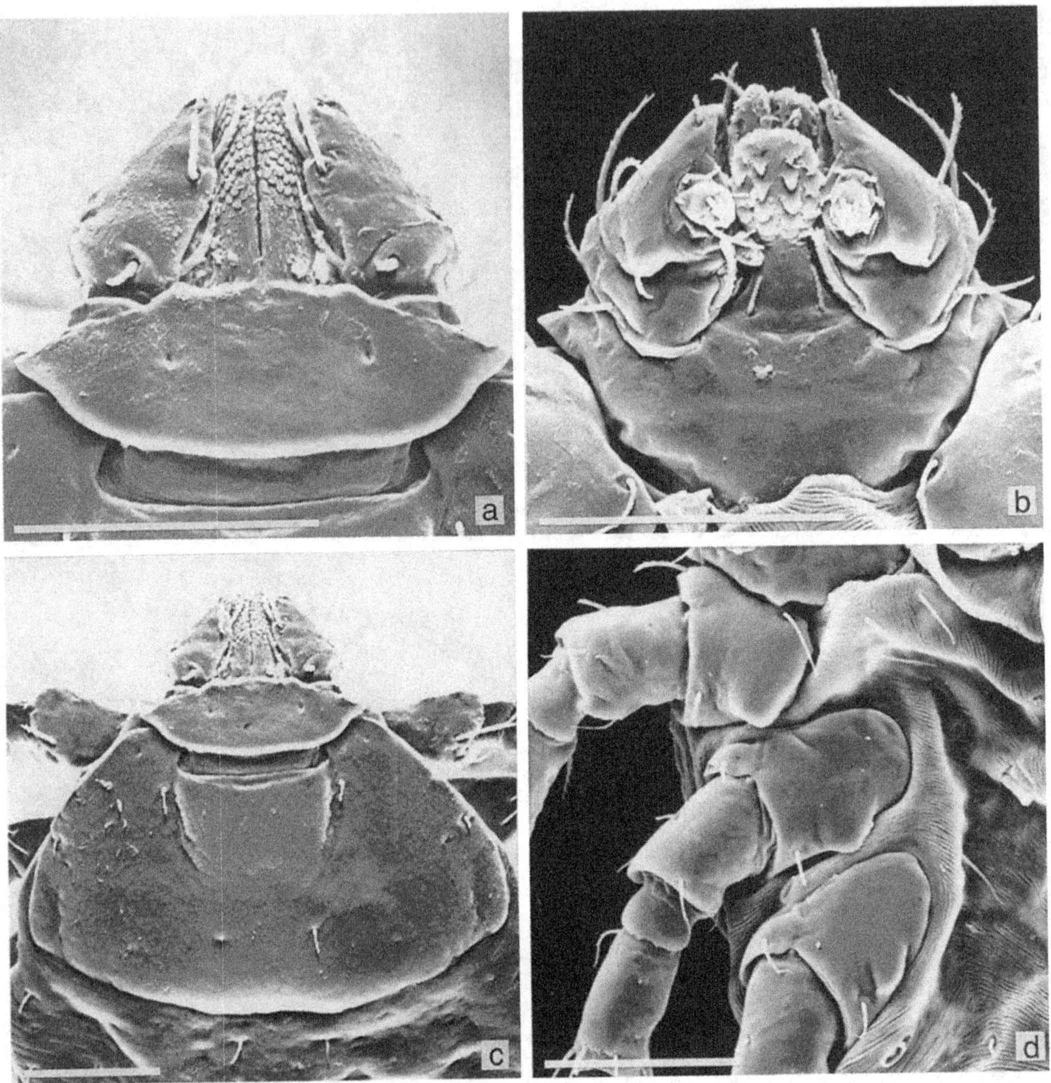

Figure 265. *Rhipicephalus schulzei* (RML 51582, collected on little souslik (*Spermophilus pygmaeus*), Western Kazakhstan, June 22, 1931, from the collection of Dr Paul Schulze). Larva: (a) capitulum, dorsal; (b) capitulum, ventral; (c) scutum; (d) coxae. Scale bars represent 0.10 mm. SEMs by P. Hill.

depressed, rugose; raised external cervical margins not always marked by punctations. Background of very fine punctations over entire scutum gives it a smooth impunctate appearance. Ventrally genital aperture U-shaped and short.

Nymph (Fig. 264)

Only one specimen available for measurement. Capitulum broader than long, length × breadth 0.26 mm × 0.28 mm. Basis capituli 2.5 times as broad as long, lateral angles moderately acute, not overlapping scapulae in specimen examined. Palps pointed apically, external margins very slightly convex. Scutum slightly broader than long, length × breadth 0.51 mm × 0.53 mm. Eyes more than halfway back, slightly bulging. Scutal margin recedes markedly posterior to eyes. Cervical grooves slightly diverging, cervical fields inapparent. Ventrally coxae I each with two well-developed spurs; coxae II to IV each with an external spur, decreasing progressively in size.

Larva (Fig. 265)
Capitulum much broader than long, length × breadth ranging from 0.12 mm × 0.15 mm to 0.13 mm × 0.16 mm. Basis capituli three times as broad as long, with diverging sharply-pointed lateral angles. Palps constricted proximally, pointed apically, external margins slightly convex. Scutum much broader than long, length × breadth ranging from 0.20 mm × 0.34 mm to 0.24 mm × 0.36 mm; posterior margin broadly curved. Eyes at widest point, slightly bulging. Cervical grooves short parallel troughs. Ventrally coxae I each with a large triangular spur, coxae II each with a small triangular spur, coxae III each with a posterior thickening, but no obvious spur.

Notes on identification
A species closely related to *R. pumilio,* but differs in the male in not having internal cusps on the adanal plates, and in the female in the configuration of the capitulum.

Hosts

A three-host species. All life stages inhabit the burrows of its hosts, which are various species of sousliks, the primary host being the Tien Shan souslik, but it will feed on other small hosts and on their predators (Table 69). Pomerantsev (1950) lists the primary host as the little souslik, followed by the large-toothed souslik, common hamster, Eversmann's hamster, great gerbil, house mouse, cattle, sheep, goats, camels, dogs and humans. Larval emergence occurs from May to August; nymphs during July and August; adults in July. The pattern of seasonal abundance suggests a life cycle extending over two years. Nelzina & Danilova (1960) studied the life cycle of *R. schulzei* and concluded that there are two phases in the annual cycle of both the tick and its host. The first phase is when the ticks are active from April to July and the second is a non-active 'hibernating' phase from August to March; these phases correspond to the periods of activity and hibernation of its souslik host.

Zoogeography

Rhipicephalus schulzei is a xerophytic species found in semi-desert and dry steppe habitats. It is present in southern Russia in the areas of eastern Transcaucasia, the northern areas of the Terek River in Daghestan, and in the Astrakhan Oblast, the Aral Sea area of western Kazakhstan, and Uzbekistan (Pomerantsev, 1950; Nyalkovskaya, 1966) (Map 74) It has recently been reported in Xinjiang Province, China (Yu, Ye & Gong, 1997).

Disease relationships

Little is known of the disease relationships of this tick. Being a burrow inhabitor and rarely parasitizing humans, *R. schulzei* is probably not an important vector of any human pathogen. However, Nelzina *et al.* (1960) showed that by rubbing ticks or their faeces containing the plague bacillus into the skin of guinea pigs, they were able to transmit the disease. Consequently *R. schulzei* could be a maintenance vector for one or more tickborne diseases.

REFERENCES

Nelzina, E.N. & Danilova, G.M. (1960). *Rhipicephalus schulzei* Ol. (Ixodidae) – a burrow inhabiting parasite of susliks. *Meditsinskaya Parazitologiya, Moskva*, **29**, 291–300. [In Russian, English summary].

Nelzina, E.N., Pylenko, M.S., Chudesova, V.P., Kondrashkina, K.I. & Bykov, L.T. (1960). The role of *Rhipicephalus schulzei* in natural foci of plague. (Communication 1. Localization of *Bacillus pestis* in the tick body). *Meditsinskaya Parazitologiya, Moskva*, **29**, 202–7. [In Russian, English translation T121, NAMRU-3, Cairo.]

Nyalkovskaya, S.A. (1966). *Rhipicephalus schulzei* Ol. in Daghestan. In *Tezisy Dokladov Pervoe Akarologicheskoe Soveshchanie*, ed. B.E. Bykhovsky, pp. 138–9. Moscow and Leningrad: Akademiya Nauk SSSR. [In Russian, English translation T385, NAMRU-3, Cairo.]

Olenev, N.O. (1929). The study of the Ixodoidea of our country. *Vestnik Sovremennoy Veterinarii*, 5, 191-3. [In Russian].

Pomerantsev, B.I. (1950). Ixodid ticks (Ixodidae). Fauna SSSR, Paukoobraznye, n.s. (41), 4(2), 224 pp. [In Russian; English translation by A. Elbl, edited by G. Anastos. The American Institute of Biological Sciences, Washington, DC, 199 pp.]

Yu, X., Ye, R.-Y. & Gong, Z.-D. (1997). *The Tick Fauna of Xinjiang*. Urumqi, China: Xinjiang Scientific, Technological and Medical Publishing House.

Also see the following Basic Reference (p. 13): Filippova (1997).

11

Host/parasite list for the non-Afrotropical *Rhipicephalus* species

Unless otherwise stated the records in the host/parasite checklist refer to adult ticks. The number in brackets appearing after a tick's name represents the number of collections made from the host species under which it is listed. When a tick's name appears in **bold type** this indicates that the animal under which it is listed is a preferred host of the adults. Large numbers of domestic cattle, sheep and dogs as well as Indian gerbils and Tolai hares have been examined for ticks within the distribution range of the non-Afrotropical *Rhipicephalus* species. The numbers of collections from these animals could give a false impression of host preference.

The host groupings under which the tick collections are recorded in the host/parasite checklist follow the same sequence as those in the host record tables accompanying the individual tick species accounts, namely domestic animals, wild animals, birds and humans.

DOMESTIC ANIMALS

Bos indicus/taurus **Cattle**

R. bursa (40)
R. haemaphysaloides (11)
R. pilans (5)
R. pumilio (2)
R. rossicus (8)
R. schulzei (1)
R. turanicus (150 +)

Bubalus bubalis **Water buffaloes**

R. haemaphysaloides (2)
R. pilans (12)
R. turanicus (3)

Ovis aries **Sheep**

R. bursa (91)
R. haemaphysaloides (5)
R. pumilio (1)
R. schulzei (1)
R. turanicus (many)

Capra hircus **Goats**

R. bursa (35, including immatures)
R. haemaphysaloides (8)
R. leporis (1)
R. pumilio (1)
R. rossicus (1)
R. schulzei (1)
R. turanicus (many)

Camelus dromedarius **Camels**

R. bursa (1)
R. haemaphysaloides (2)
R. pumilio (1)
R. rossicus (1)
R. schulzei (1)
R. turanicus (15)

Equus caballus **Horses**

R. bursa (10)
R. haemaphysaloides (4)
R. pilans (8, including immatures)

Horses (*cont.*)
R. pumilio (1)
R. rossicus (2)

Equus asinus **Donkeys**
R. bursa (1)
R. haemaphysaloides (2)
R. pumilio (1)
R. turanicus (1)

Sus scrofa **Pigs**
R. haemaphysaloides (4)
R. pilans (4)
R. pumilio (1)
R. turanicus (1)

Pigs (feral)
R. pusillus (2)

Canis familiaris **Dogs**
R. bursa (1)
R. haemaphysaloides (6)
R. leporis (2)
R. pilans (6, including immatures)
R. pumilio (2)
R. pusillus (3)
R. rossicus (3)
R. schulzei (1)
R. turanicus (172)

Felis catus **Cats**
R. turanicus (44, including immatures)

WILD ANIMALS

CLASS MAMMALIA

Order Insectivora

Family Erinaceidae
Atelerix algirus **Algerian hedgehog**
R. pusillus (1)

Erinaceus concolor **East European hedgehog**
R. rossicus (1)
R. turanicus (5)

Erinaceus europaeus **West European hedgehog**
R. turanicus (4)

Hemiechinus aethiopicus **Desert hedgehog**
R. leporis (2)
R. turanicus (3)

Hemiechinus auritus **Long-eared hedgehog**
R. leporis (3)
R. pumilio (2, immatures)
R. turanicus (9, including immatures)

Hemiechinus hypomelas **Brandt's hedgehog**
R. leporis (2)
R. pumilio (1)
R. turanicus (2)

Hemiechinus micropus **Indian hedgehog**
R. turanicus (7)

'Hedgehog'
R. pusillus (1)

Hylomys suillus **Lesser moonrat**
R. pilans (1, immatures)

Family Soricidae
Crocidura nigripes **Sulawesi white-toothed shrew**
R. pilans (2, immatures)

Crocidura russula **Common European white-toothed shrew**
R. turanicus (1, immatures)

Crocidura zarudnyi **'White-toothed shrew'**
R. turanicus (1, immatures)

Suncus murinus **House shrew**
R. haemaphysaloides (2, immatures)
R. pilans (2)

Order Chiroptera

Family Pteropodidae
Rousettus leschnaulti **Leschnault's rousette**
R. pilans (1, immatures)

Family Vespertilionidae

Eptesicus serotinus **Serotine**
R. rossicus (1, immatures?)

Order Carnivora

Family Canidae

Canis aureus **Golden jackal**
R. haemaphysaloides (3)
R. leporis (3)
R. pumilio (1)
R. turanicus (7)

Canis lupus **Wolf**
R. pumilio (1)

Cuon alpinus **Dhole (Red dog)**
R. haemaphysaloides (1)

Vulpes bengalensis **Bengal fox**
R. ramachandrai (1)

Vulpes vulpes **Red fox**
R. leporis (6, including nymphs)
R. pumilio (1)
R. pusillus (5)
R. turanicus (13)

Family Felidae

Catopuma temminckii **Asian golden cat**
R. haemaphysaloides (1)

Felis chaus **Jungle cat**
R. turanicus (2)

Panthera pardus **Leopard**
R. haemaphysaloides (2)

Panthera tigris **Tiger**
R. haemaphysaloides (1)
R. pilans (1)

Family Mustelidae

Meles meles **Eurasian badger**
R. turanicus (2)

Mustela eversmannii **Steppe polecat**
R. leporis (1)
R. schulzei (2)

Mustela nivalis **Weasel**
R. pusillus (1)
R. turanicus (1, immatures)

Mustela putorius **Polecat**
R. pumilio (1)

Vormela peregusna **Marbled polecat**
R. pumilio (1)

Family Viverridae

Paradoxurus hermaphroditus **Common palm civet**
R. pilans (1)

'Genet'
R. pusillus (1)

Order Perissodactyla

Family Equidae

Equus hemionus **Asian wild ass**
R. pumilio (1)

Order Artiodactyla

Family Cervidae

Axis axis **Spotted deer**
R. haemaphysaloides (1)

Cervus timorensis **Timor deer**
R. pilans (2, including immatures)

Cervus unicolor **Sambar**
R. haemaphysaloides (2)
R. pilans (1)

Muntiacus muntjak **Indian muntjac**
R. haemaphysaloides (3)

'Deer'
R. pusillus (1)

Family Bovidae

Gazella gazella **Mountain gazelle**
R. bursa (1)

Gazella subgutturosa **Goitred gazelle**
R. pumilio (1)

Hemitragus hylocrius **Nilgiri tahr**
R. haemaphysaloides (1)

Hemitragus jemlahicus **Himalayan tahr**
R. haemaphysaloides (1)

Naemorhedus sumatraensis **Common goral**
R. haemaphysaloides (1)

Ovis ammon **Argali**
R. bursa (2)
R. turanicus (3)

Order Rodentia

Family Sciuridae

Funambulus tristriatus **Jungle striped squirrel**
R. ramachandrai (1)

Marmota **sp.** **'Marmot'**
R. haemaphysaloides (1)

Spermophilus erythrogenys **Red-cheeked souslik**
R. schulzei (1)

Spermophilus fulvus **Large-toothed souslik**
R. schulzei (3)

Spermophilus pygmaeus **Little souslik**
R. schulzei (7, including immatures)

Spermophilus relictus **Tien Shan souslik**
R. schulzei (4)

Spermophilus undulatus **Long-tailed *souslik***
R. schulzei (1)

Family Dipodidae

Allactaga major **Great jerboa**
R. schulzei (1)

Stylodipus telum **Thick-tailed jerboa**
R. schulzei (1)

Family Muridae

Arvicola terrestris **European water vole**
R. rossicus (1)

Microtus guentheri **Günther's vole**
R. turanicus (2, immatures)

Microtus socialis **Social vole**
R. schulzei (1)

Ondatra zibethicus **Muskrat**
R. pusillus (1)

Allocricetulus eversmanni **Eversmann's hamster**
R. schulzei (1)

Cricetus cricetus **Common hamster**
R. schulzei (1)

Cricetulus migratorius **Grey hamster**
R. turanicus (7, immatures)

Meriones hurrianae **Indian desert gerbil**
R. ramachandrai (2, immatures)
R. turanicus (2)

Meriones libycus **Libyan jird**
R. turanicus (5, immatures)

Meriones meridianus **Mid-day gerbil**
R. pumilio (1, immatures)
R. schulzei (1)

Meriones persicus **Persian jird**
R. turanicus (2, immatures)

Meriones tamariscinus **Tamarisk gerbil**
R. pumilio (1, immatures)
R. schulzei (1)

Meriones tristrami **Tristram's jird**
R. turanicus (1, immatures)

Rhombomys opimus **Great gerbil**
R. leporis (1)
R. pumilio (1)
R. schulzei (2)
R. turanicus (1, immatures)

Tatera indica **Indian gerbil**
R. haemaphysaloides (1, immatures)
R. ramachandrai (34, including immatures)
R. turanicus (1, immatures)

Apodemus mystacinus **Broad-toothed mouse**
R. turanicus (1, immatures)

Apodemus sylvaticus **Wood mouse**
R. bursa (7, immatures)

Bandicota bengalensis **Lesser bandicoot rat**
R. ramachandrai (2)
R. turanicus (1, immatures)

Bandicota indica **Greater bandicoot rat**
R. haemaphysaloides (1, nymph)
R. pilans (2, immatures)

Bunomys penitus **Sulawesi rat**
R. pilans (1, immatures)

Golunda ellioti **Indian bush rat**
R. haemaphysaloides (1, immatures)

Maxomys bartelsii **Oriental spiny rat**
R. pilans (2, immatures)

Millardia gleadowi **Sand-coloured rat**
R. ramachandrai (2, immatures)

Millardia meltada **Soft-furred field rat**
R. ramachandrai (2, immatures)

Mus booduga **Little Indian field mouse**
R. haemaphysaloides (4, immatures)

Mus musculus **House mouse**
R. haemaphysaloides (15, immatures)
R. pusillus (1)
R. schulzei (1)
R. turanicus (3, immatures)

Mus platythrix **Indian brown spiny mouse**
R. haemaphysaloides (1, immatures)

Mus saxicola
R. haemaphysaloides (1, immatures)

Nesokia indica **Short-tailed bandicoot rat**
R. turanicus (3, immatures)

Niviventer fulvescens **Chestnut rat**
R. pilans (1, immatures)

Rattus argentiventer **Ricefield rat**
R. pilans (3, immatures)

Rattus blandfordi **Blandford's rat**
R. haemaphysaloides (3, immatures)

Rattus exulans **Polynesian rat**
R. pilans (24, immatures)

Rattus rattus **Black rat**
R. haemaphysaloides (8, immatures)
R. turanicus (3, immatures)

Rattus tanezumi
R. pilans (3, immatures)

Rattus tiomanicus **Malaysian field rat**
R. pilans (2, immatures)

Rattus **sp.** **'Rat'**
R. pilans (2, immatures)

'Orchard rat'
R. pusillus (1)

Family Gliridae

Eliomys melanurus **'Dormouse'**
R. pusillus (1)
R. turanicus (1, immatures)

Eliomys quercinus **Garden dormouse**
R. bursa (1, immatures)

'Rodents'
R. rossicus (several records, immatures)

Order Lagomorpha

Family Ochotonidae

Ochotona pallasi **Pallas's pika**
R. rossicus (2)

Family Leporidae

Lepus capensis **Cape hare**
R. leporis (5)
R. turanicus (5)

Lepus nigricollis **Indian hare**
R. haemaphysaloides (1)
R. turanicus (8, including immatures)

Lepus tolai **Tolai hare**
R. leporis (12, including immatures)
R. pumilio (8, including immatures)

Lepus yarkandensis **Yarkand hare**
R. pumilio (1, immatures)

Lepus **spp.** **'Hare'**
R. haemaphysaloides (1)
R. leporis (3)

Oryctolagus cuniculus **European rabbit**
R. bursa (3)
R. pusillus (many, including immatures)
R. rossicus (1)
R. turanicus (2)

'Rabbit'
R. pumilio (1)

BIRDS

CLASS AVES

Order Cuculiformes

Family Cuculidae

Centropus sinensis **Crow pheasant**

R. haemaphysaloides (1, immatures?)

Order Passeriformes

Family Timaliidae

Chrysomma sinense **Oriental yellow-eyed babbler**

R. haemaphysaloides (1, immatures?)

HUMANS

R. bursa (2)
R. haemaphysaloides (9)
R. pilans (2)
R. pumilio (2)
R. rossicus (2)
R. schulzei (1)
R. turanicus (13)

12

Species groups based on the immature stages

The oft-quoted saying 'a picture is worth a thousand words' is particularly apt when applied to the immature stages of ticks of the genus *Rhipicephalus*. Their identification to species level is especially difficult as several of them are so similar in appearance. For this reason we have not attempted to produce keys for their identification. Instead we have included a series of figures in which line drawings of the capitula of nymphs and larvae of morphologically similar species have been grouped. In some contentious cases, but not all, we have been influenced by the morphology of the adults in deciding on our groupings. The diagnostic characters on the capitula of the various species within a group are easily visible in these figures and should assist readers to identify them. In addition brief descriptions of these characters have been given for each group as well as summaries of the geographic distributions of the individual species.

The classification proposed now represents a revision and extension of that initially suggested by Walker (1961). We do not, however, wish to give readers the impression that we regard the present groups as being immutable. Some do appear to be naturally well founded, e.g. the *R. evertsi* group, comprising *R. bursa*, *R. evertsi* and *R. glabroscutatum*. These three species are not only very much alike morphologically at the immature stage but they are the only species in the genus at present known to have a two-host, rather than a three-host, life cycle. In other cases, though, our placing of individual species in this group or that may be regarded as questionable, e.g. the inclusion of *R. armatus* in the *R. appendiculatus* group. Apart from the few entities presently consigned to the 'ragbag' of the miscellaneous group, the species that we have found most difficult to identify and to group are those with more-or-less pointed palps, i.e. particularly those presently included in the *R. simus*, *R. follis* and *R. capensis* groups. We feel that these are the groups whose composition is most likely to be revised in the future.

Another important characteristic that is apparently group-related in this genus is the host relationships of the various species. In general the 'pointy-palp' nymphs and larvae feed on small mammals, including rodents, not on the hosts utilized by the adults. There is also some evidence that they may be very specific in their choice of small mammal hosts. Immatures of the *pravus* group apparently favour various species of macroscelids, the elephant shrews, though the immature stages of *R. kochi* have been collected from the same large mammals as the adults. Numerous nymphs and larvae of both the *appendiculatus* and *evertsi* groups usually feed on the same hosts as the adults, though occasional specimens have been collected from small mammals. It is noteworthy that African hares and rabbits, i.e. the leporids, serve as hosts of the immature stages of species from nearly all the groups in the genus.

These host preferences are important when the potential of different species of *Rhipicephalus* as vectors of various pathogenic organisms is being considered.

Future studies of our 'groups' could provide a firmer basis for the erection of subgenera. The nominate subgenus *Rhipicephalus* encompasses the majority of *Rhipicephalus* species worldwide, but the subgenera *Digeneus, Pomerantzevia* and *Pterygodes* may, indeed, be valid. To this date, however, the erection of subgenera has been based on characters found in adult ticks only. Without the inclusion of immature characteristics we question their validity.

The illustrations comprising the various species groups have been arranged alphabetically, commencing with those of the nymphs. Whenever a species within the group is particularly well-known the group has been named after this species. In the two cases in which no well-known species is present the group has been named after the species illustrated first alphabetically.

The Afrotropical species groups are presented in the following sequence: *R. sanguineus, R. simus, R. follis, R. capensis, R. pravus, R. appendiculatus, R. evertsi* and a group containing species that we are unable to place comfortably in any of the other groups. For the species occurring outside Africa the *R. sanguineus* group is presented first followed by the *R. haemaphysaloides* group.

REFERENCE

Walker, J.B. (1961). Some observations on the classification and biology of ticks belonging to the genus *Rhipicephalus*, with special reference to the immature stages. *East African Medical Journal*, 38, 232–8.

THE AFROTROPICAL *RHIPICEPHALUS SANGUINEUS* GROUP

This group includes the following species (Fig. 266):

R. camicasi: north-eastern corner of Africa, Yemen Arab Republic, Saudi Arabia, Jordan and Lebanon.

R. guilhoni: East to West Africa between 6 °N and 18 °N.

R. sanguineus: worldwide, between approximately 50 °N and 30 °S.

R. sulcatus: widespread in West, East, Central and southern Africa.

R. turanicus: mainly in North, East and southern Africa, and in the Mediterranean countries, Europe, Russia and India.

Nymphs

(a) Basis capituli with pointed lateral angles from mid-length to two-thirds back.

(b) Palps tapering, apices narrowly to fairly broadly rounded, inclined inwards.

Larvae

(a) Basis capituli with pointed lateral angles at about mid-length.

(b) Palps tapering, apices narrowly rounded to pointed, inclined inwards.

Figure 266 (*opposite*). The Afrotropical *Rhipicephalus sanguineus* species group: capitula of nymphs (N) and larvae (L).

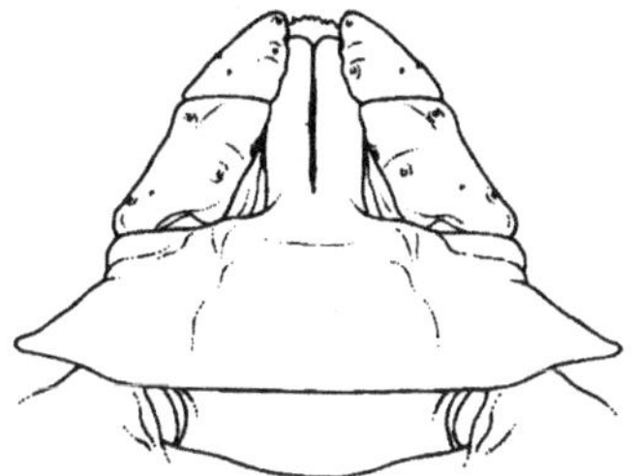
R. camicasi N.

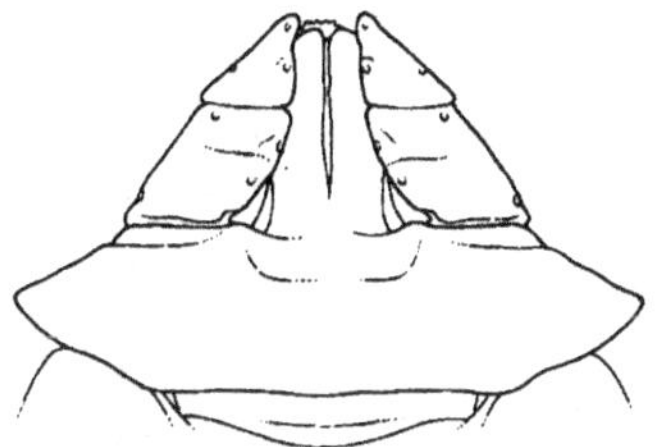
R. guilhoni N.

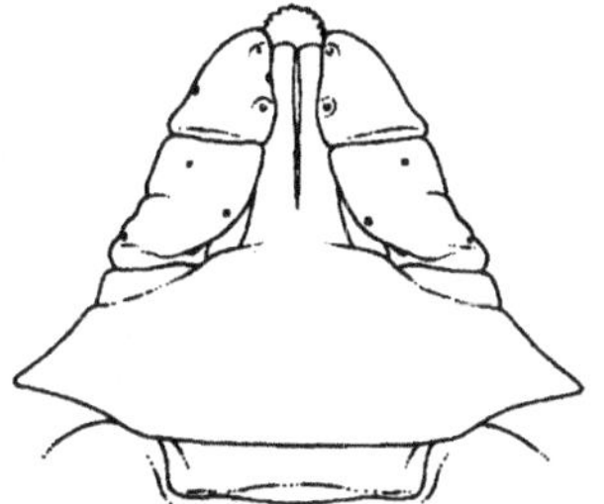
R. sanguineus N.

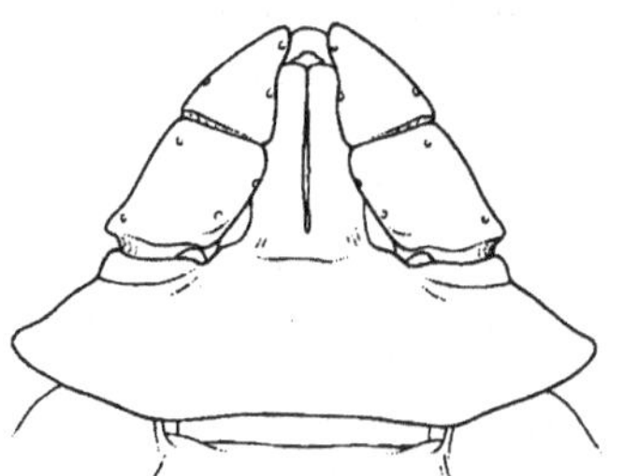
R. sulcatus N.

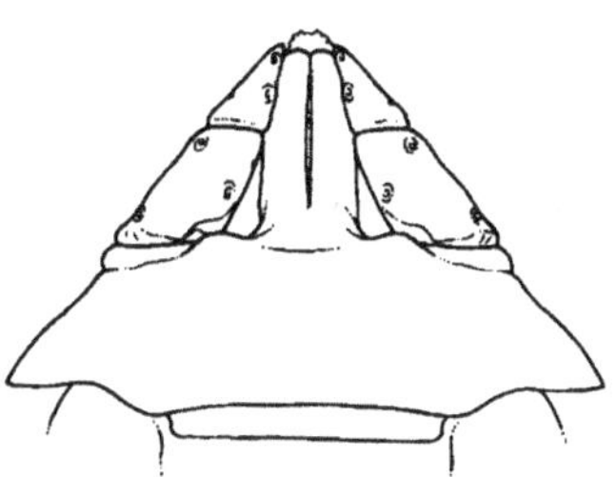
R. turanicus N. (Zambia)

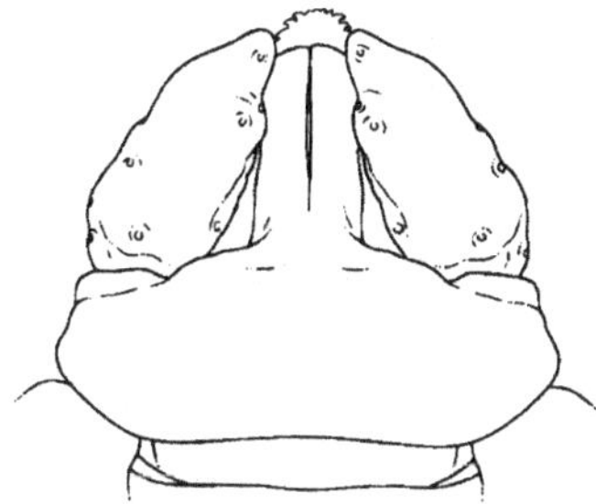
R. camicasi L.

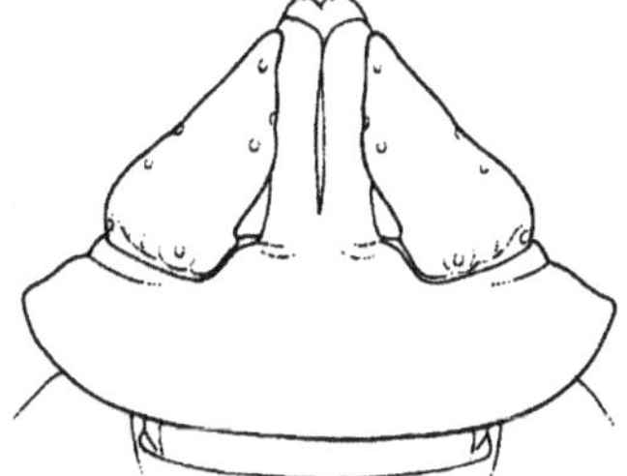
R. guilhoni L.

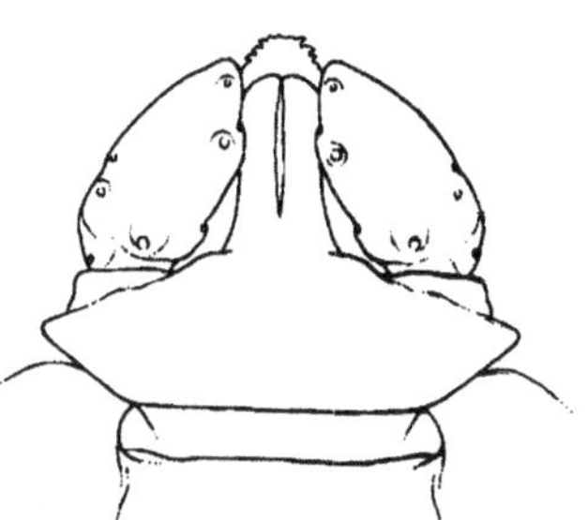
R. sanguineus L.

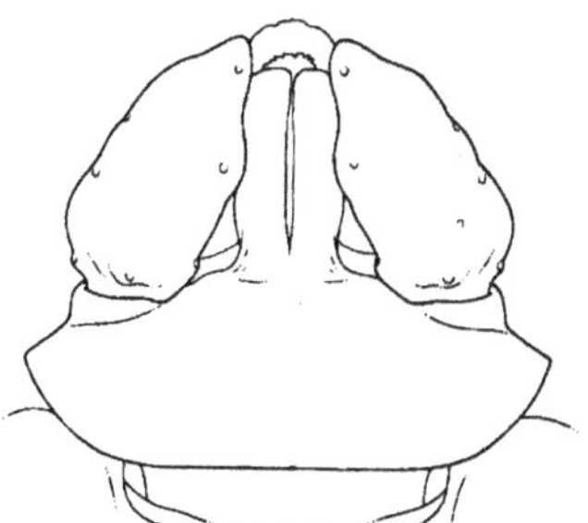
R. sulcatus L.

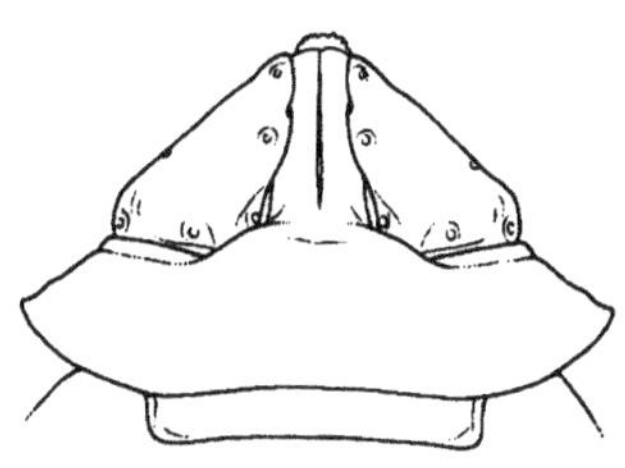
R. turanicus L. (Zambia)

RHIPICEPHALUS SIMUS GROUP

This group contains the following species (Figs 267 and 268):

R. distinctus: East, Central and southern Africa.
R. muhsamae: mainly West Africa.
R. planus: East Africa and eastern Central Africa.
R. praetextatus: north-eastern to East Africa.
R. senegalensis: from West Africa to Uganda in the east, mainly between 0° and 12 °N.
R. simpsoni: West, East and Central Africa southwards to eastern southern Africa.
R. simus: southern Africa excluding the more arid western regions.
R. zumpti: Mozambique southwards to eastern South Africa.

Nymphs

(a) Basis capituli with sharply pointed lateral angles; posterolateral margins usually concave over scapulae, posterior margin typically straight; ventrally spurs on posterior margin.
(b) Palps long with external margins straight to mildly curved, tapering, inclined inwards.

Larvae

(a) Basis capituli hexagonal, at least three times as broad as long; rounded to pointed lateral angles at about mid-length.
(b) Palps tapering, inclined inwards.

RHIPICEPHALUS FOLLIS GROUP

This group comprises the following species (Figs 269 and 270):

R. follis: southern and eastern southern Africa.
R. gertrudae: Namibia and south-western and central South Africa.
R. hurti: high altitude areas in south-western Tanzania, Kenya, Uganda and Rwanda.
R. jeanneli: high altitude areas in Kenya, Tanzania, Uganda, Rwanda and Burundi.
R. lounsburyi: south-eastern and south-western South Africa.
R. lunulatus: widely distributed in the Afrotropical region.
R. neumanni: southern Namibia, Northern and Western Cape Provinces, South Africa.
R. tricuspis: western Democratic Republic of Congo, central southern Africa.

Nymphs

(a) Basis capituli with long sharp lateral angles overlapping scapulae; cornua may or may not be present; ventrally well-developed spurs on posterior margin.
(b) Palps slender, tapering, apices narrowly rounded (somewhat broader in *R. jeanneli* and *R. lounsburyi*), inclined inwards.

Larvae

(a) Basis capituli hexagonal, sharply pointed to bluntly rounded lateral angles at about mid-length.
(b) Palps broad proximally, tapering to narrowly rounded apices (broader in *R. gertrudae*), inclined inwards.

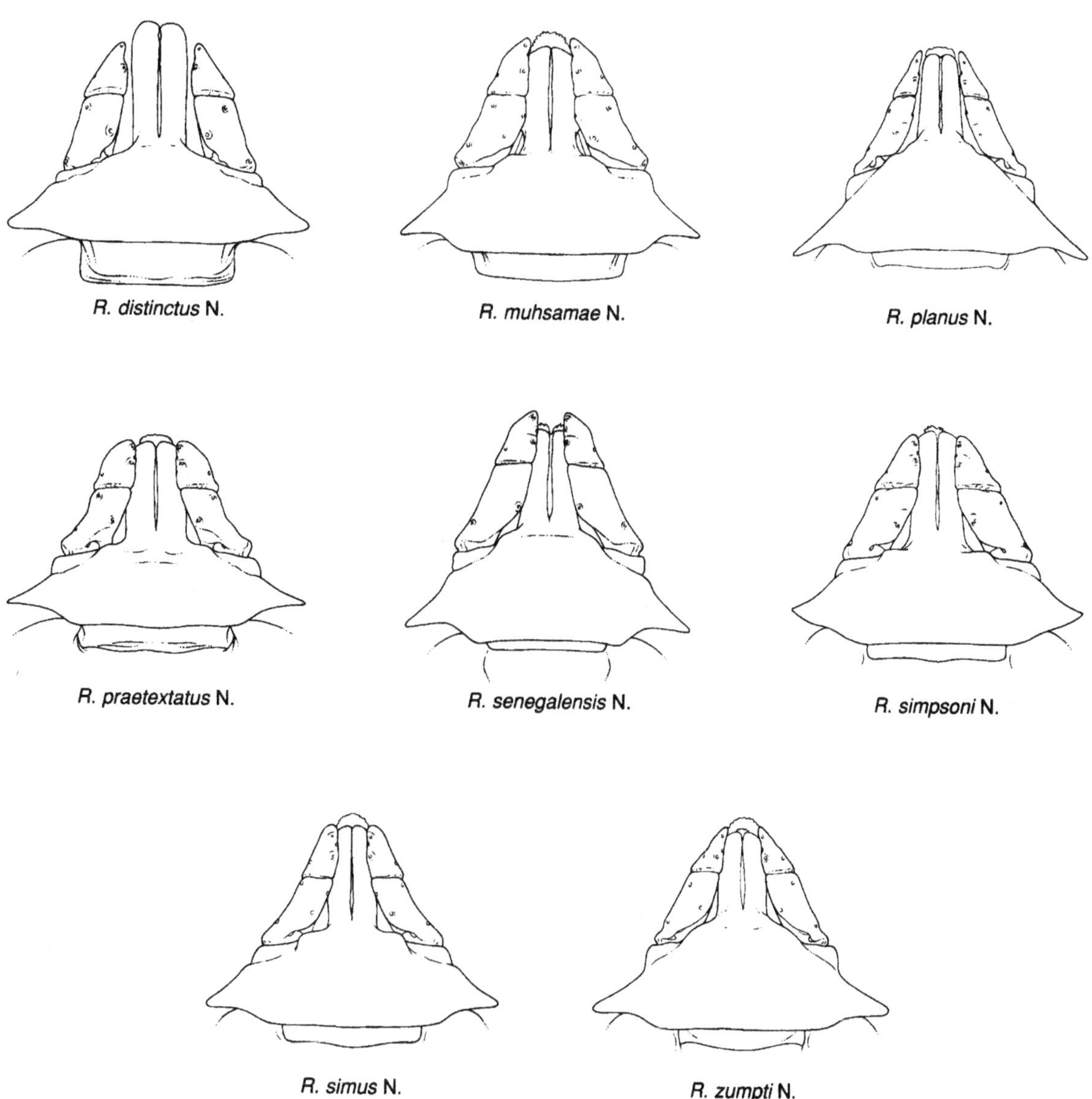

Figure 267. The *Rhipicephalus simus* species group: capitula of nymphs (N).

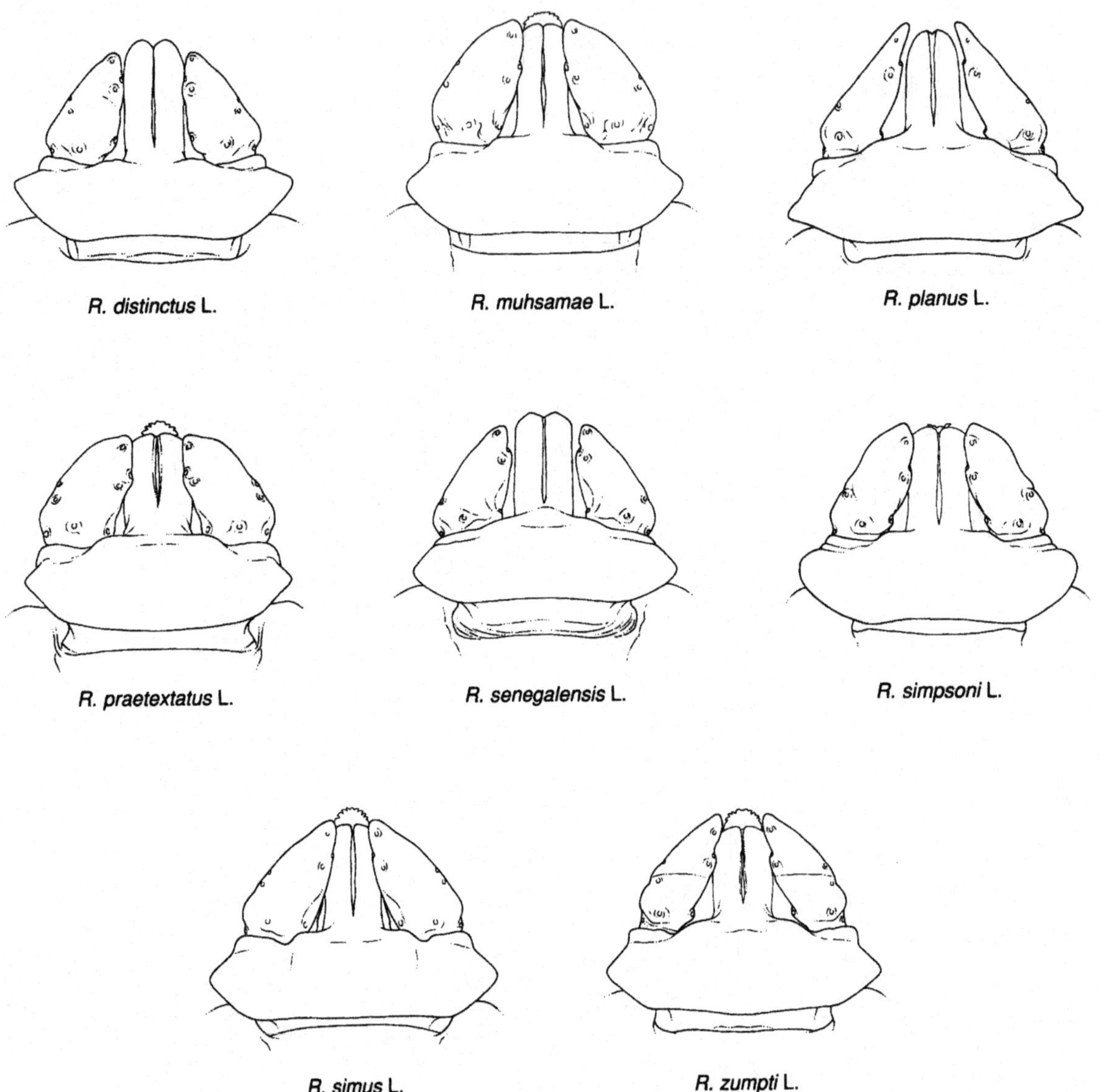

Figure 268. The *Rhipicephalus simus* species group: capitula of larvae (L).

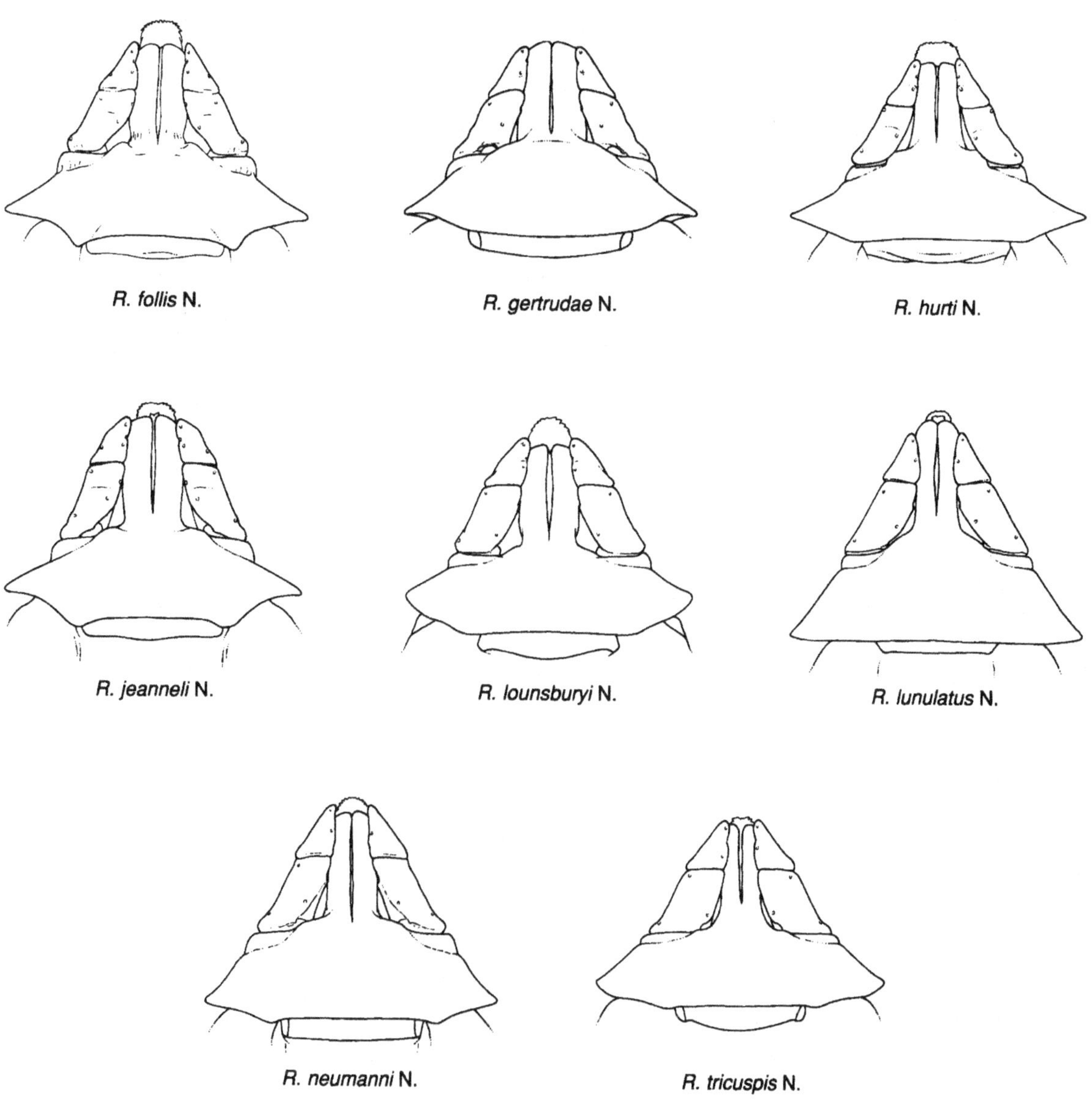

Figure 269. The *Rhipicephalus follis* species group: capitula of nymphs (N).

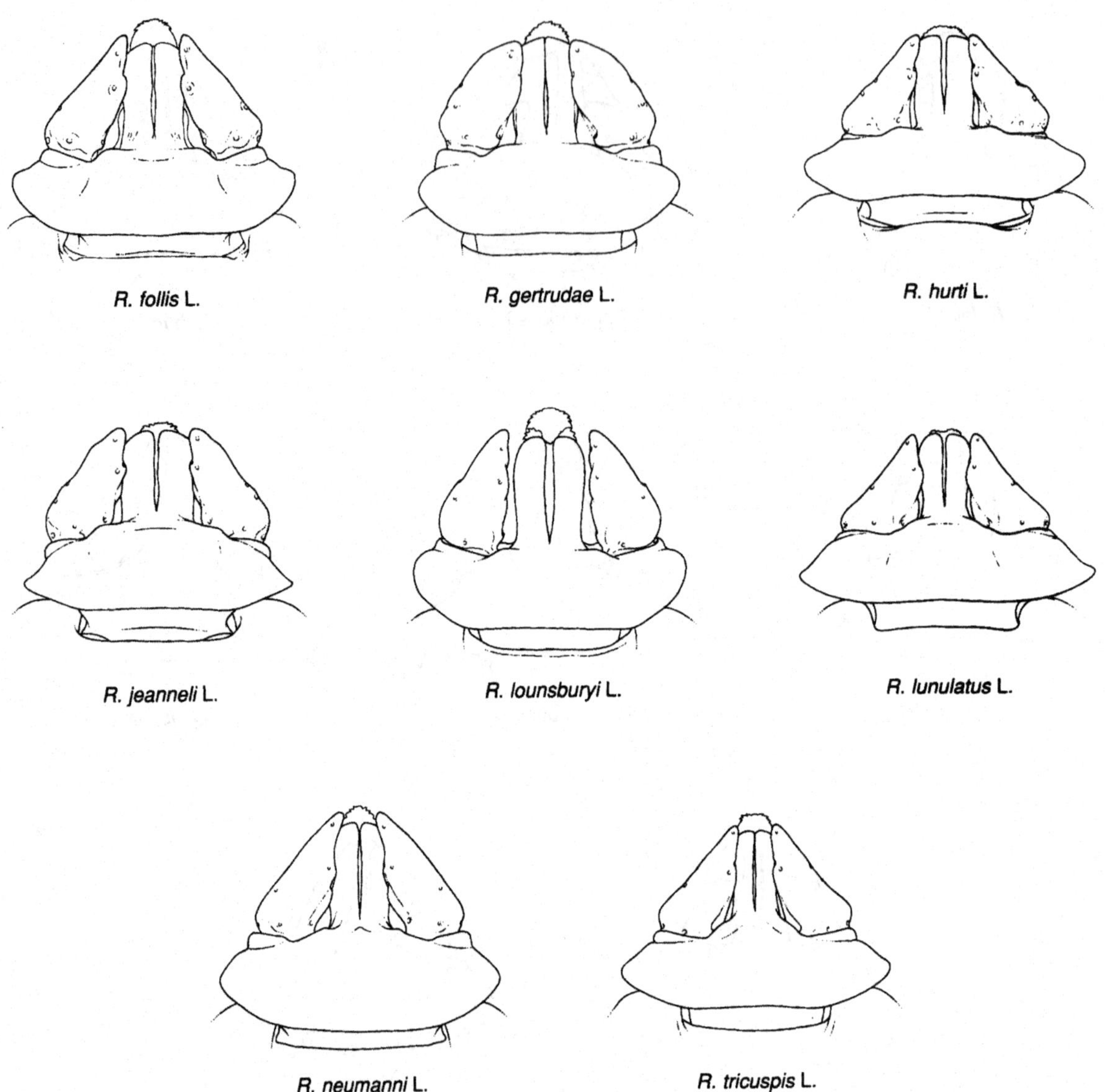

Figure 270. The *Rhipicephalus follis* species group: capitula of larvae (L).

RHIPICEPHALUS CAPENSIS GROUP

This group contains the following species (Fig. 271):

R. capensis: south-western South Africa.
R. compositus: mainly East and Central Africa from Kenya in the north to Zimbabwe in the south.
R. longus: across Africa approximately between 6 °N and 12 °S.
R. pseudolongus: West Africa eastwards to Uganda.

Nymphs

(a) Basis capituli with long sharp lateral angles overlapping scapulae; cornua present (somewhat reduced in *R. longus*); well-developed spurs on ventral posterior margin.
(b) Palps slender, tapering gradually, inclined inwards.

Larvae

(a) Basis capituli hexagonal, lateral angles at about mid-length, bluntly pointed.
(b) Palps broad proximally, tapering to broadly rounded apices, inclined inwards.

RHIPICEPHALUS PRAVUS GROUP

This group encompasses the following species (Fig. 272):

R. arnoldi: south-western Zimbabwe and central South Africa.
R. exophthalmos: mainly Namibia and southern South Africa.
R. kochi: south-eastern Kenya, Tanzania and Central Africa southwards to north-eastern South Africa.
R. oculatus: southern Namibia and southern and central South Africa.
R. pravus: East Africa
R. warburtoni: Free State Province, South Africa.

Nymphs

(a) Basis capituli hexagonal, three times as broad as long; dorsally cornua, and ventrally spurs, absent.
(b) Palps constricted proximally, becoming broader then tapering slightly to rounded tips.

Larvae

(a) Basis capituli quadrangular.
(b) Palps constricted proximally, internal margins straight and external margins convex giving the palps a more-or-less bulbous appearance, tips broadly rounded.

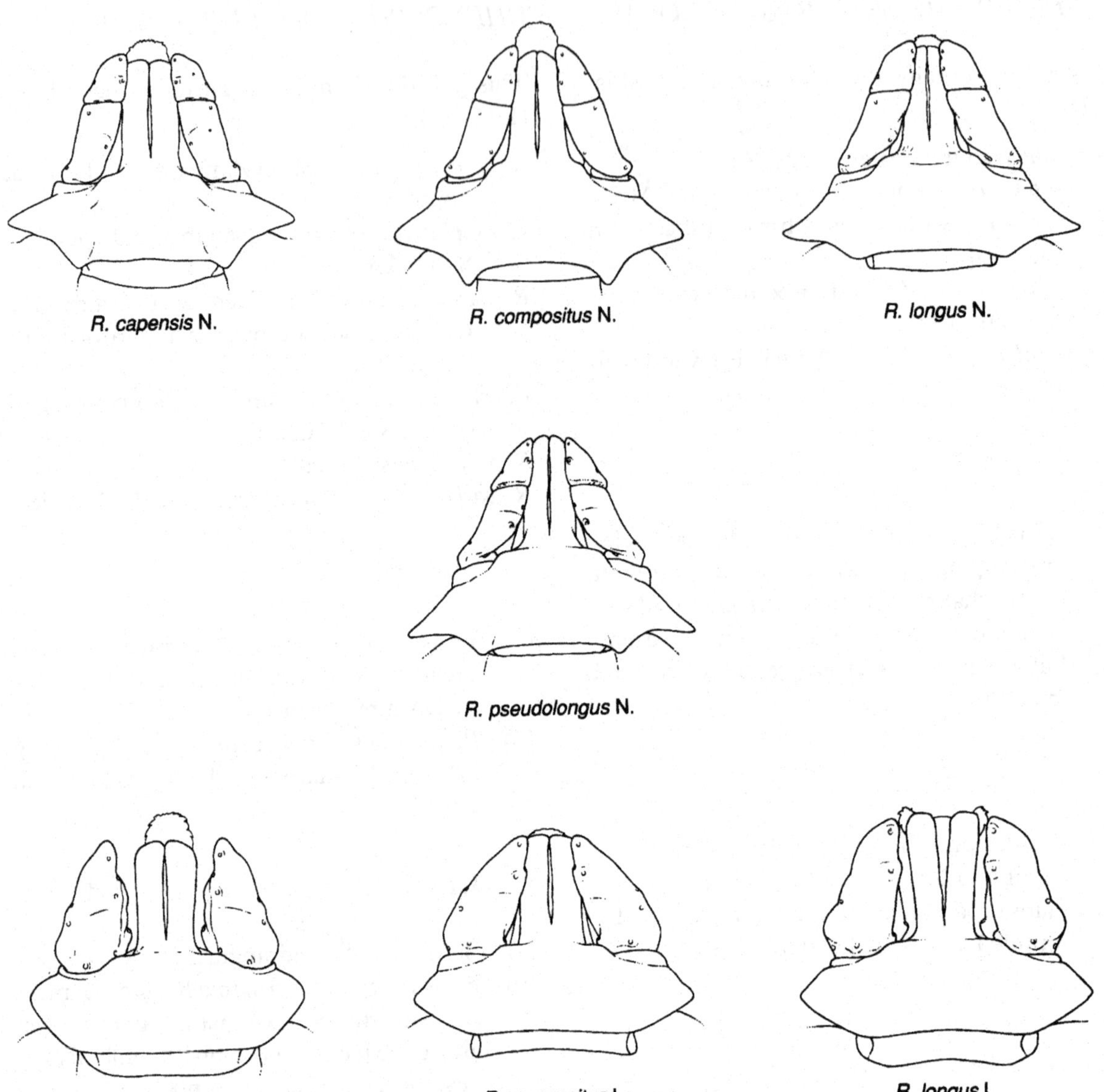

Figure 271. The *Rhipicephalus capensis* species group: capitula of nymphs (N) and larvae (L).

Figure 272 (*opposite*). The *Rhipicephalus pravus* species group: capitula of nymphs (N) and larvae (L).

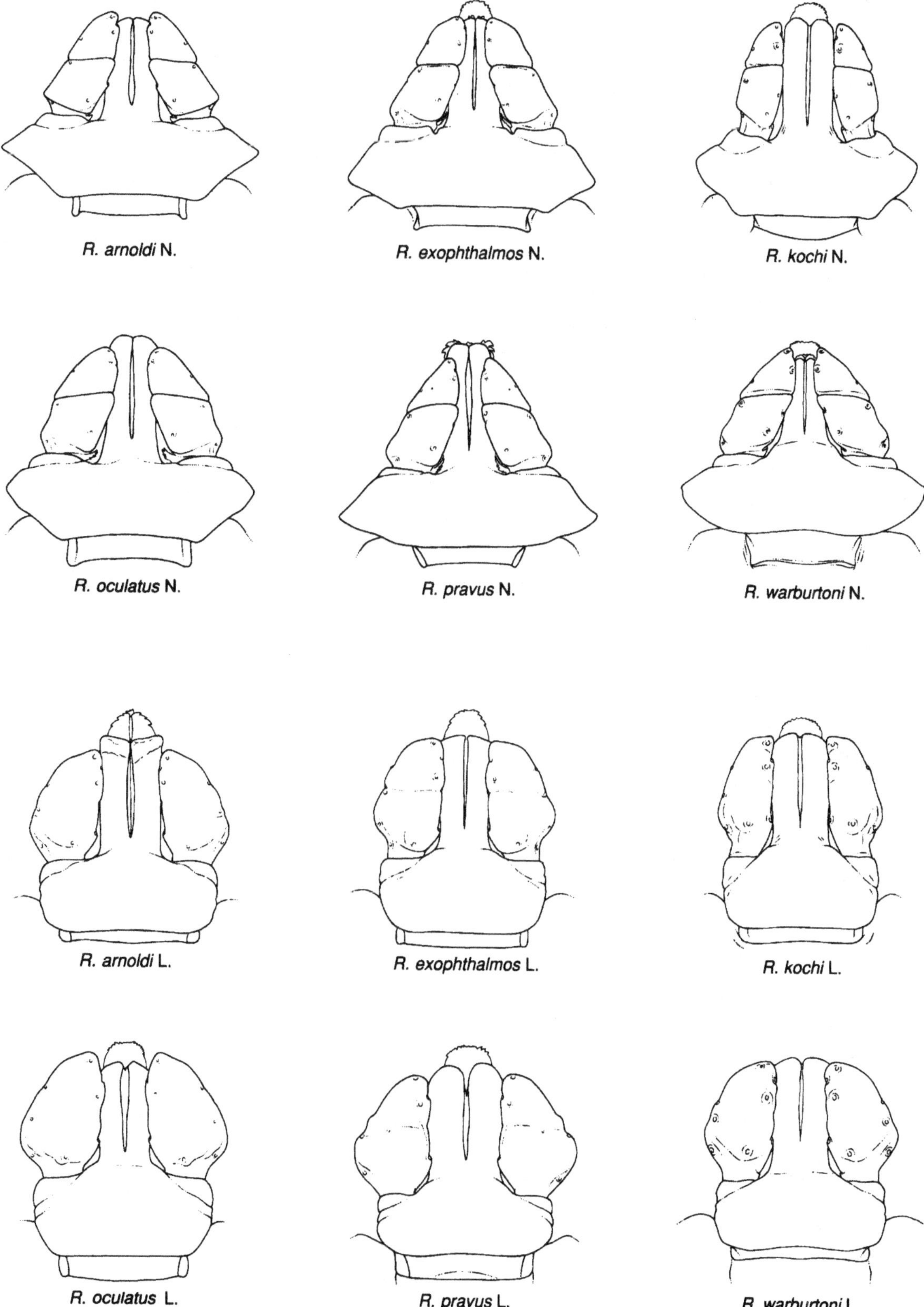
R. arnoldi N.
R. exophthalmos N.
R. kochi N.
R. oculatus N.
R. pravus N.
R. warburtoni N.
R. arnoldi L.
R. exophthalmos L.
R. kochi L.
R. oculatus L.
R. pravus L.
R. warburtoni L.

RHIPICEPHALUS APPENDICULATUS GROUP

This group comprises the following species (Figs 273 and 274):

R. appendiculatus: East and Central Africa southwards to eastern southern Africa.
R. armatus: dry areas in eastern Uganda, Kenya, Somalia and Ethiopia.
R. carnivoralis: mainly Kenya, Uganda and Tanzania.
R. duttoni: almost exclusively western Angola.
R. humeralis: south-eastern Somalia and Kenya and north-eastern Tanzania.
R. maculatus: eastern East and southern Africa.
R. muehlensi: eastern Kenya and Tanzania southwards to north-eastern South Africa.
R. nitens: southern to south-western South Africa.
R. pulchellus: mainly east of the Rift Valley from Eritrea southwards to north-eastern Tanzania.
R. sculptus: Tanzania, Malawi and Zambia.
R. zambeziensis: mainly south-western Tanzania, southern Zambia, Zimbabwe, northern Namibia, south-eastern Botswana and northern South Africa.

Nymphs

(a) Basis capituli hexagonal; lateral angles sharp, anterior to mid-length, pointing either forwards or sideways; cornua absent; ventrally short spurs or indications of spurs on posterior margin.
(b) Palps constricted proximally, broad, apices bluntly rounded.

Larvae

(a) Basis capituli quadrangular to hexagonal with short lateral angles.
(b) Palps broad with bluntly rounded apices.

RHIPICEPHALUS EVERTSI GROUP

This group comprises the following species (Fig. 275):

R. bursa: Mediterranean countries.
R. evertsi (includes subspecies *evertsi* and *mimeticus*): widespread in the Afrotropical region
R. glabroscutatum: Eastern and Western Cape Provinces, South Africa.

Nymphs

(a) Basis capituli with hypostome longer than palps; dorsally cornua, and ventrally spurs, absent.
(b) Palps cylindrical, not constricted proximally.

Larvae

(a) Basis capituli quadrangular; dorsally cornua and ventrally spurs absent.
(b) Palps cylindrical.

Figure 273 (*opposite*). The *Rhipicephalus appendiculatus* species group: capitula of nymphs (N).

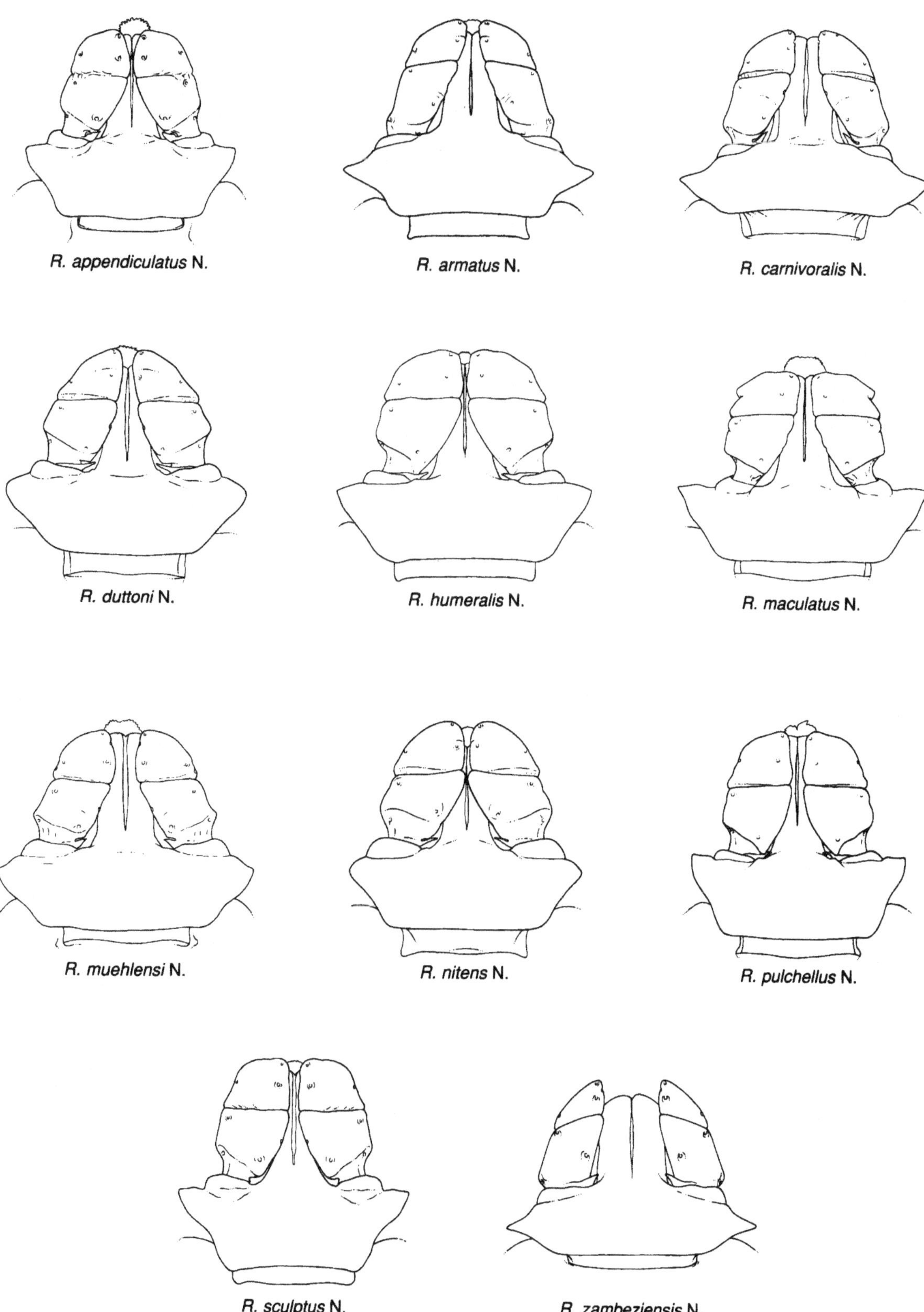

R. appendiculatus N.
R. armatus N.
R. carnivoralis N.
R. duttoni N.
R. humeralis N.
R. maculatus N.
R. muehlensi N.
R. nitens N.
R. pulchellus N.
R. sculptus N.
R. zambeziensis N.

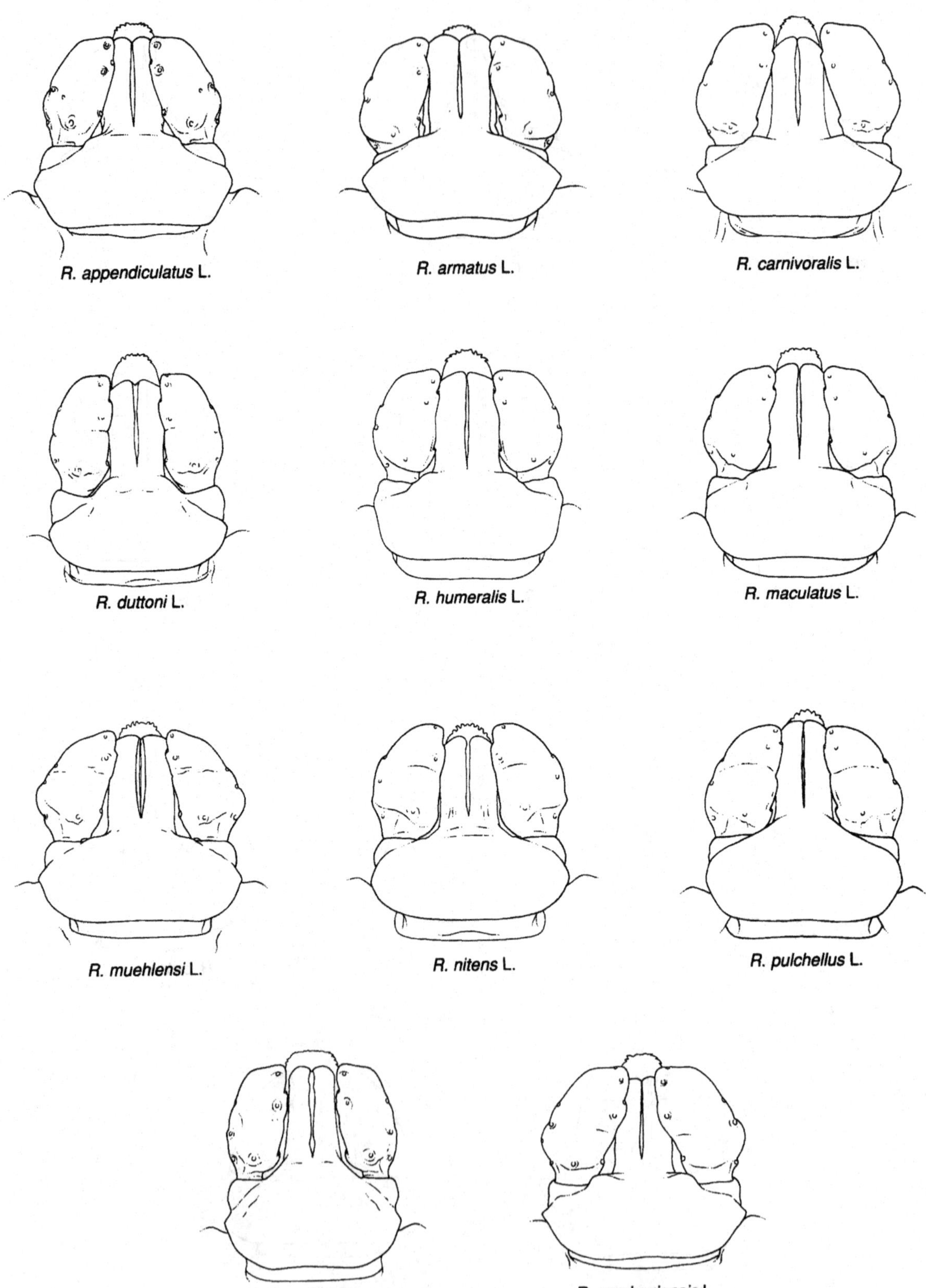
R. appendiculatus L.
R. armatus L.
R. carnivoralis L.
R. duttoni L.
R. humeralis L.
R. maculatus L.
R. muehlensi L.
R. nitens L.
R. pulchellus L.
R. sculptus L.
R. zambeziensis L.

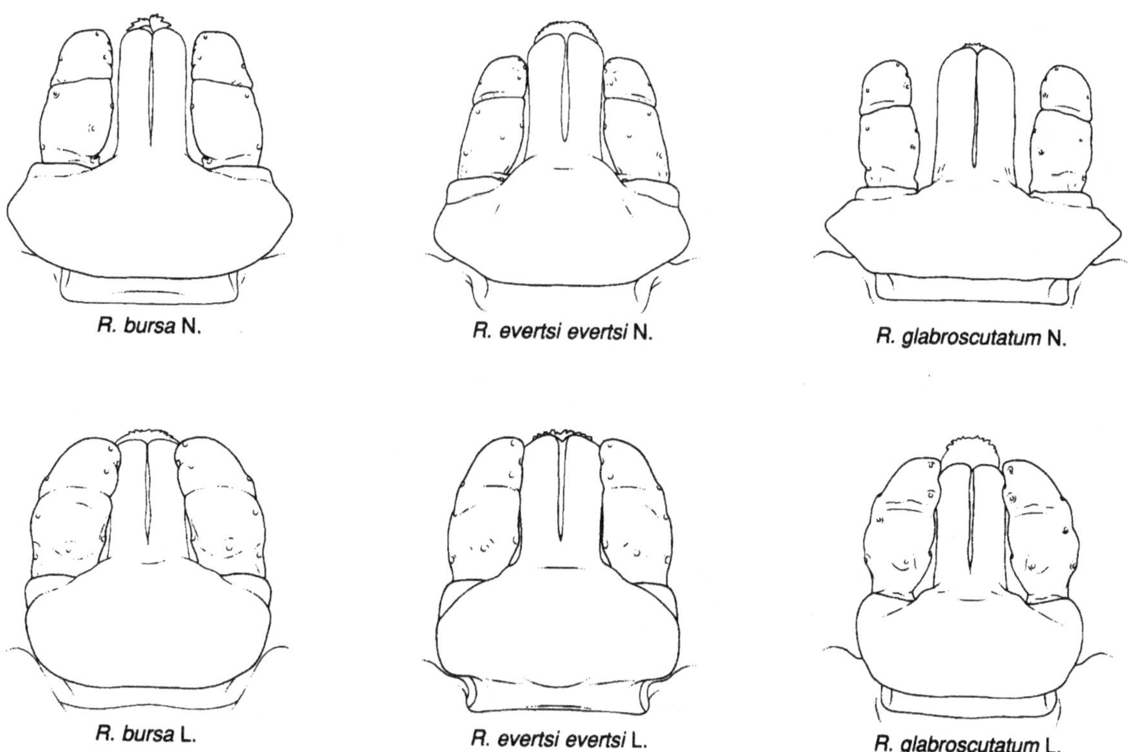

Figure 275 (*above*). The *Rhipicephalus evertsi* species group: capitula of nymphs (N) and larvae (L).

Figure 274 (*opposite*). The *Rhipicephalus appendiculatus* species group: capitula of larvae (L).

OTHER SPECIES

There are a few species that we have been unable to assign satisfactorily in our minds to any of the foregoing groups (Fig. 276). These are:

R. cuspidatus: extending from West Africa to Sudan.

R. longicoxatus: southern Kenya and northern and central Tanzania.

R. theileri: central Namibia, northern Botswana, north-western, northern and central South Africa.

R. fulvus: north-eastern Algeria, Tunisia, Niger and Chad.

We have considered placing *R. cuspidatus* in the *R. evertsi* group and *R. theileri* in the *R. appendiculatus* group but decided against it. Future research will show whether this and other decisions we have made on the species groupings are taxonomically and evolutionarily defensible.

THE NON-AFROTROPICAL *RHIPICEPHALUS SANGUINEUS* GROUP

This group includes the following species (Figs 277 and 278):

R. camicasi: Yemen Arab Republic, Saudi Arabia, Jordan, Lebanon and the north-eastern corner of Africa.

R. leporis: Afghanistan, Iran, Kazakhstan and Turkmenistan.

R. pumilio: southern Russia, Turkmenistan, Tajikistan, Kashmir and Mongolia.

R. pusillus: western Mediterranean countries.

R. rossicus: eastern Europe, Sinai Peninsula, western Kazakhstan and Xinjiang Province, China.

R. sanguineus: worldwide, between approximately 50 °N and 30 °S.

R. schulzei: southern Russia.

R. turanicus: Mediterranean countries, Europe, Russia, India and in North, East and southern Africa.

Nymphs

(a) Basis capituli hexagonal, lateral angles at about mid-length, pointed; with the exception of *R. pusillus*, which has slight cornua, posterior margin straight.

(b) Palps with external margins straight to mildly convex, tapering gradually to rapidly, apices narrowly to bluntly rounded, inclined inwards.

Larvae

(a) Basis capituli hexagonal, lateral angles obtuse and bluntly rounded to acute and narrowly pointed, at about mid-length or slightly more forward.

(b) Palps tapering, apices narrowly rounded to pointed, inclined inwards.

Figure 276 (*opposite*). The *Rhipicephalus* species not placed in a group: capitula of nymphs (N) and larvae (L).

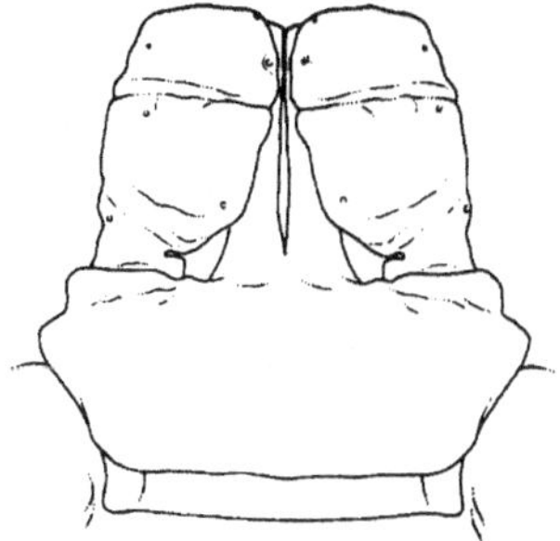

R. cuspidatus N.

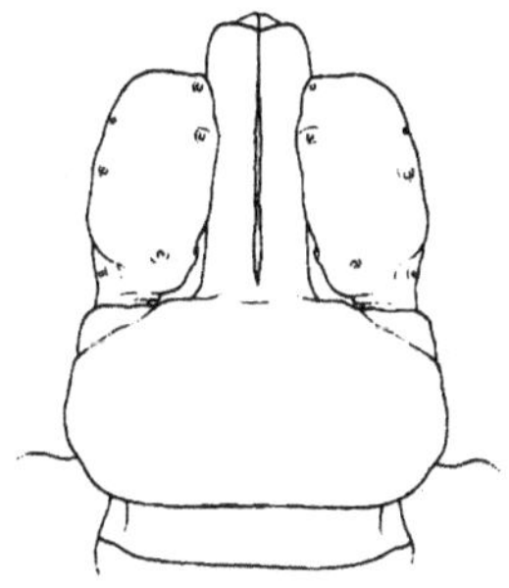

R. cuspidatus L.

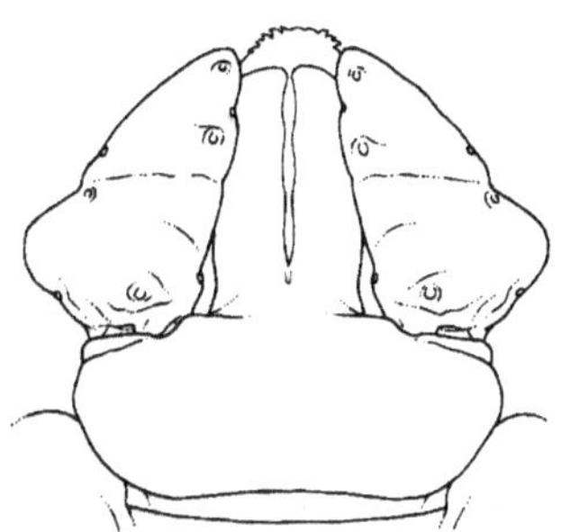

R. longicoxatus L.

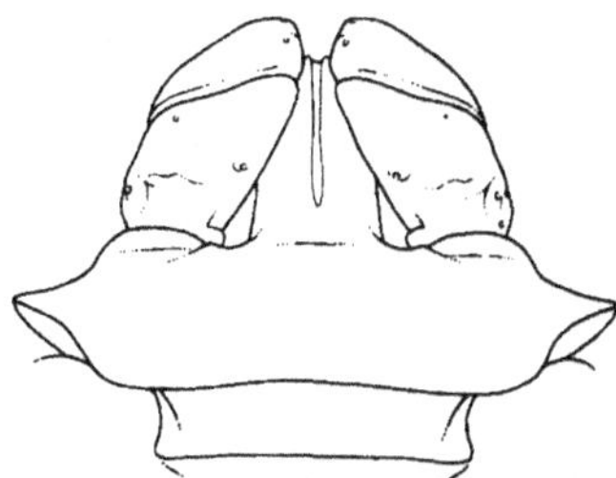

R. theileri N.

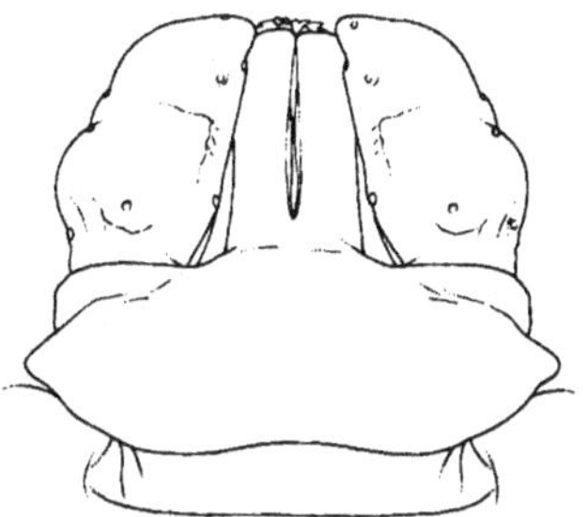

R. theileri L.

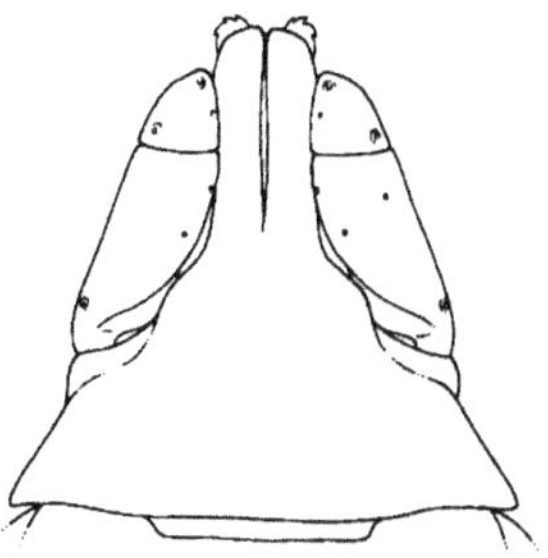

R. fulvus N.

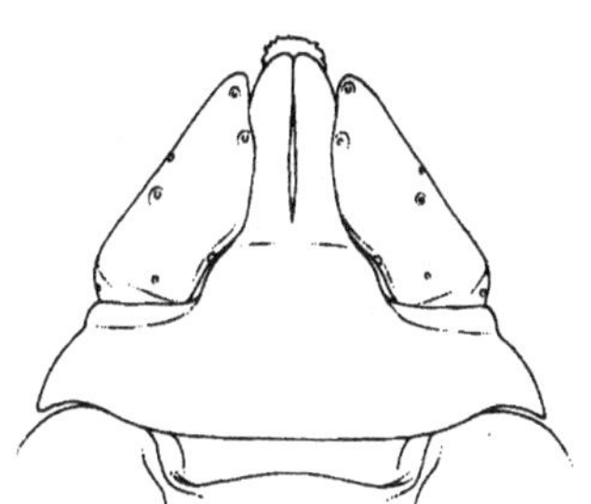

R. fulvus L.

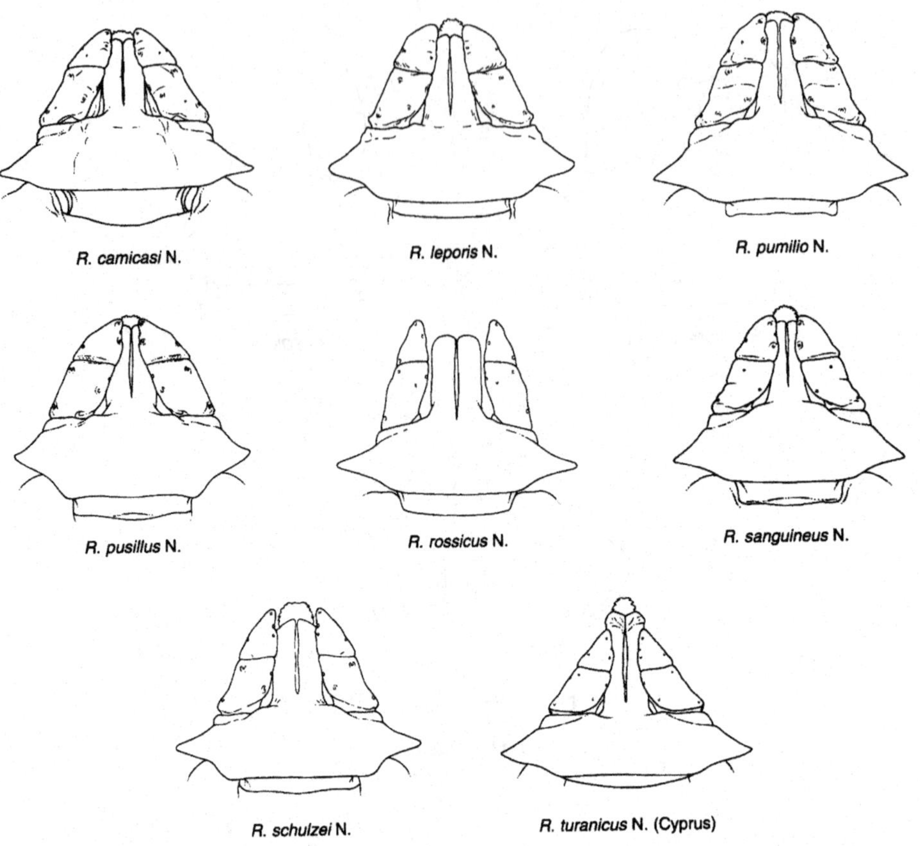

Figure 277. The non-Afrotropical *Rhipicephalus sanguineus* species group: capitula of nymphs (N).

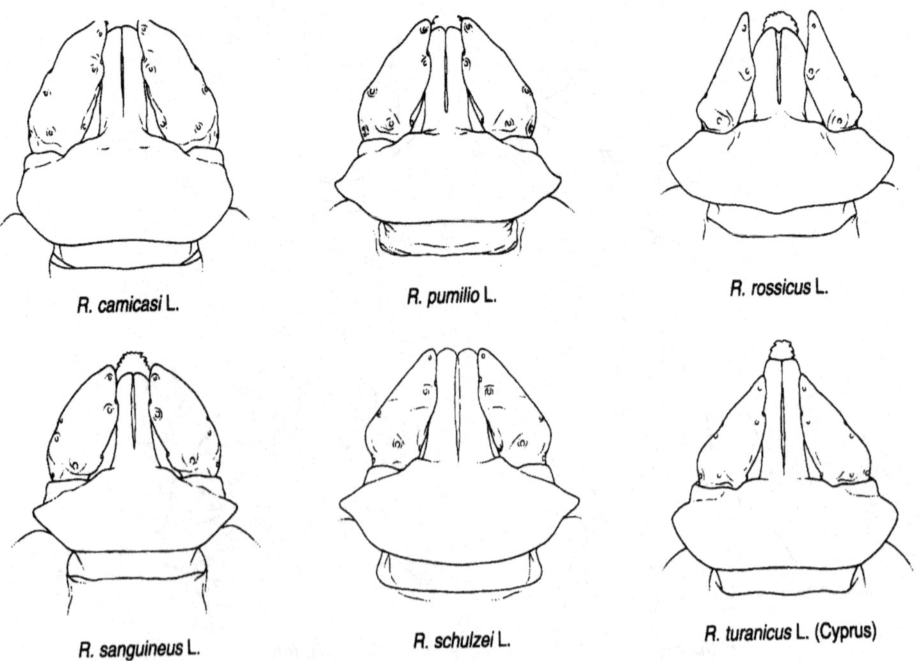

Figure 278. The non-Afrotropical *Rhipicephalus sanguineus* species group: capitula of larvae (L).

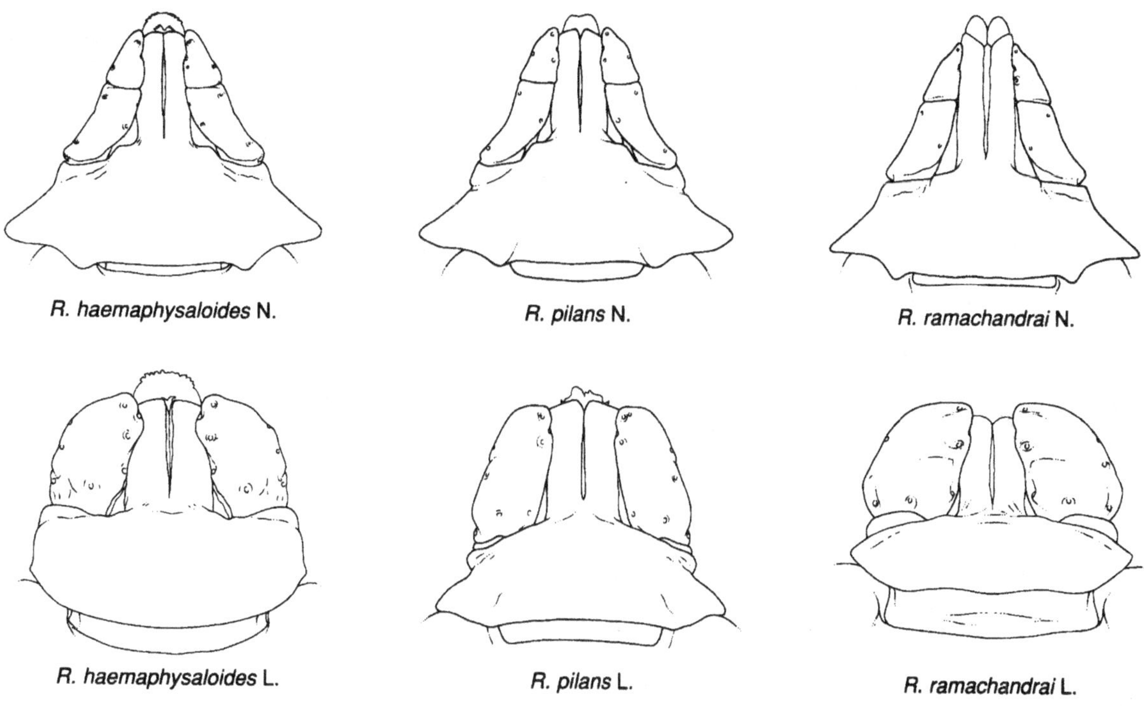

Figure 279. The *Rhipicephalus haemaphysaloides* species group: capitula of nymphs (N) and larvae (L).

RHIPICEPHALUS HAEMAPHYSALOIDES GROUP

This group encompasses the following species (Fig. 279):

R. haemaphysaloides: South-east Asia and Indonesia.
R. pilans: the Philippines and Indonesia.
R. ramachandrai: India, Nepal and Pakistan.

The morphological disparity between the larvae of this group compared with the uniform appearance of the nymphs suggests that some of the specimens have perhaps not been correctly matched either with each other or with the adults.

Nymphs

(a) Basis capituli hexagonal, pointed to bluntly rounded lateral angles at about posterior third of its length, cornua present dorsally and spurs ventrally.
(b) Palps slender, external margins concave, tapering, apices narrowly rounded, inclined inwards.

Larvae

(a) Basis capituli quadrangular to hexagonal, lateral angles obtuse to acute, posterior margin convex to concave.
(b) Palps broadly cylindrical to broadly conical with bluntly rounded apices.

13

The transmission of tick-borne diseases of animals and humans by *Rhipicephalus* species

The format of the tables given below is based on those compiled by Neitz (1956a, b) and Walker (1974). Further information is given in Hoogstraal (1967, 1973, 1979), Yunker (1970), Friedhoff (1988) and Coetzer, Thomson & Tustin (1994). No attempt has been made to cite all the references to studies on the transmission by the different *Rhipicephalus* spp. vectors of the organisms or conditions listed; such an undertaking lies outside the scope of this work. Usually only the first or an early account of the transmission of an organism or condition by a particular *Rhipicephalus* sp. is quoted, unless this is very incomplete. In such cases additional references are given.

Notes to Table 70

* Indicates the most important vector, if known.

† Indicates laboratory transmission, sometimes in splenectomized animals subject to very high tick infestations. Such vectors could be of little or no importance in the field.

L, larva.

N, nymph.

A, adult.

E, egg.

Table 70. *Transmission of tick-borne diseases of animals*

Disease and causative agent	Animals affected	Tick vectors	Number of hosts	Acquisition [x] and transmission [)] of disease: L	N	A	E	L	N	A	References
East Coast fever *Theileria parva parva*	Cattle, African buffalo	* *R. appendiculatus*	3	X—)	X—)						Lounsbury (1904, 1906), Lewis (1943), De Vos (1981)
		† *R. capensis*	3		X—)						Lounsbury (1906)
		† *R. carnivoralis*	3		X—)						Brocklesby, Bailey & Vidler (1966)
		† *R. compositus* (listed as *R. ayrei*, see p. 128)	3	X—)	X—)						Wilson (1951)
		R. evertsi evertsi	2	X—	X—)						Lounsbury (1906)
		† *R. jeanneli*	3	Stages involved not stated							Wilson (1953)
		† *R. nitens*	3	X—)							Lounsbury (1906)
		† *R. pravus* (listed as *R. neavei*, see p. 358)	3	X—)	X—)						Lewis, Piercy & Wiley (1946)
		† *R. pulchellus*	3		X—)						Brocklesby (1965)
		† *R. simus*	3		X—)						Lounsbury (1906)
		R. zambeziensis	3		X—)						Lawrence, Norval & Uilenberg (1983)

Table 70. *(cont.)*

Disease and causative agent	Animals affected	Tick vectors	Number of hosts	Acquisition [x] and transmission [)] of disease L N A E L N A	References
Corridor disease *Theileria parva lawrencei*	Cattle, African buffalo	* *R. appendiculatus*	3	X—) X—)	Neitz (1955), De Vos (1981)
		R. duttoni	3	Stages involved not stated	Da Graça & Serrano (1971)
		† *R. simus*	3	X—)	Neitz (1962)
		* *R. zambeziensis*	3	X—) X—)	Lawrence, Norval & Uilenberg (1983)
Zimbabwe theileriosis *Theileria parva bovis*	Cattle	* *R. appendiculatus*	3	X—)	Matson & Hill (1967), Fivaz, Norval & Lawrence (1989)
		R. zambeziensis	3	X—) X—)	Lawrence, Norval & Uilenberg (1983)
Benign bovine theileriosis *Theileria taurotragi*	Eland, cattle	*R. appendiculatus*	3	X—) X—)	Lawrence & MacKenzie (1980)
		R. pulchellus	3	X—)	Brocklesby (1965)
		R. zambeziensis	3	X—) X—)	Lawrence, Norval & Uilenberg (1983)
Ovine theileriosis *Theileria separata* (as *Theileria ovis*)	Sheep	*R. bursa*	2	X—X—)	Neitz (1972)

		* *R. evertsi evertsi*	2	X—X—)	Jansen & Neitz (1956), Uilenberg & Schreuder (1976)
		† *R. evertsi mimeticus*	2	X X—)	Neitz (1972)
Bovine babesiosis *Babesia bigemina*	Cattle	*R. bursa*	2	X—X—) X———)—)—)	Brumpt (1931), Sergent *et al.* (1931)
		† *R. evertsi evertsi*	2	X———)	Büscher (1988)
Bovine babesiosis *Babesia bovis*	Cattle	*R. bursa*	2	X—X—)	Sergent *et al.* (1945, cited by Neitz, 1956a)
Equine babesiosis *Babesia caballi*	Horses, donkeys, mules	† *R. bursa*	2	X—X—)	Enigk (1943)
		R. evertsi evertsi	2	X—)	De Waal & Potgieter (1987)
		† *R. turanicus* (listed as *R. sanguineus*, see p. 462)	3	X—) X——————)	Enigk (1943)
Equine babesiosis *Babesia equi* (Now called **Equine theileriosis** and *Theileria equi*. This is controversial. See the last three references on p. 622.)	Horses, donkeys, mules, zebras	*R. bursa*	2	X—X—)	Kurchatov & Markov (1940, cited by Neitz, 1956a)
		* *R. evertsi evertsi*	2	X—X—)	Theiler (1906), De Waal (1983–88, cited by De Waal & Van Heerden, 1994)
		R. evertsi mimeticus	2	X—X—)	Potgieter, De Waal & Posnett (1992)
		† *R. turanicus* (listed as *R. sanguineus*, see p. 389)	3	X—) X—)	Enigk (1943)

Table 70. *(cont.)*

Disease and causative agent	Animals affected	Tick vectors	Number of hosts	Acquisition [x] and transmission [)] of disease L N A E L N A	References
Porcine babesiosis *Babesia trautmanni*	Pigs, warthog, bushpig	*R. lunulatus*	3	Stages involved not stated	Tendeiro (1952)
		† *R. simus*	3	X————)—)	De Waal, López-Rebollar & Potgieter (1992)
		† *R. turanicus*	3	X————)—)	López-Rebollar & De Waal (1994)
Ovine and caprine babesiosis *Babesia motasi*	Sheep, goats	*R. bursa*	2	X—X—)	Rastegaïeff (1933)
Ovine and caprine babesiosis *Babesia ovis*	Sheep, goats	* *R. bursa*	2	X—X—) X————)	Rastegaïeff (1933), Büscher, Friedhoff & El-Allawy (1988)
Canine babesiosis *Babesia canis*	Dogs	*R. sanguineus*	3	X—) X—) X———)—)—)	Christophers (1907), Shortt (1936), Liebisch & Gillani (1979)
		† *R. turanicus*	3	)) Stage in which infection acquired not stated	Achuthan, Mahadevan & Lalitha (1980)
Canine babesiosis *Babesia gibsoni*	Dogs, wolf, jackal, fox	† *R. sanguineus*	3	X—)	Sen (1933)

Canine hepatozoonosis *Hepatozoon canis*	Dogs	† *R. sanguineus*	3	X—)	These infected adult ticks fed to dogs	Nordgren & Craig (1984)
Malignant bovine anaplasmosis *Anaplasma marginale*	Cattle	*R. bursa*	2	X—X—)		Brumpt (1931)
		R. evertsi evertsi	2	X—)	Intrastadial transmission by adult ticks	Potgieter (1981)
		R. simus	3	X—) X————) X—)		Theiler (1912), Potgieter (1981)
				X—)	Intrastadial transmission by adult ticks	
Benign bovine anaplasmosis *Anaplasma centrale*	Cattle	*R. simus*	3	X—)		Potgieter & Van Rensburg (1987)
Ovine and caprine anaplasmosis *Anaplasma ovis*	Sheep, goats	*R. bursa*	2	X—X—)		Rastegaïeff (1933)
Canine haemobart-onellosis *Haemobartonella canis*	Dogs	† *R. sanguineus*	3	X—) X————)—)		Seneviratna, Weerasinghe & Ariyadasa (1973)
Bovine ehrlichiosis *Ehrlichia bovis*	Cattle	⋆ *R. appendiculatus*	3	X—)		Matson (1967), Norval (1979)
		† *R. nitens*	3	X—)		Stoltsz (1994)
		† *R. zambeziensis*	3	X—) X—)		Stoltsz (1994)

Table 70. *(cont.)*

Disease and causative agent	Animals affected	Tick vectors	Number of hosts	Acquisition [x] and transmission [)] of disease L N A E L N A	References
Ovine ehrlichiosis *Ehrlichia ovina*	Sheep	† *R. bursa*	2	Tick emulsion injected into sheep	Donatien & Lestoquard (1937)
		† *R. evertsi evertsi*	2	X—X—)	Neitz (1952, cited by Neitz, 1956b)
Canine ehrlichiosis *Ehrlichia canis*	Dogs	* *R. sanguineus*	3	X—)—) X—)	Groves *et al.* (1975)
Rocky mountain spotted fever *Rickettsia rickettsi*	Dogs	*R. sanguineus*	3	Stages involved not stated	Greene & Breitschwerdt (1990)
Nairobi sheep disease (Bunyaviridae)	Sheep, goats	* *R. appendiculatus*	3	X—) X———) X—)	Montgomery (1917), Daubney & Hudson (1931), Davies & Mwakima (1982)
		R. hurti or *R. jeanneli* (listed as *R. bursa*, see p. 227)	3	X—)	Daubney & Hudson (1934)
		Probably *R. praetextatus* (listed as *R. simus*, see p. 425)	3	Stages involved not stated	Lewis (1949)
		* *R. pulchellus*	3	X—) X———)—)	Pellegrini (1950)

Kisenye sheep disease (Bunyaviridae)	Sheep	*R. appendiculatus*	3	X—)	Bugyaki (1955)
Louping ill (Flaviviridae)	Sheep, cattle	†*R. appendiculatus*	3	X—) X—)	Alexander & Neitz (1935)
Spirochaetosis *Borrelia theileri*	Cattle, sheep, goats, horses, donkeys, mules	**R. evertsi evertsi*	2	X—X—) X———)	Theiler (1909)
Filarid nematodosis *Dipetalonema dracunculoides*	Dogs	†*R. sanguineus*	3	X—)	Olmeda-García, Rodríguez-Rodríguez & Rojo-Vázquez (1993)
Tick paralysis	Sheep	*R. bursa*	2	)	Anonymous (1984)
	Lambs	*R. evertsi evertsi*	2	)	Clark (1938), Hamel & Gothe (1978)
	Dogs (facial paralysis)	*R. evertsi evertsi*	2	)	Norval (1981)
	Rabbits	*R. exophthalmos*	3	)	Neitz (1970, cited by Keirans *et al.*, 1993)
	Calves, lambs	*R. lunulatus* (as *R. tricuspis*)	3	)	Theiler (1962), Lawrence & Norval (1979)

Table 70. *(cont.)*

Disease and causative agent	Animals affected	Tick vectors	Number of hosts	Acquisition [x] and transmission [)] of disease							References
				L	N	A	E	L	N	A	
Tick paralysis (cont.)	Cattle	*R. praetextatus* (as *R. simus)*	3			)					Pegram, Hoogstraal & Wassef (1981)
	Dogs	*R. sanguineus*	3			)					Viloria (1954)
	Calves, lambs	*R. simus*	3			)					Norval & Mason (1981)
	Goat kids	*R. warburtoni* (as *R. pravus* group)	3			)					Fourie, Horak & Marais (1988)
Tick toxicosis	Cattle, antelope	*R. appendiculatus*	3			)					Thomas & Neitz (1958), Van Rensburg (1959), Lightfoot & Norval (1981)
	Springbok	*R. nitens*	3			)					I.G.H. (unpublished data)

Note: * Indicates the most important vector, if known.
† Indicates laboratory transmission, sometimes in splenectomized animals subject to very high tick infestations. Such vectors could be of little or no importance in the field.
L, larva.
N, nymph.
A, adult.
E, egg.

REFERENCES

Achuthan, H.N., Mahadevan, S. & Lalitha, C.M. (1980). Studies on the developmental forms of *Babesia bigemina* and *Babesia canis* in ixodid ticks. *Indian Veterinary Journal*, **57**, 181–4.

Alexander, R.A. & Neitz, W.O. (1935). The transmission of Louping ill by ticks (*Rhipicephalus appendiculatus*). *Onderstepoort Journal of Veterinary Science and Animal Industry*, **5**, 15–33.

Anonymous (1984). *Ticks and Tick-borne Disease Control. A Practical Field Manual. Volume I. Tick Control.* Rome: Food and Agricultural Organization of the United Nations.

Brocklesby, D.W. (1965). Evidence that *Rhipicephalus pulchellus* (Gerstäcker 1873) may be a vector of some piroplasms. *Bulletin of Epizootic Diseases of Africa*, **13**, 37–44.

Brocklesby, D.W., Bailey, K.P. & Vidler, B.O. (1966). The transmission of *Theileria parva* (Theiler, 1904) by *Rhipicephalus carnivoralis* Walker, 1965. *Parasitology*, **56**, 13–14.

Brumpt, E. (1931). Transmission d'*Anaplasma marginale* par *Rhipicephalus bursa* et par *Margaropus*. *Annales de Parasitologie Humaine et Comparée*, **9**, 4–9.

Bugyaki, L. (1955). La 'maladie de Kisenyi', du mouton due à un virus filtrable et transmise par des tiques. *Bulletin Agricole du Congo Belge*, **46**, 1455–62.

Büscher, G. (1988). The infection of various tick species with *Babesia bigemina*, its transmission and identification. *Parasitology Research*, 74, 324–30.

Büscher, G., Friedhoff, K.T. & El-Allawy, T.A.A. (1988). Quantitative description of the development of *Babesia ovis* in *Rhipicephalus bursa* (hemolymph, ovary, eggs). *Parasitology Research*, **74**, 331–9.

Christophers, S.R. (1907). Preliminary note on the development of *Piroplasma canis* in the tick. *British Medical Journal*, **No. 1 for 1907**, 76–8.

Clark, R. (1938). A note on paralysis in lambs caused apparently by *Rhipicephalus evertsi*. *Journal of the South African Veterinary Medical Association*, **9**, 143–5.

Coetzer, J.A.W., Thomson, G.R. & Tustin, R.C. (eds) (1994). *Infectious Diseases of Livestock with Special Reference to Southern Africa. Vols 1 and 2.* Cape Town: Oxford University Press.

Da Graça, H.M. & Serrano, F.M.H. (1971). Contribuição para o estudo da Theileriose cincerina maligna dos bovinos, em Angola. *Acta Veterinária, Nova Lisboa*, **7**, 1–8.

Daubney, R. & Hudson, J.R. (1931). Nairobi sheep disease. *Parasitology*, **23**, 507–24.

Daubney, R. & Hudson, J.R. (1934). Nairobi sheep disease: natural and experimental transmission by ticks other than *Rhipicephalus appendiculatus*. *Parasitology*, **26**, 496–509.

Davies, F.G. & Mwakima, F. (1982). Qualitative studies of the transmission of Nairobi sheep disease virus by *Rhipicephalus appendiculatus* (Ixodoidea, Ixodidae). *Journal of Comparative Pathology*, **92**, 15–20.

De Vos, A.J. (1981). *Rhipicephalus appendiculatus*: cause and vector of diseases in Africa. *Journal of the South African Veterinary Association*, **52**, 315–22.

De Waal, D.T., López-Rebollar, L.M. & Potgieter, F.T. (1992). The transovarial transmission of *Babesia trautmanni* by *Rhipicephalus simus* to domestic pigs. *Onderstepoort Journal of Veterinary Research*, **59**, 219–21.

De Waal, D.T. & Potgieter, F.T. (1987). The transstadial transmission of *Babesia caballi* by *Rhipicephalus evertsi evertsi*. *Onderstepoort Journal of Veterinary Research*, **54**, 655–6.

De Waal, D.T. & Van Heerden, J. (1994). Equine babesiosis. In *Infectious Diseases of Livestock with Special Reference to Southern Africa*, Vol.1, ed. J.A.W. Coetzer, G.R. Thomson & R.C. Tustin, pp. 295–304. Cape Town: Oxford University Press.

Donatien, A. & Lestoquard, F. (1937). État actuel des connaissances sur les rickettsioses animales. *Archives de l'Institut Pasteur d'Algérie*, **15**, 142–87, 6 pls.

Enigk, K. (1943). Die Überträger des Pferdepiroplasmose, ihre Verbreitung und Biologie. *Archiv für Wissenschaftliche und Praktische Tierheilkunde*, **78**, 209–40.

Fivaz, B.H., Norval, R.A.I. & Lawrence, J.A. (1989). Transmission of *Theileria parva bovis* (Boleni strain) to cattle resistant to the brown ear tick *Rhipicephalus appendiculatus* (Neumann). *Tropical Animal Health and Production*, **21**, 129–34.

Fourie, L.J., Horak, I.G. & Marais, L. (1988). An undescribed *Rhipicephalus* species associated with field paralysis of Angora goats. *Journal of the South African Veterinary Association*, **59**, 47–9.

Friedhoff, K.T. (1988). Transmission of *Babesia.* In *Babesiosis in domestic Animals and Man,* ed. M. Ristic, pp. 23–52. Boca Raton, FL: CRC Press.

Greene, C.E. & Breitschwerdt, E.B. (1990). Rocky Mountain spotted fever and Q fever. In *Infectious Diseases of the Dog and Cat,* ed. C.E. Greene, pp. 419–33. Philadelphia: W.B. Saunders Company.

Groves, M.G., Dennis, G.L., Amyx, H.L. & Huxsoll, D.L. (1975). Transmission of *Ehrlichia canis* to dogs by ticks (*Rhipicephalus sanguineus*). *American Journal of Veterinary Research,* **36**, 937–40.

Hamel, H.D. & Gothe, R. (1978). Influence of infestation rate on tick-paralysis in sheep induced by *Rhipicephalus evertsi evertsi* Neumann, 1897. *Veterinary Parasitology,* **4**, 183–91.

Jansen, B.C. & Neitz, W.O. (1956). The experimental transmission of *Theileria ovis* by *Rhipicephalus evertsi. Onderstepoort Journal of Veterinary Research,* **27**, 3–6.

Keirans, J.E., Walker, J.B., Horak, I.G. & Heyne H. (1993). *Rhipicephalus exophthalmos* sp. nov., a new tick species from southern Africa, and a rediscription of *Rhipicephalus oculatus* Neumann, 1901, with which it has hitherto been confused (Acari : Ixodida : Ixodidae). *Onderstepoort Journal of Veterinary Research,* **60**, 229–46.

Lawrence, J.A. & MacKenzie, P.K.I. (1980). Isolation of a non-pathogenic theileria of cattle transmitted by *Rhipicephalus appendiculatus. Zimbabwe Veterinary Journal,* **11**, 27–35.

Lawrence, J.A. & Norval, R.A.I. (1979). A history of ticks and tick-borne diseases of cattle in Rhodesia. *Rhodesian Veterinary Journal,* **10**, 28–40.

Lawrence, J.A., Norval, R.A.I. & Uilenberg, G. (1983). *Rhipicephalus zambeziensis* as a vector of bovine theileriae. *Tropical Animal Health and Production,* **15**, 39–42.

Lewis, E.A. (1943). East Coast fever and the African buffalo, the eland and the bushbuck. *East African Agricultural Journal,* **9**, 90–2.

Lewis, E.A. (1949). *Nairobi sheep disease.* Report of the Veterinary Department, Kenya, for 1947, pp. 45, 51.

Lewis, E.A., Piercy, S.E. & Wiley, A.J. (1946). *Rhipicephalus neavei* Warburton, 1912 as a vector of East Coast fever. *Parasitology,* **37**, 60–4.

Liebisch, A. & Gillani, S. (1979). Experimentelle Übertragung der Hundebabesiose (*Babesia canis*) durch einheimische deutsche Zeckenarten: I. Die braune Hundezecke (*Rhipicephalus sanguineus*). *Deutsche Tierärztliche Wochenschrift,* **86**, 149–53.

Lightfoot, C.J. & Norval, R.A.I. (1981). Tick problems in wildlife in Zimbabwe. I. The effects of tick parasitism on wild ungulates. *South African Journal of Wildlife Research,* **11**, 41–5.

López-Rebollar, L.M. & De Waal, D.T. (1994). Tick vectors of *Babesia trautmanni* in domestic pigs in South Africa. *8th International Congress of Parasitology, October 10–14, 1994, Abstracts, Vol. 1,* 231. Izmir: Turkey.

Lounsbury, C.P. (1904). Transmission of African Coast fever. *Agricultural Journal, Cape of Good Hope,* **24**, 428–32.

Lounsbury, C.P. (1906). Ticks and African Coast fever. *Agricultural Journal, Cape of Good Hope,* **28**, 634–54.

Matson, B.A. (1967). Theileriosis in Rhodesia: I. A study of diagnostic specimens over two seasons. *Journal of the South African Veterinary Medical Association,* **38**, 93–102.

Matson, B.A. & Hill, R.R. (1967). Recent advances in the study of theileriosis in Rhodesia. *Rhodesia Agricultural Journal,* **64**, 88–92.

Mehlhorn, H. & Schein, E. (1984). The piroplasms: life cycle and sexual stages. *Advances in Parasitology,* **23**, 37–103.

Mehlhorn, H. & Schein, E. (1998). Redescription of *Babesia equi* Laveran, 1901 as *Theileria equi* Mehlhorn, Schein 1998. *Parasitology Research,* **84**, 467–75.

Montgomery, E. (1917). On a tick-borne gastro-enteritis of sheep and goats occurring in British East Africa. *Journal of Comparative Pathology and Therapeutics,* **30**, 28–57.

Neitz, W.O. (1955). Corridor disease: a fatal form of bovine theileriosis encountered in Zululand. *Bulletin of Epizootic Diseases of Africa,* **3**, 121–3.

Neitz, W.O. (1956a). Classification, transmission, and biology of piroplasms of domestic animals. *Annals of the New York Academy of Sciences,* **64**, 56–111.

Neitz, W.O. (1956b). A consolidation of our knowledge of the transmission of tick-borne diseases. *Onderstepoort Journal of Veterinary Research,* **27**, 115–63.

Neitz, W.O. (1962). Review of recent developments in the protozoology of tick-borne diseases. In *Report of the Second Meeting of the FAO/OIE*

Expert Panel on Tick-Borne Diseases of Livestock, Appendix D pp. 34–5. Cairo: Food and Agricultural Organization of the United Nations.

Neitz, W.O. (1972). The experimental transmission of *Theileria ovis* by *Rhipicephalus evertsi mimeticus* and *R. bursa*. *Onderstepoort Journal of Veterinary Research*, **39**, 83–5.

Nordgren, R.M. & Craig, T.M. (1984). Experimental transmission of the Texas strain of *Hepatozoon canis*. *Veterinary Parasitology*, **16**, 207–14.

Norval, R.A.I. (1979). Tick infestations and tick-borne diseases in Zimbabwe Rhodesia. *Journal of the South African Veterinary Association*, **50**, 289–92.

Norval, R.A.I. (1981). The ticks of Zimbabwe. III. *Rhipicephalus evertsi evertsi*. *Zimbabwe Veterinary Journal*, **12**, 31–5.

Norval, R.A.I. & Mason, C.A. (1981). The ticks of Zimbabwe. II. The life cycle, distribution and hosts of *Rhipicephalus simus* Koch, 1844. *Zimbabwe Veterinary Journal*, **12**, 2–9.

Olmeda-García, A.S., Rodríguez-Rodríguez, J.A. & Rojo-Vázquez, F.A. (1993). Experimental transmission of *Dipetalonema drancunculoides* (Cobbold 1870) by *Rhipicephalus sanguineus* (Latreille 1806). *Veterinary Parasitology*, **47**, 339–42.

Pegram, R.G., Hoogstraal, H. & Wassef, H.Y. (1981). Ticks (Acari: Ixodoidea) of Ethiopia. I. Distribution, ecology and host relationships of species infesting livestock. *Bulletin of Entomological Research*, **71**, 339–59.

Pellegrini, D. (1950). La gastro-enterite emorragica delle pecore. Esperimenti di transmissione col '*Rhipicephalus pulchellus*'. *Bolletino della Societa Italiana di Medicina e Igiene Tropicale*, **10**, 164–70.

Potgieter, F.T. (1981). Tick transmission of anaplasmosis in South Africa. In *Proceedings of an International Conference on Tick Biology and Control*, ed. G.B. Whitehead & J.D. Gibson, pp. 53–6. Grahamstown: Rhodes University.

Potgieter, F.T., De Waal, D.T. & Posnett, E.S. (1992). Transmission and diagnosis of equine babesiosis in South Africa. *Memórias do Instituto Oswaldo Cruz, Rio de Janeiro*, **87**, (Suppl III), 139–42.

Potgieter, F.T. & Van Rensburg, L. (1987). Tick transmission of *Anaplasma centrale*. *Onderstepoort Journal of Veterinary Research*, **54**, 5–7.

Rastegaïeff, E.F. (1933). Zur Frage der Überträger der Schafpiroplasmosen in Azerbaidschan (Transkaukasien). *Archiv für Wissenschaftliche und Praktische Tierheilkunde*, **67**, 176–86.

Sen, S.K. (1933). The vector of canine piroplasmosis due to *Piroplasma gibsoni*. *Indian Journal of Veterinary Science*, **3**, 356–63.

Seneviratna, P., Weerasinghe, N. & Ariyadasa, S. (1973). Transmission of *Haemobartonella canis* by the dog tick *Rhipicephalus sanguineus*. *Research in Veterinary Science*, **14**, 112–14.

Sergent, E., Donatien, A., Parrot, L. & Lestoquard, F. (1931). Transmission héréditaire de *Piroplasma bigeminum* chez *Rhipicephalus bursa*. Persistence du parasite chez des tiques nourries sur des chevaux. *Bulletin de la Société de Pathologie Exotique*, **24**, 195–8.

Shortt, H.E. (1936). Life-history and morphology of *Babesia canis* in the dog-tick *Rhipicephalus sanguineus*. Part I. The life-cycle of *B. canis* in *R. sanguineus*. Part II. Cytology of the different stages in the life-cycle of *B. canis* in *R. sanguineus*. *Indian Journal of Medical Research*, **23**, 885–920.

Stoltsz, W.H. (1994). Transmission of *Ehrlichia bovis* by *Rhipicephalus* spp. in South Africa. *Journal of the South African Veterinary Association*, **65**, 159.

Tendeiro, J. (1952). Infestação natural do porco da Guiné pela *Babesia trautmanni* (Knut e Du Toit). *Boletim Cultural da Guiné Portuguesa*, **7**, 359–64.

Theiler, A. (1906). Transmission of equine piroplasmosis by ticks in South Africa. *Journal of Comparative Pathology and Therapeutics*, **19**, 283–92.

Theiler, A. (1909). Transmission des spirilles et des piroplasmes par différentes espèces de tiques. *Bulletin de la Société de Pathologie Exotique*, **2**, 293–4.

Theiler, A. (1912). The transmission of gall-sickness by ticks. *Agricultural Journal of the Union of South Africa*, **3**, 173–81.

Theiler, G. (1962). *The Ixodoidea Parasites of Vertebrates in Africa South of the Sahara (Ethiopian Region)*. Project S9958. Report to the Director of Veterinary Services, Onderstepoort. 260 pp. Mimeographed.

Thomas, A.D. & Neitz, W.O. (1958). Rhipicephaline tick toxicosis in cattle: its possible aggravating effects on certain diseases. *Journal of the South African Veterinary Medical Association*, **29**, 39–50.

Uilenberg, G. (1986). Highlights in recent research on

tick-borne diseases of domestic animals. *Journal of Parasitology*, 72, 485–91.

Uilenberg, G. & Schreuder, B.E.C. (1976). Further studies on *Haematoxenus separatus* (Sporozoa, Theileriidae) of sheep in Tanzania. *Revue d'Élevage et de Médecine Vétérinaire des Pays Tropicaux*, **29** (nouvelle série), 119–26.

Van Rensburg, S.J. (1959). Haematological investigations into the rhipicephaline tick toxicosis syndrome. *Journal of the South African Veterinary Medical Association*, **30**, 75–95.

Viloria, D. (1954). Paralisis por garrapatas en caninos. *Revista de Medicina Veterinaria y Parasitologia Maracay*, **13**, 67–70.

Walker, J.B. (1974). *The Ixodid Ticks of Kenya. A Review of present Knowledge of their Hosts and Distribution.* London: Commonwealth Institute of Entomology.

Wilson, S.G. (1951). *Report by Chief Field Zoologist. Ticks and Tick-borne Diseases. Tick Collections.* Report of the Department of Veterinary Services Kenya for 1951, 1–10. Mimeographed.

Wilson, S.G. (1953). A survey of the distribution of tick vectors of East Coast fever in East and Central Africa. In *Proceedings of the 15th International Veterinary Congress, Stockholm*, **1**, 287–90.

Table 71. *Transmission of tick-borne diseases of humans*

Disease and causative agent	Tick vectors	Number of hosts	Mode of infection or isolation of causative agent	References
Tick typhus *Rickettsia conori*	*R. appendiculatus*	3	Injection of tick emulsion, and rickettsiae isolated from ticks	Pijper & Dau (1934), Gear (1954)
	R. evertsi evertsi	2	Rickettsiae isolated from ticks	Gear (1954)
	R. pulchellus	3	Antibodies to *R. conori* in guinea-pigs after tick feeding	Philip *et al.* (1966)
	R. pusillus	3	Antibodies to *R. conori* detected in European rabbits parasitized by *R. pusillus*	Ciceroni *et al.* (1988)
	R. sanguineus	3	Stage to stage biological transmission	Roberts & Tonking (1933), Neitz, Alexander & Mason (1941)
	Probably *R. praetextatus* (as *R. simus*)	3	Injection of tick emulsion	Heisch, McPhee & Rickman (1957)
Rocky mountain spotted fever *Rickettsia rickettsi*	*R. sanguineus*	3	Rickettsiae isolated from ticks	Hoogstraal (1967)
Siberian tick typhus *Rickettsia sibirica*	*R. sanguineus*	3	Rickettsiae isolated from ticks	Hoogstraal (1967)
	R. turanicus	3	Rickettsiae isolated from ticks	Berdyev (1980)
A tick typhus or fever *Rickettsia slovaca*	*R. pusillus*	3	Antibodies to *R. slovaca* detected in European rabbits parasitized by *R. pusillus*	Ciceroni *et al.* (1988)
Astrakhan fever *Rickettsia* sp.	*R. pumilio*	3	Rickettsiae isolated from ticks	Eremeeva *et al.* (1994)

Table 71. *(cont.)*

Disease and causative agent	Tick vectors	Number of hosts	Mode of infection or isolation of causative agent	References
Q-fever *Coxiella burneti*	*R. bursa*	2	Rickettsiae isolated from ticks	Babudieri (1959)
	R. cuspidatus	3	Rickettsiae isolated from ticks	Tendeiro (1954)
	R. pumilio	3	Rickettsiae isolated from ticks	Hoogstraal (1979)
	R. rossicus	3	Rickettsiae isolated from ticks	Hoogstraal (1979)
	R. sanguineus	3	Injection of tick faeces, and stage to stage biological transmission	Smith (1941)
	R. senegalensis (as *R. simus senegalensis*)	3	Rickettsiae isolated from ticks	Hoogstraal (1956)
	R. turanicus	3	Rickettsiae isolated from ticks	Berdyev (1980)
	Probably *R. praetextatus* (as *R. simus*)	3	Injection of tick emulsion	Heisch (1960), Heisch *et al.* (1962)
Plague *Yersinia pestis*	*R. schulzei*	3	Rubbing ticks or their faeces into the skin	Nelzina *et al.* (1960)
Lyme disease *Borrelia burgdorferi*	*R. bursa*	2	Spirochaetes isolated from ticks	Georgieva *et al.* (1993)
	R. sanguineus	3	Spirochaetes isolated from ticks	Georgieva *et al.* (1993)
Tularaemia *Francisella tularensis*	*R. pumilio*	3	Bacteria isolated from ticks	Hoogstraal (1979)
	R. rossicus	3	Bacteria isolated from ticks	Borodin, Samsonova & Koroleva (1958)
	R. sanguineus	3	Survival of causative organism in ticks from larval to adult stage	Parker (1933)
Crimean-Congo haemorrhagic fever (Bunyaviridae)	*R. bursa*	2	Virus isolated from ticks	Antoniadis & Casals (1982)

	R. evertsi evertsi	2	Virus isolated from ticks	Swanepoel *et al.* (1983)
	R. evertsi mimeticus	2	Nymphs inoculated and transmitted virus as adults	Shepherd *et al.* (1989)
	R. pulchellus	3	Virus isolated from ticks	Wood *et al.* (1978)
	R. pumilio	3	Virus isolated from ticks	Hoogstraal (1979)
	R. rossicus	3	Virus isolated from ticks	Kondratenko *et al.* (1970)
	R. sanguineus	3	Virus isolated from ticks	Hoogstraal (1979)
	R. schulzei	3	Virus isolated from ticks	Smirnova *et al.* (1991)
	R. turanicus	3	Virus isolated from ticks	Hoogstraal (1979)
Dugbe virus (Bunyaviridae)	*R. pulchellus*	3	Virus isolated from ticks	Wood *et al.* (1978)
Thogoto virus (Bunyaviridae)	*R. bursa*	2	Virus isolated from ticks	Albanese *et al.* (1972)
	R. sanguineus	3	Virus isolated from ticks	Filipe & Calisher (1984)
Kyasanur Forest disease (Flaviviridae)	*R. haemaphysaloides*	3	Stage to stage transmission of the virus to laboratory rodents	Bhat *et al.* (1978)
Louping ill (Flaviviridae)	*R. appendiculatus*	3	Stage to stage transmission of the virus to sheep	Alexander & Neitz (1935)
West Nile virus (Flaviviridae)	*R. bursa*	2	Virus isolated from ticks	Hoogstraal (1979)
	R. rossicus	3	Virus survived experimentally in ticks from the larval to the adult stage	Hoogstraal (1979)
	R. turanicus	3	Virus isolated from ticks	Hoogstraal (1979)

REFERENCES

Albanese, M., Bruno-Smiraglia, C., Di Cuonzo, G., Lavagnino, A. & Srihongse, S. (1972). Isolation of Thogoto virus from *Rhipicephalus bursa* ticks in western Sicily. *Acta Virologica*, **16**, 267.

Alexander, R.A. & Neitz, W.O. (1935). The transmission of Louping ill by ticks (*Rhipicephalus appendiculatus*). *Onderstepoort Journal of Veterinary Science and Animal Industry*, **5**, 15–33.

Antoniadis, A. & Casals, J. (1982). Serological evidence of human infection with Congo-Crimean hemorrhagic fever virus in Greece. *American Journal of Tropical Medicine and Hygiene*, **31**, 1066–7. (Brief communication).

Babudieri, B. (1959). Q fever: a zoonosis. *Advances in Veterinary Science*, **5**, 81–182.

Berdyev, A. (1980). *Ecology of Ixodid Ticks of Turkmenistan and their Importance in Natural Focal Disease Epizootiology*. Ashkhabad, Turkmenistan: Ylym. [In Russian].

Bhat, H.R., Naik, S.V., Ilkal, M.A. & Banerjee, K. (1978). Transmission of Kyasanur Forest disease virus by *Rhipicephalus haemaphysaloides* ticks. *Acta Virologica*, **22**, 241–4.

Borodin, V.P., Samsonova, A.P. & Koroleva, A.P. (1958). Two cases of allergic reaction in persons inoculated against tularemia, resulting from bites by infected ticks of the species *Rhipicephalus rossicus*. *Zhurnal Mikrobiologii, Epidemiologii i Immunobilogii, Moskova*, **29**, 117–18. [In Russian].

Ciceroni, L., Pinto, A., Rossi, C., Khoury, C., Rivosecchi, L., Stella, E. & Cacciapuoti, B. (1988). Rickettsiae of the spotted fever group associated with the host-parasite system *Oryctolagus cuniculus/Rhipicephalus pusillus*. *Zentrablatt für Bakeriologie und Hygiene*, Series A, **269**, 211–17.

Eremeeva, M.E., Beati, L., Markova, V.A., Fetisova, N.F., Tarasevich, I.V., Balayeva, N.M. & Raoult, D. (1994). Astrakhan fever rickettisae: antigenic and genotypic analysis of isolates obtained from human and *Rhipicephalus pumilio* ticks. *American Journal of Tropical Medicine and Hygiene*, **51**, 697–706.

Filipe, A.R. & Calisher, C.H. (1984). Isolation of Thogoto virus from ticks in Portugal. *Acta Virologica*, **28**, 152–5.

Gear, J. (1954). The rickettsial diseases of southern Africa. A review of recent studies. *South African Journal of Clinical Science*, **5**, 158–75.

Georgieva, G., Manev, Kh., Georgieva, V., Matev, G. & Karanikolova, N. (1993). Ixodid ticks infected with *Borrelia* in two different eco- geographical regions. *Infectology*, **30**, 38–40. [In Bulgarian] (CAB Abstracts 1995).

Heisch, R.B. (1960). The isolation of *Rickettsia burneti* from *Lemniscomys* sp. in Kenya. *East African Medical Journal*, **37**, 104.

Heisch, R.B., Grainger, W.E., Harvey, A.E.C. & Lister, G. (1962). Feral aspects of rickettsial infections in Kenya. *Transactions of the Royal Society of Tropical Medicine and Hygiene*, **56**, 272–86.

Heisch, R.B., McPhee, R. & Rickman, L.R. (1957). The epidemiology of tick-typhus in Nairobi. *East African Medical Journal*, **34**, 459–77.

Hoogstraal, H. (1956). *African Ixodoidea. I. Ticks of the Sudan (with Special Reference to Equatoria Province and with Preliminary Reviews of the Genera* Boophilus, Margaropus, *and* Hyalomma). Research report NM 005 050. 29.07, 1101 pp. Washington DC: Department of the Navy, Bureau of Medicine and Surgery.

Hoogstraal, H. (1967). Ticks in relation to human diseases caused by *Rickettsia* species. *Annual Review of Entomology*, **12**, 377–420.

Hoogstraal, H. (1973). Viruses and ticks. In *Viruses and Invertebrates*, ed. A.J. Gibbs, pp. 349–90. Amsterdam and London: North Holland Publishing Co.

Hoogstraal, H. (1979). The epidemiology of tick-borne Crimean-Congo hemorrhagic fever in Asia, Europe, and Africa. *Journal of Medical Entomology*, **15**, 307–417.

Kondratenko, V.F., Blagoveshchenskaya, N.M., Butenko, A.M., Vishnivetskaya, L.K., Zarubina, L.V., Milyutin, V.N., Kuchin, V.V., Novikova, E.N., Rabinovich, V.D., Shevchenko, S.F. & Chumakov, M.P. (1970). Results of virological investigation of ixodid ticks in Crimean hemorrhagic fever focus in Rostov Oblast. In *Crimean Hemorrhagic Fever*, ed. M.P.Chumakov, pp. 29–35. Materialy III Oblastnoy Nauchno-Prakticheskoy Konferentsii, Rostov-na-Donu, May 1970. [In Russian, Translation T524, NAMRU-3, Cairo].

Neitz, W.O., Alexander, R.A. & Mason, J.H. (1941).

The transmission of tick-bite fever by the dog tick *Rhipicephalus sanguineus*, Latr. *Onderstepoort Journal of Veterinary Science and Animal Industry*, **16**, 9–17.

Nelzina, E.N., Pylenko, M.S., Chudesova, V.P., Kondrashkina, K.I. & Bykov, L.T. (1960). The role of *Rhipicephalus schulzei* in natural foci of plague. (Communication 1. Localization of *Bacillus pestis* in the tick body). *Meditsinskaya Parazitologiya, Moskva*, **29**, 202–7. [In Russian, English translation T121, NAMRU-3, Cairo].

Parker, R.R. (1933). Recent studies of tick-borne diseases made at the United States Public Health Service Laboratory at Hamilton, Montana. *Proceedings of the 5th Pacific Science Congress*, **6**, 3367–74.

Philip, C.B., Hoogstraal, H., Reiss-Gutfreund, R.J. & Clifford, C.M. (1966). Evidence of rickettsial disease agents in ticks from Ethiopian cattle. *Bulletin of the World Health Organisation*, **35**, 127–31.

Pijper, A. & Dau, H. (1934). Die fleckfieberartigen Krankheiten des südlichen Afrika. *Zentralblatt für Bakteriologie*, **133**, 7–22.

Roberts, J.I. & Tonking, H.D. (1933). A preliminary note on the vector of tropical typhus in Kenya. *East African Medical Journal*, **9**, 310–15.

Shepherd, A.J., Swanepoel, R., Cornel, A.J. & Mathee, O. (1989). Experimental studies on the replication and transmission of Crimean-Congo hemorrhagic fever virus in some African tick species. *American Journal of Tropical Medicine and Hygiene*, **40**, 326–31.

Smirnova, S.E., Karavanov, A.S., Zimina, Yu. V. & Sedova, A.G. (1991). Study of the vectorial capacity of ticks in the transmission of Crimean hemorrhagic fever. *Meditsinskaya Parazitologiya i Parazitarnye Bolezni*, **No.1**, 32–4. [In Russian] (CAB Abstracts 1990–1991).

Smith, D.J.W. (1941). Studies in the epidemiology of Q fever. 8. The transmission of Q fever by the tick *Rhipicephalus sanguineus*. *Australian Journal of Experimental Biology and Medical Science*, **19**, 133–6.

Swanepoel, R., Struthers, J.K., Shepherd, A.J., McGillivray, G.M., Nel, M.J. & Jupp, P.G. (1983). Crimean-Congo hemorrhagic fever in South Africa. *American Journal of Tropical Medicine and Hygiene*, **32**, 1407–15.

Tendeiro, J. (1954). Posição actual do problema da febre Q. *Revista de Ciencias Veterinarias*, **49** (350), 283–311.

Wood, O.L., Lee, V.H., Ash, J.S. & Casals, J. (1978). Crimean-Congo hemorrhagic fever, Thogoto virus, Dugbe and Jos viruses isolated from ixodid ticks in Ethiopia. *American Journal of Tropical Medicine and Hygiene*, **27**, 600–4.

Yunker, C. (1970). Worldwide research on human and animal diseases caused by tickborne viruses. *Miscellaneous Publications of the Entomological Society of America*, **6**, 330–8.

Index

Page references to species under the *Rhipicephalus* entry in **bold** are the primary references.
Page references to hosts in **bold** refer to host/parasite lists.
Page references in *italics* refer to maps.